2020 22nd European Conference on Power Electronics and Applications (EPE'20 ECCE Europe)

Lyon, France
7-11 September 2020

Pages 2059-2742

IEEE Catalog Number: CFP20850-POD
ISBN: 978-1-7281-9807-1

Copyright © 2020, EPE Association
All Rights Reserved

*** *This is a print representation of what appears in the IEEE Digital Library. Some format issues inherent in the e-media version may also appear in this print version.*

IEEE Catalog Number: CFP20850-POD
ISBN (Print-On-Demand):   978-1-7281-9807-1
ISBN (Online):   978-9-0758-1536-8

Additional Copies of This Publication Are Available From:

Curran Associates, Inc
57 Morehouse Lane
Red Hook, NY 12571 USA
Phone: (845) 758-0400
Fax: (845) 758-2633
E-mail: curran@proceedings.com
Web: www.proceedings.com

2020 22nd European Conference on Power Electronics and Applications (EPE'20 ECCE Europe)

Lyon, France
7-11 September 2020

Pages 2059-2742

IEEE Catalog Number: CFP20850-POD
ISBN: 978-1-7281-9807-1

TABLE OF CONTENTS

VALIDATION OF THERMAL STRESS MODELING IN PV INVERTERS UNDER MISSION PROFILE OPERATION 1
Ariya Sangwongwanich, Huai Wang, Frede Blaabjerg

ON THE LIMITATIONS OF USING A LTI MODELLING APPROACH FOR CONTROL TUNING OF VSC-HVDC SYSTEMS 9
Pablo Briff, Julián Freytes, Guillaume De-Preville, Jiaqi Li, Omar Jasim

A VOLTAGE CONTROL METHOD FOR POWER DISTRIBUTION LINES UTILIZING DISPERSED CUSTOMER RESOURCES 19
Hiroki Ishihara, Kaho Nada, Miwako Tanaka, Sadayuki Inoue, Akiko Kuwata, Tomihiro Takano

PERFORMANCE COMPARISON BETWEEN SIC AND SI INVERTER MODULES IN AN ELECTRICAL VARIABLE TRANSMISSION APPLICATION 27
Mauricio Dalla Vecchia, Simon Ravyts, Florian Verbelen, Jeroen Tant, Peter Sergeant, Johan Driesen

SEAMLESS INTEGRATION OF FEEDFORWARD AND FEEDBACK CONTROL OF BALANCE OF ARM CAPACITOR VOLTAGES IN STATCOMS BASED ON CHAIN LINKS OF H BRIDGE MODULES 37
D. Basic, N. Lapassat

ASYNCHRONIZED ELECTROMECHANICAL CONVERTER IN THE ELECTRICAL SUPPLY SYSTEM OF POWERFUL ENERGY CONSUMERS 47
Aleksey G. Vorontsov, Mikhail V. Pronin, Anastasiia D. Stotckaia, Vasiliy V. Glushakov, Pavel V. Sokur

SYMMETRIC AND ASYMMETRIC OPERATING MODES OF HYBRID CASCADE FREQUENCY CONVERTERS 56
Aleksey G. Vorontsov, Vasiliy V. Glushakov, Mikhail V. Pronin, Anastasiia D. Stotckaia

SYSTEM FREQUENCY DYNAMIC RESPONSE OF A NOVEL, SELF-SYNCHRONIZING INVERTER IN A HIGH RENEWABLE PENETRATION GRID 65
Christian Perenyi, Moath Alqatamin, Thibaut Harzig, Michael McIntyre, Brandon M. Grainger

ROTOR POSITION ESTIMATION WITH HALL-EFFECT SENSORS IN BEARINGLESS DRIVES 75
Patricio Peralta, Jacopo Leo, Yves Perriard

NON-UNIT ROCOV SCHEME FOR PROTECTION OF MULTI-TERMINAL HVDC SYSTEMS 85
María José Pérez-Molina, Pablo Eguia, Marene Larruskain, Garikoitz Buigues, Esther Torres

MODELLING OF CONVERTER SYSTEMS PARALLELED VIA INTERPHASE TRANSFORMERS IN CYCLIC CASCADE TOPOLOGY AND OPTIMIZATION OF PWM CARRIER SHIFTS 95
D. Basic, H. Baërd, S. Siala

MEASUREMENT AND CALCULATION METHOD OF WIRELESS POWER TRANSFER COIL EQUIVALENT SERIES RESISTANCE UNDER THE VEHICLE 105
Norihito Kimura, Hiroaki Yuasa

DESIGN OF A CIRCUMSCRIBING POLYGON WIDE BANDGAP BASED INTEGRATED MODULAR MOTOR DRIVE TOPOLOGY WITH THERMALLY DECOUPLED WINDINGS AND POWER CONVERTERS 115

Abdalla Hussein Mohamed, Hendrik Vansompel, Peter Sergeant

LIMITS OF ENHANCED DESATURATION DETECTION METHOD WITH ADAPTIVE BLANKING FOR GAN HEMTS 124

Jan Schmitz, Markus Meißner, Steffen Bernet

CURRENT CONTROL OF A GRID-CONNECTED SINGLE-PHASE VOLTAGE-SOURCE INVERTER WITH LCL FILTER 134

Alfonso Parreño Torres, Fco. Javier López-Alcolea, Pedro Roncero-Sánchez, Javier Vázquez, Emilio J. Molina-Martínez, Felix García-Torres

FOUR SWITCH BUCK/BOOST CONVERTER FOR DC MICROGRID APPLICATIONS 143

Matthias Schulz, Nico Schleippmann, Kilian Gosses, Bernd Wunder, Martin März

STABILITY INVESTIGATION OF THREE-PHASE GRID-TIED PV INVERTERS WITH IMPEDANCE-BASED METHOD 153

Zhiqing Yang, Wanchao Gou, Xian Luo, Chirag Shah, Nurhan Rizqy Averous, Rik W. De Doncker

STABILITY INVESTIGATION OF LARGE-SCALE PV PARKS WITH EIGENVALUE-BASED METHOD 163

Zhiqing Yang, Christian Bendfeld, Jin Qiang, Benedict Mortimer, Rik W. De Doncker

COMPACT CORE LOSS MODEL BASED ON AN EFFECTIVE FREQUENCY FOR ARBITRARY CORE EXCITATIONS INCLUDING DC-BIAS 173

Erika Stenglein, Manfred Albach, Thomas Dürbaum

ASSESSMENT OF AGING AND PERFORMANCE DEGRADATION OF SUPERCAPACITORS INTEGRATED INTO A MODULAR MULTILEVEL CONVERTER 183

F. Errigo, L. Chédot, F. Morel, P. Venet, A. Sari, A. Hijazi, R. A. Peña

SEPARATION OF MAGNETIC FLUX DENSITY TRAJECTORIES INTO SUBLOOPS FOR THE PREDICTION OF HYSTERESIS LOSS 193

Erika Stenglein, Manfred Albach, Thomas Dürbaum

INFLUENCE OF GENERALIZED DISCONTINUOUS PULSE WIDTH MODULATION (GDPWM) ON THE DC-LINK CURRENT AND VOLTAGE RIPPLE IN BATTERY-FED PWM INVERTER SYSTEMS 203

Panagiotis Mantzanas, Alexander Bucher, Daniel Kuebrich, Alexander Pawellek, Christian Hasenohr, Harald Hofmann, Thomas Duerbaum

AUTOMATED DESIGN METHOD FOR SINE WAVE FILTERS IN MOTOR DRIVE APPLICATIONS WITH SIC-INVERTERS 213

Thorben Schobre, Regine Mallwitz

A SYMMETRICAL BOOST CONVERTER WITH REDUCED COMMON-MODE LEAKAGE CURRENTS FOR EV APPLICATIONS 223

Caniggia Viana, Netan Yakop, Damien Frost, Peter Lehn

MODELING AND ANALYSIS OF CONDUCTED EMI ON FLYBACK CONVERTER USING POWER MANAGEMENT IC WITH CHAOTIC SUPPRESSION EMI 231

Diao Jiaqi, Yang Ru, Liu Zuolian, Yang Hong, Jie Hai

HIGH PERFORMANCE DRIVE INVERTER FOR AN ELECTRIC TURBO COMPRESSOR IN FUEL CELL APPLICATIONS........... 241

N. Langmaack, G. Tareilus, R. Mallwitz

DEVELOPMENT OF AN ALGORITHM FOR THE AUTOMATION OF THE MODELLING PROCESS OF POWER CONVERTERS........... 251

Jon Anzola, Iosu Aizpuru, Asier Arruti

A NOVEL FULLY DISTRIBUTED COST OPTIMAL CONTROL METHOD FOR DC MICROGRID........... 260

Qingping Xia, Hua Han, Yao Liu, Zhangjie Liu, Yao Sun, Mei Su

MEASUREMENT OF DYNAMIC ON-STATE RESISTANCE OF HIGH-VOLTAGE GAN-HEMTS UNDER REAL APPLICATION CONDITIONS........... 266

Benedikt Kohlhepp, Carsten Kuring, Stefan Peller, Daniel Kübrich

ANALYSIS OF DC-SIDE FAULT RESPONSE OF MMCS WITH CONTROLLED FAULT BLOCKING CAPABILITY FOR DIFFERENT TRANSMISSION LINE TYPES........... 276

Willem Leterme, Paul D. Judge, Tim C. Green

A HYBRID SERIES-PARALLEL MICROGRID AND ITS LOW-DEPENDENT COMMUNICATION CONTROL........... 285

Lang Li, Yao Sun, Hua Han, Mei Su

ADAPTIVE VOLTAGE CONTROL OF ISLANDED RES-BASED RESIDENTIAL MICROGRID WITH INTEGRATED FLYWHEEL/BATTERY HYBRID ENERGY STORAGE SYSTEM........... 292

Linda Barelli, Gianni Bidini, Ermanno Cardelli, Dana-Alexandra Ciupageanu, Andrea Ottaviano, Dario Pelosi, Simone Castellini, Gheorghe Lazaroiu

AN IMPROVED λ -CONSENSUS CONTROL METHOD FOR DC MICROGRIDS........... 302

Siqi Fu, Yao Sun, Zhangjie Liu, Hua Han, Mei Su

DECREASE OF POWER ELECTRONIC SWITCHING LOSSES USING VARIABLE SWITCHING EVENTS........... 307

Hannes Ramm, Michael Homann, Torben A. Schulze, Faical Turki, Heiko Rabba

OPTIMIZATION OF MEDIUM-FREQUENCY TRANSFORMERS WITH LARGE CAPACITY AND HIGH INSULATION REQUIREMENT........... 317

Xuan Guo, Chi Li, Zedong Zheng, Yongdong Li

IMPROVED SOC BALANCING AND ACTIVE POWER SHARING CONTROL METHOD IN HIGHLY RESISTIVE LINE MICROGRID........... 326

Yuanhao Zhu, Hua Han, Guangze Shi, Zhangjie Liu, Yao Sun, Mei Su

TECHNO-ECONOMIC ANALYSIS OF SECOND-LIFE LITHIUM-ION BATTERIES INTEGRATION IN MICROGRIDS........... 332

Camille Birou, Xavier Roboam, Hugo Radet, Fabien Lacressonnière

DESIGN, MODELLING, AND TEST OF A SOLID-STATE MAIN BREAKER FOR HYBRID DC CIRCUIT BREAKER........... 342

Jiawen Xi, Xiaoze Pei, Xianwu Zeng, Liyong Niu

MODEL PREDICTIVE CONTROL FOR THREE-PHASE SPLIT-SOURCE INVERTER........... 352

Youssuf Elthokaby, Islam Mohamed, Naser Abdel-Rahim

HARDWARE IMPLEMENTATION STUDY OF VARIABLE SPEED WIND-TURBINE-DFIG IN STAND-ALONE MODE 362
    Fayssal Amrane, Bruno Francois, Azeddine Chaiba

INFLUENCE OF WIRE-BONDING LAYOUT ON RELIABILITY IN IGBT MODULE 370
    Lubin Han, Lin Liang, Wei Xin, Fang Luo

RAIL POTENTIAL CALCULATION MODEL FOR DC RAILWAY POWER SUPPLY EQUIPPED WITH VOLTAGE LIMITING DEVICE 377
    Shota Kimura, Tsutomu Miyauchi, Kenji Oguma, Hirotaka Takahashi, Keiko Teramura

HOMOGENIZATION OF CURRENT DISTRIBUTION IN PARALLEL CONNECTION OF INTERLEAVED WINDING LAYERS OF HIGH-FREQUENCY TRANSFORMERS BY OPTIMIZING DISTANCE BETWEEN WINDING LAYERS 386
    Ryo Murata, Tomohide Shirakawa, Kazuhiro Umetani, Eiji Hiraki, Hiroto Mizutani, Takaaki Takahara, Osamu Mori

REAL-TIME PARAMETERS IDENTIFICATION OF LITHIUM-ION BATTERIES MODEL TO IMPROVE THE HIERARCHICAL MODEL PREDICTIVE CONTROL OF BUILDING MICROGRIDS 396
    Daniela Yassuda Yamashita, Ionel Vechiu, Jean-Paul Gaubert

IMPACT OF DC FAULT BLOCKING CAPABILITY ON THE SIZING OF THE DC-DC MODULAR MULTILEVEL CONVERTER 406
    J. D. Paez, F. Morel, S. Bacha, Piotr Dworakowski, D. Frey

OPTIMIZATION OF HIGH FREQUENCY MAGNETIC DEVICES WITH CONSIDERATION OF THE EFFECTS OF THE MAGNETIC MATERIAL, THE CORE GEOMETRY AND THE SWITCHING FREQUENCY 416
    Sobhi Barg, Muhammad Farhan Alam, Kent Bertilsson

REAL TIME CONTROL HARDWARE IN THE LOOP TEST OF A NOVEL MVDC SOLID-STATE BREAKER 424
    Alessio Clerici, Riccardo Chiumeo, Chiara Gandolfi

IGBT LIFETIME ESTIMATION IN A MODULAR MULTILEVEL CONVERTER FOR BIDIRECTIONAL POINT-TO-POINT HVDC APPLICATION 433
    Diego Velazco, Guy Clerc, Emmanuel Boutleux, François Wallart, Laurent Chédot

OPTIMIZATION DESIGN FOR SIC DRIFT STEP RECOVERY DIODE (DSRD) 443
    Xiaoxue Yan, Lin Liang, Ziyue Wang, Guoqiang Tan

DISCRETE SUPER-TWISTING SLIDING MODE CURRENT CONTROLLER FOR INDUCTION MOTOR DRIVES 450
    Tianqing Wang, Bo Wang, Yong Yu, Yangming Zhu, Dianguo Xu

NEW GRID-CONNECTED MULTILEVEL BOOST CONVERTER TOPOLOGY WITH INHERENT CAPACITORS VOLTAGE BALANCING USING MODEL PREDICTIVE CONTROLLER 460
    Rasoul Shalchi Alishah, Kent Bertilsson, Frede Blaabjerg, Mohd. Ali Jagabar Sathik, Ali Yahya Rezaee

DCM OPERATION OF SINGLE-SWITCH HIGH STEP-UP DC-DC CONVERTER WITH THREE-WINDING COUPLED INDUCTOR 467
    Masataka Minami, Genki Hase

POWER LOSSES CALCULATION FOR MEDIUM VOLTAGE DC/DC CURRENT-FED SOLID STATE TRANSFORMER FOR BATTERY GRID-CONNECTED 471
  E. K. Hussain, Mohammad Abusara, S. M. Sharkh

MODELLING AND EXPERIMENTAL VALIDATION OF A POLE-TO-GROUND PROTECTION DEVICE IN LOW VOLTAGE DC MICROGRIDS 480
  L. Hallemans, G. Govaerts, G. Van Den Broeck, S. Ravyts, M. M. Alam, P. Van Tichelen, J. Driesen

DESIGN OF A DUAL ACTIVE BRIDGE CONVERTER FOR ON-BOARD VEHICLE CHARGERS USING GAN AND INTO TRANSFORMER INTEGRATED SERIES INDUCTANCE 490
  K. Siebke, M. Giacomazzo, R. Mallwitz

AN EXPERIMENTAL ANALYSIS OF CIRCULATING CURRENT CONTROL CIRCUIT FOR OUTPUT POWER FROM VIBRATION GENERATOR FOR VIBRATION INCLUDING THE THIRD HARMONICS 498
  Masataka Minami, Akito Nakagaki, Genki Hase

IMPLEMENTATION OF CONTROL STRATEGY FOR STEP-DOWN DC-DC CONVERTER BASED ON PIEZOELECTRIC RESONATOR 503
  Mustapha Touhami, Ghislain Despesse, François Costa, Benjamin Pollet

THERMAL IMPEDANCES AND TEMPERATURE SENSORS: A COMBINED APPROACH FOR A NOVEL THERMAL MODEL OF POWER SEMICONDUCTORS 512
  Maria De Lauretis, Jonas Millinger, Erik Baker, Martin Karlsson, Diane -Perle Sandik

A 3A LOW VOLTAGE LASER DIODE DRIVER IC IN A CMOS TECHNOLOGY FOR AN ITOF-BASED 3D IMAGE SENSOR 522
  Romain David, Bruno Allard, Xavier Branca, Charles Joubert

COMPARISON OF DECOUPLING TECHNIQUES VIA DISCRETE LUENBERGER STYLE OBSERVER FOR VOLTAGE ORIENTED CONTROL 532
  Gyanendra Kumar Sah, Michael Schütt, Hans-Günter Eckel

VARIABLE SWITCHING POINT PARALLEL PREDICTIVE CURRENT CONTROL (VSP3CC) FOR INDUCTION MOTOR 542
  Qing Chen, Ralph Kennel

OPERATION OF AN EXTERNALLY EXCITED SYNCHRONOUS MACHINE WITH A HYBRID MULTILEVEL INVERTER 551
  C. Terbrack, J. Stöttner, C. Endisch

A FACILITY FOR MIXED FLOWING GAS TESTING OF AND EXPERIMENTATION WITH POWER ELECTRONIC COMPONENTS AND SYSTEMS 563
  Juuso Rautio, Janne Jäppinen, Tommi J. Kärkkäinen, Markku Niemelä, Pertti Silventoinen, Mika Kiviniemi, Joonas Leppänen, Jonny Ingman

IMPACT OF IMPLEMENTATION OF AUXILIARY BIAS-WINDINGS ON CONTROLLABLE INDUCTORS FOR POWER ELECTRONIC CONVERTERS 571
  Jonas Pfeiffer, Pierre Küster, Yeliz Erenler, Ziyad H. S. Qashlan, Peter Zacharias

APPROXIMATED SLIDING-MODE CONTROL OF PARALLEL-CONNECTED GRID INVERTERS 581
  Albrecht Gensior

EQUIVALENT MODEL AND CONTROL OF A NEUTRAL POINT SUPPLY SYNRM DRIVE 590
Xiaokang Zhang, Jean-Yves Gauthier, Xuefang Lin-Shi

IMPROVEMENTS ON SIGNAL-TO-NOISE RATIO IN FEEDBACK MEASUREMENT IN
DC/DC CONVERTERS 598
Fernando Davalos Hernandez, Federico Ibanez, Sebastian Gutierrez, Wilmar Martinez

APPROACH OF AN ACTIVE DEVICE PROTECTION FOR DRIVE INVERTERS AGAINST
SHORT CIRCUIT FAULTS IN AN OPEN INDUSTRIAL DC GRID 608
Simon Puls, Urs Obernolte, Martin Ehlich, Holger Borcherding

A NEW DESIGN OF AN AIR CORE TRANSFORMER FOR ELECTRIC VEHICLE ON-
BOARD CHARGER 618
Valentin Rigot, Tanguy Phulpin, Daniel Sadarnac, Jihen Sakly

ENABLING FOIL WINDINGS OF MEDIUM-FREQUENCY TRANSFORMERS FOR HIGH
CURRENTS 627
Thomas B. Gradinger, Uwe Drofenik, Filip Grecki

A HIGH-EFFICIENCY WIRELESS POWER TRANSFER SYSTEM FOR UNMANNED
AERIAL VEHICLE CONSIDERING CARBON FIBER BODY 637
Kai Song, Peng Zhang, Zhengxin Chen, Guang Yang, Jinhai Jiang, Chunbo Zhu

ANALYTICAL COMPUTATION OF NORMAL AND FAULT-TOLERANT ACTIVE SHORT
CIRCUIT OPERATION OF ANISOTROPIC SYNCHRONOUS DOUBLE STAR MACHINES 644
Michael Gleissner, Johannes Häring, Wolfgang Wondrak, Mark-M. Bakran

FULL-SILICON 98.7% EFFICIENT THREE-PHASE FIVE-LEVEL 3-PORT UPS
ARCHITECTURE WITH WIDE VOLTAGE RANGE BATTERY BASED ON MULTIPLEXED
TOPOLOGY 654
Kepa Odriozola, Thierry A. Meynard, Alain Lacarnoy

ON-GRID/OFF-GRID DC MICROGRID OPTIMIZATION AND DEMAND RESPONSE
MANAGEMENT 667
Wenshuai Bai, Manuela Sechilariu, Fabrice Locment

SHEDDING AND RESTORATION ALGORITHMS FOR AN EV CHARGING STATION TO
MAXIMIZE AVAILABLE POWER 677
Dian Wang, Fabrice Locment, Manuela Sechilariu

EFFICIENCY AND COST COMPARISON OF B6 AND HYBRID ANPC CONVERTERS FOR
TRACTION DRIVES 686
Johannes Häring, Michael Gleissner, Wolfgang Wondrak, Mark-M. Bakran

DESIGN AND CONTROL OF A KE (KINETIC ENERGY) - COMPENSATED
GRAVITATIONAL ENERGY STORAGE SYSTEM 696
Alfred Rufer

A NOVEL POWER FLOW CONTROL STRATEGY FOR HETEROGENEOUS BATTERY
ENERGY STORAGE SYSTEMS BASED ON PROGNOSTIC ALGORITHMS FOR
BATTERIES 707
Markus Muehlbauer, Samantha Klier, Herbert Palm, Oliver Bohlen, Michael A. Danzer

AN IGCT-BASED MULTI-FUNCTIONAL MMC SYSTEM WITH COMMUTATION AND
SWITCHING 718
Chaoqun Xu, Mingzhu Guo, Biao Zhao, Bojin Tang, Zhanqing Yu, Dongling Zhai, Chunpin
Ren

COMMON-MODE NOISE MODELLING AND RESONANT ESTIMATION IN A THREE-PHASE MOTOR DRIVE SYSTEM: 9-150 KHZ FREQUENCY RANGE 726
Hansika Rathnayake, Amir Ganjavi, Firuz Zare, Dinesh Kumar, Pooya Davari

POLYNOMIAL MULTI-VARIABLE CONTROL STRATEGY FOR FLUX BALANCING IN DUAL ACTIVE BRIDGE CONVERTER 736
Pierre-Baptiste Steckler, Jean-Yves Gauthier, Xuefang Lin-Shi, François Wallart

ENHANCED POWER SYSTEM DAMPING ESTIMATION VIA OPTIMAL PROBING SIGNAL DESIGN 745
S. Boersma, X. Bombois, L. Vanfretti, V. Peric, J-C. Gonzalez-Torres, R. Segur, A. Benchaib

IMPROVED HIGH STEP-UP BOOST-BASED DC/DC CONVERTER WITH BUILT-IN TRANSFORMER AND ACTIVE CLAMP FOR DC MICROGRIDS 755
Konstantinos Zaoskoufis, Emmanuel C. Tatakis

ELIMINATION/MITIGATION OF OUTPUT VOLTAGE HARMONICS FOR MULTILEVEL CONVERTERS OPERATED AT FUNDAMENTAL SWITCHING FREQUENCY USING MATLAB'S GENETIC ALGORITHM OPTIMIZATION 765
Anton Kersten, Manuel Kuder, Arthur Singer, Weiji Han, Torbjörn Thiringer, Thomas Weyh, Richard Eckerle

EVALUATION OF DRIVE TOPOLOGIES FOR MACRO SCALE SYNCHRONOUS ELECTROSTATIC MACHINES 777
Peter Killeen, Daniel C. Ludois

DECENTRALIZED VOLTAGE REGULATION IN ISLANDED DC MICROGRIDS IN THE PRESENCE OF DISPATCHABLE AND NON-DISPATCHABLE DC SOURCES 787
Mohammadreza Nabatirad, Reza Razzaghi, Behrooz Bahrani

AN ULTRA-FAST GATE DRIVER WITH OVER CURRENT PROTECTION FOR GAN POWER TRANSISTORS 797
Qingqing Nie, Han Peng, Yong Kang

A NEW GAN HYBRID RESONANT-CLAMPING GATE DRIVER FOR HIGH FREQUENCY SIC MOSFETS 804
Ziyue Dang, Han Peng, Hao Peng, Yong Kang, Yu Chen, Xudan Liu, Maojun He

MAINTENANCE SCHEDULING IN POWER ELECTRONIC CONVERTERS CONSIDERING WEAR-OUT FAILURES 810
Saeed Peyghami, Frede Blaabjerg, Jose Rueda Torres, Peter Palensky

AC/DC DYNAMIC INTERACTIONS OF MMC-HVDC IN GRID-FORMING FOR WIND-FARM INTEGRATION IN AC SYSTEMS 820
Rayane Mourouvin, Kosei Shinoda, Jing Dai, Abdelkrim Benchaib, Seddik Bacha, Didier Georges

A DESIGN OF SOLID STATE POWER CONTROLLER FOR A BIDIRECTIONAL DC-DC CONVERTER IN AN AERONAUTIC CONTEXT 829
Hassan Cheaito, Bruno Allard, Guy Clerc, Joris Pallier, Pascal Pommier-Petit

A NEW APPROACH OF RESONANT CONVERTER USING LARGE AIR GAP TRANSFORMER 835
Michael Finkenzeller, Monika Poebl, Thomas Komma

REDUCED CAPACITOR SIZE AND ON-STATE LOSSES IN ADVANCED MMC
SUBMODULE TOPOLOGIES 843
 Christopher Dahmen, Rainer Marquardt

STABILITY AND ROBUSTNESS ANALYSIS OF FRACTIONAL PROPORTIONAL
RESONANT CONTROLLERS IN CURRENT-CONTROLLED VOLTAGE-SOURCE-
INVERTERS 853
 Daniel Heredero-Peris, Cristian Chillón-Antón, Daniel Montesinos-Miracle

EMPLOYING VIRTUAL SYNCHRONOUS GENERATOR WITH A NEW CONTROL
TECHNIQUE FOR GRID FREQUENCY STABILIZATION 863
 *Meysam Saeedian, Bahman Eskandari, Kumars Rouzbehi, Shamsodin Taheri, Edris
Pouresmaeil*

A HYBRID PULSE WIDTH MODULATION TECHNIQUE WITH TEMPERATURE
CONTROL FOR MODULAR MULTILEVEL CONVERTERS 871
 Ara Bissal, Waqas Ali, Rob Leedham, Mark Snook, Ibrahim Elsabrouty, Ilknur Colak

DESIGN FLOW OF A COMPACT HIGH-FREQUENCY DC/DC CONVERTER WITH
OPTIMUM AVERAGE EFFICIENCY IN A WIDE OPERATION RANGE 880
 Maximilian Nitzsche, Matthias Zehelein, Julian Weimer, Dominik Koch, Jörg Roth-Stielow

ANALYSIS OF THE TRANSFORMER MODULARIZATION FOR HIGH FREQUENCY
ISOLATED HIGH VOLTAGE GENERATOR WITH THE SILICON CARBIDE DEVICES 892
 Saijun Mao, Popovic Jelena, Jan Abraham Ferreira

IMPROVED DIRECT-MODEL PREDICTIVE CONTROL WITH A SIMPLE DISTURBANCE
OBSERVER FOR DFIGS 900
 Mohamed Abdelrahem, Christoph Hackl, José Rodríguez, Ralph Kennel

MODELING OF SIC-MOSFET CONVERTER LEG INCLUDING PARASITICS OF PRINTED
CIRCUIT BOARD LAYOUT AND DEVICE PACKAGING 909
 M. Pulvirenti, L. Salvo, A. G. Sciacca, G. Scelba, M. Cacciato

PERFORMANCE ANALYSIS OF RL DAMPER IN GAN-BASED HIGH-FREQUENCY
BOOST CONVERTER 919
 A. Gutierrez, E. Marcault, C. Alonso, D. Tremouilles

RAPID IMPEDANCE ESTIMATION ALGORITHM FOR MITIGATION OF
SYNCHRONIZATION INSTABILITY OF PARALLELED CONVERTERS UNDER GRID
FAULTS 927
 Mads Graungaard Taul, Robert Eric Betz, Frede Blaabjerg

ADAPTIVE THERMAL CONTROL FOR MOSFET-BASED MODULAR MULTILEVEL
CONVERTER 937
 Tianxiang Yin, Lei Lin, Chen Xu

ELECTRIC IMPULSE TECHNOLOGY – BREAKING ROCK 944
 Matthias Voigt, Erik Anders, Franziska Lehmann, Margarita Mezzetti, Frank Will

IMPACT OF COMBINED THERMO-MECHANICAL AND ELECTRO-CHEMICAL STRESS
ON THE LIFETIME OF POWER ELECTRONIC DEVICES 954
 Felix Hoffmann, Stefan Schmitt, Nando Kaminski

CURRENT CONTROL AND FPGA–BASED REAL–TIME SIMULATION OF GRID–TIED
INVERTERS 962
 Sabin Carpiuc, Matthias Schiesser, Carlos Villegas

IMPACT OF CONTROL LOOPS ON THE LOW-FREQUENCY PASSIVITY PROPERTIES OF GRID-FORMING CONVERTERS 969

Mebtu Beza, Massimo Bongiorno, Anant Narula

GRID IMPEDANCE ESTIMATION WITH OVERSAMPLING FOR GRID-CONNECTED CONVERTERS 979

Niklas Himker, Robin Strunk, Axel Mertens

LOW SPEED SENSORLESS CURRENT CONTROL FOR PMSM WITH SEARCH-BASED OBSERVER (SBO) 989

K. Scicluna, C. Spiteri Staines, R. Raute

INSIGHT INTO THE PECULIARITIES OF OPTIMIZED PULSE PATTERNS FOR PERMANENT-MAGNET SYNCHRONOUS MACHINES 998

Georgios Darivianakis, Ioannis Tsoumas

INVESTIGATING THE EFFECT OF DIFFERENT PARAMETERS ON HARMONICS AND EMI EMISSIONS AT THE FREQUENCY RANGE OF 0–9 KHZ 1006

Amir Ganjavi, Hansika Rathnayake, Firuz Zare, Dinesh Kumar, Amin Abbosh, Pooya Davari

FIVE-LEVEL NESTED INVERTER WITH NEUTRAL POINT CONNECTION 1016

Juhamatti Korhonen, Aleksi Mattsson, Heikki Järvisalo, Pertti Silventoinen, William Giewont, Dan Isaksson

ELECTRIC SPRING-BASED SMART WATER HEATER FOR LOW VOLTAGE MICROGRIDS 1025

Alexander Micallef, Racquel Ellul, John Licari

ENERGY-BALANCING OF A MODULAR MULTILEVEL CONVERTER USING AN ONLINE TRAJECTORY PLANNING ALGORITHM 1030

Qiuye Gui, Jan Lasse Gnärig, Hendrik Fehr, Albrecht Gensior

CAPACITOR SIZE COMPARISON ON HIGH-POWER DC-DC CONVERTERS WITH DIFFERENT TRANSFORMER WINDING CONFIGURATIONS ON THE AC-LINK 1040

Babak Khanzadeh, Torbjörn Thiringer, Yuhei Okazaki

DYNAMIC CHARACTERISTICS VERIFICATION OF LINEAR INDUCTION MOTOR BY SIMULTANEOUS PROPULSION AND LEVITATION CONTROL 1047

Shota Nakatani, Daichi Okamori, Toshimitsu Morizane, Hideki Omori

'IG,VGS' MONITORING FOR FAST AND ROBUST SIC MOSFET SHORT-CIRCUIT PROTECTION WITH HIGH INTEGRATION CAPABILITY 1057

Yazan Barazi, François Boige, Nicolas Rouger, Jean-Marc Blaquiere, Frédéric Richardeau

FAULT-TOLERANT CONTROL OF SERIES CONNECTABLE MODULAR FULL-BRIDGE INVERTER MITIGATING OPEN SWITCH FAULTS 1067

Juris Arrozy, Darian V. Retianza, Jorge L. Duarte, Henk Huisman

DESIGN AND CONTROL OF A MODULAR POWER ELECTRONIC BACK-TO-BACK CONVERTER FOR WAVE ENERGY HARVESTING APPLICATIONS 1076

Mattia Mantellini, Riccardo Morici, Marcos Blanco, Marcos Lafoz, Gustavo Navarro, Jorge Torres, Jorge Najera, Miguel Santos

INTELLIGENT HIGH CURRENT SENSOR FOR VARIOUS FREQUENCY 1086

Bohumil Skala, Vladimir Kindl, Pavel Turjanica, Ales Vobornik, Libor Polacek, Josef Stengl, Vladimir Pavlicek, Jiri Fort

FAIL-SAFE SWITCHING-CELLS ARCHITECTURES BASED ON MONOLITHIC ON-CHIP FUSE 1096
Amirouche Oumaziz, Emmanuel Sarraute, Frédéric Richardeau, Abdelhakim Bourennane

HOW GOOD ARE THE DESIGN TOOLS IN POWER ELECTRONICS? 1106
Thomas Lagier, Piotr Dworakowski, Laurent Chédot, François Wallart, Bruno Lefebvre, Jose Maneiro, Juan Páez, Philippe Ladoux, Cyril Buttay

ANALYSIS OF THE IMPACT OF MANUFACTURING DISSYMMETRY ON CURRENT DISTRIBUTION FOR MAGNETICALLY COUPLED INTERLEAVED INVERTERS 1118
Rita Mattar, Mickael Petit, Eric Monmasson, Stéphane Lefebvre, Christelle Saber, Cyrille Gautier, Marwan Ali

POWER FLOW CONTROL USING A BIDIRECTIONAL Z-SOURCE INVERTER–BASED STATIC SYNCHRONOUS SERIES COMPENSATOR 1128
Xuejiao Pan, Han Huang, Li Zhang

INVESTIGATION OF HARMONICS CONTENT IN PWM NATURAL AND REGULAR SAMPLING INCLUDING DEAD TIME AND LOAD CURRENT PHASE 1138
Tonny Wederberg Rasmussen, Anushruti Vashishtha, Ankit Jotwani

USING A WEB SCRAPING ALGORITHM FOR COMPONENT MODEL GENERATION IN MULTIOBJECTIVE OPTIMIZATION OF POWER ELECTRONIC APPLICATIONS 1148
Marcel Gladen, Volker Staudt

IMPACT ON THE ELECTRICAL CHARACTERISTICS, WAVEFORMS AND LOSSES OF THE ZERO-SEQUENCE INJECTION ON THE MODULAR MULTILEVEL CONVERTER 1158
Francois Gruson, Pierre Vermeersch, Philippe Delarue, Philippe Le Moigne, Frédéric Colas, Haibo Zhang, Moez Belhaouane, Xavier Guillaud

WIDE BANDWIDTH CURRENT SENSOR FOR COMMUTATION CURRENT MEASUREMENT IN FAST SWITCHING POWER ELECTRONICS 1168
Philipp Ziegler, Nathan Tröster, Dimitri Schmidt, Johannes Ruthardt, Manuel Fischer, Jörg Roth-Stielow

A SERIES–PARALLEL-TYPE RESONANT CIRCUIT WIRELESS POWER TRANSFER SYSTEM WITH A DUAL ACTIVE BRIDGE DC–DC CONVERTER 1177
Kohei Sugiyama, Taishi Kitamura, Shuto Uwai, Takahiro Yano, Yoshitaka Kawabata

STRAY VOLTAGE CAPTURE FOR ROBUST AND ULTRA-FAST SHORT CIRCUIT DETECTION IN POWER ELECTRONICS WITH HALF-BRIDGE STRUCTURE: THE LIMITATION AND IMPLEMENTATION 1186
Darian Verdy Retianza, Jeroen Van Duivenbode, Henk Huisman

ON THE INFLUENCE OF THE STATOR WINDING TOPOLOGY ON THE ELECTROMAGNETIC EMISSIONS OF FRACTIONAL HORSEPOWER BLDC MOTORS 1196
Felix Krall, Annette Muetze

IMPACT OF SILICON CARBIDE DEVICES IN 2 MW DFIG BASED WIND ENERGY SYSTEM 1205
Antxon Arrizabalaga, Aitor Idarreta, Mikel Mazuela, Iosu Aizpuru, Unai Iraola, José Luis Rodriguez, Daniel Labiano, Ibrahim Alisar

SMALL-SIGNAL STABILITY OF HVDC SYSTEM COMPRISING DC REACTORS 1215
Kosei Shinoda, Abdelkrim Benchaib, Jing Dai

MODEL PREDICTIVE CONTROL FOR THE REDUCTION OF DC-LINK CURRENT RIPPLE IN TWO-LEVEL THREE-PHASE VOLTAGE SOURCE INVERTERS 1224
Junzhong Xu, Fei Gao, Thiago Batista Soeiro, Linglin Chen, Luca Tarisciotti, Houjun Tang, Pavol Bauer

CARRIER-BASED MODULATED MODEL PREDICTIVE CONTROL FOR VIENNA RECTIFIERS 1233
Junzhong Xu, Fei Gao, Thiago Batista Soeiro, Linglin Chen, Luca Tarisciotti, Houjun Tang, Pavol Bauer

NEW HIGH-EFFICIENCY POWER GENERATION USING POSITION SENSOR-LESS PERMANENT MAGNET SYNCHRONOUS GENERATOR 1243
Somi Takeuchi, Hiroyuki Takahashi, Shota Yamada, Yoshitaka Kawabata

ACTIVE CLAMPING METHOD FOR SIC MOSFET HIGH POWER MODULES - BENEFITS AND LIMITS 1252
Robert W. Maier, Mark-M. Bakran

PREDICTIVE TORQUE CONTROL OF INDUCTION MACHINE WITH AN ADAPTIVE OBSERVER FOR TRAJECTORY PLANNING OF SERVO PRESS 1262
Qi Li, Jianbo Gao, Qiwu Wang, Ralph Kennel

FUTURE GRID STABILITY, A COST COMPARISON OF GRID-FORMING AND SYNCHRONOUS CONDENSER BASED SOLUTIONS 1270
Thibault Prevost, Guillaume Denis, Clementine Coujard

DEMONSTRATION OF THE SHORT-CIRCUIT RUGGEDNESS OF A 10 KV SILICON CARBIDE BIPOLAR JUNCTION TRANSISTOR 1279
Besar Asllani, Hervé Morel, Pascal Bevilacqua, Dominique Planson

LOSS MINIMIZATION OF TRACTION SYSTEMS IN BATTERY ELECTRIC VEHICLES USING VARIABLE DC-LINK VOLTAGE TECHNIQUE — EXPERIMENTAL STUDY 1289
Libo Liu, Boyang Li, Gunther Götting, Yusheng Xiang, Qusay Salem, Muhammad Hamid, Jian Xie

DIRECT MULTIVARIABLE CONTROL FOR MMC: DIGITAL SIGNAL PROCESSING AND EXPERIMENTAL RESULTS 1297
Daniel Dinkel, Claus Hillermeier, Rainer Marquardt

STATE OF CHARGE CONTROL FOR A FREQUENCY-SUPPORTING STORAGE SYSTEM BASED ON AN AUTO-REGRESSIVE FREQUENCY FORECAST 1306
A. Bolzoni, R. Todd, Q. Zhu, A. J. Forsyth

DESIGN OF A WIDE INPUT VOLTAGE RANGE CURRENT-FED DC/DC CONVERTER WITHIN A REDUCED DUTY-CYCLE RANGE 1316
Michael Gerstner, Martin Maerz, Armin Dietz

AN IMPROVED CONTROL STRATEGY FOR RENEWABLE ENERGY SOURCES (RES) BASED DC MICROGRID WITH ENHANCED SYSTEM STABILITY AND CONTROL PERFORMANCE 1326
Muhammad Adnan Mumtaz, Zheng Yan

TRANSIENT VOLTAGE DIP MITIGATION SYSTEM BASED ON HYBRID MODULAR MULTILEVEL CONVERTERS 1336
Manuel Colmenero, Francisco R. Blanquez, Karsten Kahle

A LOSS-COMPENSATED CONTROL SCHEME FOR SIC-BASED DUAL ACTIVE BRIDGE CONVERTER 1346

Ishan Pendharkar, Tobias Strittmatter, Paula Diaz Reigosa, Nicola Schulz

EXPERIMENTAL HYBRID AC/DC-MICROGRID PROTOTYPE FOR LABORATORY RESEARCH 1354

Enrique Espina, Claudio Burgos-Mellado, Juan S. Gomez, Jacqueline Llanos, Erwin Rute, Alex Navas F., Manuel Martínez-Gómez, Roberto Cárdenas, Doris Sácz

EXPERIMENTAL AND NUMERICAL CHARACTERIZATION OF PCB-EMBEDDED POWER DIES USING SOLDERLESS PRESSED METAL FOAM 1363

S. Bensebaa, M. Berkani, S. Lefebvre, M. Petit, N. Schmitt

FEASIBILITY STUDY OF A SUPERCONDUCTING POWER FILTER FOR HVDC GRIDS 1373

Loïc Quéval, Olivier Despouys, Frédéric Trillaud, Bruno Douine

POWER DECOUPLING METHOD OF DC TO SINGLE-PHASE AC CONVERTER USING FLYING CAPACITOR DC/DC CONVERTER WITH BOUNDARY CURRENT MODE 1380

Hiroki Watanabe, Keisuke Kusaka, Jun-Ichi Itoh

AN ARCHITECTURE FOR LEVEL-3 EV BATTERY CHARGER STATIONS USING INTEGRATED SOLID STATE TRANSFORMER (I-SST) 1390

Erick I. Pool-Mazun, Prasad Enjeti, Gerardo Escobar, Ira Pitel

LQR AND H-INFINITY CONTROL OF VOLTAGE SOURCE INVERTERS FOR AC MICROGRIDS 1400

Tenorio Jorge, Jose Miguel Ramirez Scarpetta, Fabio Andrade

FAMILY OF SPLITTING CURRENT SINGLE-LOOP CONTROL FOR *LCL*- TYPE GRID-CONNECTED INVERTER 1410

Yuying He, Xuehua Wang, Xinbo Ruan, Guoxing Su, Fuxin Liu

ANALYSIS AND DESIGN OF HIGH-POWER SINGLE-STAGE THREE-PHASE DIFFERENTIAL-BASED FLYBACK INVERTER FOR PHOTOVOLTAIC APPLICATIONS 1417

Ahmed Ismail M. Ali, Mahmoud A. Sayed, Takaharu Takeshita

INVESTIGATION OF IMPROVEMENT OF MODELING PRECISION FOR CONDUCTED NOISE ON ISOLATED AC/DC CONVERTER USING SIC DEVICES 1425

Kazuki Kuwana, Kohei Mitani, Wataru Kitagawa, Takaharu Takeshita

PASSIVITY-BASED DESIGN FOR THE PLUG-AND-PLAY SINGLE-LOOP CONTROLLED LCL-FILTERED INVERTER 1435

Yuying He, Xuehua Wang, Xinbo Ruan, Yixiao Ma, Fuxin Liu

CHARACTERISTICS OF AN INTEGRATED MOTOR CONTROLLED INDEPENDENTLY BY MULTI-INVERTERS TO ACHIEVE HIGH EFFICIENCY AND A WIDE SPEED RANGE 1442

Kazuto Sakai, Yano Hideaki

AN ISOLATED MEDIUM-VOLTAGE AC-DC CONVERTER USING LEVEL-SHIFTED PWM CONTROL OF A MODULAR MATRIX CONVERTER 1450

Kohei Budo, Takaharu Takeshita

DETAILED SIMULATION MODEL OF AN ASYMMETRICAL HALF-BRIDGE PWM CONVERTER WITH SYNCHRONOUS RECTIFICATION INCLUDING PARASITIC ELEMENTS 1460

Benedikt Kohlhepp, Valentin Zeller, Markus Barwig, Thomas Dürbaum

ELECTRICAL PROPERTY VARIABILITY OF GAN TRANSISTORS IN PARALLEL AND THEIR IMPACT ON FAST SWITCHING OPERATIONS 1470

Thilini Wickramasinghc, Bruno Allard, Réne Escofficr, Marc Plissonnicr

A COMPARISON BETWEEN DIFFERENT MODELS OF THE MODULAR MULTILEVEL CONVERTER 1479

Rafael Coelho-Medeiros, Bogdan Džonlaga, Jean-Claude Vannier, Jing Dai, Loic Queval, Philippe Egrot

PACKAGING TECHNOLOGY FOR THE IMPROVEMENT OF POWER CYCLING CAPABILITY OF HVIGBTS 1489

Kenji Hatori, Keiichi Nakamura, Nobuhiko Tanaka, Yasuhiro Sakai, Norikazu Sakai, Kenji Ota, Takeshi Higashihata, Eckhard Thal, Nils Soltau

A BIDIRECTIONAL DAB-LLC DCX TO ACHIEVE VOLTAGE REGULATION AND WIDE ZVS RANGE CAPABILITY 1498

Yuefeng Liao, Tao Peng, Mei Su, Yao Sun, Weijing Xiong, Guo Xu

SALIENCY SELECTION FOR SEARCH-BASED AC MACHINE LOW AND ZERO SPEED ESTIMATION METHODS 1506

K. Scicluna, C. Spiteri Staines, R. Raute

GENETIC ALGORITHM BASED MULTI OBJECTIVE OPTIMIZATION FOR INDUCTOR DESIGN 1515

Thorben Schobre, Raquel González Aríztegui, Regine Mallwitz

DIGITAL SMART DRIVER FOR SIC MOSFETS 1524

Nerea Arandia, José Ignacio Garate, Jon Mabe, Ander Ordoño

FASTER SWITCHING WITH LESS OVERVOLTAGE - OPERATING A SIC-MOSFET AT ITS SPEED LIMIT 1533

Pablo Rodriguez De Mora, Mark-M. Bakran

THE ENERGY RING TO SUPPLY THE EXPOELECTRIC'18 SHOW WITH RENEWABLE ENERGY SOURCES AND ELECTRIC VEHICLES 1542

Cristian Chillón-Antón, Daniel Heredero-Peris, Francesc Girbau-Llistuella, Paula González-Fontderubinat, Marc Llonch-Masachs, Daniel Montesinos-Miracle, Oriol Gomis-Bellmunt

IMPEDANCE-BASED MODELING OF A THREE-LEVEL CONVERTER UNDER BALANCED AND UNBALANCED CONDITION FOR THE STABILITY ANALYSIS OF BIPOLAR LVDC GRIDS 1551

T. Roose, G. Van Den Broeck, M. M. Alam, J. Beerten

LCL FILTER DESIGN FOR THREE PHASE AC-DC CONVERTERS CONSIDERING SEMICONDUCTOR MODULES AND MAGNETICS COMPONENTS PERFORMANCE 1561

Marco Stecca, Thiago Batista Soeiro, Laura Ramirez Elizondo, Pavol Bauer, Peter Palensky

SWITCHING BEHAVIOR AND COMPARISON OF 600V SMD WIDE BANDGAP POWER DEVICES 1569

Markus Meißner, Jan Schmitz, Steffen Bernet

ANALYSIS OF THE COUPLING BETWEEN THE OUTER AND INNER CONTROL LOOPS OF A GRID-FORMING VOLTAGE SOURCE CONVERTER 1579

T. Qoria, F. Gruson, F. Colas, X. Kestelyn, X. Guillaud

INFLUENCE OF DIFFERENT PULSE-WIDTH MODULATION METHODS ON MAGNET LOSSES IN PERMANENT MAGNET SYNCHRONOUS MACHINES 1589
Narciso G. Marmolejo, Xiaohu Tang, Martin Doppelbauer

RESONANT DC/DC CONVERTER WITH CLASS ϕ_2 INVERTER AND CLASS DE RECTIFIER BASED ON GAN HEMT 1599
Cai Si-Yuan, He Jun-Ping, Li Zi-Fan

FOUR-LEVEL INVERTER WITH VARIABLE VOLTAGE LEVELS FOR HARDWARE-IN-THE-LOOP EMULATION OF THREE-PHASE MACHINES 1605
Manuel Fischer, Johannes Ruthardt, Vasken Ketchedjian, Philipp Ziegler, Maximilian Nitzsche, Jörg Roth-Stielow

POWDER INJECTION MOLDING IN THE FABRICATION OF SOFT FERRITE MATERIAL FOR POWER ELECTRONICS 1613
J-S Ngoua-Teu, U. Soupremanien, P. Sallot, G. Delette, M. Bohnke

MODULATION SCHEME WITH COMMON MODE AND DIFFERENTIAL MODE VOLTAGE ELIMINATION FOR A FIVE LEVEL INVERTER FED OPEN END WINDING INDUCTION MOTOR DRIVE 1619
Greeshma Nadh, Durga Nair S., Arun Rahul S.

A FAST AND ROBUST MODEL OF DUAL-ACTIVE BRIDGE CONVERTERS IN REAL-TIME SIMULATION 1627
Ming Jia, Philipp Joebges, Rik W. De Doncker

DUAL INTERLEAVED 3.6 KW LLC CONVERTER OPERATING IN HALF-BRIDGE, FULL-BRIDGE AND PHASE-SHIFT MODE AS A SINGLE-STAGE ARCHITECTURE OF AN AUTOMOTIVE ON-BOARD DC-DC CONVERTER 1638
Philipp Rehlaender, Sergey Tikhonov, Frank Schafmeister, Joachim Bocker

SWITCHING LOSS ESTIMATION USING A VALIDATED MODEL OF 650 V GAN HEMTS 1648
Joao Oliveira, Florent Loiselay, Hervè Morel, Dominique Planson

REDUCTION OF CONDUCTION LOSSES IN RESONANT CONVERTERS BY CONNECTING THREE SINGLE-PHASE INVERTERS TO A COMMON GENERATOR 1658
Sergio Tárraga, John Paul Mayorga, Esther De Jódar, José Villarejo

COMPARISON OF DIFFERENT LOW VOLTAGE MULTILEVEL CONVERTER TOPOLOGIES FOR DISTRIBUTED POWER GENERATION 1666
Ingmar Kaiser, Hans-Günter Eckel

LOSS DISTRIBUTION COMPARISON OF VARIABLE AND FIXED INDUCTOR DAB CONVERTERS 1675
Erik Smailus, Gerd Griepentrog, Markus Pfeifer, Marcel Lutze

DESIGN BY OPTIMIZATION OF MULTIPHASE INVERTER FOR ELECTRIC VEHICLE DRIVE 1685
Nasreddine Kesbia, Jean-Luc Schanen, Hadi Alawieh, Lauric Garbuio, Yvan Avenas

OPTIMAL TORQUE/SPEED CHARACTERISTICS OF A FIVE-PHASE SYNCHRONOUS MACHINE UNDER PEAK OR RMS CURRENT CONTROL STRATEGIES 1693
Tiago José Dos Santos Moraes, Hailong Wu, Eric Semail, Ngac Ky Nguyen, Duc Tan Vu

COMPARATIVE STUDY OF TWO CONTROL TECHNIQUES OF REGENERATIVE
BRAKING POWER RECOVERING INVERTER BASED DC RAILWAY SUBSTATION 1700
Youssef Krim, Khaled Almaksour, Hervé Caron, Tony Letrouvé, Christophe Saudemont,
Bruno Francois, Benoit Robyns

JUNCTION TEMPERATURE CONTROL STRATEGY FOR LIFETIME EXTENSION OF
POWER SEMICONDUCTOR DEVICES 1709
Johannes Ruthardt, Hendrik Schulte, Philipp Ziegler, Manuel Fischer, Maximilian Nitzsche,
Jörg Roth-Stielow

HIGH DYNAMIC POWER BALANCING FOR DUAL TWO-LEVEL INVERTERS DURING
HIGH-SPEED MACHINE OPERATION 1718
Johannes Büdel, Johannes Teigelkötter, Alexander Stock, Christian Herkommer, Kai
Kuhlmann

CHARGING HIGH VOLTAGE CAPACITORS IN PULSED POWER APPLICATIONS WITH A
CAPACITOR DIODE VOLTAGE MULTIPLIER OF REDUCED SIZE AND LOWER RIPPLE
CURRENTS 1727
Tristan Weinert, Wolfgang Oberschelp, Günter Schröder

REVIEW OF OPTIMIZATION METHODS FOR THE DESIGN OF POWER ELECTRONICS
SYSTEMS 1737
Mylène Delhommais

A FLEXIBLE POWER CROSSBAR-BASED ARCHITECTURE FOR SOFTWARE-DEFINED
POWER DOMAINS 1747
Francesco Di Gregorio, Gilles Sassatelli, Abdoulaye Gamatié, Arnaud Castelltort

IMPACT OF GRID-FORMING CONTROL ON THE INTERNAL ENERGY OF A MODULAR
MULTILEVEL CONVERTER 1756
Ebrahim Rokrok, Taoufik Qoria, Antoine Bruyere, Bruno Francois, Haibo Zhang, Moez
Belhaouane, Xavier Guillaud

COMBINING MULTIPLE TEMPERATURE-SENSITIVE ELECTRICAL PARAMETERS
USING ARTIFICIAL NEURAL NETWORKS 1766
Daniel Herwig, Torben Brockhage, Axel Mertens

SINGLE-PHASE MEASUREMENT OF THE OUTPUT IMPEDANCE OF THE FOUR-
QUADRANT CASCADED H-BRIDGE CONVERTER CELL USING WIDEBAND SIGNALS 1776
Marko Petkovic, Dražen Dujic

A NOVEL THREE-PHASE PFC DIODE RECTIFIER BY LC NETWORK CIRCUITS FOR
HIGH FREQUENCY GENERATOR 1786
Shin-Ichi Motegi, Yasuyuki Nishida

FREQUENCY-DOMAIN SIMULATION OF POWER ELECTRONIC SYSTEMS BASED ON
MULTI-TOPOLOGY EQUIVALENT SOURCES MODELLING METHOD 1793
Stephane Vienot, Arnaud Videt, Nadir Idir, Lamine Kone, Sébastien Weiss, Frederic Lafon

MODULAR MULTILEVEL CONVERTER WITH DISTRIBUTED GALVANIC ISOLATION: A
DECENTRALIZED VOLTAGE BALANCING ALGORITHM WITH SMART GATE DRIVERS 1803
Darbas Corentin, Ginot Nicolas, Olivier Jean-Christophe, Poitiers Frédèric

COMPARISON AND OPTIMIZATION OF MAGNETICALLY COUPLED AND NON-
COUPLED MAGNETIC DEVICES IN INTERLEAVED OPERATION 1813
Peter Zacharias, Alejandro Aganza-Torres

EXPERIMENTAL TUNING AND DESIGN GUIDELINES OF A DYNAMICALLY
RECONFIGURED WEIGHTING FACTOR FOR THE PREDICTIVE TORQUE CONTROL OF
AN INDUCTION MOTOR 1823
Ilker Sahin, Ozan Keysan, Eric Monmasson

COMPENSATION OF TEMPERATURE DEPENDENCE IN A MODULE PARASITIC BASED
CURRENT MEASUREMENT SYSTEM 1831
Frank Lautner, Mark-M. Bakran

DEVELOPMENT AND IMPLEMENTATION OF A LOW-COST RESEARCH PLATFORM
FOR CONTROL APPLICATIONS FOR INVERTER-BASED GENERATORS 1841
Jesus D. Vasquez Plaza, Juan F. Patarroyo-Montenegro, Fabio Andrade

CONTROL OF PARALLEL CONNECTED VOLTAGE SOURCE INVERTERS IN A
MICROGRID FOR EXPERIMENTAL TESTING 1850
*Jesus D. Vasquez-Plaza, Jorge Tenorio, J. M. Ramírez-Scarpetta, Jose Alex Restrepo, Fabio
Andrade*

OPTIMIZATION STRATEGY FOR THE SIZING OF PASSIVE MAGNETIC COMPONENTS 1858
*Guillaume Devos, Maya Hage-Hassan, Philippe Dessante, Cyrille Gautier, Adrien Mercier,
Eric Labouré*

EXPLOITING A MULTI-PORT TRANSFORMER FOR MINIMAL DC-LINK CAPACITANCE
FOR AN AUTOMOTIVE ONBOARD CHARGER 1866
Franz Vollmaier, Alexander Connaughton, Thomas Langbauer, Klaus Krischan

DESIGN AND OPTIMIZATION OF HIGH-EFFICIENCY 1W 500V-12V ISOLATED LOW-
COST DC/DC CONVERTER 1874
Etienne Foray, Christian Martin, Bruno Allard

CHALLENGES IN CALIBRATING AN UNCONVENTIONAL PARTIAL DISCHARGE
MEASUREMENT SYSTEM FOR PULSED VOLTAGES 1885
Markus Fürst, Mark-M. Bakran

ELECTROTHERMAL MODELING OF GAN POWER TRANSISTOR FOR HIGH
FREQUENCY POWER CONVERTER DESIGN 1895
*Loris Pace, Florian Chevalier, Arnaud Videt, Nicolas Defrance, Nadir Idir, Jean-Claude De
Jaeger*

MODELING AND FAULT DETECTION IN PHOTOVOLTAIC SYSTEMS USING THE I-V
SIGNATURE 1905
*Abdelhadi Benzagmout, Thierry Talbert, Olivier Fruchier, Thierry Martire, Philippe
Alexandre, Carolina Penin*

EFFICIENCY REQUIREMENTS FOR PASSIVELY COOLED CONVERTERS WITH
THERMAL MEASUREMENT BASED 3D-FEM SIMULATION 1915
Julian Weimer, Dominik Koch, Maximilian Nitzsche, Matthias Zehelein, Ingmar Kallfass

GENERIC CONTROL LAW FOR DC AND AC MACHINES 1923
Pierre-Philippe Robet, Maxime Gautier, Yannick Aoustin

A HIGH PERFORMANCE 48-TO-8 V MULTI-RESONANT SWITCHED-CAPACITOR
CONVERTER FOR DATA CENTER APPLICATIONS 1934
Rose A. Abramson, Zichao Ye, Robert C. N. Pilawa-Podgurski

SISO CONTROL STRATEGY OF RESONANT DUAL ACTIVE BRIDGE WITH A TUNED CLC NETWORK 1944

Meiqi Wang, Bo Yang, Lie Xu, Jing Li, David Gerada, Chunyang Gu, He Zhang, Chris Gerada, Yongdong Li

IMPACT OF STEADY-STATE GRID-FREQUENCY DEVIATIONS ON THE PERFORMANCE OF GRID-FORMING CONVERTER CONTROL STRATEGIES 1952

Anant Narula, Massimo Bongiorno, Mebtu Beza, Jan R Svensson, Xavier Guillaud, Lennart Harnefors

A GENERAL METHOD TO DAMP WIND TURBINE SSR WITH DIFFERENT TRANSMISSION SYSTEMS 1962

Ignacio Vieto, Jian Sun

A TEST SCHEME FOR THE COMPREHENSIVE QUALIFICATION OF MMC SUBMODULE BASED ON 10 KV SIC MOSFETS UNDER HIGH DV/DT 1972

Xingxuan Huang, Shiqi Ji, Dingrui Li, Cheng Nie, William Giewont, Leon M. Tolbert, Fred Wang

PWM GAIN LINEARIZATION ALGORITHM FOR MEDIUM VOLTAGE SOURCE INVERTER 1982

Hamza El Jihad, Sami Siala, Elise Savarit

AUTO-COMMISSIONING OF ACOUSTIC CONTROL OF IM DRIVE USING BAYESIAN OPTIMIZATION 1992

Michal Kroneisl, Václav Šmídl

EXPERIMENTAL EMI STUDY OF A 3-PHASE 100KW 1200V DUAL ACTIVE BRIDGE CONVERTER USING SIC MOSFETS 2000

Hadiseh Geramirad, Florent Morel, Piotr Dworakowski, Philippe Camail, Bruno Lefebvre, Thomas Lagier, Christian Vollaire

MODELING OF A DAB UNDER PHASE-SHIFT MODULATION FOR DESIGN AND DM INPUT CURRENT FILTER OPTIMIZATION 2010

Glauber De Freitas Lima, Yves Lembeye, Fabien Ndagijimana, Jean-Christophe Crebier

ACTIVE CURRENT AND ENERGY CONTROL FOR THE QUASI-THREE-LEVEL OPERATION MODE OF AN EXTENDED MODULAR MULTILEVEL CONVERTER TOPOLOGY 2020

Malte Lorenz, Jakub Kucka, Axel Mertens

TORQUE RIPPLE REDUCTION TECHNIQUE FOR A SWITCHED RELUCTANCE MOTOR 2029

Krzysztof Jackiewicz, Arkadiusz Kaszewski, Andrzej Stras, Bartlomiej Ufnalski, Tomasz Balkowiec

EXPERIMENTAL VALIDATION OF THE PERFORMANCES OF AN INVERTER SIZED WITH OPTIMIZATION METHODS 2039

Adrien Voldoire, Jean-Luc Schanen, Jean-Paul Ferrieux, Alexis Derbey, Cyrille Gautier, Marwan Ali

INFLUENCE OF SYSTEM PARAMETERS IN VARIABLE SPEED AC-INDUCTION MOTOR DRIVES ON PARASITIC ELECTRIC BEARING CURRENTS 2049

Martin Weicker, Guilherme Bello, Dennis Kampen, Andreas Binder

PLASMA IMPACT ON OVERVOLTAGE SHORT-CIRCUIT FAILURES IN ANPC CONVERTERS 2059

David Hammes, Sidney Gierschner, Dietmar Krug, Hans-Günter Eckel

NOVEL SOFT-SWITCHING INTERLEAVED BOOST CONVERTERS FOR RENEWABLE ENERGY CONVERSION SYSTEMS 2068

Madhuchandra Popuri, V. V. Subrahmanya Kumar Bhajana, Pavel Drabek, Manoj Kumar Maharana

POWER DENSITY OF PLANAR TRANSFORMERS DESIGNED WITH COMMERCIAL STANDARD CORES 2078

Reda Bakri, Xavier Margueron, Jean Sylvio Ngoua Teu Magambo, Philippe Le Moigne, Nadir Idir

EFFECTS OF PV PANEL AND BATTERY DEGRADATION ON PV-BATTERY SYSTEM PERFORMANCE AND ECONOMIC PROFITABILITY 2088

Monika Sandelic, Ariya Sangwongwanich, Frede Blaabjerg

FULL SENSORLESS OPERATION OF INDUCTION MACHINES BASED ON ONLINE IDENTIFICATION OF SALIENCIES USING HARMONIC COMPENSATION LUTS IN TRACTION APPLICATIONS 2098

E. Rodriguez Montero, M. Vogelsberger, T. Wolbank

MITIGATING DRAIN SOURCE VOLTAGE OSCILLATION WITH LOW SWITCHING LOSSES FOR SIC POWER MOSFETS USING FPGA-CONTROLLED ACTIVE GATE DRIVER 2106

Zheming Li, Robert W. Maier, Mark-M. Bakran

ONLINE TRAJECTORY PLANNING DURING LOW-VOLTAGE FRT OF A MODULAR MULTILEVEL CONVERTER 2116

Hendrik Fehr, Albrecht Gensior

EVALUATING FREQUENCY STABILITY WITH CONSIDERATION OF LOAD TYPE IN DIFFERENT SHARE OF RENEWABLES AND EMULATED INERTIA IN CASE OF SYSTEM SPLIT 2126

Nastaran Fazli, Sidney Gierschner, Hans-Günter Eckel

DISCRETE-TIME DIRECT POLE PLACEMENT FOR STABILITY ENHANCEMENT OF LCL-FILTERED INVERTERS IN THE SYNCHRONOUS-REFERENCE FRAME 2135

Pei Cai, Xiaohua Wu, Yongheng Yang, Wenli Yao, Weilin Li, Frede Blaabjerg

ON THE SWITCHING-INDUCED DC-LINK VOLTAGE RIPPLE IN THREE-LEVEL CONVERTERS WITH A NEUTRAL POINT 2145

Ioannis Tsoumas, Tobias Geyer

EFFECT OF PASSIVE INVERTER OUTPUT MOTOR FILTERS ON DRIVE SYSTEMS 2153

Dennis Kampen, Martin Weicker

IMPACT OF THE NEUTRAL POINT POTENTIAL RIPPLE ON THE GRID SIDE HARMONICS OF A 3LNPC BACK-TO-BACK CONVERTER EMPLOYED IN A MEDIUM VOLTAGE WECS 2163

Ioannis Tsoumas

TWO-LAYER GENETIC ALGORITHM FOR THE CHARGE SCHEDULING OF ELECTRIC VEHICLES 2172

Nikolaos T. Milas, Dimitris A. Mourtzis, Panagiotis I. Giotakos, Emmanuel C. Tatakis

SIX-PHASE PMSM DRIVE INVERTER TESTING ON A HIGH PERFORMANCE POWER HARDWARE-IN-THE-LOOP TESTBED 2182

Yasser Rahmoun, Patrick Winzer, Alexander Schmitt, Horst Hammerer

AN IMPROVED BIDIRECTIONAL HYBRID SWITCHED INDUCTOR CONVERTER........................ 2192

Dan Hulea, Mihaita Gireada, Danut Vitan, Octavian Cornea, Nicolae Muntean

HYBRID MULTIPLE CHOPPER CELLS OF PWM AND SQUARE-WAVE OPERATION FOR
SOLID-STATE TRANSFORMER 2200

Naoto Kikuchi, Jun-Ichi Itoh, Keisuke Kusaka, Hoai Nam Le

A NEW ZVS ZONE IDENTIFICATION FOR DUAL ACTIVE BRIDGE WITH A GENERAL
MODULATION OBJECTIVE.. 2210

Suman Maharana, Dipankar De, Alberto Castellazzi

SINGLE-STAGE BOOST MODULAR MULTILEVEL CONVERTER (BMMC) FOR ENERGY
STORAGE INTERFACE.. 2220

Ahmed Abdelhakim, Frede Blaabjerg, Hans-Peter Nee

LOW VOLTAGE GAN-BASED GATE DRIVER TO INCREASE SWITCHING SPEED OF
PARALLELED 650 V E-MODE GAN HEMTS 2230

Raffael Risch, Jürgen Biela

GATE STRESSES AND THRESHOLD VOLTAGE INSTABILITY IN NORMALLY-OFF GAN
HEMTS 2241

Jose Ortiz Gonzalez, Burhan Etoz, Olayiwola Alatise

NEW ENERGY MANAGEMENT ALGORITHM BASED ON FILTERING FOR ELECTRICAL
LOSSES MINIMIZATION IN BATTERY-ULTRACAPACITOR ELECTRIC VEHICLES 2251

*Bakou Traoré, Moustapha Doumiati, Cristina Morel, Jean-Christophe Olivier, Ousmane
Soumaoro*

MECHANISTIC POWER MODULE DEGRADATION MODELLING CONCEPT WITH
FEEDBACK.. 2258

Martin Bendix Fogsgaard, Paula Diaz Reigosa, Francesco Iannuzzo, Michael Hartmann

EXPERIMENTAL VALIDATION AND COMPARISON OF A SIC MOSFET BASED 100 KW
1.2 KV 20 KHZ THREE-PHASE DUAL ACTIVE BRIDGE CONVERTER USING TWO
VECTOR GROUPS 2265

*Thomas Lagier, Piotr Dworakowski, Cyril Buttay, Philippe Ladoux, Andrzej Wilk, Philippe
Camail, Elissa Cresenta Anak Justin*

IMPEDANCE ANALYSIS OF AN AUTOMOTIVE DC BUS.. 2274

Michael Schlüter, Marius Gentejohann, Sibylle Dieckerhoff

A NEW DUAL-MODE MPPT ALGORITHM APPLIED TO A QUADRATIC CONVERTER IN
A SOLAR ENERGY SYSTEM 2284

Ahmad Ghamrawi, Jean-Paul Gaubert, Driss Mehdi

THERMAL MODEL DEVELOPMENT FOR SIC MOSFETS ROBUSTNESS ANALYSIS
UNDER REPETITIVE SHORT CIRCUIT TESTS 2293

M. Pulvirenti, D. Cavallaro, N. Bentivegna, S. Cascino, E. Zanetti, M. Saggio

COMPENSATION OF THE RADIAL AND CIRCUMFERENTIAL MODE 0 VIBRATION OF A
PERMANENT MAGNET ELECTRIC MACHINE BASED ON AN EXPERIMENTAL
CHARACTERISATION... 2303

Jan Andresen, Stephan Vip, Axel Mertens, Sebastian Paulus

MEASUREMENT BASED MODEL FOR THE CALCULATION OF CURRENT DISTRIBUTIONS BETWEEN PARALLELED POWER SEMICONDUCTORS DURING HIGH CURRENT OPERATION 2312
Julian Da Cunha

DUAL-LOOP CONTROL SCHEME WITH OPTIMIZED TYPE-III CONTROLLER BASED ON GENETIC ALGORITHM FOR 6-PHASE INTERLEAVED CONVERTER IN ELECTRIC VEHICLE DRIVETRAINS 2320
Dai-Duong Tran, Sajib Chakraborty, Thomas Geury, Joeri Van Mierlo, Mohamed El Baghdadi, Omar Hegazy

HIGH SENSITIVITY CURRENT TRANSFORMER WITH LOW SETTLING TIME, FOR MAGNIFIED AC CURRENT MEASUREMENTS IN PULSED APPLICATIONS 2331
Georgios Tsolaridis, Pascal Seiler, Juergen Biela

LOSS SEPARATION IN HARD- AND SOFT-SWITCHING GAN HEMTS OPERATED IN A 10 KW ISOLATED DC/DC CONVERTER 2341
Jan Böcker, Sören Heucke, Sibylle Dieckerhoff

A SWITCHED-MODE POWER AMPLIFIER FOR ION ENERGY CONTROL IN PLASMA ETCHING 2350
Qihao Yu, Erik Lemmen, Korneel Wijnands, Bas Vermulst

EXPLORING THE BOUNDARIES AND EFFECTS OF THE DISCONTINUOUS CONDUCTION MODE IN H-BRIDGE INVERTER WITH DEAD-TIME 2358
Qihao Yu, Erik Lemmen, Korneel Wijnands, Bas Vermulst

FIGURES-OF-MERIT AND CURRENT METRIC FOR THE COMPARISON OF IGCTS AND IGBTS IN MODULAR MULTILEVEL CONVERTERS 2366
Arthur Boutry, Cyril Buttay, Dong Dong, Rolando Burgos, Bruno Lefebvre, Florent Morel, Colin Davidson

ZERO-CURRENT SWITCHING WITH LC RESONANT TANK CIRCUIT AND CAPACITOR ISOLATION DC-DC CONVERTER 2376
Hideki Jonokuchi, Osamu Nakashima, Daichi Hiwatari, Hiroshi Hirayama

A FULL STATE-VARIABLE PREDICTIVE CONTROL OF BI-DIRECTIONAL BOOST CONVERTERS WITH GUARANTEED STABILITY 2386
Yu Li, Zhenbin Zhang, Ralph Kennel

SYSTEM-LEVEL RELIABILITY ANALYSIS OF A REPAIRABLE POWER ELECTRONIC-BASED POWER SYSTEM CONSIDERING NON-CONSTANT FAILURE RATES 2393
Amirali Davoodi, Yongheng Yang, Tomislav Dragicevic, Frede Blaabjerg

AN EFFICIENCY ANALYSIS OF A FERRITE MAGNET ASSISTED SYNCHRONOUS RELUCTANCE MACHINE FOR LOW POWER DRIVES INCLUDING FLUX WEAKENING 2403
Matthias Hofer, Mario Nikowitz, Thomas Kirowitz, Manfred Schrödl

HIGH PERFORMANCE LQR CONTROL OF MODULAR MULTILEVEL CONVERTERS WITH SIMPLE CONTROL STRUCTURE AND IMPLEMENTATION 2409
Min Jeong, Simon Fuchs, Jürgen Biela

FAULT DETECTION AND CLASSIFICATION BASED ON DEEP LEARNING IN LVDC OFF-GRID SYSTEM 2419
Iurii Demidov, Antti Pinomaa, Andrey Lana, Olli Pyrhönen

AN INPUT-SERIES OUTPUT-INDEPENDENT FULL-BRIDGE DUAL ACTIVE BRIDGE CONVERTER WITH SOFT-SWITCHING CHARACTERISTICS FOR CHARGING AND BALANCING ELECTRIC VEHICLE BATTERY STACKS 2429
Alex V. Mirtchev, Emmanuel C. Tatakis

A METHOD TO SEARCH GLOBAL MAXIMA BY PERMANENT MONITORING OF VOLTAGE AND CURRENT OF EACH PV PANEL 2439
Shailendra Rajput, Moshe Averbukh

SURVEY AND COMPARISON OF 1D/2D ANALYTICAL MODELS OF HF LOSSES IN LITZ WIRE.. 2446
Qingchao Meng, Jürgen Biela

HIGH-FREQUENCY SIC-BASED MEDIUM VOLTAGE QUASI-2-LEVEL FLYING CAPACITOR DC/DC CONVERTER WITH ZERO VOLTAGE SWITCHING.......................... 2457
Rafal Kopacz, Przemyslaw Trochimiuk, Grzegorz Wrona, Jacek Rabkowski

SMART FUEL CELL MODULE (6.5 KW) FOR A RANGE EXTENDER APPLICATION 2467
Pascal Bazin, Bruno Beranger, Jacques Ecrabey, Laurent Garnier, Sylvain Mercier

IMPACT OF THE INITIAL TRANSIENT INTERRUPTION VOLTAGE (ITIV) ON THE DESIGN AND OPERATION OF HYBRID CURRENT-INJECTION DC CIRCUIT BREAKERS............ 2475
Andreas Jehle, Jürgen Biela

FOUR QUADRANT BUS-TIE SWITCH FOR PROTECTION OF SHIPBOARD POWER SYSTEMS 2486
Gabriele Ulissi, Seong-Yong Lee, Drazen Dujic

ESTIMATION OF AN UNBALANCED GRID IMPEDANCE USING A THREE-PHASE POWER CONVERTER 2495
Jarno Kukkola, Ville Pirsto, Mikko Routimo, Marko Hinkkanen

FAULT DIAGNOSIS OF HVDC TRANSMISSION SYSTEM USING WAVELET ENERGY ENTROPY AND THE WAVELET NEURAL NETWORK 2505
Cuicui Liu, Feng Wang, Fang Zhuo, Ziqian Zhang

REDUCING THE ENERGY STORAGE REQUIREMENTS OF MODULAR MULTILEVEL CONVERTERS WITH OPTIMAL CAPACITOR VOLTAGE TRAJECTORY SHAPING...................... 2513
Simon Fuchs, Min Jeong, Jürgen Biela

LEAKAGE INDUCTANCE MODELLING OF TRANSFORMERS: ACCURATE AND FAST MODELS TO SCALE THE LEAKAGE INDUCTANCE PER UNIT LENGTH... 2524
Richard Schlesinger, Jürgen Biela

A GAN-BASED DC/DC CONVERTER FOR E-VEHICLES APPLICATIONS 2535
Eduardo F. De Oliveira, Sebastian Sprunck, Jonas Pfeiffer, Peter Zacharias

THEORY OF INFLUENCING THE BREATHING MODE AND TORQUE PULSATIONS OF PERMANENT MAGNET ELECTRIC MACHINES WITH HARMONIC CURRENTS 2545
Jan Andresen, Stephan Vip, Axel Mertens, Sebastian Paulus

POWER HARDWARE IN THE LOOP SYSTEM BASED ON INTERLEAVED CONVERTER AND FPGA - APPLICATION TO DC AND AC SIDE EMULATION FOR PHOTOVOLTAIC INVERTER TESTING.. 2554
R. Kadri, R. Bakri, A. Omrane, F. Colas, F. Delpech

IMPLEMENTATION OF TAPIR SWITCHING CELLS WITH INTEGRATED DIRECT AIR-COOLING FOR SIC POWER DEVICES 2564
Wendpanga Fadel Bikinga, Kouceila Alkama, Bachir Mezrag, Jean Michel Guichon, Yvan Avenas

EFFECT OF UNIPOLAR AND BIPOLAR SPWM ON THE LIFETIME OF DC-LINK CAPACITORS IN SINGLE-PHASE VOLTAGE SOURCE INVERTERS 2573
Silpa Baburajan, Saeed Peyghami, Dinesh Kumar, Frede Blaabjerg, Pooya Davari

TRANSIENT THERMAL MODELS OF CAPACITORS AND INDUCTORS FOR SYSTEM OPTIMIZATION 2583
Vasilios Karaventzas, Juergen Biela, Felix Rodriguez Mateos

ENERGY MANAGEMENT FOR ISOLATED RENEWABLE-POWERED MICROGRIDS USING REINFORCEMENT LEARNING AND GAME THEORY 2594
Rui Hu, Alexis Kwasinski

ALL-GAN BIDIRECTIONAL ANPC-BASED RESONANT DC-DC CONVERTER 2603
Tino Kahl, Laurenz Wernicke, Sibylle Dieckerhoff, Christopher Fromme, Marvin Tannhäuser, Ag Siemens

LIFETIME ESTIMATION AND DIMENSIONING OF THE MACHINE-SIDE CONVERTER FOR PUMPING-CYCLE AIRBORNE WIND ENERGY SYSTEM 2613
Bakr Bagaber, Patrick Junge, Axel Mertens

A DESIGN OF HIGH-POWER INVERTER CIRCUIT INCLUDING GAN POWER DEVICES 2623
Takashi Sawada, Hiroshi Tadano, Koji Shiozaki

SPEED SENSORLESS COMMISSIONING OF RESONATING MECHANICAL SYSTEM IN ELECTRIC DRIVES 2630
A. Putkonen, N. Nevaranta, O. Liukkonen, M. Niemelä, O. Pyrhönen

CONTROL OF A TWO-STAGE, SINGLE-PHASE GRID-TIED, GAN BASED SOLAR MICRO-INVERTER 2638
Anthony Bier, Van Sang Nguyen, Stéphane Catellani, Jérémy Martin

A DC/DC BUCK-BOOST CONVERTER CONTROL USING SLIDING SURFACE MODE CONTROLLER AND ADAPTIVE PID CONTROLLER 2648
Bassem Saleh, Ahmed Teirelbar, Amr Wasfi

SENSORLESS NEUTRAL POINT VOLTAGE STABILIZATION IN THREE-PHASE FOUR-WIRE CONVERTERS 2656
Xinwei Xu, Gabriel Tibola, Jorge L. Duarte

BIDIRECTIONAL ISOLATED RIPPLE CANCEL TRIPLE ACTIVE BRIDGE DC-DC CONVERTER 2666
Takahiro Ohta, Pin-Yu Huang, Yuichi Kado

DESIGN OF THE SPEED SENSORLESS FIELD ORIENTED CONTROL SYSTEM FOR INDUCTION MOTORS CONSIDERING SUDDEN CHANGE OF THE ROTOR SPEED 2675
Yoshiki Sakurazawa, Osamu Yamazaki, Kazuaki Yuki, Yosuke Nakazawa, Kenji Natori, Keiichiro Kondo

EFFICIENCY POTENTIAL OF SOLID-STATE PULSE MODULATORS USING SIC DEVICES 2684
Spyridon Stathis, Michael Jaritz, Sebastian Blume, Jürgen Biela

EFFICIENT AND SCALABLE POWER CONTROL IN MULTI-PORT ACTIVE-BRIDGE CONVERTERS 2695

Soleiman Galeshi, David Frey, Yves Lembeye

COMPARISON OF PRESS-PACK AND WIRE-BONDING TECHNOLOGIES FOR SIC MOSFETS UNDER SHORT-CIRCUIT CONDITIONS 2704

Ran Yao, Francesco Iannuzzo, Amir Sajjad Bahman, Hui Li

ERROR INDUCED BY THE OPTICAL PATH OF A HIGH ACCURACY AND HIGH BANDWIDTH OPTICAL CURRENT MEASUREMENT SYSTEM 2712

Stefan Rietmann, Jürgen Biela

ANALYSIS OF THE RMS CURRENT STRESS ON THE DC LINK CAPACITORS OF THE FOUR PHASE 3-LEVEL T-TYPE VOLTAGE SOURCE CONVERTER 2723

Zoran Miletic, Werner Tremmel, Roland Bründlinger, Johannes Stöckl, Petar J. Grbovic

AN ADAPTIVE DROOP CONTROL METHOD FOR INTERLINK CONVERTER IN HYBRID AC/DC MICROGRIDS 2733

Mohammad S. Golsorkhi, Rasool Heydari, Mehdi Savaghebi

SIMPLIFIED CALCULATION OF PARASITIC ELEMENTS AND MUTUAL COUPLINGS OF WIDE-BANDGAP POWER SEMICONDUCTOR MODULES 2743

Mohammad Ali, Jens Friebe, Axel Mertens

VARIABLE-SPEED-DRIVE-BASED SENSORLESS ESTIMATION OF PUMP SYSTEM RESERVOIR FLUID LEVEL 2753

Santeri Pöyhönen, Aleksi Simola, Jero Ahola

ANALYSIS OF SWITCHING PERFORMANCE AND EMI EMISSION OF SIC INVERTERS UNDER THE INFLUENCE OF PARASITIC ELEMENTS AND MUTUAL COUPLINGS OF THE POWER MODULES 2763

Mohammad Ali, Jan-Kaspar Müller, Jens Friebe, Axel Mertens

WIRE-WOUND MULTI-PHASE STATOR BASED EMEH WITH MPPT SELF-POWERED ENERGY MANAGEMENT SYSTEM 2773

Mahmoud Shousha, Dragan Dinulovic, Talha Zafar, Michael Brooks, Martin Haug

COMPARISON OF OPTIMIZED MOTOR-INVERTER SYSTEMS USING A STACKED POLYPHASE BRIDGE CONVERTER COMBINED WITH A 3-, 6-, 9-, OR 12-PHASE PMSM 2780

Thilo Bringezu, Jürgen Biela

DESIGN OF A PULSE MODULATOR BASED ON TRANSMISSION LINES FOR GENERATING FAST CURRENT PULSES FOR PLASMA DRILLING 2791

Oliver Keel, Melissa Artiglia, Juergen Biela

ANALYSIS OF CURRENT IN PULSATING DC LINK CONVERTER WITH ZERO VOLTAGE TRANSITION 2802

Daniele Marciano, Giovanni Busatto, Carmine Abbate, Annunziata Sanseverino, Davide Tedesco, Francesco Velardi

SIGNAL INJECTION FOR SENSORLESS CURRENT SHARING WITH EXPERIMENTAL VERIFICATION ON 1 MHZ GAN PROTOTYPE 2812

N. Boškovic, J. Duarte, E. A. Lomonova

MODELLING AND ANALYSIS OF SENSORLESS CURRENT SHARING APPROACH 2820

N. Boškovic, J. Duarte

PWM-INDUCED HARMONIC POWER IN 75 KW IM DRIVE SYSTEM 2829
Lassi Aarniovuori, Hannu Kärkkäinen, Markku Niemelä, Juha Pyrhönen

PROPOSAL OF BOOST CONVERTER WITHOUT REACTOR USING OPEN-ENDED
WINDING PMSM FOR PHOTOVOLTAIC PUMP SYSTEM.. 2838
Akihiro Okazaki, Sari Maekawa

THE PROPOSAL OF DISCRIMINATING STABLE CONTROL BANDWIDTH USING ANN IN
SENSORLESS SPEED CONTROL SYSTEM FOR PMSM.. 2844
Ami Tanaka, Sari Maekawa

COST FUNCTION DESIGN FOR STABILITY ASSESSMENT OF MODULATED MODEL
PREDICTIVE CONTROL 2851
*Jordan P. Zucuni, Fernanda Carnielutti, Humberto Pinheiro, Margarita Norambuena, Jose
Rodriguez*

A ROBUST FUZZY-BASED CONTROL TECHNIQUE FOR WIND FARM TRANSIENT
VOLTAGE STABILITY USING SVC AND STATCOM: COMPARISON STUDY 2860
*Reza Ebrahimi, Vahid Eslampanah, Hossein Madadi Kojabadi, Mohammadreza Azizian,
Naser Nourani Esfetanaj, Dao Zhou*

TEMPERATURE EVOLUTION AS AN EFFECT OF WIRE-BOND FAILURES IN A MULTI-
CHIP IGBT POWER MODULE.. 2865
N. Degrenne, R. Delamea, S. Mollov

COST OF ENERGY ASSESSMENT OF WIND TURBINE CONFIGURATIONS 2873
Catalin Dincan, Philip Kjær, Lars Helle

ENERGY MANAGEMENT IN A MULTI-SOURCE SYSTEM USING ISOLATED DC-DC
RESONANT CONVERTERS.. 2881
M. Arazi, A. Payman, M. B. Camara, B. Dakyo

LONG-TERM CLIMATE IMPACT ON IGBT LIFETIME... 2888
Martin Vang Kjaer, Yongheng Yang, Huai Wang, Frede Blaabjerg

COMMUNICATION-FREE SECONDARY FREQUENCY AND VOLTAGE CONTROL OF
VSC-BASED MICROGRIDS: A HIGH-BANDWIDTH APPROACH 2898
*Rasool Heydari, Mohammad S. Golsorkhi, Mehdi Savaghebi, Tomislav Dragicevic, Frede
Blaabjerg*

OFFSHORE WIND FARM LAYOUT OPTIMIZATION CONSIDERING WAKE EFFECTS 2907
Asma Dabbabi, Salvy Bourguet, Rodica Loisel, Mohamed Machmoum

SMALL-SIGNAL STABILITY ANALYSIS OF SMART GRIDS CONSIDERING HIGH
PENETRATION OF POWER ELECTRONICS CONVERTERS AND ENERGY MARKETS 2917
Javiera Meneses, Patricio Mendoza-Araya

COMPONENT-LEVEL RELIABILITY ASSESSMENT OF A DIRECT-DRIVE PMSG WIND
POWER CONVERTER CONSIDERING LONG-TERM AND SHORT-TERM THERMAL
CYCLES.. 2928
Shuaichen Ye, Dao Zhou, Frede Blaabjerg

A SUBMODULE IMPLEMENTATION FOR PARALLEL CONDUCTION OF DIODES IN
MODULAR MULTILEVEL CONVERTERS... 2938
Martin Geske, Duro Basic, Christian Keller, Thomas Brückner

EVALUATION OF THE I_{MAX}-F_{SW}-DV/DT TRADE-OFF OF HIGH VOLTAGE SIC MOSFETS BASED ON AN ANALYTICAL SWITCHING LOSS MODEL 2946

Anliang Hu, Jürgen Biela

PROTECTION MEASURES FOR MODULAR MULTILEVEL CONVERTERS IN CASE OF DC SHORT-CIRCUIT FAULTS 2957

Martin Geske, Duro Basic, Roland Jakob, Christian Keller, Thomas Brückner

INVESTIGATION ON PARALLEL OPERATION OF TWO MMC-HVDC LINKS IN GRID FORMING CONNECTED TO AN EXISTING NETWORK 2967

H. Saad, P. Rault, S. Dennetière

MODELLING AND EXPERIMENTAL VALIDATION OF A LABORATORY-SCALED HVDC CABLE EMULATOR TESTED IN AN MMC-BASED PLATFORM 2977

Enric Sánchez-Sánchez, Adrià Junyent-Ferré, Eduardo Prieto-Araujo, Oriol Gomis-Bellmunt, Tim Green

DAISY CHAIN PN CELL FOR MULTILEVEL CONVERTER USING GAN FOR HIGH POWER DENSITY 2987

Faheem Ahmad, Asger Bjørn Jørgensen, Szymon Michal Beczkowski, Stig Munk-Nielsen

GRID-FREQUENCY VIENNA RECTIFIER AND ISOLATED CURRENT-SOURCE DC-DC CONVERTERS FOR EFFICIENT OFF-BOARD CHARGING OF ELECTRIC VEHICLES 2996

Jacek Rabkowski, Andrei Blinov, Denys Zinchenko, Grzegorz Wrona, Mariusz Zdanowski

UNIDIRECTIONAL THYRISTOR-BASED DC-DC CONVERTER FOR HVDC CONNECTION OF OFFSHORE WIND FARMS 3006

Pierre Le Métayer, Piotr Dworakowski, Jose Maneiro

INDUCTOR SIZE EVALUATION OF AN ELECTROMAGNETIC INTERFERENCE FILTER FOR A TWO-LEVEL POWER FACTOR CORRECTION RECTIFIER USING DIFFERENT MODULATION TECHNIQUES 3015

Mohammad Najjar, Alireza Kouchaki, Morten Nymand

EVALUATION OF MMCS FOR HIGH-POWER LOW-VOLTAGE DC-APPLICATIONS IN COMBINATION WITH THE MODULE LLC-DESIGN 3024

Roland Unruh, Frank Schafmeister, Joachim Böcker

IRON LOSS CHARACTERISTICS OF MNZN FERRITES UNDER GAN INVERTER EXCITATION IN THE MHZ ORDER 3034

Wilmar Martinez, Camilo Suarez, Federico Ibanez

VIBRATION SUPPRESSION AND CONTROL PARAMETER DESIGN OF A SENSORLESS PMSM ROTARY COMPRESSOR DRIVE 3044

Tao Li, Chaohui Liang

3D PCB PACKAGE FOR GAN INVERTER LEG WITH LOW EMC FEATURE 3054

Pawel B. Derkacz, Jean-Luc Schanen, Pierre-Olivier Jeannin, Piotr Musznicki, Piotr J. Chrzan, Mickael Petit

ESTIMATION OF THE WINDING LOSSES OF MEDIUM FREQUENCY TRANSFORMERS WITH LITZ WIRE USING AN EQUIVALENT PERMEABILITY AND CONDUCTIVITY METHOD 3064

Mohammad Kharezy, Morteza Eslamian, Torbjörn Thiringer

IMPROVEMENT OF DRIVING EFFICIENCY OF PMSM BY USING MODIFIED TRAPEZOIDAL MODULATING SIGNAL 3071
Kento Betto, Satoshi Joryo, Toshimitsu Morizane

DESIGN AND CONTROL OF A VIRTUAL DC-LINK FOR A FULL GAN-BASED SINGLE PHASE CONVERTER WITH HIGH POWER DENSITY 3081
Yugandhara H. Wankhede, Leon Fauth, Jens Friebe

USING BOTH THE CIRCULATING CURRENTS AND THE COMMON-MODE VOLTAGE FOR THE BRANCH ENERGY CONTROL OF MODULAR MULTILEVEL CONVERTERS 3091
Rebecca Dierks, Jakub Kucka, Axel Mertens

ANALYTICAL HARMONIC CURRENT MODEL FOR A PERMANENT MAGNET ASSISTED SYNCHRONOUS RELUCTANCE MOTOR (PMA-SYNRM) FED BY PWM INVERTER 3101
Jessica Neumann, Carole Hénaux, Maurice Fadel, Etienne Founier, Dany Prieto, Mathias Tientcheu Yamdeu

GENERALIZED SMALL-SIGNAL AVERAGED SWITCH MODEL ANALYSIS OF A WBG-BASED INTERLEAVED DC/DC BUCK CONVERTER FOR ELECTRIC VEHICLE DRIVETRAINS 3111
Sajib Chakraborty, Dai-Duong Tran, Joeri Van Mierlo, Omar Hegazy

ADAPTIVE PREDICTIVE-DPC FOR LCL-FILTERED GRID CONNECTED VSC WITH REDUCED NUMBER OF SENSORS 3119
Hosein Gholami-Khesht, Pooya Davari, Frede Blaabjerg

FPGA IMPLEMENTATION OF MODIFIED SPACE VECTOR MODULATION (SVM) FOR HIGH-FREQUENCY HYBRID ACTIVE NEUTRAL-POINT-CLAMPED (NPC) POWER FACTOR CORRECTION RECTIFIER 3129
Mohammad Najjar, Alireza Kouchaki, Morten Nymand

ENHANCED FLUX CONTROL INCLUDING A CLOSED LOOP VOLTAGE CONTROLLER TO OPTIMIZE THE VOLTAGE USAGE AND THE TORQUE COMPUTATION FOR A 48V IPMSM 3137
Felix Bertele, Ulrich Ammann, Christoph Cheshire, Tobias Röser

EXTENDED BOOST PV INVERTER TOPOLOGY FOR THE REDUCTION OF COMMON-MODE LEAKAGE CURRENT IN THREE-PHASE APPLICATIONS 3146
Georgios I. Orfanoudakis, Eftychios Koutroulis, Michael A. Yuratich, Suleiman M. Sharkh

A ROBUST CONTROL DESIGN TO REAL-TIME CONDITIONS AND MODELLING OF A MICROGRID 3156
Iréna Horvatic, Delphine Riu, Moataz Elsied, Sébastien Benjamin

DESIGN OF MODULAR LOW-PROFILE FREQUENCY CONVERTER FOR MULTI-MOTOR MANIPULATORS 3166
Tomas Glasberger, Zdenek Kehl, Tomas Kosan, Jan Molnar

STUDY OF THE CONTROL OF A NEW AC VOLTAGE STABILIZER USING LINEAR CONTROLLER WITH REFERENCE FRAME TRANSFORMATION 3172
Bunthern Kim, Etienne Boulaud, Emile Boisaubert, Sokchea Am, Phok Chrin

HYBRID ENERGY STORAGE SYSTEM FOR MVDC-GRIDS 3179
Florian Mahr, Johann Jaeger, Stefan Henninger, Hubert Rubenbauer

A COMBINED MODEL FOR OPTIMAL POWER FLOW APPLIED TO MT-HVDC SYSTEMS 3189
Fernando Torres, Javier Muñoz, Fredy Muñoz, Claudio Roa

CHARACTERIZATION OF LITHIUM ION SUPERCAPACITORS..3198
   Zeyang Geng, Felix Mannerhagen, Torbjöm Thiringer

GREY WOLF OPTIMIZER BASED PREDICTIVE TORQUE CONTROL FOR ELECTRIC
VEHICLE APPLICATIONS...3205
   Ali Djerioui, Azeddine Houari, Mohamed Machmoum, Malek Ghanes, Tedjani Mesbahi,
   Mohamed Fouad Benkhoris

OPERATION PRINCIPLE AND PERSPECTIVE PERFORMANCES OF METAL OXIDE
VACUUM FIELD EFFECT TRANSISTOR - MOVFET..3210
   Davide Patti, G. Busatto, G. Golluccio, D. Marciano, A. Sanseverino, F. Velardi

IMPROVED METHODOLOGY FOR PREDICTING CORRELATED COLOR TEMPERATURE
IN MIXED LED LIGHTING SOURCES 3217
   Thais E. Bolzan, Bruno F. Almeida, Renan R. Duarte, Vitor C. Bender, Rafael A. Pinto

DC MICROGRID CONCEPT FOR MINE ENVIRONMENT..3227
   Jooa Pursiainen, Jenni Rekola, Raimo Juntunen, Mikko Valtee, Pasi Peltoniemi

A COMPARISON OF TWO-STAGE INVERTER AND QUASI-Z-SOURCE INVERTER FOR
HYBRID ENERGY STORAGE APPLICATIONS 3237
   V. Castiglia, R. Miceli, F. Blaabjerg, Y. Yang

STATE ESTIMATION FOR MEDIUM AND LOW VOLTAGE DISTRIBUTION GRIDS
BASED ON NEAR REAL-TIME GRID MEASUREMENTS AND DELAYED SMART
METERS DATA 3247
   Mohammad Rayati, Thomas Pidancier, Mauro Carpita, Mokhtar Bozorg

GROUND FAULT ACTIVE COMPENSATION IN EMULATED DISTRIBUTION GRID OF 10
KV 3257
   Tomáš Komrska, Antonín Glac, Jakub Talla, Bohumil Skala, Jan Štepánek, Lubeš Streit,
   Zdenek Peroutka

MODELING OF A POWER TRANSFORMER INCLUDING HIGHER ORDER RESONANCES 3263
   Lukas Reißenweber, Alexander Stadler

A COMPARISON OF TWO STATE-SPACE MODELS OF AN INDUCTION MACHINE
CONSIDERING DIFFERENT SETS OF WINDING DISTRIBUTION HARMONICS.............................3272
   Julien Cordier, Stefan Klass, Ralph Kennel

PERFORMANCE IMPROVEMENT FOR PLUG-IN REVERSE CONDUCTING IGBTS
THROUGH GATE-VOLTAGE OBSERVATION..3282
   Daniel Lexow, Hans-Günter Eckel

DIFFERENTIAL FLATNESS FOR SMOOTH TRANSITION BETWEEN GRID-CONNECTED
AND STANDALONE MODE OF THREE-PHASE INVERTER..3289
   Abdelhakim Saim, Azeddine Houari, Mourad Ait-Ahmed, Mohamed Machmoum, Josep. M
   Guerrero

DIFFERENTIAL MODEL EMI FILTER ANALYSIS FOR INTERLEAVED BOOST PFC
CONVERTERS CONSIDERING OPTIMAL PHASE SHIFTING...3295
   Naser Nourani Esfetanaj, Yamen Saad, Omar Ahmed Sakaria, Huai Wang, Pooya Davari

MODULAR HYBRID DC BREAKER-BASED ADAPTIVE AUTO-RECLOSING METHOD
FOR MMC-HVDC SYSTEMS 3305
   Hossein Iman-Eini, M. Langwasser, L. Camurca, Marco Liserre

MULTISTEP MPC OF DUAL INVERTER FOR SWITCHING LOSSES OPTIMIZATION 3314
Martin Votava, Tomas Glasberger, Zdenek Peroutka

A HIGH-EFFICIENCY CONTROL OF A DOUBLE-INPUT CONVERTER FOR RENEWABLE
ENERGIES AND HYBRID VEHICLES 3321
Mario Marchesoni, Massimiliano Passalacqua, Luis Vaccaro

DEAD-TIME INFLUENCE ON FAST SWITCHING PULSED POWER CONVERTERS
DESIGN - A HIGH CURRENT APPLICATION FOR ACCELERATOR'S MAGNETS 3330
Ludovic Horrein, Jean-Marc Cravero, Philippe Delarue, Alain Bouscayrol, Davide Aguglia,
Carmen Ortega-Perez

DYNAMIC CHARACTERIZATION OF A SIC-MOSFET HALF BRIDGE IN HARD- AND
SOFT-SWITCHING AND INVESTIGATION OF CURRENT SENSING TECHNOLOGIES 3340
Janine Ebersberger, Jan-Kaspar Müller, Axel Mertens

POWER SUPPLY DESIGN CONSIDERATIONS FOR 400HZ AIRCRAFT APPLICATIONS 3348
Bilal Ahmad, Jorma Kyyrä, Juha Mäkelä

DC CAPACITOR VOLTAGE FEEDBACK METHOD FOR A PEAK VOLTAGE
SUPPRESSION CONTROL WITH MULTIPLE LEG-SHORT-CIRCUITS USING SIC-
MOSFETS EMPLOYED IN POWER CONVERTERS 3358
Tomoyuki Mannen, Takanori Isobe, Keiji Wada

INVESTIGATION OF BOND WIRE LIFT-OFF BY ANALYZING THE CONTROLLER
OUTPUT VOLTAGE HARMONICS FOR THE PURPOSE OF CONDITION MONITORING 3366
Firat Yüce, Marc Hiller

FRUGAL INNOVATION FOR SUSTAINABLE RURAL ELECTRIFICATION 3376
Bunthern Kim, Phok Chrin, Maria Pietrzak-David, Pascal Maussion

A CURRENT-MODULUS DERIVATIVE-BASED PROTECTION METHOD IN A FLEXIBLE
DC GRID 3385
Jianquan Liao, Niancheng Zhou, Qianggang Wang

COMPARATIVE ASSESSMENT OF VOLTAGE MODULATION METHODS FOR
ASYMMETRIC SIX-PHASE MACHINES 3393
R. S. Kanchan, Omer Ikram Ul Haq, Luca Peretti

SIMULATION AND MEASUREMENT-BASED ANALYSIS OF EFFICIENCY
IMPROVEMENT OF SIC MOSFETS IN A SERIES-PRODUCTION READY 300 KW / 400 V
AUTOMOTIVE TRACTION INVERTER 3403
A. Nisch, M. Heller, W. Wondrak, A. Bucher, C. Hasenohr, K. Kefer, B. Lunz, A. Pawellek, A.
Smit, M. Gärtner, N. Twardon, U. Kirchenberger

VALIDITY OF POWER CYCLING LIFETIME MODELS FOR MODULES AND EXTENSION
TO LOW TEMPERATURE SWINGS 3413
Josef Lutz, Christian Schwabe, Guang Zeng, Lukas Hein

ROADMAP FOR DC 3422
Pavol Bauer

THE ROLE OF COLLABORATIVE RESEARCH TO SUPPORT INNOVATION FOR CLEAN
ENERGY TRANSITION 3424
Hubert De La Grandiere

THOMAS EDISON VINDICATED — THE RESURGENCE OF DC IN MV AND HV POWER GRIDS 3425

Colin Davidson

INTEGRATION OF ELECTRIC MOBILITY IN THE FRENCH PUBLIC ELECTRICITY DISTRIBUTION NETWORK 3426

Anne-Sophie Cochelin

A CRITICAL ROLE FOR R&I FOR CLEAN ENERGY FOR THE EU GREEN AND DIGITAL RECOVERY 3427

Hélène Chraye

Author Index

Plasma Impact on Overvoltage Short-Circuit Failures in ANPC Converters

David Hammes[1], Sidney Gierschner[1], Dietmar Krug[2], Hans-Günter Eckel[1]

UNIVERSITY OF ROSTOCK[1]
Albert-Einstein-Str. 2
18059 Rostock, Germany
E-Mail: david.hammes@uni-rostock.de

SIEMENS AG[2]
Process Industries and Drives Division
Vogelweiherstr. 1-15
90441 Nuremberg, Germany

Keywords

«Fault tolerance», «Multilevel converters», «Robustness», «IGBT», «Diode»

Abstract

Three-level converters experience more complex short-circuit patterns than the standard four types of semiconductor failures for IGBT and / or diode. In literature, there is a special pattern described where three semiconductors in series suffer due to an overvoltage protection that interferes. This happens for saving one of the switches from or against overvoltage in certain situations. However, it is possible that plasma, the free charge carriers inside the semiconductor, is still inside a fourth one as a result of previous switching actions. This allows a second parallel path and impacts the first one. Besides, the plasma itself dissipates over time, which further changes the trend of voltages and currents for the short-circuit situation.

Introduction

The three-level Active-Neutral-Point-Clamped (ANPC) converter [1], as a subvariant of the NPC type [2], offers advantages over a standard two-level converter like every multilevel converter. The DC-link voltage (V_{DC}) can be doubled for the three-level type if the blocking voltage class (V_{CES}) of the semiconductors is the same, which increases twice the rated power of the converter itself. As an alternative, V_{DC} can be kept constant, but V_{CES} of the switches is reduced. This can be accompanied by an increase in the switching frequency as an advantageous example. In both ways, V_{DC} of the three-level ANPC converter is then significantly above V_{CES}. Further positive or negative aspects between two- and three-level converters are well-described in literature, for example in [3]. While the rise of V_{DC} above the nominal operating voltage of the semiconductor is beneficial, it offers in some situations a special kind of short-circuit (SC) failure. These so-called overvoltage failures have been first described for NPC converters in [4] and for ANPC types in [5] as well as in [6]. The cause, as it will be described in detail later on, is a loss of one semiconductor, which can result in some cases in a voltage drop of the whole V_{DC} across another one. If this occurs, overvoltage protection interferes but causes as a secondary effect these specific short-circuit faults. A detailed analyses of overvoltage SC occurrences has been done in [7]. This source distinguishes between a failure that affects three semiconductors in series (type A) and in parallel to these three a fourth switch (type B).

Bipolar semiconductors, like the widely used high-voltage IGBTs and PiN-diodes, have charge carriers (electrons and holes) inside during conducting. If the turn-OFF process does not remove this plasma, it can recombine over time, or offers possibilities for further short-circuit scenarios [5]. Therefore, a third subversion (type C) of the overvoltage short circuit is also available, which is described for the first time in this paper. In type B, the fourth switch is in ON-state before the fault. Now in type C, this switch is in OFF-state (as for A) but has still charge carriers inside. This leads to different behavior than in the other two variants. Besides, the plasma recombines during a specific period, which will alter the fault occurrence at different times of appearance.

As it is stated in [7], the overvoltage short-circuit failures can be built out of the single semiconductor failures of the two-level converter in theory. The first case (SC I) for the IGBT occurs when it switches onto an existing short circuit as described, for example in [8]. The behavior is different for the second fault (SC II) [9]. There, the IGBT is in ON-state when the fault occurs. The third type (SC III) affects the free-wheeling diode as well as the corresponding IGBT [10]. It occurs when the anti-parallel IGBT is in ON-state, but its diode conducts the load current (I_0) when the failure happens. It is called type four (SC IV) if this IGBT is not turned ON before the fault, but the diode is still conducting I_0 [11]. Two more short circuits are possible by changing from two-level to three-level ANPC converter. The inner four elements (IGBTs or diodes) on the positions 12, 21, 31 and 32 (see Fig. 1a) can be switched OFF in some switching situations without taking blocking voltage, as described in [5]. Therefore, free charge carriers are still stored in either IGBT and / or diode. A loss of a corresponding opposing switch in a half-bridge removes this plasma rapidly. The failure is called SC V for the IGBT [12] and SC IV ZC (zero [load] current) for the diode as an extension of SC IV [13]. These last two types are the key component for the overvoltage short-circuit failures of type C. Without them, only types A and B would exist. Therefore, an ANPC converter with MOSFETs in synchronous mode (the internal bipolar diode is nearly not used) cannot suffer under the third subcategory. Every converter with bipolar elements (PIN-diode and / or IGBT) will show this behavior in some instances during overvoltage short-circuit faults. However, the series and parallel short-circuit failures in multilevel converter differ in some aspects in their trends for voltage and current from the single SC types (I - V) [7]. Nevertheless, as a first assumption, it is valid to take these five types to describe the more complex patterns [5].

All measurements in this paper have been done in a test bench equipped with 3.3 kV / 1.5 kA Trench / Fieldstop IGBTs, including Emitter-Controlled free-wheeling diodes in the module. The DC-link voltage across one half-bridge is kept at 1.8 kV, the whole V_{DC} is then 3.6 kV. The nominal current through the device before the fault is 1.5 kA. The values for $L_{\sigma1}$ and $L_{\sigma3}$ are 120 nH respectively 90 nH, $L_{\sigma2}$ is 20 nH, and both $L_{\sigma4}$ have 50 nH each, see also Fig. 1a. The used test bench is derived from a real ANPC converter, and the stray inductances are not distributed equally, as stated. It is unimportant if the direction of the load current is positive or negative (here it is always set to positive). This is because the failure situations are symmetrical to each other, depending on I_0. As a result, the number of failures, that has to be investigated, can be reduced significantly [4].

Formation of overvoltage short circuit with plasma

The general concept to have charge carriers inside the inner switches after their turn-OFF (11, 12, 31 & 32) is quite simple [13]. The initial point is a zero-voltage output for the phase of the ANPC converter. It is unimportant if the upper (via 12 & 31) or the lower path (via 21 & 32, as displayed in Fig. 1a) is chosen. The result regarding the plasma is, in the end, the same, only the failure case changes slightly. This means, that either a diode or an IGBT has plasma inside and therefore SC V or SC IV ZC occurs. As a next step, a commutation to the opposing positive or negative voltage vector must occur, as displayed in Fig. 1b to + V_{DC} with turning ON T_{11} & T_{31}. Therefore, the plasma is forced out of D_{32} due to the applied voltage. However, the voltage between collector and emitter of T_{21} stays at approximately zero volts, and the charge carriers are not removed like in a normal turn-OFF process (indicated with superscript "PL") [13]. A breakthrough of the outer semiconductor (T_{22} / D_{22} for the displayed situation) will lead to a change of the emitter potential from 0 V towards - V_{DC}. How fast this occurs depends on the overall time of the arising SC V for T_{21} [12]. During its period, a positive and negative current slope (di / dt) exists, which, in combination with the stray inductances ($L_{\sigma2} + L_{\sigma3}$), causes a voltage drop or overvoltage. This interferes with the voltage drop for the other affected switches in this case. However, the plasma short-circuit faults for high-voltage devices end in some microseconds [12]. Afterward, the di / dt is zero, and the potential reaches finally -V_{DC}. As a consequence, the voltage drop across T_{32} exceeds its V_{CES} of 3.3 kV for 300 V with an overall DC-link voltage of 3.6 kV (in the used testbench). This point does not even consider that during highly dynamic processes like a short circuit, the voltage drop will oscillate and, therefore, even exceed these levels.

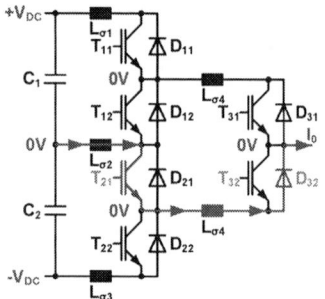
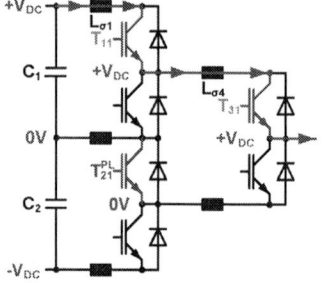
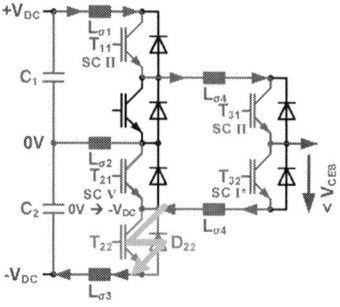

Fig. 1a: Initial state with zero output voltage & pos. I_0 of one phase, T_{21} & T_{32} are in ON-state (marked in green)

Fig. 1b: Commutating to pos. output (T_{11} & T_{31} are in ON-state), zero voltage drop across T_{21}, plasma remains

Fig. 1c: Intervention of AGC for T_{32} prevents overvoltage after loss of T_{22}, allows two short-circuit paths

Therefore, as mentioned in different sources like [4], the voltage drop across the affected IGBT (V_{CE}) has to be clamped to protect against overvoltage breakthrough. An Active Gate Clamping (AGC) circuit [14], which turns ON the device if V_{CE} rises above a predefined level, is used. For this, a standard solution with transil voltage suppressor diodes has been used. Mandatory is that this protection is fast enough to prevent the devastating overvoltage with the beginning of the fault [7]. For this testbench, the threshold value of the AGC ($V_{th,AGC}$) is set to a safety margin of roundabout 2.4 kV. This solution prevents the loss of T_{32}. Then, this IGBT conducts current to clamp its blocked voltage below V_{CES}. However, it allows a second short-circuit path from the positive (+V_{DC}) to the negative potential (-V_{DC}), the so-called overvoltage short circuit [7]. This is in parallel to the existing SC V for T_{21} in theory, see Fig. 1c. In reality, the two events do overlap but are also staggered in time. Moreover, a plasma short circuit happens during several hundred nanoseconds [12]. A failure of the types I to IV is in the range of 10 µs or more [10].

The current slope of plasma short circuits is determined by the stray inductances of this circuit and the removal of the charge carriers [13]. The second path has a rise of the short-circuit current (I_{SC}), which is determined by the voltage drop ($V_{L\sigma}$) across the stray inductances ($L_{\sigma1} + L_{\sigma3} + 2 \times L_{\sigma4}$). This voltage is composed of the whole DC-link voltage minus the actual V_{CE} of T_{32}. It is defined by $V_{th,AGC}$ and leads to approximately a $V_{L\sigma}$ of 1.2 kV for a V_{DC} of 3.6 kV. A value which is only 2 / 3 of the 1.8 kV, which a single half-bridge short-circuit failure would typically experience. The measurements of Fig. 2 show the trend for the impact of the charge carriers on the overvoltage short circuit in the situation of Fig. 1c. In theory, T_{21} faces an SC V, and T_{11} plus T_{31} suffer both from an SC II.

The switch T_{31} faces a kind of SC I (indicated with superscripted *) due to the interference of the overvoltage protection [7]. Some parts of this short-circuit trend have similarities, others not. A more detailed view of this particular problem is given in [7]. The load current through the IGBT T_{21}, which will suffer under the SC V, has been stopped 15 µs before the failure occurs. This happens due to the change of the output voltage vector from zero to $-V_{DC}$. This period between shutdown and fault occurrence is labeled as t_{DELAY} [13].

At 0 µs, T_{22} breaks through at once, and the SC V for T_{21} arises, see Fig. 2a. The drop of 1.8 kV across $L_{\sigma1} + L_{\sigma2}$ defines the rising current slope. This leads to the plasma removing, which defines the voltage slope across this IGBT, and mutual interference happens [12]. The voltage across T_{32} is also significantly increased with the rise of V_{CE} of T_{21}. This happens because a change of the emitter potential of T_{21} also interferes with the one of T_{32}, as can be seen in Fig. 1c and Fig. 2a. The voltage of T_{32} rises until its AGC reacts at ~ 0.5 µs and does a clamping of a maximum dynamic value of ~ 3.0 kV. The IGBT starts to conduct and therefore allows the second short-circuit path. At the same time, the voltage across T_{21} is still not at its maximum. This difference in time is explainable because the voltage across T_{32} is already at half of V_{DC} (1,8 kV) before the fault (the starting condition is then different).

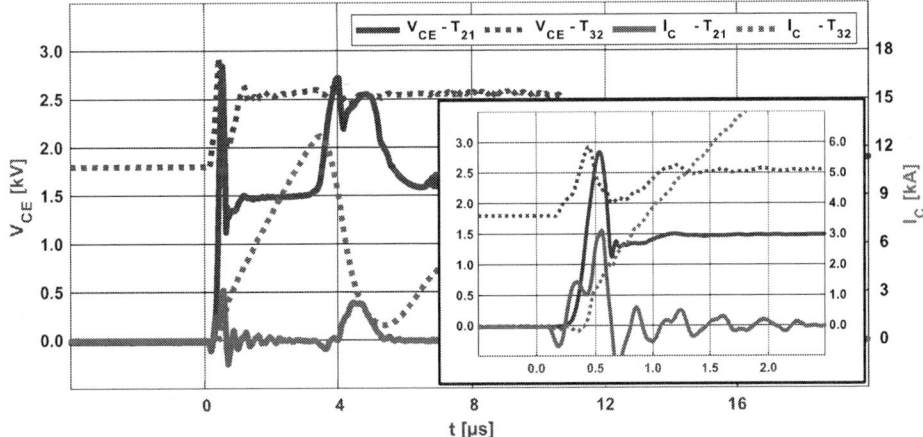

Fig. 2a: Measurement: SC V ‖ I* + 2x II, SC I* for T_{32} & SC V for T_{21} at $\sum V_{DC}$ = 3.6 kV, I_0 = 1.5 kA, t_{DELAY} = 15 µs

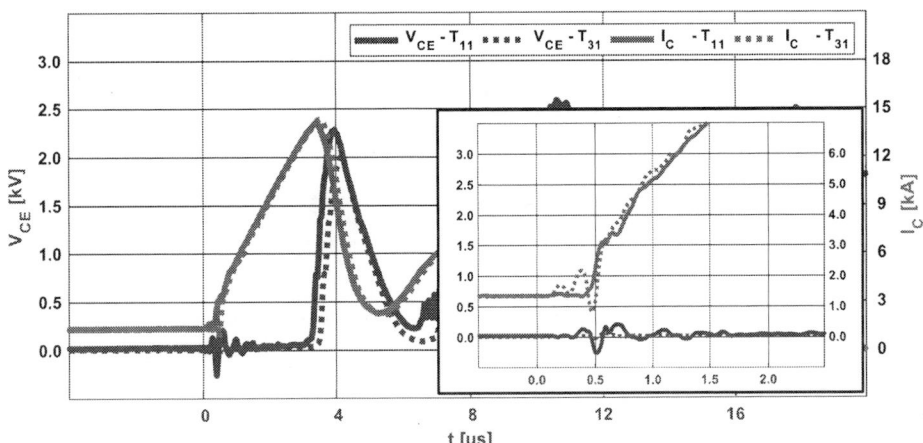

Fig. 2b: Measurement: SC V ‖ I* + 2x II, SC II for T_{11} & T_{31} at $\sum V_{DC}$ = 3.6 kV, I_0 = 1.5 kA, t_{DELAY} = 15 µs

A first short-circuit current peak for SC V occurs at approximately 0.3 µs with 1.4 kA. Then, it declines for some nanoseconds until it rises again towards 3.1 kA. The second maximum could be seen at first glance of interference of the voltage clamping for T_{21} before 0.5 µs. However, the gate-emitter voltage (V_{GE}) of T_{21} could not be measured due to probe limitation to prove this theory. Moreover, the trend of V_{CE} for this IGBT shows no real voltage clamping effect and is still below the value, where T_{32} shows the voltage reduction by its AGC. Another more likely explanation is a superimposed high-frequency oscillation on the current measurement, which is visible from 0.6 µs until 2.0µs in the zoom of Fig. 2a.

The I_{SC} starts flowing through T_{22}, T_{31} and T_{32} at 0.5 µs when the AGC turns ON T_{32}. Now the two parallel SC paths exist, which form a kind of complex current divider [7]. This also explains the fast rise from zero towards 3.0 kA of the short-circuit current for T_{11} and T_{31}, see Fig. 2b at 0.5 µs. Afterwards, the SC V ends at ~ 0.6 µs. The parallel type is reduced somehow to the series overvoltage case of type A. The di / dt for the SC II of T_{11} and T_{31} is declining notably. It is now only depending on the stray inductances in the remaining short-circuit path until the IGBTs desaturate. The rest of the trends for voltage and current of all affected IGBTs are nearly similar to the type A as expected. The interference of the plasma SC V seems to be small, the current peak is 'just' 22 % of the value for SC II of T_{11} (compare Fig. 2a with b). However, the impact on the dv / dt, and therefore the interference with V_{CE} of T_{32}, is substantial. The following chapter delivers a more in-depth analysis between the differences of types A, B and C.

Comparison of the overvoltage short circuits with and without plasma

A comparison is made to clarify the differences between the effect of the remaining charge carriers on the overvoltage short circuit. The conditions are as before (3.6 kV, 1.5 kA, and 15 µs t_{DELAY} for type C). The starting point for all three types is the same except for T_{21} (A – T_{21} in OFF-state / B - T_{21} in ON-state / C – T_{21} has plasma inside). First, an analyze of the early stage up to 2.5 µs after the breakthrough of T_{22} is made, see Fig. 3.

As it is visible in the measurement for SC V ∥ SC I* + 2x II (C - Fig. 3c), the beginning is different from SC I* + 2x II (A - Fig. 3a) or SC II ZC ∥ I* + 2x II (B - Fig. 3b). T_{21} shows the removal of charge carriers for several hundred nanoseconds, as described before. During this time, the IGBT takes over V_{DC}, but with a lower dv/dt of ~ 45 % in comparison to the capacitive voltage jump in type A. However, the peak value of V_{CE} for T_{21} is the same percentage higher. In type A, the overvoltage is caused by the jump and the subsequent oscillations. For type C, the negative current slope causes the overvoltage. The plasma charge of SC V is despite the superimposed oscillations still visible if SC V ∥ SC I* + 2x II is set in comparison to SC I* + 2x II. The charge carriers inside T_{21} prevents the oscillations of the capacitive voltage jumps of type A. This effect has been already seen in the investigations for the plasma failures, compare this to [13].

However, the differences with or without SC V are quite small, except for the reduction of the dv / dt for T_{21} and, as a result of this, also for T_{32}. This reduced voltage slope for T_{21} is an advantage for the protection of T_{31} or T_{32} against critical overvoltage. The AGC protection does not have to interfere as fast as for the cases of type A. After 1.0 µs, the trends for type A and C are nearly identical. Another small difference is the rise of the SC current for T_{32}. Because the AGC interferes delayed for the plasma type, the short-circuit occurrence for T_{11}, $T_{31,}$ and T_{32} happens several nanoseconds later.

Both types (A & C) differ a lot in comparison to B when T_{21} is in ON-state before the failure, see Fig. 3b. There, T_{21} first conducts an SC II without zero current before the failure (abbreviated as ZC) as described in [7]. The voltage across T_{21} is close to zero (a forward voltage recovery is visible until 0.5 µs, which is characteristic for SC II ZC) as long as the IGBT does not desaturate at several kiloamperes. Before the fault, the collector of the IGBT T_{32} lies on + 1.8 kV, whereas its emitter is fixed on 0 kV. The ~ 16 kA / µs as slope of the SC II ZC for T_{21} causes a voltage drop of 320 V at $L_{\sigma2}$. Therefore, the emitter of T_{32} is moved towards – 0.32 kV. The overall voltage change is minimal (from 1.8 kV to ~ 2.1 kV) and below the threshold voltage of its AGC.

The desaturation of T_{21} at 1.2 µs leads to a voltage rise across T_{32} above $V_{th,AGC}$. The voltage clamping turns ON the IGBT, and the second short-circuit path starts to form at 1.3 µs. The comparison shows that the status of T_{21} before the failure has a measurable impact on the three types. The charge carriers inside the IGBT reduces the harshness of the voltage rise across T_{32}. As a result, the AGC no longer has to react so fast, which reduces the requirement of it as well.

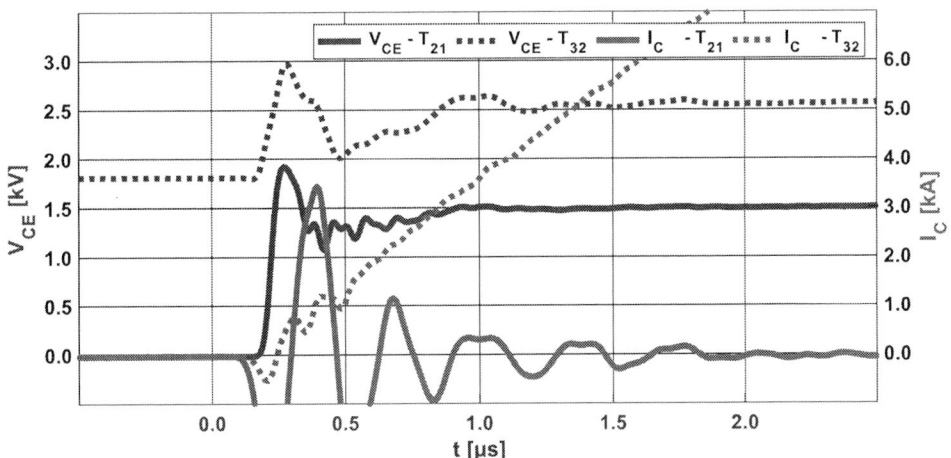

Fig. 3a: Measurement: SC I* + 2x II, T_{21} in OFF-state (type A), capacitive voltage jump for T_{21}

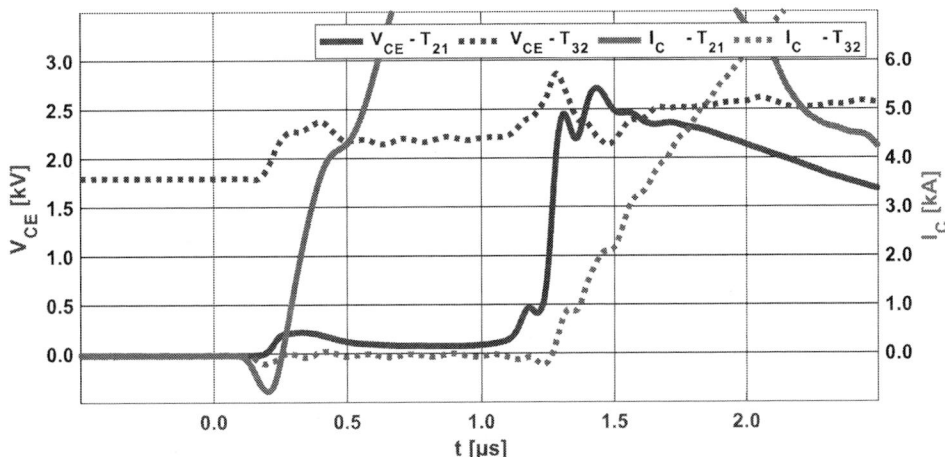

Fig. 3b: Measurement: SC II ZC || I* + 2x II, T_{21} in ON-state (type B), SC II ZC for T_{21}

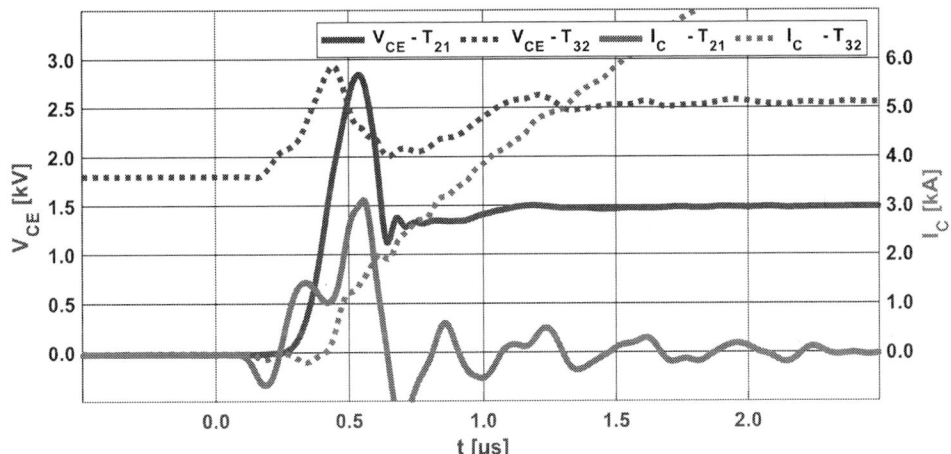

Fig. 3c: Measurement: SC V || I* + 2x II, T_{21} with plasma (type C), SC V for T_{21}

Time impact on the plasma overvoltage short-circuit case

The impact of the time delay on the charge carriers inside T_{21} is visible, see Fig. 4. The increase of t_{DELAY} from 15 µs to 20 µs is already enough to reduce the plasma in T_{21}. The peak current of the charge carrier extraction over time should, according to theory, be reduced as well [13]. This is not measurable here, because oscillations are overlaying this effect, see Fig. 4. A time delay in the range of 50 µs or higher will vanish the plasma inside the used IGBT due to recombination. Then type C is, in fact, a type A, which is again visible by the not damped oscillations in the measurement.

How long the full recombination will take depends on the amount of stored charge carriers before the shutdown, the voltage rating of the semiconductor and its recombination rate [15]. Unfortunately, the test bench itself doesn't allow a smaller delay time then 15 µs. Everything below would interfere with the blocking time of the semiconductors and cause an unwanted short-circuit situation.

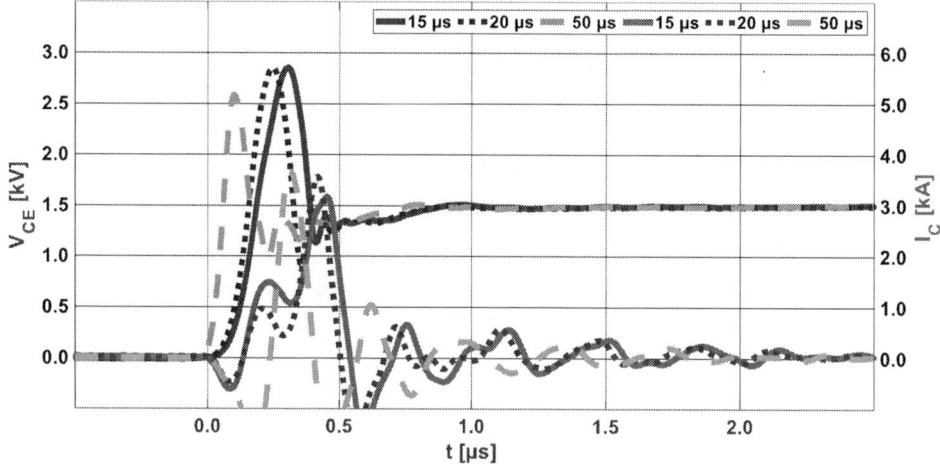

Fig. 4: Measurement: Variation of t_{DELAY} which changes the amount of plasma inside T_{21} (SC V) before failure

As already stated before, the less plasma is inside T_{21}, the earlier the IGBT takes over voltage. Moreover, the voltage slope will increase slowly, see Fig. 4. Other effects, except the transformation from type C to A with increasing of t_{DELAY}, are not detectable, as expected.

Further overvoltage failures with plasma inside a diode

The initial situation for SC V || I* + 2x II can be mirrored. Then, in contrast to Fig. 1 a - c, the zero-voltage vector of the phase is going through D_{12} and T_{31} while positive output load current. The transformation then goes to the negative phase output by turning-ON T_{22} and T_{32}. Therefore, charge carriers are still inside D_{12}. A loss of the opposing T_{11} will lead to an SC IV ZC for this diode [13]. The voltage across D_{12} is labeled with V_{CA} (cathode-anode voltage) and the current through it is the inverse of the forward current (I_F) in the measurement. This is done to have the direction of the reverse recovery effect of SC IV ZC in the positive direction, see Fig. 5a. In parallel to this is the series connection of SC I* (T_{31}) plus two times SC III (T_{22} & T_{32}). As the measurements in Fig. 5 show, the occurring failure scenario is similar to the SC I + 2x III except for the beginning.

The charge carriers inside the diode are minimal. The peak of the reverse recovery current for D_{12} is only ~ 150 A, see Fig. 5a. This value is marginal in contrast to the before described SC V with its maximum of roundabout 3 kA, compare this to Fig. 2. The whole current measurement is in addition overlaid again with oscillations, which interferes with the results. However, if a comparison with this failure case with and without plasma inside D_{12} is done (type A versus C), the indirect effect of the charge carriers is measurable. Without plasma, the voltage slope is higher and starts earlier than with plasma. The same effect, but better visible, is also existing in the before discussed SC V || I* + 2x II. Therefore, the current measurement is not entirely correct because of the disturbances. As before, the charge carriers inside the diode recombine quite fast, as it happened before for the described overvoltage failure with SC V. In direct comparison, the SC V in the overvoltage failures of type C shows a deeper impact than the SC IV ZC. However, this is always depending on the used IGBTs and diodes in the module. Changing the relation between these two items more into the favor of the diode, the effect of the plasma will be, of course, higher for the SC IV ZC || I* + 2x III.

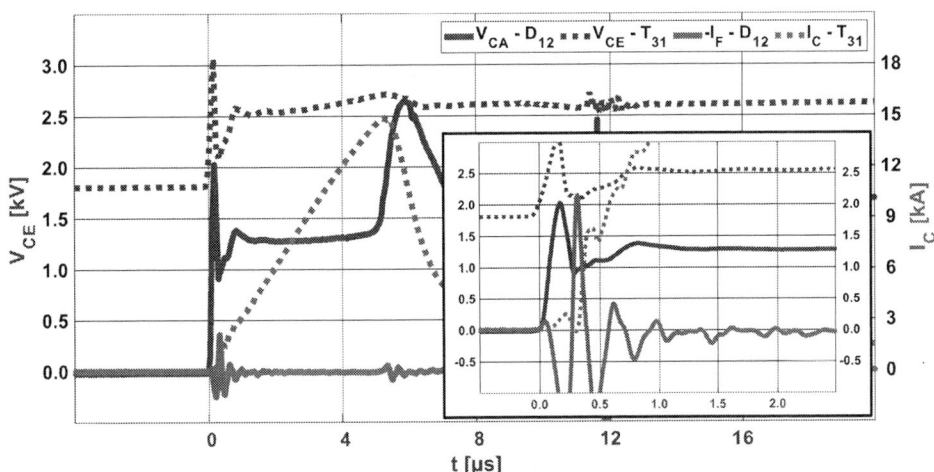

Fig. 5a: Measurement: SC IV ZC ‖ I* + 2x III, SC I* for T_{31} & SC IV ZC for D_{21} at $\sum V_{DC} = 3.6$ kV, $I_0 = -1.5$ kA, $t_{DELAY} = 15$ µs

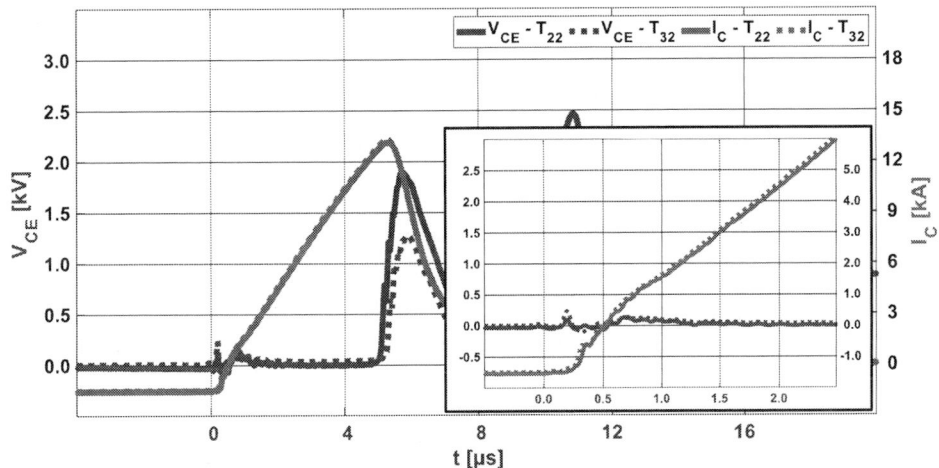

Fig. 5b: Measurement: SC IV ZC ‖ I* + 2x III, SC III for T_{22} & T_{32} at $\sum V_{DC} = 3.6$ kV, $I_0 = -1.5$ kA, $t_{DELAY} = 15$ µs

The initial situation for the series failure of two times SC III can be slightly changed [7]. The first derivation is that T_{22} is now in OFF-state. Then this switch suffers under an SC IV failure because only the conducting diode reacts. The SC IV has as a side effect that the SC III for T_{32} is not occurring [7]. If also T_{32} is meanwhile in OFF-state, two SC IV will then happen in a series combination next to the SC I* of T_{31}. Technically, the charge carriers inside D_{12} before the failure occurrence is the same in both cases as for SC IV ZC ‖ I* + 2x III. The SC IV ZC for D_{12} should also not be affected if T_{22} and T_{32} suffer under an SC III and / or an SC IV. These basic short-circuit cases take over the blocking voltage only after several microseconds are already past, be it through the desaturation process (SC III) or the reverse recovery effect (SC IV) [16]. However, the plasma failures, either SC IV ZC [13] or SC V [12], occur during some hundred nanoseconds, as also the here shown measurements indicate. As a result, it is unimportant, which short-circuit case happens for T_{22}, respectively T_{32}. The trend for D_{12} should be the same.

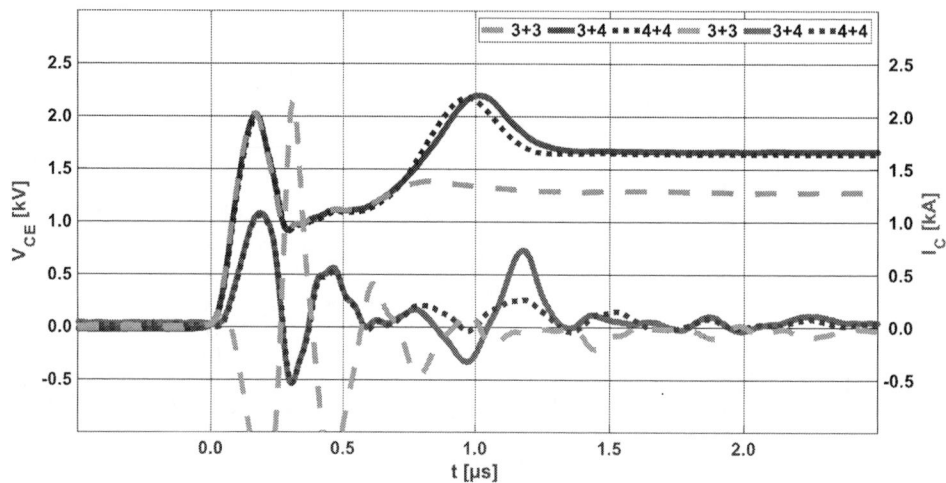

Fig. 6: Measurement: Comparison of the trends for D_{12} (SC IV ZC) in the variation of the failure cases for T_{22} & T_{32}

The trend for the cathode-anode voltage across D_{12} during the removing of the charge carriers is in all three variations (SC IV ZC || I* 2x III / SC IV ZC || I* III + IV / SC IV ZC || I* 2x IV) the same, as the comparison of Fig. 6 shows. After 0,5 µs, when the plasma is already removed, the case with 2x SC III differs from the other two. The parallel happening of SC IV for D_{22} (and D_{32}) forces D_{12} to block the whole DC-link voltage of nearly 1.8 kV. This particular failure has a time delay between failure occurrence and starting to block voltage as a result of the reverse recovery process [16]. In this period, the forward current through the diode is reduced from 1.5 kA to zero. In the measurement above, the zero crossing and following removal of charge carriers happen after the 0.5 µs. In contrast to this, the desaturation of an SC III for T_{22} or T_{32} takes more microseconds until it occurs. Therefore, the increase of V_{CA} of D_{12} comes much later, here it is at approximately 5.0 µs (not displayed).

However, the current trend for the case with SC IV ZC || I* 2x III differs from the beginning, see Fig. 6. This is contrary to the mentioned theory. The explanation lies in the described impact of the oscillations on this failure case for D_{12}. Without these, the trend of the reverse recovery during SC IV ZC should be the same as for the other two displayed cases. As proof of this theory stands the voltage trends. All of them are the same during plasma removal. Thus, the current trends should also be the same for all three failure versions if the oscillations would not interfere.

Conclusion

In addition to the already known overvoltage short-circuit cases for ANPC converters, the occurrence of remaining charge carriers in a fourth switch can change the trend for current and voltage in some degrees. The most notable change is a declined voltage slope for the switch, which will face the whole DC-link voltage. This reduces the stress for the protection circuit of that switch. The plasma failure itself usually is not harmful to either IGBT or diode, but it can decrease the difficulty of the overvoltage failures.

However, the effect of the SC V, respectively SC IV ZC, is just measurable for some microseconds. Moreover, the charge carriers recombine in the period up to 50 µs with the used 3.3 kV module at nominal load current before the turn-OFF. Afterwards, the overvoltage short-circuit cases do not have any plasma impact. They act then as the already described "normal" failures.

References

[1] T. Brückner and S. Bernet, "Loss Balancing in Three-Level Voltage Source Inverters applying Active NPC Switches," in *IEEE Annual Power Electronics Specialists Conference*, Vancouver, Canada, 2001.

[2] A. Nabae, I. Takahashi and H. Akagi, "A New Neutral-Point-Clamped PWM Inverter," *IEEE Transactions on Industry Applications*, 1981.

[3] G. Sinha, C. Hochgraf, R. Lasseter, D. Divan and T. Lipo, "Fault protection in a multilevel inverter implementation of a static condenser," in *30th IEEE Industry Applications Conference (IAS)*, Orlando, USA, 1995.

[4] M. Sprenger, R. Alvarez, M. Tannhaeuser and S. Bernet, "Experimental investigation of short-circuit failures in a three level neutral-point-clamped voltage-source converter phase-leg with IGBTs," in *Energy Conversion Congress and Exposition (ECCE)*, Denver, USA, 2013.

[5] D. Hammes, J. Fuhrmann, S. Gierschner, M. Beuermann and H.-G. Eckel, "Short-Circuit Behavior of the Three-Level Advanced-Active-Neutral-Point-Clamped Converter," in *Power Electronics and Applications (EPE)*, Warsaw, Poland, 2017.

[6] M. Gleissner, T. Bertelshofer and M.-M. Bakran, "Driver integrated fault-tolerant reconfiguration after short-on failures of a SiC MOSFET ANPC inverter phase," in *PCIM Europe*, Nuremberg, Germany, 2018.

[7] D. Hammes, J. Fuhrmann, S. Gierschner, D. Krug and H.-G. Eckel, "Overvoltage Short-Circuit Failures in Three-Level ANPC Converters," in *European Conference on Power Electronics and Applications (EPE)*, Genova, Italy, 2019.

[8] T. Rogne, N. Ringheim, B. Odegard, J. Eskedal and T. Undeland, "Short-circuit capability of IGBT (COMFET) transistors," in *Industry Applications Society Annual Meeting*, Pittsburgh, USA, 1988.

[9] H.-G. Eckel and L. Sack, "Experimental investigation on the behaviour of IGBT at short-circuit during the on-state," in *Industrial Electronics, Control and Instrumentation, IECON*, Bologna, Italy, 1994.

[10] J. Lutz, R. Dobler, J. Mari and M. Menzel, "Short circuit III in high power IGBTs," in *Power Electronics and Applications (EPE)*, Barcelona, Spain, 2009.

[11] S. Pierstorf and H.-G. Eckel, "Short-circuit behavior of diodes in voltages," in *PCIM Europe*, Nuremberg, Germany, 2012.

[12] J. Fuhrmann, S. Klauke and H.-G. Eckel, "Passive IGBT turn-off during short-circuit type V," in *PCIM Europe*, Nuremberg, Germany, 2016.

[13] D. Hammes, J. Fuhrmann, R. Schrader, S. Gierschner, D. Krug and H.-G. Eckel, "Plasma-induced Diode Short-Circuit in Neutral-Point-Clamped Converters," in *PCIM Europe*, Nuremberg, Germany, 2018.

[14] A. Volke and M. Honrkamp, IGBT Modules - Technologies, Driver and Application, Munich, Germany: Infineon Technologies AG, 2012.

[15] J. Lutz, Halbleiter-Leistungsbauelemente, Berlin Heidelberg Chemnitz, Deutschland: Springer Vieweg, 2012.

[16] J. Fuhrmann, D. Hammes and H.-G. Eckel, "Short-circuit behavior of high-voltage IGBTs," in *IECON Proceedings (Industrial Electronics Conference)*, Florence, Italy, 2016.

Novel Soft-Switching Interleaved Boost Converters for Renewable Energy Conversion Systems

Madhuchandra Popuri[1], V.V.Subrahmanya Kumar Bhajana[2], Pavel Drabek[3],
Manoj Kumar Maharana[1]

[1]School of Electrical Engineering, KIIT University,Bhubaneswar, India
[2]School of Electronics Engineering, KIIT University, Bhubaneswar, India
[3]Regional Innovation Centre for Electrical Engineering, University of West Bohemia, Pilsen, Czech Republic
E-Mail: madhuflo@gmail.com, bvvs.kumarfet@kiit.ac.in, drabek@ieee.org, mkmfel@kiit.ac.in

Acknowledgements

This research has been supported by the Ministry of Education, Youth and Sports of the Czech Republic under the project OP VVV Electrical Engineering Technologies with High-Level of Embedded Intelligence CZ.02.1.01/0.0/0.0/18_069/0009855 and project No. SGS-2018-009

Keywords

« Boost Converters », « », « Interleaved Converter», « Soft-Switching », «Zero Voltage Switching (ZVS)»,« Zero Current Switching ».

Abstract

This paper proposes new soft-switching interleaved boost converters (SSIBC) for photovoltaic power generation systems. The SSIBCs presented comprises with additional dual and single auxiliary cells. The main active switches of these converters are able to provide zero voltage switching (ZVS) and zero current switching (ZCS) operation by means of dual auxiliary and single auxiliary cells, respectively. The proposed converters are capable to reduce turn on and turn off switching power losses with their soft-switching operation and consequently, they mitigate the overall efficiency. In addition, the auxiliary switches also achieve soft-turn on and turn off operation, during the total soft-switching region of main switches. The detailed principle of operation and simulation evaluations presented in this paper validate the effectiveness and soft-switching ability of the proposed interleaved boost converters.

Introduction

Recently, the boost converters are playing a significant role in renewable energy sources such as fuel cells, super capacitors and photovoltaic modules, to provide desired high output voltage and high powers. Especially, interleaved boost converters (IBC) are being used for high power applications. To reduce the input inductor current ripples and also to extend the voltage conversion ratios, the boost converters are usually connected in parallel. Early days of research was focused on interleaved converters with different controlling techniques such as voltage mode, current mode [1-2], sliding mode controls [3] which are implemented to ensure interleaving stability and inductor tolerances. On the other hand, interleaving structures are realized with additional inductor and capacitors to their main switches between two boost converters to achieve zero voltage transition (ZVT) [4] and zero voltage switching [5]. Nevertheless, those converters may not be operated at high power levels due to the limitation on main switch conduction. Aiming at photovoltaic distribution systems, another soft-switching converter is reported in [6] with additional inductor, capacitor and diodes connected with each module. Hence, this converter achieves zero voltage switching during turn on commutation and it cannot be useful for high gain applications. For high gain applications, utilization of auxiliary inductors for a three phase IBC [7] and coupled inductors in IBC [8] have already been initiated. Since, regardless of voltage gain, these converters achieve soft-switching by means of one parallel capacitor each with their main switches, some high current peaks appear during turn off. Furthermore, coupled inductor integrated with an active clamp circuit is incorporated in IBC [9] to extend the conversion ratio and also to ensure soft-switching. Interleaved converter utilizes dual auxiliary resonant networks [10], which are indeed to guarantee the zero voltage switching and operated at high powers, as well. In same manner, dual auxiliary resonant networks [11] are also used in IBC, which is having dual soft-switching characteristics such as turn on under zero current switching and turn off with zero voltage switching operations, respectively.

To decrement the auxiliary component count, several IBCs were improved with a single auxiliary circuit [12-13]. Therefore, the zero voltage zero current switching (ZVZCS) operations are achieved at low switching frequency and low output power. Then, a low power, high switching frequency IBC [14] is implemented with an auxiliary circuit. This has achieved zero current transition (ZCT) condition for both main and auxiliary switching devices. To reduce turn off current tails across IGBTs, IBCs [15-16] are developed with a resonance clamp cell, which provides zero current switching (ZCS) turn off. Consequently, ZCS turn off commutation is obtained and an intermittent current flows through the main IGBTs throughout the conduction period. In recent past, research was also focused on IBC with a one auxiliary resonant cell in addition to a coupled inductor [17] to obtain ZVT turn on operation. Though, this converter achieves the ZVT turn on with lesser voltage gain, auxiliary switch stress is increased. In an IBC, if the duty cycle of the main switches is less than 50%, a single auxiliary resonant circuit [18] will be sufficient to achieve the zero current switching (ZCS) turn on.

The previous existing IBCs are lacking in obtaining desired gain, high output power and better efficiency. Thus, it is of interest to design new soft-switching interleaved boost converter (SSIBC) with significant benefits such as soft-switching (ZVS and ZCS), high voltage gain and high efficiency. The main attention of this paper is to design new soft-switching interleaved converters (SSIBC) for the applications in renewable energy sources like fuel cell, super-capacitor and photovoltaic systems. Mainly two different SSIBC topologies are proposed in this paper, one is improved SSIBC with dual auxiliary cell and another with a single auxiliary cell. The significant benefit of dual auxiliary cell based SSIBC-1 is that zero voltage switching (ZVS) turn-on operation can be obtain for all active power switches. On the other hand, the second topology SSIBC-2 has zero current switching (ZCS) turn-on/turn-off operations achieved for the main IGBTs. The description and operating principles are discussed in the following section. Section 3 presents the performance of proposed SSIBCs with simulation results.

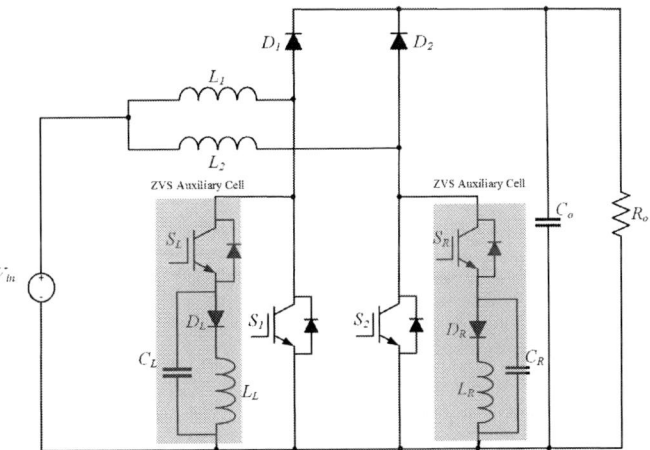

Fig. 1: Proposed topology-1 soft-switching interleaved boost converter (SSIBC-1)

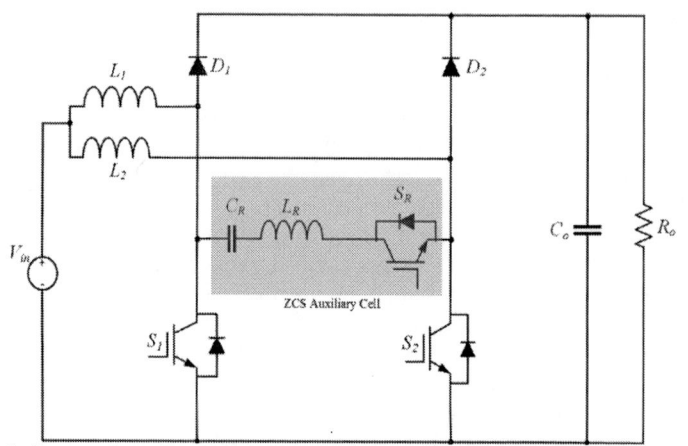

Fig. 2: Proposed topology-2 soft-switching interleaved boost converter (SSIBC-2)

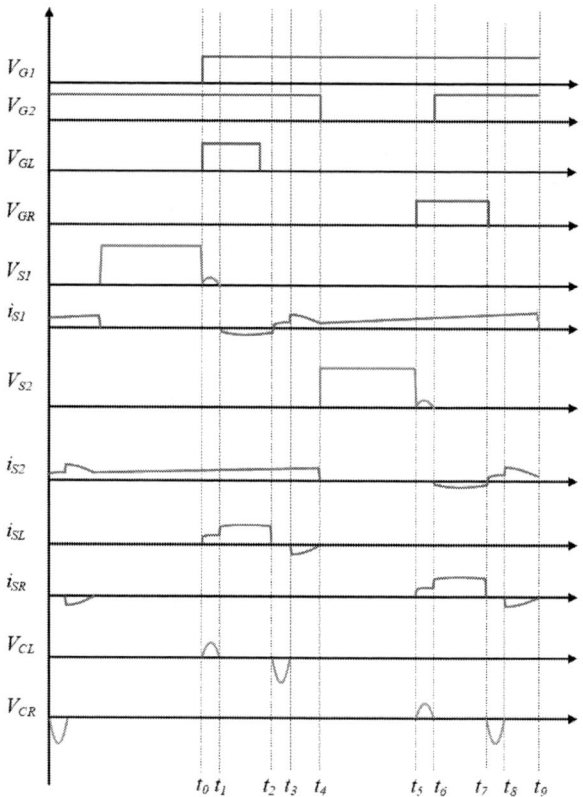

Fig. 3: Key waveforms: SSIBC-1

Description and Operation Principles

This paper proposes new topologies that obtain soft commutation during turn on and turn off processes with the help of dual auxiliary resonant and single auxiliary circuits, respectively. Fig.1 shows the first variant of SSIBC-1 with dual auxiliary resonant circuit. Except auxiliary circuit components, the conventional configuration of IBC is same in both the topologies, in general, which consists of two main switches S_1 and S_2, two input inductors L_1 and L_2 and two output diodes D_1 and D_2. The dual auxiliary resonant circuit adopted separately in left and right legs comprises two additional IGBTs S_L

and S_R, two resonant inductors L_L and L_R, two resonant capacitors C_L and C_R and two diodes D_L and D_R. Fig.2 shows the proposed topology SSIBC-2 improved with only one auxiliary resonant circuit connected between the two legs of the converter. The auxiliary circuit consists of a MOSFET, S_R, a resonant inductor, L_R and a capacitor, C_R. The operations of both the topologies are explained with the aid of key waveforms illustrated in Fig.3 and Fig.6, respectively.

Operation Intervals of SSIBC-1

The topology SSIBC-1 operations are divided in to six intervals from t_0-t_6 and the overall time intervals simply classified into four operating modes as resonant, freewheeling, non-resonant and energy accumulating intervals. The equivalent current flow schematics of SSIBC-1 are shown in Fig.4 (a, b, c, d) & Fig.5 (a, b, c, d).

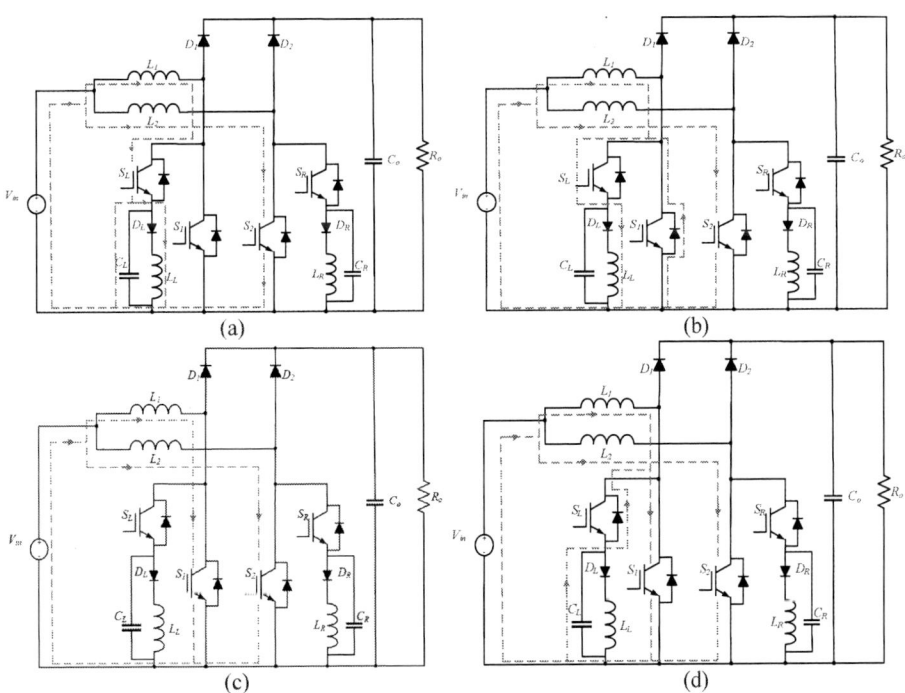

Fig. 4: Equivalent current flow schematics of SSIBC-1 (a) Interval (t_0-t_1) (b) Interval (t_1-t_2) (c) Interval (t_2-t_3) (d) Interval (t_3-t_4)

Resonant Interval (t_0-t_1): During the first interval (t_0-t_1), the main IGBT S_2 is already in conduction from the previous interval and the auxiliary IGBT S_L is turned on at t_0 to ensure soft-switching condition to the IGBT S_1. The current in the inductor L_L increases gradually, capacitor, C_L is charged up to 1/4th of V_{in} and at end of first interval C_L gets discharged completely. When the voltage in C_L reduces to zero, the voltage across S_1 becomes zero in sinusoidal manner. Hence, the ZVS condition is achieved for the S_1. Therefore, interval t_0-t_1 can be defined as resonant interval.

Freewheeling Interval (t_1-t_2): From the beginning of interval t_1-t_2, the body diode of S_1 is forward biased to create a path for the resonant current. At t_2, the diode of S_1 stops conducting. So, t_1-t_2 interval can be defined as freewheeling interval.

Non-resonant Interval (t_2-t_3): During the interval t_2-t_3, the current in the auxiliary IGBT S_L becomes zero; hence, the ZCS condition is achieved for auxiliary IGBT S_L. The current through S_1 is equal to the input current and its body diode is forward biased at t_3.

Energy Accumulating Interval (t_4-t_9): From t_4-t_9, the main IGBT S_1 conducts and energy is accumulated in input inductor L_1.

At t_5, the auxiliary IGBT S_R is turned on to obtain ZVS operation in S_2. The principle of operation repeats from t_5-t_9 similar to that of t_0-t_4.

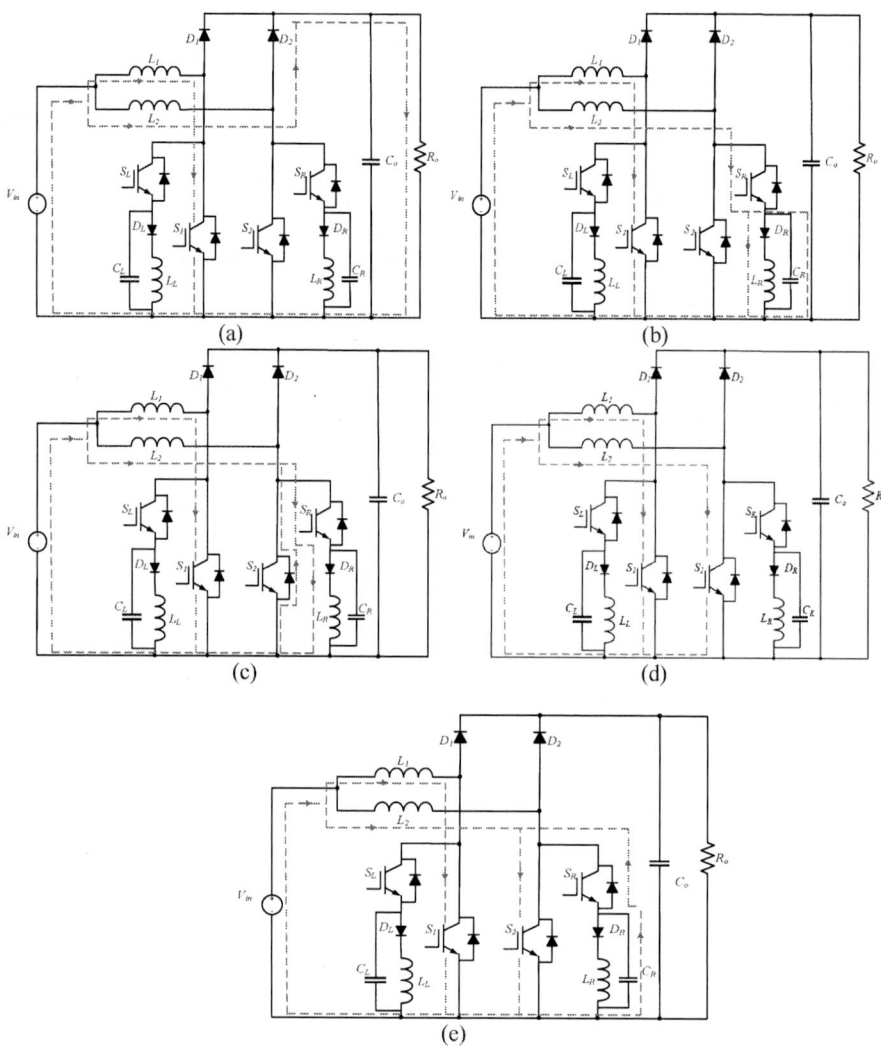

Fig. 5: Equivalent current flow schematics of SSIBC-1 (Contd) (a) Interval (t_4-t_5) (b) Interval (t_5-t_6) (c) Interval (t_6-t_7) (d) Interval (t_7-t_8) (e) Interval (t_8-t_9)

Operation Intervals of SSIBC-2

The operation principles of second topology are analyzed with the help of key waveform shown in Fig.6. The operation is divided in to five intervals.

Interval (t_0-t_1): During the first interval t_0-t_1, the auxiliary MOSFET S_R is turned on at t_0 to achieve ZCS turn on conditions for IGBT S_1. Since, the inductor L_R current decreases in reverse direction and capacitor is being discharged from $+V_0$ and gets charged to $-V_0$. Throughout this interval, the anti-parallel diode of IGBT S_1 conducts in order to allow the resonant tank current and at same instant, resonant peak current flows through S_R.

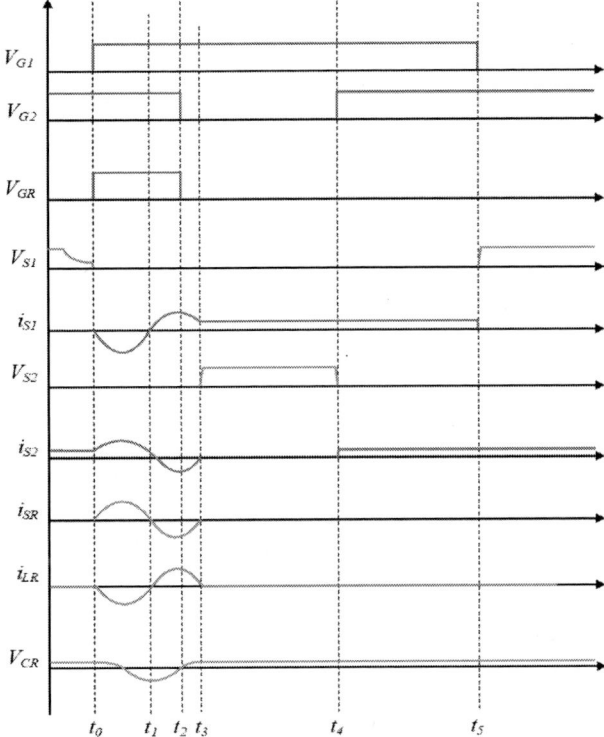

Fig. 6: Key waveforms: SSIBC-2

Interval (t_1-t_2): After a short duration from t_1, the current through S_2 reduces to zero and its body diode allows the resonant current. Therefore ZCS turn off condition is achieved for S_2. From t_1-t_3, the body diode of main MOSFET S_2 and auxiliary MOSFET S_R starts conducting to create a path to allow resonant current and at t_2, the auxiliary IGBT S_R is turned off.

Interval (t_2-t_3): From t_2-t_3, there is a resonant peak current flow through the S_1 and capacitor, C_R is being charged to $+V_0$ again. During this interval, the ZCS turn-off condition is achieved for the auxiliary MOSFET S_R and ZCS turn on operation is achieved for S_2.

Interval (t_3-t_4): During t_3-t_4, the S_1 is in conducting state, L_1 accumulates the energy and output power will be delivered by L_2.

Interval (t_4-t_5): During t_4-t_5, the IGBT S_2 will be conducting and hence energy will be accumulated by L_2.

Simulation Results

The proposed topologies are modeled by using MATLAB and simulations results. The parameters and components considered for simulations are as mentioned in table 1.

Table I: Simulation Parameters and Components

Parameters	Value	Components	Value
Input voltage (V_{in})	100 V	Resonant inductors (L_L, L_R)	8 μH
Output voltage (V_o)	400 V	Resonant capacitors (C_L, C_R)	10 nF
Switching frequency (f_{sw})	100 kHz	Output capacitor (C_o)	470 μF
Output power (P_o)	1 kW	Input inductors (L_1, L_2)	100 μH

The design simulations of the proposed topology, SSIBC-1 are carried out using MATLAB Simulink. It can be seen from Fig.7 that, the ZVS turn on is achieved for the both the switching devices S_1 and S_2. The voltage and current waveforms of S_L and S_R are shown in Fig.8. It is observed that the auxiliary switches S_L and S_R are turned off at zero current. It can also be seen from Fig.8, there are negligible resonant voltage ripples through the auxiliary switches. These voltage ripples may not affect the overall performance of the converter. The obtained simulation results are validated with theoretical expectations and it is proved that ZVS turn on feature is obtained without increasing additional losses.

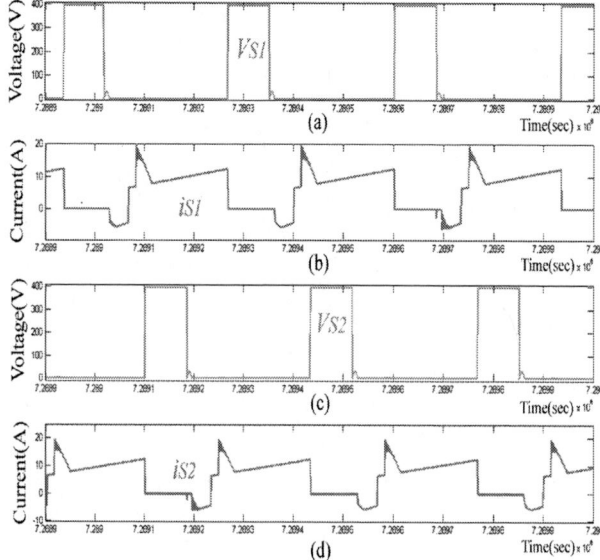

Fig.7: Simulated Results: SSIBC-1 (a) V_{S1} (b) i_{S1} (c) V_{S2} (d) i_{S2}

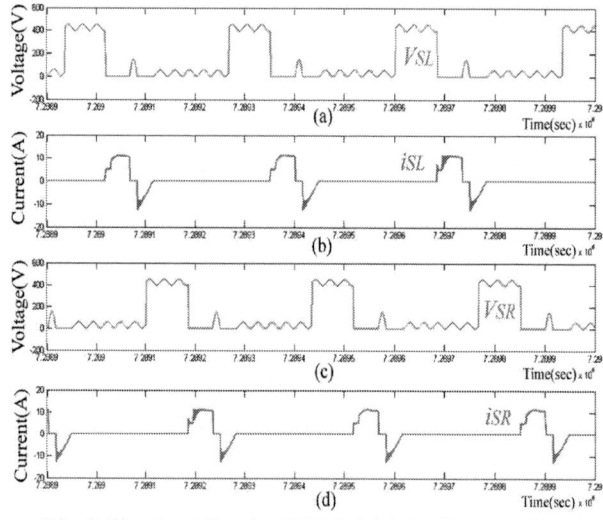

Fig. 8: Simulated Results: SSIBC-1 (a) V_{SL} (b) i_{SL} (c) V_{SR} (d) i_{SR}

Similarly, the design simulations of the proposed second topology SSIBC-2 are performed using MATLAB Simulink. Fig.9 shows the resonant inductors L_R, L_L current waveforms and voltage waveforms of C_L, C_R. Fig.10 shows the ZCS turn on commutation of S_1 and S_2 is commutated with ZCS turn off. The auxiliary switch, S_R is turned on and turned off at zero current. Soft-switching operation of the switches is obtained with the help of a single auxiliary cell, where as in SSIBC-1 two auxiliary cells are utilized. Fig.11 shows the voltage and current waveforms of auxiliary switch, S_R.

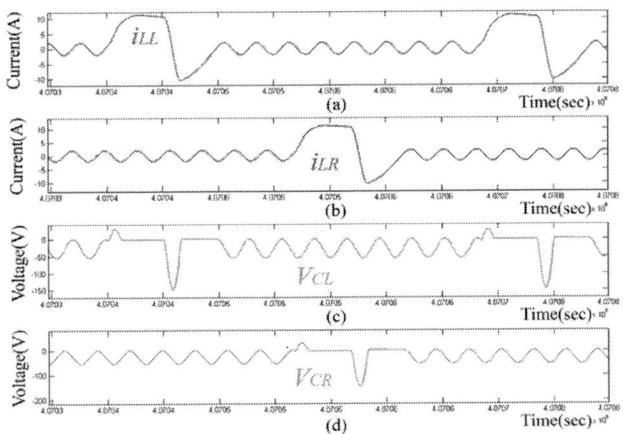

Fig. 9: Simulated Results: SSIBC-1 (a) i_{LL} (b) i_{LR} (c) V_{CL} (d) V_{CR}

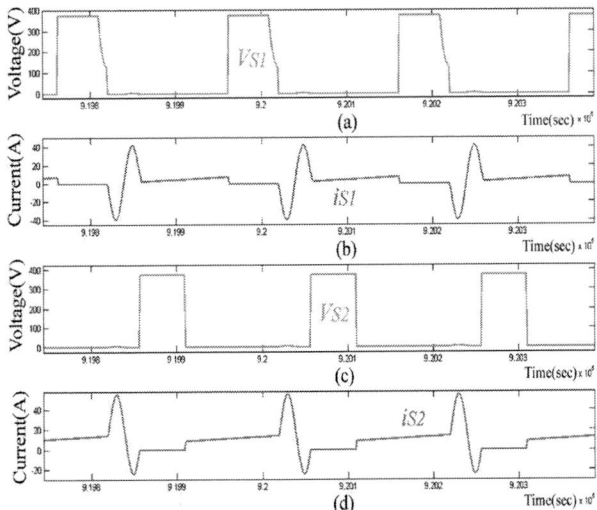

Fig. 10: Simulated Results: SSIBC-2 (a) V_{S1} (b) i_{S1} (c) V_{S2} (d) i_{S2}

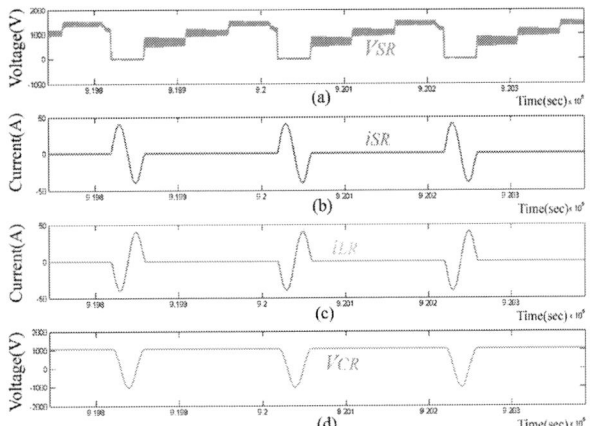

Fig. 11: Simulated Results: SSIBC-2 (a) V_{SR} (b) i_{SR} (c) i_{LR} (d) V_{CR}

It is observed that the switch, S_R is turned on and turned off with ZCS. The peak current of i_{LR} is exactly equal to S_R current and average voltage is half the input voltage across resonant capacitor is V_{CR}. The simulation evaluations of SSIBC-2 prove the characteristics of ZCS turn on and ZCS turn off for the main switches S_1 and S_2.

Conclusion

This paper proposed new soft-switching interleaved boost converter topologies for renewable energy conversion systems. The proposed topologies are described and it operation principles are discussed and validates with simulations analysis on 100 V/400V converters with 100 kHz switching frequency and maximum output power 1 kW . The conventional IBCs are improved with dual auxiliary cells, which provide ZVS turn on operation and another version was improved with a single auxiliary cell, which provides ZCS turn on/ turn off operations. These proposed SSIBCs are one of the alternate candidates for high power, high gain applications with improved efficiencies. The experimental evaluations for these topologies are in process.

References

[1] R. Giral, L. Martinez-Salamero and S. Singer .: Interleaved converters operation based on CMC, IEEE Transactions on Power Electronics,Vol.14,no.4,pp.643-652,July1999

[2] L. Huber, B. T. Irving and M. M. Jovanovic.: Open-Loop Control Methods for Interleaved DCM/CCM Boundary Boost PFC Converters, IEEE Transactions on Power Electronics, Vol. 23, no. 4, pp. 1649-1657, July 2008.

[3] R. Giral, L. Martinez-Salamero, R. Leyva and J. Maixe.: Sliding-mode control of interleaved boost converters, IEEE Transactions on Circuits and Systems I: Fundamental Theory and Applications, Vol. 47, no. 9, pp. 1330-1339, Sept.2000.

[4] Y. Hsieh, T. Hsueh and H. Yen.: An Interleaved Boost Converter With Zero-Voltage Transition, IEEE Transactions on Power Electronics, Vol.24,no.4,pp.973-978,April 2009.

[5] M. Veerachary and M. N. V. Bhavana.: Analysis and design of zero-voltage switching high gain interleaved boost converter,2014 6th IEEE Power India International Conference (PIICON), Delhi, 2014, pp. 1-6. 5-7 Dec. 2014.

[6] D. Jung, Y. Ji, S. Park, Y. Jung and C. Won.: Interleaved Soft-Switching Boost Converter for Photovoltaic Power-Generation System,IEEE Transactions on Power Electronics, Vol. 26, no. 4, pp. 1137-1145, April 2011.

[7] J. Yi, W. Choi and B. Cho.: Zero-Voltage-Transition Interleaved Boost Converter With an Auxiliary Coupled Inductor, IEEE Transactions on Power Electronics, Vol. 32, no. 8, pp. 5917-5930, Aug. 2017.

[8] T. Rahimi, S. H. Hosseini, M. Sabahi, M. Abapour and G. B. Gharehpetian.: Three-phase soft-switching-based interleaved boost converter with high reliability, IET Power Electronics, Vol. 10, no. 3, pp. 377-386, 2017.

[9] G. Yao, A. Chen and X. He.: Soft Switching Circuit for Interleaved Boost Converters, IEEE Transactions on Power Electronics, Vol. 22, no. 1, pp. 80-86, Jan. 2007.

[10] M. Packnezhad, H. Farzanehfard and E. Adib.: Integrated soft switching cell and clamp circuit for interleaved high-step-up converters, IET Power Electronics, Vol. 12, no. 3, pp. 430-437, 2019.

[11] M. Pahlevaninezhad, P. Das, J. Drobnik, P. K. Jain and A. Bakhshai.: A ZVS Interleaved Boost AC/DC Converter Used in Plug-in Electric Vehicles, IEEE Transactions on Power Electronics, Vol. 27, no. 8, pp. 3513-3529, Aug. 2012.

[12] K. Chao and M. Yang.: High step-up interleaved converter with soft-switching using a single auxiliary switch for a fuel cell system, IET Power Electronics, Vol. 7, no. 11, pp. 2704-2716, 2014.

[13] Y. Chen, S. Shiu and R. Liang.: Analysis and Design of a Zero-Voltage-Switching and Zero-Current-Switching Interleaved Boost Converter, IEEE Transactions on Power Electronics, Vol. 27, no. 1, pp. 161-173, Jan. 2012.

[14] M. Rezvanyvardom, E. Adib, H. Farzanehfard and M. Mohammadi.: Analysis, design and implementation of zero-current transition interleaved boost converter, IET Power Electronics, Vol. 5, no. 9, pp. 1804-1812, November 2012.

[15] C. Wang, C. Lin, S. Hsu, C. Lu and J. Li.: Analysis, design and performance of a zero-current switching pulse-width-modulation interleaved boost dc/dc converter, IET Power Electronics, Vol. 7, no. 9, pp. 2437-2445, September 2014.

[16] H. Bahrami, E. Adib, S. Farhangi, H. Iman-Eini and R. Golmohammadi.: ZCS-PWM interleaved boost converter using resonance-clamp auxiliary circuit, IET Power Electronics, Vol. 10, no. 3, pp. 405-412, 10 3 2017.

[17] Y. Chen, Z. Li and R. Liang.: A Novel Soft-Switching Interleaved Coupled-Inductor Boost Converter With Only Single Auxiliary Circuit, IEEE Transactions on Power Electronics, Vol. 33, no. 3, pp. 2267-2281, March 2018.

[18] M. Rezvanyvardom, A. Mirzaei and S. Rahimi.: New interleaved fully soft switched pulse width modulation boost converter with one auxiliary switch, IET Power Electronics, Vol. 12, no. 5, pp. 1053-1060, 2019.

Power density of planar transformers designed with commercial standard cores

Reda Bakri[1,2], Xavier Margueron[1], Jean Sylvio Ngoua Teu Magambo[3], Philippe Le Moigne[1], Nadir Idir[1]

[1] Univ. Lille, Arts et Metiers Institute of Technology, Centrale Lille, Yncrea Hauts-de-France, ULR 2697 - L2EP, F-59000 Lille, France

[2] Univ. Lille, Arts et Metiers Institute of Technology, Centrale Lille, Yncrea Hauts-de-France, HESAM Université, ULR 2697 - L2EP, F-59000 Lille, France

[3] SAFRAN, Electrical Systems & Electronics Dept. SafranTech, F-78772 Magny les Hameaux, France

E-Mail: reda.bakri@ensam.eu

Keywords

« Passive component integration », «transformer», « Thermal design », «High power density systems »

Abstract

This paper presents a methodology to evaluate and analyze the volumetric power density of planar magnetics used in power electronics converters. The power density is computed for various EE and E/PLT cores considering optimal configurations for the planar transformers' design and for its cooling heatsink. The analysis is performed for three cooling configurations: natural convection without heatsink, single sided cooled component with one heatsink, and double sided cooled with two heatsinks. This study can be very useful for designers to evaluate their design specifications and to adapt their technological choices to achieve the desired planar magnetics' characteristics.

Introduction

Magnetic components are essential in power electronics converters (PECs). They ensure a number of vital functions such as filtering, storing and transferring energy, or providing galvanic isolation. However, these components occupy a considerable volume inside PECs and involve a non-negligible part of power losses.

For embedded systems, transportation or handheld equipment, more compact and more efficient components are required, hence the need of high-density components. To reach this goal, increasing operating frequencies is unavoidable [1].

Even if wide band gap semiconductors like Silicon Carbide (SiC) and Gallium Nitride (GaN) components allow to reach switching frequencies beyond the MHz [2]-[3], passive components are a serious issue to operate at such frequencies. They limit the opportunities towards power electronics integration.

In this context, planar magnetic components (PMCs) present an interesting solution towards high density and integrated PECs, thanks to some of its advantageous properties compared to wound components [4]-[5]:

- Lower thickness
- Less high frequency (HF) losses
- Good thermal behavior
- Good reproducibility

For their first applications, custom cores were used to manufacture planar magnetics as presented in [6]-[7]. Then, standard planar magnetic cores have been introduced in the market [8]-[9]. Nowadays, standard cores are cost effective and very used in power converter designs. The power handled by such cores is linked to their size, winding properties, frequency and core materials. In a design with volumetric power density constraints, it is necessary to know the capabilities of these various available cores in order to guide the designers in their choices.

On the one hand, increasing operating frequencies allows to reduce component's volume. On the other hand, it can drastically increase the power losses and so the loss density. As a consequence, the component's thermal constraints also raise and a particular attention must be payed to the cooling system and its incidence on the power density.

In order to guide PMCs designers on their technological choices, this paper presents a methodology to study the power density of various standard EE and E/PLT planar cores, for three cooling modes. The paper is organized as follows: In a first section, the hypotheses of the study are presented. Then, in the second section, the methodology and the different models are detailed. Finally, results are dressed in order to analyze the volumetric power density behavior for various EE and E/PLT planar cores [8]-[9].

Study hypotheses

The objective of the present study is to investigate theorical limits of the volumetric power density for HF planar transformers. These limits represent the maximal power density that can be achieved with a planar transformer based on standard planar core in optimal conditions where the transformer handled power is maximized by taking some hypotheses as addressed in this section.

Three cooling configurations are investigated as presented in Fig. 1: natural convection cooling (Fig.1a), single-sided cooling with a heat sink attached to one external surface of the core (Fig.1b) and a double-sided cooling with two heat sinks attached to the core (Fig.1c). To maximize the extracted heat, the transformer and heatsinks are placed vertically as shown in Fig.1.

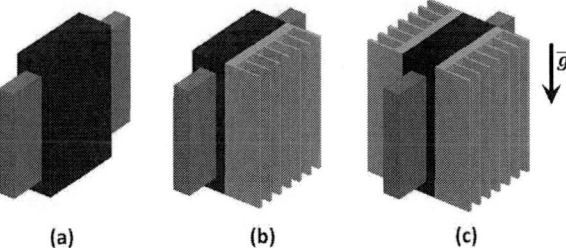

| (a) | (b) | (c) |

Fig. 1: Studied cooling configurations: (a) free convection without heatsink, (b) single-sided cooled and (c) double-sided cooled.

The following hypotheses are considered for the study:
- Copper and insulator layers occupy the total available core window area as shown in Fig. 2a
- The primary and secondary windings have the same number of layers
- Copper and insulator thicknesses (e_{cu} and e_{is}) are considered to be the same for all core sizes
- Maximal interleaving of primary and secondary layers is considered to minimize HF losses
- Copper and core losses are equal if the core is not saturated
- Losses are totally evacuated by the transformer's external area (Fig.2b) by thermal convection and thermal radiation
- Thermal conduction inside component is neglected
- Thermal constraints are fixed: The ambient temperature is set to 40°C and the maximum operation temperature of the transformer is 100°C.

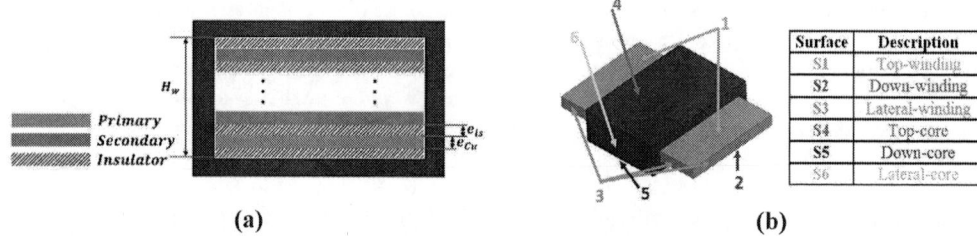

Fig. 2: (a) planar transformer core window and (b) transformer cooling external surfaces

These hypotheses allow some transformer design with the highest power possible. It also simplifies the mathematical models used to elaborate power density evaluating methodology as it will be explained in the next section.

Power density computation methodology

The global flowchart of the power density is presented in Fig.3. First, design specifications are fixed like ambient temperature (T_{amb}), maximal component temperature (T_{max}), copper (e_{cu}) and insulator (e_{is}) thicknesses, core geometrical and material properties and frequency. Then, based on these specifications, the allowable losses are computed. These losses are spited into copper and core losses. Then admissible voltage is deduced from copper losses and admissible current is computed from copper losses.

From admissible voltage and current the amount of power that can be handled by the transformer is computed, the last one is, then, divided by the volume of the transformer and the attached heatsinks if it is the case to evaluate the power density. evaluate the volumetric power density.

The computation of the allowable power, the modeling of the heatsink and the evaluation of the power density are detailed below.

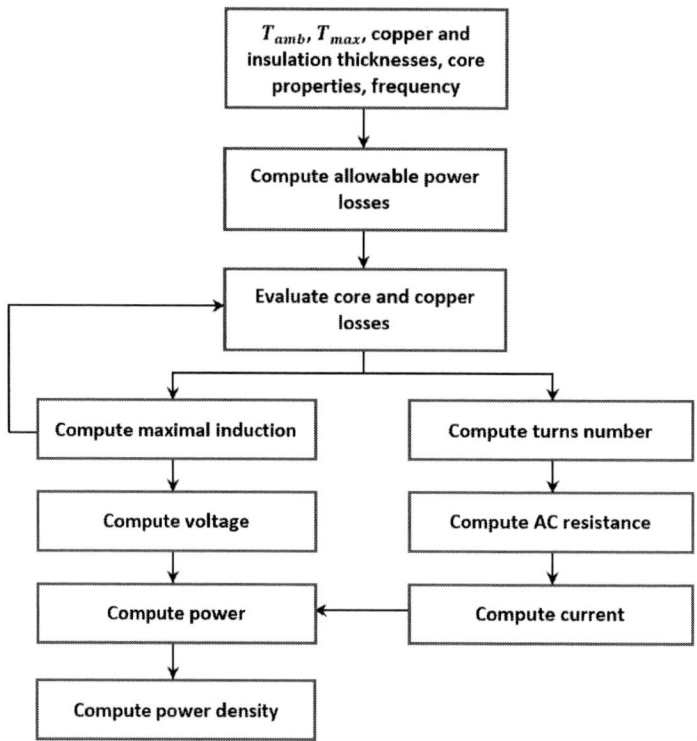

Fig. 3: Planar transformer's volumetric power density computation flowchart

Allowable loss computation

Allowable losses are computed from the ambient (T_{amb}) and the maximal temperature (T_{max}) using a reverse thermal model that compute power losses from these two temperatures. As mentioned before, it is assumed that all the power losses inside the component are evacuated by convection and radiation from the component's external surfaces. This assumption can be justified by the good thermal conductivity of the component's materials (copper, ferrite), associated to the low level of losses that does not generate a high heat flux density inside the component.

In this case, the power losses are the sum of the heat leaving through all the external surfaces of the component by thermal convection and radiation [10] as expressed by (1):

$$P_{loss\_adm} = \sum_{i=1}^{6} S_i h_i (T_{max} - T_{amb}) + \sum_{i=1}^{6} S_i \varepsilon_i \sigma (T_{max}^{4} - T_{amb}^{4}) \tag{1}$$

With h_i the convection heat exchange coefficients for the different external surfaces of the transformer (Fig.2b), S_i the transformer external surfaces (Fig.2 b), ε_i the external surfaces emissvities and σ Plank constant.

Coefficients h_i depend on the surface orientation, its temperature T and the ambient temperature T_a. Expressions for computing those coefficients are given in Table I [10]. Regarding emmessivities, for winding external area it is fixed to $\varepsilon_i = 0.5$, and $\varepsilon_i = 0.9$ for the core.

Table I: Convection heat transfer coefficient for different configurations

Configuration	Geometry	h [W.m⁻².K⁻¹]
Horizontal face top heating		$h = 1.32 \left(\dfrac{T - T_a}{L}\right)^{0.25}$
Vertical face		$h = 1.42 \left(\dfrac{T - T_a}{H}\right)^{0.25}$
Vertical face bottom heating		$h = 0.66 \left(\dfrac{T - T_a}{L}\right)^{0.25}$

Heatsink modeling

In the case where a heat sink is attached to the planar core's external face, an equivalent heat exchange coefficient can be computed from the heatsink modeling. From this study perspective, an optimal heatsink is designed to maximize the heat exchange coefficients. The considered heat sink and its geometrical parameters are shown in Fig.4.

The heat sink base's length L and width W are the same as those of the planar core where they are set. The height and the thickness of the fins are noted H and t respectively. The spacing between them is noted s.

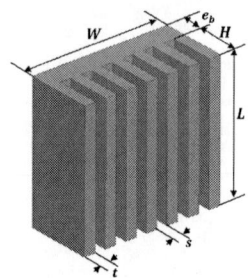

Fig. 1. Heatsink dimensions

In order to neglect the thermal conduction, effect the heatsink base height (e_b) is reduced. Taken into account manufacturing constraints, the base height is fixed to 5mm while the thickness of the fins is set to 1mm to define a manufacturable heatsinks.

The optimal fins' spacing, that allow a maximal heat exchange coefficient is given in (2). S_{opt} is a function of the Rayleigh dimensionless number Ra and the length of the heatsink L [11]:

$$S_{opt} = 2.714 \frac{L}{R_a^{1/4}} \qquad (2)$$

The Rayleigh dimensionless number expression is:

$$R_a = \frac{g \, \beta C_p}{\mu \, k} \Delta T \, L^3 \qquad (3)$$

with g the Gravity acceleration [m.s^{-2}], β the air's expansion coefficient [K^{-1}], C_p the air specific heat [J kg^{-1} K^{-1}], μ the dynamic velocity [Pa.s] and k the air's thermal conductivity [W·m^{-1}·K^{-1}]

Based on (2), the heatsink optimal exchange coefficient can be computed by [11]:

$$h_{hs\_opt} = 1,31 \frac{k_{alu}}{S_{opt}} \qquad (4)$$

The heatsink heat exchange coefficient is computed for a given heatsink external area. It must be reported to the core base surface as shown in equation (5):

$$h_{eq} = h_{hs} . \frac{S_{heatsink}}{S_{bcore}} \qquad (5)$$

with $S_{heatsink}$ the total external surface of the heatsink and S_{bcore} the core base surface where the heatsink is attached.

The obtained value h_{eq} represents the heatsink in equation (1). The losses computed with this equation will be used to compute the maximal allowable power on the component for all the cores and for the three cooling configurations.

Maximal allowable power computation and volumetric power density

In order to evaluate the planar transformer maximal admissible power, the admissible voltage and current are needed. Therefore, the number of primary and secondary turns N must be determined. This parameter is evaluated from the core window height H_w, the conductors' thickness e_{cu} and the insulator thickness e_{is} by assuming a maximal filling of the core window (Fig.2) as expressed by (6):

$$H_W \approx 2N * e_{cu} + (2N + 1)e_{is} \tag{6}$$

The number of turns N can be deduced from the last equation taking into account that it must be integer:

$$N = INT\left[\frac{H_w - e_{is}}{2(e_{cu} + e_{is})}\right] \tag{7}$$

For an optimal design, the computed allowable losses are equitably shared between copper losses P_{cu} and core losses P_{core} (8):

$$P_{cu} = P_{core} = \frac{P_{loss\_adm}}{2} \tag{8}$$

Then, the maximal induction is computed from core losses using the Steinmetz equation [12]:

$$P_{core} = k_c f^\alpha B_{max}{}^\beta V_{core} \tag{9}$$

with B_{max} the maximal magnetic induction [T], α, β and k_c the Steinmetz parameters of the used magnetic material.
V_{core} the core volume
From (6), the maximal induction can be expressed as:

$$B_{max} = \left(\frac{P_{core}}{k_c V_{core} f^\alpha}\right)^\beta \tag{10}$$

Then, by applying Faraday's law, the allowable voltage (V_{adm}) can be computed:

$$V_{adm} = \frac{2\pi}{\sqrt{2}} N.f.S_c.B_{max} \tag{11}$$

If the maximal induction computed in (10) exceeds the core material saturation induction, then the maximal induction is set to the saturation induction value:

$$B_{max} = B_{sat} \tag{4}$$

In this case, core losses also are revaluated by Steinmetz equation based on the saturation induction value. The remained part of the allowable losses is affected to copper losses.

$$P_{cu} = P_{loss\_adm} - P_{core} \tag{12}$$

The maximal allowable current (I_{adm}) can be computed from copper losses and winding resistance calculated with the Dowell's expression [13]:

$$I_{adm} = \sqrt{\frac{P_{cu}}{R_{ac}}} \tag{13}$$

with R_{ac} the Dowell's HF winding resistance (14):

$$R_{ac} = R_{dc}\left[\frac{sinh(2X) + sin(2X)}{cosh(2X) - cos(2X)} + 2X\left(\frac{(m^2 - 1)}{3}\frac{s\ sinh(X) - sin(X)}{cosh(X) + cos(X)}\right)\right] \tag{14}$$

with X the ratio between conductors' thickness and the skin depth δ (16) and R_{dc} is the low frequency resistance given in (16).

$$X = \frac{e_{cu}}{\delta} \tag{15}$$

$$R_{dc} = \frac{l_m\,\rho}{S} \tag{16}$$

with S the conductor section [m^2], ρ the copper resistivity [$\Omega.m$] and l_m the winding mean length [m] that can be estimated from core dimensions (Fig.5):

$$l_m = 2G + 3B - C \tag{17}$$

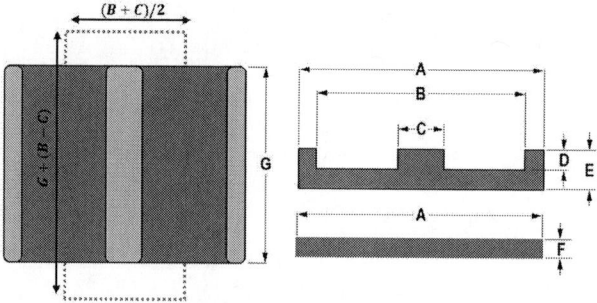

Fig. 2. Magnetic core dimensions and mean winding length

When the maximal allowable voltage and current are known, the maximal admissible power can be computed by the product of the two quantities (18):

$$P_{adm} = V_{adm}.\,I_{adm} \tag{18}$$

Finally, the volumetric power density is evaluated by dividing the maximal allowable power by the transformer volume ($V_{transfo}$) with the attached heatsinks ($V_{heatsink}$) in this case.

$$\rho_{puiss} = \frac{P_{adm}}{V_{transfo} + V_{heatsink}} \tag{19}$$

Transformer volume (20) and heatsink volume (21) can be calculated from geometrical parameters presented in Fig.5 and Fig.4.

$$V_{transfo} = (E + F) * A * G + l_{moy} * \frac{(B - C)}{2} \tag{20}$$

$$V_{radiateur} = L * W * (H + e_b) \tag{21}$$

Results and discussion

The used methodology is established and implemented in Matlab software. All the standard EE and E/PLT planar cores are tested. The impact of their cooling on the volumetric power density is deduced form this methodology. In addition, the impact of conducting and insulating layers' thicknesses is investigated.

Fig.6.a shows the power density evolution for EE64 planar core. Three conductor thickness are considered: 0.075, 0.15 and 0.2 mm. The insulating thickness is fixed to 0.2mm for all the studied cases. According to the plots, the power density presents a maximal value that is reached for different frequencies depending on the conductors' thickness. The thinner is the conductor, the highest is the power density and the highest is the frequency where the maximum is reached. This can be explained by the HF effects on the conductors that generate less losses for thinner conductors.

As regards Fig.6.b, the power density evolution is plotted with frequency for various insulator thicknesses (0.1, 0.2 and 0.3 mm) for the same conductors' thickness (0.2mm). In this case, the maximal volumetric power density is reached for the same frequency, which is linked to the conductor thickness that is the same for the three configurations. The value of the maximal power density is larger for the small insulator thickness because it allows better filling of the window's core.

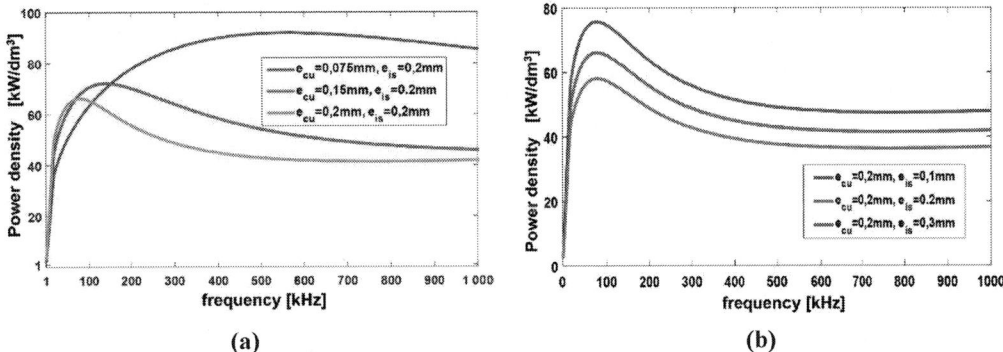

          (a)   **(b)**

Fig. 3. power density evolution for EE64 planar core: (a) for various copper thicknesses and (b) for various insulator thicknesses

The power density evolution with frequency for all the EE and E/PLT cores sizes without heatsink is shown in Fig.7. EE type cores present a higher power density compared to E/PLT cores. This can be explained by the core's window area in the case of EE cores which is two times bigger compared to E/PLT cores, with a less volume difference between the two types of cores. Thus, for a designer it is more attractive to choose an EE core if it is possible instead of an E/PLT type regarding the power density criteria.

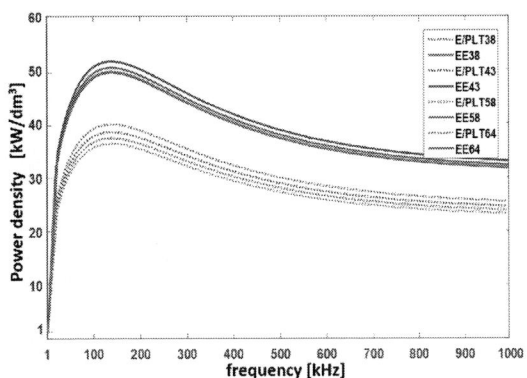

Fig. 7. power density evolution for the studied E/PLT and EE planar cores without heatsink, for e_{cu}=0.2, e_{is}=0.2.

The effect of cooling is illustrated in Fig.8 for E/PLT cores (sizes 38, 43 ,58 and 64) and Fig.9 for EE ones. For these plots, a conductors' thickness of 0.2mm and 0.1mm for insulating layers are considered.

The single-sided cooling presents the better volumetric power density for all the cores. The double-sided cooling presents more power capability. However, its volumetric power density is penalized by the added volume of the second heatsink. For the largest core sizes (E/PLT64 and EE64) it becomes worst compared to the case without heat sink (Fig.8d and Fig.9d).

Conclusion

This paper suggests a methodology for studying volumetric power density of HF planar transformers under thermal constraints for three cooling types: natural convection without heatsink, single sided cooled component with one heatsink and double-sided cooled component with two heatsinks. The approach was applied to study the theorical limits for planar transformers, based on the available commercial E/PLT and EE cores.

The results show that it is more attractive to choose an EE core if it is possible instead of an E/PLT type, because they present a higher power density. The single-sided cooling seems to be the best choice for power density maximization. Indeed, the double-sided cooling can be worse than free air cooling especially for the bigger core sizes.

Even if the limits of power density computed in this paper are difficult to obtain in practice, such study can be used to accelerate a design process by eliminating cores that cannot reach the desired power density for a given frequency and cooling condition. It can also help in adapting technological choices to the power density specification in the early design stage.

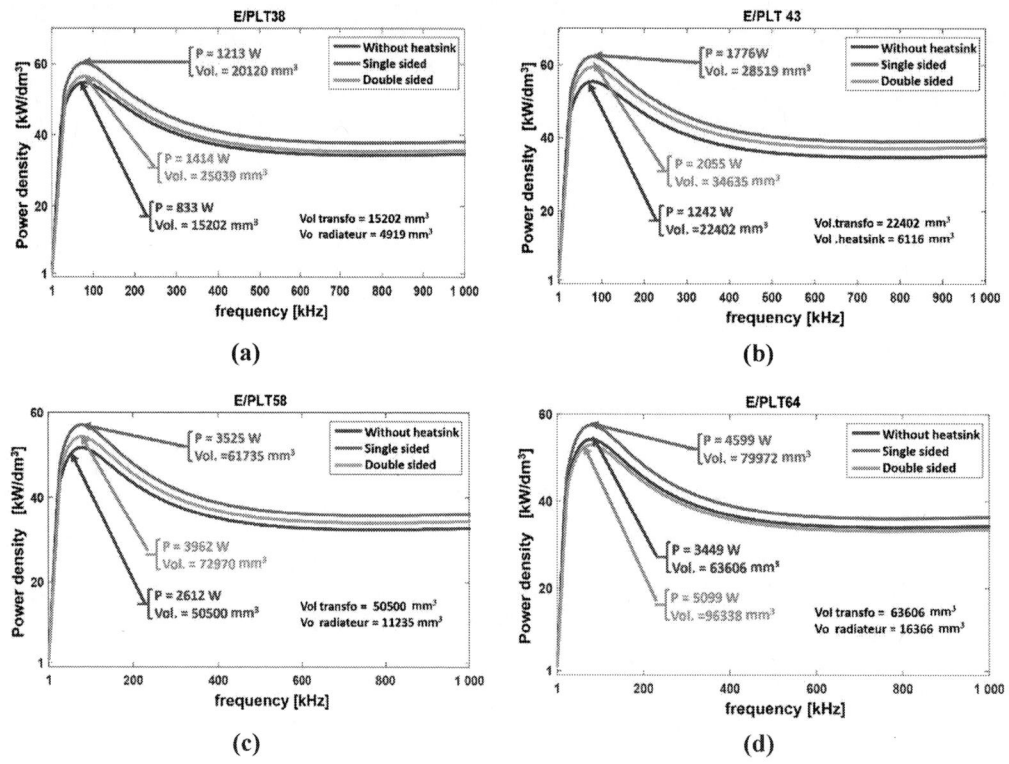

Fig. 8: Studied evolution of power density with frequency for three cooling configurations (E/PLT cores): (a) E/PLT 38, (b) E/PLT 43, (c) E/PLT 58, (c) E/PLT 64

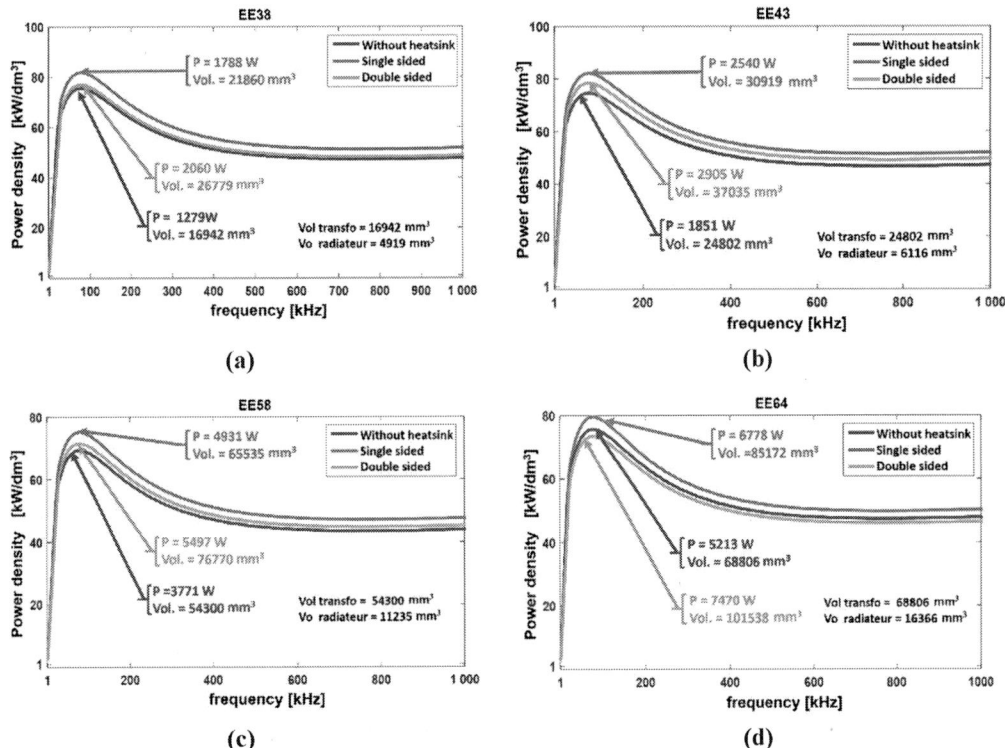

Fig. 9: Studied evolution of power density with frequency for three cooling configurations (EE cores): (a) EE 38, (b) EE 43, (c) EE 58, (c) EE 64

References

[1] Kolar J. W., Drofenik U., Biela J. et al.: PWM Converter Power Density Barriers, Proc. 2007 Power Conversion Conference – Nagoya, Japan, 2007, pp. 9-29

[2] Millán .J, Godignon .P, Perpiñà .X, Pérez-Tomás .A, and Rebollo .J.: A Survey of Wide Bandgap Power Semiconductor Devices, IEEE Transactions on Power Electronics, vol. 29, no. 5, pp. 2155–2163, May 2014

[3] Mantooth .H. A, Glover .M. D, and Shepherd.P.: Wide Bandgap Technologies and Their Implications on Miniaturizing Power Electronic Systems, IEEE Journal of Emerging and Selected Topics in Power Electronics, vol. 2, no. 3, pp. 374–385, Sep. 2014

[4] Z. Ouyang and M. A. E. Andersen.: Overview of Planar Magnetic Technology—Fundamental Properties, IEEE Transactions on Power Electronics, vol. 29, no. 9, pp. 4888–4900, Sep. 2014

[5] Ngoua Teu Magambo J. S. Bakri R., Margueron X. et al.: Planar Magnetic Components in More Electric Aircraft: Review of Technology and Key Parameters for DC–DC Power Electronic Converter, IEEE Transactions on Transportation Electrification, 2017, 3, (4), pp. 831-842

[6] Sayani .M. P, Skutt .G. R, and Venkatraman.P. S.: Electrical and thermal performance of PWB transformers, in [Proceedings] APEC '91: Sixth Annual Applied Power Electronics Conference and Exhibition, 1991, pp. 533–542

[7] van der Linde .D, Boon.C. A. M, and Klaassens.J. B.: Design of a high-frequency planar power transformer in multilayer technology, IEEE Transactions on Industrial Electronics, vol. 38, no. 2, pp. 135–141, Apr. 1991

[8] Ferroxcube: Soft Ferrites and Accessories Data Handbook. Available online on www.ferroxcube.com, 2013.

[9] Magnetics: Ferrite core catalogue. Available online on https://www.mag-inc.com/getattachment/Products/Ferrite-

[10] Incropera F .: Fundamentals of Heat and Mass Transfer », New York: John Wiley and Sons, 1985.

Cores/Ferrite-Shapes/Learn-More-about-Ferrite-Shapes/Magnetics-Ferrite-Catalog-2017.pdf?lang=en-US, 2017.

[11] Kraus A D and A. Bar-Cohen A.: Design and analysis of heat sinks. New York: Wiley, 1995.

[12] Brittain.J.E.: A steinmetz contribution to the AC power revolution.: Proceedings of the IEEE, vol. 72, no. 2, pp. 196–197, Feb. 1984

[13] Dowell.P. L.: Effects of eddy currents in transformer windings, Proceedings of the Institution of Electrical Engineers, vol. 113, no. 8, pp. 1387–1394, Aug. 1966

Effects of PV Panel and Battery Degradation on PV-Battery System Performance and Economic Profitability

Monika Sandelic, Ariya Sangwongwanich, Frede Blaabjerg
Department of Energy Technology, Aalborg University, Aalborg, Denmark
Email: mon@et.aau.dk, ars@et.aau.dk, fbl@et.aau.dk

Keywords

≪Batteries≫, ≪Photovoltaic≫, ≪IGBT≫, ≪Thermal stress≫, ≪Energy system management≫

Abstract

The performance of the photovoltaics (PV) systems with integrated battery energy storage is influenced by key-components degradation over time. In long-term operation, this aspect can affect the overall economic profitability of the system. In this paper, the impacts of the degradation and the lifetime of the key components of the PV-battery system on economic profitability are investigated. The influences of PV panel and battery degradation on the electrical loading and lifetime of the power converters are also demonstrated. A case study reveals that the economic profitability reduces more than 35% due to the degradation over time in service. It is concluded that the performance degradation needs to be carefully considered when analysing the economic profitability of the PV-battery systems.

Introduction

In recent years, PV systems are increasingly being coupled with battery energy storage systems, especially on a residential level [1]. This is mainly due to the decreasing price of PV modules and the subsequent cost of PV energy, as well as the decreasing price of the battery packs [2]. Those cost reductions are the key enablers for even larger integration of PV-battery systems globally. Since the economic profitability plays an important role in the integration of PV-battery systems, accurate predictions of the associated costs and profit are necessary in the planning stage. In that case, the lifetime of the key components needs to be taken into account, since their failure can strongly affect the operation and maintenance cost. With current industry developments, lifetime of PV panels can reach up to 25-30 years [3], while the lifetimes of power electronics and battery are typically much lower [4], [5]. Thus, those components need to be replaced several times during the entire operation, imposing an additional cost of the replacement. Another important aspect that needs to be considered is the degradation of PV panels and battery unit during their time in service. In fact, the degradation of those components influences their power capability, which has an impact on the overall system performance and subsequent economic profitability [6], [7].

Previous research focused on the influence of PV or battery degradation solely on lifetime. For instance, the influence of PV panel degradation on lifetime and reliability of PV inverters has been investigated in [4]. From a battery perspective, in [8], the authors investigated how battery lifetime is influencing its capability of providing certain service. The impact of degradation and lifetime of components has been taken into account in economic assessments, as discussed in [9] and [10]. However, in those cases, the component lifetime is assumed based on the statistical data when calculating the replacement cost and the economic profitability. This assumption does not reflect the performance of PV-battery system in real operation, where the lifetime of the components may vary considerably according to their operating condition (e.g., mission profile) [11]. Thus, a more realistic interval replacement of the components and the associated cost should be considered based on their lifetime/degradation model. Moreover, the

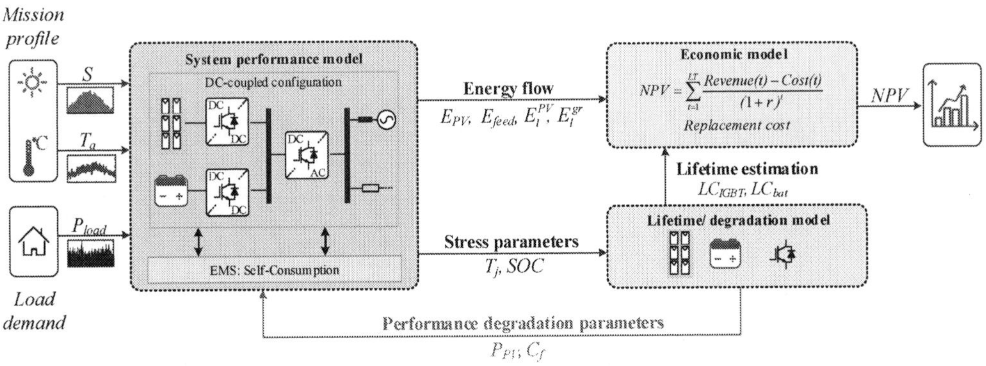

Fig. 1: Diagram showing interactions between three developed models of PV-battery system. The external input is a mission profile consisting of the solar irradiance S, ambient temperature T_a, and load demand P_{load}. The output is net present value NPV. Stress parameters are state-of-charge SOC, and junction temperature T_j. Lifetime estimation parameters are lifetime consumption of IGBT LC_{IGBT}, and battery LC_{bat}. Degradation parameters are capacity fade C_f, and available PV power P_{PV}.

degradation of components results in their limited performance (lower power capability) but changes also the loading of power converters and their lifetime. Furthermore, the degradation of PV and battery affect also the economic profitability of the system due to a continuing decrease in the PV energy yield and battery capacity. This information, together with information about the replacement time and associated replacement cost, need to be considered in an economic analysis. In that way, a more accurate estimation of PV-battery system performance, lifetime, and economic profitability can be obtained. This comprehensive analysis can then provide a more accurate insight during the planning stage, which has not been investigated so far. This can help in leading to a further improvements in terms of efficiency and cost-effectiveness of the PV-battery systems.

In this paper, the impact of PV panel and battery degradation is taken into account when estimating the cost associated with the component replacement (e.g., lifetime) as well as the long-term economic profitability analysis of the PV-battery system. The rest of the paper is organized as follows. The PV-battery system modelling is outlined in Section II. This includes the performance, lifetime/degradation, and economic models of the PV-battery system. The interaction among the models and inclusion of the degradation influence on the system performance and profitability is elaborated in Section III. In Section IV, a case study of PV-battery system in Germany is presented. Finally, concluding remarks are given in Section V.

PV-Battery System Modelling

In the following, three different models are considered to investigate the influence of the degradation on the economic profitability of the PV-battery system. First, a performance model is used to obtain information about power generation and energy distribution in the system. Afterwards, lifetime/degradation model is employed to estimate the component lifetime and thereby its replacement times. Finally, the economic model is used to analyse the economic profitability of PV-battery system. An overview of the model and connections among different parts are shown in Fig. 1.

Performance Model

A single-phase PV-battery system shown in Fig. 2(a) is considered in this study. The system is connected in DC-coupled configuration, where the battery unit is connected at the DC side. Power electronic interface consists of two DC/DC converters, i.e., PV converter and battery converter, as well as a DC/AC inverter connected to the AC grid and the load.

By means of the performance model, it is necessary to determine the energy generation and distribution

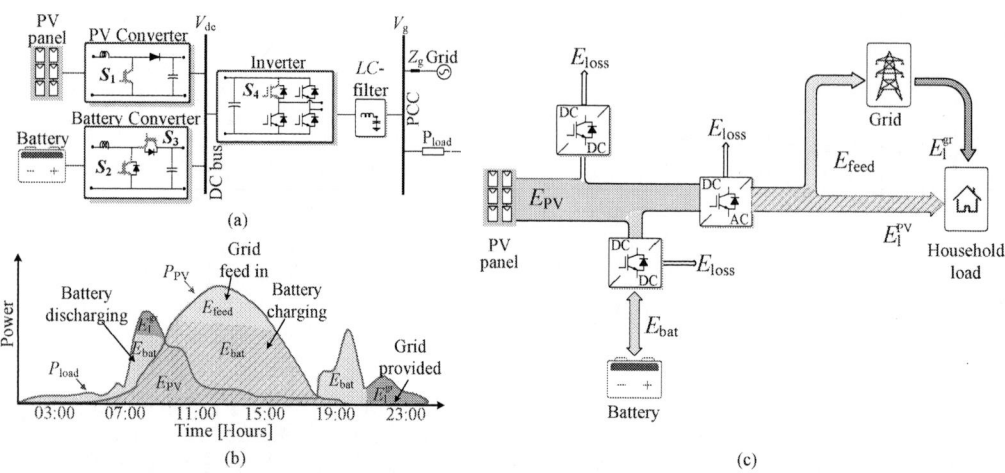

Fig. 2: Performance model of the PV-battery system: (a) DC-coupled configuration, (b) One-day operation based on the self-consumption principles and (c) energy flow of the system corresponding to a one-day operation based on the self-consumption. P_{PV} and E_{PV} are PV power generation and associated energy over investigated time interval respectively, while P_{load} is power demand of household load. E_l^{gr} and E_l^{PV} are energy delivered to the load by grid and PV-battery system respectively. E_{feed} is the excess PV-battery system energy fed to the grid, E_{bat} is energy stored or released by the battery unit and E_{loss} represent the converter losses.

which are later used in the economic analysis. The energy distribution and the loading of the power converters are based on the energy management strategy of the system. One of the most common energy management strategies in residential grid-connected PV-battery systems is self-consumption. Its key target is to maximize the use of PV power generation locally (e.g., internal household loads). The self-consumption operation is illustrated in Fig. 2(b), where a typical one-day PV power P_{PV} and load profiles P_{load} are considered. By means of the aforementioned control strategy and by following the rule of maximizing the internal load supply, the energy distribution in the system is illustrated in Fig. 2(c). As shown, the load E_{load} is supplied by energy from PV-battery system E_l^{PV} and grid E_l^{gr} defined in a way ensuring the priority of internal load supply. To further elaborate this, two periods during the day are examined. The first period is between 11:00 and 13:00 hours where the PV panels energy is higher than load demand, i.e., $E_{PV} > E_{load}$. In such case, the energy yield from the PV panel is firstly used to supply the load and the remaining part is stored in the battery E_{bat}. However, due to battery capacity limitations, not all excess energy can be stored in the battery and it is fed to the grid E_{feed}. The second period is between 19:00 and 23:00 hours where the load demand is higher than PV production, i.e., $E_{load} > E_{PV}$. In such case, the previously stored energy of the battery unit E_{bat} is used to cover the load demand. If it is not sufficient, additional energy is delivered from the grid, E_l^{gr}.

Similarly, a set of parameters determined in the performance model is connected to the lifetime/degradation model. Those are the stress parameters of power converters and battery resulting from the operating conditions. The stress associated with the power converters is represented though the junction temperature T_j of its devices e.g., Insulated-Gate Bipolar Transistors (IGBTs) [12]. In order to evaluate this stress parameter, the electrical loading of the power converters is mapped to the thermal stress of the associated IGBTs. A detailed procedure (implemented in this system) is provided in [12]. Stress parameter associated with the battery unit is its state-of-charge SOC [13]. To monitor this parameter, a SOC estimator needs to be implemented. In this analysis, the Coulomb counting method is used, where the battery current is required as an input.

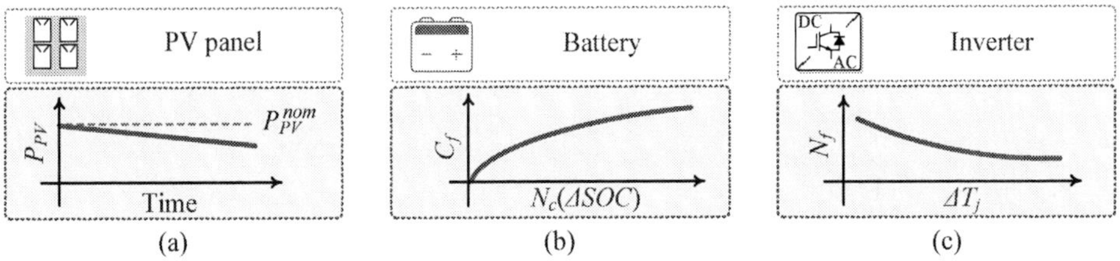

Fig. 3: Lifetime model characteristics diagram of: (a) PV panel, (b) battery, and (c) IGBT of DC/AC inverter.

Lifetime/Degradation model

In the following, lifetime models of the system components are presented. The overview of the lifetime characteristics is given in Fig. 3.

PV Panels

In general, the project lifetime of the PV-battery systems is defined based on the expected lifetime of the PV panels (being the component with the highest lifetime). However, they degrade during their time in service, which is reflected in their decreased performance. In order to account for such behavior, a degradation model of PV panels is defined. It is based on a linear approximation of the output power reduction over time [14], as illustrated in Fig. 3(a). The expression for PV power output with included degradation rate is given in (1), where t represents the year in service and P_{PV}^{nom} is nominal PV power at beginning-of-life (BOL).

$$
P_{PV} = \left(-\frac{0.05}{100} \cdot t + 1 \right) \cdot P_{PV}^{nom}
\tag{1}
$$

Battery

In order to determine the lifetime of the battery unit, it is necessary to monitor its performance-limiting factor - capacity. With time in service, battery capacity gradually reduces, limiting battery availability. Hence, by means of the lifetime model based on [15] and illustrated in Fig. 3(b), the capacity fade C_f resulting from the battery operation is defined as:

$$
C_f = a_{cyc} \cdot N_c (\Delta SOC)^{b_{cyc}} + a_{idl} \cdot k_T \cdot k_V \cdot t_l
\tag{2}
$$

$$
LC_{bat} = \sum_i C_{fi}
\tag{3}
$$

Parameter a_{cyc} is a function of the mean state-of-charge SOC_m and the cycle amplitude ΔSOC. $N_c(\Delta SOC)$ represents the equivalent full cycles determined based on the Wöhler function and its associated parameters a_w and b_w. k_T and k_V are the Arrhenius and Tafel expression respectively and t_l is the idling time of the battery. The rest of the parameters are provided in Table I.

Battery lifetime consumption (LC) for certain operating conditions LC_{bat} is defined in (3). At BOL, LC_{bat} equals 0, meaning that the battery nominal capacity is entirely available. When LC_{bat} accumulates to 1, battery capacity is reduced to 80% of the nominal capacity. This condition represents the battery's end-of-life (EOL) and the component replacement should take place.

Power Electronics Interface

According to the previous study in [16], IGBTs and capacitors are the components that are most prone to failures in the power electronic interface. In this paper, the lifetime assessment of IGBTs is considered in

Table I: Battery lifetime model parameters [7].

Parameter	Value	Description
b_{cyc}	1	Cycle aging parameter
a_{idl}	6.6269×10^{-4}	Idle aging parameter
a_w	151245.25	Wöhler parameter
b_w	-0.968423	Wöhler parameter

Table II: IGBT lifetime model parameters [10].

Parameter	Value	Constant	Value
		β_1	-4.416
I	10 A	β_2	1285
V	6 V	β_3	-0.463
D	300 μm	β_4	-0.716
K	2.03×10^{14}	β_5	-0.761
		β_6	-0.5

the case study. The temperature-related lifetime model, which represents the dominant wear-out failure mode (i.e., bond wire lift-off) is used. The expression for the number of cycles to failure N_f is based on the model developed in [17], outlined in (4) and illustrated in Fig. 3(c). The stress conditions of the mean junction temperature T_{jm} the cycle amplitude ΔT_j and the cycle period t_{on} are considered. The parameters of the lifetime model are provided in Table II. A number of cycles for certain operating conditions n_i is extracted from the mission profile and used together with the number of cycles to failure N_f to calculate LC_{IGBT} following (5).

$$N_f = K \cdot (\Delta T_j)^{\beta_1} \cdot e^{\frac{\beta_2}{T_{jm}+273}} \cdot (t_{on})^{\beta_3} \cdot I^{\beta_4} \cdot V^{\beta_5} \cdot D^{\beta_6} \tag{4}$$

$$LC_{IGBT} = \sum_i \frac{n_i}{N_{fi}} \tag{5}$$

LC_{IGBT} equals 0 at BOL and it accumulates to 1 at EOL. When the converter unit reaches EOL, a replacement and its associated cost are assigned. In the system under study, 4 different IGBTs are employed in the power electronics interface (i.e., S_1-S_4 as highlighted in Fig. 2(a)). It is considered that a converter is failed when at least one of the IGBTs being part of that converter is failed.

Economic Model

The main aim of the economic model is to determine the profitability of the PV-battery system. In order to do so, it is necessary to determine the cash inflow, i.e., revenue, and the cash outflow, i.e., cost of the system. Those two parameters are used for representing the performance and lifetime of the system in the economic aspects. Once cash inflow and outflow are defined, an adequate profitability metrics needs to be chosen.

For residential grid-connected PV-battery systems, revenue is typically generated in two ways. Firstly, it accounts for the savings in the cost of electricity needed to cover load consumption (i.e., the difference between the cost of electricity supplied from the grid and the electricity generated from the PV), R_{sav}. Secondly, it accounts for earnings from selling the excess electricity to the grid (e.g., feed-in tariff), R_{feed}. The total revenue R_{tot} is then the sum of the two, where the mathematical expressions are given in (6)-(8):

$$R_{sav} = E_{load} \cdot C_{gr} - E_l^{PV} \cdot LCOE - E_l^{gr} \cdot C_{gr} \tag{6}$$

$$R_{feed} = E_{feed} \cdot C_{feed} \tag{7}$$

$$R_{tot} = R_{sav} + R_{feed} \tag{8}$$

where C_{gr} is grid electricity price, C_{feed} is feed-in tariff of the PV-battery system and $LCOE$ is system's levelized cost of energy that is calculated by using [18]. E_{load}, E_l^{PV}, E_l^{gr} and E_{feed} are energy parameters defined in the performance model.

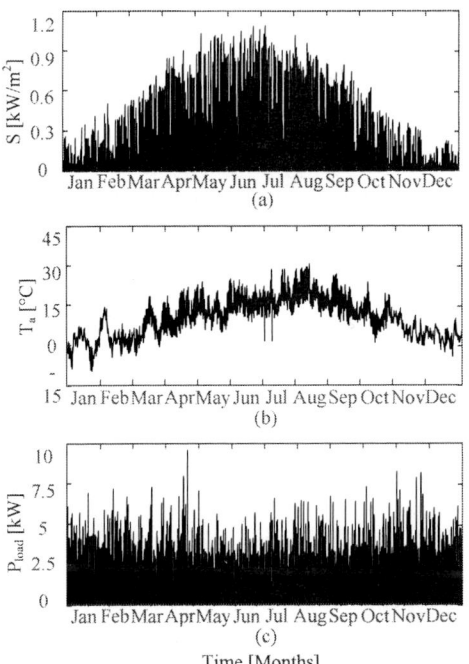

Fig. 4: A One-year mission profile of the PV-battery system: (a) solar irradiance, (b) ambient temperature, and (c) load demand.

Table III: Configuration parameters of the PV-battery system.

Parameter	Value
PV panel rated power	5.5 kW
Battery energy capacity	5.5 kWh
PV converter rated power	6 kW (3 kW x 2 units)
Battery converter rated power	3 kW
PV inverter rated power	6 kW
DC-link voltage	$v_{dc}^* = 450$ V
Grid nominal voltage (RMS)	$V_g = 230$ V

Table IV: Economic parameters of the PV-battery installation in Germany.

Parameter	Value	Description
C_{cpt}^{PV}	1550 USD/kW	PV investment cost
C_{cpt}^{bat}	400 USD/kWh	Battery investment cost
$C_{o\&m}^{PV}$	14.15 USD/kW	PV O&M cost
$C_{o\&m}^{bat}$	7 USD/kWh	Battery O&M cost
C_{rep}^{conv}	18% of C_{cpt}^{PV}	Converter replacement cost
C_{rep}^{bat}	20% of C_{cpt}^{bat}	Battery replacement cost
C_{gr}	0.32 USD/kWh	Grid electricity cost
C_{feed}	0.14 USD/kWh	Feed-in tariff
r	3%	Discount rate
LT	25 years	PV-battery project lifetime

Cost covers the capital, and the operation and maintenance cost. The capital cost consists of the hardware and software cost as well as the installation cost, and represents the initial investment. The operation and maintenance cost also accounts for the replacement cost of the components [19]. In this case, it is affected by the lifetime of the components discussed previously, where the replacement is needed when the component reaches EOL before the end of the project LT (e.g., 25 years).

The system cost can, therefore, be expressed as follows:

$$C_{tot} = C_{cpt}^{PV} + C_{cpt}^{bat} + C_{o\&m}^{PV} + C_{o\&m}^{bat} + C_{rep}^{conv} + C_{rep}^{bat} \tag{9}$$

The capital cost of the PV panels and battery are C_{cpt}^{PV} and C_{cpt}^{bat}, respectively and the operation and maintenance cost are denoted with $C_{o\&m}^{PV}$ and $C_{o\&m}^{bat}$. The replacement cost of power converters and battery are C_{rep}^{conv} and C_{rep}^{bat} respectively.

The economic profitability of the PV-battery system is determined by the net present value (NPV) defined as follows:

$$NPV = \sum_{t=1}^{LT} \frac{R_{tot}(t) - C_{tot}(t)}{(1+r)^t} \tag{10}$$

where r represents the discount rate and LT is the project lifetime. *NPV* evaluates if the project earnings throughout the system operation time are sufficient to cover the investments. The final value of the *NPV* gives information about project profitability. In fact, if the final *NPV* value is lower than zero, the project is not profitable, meaning that the revenue generated throughout the years in operation is not sufficient to cover all the project-associated costs.

Case Study

In order to investigate the impact of the PV and battery degradation, two case studies, with and without considering degradation, are analysed. The analysis is performed based on the diagram shown in Fig. 1. For both cases, the input to the system is mission profile which considers the environmental conditions at the installation site. Those influence the power generation and energy distribution as well as loading and stress conditions of system components (which are evaluated by means of the performance model). The two cases differ based on the way the output information from the lifetime/degradation model is used, which is elaborated in the following.

Case I: Without Considering Degradation

In the first case study, the lifetime of the components is only taken into account in the economic model to determine the replacement cost. In fact, it only evaluates whether or not the unit is failed and how many times it needs to be replaced. This is shown in Fig. 1, where the output of the lifetime model (i.e., LC of power converters and battery, LC_{IGBT} and LC_{bat}) is only connected to the economic model. Calculated replacement cost is the additional cost of the system under operation and it is used in the determination of project NPV. In this case, it is assumed that the system performance is not affected by the component degradation (e.g., the PV power production and loading conditions are similar for each year in operation).

Case II: Considering Degradation

In the second case, the degradation of the components is taken into account not only in the economic analysis but also to assess the system performance. This influence is shown in Fig. 1 as the feedback from the lifetime model to the performance model (marked in red). The LC of each component is calculated after each year in service and PV power capability and battery capacity are updated based on the degradation level calculated from the lifetime/degradation model. This means that the PV-battery system will not have the same performance in the first and the years to follow. As a result, the economic analysis will not only be impacted by the replacement cost (as in the first case), but also by the reduced revenue due to limited power capability of the PV and battery units.

Mission Profile and System Configuration

To investigate the two aforementioned cases, a PV-battery installation site in Germany is considered. The configuration and economic parameters are provided in Table III and Table IV, respectively. The PV power rating is 5.5 kW and battery capacity is 3 kW/5.5 kWh. To perform the evaluation for 25 years period, the one-year mission profile shown in Fig. 4 is repetitively used until the project LT is reached.

Impact of Degradation on the System

In the following, the results of the degradation impact on the performance and the economic profitability are provided. The impact of the degradation on the loading and lifetime of the power converters is shown based on a one-day operation profile and LC curves for LT time span. Furthermore, the NPV of the two aforementioned cases is determined and conclusions regarding the influence of degradation on the economic profitability are drawn.

Stress Profiles

One-day operation is investigated for the Case I and Case II after 25 years of operation. The PV power as well as the battery and power electronics stress profiles are as shown in Fig. 5. It can be seen that there is a substantial difference in PV power profile for the two cases due to the PV panel degradation. The reduction in PV power generation results in a decrease in the battery and power electronics loading and associated stress. This can be observed from the junction temperature profiles of IGBTs, where the largest difference is observed for the IGBT S_4. The reduction in stress of the components due to degradation is also further considered for the whole operation time of 25 years in the following.

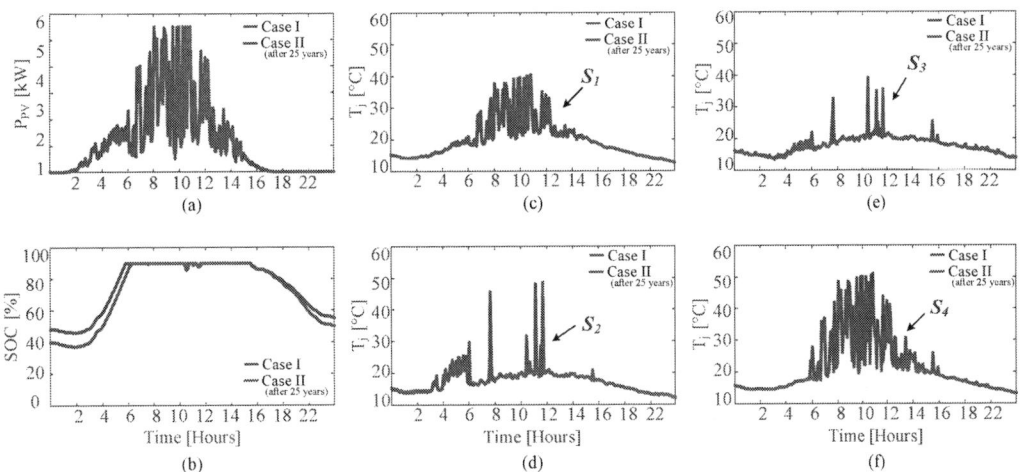

Fig. 5: Impact of PV panel and battery degradation on the one-day operation of PV-battery system: (a) PV power P_{PV}, (b) battery SOC, (c) junction temperature of the IGBT S_1, (d) junction temperature of the IGBT S_2, (e) junction temperature of the IGBT S_3, (f) junction temperature of the IGBT S_4.

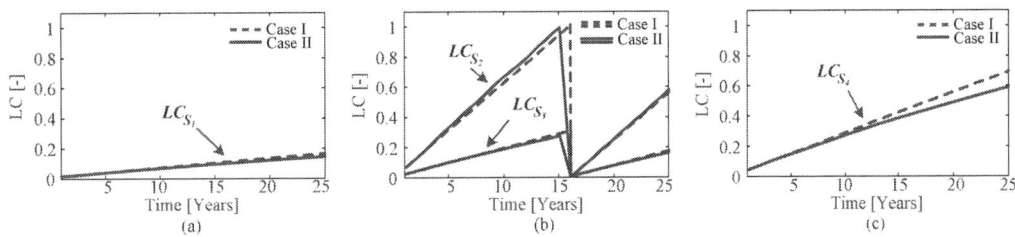

Fig. 6: Lifetime consumption of power electronic components: (a) LC of the IGBT S_1 (PV converter), (b) LC of the IGBTs S_2 and S_3 (Battery converter), and (c) LC of the IGBT S_4 (Inverter).

Lifetime Results

The results for the two cases are shown in Fig. 6, where it can be seen that the LC in Case II is lower for all IGBTs except for S_2. The IGBT S_2 is a part of the battery converter and it is highly loaded during battery discharging. In Case II, due to the reduced PV power generation, battery is required to discharge more often which is then resulting in a higher LC of the IGBT S_2. Similarly to the results shown for one-day operation (see Fig. 5), the highest difference in LC is seen for the IGBT S_4. This difference is a direct result of PV panel degradation where the reduction in its power capability strongly influenced the loading of the DC/AC inverter. Over the years in operation, the power delivered to the load is gradually being reduced, all resulting in a decreased lifetime consumption of IGBT S_4.

Based on the aforementioned results, it can be concluded that neglecting PV panel and battery degradation could lead to a higher LC of most power converters. Thus, without considering the degradation impact, it can result in over-design of the components in the planning stage of the PV-battery systems.

Economic Evaluation

The results of the economic analysis are shown in Fig. 7(a). The final *NPV* after 25 years in service accounts for 5370 USD for the Case I. This final NPV value indicates that the project is profitable. For Case II, the final *NPV* is 3407 USD, which is a reduction of profit for 1963 USD compared to Case I. Furthermore, it can be observed that the PV panel and battery degradation is having a gradual decrease of *NPV* over the years of operation. On the contrary, the lifetime and replacement of the power electronics

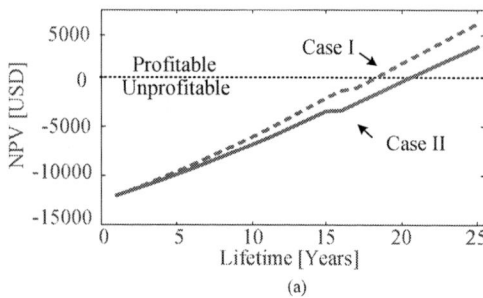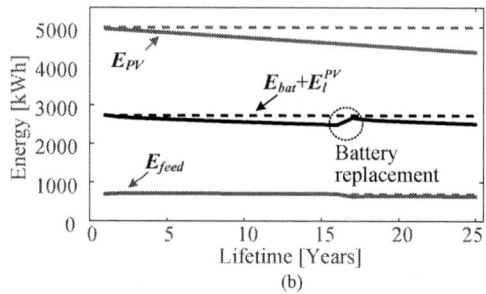

Fig. 7: Case study results: (a) net present value, (b) energy distribution, where solid line represents the results for the Case II (with included degradation) and dashed line Case I (without degradation).

and battery components significantly influence the *NPV* in years in which the replacement occurs.

The two effects (degradation and lifetime) can be, further on, studied through energy generation and distribution over lifespan of the project, *LT*. These results are shown in Fig. 7(b), where the PV generation is linearly decreasing over time with the largest difference in the first and the last year of project lifetime, e.g., 25 years. The self-consumed energy of the PV-battery system is influenced by both PV and battery degradation and it is decreasing until battery replacement takes place (around 15 years). After the battery replacement, a smaller difference in the energy level is seen compared to the first year of operation. The lowest difference in energy level is seen for the feed-in energy. However, this energy is in general lower than the aforementioned two. Therefore, changes in all three energy levels over time strongly influence the economic profitability of the system.

Conclusion

In this paper, the impact of PV panel and battery degradation on the performance and associated cost-effectiveness of the PV-battery system are investigated. The economic profitability of the system is analysed where both the replacement cost of the components (i.e., power converters and battery unit) and the reduction in revenue due to the reduced performance (PV panel and battery degradation) are included. The case study results show that the reduction of the PV and battery power capability influences the performance of the overall system. Furthermore, the impact of degradation on the loading of the power electronic interface units and their lifetime is shown. The results of economic analysis have shown that the degradation of components accounts for more than 35% reduction in the final *NPV*. Hence, it is concluded that the degradation of units needs to be taken into account in order to obtain accurate results about the PV-battery project profitability.

References

[1] R. Akshat, "100,000 homes in Germany now have battery-storage systems connected to the grid," Quarty, 2018. [Online]. Available: https://qz.com /. [Accessed: 21-Oct-2019].

[2] REN21, "Renewables 2019: Global Status Report (GRS)," 2020. [Online]. Available: http://www.ren21.net. [Accessed: 03-Jul-2020].

[3] Energy Informative, "Solar Panel Warranty Comparison," [Online]. Available: https://energyinformative.org/solar-panel-warranty-comparison/. [Accessed: 03-Jul-2020].

[4] A. Sangwongwanich, Y. Yang, D. Sera and F. Blaabjerg, "Lifetime Evaluation of Grid-Connected PV Inverters Considering Panel Degradation Rates and Installation Sites," in *IEEE Trans. Power Electron.*, vol. 33, no. 2, pp. 1225-1236, Feb. 2018.

[5] J. Li, R. Xiong, H. Mu, B. Cornélusse, P. Vanderbemden, D. Ernst and W. Yuan, "Design and real-time test of a hybrid energy storage system in the microgrid with the benefit of improving the battery lifetime," in *Appl. Energy*, vol. 218, pp. 470-478, May 2018.

[6] National Renewable Energy Laboratory, "The Role of Reliability and Durability in Photovoltaic System Economics," Tech. Rep. No. NREL/PR-6A20-73751, 2019.

[7] B. Xu, A. Oudalov, A. Ulbig, G. Andersson and D. S. Kirschen, "Modeling of Lithium-Ion Battery Degradation for Cell Life Assessment," in *IEEE Trans. Smart Grid*, vol. 9, no. 2, pp. 1131-1140, March 2018.

[8] G. Angenendt, S. Zurmühlen, R. Mir-Montazeri, D. Magnor and D. Uwe Sauer, "Enhancing Battery Lifetime in PV Battery Home Storage System Using Forecast Based Operating Strategies," in *Energy Procedia*, vol. 99, pp. 80–88, Nov. 2016.

[9] C. Galilea, J. Pascual, A. Berrueta, A. Ursua and L. Marroyo, "Economic analysis of residential PV self-consumption systems with Li-ion batteries under different billing scenarios," *Proc. of EEEIC/I&CPS Europe*, Genova, Italy, 2019, pp. 1-6.

[10] P. C. Palavicino, Y. Wu, M. Smuda, W. Choi, W. Lee and B. Sarlioglu, "Methodology for Evaluating Potential Benefits and Economic Value of Residential Photovoltaic and Battery Energy Storage System," *Proc. of ITEC*, Long Beach, CA, 2018, pp. 725-730.

[11] M. Musallam, C. Yin, C. Bailey and M. Johnson, "Mission Profile-Based Reliability Design and Real-Time Life Consumption Estimation in Power Electronics," in *IEEE Trans. Power Electron.*, vol. 30, no. 5, pp. 2601-2613, May 2015.

[12] P. D. Reigosa, H. Wang, Y. Yang and F. Blaabjerg, "Prediction of Bond Wire Fatigue of IGBTs in a PV Inverter Under a Long-Term Operation," in *IEEE Trans. Power Electron.*, vol. 31, no. 10, pp. 7171-7182, Oct. 2016.

[13] M. Sandelic, A. Sangwongwanich and F. Blaabjerg, "Reliability Evaluation of PV Systems with Integrated Battery Energy Storage Systems: DC-Coupled and AC-Coupled Configurations," in *Electronics*, vol. 8, no.9, pp. 1059-1077, August 2019.

[14] D. C. Jordan, T. J. Silverman, B. Sekulic and S. R. Kurtz, "PV degradation curves: non-linearities and failure modes," in *Prog. Photovolt*, vol. 25, no. 7, pp. 583– 591, July 2017.

[15] G. Angenendt, S. Zurmühlen, H. Axelsen and D. Uwe Sauer, "Comparison of different operation strategies for PV battery home storage systems including forecast-based operation strategies," in *Appl. Energy*, vol. 229, pp. 884-899, Nov. 2018.

[16] H. Wang, M. Liserre, F. Blaabjerg, P. de Place Rimmen, J. B. Jacobsen, T. Kvisgaard, J. Landkildehus, "Transitioning to Physics-of-Failure as a Reliability Driver in Power Electronics," in *IEEE Trans. Emerg. Sel. Topics Power Electron.*, vol. 2, no. 1, pp. 97-114, March 2014.

[17] R. Bayerer, T. Herrmann, T. Licht, J. Lutz and M. Feller, "Model for Power Cycling lifetime of IGBT Modules - various factors influencing lifetime," *Proc. of CIPS*, Nuremberg, Germany, 2008, pp. 1-6.

[18] International Renewable Energy Agency, "Cost and competitiveness indicators: Rooftop solar PV," 2017. [Online]. Available: https: //www.irena.org/

[19] National Renewable Energy Laboratory, "PV O&M Cost Model and Cost Reduction," Tech. Rep. No. NREL/PR-7A40-68023, 2017.

Full Sensorless Operation of Induction Machines based on Online Identification of Saliencies using Harmonic Compensation LUTs in Traction Applications

E. Rodriguez Montero[1], M. Vogelsberger[2], T. Wolbank[3]
VIENNA UNIVERSITY OF TECHNOLOGY[1,3]
Gusshausstrasse 27/29 E370-2, 1040
Vienna, Austria.
Tel.: +43 (1) 58801 – 370231[1]
Tel.: +43 (1) 58801 – 370226[3]
eduardo.montero@tuwien.ac.at[1]
thomas.wolbank@tuwien.ac.at[3]
BOMBARDIER TRANSPORTATION AUSTRIA GMBH
PRODUCT AND R&D MGMT. DRIVES
BT COO_EQUIPMENT[2]
Hermann Gebauer Straße 5, 1220
Vienna, Austria.
Tel.: +43 (1) 25110 – 599[2]
markus.vogelsberger@rail.bombardier.com[2]

Acknowledgements

This research is supported by Austrian Research Promotion Agency (FFG) under the Bridge-program, grant no. 858502.

The authors want to thank Bombardier Transportation- COO_Equipment, especially Mr. H. Mannsbarth (global head of BT_COO Equipment_Module Center Bogies & Drive) for research/development funding and project supervision.

Special thanks for his generous support and technical feedbacks goes to Mr. M. Bazant (head of DRIVES_ Product- and R&D_Management in BT_COO Equipment_Module Center Bogies & Drives) and Mr. G.Hilpert (BT_COO_Equipment_ Energy&Motion _Head of Platform Mgmt.).

Furthermore the authors want to thank colleagues from Bombardier Transportation BT_COO_Equipment (Mr. Ch. Wirth, Mr. U. Sorg, Mr. Chr. Guntermann, Dr. J. Y. Favez, Mr. G. Harasleben, Mr. M. Ganster, Mr. W. Cepak and Mr. E. Moser), as well as Prof. Dr. H. Ertl (TU-Wien) for all their feedback.

The authors further are very indebted to LEM Company (especially Mr. A. Hürlimann/Chairman of Board of Directors, Dr. W. Teppan & Mr. J. Burk) for the cooperation and generous support.

Keywords

AC machine, Estimation technique, Sensorless control

Abstract

Speed sensorless control of induction machines (IM) around zero frequency is distinctively carried out by external injection, whose objective is to extract rotor-related information by evaluating the actual state of the machine saliencies. Commonly, IM saliencies are identified during an offline encoder-based test where each saliency signal component is estimated regarding its amplitude and phase shift for each required torque point. It is then, during the sensorless online operation, where the identified saliencies are eliminated from the total saliency vector, leading to a single saliency component, which is used for sensorless control scheme. The offline process needed for saliency identification requires time, a shaft encoder (during offline test) and cannot deal with online parameter variations. To deal with such drawbacks, this paper proposes a fully sensorless saliency identification method that runs without position sensor even in the identification phase and can be updated during online operation.

Introduction

Sensorless Field Oriented Control (FOC) of IM requires a rotor flux position as feedback signal for common torque or speed control. This can be calculated without any further sensors by means of models that estimate the flux position based on the actual current and fundamental wave reference voltage. However, the actual machine voltage strongly differs from the reference voltage at very low speeds due to the nonlinear switching of the inverter. Thus, stable zero speed sensorless FOC can only be accomplished by the so-called external signal injection methods. These methods can be sorted into high frequency injection [10-11] and voltage step excitation [6-7], where the second group is often preferred as it involves no current control interference and no additional saliencies due to inverter non-linear switching.

Voltage step excitation strategies draw on the machine inherent saliencies that are assessable at the leakage inductance, which is inversely related to the time derivative of the line current. Therefore, it is sufficient to evaluate the current response to a voltage step caused by the inverter in order to access the saliency information. However, two additional effects are contained in the identified value of the leakage inductance. First, the voltage step applied to the inductance is not constant due to the back EMF and stator resistance and, second, an average inductance value is always present, accounting for the symmetric value of the leakage inductance that would exist if there were no spatial asymmetries.

In order to obtain a saliency position vector without any other effect, different excitation strategies have been presented, tending to integrate the voltage steps into the SVPWM switching pattern [6-7]. The output of all of them is a vector, where all saliency harmonics are superimposed, being only one of them the sensorless control signal. It is then the challenge of the signal processing and signal analysis, to identify and separate all signals so that only one remains for the control purpose. Several methods have proposed in literature to identify and eliminate the undesired saliency harmonics. A few strategies are based on advanced filtering [1-6], but do not address the problem of frequency overlap. On the other hand, saliency harmonic compensation LUTs (Look-Up Table), as in [7-10], are based on the individual identification of each saliency regarding amplitude and phase shift as a function of the torque. They are able to work at spectral overlap, but have traditionally been identified during a time-consuming offline pre-commissioning test for which an encoder was required. Besides, they do not adapt during online operation, which is indispensable in case of parameter modification. Particularly for railway applications, it is crucial that the saliency identification LUTs can be updated online and filled up from zero, even when the train is on the rails. After identifying the saliencies, normal sensorless operation is done by accessing these LUTs with the actual torque current component.

This paper proposes a fully sensorless saliency identification method that will be performed on a test bench, but could be theoretically carried out with the motor mounted in the train. At first instant, a constant speed stator current is supplied to the machine, which enables the initial identification of the offset, saturation and intermodulation. After this initial saliency identification, sensorless FOC follows, updating the saliency identified values during online operation. The saliency parameters (angle and amplitude), which are stored and updated in a LUT for online access, are calculated based on a minimization function [10]. The excitation strategy used is explained in [7]. The slotting harmonic is finally selected as sensorless information to calculate the rotor flux position.

Saliency-offset vector acquisition

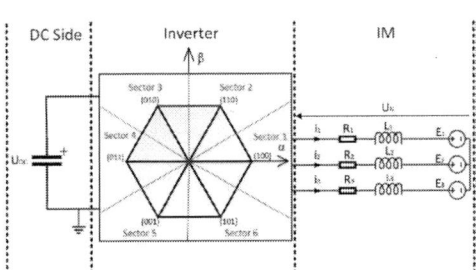

Fig. 1: Typical figure caption

Fig.1 shows the DClink-inverter-machine system. U denotes voltage, i current, R phase resistance, L inductance, E phase back electromotive force (EMF), *123* the three phase sub-indices and N an arbitrary phase.

Considering the IM model of Fig. 1, under any ijk inverter state, the leakage inductance can be related to the current derivate as shown in (1), U_N taking the values of $\pm 2/3 U_{DC}$, $\pm 1/3 U_{DC}$ or 0 depending on the phase and switching state.

$$(U_N - R_N \cdot i_N - E_N) \cdot \frac{1}{L_N} = \frac{di_N{}^{(ijk)}}{dt} \tag{1}$$

Besides, the leakage inductance can be expressed as a summation of a constant value L_0 and a sinusoidal term for each spatial anisotropy present in the machine. The mean value L_0 represents the symmetrical value of the inductance that would exist, if there were no machine saliencies present. For clarity reasons, only one arbitrary saliency with L_m amplitude and γ angle is represented in (2).

$$L_N = L_{0,N} + L_m \cdot \sin(\gamma + (N-1) \cdot 2 \cdot \frac{\pi}{3}) \tag{2}$$

However, there are several saliencies in induction machines. Saturation, slotting and intermodulation are usually present, and are related with the machine geometry or flux as shown below.

$$L_{m,sat} = f(2 \cdot \theta_{Flux}) \tag{3}$$

$$L_{m,slot} = f(N_{RS} \cdot \theta_{Rotor}) \tag{4}$$

$$L_{m,inter} = \pm f(L_{N,sat}) - f(L_{N,slot}) \tag{5}$$

Where N_{RS} is the number of rotor slots.

As proposed in [7], the current responses of only one active and one inactive inverter states are sufficient to obtain a saliency-offset vector, containing both, the constant value as well as the sinusoidal saliency-related values mentioned above. This enables the integration of the excitation sequence into the SVPWM.

Assuming the fundamental reference voltage lies within Sector 1 (Fig. 1), a different saliency sinusoid plus offset is obtained by means of (1) – (2) for each phase (phases 1-2-3). Note, for clarity purposes only one saliency is expressed in (6)-(9).

$$\frac{di_1{}^{(100)}}{dt} - \frac{di_1{}^{(000)}}{dt} = \frac{2 \cdot L_{0,A} - L_m \cdot \sin(\gamma_m)}{cst} \tag{6}$$

$$\frac{di_2{}^{(100)}}{dt} - \frac{di_2{}^{(000)}}{dt} = \frac{-L_{0,A} - L_m \cdot \sin(\gamma_m + 4 \cdot \pi/3)}{cst} \tag{7}$$

$$\frac{di_3{}^{(100)}}{dt} - \frac{di_3{}^{(000)}}{dt} = \frac{2 \cdot L_{0,A} - L_m \cdot \sin(\gamma_m + 2 \cdot \pi/3)}{cst} \tag{8}$$

Where cst is a constant term at constant flux and torque.

In order to obtain a saliency-offset vector, currents are combined as in (9)

$$\frac{di^{(ijk)}}{dt} = \frac{di_1^{(ijk)}}{dt} + \frac{di_2^{(ijk)}}{dt} \cdot e^{i\left(2 \cdot \frac{\pi}{3}\right)} + \frac{di_3^{(ijk)}}{dt} \cdot e^{i\left(4 \cdot \frac{\pi}{3}\right)} \tag{9}$$

The saliency-offset vector for each sector defined in Fig. 1 is presented in Table I

Table I: Sector saliency-offset vector using one active and one inactive inverter state.

INVERTER SECTOR	SLOPE COMBINATION	SALIENCY-**OFFSET** VECTOR
S1	$\dfrac{di_{100} - di_{111}}{dt}$	$\dfrac{\mathbf{L_{0,S1}} - \frac{1}{2} \cdot L_m \cdot e^{i \cdot \gamma_m}}{cst}$
S2	$\dfrac{di_{110} - di_{111}}{dt}$	$\dfrac{-\mathbf{L_{0,S2}} \cdot \mathbf{e}^{i \cdot 4\frac{\pi}{3}} - \frac{1}{2} \cdot L_m \cdot e^{i\left(\gamma_m + 2 \cdot \frac{\pi}{3} + \pi\right)}}{cst}$
S3	$\dfrac{di_{010} - di_{111}}{dt}$	$\dfrac{\mathbf{L_{0,S3}} \cdot \mathbf{e}^{i \cdot 2 \cdot \frac{\pi}{3}} - \frac{1}{2} \cdot L_m \cdot e^{i\left(\gamma_m + 4\frac{\pi}{3}\right)}}{cst}$
S4	$\dfrac{di_{011} - di_{111}}{dt}$	$\dfrac{-\mathbf{L_{0,S4}} - \frac{1}{2} \cdot L_m \cdot e^{i(\gamma_m + \pi)}}{cst}$
S5	$\dfrac{di_{001} - di_{111}}{dt}$	$\dfrac{\mathbf{L_{0,S5}} \cdot \mathbf{e}^{i \cdot 4\frac{\pi}{3}} - \frac{1}{2} \cdot L_m \cdot e^{i\left(\gamma_m + 2\frac{\pi}{3}\right)}}{cst}$
S6	$\dfrac{di_{101} - di_{111}}{dt}$	$\dfrac{-\mathbf{L_{0,S6}} \cdot \mathbf{e}^{i \cdot 2 \cdot \frac{\pi}{3}} - \frac{1}{2} \cdot L_m \cdot e^{i\left(\gamma_m + 4 \cdot \frac{\pi}{3} + \pi\right)}}{cst}$

Offset estimation strategy

As Table I shows, the saliency-offset vector is composed by an offset term (bold), and the saliency term, where only one saliency is written for clarity reasons. Since the offset amplitude is up to hundreds of times bigger than the saliency amplitude, the offset estimation and accurate elimination is crucial in order to be able to evaluate the saliencies information. After the offset elimination, all what remains is a saliency vector where all saliencies are superposed. The offset elimination of the saliency-offset vector is explained in [7]. It is based on spatial rotations and a signal displacement, and can be carried out during online operation. It requires half an electrical revolution, and delivers the vector offsets of all sectors. After the sector- and load-dependent offset elimination, a saliency vector is obtained with all present saliencies superimposed.

Saliency harmonic compensation LUT

The saliency vector consists of three main saliencies superimposed. These are saturation harmonic, which rotates with twice the electrical frequency; the slotting harmonic, whose frequency is a factor of the mechanical frequency times the rotor slot number; and the intermodulation, which rotates with the - slotting +- saturation frequency.

Throughout this work, the slotting harmonic will be used as the base to calculate the rotor flux angle for sensorless control. In this sense, the intermodulation and saturation components will be eliminated applying the harmonic compensation LUT method. It consists of an identification of the machine saliencies regarding phase shift and amplitude as a function of the machine load, and a further elimination of all saliencies expect one (slotting).

The LUT is a 2D Look-up Table where the saliencies, as explained in [10], are characterized as a vector regarding amplitude and phase shift, by means of a function f_1 that minimizes the angle and amplitude deviation of an ideal vector from the saliency vector.

It is summarized as follows, x_1 θ_1 being amplitude and angle respectively, N sample amount and SV saliency-vector:

$$C_1(x_1, \theta_1) = \frac{1}{N} \sum_{i=1}^{N} abs[\overline{SV} - f_1] \tag{10}$$

$$f_1 = x_1 \cdot e^{j(\gamma - \theta_1)} \tag{11}$$

Full sensorless identification and controlled operation

This section explains the proposed method for a sensorless identification and FOC.

The procedure, completely sensorless, begins by supplying the machine with fixed amplitude and a very low speed stator current to reach rated magnetization. After a few rotor time constants, the saliency-offset vector is acquired for a half electrical period using the one-active inverter state excitation and signal processing of Table I. After this, the offset is estimated, stored and eliminated using [7] signal processing. Immediately after, the saliency identification method of [10] is used to estimate only the saturation harmonic, as the intermodulation is not evaluable yet due to the lack of sensorless mechanical position. The estimated saturation is then subtracted from the saliency vector, which means the slotting is the biggest remaining component in the vector. Thus, a sensorless mechanical angle is now available, used to estimate the intermodulation saliency parameters.

Once the initial offset and the saturation and intermodulation saliency parameters are known at rated flux, the control switches to FOC at rated flux. At constant low speed (controlled by a coupled load machine), the process continues by estimating again the offset, saturation and intermodulation values at no-load. These values are stored and updated in the LUT and eliminated from the saliency-offset vector. In a next stage, the load is increased in a small step and, after half an electrical period, the offset, intermodulation and saturation are estimated, stored and subtracted from the saliency-offset vector. This can either be done in a test bench or with the motor mounted in the train. The load is afterwards gradually increased in small steps, estimating, storing and eliminating the offset, saturation and intermodulation, which leads to a load-dependent LUT having the offset, saturation and intermodulation as outputs.

During online sensorless operation, the offset and saliencies can be estimated and updated in the LUT if load is kept constant during at least half an electrical period.

Flowchart of the method

The above explained sensorless identification procedure is shown in the flowchart of Fig. 2.

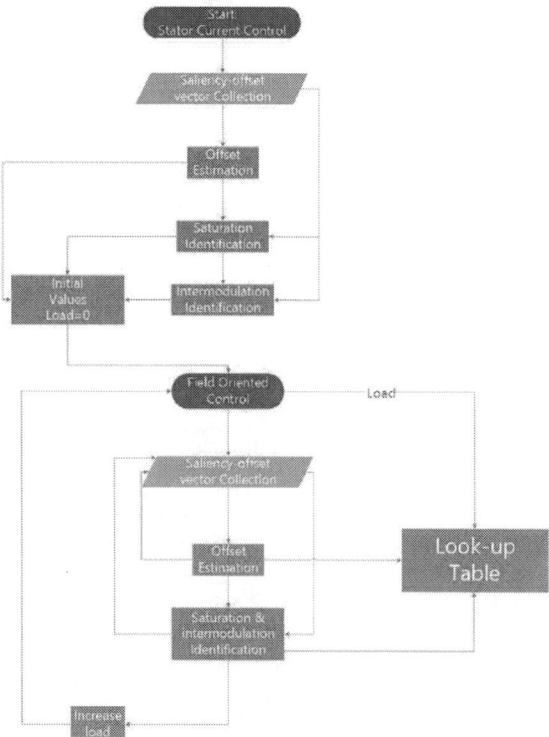

Fig. 2. Flowchart of proposed method for load-dependent harmonic identification and compensation.

Limitations and contribution of the method

The method described above has certain conditions and limitations, presented below:
- For both offset and saliency identification, a minimum of a half electrical period is required during commissioning as well as during online parameter adaptation.
- The slotting harmonic magnitude is higher than the intermodulation at no-load.
- Load steps need to be small enough so that the offset, saliency and intermodulation parameters from the previous load level still ensure stable sensorless control.

Conversely, the proposed identification procedure brings some additional advantages compared to literature:
- No position sensor is required at any stage of the process.
- Computational time is decreased to half an electrical period.
- Can be used during online operation in order to adapt LUT values to parameter variations.
- Does not require a test bench: might be performed with the motor mounted on the train.

Full sensorless saliency experimental results

The proposed sensorless identification procedure is tested on an induction motor to prove its practical applicability, performing constant speed stator current (I_S) control initially and later FOC. During the first experiment, the LUT for saliency harmonic compensation is filled up without any position sensor from start-up. During the second experiment, online operation at different load points is shown together with the mechanical deviation angle (note that encoder is used only for angle comparison and not during the proposed process).

Offset and saliency identification from start-up.

Next figures shows the offset, saturation and intermodulation results using the sensorless identification proposed from start-up. During the first part, a stator current with constant low speed is supplied to the motor. After the identification of offset, saturation and intermodulation, the control mode switches to FOC, and load is increased in steps of 5% up to 50% rated load.

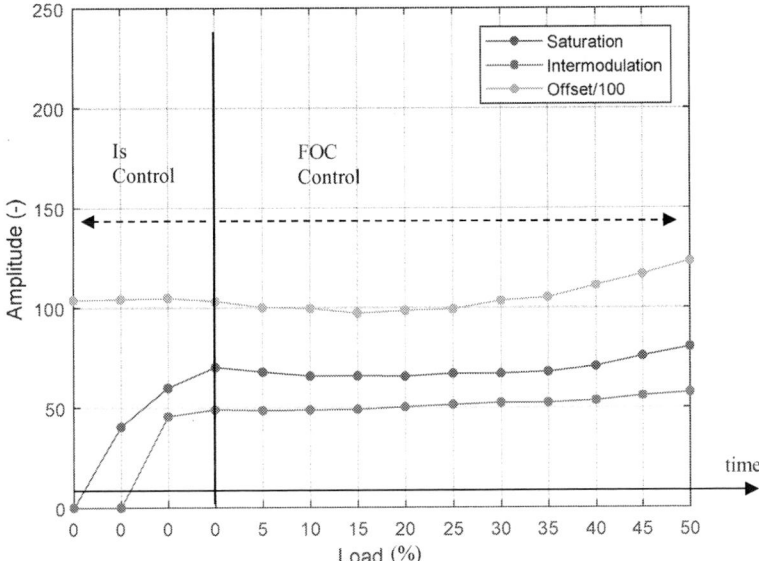

Fig. 3: Sensorless offset and saliency identification from start-up to 50% load for sector 1 switching sector (See Fig. 1).

As observed, after offset estimation, saturation can be estimated. After that, intermodulation identification is possible, which yields in saliency vector where only slotting is present, allowing FOC.

5.2. Sensorless FOC.

Next figure shows the saliencies and offset values for positive and negative loads during the online commissioning, covering negative and positive load at constant speed of 10 rpm.

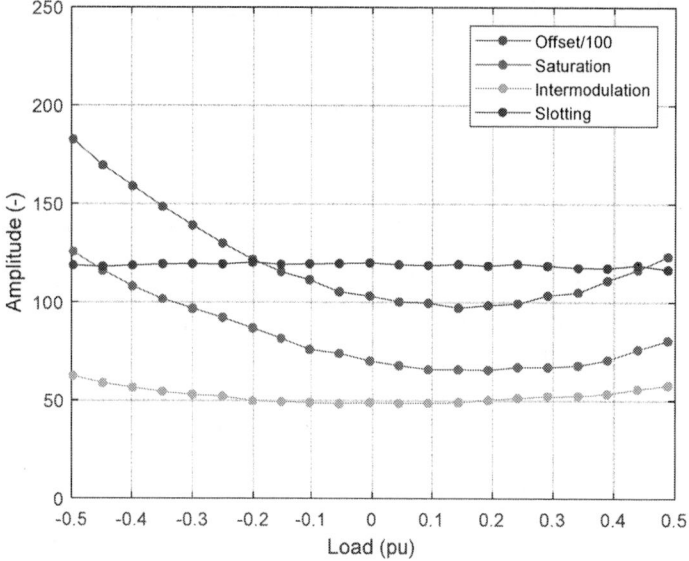

Fig. 4: Saliency & offset values during sensorless online operation from -50 to 50 % load for sector 1.

Next figure shows sensorless rotor angle and the angle deviation (sensorless vs. encoder) during sensorless FOC using the saliency values identified in Fig. 4. The deviation signal has been set to zero at t=0s in the right diagram.

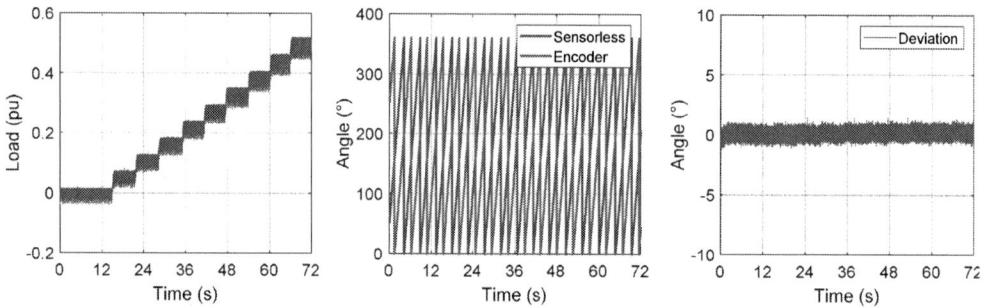

Fig. 5: Torque and sensorless position during the proposed procedure.

As observed in previous figure, the sensorless angle presents a very low deviation compared to the encoder angle (below +-2°).

Conclusion

The identification of machine saliencies by harmonic compensation LUT in voltage-step-excitation-based sensorless drives has typically required a time consuming offline commissioning phase on a test bench equipped with a shaft encoder, and is not able to adapt to parameter variations. A new saliency identification procedure based on harmonic compensation LUT for 1-active SVPWM-integrated excitation is proposed. It works fully sensorless and only requires half an electrical period of constant load operation. Furthermore, it can also be carried out when the motor is mounted in the rail vehicle.

Besides, it can be performed during online operation, what enhances the ability of the control to adapt to potential machine parameter variation. Experimental results from an induction motor proved the applicability of the method and high accuracy of sensorless slotting-based angle.

References

[1] Q. Gao, G. M. Asher, M. Sumner, P. Makys, "Position Estimation of AC Machines over a Wide Frequency Range Based on Space Vector PWM Excitation," in *IEEE Transactions on Industry Application*, vol. 43, pp. 1001-1011. July, 2007.

[2] Y. D. Yoon and S. K. Sul, "Sensorless control for induction machines based on square-wave voltage injection," *IEEE Trans. Power Electron.*, vol. 29, no. 7, pp. 3637–3645, Jul. 2014

[3] P. Garcia, F. Briz, D. Raca, and R. D. Lorenz, "Saliency-tracking-based sensorless control of AC machines using structured neural networks," in *IEEE Transactions on Industry Applications*, vol. 43, pp. 77-86, 2007.

[4] S. Damkhi, M. Said, N. Said, "Slotting effects and high frequency signal injection for induction machine rotor speed estimation", International Conference and Exposition on Electrical and Power Engineering (EPE), pp. 401-408, 2012.

[5] Z. Chen, F. Wang, G. Luo, Z. Zhang, and R. Kennel, "Secondary saliency tracking-based sensorless control for concentrated winding SPMSM," in *IEEE Transactions in Industrial Informatics*, vol. 12, no. 1, pp. 201–210, Feb. 2016.

[6] Q. Gao, G.M. Asher, M. Sumner: "Sensorless Position and Speed Control of Induction Motors Using High-Frequency Injection and Without Offline Precommissioning", in *IEEE Transactions on Industrial Electronics*, Volume 54, Issue 5, pp. 2474 – 2481, Oct. 2007.

[7] E. Rodriguez Montero, M. A. Vogelsberger, and T. Wolbank, "Robust Signal Offset Identification for Sensorless Control of Induction Machines at Rated Load using One-Active Modulating Pulse Excitation," presented at *IEEE Energy Conversion Congress and Expo (ECCE)*, 2019, Baltimore, USA.

[8] A. Yousefi-Talouki, P. Pescetto, G. Pellegrino: "Sensorless Direct Flux Vector Control of Synchronous Reluctance Motors Including Standstill, MTPA and Flux Weakening", IEEE Transactions on Industry Applications, Volume: PP, Issue: 99, 2017

[9] F. Briz, M.W. Degner, P. Garcia, and J. M. Guerrero, "Rotor position Estimation of AC Machines Using the Zero Sequence Carrier Signal Voltage", *Proc. IEEE-IAS Annual Meeting*, Seattle, Oct. 2004, pp. 1305-1312.

[10] W. Fahrner, M. A. Vogelsberger and T. Wolbank, "A new technique to identify induction machine rotor parameters during dynamic operation and low speed," in *IEEE 18th International Power Electronics and Motion Control Conference*, pp. 471-476, 2018, Budapest, Hungary.

[11] P. Pescetto, G. Pellegrino, "Automatic Tuning for Sensorless Commissioning of Synchronous Reluctance Machines Augmented With High-Frequency Voltage Injection," *IEEE Transactions on Industrial Electronics*, vol. 54, pp. 4485-4493, 2018.

[12] D. Raca, P. Garcia, D. D. Reigosa, F. Briz, and R. D. Lorenz, "Carriersignal selection for sensorless control of pm synchronous machines at zero and very low speeds," *IEEE Transactions on Industry Applications*, vol. 46, pp. 167–178, 2010.

Mitigating Drain Source Voltage Oscillation with Low Switching Losses for SiC Power MOSFETs Using FPGA-Controlled Active Gate Driver

Zheming Li, Robert W. Maier, Mark-M. Bakran
University of Bayreuth, Department of Mechatronics
Universitätsstraße 30
Bayreuth, Germany
Tel.: +49 (0) 921 55 7811
E-Mail: zheming.li@uni-bayreuth.de

Acknowledgements

We want to thank Dr. Domes and Dr. Niedernostheide from Infineon Technologies AG for their help.

Keywords

«Silicon Carbide (SiC)», «MOSFET», «Field Programmable Gate Array (FPGA)», «EMC/EMI».

Abstract

In order to improve the switching performance of SiC MOSFETs at turn-off, the drain-source voltage oscillation should be mitigated with low switching losses. To achieve this improvement, an approach, which uses an FPGA-controlled active gate driver with two level switchable gate resistances, is investigated and presented in this paper. To ensure the performance of this approach for varying operating points in a wide range, three methods are shown and compared to find the best solution.

Introduction

Wide-bandgap-semiconductors like SiC MOSFETs are increasingly employed in many power electronic applications [1, 2], due to their outstanding advantages such as low switching losses and conduction losses. However, the performance of SiC MOSFETs can be limited by parasitic parameters or electromagnetic compatibility (EMC) problems. For example, the switching speed at turn-off can be limited by the drain-source overvoltage caused by commutation circuit inductance and the following drain-source voltage (U_{DS}) oscillation [3-6], which causes EMC-problems related to switching. This work aims to improve the switching performance of SiC MOSFETs at turn-off with a new control method, which combines mitigating drain-source voltage oscillation and low switching losses.

In this work, an approach, which uses an active gate driver (AGD) to switch between different gate resistances, is chosen. The switching speed at turn-off can be first increased in the dU_{DS}/dt-phase using a small gate resistance (boost gate resistance, $R_{G, boost}$) to reduce switching losses and then reduced again in the dI_D/dt-phase with the help of a high gate resistance (normal gate resistance, $R_{G, normal}$) to suppress drain-source overvoltage and oscillation. Generally, there are two approaches to control the switching between different gate resistances. One approach is to use specific signals as indicators for gate resistance switching, such as drain-source voltage (U_{DS}), drain current (I_D) in [7, 8] and gate-source voltage (U_{GS}) in [9]. When the indicator signal reaches a certain value (trigger value), the boost process is ended. Another approach is to set a certain on-time (boost-time, t_B) for the boost gate resistance using AGDs. A field programmable gate array (FPGA), which can be programmed to fulfill various logic functions with a high resolution, can realize the boost-time-setting function.

The methods, which use AGDs to improve the switching performance of SiC MOSFETs, have been widely discussed in recent times, as for example presented in [7-13]. However, these methods were only applied to several operating points in these papers. Whether good switching performance can be realized for varying operating points in a wide range, was not verified. It will be shown that a certain trigger value of the indicator signal or a certain boost-time for the boost gate resistance can only ensure good switching performance for a small range of operating condition. Therefore, methods should be found to ensure good switching performance for varying operating points in the application. In this work, only the turn-off process is investigated.

Experimental setup and realization of the approach

The switching behavior of SiC MOSFET is investigated by the double pulse experiment, which in this work is performed with scaled single chip measurements, as shown in [14]. Hence, the commutation circuit inductance and the gate inductance were scaled from a single chip up to the whole module in the performed tests. The drain-source breakdown voltage of the SiC MOSFET is 1700 V, the rated current of a single chip is 37.5 A. The minimum time resolution of the used FPGA is 5 ns. Hence, the minimum boost-time increment in the performed tests is 5 ns.

Experimental Setup

The range of the applied values for the DC-link voltage (U_{DC}), drain current (I_D), junction temperature (T_j), gate driver voltage (U_{DR}) is shown in Table I. The normal value of the gate driver voltage is -5 V/+15 V. For the threshold voltage variation of a SiC MOSFET a typical value of 0.5 V has been assumed. Besides, the gate driver voltage changes about 0.15 V due to aging in the lifetime. Thus, the gate driver voltage has a setting range of -5.7 V/+14.3 V to -4.3 V/+15.7 V to investigate both of the threshold voltage variation and gate driver voltage variation. In this work, some tests were performed with a short lead between gate driver and gate of the chips, whose gate inductance (L_G) is negligible. However, the other tests were carried out with a gate inductance of 7 µH, which is scaled from a single chip up to the whole module, in order to simulate the switching behavior in the real application. The commutation inductance was also scaled up to 1120 nH and kept constant in the investigation.

Table I: Range of the test conditions

	U_{DC}/V	I_D/A	$T_j/°C$	U_{DR}/V
Minimum value	600	0	-40	-5.7 V/+14.3 V
Normal value			25	-5 V/+15 V
Maximum value	1200	112	150	-4.3 V/+15.7 V

Working principle of the system

In order to realize the approach, two diodes and three output stages were used as shown in Fig. 1 (a). The control signal and the output voltage of the normal-driver $U_{DR, normal}$, the turn-on boost-driver $U_{DR, boost-on}$ and the turn-off boost-driver $U_{DR, boost-off}$ during turn-off phase are shown in Fig. 2 (b). The output voltage of the turn-on boost-driver is kept on low level during turn-off phase to ensure that no current flows and generates driver losses. The output voltage of the turn-off boost-driver is normally of high level. It only changes to low level for a certain boost-time t_B, when the falling edge of the control signal arrives, which ensures that the turn-off boost-driver only works at turn-off phase with the help of the diode. Only the turn-off process was investigated in this paper.

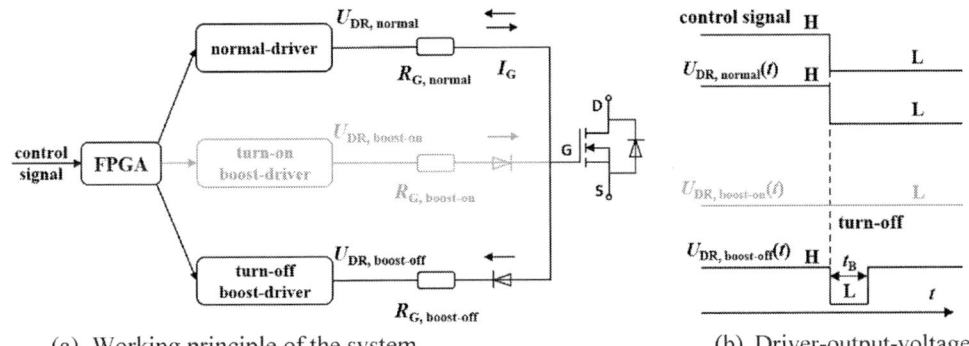

(a) Working principle of the system   (b) Driver-output-voltage

Fig. 1: Working principle of the system

Demonstration of working principle

A comparison between the boost-switching and a pure ohmic switching is shown in Fig. 2. The pure ohmic switching with low switching losses is shown in dashed lines with a maximum drain-source voltage of 1700 V to fully utilize the device blocking voltage, but its drain-source voltage oscillation is severe. Its switching loss (E_{off}) is 49.8 mJ. The pure ohmic switching with a small oscillation is shown in dotted lines, its switching loss is 185.6 mJ. The boost-switching is shown in solid lines with a similar oscillation as the pure ohmic switching with a small oscillation, its switching loss is 48.8 mJ, even smaller than the pure ohmic switching with low switching losses. The output voltage of the turn-off boost-driver is shown in green.

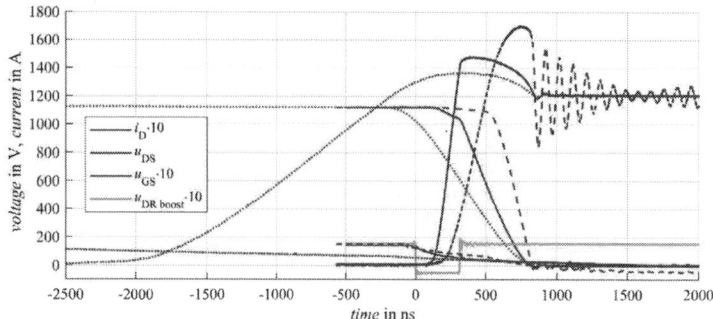

Fig. 2: Comparison between boost-switching and pure ohmic switching at the test condition: U_{DC} = 1200 V, I_D = 112 A, T_j = 25°C, U_{DR} = -5 V/+15 V, with negligible gate inductance

Influence of the boost-time on the boost-switching performance

To investigate the influence of the boost-time on the boost-switching performance, a series of tests with different boost-times were carried out at the same test condition: U_{DC} = 600 V, I_D = 112 A, T_j = 25°C, U_{DR} = -5 V/+15 V, with negligible gate inductance. In order to evaluate the drain-source voltage oscillation quantitatively, the magnitude of the oscillation (U_{DS}-oscillation) is defined as |U1 – U2|, as shown in Fig. 3. The relationship between the drain-source voltage oscillation and the turn-off switching losses E_{off} is shown in Fig. 4. Each point represents a test with a certain boost-time. It can be seen that with increasing boost-time the turn-off switching losses decrease monotonously and the oscillation increases monotonously. This means that a trade-off between the turn-off switching losses and the drain-source voltage oscillation exists with varying boost-time. It can be also seen that the green marked test, which boost-time is the maximum boost-time with an oscillation of no more than 50 V achieves low switching losses and a small oscillation at the same time. Therefore, the maximum boost-time with a drain-source voltage oscillation of no more than 50 V is defined as "optimum boost-time", the corresponding boost-switching is called "optimum boost-switching". An oscillation of no more than 50 V is considered oscillation free.

The curves of the pure ohmic switching (purple marked test in Fig. 3 and Fig. 4) are shown in dash-dotted lines, the boost-switching with a boost time of 150 ns (blue marked test) in solid lines, 245 ns (green marked test) in dashed lines and 285 ns (red marked test) in dotted lines in Fig. 3. It can be seen that the drain-source voltage curves of pure ohmic switching and 150 ns boost-switching differs from each other only before t = 150 ns, afterwards they overlap with each other. Their drain current curves overlap with each other. The switching loss of pure-ohmic switching is 69.4 mJ, the switching loss of 150 ns boost-switching is 65.9 mJ. It means that 150 ns boost-time is not enough for this operating point. Their oscillation magnitude is almost the same and smaller than 50 V. The 245 ns boost-switching was found to be the optimum boost-switching for this operating point with an oscillation of 40 V, the switching loss is 24.0 mJ. As the drain-source breakdown voltage is 1700 V and the maximum drain-source voltage in operation is 1200 V, the drain-source breakdown voltage cannot be exceeded with an overvoltage of no more than 500 V. The 285 ns boost-switching has an overvoltage of 500 V. Its oscillation magnitude is 259 V, its switching loss is 18.9 mJ. The switching losses decrease not much compared to optimum boost-switching, but the oscillation is much higher.

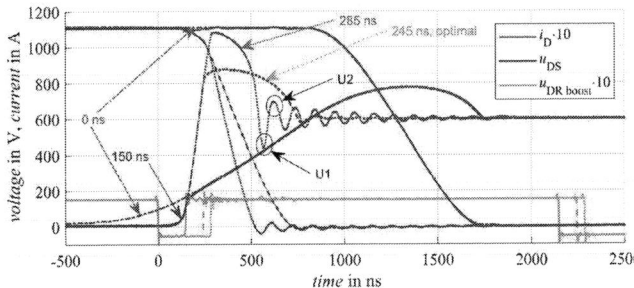 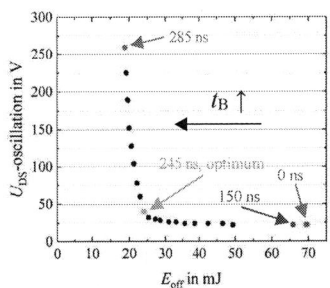

Fig. 3: Boost-switching behaviors with different boost-times at the test condition: $U_{DC} = 600$ V, $I_D = 112$ A, $T_j = 25°C$, $U_{DR} = -5$ V/+15 V, with negligible gate inductance

Fig. 4: Relationship between U_{DS}-oscillation ($|U1 - U2|$) and switching losses (E_{off})

In conclusion, for a certain operating point, the boost-time should be set as the corresponding optimum boost-time to achieve the best turn-off boost-switching performance.

Influence of operation parameters on the boost-switching performance

It has been shown that the actual boost-time must match the optimum boost-time of the test condition to ensure good boost-switching performance. Therefore, the influence of operation parameters on the boost-switching performance can be measured by the influence of operation parameters on optimum boost-time. In this part of the investigation, all the operation parameters were kept constant except for the investigated parameter. First of all, the relationship between the optimum boost-time $t_{B, opt}$ and the investigated parameter was found. The parameter value, which corresponds to the maximum optimum boost-time, is named as C_{max}. The parameter value corresponding to the minimum optimum boost-time is named as C_{min}. Then the influence of this parameter on boost-switching performance was evaluated. The worst case for switching losses occurred, when the boost-switching with C_{max} was performed with the minimum optimum boost-time. The worst case for oscillation occurred, when the boost-switching with C_{min} was performed with the maximum optimum boost-time. It has been shown in Fig. 4 that, the boost-time difference between the boost-time for boost-switching with 500 V drain-source overvoltage 285 ns and the optimum boost-time 245 ns is 40 ns. Hence, if the optimum boost-time variation due to this parameter is more than 40 ns, the corresponding measurement for oscillation worst case should not be performed, in order to ensure that the drain-source breakdown voltage cannot be exceeded.

Influence of the DC-link voltage on the boost-switching performance

In order to investigate the influence of the DC-link voltage on the boost-switching performance, three tests were performed with corresponding optimum boost-time at the test condition: $I_D = 112$ A, $T_j = 25°C$, $U_{DR} = -5$ V/+15 V, with negligible gate inductance with different DC-link voltages of 1200 V, 900 V, 600 V. The curves of optimum boost-switching with $U_{DC} = 1200$ V are shown in solid lines, with $U_{DC} = 900$ V in dashed lines and with $U_{DC} = 600$ V in dotted lines in Fig. 5 (a). Their optimum boost-time is 310 ns, 280 ns and 245 ns respectively. The relationship between optimum boost-time and DC-link voltage is shown in Fig. 5 (b). It can be seen that, the optimum boost-time increases linearly with increasing DC-link voltage. The optimum switching loss for $U_{DC} = 1200$ V is 48.8 mJ. If the influence of the DC-link voltage is not considered, the worst case for switching losses occurred, when the boost-switching for $U_{DC} = 1200$ V is performed with 245 ns. The corresponding switching loss is 122.6 mJ, which leads to a 151% increase compared to the optimum switching loss. The optimum boost-time variation due to DC-link voltage is 65 ns, more than 40 ns, which means a severe oscillation in a worst case. Therefore, the influence of the DC-link voltage is non-negligible.

Influence of the drain current on the boost-switching performance

Three tests were performed with optimum boost-time at the test condition: $U_{DC} = 600$ V, $T_j = 25°C$, $U_{DR} = -5$ V/+15 V, with negligible gate inductance with different drain currents of 112 A, 37.5 A, 10 A. The curves of boost-switching with $I_D = 112$ A are shown in solid lines, with $I_D = 37.5$ A in dashed lines and with $I_D = 10$ A in dotted lines in Fig. 6 (a). Their optimum boost-time is successively 245 ns, 280 ns, 325 ns. The relationship between optimum boost-time and drain-current is shown in

Fig. 6 (b). It can be seen that the optimum boost-time decreases logarithmically with increasing drain current. The optimum switching loss for $I_D = 10$ A is 0.5 mJ. For the switching loss worst case, boost-switching with $I_D = 10$ A with 245 ns, the switching loss is 3.3 mJ, which leads to a 560% increase of switching loss. The optimum boost-time variation due to drain current is 80 ns, more than 40 ns, which means a severe oscillation in a worst case. Hence, the impact of the drain current muss be considered.

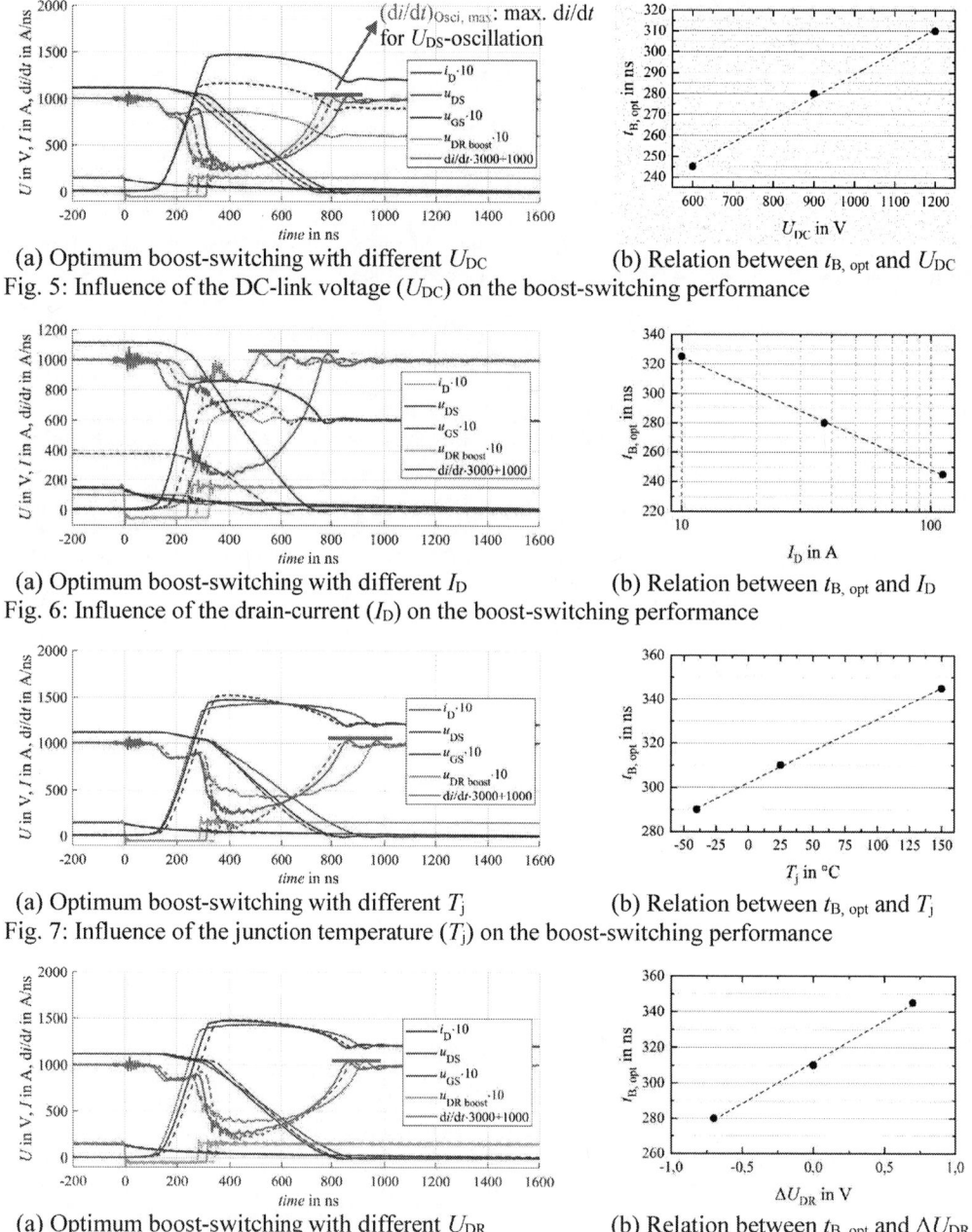

(a) Optimum boost-switching with different U_{DC} (b) Relation between $t_{B, opt}$ and U_{DC}

Fig. 5: Influence of the DC-link voltage (U_{DC}) on the boost-switching performance

(a) Optimum boost-switching with different I_D (b) Relation between $t_{B, opt}$ and I_D

Fig. 6: Influence of the drain-current (I_D) on the boost-switching performance

(a) Optimum boost-switching with different T_j (b) Relation between $t_{B, opt}$ and T_j

Fig. 7: Influence of the junction temperature (T_j) on the boost-switching performance

(a) Optimum boost-switching with different U_{DR} (b) Relation between $t_{B, opt}$ and ΔU_{DR}

Fig. 8: Influence of the gate driver voltage (U_{DR}) on the boost-switching performance

Influence of the junction temperature on the boost-switching performance

Three tests were performed with optimum boost-time at the test condition: $U_{DC} = 1200$ V, $I_D = 112$ A, $U_{DR} = -5$ V/+15 V, with negligible gate inductance with different junction temperatures of -40°C, 25°C, and 150°C. The curves of boost-switching with $T_j = -40$°C are shown in dotted lines, with 25°C in solid lines and with 150°C in dashed lines in Fig. 7 (a). Their optimum boost-time is successively 290 ns, 310 ns, 345 ns. The relation between optimum boost-time and junction temperature is shown in Fig. 7 (b). It can be seen that the optimum boost-time increases linearly with increasing junction temperature. The optimum switching loss for $T_j = 150°C$ is 42.7 mJ. For the switching loss worst case, boost-switching with $T_j = 150°C$ with 290 ns, the switching loss is 83.3 mJ, 95% increase compared to 42.7 mJ. The optimum boost-time variation due to junction temperature is 55 ns, which means a severe oscillation in a worst case. Thus, the impact of the junction temperature could not be neglected.

Influence of the gate driver voltage on the boost-switching performance

Three tests were performed with optimum boost-time at the test condition: $U_{DC} = 1200$ V, $I_D = 112$ A, $T_j = 25°C$, with negligible gate inductance with different gate driver voltages (U_{DR}) of -5.7 V/+14.3 V, -5 V/+15 V and -4.3 V/+15.7 V. The curves of boost-switching with $U_{DR} = -5.7$ V/+14.3 V are shown in dotted lines, with -5 V/+15 V in solid lines and with -4.3 V/+15.7 V in dashed lines in Fig. 8 (a). Their optimum boost-time is respectively 280 ns, 310 ns, 345 ns. The relationship between optimum boost-time and gate driver voltage difference compared to -5 V/+15 V is shown in Fig. 8 (b). It can be seen that the optimum boost-time increases linearly with increasing gate driver voltage. The optimum switching loss for $U_{DR} = -4.3$ V/+15.7 V is 46.5 mJ. For the switching loss worst case, boost-switching with $U_{DR} = -4.3$ V/+15.7 V with 280 ns, the switching loss is 112.6 mJ, 142% increase compared to 46.5 mJ. The optimum boost-time variation due to gate driver voltage is 65 ns, which means a severe oscillation in a worst case. Therefore, the impact of the gate driver voltage must be considered.

Methods to realize an optimum boost-switching for varying operating points

It has been proven that the optimum boost-time is sensitive to the change of operation parameters. It is impossible to set a constant boost-time to ensure good switching performance for the whole range of working conditions. In order to realize an optimum boost-switching for varying operating points as much as possible automatically, three methods are shown and compared to find the best solution: "Trigger Method", "Fully-controlled Method" and "Tracking Method".

"Trigger Method"

The trigger method is similar as the methods shown in [7-9]. A specific signal can be used as indicator for gate resistance switching. When the indicator signal reaches the trigger value, the boost process is terminated. The chosen indicator signal in this work is the drain-source voltage (U_{DS}).

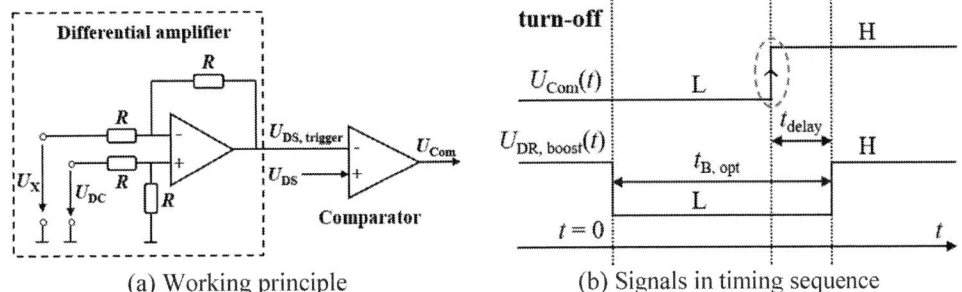

(a) Working principle   (b) Signals in timing sequence

Fig. 9: Working principle of the "Trigger Method"

The working principle of the trigger method is shown in Fig. 9 (a). The trigger value $U_{DS, trigger}$ is set by the output voltage of the differential amplifier, which is the voltage difference between the DC-link voltage U_{DC} and a pre-defined voltage U_X. The drain-source voltage and the trigger value are compared in the comparator. The output voltage of comparator U_{Com} and the output voltage of boost-driver $U_{DR, boost}$ are shown in Fig. 9 (b) in timing sequence. When the turn off process is initiated, the boost-switching starts. When the positive edge of the comparator output voltage is detected, the boost-switching stops after a systematic delay time t_{delay}. Therefore, the comparator output voltage should change to high level at the time point "$t_{B, opt} - t_{delay}$", in order to realize an optimum boost-switching. The trigger value $U_{DS, trigger}$ should be set as the drain-source voltage at "$t_{B, opt} - t_{delay}$", $U_{DS} (t_{B, opt} - t_{delay})$. For a certain working point, when the voltage U_X, which can be considered as the drain-source voltage variation in t_{delay}, is set as the difference between DC-link voltage U_{DC} and the $U_{DS}(t_{B, opt} - t_{delay})$, the optimum boost-switching can be achieved. This relation is described in the equation (1).

$$U_X = U_{DC} - U_{DS,trigger} = U_{DC} - U_{DS}(t_{B,opt} - t_{delay}) \qquad (1)$$

In order to ensure low drain-source voltage oscillation for all the operating points at boost-switching, the voltage U_X should be set as the maximum U_X for the whole range of working conditions, which leads to an increase of switching losses because of the earlier boost stop compared to the optimum boost-switching. Therefore, the performance of the trigger method should be evaluated from the perspective of switching losses. For this evaluation, a series of tests at different test conditions with negligible gate inductance were performed at optimum boost-switching. The DC-link voltage is varied from 600 V to 1200 V, the drain current from 10 A to 112 A, the junction temperature from 25°C to 150°C, the gate driver voltage from -5.7 V/+14.3 V to -4.3 V/+15.7 V, which covers nearly all the range of the working conditions. The maximum U_X of these tests is 536 V. The delay time t_{delay} is estimated as 60 ns. The process of the trigger method is simulated at the test condition: $U_{DC} = 600$ V, $I_D = 112$ A, $T_j = 25$°C, $U_{DR} = -5$ V/+15 V, with negligible gate inductance and shown in Fig. 10 (a). The optimum boost-time $t_{B, opt}$ is 245 ns. According to the equation (1), the trigger value $U_{DS, trigger}$ should be $U_{DC} - U_X = 600$ V - 536 V = 64 V. After a delay time of 60 ns, the boost process stops with an actual boost time of 196.5 ns, smaller than 245 ns. A test was then performed with a boost-time of 195 ns at this test condition, its switching behavior is shown in Fig. 10 (b). The switching loss is 49.4 mJ, which leads to a 105.8% increase of switching loss compared to the optimum switching loss 24.0 mJ. Hence, the realized switching losses through the trigger method are much higher than the optimum switching losses. The trigger method is therefore not considered to be efficient.

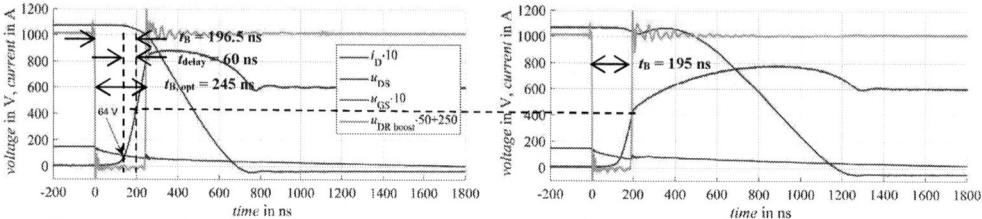

(a) Process simulation of the trigger method   (b) Boost-switching with $t_B = 195$ ns

Fig. 10: Performance evaluation of the trigger method at the test condition: $U_{DC} = 600$ V, $I_D = 112$ A, $T_j = 25$°C, $U_{DR} = -5$ V/+15 V, with negligible gate inductance

"Fully-controlled Method"

With the fully-controlled method, a matrix or an equation should be derived to describe the relation between test conditions and the corresponding optimum boost-time. Some parameters such as gate inductance, threshold voltage can be characterized before the operation, the other parameters like DC-link voltage, drain current, junction temperature should be fed back to FPGA in real time in operation to obtain the optimum boost-time for this time point. If the derived matrix or equation is precise enough, an optimum boost-switching can be realized for all the working points theoretically.

It has been shown in Fig. 5 (b), Fig. 6 (b), Fig. 7 (b) and Fig. 8 (b) that the relationship between optimum boost time $t_{B, opt}$ and DC-link voltage is linear, between $t_{B, opt}$ and drain current logarithmic, between $t_{B, opt}$ and temperature linear, between $t_{B, opt}$ and gate driver voltage linear. In order to derive an equation to describe the relationship between optimum boost-time and test conditions, the test condition $U_{DC} = 1200$ V, $I_D = 75$ A, $T_j = 25$°C, $U_{DR} = -5$ V/+15 V, with negligible gate inductance is defined as the standard test condition. Its optimum boost-time is 325 ns. A series of tests with different test conditions around the standard test condition were performed at optimum boost-switching. Their optimum boost-times and test conditions were recorded and used to derive the equation. Finally, the following equation (2) for tests with negligible gate inductance and equation (3) for tests with a gate inductance of 7 μH were derived. $t_{B, cal}$ means the calculated boost-time. The interactions between operation parameters are not considered in these two equations.

$$t_{B,cal} = 325\ ns + 0.12 \cdot (U_{DC} - 1200\ V) + 41.11 \cdot \ln\left(\frac{75\ A}{I_D}\right) + 46.43 \cdot \Delta U_{DR} + 0.32 \cdot$$
$$(T_j - 25°C),\ with\ negligible\ gate\ inductance \tag{2}$$

$$t_{B,cal} = 325\ ns + 0.12 \cdot (U_{DC} - 1200\ V) + 41.11 \cdot \ln\left(\frac{75\ A}{I_D}\right) + 46.43 \cdot \Delta U_{DR} + 0.32 \cdot$$
$$(T_j - 25°C) + 60\ ns,\ with\ a\ gate\ inductance\ of\ 7\ \mu H \tag{3}$$

In order to verify the accuracy of the two equations, a comparison between optimum boost-time $t_{B,\,opt}$ and calculated boost-time $t_{B,\,cal}$ is performed and shown in Fig. 11. Each point represents a certain test. It can be seen in Fig. 11 (a) that for the tests with negligible gate inductance, the calculated boost-time and the optimum boost-time match well with each other. The difference between them varies from -7 ns to +13 ns. The corresponding tests are shown in blue and red. The accuracy of equation (2) is proven to be good. However, it can be seen in Fig. 11 (b) that for the tests with $L_G = 7\ \mu H$, the difference between the calculated boost-time and the optimum boost-time is much higher and varies from -7 ns to +35 ns. The corresponding tests are shown in blue an red. The accuracy of equation (3) is proven to be insufficient. In order to improve the accuracy of equation (3), the interactions between operation parameters should be taken into consideration, which will increase the complexity of the equation. Moreover, more tests need to be carried out according to a more detailed test design.

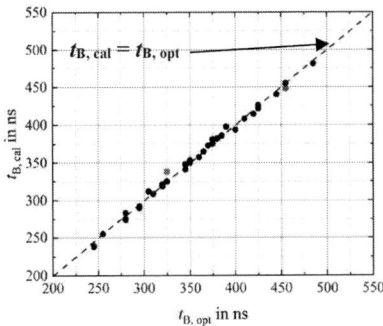 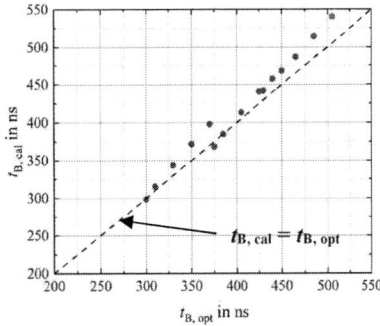

(a) Comparison of tests with negligible L_G (b) Comparison of tests with $L_G = 7\ \mu H$

Fig. 11: Comparison between optimum boost-time $t_{B,\,opt}$ and calculated boost-time $t_{B,\,cal}$

Besides the accuracy of the equations, the fully-controlled method faces some other problems. Firstly, only four operation parameters, DC-link voltage, drain current, junction temperature and gate driver voltage are considered. However, some other parameters may also have strong impact on the optimum boost-time. To investigate the influences of these parameters, the necessary number of tests increases exponentially. Secondly, the matrix or equation of the fully-controlled method is derived from a certain application, certain chip or module, certain optimum boost-switching criterion (50 V U_{DS}-oscillation), etc. A variation of these conditions leads to an unavoidable new derivation of the matrix or equation, which needs a number of new tests. Thirdly, the non-pre-characterizable parameters must be fed back to FPGA in real time in operation, which makes the implementation more difficult. Fourthly, the measurement inaccuracy of the operation parameters in operation must be considered. Hence, a more advanced method compared to the fully-controlled method needs to be found.

"Tracking Method"

The optimum boost-time varies with test conditions. If the actual boost-time is adjusted by the FPGA for each boost-switching process to match the varying optimum boost-time, optimum boost-switching can be realized for all the operating points theoretically. This optimum boost-time tracking function is realized by the tracking method, which will be explained in this section.

It has been shown in Fig. 4 that, the drain-source voltage oscillation increases monotonously with increasing boost-time. If the drain-source voltage oscillation of a turn-off process is less than 50 V, the corresponding boost-time must be less than the optimum boost-time. The working principle of the optimum boost-time tracking function is shown in Fig. 12. The initial value of the actual boost-time

$t_{B, 1}$ is set as the minimum optimum boost-time $t_{B, min}$ of all the operating points, in order to ensure a low drain-source voltage oscillation at the beginning of the boost-time tracking. After the dynamic turn-off of a boost-switching, it is verified whether the oscillation-criterion, which is the occurrence of a drain-source voltage oscillation of more than 50 V is fulfilled in this turn-off process. If so, it means that the actual boost-time is already longer than the optimum boost-time. Thus, the actual boost-time subtracts the boost-time increment Δt_B, which can ensure the best boost-switching performance with the tracking method. The minimum time resolution of the used FPGA is 5 ns. It is assumed that the optimum boost-time increment in this work is 5 ns. If the oscillation-criterion is not fulfilled, it means that the actual boost-time is not yet long enough. Hence, the actual boost-time adds 5 ns. The updated actual boost-time will be used for the next turn-off process.

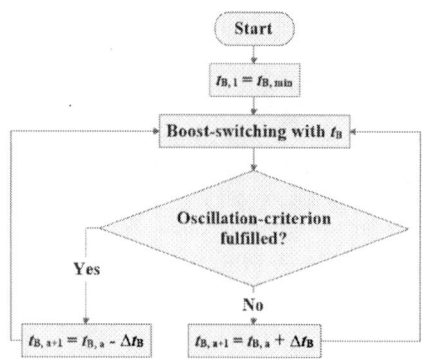

(a) Working principle

- t_B: actual boost-time
- $t_{B, a}$: t_B for „a"-th turn-off process
- $t_{B, opt}$: optimum boost-time
- $t_{B, min}$: minimum $t_{B, opt}$ for all operation points
- Δt_B: boost-time increment (5 ns in this work)
- Oscillation-criterion: drain-source voltage oscillation more than 50 V

(b) Parameter definitions

Fig. 12: Flowchart description of the optimum boost-time tracking function

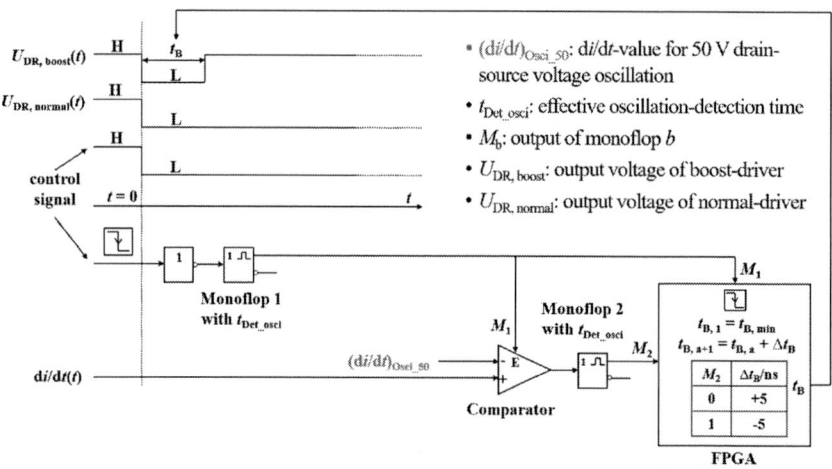

- $(di/dt)_{Osci\_50}$: di/dt-value for 50 V drain-source voltage oscillation
- t_{Det_osci}: effective oscillation-detection time
- M_b: output of monoflop b
- $U_{DR, boost}$: output voltage of boost-driver
- $U_{DR, normal}$: output voltage of normal-driver

Fig. 13: Exemplary time signals of important signals and exemplary implementation of the tracking method in a turn-off process

The oscillation-criterion is defined as the occurrence of a drain-source voltage oscillation of more than 50 V. However, the drain-source voltage will not be measured in the real application. Therefore, an alternative indicator should be found to evaluate the drain-source voltage oscillation. The di/dt-signals are shown in Fig. 5 (a) - Fig. 8 (a) in red. It can be seen that a positive di/dt-peak, which is referred to as $(di/dt)_{Osci, max}$ in Fig. 5 (a), occurs simultaneously as the first minimum of the U_{DS}-oscillation. It can be seen that $(di/dt)_{Osci, max}$ is the same at different test conditions at optimum boost-switching. Thus, $(di/dt)_{Osci, max}$ can be used as an indicator to estimate whether a boost-switching is optimum or not.

The working principle of the tracking method is shown in Fig. 13. t_{Det_osci} is the effective oscillation-detection time, which is long enough to detect oscillation for all operating points. The pulse duration in monoflop 1 is set as $t_{\text{Det\_osci}}$ to enable the comparator for oscillation detection. M_b is the output of monoflop b. The di/dt-value for 50 V drain-source voltage oscillation is referred to as (di/dt)$_{\text{osci\_50}}$ and used as reference value in the comparator. When a turn-off process is initiated, the comparator is enabled by monoflop 1 with $t_{\text{Det\_osci}}$. The di/dt-signal is compared with (di/dt)$_{\text{osci\_50}}$ in the comparator to estimate whether the oscillation-criterion is fulfilled. When the signal level of M_1 changes to zero, the oscillation must have already occurred. The actual boost-time is updated at this time according to the look-up-table and then will be used for the next turn-off process.

The tracking method seems to be the best one among these three methods due to its simple working principle, short time-consumption for pre-characterization and good compatibility for different semiconductors, oscillation-criterions, applications etc. This promising method will be investigated in further work. It should be proven that the actual boost-time can keep track of the varying optimum boost-time with the tracking method. Afterwards, the tracking method should be realized with a consideration of its robustness against external influences.

Conclusion

With the proposed AGD with two level switchable gate resistances, low switching losses and low drain-source voltage oscillation can be achieved at the same time at turn-off. The optimum boost-time varies strongly with operating conditions. In order to realize an optimum boost-switching for varying operating points, the trigger method, the fully-controlled method and the tracking method are investigated and compared. The tracking method is found to be the most promising method. The traceability of the tracking method and its realization will be investigated in future work.

References

[1] K. O. Armstrong, S. Das and J. Cresko: Wide bandgap semiconductor opportunities in power electronics, 2016 IEEE 4th Workshop on Wide Bandgap Power Devices and Applications (WiPDA), pp. 259-264, 2016.

[2] L. Middelstaedt and A. Lindemann: Strategy for reducing oscillations in power electronic circuits using gate control, PCIM Europe 2018, pp. 1714-1720, 2018.

[3] L. Middelstaedt and A. Lindemann: Optimization of critical oscillations within a boost converter based on an analytical model, EPE'16 ECCE Europe, pp. 1-9, 2016.

[4] F. Wolfgang: Simple slew-rate control technique cuts switching losses, PCIM Europe 2019, pp. 352-356.

[5] P. Hofstetter and Mark-M. Bakran: Mitigating drain source voltage oscillation for SiC power MOSFETs in order to reduce electromagnetic interference, EPE'19 ECCE Europe, pp. 1-10, 2019.

[6] E. U. Krafft, B. Laska, A. Nagel and J. Weigel: A new standard IGBT housing for high-power converters, EPE'15 ECCE Europe, pp. 1-11, 2015.

[7] P. Nayak and K. Hatua: Parasitic inductance and capacitance-assisted active gate driving technique to minimize switching loss of SiC MOSFET, IEEE Transactions on Industrial Electronics Vol. 64 no. 10, pp. 8288-8298, 2017.

[8] P. Nayak and K. Hatua: Active gate driving technique for a 1200 V SiC MOSFET to minimize detrimental effects of parasitic inductance in the converter layout, IEEE Transactions on Industry Applications Vol. 54 no. 2, pp. 1622-1633, 2018.

[9] A. Paredes, H. Ghorbani, V. Sala, E. Fernandez and L. Romeral: A new active gate driver for improving the switching performance of SiC MOSFET, 2017 IEEE APEC, pp. 3557-3563, 2017.

[10] M. V. Krishna and K. Hatua: An easily implementable gate charge controlled active gate driver for SiC MOSFET, IECON 2018 - 44th Annual Conference of the IEEE Industrial Electronics Society, pp. 999-1004.

[11] J. Gottschlich and Rik W. De Doncker: A programmable gate driver for power semiconductor switching loss characterization, IEEE PEDS 2015, pp. 456-461, 2015.

[12] S. Acharya, X. She, F. Tao, T. Frangieh, M. H. Todorovic and R. Datta: Active gate driver for SiC-MOSFET-based PV inverter with enhanced operating range, IEEE Transactions on Industry Applications, Vol. 55 no. 2, pp. 1677-1689, 2019.

[13] V. K. M and K. Hatua: Current controlled active gate driver for 1200 V SiC MOSFET, PEDES, 2016.

[14] R. W. Maier and Mark-M. Bakran: Switching SiC MOSFETs under conditions of a high power module, EPE'18 ECCE Europe, pp. 1-10, 2018.

Online Trajectory Planning During Low-Voltage FRT of a Modular Multilevel Converter

Hendrik Fehr and Albrecht Gensior
TECHNISCHE UNIVERSITÄT DRESDEN
Helmholtzstr. 9, 01069
Dresden, Germany
Phone: +49 (0) 351-46332436
Email: hendrik.fehr@tu-dresden.de
URL: https://www.tu-dresden.de

Acknowledgments

This work was supported by *Deutsche Forschungsgemeinschaft*, DFG, grant GE 2502/4-2.

Keywords

≪Modular Multilevel Converters (MMC)≫ ≪Converter control≫ ≪Fault ride-through≫

Abstract

Balancing the modular multilevel converter arm energies in case of a sudden grid voltage sag is a challenging open-loop control problem, because it requires the solution of a differential equation. Unlike existing numerical approaches, an analytical solution was obtained. Ride-through experiments are conducted on a low-voltage MMC test bench using a grid emulator. In contrast to feedback balancing alone, a faster return of the cell voltages into the region of stationary operation is achieved.

Introduction

This paper focuses on the open-loop control of the Modular Multilevel Converter (MMC) arm energies during large-signal transients in order to add feed-forward currents supporting the well-proven feedback-based [1–3] energy balancing. Here, the low-voltage fault ride-through (FRT) of grid-connected MMCs is considered [4,5] for the circuit shown in Fig. 1. In addition to the common FRT requirements concerning the ac and dc terminal behavior, an MMC based application faces the challenge to quickly reach a balanced operation of the cell voltages [8–10]. This paper focuses on a trajectory planning to solve the latter.

The use of a trajectory planning to add a feed forward control has been proposed in [11] for a single-phase MMC and in [12–14] for a three-phase application. However, adopting the single-phase solution is inconvenient, because the respective models are incompatible. The three-phase trajectory plannings [12, 13] have limited practical merit since they lack an analytical solution implying higher computational cost. Moreover, the existing three-phase solutions assume the energies to start the planned transition from a balanced condition, which is violated in case of an FRT. The present work reports on the implementation of an analytical solution that overcomes these drawbacks. The approaches in [11–14] are based on a *detour* in the nominal trajectories in order to gain influence on the subsystem which cannot be considered in the planning. This is similar to the approach in [15], which the authors only recently became aware of.

The paper is organized as follows: An MMC model is introduced and the trajectory planning is explained in the next two sections. Subsequently, its implementation on a low-voltage test bench is shown and the obtained experimental results are presented. A conclusion is given at the end.

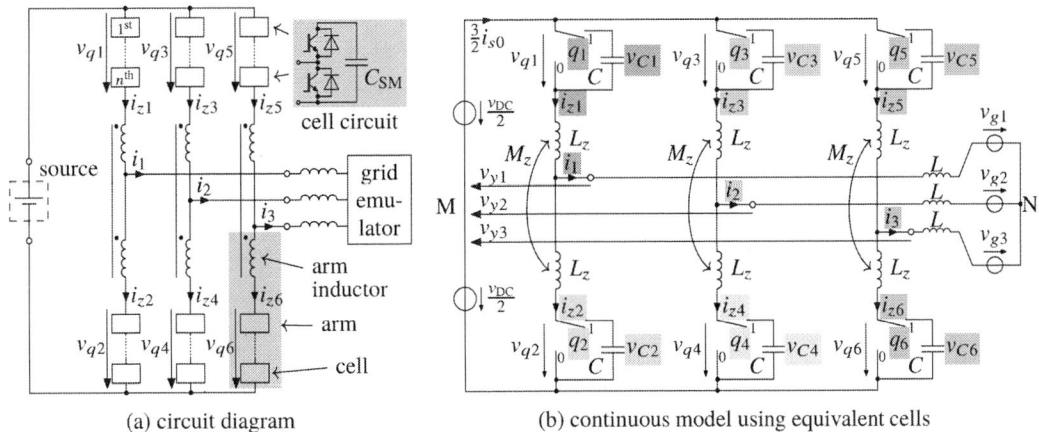

(a) circuit diagram (b) continuous model using equivalent cells

Fig. 1: Circuit (a) and continuous model (b) of an MMC. The cells of each arm are represented by equivalent cells [6, 7] and their duty cycles $q_k \in [0, 1]$, $k = 1, \ldots, 6$ are used as control inputs. Parameters: $C_{SM} = 375\,\mu F$, $n = 6$, $C = C_{SM}/n$, $L_z = 1.2\,mH$, $M_z = 0.94\,mH$, $L = 15\,mH$, $v_{DC} = 650\,V$.

Table I: Transformations of the arm energies (1), currents, and voltages of the circuit in Figure 1b.

definition	description
$g_0 = \frac{1}{3}(1, 1, 1)$	zero sequence component projection
$\underline{g}_{dq} = \frac{1}{3}e^{-j\theta}(2, -1 + j\sqrt{3}, -1 - j\sqrt{3})$	3ph variables to rotating frame projection
$e_{s0} = 2g_0\left[(e_{z1}, e_{z3}, e_{z5})^T + (e_{z2}, e_{z4}, e_{z6})^T\right]$	stored energy (scaled)
$e_{d0} = 2g_0\left[(e_{z1}, e_{z3}, e_{z5})^T - (e_{z2}, e_{z4}, e_{z6})^T\right]$	vertical difference (scaled)
$\underline{e}_s = 2\underline{g}_{dq}\left[(e_{z1}, e_{z3}, e_{z5})^T + (e_{z2}, e_{z4}, e_{z6})^T\right]$	complex energy sum in rotating frame
$\underline{e}_d = 2\underline{g}_{dq}\left[(e_{z1}, e_{z3}, e_{z5})^T - (e_{z2}, e_{z4}, e_{z6})^T\right]$	complex energy difference in rotating frame
$i_{s0} = g_0\left[(i_{z1}, i_{z3}, i_{z5})^T + (i_{z2}, i_{z4}, i_{z6})^T\right]$	(scaled) dc current
$\underline{i}_s = \underline{g}_{dq}\left[(i_{z1}, i_{z3}, i_{z5})^T + (i_{z2}, i_{z4}, i_{z6})^T\right]$	complex circulating current in rotating frame
$\underline{i} = \underline{g}_{dq}\left[(i_{z1}, i_{z3}, i_{z5})^T - (i_{z2}, i_{z4}, i_{z6})^T\right]$	complex output current in rotating frame
$\underline{v}_y = \underline{g}_{dq}(v_{y1}, v_{y2}, v_{y3})^T$	complex output voltage in rotating frame
$v_{y0} = g_0(v_{y1}, v_{y2}, v_{y3})^T$	common-mode voltage referred to dc midpoint M

MMC Model

The MMC arm energies

$$e_{zk} = \frac{1}{2}Cv_{Ck}^2 + \frac{1}{2}(L_z + M_z)i_{zk}^2 - \frac{1}{4}M_z i_p^2, \quad p = \left\lceil \frac{k}{2} \right\rceil, \quad k = 1, \ldots, 6 \tag{1}$$

are represented using transformed coordinates, based on the usual sum and difference between upper and lower arms and a park transform, explained in Table I. This leads to the model [16]

$$\dot{e}_{s0} = v_{DC}i_{s0} - \mathrm{Re}\left(\underline{i}\,\underline{v}_y^*\right) \qquad \text{twice the total stored energy} \tag{2a}$$

$$\dot{e}_{d0} = -2v_{y0}i_{s0} - \mathrm{Re}\left(\underline{i}_s^*\underline{v}_{y\Delta}\right) \qquad \text{twice the vertical energy difference} \tag{2b}$$

$$\underline{\dot{e}}_s = v_{DC}\underline{i}_s - e^{-j3\theta}\underline{v}_y^*\underline{i}^* - 2\underline{i}\,v_{y0} - j\omega\underline{e}_s \qquad \text{complex sum in rotating frame} \tag{2c}$$

$$\underline{\dot{e}}_d = v_{DC}\underline{i} - e^{-j3\theta}\underline{i}_s^*\underline{v}_{y\Delta}^* - 2\underline{i}_s v_{y0} - 2i_{s0}\underline{v}_{y\Delta} - j\omega\underline{e}_d. \qquad \text{complex difference in rotating frame} \tag{2d}$$

The complex variables are defined in a rotating frame with angle θ and angular frequency $\omega = \dot{\theta}$, given by a phase-locked loop (PLL) in grid-side applications. The energies e_{d0}, $\underline{e}_s$, and $\underline{e}_d$ represent the energy deflections w.r.t. the equally distributed stored energy e_{s0}. The output voltage and the common-mode voltage are denoted by $\underline{v}_y$ and v_{y0}, respectively. And $\underline{v}_{y\Delta} = \underline{v}_y - M_z(j\omega\underline{i} + \frac{\mathrm{d}}{\mathrm{d}t}\underline{i})$ denotes the output voltage subtracted by a voltage drop caused by the mutual inductance $M_z \in [-L_z, L_z]$. In contrast to the previously published trajectory planning algorithms in [11, 16], the output current $\underline{i}$, circulating current $\underline{i}_s$ and dc current i_{s0} are not restricted by a continuity requirement of a differential equation here.

In the context of energy balancing and trajectory planning, the output current $\underline{i}$ and the output voltage $\underline{v}_y$ are known parameters to provide for the separation [12] of the output current controller. Also the dc voltage v_{DC} and the common-mode voltage v_{y0} are considered as known parameters. Now, the dc and the circulating current can be obtained from the stored energy e_{s0} and the the complex energy $\underline{e}_s$ as

$$i_{s0} = \frac{\dot{e}_{s0} + \mathrm{Re}\left(\underline{i}\,\underline{v}_y^*\right)}{v_{DC}} \tag{3a}$$

$$\underline{i}_s = \frac{\dot{\underline{e}}_s + e^{-j3\theta}\underline{v}_y^*\underline{i}^* + 2\underline{i}\,v_{y0} + j\omega\underline{e}_s}{v_{DC}}. \tag{3b}$$

The so called parametrization (3) is an inversion of the subsystems (2a) and (2c). This fixes the left hand sides of the remaining subsystems (2b) and (2d) as well, because they constitute the internal dynamics of the energy system (2) w.r.t. the output $y = (e_{s0}, \underline{e}_s)$. In other words, a solution of the energy system (2) is given by trajectories for the energies e_{s0} and $\underline{e}_s$ and initial values for the energies e_{d0} and $\underline{e}_d$, because the parametrization (3) provides the required information to solve the subsystems (2b) and (2d). The energies e_{d0} and $\underline{e}_d$ cannot be determined by trajectories independent from the trajectories for e_{s0} and $\underline{e}_s$, posing a challenge for the trajectory planning whose solution is explained in the next section.

Although the parametrization (3) increases the order of the internal dynamics in contrast to the solution in [12], it covers a wider operating range, including zero ac power.

Stationary operation of a grid-connected MMC is characterized by constant terminal variables, denoted

$$\underline{i} = \underline{I}, \qquad \underline{v}_y = \underline{V}_y = \underline{V}_g + j\omega L\underline{I}, \qquad i_{s0} = I_{s0} = \mathrm{Re}\left(\underline{V}_y^*\underline{I}\right)/V_{DC}, \qquad v_{DC} = V_{DC}, \tag{4}$$

in which $\underline{V}_g$ denotes the grid voltage and L the filter inductance. The common-mode voltage is assumed to be

$$v_{y0} = \mathrm{Re}\left(\underline{V}_{y0}^{[3]}e^{j3\omega t}\right), \tag{5}$$

i.e. the phase and amplitude of the third harmonic is determined by the complex parameter $\underline{V}_{y0}^{[3]}$. During symmetric stationary operation, the MMC energies (2) evolve according to the stationary operating regime

$$e_{s0}^{\mathrm{stat}} = E_{s0}, \qquad e_{d0}^{\mathrm{stat}} = \mathrm{Re}\left(\underline{E}_{d0}^{[3]}e^{j3\omega t}\right), \qquad \underline{e}_s^{\mathrm{stat}} = \underline{E}_s^{[3]}e^{j3\omega t} + \underline{E}_s^{[-3]}e^{-j3\omega t}, \qquad \underline{e}_d^{\mathrm{stat}} = \underline{E}_d, \tag{6}$$

where the constants $\underline{E}_{d0}^{[3]}$, $\underline{E}_s^{[-3]}$, $\underline{E}_s^{[3]}$, and $\underline{E}_d$ depend on the parameters $\underline{I}$, $\underline{V}_y$, V_{DC}, $\underline{V}_{y0}^{[3]}$ introduced in (4) and (5) and on the voltage

$$\underline{V}_{y\Delta} = \underline{V}_y - M_z j\omega\underline{I} \tag{7a}$$

and the modulation index

$$m = \left|\underline{V}_y\right|\frac{2}{V_{DC}} \tag{7b}$$

as shown in Table II.

Table II: Coefficients of the stationary operating regime (6) with the common-mode voltage (5), depending on the settings (4) and the definitions (7).

$j\omega\underline{E}_{d0}^{[3]}$	$j\omega\underline{E}_s^{[-3]}$	$j\omega\underline{E}_s^{[3]}$	$j\omega\underline{E}_d$	$\underline{V}_{y0}^{[3]}$	m_3
$-\frac{2}{3}I_{s0}\underline{V}_{y0}^{[3]}$	$\frac{1}{2}I\underline{V}_{y0}^{[3]*} + \frac{1}{2}\underline{V}_y^*\underline{I}^*$	$-\frac{1}{4}I\underline{V}_{y0}^{[3]}$	$V_{\mathrm{DC}}\underline{I} - 2I_{s0}\underline{V}_{y\Delta}$	$-\frac{1}{2}m_3 V_{\mathrm{DC}}\mathrm{e}^{3j\arg\underline{V}_y}$	$\max(0.95 - m, \frac{1}{6})$

Trajectory Planning

Starting at $t = 0$ from an initial state of the system (2), the goal of trajectory planning is to specify a transition that reaches a desired state at $t = T$, i.e. the transition time $T > 0$. The desired state is given by the regime (6), such that the MMC is able to continue a balanced stationary operation for $t > T$. Only a subset of the energies (2) can be specified unambiguously [16]. Here, nominal trajectories for the stored energy e_{s0} and the complex energy sum $\underline{e}_s$ are constructed from two parts: The first part cross-fades between the initial state e_{s0}^{init}, $\underline{e}_s^{\mathrm{init}}$ and the desired operation, respectively, while the second part, the *detour*, only affects the trajectories during the transfer, leaving start and end unchanged. In particular

$$e_{s0}^{\mathrm{d}} = \underbrace{e_{s0}^{\mathrm{init}} + (E_{s0} - e_{s0}^{\mathrm{init}})\frac{t}{T}}_{\text{fixed transition}} + \underbrace{D_{s0}(k_1)}_{\text{tunable detour}} \tag{8a}$$

$$\underline{e}_s^{\mathrm{d}} = \underbrace{\underline{E}_s^{[3]}\mathrm{e}^{j3\omega t} + \underline{E}_s^{[-3]}\mathrm{e}^{-j3\omega t} + (\underline{e}_s^{\mathrm{init}} - \underline{E}_s^{[3]} - \underline{E}_s^{[-3]})(1 - \frac{t}{T})}_{\text{fixed transition}} + \underbrace{\underline{D}_s(k_1, \underline{K})}_{\text{tunable detour}} . \tag{8b}$$

The detour is given by

$$D_{s0}(k_1) = k_1 \mathrm{Re}\left(-j\underline{V}_{y0}^{[3]}\mathrm{e}^{3j\omega t}\right) \sin\frac{\pi t}{T} \tag{9a}$$

$$\underline{D}_s(k_1, \underline{K}) = j\left[\underline{K}\,\underline{V}_{y0}^{[3]}\mathrm{e}^{j2\omega t} - \underline{K}^*\,\underline{V}_{y\Delta}^*\mathrm{e}^{-j2\omega t} - \underline{K}\,\underline{V}_{y0}^{[3]*}\mathrm{e}^{-j4\omega t} - \lambda\underline{V}_{y\Delta}k_1\right]\sin\frac{\pi t}{T} \tag{9b}$$

with the choice $\lambda = 942.5$ and its parameters are normalized as

$$k_1 = \frac{\tilde{k}_1}{Q_{d0}\lambda} \quad\text{and}\quad \underline{K} = \frac{\tilde{\underline{K}}}{Q_d} \quad\text{with}\quad Q_{d0} = |\underline{V}_{y\Delta}|^2 + |\underline{V}_{y0}^{[3]}|^2 \quad\text{and}\quad Q_d = |\underline{V}_{y\Delta}|^2 + 2|\underline{V}_{y0}^{[3]}|^2. \tag{10}$$

Before the calculation of k_1 and $\underline{K}$ is explained in detail, the idea of the detour is demonstrated on the example of a grid voltage sag by comparing the shaded nominal transfers in Fig. 2. The transfers are assumed to immediately start after the voltage sag at $t = 0$ and span the time $0 \leq t \leq T$. On the left side, the detour is disabled by setting $k_1 = 0$, $\underline{K} = 0$ and on the right side, the detour parameters are calculated as will be explained below. As visible on the left side in Figs. 2a and c, the fixed transitions in the nominal trajectories (8) already transfer the nominal stored energy e_{s0}^{d} and the nominal complex energy $\underline{e}_s^{\mathrm{d}}$ from their initial values e_{s0}^{init}, $\underline{e}_s^{\mathrm{init}}$ into the desired regime given by E_{s0}, $\underline{E}_s^{[-3]}$, and $\underline{E}_s^{[3]}$. The corresponding nominal currents i_{s0}^{d} and $\underline{i}_s^{\mathrm{d}}$, shown in Figs. 2e and 2f, are obtained by inserting the nominal trajectories (8) and their derivatives into the parametrization (3), leading to

$$i_{s0}^{\mathrm{d}} = \frac{\dot{e}_{s0}^{\mathrm{d}}}{V_{\mathrm{DC}}} + I_{s0} \tag{11a}$$

$$\underline{i}_s^{\mathrm{d}} = \frac{\dot{\underline{e}}_s^{\mathrm{d}} + \mathrm{e}^{-j3\omega t}\underline{V}_y^*\underline{I}^* + 2\underline{I}\,\mathrm{Re}\left(\underline{V}_{y0}^{[3]}\mathrm{e}^{j3\omega t}\right) + j\omega\underline{e}_s^{\mathrm{d}}}{V_{\mathrm{DC}}}. \tag{11b}$$

The nominal energy difference e_{d0}^{d} and the nominal complex energy $\underline{e}_d^{\mathrm{d}}$ can be obtained by inserting the nominal currents (11) into the corresponding subsystems of the energy system (2), leading to the

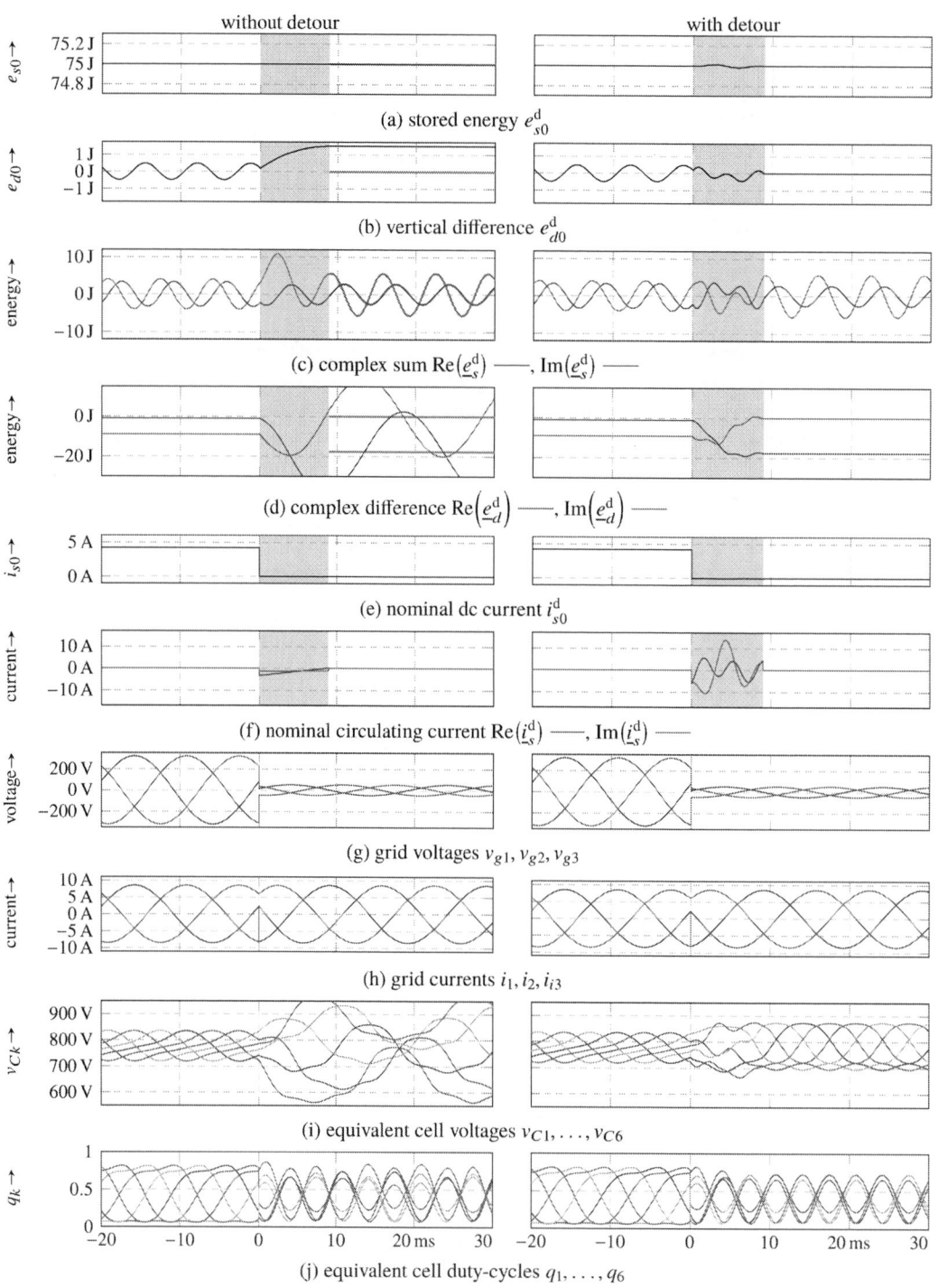

Fig. 2: Example trajectories: without (left) and with detour (right) for a grid voltage sag starting at 0 ms with 0.15 p. u. remaining voltage. The output current $\underline{i}$ is assumed to immediately change from $(-8.5 + 0\mathrm{j})$ A to $(0 - 8.5\mathrm{j})$ A. The dc voltage is assumed to be constant at $v_{DC} = 650$ V. The blue shading indicates the range $0 < t < T$ of the transition, with $T = 9$ ms. The gray traces on the left axis of (a) to (d) indicate the designated stationary regime e_{d0}^{stat} and $\underline{e}_d^{stat}$.

differential equations

$$\dot{e}_{d0}^{\mathrm{d}} = -2i_{s0}^{\mathrm{d}}\mathrm{Re}\left(\underline{V}_{y0}^{[3]}e^{j3\omega t}\right) - \mathrm{Re}\left(\underline{V}_{y\Delta}\,\underline{i}_{s}^{\mathrm{d}*}\right) \tag{12a}$$

$$\underline{\dot{e}}_{d0}^{\mathrm{d}} = V_{\mathrm{DC}}\underline{I} - e^{-j3\omega t}\underline{i}_{s}^{\mathrm{d}*}\underline{V}_{y\Delta}^{*} - 2\underline{i}_{s}^{\mathrm{d}}\mathrm{Re}\left(\underline{V}_{y0}^{[3]}e^{j3\omega t}\right) - 2i_{s0}^{\mathrm{d}}\underline{V}_{y\Delta} - j\omega\underline{e}_{d}. \tag{12b}$$

However, without detour, the nominal trajectories e_{d0}^{d} and $\underline{e}_{d}^{\mathrm{d}}$ do not reach their desired regime at the end of the transfer, because they were not yet considered in the planning process. This is visible from the large deviation in Figs. 2b and d. In order to include these energies in the planning, the detour (9) is designed such that it invokes adjustable dc and circulating current components in (11), without changing the start and end of the other energies (8), i.e. D_{s0} and $\underline{D}_{s}$ are zero at $t = 0$ and $t = T$ independent from k_1 and $\underline{K}$. Now, with proper values of the detour parameters k_1, $\underline{K}$ the desired operating regime for the nominal energies e_{d0}^{d} and $\underline{e}_{d}^{\mathrm{d}}$ can be reached as well, as visible on the right side in Fig. 2.

The detour parameters k_1, $\underline{K}$ can be obtained by setting equal the final values of e_{d0}^{d} and $\underline{e}_{d}^{\mathrm{d}}$ and the values given by the desired operating regime, respectively. That means, at the end of the transitions, i.e. at $t = T$, the requirement is $e_{d0}^{\mathrm{d}} = e_{d0}^{\mathrm{stat}}$ and $\underline{e}_{d}^{\mathrm{d}} = \underline{e}_{d}^{\mathrm{stat}}$. The nominal trajectories e_{d0}^{d} are obtained by solving the differential equation (12) by integration, while e_{d0}^{stat} and $\underline{e}_{d}^{\mathrm{stat}}$ are given by the desired regime (6), after inserting $t = T$. This results in

$$\int_{0}^{T} -2i_{s0}^{\mathrm{d}}\mathrm{Re}\left(\underline{V}_{y0}^{[3]}e^{j3\omega t}\right) - \mathrm{Re}\left(\underline{V}_{y\Delta}\,\underline{i}_{s}^{\mathrm{d}*}\right)\,dt = \mathrm{Re}\left(\underline{E}_{d0}^{[3]}e^{j3\omega T}\right) \tag{13a}$$

$$e^{-j\omega T}\underline{e}_{d}^{\mathrm{init}} + e^{-j\omega T}\int_{0}^{T}\left[V_{\mathrm{DC}}\underline{I} - e^{-j3\omega t}\underline{i}_{s}^{\mathrm{d}*}\underline{V}_{y\Delta}^{*}\right.$$
$$\left. - 2\underline{i}_{s}^{\mathrm{d}}\mathrm{Re}\left(\underline{V}_{y0}^{[3]}e^{j3\omega t}\right) - 2i_{s0}^{\mathrm{d}}\underline{V}_{y\Delta}\right]e^{-j\omega t}\,dt = \underline{E}_{d}. \tag{13b}$$

Carrying out the integration analytically is possible because the remaining time dependencies from the nominal trajectories (8), the detours (9), and the nominal currents (11) boil down to linear and complex exponential functions and combinations thereof. This leads to a linear system of the form

$$\boldsymbol{A}\left(k_1, \mathrm{Re}(\underline{K}), \mathrm{Im}(\underline{K})\right)^{\mathsf{T}} = \boldsymbol{b}, \quad \boldsymbol{A} \in \mathbb{R}^{3\times3} \quad \boldsymbol{b} \in \mathbb{R}^{3}, \tag{14}$$

whose solution, assuming the matrix $\boldsymbol{A}$ is regular, provides the detour parameters k_1, $\underline{K}$. Solving (14) in the real time system was done by the product $\boldsymbol{A}^{-1}\boldsymbol{b}$ in which the separate inversion of $\boldsymbol{A}$ requires 27 multiplications and 14 additions or subtractions.

The block diagrams in Fig. 3 depict the implemented trajectory planning and its integration into an MMC control scheme. The *trajectory planning* block in Fig. 3a implements the calculations described above and is updated upon request from the *replan trigger* block. Planning triggers occur when a change is detected in the output voltage $\underline{V}_y$ (calculated with a load model), or when the energies deviate too far from their nominal trajectories, or once the end of the current transition is reached. The respective initial values and parameters are retained between the planning instants and used by the *trajectory generator* block to calculate the trajectories at the current time t. In order to prevent overflows, the time t is restarted at each trigger event.

The nominal currents are used as feed-forward signals for the current controlled MMC, as shown in Fig. 3b. In order to compensate for modeling inaccuracies, a *balancing feedback* block, similar to the ones from [17], is used and its output is added to the nominal dc and circulating current. The balancing feedback depends on the energy errors w.r.t. the nominal trajectories, as given by

$$i_{s0}^{\mathrm{b}} = k_{s0}\frac{e_{s0}^{\mathrm{d}} - e_{s0}}{V_{\mathrm{DC}}} + 2f_{d0}v_{y0}\left|\underline{V}_{y0}^{[3]}\right|^{-1} \tag{15a}$$

$$\underline{i}_{s}^{\mathrm{b}} = k_{s}\frac{\underline{e}_{s}^{\mathrm{d}} - \underline{e}_{s}}{V_{\mathrm{DC}}} + f_{d0}\,e^{j\arg\underline{V}_{y}} + \underline{f}_{d}^{*}e^{-j\arg\underline{V}_{y}}e^{-2j\omega t} + 2\underline{f}_{d}v_{y0}\left|\underline{V}_{y0}^{[3]}\right|^{-1}, \tag{15b}$$

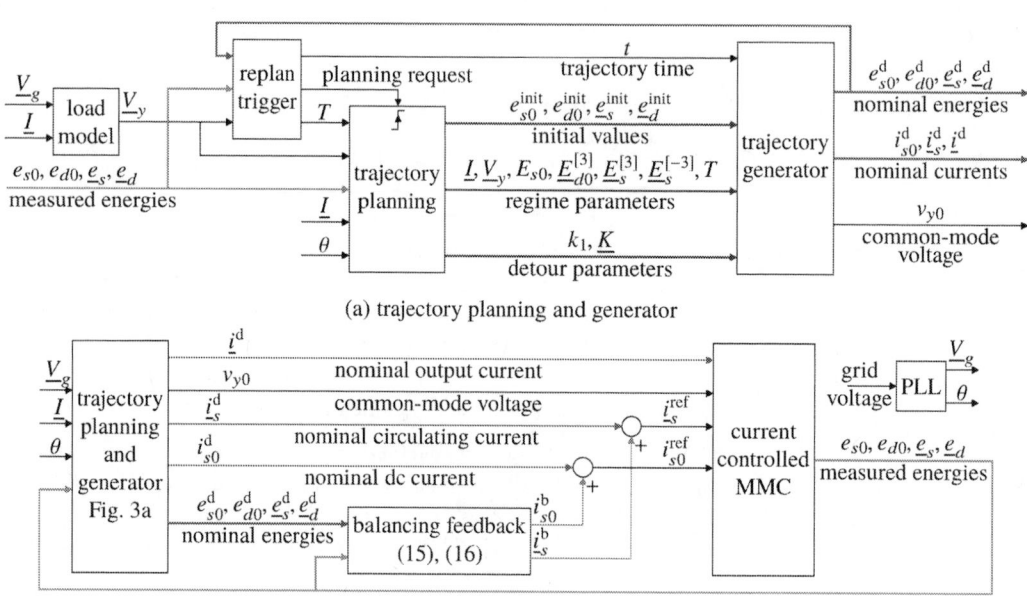

(a) trajectory planning and generator

(b) control scheme with trajectory planning

Fig. 3: Block diagrams of the trajectory planning and generator (a) and the control scheme (b).

in which

$$f_{d0} = -k_{d0}\frac{e_{d0}^{\mathrm{d}} - e_{d0}}{N}, \quad \underline{f}_d = -k_d\frac{\underline{e}_d^{\mathrm{d}} - \underline{e}_d}{N}, \quad N = \left|\underline{V}_y\right| + 2\left|\underline{V}_{y0}^{[3]}\right|. \tag{16}$$

Experimental Results

In order to demonstrate the advantages of the trajectory planning, a low-voltage MMC is exposed to a sudden voltage sag with a remaining voltage of approximately $0.15\,\mathrm{p.\,u.}$, while operating as an inverter. The dc source is a three-level voltage source converter connected to a permanent magnet synchronous machine running as generator and driven by a dc motor. The measurement results and the main components of the test bench in are shown in Figs. 4, 5, and 6. Fig. 4 and 5 compare the two cases:

left column: planning disabled The energy references e_{s0}^{d}, e_{d0}^{d}, $\underline{e}_s^{\mathrm{d}}$, and $\underline{e}_d^{\mathrm{d}}$ are calculated from the stationary regime (6) skipping the transition phase. This leaves the energy control solely to the balancing feedback, because the feed-forward currents i_{s0}^{d} and $\underline{i}_s^{\mathrm{d}}$ reduce to their respective stationary values. A change in the background shading indicates an update of the stationary regime.

right column: planning enabled The energy references e_{s0}^{d}, e_{d0}^{d}, $\underline{e}_s^{\mathrm{d}}$, and $\underline{e}_d^{\mathrm{d}}$ are given by the real-time trajectory planning. This entails corresponding feed-forward currents i_{s0}^{d} and $\underline{i}_s^{\mathrm{d}}$ that steer the energies into the designated stationary regime. The task of the balancing feedback boils down to rejecting disturbances mainly originating from modeling simplifications. An updated trajectory is indicated by alternating the background shading.

The voltage sag at $t = 0$ triggers parameter updates (left column) or planning requests (right column), as the blue background shading in Fig. 4 reveals. In both cases, adaptation to the new conditions takes until the third request, after which the updates fall back to the minimum rate, given by the setting $T = 9.2\,\mathrm{ms}$. In contrast to the standard balancing (left), the trajectory planning (right) improves the control performance w.r.t. the respective energy references shown in gray in Fig. 4. As a result, the equivalent cell voltages stay closer together and reach the balanced operation faster as visible in Fig. 5e. A lower balancing current $\underline{i}_s^{\mathrm{b}}$

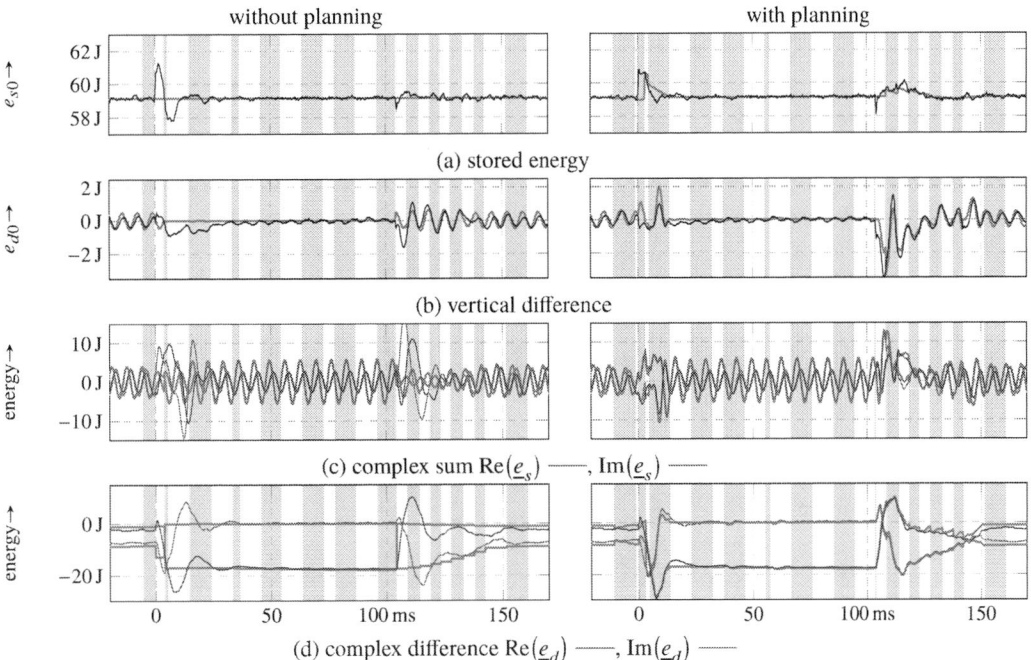

Fig. 4: Measurements of an FRT event without (left) and with (right) online trajectory planning.

on the right in Fig. 5b indicates a reduced effort of the balancing feedback during transients. However, faster restoration of the stationary regime requires increased arm currents, as visible from Fig. 5f.

The voltage and current at the dc terminal and the grid side, shown in Fig. 5c, d, g, and h, respectively, exhibit very similar behavior in both cases, because the grid current is not used by the balancing regime and the additional dc current component is very small. The usual rise of the dc voltage v_{DC} to a chopper level during the voltage sag is missing, because the source power is already reduced on the generator side depending on the MMC ac terminal power.

Conclusion

An efficient real-time trajectory planing was designed and implemented that provides consistent energy references and feed-forward currents to support the feedback-based MMC energy balancing during large-signal transients caused by e.g. sudden grid voltage sags. Thanks to the proposed analytical solution of the integrals and the detour parameters, the resulting implementation clearly outperforms existing numerical trajectory plannings. Experimental results demonstrate improved energy control performance compared to feedback balancing alone. Implementing the support of asymmetric fault conditions is left for the future.

References

[1] J. Pou, et al. Circulating Current Injection Methods Based on Instantaneous Information for the Modular Multilevel Converter. *IEEE Trans. Ind. Electron.*, 62(2):777–788, Feb. 2015.

[2] M. Jankovic, et al. Arm-Balancing Control and Experimental Validation of a Grid-Connected MMC With Pulsed DC Load. *IEEE Trans. Ind. Electron.*, 64(12):9180–9190, Dec. 2017.

[3] G. Rizzoli, et al. Decoupled Control of the Arms of a Modular Multilevel Converter with Orthogonal Reference Signals. In *EPE'19 ECCE Europe*, Sep. 2019.

[4] M. Wang, et al. Application of modular multilevel converter in medium voltage high power permanent magnet synchronous generator wind energy conversion systems. *IET Renewable Power Generation*, 10(6):824–833, 2016.

Online Trajectory Planning During Low-Voltage FRT of a Modular Multilevel Converter

FEHR Hendrik

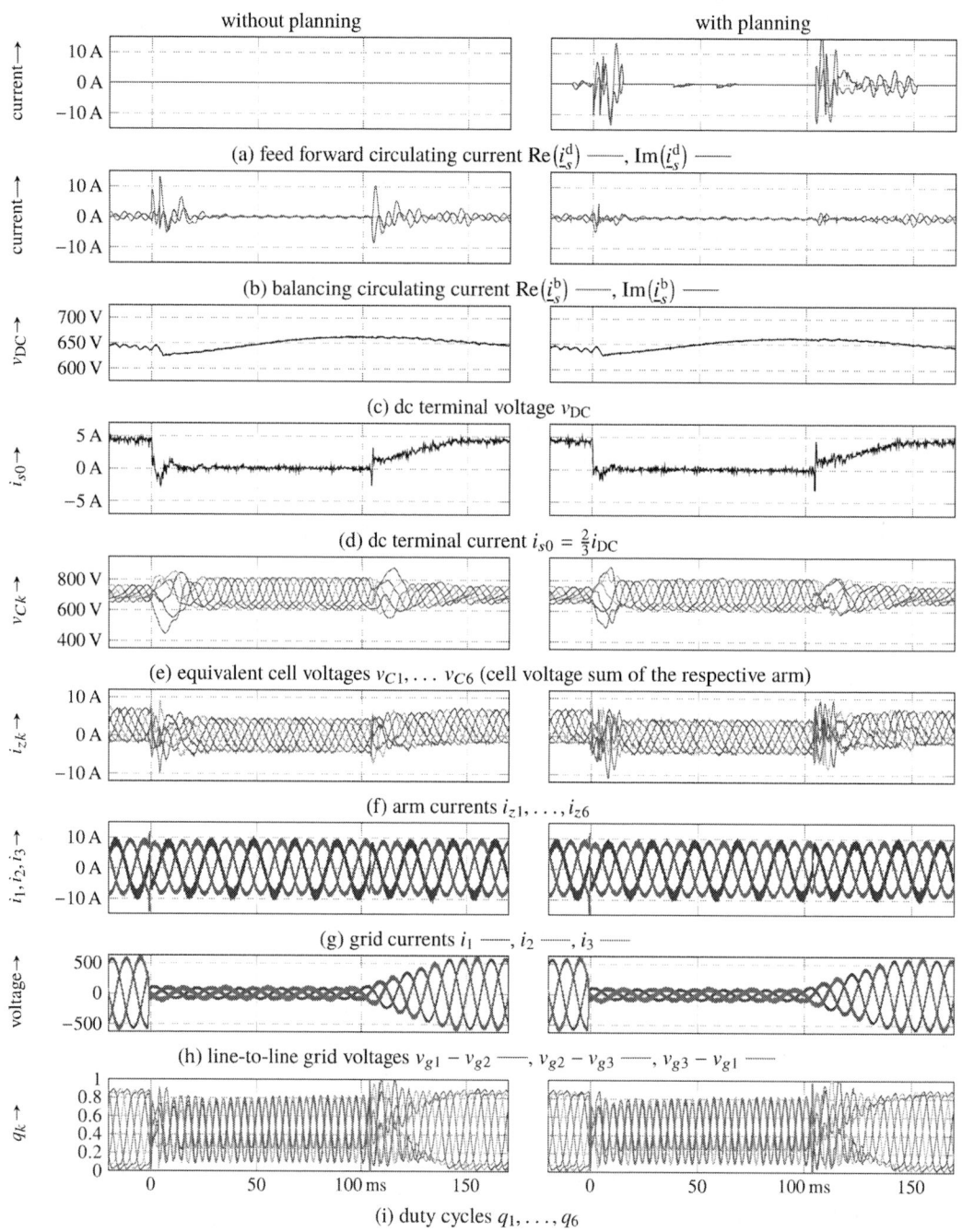

Fig. 5: Measurements of an FRT event without (left) and with (right) online trajectory planning.

(a)   (b)   (c)

Fig. 6: Test bench for the experiments in Figs. 4 and 5: view of the grid side MMC cabinet (a), grid emulator (b), dc machine and permanent magnet synchronous generator (c).

[5] K. Oguma and H. Akagi. Low-voltage-ride-through (LVRT) control of an HVDC transmission system using two modular multilevel DSCC converters. *IEEE Trans. Power Electron.*, 32(8):5931–5942, Aug. 2017.

[6] S. Rohner, J. Weber, and S. Bernet. Continuous model of Modular Multilevel Converter with experimental verification. In *ECCE*, pages 4021–4028, Phoenix, USA, Sep. 2011.

[7] H. Bärnklau, A. Gensior, and S. Bernet. Derivation of an Equivalent Submodule per Arm for Modular Multilevel Converters. In *EPE-PEMC ECCE Europe*, pages LS2a.2–1–LS2a.2–5, Novi Sad, Serbia, 2012.

[8] S. Cui, et al. A Comprehensive AC-Side Single-Line-to-Ground Fault Ride Through Strategy of an MMC-Based HVDC System. *IEEE J. Emerg. Sel. Topics Power Electron.*, 6(3):1021–1031, 2018.

[9] Q. Tu, et al. Suppressing dc voltage ripples of mmc-hvdc under unbalanced grid conditions. *IEEE Trans. Power Del.*, 27(3):1332–1338, 2012.

[10] X. Shi, et al. Steady-state analysis of modular multilevel converter (mmc) under unbalanced grid conditions. In *IEEE Applied Power Electronics Conference and Exposition (APEC)*, pages 2637–2644, 2016.

[11] H. Fehr, A. Gensior, and M. Müller. Analysis and trajectory tracking control of a modular multilevel converter. *IEEE Trans. Power Electron.*, 30(1):398–407, Jan. 2015.

[12] H. Fehr and A. Gensior. Improved Energy Balancing of Grid-Side Modular Multilevel Converters by Optimized Feedforward Circulating Currents and Common-Mode Voltage. *IEEE Trans. Power Electron.*, 33(12):10903–10913, Dec. 2018.

[13] H. Fehr. *Beiträge zur Modulation, Modellbildung und Energieregelung von modularen Mehrpunktstromrichtern (M2C)*. (in German), Dr.-Ing. dissertation, Technische Universität Dresden, Jul. 2019.

[14] Q. Gui, et al. Energy-Balancing of a Modular Multilevel Converter Using an Online Trajectory Planning Algorithm. In *EPE'20 ECCE Europe*, Sep. 2020.

[15] K. Graichen, V. Hagenmeyer, and M. Zeitz. A new approach to inversion-based feedforward control design for nonlinear systems. *Automatica*, 41(12):2033 – 2041, 2005.

[16] H. Fehr and A. Gensior. Improved Energy Balancing of Grid Side Modular Multilevel Converters by Optimized Feed-Forward Circulating Currents and Common-Mode Voltage. *IEEE Trans. Power Electron.*, 33(12):10903–10913, Dec. 2018.

[17] A. Gensior and H. Fehr. Modeling and Energy Balancing Control of Modular Multilevel Converters using Perturbation Theory for Quasi-Periodic Systems. *IEEE Trans. Power Electron.*, to be published.

Evaluating frequency stability with consideration of load type in different share of renewables and emulated inertia in case of system split

Nastaran Fazli, Sidney Gierschner, Hans-Günter Eckel
Institute of Electrical Power Engineering
University of Rostock
Rostock, GERMANY
Phone: +49 381 498-7115
Email: nastaran.fazli@uni-rostock.de

Acknowledgments

This paper was made within the framework of the research project "Netz-Stabil" and financed by the European Social Fund (ESF/14-BM-A55-0015/16). This paper is part of the qualification programme "Promotion of Young Scientists in Excellent Research Associations-Excellence Research Programme of the State of Mecklenburg Western Pomerania".

Keywords

≪Wind Energy≫, ≪Converter Control≫, ≪Voltage Source Inverters (VSI)≫, ≪Inertia Support≫

Abstract

Nowadays, with increasing the share of renewable generation units, a modern power system encounters stability challenges due to the shrunken inertia. Yet the majority of installed wind turbines as an alternative for synchronous generator do not perform a similar role regarding inertia provision and grid support, which sets a limit for higher penetration of them into a power system. As a solution, new control for converter-based generation units is introduced, enabling them to be a grid supporting voltage control system, which reacts seamlessly to the variation of the grid parameters. Hence, this paper aims to evaluate the required minimum amount of inertia to keep stability after a system split. To this aim, a comparison will be drawn between the synchronous generator and these grid-supporting voltage control power sources in combination with the installed capacity of grid feeding wind turbines. Besides, the tests will be done in the presence of different type of loads, as the effect of converter-fed loads is worth investigating in the evaluation of the border for the share of renewables.

Introduction

From the early 2000s, policies and policymaker's aim in the European countries was only an unconditional expansion of renewable energy generation as the main priority to deal with the problem of emission of greenhouse gases. This ended with the current power system condition with a high share of renewables [1].

There is the actual trend for having a high share of non-synchronous generation and decreasing the percentage of synchronous generation (SG). This has made a significant change in the power system's condition regarding frequency stability and voltage support. Both were conventionally the tasks of the generation units with SG. A survey from 11 TSOs categorizes the most concerning issues in the power system from top to down considering severity and probability, namely frequency stability and voltage stability. The highest ranking among them belongs to the frequency stability issues due to the decrease of inertia in the power system [2].

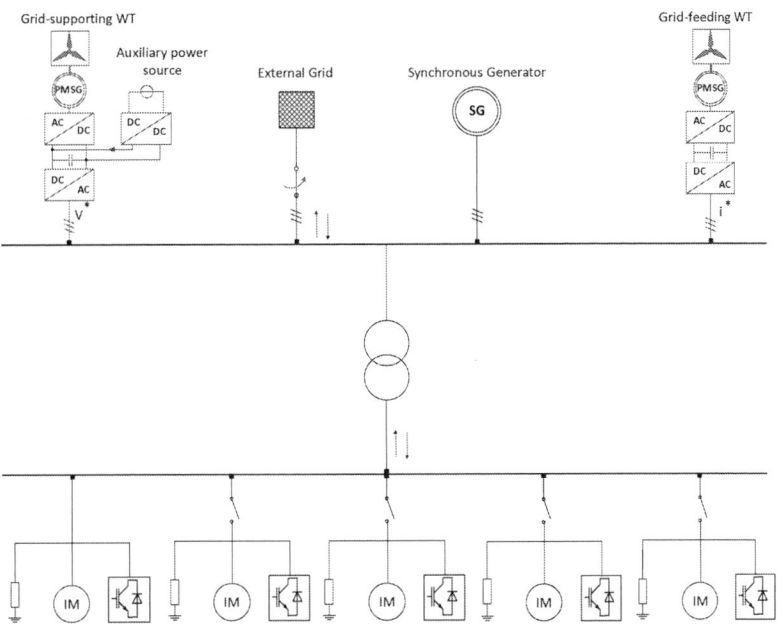

Fig. 1: Model of simulated power system

On the other hand, increasing transmission capacities lead to an increasing number of possible imbalances in the case of a system split, which directly affects frequency stability [3]. This will force new challenges that require new policies and inertia re-evaluation in the power system to keep security in the system split.

The system split can occur after an outage in transmission lines, which can violate voltage and frequency depending on the depth of unbalance. SG and wind turbines (WTs) with grid-supporting features can support the grid on such occasions. However, all types of generation, including non-synchronous, have a natural delay for primary control. Therefore, a high Rate of Change of Frequency (RoCoF) value can make the system prone to blackouts. How high the RoCoF is, depends on the amount of imbalance and also the share of generation based on power electronics that their increasing share reduces the inertia constant of the system. A severe frequency gradient can put mechanical stress on the prime mover and give no time for reaction of reserve containment, which results in misfunction of load shedding relays (LSR) and cascaded trips. The expected data in 2030 from the European Network of Transmission System Operators for Electricity (ENTSO-E) illustrates that the inertia constant in Germany drops to H=0.9 s for 90% of the hours per year [3]-[4]. According to this data and in the absence of generation units with intrinsic inertia, the same imbalance can cause a severe gradient of frequency, making the system uncoverable.

Another major issue in the power system nowadays is increasing the share of loads with constant power consumption like converter-fed drives. As a precise load estimation is not possible for TSOs, a rough estimation shows that migration toward 70% power-electronic loads (PE-load) is expected in the near future. This draws attention to their existence. Converter-fed drives have no inherent inertia and their consumption stays constant during frequency dips and voltage variations to an extend. While loads like directly-connected drives perform a self-regulating effect that the power system can benefit from it. In [5]-[7], an islanding scenario with increments or decrements of passive RL or RC loads is investigated. These passive loads have mainly voltage-dependent consumption, and they will eventually cause active power imbalance as well, which has to be dealt with. In a power system, different shares and types of load exist. In addition to passive loads, investigating induction machines and converter-fed loads with variations of their share can be beneficial for frequency stability studies and LSR performance in an islanded grid.

In such a condition, different solutions and countermeasures can be taken to secure the power system. One of them is to use the capacity of installed WTs for inertia emulation, but this is accompanied by a high degree of difficulty, developing the existing capacity, and also has a level of uncertainty as the wind speed is constantly changing. In some operating points, the provision of inertia is very low or almost impossible without the risk of falling below cut-in speed and losing the existing capacity [8]. Another technology is fly-wheel inertia storage as an alternative, although the cost can escalate due to losses of the fly-wheel [6]. To reduce the uncertainty associated with renewable sources' nature, an auxiliary power source connected to the grid-side converter of WTs with a grid-supporting control can support the grid without the risk of disconnection. Usually, during sudden load changes, high-density power storage (like supercapacitors) as an auxiliary power source can be beneficial to reduce the frequency nadir and RoCoF. Also, in case inertia and primary provision are demanded, battery storage directly connected to the dc-ac converter of WT can serve the power system with ancillary services. In this paper, an auxiliary power source (APS) is connected to the grid-side converter of a WT. The converter's provided control is modeling this system like a voltage source, which can sense the variation of frequency and voltage of the grid and react to it in time. As mentioned, this paper focuses on increasing the share of converter-fed loads as well.

In the following sections, first, the criteria for the islanding system will be explained. Next, the control details for grid-supporting WT (GS-WT) and grid-feeding WT (GF-WT) will be presented. Then, the simulation results for the test scenarios between emulated inertia and ancillary service of the WT equipped to the APS and SG with considering the effects of loads will be shown.

Test System Criteria and Scenarios

The following criteria and scenarios are considered for the simulation of this power system , as shown in Fig.1. The generation units consist of a combination of a conventional generation plant with SG and two WTs with two different types of control on their grid-side converter. For the test, it is essential to see the frequency stability's borders, which sits within the criteria based on the grid codes.

The islanding incident in this paper is chosen for all scenarios according to a 5% import deficit. This is similar to the historical data of the last system split in continent Europe in 2006 in Germany, which experienced import interruption of 5% in the western area (area 1) and led to under-frequency and separation from the rest of the synchronous area [3]. Droop control for SG and GS-WT is designed in accordance with the ENTSO-E code, which is 5% droop coefficient. Reactive power droop is also chosen 5% for the Automatic Voltage Regulator (AVR) of the SG and the GS-WT droop loop. For this test system, the acceleration time constant of the system can be calculated according to (1) and (2)[3].

$$T_N = \frac{\sum_1^n T_{A,SG,n} * P_{Nom,SG,n}}{P_{Load}} \qquad (1)$$

$$T_A = \frac{J * \omega_N^2}{P_{SG}} \qquad (2)$$

In (1), T_A is the time constant of the SG in seconds, $P_{Nom,SG,n}$ is the nominal power of number nth SG in the power system, and P_{Load} is the total load power. The value of T_N is the acceleration time constant of the power system in seconds, which indicates the strength of the power system associated with the amount of inertia. According to (1) and (2), in a 100% share of SG/GS-WT, $T_N = 10$s. For different shares of SG/GS-WT, the inertia constant is scaled based on the penetration level of the SG/GS-WT to the total load. The grid-feeding control on power converters of typical WTs will not participate in inertia as the converter decouples the generator side from the variation of the grid parameters. After islanding,the unbalanced value and the inertia determine what the gradient of the frequency will be according to (3).

EPE'20 ECCE Europe

Assigned jointly to the European Power Electronics and Drives Association & the Institute of Electrical and Electronics Engineers (IEEE)

$$RoCoF_{max} = \Delta P_{imbalance} * \frac{f_0}{T_N} \tag{3}$$

In (3), ΔP is the unbalanced value relative to the initial load demand in the per-unit system, and f_0 is the initial frequency before islanding. Theoretically, according to (3), with this time constant in the system, maximum and minimum RoCoF for ΔP= 5% and T_N = [10 - 0.1 s] stay in the range of [0.25 - 25 Hz/s] [3]. But in practice, the dynamics of the loads and fast frequency support of the GS-WT are playing a decisive role in max df/dt value.

The load configuration is also shown in Fig.1. The loads, which are used in this power system, are divided into three different groups. There are resistive loads with voltage-dependent power consumption and directly-connected drives like induction machines, which have a frequency-dependent power consumption. The last group is converter-connected drives. Their effect is considered of the essence as their consumption has no dependency on frequency.

As the grid code proposes, four load groups are implemented in this power system. Each has a share of 12.5% of the total load and is equipped with an LSR. The LSRs are triggered as the frequency reaches to specified thresholds of [49, 48.8, 48.6, 48.4 Hz]. The 5th group of load forms the rest 50% of the total load, which has no LSR. The time delay of LSR in this paper is 100 ms for detection and 70 ms for the disconnection of each switch after detection.

Control of Grid-Supporting WT equipped to APS

In a GS-WT, inertial support during frequency deviation can be provided. As shown in Fig.1, in the production side, apart from the SG, a grid-feeding state of the art WT is implemented to model the behavior of a power system dominated by renewables. Also, a GS-WT equipped with an APS with grid-supporting functionalities is modeled into the sample power system. In such a power system with a share of SG, GF-WT, GS-WT and loads, the relationship according to (4) exists.

$$P_{WT} + P_{SG} - P_L = \pm \Delta P \tag{4}$$

In an islanding scenario, the $\pm \Delta P$ that could conventionally be provided only by SG, can be provided by GS-WT as well. A GS-WT, similar to an SG depending on its share, can provide ancillary services to ease frequency/voltage deviation. In this power system, APS is providing this additional power demand. Therefore, $\pm \Delta P$ command is given to the APS control.

$$\pm \Delta P = P_{APS} \tag{5}$$

This command can be expressed as (6).

$$P_{APS} = J * \frac{d\omega}{dt} + D_p * \Delta\omega \tag{6}$$

The APS power demand is enabled using (5) to obtain the amount of current (i_{APS}) to be released from APS. This current is converted into PWM to switch the buck-boost converter either to release or to absorb power (7).

$$i_{APS} = \pm \Delta P / v_{APS} \tag{7}$$

A PI controller is implemented to control the converter's duty cycle to make sure the power command is achieved. By comparing the duty cycle with a triangular carrier signal (frequency of 5 kHz), the switching signals for IGBTs are obtained.

In the grid-side control of GS-WT, adjusting the voltage at the Point of Common Coupling (PCC) in the islanding situation is done by a reactive droop loop. The inertia constant is chosen in the same manner as for SG to keep a T_N = 10 s in case of 100% GS-WT with APS share. The power is limited via the frequency loop. The droop factor is chosen as for the SG, 5%, to be comparable to the frequency excursion scenarios with SG.

Control of Grid-feeding Converter

A voltage-oriented control (VOC) for the grid-side converter of the WT is realized via a cascaded control structure. Based on this control strategy, power converters behave as a current source, and the current references determine the amount of active and reactive power exchange to the grid [9]. The operation in this control is limited to the grid-connected mode and requires grid information to be synchronized. Hence, a fast phase-locked-loop (PLL) is implemented for control. Therefore, it has to be always an SG or GS-WT which forms the grid.

Scenario I and Simulation Results

In this scenario, the load is composed of a 10% resistive load, 80% directly-connected drives and 10% converter-fed drives. The share of GF-WT is changing in the steps of 10%. The total amount of generation is 95% of the total load in the time of disconnection for active power, which means that the rest 5% active power is provided via the external grid. However, a total capacity of 96% of the total load can be provided, which is similar to the available 1% reserve according to data of ENTSO-E [10] .

At t=0 s, the external grid is disconnected and interrupts the import. The reactive power exchange between the external grid and the loads is zero as the total reactive power is transferred from SG/GS-WT in each power share. Fig. 2 shows the simulation results for scenario I.

Each color shows the share of SG/GS-WT in percent in the system. As there is no reference scenario to define minimum required inertia, hence, in this test, the results are compared to show the maximum share of grid-feeding WTs based on how fast load shedding is triggered and the RoCoF value of each share.

As it is shown in Fig. 2(a) and (b), RoCoF is different in each SG/GS-WT share as it will change the total acceleration time constant of the power system. In 0% share of SG or GS-WT, the RoCoF is so high, and frequency drops to a low nadir that the first step of LS is inevitable. Such a high RoCoF will neither allow the self-regulating effect nor the primary frequency control to stabilize the frequency from falling below 49 Hz.

In tests 2 to 4, with the higher shares of a generation with inertia provision, frequency stability is feasible . However, RoCoF and nadir have a slight difference in each test, as shown in 2(a). Although this difference is more noticeable 2(b). The reason lies in the time constant of the governor of the SG. In a GS-WT, however, the primary frequency control is applied as soon as $\Delta\omega$ is measured in the frequency control loop. This resulted in different steady-state frequency values.

The RoCoF comparison of these four tests is illustrated in Fig. 2 (c) and (d). According to the graphs, the worst gradient of frequency is measured when the share of SG or GS-WT is 0%. In test 1 with SG and GF-WT combination, this value is -0.63 [$Hz\,s^{-1}$], and with GS-WT and GF-WT, it is reduced slightly to -0.57 [$Hz\,s^{-1}$]. The RoCoF measurement is designed to calculate RoCoF over a 500 ms time frame according to ENTSO-E criteria [11].

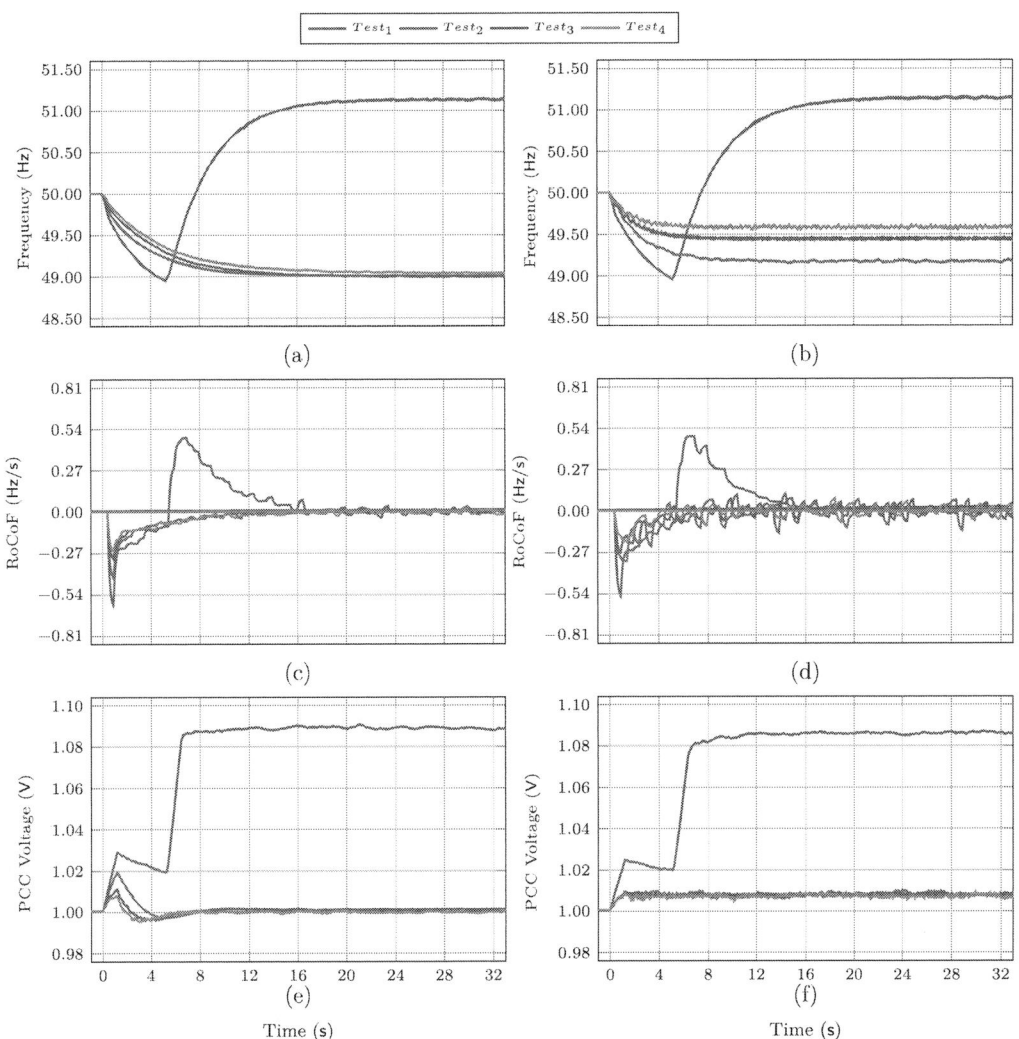

Fig. 2: Scenario I : Frequency, RoCoF and PCC voltage with 10% resistive - 80% directly-connected drives - 10% converter-fed drives, left) generation composed of GF-WT and SG, right) generation composed of GF-WT and GS-WT. Test description: test 1) 0% SG/GS-WT and 95% GF-WT, test 2) 10% SG/GS-WT and 85% GF-WT, test 3) 20% SG/GS-WT and 75% GF-WT, test 4) 30% SG/GS-WT and 65% GF-WT

The voltage at PCC is also shown in Fig. 2(e) and (f). The trend is following a slightly different path between tests with SG and GS-WT. After disconnection of the external grid and with a deceleration of directly-connected drives due to active power imbalance, the reactive power consumption of drives is also reduced as they differ from the nominal operating point. Due to higher reactive power exchange between generation units and loads, the voltage in all tests increases at the beginning and will be regulated either by AVR or by the reactive droop loop of GS-WT. In test 1 in both cases, the voltage increased further as a result of load shedding. In Fig. 2(e), AVR could take back the voltage after almost 8 seconds. However, in test 1 in Fig. 2(f), due to the faster regulating of active power, the voltage settled down after 2-3 seconds in a steady-state, which is higher than the nominal value.

Scenario II and Simulation Results

To evaluate the minimum required inertia to keep frequency robustness, the effect of constant power loads in such system splits is considered and compared to the first scenario with 10% converter-fed loads. Therefore, in this test, the share of directly-connected drives diminished to 70%, and converter-fed drives share increased to 20% of the total load power. simulation results are shown in Fig. 3(a) to (f).

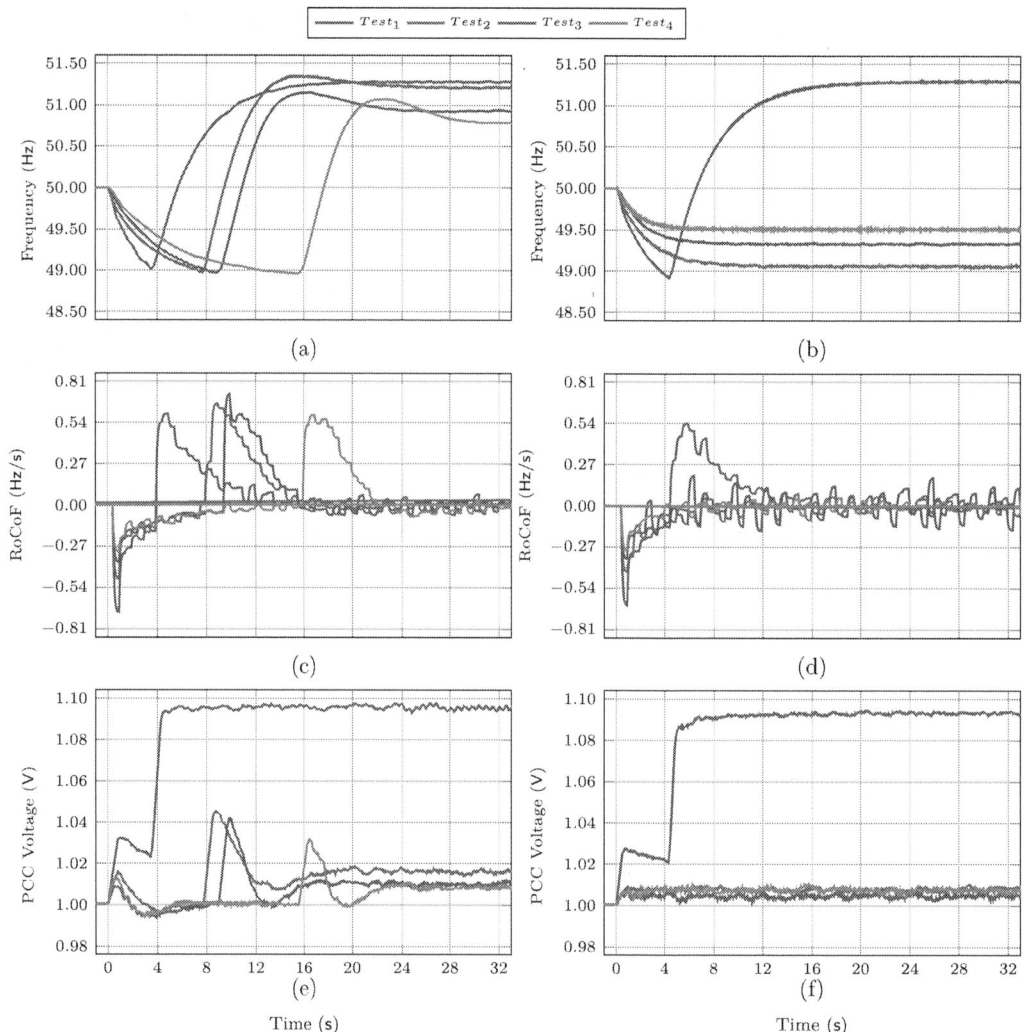

Fig. 3: Scenario II : Frequency, RoCoF and PCC voltage with 10% resistive - 70% directly-connected drives - 20% converter-fed drives, left) generation composed of GF-WT and SG, right) generation composed of GF-WT and GS-WT. Test description: test 1) 0% SG/GS-WT and 95% GF-WT, test 2) 10% SG/GS-WT and 85% GF-WT, test 3) 20% SG/GS-WT and 75% GF-WT, test 4) 30% SG/GS-WT and 65% GF-WT

As shown, the same trend as in scenario I is followed by different shares of SG or GS-WT. Although, the highest RoCoF related to the share of 0% SG/GS-WT is increased in comparison to scenario I with a lower share of PE- loads. As shown in Fig. 3(a), all shares of GF-WT will reach to load shedding which indicates that with this share of the load, even with a higher proportion of SG (up to 30%), the frequency stability after a 5% imbalance cannot be achieved and load shedding will occur. On the other hand, in Fig.3(b), the RoCoF is increased in all tests compared to scenario I, but it will already show a

Table I: Frequency comparison in scenarios I and II in different tests

		SG with GF-WT			GS-WT with GF-WT		
		RoCoF (Hz/s)	Nadir (Hz)	LS time (s)	RoCoF (Hz/s)	Nadir (Hz)	LS time (s)
Scenario I	Test 1	-0.63	49.00	5.20	-0.57	49.00	5.40
	Test 2	-0.44	49.01	-	-0.32	49.17	-
	Test 3	-0.31	49.03	-	-0.27	49.44	-
	Test 4	-0.25	49.05	-	-0.19	49.50	-
Scenario II	Test 1	-0.70	49.00	3.50	-0.65	49.00	4.35
	Test 2	-0.48	49.00	7.74	-0.43	49.06	-
	Test 3	-0.37	49.00	8.90	-0.34	49.30	-
	Test 4	-0.30	49.00	15.50	-0.29	49.51	-

better performance than Fig. 3(a). It was stated already that the time constant of the dc-dc converter to deliver this primary power to the grid from the beginning of islanding is very small in contrast to the governor's time constant is much higher. This could capture the frequency in a higher nadir and prevent load shedding in tests 2-4.

The RoCoF measurement graphs are plotted in Fig. 3(c) and (d). Table I shows a summary of the results related to the frequency gradient, nadir and load shedding time in each test in scenario I and II. According to it, only by increasing 10% more PE-loads, the gradient of frequency increases to an extend, that load shedding is happening in SG and GF-WT systems in all shares. Also, in the GS-WT and GF-WT, the value of RoCof increased in all four tests comparing to scenario I, the nadir reduced, and LSR is triggered faster. Looking at this data shows the limitation upon further share of renewables without inertia as the loads are migrating towards PE-loads. Using renewable generation with intrinsic inertia can help the power system to increase in the share of GF-WTs.

Voltages at PCC are also shown in Fig. 3(e) and (f). In the combination of SG and GF-WT, in test 1, the voltage settled in a higher steady-state value as the directly-connected loads were shrunken. In the rest of the tests, the voltage has risen in each in the moment of load shedding. Similar to scenario I, for GS-WT and GF-WT combination, the voltages reached faster to a steady-state.

Conclusion

In this paper, the frequency comparison in different shares of grid-feeding WTs and converter-fed loads was investigated. As the results showed, for the same islanding scenario, there were limits for further share of grid-feeding WTs, which means that going toward a 100% inverter-dominated power system is only possible when at least 10% of the total load power is generated by an inverter system, forming the grid and has intrinsic inertia.

Also, in changing the proportion of the loads, a 10% increase of PE-loads showed a noticeable difference in the RoCoF and nadir value in the results, which requires further investigation for studying the required future minimum inertia for a 100% inverter-dominated power system.

References

[1] S. Karamitsos, A. Canelhas, M. Bazargan, R. Ierna, B. Marshall and S. Kelly , "Inverter Dominated UK Grid", CIGRE Session Technical Programme, Paris, France 2018

[2] V. N. Sewdien et al., "Effects of Increasing Power Electronics on System Stability: Results from MIGRATE Questionnaire," 2018 International Conference and Utility Exhibition on Green Energy for Sustainable Development (ICUE), Phuket, Thailand, 2018

[3] ENTSO-E, "Frequency Stability Evaluation Criteria for the Synchronous Zone of Continental Europe: Requirements and impacting factors," Brussels, Belgium, March 2016

[4] ENTSO-E, "High Penetration of Power Electronic Interfaced Power Sources (HPoPEIPS): ENTSO-E Guidance document for national implementation for network codes on grid connection," Brussels, Belgium, March 2017

[5] M. Nuschke, B. Winter, D. Strauß-Mincu and B. Engel, "Power System Stability Analysis for System-split Situations with Increasing Shares of Inverter based Generation", Conference on Sustainable Energy Supply and Energy Storage Systems (NEIS), Hamburg, Germany, 2019

[6] D. Duckwitz, "Power System Inertia: Derivation of Requirements and Comparison of Inertia Emulation Methods for Converter-based Power Plants", Dissertation, Kassel, Germany, 2019

[7] N. S. Hasan, N. Rosmin, N. M. Nordin and M. Y. Hassan, "Virtual inertial support extraction using a super-capacitor for a wind-PMSG application," in IET Renewable Power Generation, vol. 13, no. 10, pp. 1802-1808, 29 7 2019

[8] A. Gloe, C. Jauch, B. Craciun and J. Winkelmann, "Continuous provision of synthetic inertia with wind turbines: implications for the wind turbine and for the grid," in IET Renewable Power Generation, vol. 13, no. 5, 2019

[9] T. Qoria, T. Prevost, G. Denis, F. Gruson, F. Colas and X. Guillaud, "Power Converters Classification and Characterization in Power Transmission Systems," 21th European Conference on Power Electronics and Applications (EPE'19 ECCE Europe), Genova, Italy, 2019

[10] ENTSO-E, "Continental Europe Operation Handbook Policy P1: Load- Frequency Control and Performance," 19 March 2009. [Online]

[11] ENTSO-E, "Rate of Change of Frequency (RoCoF) withstand capability," Brussels, Belgium, 2018

Discrete-time Direct Pole Placement for Stability Enhancement of LCL-Filtered Inverters in the Synchronous-Reference Frame

Pei Cai[1], Xiaohua Wu[1], Yongheng Yang[2], Wenli Yao[1], Weilin Li[1], Frede Blaabjerg[2]

[1]School of Automation, Northwestern Polytechnical University
127 Youyi West Road, 710000
Xi'an, Shaanxi, 710072, P.R.China
[2]Department of Energy Technology, Aalborg University
Pontoppidanstraede 101, 9220
Aalborg, Denmark
E-Mail: caipei1992@mail.nwpu.edu.cn, wxh@nwpu.edu.cn, yoy@et.aau.dk,
yaowl@nwpu.edu.cn, wli907@nwpu.edu.cn, fbl@et.aau.dk

Keywords

«Digital control», «Active damping», «Robustness», «Regulators», «Voltage Source Converter (VSC)».

Abstract

For LCL-filtered grid-tied inverters, direct pole-placement techniques can combine the fundamental-frequency current regulation and resonance damping simultaneously. As known, for state-space-based direct pole-placement, to purse sensorless operations, state observers have to be employed, which means heavy computation burden and might limit the switching and sampling frequencies. While for transfer-function-based one, the number of necessary variables decreases and meanwhile it saves the computation load from observers. Therefore, the latter is more suitable in practice. Based on this, this paper extends and summarizes this approach for better accessibility, and it shows that, compared to previous applications in the stationary frame, the transfer-function-based discrete-time domain (z-domain) direct pole placement in the synchronous frame presents better robustness against grid impedance variations, improved phase and gain margins, and simplified feed-forward compensation, presuming the cross coupling is ignored. Simulations have verified the analysis and design.

Introduction

LCL-filtered grid-connected inverters play a key role in interfacing renewable energy resources to the power grid [1]–[3]. The employment of LCL-type filters effectively attenuates switching harmonics caused by pulse width modulation (PWM) behaviors of inverters, with low volume and cost compared to their counterparts like L-type filters [2]. However, instability may arise in LCL systems due to almost zero impedance at the resonance frequency of LCL filters.

Therefore, various damping solutions to the resonance have been proposed in the literature [4]–[10]. Passive damping methods, which add resistors into LCL branches, can provide reliable damping at a wide frequency range but it is with inevitable power losses [5], especially in high power applications. In contrast, active damping strategies are realized without any additional passive components. From the point of stability of control loops, LCL resonance is damped by auxiliary-variable-involved loop-gain modifications in both phase and magnitude [4], [9], [11]–[14]. For sensorless purpose [15], the capacitor voltage [12], [16] or PCC (point of common connection) voltage [11], [17] utilized for the synchronization, grid-side [18] or inverter-side current [19] for closed-loop control are effective variables to achieve the active damping. In addition, active damping design is expected to be separated from the fundamental current regulation [20], [21], though in certain cases a comprehensive design had to be considered for robustness against the parameter variations, by root locus [22], or combined stability margin constraints graphically [23], pre-defined pole placements [18], and LQR [15]. In [18], pole locations of inner damping loop rather than the whole loop was defined, and thus, the closed-loop response needs to be carefully checked after the design.

Actually, the pole placement can be promising to achieve fundamental-frequency current regulation and resonance damping simultaneously, i.e., the so-called direct pole placement [24], [25], including state-space-based and transfer-function-based ones. As for the former, [24] measured all necessary variables to implement the state-space-based pole placement, which, however, was undesired in the sensorless trending and can be improved by using state observers [25]–[27]. Meanwhile, the cross-coupling ignored in the design process in synchronous coordinates might decrease the design accuracy. Then, without considering time delays into system modeling but with compensation for simplicity, [25] presented an analytical tuning procedure for both controller and observer as functions of physical system parameters and design specifications. Moreover, the cross-coupling between d- and q-axis was weakened thanks to specific pole locations. In addition, [26] extended it by taking time delays into system modeling (means higher system orders) for real-time computation in synchronous coordinates. Furthermore, [27] achieved zero steady-state errors for both positive- and negative-sequences currents under unbalanced grid conditions in the stationary frame, yielding fast dynamics and negligible overshoots. However, a common issue for the state-space-based direct pole placement is that, due to the involvement of state observers to estimate variables to reduce the number of sensors, the computation burden in digital controllers rises significantly.

On the contrary, when transfer-function-based direct pole placement in the discrete-time domain (z-domain) is adopted, as proven in [28], the computational load is five times less than that in [27], which indicates more suitable implementations especially with high switching frequencies. In [28], a z-domain-based realization in the alpha-beta-zero (αβ0) frame was proposed, but with a fifth-order controller and a limited phase margin (PM). Meanwhile, the reference signals were actually obtained via a separating matrix to separate real- and imaginary-numbers due to a complex-number feedforward. In light of the above, this paper extends and summarizes the z-domain pole-placement technique for better accessibility, that is, easier to implement; and then, investigates its application in the synchronous frame (dq0). It is found that both the PM and gain margin (GM) rise and the order of the controller is reduced; furthermore, the feed-forward part is simplified and easy to maintain the DC gain. It means the complex-number separating matrix for an unaffected line-frequency gain in αβ0 in [28] is avoided. Simulations in MATLAB/Simulink have verifies the analysis and design.

The system is described and modeled in Section II. The discrete-time pole placement algorithm is summarized in Section III, followed by the applications of that in grid-side current feedback (GCF) scheme and the case study in Sections IV and V, respectively. Simulation results of are presented in Sections VI. Sections VII gives conclusions.

System Description and Modeling

Fig. 1(a) shows the phase-a circuit diagram of a balanced three-phase grid-connected inverter system with an LCL filter. The grid voltage e_g is measured for synchronization, and L_g represents the grid impedance. Fig. 1(b) depicts the GCF control loop in the z-domain and in the dq0 reference frame, where the cross-coupling between d- and q-axis is neglected for design [24]. $G_{pp}(z) = G_c(z)/(z-1)$ is the pole-placement controller to be designed. K_{PWM} is the inverter PWM gain. $G_{i2}(s)$ is the system plant in the s-domain as:

$$G_{i2}(s) = \frac{1}{sL_1(L_2+L_g)C}\frac{1}{s^2+\omega_r^2}.$$

(1)

After discretizing by the zero-order hold (ZOH) method, it gives $G_{i2}(z)$ in the z-domain as:

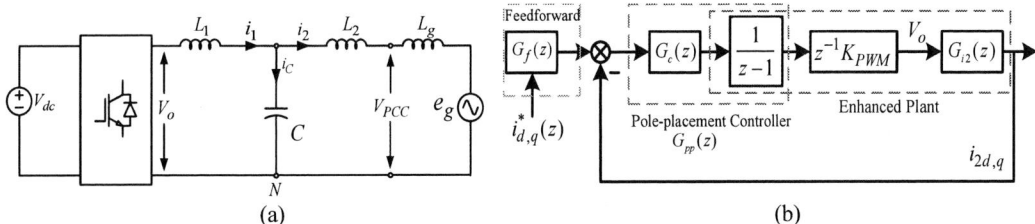

(a)   (b)

Fig. 1. Single-phase LCL-filtered grid-tied inverter: (a) circuit diagram and (b) current control loop (K_{PWM}: PWM gain, V_o: inverter output voltage, V_{PCC}: point of common coupling).

$$G_{i2}(z) = \frac{\omega_r T\left[z^2 - 2z\cos(\omega_r T) + 1\right] - \sin(\omega_r T)(z-1)^2}{\omega_r (z-1)\left[z^2 - 2z\cos(\omega_r T) + 1\right]},$$ (2)

where $\omega_r = [(L_1+L_2+L_g)/(L_1(L_2+L_g)C)]^{1/2}$. Furthermore, the total time delay considered in this paper is $1.5T$, where T is sampling period, and calculation delay is one T, i.e., z^{-1}, and ZOH represents $0.5T$, respectively. Finally, the extra presence of $1/(z-1)$, together with the enhanced plant and feed-forward transfer function $G_f(z)$ will be explained in the following.

Discrete-Time Pole Placement Algorithm

The pole placement is a problem that one can synthesize a controller such that the closed-loop poles are in predefined locations [29]. Different form the state-space method [25]–[27], the method used in this paper is a polynomial approach. [29] proposed a concise pole placement algorithm, which was also adopted in [28]. Here, it will be briefly reviewed and extended to be a general method. That is, additional poles or zeros can be integrated into the original plant model to produce an enhanced plant; meanwhile, the constraints for choosing the controller degree are established for better accessibility.

For a system plant model $G(z) = B(z)/A(z)$ and a controller $G_c(z) = P(z)/L(z)$, given as:

$$\begin{cases} A(z) = a_n z^n + \cdots + a_1 z + a_0, & B(z) = b_m z^m + \cdots + b_1 z + b_0 \\ L(z) = l_{n_l} z^{n_l} + \cdots + l_1 z + l_0, & P(z) = p_{n_p} z^{n_p} + \cdots + p_1 z + p_0 \end{cases},$$ (3)

where their degrees are defined as the respective highest order. The characteristics polynomial is thus given as:

$$A_{CL}(z) = A(z) \cdot L(z) + B(z) \cdot P(z),$$ (4)

where $A_{CL}(z)$ is pre-defined closed-loop poles.

Considering an extended case, where either additional poles or zeros are added into the original plant $G(z)$ and denoting the additional pole polynomial as $A_{ad}(z)$ and additional zero polynomial as $B_{ad}(z)$, and their degrees are t and k, respectively. The enhanced plant is obtained as $\{B_{ad}(z)B(z)\}/\{A_{ad}(z)A(z)\}$, where the degree of the numerator and the denominator is $m+k$ and $n+t$, respectively. The characteristics polynomial can then be obtained as:

$$A_{CL}(z) = A(z) \cdot A_{ad}(z) \cdot L(z) + B(z) \cdot B_{ad}(z) \cdot P(z) = a_{n_c}^c z^{n_c} + \cdots + a_1^c z + a_0^c.$$ (5)

Then, according to the Sylvster's theorem [29], if $A(z)A_{ad}(z)$ and $B(z)B_{ad}(z)$ are relatively prime, i.e., they have no common factors, there is a non-singular matrix S, whose elements are obtained by equating the coefficient of the same order of (5), making the controller vector U solvable, that is, $SU = V$, where $U = [l_{n_l}, \ldots, l_0, p_{n_p}, \ldots, p_0]^T$, representing the unknown coefficients of the controller, and here $V = [a_{n_c}^c, \ldots, a_0^c]^T$, coefficients of the closed-loop poles.

Note that the number of unknown coefficients and that of equations should be equal to ensure that S is a square matrix. Thus, three constraints (6)-(8) are built to choose the degree of $P(z)$ and $L(z)$.

- The number of unknown coefficients is equal to that of equations:

$$n_l + n_p + 2 = n + t + n_l + 1 \ . \tag{6}$$

- For the pole-placement controller $G_{pp}(z) = B_{ad}(z)/A_{ad}(z)*P(z)/L(z)$, to guarantee its causality, the degree of its denominator is larger than that of its numerator:

$$n_l + t \geq n_p + k \tag{7}$$

- For the open-loop transfer function, the degree of the denominator $B(z)B_{ad}(z)L(z)$ is larger than that of the numerator $A(z)A_{ad}(z)P(z)$. Actually, in the z-domain, the degree of the denominator always exceeds that of the numerator in the system plant [5], which implies $n > m$ and from (7), (8) always holds.

$$n_c = n + t + n_l \geq m + k + n_p \tag{8}$$

Solving (6)-(8) gives (9). Based on (9), if the minimum degree of n_l is desired, there is $n_l = n + k - 1$. In this way, the degrees of n_p and n_l can be determined and the generalized equation $SU=V$ is written as (10), where the subscript of b should be within $[0, m + k]$; otherwise b is zero, where $m + k$ is the degree of the numerator of the enhanced plant. By solving $U=S^{-1}V$, one can get $G_c(z)= P(z)/L(z)$, as a part of the full controller $G_{pp}(z)$. The generalized results are then summarized in Table I.

$$\begin{cases} n_p = n+t-1 \\ n_l \geq m+k-1 \Rightarrow \\ n_l \geq n+k-1 \end{cases} \begin{cases} n_p = n+t-1 \\ n_l \geq n+k-1, \text{with } n \geq m \end{cases} \tag{9}$$

$$\begin{bmatrix} a_{n+t} & & & b_{n+k} & & \\ a_{n+t-1} & \ddots & & b_{n+k-1} & \ddots & \\ \vdots & & \ddots & \vdots & & \ddots \\ \vdots & & \ddots & \vdots & \vdots & \ddots & \vdots \\ a_0 & & \ddots & a_{n+k-1} & b_{k-t} & \ddots & b_{n+k-1} \\ & \ddots & & \vdots & \vdots & \ddots & \vdots \\ & & \ddots & \vdots & & \ddots & \vdots \\ & & & a_0 & & & b_0 \end{bmatrix} \begin{bmatrix} l_{n_l} \\ \vdots \\ l_0 \\ \hline p_{n_p} \\ \vdots \\ p_0 \end{bmatrix} = \begin{bmatrix} a_{n_c}^c \\ a_{n_c-1}^c \\ \vdots \\ \vdots \\ a_1^c \\ a_0^c \end{bmatrix}, \text{ denoted by } SU = V \tag{10}$$

TABLE I: GENERALIZED RESULTS OF DISCRETE-TIME POLE-PLACEMENT TECHNIQUES

Name	Polynomial	Name	Polynomial
Original Plant num. $B(z)$	$b_m z^m + \cdots + b_1 z + b_0$	Original Plant den. $A(z)$	$a_n z^n + \cdots + a_1 z + a_0$
Additional zeros $B_{ad}(z)$	$b_k^{ad} z^k + \cdots + b_1^{ad} z + b_0^{ad}$	Additional poles $A_{ad}(z)$	$a_t^{ad} z^t + \cdots + a_1^{ad} z + a_0^{ad}$
Controller num. $P(z)$	$p_{n_p} z^{n_p} + \cdots p_0$, $n_p = n + t - 1$	Controller den. $L(z)$	$l_{n_l} z^{n_l} + \cdots + l_0$, $n_l \geq n + k - 1$
Enhanced plant num.	$B_{ad}(z)B(z)$, $m + k$	Enhanced plant den.	$A_{ad}(z)A(z)$, $n + t$
Full controller num.	$B_{ad}(z)P(z)$, $n_p + t$	Full controller den.	$A_{ad}(z)L(z)$, $n_l + t$
Closed-loop poles $A_{CL}(z)$	$A_{CL}(z) = A(z) \cdot A_{ad}(z) \cdot L(z) + B(z) \cdot B_{ad}(z) \cdot P(z) = a_{n_c}^c z^{n_c} + \cdots + a_1^c z + a_0^c$, $n_c = n + t + n_l$		

Applications to Grid-Current Feedback Scheme in dq0 Frame

Referring to Fig. 1(b), in this section, the abovementioned discrete-time pole-placement method is used to regulate the fundamental-frequency current and dampen the LCL resonance simultaneously. To achieve zero steady-state errors in the dq0 frame, there is at least one pole at $z = 1$ in the z-domain, like what proportional-integral (PI) controllers do, to regulate the DC component following the reference and reject the disturbance from the grid voltage e_g. Therefore, there is a $1/(z-1)$ as an additional pole in Fig. 1(b). Similarly, a pair of internal-model resonant poles in [28] are added to eliminate the steady-state error in the αβ0 frame. Subsequently, the original plant $G_{OP}(z)$ is given as:

$$G_{OP}(z) = \frac{K_{PWM}}{z} \frac{\omega_r T \left[z^2 - 2z\cos(\omega_r T) + 1 \right] - \sin(\omega_r T)(z-1)^2}{\omega_r (z-1)\left[z^2 - 2z\cos(\omega_r T) + 1 \right]}, \tag{11}$$

of which the numerator degree is $m = 2$ and the denominator degree is $n = 4$. Adding no additional zeros but one additional pole to (11), the enhanced plant is derived as:

$$G_{EP}(z) = \frac{G_{OP}}{(z-1)}, \tag{12}$$

which gives $k = 0$ and $t = 1$. Then, according to Table I, comparisons of full controller degree between the αβ0 and the dq0 frame are listed in Table II. It can be seen that the order of pole-placement controller $G_{pp}(z)$ is reduced.

TABLE II: COMPARISONS OF DEGREES BETWEEN AB0 AND DQ0 FRAME

	m	n	t	k	n_l	n_p	n_c
αβ0 [28]	2	4	2	0	3	5	9
dq0	2	4	1	0	3	4	8

TABLE III: CLOSED-LOOP POLE LOCATIONS

Desired closed-loop poles $n_c = n + t + n_l = 8$

- Five poles at $z = 0$;
- One pole at $z = e^{-\omega_{dom} T}$;
- A pair of conjugated poles at
  $z = e^{-\xi \omega_r T} \cdot e^{\pm \sqrt{1-\xi^2} \omega_r T \cdot i}$.

Case Study

With parameters defined in Table IV, the desired pole locations are given in Table III, which shows that it is equivalent to a first-order system with the dominant frequency f_{dom} being 230 Hz [28] and ξ being 0.7 for a pair of conjugated poles, for a fair comparison between the dq0 in this paper and αβ0 in [28]. The original poles at the resonant frequency f_r are now placed at a location with the same natural frequency and the damping ratio 0.7 to reduce the control effort [25], [26], [28], [30].

$$
\begin{bmatrix}
a_5 & 0 & 0 & 0 & 0 & 0 & 0 & 0 & 0 \\
a_4 & a_5 & 0 & 0 & 0 & 0 & 0 & 0 & 0 \\
a_3 & a_4 & a_5 & 0 & b_2 & 0 & 0 & 0 & 0 \\
a_2 & a_3 & a_4 & a_5 & b_1 & b_2 & 0 & 0 & 0 \\
a_1 & a_2 & a_3 & a_4 & b_0 & b_1 & b_2 & 0 & 0 \\
a_0 & a_1 & a_2 & a_3 & 0 & b_0 & b_1 & b_2 & 0 \\
0 & a_0 & a_1 & a_2 & 0 & 0 & b_0 & b_1 & b_2 \\
0 & 0 & a_0 & a_1 & 0 & 0 & 0 & b_0 & b_1 \\
0 & 0 & 0 & a_0 & 0 & 0 & 0 & 0 & b_0
\end{bmatrix}
\begin{bmatrix}
l_3 \\ l_2 \\ l_1 \\ l_0 \\ p_4 \\ p_3 \\ p_2 \\ p_1 \\ p_0
\end{bmatrix}
=
\begin{bmatrix}
a_8^c \\ a_7^c \\ a_6^c \\ a_5^c \\ a_4^c \\ a_3^c \\ a_2^c \\ a_1^c \\ a_0^c
\end{bmatrix}
\tag{13}
$$

Then, notice that there is no common factor between the numerator and denominator in (12), thereby, the Sylvster's theorem is met and controller coefficients can be calculated by (13) where l is for the denominator and p for the numerator, as seen in Table II. In such a way, $G_c(z)$, a part of the pole-placement controller $G_{pp}(z) = G_c(z)/(z-1)$ can be obtained, and the open-loop transfer function (OLTF) is $G_{EP}(z)G_c(z)$. For stability analysis, the closed-loop transfer function (CLTF) is derived as (14). Zero-pole maps and Bode diagrams are plotted in Fig. 2 based on (12), OLTF and (14). Meanwhile, parameter variations of grid impedance L_g are also included for robustness evaluation.

$$CLTF = \frac{i_2}{i_2^*} = \frac{OLTF}{1+OLTF} = \frac{G_c G_{EP}}{1+G_c G_{EP}} \tag{14a}$$

$$CLTF_{com} = \frac{G_f G_c G_{EP}}{1+G_c G_{EP}} \tag{14b}$$

$$G_f = \frac{z(1-z_0)}{z-z_0} \tag{15}$$

In Fig. 2(a), poles and zeros are shown to verify that there is no common factor in the numerator and the denominator. Note that one zero outside the unit circle was ignored. Desired poles listed in Table III are drawn in Fig. 2(b). Similarly, unimportant outside zeros are ignored but there is one undesired zero z_0 below the dominant frequency, which will reduce the system damping and induce overshoots. To eliminate its side effects, a feed-forward compensation is added before the reference signal, as illustrated in Fig. 1(b). The compensated CLTF and feed-forward compensation is given by (14b) and (15). Here, $(1-z_0)$ is a real-number gain to keep the DC gain unaffected, which is vital in the dq0 frame. While in the $\alpha\beta0$ frame in [28], the gain at the fundamental frequency f_0 should be retained, and thus G_f becomes a complex number at f_0; in this way, the reference $I_{2\alpha}^* + I_{2\beta}^* \cdot j$ multiplied by a complex number G_f is cross-coupled. Therefore, an additional matrix to separate the real and the imaginary part has to be inserted into the feed-forward path. It means the calculation burden in digital processors increases. The effect of compensation is shown in Fig. 2(e) and Fig. 3(a).

The OLTF is depicted in Fig. 2(d). It can be seen that the PM and the GM in the $\alpha\beta0$ frame are 18.1 deg and 2.31 dB, respectively, while in the dq0 frame they reach 31.2 deg and 3.81 dB, which implies better relative stability and robustness against the parameter variations. This point is indicated in Fig. 2(c), where L_g varies from 0 to 30 mH (0.6 p.u.). That is, L_g boundary in [28] is about 15 mH (0.3 p.u.) but in the dq0 frame, even if it is over 30 mH, the system is still stable.

Fig. 2(f) shows different DC gains (indicating by 1 Hz) of the pole-placement controller with and without an additional pole $1/(z-1)$, and of the traditional proportional-integral (PI) controller, which is used to demonstrate the necessity of involving an pole at $z = 1$. There is a pole at $z = 1$ in the system plant (11) due to the ZOH discretization, but it is not sufficient for the DC gain, i.e., −29.1 dB. Although an additional pole at $z = 1$ increases the degree of the full controller, it can increase the DC gain, to 8.93 dB in this case compared to -3.47 dB with a PI, which is beneficial to the steady-state accuracy.

TABLE IV: PARAMETER DEFINITIONS

Name	Symbol	Value
Inverter Side Inductor	L_1	3.75 mH (0.07p.u.)
Grid Side Inductor	L_2	3.75 mH (0.07p.u)
Grid Impedance	L_g	15 mH (0.3 p.u.)
Filter Capacitor	C	15 μF (0.07 p.u.)
Grid Voltage(1φ-RMS)	V_g	230 V/50 Hz
Sampling Frequency	f_s	5 kHz
Switching Frequency	f_{sw}	2.5 kHz
DC Bus Voltage	V_{dc}	730 V

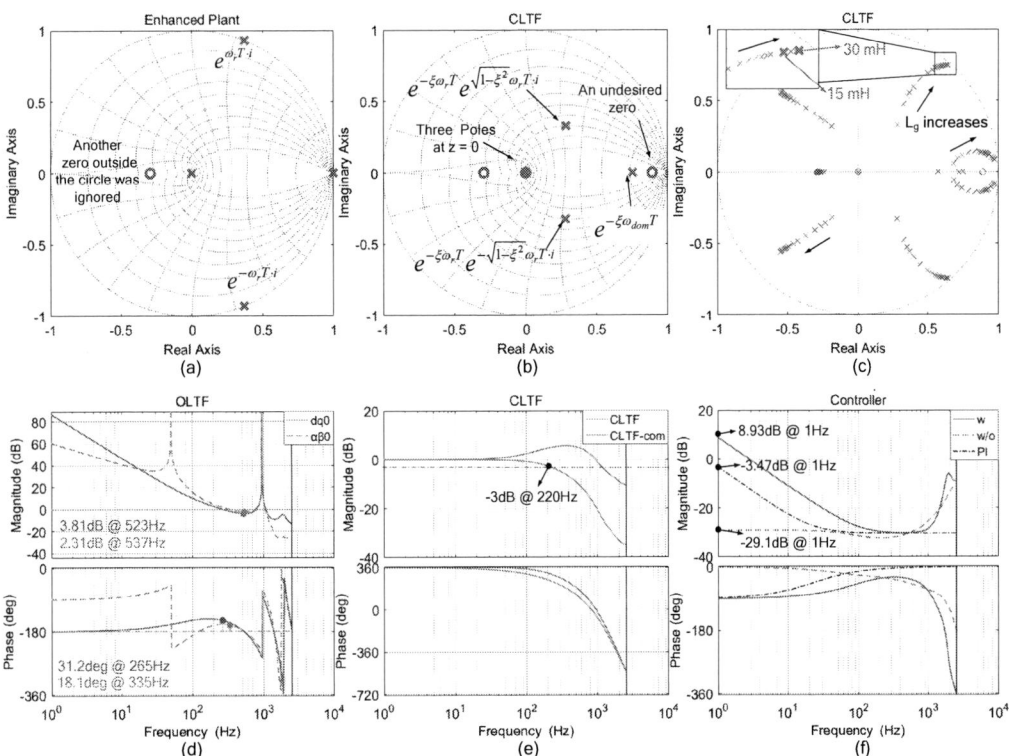

Fig. 2. Pole-zero maps of (a) the enhanced plant, (b) desired closed-loop poles, and (c) parameter variations of L_g. Bode diagrams of (d) open-loop transfer function, (e) closed-loop transfer function, and (f) controllers.

Simulation Verifications

Simulations are carried out in MATLAB/SIMULINK. As seen in Table IV, asymmetric regular sampling is adopted. A three-phase LCL-filtered grid-tied inverter is connected to a three-phase balanced grid. Parameters are the same as those in [28] and f_r / f_s is 0.19, larger than 1/6. In Fig. 3(a), step responses of the closed-loop transfer function CLTF with and without feed-forward $G_f(z)$ are compared. As seen, the feed-forward compensation reduces the percentage overshoot (PO, 82%) and also the setting time to some extent. In Fig. 3(b), when the d-axis reference steps at 0.1s from 0 to 8 A, the PO is about 63% and the setting time about 4.24 ms without compensation, while the overshoot almost disappears as a first-order system and the setting time is 2.81 ms with compensation, which is similar to the analytical step response in Fig. 3(a). The corresponding three-phase current are depicted in Fig. 3(f).

For the robustness verification related to Fig. 2(c) and (e), $L_g = 15$ mH (0.3 p.u. and f_r / f_s is 0.147) and 30 mH (0.6 p.u.) are shown in the following plots. As seen in Fig. 3(c) and 3(g) with $L_g = 15$ mH in the dq0 frame the system is stable, while in the αβ0 frame in Fig. 3(e), the stability boundary [28] has already been reached. Furthermore, when $L_g = 30$ mH in the dq0 frame, the stability is still guaranteed as shown in Fig. 3(d) and 3(h), but the oscillation occurs with poles approaching the unit circle, which can be observed in Fig. 2(c). Therefore, based on the same parameters, the pole-placement controller in the dq0 frame presents better robustness against grid impedance variations.

Conclusion

This paper has investigated transfer-function-based z-domain pole-placement techniques in the dq0 frame for LCL-filtered grid-tied inverters, and extended and summarized it for better accessibility.

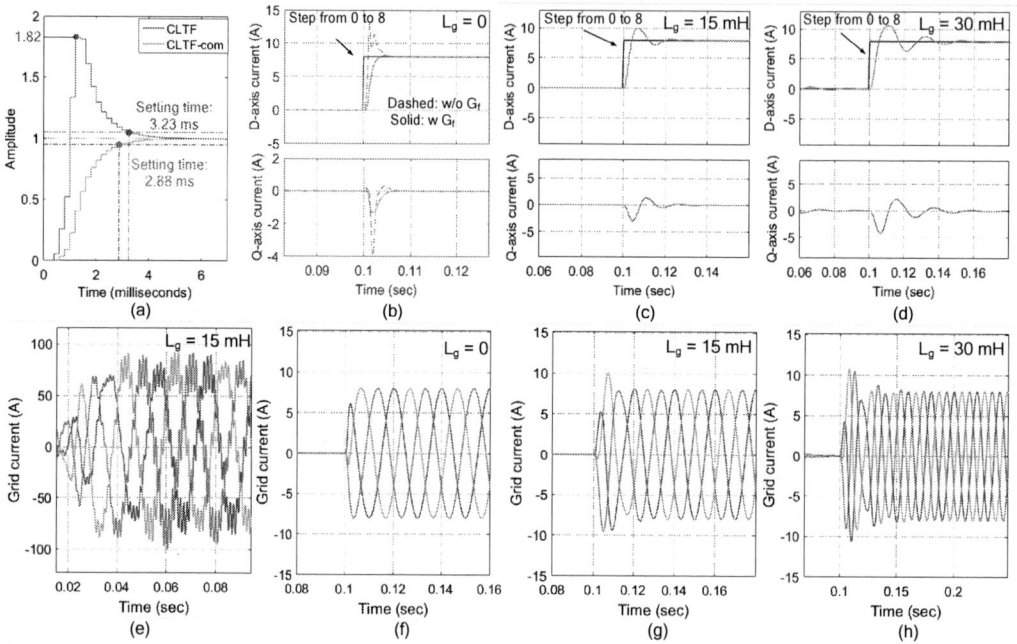

Fig. 3. (a) Pole-zero maps of the step response of the CLTF with (red) and without (blue) feed-forward compensation. Three-phase grid current in the dq0 frame with (b) $L_g = 0$, (c) $L_g = 15$ mH, (d) $L_g = 30$ mH. Three-phase grid current in the αβ0 frame with $L_g = 15$ mH in (e). And three-phase grid current in the dq0 frame with (f) $L_g = 0$, (f) $L_g = 15$ mH, and (g) $L_g = 30$ mH.

Presuming that the cross-coupling is ignored, compared to the z-domain pole-placement technique in the αβ0 frame, it presents better robustness in the dq0 frame against varying grid impedance and improved PM and GM; meanwhile, the feed-forward compensation for eliminating undesired zeros is simpler and the complex-number separating matrix in the αβ0 for unaffected line-frequency gain is avoided. Simulations have verified the analysis and design.

References

[1] F. Blaabjerg, R. Teodorescu, M. Liserre, and A. V. Timbus, "Overview of Control and Grid Synchronization for Distributed Power Generation Systems," *IEEE Trans. Ind. Electron.*, vol. 53, no. 5, pp. 1398–1409, Oct. 2006.

[2] M. Liserre, A. Dell'Aquila, and F. Blaabjerg, "Stability improvements of an LCL-filter based three-phase active rectifier," *in Proc. IEEE Power Electron. Spec. Conf.*, 2002, vol. 3, pp. 1195–1201.

[3] J. Dannehl, C. Wessels, and F. W. Fuchs, "Limitations of Voltage-Oriented PI Current Control of Grid-Connected PWM Rectifiers With LCL Filters," *IEEE Trans. Ind. Electron.*, vol. 56, no. 2, pp. 380–388, Feb. 2009.

[4] D. Pan, X. Ruan, C. Bao, W. Li, and X. Wang, "Capacitor-Current-Feedback Active Damping With Reduced Computation Delay for Improving Robustness of LCL-Type Grid-Connected Inverter," *IEEE Trans. Power Electron.*, vol. 29, no. 7, pp. 3414–3427, Jul. 2014.

[5] R. Peña-Alzola, M. Liserre, F. Blaabjerg, R. Sebastián, J. Dannehl, and F. W. Fuchs, "Analysis of the Passive Damping Losses in LCL-Filter-Based Grid Converters," *IEEE Trans. Power Electron.*, vol. 28, no. 6, pp. 2642–2646, Jun. 2013.

[6] R. N. Beres, X. Wang, F. Blaabjerg, M. Liserre, and C. L. Bak, "Optimal Design of High-Order Passive-Damped Filters for Grid-Connected Applications," *IEEE Trans. Power Electron.*, vol. 31, no. 3, pp. 2083–2098, Mar. 2016.

[7] J. L. Agorreta, M. Borrega, J. López, and L. Marroyo, "Modeling and Control of N-Paralleled Grid-Connected Inverters With LCL Filter Coupled Due to Grid Impedance in PV Plants," *IEEE Trans. Power Electron.*, vol. 26, no. 3, pp. 770–785, Mar. 2011.

[8] J. Dannehl, M. Liserre, and F. W. Fuchs, "Filter-Based Active Damping of Voltage Source Converters With LCL Filter," *IEEE Trans. Ind. Electron.*, vol. 58, no. 8, pp. 3623–3633, Aug. 2011.

[9] W. Yao, Y. Yang, X. Zhang, F. Blaabjerg, and P. C. Loh, "Design and Analysis of Robust Active Damping for LCL Filters Using Digital Notch Filters," *IEEE Trans. Power Electron.*, vol. 32, no. 3, pp. 2360–2375, Mar. 2017.

[10] J. Wang, J. D. Yan, L. Jiang, and J. Zou, "Delay-Dependent Stability of Single-Loop Controlled Grid-Connected Inverters with LCL Filters," *IEEE Trans. Power Electron.*, vol. 31, no. 1, pp. 743–757, Jan. 2016.

[11] X. Li, J. Fang, Y. Tang, and X. Wu, "Robust Design of LCL Filters for Single-Current-Loop-Controlled Grid-Connected Power Converters With Unit PCC Voltage Feedforward," *IEEE J. Emerg. Sel. Top. Power Electron.*, vol. 6, no. 1, pp. 54–72, Mar. 2018.

[12] B. Liu, Q. Wei, C. Zou, and S. Duan, "Stability Analysis of LCL-type Grid-connected Inverter under Single-loop Inverter-side Current Control with Capacitor Voltage Feedforward," *IEEE Trans. Ind. Inform.*, vol. 14, no. 2, pp. 691–702, Feb. 2018.

[13] W. Yao, Y. Yang, Y. Xu, F. Blaabjerg, S. Liu, and W. Gary, "Phase Reshaping via All-Pass Filters for Robust LCL-Filter Active Damping," *IEEE Trans. Power Electron.*, vol. 35, no. 3, pp. 3114-3126, March 2020.

[14] D. Pan, X. Ruan, and X. Wang, "Direct Realization of Digital Differentiators in Discrete Domain for Active Damping of LCL-Type Grid-Connected Inverter," *IEEE Trans. Power Electron.*, vol. 33, no. 10, pp. 8461-8473, Oct. 2018.

[15] M. Su *et al.*, "Single-Sensor Control of LCL-Filtered Grid-Connected Inverters," *IEEE Access*, vol. 7, pp. 38481–38494, 2019, doi: 10.1109/ACCESS.2019.2906239.

[16] M. T. Faiz, M. M. Khan, J. Xu, M. Ali, S. Habib, K. Hashmi, H. Tang, "Capacitor Voltage Damping Based on Parallel Feedforward Compensation Method for LCL Filter Grid-Connected Inverter," *IEEE Trans. Ind. Appl.*, vol. 56, no. 1, pp. 837-849, Jan.-Feb. 2020.

[17] M. Lu, A. Al-Durra, S. M. Muyeen, S. Leng, P. C. Loh, and F. Blaabjerg, "Benchmarking of Stability and Robustness against Grid Impedance Variation for LCL-Filtered Grid-Interfacing Inverters," *IEEE Trans. Power Electron.*, vol. 33, no. 10, pp. 9033-9046, Oct. 2018.

[18] J. Xu, S. Xie, and T. Tang, "Active Damping-Based Control for Grid-Connected LCL -Filtered Inverter With Injected Grid Current Feedback Only," *IEEE Trans. Ind. Electron.*, vol. 61, no. 9, pp. 4746–4758, Sep. 2014.

[19] Y. Tang, P. C. Loh, P. Wang, F. H. Choo, and F. Gao, "Exploring Inherent Damping Characteristic of LCL-Filters for Three-Phase Grid-Connected Voltage Source Inverters," *IEEE Trans. Power Electron.*, vol. 27, no. 3, pp. 1433–1443, Mar. 2012.

[20] S. G. Parker, B. P. McGrath, and D. G. Holmes, "Regions of Active Damping Control for LCL Filters," *IEEE Trans. Ind. Appl.*, vol. 50, no. 1, pp. 424–432, Jan. 2014.

[21] D. G. Holmes, T. A. Lipo, B. P. McGrath, and W. Y. Kong, "Optimized Design of Stationary Frame Three Phase AC Current Regulators," *IEEE Trans. Power Electron.*, vol. 24, no. 11, pp. 2417–2426, Nov. 2009.

[22] X. Wang, F. Blaabjerg, and P. C. Loh, "Grid-Current-Feedback Active Damping for LCL Resonance in Grid-Connected Voltage-Source Converters," *IEEE Trans. Power Electron.*, vol. 31, no. 1, pp. 213–223, Jan. 2016.

[23] C. Bao, X. Ruan, X. Wang, W. Li, D. Pan, and K. Weng, "Step-by-Step Controller Design for LCL-Type Grid-Connected Inverter with Capacitor-Current-Feedback Active-Damping," *IEEE Trans. Power Electron.*, vol. 29, no. 3, pp. 1239–1253, Mar. 2014.

[24] J. Dannehl, F. W. Fuchs, and P. B. Thøgersen, "PI State Space Current Control of Grid-Connected PWM Converters With LCL Filters," *IEEE Trans. Power Electron.*, vol. 25, no. 9, pp. 2320–2330, Sep. 2010.

[25] J. Kukkola and M. Hinkkanen, "Observer-Based State-Space Current Control for a Three-Phase Grid-Connected Converter Equipped With an LCL Filter," *IEEE Trans. Ind. Appl.*, vol. 50, no. 4, pp. 2700–2709, Jul. 2014.

[26] J. Kukkola, M. Hinkkanen, and K. Zenger, "Observer-Based State-Space Current Controller for a Grid Converter Equipped With an LCL Filter: Analytical Method for Direct Discrete-Time Design," *IEEE Trans. Ind. Appl.*, vol. 51, no. 5, pp. 4079–4090, Sep. 2015.

[27] D. Pérez-Estévez, J. Doval-Gandoy, A. G. Yepes, and Ó. López, "Positive- and Negative-Sequence Current Controller With Direct Discrete-Time Pole Placement for Grid-Tied Converters With LCL Filter," *IEEE Trans. Power Electron.*, vol. 32, no. 9, pp. 7207–7221, Sep. 2017.

[28] D. Pérez-Estévez, J. Doval-Gandoy, A. G. Yepes, Ó. López, and F. Baneira, "Enhanced Resonant Current Controller for Grid-Connected Converters With LCL Filter," *IEEE Trans. Power Electron.*, vol. 33, no. 5, pp. 3765–3778, May 2018.

[29] G. C. Goodwin, S. F. Graebe, and M. E. Salgado, *Control system design*. Englewood Cliffs, NJ: Prentice–Hall, 2001.

[30] G. F. Franklin, J. D. Powell, and M. L. Workman, *Digital control of dynamic systems*, 3rd ed. Addison-wesley Menlo Park, CA, 1998.

On the switching-induced dc-link voltage ripple in three-level converters with a neutral point

Ioannis Tsoumas
ABB System Drives
Austrasse
5300 Turgi, Switzerland
E-Mail: ioannis.tsoumas@ch.abb.com

Tobias Geyer
ABB System Drives
Austrasse
5300 Turgi, Switzerland
E-Mail: tobias.geyer@ch.abb.com

Keywords

«Voltage Source Converter», «Pulse Width Modulation», «Harmonics».

Abstract

The operation of a three-level (3L) converter with a neutral point (NP) employed in a medium voltage (MV) wind energy conversion system (WECS) is investigated. In particular, the voltage ripple of the upper or the lower half of the dc-link and its dependence on the relative phase of the carriers in a carrier-based pulse width modulation (CB-PWM) scheme is investigated. It is shown that the carrier phase has no influence on the half dc-link voltage ripple.

Introduction

The peak value of the voltage of each half of the dc-link is an important design related quantity of a 3L converter with a neutral point, because its value must not exceed the maximum blocking voltage of the semiconductors. When carrier-based pulse width modulation (CB-PWM) is employed for the converter modulation the ripple of the above voltage can be separated in two groups of frequency components: low frequency components related to the average value of the output voltage within one switching period (switching is not taken into account) [1] and high frequency components related to the switching. The former depend on the CB-PWM reference and the current displacement factor and they have been examined in detail in previous works [1]-[3], where a special CB-PWM reference to suppress the low frequency ripple has been derived.

In this work we focus on the latter, the frequency components related to the switching. Exact knowledge of the characteristics of the above components is of paramount importance, since they contribute to the peak value of half the dc-link voltage. Knowledge of their dependence on the CB-PWM reference and the carrier configuration, as well as on the modulation index and the current displacement factor is necessary for a correct selection of the dc-link capacitance value.

Mathematical modelling of the converter

The circuit diagram of a 3L voltage source converter with an NP is shown in Fig. 1. The dc-link capacitances C are considered identical but the instantaneous voltages of the upper and lower dc-link halves, $v_{\text{dc,h}}(t)$ and $v_{\text{dc,l}}(t)$, may be different. The integer variable $u_x \in \{-1,0,1\}$ denotes the switch position in the phase leg x, with $x \in \{a, b, c\}$. The variable will henceforth be referred to as switching function. The voltages of the three phases referred to the NP for different values of the switching function are given by

$$v_x(t) = \begin{cases} v_{\text{dc,h}}(t), & u_x(t) = 1 \\ 0, & u_x(t) = 0 \\ v_{\text{dc,l}}(t), & u_x(t) = -1 \end{cases} . \tag{1}$$

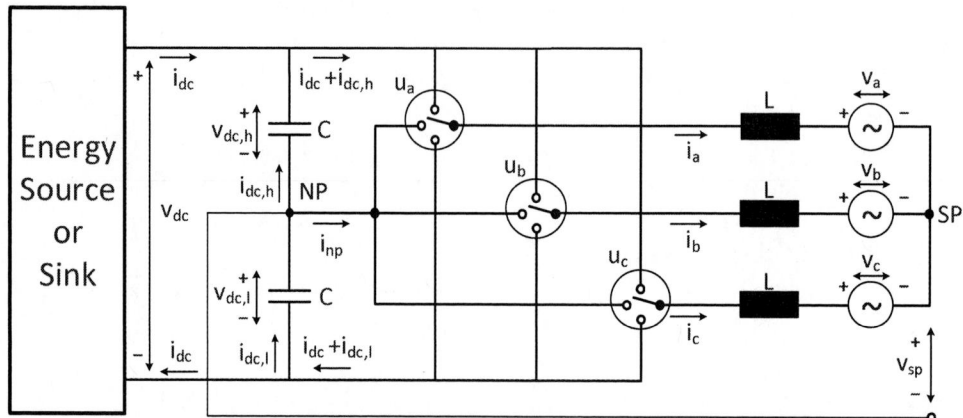

Fig. 1: Circuit diagram of a 3L voltage source converter with a neutral point and ideal switches supplying a 3-phase AC system [4].

The total instantaneous dc-link voltage is given by

$$v_{dc}(t) = v_{dc,h}(t) + v_{dc,l}(t) \tag{2}$$

and the NP potential by

$$v_{np}(t) = \frac{1}{2}\left[v_{dc,l}(t) - v_{dc,h}(t)\right]. \tag{3}$$

The NP potential evolves as a function of the NP current $i_{np}(t)$ according to

$$\frac{dv_{np}}{dt} = \frac{1}{2}\left[\frac{dv_{dc,l}(t)}{dt} - \frac{dv_{dc,h}(t)}{dt}\right] = -\frac{1}{2}\left[\frac{1}{C}i_{dc,l}(t) - \frac{1}{C}i_{dc,h}(t)\right] =$$

$$-\frac{1}{2C}\left[i_{dc,l}(t) - i_{dc,h}(t)\right] = -\frac{1}{2C}i_{np}(t),$$

which leads to

$$v_{np}(t) = -\frac{1}{2C}\int_0^t i_{np}(\tau)d\tau + v_{np}(0). \tag{4}$$

To calculate v_{np} as a function of the angle in (4) we make a change of variable from t to $\theta = \omega t$

$$v_{np}(\theta) = -\frac{1}{2\omega C}\int_0^\theta i_{np}(\zeta)d\zeta + v_{np}(0).$$

The NP current can be calculated from the AC currents as follows:

$$i_{np}(t) = \sum_{x\in\{a,b,c\}} (1 - |u_x(t)|)\, i_x(t), \quad \text{with } x \in \{a, b, c\}. \tag{5}$$

If $\sum_{x\in\{a,b,c\}} i_x(t) = 0$ (star connection of the phases or symmetrical currents) we get from (5)

$$i_{np}(t) = -\sum_{x\in\{a,b,c\}} |u_x(t)| i_x(t). \tag{6}$$

The phase currents appear on the upper and the lower rail of the dc-link if the corresponding switching function u_x is equal to 1 and −1 respectively. Thus, they can be modelled as follows [4]:

$$i_{dc}(t) + i_{dc,h}(t) = \sum_{x\in\{a,b,c\}} u_x(t)\frac{u_x(t)+1}{2}i_x(t) =$$

$$\frac{1}{2}\sum_{x\in\{a,b,c\}} |u_x(t)| i_x(t) + \frac{1}{2}\sum_{x\in\{a,b,c\}} u_x(t)i_x(t) \tag{7.a}$$

and

$$i_{dc}(t)+i_{dc,l}(t) = -\sum_{x\in\{a,b,c\}} u_x(t)\frac{u_x(t)-1}{2} i_x(t) =$$
$$-\frac{1}{2}\sum_{x\in\{a,b,c\}} |u_x(t)|i_x(t) + \frac{1}{2}\sum_{x\in\{a,b,c\}} u_x(t)i_x(t). \quad (7.b)$$

If we subtract i_{dc} from the above currents, the currents in the dc-link capacitors can be calculated:

$$i_{dc,h}(t) = \frac{1}{2}\sum_{x\in\{a,b,c\}} u_x(t)i_x(t) + \frac{1}{2}\sum_{x\in\{a,b,c\}} |u_x(t)|i_x(t) - i_{dc}(t) \quad \text{and} \quad (8.a)$$

$$i_{dc,l}(t) = \frac{1}{2}\sum_{x\in\{a,b,c\}} u_x(t)i_x(t) - \frac{1}{2}\sum_{x\in\{a,b,c\}} |u_x(t)|i_x(t) - i_{dc}(t) . \quad (8.b)$$

The capacitor voltages are given by

$$v_{dc,h}(t) = -\frac{1}{C}\int_0^t i_{dc,h}(\tau)d\tau + v_{dc,h}(0) =>$$

$$v_{dc,h}(\theta) = -\frac{1}{\omega C}\int_0^\theta \left[\frac{1}{2}\sum_{x\in\{a,b,c\}} |u_x(\zeta)|i_x(\zeta) + \frac{1}{2}\sum_{x\in\{a,b,c\}} u_x(\zeta)i_x(\zeta) - i_{dc}(\zeta)\right]d\zeta + v_{dc,h}(0) \quad (9.a)$$

and

$$v_{dc,l}(t) = -\frac{1}{C}\int_0^t i_{dc,l}(\tau)d\tau =>$$

$$v_{dc,l}(\theta) = -\frac{1}{\omega C}\int_0^\theta \left[-\frac{1}{2}\sum_{x\in\{a,b,c\}} |u_x(\zeta)|i_x(\zeta) + \frac{1}{2}\sum_{x\in\{a,b,c\}} u_x(\zeta)i_x(\zeta) - i_{dc}(\zeta)\right]d\zeta + v_{dc,l}(0). \quad (9.b)$$

Finally, from (2) and (9) the total dc-link voltage can be calculated:

$$v_{dc}(t) = v_{dc,h}(t) + v_{dc,l}(t) =>$$

$$v_{dc}(\theta) = \frac{1}{\omega C}\int_0^\theta \left[-\sum_{x\in\{a,b,c\}} u_x(\zeta)i_x(\zeta) + 2i_{dc}(\zeta)\right]d\zeta + v_{dc}(0) . \quad (10)$$

By reformulating (2) and (3) the capacitor voltages can be also expressed as

$$v_{dc,h}(t) = \frac{v_{dc}(t)}{2} - v_{np}(t) , \quad (11.a)$$

$$v_{dc,l}(t) = \frac{v_{dc}(t)}{2} + v_{np}(t) . \quad (11.b)$$

As can be seen from (11) the voltage of each half of the dc-link is equal to the sum or the difference of half the total dc-link voltage and the NP potential.

Harmonics of the switching function

The switching that takes place in a 3L converter generates harmonic components at multiples of the carrier frequency and at their respective sidebands. The two carriers can be arranged either in phase disposition (PD) or in phase opposition-disposition (POD). With PD the two level-shifted carriers are in phase, whereas with POD the carriers have a phase difference of π. Analytical solutions of the switching function harmonics caused by modulation have been calculated in [5] for both cases when a sinusoidal reference is considered for the CB-PWM.

In the case of PD the switching function in phase a can be expressed as the following sum of harmonics:

$$u_a = \sum_{m\in\mathbb{N}} \frac{1}{2m-1} C_{2m-1} \cos[(2m-1)\omega_c t] + \frac{2}{\pi}\sum_{m\in\mathbb{N}} \frac{1}{2m}\sum_{n\in\mathbb{Z}} C_{2m,2n+1} \cos[2m\omega_c t +$$
$$(2n+1)\omega_1 t] + \frac{8}{\pi^2}\sum_{m\in\mathbb{N}} \frac{1}{2m-1}\sum_{n\in\mathbb{Z}\setminus\{0\}} C_{2m-1,2n} \cos[(2m-1)\omega_c t + 2n\omega_1 t], \quad (12)$$

where ω_c is the carrier angular frequency and ω_1 is the fundamental angular frequency. The coefficients C_{2m-1}, $C_{2m,2n+1}$, $C_{2m-1,2n}$, have been analytically calculated for a sinusoidal reference and natural sampling in [5], but not for a reference that includes a CM component. As it will become evident from the subsequent numerical analysis the coefficients C_{2m-1}, $C_{2m,2n+1}$, $C_{2m-1,2n}$ depend on the modulation index. The latter is defined by the ratio of the amplitude of the fundamental harmonic to half the dc-link voltage:

$$m_{3L} = \frac{\hat{v}_{a1}}{v_{dc}/2} \quad (13)$$

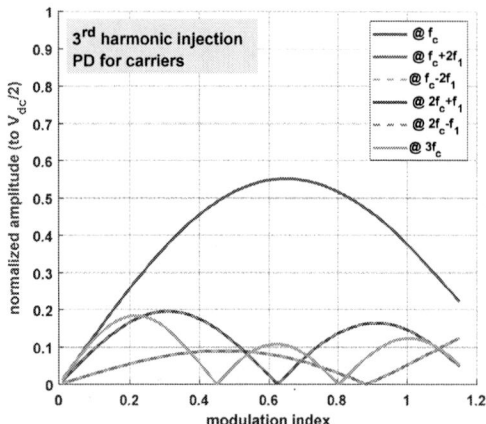

Fig. 2: Switching function harmonics for PD.

Fig. 3: Absolute switching function harmonics for PD.

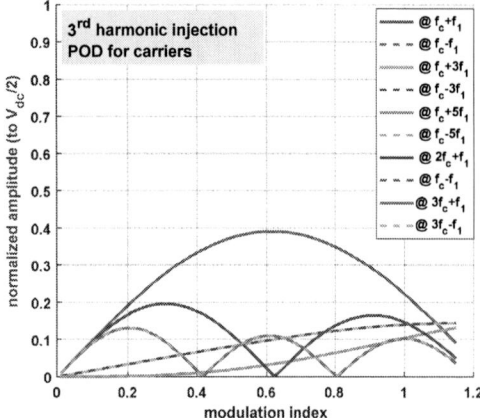

Fig. 4: Switching function harmonics for POD.

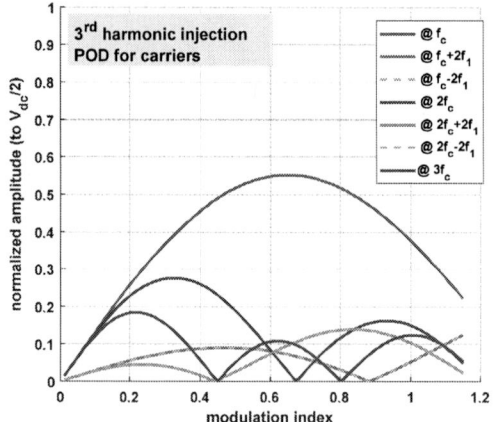

Fig. 5: Absolute switching function harmonics for POD.

It can be observed in (12) that harmonic components are present at multiples of the carrier frequency. This is exactly the feature that gives PD its superior performance when it comes to current ripple, since the above harmonics are common-mode (CM) harmonics and do not appear in the line-to-line or the phase-to-star point voltage. In addition, one can observe odd sidebands around even carrier multiples and even sidebands around odd carrier multiples. As with all CB-PWM strategies, the triplen sideband harmonics (i.e. multiples of three of the fundamental) are CM harmonics and cancel in the line-to-line voltage [5].

The absolute value of the switching function can be expressed as the following sum of harmonics:

$$|u_a| = \sum_{m \in \mathbb{N}} \frac{1}{2m} C_{2m} \sin[2m\omega_c t] + \sum_{m \in \mathbb{N}} \frac{1}{2m-1} \sum_{n \in \mathbb{Z}} C_{2m-1,2n+1} \sin[(2m-1)\omega_c t + (2n+1)\omega_1 t] + \sum_{m \in \mathbb{N}} \frac{1}{2m} \sum_{n \in \mathbb{Z} \setminus \{0\}} C_{2m,2n} \sin[2m\omega_c t + 2n\omega_1 t]. \tag{14}$$

In the case of POD the switching function can be expressed as the following sum of harmonics:

$$u_a = \frac{2}{\pi} \sum_{m \in \mathbb{N}} \frac{1}{m} \sum_{n \in \mathbb{Z}} C_{m,2n+1} \cos[m\omega_c t + (2n+1)\omega_1 t]. \tag{15}$$

The absolute value of the switching function can be expressed as the following sum of harmonics:

$$|u_a| = \sum_{m \in \mathbb{N}} \frac{1}{m} \sum_{n \in \mathbb{Z}} C_{m,2n+1} \sin[m\omega_c t + (2n+1)\omega_1 t]. \tag{16}$$

Consider CB-PWM with a sinusoidal CM component at three times the fundamental frequency that extends the linear region of the modulator. This technique is known as "third harmonic injection" [5]. Numerical analysis via FFT shows in the case of PD a strong CM harmonic component in the switching function at the switching frequency (cf. Fig. 2), which takes its maximum value close to a modulation index of $m_{3L} \approx 0.65$. This component does not appear in the harmonics of the absolute value of the switching function (cf. Fig. 3). When POD is applied the opposite happens: the large CM harmonic components appear only in the harmonics of the absolute value of the switching function (cf. Fig. 4 and Fig. 5).

The presence or lack of a CM harmonic component at the carrier frequency is of paramount importance for the amplitude of the NP potential ripple and of the total dc-link ripple, as it will become clear from the harmonic transfer rules and the numerical analysis presented in the next sections.

Harmonics transfer rules from the ac to the dc side

As shown in the previous section the total dc-link voltage and the NP potential can be calculated from the convolution of the phase currents with the switching functions and their absolute value respectively. In such calculations sums of the three-phase quantities appear. In [4] symmetrical operating conditions were assumed and the following rules have been derived for the resulting dc-link components in the frequency domain:

1. If the symmetrical components of the current harmonic and the switching function harmonic describe either a positive-sequence system or a negative sequence system, then their convolution onto the DC side will result in a single harmonic.

2. If the current harmonic and the switching function harmonic have the same/different phase sequence, the harmonic order of the resulting dc-link harmonics will be the difference/sum of the individual AC-side harmonic orders.

3. If the symmetrical components of either the current harmonic or the switching function harmonic describe a zero-sequence system, their convolution will not result in any dc-link harmonics.

The above harmonic transfer rules, especially the last one, are very important for the understanding of the results that are presented in the next section. They show that only some sums and differences of frequencies appear in the dc-link harmonic components whereas others cancel out when the sum of the three phases is considered.

Harmonics of the dc-link voltage ripple

Consider a 3.3 kV converter rated at 8.5 MVA with a total dc-link voltage of 5 kV for a WECS. Characteristic numerical results for the harmonics of the NP potential, total dc-link voltage and upper (dc-link) capacitor voltage are presented in Fig. 6 - Fig. 7, Fig. 8 - Fig. 9, and Fig. 10-Fig. 11 respectively. Naturally sampled CB-PWM has been considered. The selected carrier frequency $f_c = 350$ Hz for the investigation lies in the range of typical carrier frequencies of converters in this power range; the fundamental frequency (of the generator) is $f_1 = 10$ Hz. The current ripple has been neglected in the analysis, since it is very small for such a carrier to fundamental frequency ratio. The CB-PWM reference includes a CM component of three times the fundamental frequency that extends the linear region of the modulator.

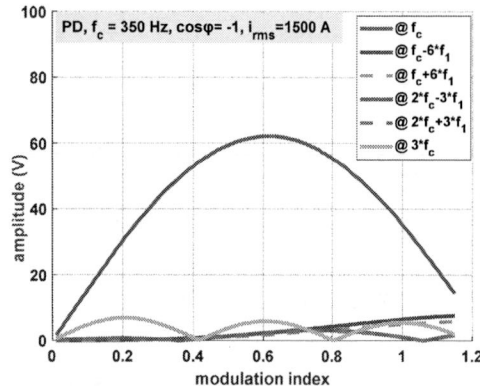

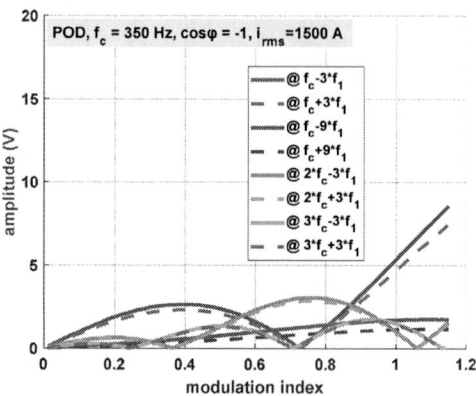

Fig. 6: Harmonics of the NP potential for PD.

Fig. 7: Harmonics of the NP potential for POD.

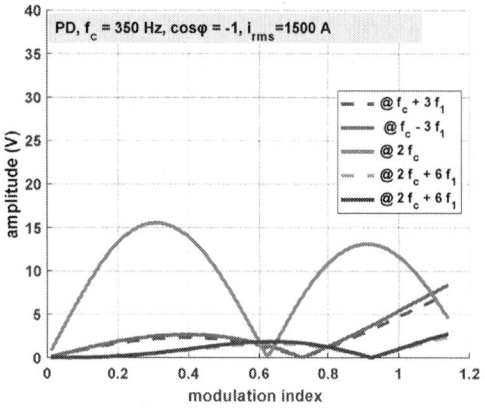

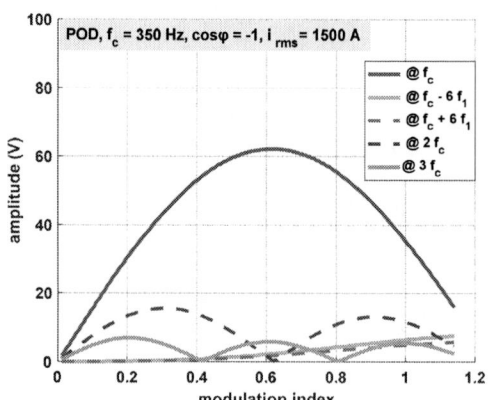

Fig. 8: Harmonics of the total dc-link voltage for PD.

Fig. 9: Harmonics of the total dc-link voltage for POD.

In Fig. 6 - Fig. 7 one can observe that if PD is used a strong harmonic component at the carrier frequency is present in the NP potential, which doesn't appear when POD is employed. This happens because in POD a strong CM component at the carrier frequency appears in the absolute value of the switching function (cf. Fig. 5). The latter does not cause any component on the dc-side according to the harmonic transfer rule No. 3 of the previous section. The strongest harmonic components of the NP ripple in the case of POD appear at $f_c \pm 3f_1$ and at $f_c \pm 9f_1$. Their amplitudes are significantly smaller than those at the carrier frequency in the case of PD.

Regarding the harmonics on the total dc-link voltage, it can be clearly seen that PD produces lower harmonics compared to POD (cf. Fig. 8 and Fig. 9). In POD a strong harmonic component at the carrier frequency appears, whose amplitude gets its maximum value for a modulation index of approximately 0.62. Because of this strong harmonic component POD is suboptimal when it comes to dc-link voltage ripple. In PD no strong component close to the carrier frequency appears, because the harmonic of the switching function at the carrier frequency is a CM harmonic (see again rule No. 3 of previous section).

For the semiconductors the sum or the difference of the previously discussed quantities is critical, i.e., the voltage harmonics of the upper (or lower) dc-link capacitor. This voltage contains the harmonics from both the NP potential and the total dc-link voltage. Because of that the voltage harmonics have the same frequency and amplitude regardless of the selection of PD or POD. An inspection of Fig. 10 and Fig. 11 proves that. In fact, for all possible phase displacements between the two carriers the frequencies and the amplitudes of the harmonics are the same.

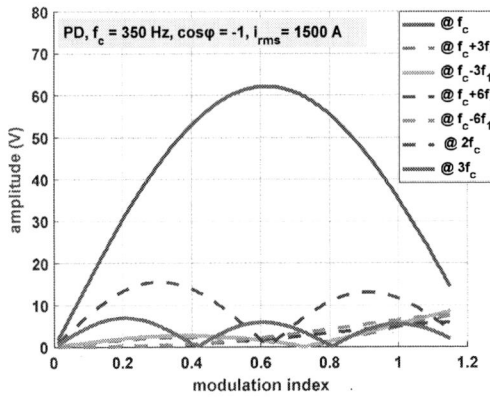

Fig. 10: Harmonics of the upper half of the dc-link voltage for PD.

Fig. 11: Harmonics of the upper half of the dc-link voltage for POD.

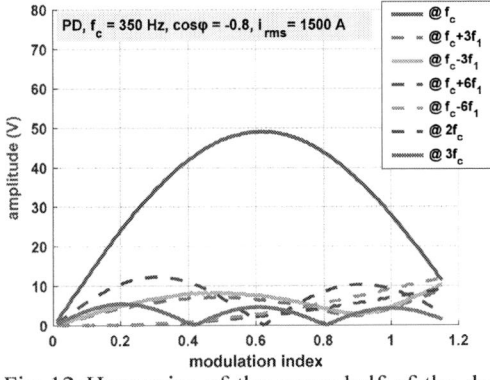

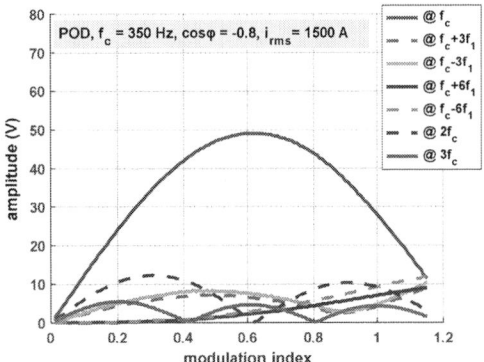

Fig. 12: Harmonics of the upper half of the dc-link voltage for PD.

Fig. 13: Harmonics of the upper half of the dc-link voltage for POD.

For a given carrier frequency and dc-link capacitance their amplitudes depend only on the current displacement factor. Their amplitudes decrease slightly with decreasing displacement factor as can be seen when comparing Fig. 12 with Fig. 13.

Conclusion

The switching related ripple of the dc-link voltage of a carrier-based pulse width modulated three-level converter has been investigated. Of special interest is the voltage ripple of each half of the dc-link. Its peak value must not exceed the maximum blocking voltage of the semiconductors. Numerical analysis has shown that although the harmonics of the NP potential and the harmonics of the total dc-link voltage depend on the relative phase of the two carriers, the harmonics of the upper and the lower half of the dc-link voltage are unaffected by the carrier phase. For a given carrier frequency and dc-link capacitance their amplitudes are influenced only be the displacement factor of the fundamental current component.

The presented analysis is valid for high ratios between the carrier frequency and the fundamental frequency, as it is the case for gearless wind energy generation systems. In such a case the current ripple doesn't contribute significantly to the dc-link voltage ripple and can be neglected.

References

[1] Ogasawara S. and Akagi H., Analysis of variation of neutral point potential in neutral-point-clamped voltage source PWM inverters, IEEJ Transactions on Industry Applications, November 1993, pp. 965-970.

[2] Song Q., Lui W., Yu Q. and Wang X., A neutral-point potential balancing algorithm for three-level NPC inverters using analytically injected zero-sequence voltage, 18th Annual IEEE Applied Power Electronics Conference and Exposition (APEC), February 2003, pp. 228-233.

[3] Wang Ch. and Li Y., Analysis and calculation of zero-sequence voltage considering neutral-point potential balancing in three-level NPC converters, IEEE Transactions on Industrial Electronics, July 2010, pp. 2262-2271.

[4] Scheuer G., Investigation of the 3-level voltage source inverter for flexible AC transmission systems exemplified on a static VAr compensator, Dissertation, ETH Zurich 1997.

[5] Holmes D.G. and Lipo T.A, Pulse width modulation for power converters, IEEE Press, 2003.

Effect of passive inverter output motor filters on drive systems

Dr. Dennis Kampen / Martin Weicker
BLOCK TRANSFORMATOREN-ELEKTRONIK GMBH / TECHNISCHE UNIVERSITÄT
DARMSTADT
Max-Planck-Str. 36-46 / Landgraf-Georg-Str. 4
27283 Verden (Aller) / 64283 Darmstadt, Germany
Tel.: +49 4231 678-178 / +49 6151 16-24191
E-Mail: dennis.kampen@block.eu / mweicker@ew.tu-darmstadt.de
URL: http://www.block.eu / http://www.ew.tu-darmstadt.de

Acknowledgements

This research is based on the work and results of the industry group bearing currents, offered by the Technical University of Darmstadt.

Keywords

«passive filter», «insulation», «EMC/EMI», «industrial application».

Abstract

The evaluation shows the effect of different on the market available inverter output filters on important drive system characteristics. Each filter is explained and the filter effect on motor insulation, inverter losses, cable losses, motor losses, audible noise, EMI, bearing currents and costs is compared. This paper offers a comprehensive overview and might help drive system designers in choosing the most suitable passive filter solution.

Introduction

Using inverters in drive systems comes with many drawbacks. Inverter caused bearing currents can lead to severe damage of electrical motors or connected load machines. The currents can be divided into EDM-, rotor-to-ground and circulating bearing currents. To reduce these currents, there are some options like using insulated bearings or grounding of the motor shaft, which might protect the bearings [1], [9]-[16], but with no effect on other inverter caused drawbacks like insulation faults, EMI or additional inverter, cable and motor losses. Passive filters at the output of inverters are a state-of-the-art solution for different inverter caused drawbacks. In this evaluation, the effect of the different on the market available filter types on motor insulation, inverter losses, cable losses, motor losses, audible noise, EMI, bearing currents and costs is presented in an overview.

Passive motor filter types

In this section, the effect of different motor filters on the motor terminal voltage u_{DM} (differential mode motor voltage) and on the motor star point-to-ground voltage u_{CM} (common mode motor voltage). Also, general information about the filters are given.

Common mode ring cores

Ring cores, i.e. nanocrystalline or ferrite cores, at the inverter output have an excellent effect on EMI. In addition, if probably designed, they have an eliminating effect on rotor-to-ground and circulating bearing currents. However, the design of the ring cores depends strongly on the drive system setup and allows no standard filter solution. Saturation effects of the cores by high leakage currents must be avoided by choosing a good compromise between permeability and core cross section. The resonance frequency, typically 80 kHz - 300 kHz, of this filter depends on permeability, core cross section, number of windings and on the parasitic cable/motor capacitance. An excitation of this resonance must be avoided by choosing the right components depending on cable and cable length. Otherwise the motor is penetrated even worse, than without filter [2]. The common mode motor voltage u_{CM} overshot depends on the ohmic damping, provided mainly by the core losses. Depending on the installation, several nanocrystalline ring cores are required. Additionally, the ring cores have negligible effects on voltage waveforms, Fig. 2 and Fig. 3, and are therefore not reducing insulation faults in the motor or EDM bearing currents. In addition, their effect on inverter, cable or motor losses is very low. An advantage of the ring cores is that they do not generate a motor voltage drop and they can be used in drives with high dynamic control because nearly no differential mode inductance is provided.

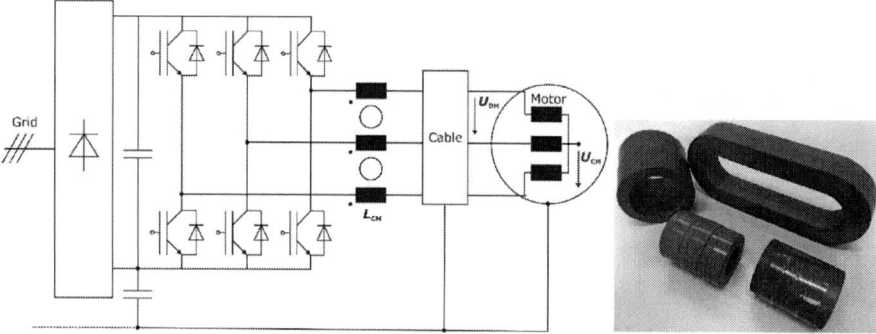

Fig. 1 Setup with common mode ring cores

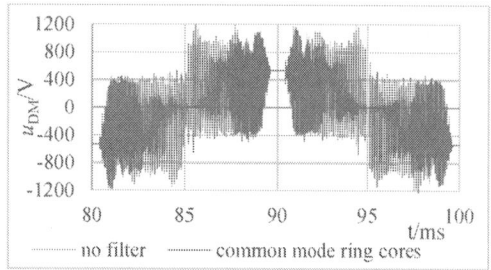

Fig.1 Differential mode motor voltage u_{DM}:
Nearly no voltage shaping effect by ring cores.

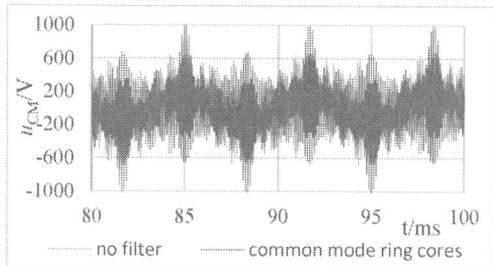

Fig.2 Common mode motor voltage u_{CM}:
Very small voltage shaping effect by ring cores.

Differential mode motor choke

The cheapest filter solution is a three-phase differential mode motor choke. Its inductance reduces parasitic cable capacitance caused inverter output peak currents and losses in the inverter. By that, longer motor cables can be used. In addition, its inductance together with the parasitic cable capacitance forms a differential mode low pass filter with a resonance frequency of typically 80 kHz - 500 kHz and reduces dv/dt of the PWM pulses. The peak voltage overshot is reduced mainly by the iron losses of the electrical steel core, Fig.4 and Fig.5. By that the motor insulation is protected between the phases. Also circulating and rotor-ground bearing currents are partly reduced by the stray inductance of the choke.

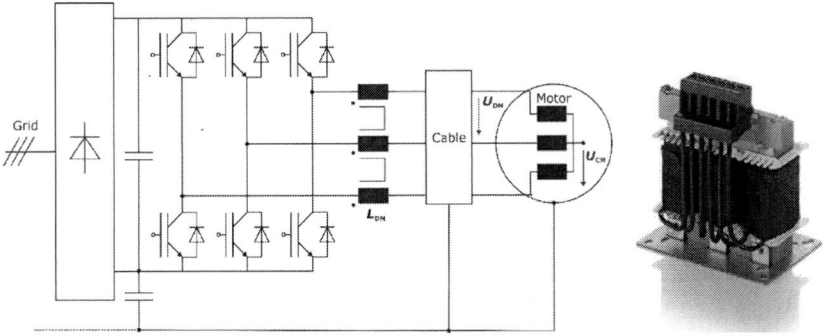

Fig. 3 Setup with differential mode motor choke

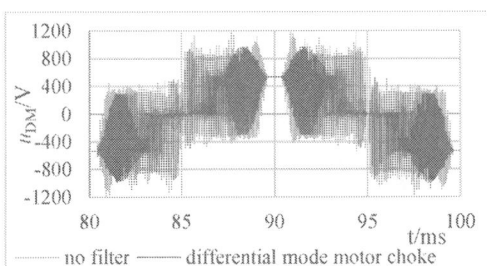

Fig.4 Differential mode motor voltage u_{DM}: Medium voltage shaping effect (dv/dt and peak voltage) with filter.

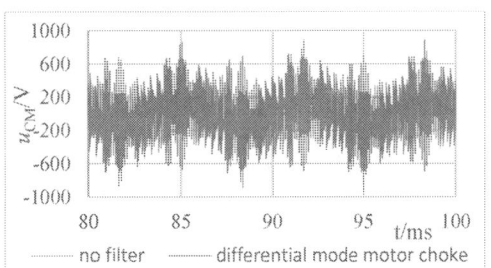

Fig.5 Common mode motor voltage u_{CM}: Nearly no effect by filter.

Differential mode dv/dt-filter

Differential mode dv/dt-filters consist of a differential mode inductance with parallel ohmic damping resistors and a capacitance. The filter effect is independent from cable length. The resonance frequency of this filter set higher than the switching frequency of the inverter, typically 80 kHz - 500 kHz. The damping resistors set the voltage overshot at the motor terminals, Fig.7. Like the motor choke, differential mode dv/dt-filters only offer insulation protection between the phases and not against ground, Fig.8. The benefit of a dv/dt-filter in comparison to a motor choke is, that the filter effect is decoupled from the cable and the *dv/dt* and voltage overshoot can be defined in advance to meet different standard limits, i.e. [3] or [4]. The filter generates quite high ohmic losses but has a beneficial effect on inverter and cable losses by damping parasitic capacitance charging currents. EMI or bearing current reduction is offered by this solution only partly by the stray inductance of the choke.

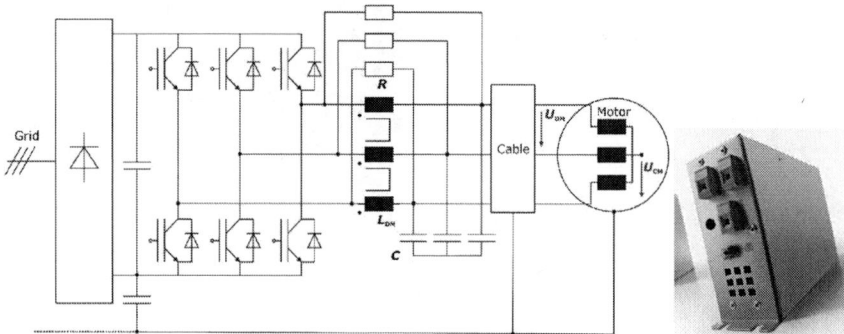

Fig. 6 Setup with differential mode dv/dt-filter

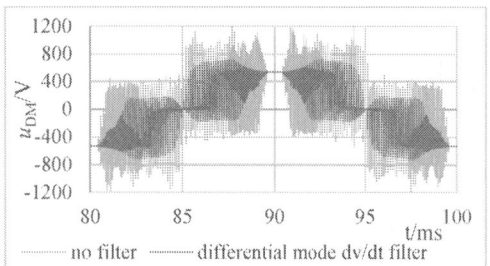

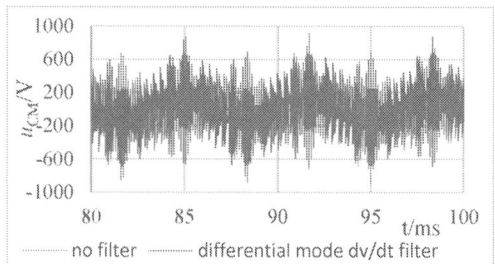

Fig.7 Differential mode motor voltage u_{DM}: Strong voltage shaping effect (*dv/dt* and peak voltage) with filter.

Fig.8 Common mode motor voltage u_{CM}: Nearly no effect by filter.

Dual mode (differential and common mode) dv/dt-filter

This filter type offers a common mode low pass in addition to differential mode dv/dt-filters.
The dual mode, or allpole, dv/dt-filters offer a common mode inductance and might also have a Y-connection to ground or DC link, so that also a common mode low pass filter is provided. By that, the shape of the common mode voltage at the motor can really be affected in a defined way. Excellent EMI reduction and damping of rotor-to-ground and circulating bearing currents like using a couple of nanocrystalline ring cores is provided in addition to the characteristics of a differential mode dv/dt-filter.

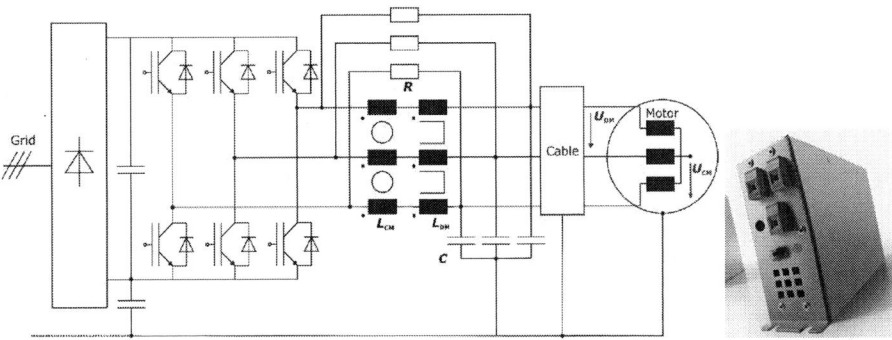

Fig. 9 Setup with dual mode dv/dt-filter

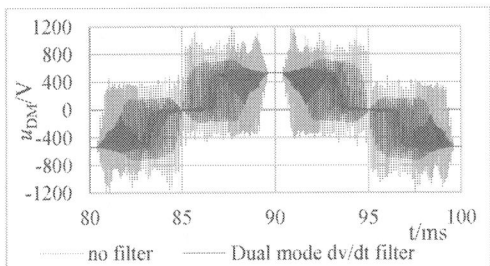

Fig.10 Differential mode motor voltage u_{DM}: Strong voltage shaping effect (*dv/dt* and peak voltage) with filter.

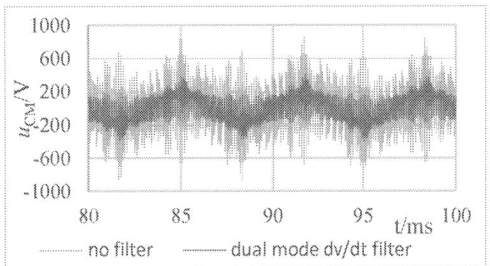

Fig.11 Common mode motor voltage u_{CM}: Strong voltage shaping effect (*dv/dt* and peak voltage) with filter.

Differential mode sine filters

Differential mode sine filters have been used as a high-end filter solution for motor protection in the past. The inductance is built as a differential mode three phase choke. The capacitances are set in X-connection with no connection to ground or DC-link. In comparison to dv/dt-filters, the inductance and capacitance values are higher, and no additional ohmic damping is required. At the motor terminals, a pure sine wave form between the phases is generated. The filter resonance frequency is set smaller than the switching frequency of the inverter, typically between 500 Hz and 5 kHz for IGBT Inverters. Because of that, differential mode sine filters only offer insulation protection between the phases and not against ground. EMI or bearing current reduction is offered by this solution only partly by the stray inductance of the choke. The filter eliminates PWM and high frequency losses in the motor cable and in the motor. Also, inverter losses are reduced. Because of the high inductance value, a voltage drop of up to 8% nominal voltage at nominal current will occur and this filter might be used in uncontrolled or low dynamic controlled drives.

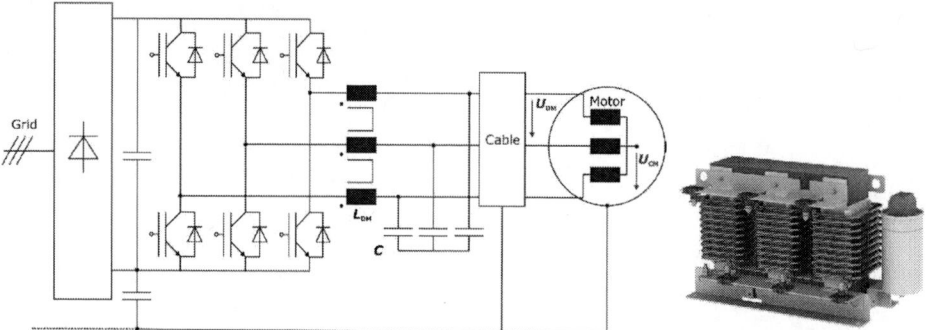

Fig. 12 Setup with sine filter

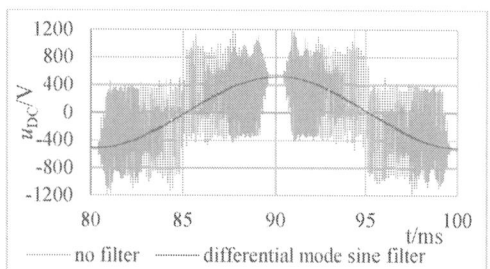

Fig.13 Differential mode motor voltage u_{DM}: Pure sine wave voltage.

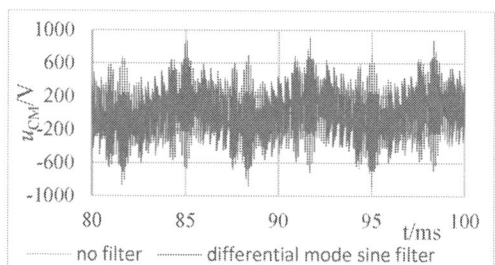

Fig.14 Common mode motor voltage u_{CM}: Nearly no effect by filter.

Dual mode (Differential and Common mode) sine filters

Dual mode or allpole sine filters have occurred on the market since some years and offer real high-end motor protection for differential and common mode, [5], [6]. Common mode currents are led back to the DC-link directly (optimal) or to ground (if no DC link connection available). No inverter caused bearing currents or motor insulation faults occur and high EMI reduction is offered by this solution. Total drive system loss is reduced [6]. EMI and leakage currents are excellently damped. Unshielded and very long motor cable might be used, and the grid side EMI filter can be built up with less or smaller components.

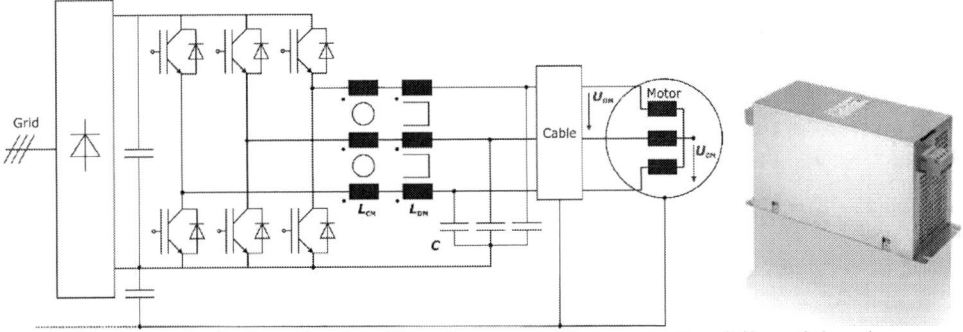

Fig. 15 Setup with dual mode sine filter. Resonance frequency for differential and common mode typically 500 Hz - 5 kHz for IGBT inverters.

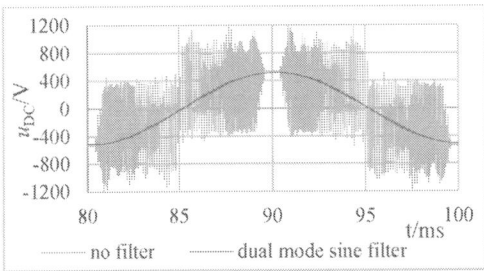

Fig.16 Differential mode motor voltage u_{DM}: Pure sine wave voltage.

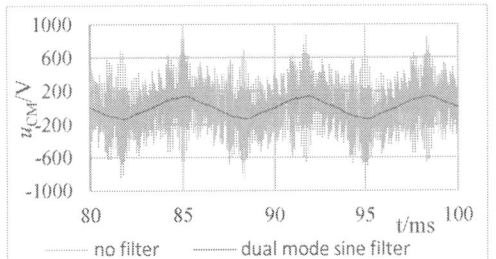

Fig.17 Common mode motor voltage u_{CM}: No switching frequency voltage components any more by filter.

Effects of the motors filters on different drive system characteristics

Motor insulation

All differential mode filters offer insulation protection. The sine filters generate pure sine voltage between the phases and eliminated inverter caused insulations failures and dv/dt-filters only shape the PWM voltage pulse so that rise time and voltage overshot stays within limits. Differential mode insulations failures from experience are the dominant failure mechanism, mostly in the winding heads of the machine or within the slots of a distributed winding. But in some cases, also common mode insulations fault, i.e. winding to slot or star point to housing, occur and a common mode filter is solving the problem.

Bearing currents

For smaller motors, i.e. <200 mm shaft height EDM currents dominate. Only the dual mode sine filters eliminate these currents because the common mode voltage is influenced strongly. For bigger motors, i.e. >200 mm shaft height, rotor-to-ground and circulating currents dominate. Here all filters have a positive damping effect. Excellent damping is provided by a matched amount of nanocrystalline cores or a dual mode sine filter [1], [8].

Motor voltage drop

While ring cores do not have a differential mode voltage drop, motor chokes or dv/dt-filters have voltage drop of <1% of the nominal motor voltage. Sine filters generate motor voltage drops up to 8%, depending on for what switching frequency the filter is designed.

Dynamic drive control

If the maximum control frequency and the switching frequency are close to each other, then the design of a sine filter is not possible. Therefore, in high dynamic controlled drives a sine filters might not be used.

Losses

If a passive filter is inserted into a drive system, then filter losses are added to the system losses. On the other hand, all filters, especially the dual mode sine filter, reduces the inverter, cable and motor losses. This especially in drive systems with long cables and partial load increases total system efficiency drastically [7].

EMI

Conducted and radiated EMI is improved by the ring cores and the two dual mode filters. Especially with a dual mode sine filter, extreme long unshielded cables are possible.

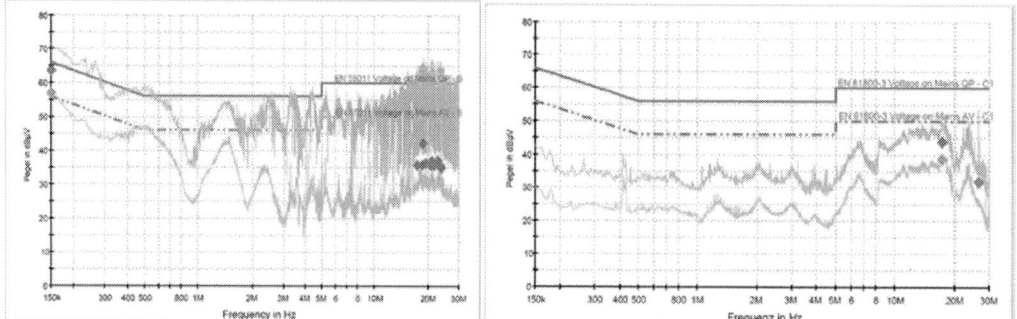

Fig.18 Conducted EMI of a drive system without any filter (left) and with dual mode sine filter (right). 3x400 V/7,5 kW inverter and motor, 100 m shielded cable length. Limit lines according to EN 61800-3 C1. Superior EMI behavior with a dual mode sine filter.

Audible noise

Passive filters produce additional noise in the drive system mainly due to magnetostriction in the magnetic cores. On the other hand, both sine filters eliminated PWM caused noise in the motors.

Costs

The cost for the passive filters varies from a cheap motor choke to a high-end dual mode sine filter with 4 times higher costs. On the other hand, cost savings for the drive system can be achieved using passive motor filters.

- Increased lifetime of motors and other system components lead to lower maintenance and failure costs.
- Unshielded cables might be used with a dual mode sine filter.
- Reduced system losses and therefore reduced energy costs.
- If differential mode output filters are used, inverter losses are decreased up to 30% depending on operating point [7]. This eliminates the necessity of using one size up inverter with longer cables.

Overview of the effect of each filter on various system characteristics

The following table gives an overview on the impact of each motor filter on different important parameters of a drive system:

Table I Overview on the impact of each motor filter

motor filter type	conducted & radiated EMI	inverter losses	motor cable losses	DM voltage drop over filter**	motor insulation	motor audible noise	motor losses	Bearing currents Circulating and rotor-to-ground currents (dominant in motors > 200 mm shaft height)	EDM-currents (dominant in motors <200 mm shaft height)	Costs & Size
Common mode ring cores*	XX			0%				XXX		100%
Differential mode motor choke		X	X	<1%	X			X		70%
Differential mode dv/dt-filter		X	XX	<1%	X			X		150%
Dual mode dv/dt-filter	XX	X	XX	<1%	X			XXX		200%
Differential mode sine filters		XX	XX	<8%	XX	XXX	XXX	X		250%
Dual mode sine filters	XXX	XXX	XXX	<8%	XXX	XXX	XXX	XXX	XXX	300%

X-beneficial impact

XX-high beneficial impact

XXX- very high beneficial impact

*depending on installation several nanocrystalline ring cores are required to achieve the listed effects

**assuming filter design for 2 kHz - 5 kHz switching frequency. Voltage drop only for motor frequency voltage component.

Conclusion

There are a lot of different passive motor filters available on the market affecting different drive system characteristics in different expressions. Some the filters only affect the dv/dt of the motor voltage and some generate a real sinus wave form. The cost and size difference between these filters might have a factor of 5 to 10. There are filters available, which affect only common or differential mode voltages. If a filter is required i.e. to protect the motor, also the beneficial effect on other system parameters like total drive system losses, EMI or acoustical noise should be considered. Therefor a guidance in selecting the right one is required by drive system designers.

This paper offers a comprehensive overview and might help drive system designers in choosing the most suitable passive filter solution.

References

[1] A. Muetze, A. Binder, „Don't lose your bearings – Mitigation techniques for bearing currents in inverter-supplied drive systems," IEEE Industry Applications Magazine (IAS), vol. 12, no. 4, pp. 22 – 31, 2006.

[2] B. Weis, „Erhöhte Spannungsbelastung von Motoren durch Umrichter-Ausgangsfilter", SPS/IPC/DRIVES, Nürnberg, 2009.

[3] IEC TS 60034-25:2014, 2014.

[4] NEMA MG-1: Motors and Generators, 2011 .

[5] D. Kampen, "Efficiency of Motor Side Common Mode (CM) Filtering Techniques for PWM Inverters," PCIM Europe, Nürnberg, 2012.

[6] N. Hanigovszki, "EMC Output Filters for Adjustable Speed Drives" Disseration, Aalborg, Denmark, 2005.

[7] M. Burger, D. Kampen, "Drive system loss reduction by allpole sine filters", in Proc. of the International Exhibition and Conference for Power Electronics (PCIM) Europe, 10. - 12. May, Nuremberg, 2016.

[8] M. Weicker, G. Bello, D. Kampen, A. Binder, „Influence of system parameters in variable speed AC-induction motor drives on parasitic electric bearing currents," in Proc. of the 22nd European Conference on Power Electronics and Applications (EPE'20 ECCE) Europe, 07. – 11. September, 2020, Virtual, submitted 2020.

[9] V. Hausberg, H. O. Seinsch, "Schutzmaßnahmen gegen Lagerschäden umrichtergespeister Motoren", Electrical Engineerig, vol. 82, pp. 339-345, 2000.

[10] A. Muetze, A. Binder, "Practical Rules for Assessment of Inverter-Induced Bearing Currents in Inverter-Fed AC Motors up to 500 kW," *IEEE Transactions on Industrial Electronics*, vol. 54, pp. 1614-1622, 2007.

[11] F. J. T. E. Ferreira, M. V. Cistelecan und A. T. de Almeida, "Evaluation of Slot-Embedded Partial Electrostatic Shield for High-Frequency Bearing Current Mitigation in Inverter-Fed Induction Motors," *IEEE Transactions on Energy Conversion*, vol. 27, pp. 382-390, 2012.

[12] D. Kampen, F. Quattrone, B. Ponick, "Was wirklich hilft - Filter zur Beseitigung Frequenzumrichter-bedingter Lagerströme," Antriebstechnik, vol. 52(9), pp. 56-58, 2013.

[13] Q. Jing, B. Baodong, W. Yu und L. Weifeng, "Research on electrostatic shield for discharge bearing currents suppression in variable-frequency motors," 17th International Conference on Electrical Machines and Systems (ICEMS), 2014, pp. 139-143.

[14] D. C. Ludois, J. K. Reed, "Brushless Mitigation of Bearing Currents in Electric Machines via Capacitively Coupled Shunting," Proc. of International Conference on Electrical Machines (ICEM), 02. – 05. September, 2014, pp. 2336 - 2342.

[15] DIN EN 60034-17:2005, Rotating electrical machines - Part 17: Cage induction motors when fed from converters – Application guide (IEC 2/1348/DTS:2005), Geneva, Switzerland.

[16] DIN EN 60034-24:2011, Rotating electrical machines – Part 24: Online detection and diagnosis of potential failures at the active parts of rotating electrical machines and of bearing currents – Application guide (IEC/TS 60034-24:2009), Geneva, Switzerland.

Impact of the neutral point potential ripple on the grid side harmonics of a 3LNPC back-to-back converter employed in a medium voltage WECS

Ioannis Tsoumas
ABB System Drives
Austrasse, 5300
Turgi, Switzerland
E-Mail: ioannis.tsoumas@ch.abb.com

Keywords

«Wind generator systems», «Interharmonics», «Voltage Source Converter», «Pulse Width Modulation».

Abstract

The operation of a back-to-back three-level neutral point clamped (3LNPC) converter employed in a medium voltage (MV) wind energy conversion system (WECS) is investigated. In particular, the interharmonics that appear on the grid side due to low frequency oscillations of the neutral point (NP) potential of the converter.

Introduction

In Fig. 1 the system under investigation is shown. It consists of a wind turbine, a synchronous generator with two windings, two back-to-back 3LNPC converters with separate dc-links and a wye/delta-wye transformer, which connects the inverters to the grid. The two rectifiers on the generator side are modulated with carrier-based pulse width modulation (CB-PWM), whereas on the grid-side inverters optimized pulse patterns (OPPs) are employed.

Due to the low fundamental frequency on the generator side in case of gearless systems, which can take values even below 10 Hz, strong low frequency oscillations of the neutral point (NP) potential appear in the dc-link of the 3LNPC converters. These low frequency oscillations are transferred to the transformer primary, where they appear as voltage interharmonics. Since the related standards, e.g. [1]-[2], are very stringent on the maximum acceptable values of interharmonics it is of paramount importance to be able to predict their amplitude and design the system in such a way that the limits set by the standards are not violated.

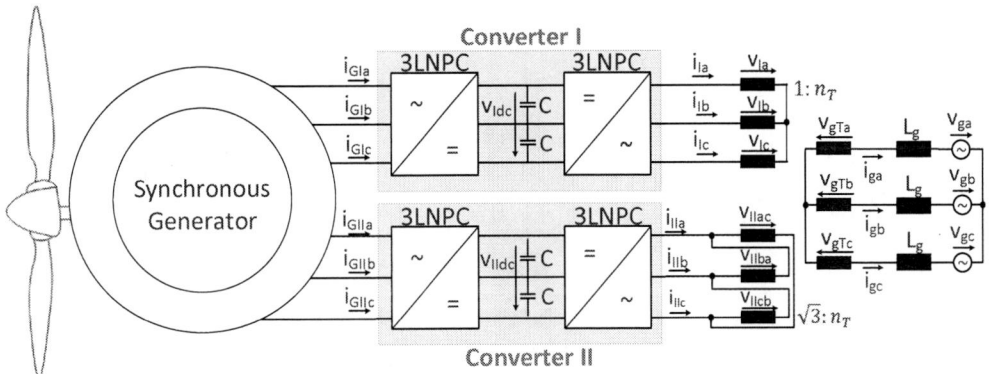

Fig. 1: Schematic diagram of the system under investigation.

NP potential ripple caused by the generator side rectifier

Switching function of the CB-PWM on the generator side rectifier

For the low frequency NP harmonics one needs to consider only the average value of the CB-PWM output waveform (switching function) in a switching period, i.e. the CB-PWM reference [4]-[6]. It is assumed that the modulator is operated in the linear modulation region, which means that the values of the reference do not exceed the positive and negative peaks of the carrier. The CB-PWM can be described for the three phases as follows:

$$u_a(\vartheta) = u_{a\sin} + u_0 = m_G \cdot \sin(\vartheta + \vartheta_G) + u_0 \qquad , \tag{1}$$

$$u_b(\vartheta) = u_{b\sin} + u_0 = m_G \cdot \sin\left(\vartheta + \vartheta_G - \frac{2\pi}{3}\right) + u_0 \quad , \tag{2}$$

$$u_c(\vartheta) = u_{a\sin} + u_0 = m_G \cdot \sin\left(\vartheta + \vartheta_G + \frac{2\pi}{3}\right) + u_0 \quad , \tag{3}$$

where $u_{a\sin}, u_{b\sin}, u_{c\sin}$ are the sinusoidal references, u_0 is a common mode (CM) component which is added to them, $\vartheta = \omega_G t$ and ϑ_G is the initial phase angle. The term m_G represents the modulation index of the generator-side rectifier, defined by

$$m_G = \frac{\hat{v}_{G1}}{v_{dc}/2} \ , \tag{4}$$

where $\hat{v}_{G1}$ is the amplitude of the fundamental phase-to-NP voltage component and v_{dc} is the total dc-link voltage.

The purpose of adding a CM component is twofold [3]: to extend the linear region of the modulator up to a modulation index of $2/\sqrt{3}$ (CB-PWM equivalent of the space vector modulation often referred to as "min-max") as well as to transfer some harmonic energy to higher frequencies in order to decrease the current ripple of an inductive load. The CM component in the CB-PWM reference can be also used for other purposes, e.g. for the suppression of the NP potential ripple in a 3LNPC converter [4].

In [5] an accurate analytical approach for the calculation of the required CM component in order to suppress the NP ripple has been introduced. However, the analytical calculations presented in [4] and [5] are not suitable for real time implementation, among others because the displacement factor must be known accurately and revision of the solutions may be necessary. To overcome this, a straightforward approach has been proposed in [6], which shows that two or three characteristic corner points are enough to fully describe the transfer function $i_{np} = f(u_0)$. From the inverse transfer function $u_0 = f^{-1}(i_{np})$ one can calculate the necessary CM component u_0 to achieve a desired NP current i_{np}, that eliminates the

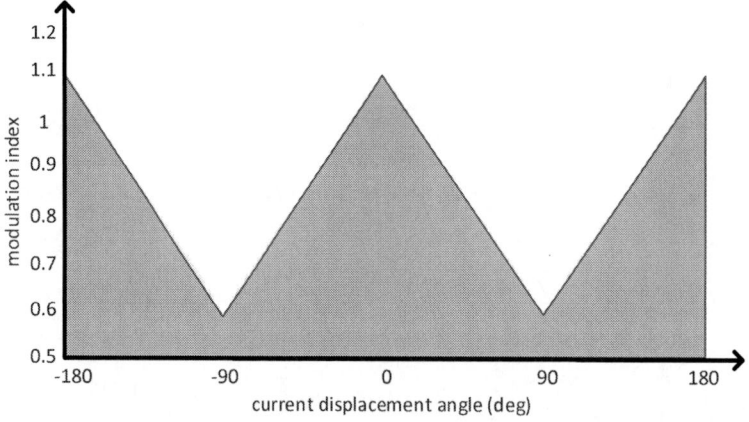

Fig. 2: NP potential AC ripple fully controllable region as a function of the modulation index and the current displacement angle [5].

low frequency NP ripple. Such an approach can be used for controlling both the ripple as we well as the drift (average value) of the NP potential [7]. To eliminate the low frequency oscillating component a commanded NP current equal to $i_{np}^* = 0$ (zero average NP current) is necessary.

Nevertheless, it is not always possible to add the desired CM component for the NP ripple suppression, since depending on the modulation index m_G and the current displacement angle φ_G the above addition may lead to a CB-PWM reference with an absolute value higher than 1 (overmodulation). To avoid this the CM component is clamped in such a case to a value for which the CB-PWM reference is either 1 or −1. The fully controllable region for which no such clamping is necessary has been numerically investigated in [5] and it is depicted with a grey area in Fig. 2. In the non-fully controllable region a low frequency NP potential oscillation will always appear.

Harmonic transfer rules from the AC side to the DC side

The neutral point potential can be calculated from the convolution of the phase currents with the absolute value of the switching function. In such a calculation sums of the three-phase quantities appear. In [8] symmetrical operating conditions were assumed and the following rules have been derived for the resulting dc-link components in the frequency domain:

- If the symmetrical components of the current harmonic and the switching function harmonic describe either a positive-sequence system or a negative sequence system, then their convolution onto the DC side will result in a single harmonic.

- If the current harmonic and the switching function harmonic have the same (different) phase sequence, the harmonic order of the resulting dc-link harmonics will be the difference (sum) of the individual AC-side harmonic orders.

- If the symmetrical components of either the current harmonic or the switching function harmonic describe a zero-sequence system, their convolution will not result in any dc-link harmonics.

The above rules are very important for the understanding of the equation presented in the next section because they explain why only some sums and differences of frequencies appear in the NP ripple whereas others cancel out.

NP ripple harmonics caused by the CB-PWM of the rectifier

Since in low speed generators the current ripple is very small (due to the high ratio of carrier to fundamental frequency) only the fundamental harmonic of the current is considered for the calculation of the NP ripple, with a phase displacement of φ_G with respect to the sinusoidal reference:

$$i_{GIa}(\vartheta) = I_{GI} \cdot \sin(\vartheta + \vartheta_G + \varphi_G) \quad , \tag{5}$$

$$i_{GIb}(\vartheta) = I_{GI} \cdot \sin\left(\vartheta + \vartheta_G + \varphi_G - \frac{2\pi}{3}\right), \tag{6}$$

$$i_{GIc}(\vartheta) = I_{GI} \cdot \sin\left(\vartheta + \vartheta_G + \varphi_G + \frac{2\pi}{3}\right). \tag{7}$$

The NP potential is calculated by [8]

$$v_{Inp}(\theta) = -\frac{1}{2\omega C} \int_0^\vartheta i_{np}(\zeta) d\zeta + v_{np}(0), \tag{8}$$

where
$$i_{Inp}(\zeta) = \sum_{x \in \{a,b,c\}} (1 - |u_x(\zeta)|) \, i_x(\zeta), \qquad \text{with } x \in \{a, b, c\}. \tag{9}$$

From (5)-(9) the harmonics of the NP ripple yield

$$v_{GInp}(\vartheta) = \frac{3I_G}{4\omega_G C} \sum_{v \in 6k+3, k \in \mathbb{N}_0} \frac{1}{v} \Big[(b_{v+1} - b_{v-1}) \cos(\varphi_g) \cos[v(\vartheta + \vartheta_G)] +$$
$$(b_{v+1} + b_{v-1}) \sin(\varphi_g) \sin[v(\vartheta + \vartheta_G)] \Big] + v_{GInp}(0), \tag{10}$$

where the coefficients b_{v+1} and b_{v-1} represent the amplitudes of the harmonics of the absolute value of the CB-PWM reference (1)-(3) with orders $2, 4, 8, 10, 14, 16, \ldots, I_G$ is the peak value of the fundamental component of the generator current, ω_G is the generator fundamental frequency and C is the capacitance of each dc-link half. As can be seen in (10) only harmonics with orders $3, 9, 15, 21, \ldots$ appear in the NP potential. This is expected if the harmonic transfer rules mentioned in the previous section are considered. The amplitude of the coefficients b_{v+1} and b_{v-1} depends on the modulation index and on the CM component u_0 in (1)-(3). It can be shown that for the typical CM components (e.g. 3rd harmonic and "min-max") the NP ripple is sufficiently approximated by considering only the 3rd and the 9th harmonic. i.e. for $k = 0, 1$ in (10).

Grid Side Harmonics

The voltage harmonics at the output of the grid side converters can be computed from

$$v_x(t) = u_x(t) \cdot \frac{v_{dc}(t)}{2} - |u_x(t)| \cdot v_{np}(t) \text{ , with } x \in \{a, b, c\}. \tag{11}$$

The resulting voltage harmonics can be sorted into three categories:

1. The harmonics that arise from the convolution of the switching function u_x with the average value of the total dc-link voltage V_{dc} (divided by 2). These are the harmonics of the switching function u_x of the inverter multiplied by a constant factor. The product of $V_{dc}/2$ with the fundamental component of u_x yields the desired fundamental voltage component.

2. The harmonics that appear due to the convolution of u_x with the harmonics of the total dc-link voltage v_{dc} divided by two. Some of the latter harmonics come from the rectifier and some other from the inverter.

3. The harmonics that appear due to the convolution of $|u_x|$ with the harmonics of the NP potential v_{np}. Some of the latter harmonics come from the rectifier and some other from the inverter.

In this work we focus on the 3rd part and more specifically on grid side harmonics related to the low frequency harmonics of the NP potential caused by the generator side rectifier. As shown in (11) the NP potential ripple is multiplied by $|u_x|$ and generates voltage harmonics at the transformer secondary windings. For the harmonics at the transformer primary the sum of the harmonics of the two converters must be considered as follows:

$$v'_{gnpG\,x} = \frac{n_T}{2}\left(|u_{gI\,x}| \cdot v_{GInp} + \frac{1}{\sqrt{3}}|u_{gII\,xy}| \cdot v_{GIInp}\right), \text{ with } x \in \{a, b, c\} \text{ and } y \in \{c, a, b\}. \tag{12}$$

The parameter n_T is the transformation ratio of the transformer. In (12) $u_{gI\,x}$ and $u_{gII\,x}$ are the switching functions of the grid side inverter of the converter line I and II respectively and v_{GInp} and v_{GIInp} is the NP potential ripple caused by the rectifier on the generator side of the converter I and II respectively. Before calculating the grid side harmonics from (12) we express the NP potential ripple of converter I and II as

$$v_{GInp}(\vartheta) = \sum_{\mu \in 6k+3, k \in \mathbb{N}_0} C_\mu \, \sin[\mu(\vartheta + \vartheta_G) + \varphi_{Gnp}] \tag{13.a}$$

$$\text{and} \quad v_{GIInp}(\vartheta) = \sum_{\mu \in 6k+3, k \in \mathbb{N}_0} C_\mu \, \sin[\mu(\vartheta + \vartheta_G + \Delta\vartheta_G) + \varphi_{Gnp}], \tag{13.b}$$

where C_μ is the amplitude of the NP ripple harmonic, ϑ_G is the initial phase of the reference voltage of converter I which appears in (5)-(6) and $\Delta\vartheta_G$ is the phase difference between the CB-PWM references of converter II and I. The amplitude C_μ and the phase φ_{Gnp} can be calculated from (10) for $\mu = v$. The phase φ_{Gnp} can vary between $-\pi/2$ and 0 depending on the displacement factor of the phase current. Only the harmonics that arise from the CB-PWM reference are considered. The switching is not considered, since we are only interested in the low frequency ripple of the NP potential.

For the calculation of the harmonics of the absolute switching function of the grid side inverter we consider that it is modulated by OPPs. If we consider a quarter wave symmetry for the OPPs the Fourier approximation of the absolute value of the switching function of e.g. phase a is the following:

$$|u_a(\theta)| = \frac{a_0}{2} + \sum_{\nu \in 2k, k \in \mathbb{N}} a_\nu \cos(\nu\theta), \tag{14.a}$$

where $\quad a_\nu = -\frac{1}{\nu}\frac{4}{\pi}\sum_{i=1}^{d} \Delta u_i \sin\nu\theta_i \,, \ \nu > 0 \qquad$ and $\tag{14.b}$

$$a_0 = -\frac{4}{\pi}\sum_{i=1}^{d}\Delta u_i \theta_i + 2\sum_{i=1}^{d}\Delta u_i \tag{14.c}$$

where $\theta \in [0, \pi/2]$ and d is the OPP pulse number. The latter is an integer number that expresses the ratio between switching and fundamental frequency.

From (12), (13) and (14) we have for the grid side harmonics in phase a:

$$v'_{\text{gnpG a}} = \frac{n_T}{2}\left\{\sum_{\nu \in 6k+3 \,\pm 1, k \in \mathbb{N}_0} a_\nu \cos(\nu\theta) \cdot \sum_{\mu \in 6k+3, k \in \mathbb{N}_0} C_\mu \, \sin[\mu(\vartheta + \vartheta_G) + \varphi_{\text{Gnp}}]\right.$$

$$\left. + \sum_{\nu \in 6k+3 \,\pm 1, k \in \mathbb{N}_0} a_\nu \, \xi \sin\left[\nu\left(\theta + \frac{\pi}{2}\right)\right] \cdot \sum_{\mu \in 6k+3, k \in \mathbb{N}_0} C_\mu \, \sin[\mu(\vartheta + \vartheta_G + \Delta\vartheta_G) + \varphi_{\text{Gnp}}]\right\}, \tag{15.a}$$

$$\text{where } \xi = \begin{cases} 1, & \nu \in 2 + 6k, \ k \in \mathbb{N}_0 \\ -1, & \nu \in 4 + 6k, \ k \in \mathbb{N}_0 \end{cases}. \tag{15.b}$$

In (15) the CM components of the absolute value of the switching function with orders $6k$, $k \in \mathbb{N}_0$, have been omitted, since their multiplication with v_{GInp} and v_{GIInp} yields CM components that don't appear at the transformer windings.

Using well known trigonometric identities for sums of sine and cosine functions (14) yields

$$v'_{\text{gnpG a}} = \frac{n_T}{4}\left\{\sum_{\nu \in 6k+3 \,\pm 1, k \in \mathbb{N}_0}\sum_{\mu \in 6k+3, k \in \mathbb{N}_0} a_\nu \, C_\mu \left\{\sin\left[\left(-\nu + \mu\frac{f_G}{f_g}\right)\theta + \mu\,\vartheta_G + \varphi_{\text{Gnp}}\right]\right.\right.$$

$$\left. + \xi \cos\left[\left(-\nu + \mu\frac{f_G}{f_g}\right)\theta + \mu\vartheta_G + \varphi_{\text{Gnp}} - \nu\frac{\pi}{2} + \mu\,\Delta\vartheta_G\right]\right\}$$

$$+ \sum_{\nu \in 6k+3 \,\pm 1, k \in \mathbb{N}_0}\sum_{\mu \in 6k+3, k \in \mathbb{N}_0} a_\nu \, C_\mu \left\{\sin\left[\left(\nu + \mu\frac{f_G}{f_g}\right)\theta + \mu\,\vartheta_G + \varphi_{\text{Gnp}}\right]\right.$$

$$\left.\left. - \xi \cos\left[\left(\nu + \mu\frac{f_G}{f_g}\right)\theta + \mu\vartheta_G + \varphi_{\text{Gnp}} + \nu\frac{\pi}{2} + \mu\,\Delta\vartheta_G\right]\right\}\right\}. \tag{16}$$

The above equation gives the voltage harmonic components on the grid side for the general case of an arbitrary value for $\Delta\vartheta_G$. For the system under investigation we typically have either $\Delta\vartheta_G = 0$ or $\Delta\vartheta_G = \pm\frac{\pi}{6}$, where the sign depends on which generator winding is connected to the 2nd converter line (the one connected to the transformer delta). The two latter cases for $\Delta\vartheta_G$ are examined separately in the next two subsections.

Case 1: $\Delta\vartheta_G = +\frac{\pi}{6}$

In this case we have a phase of $3 \cdot \Delta\vartheta_G = \frac{\pi}{2}$ for the 3rd harmonic of the NP potential ripple. Then (16) yields for $\mu = 3$

$$v'_{\text{gnpG a,3}} = \frac{n_T}{4}\left\{C_3 \sum_{\nu \in 2k, k \in \mathbb{N}} a_\nu \left\{\sin\left[\left(-\nu + 3\frac{f_G}{f_g}\right)\theta + 3\,\vartheta_G + \varphi_{\text{Gnp}}\right]\right.\right.$$

$$\left. + \xi \cos\left[\left(-\nu + 3\frac{f_G}{f_g}\right)\theta + 3\vartheta_G + \varphi_{\text{Gnp}}(-\nu + 1)\frac{\pi}{2}\right]\right\}$$

$$+ C_3 \sum_{\nu \in 2k, k \in \mathbb{N}} a_\nu \left\{\sin\left[\left(\nu + 3\frac{f_G}{f_g}\right)\theta + 3\,\vartheta_G + \varphi_{\text{Gnp}}\right]\right.$$

$$- \xi \, \cos\left[\left(v + 3\frac{f_G}{f_g}\right)\theta + 3\vartheta_G + \varphi_{Gnp} + (v+1)\frac{\pi}{2}\right]\Big\}\Big\}. \tag{17}$$

Recalling that $\cos\left(\theta \pm \frac{\pi}{2}\right) = \mp \sin(\theta)$ we finally get

$$v'_{gnpG\,a,\,3} = \frac{n_T}{2}\,C_3 \left\{ \sum_{v\,\in\,2k\cdot6+3\,\pm1,k\in\mathbb{N}_0} a_v \sin\left[\left(-v+3\frac{f_G}{f_g}\right)\theta + 3\,\vartheta_G + \varphi_{Gnp}\right] \right.$$

$$\left. + \sum_{v\,\in\,(2k+1)\cdot6+3\,\pm1,k\in\mathbb{N}_0} a_v \sin\left[\left(v+3\frac{f_G}{f_g}\right)\theta + 3\,\vartheta_G + \varphi_{Gnp}\right] \right\}. \tag{18}$$

In (18) we observe that for the harmonics of $|u_{gI\,x}|$ with orders 2, 4, 14, 16, 26, 28, ... only the left frequency component of the convolution of $|u_{gI\,x}|$ and v_{GInp} with frequency $\left|-v+3\frac{f_G}{f_g}\right| f_G$ appears at the transformer primary winding, whereas for the harmonics with orders 8, 10, 20, 22, 32, 34, ... only the right component with frequency $\left(v+3\frac{f_G}{f_g}\right) f_G$ appears. The same will happen for the 15th harmonic of the NP potential ripple, because we have the same phase displacement of the NP potential frequency component $15 \cdot \Delta\vartheta_G = \frac{\pi}{2}$. The opposite will happen with the 9th harmonic of the NP potential ripple because $9 \cdot \Delta\vartheta_G = -\frac{\pi}{2}$. The above can be summarized in the following equation for the harmonics at the transformer primary:

$$v'_{gnpG\,a} = \frac{n_T}{2} \sum_{\mu\,\in\,2k\cdot6+3,k\in\mathbb{N}_0} C_\mu \left\{ \sum_{v\,\in\,2k\cdot6+3\,\pm1,k\in\mathbb{N}_0} a_v \sin\left[\left(-v+\mu\frac{f_G}{f_g}\right)\theta + \mu\,\vartheta_G + \varphi_{Gnp}\right] \right.$$

$$\left. + \sum_{v\,\in\,(2k+1)\cdot6+3\,\pm1,k\in\mathbb{N}_0} a_v \sin\left[\left(v+\mu\frac{f_G}{f_g}\right)\theta + \mu\,\vartheta_G + \varphi_{Gnp}\right] \right\}$$

$$\frac{n_T}{2} \sum_{\mu\,\in\,(2k+1)\cdot6+3,k\in\mathbb{N}_0} C_\mu \left\{ \sum_{v\,\in\,(2k+1)\cdot6+3\,\pm1,k\in\mathbb{N}_0} a_v \sin\left[\left(-v+\mu\frac{f_G}{f_g}\right)\theta + \mu\,\vartheta_G + \varphi_{Gnp}\right] \right.$$

$$\left. + \sum_{v\,\in\,2k\cdot6+3\,\pm1,k\in\mathbb{N}_0} a_v \sin\left[\left(v+\mu\frac{f_G}{f_g}\right)\theta + \mu\,\vartheta_G + \varphi_{Gnp}\right] \right\}. \tag{19}$$

Case 2: $\Delta\vartheta_G = -\frac{\pi}{6}$

In this case we get for the grid harmonics

$$v'_{gnpG\,a,} = \frac{n_T}{2} \sum_{\mu\,\in\,2k\cdot6+3,k\in\mathbb{N}_0} C_\mu \left\{ \sum_{v\,\in\,(2k+1)\cdot6+3\,\pm1,k\in\mathbb{N}_0} a_v \sin\left[\left(-v+\mu\frac{f_G}{f_g}\right)\theta + \mu\,\vartheta_G + \varphi_{Gnp}\right] \right.$$

$$\left. + \sum_{v\,\in\,2k\cdot6+3\,\pm1,k\in\mathbb{N}_0} a_v \sin\left[\left(v+\mu\frac{f_G}{f_g}\right)\theta + \mu\,\vartheta_G + \varphi_{Gnp}\right] \right\}$$

$$\frac{n_T}{2} \sum_{\mu\,\in\,(2k+1)\cdot6+3,k\in\mathbb{N}_0} C_\mu \left\{ \sum_{v\,\in\,2k\cdot6+3\,\pm1,k\in\mathbb{N}_0} a_v \sin\left[\left(-v+\mu\frac{f_G}{f_g}\right)\theta + \mu\,\vartheta_G + \varphi_{Gnp}\right] \right.$$

$$\left. + \sum_{v\,\in\,(2k+1)\cdot6+3\,\pm1,k\in\mathbb{N}_0} a_v \sin\left[\left(v+\mu\frac{f_G}{f_g}\right)\theta + \mu\,\vartheta_G + \varphi_{Gnp}\right] \right\}. \tag{20}$$

From (20) it is evident that in this case the opposite to $\vartheta_G = +\frac{\pi}{6}$ happens. For example, regarding the 3rd NP ripple harmonic one can make the following observation: for the harmonics of $|u_{gI\,x}|$ with orders 2, 4, 14, 16, 26, 28, ... only the right frequency component of the convolution of $|u_{gI\,x}|$ and v_{GInp} with frequency $\left|-v+3\frac{f_G}{f_g}\right| f_G$ appears at the transformer primary winding, whereas for the harmonics with orders 8, 10, 20, 22, 32, 34, ... only the left component with frequency $\left(v+3\frac{f_G}{f_g}\right) f_G$ appears.

Harmonic transfer rules from the DC side to the transformer primary

The results presented in (17) - (20) can be summarized in the following harmonic transfer rules from the DC side of the converters to the resulting voltage on the transformer primary winding:

- Regardless of the value of $\Delta\vartheta_G$ the NP ripple harmonics convoluted with the harmonics with order $6k$, $k\epsilon\mathbb{N}_0$, of $\left|u_{gI\,x}\right|$ do not appear on the grid side, because the products of (12) yield CM harmonics.

- For $\Delta\vartheta_G = \frac{\pi}{6}$ and $\mu \in 2k \cdot 6 + 3$, $k\epsilon\mathbb{N}_0$ $(3, 15, \ldots)$ the following frequency components appear on the grid side:

  - For $\nu \in 2k \cdot 6 + 3$, $k\epsilon\mathbb{N}_0$ $(2, 4, 14, 16, \ldots)$ frequency components with orders $\left|-\nu + \mu \frac{f_G}{f_g}\right|$ appear on the grid side.

  - For $\nu \in (2k + 1) \cdot 6 + 3$, $k\epsilon\mathbb{N}_0$ $(8, 10, 20, 22, \ldots)$ frequency components with orders $\left(\nu + \mu \frac{f_G}{f_g}\right)$ appear on the grid side.

- For $\Delta\vartheta_G = \frac{\pi}{6}$ and $\mu \in (2k + 1) \cdot 6 + 3$, $k\epsilon\mathbb{N}_0$ $(9, 21, \ldots)$ the following frequency components appear on the grid side:

  - For $\nu \in 2k \cdot 6 + 3$, $k\epsilon\mathbb{N}_0$ $(2, 4, 14, 16, \ldots)$ frequency components with orders $\left(\nu + \mu \frac{f_G}{f_g}\right)$ appear on the grid side.

  - For $\nu \in (2k + 1) \cdot 6 + 3$, $k\epsilon\mathbb{N}_0$ $(8, 10, 20, 22, \ldots)$ frequency components with orders $\left|-\nu + \mu \frac{f_G}{f_g}\right|$ appear on the grid side.

- For $\Delta\vartheta_G = -\frac{\pi}{6}$ and $\mu \in 2k \cdot 6 + 3$, $k\epsilon\mathbb{N}_0$ $(3, 15, \ldots)$ the following frequency components appear on the grid side:

  - For $\nu \in 2k \cdot 6 + 3$, $k\epsilon\mathbb{N}_0$ $(2, 4, 14, 16, \ldots)$ frequency components with orders $\left(\nu + \mu \frac{f_G}{f_g}\right)$ appear on the grid side.

  - For $\nu \in (2k + 1) \cdot 6 + 3$, $k\epsilon\mathbb{N}_0$ $(8, 10, 20, 22, \ldots)$ frequency components with orders $\left|-\nu + \mu \frac{f_G}{f_g}\right|$ appear on the grid side.

- For $\Delta\vartheta_G = -\frac{\pi}{6}$ and $\mu \in (2k + 1) \cdot 6 + 3$, $k\epsilon\mathbb{N}_0$ $(9, 21, \ldots)$ the following frequency components appear on the grid side:

  - For $\nu \in 2k \cdot 6 + 3$, $k\epsilon\mathbb{N}_0$ $(2, 4, 14, 16, \ldots)$ frequency components with orders $\left|-\nu + \mu \frac{f_G}{f_g}\right|$ appear on the grid side.

  - For $\nu \in (2k + 1) \cdot 6 + 3$, $k\epsilon\mathbb{N}_0$ $(8, 10, 20, 22, \ldots)$ frequency components with orders $\left(\nu + \mu \frac{f_G}{f_g}\right)$ appear on the grid side.

The above grid harmonics will have orders which are not integer multiples of the grid side fundamental (with only exception a generator fundamental frequency of $f_g/3$). They are thus interharmonics. From the harmonic transfer rules, it is evident that by appropriate selection of $\Delta\vartheta_G$ one can eliminate specific interharmonics, either on the left or on the right side of the corresponding harmonic of the absolute value of the grid side inverter switching function.

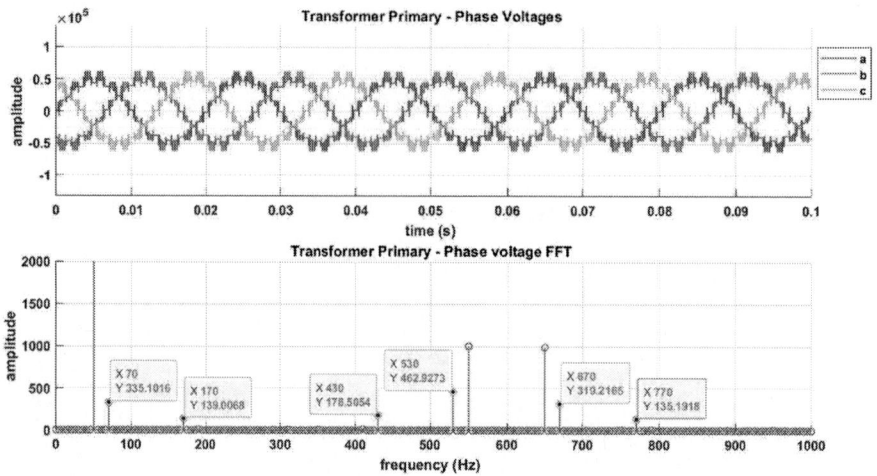

Fig. 3: Grid side voltage waveform and harmonics for $\Delta\vartheta_G = +\frac{\pi}{6}$, $f_G = 10\ Hz$ and $f_g = 50\ Hz$.

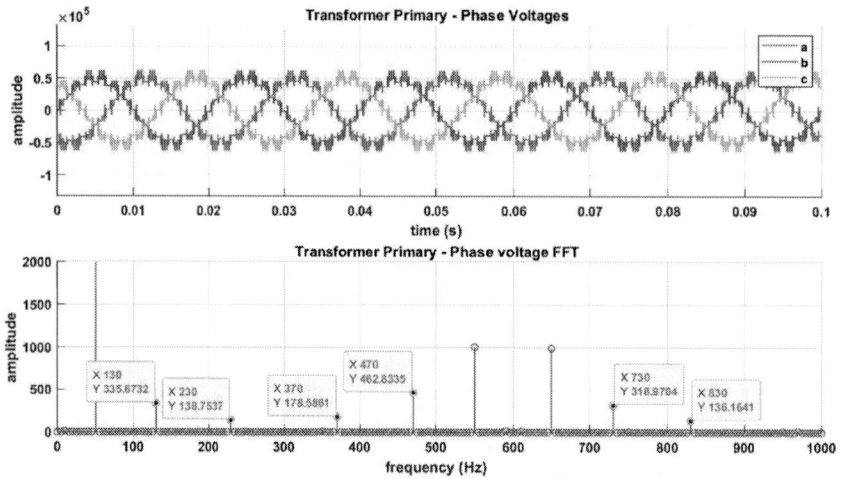

Fig. 4: Grid side voltage waveform and harmonics for $\Delta\vartheta_G = -\frac{\pi}{6}$, $f_G = 10\ Hz$ and $f_g = 50\ Hz$.

The standards are very stringent when it comes to the amplitude limits of voltage interharmonics. For example the limits defined in [1] are 0.2% of the fundamental for interharmonics with a frequency below twice the fundamental and 0.5% above twice the fundamental and up to 2.5 kHz. Considering that, it would be convenient to select $\Delta\vartheta_G = -\pi/6$. Then according to the afore mentioned harmonic transfer rules interharmonics with a frequency below twice the fundamental will not appear and the very stringent limit of 0.2% will not have to be considered.

Numerical Results

Characteristic numerical results on the grid side harmonics for two different values of the phase difference $\Delta\vartheta_G$ between the windings of the generator are shown in Fig. 3 and Fig. 4. As predicted by the harmonic transfer rules of the previous section, for a specific even multiple of the fundamental only the left or the right-side band interharmonic appears when $\Delta\vartheta_G = \pm\frac{\pi}{6}$, the second is cancelled out. In addition, as also predicted by the harmonic transfer rules, the left side interharmonic swaps to the right side and vice versa when $\Delta\vartheta_G$ changes form $+\frac{\pi}{6}$ to $-\frac{\pi}{6}$.

Conclusion

A double converter system with separate dc-links and connection to the grid via wye/delta-wye transformer has been investigated. A general formula for the calculation of the grid side interharmonics caused by the converter low frequency NP potential oscillation has been derived. Based on this formula harmonic transfer rules have been established, that determine the relationship between NP potential ripple and the grid side interharmonics.

Their amplitude depends on two factors: the amplitude of the harmonics of the NP ripple and the amplitude of the harmonics of the absolute value of the grid side switching function. Furthermore, depending on the relative phase of the currents in the two generator windings some of the interharmonics can be cancelled out.

The analytical analysis enables an appropriate system design, e.g. appropriate selection of the relative phase of the currents in the generator windings, of the dc-link capacitance value and of the grid side modulation, in order to minimize the number of harmonics that appear and meet the requirements of the most stringent standards on the remaining ones.

The only way to eliminate all of them is to employ an appropriate modulation method in the rectifier on the generator side that eliminates the low frequency NP ripple. Such a methods appear in the literature, but they are not effective in high modulation indices and low power factors. A modulation approach that overcomes these shortcomings will be presented in a future work.

References

[1] IEEE Standard 519: IEEE Recommended Practice and Requirements for Harmonic Control in Electric Power Systems, 2014 (Revision of IEEE Std 519-1992).

[2] IEC/TR 61000-3-6: Limits – Assessment of emission limits for the connection of distorting installations to MV, HV and EHV power systems, Edition 2.0, 2008.

[3] Holmes D.G. and Lipo T.A, Pulse width modulation for power converters, IEEE Press, 2003.

[4] Ogasawara S. and Akagi H., Analysis of variation of neutral point potential in neutral-point-clamped voltage source PWM inverters, IEEJ Transactions on Industry Applications, November 1993, pp. 965-970.

[5] Song Q., Lui W., Yu Q. and Wang X., A neutral-point potential balancing algorithm for three-level NPC inverters using analytically injected zero-sequence voltage, 18th Annual IEEE Applied Power Electronics Conference and Exposition (APEC), February 2003, pp. 228-233.

[6] Wang Ch. and Li Y., Analysis and calculation of zero-sequence voltage considering neutral-point potential balancing in three-level NPC converters, IEEE Transactions on Industrial Electronics, July 2010, pp. 2262-2271.

[7] Chen J, Schröder S., Duro B., and Roesner R., A neutral-point balancing controller for a three-level inverter with full power-factor range and low distortion, IEEE Transactions on Industry Applications, January/February 2013, pp. 138-147.

[8] Scheuer G., Investigation of the 3-level voltage source inverter for flexible AC transmission systems exemplified on a static VAr compensator, Dissertation, ETH Zurich 1997.

Two-Layer Genetic Algorithm for the Charge Scheduling of Electric Vehicles

Nikolaos T. Milas[1], Dimitris A. Mourtzis[2], Panagiotis I. Giotakos[3], Emmanuel C. Tatakis[1]

[1]UNIVERSITY OF PATRAS, Laboratory of Electromechanical Energy Conversion, Electrical and Computer Engineering Department, Rion-Patras, Greece 26504, Tel.: +30.2610.996414, E-Mail: nmilas@ece.upatras.gr, e.c.tatakis@ece.upatras.gr, URL: http://lemec.ece.upatras.gr

[2]UNIVERSITY OF PATRAS, Laboratory for Manufacturing Systems and Automation (LMS), Mechanical Engineering and Aeronautics Department, Rion-Patras, Greece 26504, Tel.: +30.2610.910160, E-Mail: mourtzis@lms.mech.upatras.gr URL: http://lms.mech.upatras.gr

[3]FOUNDATION FOR RESEARCH AND TECHNOLOGY HELLAS(FORTH), Institute of Chemical Engineering Sciences (ICEHT), Laboratory of Energy Processes, Stadiou Street Platani, Patra, Greece 26504, Tel.:+302610965238, E-mail: pgiotakos@forth.iceht.gr URL: www.iceht.forth.gr

Acknowledgements

This research is co-financed by Greece and the European Union (European Social Fund- ESF) through the Operational Programme «Human Resources Development, Education and Lifelong Learning» in the context of the project "Strengthening Human Resources Research Potential via Doctorate Research" (MIS-5000432), implemented by the State Scholarships Foundation (IKY).

Keywords

«Electric vehicle», «Charge Scheduling», «Genetic Algorithm», «Optimisation», «OPC-UA»

Abstract

The advent of Electric Vehicles (EV) may introduce disturbances to the operation of the Power Grid, due to the great demands of electric power that is required during the simultaneous charging of large EV fleets. Towards this end, novel approaches for the management of the EV charging have been proposed in recent literature. Nevertheless, the implementation of a framework that allows flexibility in the definition of the decision-making objectives, along with user-defined criteria is still a challenge. Towards addressing this challenge, a framework for the smart charging of EVs is presented in this paper. The smart charging is facilitated by a two-layer Genetic Algorithm that operates in Charging Stations with various types of chargers that are connected to multiple charging points in a resource-sharing manner. The benefits of the proposed approach are the fast optimisation time, the inclusion of user-defined criteria, and the extraction of feasible solutions considering the availability of the chargers in the station. The communications between the EV and the Charging Station are facilitated by the Open Platform Communications–Unified Architecture (OPC-UA) standard. The proposed algorithm succeeds into finding competitive solutions even during charging scenarios with conflicting criteria.

Introduction

The degradation of the environment dictates the adoption of non-conventional means of transportation. Towards this end, the Electric Vehicles (EVs) are eligible candidates to take the place of conventional vehicles, especially for urban transportations. The subject of the charge scheduling of EVs has been addressed in the recent literature, where a significant number of publications aim to provide solutions to various aspects of the problem [1]. Both analytical methods and heuristics have been employed to optimise the objective function of each problem. The analytical optimisation methods have advantages when the problem is explicitly defined [2], while the heuristics are applied when the problem results into a high number of alternatives, when the formulation is opaque, or when multiple local optima exist [3]. The major challenges that are yet to be addressed are the inclusion of user-defined preferences, the inclusion of multiple types of chargers, the maximisation of the utilisation of the resources in a Charging Station (CS) via resource sharing, and the introduction of communication mechanisms.

During the last years, there is a number of scientific works that deal with the subject of charge scheduling. A brief literature review that is specified on existing challenges on the subject, is presented in this section. The classification of the charge scheduling problems as centralized or decentralised has been performed in [1], while the classification based on the application type of the scheduling has been performed in [4]. The single aggregator architecture, which considers only one entity to manage the charge scheduling of all EVs in a specific area provides limited scalability. Aiming to address this limitation, distributed mechanisms that operate in a hierarchical structure have been proposed in [2], [5]. In the proposed architectures of [2], [5], an optimisation mechanism among the Smart Grid and the CS Operators results into setpoints that are the inputs for the low-level scheduling that considers the criteria set by the CS operators and the EV owners. A method that can result into high utilization of the resources in a CS is the resource-sharing [6]-[8]. According to this concept, the chargers that are installed in a CS can be connected to multiple charging spots via industrial automation. As a result, when a charging task is completed the charger is automatically reallocated to a charging spot minimizing the idle times. In addition, the satisfaction of the specific requirements of the driver, such as the charge completion time and the requirement for reduced cost, are optimisation criteria that might be conflicting compared to the criteria of the CS and the Grid Operator [9], [10]. Another issue of crucial importance is the conclusion to optimal or near-optimal solutions in a timely manner. Therefore, optimisation algorithms based on heuristics ([3], [4], [5]) and game theoretic approaches ([11], [12]) have been proposed to mitigate the computational time of traditional optimisation approaches.

To address these challenges and support the prevalence of EVs, a method for the charge scheduling of EVs enabled by a two-layer Genetic Algorithm (GA) is introduced in this paper. The method is applied for each EV individually to generate a feasible charging profile based on the availability of the chargers in a CS, considering also the preferences of the driver and the operator of the CS. Using the GA, the optimisation criteria are decoupled from the search in the solution space. As a result, the optimisation criteria can be modified according to the specific requirements of each application. The communications between the EV and the CS are facilitated by OPC-Unified Architecture (OPC-UA) [13].

Overall Architecture of the Framework

In this paper, a framework is introduced to facilitate managed charging of EVs enabled by Vehicle-to-Grid (V2G) communications. This work completes the one presented in [13] by focusing on the charge scheduling aspect. Through the OPC-UA-enabled communications, the EV requests the charging capabilities of the nearby charging stations to select the most appropriate, based on the driver's requirements for electric energy [5]. These requirements are concluded, based on the information that are generated by the battery management system (BMS) and the transportation needs of the driver, along with the priorities set by the latter. The decision-making procedure for the charging of the EVs is twofold. On the side of the charging station, a charge scheduling for the EVs that request charging services is performed, aiming to satisfy the priorities set by the EV owners and the station operator. On the side of the EV, the available charging stations should be ranked based on the quality of the services that they provide [13]. This approach aims to decouple the criteria that are considered by the EV and its driver, from the criteria that are considered by the charging station operator.

The overall architecture of the proposed framework is presented in Fig. 1. In the first step, the EV communicates the charging requirements to the Charging Station. This information refers to the charging energy in order to satisfy the transportation needs of the driver. In addition, the nominal battery voltage and current, along with the maximum battery current are taken into consideration during the decision-making, in order to preserve the health of the battery in high levels. Furthermore, the CS is informed about the preferred completion time of the charging task, based on the schedule of the driver. Finally, the charging priorities of the driver in form of weights are also communicated to the CS, to facilitate the decision-making. When the CS receives the requirements from the EV, its resource management system executes a Genetic Algorithm (GA) and offers a charging solution. The proposed algorithm operates in discrete timeslots instead of operating in continuous time. If the EV accepts the offer, the resources are reserved, and they are considered occupied.

The requirements for the operation of the proposed framework are as follows. Each power socket has to be connected to many DC chargers via power relays, to enable the allocation of different chargers to a charging task (i.e. EV), towards finding the optimal sequence. In small charging stations, the low complexity allows for all chargers to be connected to all power sockets. In the occasion of large charging stations that also offer parking services, the connection of all chargers to all parking slots may require the use of thousand switches and increase the complexity. In this case, the chargers can be grouped into clusters and the CS searches for the optimal solution in all clusters in parallel. In addition, both the EV and the CS should communicate using the same standard (in this paper OPC-UA is selected) to exchange the required information.

The benefits of the proposed approach are the resource sharing, the fast identification of a near-optimal solution, the preservation of the battery health, and the inclusion of user-defined priorities. Furthermore, the use of a global optimisation method, such as the GA, enables the decoupling of the optimisation objectives from the search mechanism of the solution space. As a result, the optimisation objectives can be modified based on the specific requirements that may emerge.

Mathematical Formulation of the Problem

In this section, the mathematical formulation of the decision-making objectives is presented. These objectives are presented by their corresponding non-negative criteria as follows. The charge completion time criterion (Eq. 1) is introduced to ensure the timely completion of the charging task based on the schedule of the driver. It is not a constraint, as the algorithm allows for small delays of the charging task if this can conclude to an overall better solution. If the charging task is finished before the time specified by the driver, this criterion has its maximum score. If a delay occurs, the value of the criterion is decreased as a linear function of the delay time. The maximum delay of the task is given with the variable tolerance.

The price criterion refers to the total price of the charging task (Eq. 2). The calculated price is normalised in the range [0, 1] based on the minimum and the maximum price of the timeframe under evaluation. The price criterion has its maximum value if the EV charges with the lowest price constantly during the timeframe under evaluation. On the contrary, it takes zero value when the EV charges with the highest price constantly during the timeframe under evaluation.

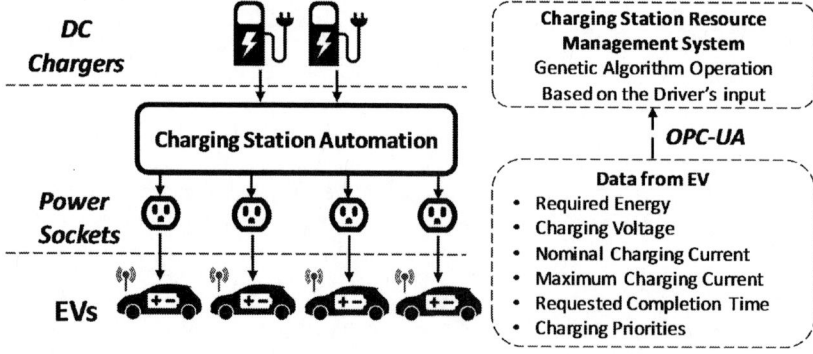

Fig. 1: Overall Architecture of the Framework

The third criterion refers to the utilisation of the resources of the CS. The utilisation of resources during a charging task is calculated by dividing the charging current with the nominal current of a charger during each charging timeslot (Eq. 3). Non-charging timeslots are not considered during the calculation of this criterion, as the proposed solution does not occupy resources during the non-charging timeslots.

The fourth and the fifth criteria are related to the preservation of the battery health, and they refer to the average charging current and the maximum charging current during the charging task. The average charging current is calculated during charging timeslots (non-charging timeslots are not considered) as presented in Eq. 4. Afterwards, the average current criterion (Eq. 5) takes its maximum value if this current is less or equal than the nominal battery charging current. If the average current exceeds the nominal battery current the criterion is linearly decreased (Eq. 5). This criterion takes its lowest value (zero value) if the average charging current exceeds a pre-defined tolerance. A similar procedure applies to the criterion that refers to the maximum charging current (Eq. 6). The fitness function (Eq. 7) is a weighted sum of the criteria that are presented in Eq. 1-6.

$$
complTime_{criterion} = \begin{cases} 1, & complTime \leq PreferredComplTime \\ 1 - \dfrac{complTime - PreferredComplTime}{Tolerance}, & complTime > PreferredComplTime \end{cases} \tag{1}
$$

$$
price_{criterion} = 1 - normalisedPrice \tag{2}
$$

$$
utilisation_{criterion} = \frac{\sum_{i=1}^{timeslots} \frac{I_{i,charging}}{I_{i,chargerRated}}}{timeslots} \tag{3}
$$

$$
Iavg = \frac{\sum_{i=1}^{timeslots} I_{i,charging}}{timeslots} \tag{4}
$$

$$
Iavg_{criterion} = \begin{cases} 1, & Iavg \leq NomCurrent \\ 1 - \dfrac{Iavg - NomCurrent}{MaxAvgCurrent - NomCurrent}, & Iavg > NomCurrent \end{cases} \tag{5}
$$

$$
Imax_{criterion} = \begin{cases} 1, & Imax \leq NomCurrent \\ 1 - \dfrac{Imax - NomCurrent}{MaxBatteryCurrent - NomCurrent}, & Imax > NomCurrent \end{cases} \tag{6}
$$

$$
f_{fitness} = (w_{complTime} \cdot complTime_{criterion}) + (w_{price} \cdot price_{criterion}) + (w_{util} \cdot utilisation_{criterion}) + (w_{Iavg} \cdot Iavg_{criterion}) + (w_{Imax} \cdot Imax_{criterion}) \tag{7}
$$

Where,

$complTime_{criterion}$: A value in the [0, 1] which corresponds to the charge completion time criterion.

$complTime$: The charge completion time in timeslots for a specific alternative.

$PreferredComplTime$: The preferred by the driver charge completion time in timeslots.

$Tolerance$: The acceptable tolerance, in timeslots, for the delay of the charging task.

$price_{criterion}$: A value in the [0, 1] which corresponds to the overall charging price criterion.

$normalisedPrice$: The charging price normalised in the [0, 1] interval. The 1 denotes charging constantly with the highest price, while 0 denotes charging constantly with the lowest price.

$utilisation_{criterion}$: A value in the [0, 1] which corresponds to the utilisation criterion.

$timeslots$: The charging timeslots. The timeslots when the EV is parked but not charged are not included in this value.

$I_{i,charging}$: The charging current during the ith timeslot.

$I_{i,chargerRated}$: The rated current of the selected charger during the ith timeslot.

$Iavg$: The average charging current of a charging task.

$Iavg_{criterion}$: A value in the [0 ,1] which corresponds to the average charging current criterion.

$NomCurrent$: The nominal battery charging current.

$MaxAvgCurrent$: The maximum tolerance of the average charging current of a task.

$Imax_{criterion}$: A value in the [0 ,1] which corresponds to the maximum charging current criterion.

$Imax$: The maximum current during a charging task.

$MaxBatteryCurrent$: The maximum battery current that is permitted by the battery manufacturer.

$w_{complTime}, w_{price}, w_{util}, w_{Iavg}, w_{Imax}$: The weight factors for each criterion.

Two-Layer Genetic Algorithm

The nature of the charge scheduling problem results into many alternatives that are not possible to be evaluated in a timely manner. In addition, the requirement for an acceptable near-optimal solution in a short time suggests the usage of heuristics towards the solution of the problem [14]. Considering that a multicriteria decision-making problem typically has multiple optima, the selected optimisation algorithm has to be able to escape local optima. As a final requirement, the inclusion of user-defined weights, along with the requirement for encapsulation of new criteria, besides the ones that are already considered, suggest the usage of a global optimisation method, where the operation of the optimisation is decoupled from the objective function. In this context, the Genetic Algorithm has been selected as the most eligible candidate for the optimisation of the charge scheduling of EVs in a CS.

The formulation of the problem divides the charging time in timeslots of five minutes duration each. During each timeslot, the type of the charger and the charging current have to be selected following the optimisation criteria. As a result, the formation of the alternatives follows the one that is depicted in Fig. 2. Towards the optimisation of the fitness function, the proposed algorithm evaluates charging the EV with various types of chargers. A main advantage of the approach that is demonstrated in this paper, is the inclusion of available resources only. As a result, the generated solutions are feasible.

As depicted in Fig. 2, for each timeslot, the algorithm has to select the charging current based on the availability of the chargers. Each charger in the CS is given an identification number. Assuming that the CS under evaluation has two types of chargers and more specifically three "Normal" chargers and two "Fast" chargers, the identification is performed as follows. Number 1 models the non-charging timeslot. This means that if the charger No. 1 is selected for a timeslot, the EV is not being charged in that specific timeslot. In addition, if one of the numbers 2, 3, 4 is selected, the EV charges using a normal charger. Moreover, if one of the numbers 5, 6 is selected the EV charges using a fast charger. Aiming to simplify the operation of the algorithm and reduce the number of alternatives, and hence the computational time, during each timeslot the algorithm identifies the availability of the chargers and samples the charging current from probabilistic distributions that include only the feasible charging currents.

As it can be observed from Fig. 2 each chromosome consists of two layers. The top layer refers to the selected charger, while the lower layer refers to the charging current. It is obvious that the two layers are not decoupled. As a result, the GA operations have to consider the relationship between the two layers. During the first step of the operation of the algorithm, three initial solutions are generated as starting points. The first initial solution refers to the constant charging with the maximum current of the normal charger. The second initial solution refers to the constant charging with the maximum current of the fast charger. The third initial solution refers to the constant charging with the current that is calculated using the preferences of the driver. The initial solutions support the convergence of the algorithm and ensure that a solution exists, even if the algorithm does not succeed finding a competitive one.

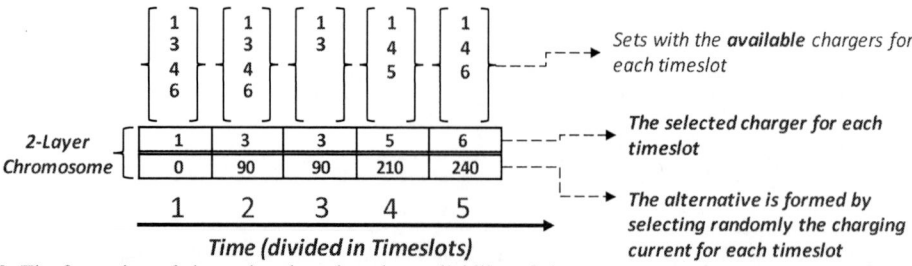

Fig. 2: The formation of alternatives based on the availability of chargers

After the generation of the initial solutions, the GA generates randomly solutions, based on the availability of the chargers, to complete the population until the first generation is created. The probabilistic distributions that are used for the creation of the population are Beta distributions (Eq. 8, Eq. 9) with parameters that provide samples based on the current that is supported by the available chargers. The (a, b) parameters for the Beta distributions modify the shape of the distribution according to the specific charging requirements. As a result, during the real-life implementation of the algorithm, the shapes of the distributions can be modified following the statistical information related to the interarrival of the charging tasks and the constraints of the Smart Grid. The current implementation of the algorithm supports the usage of two types of chargers, namely the "Normal" and the "Fast" charger. Therefore, three scenarios exist during the sampling of alternatives, i.e. only "Normal" chargers are available, only "Fast" chargers are available, and both "Normal" and "Fast" chargers are available. Following an experimental validation with different charging scenarios that consider a normal distribution on the energy requirement of the drivers, the selected Beta distributions are presented in Fig. 3. The uniform distribution (a=1 and b=1) is selected for the occasions of "Fast", and "Normal" and "Fast" charger availability (Fig. 3). A left-skewed beta distribution (a=7, b=6) is selected for the occasion of only "Normal" chargers (Fig. 4).

$$f(x|a,b) = \frac{1}{B(a,b)} x^{a-1}(1-x)^{b-1} I_{(0,1)}(x) \qquad (8)$$

$$B(a,b) = \int_0^1 t^{a-1}(1-t)^{b-1} dt \qquad (9)$$

Where,

$f(x|a,b)$: the probability density function of the beta distribution [15].

$B(a,b)$: the Beta function [16].

$I_{(0,1)}(x)$: the indicator function, which ensures that only values of x in the range (0,1) have nonzero probability [15].

Afterwards, the crossover operation is performed. If the offsprings have better fitness values than their parents, they are inserted in the population, otherwise they are discarded. The crossover operation is followed by the mutation operation. During the mutation operation, random genes are selected for mutation. The mutation operation generates a charger-current alternative for a specific timeslot based on the availability of the chargers. The mutation operation contributes towards improving the fitness function of a specific alternative. In the context of this paper, a new operation is introduced. This operation is called the "swap" operation and randomly swaps the position of two genes in the alternative. To perform this operation and ensure the feasibility of the result, the algorithm has to verify that the availability of chargers in those two timeslots. It is remarked that the merits of the swap operation are observable in the situations when the price constantly variates during short durations (during the 1-2 hours that are required for a typical normal charging task). The flowchart depicting the operations of the GA is presented in Fig. 5.

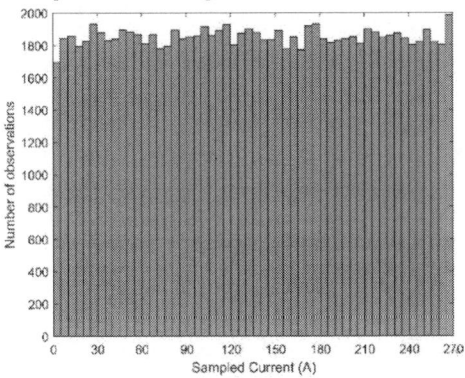

Fig. 3: Beta distribution for "Fast", and both "Normal" and "Fast" charger availability.

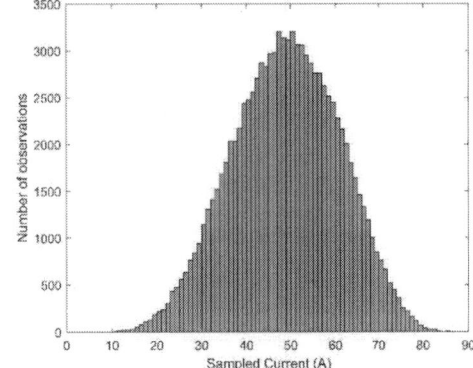

Fig. 4: Beta distribution for "Normal" charger availability.

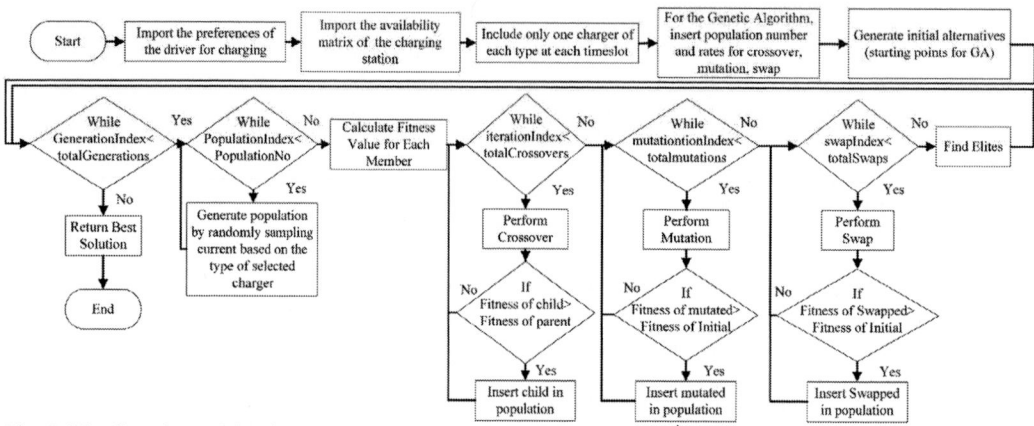

Fig. 5: The flowchart of the algorithm

OPC-Unified Architecture (OPC-UA) communications

This framework is enabled by the OPC-UA standard that has found great attention in the industrial communications [17] and is identified as a promising candidate for the smart grid communications [13], [18]. The application of the OPC-UA standard in this case study can be the basis for further developments and subsystem integrations on the subject under investigation. Additionally, the proposed framework enables the interoperability of different systems during runtime, without needing complicated software adaptors. The OPC-UA standard enables the inclusion of the subsystems of an EV into a common model. As a result, heterogenous systems can be integrated. The semantics that are included in the standards that are defined by different organisations can be incorporated during the creation of the model. Both the EV and the CSs are servers and clients for the OPC-UA communications. As servers, they expose their information model to the rest of the systems that exist in a smart grid. As clients, they consume the information that are contained in the information models of other systems.

The client can browse through the server to gather the required information and understand its content through metadata and semantic representation. Using this feature of the OPC-UA, systems of different vendors can be synchronised during runtime by exploiting semantic data. For example, in the case of the charging of the EV, the information models of the charging station should include the corresponding semantics in the description of the variables, confronting to acceptable standards [19]. As a result, the variable that reflects the energy price should have the tags "price", "cost", "tariff" etc in its description, following a common vocabulary without needing explicitly identical information models. Therefore, interoperability among the EVs and the charging stations with different information models will be achieved. The sequence diagram for the OPC-UA communications is presented in Fig. 6 and the information model of the charging station is depicted in Fig. 7 as a Unified Modelling Language (UML) following the appropriate transformation rules [17], [20].

The OPC-UA client and server on the side of the EV is implemented on a Raspberry Pi 3 microcomputer, while its counterpart referring to the charging station can be implemented either on dedicated or Cloud server. In the context of this paper, the OPC-UA server is hosted on the same server as the optimisation module. The server operates using the Intel i7 8700 (6 cores, 12 threads, 12 MB cache, base frequency 3.2 GHz, and maximum frequency 4.6 GHz). The Raspberry Pi 3 microcomputer communicates with the subsystems of the EV via CAN bus. The real-life implementation of the OPC-UA V2G communications require the EVs to be equipped cellular communication systems, a feature that is present in modern vehicles.

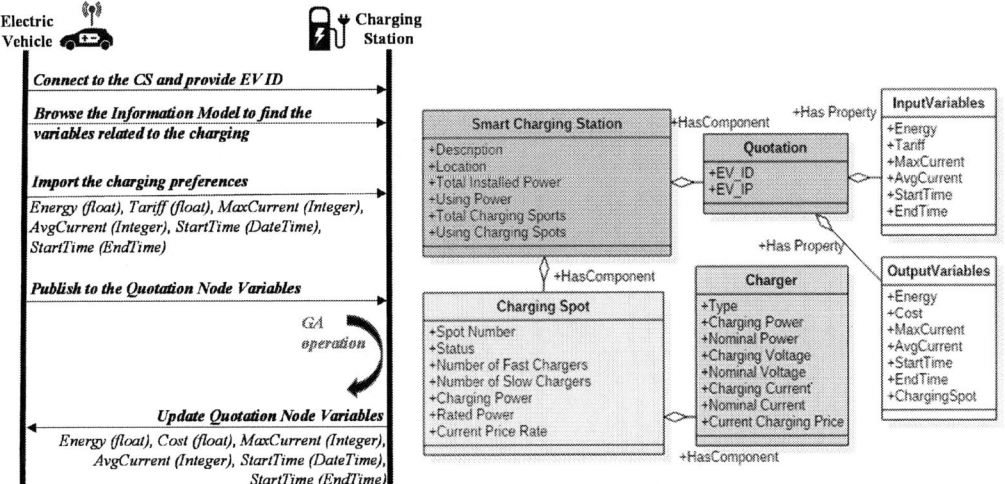

Fig. 6: Sequence diagram for the OPC-UA communications

Fig. 7: OPC-UA Information Model for the Charging Station

Case study and results

In this section of the paper, the operation of the proposed algorithm is evaluated, and the results are presented using a software implemented in MATLAB. The case study refers to a CS including three Normal and two Fast chargers of 90A and 300A maximum charging current, respectively. The availability of the chargers is created randomly for 20 timeslots. Each timeslot has a fixed value of 5 minutes. For the case study, various experiments have been performed with EVs for urban transportations [21] that have a nominal battery voltage of 86.4 V, a maximum current of 270 A, and a total capacity of 180 Ah. To perform experiments that correspond to 100 randomly generated charging tasks, the charging requirements are generated as follows. The charging tasks require an increase in the charge of the EV batteries that is sampled from a normal distribution in the interval [50, 160] Ah and the driver's preferred charge duration is sampled from a normal distribution in the interval [6, 15] timeslots (i.e. [30, 75] minutes). The weights for the criteria are fixed to 0.1 for the completion time, 0.2 for the price, 0.3 for the utilisation, 0.3 for the average charging current, and 0.1 for the maximum charging current.

Using these charging requirements, the GA operates in populations of 500 members for 100 generations. The crossover rate is 0.8 and the mutation rate is 0.4. The number of elites is 10. The selection of the population size and number of generations is made with the goal to finalise all the iterations of the algorithm within 10 seconds for a charging task of 1-hour duration (12 timeslots). The experiments are performed on a computer with the Intel i7 8700 (6 cores, 12 threads, 12 MB cache, base frequency 3.2 GHz, and maximum frequency 4.6 GHz). The crossover and the mutation parameters are set to high values to enhance the search of local optima. The selection of 10 elites enables sampling 490 alternatives from the pool of alternatives, to create the population during the beginning of each generation. Hence, the algorithm has increased chances to escape from trapping in local optima.

The results are presented in Fig. 8, and Fig. 9. In Fig. 8, the improvement of the fitness function at the end of each generation is presented for one experiment. In Fig. 9, the fitness function of the best solution for each one of the 100 experiments is depicted.

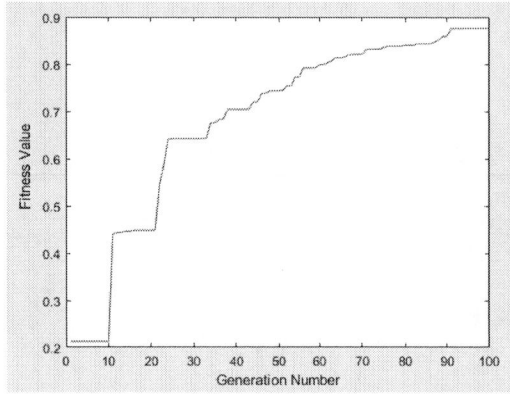 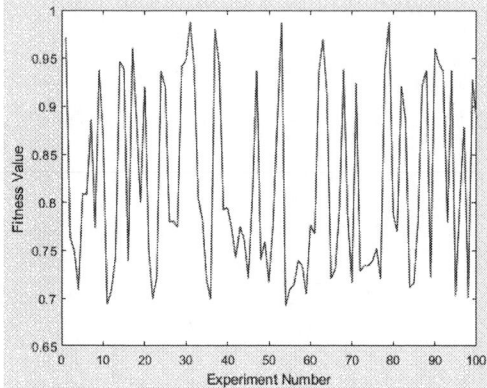

Fig. 8: The improvement of the fitness function over the generations

Fig. 9: The fitness function of the best solution for each one of the 100 experiments

In Fig. 9, it can be observed that the fitness function takes values between 70% and 98%. The low values of the fitness function correspond to charging tasks with conflicting requirements, i.e. simultaneous requirement for low battery current, high charging energy, and short completion time. Even in such scenarios, the proposed algorithm can identify near-optimal solutions with accuracy of 0.4 % - 4.63 % compared to the exhaustive search in the solution space. As aforementioned, the proposed algorithm operates in fixed timeslots of 5-minute duration each. The computing requirements for the operation of the GA for charging tasks that have a duration of 4 timeslots (20 minutes) to 25 timeslots (125 minutes) are presented in Table I.

Table I: Computing requirements for the GA operation

No of Timeslots	4	7	10	13	16	19	22	25
CPU Time (s)	2.90	4.26	5.95	7.54	9.04	10.79	12.46	12.94

The results of Table I indicate that the proposed algorithm is an eligible approach for the generation of charging schedules for EVs. The computing requirements of the method can be reduced by implementing the algorithm in C++, rather than in MATLAB that is currently developed for the evaluation of the proof-of-concept.

Conclusion

In this paper, a Two-Layer Genetic Algorithm for the charge scheduling of EVs is introduced. The proposed scheduling mechanism optimises the charging of each vehicle separately in order to provide personalised solutions. Meta-Heuristic optimisation is used to decouple the decision-making objectives from the solution searching mechanism. As a result, the decision-making objectives can be modified based of the specific requirements of each application. Moreover, this criteria-based decision-making allows the introduction of the preferences of the users via dynamically selected weight factors. In the CS, the method allows the usage of normal and fast chargers and optimises the utilisation of the resources via a resource-sharing mechanism. The algorithm establishes near-optimal solutions without being time consuming. This advantage supports the application of the proposed framework in real-life scenarios, when the driver requires timely responses from the charging stations. Another benefit of the proposed framework is the usage of OPC-UA to facilitate the communications between the CS and the EV. The OPC-UA enables the integration of heterogeneous systems using semantic representation of information.

During the design of the proposed framework, the applicability in real-life scenarios was of high priority. As a result, the decision-making procedure within the framework is intended to consider the specific preferences of the drivers and the CS operators. The criteria for the decision-making and their priorities can be modified during the operation of the algorithm to satisfy the requirements that may emerge.

References

[1] J. C. Mukherjee and A. Gupta, "A Review of Charge Scheduling of Electric Vehicles in Smart Grid," in *IEEE Systems Journal*, vol. 9, no. 4, pp. 1541-1553, Dec. 2015.

[2] X. Chen, K. Leung, A. Y. S. Lam and D. J. Hill, "Online Scheduling for Hierarchical Vehicle-to-Grid System: Design, Formulation, and Algorithm," in *IEEE Trans. on Vehicular Technology*, vol. 68, no. 2, pp. 1302-1317, Feb. 2019.

[3] H. Wu, G. K. H. Pang, K. L. Choy and H. Y. Lam, "An Optimization Model for Electric Vehicle Battery Charging at a Battery Swapping Station," in *IEEE Trans. on Vehicular Technology*, vol. 67, no. 2, pp. 881-895, Feb. 2018.

[4] O. Sassi and A. Oulamara, "Electric vehicle scheduling and optimal charging problem: complexity, exact and heuristic approaches," in *Int. Journal of Production Research*, vol. 55, no. 2, pp. 519–535, Jun. 2016.

[5] T. Mao, W. Lau, C. Shum, H. S. Chung, K. Tsang and N. C. Tse, "A Regulation Policy of EV Discharging Price for Demand Scheduling," *in IEEE Trans. on Power Systems*, vol. 33, no. 2, pp. 1275-1288, March 2018.

[6] H. Zhang, Z. Hu, Z. Xu and Y. Song, "Optimal Planning of PEV Charging Station With Single Output Multiple Cables Charging Spots," in *IEEE Trans. on Smart Grid*, vol. 8, no. 5, pp. 2119-2128, Sept. 2017.

[7] N. Tucker and M. Alizadeh, "An Online Admission Control Mechanism for Electric Vehicles at Public Parking Infrastructures," in *IEEE Trans. on Smart Grid*, vol. 11, no. 1, pp. 161-170, Jan. 2020.

[8] S. Wang, S. Bi, Y. A. Zhang and J. Huang, "Electrical Vehicle Charging Station Profit Maximization: Admission, Pricing, and Online Scheduling," in *IEEE Trans. on Sustainable Energy*, vol. 9, no. 4, pp. 1722-1731, Oct. 2018.

[9] Y. Wang, Z. Su, Q. Xu, T. Yang and N. Zhang, "A Novel Charging Scheme for Electric Vehicles With Smart Communities in Vehicular Networks," in *IEEE Trans. on Vehicular Technology*, vol. 68, no. 9, pp. 8487-8501, Sept. 2019.

[10] Y. Cao, T. Wang, O. Kaiwartya, G. Min, N. Ahmad and A. H. Abdullah, "An EV Charging Management System Concerning Drivers' Trip Duration and Mobility Uncertainty," in *IEEE Trans. on Systems, Man, and Cybernetics: Systems*, vol. 48, no. 4, pp. 596-607, April 2018.

[11] M. E. Kabir, C. Assi, M. H. K. Tushar and J. Yan, "Optimal Scheduling of EV Charging at a Solar Power-Based Charging Station," in *IEEE Systems Journal*, pp. 1–11, 2020.

[12] R. F. Atallah, C. M. Assi, W. Fawaz, M. H. K. Tushar and M. J. Khabbaz, "Optimal Supercharge Scheduling of Electric Vehicles: Centralized Versus Decentralized Methods," in *IEEE Trans. on Vehicular Technology*, vol. 67, no. 9, pp. 7896-7909, Sept. 2018.

[13] N. T. Milas and E. C. Tatakis, "Charging Station Selection through the Analytic Hierarchy Process enabled by OPC-UA for Vehicle-to-Grid Communications," 2019 *21st European Conference on Power Electronics and Applications (EPE'19 ECCE Europe)*, Genova, 2019, pp. 1-10.

[14] D. Mourtzis, M. Doukas, and F. Psarommatis, "Design of manufacturing networks for mass customisation using an intelligent search method," *International Journal of Computer Integrated Manufacturing*, vol. 28, no. 7, pp. 679–700, Mar. 2014.

[15] MATLAB documentation, "Beta Distribution," [Online]. https://uk.mathworks.com/help/stats/beta-distribution.html Accessed 12-06-2020.

[16] MATLAB documentation, "beta," [Online]. https://uk.mathworks.com/help/matlab/ref/beta.html Accessed 12-06-2020.

[17] D. Mourtzis, N. Milas, and N. Athinaios, "Towards Machine Shop 4.0: A General Machine Model for CNC machine-tools through OPC-UA," in Procedia CIRP, vol. 78, pp. 301–306, 2018.

[18] S. Rohjans, D. Fensel and A. Fensel, "OPC UA goes semantics: Integrated communications in smart grids," *ETFA2011*, Toulouse, 2011, pp. 1-4.

[19] S. Sučić, S. Rohjans and W. Mahnke, "Semantic smart grid services: Enabling a standards-compliant Internet of energy platform with IEC 61850 and OPC UA," *Eurocon 2013*, Zagreb, 2013, pp. 1375-1382.

[20] B. Lee, D.-K. Kim, H. Yang, and S. Oh, "Model transformation between OPC UA and UML," in *Computer Standards & Interfaces*, vol. 50, pp. 236–250, Feb. 2017.

[21] N. T. Milas, E. C. Tatakis and E. D. Mitronikas, "Investigation of the operation of an electric city car equipped with electronic differential using CAN-enabled monitoring," *2017 Panhellenic Conference on Electronics and Telecommunications (PACET)*, Xanthi, 2017, pp. 1-4.

Six-Phase PMSM Drive Inverter Testing on a High Performance Power Hardware-in-the-Loop Testbed

Yasser Rahmoun, Patrick Winzer, Alexander Schmitt, Horst Hammerer
AVL SET GmbH
Franz-Josef-Spiegler-Str. 5
D- 88239 Wangen/Allgäu
Phone: +49 (0) 7522 91609-0
Fax: +49 (0) 7522 91609-299
Email: Yasser.Rahmoun@avl.com
URL: https://www.avl-set.com/

Keywords

≪Power Hardware-in-the-Loop≫, ≪Six-phase PMSM≫, ≪FEA≫, ≪Inverter testing≫, ≪Electrical machine emulator≫

Abstract

This paper presents for the first time a six-phase Power Hardware-In-the-Loop (PHIL) testbench, featuring electromagnetic effects such as magnetic coupling and spatial harmonics. The proposed PHIL testbed can be used for testing six-phase Permanent Magnet Synchronous Machine (PMSM) drive inverters for electric vehicles under realistic operation conditions. This paper describes the developed six-phase PMSM model and its parameterization, as well as the structure of the PHIL testbed. Finally, measurements and simulation results are compared and illustrated.

Introduction

Modern drive inverters of electric vehicles are becoming increasingly more complex; however, the development time and costs must be reduced. This conflict of objectives can only be solved using new development methods combined with new dedicated development tools for each component of the electric vehicle. This allows the parallel development of the components and consequently accelerates the complete development process of the vehicle [1].

Drive inverter testing on conventional testbeds (see Fig. 1) has a multitude of limitations, and therefore new tools are required. One of these limitations is that the tests can be only conducted when the real motor is already available, which is usually not the case as the motor must be developed simultaneously with the inverter. Moreover, fabrication tolerances of the electric motor lead to varying behavior of the Unit Under Test (UUT) on different motor testbeds at identical operation conditions, which make the tests on these testbeds not 100% reproducible.

To overcome the afore-mentioned limitations and to cope with the new test demands of the e-mobility industry, Power Hardware-in-the-Loop (PHIL) testbeds (see Fig. 2) are often employed. With these kind of testbeds drive inverter testing can start at a very early development stage even if the real motor is still not available. The reason behind this is that PHIL testbeds are capable of emulating electric motors based on their parameters, which can be either measured or calculated by means of Finite Element Analysis (FEA). Furthermore, PHIL tests are 100% reproducible, since firstly, the onetime measured or calculated parameters of an e-motor can be used to parametrize multiple testbenches and secondly, the tolerances of such testbeds are only determined by the accuracy of the used data acquisition. Therefore, the PHIL testbed solves some critical issues by the testing of drive inverters and dramatically accelerates their

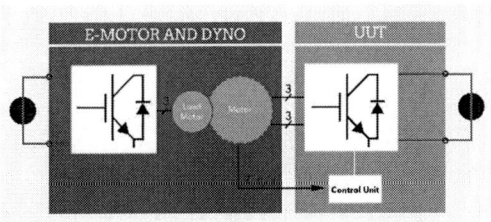

Fig. 1: Conventional testbed for six-phase inverters.
T is the torque, n is the mechanical rotor speed and γ is the mechanical rotor position.

development process. In general, a PHIL testbed comprises an Electrical Machine Emulator (EME) and two Power Supply Units (PSU) for both EME and UUT (see Fig. 2). The EME, which is the dominant part of the testbed, consists of the following three main components [2]:

1. A real-time processing system, which calculates the physical behavior of the emulated machine in real time based on its electrical and mechanical models.
2. A high switching frequency emulation converter, which imitates the electrical behavior of the machine in real time.
3. An ohmic inductive coupling network, which ensures a control path between the emulation converter and inverter.

The EME is supplied by a galvanically isolated DC PSU as the real machine windings are also galvanically isolated [3] [4].

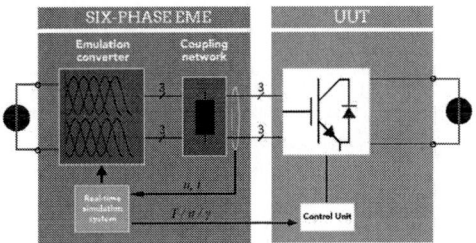

Fig. 2: PHIL testbed for six-phase inverters.
u is the measured line voltages and i is the measured phase currents.

Since the benefits of PHIL are not limited to a specific machine type, it can be applied to six-phase Permanent Magnet Synchronous Machines (PMSM). This machine type is gaining more and more attention in the field of e-mobility. This is due to firstly, the redundancy in its design, which makes it more reliable than the three-phase PMSM as it has two galvanically isolated stator winding systems and can continue to deliver power even if one of its stator winding systems fails. This is an important safety feature for autonomous driving vehicles, where the autonomous driving system can still get the vehicle off the road in the case of such faults [5]. Secondly, the six-phase PMSM has minor flux linkage harmonics in comparison with the three-phase PMSM and therefore, it is quieter, has better efficiency and has better torque quality [6]. Finally, the inverters of six-phase PMSM drives can be designed for either half the voltage or half the current at a given motor output power as compared to three-phase PMSM drive inverters, thereby potentially achieving cost or performance advantages [6].

On the other hand, the six-phase PMSM is more complex and needs a more sophisticated control scheme, because of the coupling between its stator winding systems. To overcome these drawbacks and to cover the test demand of six-phase traction inverters a high performance six-phase PHIL testbed was developed.

The goal of this contribution is to demonstrate the ability of the developed six-phase PHIL testbench to correctly emulate the highly sophisticated electromagnetic characteristics of a six-phase PMSM and finally to be used to test and evaluate the control scheme of a six-phase drive inverter. In the following section a theoretical six-phase PMSM electrical model is obtained. After that, the six-phase PMSM design is discussed and FEA data is derived. The subsequent section explains the developed six-phase PHIL testbed structure. Finally, real measurements of the emulated six-phase PMSM driven by a test inverter are presented and compared to simulation results.

Theoretical Description of the Emulated Six-Phase PMSM

Modeling of the Six-Phase PMSM in dq Domain

The six-phase PMSM is modeled in dq domain to simplify and decrease the calculation efforts of the model. It consists of two identical, galvanically isolated but magnetically coupled three phase systems. The modeled six-phase PMSM has a phase displacement $\alpha_{s1,s2}$ between its two stator winding systems and the star points of the two systems are not connected. Fig. 3 and Fig. 4 represent the six-phase PMSM equivalent circuits in UVW and dq domains respectively.

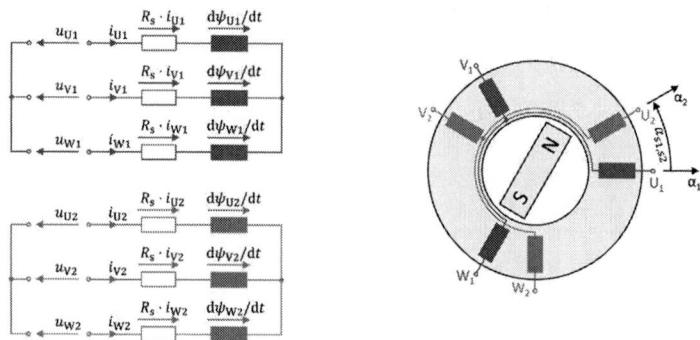

Fig. 3: Six-phase PMSM representation in UVW domain

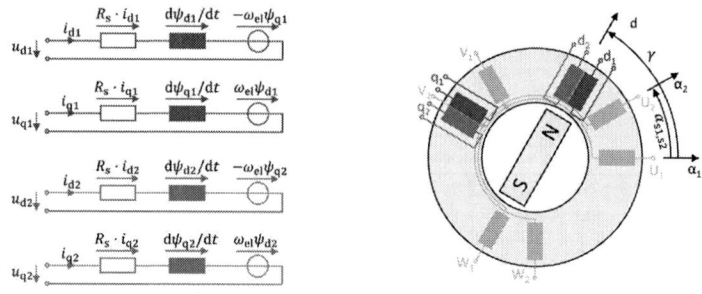

Fig. 4: Six-phase PMSM representation in dq domain

The voltage equations of the six-phase PMSM in UVW domain are given by Equation (1) and Equation (2), where R_s is the stator phase resistance, ψ is the flux linkage, u is the voltage, i is the current and t is the time.

$$u_{U1} = R_s \cdot i_{U1} + \frac{d\psi_{U1}}{dt}$$
$$u_{V1} = R_s \cdot i_{V1} + \frac{d\psi_{V1}}{dt} \qquad (1)$$
$$u_{W1} = R_s \cdot i_{W1} + \frac{d\psi_{W1}}{dt}$$

$$u_{U2} = R_s \cdot i_{U2} + \frac{d\psi_{U2}}{dt}$$
$$u_{V2} = R_s \cdot i_{V2} + \frac{d\psi_{V2}}{dt} \qquad (2)$$
$$u_{W2} = R_s \cdot i_{W2} + \frac{d\psi_{W2}}{dt}$$

By transforming the previously mentioned Equation (1) and Equation (2) in dq domain we get Equation (3).

$$u_{d1} = R_s \cdot i_{d1} + \frac{d\boldsymbol{\psi_{d1}}}{dt} - \omega_{el} \cdot \boldsymbol{\psi_{q1}}$$

$$u_{q1} = R_s \cdot i_{q1} + \frac{d\boldsymbol{\psi_{q1}}}{dt} + \omega_{el} \cdot \boldsymbol{\psi_{d1}}$$

$$u_{d2} = R_s \cdot i_{d2} + \frac{d\boldsymbol{\psi_{d2}}}{dt} - \omega_{el} \cdot \boldsymbol{\psi_{q2}} \tag{3}$$

$$u_{q2} = R_s \cdot i_{q2} + \frac{d\boldsymbol{\psi_{q2}}}{dt} + \omega_{el} \cdot \boldsymbol{\psi_{d2}}$$

In this paper, bold print is used to highlight parameters which are dependent on the current components $i_{d1}, i_{q1}, i_{d2}, i_{q2}$ and the electrical rotor position γ_{el}. Equation (4) shows an example for $\boldsymbol{\psi_{d1}}$.

$$\boldsymbol{\psi_{d1}} = \psi_{d1}(i_{d1}, i_{q1}, i_{d2}, i_{q2}, \gamma_{el}) \tag{4}$$

In general, the derivatives of flux linkages noted in bold in Equation (3) can be written as total differentials. Equation (5) shows exemplarily the total differential of $\boldsymbol{\psi_{d1}}$.

$$\frac{d\boldsymbol{\psi_{d1}}}{dt} = \frac{\partial\boldsymbol{\psi_{d1}}}{\partial i_{d1}} \cdot \frac{di_{d1}}{dt} + \frac{\partial\boldsymbol{\psi_{d1}}}{\partial i_{q1}} \cdot \frac{di_{q1}}{dt} + \frac{\partial\boldsymbol{\psi_{d1}}}{\partial i_{d2}} \cdot \frac{di_{d2}}{dt} + \frac{\partial\boldsymbol{\psi_{d1}}}{\partial i_{q2}} \cdot \frac{di_{q2}}{dt} + \frac{\partial\boldsymbol{\psi_{d1}}}{\partial \gamma_{el}} \cdot \frac{d\gamma_{el}}{dt}$$
$$\tag{5}$$

The partial derivatives of $\boldsymbol{\psi_{d1}}$ can be written as follows:

$$\frac{\partial\boldsymbol{\psi_{d1}}}{\partial i_{d1}} = \boldsymbol{L_{d1,d1}}, \frac{\partial\boldsymbol{\psi_{d1}}}{\partial i_{q1}} = \boldsymbol{L_{d1,q1}}, \frac{\partial\boldsymbol{\psi_{d1}}}{\partial i_{d2}} = \boldsymbol{L_{d1,d2}}, \frac{\partial\boldsymbol{\psi_{d1}}}{\partial i_{q2}} = \boldsymbol{L_{d1,q2}}, \frac{\partial\boldsymbol{\psi_{d1}}}{\partial \gamma_{el}} = \boldsymbol{\Lambda_{d1,el}}$$
$$\tag{6}$$

Where:

$\boldsymbol{L_{d1,d1}}$ is the differential self inductance of d1 axis.

$\boldsymbol{L_{d1,q1}}$ is the differential mutual inductance between d1- and q1-axes.

$\boldsymbol{L_{d1,d2}}$ is the differential mutual inductance between d1- and d2-axes.

$\boldsymbol{L_{d1,d2}}$ is the differential mutual inductance between d1- and q2-axes.

$\boldsymbol{\Lambda_{d1,el}}$ is the differential electric angular dependency of the flux linkage in d1-axis.

In a similar manner it is possible to write the six-phase PMSM voltage equations given in Equation (3) as presented in Equation (7).

$$u_{d1} = R_s \cdot i_{d1} + \boldsymbol{L_{d1,d1}} \cdot \frac{di_{d1}}{dt} + \boldsymbol{L_{d1,q1}} \cdot \frac{di_{q1}}{dt} + \boldsymbol{L_{d1,d2}} \cdot \frac{di_{d2}}{dt} + \boldsymbol{L_{d1,q2}} \cdot \frac{di_{q2}}{dt} + \omega_{el}(\boldsymbol{\Lambda_{d1,el}} - \boldsymbol{\psi_{q1}})$$

$$u_{q1} = R_s \cdot i_{q1} + \boldsymbol{L_{q1,d1}} \cdot \frac{di_{d1}}{dt} + \boldsymbol{L_{q1,q1}} \cdot \frac{di_{q1}}{dt} + \boldsymbol{L_{q1,d2}} \cdot \frac{di_{d2}}{dt} + \boldsymbol{L_{q1,q2}} \cdot \frac{di_{q2}}{dt} + \omega_{el}(\boldsymbol{\Lambda_{el,q1}} + \boldsymbol{\psi_{d1}})$$

$$u_{d2} = R_s \cdot i_{d2} + \boldsymbol{L_{d2,d1}} \cdot \frac{di_{d1}}{dt} + \boldsymbol{L_{d2,q1}} \cdot \frac{di_{q1}}{dt} + \boldsymbol{L_{d2,d2}} \cdot \frac{di_{d2}}{dt} + \boldsymbol{L_{d2,q2}} \cdot \frac{di_{q2}}{dt} + \omega_{el}(\boldsymbol{\Lambda_{el,d2}} - \boldsymbol{\psi_{q2}})$$

$$u_{q2} = R_s \cdot i_{q2} + \boldsymbol{L_{q2,d1}} \cdot \frac{di_{d1}}{dt} + \boldsymbol{L_{q2,q1}} \cdot \frac{di_{q1}}{dt} + \boldsymbol{L_{q2,d2}} \cdot \frac{di_{d2}}{dt} + \boldsymbol{L_{q2,q2}} \cdot \frac{di_{q2}}{dt} + \omega_{el}(\boldsymbol{\Lambda_{el,q2}} + \boldsymbol{\psi_{d2}})$$
$$\tag{7}$$

Six-Phase PMSM Model Depths

Using Equation (7), three model depths are considered by the parameterization of the six-phase PMSM model, which are:
- Linear model
- Nonlinear model
- Nonlinear model, with spatial harmonics

Linear Model

The linear model considers just the magnetic coupling of stator winding systems (d1 $\leftrightarrow$ d2 and q1 $\leftrightarrow$ q2) and the potentially anisotropic geometry of a six-phase PMSM. In this model just the constant quantities for R_s, $L_{d1,d1}$, $L_{d1,d2}$, $L_{q1,q1}$, $L_{q1,q2}$ and ψ_{PM} are used to parameterize the model. By making use of the symmetry of the stator winding systems the quantities mentioned in Equation (8) can be also obtained.

$$L_{d2,d2} = L_{d1,d1}, \; L_{d2,d1} = L_{d1,d2}$$
$$L_{q2,q2} = L_{q1,q1}, \; L_{q2,q1} = L_{q1,q2} \tag{8}$$

The flux linkages of the six-phase PMSM in this model are given by Equations (9).

$$\psi_{d1} = \psi_{PM} + L_{d1,d1} \cdot i_{d1} + L_{d1,d2} \cdot i_{d2}$$
$$\psi_{q1} = L_{q1,q1} \cdot i_{q1} + L_{q1,q2} \cdot i_{q2}$$
$$\psi_{d2} = \psi_{PM} + L_{d2,d1} \cdot i_{d1} + L_{d2,d2} \cdot i_{d2}$$
$$\psi_{q2} = L_{q2,q1} \cdot i_{q1} + L_{q2,q2} \cdot i_{q2} \tag{9}$$

Nonlinear Model

The nonlinear model considers the nonlinear magnetic characteristic of iron (saturation) and the magnetic cross-coupling of d- and q-axes in addition to the linear model properties. In this model two 4-dimensional Lookup Tables (LUT) for each ψ_{d1} and ψ_{q1} (see Equation (10)) are used to parameterize the model, where ψ_{d2} and ψ_{q2} can be obtained from ψ_{d1} and ψ_{q1} basing on the symmetry of the stator winding systems.

$$\psi_{d1} = \psi_{d1}(i_{d1}, i_{q1}, i_{d2}, i_{q2})$$
$$\psi_{q1} = \psi_{q1}(i_{d1}, i_{q1}, i_{d2}, i_{q2}) \tag{10}$$

Nonlinear Model with Spatial Harmonics

The nonlinear model with spatial harmonics considers, in addition to the properties of linear and nonlinear models, the spatial harmonics effects. For the parameterization of this model two 5-dimensional LUTs for each ψ_{d1} and ψ_{q1} (see Equation (11)) are used, where ψ_{d2} and ψ_{q2} similarly to the nonlinear model can be obtained from ψ_{d1} and ψ_{q1} basing on the symmetry of the stator winding systems.

$$\psi_{d1} = \psi_{d1}(i_{d1}, i_{q1}, i_{d2}, i_{q2}, \gamma_{el})$$
$$\psi_{q1} = \psi_{q1}(i_{d1}, i_{q1}, i_{d2}, i_{q2}, \gamma_{el}) \tag{11}$$

Six-Phase PMSM Example

The used six-phase PMSM has been calculated by means of FEA. Its geometry has been derived from a three-phase PMSM geometry. The machine has buried Neodymium magnets in a U-shape, which yield anisotropic behavior. The iron saturation has been considered by the FEA. The machine's windings are distributed, in such way that there is one coil per phase and pole pair. Fig. 5a illustrates the FEA model

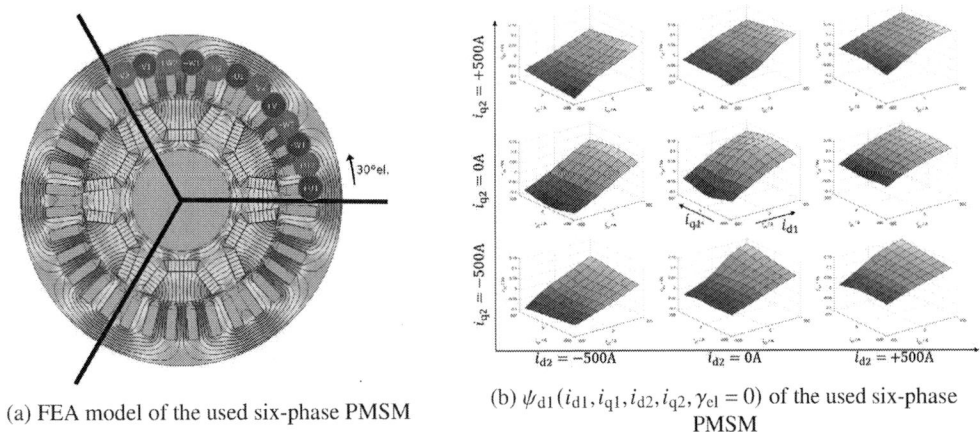

(a) FEA model of the used six-phase PMSM

(b) $\psi_{d1}(i_{d1}, i_{q1}, i_{d2}, i_{q2}, \gamma_{el} = 0)$ of the used six-phase PMSM

Fig. 5: Used six-phase PMSM example

of the analyzed six-phase PMSM. The six-phase PMSM has a number of pole pairs $p = 3$, a stator phase resistance $R_s = 12m\Omega$ and a phase displacement $\alpha_{s1,s2} = 30°$.

The resulting flux linkage maps are presented in Fig. 5b, where it can be seen that the analyzed machine exhibits various electromagnetic effects such as saturation, d/q-cross coupling, inter-stator coupling and inter-stator d/q-cross coupling.

Fig. 6a shows the flux linkage in d1 axis as function to i_{d1}, i_{q1} and γ_{el}, whereas $i_{d2} = 0$ and $i_{q2} = 0$. Fig. 6b and Fig. 6c show the partial differential of ψ_{d1} with respect to i_{d1} and i_{q1} respectively. The flux linkage of the machine has an extra dependency on rotor position. Fig. 6d shows the dependency of ψ_{d1} on γel, which is the 5th dimension of the LUTs.

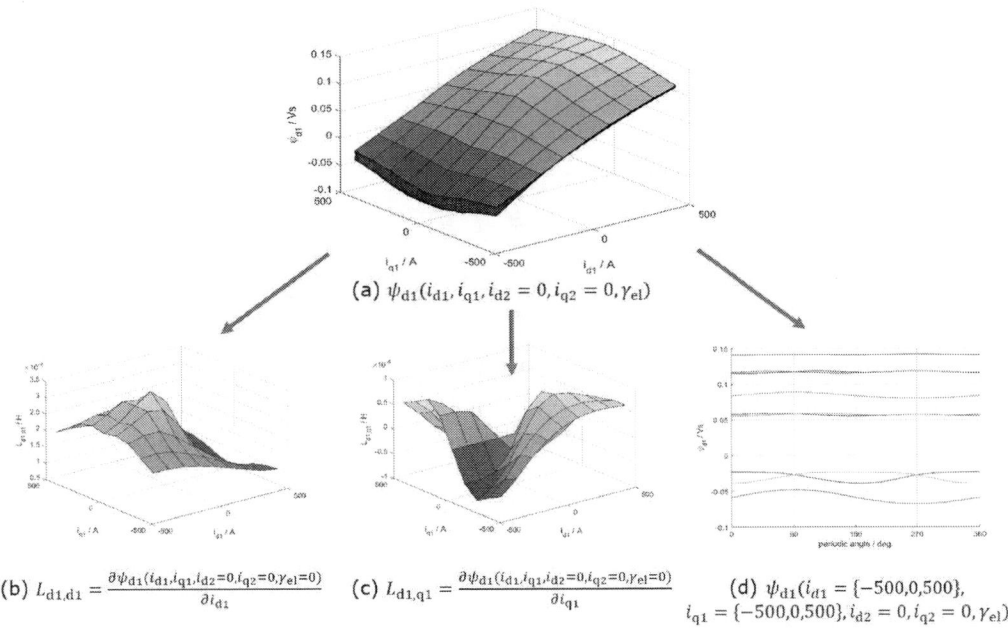

(a) $\psi_{d1}(i_{d1}, i_{q1}, i_{d2} = 0, i_{q2} = 0, \gamma_{el})$

(b) $L_{d1,d1} = \dfrac{\partial \psi_{d1}(i_{d1}, i_{q1}, i_{d2}=0, i_{q2}=0, \gamma_{el}=0)}{\partial i_{d1}}$

(c) $L_{d1,q1} = \dfrac{\partial \psi_{d1}(i_{d1}, i_{q1}, i_{d2}=0, i_{q2}=0, \gamma_{el}=0)}{\partial i_{q1}}$

(d) $\psi_{d1}(i_{d1} = \{-500, 0, 500\}, i_{q1} = \{-500, 0, 500\}, i_{d2} = 0, i_{q2} = 0, \gamma_{el})$

Fig. 6: ψ_{d1} dependencies on i_{d1}, i_{q1} and γ_{el}

Six-Phase PMSM Emulator

Based on the obtained six-phase PMSM electrical model in dq domain (see Equation (7)) a numerical fixed point six-phase PMSM model (see Fig. 7) is developed and implemented in a Field Programable Gate Array (FPGA) to minimize dead times and achieve the highest possible model calculation frequency of 3.125 MHz (with the used FPGA). Within each calculation interval the electromagnetic equations are evaluated and an interpolation of the magnetic parameters' LUTs of up to five dimensions is executed. Furthermore, the FPGA model (see Fig. 7) calculates both three-phase set voltages, which have to be applied on the coupling network of each stator system very precisely and with a minimal dead-time to ensure the correct emulation of the six-phase PMSM phase currents in both stator systems.

To ensure a precise and fast realization of the two three-phase set voltages calculated by the FPGA model, two three-phase high switching frequency multilevel inverters (one for each stator system) are employed and the multilevel inverters switching frequency is set to 800 kHz. The two multilevel inverters are connected to two ohmic inductive coupling networks (one for each stator system), which ensure a control path between the two multilevel inverters and the inverter under test.

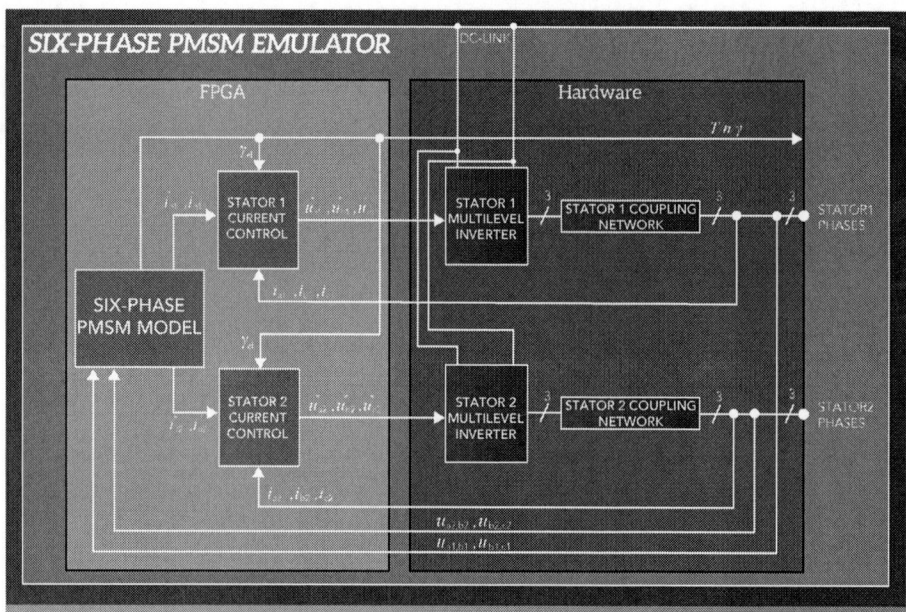

Fig. 7: Six-phase emulator functional diagram

Results

To prove the emulation quality of the developed six-phase PMSM testbench (see previous chapters) a series of tests was performed on the PHIL testbed (see Fig. 2) using the afore-mentioned parametrization for the nonlinear model with spatial harmonics. As a drive inverter, a six-phase PMSM drive inverter was utilized, which has a flux based predictive control scheme to drive the six-phase PMSM, derived from [7].

The results of the tests were then compared to simulation results of a MATLAB/Simulink® model, which represents the ideal case.

Back-EMF Test

In this test the emulated six-phase PMSM has no current flowing in its phases and the back EMF voltage amplitude is dependent on the machine permanent magnet flux and rotor speed. The voltage harmonics depend on how much the permanent magnet flux fluctuates as the machine rotates.

Fig. 8 shows the measured and simulated six-phase PMSM back-EMF line to line voltages of the first stator winding system at a rotor speed of 5000 RPM. The measured voltages are slightly filtered. Although the EME is predominantly a controlled current source (see Fig. 7), the back EMF matches the theoretical result well. The deviations between the measured and simulation results are due to the imperfection of the hardware.

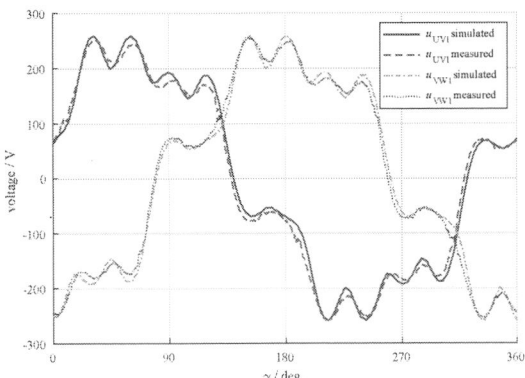

Fig. 8: Back-EMF voltage

Load Step Test

In order to present the capability of the emulator to precisely emulate the various electromagnetic characteristics of the six-phase PMSM in dynamic operation cases a dynamic current step in the q2 axis from 200 A to -200 A is performed on the PHIL testbed, while: $i_{q1\,set}$ =200 A, $i_{d1\,set}$ = $-200A$ and $i_{d2\,set}$ = $-200A$. The current step in the q2 axis is delivered at a rotor speed of 1000 RPM. Fig. 9 demonstrates the measured and simulated currents of the six-phase PMSM in dq domain during the above-mentioned current step.

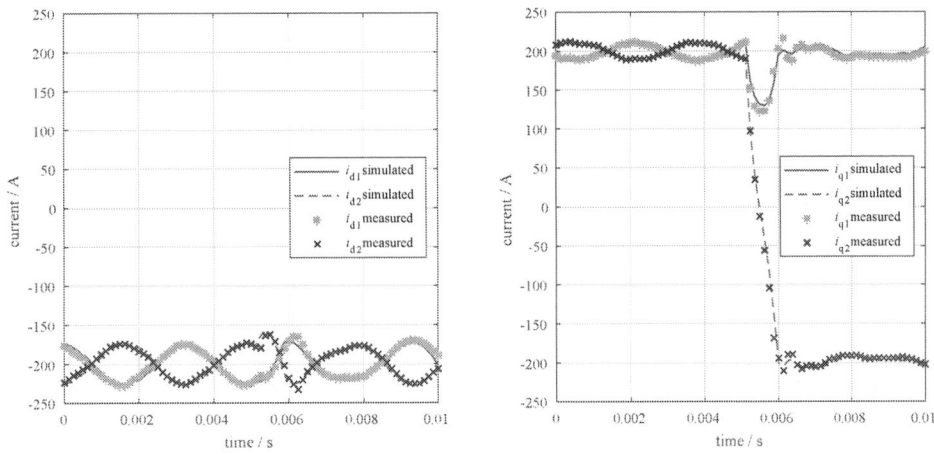

Fig. 9: Six-phase PMSM step response on q axis.

Note: Both measured and simulated results are sampled at 8 kHz sampling frequency, which is also the inverter switching frequency, hence it is not possible to see the current ripple of the inverter itself.

The results in Fig. 9 demonstrate that all magnetic effects of the six-phase PMSM, which are mentioned in the NLH model's properties, are covered by the developed six-phase PMSM testbed. The opposite phase of i_{d1} and i_{d2} respectively i_{q1} and i_{q2} is due to the 30° phase displacement between the two stator winding systems ($\alpha_{s1,s2}$), which leads to a 180° phase shift in the dominating 6th order harmonics of the flux linkages.

Active Short Circuit Test

In this test, while the emulated six-phase PMSM runs in steady state and exhibits $i_{q1} = i_{q2} = 100A$ and $i_{d1} = i_{d2} = -100A$ at a rotor speed of 500 RPM, the six-phase inverter goes into safe state and applies an Active Short Circuit (ASC) on the 3 phases of each stator winding system at the same time. Fig. 10 shows the response of the emulated six-phase PMSM in this test.

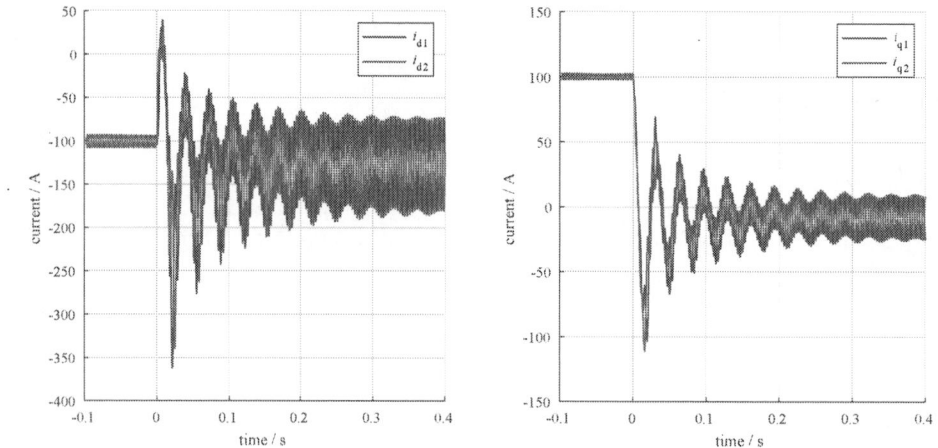

Fig. 10: Active short circuit test results (sampling rate 8 kHz)

In Fig. 11, the resulting currents in d- and q-axes are plotted against each other to show the course of the currents in both stator winding systems after the inverter applies the ASC. Fig. 11 shows that the inverter safe mode entering, at the previously mentioned operation point, could lead to phase overcurrents in the real machine, as the circle in the figure with the smaller radius represents its phase maximum current ($i_{\mathrm{max\ real\ machine}}$). On the other hand, it is very important that the used PHIL testbed provides a high enough EME phase maximum current ($i_{\mathrm{max\ EME}}$), represented by the circle with the greater radius, that it is able to emulate such high fault currents.

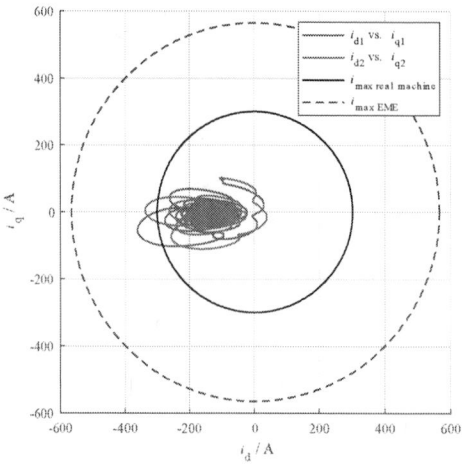

Fig. 11: Active short circuit test results in the $i_d - i_q$ plane (sampling rate 8 kHz)

Conclusion

In this contribution, a PHIL testbed for testing six-phase PMSM drive inverters has been proposed for the first time. The implemented PMSM model and its parameterization have been derived and discussed in detail, besides the structure of the proposed PHIL testbench.

The capability of the used PHIL testbed to emulate all relevant electromagnetic characteristics of the six-phase PMSM, along with the control scheme evaluation of a six-phase PMSM drive inverter have been then presented.

The test measurements have proved the high emulation quality of the developed PHIL testbench, which is due to the precise six-phase PMSM model and parameterization of the machine, the high calculation frequency of the FPGA-based machine model and the high dynamic emulation converter used to imitate the machine.

The measurements have also shown the ability of the used PHIL testbed to emulate fault scenarios, which permits the early identification of hardware limitations in the inverter and machine design.

With the proposed PHIL testbed, the development of inverters can be accelerated significantly, as many tests can be carried out even before the real motor exists using e.g. FEA data. Due to the automatization capability of the PHIL testbed, fast and reproducible test runs are possible.

References

[1] A. Schmitt, H. Hammerer, P. Winzer, and M. Schnarrenberger, "Drive inverter testing with virtual e-motors using power hardware-in-the-loop emulator concept," in AVL 8th International Symposium on Development Methodology, Nov. 2019.

[2] A. Schmitt, J. Richter, M. Braun, and M. Doppelbauer, "Power hardware-in-the-loop emulation of permanent magnet synchronous machines with nonlinear magnetics – concept & verification," in Proceedings of PCIM Europe, May 2016, pp. 393–400.

[3] A. Schmitt, M. Gommeringer, J. Kolb, and M. Braun, "A high current, high frequency modular multiphase multilevel converter for power hardware-in-the-loop emulation," in Proceedings of PCIM Europe, May 2014, pp. 1537–1544.

[4] A. Schmitt, M. Gommeringer, C. Rollbühler, P. Pomnitz, and M. Braun, "A novel modulation scheme for a modular multiphase multilevel converter in a power hardware-in-the-loop emulation system," in IECON - 41st Annual Conference of the IEEE Industrial Electronics Society, Nov. 2015, pp. 1276–1281.

[5] A. Kilic, J. Faßnacht, T. Shen, and C. Thulfaut, "Fehlertolerante Antriebsstränge für zukünftige Mobilität", MTZ-Motorteschnische Zeitschrift, Tech. Rep., 2019.

[6] E. Levi, R. Bojoi, F. Profumo, H. A. Toliyat and S. Williamson, "Multiphase induction motor drives - a technology status review," in IET Electric Power Applications, vol. 1, no. 4, pp. 489-516, July 2007, doi: 10.1049/iet-epa:20060342.

[7] P. Winzer, J. Richter and M. Doppelbauer, "Dynamic control of generalized electrically excited synchronous machines using predictive flux control," IECON 2016 - 42nd Annual Conference of the IEEE Industrial Electronics Society, Florence, 2016, pp. 2772-2777, doi: 10.1109/IECON.2016.7793666.

An Improved Bidirectional Hybrid Switched Inductor Converter

Dan Hulea[1], Mihaita Gireada[1], Danut Vitan[1], Octavian Cornea[1], Nicolae Muntean[1,2]
[1] Politehnica University of Timisoara   [2] Romanian Academy, Timisoara Branch
[1] Piata Victoriei No. 2   [2] Bv. Mihai Viteazu 24
Timisoara, Romania
Tel.: +40-256-403-450
E-Mail: dan.hulea@upt.ro; mihaita.gireada@student.upt.ro; daunt.vitan@upt.ro;
octavian.cornea@upt.ro; nicolae.muntean@upt.ro;
URL: http://www.et.upt.ro/en

Acknowledgements

This work was supported by a grant of the Romanian Ministery of Research and Innovation, CCCDI – UEFISCDI, project number PN-III-P1-1.2-PCCDI-2017- 0391 / CIA_CLIM – *"Smart buildings adaptable to the climate change effects"*, within PNCDI III.

This work was also supported by a grant of the Romanian Ministry of Research and Innovation no. 10PFE/16.10.2018, PERFORM-TECH-UPT - The increasing of the institutional performance of the Polytechnic University of Timişoara by strengthening the research, development and technological transfer capacity in the field of "Energy, Environment and Climate Change"

Keywords

«Converter circuit», «Converter control», «DC power supply», « Supercapacitor», « High voltage power converters».

Abstract

This paper proposes an improved high ratio bidirectional hybrid DC-DC converter that incorporates a switched inductor cell which helps achieve wider conversion ratios, reduced components size, and reduced semiconductor stress. The improvements for this topology, when compared to other bidirectional hybrid switched capacitor converters, consist of the elimination of inductor voltage oscillations, reduction of maximum voltage stress on transistors, minimization of parasitic switching inductances and elimination of high frequency common mode voltage. The topology is realized with two conventional half bridges, which help achieve a faster implementation, and provides the topology with the benefit of utilizing the research realized on these structures.

Introduction

High voltage conversion ratio converters are used where a large difference is present between the input and the output voltage levels. High ratio bidirectional converters are useful for storage elements with large voltage variations with respect to their stored energy, such as supercapacitors [1]. Hybrid converters make use of inductive or capacitive switching cells to achieve a wider voltage conversion ratio, lower active switch stress or smaller passive components [2], [3].

Compared to other topologies, the bidirectional hybrid switched inductor converter (BHSI) presented in [4] was proven to have good characteristics, as the ones mentioned above. A significant disadvantage was also found, present in the unidirectional topology as well [3]: large inductor voltage oscillations that translate in switch overvoltage. The scope of this paper is to propose a new, improved, BHSI configuration that overcomes this disadvantage, while introducing additional advantages.

Topology description

The new BHSI converter, presented in Fig. 1, adds an additional switch (S_{L2}) in order to make use of two half bridges, consisting of S_{H1} - S_{L1} and S_{H2} - S_{L2} transistor groups. The use of half bridges and two additional capacitors, C_1 and C_2, introduce multiple benefits, the most notable being (i) the elimination of switch voltage oscillations, (ii) minimization of switching loop parasitic inductances, (iii) simplification of PCB design, and (iv) the use of already available research on these building blocks.

The inductor voltage oscillations appear due to the parasitic capacitance of S_{L1} and S_{H2}, a phenomenon which was previously studied for the unidirectional boost hybrid switched inductor converter in [3]. As presented in the schematic, the improvements for the proposed BHSI are different than [3], because each additional capacitor is connected in parallel to one of the two half bridges, therefore it also minimizes the parasitic inductances within the switching loops (S_{H1} - S_{L1} - C_1 and S_{H2} - S_{L2} – C_2). The use of half bridges has a benefit in the PCB design because of the already available power modules, and the numerous studies based on these structures used as building blocks help for determining specific switching parameters [5]–[7].

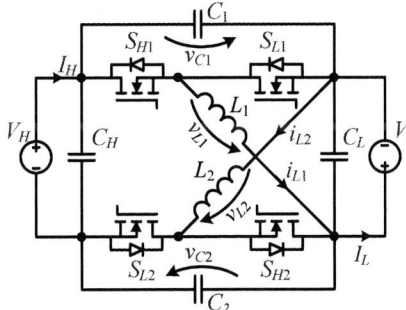

Fig. 1: The proposed BHSI

A few simplifying assumptions are made in order to explain the operation of the BHSI:
- L_1 and L_2 inductors, and their currents (i_{L1} and i_{L2}) are considered identical,
- C_1 and C_2 capacitors, and their voltages (V_{C1} and V_{C2}) are considered identical,
- Capacitors are considered large enough to maintain a constant voltage,
- All components are considered ideal,
- Steady state operation is considered.

Taking into account the simplifying assumptions, the operation of the BHSI can be explained starting from the two equivalent switching states, t_{on} and t_{off}, presented in Fig. 2. During t_{on}, S_{H1} and S_{L2} switches are turned on and S_{H2} and S_{L1} are turned off, so that L_1 and L_2 inductors are connected between the high voltage source, V_H, and the low voltage, V_L. The voltage across C_1 and C_2 is considered:

$$V_{C1} = V_{C2} = \frac{V_H + V_L}{2}, \tag{1}$$

therefore, the voltages across the inductors are:

$$v_{L1} = V_{C1} - V_L = \frac{V_H - V_L}{2}, \tag{2}$$

$$v_{L2} = V_{C2} - V_L = \frac{V_H - V_L}{2}. \tag{3}$$

During t_{off}, the two inductors are connected in parallel to V_L voltage by S_{L1} and S_{H2} transistors, which means that the inductor voltages are:

$$v_{L1} = v_{L2} = -V_L. \tag{4}$$

The main theoretical waveforms, consisting of inductor voltages and currents, input and output currents, and transistor voltages, are presented in Fig. 3. A significant improvement achieved through this topology is the limitation of the maximum voltage on each transistor to a lower value, equal to $(V_H + V_L)/2$.

By applying the inductors volt-second balance:

$$V_{L1} = V_{L2} = D \cdot \frac{V_H - V_L}{2} + (1-D) \cdot (-V) = 0, \tag{5}$$

the relation between the two input voltages is determined:

$$V_L = V_H \cdot \frac{D}{2-D}. \tag{6}$$

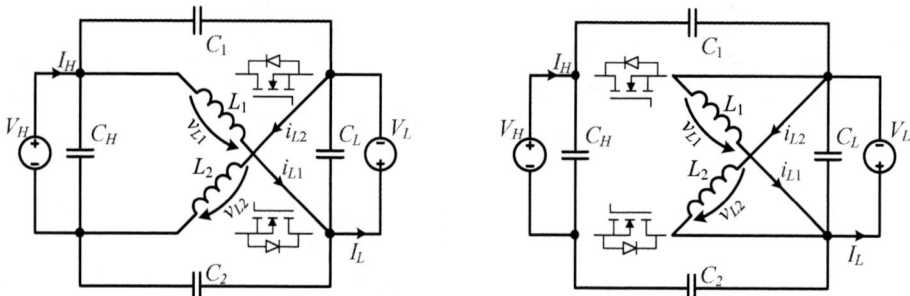

Fig. 2: The two switching states of the BHSI: t_{on} (left) and t_{off} (right)

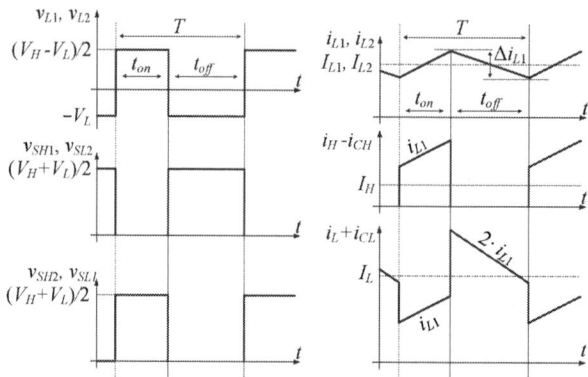

Fig. 3: Main theoretical waveforms of the improved BHSI

Oscillation elimination

Previous papers proposed methods to eliminate the voltage oscillations which appear in the unidirectional topology [3], and experimental results were obtained in the bidirectional topology with oscillations damped by a passive snubber [4]. In this family of topologies, the oscillations are caused by the parasitic capacitors of the transistors in the switching cell, C_{SL1} and C_{SH2} shown in Fig. 4, and the small differences between the switched inductances. The voltage on each transistor will oscillate around the theoretical level with amplitudes and frequencies depending on the values of the resonant circuits. Typical theoretical waveforms are presented in Fig. 5.

The improvement proposed in this paper consists of adding the C_1 and C_2 capacitors and an additional switch, which will eliminate the voltage oscillations by setting a constant voltage between the two voltage sources, V_H and V_L, with a value equal to V_{C1}, from (1). Another benefit is the elimination of the switching frequency voltage between V_H and V_L which was previously present in both unidirectional and bidirectional topologies.

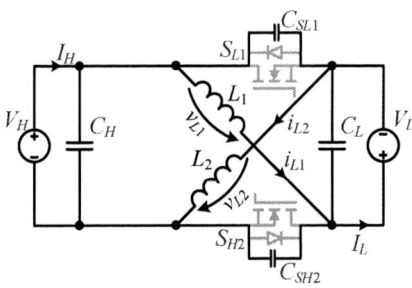

Fig. 4: Equivalent schematic during t_{on} of the unimproved BHSI [4]
(or t_{off} for the unidirectional topology [3])

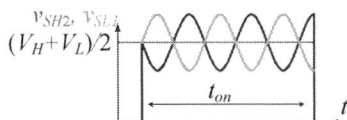

Fig. 5: Transistor voltage waveform during t_{on} state without C_1 and C_2 capacitors
(unimproved BHSI topology)

Comparison of the proposed topology to other converters

In order to compare the proposed BHSI to other topologies, the total inductor energy and the total device stress are used as metrics. To compare the total inductor energy, the inductor currents are calculated using the values of the low voltage side current (I_L) and the duty cycle:

$$I_{L1} = I_L \cdot \frac{1}{2-D} = I_L \cdot \frac{V_L + V_H}{2 \cdot V_H}. \tag{7}$$

The inductor current ripple ratio, r, is used to guarantee that the sizing of the inductors is performed similarly in different topologies. The inductor current ripple is calculated with:

$$\Delta i_{L1} = r \cdot I_{L1}. \tag{8}$$

Considering a constant value for r, the two inductors have the following sizing relation:

$$L_1 = L_2 = \frac{2 \cdot V_H \cdot V_L \cdot (V_H - V_L)}{r \cdot f_{sw} \cdot I_L \cdot (V_H + V_L)^2}. \tag{9}$$

Using (7) and (9), the L_1 inductor energy is expressed as:

$$W_{L1} = \frac{L_1 \cdot I_{L1}^2}{2} = \frac{I_L \cdot V_L \cdot (V_H - V_L)}{4 \cdot r \cdot f_{sw} \cdot V_H}, \tag{10}$$

and the total inductor energy can be calculated for any converter with:

$$W_L = \sum_{i=1}^{k} W_{Li} = \sum_{i=1}^{k} \frac{L_i \cdot I_{Li}^2}{2}. \tag{11}$$

The total active switch stress is given by:

$$S = \sum_{i=1}^{2} V_{SHi} \cdot I_{SHi} + \sum_{i=1}^{2} V_{SLi} \cdot I_{SLi}, \tag{12}$$

where the maximum voltages (V_{SHi} and V_{SLi}) and currents (I_{SHi} and I_{SLi}) for each transistor are calculated with:

$$V_{SH1} = V_{SL1} = V_{SH2} = V_{SL2} = V_{C1} = V_{C2} = \frac{V_H + V_L}{2}, \tag{13}$$

$$I_{SH1} = I_{SL1} = I_{SH2} = I_{SL2} = I_{L1} = I_{L2} = I_L \cdot \frac{V_L + V_H}{2 \cdot V_H}. \tag{14}$$

The total active switch stress for this topology is:

$$S = I_L \cdot \frac{(V_L + V_H)^2}{V_H}.$$

(15)

Even if it has an additional transistor compared to [4], it can be observed that, because of its lower voltage, the total active switch stress is the same. A comparison is presented in Fig. 6 with the relations calculated above, and it is clear that the proposed BHSI has a good conversion ratio, smaller inductors and reduced stress in the active switches.

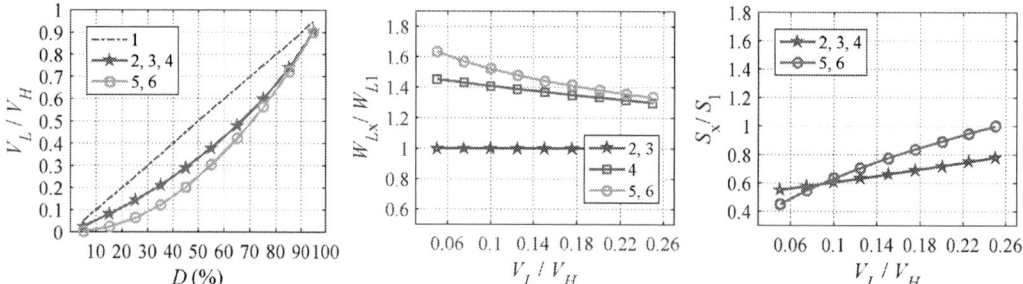

Fig. 6: Comparisons with other topologies (x = 1..6): 1. Conventional Buck/Boost; 2. proposed BHSI; 3. Converter [2]; 4. Converter [1]; 5. Conventional Quadratic [8]; 6. Converter [9]

Simulation and experimental results

The simulations results for the unimproved BHSI topology (without the additional capacitors) present inductor voltage oscillations as discussed in the previous sections, and are presented in Fig. 8. The steady state simulation results for the proposed BHSI from Fig. 8, show that the topology has no inductor voltage oscillations, compared to the conventional structure.

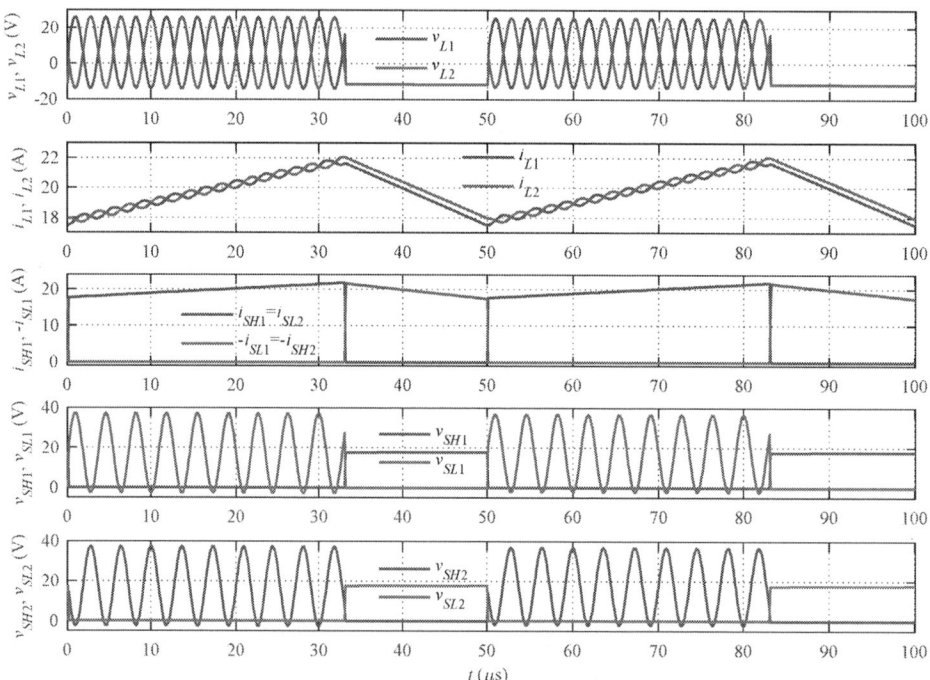

Fig. 7: Steady state simulation results of the unimproved BHSI (without C_1 and C_2)

The simulations for transient operation mode from Fig. 9 present a stable operation with a simple PI controller for i_{L1} current, with a fast transition between buck and boost operation. A small damped oscillation is present in the second inductor current, i_{L2}, which can be ignored, or it can be eliminated with other control strategies.

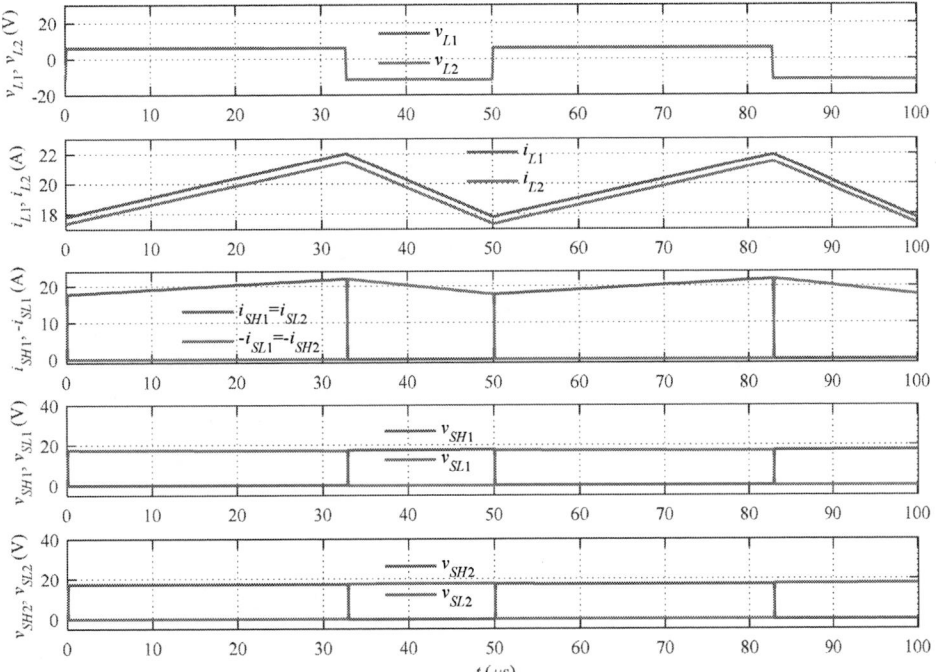

Fig. 8: Steady state simulation results of the proposed BHSI

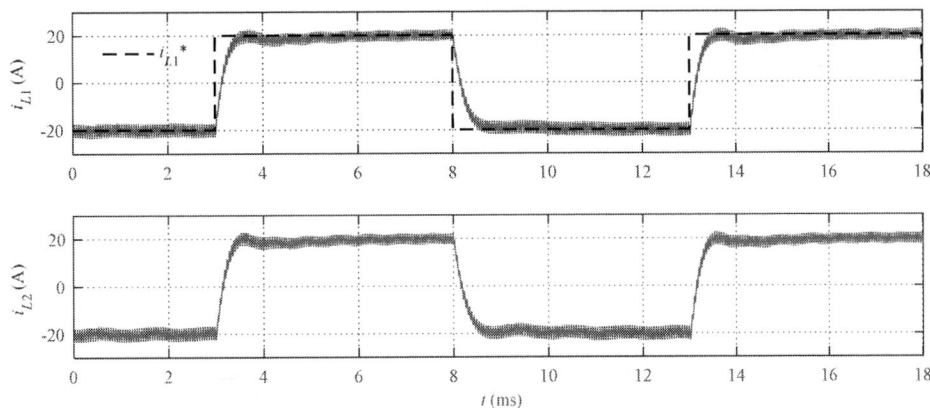

Fig. 9: Transient simulation results $i_{L1}* = \pm 20$ (A)

Steady state experimental results for the proposed BHSI are presented in Fig. 10 to Fig. 12 for both step-up and step-down operation mode. The converter was built and tested with the parameters presented in Table I, having a resistor connected at each respective output (R_{loadH} and R_{loadL}). The inductor voltages and currents presented in Fig. 10, for step-up operation (left) and step-down operation (right) present no oscillations, and are in concordance to the simulation results from Fig. 8.

The voltages on the high side branch switches, v_{SH1} and v_{SL1}, and the voltages on the low side branch switches, v_{SH2} and v_{SL2}, are presented together with the two complementary driving signals, v_{PWM_off}, v_{PWM_on}, in Fig. 11 and Fig. 12, respectively. As the voltage on the inductors do not present any oscillations, the same is valid for the transistor voltages.

Table I: BHSI prototype parameters

Element	V_H	V_L	D	R_{loadH}	R_{loadL}	f	$I_{L1} \approx I_{L2}$	Switches
Value	30	10	50	10	1	20	5.5 / 6.5	BSC040N10NS5
Unit	V	V	%	Ω	Ω	kHz	A	-

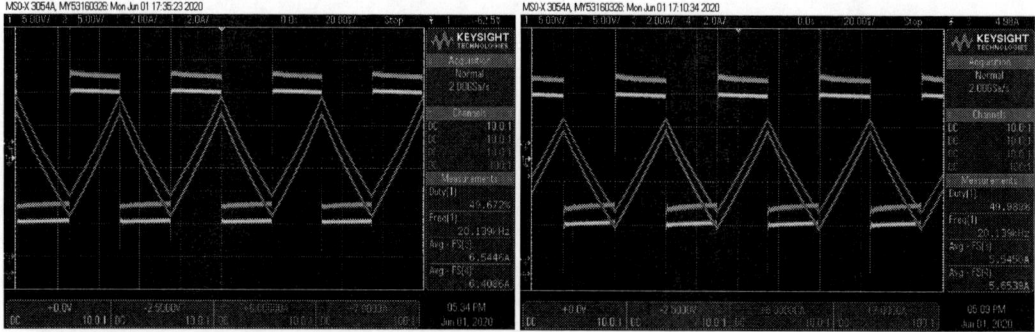

Fig. 10: BHSI experimental results for step-down operation (left) and step-up operation (right),
(ch1: v_{L1}, ch2: v_{L2}, ch3: i_{L1}, ch4: i_{L2})

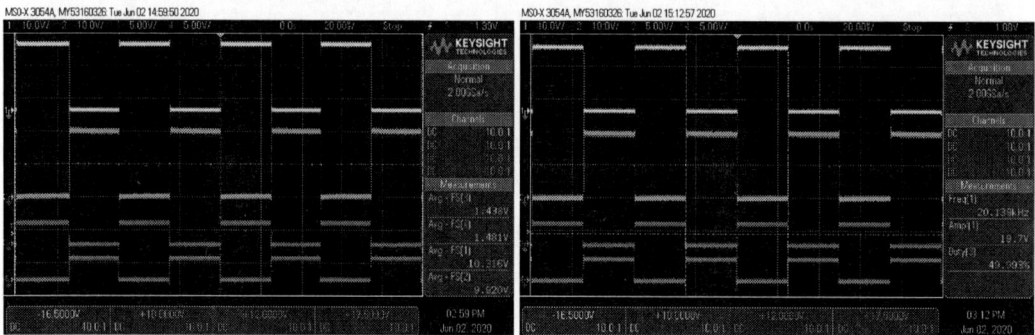

Fig. 11: BHSI experimental results for step-down operation (left), and step-up operation (right)
(ch1: v_{SH1}, ch2: v_{SL1}, ch3: v_{PWM_off}, ch4: v_{PWM_on})

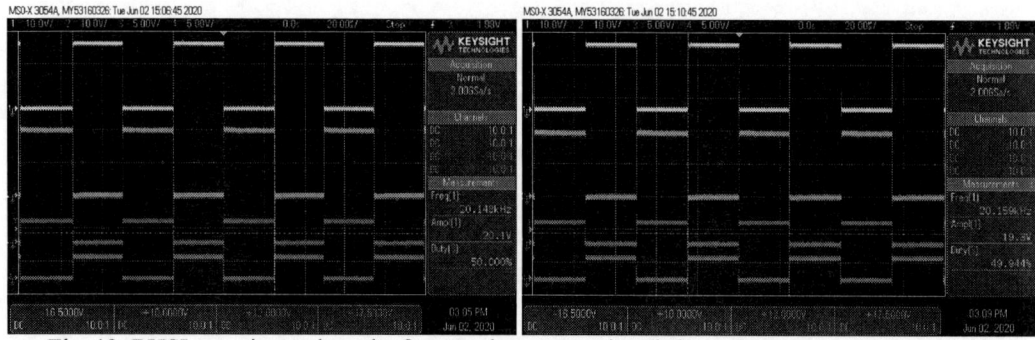

Fig. 12: BHSI experimental results for step-down operation (left), and step-up operation (right)
(ch1: v_{SH2}, ch2: v_{SL2}, ch3: v_{PWM_off}, ch4: v_{PWM_on})

Fig. 13: Prototype of the proposed BHSI converter

Conclusion

This paper presents an improved bidirectional hybrid switched inductor converter which has the advantage of no voltage oscillation in the inductors, a constant DC voltage between the inputs, reduced maximum voltage on the transistors and a reduced inductance in the switching loop. All benefits are introduced by using two half bridges (each in parallel to an additional capacitor) in a way specific to hybrid converters. The topology preserves other benefits introduced by the hybrid switched inductor converters, such as increased voltage conversion ratio, reduced passive component size and reduced stress on the transistors. The characteristics of this topology have been compared to other state of the art structures, and its advantages are highlighted. The simulation results confirm the theoretical assumptions, proving a stable operation in transient regimes, and the experimental results confirm a stable operation without voltage oscillations.

References

[1] Y. Zhang, Q. Liu, J. Li, and M. Sumner, "A Common Ground Switched-Quasi-Z -Source Bidirectional DC–DC Converter With Wide-Voltage-Gain Range for EVs With Hybrid Energy Sources," *IEEE Transactions on Industrial Electronics*, vol. 65, no. 6, pp. 5188–5200, Jun. 2018.

[2] O. Cornea, G. Andreescu, N. Muntean, and D. Hulea, "Bidirectional Power Flow Control in a DC Microgrid Through a Switched-Capacitor Cell Hybrid DC–DC Converter," *IEEE Transactions on Industrial Electronics*, vol. 64, no. 4, pp. 3012–3022, Apr. 2017.

[3] Y. Tang and T. Wang, "Study of An Improved Dual-Switch Converter With Passive Lossless Clamping," *IEEE Transactions on Industrial Electronics*, vol. 62, no. 2, pp. 972–981, Feb. 2015.

[4] D. Hulea, B. Fahimi, N. Muntean, and O. Cornea, "High Ratio Bidirectional Hybrid Switched Inductor Converter Using Wide Bandgap Transistors," in *2018 20th European Conference on Power Electronics and Applications (EPE'18 ECCE Europe)*, 2018, p. P.1-P.10.

[5] J. K. Jorgensen *et al.*, "Loss Prediction of Medium Voltage Power Modules: Trade-offs between Accuracy and Complexity," in *2019 IEEE Energy Conversion Congress and Exposition (ECCE)*, 2019, pp. 4102–4108.

[6] A. B. Jørgensen, T.-H. Cheng, D. Hopkins, S. Beczkowski, C. Uhrenfeldt, and S. Munk-Nielsen, "Thermal Characteristics and Simulation of an Integrated GaN eHEMT Power Module," in *2019 21st European Conference on Power Electronics and Applications (EPE '19 ECCE Europe)*, 2019, p. P.1-P.7.

[7] N. Fichtenbaum, M. Giandalia, S. Sharma, and J. Zhang, "Half-Bridge GaN Power ICs: Performance and Application," *IEEE Power Electronics Magazine*, vol. 4, no. 3, pp. 33–40, Sep. 2017.

[8] A. Ahmad, R. K. Singh, and R. Mahanty, "Bidirectional quadratic converter for wide voltage conversion ratio," in *2016 IEEE International Conference on Power Electronics, Drives and Energy Systems (PEDES)*, 2016, pp. 1–5.

[9] H. Ardi, A. Ajami, F. Kardan, and S. N. Avilagh, "Analysis and Implementation of a Nonisolated Bidirectional DC–DC Converter With High Voltage Gain," *IEEE Transactions on Industrial Electronics*, vol. 63, no. 8, pp. 4878–4888, Aug. 2016.

Hybrid Multiple Chopper Cells of PWM and Square-wave Operation for Solid-state Transformer

Naoto Kikuchi, Jun-ichi Itoh, Keisuke Kusaka, *Hoai Nam Le
Nagaoka University of Technology/ * Norwegian University of Science and Technology
1603-1, Kamitomioka-machi, Nagaoka, Niigata, JAPAN
Tel., Fax: +81 / (258) – 47.9533.
E-Mail: n_kikuchi@stn.nagaokaut.ac.jp
URL: http:// itohserver01.nagaokaut.ac.jp/itohlab/index.html

Keywords

«Smart grids», «Modulation strategy», «Wide bandgap devices», «ZVS converters», «Solid-state transformer».

Abstract

This paper proposes a modulation method combining PWM and square-wave drive for solid-state transformers (SSTs). Modular multilevel configurations based on input series and output parallel (ISOP) have been widely used. The advantage of ISOP configuration is that low on-resister and low-switching-loss devices are available because the applied voltage on each cell is divided by the number of cells. For this reason, SST based on the ISOP configuration is widely used in a medium-voltage system. However, this configuration has a problem that a high number of medium frequency devices, such as SiC-MOSFET, increases the cost of SST. In order to solve above problem, a reduction method of medium frequency devices are proposed. In the proposed method, one cell is driven with the PWM operation in order to compensate for the harmonic component. The other cells are driven by square-wave operation. Thus, the power factor correction (PFC) is held by different switching frequencies in each cell converter operated.

Owing to the proposed operation, the medium frequency driving devices can be replaced with low-frequency devices, such as Si-IGBT, for cost-saving. As the drawback of the proposed control, the conduction time is unbalanced in the square wave cells. Thus, this paper also proposes the sorting operation to balance the output power of cells operated with square-wave drive. From the experimental results, THD of the input current is 2.91%, the input power factor is 0.99, the maximum efficiency is 94.2% at 0.3p.u.(1.p.u.=1.0 k W).

Introduction

In recent years, the smart grid has attracted much attention due to the increasing demand for renewable resources. The solid-state transformer (SST) is known as a critical component in the smart grid system. SST consists of power converters and a medium frequency transformer for galvanic isolation. The power converter provides power flow control and compensation of the reactive power. The medium frequency transformer enables to increase the power density of the converter.

Focusing on the circuit topology of SST, various circuit topologies have been proposed in this decade from the view of efficiency, cost, and protection functions [1-20]. In [8-9], the ISOP configuration based on the PWM rectifier and dual active bridge has been proposed. The advantage of this topology is achieving soft-switching in a wide load range. However, a lot of switching devices are required for the multistage cell. Also, the DC link capacitor with a large capacity is needed in order to keep the DC-link voltage in the primly side. In [10-11], AC/AC-based ISOP configuration to reduce conversion stage and passive component has been proposed. This topology increases the number of components because the converter needs bidirectional switches. For these topologies, SiC-MOSFETs have been widely used because of them high-switching speed and low on-resistance. The use of SiC-MOSFETs has a problem with the cost of the SiC devices. The SiC devices are still more expensive than Si-IGBT, even if the demand for SiC devices is increased.

In this paper, the hybrid modulation method for ISOP is proposed in order to reduce the cost of SST. The new modulation method is that one of the cells is driven with PWM in order to compensate for the harmonic components of the input current. The other cells are driven by square-wave operation in order to correct the input power factor. It is possible to use the devices with the slower switching speed because these switches are operated at the double of the grid frequency. It will contribute to the cost-saving because Si-IGBT will be an option for these cells.

Besides, the proposed method causes the unbalance of the power loss among the cells operated with the square-wave operation due to the different conduction time of switching devices. Therefore, this paper also proposes the soring operation, which improves the unbalance of power loss among square wave cells. The new contribution of this paper is reducing the cost of SST by replacing the high-frequency wide-bandgap devices to low-cost Si devices, in which only low-frequency operation is available.

Circuit configuration

System configuration

Figure 1 shows the circuit configuration of the single-phase SST with multiple cells. Each cell has PFC stages and isolated resonant DC/DC converter. The input of the PFC stage in the cells is connected in series. The outputs of the isolated resonant DC/DC converters, which ensure the galvanic isolation [16], are connected in parallel. The input diode rectifier is common for all cells to reduce the number of components. The high-frequency operation contributes to minimizing the isolation transformer. The transformer of the isolated resonant DC/DC converter is smaller than the conventional commercial frequency transformer because the switching frequency f_{sw_re} is higher than the grid frequency. Besides, the resonant capacitor C_r is connected to the primary side of the isolated transformer in series for the zero-voltage switching (ZVS) of the DC-DC converter with open-loop control. For the ZVS operation, the leakage inductance is designed to be negligibly smaller than the excitation inductance.

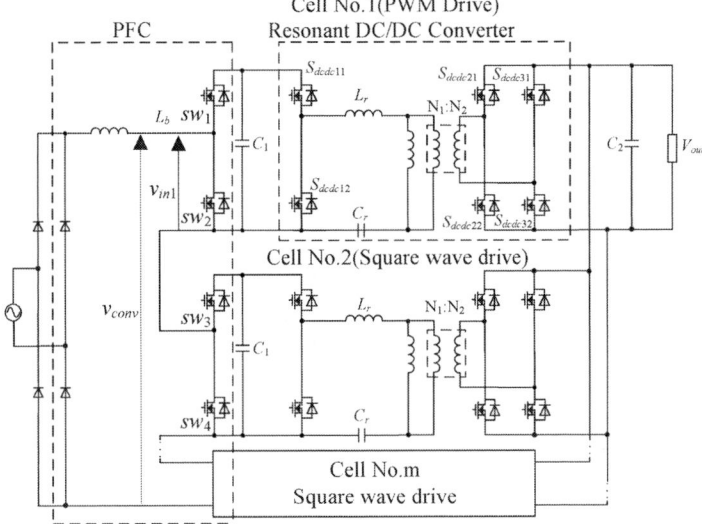

Fig. 1. Circuit configuration of the single-phase SST.
The new modulation method is applied into the PFC stage.

Conventional control of PFC stage

Figure 2 shows the block diagram of the conventional control method of the PFC stages. The conventional method controls output voltage with PWM for all cells. The overall PFC circuit controls output voltage and phase and amplitude of the input current to correct the input power factor. The phase of the inductor current is detected from the grid voltage. The output voltage of the overall cell converter is equally divided into each cell. Note that the phase-shifted carrier is used to reduce the current ripple.

The reference of the input current is a full wave rectified sine wave because the current reference is fed to the inductor on the PFC stage. The current reference is multiplied by the phase, which is generated by PLL of the grid voltage. The operation value of PI control is the total voltage of all cells. Thus, the voltage divided by the number of cells is the reference voltage for each cell. The switching timing of the PFC stage is shifted in each cell using the phase shift-carriers. Hence, the equivalent switching frequency is increasing in proportional to the number of cells. Equivalent switching frequency f_{eq} is given by

$$f_{eq} = mf_{sw\_pfc} \tag{1},$$

where f_{sw_pfc} is the switching frequency of the PFC stage in each cell.

Figure 3 shows the switching pulse generation of the secondary-side rectifier. The full-bridge converter on the secondary side operates as a synchronous rectifier. The switching pulse is the same as the pulse of the primary side. In the primary side, resonant current i_{re} is positive when S_{dcdc11}, S_{dcdc21}, S_{dcdc32} are turn-on. Note that phase of primly side resonant current is a little different in order to design resonant DC/DC converter. Similarly, S_{dcdc12}, S_{dcdc22}, S_{dcdc31} are turn-on when the resonant current is negative.

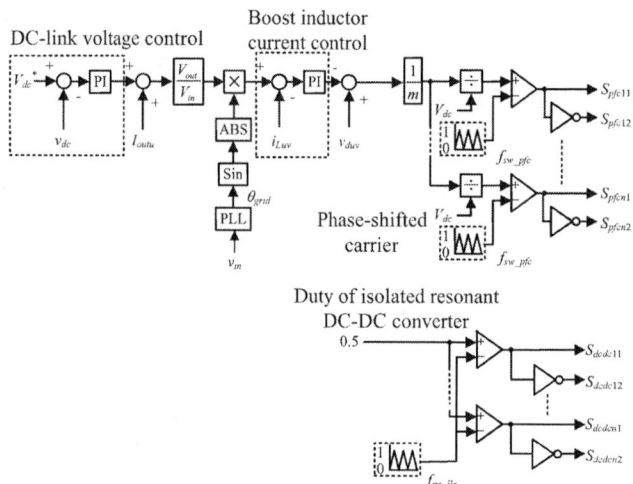

Fig. 2. Control diagram of conventional method in single-phase SST.

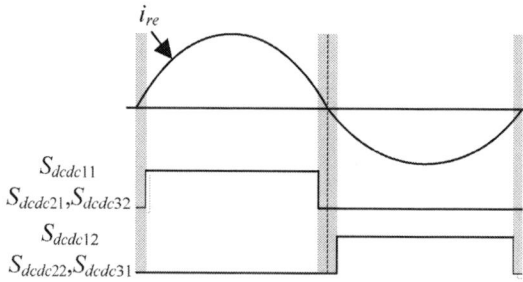

Fig. 3. Pulse generation for secondly side rectifier

Proposed modulation method of PFC stage

Control with square-wave opearion

Figure 4 shows the relationship between the input voltage and output voltage of cell #1-3. The one of cells #1 is driven by the PWM operation in order to compensate for the harmonic component of other cells. The other cells #2 and 3 are driven by the square-wave operation. The output voltage of the cell operated with square wave is determined by the comparison with the reference voltage of the PFC stage and the DC-link voltage of each cell. The switching device on the upper arm of cell #2 sw₃ turns on when the output voltage reference v_{inv} is higher than DC-link voltage of cell #2. Similarly, the switching device on the upper arm of cell #3 sw_5 turns on when the output voltage reference is higher than the sum of DC-link voltages of cell #2 and #3. Thus, the output voltage of the cell #2 and #3 is double the grid frequency.

Figure 5 shows the block diagram with the proposed modulation. The PFC stage of one cell converter is driven by PWM to compensate the harmonics component of the input current. The output value of the PI controller is the total output voltage of all the cell converters. Thus input voltage of cell #1 v_{in1} is given by

$$v_{in1} = |v_{in}| - v_{dc2}S_{sqr11} - v_{dc3}S_{sqr21} \qquad (2),$$

where v_{dc2} and v_{dc3} are the DC-link voltage of cell #2 and cell #3, respectively. Moreover, S_{sqr11} and S_{sqr21} are the switching states of the PFC stage of square-wave cells. The current controller compensates for the output voltage of the square-wave cells with feed-forward control.

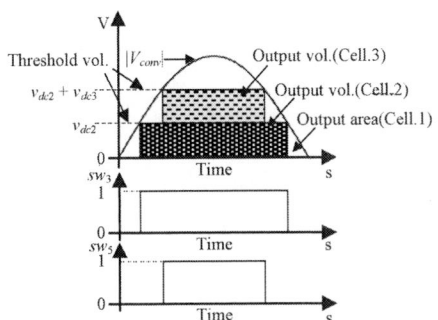

Fig. 4. Operation principle of proposed method.

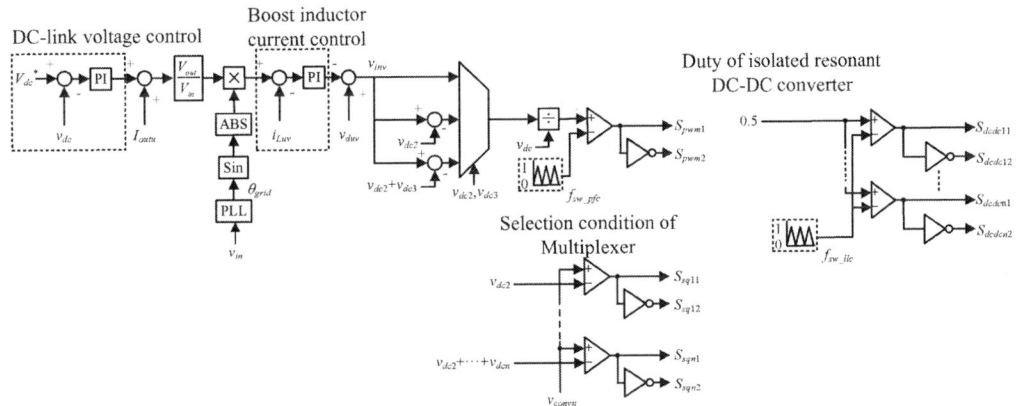

Fig. 5. Control block diagram with proposed method.
Only one cell is operated by PWM. Others output square wave.

Table I shows the comparison of the number of switches between the proposed strategy and conventional strategy. As shown in Table I, the number of SiC devices is reduced by 25% compared to the conventional strategy. Therefore, the proposed strategy is effective in improving the cost.

Thermal balancing control

In the proposed method, the output power of each cell is unbalanced because the conduction time between the upper arm and the lower arm of the

Table I. Comparison of SiC-MOSFET and Si-IGBT device between conventional strategy and proposed strategy.

Rated voltage	Number of cell	Number of switching devices	
		Conventional	Proposed
3.3 kV	6	SiC:52	SiC:40, Si:12
1.7 kV	11	SiC:92	SiC:70, Si:22
1.2 kV	16	SiC:132	SiC:100, Si:32

PFC stage is different in each cell. It may accelerate the aging of capacitors or switching devices and shorten the running time of a specific cell by concentering power loss.

Figure 6 shows the sorting operation with a PWM cell and five cells operated by square-wave drive. When the sorting operation is not worked, the output power of cell #2 is the largest in the all cell, and the output power of cell #6 is the smallest in all cell. It causes the problem that the advantage of the ISOP configuration is disappeared from the view of equally sharing load in each cell.

The proposed modulation has the freedom on the option which cell output in the long term. Therefore, the cell is selected to average the conduction time by the sorting operation. The sorting operation has five modes in five square-wave cells. First, the cell #2 output the maximum power, and #6 output the minimum power in the mode I. Second, the operation mode is shifted to mode II. The cell #6 outputs the maximum power and the cell #5 outputs the minimum power. Similarly, the mode is shifted in sequence from I to V. The average output power of each cell is balanced in 2.5 times of the output grid period.

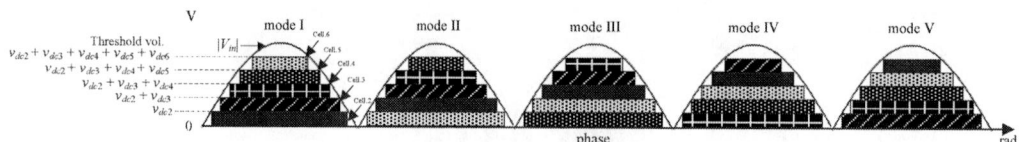

Fig. 6. Sharing input voltage in sorting operation with proposed method (m = 6).

Table II. Sorting operation of square wave cell with proposed method.

	$v_{dc2} < \|v_{in}\| < \sum_{i=3}^{5} v_{dci}$	$\sum_{i=3}^{5} v_{dci} < \|v_{in}\| < \sum_{i=2}^{5} v_{dci}$	$\sum_{i=2}^{4} v_{dci} < \|v_{in}\| < \sum_{i=2}^{5} v_{dci}$	$\sum_{i=2}^{5} v_{dci} < \|v_{in}\| < \sum_{i=2}^{6} v_{dci}$	$\sum_{i=2}^{6} v_{dci} < \|v_{in}\|$
mode I	Cell.2	Cell.3	Cell.4	Cell.5	Cell.6
mode II	Cell.6	Cell.2	Cell.3	Cell.4	Cell.5
mode III	Cell.5	Cell.6	Cell.2	Cell.3	Cell.4
mode IV	Cell.4	Cell.5	Cell.6	Cell.2	Cell.3
mode V	Cell.3	Cell.4	Cell.5	Cell.6	Cell.2

Simulation and Experimental Results

Simulation result of balancing control in conduction loss

In this section, the simulation result of the sorting operation is shown. Table III presents the simulation conditions. In this simulation, a constant voltage source of 200 V is used as the primary side DC link voltage of each cell because of evaluating the effects by eliminating the power ripple of DC-link voltage. The switching devices for cell driven by PWM is SiC-MOSFET (Rohm, SCT3040KR)[21]. The switching devices for the cells driven by square-wave is Si-IGBT (Fairchild, FGA20N120FTD)[22]. Figure 7 shows the simulation result with or without sorting operation. In this simulation, cell #1 is driven by PWM, and cell #2-6 are driven by square-wave. The conduction loss of each cell is given by

$$P_{con\_PFC} = f_{sw} \int_{t_{on}}^{t_{off}} d \cdot i_{out} v_{ce} dt \tag{3},$$

where f_{sw} is the switching frequency of PFC in each cell. d is the duty ratio of PFC. The current i_{out} is RMS value of input current. v_{ce} is the corrector-emitter voltage of IGBT. The relationship between collector-emitter voltage is approximated as

$$v_{ce} = v_0 + r_{on} i_{out} \tag{4},$$

whereas v_0 is the corrector-emitter voltage at zero current of corrector-emitter, r_{on} is approximated on-resistance, and i_{out} is the instantaneous value of the conduction current.

Fig. 7(a) shows the conduction loss and switching loss without sorting operation. In the cell #2, the conduction loss of FWD is dominant because the conduction time of the upper arm is the longest in all cells. In contrast, in cell 6, conduction loss of lower IGBT is dominant because the conduction time of the lower arm is the second-longest in all cells.

Fig. 7(b) shows the simulation result of loss analysis with sorting algorism. Meanwhile, the power loss of cells #2-6 is 16.7 W. It means that conduction loss of the upper and lower arm is balanced. Note that total loss is 83.3 W, which is equivalent to a total loss of the result without sorting operation of 83.3 W. Therefore, the loss balancing is achieved by the proposed method.

Table III. Simulation condition in sorting operation of square-wave cell.

Input voltage	V_{in}	800 V$_{rms}$
Rated output power	P_{out}	16 kVA
Rated output voltage	V_{out}	100 V
Switching frequency of PFC	f_{sw_pfc}	30 kHz
Resonant frequency	f_o	50 kHz
Number of cells	m	6
Boost inductor	L_b	4 mH
Primary side capacitor	C_1	48 µF
Resonant capacitor	C_r	204 nF
Leakage inductor	L_r	50 µH
Secondary side capacitor	C_{out}	3000 µF
Trans turns ration	$N_1:N_2$	1:1
Switching device(SiC-MOSFET)		SCT3040KR
Switching device(Si-IGBT)		FGA20N120FTD

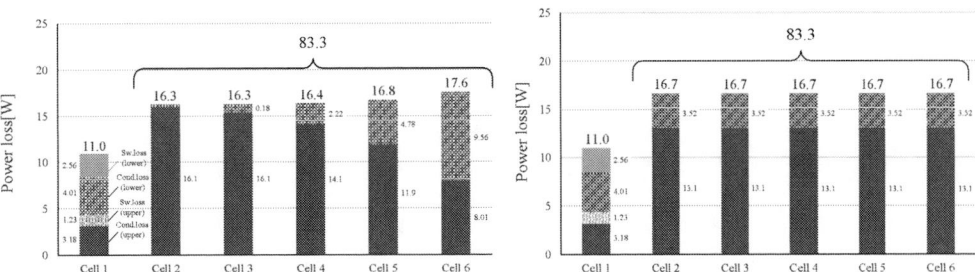

(a) Simulation result of power loss without sorting operation. (b) Simulation result of power loss with sorting operation.

Fig. 7. Simulation result of power loss in each cell.

Experimental result with conventional method

Table IV shows the parameters for the experiment. The input voltage is 200 V, the rated power is 1.0 kW, and the number of cells is three. Moreover, the prototype is operated with the conventional control block diagram, as shown in Fig. 3, and the proposed control, as shown in Fig. 5. The switching frequency of the PWM cell in the proposed control is three-times higher than the switching frequency of the conventional control for a fair comparison.

Figure 9 shows the operation waveforms of the conventional method. Fig. 9(a) shows that the input current THD is 2.91%, and the input power factor is 0.99. Fig. 9(b) shows that each cell is driven by PWM against grid voltage.

Experimental result with proposed method

Figure 10 shows the operation of the proposed method. In the proposed modulation, the switching frequency of the PFC circuits is 30 kHz. It means that the equivalent switching frequency f_{sw_eq} is 30 kHz. Whereas the switching frequency of the PWM cell in the proposed control method is 30 kHz for the fair comparison. Fig. 10(a) shows the result of the proposed modulation method. The input power factor is also 0.99. The input current THD is 2.91%. In addition, Fig. 10(b) shows the output square-wave driving in each cell with double of the grid frequency. The PWM cell compensates for the harmonic components by the square-wave cell.

Figure 11 shows that input current THD against the rotted power with the conventional and proposed method. PFC shows to compensate the harmonic component of input current. However, in light load, THD of the input current exceeded 5% with both controls. Improving the THD is future work.

Figure 12 shows the efficiency characteristic. The maximum efficiency is 94.2% at a load of 0.3 kW.. The efficiency at rated load is 84.4%.

Experimental result of the sorting operation

In this experiment, the output power is measured with or without the sorting operation. Experimental parameters are common shown in Table IV. The square-wave cell outputs in double of the grid frequency.

Figure 13 shows the operation waveform with sorting operation. The power factor of the input current is 0.99.

Figure 14 shows the experimental result of output power with or without the operation. The output power of cell #1 is 242 W, the output power of cell #2 is 485 W, and the output power of cell #3 is 233W without the sorting operation. It means that concentering power loss occurs at a specific cell. In contrast, the output power of each cell is 304 W, 329 W, and 332W, respectively with the sorting operation. It clearly shows that unbalancing output power improves.

Table IV. Experiment parameter.

Input voltage	V_{in}	200 V_{rms}
Rated output power	P_{out}	1.0 kW
Rated output voltage	V_{out}	50 V
Switching Device		SCT2080KE
Grid Freqency	f_s	50 Hz
Primary side capacitor	C_1	1500 μF
Resonant capacitor	C_r	204 nF
Leakage inductor	L_r	50 μH
Secondary side capacitor	C_2	3600 μF
Trans turns ration	$N_1{:}N_2$	1 : 1
Number of cells	m	3
Switching frequency of PFC	f_{sw_pfc}	30 kHz
Switching frequency of LLC	f_{sw_llc}	50 kHz

Fig. 8. Waveforms of ZVS operation in resonant DC/DC converter. The converter achieves ZVS of upper switches on resonant DC/DC.

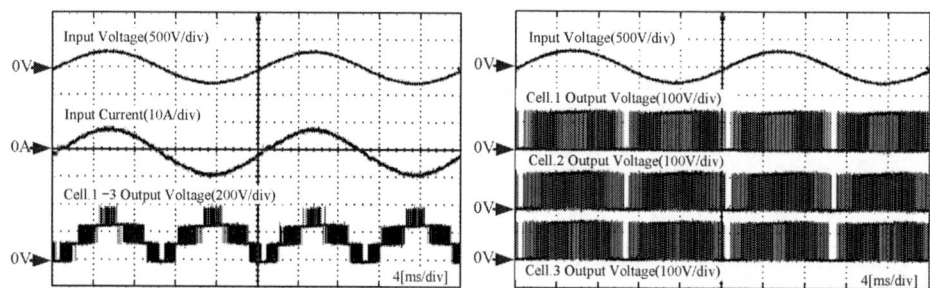

(a) Operation waveform of SST with conventional method. (b) Operation waveform of SST with conventional method.

Fig. 9. Waveforms of SST, using conventional method.

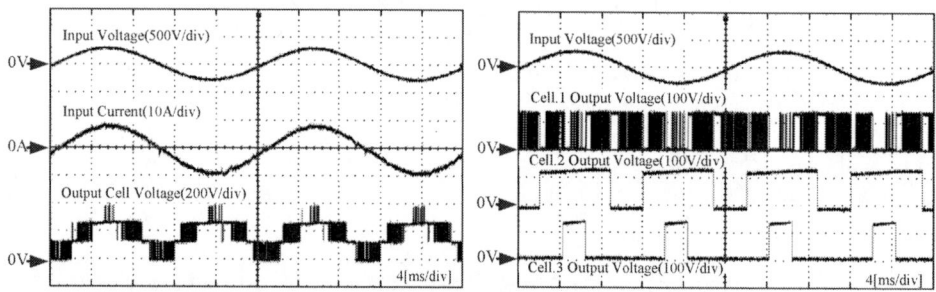

(a) Operation waveform of SST with proposed method. (b) Output voltage of each cell with proposed method.

Fig. 10. Waveforms of SST, using proposed method.

Conclusion

This paper has proposed the hybrid modulation method, which combines PWM and square-wave operation for ISOP based SST. In the proposed method, one of the cells is driven by PWM in order to compensate for the harmonic component. The other cells are driven by square-wave operation in order to share the load. Moreover, this paper proposed the sorting operation in order to improve unbalancing power in each cell when the proposed modulation is used. As the experimental result, the input current THD is 2.91% at an output power of 1.0 kW whereas the input current THD of conventional modulation is 3.18%. The efficiency of the total system is 82.0% at 1.0 kW. Besides, the output power of the cell with the square-wave operation is balanced with the sorting operation.

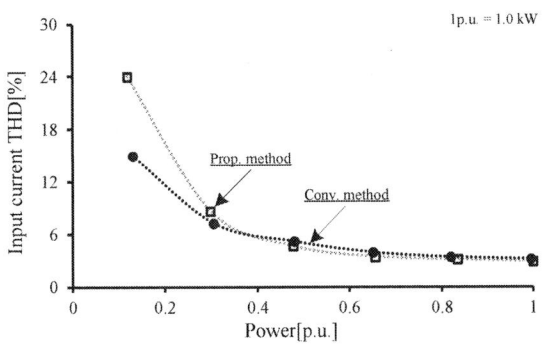

Fig. 11. Relationship between input current THD and input power with conventional and proposed modulation

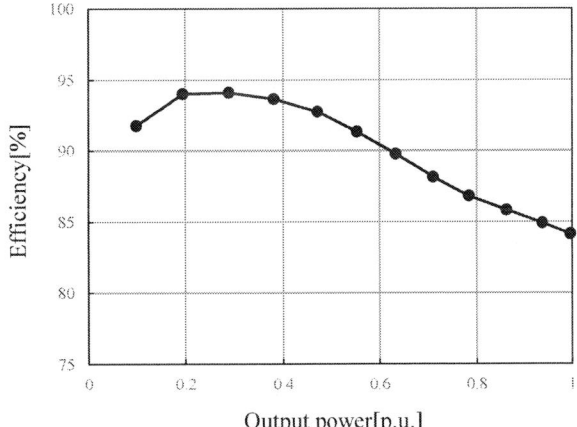

Fig. 12. Efficiency characteristic with proposed method.

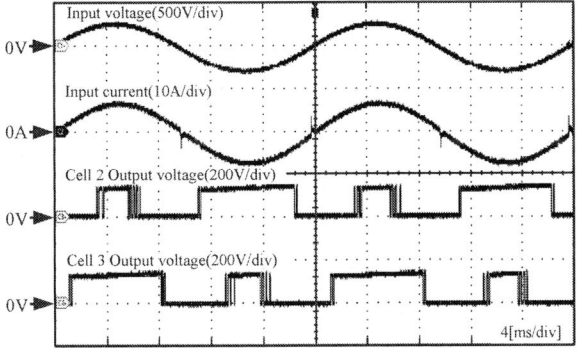

Fig. 13. Operation waveform with sorting operation.

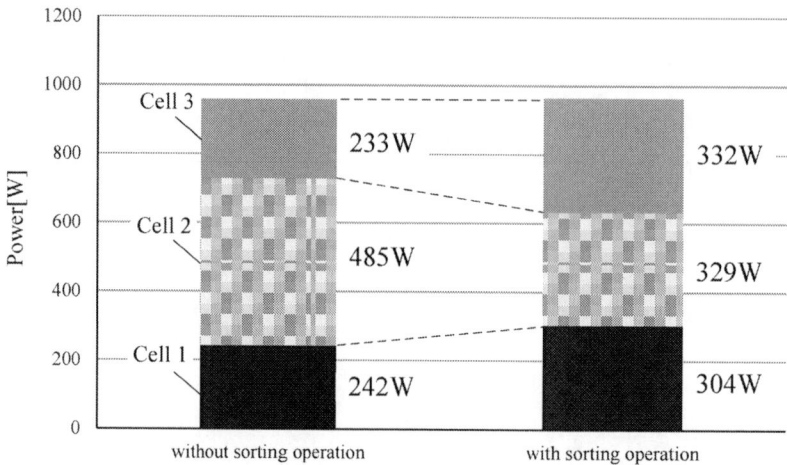

Fig. 14. Measured output power with no using sorting operation and using sorting operation.

References

[1] J. E. Huber and J. W. Kolar, "Applicability of Solid-State Transformers in Today's and Future Distribution Grids," in IEEE Transactions on Smart Grid, vol. 10, no. 1, pp. 317–326, Jan. 2019.

[2] L. Ferreira Costa, G. De Carne, G. Buticchi and M. Liserre, "The Smart Transformer: A solid-state transformer tailored to provide ancillary services to the distribution grid," in IEEE Power Electronics Magazine, vol. 4, no. 2, pp. 56–67, June 2017.

[3] A. Q. Huang and J. Baliga, "FREEDM system: Role of power electronics and power semiconductors in developing an energy internet," in Proc. 21st Int. Symp. Power Semicond. Devices ICs, pp. 9–12, Jun. 14–18, 2009.

[4] J. E. Huber and J. W. Kolar, "Volume/weight/cost comparison of a 1MVA 10 kV/400 V solid-state against a conventional low-frequency distribution transformer," in Proc. IEEE Energy Convers. Congr. Expo. (ECCE), Pittsburgh, PA, USA, pp. 4545–4552, Sep. 2014.

[5] Mizuki Nakahara, Yuki Kawaguchi, Kimihisa Furukawa, Mitsuhiro Kadota, Yuichi Mabuchi, Akihiko Kanoda, Development of a Control Method for LLC Converter Utilized for Input-Parallel-Output-Series Inverter System with Solid-State Transformers, IEEJ Journal of Industry Applications, vol. 8 , no. 4, pp. 652–659, 2019.

[6] J. E. Huber, J. Böhler, D. Rothmund and J. W. Kolar, "Analysis and cell-level experimental verification of a 25 kW all-SiC isolated front end 6.6 kV/400 V AC-DC solid-state transformer," in CPSS Transactions on Power Electronics and Applications, vol. 2, no. 2, pp. 140–148, 2017.

[7] A. Rodriguez et al., "Auxiliary power supply based on a modular ISOP flyback configuration with very high input voltage," 2016 IEEE Energy Conversion Congress and Exposition (ECCE), Milwaukee, WI, pp. 1–7, 2016.

[8] T. Zhao, G. Wang, S. Bhattacharya and A. Q. Huang, Voltage and Power Balance Control for a Cascaded H-Bridge Converter-Based Solid-State Transformer, in IEEE Transactions on Power Electronics, vol. 28, no. 4, pp. 1523–1532, April 2013.

[9] T. Isobe, H. Tadano, Z. He, and Y. Zou: "Control of Solid-State-Transformer for Minimized Energy Storage Capacitors", IEEE Energy Convers. Congr. Expo. (ECCE), pp.3809–3815, 2017.

[10] H. Qin and J. W. Kimball, "Solid-State Transformer Architecture Using AC–AC Dual-Active-Bridge Converter," in IEEE Transactions on Industrial Electronics, vol. 60, no. 9, pp. 3720–3730, Sept. 2013.

[11] J. Ramos-Ruiz, H. Krishnamoorthy and P. Enjeti, "Adding capacity to an existing electric power distribution network using a solid state transformer system," 2015 IEEE Energy Conversion Congress and Exposition (ECCE), Montreal, QC, pp. 6059–6066, 2015.

[12] S. Baek and S. Bhattacharya, "Analytical Modeling and Implementation of a Coaxially Wound Transformer With Integrated Filter Inductance for Isolated Soft-Switching DC–DC Converters", in IEEE Transactions on Industrial Electronics, vol. 65, no. 3, pp. 2245–2255, March 2018.

[13] X. Cai et al., "Fluctuation Power Control Strategy for MMC-based SST to Reduce the Submodule Capacitor Voltage Oscillation," 2019 10th International Conference on Power Electronics and Energy Conversion Congress and Exposition - Asia (ICPE 2019 -ECCE Asia), Busan, Korea (South), pp. 2430–2435, 2019.

[14] T. M. Parreiras, A. P. Machado, F. V. Amaral, G. C. Lobato, J. A. S. Brito and B. C. Filho, "Forward Dual-Active-Bridge Solid-State Transformer for a SiC-Based Cascaded Multilevel Converter Cell in Solar Applications," in IEEE Transactions on Industry Applications, vol. 54, no. 6, pp. 6353–6363, Nov.–Dec. 2018.

[15] J. E. Huber, J. Böhler, D. Rothmund and J. W. Kolar, "Analysis and cell-level experimental verification of a 25 kW all-SiC isolated front end 6.6 kV/400 V AC-DC solid-state transformer," in CPSS Transactions on Power Electronics and Applications, vol. 2, no. 2, pp. 140–148, 2017.

[16] Jun-ichi Itoh, Kazuki Aoyagi, Keisuke Kusaka, Masakazu Adachi,Development of Solid-state Transformer for 6.6-kV Single-phase Grid with Automatically Balanced Capacitor Voltage, IEEJ Journal of Industry Applications, vol.8, no.5 , p. 795–802, 2019.

[17] X. Zhang, Y. Xu, Y. Long, S. Xu and A. Siddique, "Hybrid-Frequency Cascaded Full-Bridge Solid-State Transformer," in IEEE Access, vol. 7, pp. 22118–22132, 2019.

[18] L. Zhang, J. Qin, Q. Duan and W. Sheng, "Component Sizing and Voltage Balancing of MMC-based Solid-State Transformers Under Various AC-Link Excitation Voltage Waveforms," 2019 IEEE Applied Power Electronics Conference and Exposition (APEC), Anaheim, CA, USA, pp. 371–375, 2019.

[19] T. Liu et al., "Design and Implementation of High Efficiency Control Scheme of Dual Active Bridge Based 10 kV/1 MW Solid State Transformer for PV Application," in IEEE Transactions on Power Electronics, vol. 34, no. 5, pp. 4223–4238, May 2019.

[20] Yusuke Hayashi, Yoshikatsu Matsugaki, Tamotsu Ninomiya, Capacitively Isolated Multicell DC-DC Transformer for Future Dc Distribution System, IEEJ Journal of Industry Applications, vol.6, no.4, p.268–277 2017.

[21] ROHM Co., Ltd., Japan: Datasheet of SiC Power Device (SCT3040KR).[Online]. Available: https://fscdn.rohm.com/en/products/databook/datasheet/discrete/sic/mosfet/sct3040kr-e.pdf

[22] ON Semiconductor., USA:Datasheet of Si IGBT(FGA20N120FTD).[Online]. Available:https://www.onsemi.jp/pub/Collateral/FGA20N120FTDJP-D.pdf

A New ZVS Zone Identification for Dual Active Bridge with a General Modulation Objective

Suman Maharana*, Dipankar De*, and Alberto Castellazzi+
*School of Electrical Sciences, Indian Institute of Technology Bhubaneswar, Arugul Campus,
Jatni-752050, Odisha, INDIA
Phone: +91 674 7134561
Email: sm64@iitbbs.ac.in
URL: http://www.iitbbs.ac.in/

+KUAS Engineering Department,
Kyoto University of Advanced Science,
18 Yamanouchi Gotanda-cho, Ukyo-ku, Kyoto 615-8577, JAPAN
Phone: +81-75-496-6504
Email: alberto.castellazzi@kuas.ac.jp
URL: https://www.kuas.ac.jp/

Acknowledgments

This work is supported by Department of Science and Technology, India under project grant ECR/2017/001079.

Keywords

≪High frequency power converter≫, ≪Silicon Carbide (SiC)≫, ≪Soft switching≫, ≪Modulation strategy≫, ≪DSP≫

Abstract

In this work, a performance optimization for dual active bridge is focused with a novel zero-voltage-switching-zone identification in instantaneous link current-transmitted power plane with the variation of inner and outer phase shifts. The proposed solution provides the soft switching zones for all switches along with an effective mapping of any general modulation objective like link current optimization.

Introduction

Soft switching zone identification and attempts to operate dual active bridge with zero voltage switching and/or zero current switching for entire operating range with large variations in input-to-output voltage ratio are quite common in the literature [1]-[6]. However in many applications, the dual active bridge operates with nearly constant input and output voltages (or with small voltage variations) [7]. The various soft switching operating zones with outer phase shift variation and with input-to-output voltage ratio variations become invalid in these constant input-output applications. The operating zone in such cases become very narrow as shown in the Fig. 1.

In the present work a new soft switching zone identification is carried out using power vs instantaneous link current at switching instants. The power and the instantaneous link currents are considered with fundamental and odd harmonic orders with not only outer phase shift variation but also inner phase shift variations. However circulating mode is selected which is a extreme case of narrow input variation for simplicity. The effectiveness of the analysis carried out is equally applicable in variation of k case as well.

EPE'20 ECCE Europe

Assigned jointly to the European Power Electronics and Drives Association & the Institute of Electrical and Electronics Engineers (IEEE)

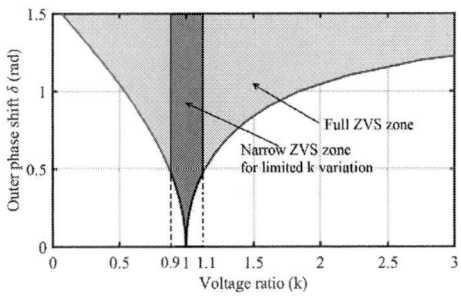

Fig. 1: ZVS zone with δ-k plane showing k variation from 0.9 to 1.1

The advantage of the method is that any general modulation scheme [8, 9] can be mapped on the identified ZVS zone to identify upto what extent that modulation scheme can be achieved without losing any soft switching for any of the eight switches in dual active bridge for full power range. Switching losses can be minimized by operating converter in soft switching region i.e. ZVS turned on and ZCS turned off. MOSFETs are generally operated at high frequency where the turn-on loss is significant than turn-off loss. Hence, ZVS analysis is preferred over ZCS [10]. The method is finally tested with optimum reactive power based modulation [11, 12] and simulation, analytical observation are presented. The results from simulation and experimental investigations are presented in this paper. The paper is organized as follows. Firstly, the soft switching zones are described for Dual active bridge under consideration. Next, the simulation and experimental results are presented and comparisons with analytically calculated data are evaluated. Finally the conclusion of the work is provided.

Description of Soft Switching Zones

To operate DAB switches in soft switching condition, one should define soft switching zones for individual switches over all power range and study the number of switches coming in soft switching zone for different power ranges. The next should be to develop a modulation strategy which can operate all switches in soft switch region for whole power range. For achieving ZVS across the switches, I_{LINK} in the high frequency (HF) AC side shown in Fig. 3 , plays a key role as its instantaneous value at the switching instant causes the switch output capacitor (C_{oss}) to discharge shown in Fig. 2. This forward biases the anti-parallel diode across the MOSFET switch which makes the voltage across the switch to zero at the switching instant. In off condition, C_{oss} charges up to full input voltage. Hence, to discharge

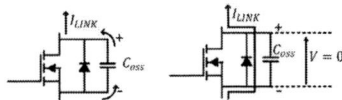

Fig. 2: MOSFET switch capacitor C_{oss} discharged by I_{LINK}

C_{oss}, two basic conditions should be satisfied to achieve ZVS across switch at switching on time [12].

(i) I_{LINK} direction should such that it discharges C_{oss} as shown in Fig. 2.

(ii) The magnitude of current at switching instant must be large enough to completely discharge the capacitor and make the body diode conducting.

Soft switching zones can be developed from this concept for all eight switches. Considering I_{LINK} waveform to be symmetrical across time-axis it can be concluded that, if one switch of a leg achieves ZVS, automatically other switch will also achieve ZVS, as they are complementary switch pairs [12]-[14]. Hence, S_4, S_2, Q_2 and Q_4 switches are taken into consideration. Soft switching condition for the same switches, satisfying the two conditions can be represented as follows.

- At S_4 switching on,

$$I_{LINK}\left(\omega t = \frac{\alpha}{2}\right) \leq -I_{mp} \tag{1}$$

- At S_2 switching on,

$$I_{LINK}\left(\omega t = \pi - \frac{\alpha}{2}\right) \geq I_{mp} \tag{2}$$

- At Q_4 switching on,

$$I_{LINK}\left(\omega t = \delta + \frac{\beta}{2}\right) \geq I_{ms} \tag{3}$$

- At Q_2 switching on,

$$I_{LINK}\left(\omega t = \pi + \delta - \frac{\beta}{2}\right) \leq -I_{ms} \tag{4}$$

Negative sign indicates the current should be in negative direction to achieve soft switching. I_{mp} and I_{ms} are minimum current required to achieve ZVS for primary and secondary (referred to primary) bridge respectively. In practice these values are very small as they directly relate with device capacitance (C_{oss}) which are very small (in pF range). Hence, these values can be approximated to zero during zone formation. $\omega t = (\frac{\alpha}{2}), (\pi - \frac{\alpha}{2}), (\delta + \frac{\beta}{2}), (\pi + \delta - \frac{\beta}{2})$ are switching on instants of S_4, S_2, Q_4, Q_2 respectively.

Considering the above conditions of soft switching (ZVS), zones of soft switching can be plotted between I_{LINK} (ωt = switching instant) and fundamental active power transfer (P) varying α (0 to π) and δ (0 to $\pi/2$) keeping β and all other parameters constant. Active power transfer equation [15] is given below.

$$P = \sum_{n=1,3,5...}^{\infty} \frac{8V_1 V_2}{n^3 \pi^2 \omega L} cos\left(\frac{n\alpha}{2}\right) cos\left(\frac{n\beta}{2}\right) sin(n\delta) \tag{5}$$

I_{LINK} equation derived from AC link voltages [15] is given as

$$I_{LINK} = \sum_{n=1,3,5...}^{\infty} \frac{4}{n^2 \pi \omega L}\{Acos(n\omega t) + Bsin(n\omega t)\} \tag{6}$$

Where

$$A = V_2' cos\left(\frac{n\beta}{2}\right) cos(n\delta) - V_1 cos\left(\frac{n\alpha}{2}\right)$$

$$B = V_2' cos\left(\frac{n\beta}{2}\right) sin(n\delta)$$

$$V_2' = \frac{V_2}{N}$$

Instantaneous current expressions of I_{LINK} at switching instant for different switches are given by substituting the switching instants in place of ωt in (6)

$$S4: \quad I_{LINK}\left(\frac{\alpha}{2}\right) = \sum_{n=1,3,5...}^{\infty} \frac{4}{n^2 \pi \omega L}\left\{Acos\left(\frac{n\alpha}{2}\right) + Bsin\left(\frac{n\alpha}{2}\right)\right\} \tag{7}$$

$$S2: \quad I_{LINK}\left(\pi - \frac{\alpha}{2}\right) = \sum_{n=1,3,5...}^{\infty} \frac{4}{n^2 \pi \omega L}\left\{Acos\left(n\pi - \frac{n\alpha}{2}\right) + Bsin\left(n\pi - \frac{n\alpha}{2}\right)\right\} \tag{8}$$

$$Q4: \quad I_{LINK}\left(\delta+\frac{\beta}{2}\right) = \sum_{n=1,3,5...}^{\infty} \frac{4}{n^2\pi\omega L}\left\{A\cos\left(n\delta+\frac{n\beta}{2}\right)+B\sin\left(n\delta+\frac{n\beta}{2}\right)\right\} \tag{9}$$

$$Q2: \quad I_{LINK}\left(\pi+\delta-\frac{\beta}{2}\right) = \sum_{n=1,3,5...}^{\infty} \frac{4}{n^2\pi\omega L}\left\{A\cos\left(n\pi+n\delta-\frac{n\beta}{2}\right)+B\sin\left(n\pi+n\delta-\frac{n\beta}{2}\right)\right\} \tag{10}$$

In this paper, DAB is presented in circulating mode configuration shown in Fig. 3. This configuration is considered for generating the soft switching zones with the parameters $f = 20kHz, V_1 = V_2 = 300V, \beta = 0, N = 1, L = 360\mu H$. $I_{mp} = I_{ms} = 0$ are taken as per ideal consideration. However actual values are $I_{mp} = I_{ms} = 0.145A$ which depends upon C_{oss} given in Table I, taken from SCT2450KE SiC MOSFET data sheet [16].

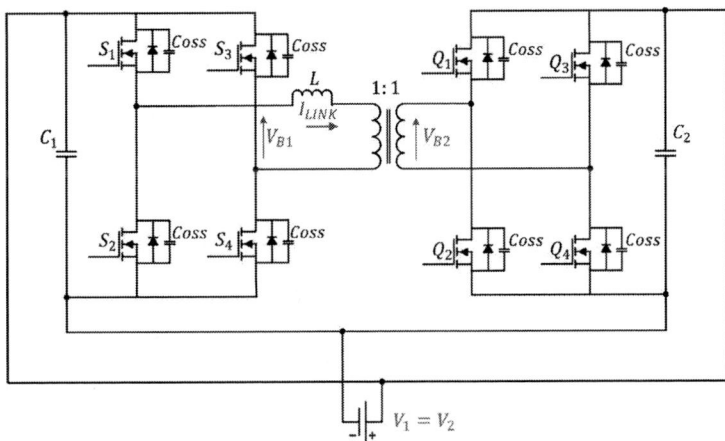

Fig. 3: DAB in circulating mode

Fig. 4(a) shows the operating lines (mapped in I_{LINK}-P plane) for five different values of α with δ changing form 0 to $\pi/2$ with the instant of the link current (I_{LINK}) is taken as that of switching instant of S_4. Fig. 4(b) repeats the same Fig. 4(a) but it shows fine variations of α which creates a full operating zone for S_4 including all harmonic components of power and link current. Red line shows the boundary of soft switching which divides the whole zone into two parts (hard switch zone and soft switch zone), according to the given condition in (1). Similarly, Fig. [4(c), 4(d)], [5(a), 5(b)] and [5(c), 5(d)] show zones for S_2, Q_4 and Q_2 respectively (which are generated according to the given conditions in (2),(3) and (4) respectively). From S_4 and S_2 zone observation, it is clear that with higher α, current stress increases rapidly for less power loading. However from Q_4 and Q_2 zone it can be observed that, switching point link current value remains constant for a power range where $\delta \leq \alpha/2$. However the constant range gets maximum at $\alpha = \pi/2$. For $\delta > \alpha/2$ case of power loading, instantaneous current value increases with increase in power.

Moreover the variation of α and δ in all the zones are independent to each other which give the total soft switching zones for any condition of power. However the general optimized modulation scheme for power transfer (OMS-III) described in [9] gives a relationship between α and δ with an optimization factor $F \in [1,2]$. The equation is given as

$$\delta = \cos^{-1}\left[\frac{1}{Fk\cos(\alpha/2)}\right] \tag{11}$$

(11) is substituted in (5), (7), (8), (9) and (10) to get optimized trajectory of soft switching on the existing zones for any value of $F \in [1,2]$. In Fig. 6, $F = 2$ (which is nothing but reactive power optimization [9])

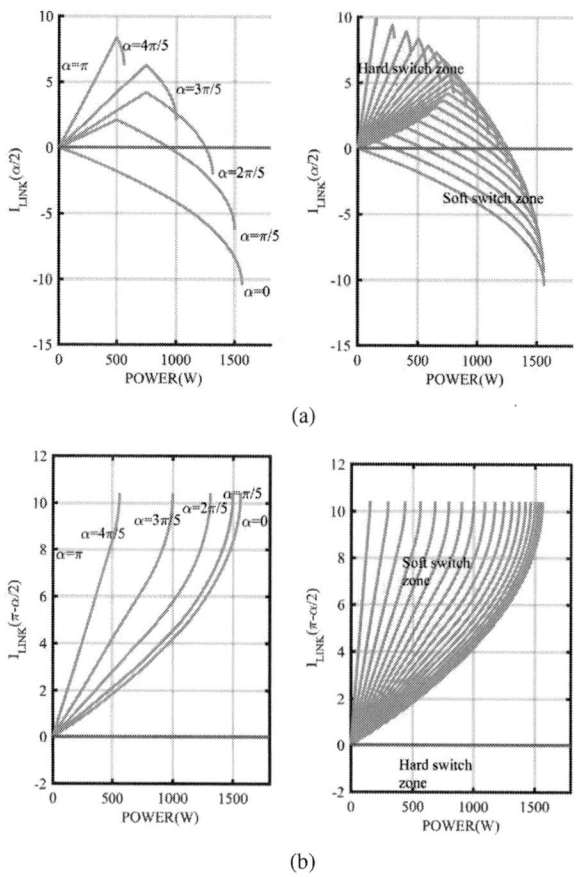

(a)

(b)

Fig. 4: S_4 soft switch zone (a), S_2 soft switch zone (b) - (left: a few illustrative curves with different α, right: with more curves)

is considered as an example for getting optimized trajectory indicated as solid blue lines. The optimized trajectory ends at maximum possible power loading with satisfying the optimization condition. After that, trajectory moves on $\alpha = 0$ line to reach the full power loading which is indicated by dotted blue lines in the same figure.

Simulation Results

For verifying the zones, each point on optimized trajectory of all zones for a power reference shown in Fig. 6, are simulated using MATLAB model. Parameters taken for simulation are given in TABLE I.

Table I: Simulation parameters

Parameters	Values
V_1	300 V
V_2	300 V
C_1, C_2	100 μF
C_{oss}	21 pF
Switching frequency	20 kHz
Inductor (L)	360 μH
Transformer turns ratio	1:1

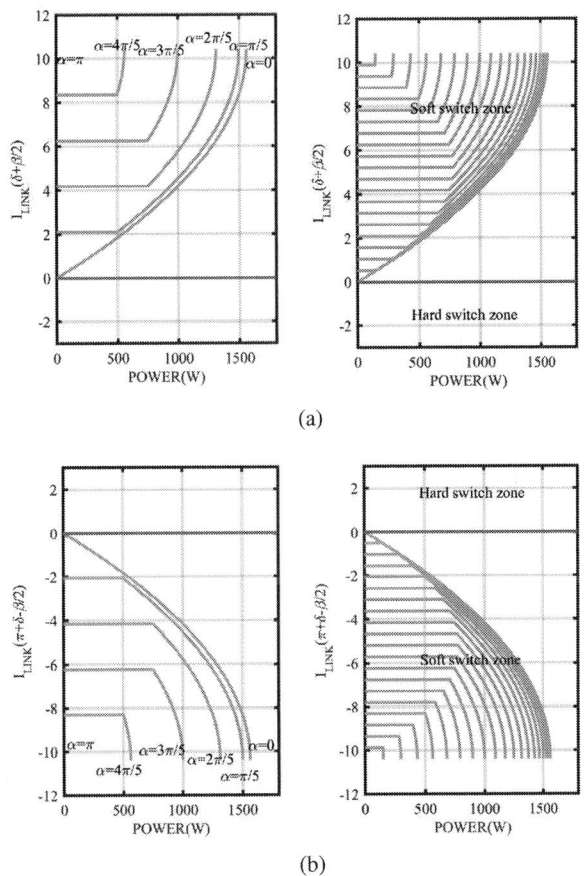

(a)

(b)

Fig. 5: Q_4 soft switch zone (a), Q_2 soft switch zone (b) - (left: a few illustrative curves with different α, right: with more curves)

The simulation verification is done at $P = 900W$ for all switch zones on reactive power optimized trajectory. At this power level $\alpha = 82.94°$ which is indicated as black curve on each zone showing constant α line with $\delta = 48.14°$, $\beta = 0°$.

I_{LINK} and switch voltages at switching instant are shown in Fig. 7 which gives instantaneous current, voltage value at switching instant and ZVS status. They are tabulated in the TABLE II.

Table II: I_{LINK} and switch voltage values at switching instants for Fig. 7

Switch	Switching instant	I_{LINK} value (A)	Switch voltage (V)	ZVS status
S_4	$\alpha/2$	5.07	300	NO
S_2	$\pi - \alpha/2$	4.48	0	YES
Q_4	$\delta + \beta/2$	5.58	0	YES
Q_2	$\pi + \delta - \beta/2$	-5.68	0	YES

Experimental Verification

Experimental setup for DAB of 1.5 kW rating is connected in circulating mode, which is taken for verification of zones, shown in Fig. 8(a). DAB gate signals are generated and controlled by DSP controller (TMS320F28335).

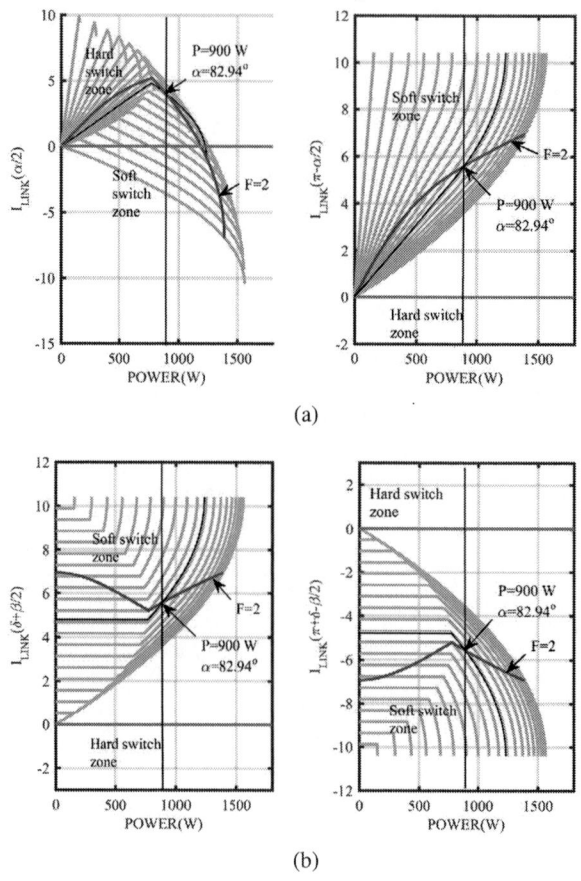

(a)

(b)

Fig. 6: Optimized performance curve on soft switching zone for S_4 (a)-left, S_2 (a)-right, Q_4 (b)-left, Q_2 (b)-right

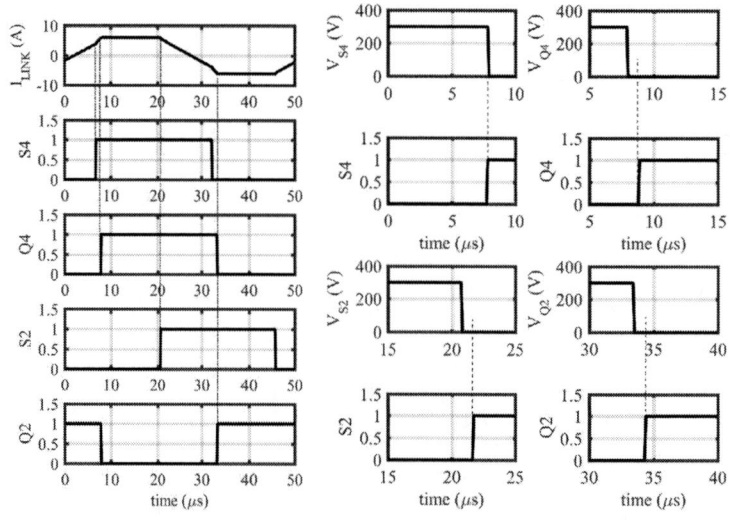

Fig. 7: Simulation result for I_{LINK} and switch voltages at switching instants for 900 W on $F = 2$

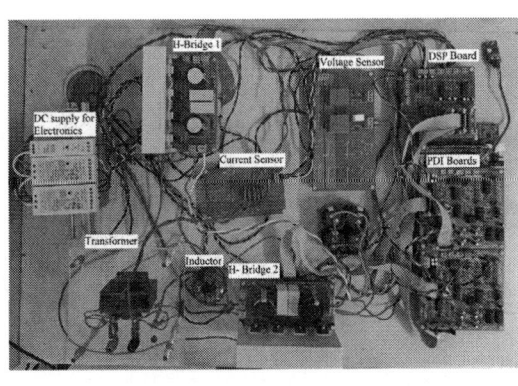

(a) Experimental set-up for circulating mode

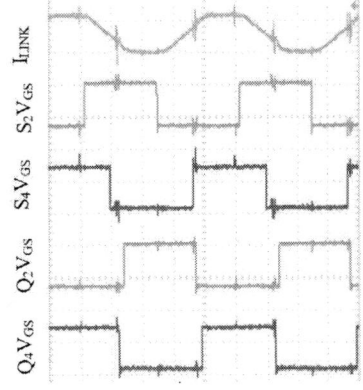

(b) Experimental results for I_{LINK} (10 A/div) and gate voltages (20 V/div) with time scale 10 μs/div

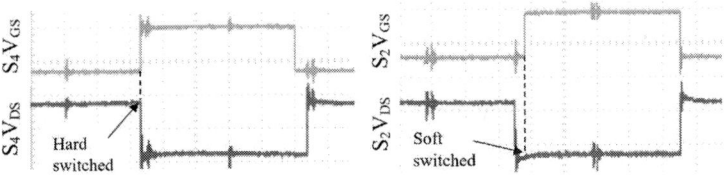

Fig. 9: Experimental results for the primary H bridge switch voltage (V_{DS}) (200 V/div) and gate voltage (V_{GS}) (20 V/div) with time scale 5 μs/div

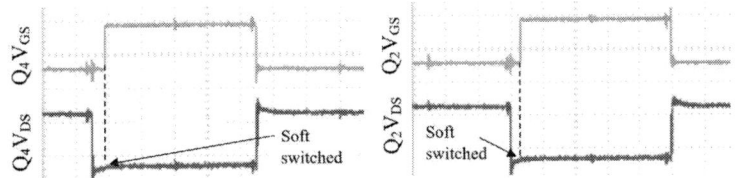

Fig. 10: Experimental results for the secondary H bridge switch voltage (V_{DS}) (200 V/div) and gate voltage (V_{GS}) (20 V/div) with time scale 5 μs/div

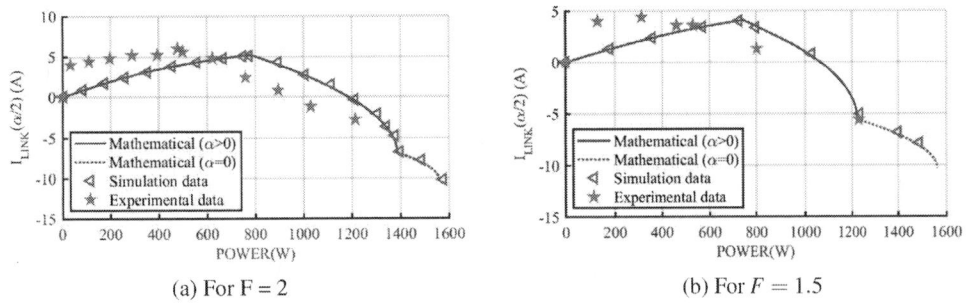

(a) For F = 2

(b) For $F = 1.5$

Fig. 11: Experimental and simulation verification of the optimized trajectories for S_4 zone

The I_{LINK} current with respect to switching instant of gate voltages V_{GS} are shown in Fig. 8(b). The current value and direction criteria for soft switching is satisfied by all three switches S_2, Q_2, Q_4 except S_4. The switch voltage (V_{DS}) with respect to switching instants are shown in the Fig. 9 and Fig. 10. From the figures it is clear that S_4 is hard switched. Moreover full operating trajectories for two different factors

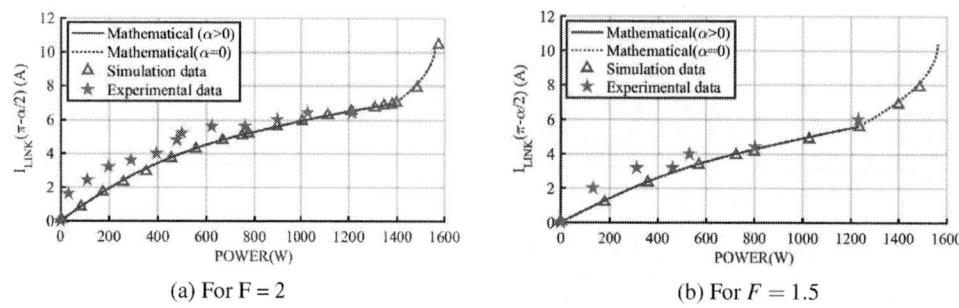

(a) For F = 2   (b) For $F = 1.5$

Fig. 12: Experimental and simulation verification of the optimized trajectories for S_2 zone

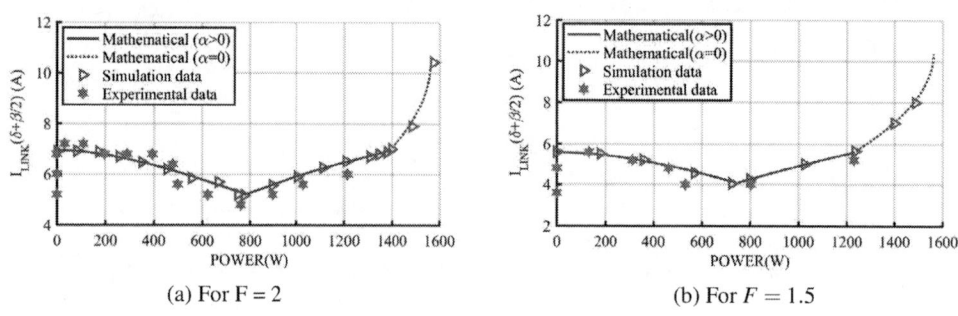

(a) For F = 2   (b) For $F = 1.5$

Fig. 13: Experimental and simulation verification of the optimized trajectories for Q_4 zone

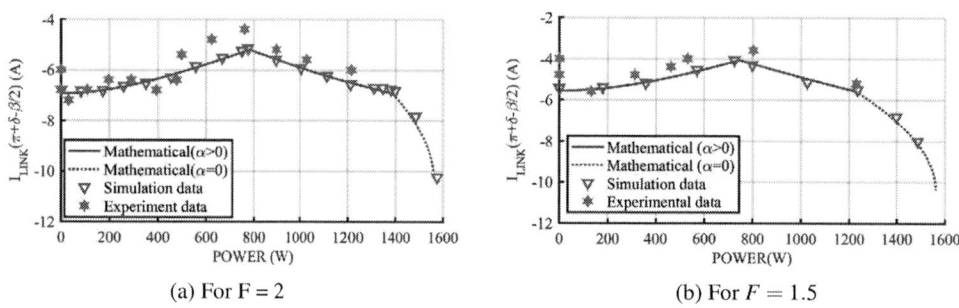

(a) For F = 2   (b) For $F = 1.5$

Fig. 14: Experimental and simulation verification of optimized trajectories for Q_2 zone

(F) for general optimized modulation on soft switching zones, verified with simulation and experimental results are presented in the figures (Fig. 11 to Fig. 14).

Fig. 11, Fig. 12, Fig. 13 and Fig. 14 shows the experimental and simulation verification data for optimized trajectories of the soft switching zones obtained from mathematical analysis for S_4, S_2, Q_4 and Q_2 respectively. Solid line indicates the optimized trajectory satisfying the particular optimized condition with $\alpha > 0$ where α decreases with increase in power and becomes zero at maximum power limit of the optimization condition. Dotted line indicates the trajectory with $\alpha = 0$ after the maximum power limit with variation of δ only.

Conclusion

This work presents a new soft switching zone identification using power vs instantaneous link current at switching instants. The proposed solution provides the soft switching zones for all switches along with an effective mapping of any general modulation objective. Soft switching and hard switching zones for individual switch in DAB for particular configuration are analyzed by all possible variations of inner phase shift α and outer phase shift δ keeping $\beta = 0$. The mathematical expressions of link current and active power considered for zone generation are included with all odd harmonic components of Fourier series. Moreover the reactive power optimized trajectory is defined on new developed soft switching zones with its maximum limit of power loading and range of soft switching capability . The zones are verified with simulation and experimental results for different factors of general optimized modulation. From the results it is clear that there is a deviation of experimental data from simulation and mathematical data. This error is due effect of dead time in the circuit and can be effectively corrected by implementing appropriate dead time compensation.

References

[1] Riedel J., Holmes D. G., McGrath B. P., and Teixeira C.: ZVS soft switching boundaries for dual active bridge DC–DC converters using frequency domain analysis, IEEE Trans. Power Electron., vol. 32, no. 4, pp. 3166–3179, Apr.2017.

[2] Jin L., Liu B., and Duan S.: ZVS operation range analysis of three-level dual active bridge DC-DC converter with phase-shift control, IEEE APEC, Tampa, FL, 2017, pp. 362-366.

[3] He X., Zhang Z., Cai Y., and Liu Y.: A variable switching frequency hybrid control for ZVS dual active bridge converters to achieve high efficiency in wide load range, IEEE APEC 2014, Fort Worth, TX, 2014, pp. 1095-1099.

[4] Fei X., Feng Z., PuQi N., and Xuhui W.: Analyzing ZVS Soft Switching Using Single Phase Shift Control Strategy of Dual Active Bridge Isolated DC-DC Converters, IEEE ICEMS, Jeju, 2018, pp. 2378-2381.

[5] Oggier G., García G. O., and Oliva A. R. : Modulation strategy to operate the dual active bridge DC-DC converter under soft switching in the whole operating range, IEEE Trans. Power Electron., vol. 26, no. 4, pp. 1228-1236, April 2011.

[6] Zhang Z., Thomsen O. C., and Andersen M. A. E.: Soft-Switched Dual-Input DC–DC Converter Combining a Boost-Half-Bridge Cell and a Voltage-Fed Full-Bridge Cell, IEEE Trans. Power Electron., vol. 28, no. 11, pp. 4897-4902, Nov. 2013.

[7] Zengin S., and Boztepe M.: Modified dual active bridge photovoltaic inverter for solid state transformer applications, IEEE ISFEE, Bucharest, 2014, pp. 1-4.

[8] Mukherjee S., Dash A., De D., and Castellazzi A.: Study of Dual Active Bridge with Modified Modulation Techniques for Harmonic Reduction in AC Link Current, 2019 IEEE ICSETS, Bhubaneswar, India, 2019, pp. 144-149.

[9] Mukherjee S., Dash A., De D., and Castellazzi A.: Trade-off in Minimization of Fundamental Link Current and Reactive Power using a Novel Online Calculation based Triple Phase Shift Modulator for Dual Active Bridge, EPE 2019 ECCE Europe, Genova, Italy, 2019.

[10] Shen Z., Burgos R., Boroyevich D., and Wang F.: Soft-switching capability analysis of a dual active bridge dc-dc converter, 2009 IEEE Electric Ship Technologies Symposium, Baltimore, MD, 2009, pp. 334-339.

[11] Wen H., and Su B.: Reactive power and soft-switching capability analysis of dual-active-bridge DC–DC converters with dual-phase-shift control, Journal Power Electron., vol. 15, pp. 18–30, Jan. 2015.

[12] Shi H., Wen H., Chen J., Hu Y., Jiang L., and Chen G.: Minimum-Reactive-Power Scheme of Dual-Active-Bridge DC–DC Converter With Three-Level Modulated Phase-Shift Control, IEEE Trans. Ind. Applicat., vol. 53, no. 6, pp. 5573-5586, Nov.-Dec. 2017.

[13] Calderon C., Barrado A., Rodriguez A., Lazaro A., Sanz M., and Olías E.: Dual active bridge with triple phase shift, soft switching and minimum RMS current for the whole operating range, IEEE IECON, Beijing, 2017, pp. 4671-4676

[14] Lagier T., and Ladoux P.: Theoretical and experimental analysis of the soft switching process for SiC MOS-FETs based Dual Active Bridge converters,IEEE SPEEDAM, Amalfi, 2018, pp. 262-267.

[15] Zhao B., Song Q., Liu W., Liu G., and Zhao Y.: Universal High-Frequency-Link Characterization and Practical Fundamental-Optimal Strategy for Dual-Active-Bridge DC-DC Converter Under PWM Plus Phase-Shift Control, IEEE Trans. Power Electron., Vol. 30, No. 12, Dec. 2015

[16] SCT2450KE, 1200V, 10A N-channel SiC Power MOSFET data-sheet, available on-line: https://www.rohm.com/products/sic-power-devices/sic-mosfet/sct2450ke-product

Single-Stage Boost Modular Multilevel Converter (BMMC) for Energy Storage Interface

Ahmed Abdelhakim[1], Frede Blaabjerg[2], and Hans-Peter Nee[3]

[1] ABB Corporate Research Center, Västerås, Sweden

[2] Aalborg University, Aalborg, Denmark

[3] KTH Royal Institute of Technology, Stockholm, Sweden

Emails: ahmed.abdelhakim@se.abb.com, fbl@et.aau.dk, hansi@kth.se

Keywords

≪Boost≫, ≪DC-AC converter≫, ≪Energy storage≫, ≪Fuel cell≫, ≪Low-voltage≫, ≪Modular multilevel converter≫, ≪Photovoltaic≫, ≪Single-stage≫, ≪Two-stage≫.

Abstract

Single-stage DC-AC power converters are gaining higher attention due to their simpler structure compared to the two-stage equivalent solution. In this paper, a single-stage DC-AC converter solution is proposed for interfacing a low voltage (LV) DC source with a higher voltage AC load or grid, where this converter has a modular structure with multilevel operation. The proposed converter, which is called boost modular multilevel converter (BMMC), comprises the boosting capability within the inversion operation, and it is mainly dedicated for interfacing LV energy storage systems, such as fuel cells and batteries, and it allows the use of LV MOSFETs (< 300 V), in order to utilize their low ON-state resistance, along with LV electrolytic capacitors. This converter is introduced and analysed in this paper, where simulation results using PLECS, considering a 10 kW three-phase BMMC, are presented in order to verify its functionality.

1 Introduction

Power electronic converters are the backbone of many applications, such as grid-tied systems [1, 2, 3], variable speed motor drives [4, 5], and uninterruptible power supply (UPS) systems [6, 7]. Each of these applications has some specific technical requirements, and this implies the utilization of different power conditioning stages (PCSs) in order to fulfil these requirements. Within these different applications, low voltage (LV) DC sources, such as fuel cells and batteries, can be utilized, and this imposes a strict need for the boosting capability or an additional regulating stage in the utilized PCS as a consequence [8, 9]. Such boosting capability might be as high as a gain of ten, which cannot be fulfilled using the conventional boost converter shown in Fig. 1(a). Thus, high gain DC-DC converters have to be utilized, such as voltage multipliers or high frequency transformer-based solutions, where the dual-active bridge- (DAB) based solution, that is shown in Fig. 1(b), is commonly used [9]. On the other hand, a low frequency transformer has to be utilized in order to interface LV 1500 V photovoltaic (PV) systems with medium-voltage (MV) networks, where this transformer is acting as a boosting stage [10]. On the other hand, higher voltage-rated devices, e.g. 600 or 1200 V, should be utilized depending on the voltage and power levels of the targeted application, and the nature of the utilized PCS. Such utilization of higher voltage-rated semiconductor devices leads to increased switching and conduction losses, which result in a deteriorated power conversion efficiency of the entire system.

Hence, having a reliable, modular, compact, redundant, and efficient PCS is a common challenge in the aforementioned applications, where multilevel converters show superior performance in terms of

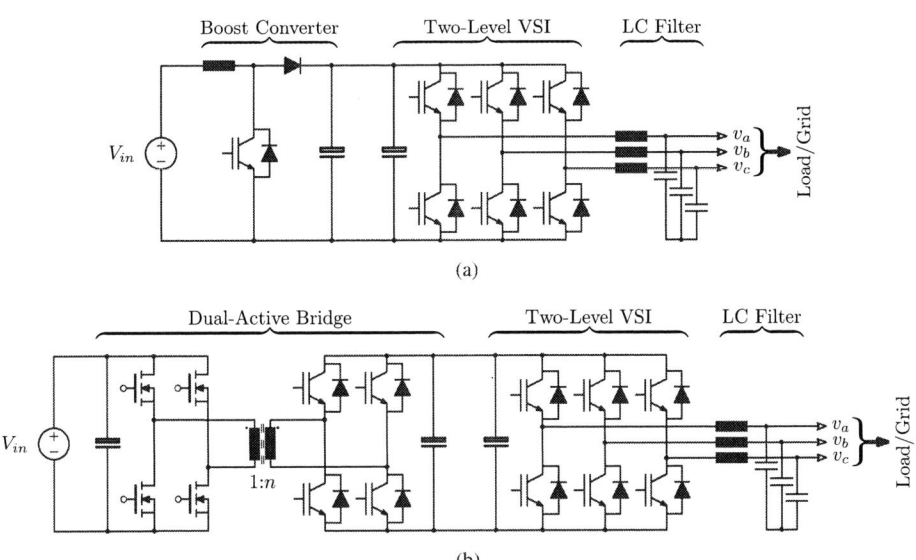

Fig. 1: State-of-the-art DC-AC power conditioning stages (PCSs) for interfacing low voltage (LV) DC sources with a higher voltage three-phase AC load or grid. (a) Boost converter-fed two-level voltage source inverter (VSI); and (b) dual-active bridge- (DAB) fed two-level VSI.

utilizing LV semiconductor devices along with LV electrolytic capacitors. Among the different multi-level converters, modular multilevel converters (MMCs), shown in Fig. 2, have demonstrated reliable, redundant, and efficient performance for numerous applications, such as high voltage DC (HVDC) systems [11, 12, 13, 14], flexible AC transmission systems (FACTS) [15, 16], MV motor drive systems [17, 18], and PV systems [19].

Over and above that, MMCs have recently been investigated for LV systems utilizing LV MOSFETs (< 300 V), for enhanced system efficiency and redundancy [20, 21, 22]. Meanwhile, it is still mandatory to utilize an additional boosting stage in order to integrate such LV DC sources, resulting in a bulkier two-stage PCS. It is worth to note that a full-bridge-based MMC solution can achieve this boosting capability within the inversion operation in a single-stage converter but it utilizes twice the number of semiconductor devices in the half-bridge-based MMC, which is seen to be bulky and complicated.

From the prior discussions and due to the seen merits behind the MMCs and the single-stage operation, this paper proposes an MMC-based single-stage DC-AC PCS solution for interfacing an LV DC source with a higher voltage AC load or grid. The proposed converter, which is called boost MMC (BMMC) and has the structure that is shown in Fig. 3, comprises the boosting capability within the inversion operation in a single-stage with a minimum number of LV semiconductor devices. It is worth to note that this converter is a combined solution between the conventional MMC configuration and the switched capacitor concept in the Marx converter [23]. As a consequence, the following merits are seen behind the proposed BMMC:

- It is single-stage DC-AC converter;
- It has a modular structure;
- It can achieve high reliability through added redundancy;
- Lower conduction losses and improved efficiency can be achieved as a consequence of utilizing LV MOSFETs;
- It has reduced voltage stresses across the different system components.

On the other hand, this converter suffers from unequal current stresses among the different semiconductor devices.

It is worth to note that the proposed BMMC can be utilized in interfacing LV 1500 V PV systems with

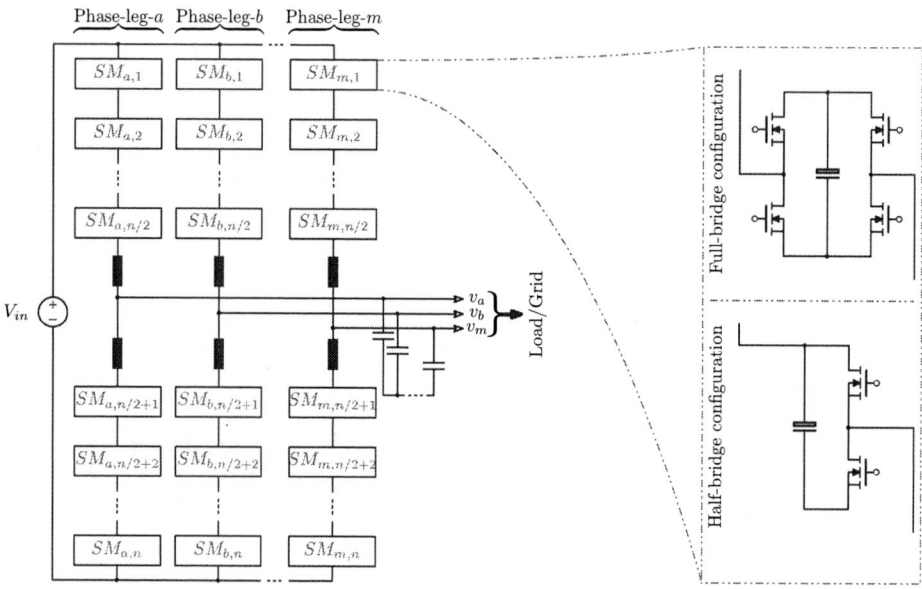

Fig. 2: General structure of the conventional modular multilevel converter (MMC), considering m phases and n sub-modules (SM) per phase-leg. Note that each SM can utilize full- or half-bridge configurations.

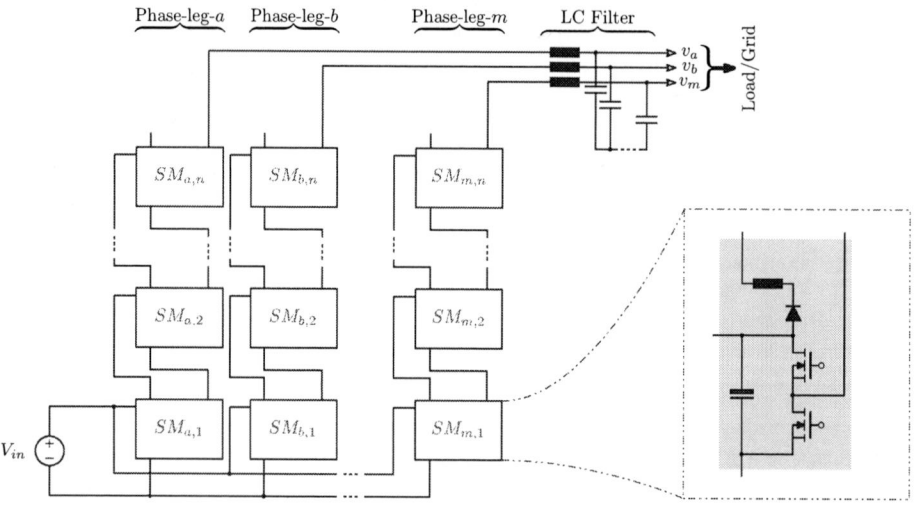

Fig. 3: General structure of the proposed boost MMC (BMMC) considering m phases and n sub-modules (SMs) per phase-leg, corresponding to $(n+1)$ levels.

MV networks, where the used LV MOSFETs in Fig. 3 can be replaced with IGBTs.

The rest of this paper is organized as follows: Section 2 introduces the structure of the proposed BMMC and describes its operation. Then, the modulation and mathematical derivation of the three-phase BMMC are then discussed in Section 3. In order to verify the functionality of the proposed BMMC and validate the introduced analysis, simulation results using PLECS are introduced in Section 4 considering a 10 kW three-phase BMMC with four SMs per phase-leg. Finally, the conclusions are drawn in Section 5.

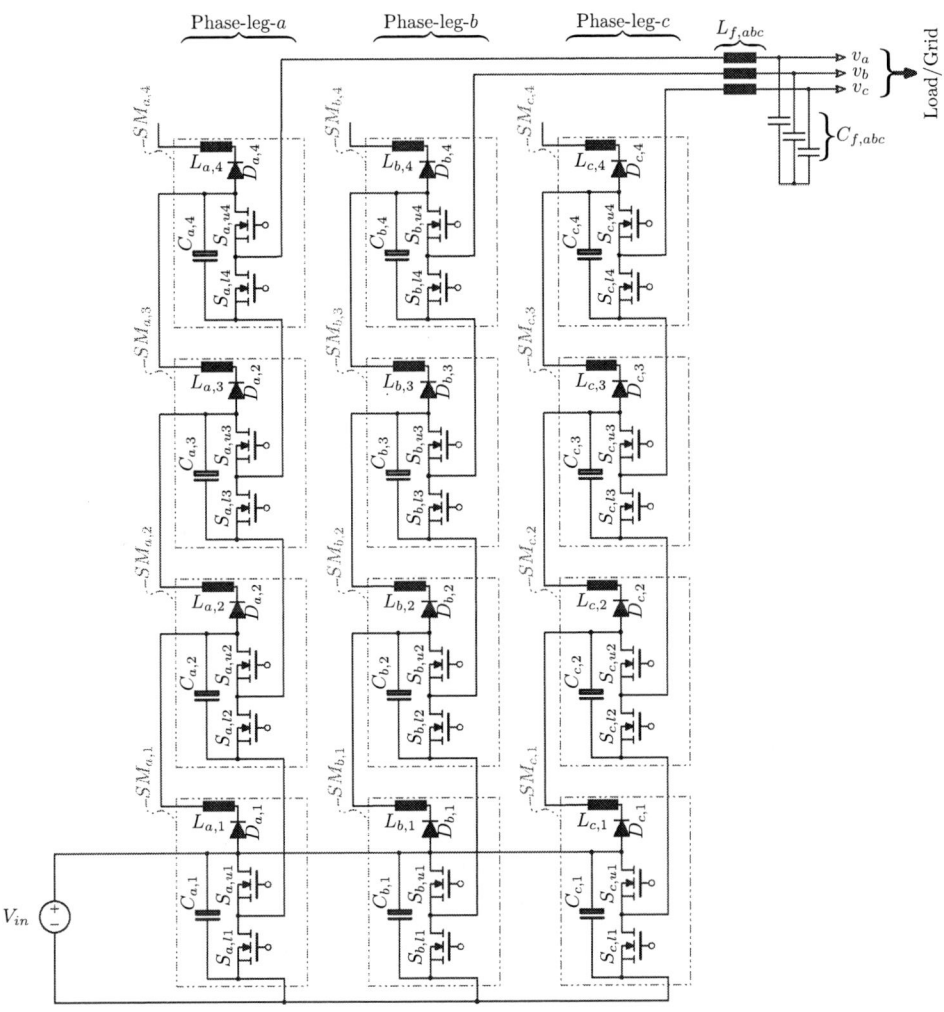

Fig. 4: Structure of the three-phase BMMC with four sub-modules (SMs) per phase-leg.

2   Structure and Operation of the Proposed BMMC

In general, the proposed BMMC can comprise any number of phase-legs in order to form a multiphase single-stage DC-AC converter, where each phase-leg comprises any number of sub-modules (SMs) to obtain the required boosting ratio and achieve the required number of voltage levels as depicted in Fig. 3, in which n SMs are shown with m phase-legs. Note that n SMs results in $(n+1)$ levels, where Fig. 4 shows the detailed structure of three-phase BMMC with four SMs per phase-leg that correspond to five-level operation.

Each SM, which is considered as the basic building block of the proposed BMMC, comprises two switches and one capacitor (C_x) to form the conventional half-bridge configuration, in addition to an inductor (L_x) that is connected in series with a diode and then connected to the positive node of the half-bridge as illustrated in Fig. 3.

Note that the utilized diode in each SM is floating, and it can be replaced with an active switch operating under zero voltage switching (ZVS) conditions for a synchronous rectification mode in order to improve the conversion efficiency. This diode is used to transfer the energy from one SM to the cascaded one. Furthermore, the utilized inductor in each SM is acting as a current limiter as it limits the current when the energy is transferred between cascaded SMs, resulting in low inductance requirements. Note that the

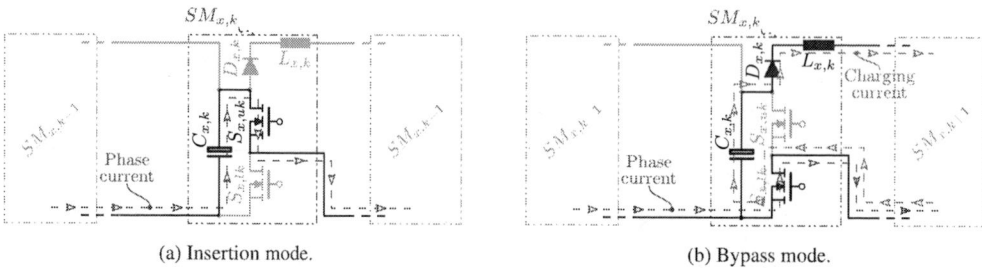

(a) Insertion mode. (b) Bypass mode.

Fig. 5: Modes of operation of any sub-module (SM) k in any phase-leg-x, denoted as $SM_{x,k}$. (a) Insertion mode and the SM capacitor energy is transferred to the load; and (b) bypass mode and the SM capacitor energy is transferred to successive cell.

floating diode and the inductor are not needed in the last or upper SM of each phase-leg. Finally, the DC input source (V_{in}) is parallel-connected to the capacitors of the first or bottom SM in each phase-leg as depicted in Fig. 3, i.e. the bottom cells are paralleled and connected to the DC input source.

In order to explain the operation of the proposed BMMC, the operation of the converter's basic building block, i.e. the utilized SM, has to be introduced first along with the energy transfer concept between the cascaded SMs. Each SM has two modes of operation as follows:

- Insertion mode which occurs when the upper switch ($S_{x,uk}$) is turned ON as shown in Fig. 5(a);
- Bypass mode which occurs when the lower switch ($S_{x,lk}$) is turned ON as shown in Fig. 5(b).

Under the insertion mode, which is shown in Fig. 5(a), the SM capacitor ($C_{x,k}$) is inserted through the upper switch ($S_{x,uk}$), and no energy is transferred to the successive SM, being x represents any phase-leg, and k represents the SM number. Meanwhile, under the bypass mode, which is shown in Fig. 5(b), $C_{x,k}$ is bypassed through $S_{x,lk}$, and the energy can be transferred to the successive SM capacitor (i.e. ($C_{x,k+1}$)) through the lower switch ($S_{x,lk}$) and SM floating diode ($D_{x,k}$) if the voltage across $C_{x,k}$ is larger than the voltage across $C_{x,k+1}$. Note that $L_{x,k}$ is limiting the rate of change of the charging current during this energy transfer period. On the other hand, if the voltage across $C_{x,k}$ is smaller than the voltage across $C_{x,k+1}$, no energy transfer will occur and $D_{x,k}$ will be reverse biased and turned OFF.

Utilizing those two modes of operation of any SM, the entire phase-leg can be controlled to transfer the energy from one SM to the successive one and obtain the desired voltage across this phase-leg at the same time. Hence, by sinusoidally modulating each phase-leg, a sinusoidal alternating voltage with an average DC component can be obtained across the entire phase-leg. Then, this DC component can be eliminated differentially between each two phase-legs, obtaining an AC component at the load side.

Since the energy is transferred from one SM to the cascaded one, a higher AC voltage can be obtained by increasing the number of SMs per phase-leg. Thus, redundant SMs can be added in order to further boost the input voltage if needed, as in fuel cells systems, in which the voltage will drop with the increase of the load power. Meanwhile, if the input voltage is sufficiently, those redundant SMs can be bypassed in order to minimize the converter losses and operate at higher modulation indices.

On the other hand, due to the energy transfer concept from one SM to the successive one, the SMs are stressed with different current levels, where the closest SM to the DC source, i.e. the bottom one, is highly stressed as it transfers the energy to all the successive SMs and the load, while the closest SM to the AC load side is experiencing the lowest stresses as it only transfers energy to the load. However, this drawback can easily be overcome with a proper design and utilization of paralleled MOSFETs that can be air cooled, resulting in lower complexity in the heat sink design. In many cases, depending on the power rating, the same semiconductors can be used, as LV MOSFETs present very low ON-state losses with low cost.

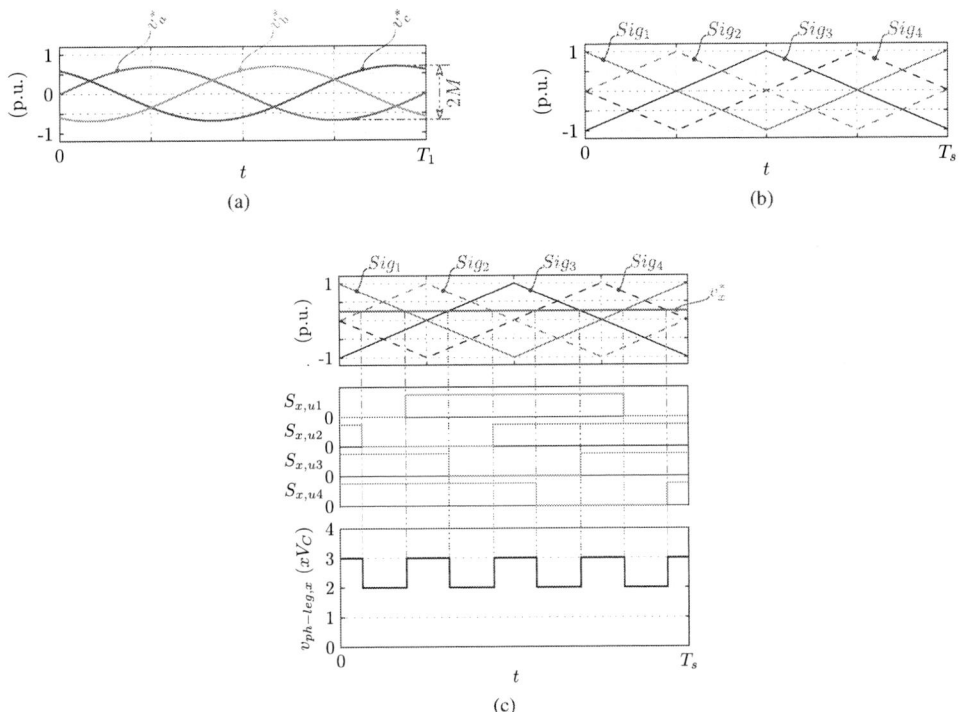

Fig. 6: Reference and carrier signals of the proposed three-phase BMMC. (a) Sinusoidal modulation reference signals for one fundamental cycle of time T_1; (b) phase-shifted carrier signals of any phase-leg-x for one switching cycle of time T_s, considering four SMs per phase; and (c) modulation of phase-leg-x for one switching cycle of time T_s, where the state of the upper switches in the different SMs are shown along with the sum of the SM capacitors' voltage, i.e. the phase-leg voltage ($v_{ph-leg,x}$). Note that x corresponds to a, b, or c, and each SM is assumed to have an average capacitor voltage of V_C.

3 Three-Phase BMMC Modulation

The three-phase BMMC, which is shown in Fig. 4 considering four SMs per phase-leg as an example, can be modulated using the traditional modulation schemes that are used with the conventional VSIs. This paper utilizes a sinusoidal modulation scheme, whose reference signals are shown in Fig. 6(a), with phase-shifted carriers, that are depicted in Fig. 6(b), in order to modulate this converter. It is worth to note that each SM has a corresponding carrier signal that is used along with the corresponding phase-leg reference signal to modulate the SM switches in a complementary manner.

Considering phase-leg-x, the different SMs can be modulated as illustrated in Fig. 6(c), in which the upper switch of each SM k, i.e. $S_{x,uk}$, is turned ON when the phase-leg reference signal v_x^* is larger than the corresponding carrier signal, otherwise it is turned OFF. For example, $S_{x,u1}$ is turned ON when v_x^* is larger than Sig_1, otherwise it is turned OFF as illustrated in Fig. 6(c). Note that the lower switch is operating complementary to the upper one. Then, the sum of the SM capacitors' voltage, i.e. the phase-leg voltage ($v_{ph-leg,x}$), which corresponds to the number of inserted SM in any phase-leg, as illustrated in Fig. 6(c). Note that $v_{ph-leg,x}$ is shown as a function of the average SM capacitor's voltage (V_C) assuming equal SM capacitors' voltage. Hence, by sinusoidally modulating each phase-leg, a sinusoidal AC voltage can be obtained across each phase-leg with a DC component, where this DC component can be eliminated differentially between the phase-legs.

Neglecting the converter losses, and the voltage ripple across each SM capacitor, and considering the parallel connection of the SM capacitors during each switching cycle over the fundamental period, the average SM capacitors' voltage (V_C) is assumed equal among the different SMs and it is given by

$$V_C \approx V_{in}, \tag{1}$$

where V_{in} is the input DC voltage. Then, the voltage across each phase-leg-x ($v_{ph-leg,x}$) can attain a maximum voltage of $\hat{v}_{ph-leg,x}$ when all the SMs are inserted, where this voltage is given by

$$\hat{v}_{ph-leg,x} = n \cdot V_C \approx n \cdot V_{in}, \tag{2}$$

being n represents the number of employed SMs per phase-leg as depicted in Fig. 3, where $n = 4$ which is the case that is shown in Fig. 4.

Then, using the sinusoidal modulation, whose reference signals are shown in Fig. 6(a), the output peak fundamental AC phase voltage ($\hat{v}_\varphi$) can be related to V_{in} by

$$\hat{v}_\varphi = M \cdot \frac{\hat{v}_{ph-leg,x}}{2} = M \cdot \frac{n \cdot V_C}{2} \approx M \cdot \frac{n \cdot V_{in}}{2}, \tag{3}$$

where M is the modulation index that is defined in Fig. 6(a).

Hence, for a certain input DC voltage, the minimum number of SMs per phase-leg, i.e. n, can be calculated in order to obtain the required output AC voltage assuming a maximum value of the modulation index. In other words, n can be calculated using

$$n \geq \frac{2 \cdot \hat{v}_\varphi}{M_{max} \cdot V_{in}}, \tag{4}$$

where M_{max} is the maximum desired modulation index under steady-state operation.

The SM capacitance (C_x) can be designed considering a maximum allowable voltage ripple, which can be estimated using

$$C_x \geq \frac{S_{con}}{3\omega \cdot \Delta V_{C,max1} \cdot \hat{v}_\varphi} \left(\frac{n \cdot M^2}{2} + 2 \right), \tag{5}$$

where S_{con} is the rated converter power in VA, ω is the fundamental frequency in rad/s, and $\Delta V_{C,max1}$ is the maximum allowable voltage ripple across each SM capacitor at the fundamental frequency. It is worth noting that this equation has been approximated using the energy balance in any phase-leg in the proposed BMMC and only considering the fundamental component in the capacitor voltage ripple, which is dominant.

On the other hand, the selection of the SM inductance (L_x) has to consider the maximum allowable current ripple in the charging current during the parallel connection of two successive SMs. The worst scenario occurs in the lower SM, where the voltage across this capacitor can be assumed constant as it is parallel-connected to the input DC source. Then, when the lower SM is parallel connected with the cascaded SM, a maximum current ripple would occur when the voltage across the cascaded SM reaches it's lowest value. In other words, the voltage difference between the two SM capacitors can reach a maximum value of $\Delta V_{C,max1}$ that will result in the highest current ripple in the inductor. Hence, the SM inductance (L_x) can be estimated based on the worst case scenario in the lower SM by

$$L_x \geq \frac{\Delta V_{C,max1}}{2 f_{sw} \cdot \Delta I_L}, \tag{6}$$

where f_{sw} is the switching frequency, and ΔI_L is the maximum allowable ripple in each inductor.

4 Simulation Results

For the sake of validating the functionality of the proposed BMMC and verifying the prior discussion and analysis, a 10 kVA three-phase BMMC with four SMs per phase-leg, whose equivalent circuit is shown

Table I: Parameters of the simulated 10 kVA three-phase BMMC with four SMs per phase-leg.

Input DC voltage (V_{in})	170 V	Output RMS phase voltage (V_φ)	220 V
Output fundamental frequency (f_1)	50 Hz	SM carrier frequency (f_{sw})	5 kHz
SM capacitance (C_x)	10 mF	SM inductance (L_x)	5 μH
Output filter capacitance (C_f)	10 μF	Output filter inductance (L_f)	50 μF

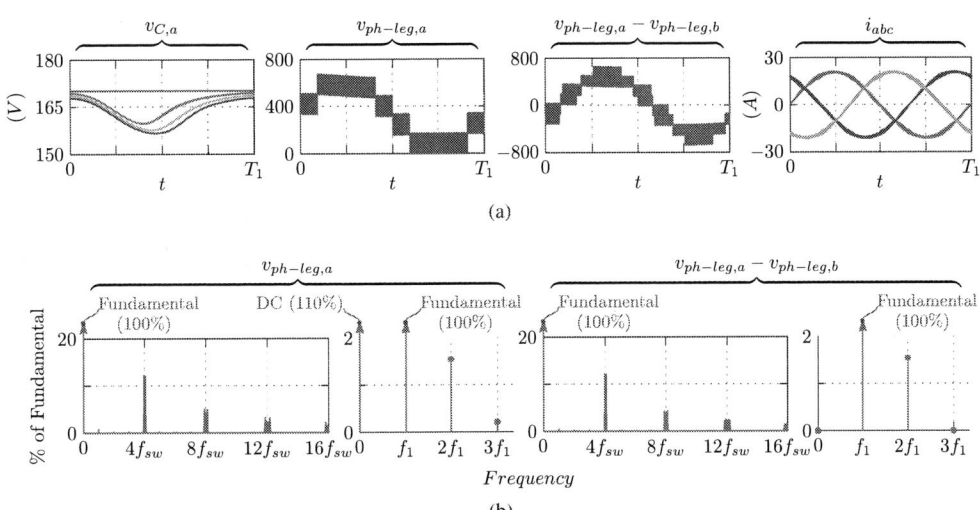

Fig. 7: Obtained open-loop simulation results of the three-phase BMMC with four sub-modules (SMs) per phase-leg. (a) SM capacitors' voltage of phase-leg-a ($v_{C,a}$), phase-leg-a voltage ($v_{ph-leg,a}$), voltage difference between phase-legs-a and b ($v_{ph-leg,a} - v_{ph-leg,b}$), and load currents ($i_{abc}$) are shown from left to right for one fundamental cycle of time T_1; and (b) frequency spectra of phase-leg-a voltage ($v_{ph-leg,a}$), and voltage difference between phase-leg-a and phase-leg-b ($v_{ph-leg,a} - v_{ph-leg,b}$). Note that $T_1 = 20$ ms and f_{sw} is the switching frequency of one SM.

in Fig. 4, is designed and simulated using PLECS. The parameters of the simulated BMMC are listed in Table I. This BMMC is feeding a resistive load of 10 kW through an LC filter, and the SM capacitance and inductance have been selected to limit the SM capacitor voltage ripple to be less than 7.4% of V_{in} and the maximum charging current ripple to be less than 250 A using (5) and (6) respectively. It is worth to note that since the capacitor voltage ripple is expected to be lower due to the approximated calculations, the current ripple shall be much lower.

The obtained simulation results are depicted in Fig. 7. Fig. 7(a) shows the SM capacitors' voltage of phase-leg-a ($v_{C,a}$), phase-leg-a voltage ($v_{ph-leg,a}$), voltage difference between phase-legs-a and b ($v_{ph-leg,a} - v_{ph-leg,b}$), and load currents ($i_{abc}$). On the other hand, the frequency spectra of phase-leg-a voltage ($v_{ph-leg,a}$), and voltage difference between phase-leg-a and phase-leg-b ($v_{ph-leg,a} - v_{ph-leg,b}$) are shown in Fig. 7(b). This figure shows that the arm voltage comprises a DC component that is taken out differentially at the load side. On top of that, the effective switching frequency at the load side is four times the SM switching frequency since a phase-shifted modulation is utilized.

In addition to that, the currents in each SM's inductor and switches in phase-leg-a are shown in Fig. 8. This figure sheds the light upon the main demerit of this converter, which is the asymmetry in the current ratings among the different SMs. This is coming from the fact that all the energy is transferred through the bottom SM and then this energy decreases gradually until it reaches the top SM. Such a demerit is not an issue for LV MOSFETs, where parallel operation of air cooled low cost SMD switches can easily be done. Finally, note that the SM capacitor voltage ripple is less than 7.4% of V_{in} as shown in Fig. 7(a),

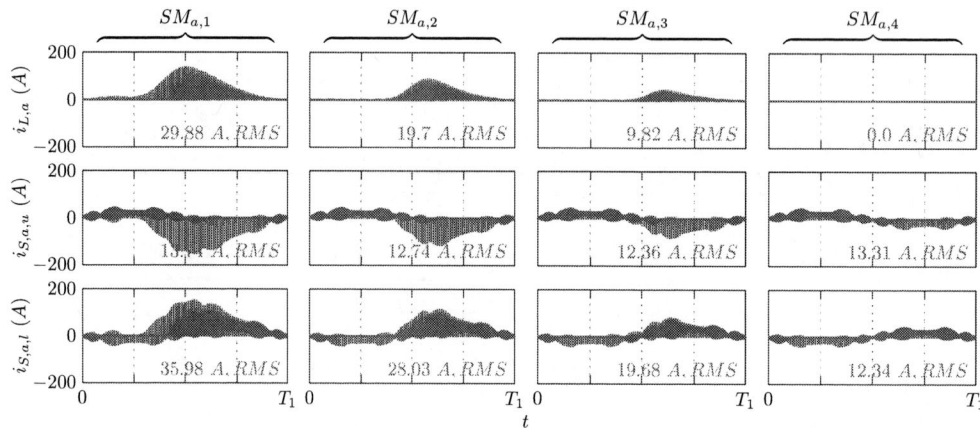

Fig. 8: Inductor and switches currents in the different sub-modules (SMs) in phase-leg-a, where $i_{L,a}$ is the inductor current, $i_{S,a,u}$ is the upper switch current, and $i_{S,a,l}$ is the lower switch current.

and the maximum SM inductor current ripple is much less than 250 A as discussed before.

5 Conclusion

A modular multilevel converter (MMC)-based single-stage DC-AC converter solution is proposed in this paper. This converter is called boost MMC (BMMC) and is used for interfacing a low voltage (LV) DC source, such as fuel cells and batteries, with a higher voltage AC load or grid. This converter has the following merits:

- Flexible boosting capability with the possibility of high voltage gains;
- Multilevel operation, where the number of levels is proportional to the boosting ratio;
- Possibility of utilizing low cost and LV MOSFETs and LV electrolytic capacitors;
- Higher efficiency due to the low ON-state resistance of the LV MOSFETs;
- High modularity and redundancy;
- Redundant SMs can be utilized in each phase-leg in order to maintain high modulation index under different conditions.

Over and above that, the proposed BMMC can be utilized to interface LV PV systems with MV AC systems utilizing IGBTs. On the other hand, this converter suffers from unequal current stresses among the different semiconductor devices. Such demerit is not an issue for LV MOSFETs, where parallel operation of air cooled low cost SMD switches can easily be achieved.

The proposed converter has been analysed and simulation results using PLECS have been introduced, where the results verify its functionality and the introduced analysis.

References

[1] T. Wu, C. Chang, L. Lin, and C. Kuo, "Power loss comparison of single- and two-stage grid-connected photovoltaic systems," *IEEE Trans. Energy Convers.*, vol. 26, no. 2, pp. 707–715, Jun. 2011.

[2] A. Abdelhakim, P. Mattavelli, V. Boscaino, and G. Lullo, "Decoupled control scheme of grid-connected split-source inverters," *IEEE Trans. Ind. Electron.*, vol. 64, no. 8, pp. 6202–6211, Aug. 2017.

[3] A. Abdelhakim, P. Mattavelli, D. Yang, and F. Blaabjerg, "Coupled-inductor-based dc current measurement technique for transformerless grid-tied inverters," *IEEE Trans. Power Electron.*, vol. 33, no. 1, pp. 18–23, Jan. 2018.

[4] I. Subotic, N. Bodo, and E. Levi, "An EV drive-train with integrated fast charging capability," *IEEE Trans. Power Electron.*, vol. 31, no. 2, pp. 1461–1471, Feb. 2016.

[5] M. Darijevic, M. Jones, and E. Levi, "An open-end winding four-level five-phase drive," *IEEE Trans. Ind. Electron.*, vol. 63, no. 1, pp. 538–549, Jan. 2016.

[6] J. Lu *et al.*, "Dc-link protection and control in modular uninterruptible power supply," *IEEE Trans. Ind. Electron.*, vol. 65, no. 5, pp. 3942–3953, May 2018.

[7] L. M. A. Caseiro, A. M. S. Mendes, and S. M. A. Cruz, "Cooperative and dynamically weighted model predictive control of a 3-level uninterruptible power supply with improved performance and dynamic response," *IEEE Trans. Ind. Electron.*, vol. 67, no. 6, pp. 4934–4945, Jun. 2020.

[8] L. Valverde, C. Bordons, and F. Rosa, "Integration of fuel cell technologies in renewable-energy-based microgrids optimizing operational costs and durability," *IEEE Trans. Ind. Electron.*, vol. 63, no. 1, pp. 167–177, Jan. 2016.

[9] A. Abdelhakim, P. Mattavelli, S. Pistollato, and G. Spiazzi, "Bidirectional dc-dc converter topologies for low-voltage battery interface: Comparative assessment," in *4th Int. Forum on Research and Technol. for Society and Ind. (RTSI)*, Sep. 2018, pp. 1–6.

[10] M. Rabiul Islam, A. M. Mahfuz-Ur-Rahman, K. M. Muttaqi, and D. Sutanto, "State-of-the-art of the medium-voltage power converter technologies for grid integration of solar photovoltaic power plants," *IEEE Trans. Energy Convers.*, vol. 34, no. 1, pp. 372–384, Mar. 2019.

[11] W. Crookes, D. Trainer, and C. Oates, "Converter," Patent WO/2010/149 200.

[12] C. C. Davidson and D. Trainer, "Modularised converter for HVDC and STATCOM," Patent WO/2011/124 260.

[13] S. Allebrod, R. Hamerski, and R. Marquardt, "New transformerless, scalable modular multilevel converters for hvdc-transmission," *IEEE Power Electron. Specialists Conf.*, pp. 174–179, 2008.

[14] A. Lesnicar and R. Marquardt, "An innovative modular multilevel converter topology suitable for a wide power range," in *IEEE Bologna Power Technol. Conf. Proc.*, vol. 3, Jun. 2003, pp. 1–6.

[15] C. Qian and M. L. Crow, "A cascaded converter-based statcom with energy storage," in *IEEE Power Eng. Society Winter Meeting*, vol. 1, Jan. 2002, pp. 544–549.

[16] D. Soto and T. C. Green, "A comparison of high-power converter topologies for the implementation of facts controllers," *IEEE Trans. Ind. Electron.*, vol. 49, no. 5, pp. 1072–1080, Oct. 2002.

[17] H. Akagi, "Classification, terminology, and application of the modular multilevel cascade converter (MMCC)," *IEEE Trans. Power Electron.*, vol. 26, no. 11, pp. 3119–3130, Nov. 2011.

[18] A. Marzoughi, R. Burgos, D. Boroyevich, and Y. Xue, "Design and comparison of cascaded h-bridge, modular multilevel converter, and 5-l active neutral point clamped topologies for motor drive applications," *IEEE Trans. Ind. Appl.*, vol. 54, no. 2, pp. 1404–1413, Mar. 2018.

[19] E. Villanueva, P. Correa, J. Rodriguez, and M. Pacas, "Control of a single-phase cascaded h-bridge multilevel inverter for grid-connected photovoltaic systems," *IEEE Trans. Ind. Electron.*, vol. 56, no. 11, pp. 4399–4406, Nov. 2009.

[20] Y. Zhong, N. M. Roscoe, D. Holliday, and S. J. Finney, "MMC with parallel-connected MOSFETs as an alternative to wide bandgap converters for LVDC distribution networks," *J. of Eng.*, vol. 2017, no. 5, pp. 149–157, May 2017.

[21] N. M. Roscoe, D. Holliday, N. McNeill, and S. J. Finney, "Lv converters: Improving efficiency and EMI using Si MOSFET MMC and experimentally exploring slowed switching," *J. of Emerg. and Sel. Topics Power Electron.*, vol. 6, no. 4, pp. 2159–2172, Dec. 2018.

[22] Y. Zhong *et al.*, "High-efficiency MOSFET-based MMC design for LVDC distribution systems," *IEEE Trans. Ind. Appl.*, vol. 54, no. 1, pp. 321–334, Jan. 2018.

[23] L. M. Redondo, J. F. Silva, P. Tavares, and E. Margato, "High-voltage high-frequency marx-bank type pulse generator using integrated power semiconductor half-bridges," in *European Conf. on Power Electron. and Appl.*, Sep. 2005, pp. 1–8.

Low Voltage GaN-Based Gate Driver to Increase Switching Speed of Paralleled 650 V E-mode GaN HEMTs

Raffael Risch and Jürgen Biela

Laboratory for High Power Electronic Systems, ETH Zurich

Email: risch@hpe.ee.ethz.ch

URL: http://www.hpe.ee.ethz.ch

Keywords

≪Gallium Nitride (GaN)≫, ≪Parallel operation≫, ≪Pulsed power≫, ≪Design≫, ≪Wide bandgap devices≫

Abstract

Using GaN HEMTs in high current applications, such as pulsed power modulators for particle accelerator systems, requires the parallelization of multiple devices. In order to achieve a dynamically balanced current distribution between the parallel devices, synchronized gate voltages are crucial. Furthermore, the high switching speeds, which are often required in pulsed power systems, requires a high driving current capability and fast rise/fall times of the gate driver. Therefore, this paper presents a gate driver, based on a low voltage GaN HEMT half bridge, for driving four paralleled 650 V e-mode GaN HEMTs in a low inductive switching cell design.

1 Introduction

Wide bandgap (WBG) semiconductor devices, such as gallium nitride (GaN) high-electron-mobility transistors (HEMTs), have lower area specific on-state resistances than conventional silicon devices. This results in smaller semiconductor dies, which have higher current density ratings and smaller device capacitances. The smaller device capacitances and the higher saturation velocities of GaN HEMTs enable faster switching speeds, which results in reduced switching losses and allows for higher switching frequencies. In addition, the high switching speeds of GaN HEMTs are also useful in pulsed power applications that require very short pulses with fast switching transients, such as the generation of transient plasmas or the driving of fast kicker magnets in particle accelerator systems. The required output voltage pulses in these applications are usually several times higher than the breakdown voltage of a single GaN HEMT. Therefore, pulse generator topologies consisting of multiple series-connected switching cells, as shown in Fig. 1(a), are used to generate the high output voltage amplitudes.

The design in this paper focusses on a single switching cell with the target specifications listed in Table I. In a single switching cell, the switches are typically arranged as chopper type half bridges (i.e. with a free-wheeling diode D_{fw}) as depicted in Fig. 1(b). The switching cell must conduct the full load current of the pulse generator,

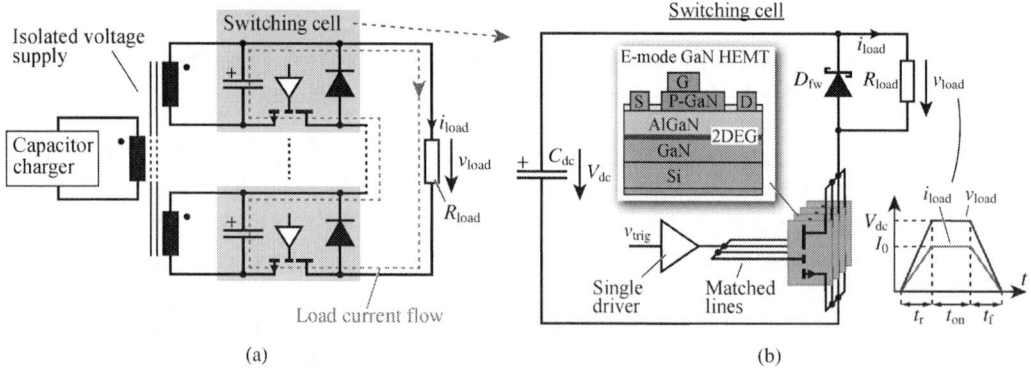

Fig. 1: (a) High-voltage pulse generator topology consisting of a series-connection of lower voltage switching cells. (b) Single switching cell implemented as a chopper type half bridge with the investigated single driver concept and parallel-connected GaN HEMTs.

Table I: Target pulse specifications of single switching cell and complete system.

	Cell	System
Pulse voltage V_{dc}	450 V	20 kV
Pulse current I_0	400 A	400 A
Rise and fall time t_r, t_f	5 ns	10 ns
Flat-top duration t_{on}	50 ns	50 ns

which in this design is several times higher than the rating of a single GaN HEMT die. Therefore, a parallelization of multiple GaN HEMT dies – usually either within a single module or within discrete packages – is required.

However, tolerances in device fabrication [1], an asymmetrical layout [2–4], and unequal gate voltages can lead to an imbalanced transient current distribution among the paralleled switches. This results in an unequal power loss share, which may lead to an over-temperature or even thermal runaway of single devices.

Several active [5] and passive approaches [6, 7] to mitigate the current imbalance for WBG devices have been presented in literature. Active methods are usually based on controlling the gate signal timing [8] or the gate drive voltages [9]. However, these methods all require separate drain current measurements and the necessary control bandwidth for the targeted nanosecond switching times is in the gigahertz range. This strongly increases the system complexity, especially when a large number of switches are paralleled. Passive methods are usually based on inserting coupled inductors, which, however, adds unwanted leakage inductance to the circuit that slows down the maximal speed. In general, these methods are not practical for designing fast, low-inductive switching cells. Therefore, a symmetrical layout and the design of a synchronized gate drive circuitry is important for synchronous switching.

Regarding the gate drive design, using separate gate drivers for each switch enables the gate drivers to be placed close to the devices and allows for low inductive gate loop designs. However, fabrication tolerances and differences in chip temperature between the gate drivers [10] can lead to potential propagation delay mismatches in the range of tens of nanoseconds and consequently to unsynchronized gate voltages. This is especially of concern in case of WBG devices with switching times in the range of only a couple of nanoseconds [11, 12].

Therefore, this paper presents the design of a fast gate driver with minimal gate voltage timing mismatch, which is not susceptible to temperature fluctuations and does not require a complex control system. The gate drive is designed to drive four parallel e-mode GaN HEMTs. In the proposed gate driver, a GaN-based amplifier stage with a high drive current is used to simultaneously drive all parallel GaN HEMTs. A high drive current becomes even more important in case of high-current rated transistors with large chip areas since a larger amount of gate charge is necessary to turn-on and turn-off the device properly. The gate voltages are distributed via signal traces of equal length, which enables a synchronization of the gate voltages.

This paper is structured as follows. First, section 2 discusses the problem of propagation delay mismatches in case of individual gate drivers. Thereafter, the component selection and the circuit modelling of the proposed single gate driver design is presented in section 3. Section 4 covers the layout design of the gate and the power loop of the switching cell. Section 5 presents the measurement results of the proposed gate driver and compares the performance with a commercial single gate driver. Finally, the conclusions are drawn in section 6.

2 Problem of Gate Voltage Timing Mismatch

In the following, the influence of mismatched gate timing and asymmetrical gate inductances on the output voltage switching speed and the transient drain current of fast switching, paralleled GaN HEMTs is discussed.

For driving multiple switches in parallel, a straightforward approach is to use individual gate drivers for each parallel-connected switch. This simplifies the circuit layout and allows gate drivers with smaller driving power to be used. The majority of commercially available gate drivers are manufactured as integrated silicon CMOS buffers, which sometimes also provide the necessary isolation and safety functionalities. However, fabrication tolerances can result in different propagation delays between these driver ICs, which can lead to delay mismatches of tens of nanoseconds according to the data sheets. Furthermore, the propagation delay of gate drivers varies with their chip temperature. Hence, temperature differences between different drivers can further increase the propagation delay mismatch. Especially in case of very fast switching GaN HEMTs, small timing mismatches can already have a significant impact on the transient drain current distribution and the overall output voltage switching speed.

In order to quantify the effect of propagation delay mismatch on the output voltage switching speed, the switching cell of Fig. 1(b) is simulated in LTSpice for two parallel 650 V/60 A e-mode GaN HEMTs with individual drivers as is indicated in Fig. 2(a). The transient drain currents of the two GaN HEMTs for a mismatch of $\Delta t = 2$ ns is shown in Fig. 2(b). The mistimed gate voltages have the effect that one switch turns on earlier than the other one. Hence, the earlier switch starts to conduct the full load current on its own. In addition, the earlier switch discharges not only its own output capacitance during turn-on but also the output capacitance of the paralleled switch. This leads to a transient overcurrent in the faster GaN HEMT and an increase in the output voltage rise time as is shown in Fig. 3(a) and Fig. 3(b).

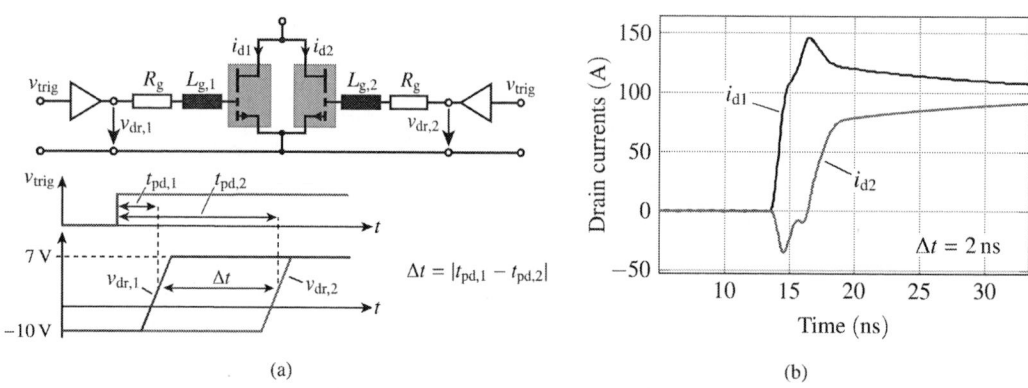

(a)   (b)

Fig. 2: (a) Simulation model of two parallel $650\,\text{V}/60\,\text{A}$ e-mode GaN HEMTs driven by separate gate drivers with unequal gate inductances and mismatched propagation delays. (b) Transient turn-on drain currents at $\Delta t = 2\,\text{ns}$ for $R_\text{g} = 2\,\Omega$.

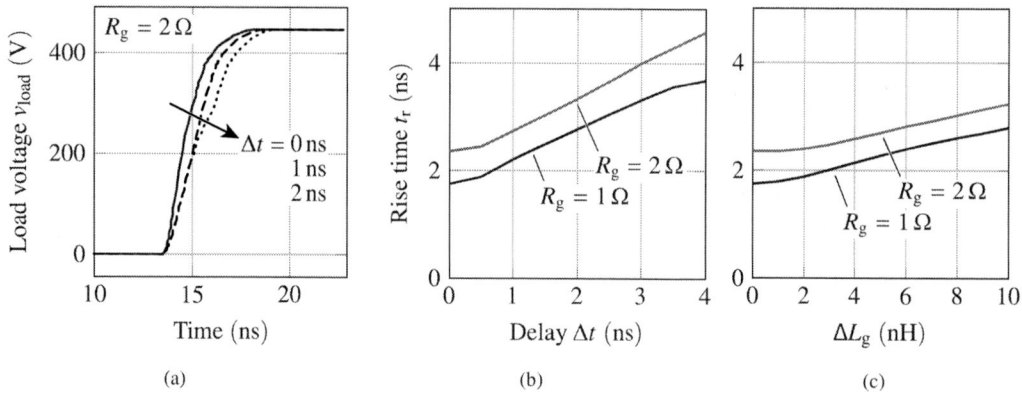

(a)   (b)   (c)

Fig. 3: (a) Turn-on transient of load voltage for different delay times. (b) Increase of turn-on times for increasing propagation delay mismatch and in (c) for increasing gate inductance mismatch.

A similar effect is caused by a mismatch of the parasitic gate inductances $L_{\text{g},i}$. A larger gate inductance – e.g. caused by a longer gate signal path – requires more time to build up its magnetic energy. Hence, the gate voltage is slightly delayed, which also results in a slower switching speed as shown in Fig. 3(c). In order to avoid the problems with the timing of separate gate drivers, in the following section a gate driver with only one driver and matched gate path lengths is presented.

3   Components and Circuit Modelling of Proposed Gate Driver

Fig. 4 shows the circuit model of the proposed gate driver including the considered parasitic elements and the switching cell in which the four parallel GaN HEMTs operate. The circuit model is used for simulation and design optimization. The parallel-connected GaN HEMTs are implemented in a chopper-type half bridge, which commonly occurs in pulse generators as basic switching cell. In pulse applications, various load conditions - resistive, capacitive, inductive as well as combinations of these - can occur, accounting for the growing variety of high-voltage pulse applications. In this paper, the focus lies on stripline kicker magnets, which are typically designed to have a broadband ohmic impedance response in order to enable very fast load voltage transients. Therefore, pulse proof chip resistors are used to emulate the kicker magnet load. The resistors are modelled by the load resistance R_{load}. In the following, the components and the equivalent circuit model of the gate driver and the switching cell are described in detail.

3.1   Gate Drive Components

In order to drive parallel GaN HEMTs at fast switching speeds, the gate driver requires a high pulse current rating and fast rise/fall times. Furthermore, a low inductive package of the driver is beneficial in order to be able to decrease the external gate resistance without exceeding the overvoltage rating of the gate structure.

Most of the currently commercially available gate driver ICs are based on silicon CMOS inverters. By cascading multiple CMOS inverters on the same substrate, relatively high current ratings are achievable. However, imple-

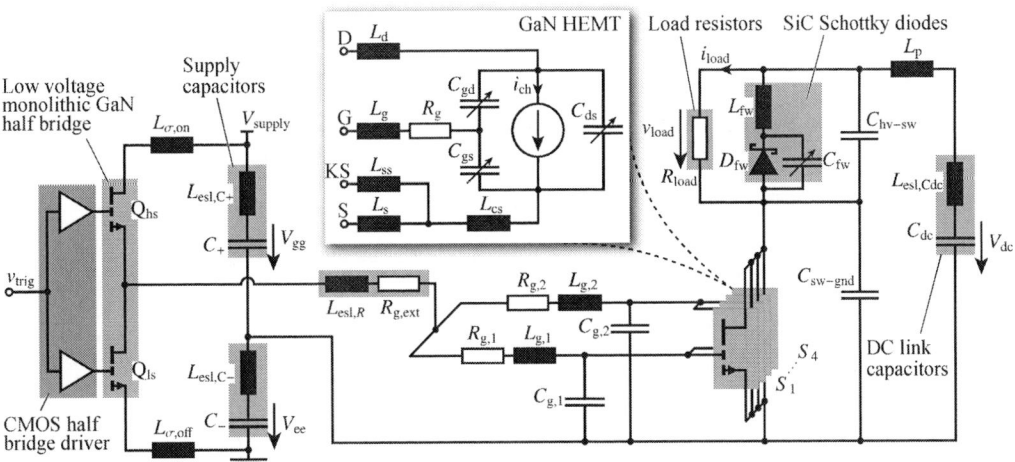

Fig. 4: Equivalent circuit modelling of the proposed gate driver and the switching cell in which the four parallel GaN HEMTs operate.

menting a gate drive circuit based on GaN would benefit from its superior material properties compared to silicon. A more compact gate driver with faster switching speeds can be achieved at the same current rating. In addition, GaN devices often already come in low inductive packages, which allows them to fully exploit their superior switching performance.

Today, GaN-based gate drivers are mainly implemented in integrated GaN power ICs. The lateral device structure of both low and high voltage GaN transistors enables a monolithic integration of the gate driver and the power transistor. This allows for a compact integrated gate loop layout, which substantially reduces the parasitics compared to the external gate loop interconnections of conventional gate drivers. However, these integrated GaN-based gate drivers usually only drive one power switch. Hence, their driving power is relatively low. Furthermore, they are not commercially available as stand-alone devices. Therefore, the proposed gate driver for driving the four parallel GaN HEMTs is based on a discrete GaN half bridge with integrated high-side (HS) and low-side (LS) switches. Both switches are n-type transistors since complementary GaN technology is not yet mature enough. A pre-driver is necessary for the GaN half bridge. This pre-driver is realized by a silicon-based CMOS half bridge driver IC. The half bridge driver already integrates the delay time control that is necessary to avoid shoot-through currents between the HS and the LS switch.

For the monolithically integrated GaN half bridge, the EPC2111 from EPC is used. It has a 30 V voltage rating, which leaves sufficient margin for the nominal gate drive voltages of the main GaN HEMTs ($+7\,\mathrm{V}/-10\,\mathrm{V}$). The IC has a nominal current rating of 16 A, whereas the HS switch Q_{hs} has a nominal on-state resistance of $14\,\mathrm{m\Omega}$ and the LS switch Q_{ls} has $6\,\mathrm{m\Omega}$ (at $25\,^{\circ}\mathrm{C}$). As further shown in Fig. 5, the GaN half bridge output stage is driven by an integrated CMOS half bridge driver (LMG1210 from Texas Instruments). It consists of separate drivers - CMOS based push-pull stages - for each GaN HEMT. The HS driver is controlled via a capacitive level shifter and supplied via the bootstrap diode D_{bst}. A sufficient dead-time between Q_{hs} and Q_{ls} is necessary to avoid detrimental shoot-through currents. The dead-time can be tuned using the external resistors R_{dlh} and R_{dhl} in the equivalent circuit. The internal drivers of the CMOS half bridge are modelled as voltage sources with the rise time shown in Fig. 5. The supply capacitors, which are charged to the gate voltages V_{gg} and V_{ee}, are modelled with their capacitances C_+ and C_- as well as their equivalent series inductances $L_{esl,C+}$ and $L_{esl,C-}$.

As will be further outlined in the next section 4, the gate driver is connected to the four GaN HEMTs via two signal traces of equal lengths. The parasitics of these two traces are taken into account with $R_{g,i}$, $L_{g,i}$ and $C_{g,i}$ for $i=1,2$.

3.2 Switching Cell Components

The parallel-connected GaN HEMTs S_1–S_4 are realized by four GS66516B enhancement-mode GaN HEMTs from GaNSystems. This considered GaN HEMT is a single-chip GaN power transistor with a nominal breakdown voltage of 650 V and a pulse current rating of 120 A. It utilizes a p-GaN layer on top of the AlGaN/GaN heterojunction channel in order to achieve a positive threshold voltage and enable an enhancement-mode operation of the device. Even though the gate exhibits a diode structure, it is not sufficiently forward biased at the nominal gate voltage so that the steady-state gate current is very small (approximately $320\,\mu\mathrm{A}$ of gate leakage current at $v_{gs}=6\,\mathrm{V}$). Therefore, the device essentially always operates in the field-effect mode where the electric field, associated with the gate voltage, controls the turn-on and turn-off behaviour of the device. As a result, the implemented GaN HEMT can be modelled with the dynamic model shown in Fig. 4. It consists of voltage-dependent capacitances C_{gs}, C_{ds}, and C_{gd}, a voltage-controlled current source i_{ch}, and an internal gate resistance R_g. Even though the GaN HEMT die is embedded in a proprietary fiberglass package and contacted by copper-filled micro vias, which significantly reduces the parasitic inductances compared to wire-bonded packages, the parasitic inductance cannot be neglected. Therefore, the parasitic inductances are included in the model by the connection inductances L_d, L_g and L_s, the Kelvin-source inductance L_{ss}, and the coupling inductance L_{cs} between the power and gate loop.

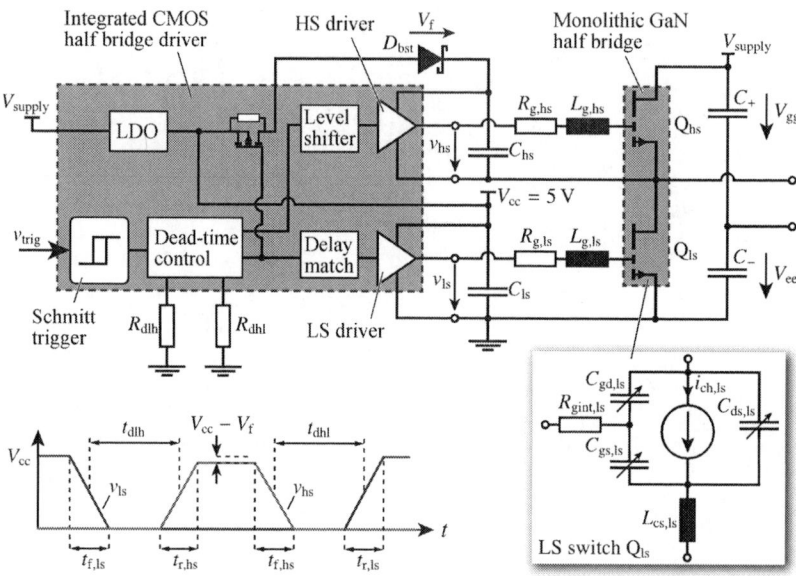

Fig. 5: Schematic of the cascaded structure of the proposed gate driver, consisting of the GaN-based output stage, which is driven by a CMOS-based input stage.

SiC Schottky diodes are connected between the drain of the GaN HEMTs and the positive supply rail in order to clamp inductively induced overvoltages. The current rating of these diodes need to be dimensioned for the full load current as they bypass the load current in case of an asynchronous switching between the series-connected switching cells described in Fig. 1(a). Five GB01SLT12-214 SiC Schottky diodes from GeneSiC in an SMD package are used. The total surge current capability of these diodes are 600 A, which leaves enough margin to the target operating current of 400 A. Their voltage-dependent junction capacitance C_{fw} is also taken into account in the model.

Ceramic SMD capacitors with a total capacitance of 2 μF are used as dc link energy storage. The capacitors are modelled by the capacitance C_{dc} and the parasitic series inductance $L_{esl,Cdc}$.

In addition to the component parasitics, also parasitics originating from the copper interconnections between the power loop components are taken into account, namely the power loop inductance L_p, and the capacitances C_{hv-sw} and C_{sw-gnd}.

4   PCB Layout Design

Due to the lower device capacitances, GaN-based switches can be operated at much higher switching speeds than comparable silicon devices. However, in order to be able to fully exploit the switching speed potential of GaN devices, special attention has to be paid to the parasitics within the circuit. Large parasitics might lead to excessive ringing, which can overstress components and consequently result in device failures. Furthermore, the output voltage switching transient is slowed down due to the additional build up of parasitic magnetic and electric energy, which results in a worse switching cell performance.

Nowadays, more and more manufacturers offer their devices in low-inductive semiconductor packages. Hence, the package parasitics are no longer the limiting factor that prevents the designer from exploiting the full switching speed potential of the devices. Instead, the influence of the layout parasitics on the circuit voltages and currents increases. Therefore, additional effort needs to be put into the design of a low inductive layout of the PCB. In the following subsections, the PCB layout and the placement of the components for the presented design is discussed in detail.

4.1   General Layout Considerations

A picture of the populated PCB is depicted in Fig. 6 and a schematic cross sectional view of the board with the main components is shown in Fig. 7(a). In general, there are three main options to design a current loop: Vertical, lateral, and *inner layer* loop design. Their layout as well as their performance with respect to the achievable minimal parasitic inductance has been thoroughly discussed in [13]. Based on [13], the inner layer loop design, in which the current return path lies on the first inner layer, features the smallest parasitic loop inductance. The main reason is that it has the smallest loop area and the magnetic field outside the loop gets cancelled due to the opposing currents on the adjacent layers. Consequently, all critical current loops in the presented design, in which fast current transients occur, are designed based on this inner layer loop design approach.

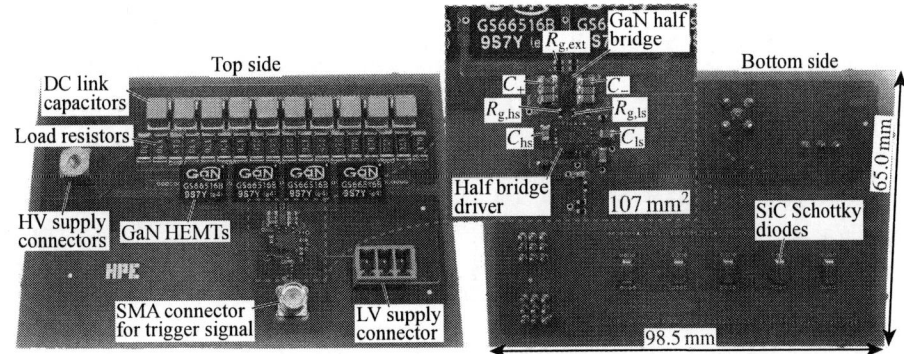

Fig. 6: Pictures of the top and bottom side of the populated PCB. The proposed gate drive circuit is shown in a zoomed view.

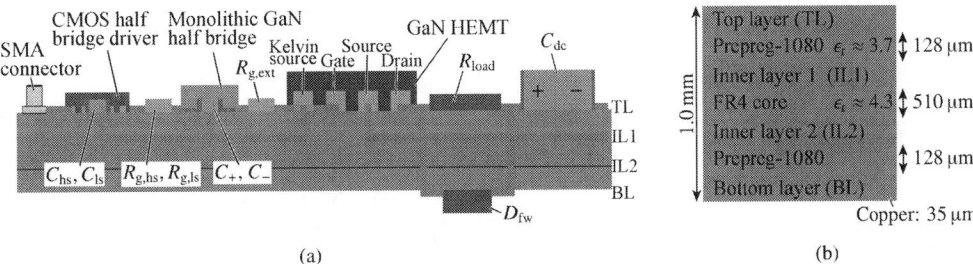

Fig. 7: (a) Cross sectional view of the PCB. (b) Layer stack configuration of the PCB.

The inner layer loop design requires that a multi-layer board is used, in which the distance between the top and first inner layer is small. The distance is determined by the thickness of the employed prepreg material and the design freedom is limited by the available layer stack configurations. In the considered design, a standard four layer PCB, as shown in Fig. 7(b), with a total thickness of 1 mm is used. The prepreg layers between the outer and the inner copper layer consists of two prepregs of 1080 type, which both have a total height of approximately 128 µm after the layers have been pressed together. In the following, the designs of the power and gate loops are described more in detail.

4.2 Gate Loop Design

The two main requirements for the gate loop design are to minimize its parasitic inductance and to have equal gate loop parasitics for each paralleled GaN HEMT. Equal gate loop parasitics ensure that the drive voltage is equally distributed among the parallel GaN HEMTs, which is especially important in case of very fast switching GaN devices where synchronized switching is key. In general, equal lengths can be achieved by using star-connected gate traces from the driver to the individual gate pads.

A major benefit of the GaN HEMTs in this design, are their dual gate drive pins. These are internally connected such that only one needs to be electrically contacted by the driver. As shown in Fig. 8(a), this allows the GaN half bridge to be placed close to the GaN HEMTs while still achieving equal gate trace lengths. As a result, there are only two gate loops necessary, a left gate loop and a right gate loop, each driving two GaN HEMTs simultaneously. Furthermore, Kelvin source pins are available for this GaN HEMT package, which reduces the coupling between power and gate loop and therefore increases the driving power.

The second design requirement is minimizing the gate loop inductance. Since the gate structure of GaN HEMTs is rather sensitive to overvoltages compared to silicon devices, minimizing the gate loop inductance is crucial in order to be able to use a small external damping resistance $R_{g,ext}$ and therefore, achieve a fast switching speed. For current loops with adjacent current conducting layers, the parasitic inductance decreases with increased trace width [14]. Hence, a small gate loop inductance can be achieved by using relatively wide signal tracks between driver and gate.

Furthermore, high frequency currents follow the path of the least inductance rather than least resistance. The gate loop is designed to have a microstrip line type structure, for which the lowest return path inductance lies directly underneath the top layer signal path. This minimizes the effective gate loop area during the fast switching transient. As an example, Fig. 8(b) illustrates the distribution of the high frequency return current during turn-off, which flows back from the gate pads to the supply capacitors C_- on the first inner layer. From the simulated current distribution, the magnetic energy density and consequently the gate loop inductance can be calculated.

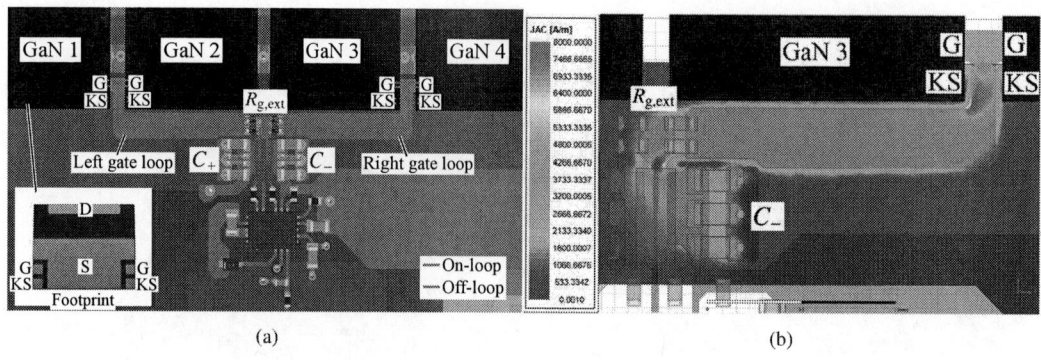

(a)   (b)

Fig. 8: (a) Top view of gate drive circuit showing the left and right gate loop. (b) Simulated (FEM) high frequency gate current distribution on the first inner layer of the right gate loop during turn-off.

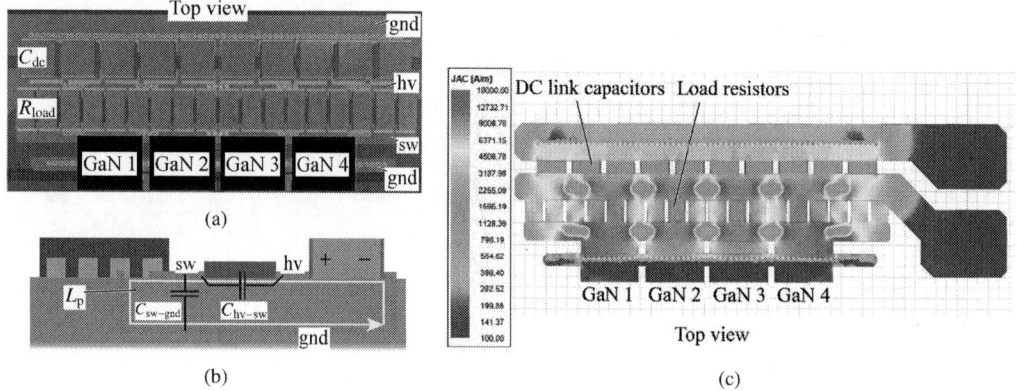

(a)   (c)

(b)

Fig. 9: (a) Top view and (b) cross sectional view of the power loop including the layout parasitics. (c) Transient load current distribution among the four parallel-connected GaN HEMTs.

4.3 Power Loop Design

A key point in the power loop design is to achieve a layout, in which each GaN HEMT has the same power loop parasitics. This ensures that transient current imbalances originating from an asymmetrical layout are minimized, which further improves the output voltage switching speed.

Whereas in power modules extra measures need to be taken in order to achieve a symmetrical layout, such as fine tuning the wire bond lengths of individual semiconductor chips, the use of discrete semiconductor devices usually simplifies this design step. In the considered design, a symmetrical layout is achieved by arranging the components in straight lines as is shown in Fig. 9(a). A top view of the simulated transient load current distribution is depicted in Fig. 9(c).

Fig. 9(b) depicts the cross sectional view of the power loop, in which the relevant parasitics are indicated. These parasitics need to be minimized in order to achieve a maximum switching speed. The parasitic power loop inductance L_p originates from the loop area of the load current. The parasitic capacitances C_{hv-sw} and C_{sw-gnd} arise from the opposite copper polygons on the top layer and the first inner layer (designated by hv, sw and gnd). These polygons essentially form parallel plate capacitors, which result in the mentioned parasitic capacitances. The third capacitance C_{hv-gnd} is neglected since the parallel connected dc link capacitors have a much larger capacitance value.

In order to accurately model the circuit, the parameters need to be extracted from the design. The component parasitics are either directly given in the data sheet or they can be extracted from impedance curves. The layout parasitics have been extracted using 3D FEM simulations of magnetic and electric energy densities. An overview of the model parameters is given in Table II. Measurements of the half bridge prototype are given in the following.

5 Hardware Measurements and Comparison with Commercial Gate Driver

This section presents measurement results of the output voltage and the gate voltage and compares these with the simulated waveforms. In addition, the achieved output voltage switching speed with the proposed GaN-based

Table II: Extracted model parameters of the power loop, the gate loop, and the CMOS driver.

Power loop

C_{gs}	512.0 pF	C_{ds}	120.0 pF	C_{gd}	5.9 pF	R_g	0.5 Ω
C_{hv-sw}	170.0 fF	C_{sw-gnd}	87.0 pH	L_p	130.0 pH	C_{dc}	2.0 µF
C_{fw}	50.0 pF	$L_{esl,Cdc}$	65.0 pH	L_d	370.0 pH	L_{ss}	0.2 nH
L_s	150.0 pH	L_{cs}	20.0 pH	L_g	3.1 nH	R_{load}	1.1 Ω
V_{dc}	450.0 V	L_{fw}	40.0 pH				

Gate loop

$L_{g,1}$	1.3 nH	$L_{g,2}$	1.3 nH	$C_{g,1}$	9.1 pF	$C_{g,2}$	9.1 pF
$R_{g,1}$	28.5 mΩ	$R_{g,2}$	28.5 mΩ	C_+	30.0 µF	C_-	30.0 µF
$L_{esl,C+}$	250.0 pH	$L_{esl,C-}$	250.0 pH	$L_{esl,R}$	100.0 pH		
$L_{\sigma,on}$	356.0 pH	$L_{\sigma,off}$	356.0 pH	V_{gg}	7.0 V	V_{ee}	−5.0 V

CMOS driver

$R_{g,hs}$	10.0 Ω	$R_{g,ls}$	10.0 Ω	$L_{g,hs}$	3.0 nH	$L_{g,ls}$	3.0 nH
$t_{r,ls}, t_{r,hs}$	2.0 ns	$t_{f,ls}, t_{f,hs}$	2.0 ns	t_{dlh}	20.0 ns	t_{dhl}	20.0 ns
V_{cc}	5.0 V	V_f	0.4 V	$R_{gint,ls}$	0.4 Ω	$R_{gint,hs}$	0.5 Ω
$C_{gd,ls}$	21.0 pF	$C_{gs,ls}$	474.0 pF	$C_{ds,ls}$	469.0 pF	$L_{cs,ls}$	100.0 pH
$C_{gd,hs}$	8.0 pF	$C_{gs,hs}$	182.0 pF	$C_{ds,hs}$	162.0 pF	$L_{cs,hs}$	100.0 pH

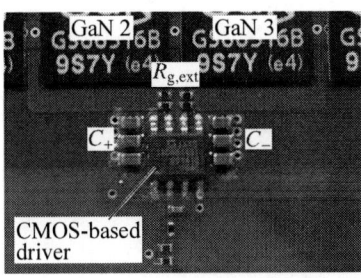

Fig. 10: Alternative gate drive design based on a commercial gate driver.

gate driver is compared to the performance of a commercially available gate driver IXDN614SI from IXYS (c.f. Fig. 10), which is fabricated in an 8-pin SOIC package and has a nominal current rating of 14 A as well as measured voltage rise and fall times of 5.2 ns, resp. 8.7 ns. This gate drive variant will be termed *SOIC* for the rest of the paper. Except from the different gate drivers, the rest of the switching cell design is exactly the same in order to allow a fair comparison.

All voltage measurements are conducted using a 500 MHz LeCroy PP008 probe connected to a LeCroy Waverunner 620Zi oscilloscope with a 500 MHz analog bandwidth (at 1 MΩ input impedance) and a sampling rate of 20 GS/s. The switching times are measured between 10 % and 90 % of the nominal voltage amplitude. The limited bandwidth of the measurement system decreases the measured rise time, which is especially of concern for the fast switching transients in this design. Therefore, when interpreting the measurement results, this error has to be taken into account. An approximate error estimation according to the formula in (1) is used.

$$t_{r,meas} = \sqrt{t_{r,signal}^2 + t_{r,probe}^2 + t_{r,osci}^2} \tag{1}$$

Here, $t_{r,meas}$ corresponds to the measured rise time that is displayed on the oscilloscope, $t_{r,signal}$ is the actual signal rise time, $t_{r,probe}$ rise time and $t_{r,osci}$ is the rise time of the oscilloscope. These rise times can be calculated from the bandwidth of the respective component by using the formula $t_r = 0.35/f_{3dB}$. No explicit current measurement system has been added to the design in order to be able to place the components closer together and to minimize the parasitics of the interconnections between the components. Instead, the current amplitude is calculated based on the measured load voltage and the load resistance. The load resistance values have to be chosen relatively small to achieve the required high load currents. Therefore, four-terminal measurements have been used for measuring the load resistance in order to increase the measurement accuracy.

Fig. 11(a) shows load voltage measurements with the proposed GaN-based gate driver. The load current has been continuously increased to finally reach the targeted operating current of 400 A. In these initial measurements, a

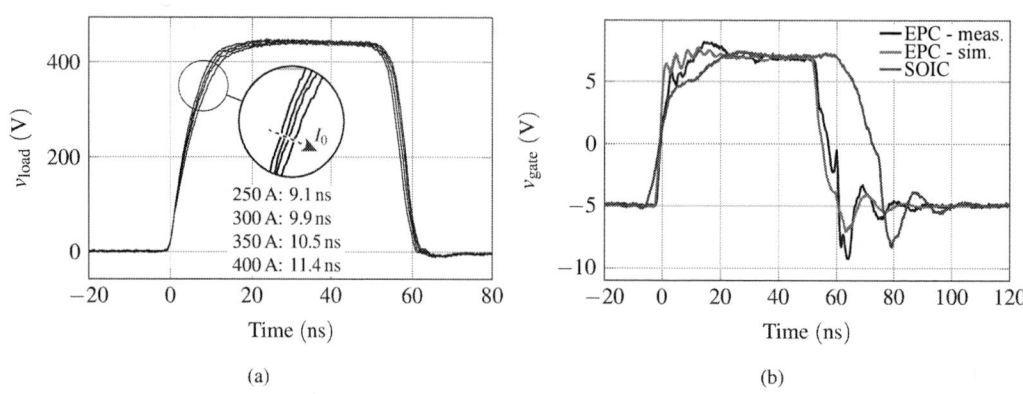

(a)

(b)

Fig. 11: (a) Load voltage measurements with the proposed gate driver using $R_{g,ext} = 1.4\,\Omega$ and increasing load current. (b) Gate voltage comparison between the commercial SOIC driver and the measured and simulated EPC design.

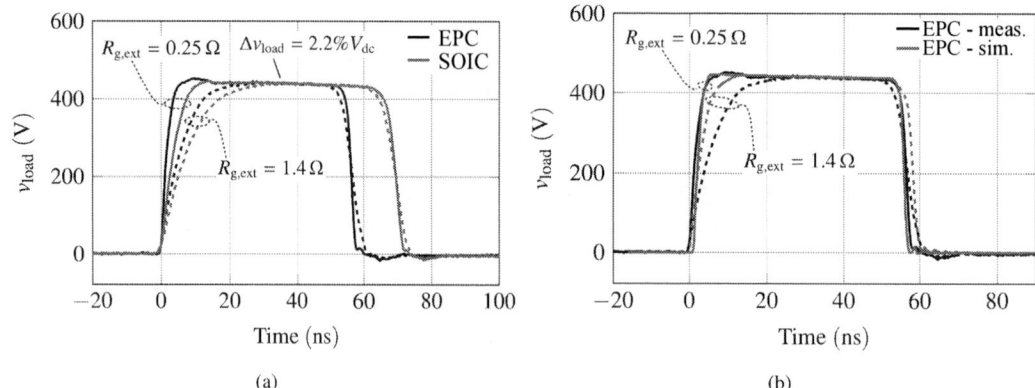

(a)

(b)

Fig. 12: (a) Comparison between the EPC design and the SOIC design. (b) Comparison between the simulated and measured load voltage curves for $R_{g,ext} = 0.25\,\Omega$ and $R_{g,ext} = 1.4\,\Omega$.

gate resistance value of $R_{g,ext} = 1.4\,\Omega$ is used. As can be seen, the rise time continuously increases with increased load current.

After the prototype has been successfully tested at the nominal operating current of 400 A, the gate resistance is continuously reduced in order to increase the switching speed performance of the prototype. The final value is $R_{g,ext} = 0.25\,\Omega$. Fig. 11(b) shows the respective gate voltage measurements of the EPC and SOIC design as well as the simulated gate voltage with drive voltages of $V_{gg} = 7\,V$ and $V_{ee} = -5\,V$.

Fig. 12(a) shows the comparison between the load voltage measurements of the two gate drive designs for the two gate resistance values $R_{g,ext} = 1.4\,\Omega$ and $R_{g,ext} = 0.25\,\Omega$. The switching times are summarized in Table III. It is clearly visible that the switching speed is increased for both designs with decreased gate resistance value. Furthermore, the EPC design performs significantly better with respect to switching speed than the commercial driver.

Fig. 12(b) shows the comparison between measured and simulated waveforms. The measured switching times are higher than the simulated values. A major contributor to this mismatch is the fact that the parameters of the GaN HEMT SPICE model have not been adjusted to the actual GaN HEMTs on the board as no characteristic device measurements of the GaN HEMTs have been conducted prior to the assembly.

Fig. 13 summarizes the achieved turn-on switching speeds in this paper and compares them with previous designs of fast switching cells in literature. The presented EPC gate drive design is approximately twice as fast as the commercial SOIC gate driver. Furthermore, the presented parallel GaN HEMT design with the proposed gate driver can achieve similar switching speeds than integrated power GaN HEMT ICs, but at current ratings that are over ten times higher.

Table III: Comparison of the measured and simulated switching times.

	Rise times (ns)				Fall times (ns)			
	EPC meas.	EPC corr.	EPC sim.	SOIC	EPC meas.	EPC corr.	EPC sim.	SOIC
$R_{g,ext} = 0.25\,\Omega$	3.4	3.25	2.5	7.0	2.7	2.51	2.1	5.0
$R_{g,ext} = 1.4\,\Omega$	11.4	11.35	6.3	11.1	6.3	6.22	4.1	7.3

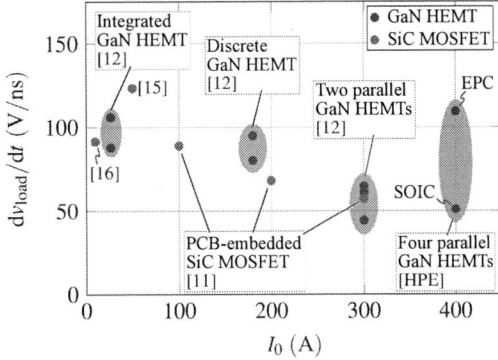

Fig. 13: Comparison of the turn-on switching speed between the proposed design in this paper and designs from literature.

6 Conclusions

This paper presents a GaN-based gate drive circuit design for four parallel-connected e-mode GaN HEMTs, which are operated in a chopper-type half bridge with resistive load. The gate voltage is generated by a single monolithically integrated GaN half bridge with high drive power and fast output switching times. The gate loop is symmetrically designed using length-matched signal traces that connect the driver to the four GaN HEMTs. This enables an equal distribution of the gate voltage which minimizes timing mismatches of the driving voltages at the gate input. Furthermore, the design of a low inductive power and gate loop is crucial in order to avoid detrimental overvoltages and to increase the output voltage switching speed of the switching cell. A circuit model of the proposed gate drive circuit has been developed and implemented in LTSpice in order to simulate all important waveforms. The parameter values of the model are extracted from the layout by using 3D FEM simulations of magnetic and electric energy densities. Finally, the GaN gate drive prototype has been tested and compared to a commercially available gate driver at the target operating point of $V_{dc} = 450\,V$ and $I_0 = 400\,A$. The GaN-based gate driver is able to drive the parallel GaN HEMTs to very high switching speeds, despite the high load current. By successively reducing the gate resistance to very low values, rise and fall times as short as 3.4 ns, resp. 2.7 ns have been achieved, which is approximately twice as fast as the commercial gate drive design.

Acknowledgements

The authors would like to thank the Deutsches Elektronen-Synchrotron (DESY) in Hamburg for their financial support of this work.

References

[1] H. Li, S. Munk-Nielsen, X. Wang, R. Maheshwari, S. Beczkowski, C. Uhrenfeldt, and W.-T. Franke, "Influences of Device and Circuit Mismatches on Paralleling Silicon Carbide MOSFETs," *IEEE Trans. Power Electron.*, vol. 31, no. 1, pp. 621–634, Jan. 2016.

[2] H. Li, W. Zhou, X. Wang, S. Munk-Nielsen, D. Li, Y. Wang, and X. Dai, "Influence of Paralleling Dies and Paralleling Half-Bridges on Transient Current Distribution in Multichip Power Modules," *IEEE Trans. Power Electron.*, vol. 33, no. 8, pp. 6483–6487, Aug. 2018.

[3] Z. Zeng, X. Zhang, and X. Li, "Layout-Dominated Dynamic Current Imbalance in Multichip Power Module: Mechanism Modeling and Comparative Evaluation," *IEEE Trans. Power Electron.*, vol. 34, no. 11, pp. 11 199–11 214, Nov. 2019.

[4] C. Zhao, L. Wang, and F. Zhang, "Effect of Asymmetric Layout and Unequal Junction Temperature on Current Sharing of Paralleled SiC MOSFETs With Kelvin-Source Connection," *IEEE Trans. Power Electron.*, vol. 35, no. 7, pp. 7392–7404, Jul. 2020.

[5] J. Ao, Z. Wang, J. Chen, L. Peng, and Y. Chen, "The Cost-Efficient Gating Drivers with Master-Slave Current Sharing Control for Parallel SiC MOSFETs," in *Proc. IEEE Conf. Expo Transp. Electrific. Asia-Pacific*, Jun. 2018.

[6] Y. Mao, Z. Miao, C. Wang, and K. D. T. Ngo, "Balancing of Peak Currents Between Paralleled SiC MOSFETs by Drive-Source Resistors and Coupled Power-Source Inductors," *IEEE Trans. Ind. Electron.*, vol. 64, no. 10, pp. 8334–8343, Oct. 2017.

[7] Z. Zeng, X. Zhang, and Z. Zhang, "Imbalance Current Analysis and Its Suppression Methodology for Parallel SiC MOSFETs with Aid of a Differential Mode Choke," *IEEE Trans. Ind. Electron.*, vol. 67, no. 2, pp. 1508–1519, Feb. 2020.

[8] S. G. Kokosis, I. E. Andreadis, G. E. Kampitsis, P. Pachos, and S. Manias, "Forced Current Balancing of Parallel-Connected SiC JFETs During Forward and Reverse Conduction Mode," *IEEE Trans. Power Electron.*, vol. 32, no. 2, pp. 1400–1410, Feb. 2017.

[9] Y. Xue, J. Lu, Z. Wang, L. M. Tolbert, B. J. Blalock, and F. Wang, "Active compensation of current unbalance in paralleled silicon carbide MOSFETs," in *Proc. IEEE Appl. Power Electron. Conf. Expo.*, Mar. 2014.

[10] M. Sack, S. Keipert, M. Hochberg, M. Greule, and G. Mueller, "Design Considerations for a Fast Stacked-MOSFET Switch," *IEEE Trans. Plasma Sci.*, vol. 41, no. 10, pp. 2630–2636, Oct. 2013.

[11] R. Risch and J. Biela, "Nanosecond switching of ohmic loads using SiC MOSFETs in ultra-low inductive PCB-packages," in *Proc. IEEE Eur. Conf. Power Electron. Appl.*, Nov. 2019.

[12] R. Risch, A. Hu, and J. Biela, "Switching speed comparison of wide bandgap semiconductors in ultra-low inductive packages for nanosecond pulse applications," in *Proc. IEEE Int. Conf. Integr. Power Electron. Syst.*, Mar. 2020.

[13] D. Reusch and J. Strydom, "Understanding the Effect of PCB Layout on Circuit Performance in a High-Frequency Gallium-Nitride-Based Point of Load Converter," *IEEE Trans. Power Electron.*, vol. 29, no. 4, pp. 2008–2015, Apr. 2014.

[14] K. Wang, L. Wang, X. Yang, X. Zeng, W. Chen, and H. Li, "A Multiloop Method for Minimization of Parasitic Inductance in GaN-Based High-Frequency DCDC Converter," *IEEE Trans. Power Electron.*, vol. 32, no. 6, pp. 4728–4740, Jun. 2017.

[15] O. Kreutzer, T. Heckel, and M. Maerz, "Using SiC MOSFET's Full Potential - Switching Faster than 200 kV/us," *Materials Science Forum*, 2016.

[16] S. Guo, L. Zhang, Y. Lei, X. Li, F. Xue, W. Yu, and A. Q. Huang, "3.38 MHz operation of 1.2kV SiC MOSFET with integrated ultra-fast gate drive," in *Proc. IEEE Workshop Wide Bandgap Power Devices Appl.*, Nov. 2015, pp. 390–395.

Gate stresses and threshold voltage instability in normally-OFF GaN HEMTs

Jose Ortiz Gonzalez, Burhan Etoz, Olayiwola Alatise
UNIVERSITY OF WARWICK
School of Engineering
Coventry, United Kingdom
Tel.: +44 / (0)2476151437
E-Mail: J.A.Ortiz-Gonzalez@warwick.ac.uk, burhan.etoz@warwick.ac.uk
O.Alatise@warwick.ac.uk

Acknowledgements

This work was supported by the UK Engineering and Physical Sciences Research Council (EPSRC) through the grant reference EP/R004366/1.

Keywords

«Gallium Nitride (GaN)», «Reliability», «Wide bandgap devices»

Abstract

This paper presents a study of gate stress and threshold voltage instability in commercially available 600/650V GaN high electron mobility transistors (HEMTs). The technologies evaluated are an ohmic gate GaN HEMT and a Schottky gate GaN HEMT. The gate leakage currents have been evaluated for two different gate contact technologies and its temperature dependency is presented. It is shown that the gate leakage current could be a temperature indicator for both technologies evaluated, with a higher temperature sensitivity in the case of the Schottky gate HEMT (showing a sixtyfold increase from 22°C to 150 °C). A novel characterization method based on the third quadrant operation of the device was applied to the two selected GaN HEMTs and the role of temperature, stress level and duration on the threshold voltage instability of GaN HEMTs has been evaluated. The method can capture both the peak shift and transient recovery. The results highlight the clear differences between both gate contact technologies with the Schottky gate HEMT exhibiting higher threshold voltage instability due to gate stress compared to the ohmic gate devices. The Schottky gate HEMT shows a positive threshold voltage shift for a gate stress voltage of 3 V, whereas at 5.5 V the shift is dependent of the stress time. For both HEMTs, the recovery transient after stress removal is accelerated with temperature.

Introduction

Wide bandgap (WBG) power devices, namely silicon carbide (SiC) and Gallium Nitride (GaN), are currently experiencing a phase of wide industrial adoption. The superior properties of WBG semiconductor materials enable low specific ON-state resistance, fast switching transients and operation at high temperatures, resulting in power devices superior to silicon-based power devices [1, 2]. WBG power devices will be fundamental for enabling more efficient power electronics systems and with a myriad WBG power devices available from different manufacturers, the time for WBG-based power electronics has come. GaN high electron mobility transistors (HEMTs) appear to be more suitable for high-frequency high-efficiency converters, at low DC link voltages, targeting the application areas in the 100-600 V range [3], whereas SiC MOSFETs are contenders for application areas targeting voltages between 600 V and 3.3 kV [3].

In order to replace Si devices, WBG power devices have to demonstrate not only a superior energy conversion performance compared to traditional Si power devices but will also have to at least match the reliability performance of Si devices. This is challenging given the decades of field operation in silicon devices. The properties of the WBG semiconductor materials add new challenges to the well-stablished qualification techniques for Si power devices [4]. Hence, it is key to develop suitable methods for evaluating the reliability of WBG power devices. Focusing on GaN, these include dynamic ON-state resistance and threshold voltage (V_{TH}) instability [5, 6]. Developing such techniques may be challenging, as the gate interface of the commercially available GaN HEMTs is more complex than the conventional insulated gate of IGBT/MOSFETs.

GaN HEMT power device structure and test prototypes

Commercially available GaN power devices are lateral devices [1]. The main feature of these power devices is the AlGaN/GaN heterojunction and band-offsets which contain a 2D electron gas (2DEG) that results in high carrier density and enhanced mobility due to reduced columbic and acoustic phonon scattering. Due to spontaneous polarization at the band discontinues of the AlGaN/GaN heterojunction, GaN HEMTS are normally-ON however they can be made normally-OFF by using a cascode configuration with a low voltage Si MOSFET as the input [7]. Using gate electrode engineering, normally-OFF GaN HEMTs can be achieved by using the depletion widths of OFF-state diodes to close the 2DEG [2]. Two main gate structures [6] have been adopted by the manufacturers: (a) a p-GaN gate on AlGaN using a Schottky Contact [8] and (b) a p-GaN gate on AlGaN using an Ohmic Contact, which is also known as Gate Injection Transistor (GIT) [9]. The two device and gate structures are shown in Fig. 1.

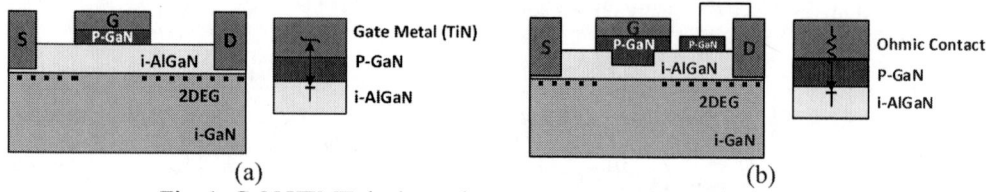

Fig. 1: GaN HEMT device and gate structures, adapted from [3, 8, 9]
(a) Schottky Gate GaN HEMT (b) Ohmic Gate GaN HEMT

The different gate structures of the normally-OFF GaN HEMTs will have a clear impact on the reliability and driving circuitry used. In the investigation presented in this paper, studies have been performed on a 600V/31A Ohmic Gate (OG) GaN HEMT from Infineon with datasheet reference IGOT60R070D1 and a 650V/30A GaN HEMT from GaN Systems, with datasheet reference GS66508T. The gate of this HEMT is identified as p-GaN in [10]. No further description has been made available by the manufacturer, as mentioned in [2, 11], but in [12] it is identified as Schottky Gate (SG) GaN HEMT.

These discrete devices are not packaged using the conventional TO-220/TO-247 package used for Si and SiC power devices hence a PCB prototype was required in order to have good access to the terminals for performing the required tests. The prototypes are shown in Fig. 2 (a) and Fig. 2(b) for the OG and SG GaN HEMT respectively. The terminals of the power device are accessible by 4 mm banana connectors, as shown in Fig. 2(c). The selected devices have a thermal pad on the top side, allowing the use of an external DC heater to set the operating temperature of the device, which is shown in Fig. 2(d).

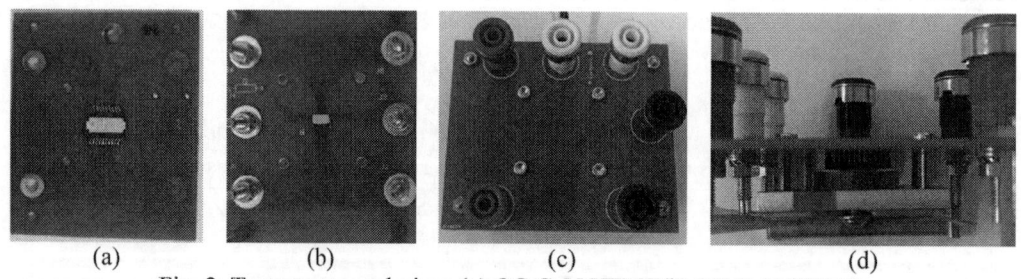

Fig. 2: Test prototype designs (a) OG GaN HEMT(b) SG GaN HEMT,
(c) Top view of the PCB prototype (d) Detail of the DC heater connection

GaN HEMT gate characteristics and impact of temperature

First, it is important to identify and compare the limitations for driving the two selected devices, including the maximum gate voltages and the impact of temperature on the gate leakage currents. Using a source measurement unit (SMU) Keithley model 2602B, positive and negative gate bias sweeps were performed for both HEMTs. The results, measured at ambient temperature (22 °C), are show in Fig. 3(a) for the SG HEMT and Fig. 3(b) for the OG HEMT. Comparing both figures, it is clearly observed how the gate leakage currents are higher for the OG HEMT.

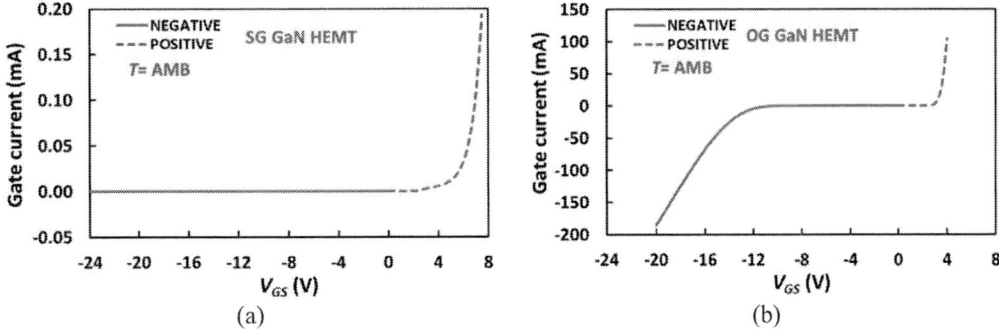

Fig. 3: Gate voltage sweep and gate leakage current at ambient temperature (22 °C)
(a) SG GaN HEMT, (b) OG GaN HEMT

Analyzing the negative gate bias, it can be concluded that it would be possible to use a negative voltage of -20 V with the SG HEMT, whereas for the OG HEMT the gate current will be approximately -200 mA in that situation. A knee voltage of around -12 V is observed in the OG HEMT, caused by an internal protection diode structure [3, 13]. In the case of the SG HEMT, the negative gate voltage can be increased up to -40 V with minimum gate leakage current, as the results in Fig. 4(a) show. For this device, increasing the positive gate voltage sweep to 10 V caused the gate leakage current to increase considerably, as shown in Fig. 4(b).

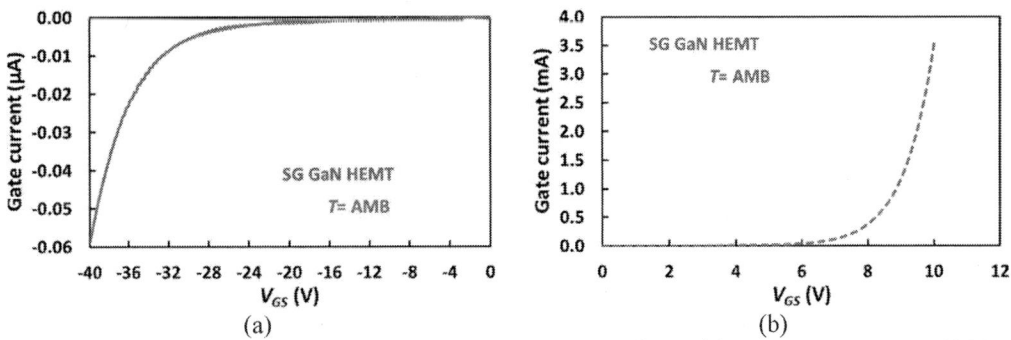

Fig. 4: SG gate HEMT (a) Negative gate sweep up to -40V (b) Positive gate sweep up to +10 V

For understanding the differences between the devices and their limitations, it is important to evaluate the impact of temperature on the gate leakage current. This has been done for both gate technologies and the results are shown in Fig. 5. The results in Fig. 5(a) show that temperature plays a fundamental role in the gate leakage currents in the SG gate HEMT. For this device, considering a gate voltage of 6 V, which is the maximum gate voltage according to the datasheet, the gate leakage current increases from 32 µA at ambient temperature to 1.09 mA at 150 °C. Analyzing the results of the OG GaN HEMT in Fig. 5(b), the temperature dependency of the gate leakage shows the expected performance of the ohmic contact gate structure (a resistor and a forward biased diode as indicated in Fig. 1(b)) with a leftwards shift of the gate leakage current with temperature.

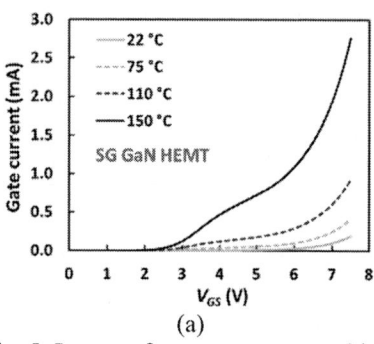 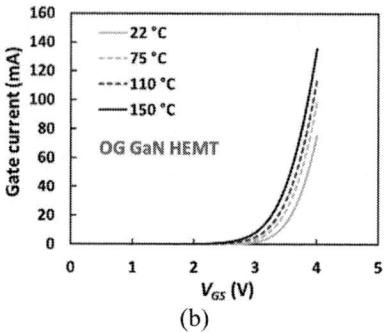

Fig. 5: Impact of temperature on positive gate voltage sweeps. (a) SG HEMT, (b) OG HEMT

This increase of the gate leakage current with temperature will have to be considered when designing gate driver circuits for GaN HEMTs, however it could also be used as temperature indicator. The calibrated relationship between the gate leakage current and temperature is shown in Fig. 6(a) and Fig. 6(b) for the SG and OG HEMT respectively. These results indicate that the gate leakage currents could be a Temperature Sensitive Electrical Parameter (TSEP) for GaN devices [14], which can be added to the TSEPs already presented in [15, 16]. In the case of the SG HEMT, for a gate voltage of 5 V, the gate leakage current increases almost 60 times, whereas in the case of the OG HEMT the increase is 2.7 times when a gate voltage of 3.5 V is used. Moreover, as already presented in [17] and shown in Fig. 6(c), the gate voltage measured at a fixed gate current I_G is a good TSEP in OG GaN HEMTs.

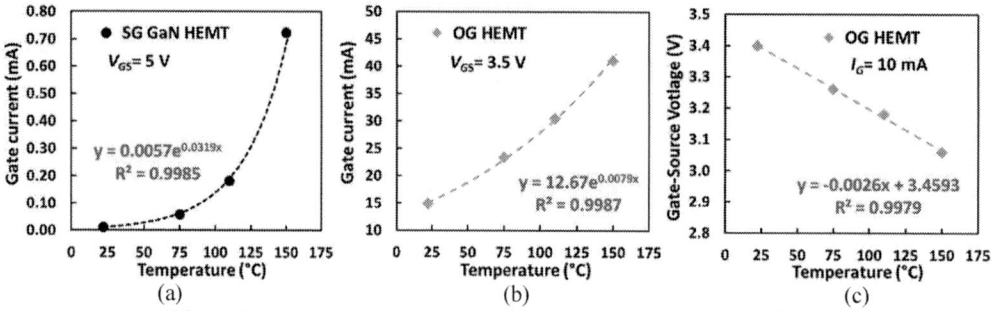

Fig. 6: Impact of temperature on gate current at fixed gate voltage
(a) SG GaN HEMT, (b) OG GaN HEMT. (c) Temperature sensitivity of V_{GS} – OG GaN HEMT

Accelerated gate stress tests in GaN HEMTs

Accelerated stress tests are performed to evaluate the reliability and lifetime of power devices, both at gate level [18] and packaging level [19]. In this paper, preliminary accelerated gate stress tests have been performed to evaluate the impact of gate biasing on the gate transfer characteristics of both GaN HEMT technologies. The stresses were performed at different stages increasing the stress level (gate voltage for the SG HEMT and gate current for the OG HEMT). Fig. 7(a) shows the measured gate current for the SG HEMT for the different cumulative gate voltage stress levels. The stress duration was 30 minutes for each stress step, with characterization at ambient temperature after recovery at V_{GS}=0 for at least 90 minutes. Fig. 7(b) shows the gate transfer characteristics, measured at ambient temperature after each stress stage, with V_{GS}=V_{DS}. Fig. 8 shows the results for the OG HEMT stress tests, consisting in different gate current applied to the device for 1000 s. The stresses for the OG HEMT were done at ambient temperature, whereas they were performed at 150 °C for the SG HEMT.

The results in Fig. 7 and Fig. 8 show that for the selected stresses there is a negative shift of V_{TH} for the SG HEMT, which is in agreement with [12], whereas in the case of the OG HEMT there is no significant shift. The negative shift in V_{TH} for the SG GaN HEMT is due to positive charge injection from the gate into the heterojunction thereby reducing the voltage necessary to from the 2DEG. Evaluating the stress itself, in the case of the OG HEMT, it can be observed that the stress voltage is dependent on the gate current, making the stress in terms of voltage complex: Increasing the stress voltage means a higher leakage current, causing a high power dissipation on the device. In the case of the SG HEMT, Fig. 7(a) shows an initial exponential rise of the gate leakage current during the stress, which reaches a steady value after 300 s. This increase can be attributed to charges being trapped in the gate stack, shown in Fig. 1. This increase of gate current at high gate stress was also reported in [12]. In the case of the OG GaN HEMT there is a modest reduction of gate voltage, which can be attributed to self-heating of the device due to the higher power dissipated.

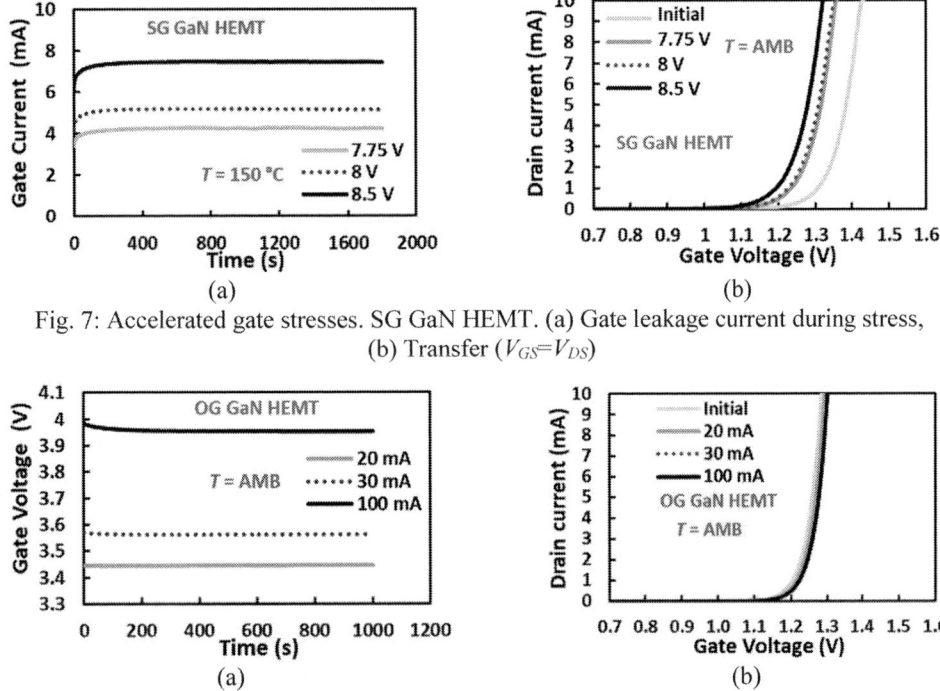

Fig. 7: Accelerated gate stresses. SG GaN HEMT. (a) Gate leakage current during stress, (b) Transfer ($V_{GS}=V_{DS}$)

Fig. 8: Accelerated gate stresses. OG GaN HEMT. (a) Gate voltage during stress (b) Transfer ($V_{GS}=V_{DS}$)

Accelerated stress tests are useful for understanding the limitations of the power devices, however these tests are time consuming hence it is paramount to define the right stress levels and durations that do not activate undesired failure mechanisms [4]. Moreover, there is a clear limitation for capturing transient and recovery phenomena. This has been proven particularly relevant for assessing threshold voltage instability in SiC power MOSFETs [20] and dynamic ON-state resistance [21, 22] in GaN HEMTs. The use of conventional methods for V_{TH} characterization in GaN is also under study [23], due to the complexity of the device structure and impact of bias history. The next section evaluates a method that can capture the dynamic threshold voltage shift caused by gate stresses in GaN HEMTs.

Third quadrant characteristics and threshold voltage instability in GaN

One of the peculiarities of GaN HEMTs is that they do not have a body diode. The reverse conduction in the third quadrant is enabled by a mechanism called self-commutated reverse conduction (SCRC) [2]. In the first quadrant, the device turns ON when a gate-source voltage exceeding $V_{GS\text{-}TH}$ is applied to the device, whereas in the third quadrant it turns ON when a gate-drain voltage higher than reverse threshold voltage ($V_{GD\text{-}TH}$) is applied. If the device is conducting in the reverse direction, $V_{GD}=V_{GS}-V_{DS}$. If higher than the threshold, the device turns ON (SCRC) and the third quadrant voltage V_{SD} is given by (1) [2].

$$V_{SD}=V_{GD\text{-}TH}-V_{GS}+I\cdot R_{SD} \tag{1}$$

Focusing on threshold voltage characterization, the value of $V_{GD\text{-}TH}$ and $V_{GS\text{-}TH}$ can be assumed equal [2] and at low currents the voltage drop in the resistance R_{SD} can be considered negligible. Analyzing (1), it can be concluded that during reverse conduction of a small sensing current at $V_{GS}=0$, V_{SD} is an indicator of V_{TH} ($V_{SD}{\approx}V_{TH}$). Measurements have been done for both SG and OG GaN HEMTs and the results are summarized in Table I. The threshold voltage was measured with $V_{GS}=V_{DS}$ and forcing a current of 10 mA.

Table 1: Measured V_{SD} and V_{TH} at ambient temperature

	V_{TH} (V) $V_{GS}=V_{DS}$ @ I = 10 mA	V_{SD} (V), $V_{GS}=0$, @ I = 10 mA
OG GaN HEMT	1.478	1.483
SG GaN HEMT	1.597	1.576

The direct relationship between V_{SD} and V_{TH} indicates that the method presented in [24, 25] could be used for monitoring V_{TH} shift in GaN HEMTs. This method is similar to the use of the body diode voltage as a TSEP in MOSFETs [14]. The test circuit for its implementation as a V_{TH} monitoring technique in GaN HEMTs is shown in Fig. 9. The circuit consists of a gate driver which is used for stressing the gate of the device under test (DUT) while a low I_{SD} current circulates through the device in reverse direction. In the case of SG HEMT the stresses are defined in voltage by adjusting the driver supply voltage V_{GG}, whereas in the case of the OG HEMTs the stresses are defined as current by using a fixed V_{GG} and adjusting the value of R_G. The circuits are shown in Fig. 9(a) and Fig. 9(b) and the operation of the method is presented in Fig. 9(c), for the SG GaN HEMT.

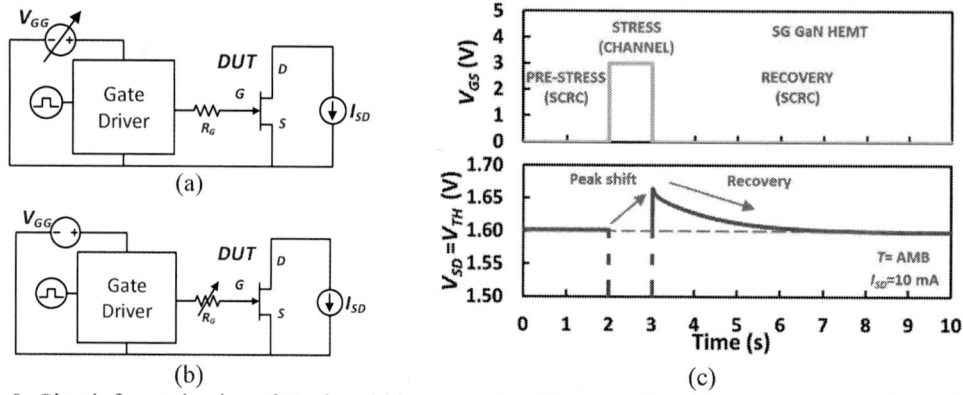

Fig. 9: Circuit for evaluation of V_{TH} instability on GaN HEMTs (a) Gate voltage stress for SG HEMT (b) Gate current stress for OG HEMT (c) Circuit operation for the SG HEMT

Depending on the gate voltage, the current circulates by means of SCRC or channel conduction. In the pre-stress stage, $V_{GS}=0$ and the current flows by means of SCRC and as mentioned previously, the measured V_{SD} can be considered equivalent to V_{TH}. During the stress phase, the device is turned ON at a determined gate stress voltage (adjustable in the circuit) and the current circulates through the channel of the DUT. After stress removal, $V_{GS}=0$ and the device operates again in the SCRC mode. The measured V_{SD} shows the peak shift of V_{TH} and recovery after stress removal. The circuit allows to study the impact of the stress duration, stress level and temperature on V_{TH} shift and recovery.

Gate stress and characterization of the Schottky gate GaN HEMT

In the case of the SG GaN HEMT, 3 stress levels have been evaluated, namely V_{GS} = 3, 5.5 and 7.5 V at both ambient and high temperature (150 °C). These stress levels were applied to the device for three different stress duration times (1, 10 and 100 s) while a current I_{SD}= 10 mA was circulating through the device. The control pulse applied to the gate driver was generated using a waveform generator model AFG3022C from Tektronix and the resulting V_{SD} voltage was captured using an oscilloscope model TDS5054B from Tektronix. The results for the stresses for the SG GaN HEMT at ambient temperature are shown in Fig. 10. The values have been normalized with respect to the measured pre-stress V_{SD}. The time t=0 was defined at the peak shift and the logarithmic time scale allows to capture the different time constants of the recovery transient. There was a recovery time of 40 minutes between each stress-characterization sequence, to allow the full recovery of V_{SD} to the pre-stress value.

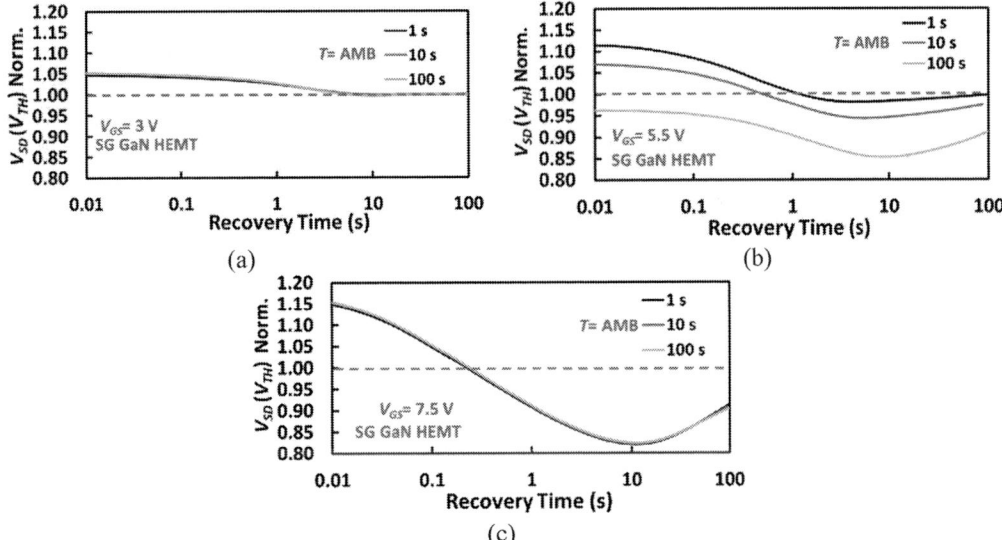

Fig. 10 Stress results for SG GaN HEMT at ambient temperature and different stress durations.
(a) Stress voltage = 3 V, (b) Stress voltage = 5.5 V (c) Stress voltage = 7.5 V

From the results in Fig 10, at ambient temperature the stress voltage has a clear impact on the measured threshold voltage shift. In the case of the 3 V stress, as shown in Fig. 10(a) and in agreement with [26], the evaluated SG HEMT shows a positive V_{TH} shift caused by trapping of electrons in the AlGaN/GaN interface [26]. For the 7.5 V stress, as presented in Fig. 10(c), an initial positive V_{TH} shift is observed (+17%), followed by a fast drop (-18%) and a long recovery transient to the pre-stress value, in the range of minutes. For these two stress levels (3 V and 7.5 V) and the evaluated stress duration range (1 s to 100 s), the stress time has no significant impact on the measured peak shift and recovery characteristics.

The situation changes for a stress voltage of 5.5 V, as the results in Fig. 10 (b) show. In this case the stress duration time plays a fundamental role on the peak shift and recovery of the SG GaN HEMT. For a stress duration of 1 s, a positive peak shift is captured (+11.8%), with a fast recovery to the pre-stress value. As the stress duration is increased, the peak shift reduces its value, with a positive peak shift of +7.3 % for the 10 s stress duration and a larger negative dip after the initial recovery phase (-5.7%). This is clearer for the 100 s stress measurement, as peak shift is now negative (-3.6%) and the negative dip after the initial fast transient has also increased (-14.7%). In [26], using a novel methodology and measuring windows of 10 µs, similar characteristics were observed for medium gate voltage stress levels. The different shifts were attributed to three different mechanisms [26]: electron trapping at the AlGaN/GaN interface, hole trapping in the AlGaN barrier and hole depletion. The traps have different time constants and this method allows to capture the transients during the recovery phase.

It is also important to evaluate the impact of temperature on the gate stress and recovery. The characterization measurements were performed at 150 °C and the results for the SG GaN HEMT are shown in Fig. 11, for stress voltages of 3, 5.5 and 7.5 V. Analyzing the results at 150 °C, the first and most obvious conclusion is that the recovery transients are accelerated for all the stress voltages evaluated. The initial drop after the peak shift is now in the range of several milliseconds, whereas the long recovery transient is in the range of seconds. Evaluating the stresses at 5.5 and 7.5 V, as shown in Fig. 11(b) and Fig .11(c), increasing the stress duration causes an increase of the dip after stress removal and there is no significant impact on the initial positive peak shift. This peak shift is higher for the 7.5 V stress (+18% for 1 s stress) than the 5.5 V stress (+13% for 1 second stress).

More relevant are the results for the 3 V stress at 150 °C, shown in Fig. 11(a). They show a trend similar to the 5.5 V stresses at ambient temperature, with a reduction of the positive peak shift as the stress duration is increased (+10.5% for 1 s stress and +8.8% for 100 s stress), as well as the presence of a negative dip (-2.5% for 100 s stress). Another interesting observation is that for a stress of 5.5 V and a duration of 100 s, at ambient temperature the peak shift is negative (-3.6%), whereas at 150 °C it is positive (+12.8%). Similar results were reported in [26], highlighting the complexity of gate stress and V_{TH} shift in GaN HEMTs. This is especially important for long duration stresses, relevant for power cycling.

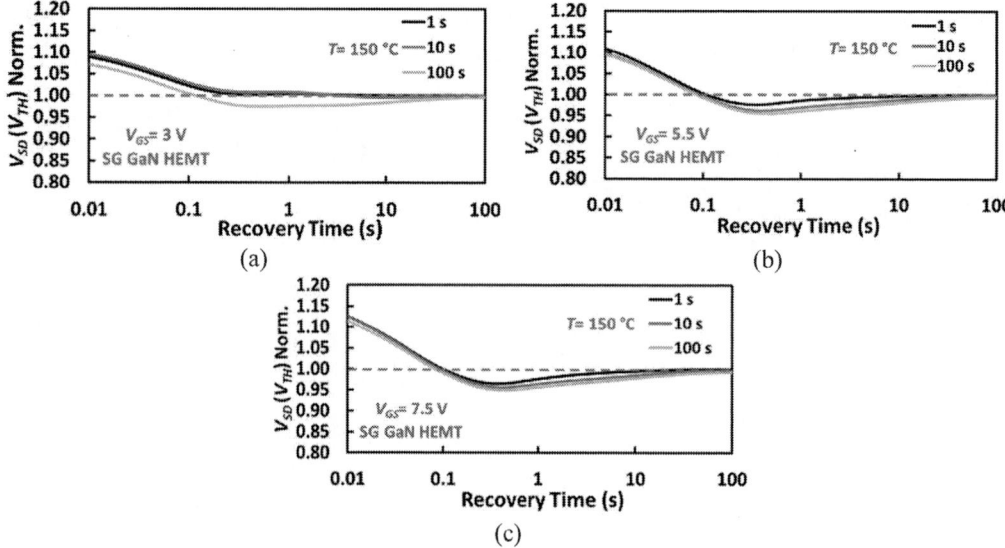

(a)   (b)

(c)

Fig. 11 Stress results for SG GaN HEMT at 150 °C and different stress durations.
(a) Stress voltage = 3 V, (b) Stress voltage = 5.5 V, (c) Stress voltage = 7.5 V

Gate stress and characterization of the ohmic gate GaN HEMT

The OG GaN HEMT was also subjected to gate stress tests. As described previously, the stresses are done in current, using the circuit shown in Fig. 9(b) and adjusting the gate resistance. Two gate current (I_G) stress levels were considered, namely I_G=5.2 mA and I_G= 73 mA, which result in gate voltage stresses of 3.2 V and 3.96 V respectively.

Compared with the SG HEMT, the results in Fig. 12 show a lower V_{TH} shift for the OG HEMT (-2.5% for I_G=5.2 mA), even for the higher gate stresses level (-4% for I_G=73 mA). This negative shift is in agreement with the results in [6]. Fig. 12(a) shows how increasing the stress level has a slight impact on the negative V_{TH} shift, which is more apparent for the high stress current, as shown in Fig. 12(b). The recovery time is in the range of seconds for both stress levels.

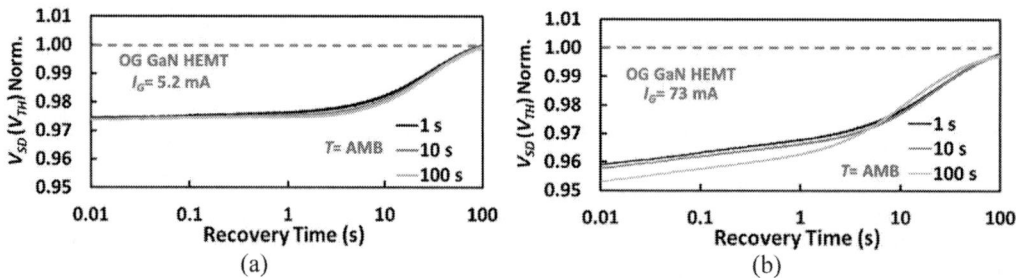

(a)   (b)
Fig. 12: OG GaN HEMT at ambient temperature and different stress durations
(a) I_G=5.2 mA, (b) I_G=73 mA

It is important to mention that at high gate currents, the high power dissipated in the gate may cause the self-heating of the device during the long gate stress pulses. This self-heating and subsequent cooling after stress removal may affect the characterization measurements at high gate current stresses, especially in the case of 73 mA, which is beyond the maximum continuous gate current of 20 mA.

It is also key to characterize the impact of temperature and this has been done for the high current stress. The results are shown in Fig. 13 and, as it was the case of the SG HEMT, at 150 °C the recovery is accelerated with no apparent V_{TH} shift. This is clearly shown in Fig. 13(b) where the complete stress sequence is shown and no apparent shift respect to the pre-stress V_{TH} is observed. From these results, the OG GaN HEMT has a superior performance regarding gate stress and threshold voltage instability.

The circuit used in these experiments is based in a voltage source gate driver and the current was adjusted by varying the gate resistance R_G in Fig.9(b). At high temperatures, the internal gate voltage drop of the OG GaN HEMT reduces, as shown in Fig. 5(b). This causes a slight change of the gate current during the stress, which at 150 °C is 78.2 mA. Using a current source gate driver may produce better results for stress and characterization of the OG HEMT, as it will enable a more precise gate current adjustment.

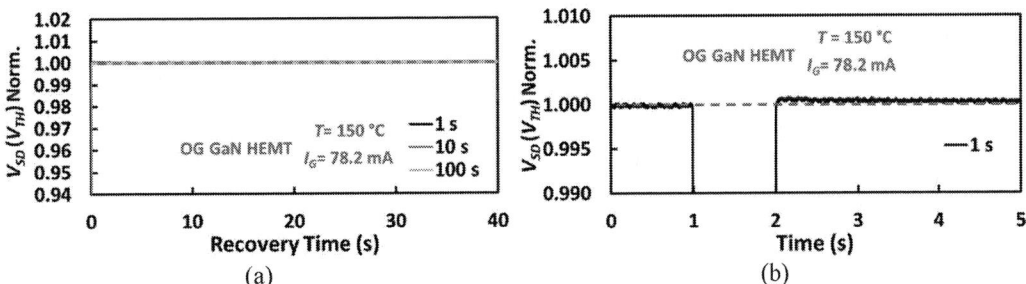

Fig. 13: Stress results for OG GaN HEMT at 150 °C. (a) Recovery for different stress durations
(b) Complete sequence for 1 s stress

From the point of view of power cycling and junction temperature determination, the impact of the gate stress on TSEPs for GaN should be evaluated, as it has been done with SiC MOSFETs in [27, 28]. This can be particularly relevant for lifetime estimation of GaN HEMTs.

Conclusion

This paper has evaluated the gate characteristics and threshold voltage instability of Schottky Gate and Ohmic Gate GaN HEMTs. The gate leakage currents have been characterized as a function of temperature and it is shown that in the case of the SG HEMT it can be a clear indicator of temperature. Accelerated gate stress tests have been performed and it is shown that the SG GaN HEMT is more susceptible to threshold voltage instability. A novel characterization technique and the impact of stress level, stress duration and temperature has been evaluated, confirming that the SG GaN HEMT has higher threshold voltage instability compared with the OG HEMT. At intermediate gate stress levels, the SG GaN HEMT shows a non-monotonic V_{TH} shift as function of the stress duration. For both gate technologies, the recovery is accelerated at high temperatures. The results presented in this paper could be fundamental for assessing the impact of gate stress on power cycling of GaN HEMTs.

References

[1] J. Millán, P. Godignon, Xavier Perpiñà , Amador Pérez-Tomás , and J. Rebollo, "A survey of wide bandgap power semiconductor devices," *IEEE Trans. on Power Electronics,* vol. 29, no. 5, pp. 2155-2163, 2014.

[2] E. A. Jones, F. F. Wang, and D. Costinett, "Review of Commercial GaN Power Devices and GaN-Based Converter Design Challenges," *IEEEJ. Emerg. Sel. Topics Power Electron.,* vol. 4, no. 3, pp. 707-719, 2016.

[3] T. Detzel, "Reliability of GaN Power Devices from the Industrial Perspective - Tutorial," presented at the 29th European Symposium on Reliability of Electron Devices, Failure Physics and Analysis - ESREF, September, 2018

[4] J. W. McPherson, "Brief history of JEDEC qualification standards for silicon technology and their applicability(?) to WBG semiconductors," in *IEEE International Reliability Physics Symposium (IRPS)*, 11-15 March 2018, pp. 3B.1-1-3B.1-8

[5] M. Meneghini *et al.*, "Reliability and failure analysis in power GaN-HEMTs: An overview," in *IEEE International Reliability Physics Symposium (IRPS)*, 2-6 April 2017, pp. 3B-2.1-3B-2.8

[6] M. Ruzzarin *et al.*, "Degradation Mechanisms of GaN HEMTs With p-Type Gate Under Forward Gate Bias Overstress," *IEEE Trans. on Electron Devices*, vol. 65, no. 7, pp. 2778-2783, 2018.

[7] X. Huang, Z. Liu, Q. Li, and F. C. Lee, "Evaluation and Application of 600 V GaN HEMT in Cascode Structure," *IEEE Trans. on Power Electronics*, vol. 29, no. 5, pp. 2453-2461, 2014.

[8] A. N. Tallarico *et al.*, "Investigation of the p-GaN Gate Breakdown in Forward-Biased GaN-Based Power HEMTs," *IEEE Electron Device Letters*, vol. 38, no. 1, pp. 99-102, 2017.

[9] Y. Uemoto *et al.*, "Gate Injection Transistor (GIT)—A Normally-Off AlGaN/GaN Power Transistor Using Conductivity Modulation," *IEEE Trans. on Electron Devices*, vol. 54, no. 12, pp. 3393-3399, 2007.

[10] GaN Systems, "GN001 Application Guide - Design with GaN Enhancement mode HEMT," 2018.

[11] E. A. Jones *et al.*, "Characterization of an enhancement-mode 650-V GaN HFET," in *IEEE Energy Conversion Congress and Exposition (ECCE)*, 20-24 Sept. 2015, pp. 400-407.

[12] J. He, G. Tang, and K. J. Chen, "V_{TH} Instability of p-GaN Gate HEMTs Under Static and Dynamic Gate Stress," *IEEE Electron Device Letters*, vol. 39, no. 10, pp. 1576-1579, 2018.

[13] E. Persson, "AN_201702_PL52_010 - CoolGaN™ Application Note," ed, 2018.

[14] Y. Avenas, L. Dupont, and Z. Khatir, "Temperature measurement of power semiconductor devices by thermo-sensitive electrical parameters: A review," *IEEE Trans. on Power Electronics*, vol. 27, no. 6, pp. 3081-3092, 2012.

[15] S. Zhu, A. Fayyaz, and A. Castellazzi, "Static and dynamic TSEPs of SiC and GaN transistors," presented at the 9th International Conference on Power Electronics, Machines and Drives (PEMD), Liverpool, 2018

[16] J. Ortiz Gonzalez, M. Hedayati, S. Jahdi, B. H. Stark, and O. Alatise, "Dynamic characterization of SiC and GaN devices with BTI stresses," *Microelectronics Reliability*, vol. 100-101, p. 113389, 2019.

[17] X. Jorda, X. Perpina, M. Vellvehi, D. Sanchez, A. Garcia, and A. Avila, "Analysis of Natural Convection Cooling Solutions for GaN HEMT Transistors," in *20th European Conference on Power Electronics and Applications (EPE'18 ECCE Europe)*, 17-21 Sept. 2018, pp. P.1-P.9

[18] S. A. Ikpe *et al.*, "Silicon-Carbide Power MOSFET Performance in High Efficiency Boost Power Processing Unit for Extreme Environments," *Additional Conferences (Device Packaging, HiTEC, HiTEN, & CICMT)*, vol. 2016, no. HiTEC, pp. 000184-000189, 2016.

[19] J. Franke, G. Zeng, T. Winkler, and J. Lutz, "Power cycling reliability results of GaN HEMT devices," in *IEEE 30th International Symposium on Power Semiconductor Devices and ICs (ISPSD)*, 13-17 May 2018, pp. 467-470

[20] K. Puschkarsky, T. Grasser, T. Aichinger, W. Gustin, and H. Reisinger, "Review on SiC MOSFETs High-Voltage Device Reliability Focusing on Threshold Voltage Instability," *IEEE Trans. on Electron Devices*, vol. 66, no. 11, pp. 4604-4616, 2019.

[21] K. Li, P. L. Evans, and C. M. Johnson, "Characterisation and Modeling of Gallium Nitride Power Semiconductor Devices Dynamic On-State Resistance," *IEEE Trans. on Power Electronics*, vol. 33, no. 6, pp. 5262-5273, 2018.

[22] P. J. Martínez, P. F. Miaja, E. Maset, and J. Rodríguez, "A Test Circuit for GaN HEMTs Dynamic R_{ON} Characterization in Power Electronics Applications," *IEEE J. Emerg. Sel. Topics Power Electron.*, vol. 7, no. 3, pp. 1456-1464, 2019.

[23] K. Murukesan, L. Efthymiou, and F. Udrea, "Gate stress induced threshold voltage instability and its significance for reliable threshold voltage measurement in p-GaN HEMT," in *IEEE 7th Workshop on Wide Bandgap Power Devices and Applications (WiPDA)*, 29-31 Oct. 2019, pp. 177-180

[24] J. A. O. González and O. Alatise, "A Novel Non-Intrusive Technique for BTI Characterization in SiC MOSFETs," *IEEE Trans. on Power Electronics*, vol. 34, no. 6, pp. 5737-5747, 2019.

[25] J. O. Gonzalez, O. Alatise, and P. Mawby, "Characterization of BTI in SiC MOSFETs Using Third Quadrant Characteristics," in *31st International Symposium on Power Semiconductor Devices and ICs (ISPSD)*, 19-23 May 2019, pp. 207-210

[26] A. Stockman, E. Canato, M. Meneghini, G. Meneghesso, P. Moens, and B. Bakeroot, "Threshold Voltage Instability Mechanisms in p-GaN Gate AlGaN/GaN HEMTs," in *31st International Symposium on Power Semiconductor Devices and ICs (ISPSD)*, 19-23 May 2019, pp. 287-290

[27] J. O. Gonzalez and O. Alatise, "Impact of the Gate Oxide Reliability of SiC MOSFETs on the Junction Temperature Estimation Using Temperature Sensitive Electrical Parameters," in *IEEE Energy Conversion Congress and Exposition (ECCE)*, 23-27 Sept. 2018, pp. 837-844

[28] F. Yang, E. Ugur, and B. Akin, "Evaluation of Aging's Effect on Temperature-Sensitive Electrical Parameters in SiC MOSFETs," *IEEE Trans. on Power Electronics*, vol. 35, no. 6, pp. 6315-6331, 2020.

New energy management algorithm based on filtering for electrical losses minimization in Battery-Ultracapacitor electric vehicles

Bakou Traoré
Université de nantes
10 Bd Jean Jeanneteau,
Angers, France
bakou.traore@etu-univ-nantes.fr

Moustapha Doumiati
ESEO School of Engineering
IREENA Lab EA 4642
10 Bd Jean Jeanneteau, Angers,
France
moustapha.doumiati@eseo.fr

Cristina Morel
ESEO School of Engineering
10 Bd Jean Jeanneteau, Angers,
France
cristina.morel@eseo.fr

Jean-Christophe Olivier
Université de Nantes, IREENA Lab EA 4642
37 Bd de l'Université, Saint Nazaire, France
jean-christophe.olivier@univ-nantes.fr

Ousmane Soumaoro
National School of Engineering (ENI-ABT)
410, Av. Vollenhoven, Bamako, Mali
samou_soumaoro@yahoo.com

Keywords

≪Energy system management≫, ≪Electric vehicle≫, ≪Batterie≫, ≪Ultra capacitors≫, ≪Adaptive control≫.

Abstract

This paper presents a new energy management algorithm based on filtering for battery-ultracapacitor electric vehicles. Compared to the passive filtering techniques, the developed strategy allows a best control of the ultracapacitor state of charge and achieves an optimization of the system electric losses. This is achieved by an online optimization of a cost function. Simulations validate the performances of the proposed method.

Introduction

To face the double challenge of reducing pollution and finding alternative means of transportation, electric vehicles (EVs) seem promising because of their advantage of not emitting atmospheric pollutants in operation. Actually, most manufacturers have their dedicated EV program and make Research and Development efforts that contribute to reveal the potential of electric vehicles.

It is well known that EVs are disadvantaged on the autonomy side because of the battery limitations. To improve the autonomy, the efficiency and the life time of the power supply of the vehicle, other sources such as fuel cell (FC) and/or ultracapacitor (UC) could be combined, thus constituting a hybrid multi-sources system [2, 3].

A multi-sources system can only be effective if it is controlled by an energy management strategy (EMS) that coordinates the power distribution between sources taking their characteristics and limitations into account. Among the energy management strategies developed in the literature aiming to preserve battery lifetime, there is in particular studies that use filtering. This approach of frequency filtering is a simple method of spectral distribution of energy between sources to protect the ones with low internal dynamics such as battery or FC [4]. Thus, the mean power is supplied by the battery or the FC and the dynamic power by sources like UC.

Unlike the methods using the optimal control and the dynamic programming that need to know in advance the entire load profile, the frequency approach has the advantage to achieve a real-time power distribution for battery-UC, fuel cell-battery and fuel cell-battery-UC hybrid power systems as in [4],[5] and [6], respectively. However, this technique has the disadvantage of not respecting the constraints on

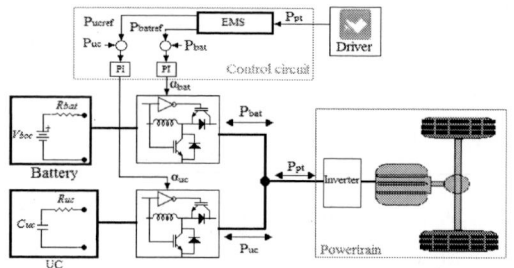

Fig. 1: System architecture.

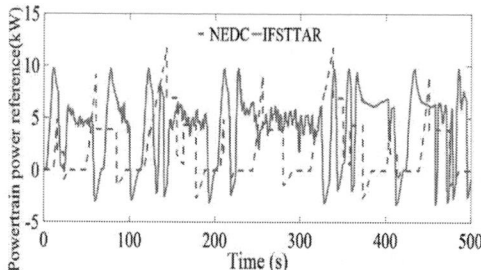

Fig. 2: NEDC and IFSTTAR power profiles

the UC state of charge [7, 4]. Thus, to solve this problem, strategies using adaptive frequency separation have been developed in [7] and [4]. Compared to the constant frequency method, these approaches adapt the filters cut-off frequency depending on the state of charge of the storage system.

This paper presents an improved EMS based on filtering for battery-UC electric vehicles with consideration of the electrical losses of the system. The battery is considered as the main source, while UC is the auxiliary one. An adaptive parameter is introduced and determined online in order to optimize a cost function depending on the system electrical losses and the UC state of charge. Reducing the system electrical losses while providing the requested load power contributes to improve the system efficiency and to preserve the battery lifetime. Indeed, the lost energy, mainly due to the sources internal resistances, is the main cause of source heating that contributes to reduce the battery lifetime [8].

Modeling

As shown in Fig. 1, the studied system is a vehicle with electric motorization powered via a DC bus by two sources: a battery and an UC. The battery is considered as the main source of energy and supply the means power, while the UC, as auxiliary source, provide the dynamic power demands. Each source is connected to the DC bus via a controlled reversible power converter.

Powertrain

Consisting essentially of the wheels, the speed reducer, the electric motor and its power converter, the powertrain must provide the required mechanical power to move the vehicle according to the instructions imposed by the driver. The required power P_{pt} to supply the powertrain of an electric vehicle of mass M, and of frontal surface S evolving at a speed v is given by (1) [4]:

$$P_{pt} = \frac{v}{\eta_{pt}} [\frac{1}{2}\rho_a SC_x v^2 + Mg\mu cos(\alpha) + Mgsin(\alpha) + M\frac{dv}{dt}], \tag{1}$$

where α, ρ_a, v_a, C_x, μ and η_{pt} are respectively, the slope of the road, the air density, the wind speed, the vehicle aerodynamic coefficient, the rolling coefficient, the gravitational acceleration and the efficiency of the powertrain.

The values of P_{pt} on a driving cycle constitute the power profile. In this work, the power profiles shown in Fig. 2 corresponding to the NEDC (New European Driving Cycle) and the IFSTTAR (Institut Français des Sciences et Technologies des Transports, de l'Aménagement et des Réseaux) driving cycles are used to represent an urban driving.

Battery and UC

In this work, a simple model of a Li-ion battery consisting of the association in series of a voltage source V_{boc} and resistance R_{bat} is used (see Fig.1) [9] and [10].. Its efficiency η_{bat} can be determined using the relation [1]:

$$\eta_{bat} = \frac{P_{bat0}}{P_0} = \frac{1}{2}(1 + \sqrt{1 - \frac{4R_{bat}P_{bat0}}{V_{boc}^2}}). \qquad (2)$$

where P_{bat0} is the battery output power and P_0 is its generated internal electrochemical power.

The UC model used in this work consists of the association of a capacitance C_{uc}, and a series resistance R_{uc} [11, 12]. (see Fig.1). In discrete time, the UC state of charge SOC_{uc} is determined at each sampling interval time kT_s by the relation :

$$\begin{cases} SOC_{uc}(k) = (E_{uc}(k) - E_{ucmin})/(E_{ucmax} - E_{ucmin}) \\ E_{uc}(k) = E_{uc}(k-1) - P_{uc}(k)T_s/(\eta_{uc} \times \eta_{cuc})^{S_u} \\ E_{uc}(0) = \frac{1}{2}C_{uc}V_{uc0}^2 \end{cases} \qquad (3)$$

where k is the discrete time index, T_s is the sampling period, η_{uc} and P_{uc} are respectively the efficiency and the power supplied or absorbed by the UC, η_{cuc} is the UC converter efficiency, S_u corresponds to the P_{uc} sign, E_{uc0} is the UC initial energy, V_{uc0} represents the initial open circuit voltage, while E_{ucmin} and E_{ucmax} are respectively the UC minimum and maximum energy.

Energy management strategy

The energy management strategy in a multi-sources system must ensure a good distribution of the power P_{pt} requested by the powertrain between the energy sources while respecting the constraints and limits of each source in terms of power, power dynamics and state of charge. In the studied system, at each sampling k of period T_s, the constraints can be summarized as follows:

$$\begin{cases} P_{pt}(k) = P_{bat}(k) + P_{uc}(k) \\ P_{batmin} \leq P_{bat}(k) \leq P_{batmax} \\ P_{ucmin} \leq P_{uc}(k) \leq P_{ucmax} \\ SOC_{ucmin} \leq SOC_{uc}(k) \leq SOC_{ucmax} \end{cases} \qquad (4)$$

where P_{bat} and P_{uc} represent respectively the power supplied by the battery and the UC through their respective converters.

The battery state of charge is not discussed here because the battery is considered as the main source, so it is supposed to be adequately charged.

As shown in Fig. 3, the energy management method developed here is based on filtering. This technique is used in several works to route the low power dynamics towards the battery and the strong dynamics towards the UC. The goal is to protect the battery from strong power dynamics.
Unlike passive filtering methods (i.e, Fig. 4) used in management strategies, the proposed method introduces an adaptive parameter k_d to modulate the amplitude of the power to be filtered. This increases

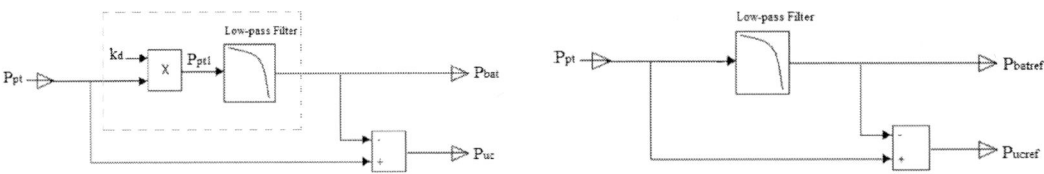

Fig. 3: Developed EMS flowchart   Fig. 4: Passive filtering EMS flowchart

the degree of freedom and allows to perform some optimization. Thus, instead of filtering all the power demanded by the powertrain, only a proportion is sent to the filter. This proportion, designated by P_{pt1}, is determined by:

$$P_{pt1}(k,k_d) = k_d \cdot P_{pt}(k). \tag{5}$$

Consequently, the battery power is calculated using the discrete model of the first order low-pass filter :

$$P_{bat}(k,k_d) = P_{pt1}(k) + (P_{bat}(k-1) - P_{pt1}(k,k_d))e^{-\omega_c \cdot T_s} \tag{6}$$

where $\omega_c(rad/s) = 2\pi f_s$, and f_s is the filter cutoff frequency defined according to the battery specifications, and is given as the ratio of the battery specific power to its specific energy [7].

The UC power is deduced by the equation :

$$P_{uc}(k,k_d) = P_{pt}(k) - P_{bat}(k,k_d) \tag{7}$$

The value of k_d is determined online in order to optimize the cost function C which takes into account the system electrical losses L_e and the SOC of the UC SOC_{uc} as given by the equation :

$$\begin{cases} C(k,k_d) = L_e(k,k_d) + K|SOC_{ucref} - SOC_{uc}(k,k_d)| \\ L_e(k,k_d) = L_{ebat}(k,k_d) + L_{euc}(k,k_d) \end{cases} \tag{8}$$

where SOC_{ucref} is the reference value of SOC_{uc}. K is a weighting constant tuned to have an appropriate regulation of the UC.

The losses considered here are the electrical losses in the sources and their converters.

For a given source i of output power P_i and its converter ci where $i \in \{bat, uc\}$, the instantaneous electrical losses L_{ei} can be expressed by the relation (9) :

$$L_{ei} = P_i[1/(\eta_i \times \eta_{ci})^{S_i} - 1], \tag{9}$$

where η_i and η_{ci} are the considered power source and its converter efficiencies. S_i is the sign of P_i: $S_i = 1$ if $P_i > 0$ and $S_i = -1$ if $P_i < 0$.
The total electrical losses of the studied system can be determined by :

$$L_e = L_{ebat} + L_{euc}, \tag{10}$$

where L_{ebat} represents the losses of the battery and its converter, and L_{euc} considers the electrical losses of the UC and its converter.

Finally, in the limit of the constraints defined by (4), for each sampling of period T_s, the algorithm scans the value of the adaptive parameter $k_d \in [0,1]$ with a step k_{ds}, and determines the reference power of each source using the set of equations :

$$\begin{cases} C_{min}(k) = min(C(k,k_d)), 0 \le K_d \le 1 \\ k_{dOpt}(k) = k_d(C = C_{min}(k)) \\ P_{batref}(k) = P_{bat}(k,k_{dOpt}) \\ P_{ucref}(k) = P_{uc}(k,k_{dOpt}) \end{cases} \tag{11}$$

k_{dOpt}, P_{batref} and P_{uctref} correspond respectively to the optimal values of k_d, P_{bat} and P_{uc} that minimize instantaneously the cost C.

The reference power P_{batref} and P_{uctref} determined using (11) are then imposed to the sources through their associated power converters using underlayer controllers. These controllers, whose designs are not detailed here, compute the appropriate value of the duty cycles for better tracking of the converter power references.

Note that although the method is iterative, it is low computational cost and thus it could be suitable for real-time implementation.

Simulations results

Matlab simulations of the developed algorithm are performed for the two power profiles shown in Fig. 2. The initial state of charge of the UC is considered equal to 50%. The sampling time $T_s = 0.001$sec and $k_{ds} = 0.01$ where used.

Fig.5 and 6 show the battery and UC power references determined by the developed algorithm for the NEDC and IFSTTAR driving cycles, respectively. For each case, the battery supplies the low power dynamics unlike the UC that assists the battery by providing the power peaks. The corresponding evolution of the UC state of charge is shown in Fig. 8. It can be noticed that it never reaches its extreme values and returns at the end of the cycle to approximately less than +10% from the reference state of charge. Indeed, the final SOC_{uc} is about 59.5% for the NEDC driving cycle and about 55% for the IFSTTAR driving cycle.

Compared to the passive filtering EMS (see Fig. 4), it appears that the proposed approach performs better than the passive filtering regarding the UC state of charge as shown in the Fig. 7. This clearly shows the influence of the adaptive parameter k_d introduced in the developed EMS. Indeed, it allows to favor the UC when needed according to its SOC. k_d especially favors the UC during regenerative braking when SOC_{uc} is too low. Fig. 9 shows the evolution of the optimal values K_{dOpt} for the IFSTTAR power profile. Although

The system electrical losses throughout the NEDC and IFSTTAR driving cycles, $\int_0^t L_e(\tau)d\tau$, are illustrated in the bar graph shown by Fig. 10. Improvements are observed when using the proposed EMS algorithm. Indeed, a reduction in the system electrical losses of about 2.4% and 3.3% are evaluated for the two considered driving cycles, respectively. These results confirm the pertinence of the proposed approach in terms of power loss minimization.

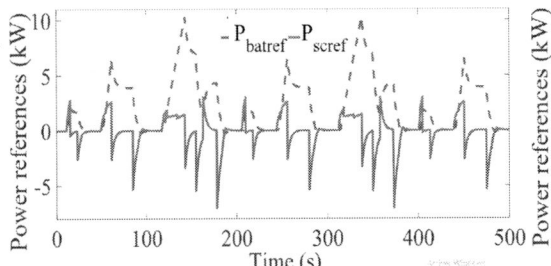

Fig. 5: Power reference of sources with the proposed EMS for NEDC profile.

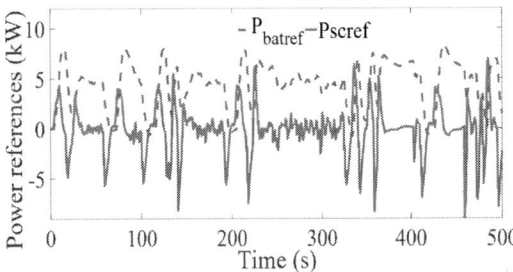

Fig. 6: Power reference of sources with the proposed EMS for IFSTTAR profile.

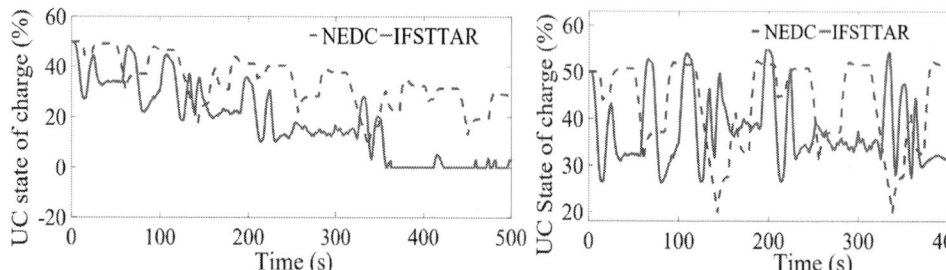

Fig. 7: UC state of charge with passive filtering EMS for NEDC and IFSTTAR driving cycle.

Fig. 8: UC state of charge with the adaptive EMS for NEDC and IFSTTAR driving cycles.

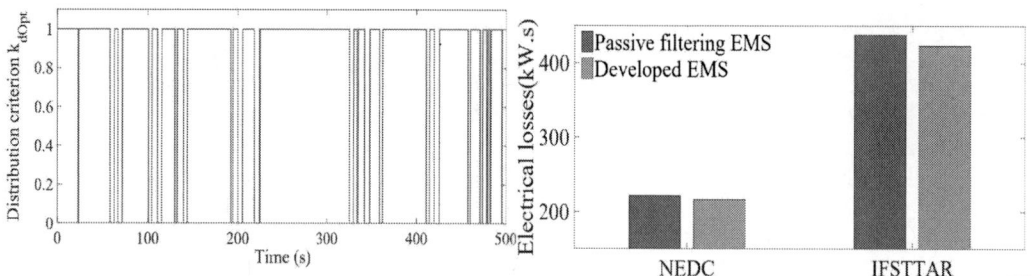

Fig. 9: k_{dOpt} for IFSTTAR profile.

Fig. 10: The electrical losses of the system for developed EMS and passive filtering EMS.

Conclusion

In this paper, an improved energy management strategy for a battery- UC electric vehicle is presented. Based on the spectral repartition of the power demand between sources, the developed energy management algorithm allows to achieve some optimization in terms of system electrical losses reduction while ensuring a good regulation of the UC state of charge. Simulations considering real power profile during urban driving cycle confirm the pertinence of the proposed approach. Although optimized compared to the classic frequency separation energy management method, the developed approach could still be subject of improvements with a more accurate model of the efficiencies of sources.

References

[1] Xie S., Hu X., Qi S., Lang K.: An artificial neural network-enhanced energy management strategy for plug-in hybrid electric vehicles, Energies2018, 2018

[2] , Snoussi J., Ben Elghali S., Benbouzid M., Mimouni M. F. :Sliding Mode Control for Frequency-Based Energy Management Strategy of Hybrid Storage System in Vehicular Application, International Symposium on Power Electronics Electrical Drives, Automation and Motion, Anacapri, Italy, 2016

[3] Sethakul P., Davat B., Hinaje M., Thounthong P., Chunkag V.: Comparative Study of Fuel-Cell Vehicle Hybridization with Battery or Supercapacitor Storage Device, IEEE Transactions on Vehicular Technology, 2009

[4] Florescu A., Munteanu I., Bratcu A. I., Bacha S.: Frequency- Separation-Based Energy management Control Strategy of Power Flows within Electric Vehicles using Ultracapacitors, IECON2012 (38th Annual Conference of the IEEE Industrial Electronics Society), Montréal, Canada, 2012

[5] Alloui H., Marouani K., Becherif M., Sid M. N.,Ben M.: A Control Strategy Scheme for Fuel Cell-Vehicle Based on Frequency Separation, IEEE ICGE, Tunisia., pp.170-175, 2014

[6] Nwesaty W. and Bratcu A. I. and Sename O.: Optimal frequency separation of power sources by multivariable LPV/H∞ control: application to on-board energy management systems of electric vehicles, 53rd IEEE Conference on Decision and Control (CDC 2014), Los Angeles, United States, 2014

[7] Snoussi J. and Ben Elghali S. and Benbouzid M. and Mimouni M. F. : Auto-adaptive filtering-based energy management strategy for fuel cell hybrid electric vehicles, Energies2018, 2018

[8] Hannan M. A., Hoque M. M., Hussein A.,Yusof Y., Ker P. J.: State-of-the-Art and Energy Management System of Lithium - Ion Batteries in Electric Vehicle Applications: Issues and Recommendations, IEEE journal, 2018

[9] Abdennadher K., Rosset C., Venet P., Rojat G.: A Real Time Predictive Maintenance System of Aluminium Electrolytic Capacitors Used in Uninterruptible Power Supplies, IEEE Industry applications society annual meeting, 2008

[10] Urbain M.,Hinajeand M., Rael S. and al: Energetical modeling of lithium ion batteries including electrode porosity effects, IEEE Transc. on Energy Conversion, IEEE Trans. On Energy Conversion,vol 25, 2010

[11] Lajnef W., Vinassa J. M., Azzopardi S., Briat O., Woirgard E., Zardini C., Accouturier J. L.: Ultracapacitors Modeling Improvement Using an Experimental Characterization Based on Step and Frequency Responses, IEEE Power Electronics Specialist Conference PESC04, Aachen, Germany, 2004

[12] Belhachemi F., Rael S., Davat B.: A physical based model of power electric double-layer supercapacitors, IEEE-IAS2000, Rome, 2000"

Mechanistic Power Module Degradation Modelling Concept with Feedback

Martin Bendix Fogsgaard
Aalborg University
Pontoppidanstræde 101
Aalborg East, Denmark
Phone: +45 28 94 27 07
Email: mbf@et.aau.dk
URL: http://www.et.aau.dk

Paula Diaz Reigosa
University of Applied Sciences and
Arts Northwestern Switzerland
Klosterzelgstrasse 2
5210 Windisch, Switzerland
Phone: +41 56 202 81 76
Email: paula.diazreigosa@fhnw.ch
URL: http://www.fhnw.ch/en

Francesco Iannuzzo
Aalborg University
Pontoppidanstræde 101
Aalborg East, Denmark
Phone: +45 99 40 33 14
Email: fia@et.aau.dk
URL: http://www.et.aau.dk

Michael Hartmann
Schneider Electric
Ruthnergasse 1
1210 Vienna, Austria
Phone: +43 (0)1 29191 2836
Email: michael.hartmann@se.com
URL: http://www.se.com

Keywords

≪Physics of Failure≫, ≪IGBT≫, ≪Lifetime Prediction≫, ≪Power Module Modelling≫, ≪Bondwire modelling≫.

Abstract

A platform will be presented based on physics-of-failure-based models. The platform gives an overview of the couplings between simulation, health monitoring and online lifetime prediction of a power module. The platform is modular and is not tied to any specific software product to make it as generally applicable as possible.

Introduction

Reliability and lifetime engineering have been relevant fields of study since their conception. After the invention of power electronics, the analyses and findings from the reliability analysis of other fields were applied to power electronics as well. As conventional energy sources are replaced with renewable sources of energy the demand for reliable power electronics increases[1].

One of the most widely used power electronics devices is the IGBT(Insulated Gate Bipolar Transistor), combining the power capability of the bipolar transistor and the fast switching of MOS(Metal-Oxide-Semiconductor) devices.

IGBTs are mature devices, and as a result of many years of improvements and study, the fatigue relevant failure mechanisms have been reduced to consisting mainly of package related failures[2]. The main degrading sections of the IGBT geometry are the solder layer attaching the semiconductor chip and the bond-wires functioning as interconnects from the top surface of the semiconductor chip.

This paper will describe the characteristic behaviour of IGBT and packaging degradation and reliability modelling. This will be presented in a large flow chart connecting cause and effect. The proposed mechanistic model is meant to serve as an overview and inspiration for power module reliability analysis.

The content of the mechanistic model will be described in this manuscript. Special focus will be placed on the empirical damage models.

The general procedure used to characterise empirical models will be described along with introductions to the relevant life testing methods for power electronics.

Finally, the mechanistic relationships will be validated by comparing the predicted wear out Vce evolution with experimental results.

Degradation Models

In manuscript [3] a range of damage models for IGBT fatigue are reviewed. One of the conclusions drawn in this paper is that cycle based damage models are limited, and time-dependent damage modelling is needed. The authors however fail to provide a complete damage modelling approach, even in subsequent papers[4].

Traditionally, empirically fitted damage equations have been used for damage modelling of a broad range of devices and constructions, physics-based simple damage models quite simply do not exist yet. Alternatively, an effort-costly approach to damage modelling can be found in the microstructural mechanical approach used for the solder layer of an IGBT in [5]. This approach was found to be able to accurately model both the crack initiation and crack propagation of the solder joint.

The traditional damage approach is used in [6], however, this approach adds a thermal model degradation feedback loop to emulate the effects of damage accumulation on the thermal characteristics of the device. The cumulative degradation of the device will in some cases greatly influence the device behaviour during device life and may have a large effect on life predictability using parameter monitoring.

The practical implementation of the traditional empirical approach to reliability modelling can be seen executed thoroughly in [7]. The authors analyse an entire DC micro-grid system with PV generation, battery storage, fuel cell and a load profile based on both a clinic and an apartment case. The final result of the paper is a reliability assessment of the entire system.

In [2] the degradation of an IGBT module is modelled through a successive series of models and physics simulations. One simulation is based on the results of the previous, and previous simulations are updated if the model requires it.

[8] offers a review of the state of the art for lifetime prediction of power electronics devices. The review reports a large number of empirical damage models and discusses degradation monitoring and power cycling methodology.

Paris Law

The Paris Law is similar in form and origin to the Coffin-Manson and Basquin equations. Instead of number of cycles to failure it predicts the crack increase per cycle. It is an empirical equation and as such requires fitting to experimental lifetime tests. The stressor input to this equation is the stress intensity factor which is a geometry weighting method for the mechanical stress.

$$\frac{da}{dN} = C(\Delta K)^p \tag{1}$$

Where a is the crack length, N is the cycle, ΔK is the stress intensity factor range and C and p are fitting parameters. It is important to note that the Paris Law only models crack propagation and not crack formulation/initiation/nucleation.

Life and degradation link

A non-linear degradation functional can be used to link the damage from one of the damage models with the degradation effects [9].

$$1 - D = \left(1 - \frac{N}{N_f}\right)^k \tag{2}$$

Table I: Table of Damage Models

Name	Equation	Stress Term(s)	Ref(s)
Coffin-Manson	$N_f = A(\Delta\varepsilon_{pl})^B$	$\Delta\varepsilon_{pl}$	[10]
Basquin	$N_f = a(\Delta\sigma_e)^{-b}$	$\Delta\sigma_e$	[11]
Modified Coffin-Manson	$N_f = A\Delta T_j^{-\alpha} exp\left(\frac{E_a}{k_B T_{jm}}\right)$	$\Delta T_j, E_a, T_{jm}$	[12]
Bayerer	$N_f = K(\Delta T_j)^{\beta_1} e^{\left(\frac{\beta_2}{T_j+273}\right)} t_{on}^{\beta_3} I^{\beta_4} V^{\beta_5} D^{\beta_6}$	$\Delta T_j, T_j, t_{on}, I, V, D$	[13]

Where D is a representation of degradation (from 0-1), N is the current number of cycles, N_f is the number of cycles to failure and k is the exponential fitting parameter.

Proposed Mechanistic Model

In the time between module assembly and first bond-wire liftoff, the bond wires on a single chip can be modelled as having identical behaviour.

This manuscript details a new power module wear-out and degradation mechanism platform. The methodology is conceptualised for a single semiconductor chip with both wire bond and die attach layers for which the degradation is modelled. The module considered in this manuscript is a Si IGBT module so it is assumed that no degradation of the semiconductor chip itself takes place.

The mechanistic model presented in this section is based on the model in fig. 1. The thermal networks is based on the thermal model from [14].

Superimposed on the material and structural representation are two networks, a blue, and a black. The blue network is the basic electric network of the micro-device, including the forward voltage of the semiconductor and bond-wire and bond-wire interface resistances.

The black network is the basic thermal network of the micro-device. Including the main thermal resistances and a simplified heat modelling approach using two heat sources, one for the chip representing the chip power loss and one for the bond-wire.

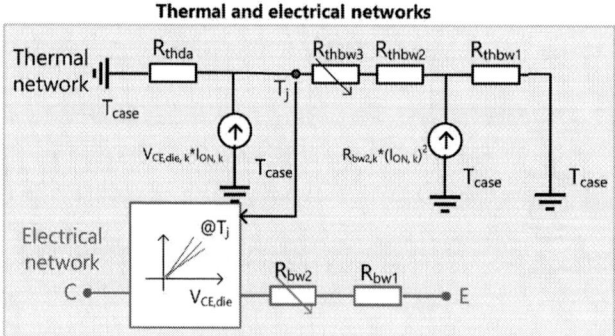

Fig. 1: Basic semiconductor thermal and electrical modelling.

Damage Functions

The damage function blocks contain the damage models needed for wire bond and die attach metallization damage calculations. The modelling platform can be updated with new damage models as long as they are capable of estimating a damage quantity which can be related to the crack propagation in the two metallization layers. A damage model can be chosen from Table I or elsewhere.

The platform itself can be implemented in both the continuous time and discrete time-domains. These functions are the green boxes in fig. 3.

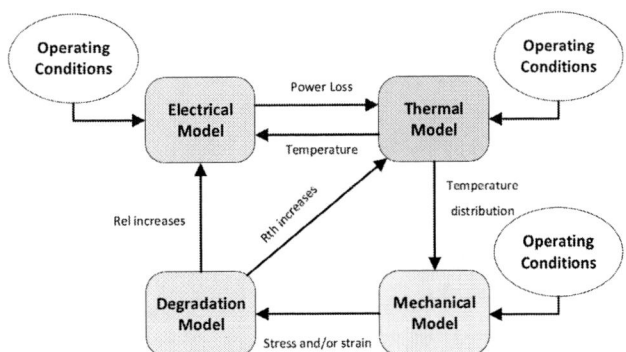

Fig. 2: Modelling Concept Flow.

Effect Functions

The effects of the degradation on the thermo-electrical model of the system are clear with the definitions of the characteristic values. These functions are the red boxes in fig. 3.

The electrical resistances change based on temperature and degradation:

$$R_{el}(T,a) = R_{0el}f(a)f(T) \tag{3}$$

Where T is temperature, a is crack length, R_{0el} is initial resistance, R_{el} is electrical resistance and f() are general formulations expressing the existence of functions describing crack and temperature dependence of the resistance.

The thermal resistance change based on degradation.

$$R_{th}(a) = R_{0th}f(a) \tag{4}$$

Where T is temperature, a is crack length, R_{0th} is initial resistance, R_{th} is thermal resistance and f() are general formulations expressing the existence of functions describing crack and temperature dependence of the resistance.

Thermal Model

The platform contains basic thermal and electrical models of the power module. The implemented thermal model contains the loop from DBC through wire bond, semiconductor die and die attach back to DBC and ambient temperature. This network is black and yellow box in fig. 3.

The split wire bond and die thermal resistances are constant, with the wire bond metallization and die attach thermal resistances being affected by the damage as described in equation 4.

Electrical Model

The electrical model included has a similar scope of the thermal model. It includes the wire bond resistance, the wire bond metallization resistance, the electrical behaviour of the semiconductor die and the die attach metallization resistance. All of the electrical resistances are affected by the local temperatures, and the *Rme* resistances are affected by the damage as described in equation 3.

Model Predicted Wear-Out Comparison

The wear-out predicted by this model is compared with the degradation reported by [15] to validate the mechanistic relationship of fig. 3.

The Vce measurement is comprised of a number of components, the main categories of which are the voltage drop across the semiconductor die and the resistive voltage drop across the interconnect path

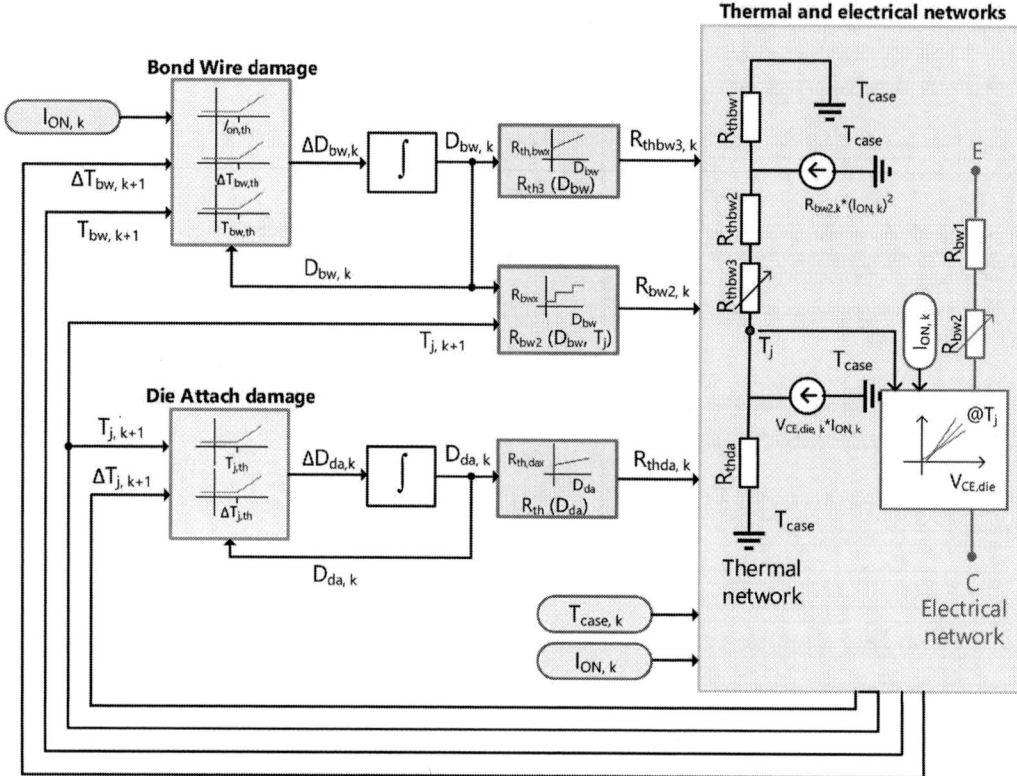

Fig. 3: Original model detailing the damage model.

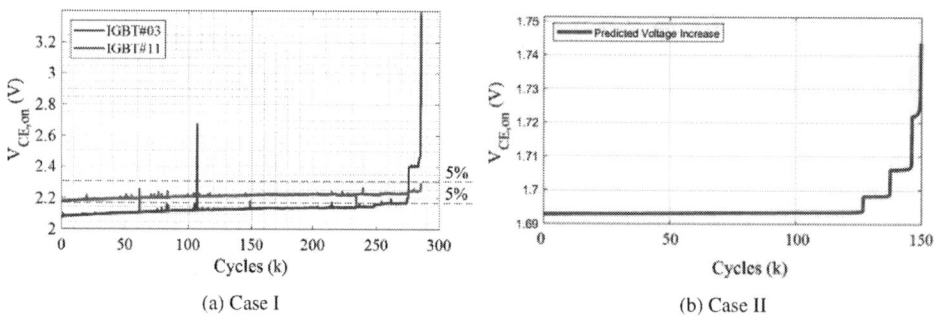

(a) Case I

(b) Case II

Fig. 4: a) Experimental wear out curve from [15]. b) Predicted wear out curve.

contained in the voltage measurement. Generally, interconnect degradation is seen in bond-wires, chip top metallization, and solder layer. Of these, the solder layer degradation has the smallest electrical effect. The resistance of the degrading bond-wires is estimated using crack propagation modelling using FEM.

A comparison of figs. 4a and 4b shows that the model is able to predict the main degradation effects experienced by the device from [15]. As the Paris Law coefficients were not fitted to the degraded device from [15] the crack propagation rate should not be considered for the comparison instead the internal timing and behaviour should be. This example implementation was done in MATLAB.

Conclusion

This manuscript presented a mechanistic damage and effect model for power electronics module reliability analysis. The model details the cause and effect of each phenomenon. General discussion is made concerning the trends of the subcomponents.

The mechanistic model is multi-physical, as it contains electrical, thermal and structural damage modelling of a power semiconductor chip. The model is formulated and presented as the basis on which reliability analysis and modelling can be conducted. The model can and should be adapted and improved according to the focus of the use case.

References

[1] Y. Yang, H. Wang, A. Sangwongwanich, and F. Blaabjerg, "Design for Reliability of Power Electronic Systems," *Power Electronics Handbook*, pp. 1423–1440, jan 2018. [Online]. Available: https://www.sciencedirect.com/science/article/pii/B9780128114070000519

[2] B. Rannestad, A. E. Maarbjerg, K. Frederiksen, S. Munk-Nielsen, and K. Gadgaard, "Converter Monitoring Unit for Retrofit of Wind Power Converters," *IEEE Transactions on Power Electronics*, vol. 33, no. 5, pp. 4342–4351, may 2018. [Online]. Available: http://ieeexplore.ieee.org/document/7953522/

[3] L. Yang, P. A. Agyakwa, and C. M. Johnson, "Physics-of-failure lifetime prediction models for wire bond interconnects in power electronic modules," *IEEE TRANSACTIONS ON DEVICE AND MATERIALS RELIABILITY*, vol. 13, Mar 2013.

[4] ——, "Calibration of a novel microstructural damage model for wire bonds," *IEEE TRANSACTIONS ON DEVICE AND MATERIALS RELIABILITY*, vol. 14, Dec 2014.

[5] V. N. Le, L. Benabou, Q. B. Tao, and V. Etgens, "Modeling of intergranular thermal fatigue cracking of a lead-free solder joint in a power electronic module," *International Journal of Solids and Structures*, vol. 106-107, pp. 1–12, 2017.

[6] B. Gao, F. Yang, M. Chen, Y. Chen, W. Lai, and C. Liu, "Thermal lifetime estimation method of IGBT module considering solder fatigue damage feedback loop," *Microelectronics Reliability*, vol. 82, no. December 2017, pp. 51–61, 2018. [Online]. Available: https://doi.org/10.1016/j.microrel.2017.12.046

[7] S. Peyghami, H. Wang, P. Davari, and F. Blaabjerg, "Mission-Profile-Based System-Level Reliability Analysis in DC Microgrids," *IEEE Transactions on Industry Applications*, vol. 55, no. 5, pp. 5055–5067, sep 2019. [Online]. Available: https://ieeexplore.ieee.org/document/8727971/

[8] A. Hanif, Y. Yu, D. Devoto, and F. Khan, "A Comprehensive Review Toward the State-of-the-Art in Failure and Lifetime Predictions of Power Electronic Devices," *IEEE Transactions on Power Electronics*, vol. 34, no. 5, pp. 4729–4746, 2019.

[9] K. S. Ravi Chandran, "A universal functional for the physical description of fatigue crack growth in high-cycle and low-cycle fatigue conditions and in various specimen geometries," *International Journal of Fatigue*, vol. 102, pp. 261–269, 2017. [Online]. Available: http://dx.doi.org/10.1016/j.ijfatigue.2017.01.046

[10] Manson and S. S, "Behavior of Materials Under Conditions of Thermal Stress," jan 1954. [Online]. Available: https://ntrs.nasa.gov/search.jsp?R=19930092197

[11] J. Bielen, J.-J. Gommans, and F. Theunis, "Prediction of high cycle fatigue in aluminum bond wires: A physics of failure approach combining experiments and multi-physics simulations," in *7th. Int. Conf. on Thermal, Mechanical and Multiphysics Simulation and*

Experiments in Micro-Electronics and Micro-Systems. IEEE, pp. 1–7. [Online]. Available: http://ieeexplore.ieee.org/document/1644022/

[12] M. Junghnel, R. Schmidt, J. Strobel, and U. Scheuermann, "Investigation on isolated failure mechanisms in active power cycle testing," *PCIM Europe 2015*, May 2015.

[13] R. Bayerer, T. Herrmann, T. Licht, J. Lutz, and M. Feller, "Model for power cycling lifetime of igbt modules - various factors influencing lifetime," Mar 2008.

[14] M. Chen, H. Wang, D. Pan, X. Wang, and F. Blaabjerg, "Thermal Characterization of Silicon Carbide MOSFET Module Suitable for High-Temperature Computationally-Efficient Thermal-Profile Prediction," *IEEE Journal of Emerging and Selected Topics in Power Electronics*, vol. 6777, no. c, pp. 1–1, 2020.

[15] M. Jiang, G. Fu, M. B. Fogsgaard, A. S. Bahman, Y. Yang, and F. Iannuzzo, "Wear-out evolution analysis of multiple-bond-wires power modules based on thermo-electro-mechanical FEM simulation," *Microelectronics Reliability*, vol. 100-101, p. 113472, sep 2019. [Online]. Available: https://www.sciencedirect.com/science/article/pii/S0026271419304640

Experimental validation and comparison of a SiC MOSFET based 100 kW 1.2 kV 20 kHz three-phase dual active bridge converter using two vector groups

Thomas LAGIER[1], Piotr DWORAKOWSKI[1], Cyril BUTTAY[1,2], Philippe LADOUX[3], Andrzej WILK[4], Philippe CAMAIL[1,2], Elissa Cresenta ANAK JUSTIN[1]

[1] SUPERGRID INSTITUTE
23 rue Cyprian
69100 Villeurbanne, France
URL: http://www.supergrid-institute.com
[2] Université de Lyon, Laboratoire Ampére, INSA Lyon, CNRS
21 bis avenue Jean Capelle
69100 Villeurbanne, France
URL: http://www.ampere-lab.fr
[3] LAPLACE, UNIVERSITÉ DE TOULOUSE, CNRS, INPT, UPS
2 rue Charles Camichel
31000 Toulouse, France
URL: http://www.laplace.univ-tlse.fr
[4] GDAŃSK UNIVERSITY OF TECHNOLOGY, FACULTY OF ELECTRICAL AND CONTROL ENGINEERING
Gabriela Narutowicza 11/12
80-233 Gdańsk, Poland

Acknowledgements

This work was supported by a grant overseen by the French National Research Agency (ANR) as part of the "Investissements d'Avenir" Program (ANE-ITE-002-01).

Keywords

High Voltage power converter, Voltage Source Converter (VSC), MOSFET, Silicon Carbide (SiC), ZVS converters.

Abstract

The Dual Active Bridge appears as a promising DC-DC converter topology when galvanic isolation and bidirectional power flow are required. Among its advantages, Zero Voltage Switching allows the switching losses to be significantly reduced. For high power applications, the three-phase topology variant may be interesting in order to reach a higher power density, especially when a three-phase transformer is implemented instead of three single-phase transformers. Moreover, the transformer vector group offers a new degree of freedom for the designers. In this paper, the authors present the experimental validation of a 1.2 kV – 100 kW – 20 kHz three-phase Dual Active Bridge converter using two medium frequency transformers and different vector groups.

Introduction

The Dual Active Bridge (DAB) topology is an interesting candidate for high power DC-DC converters when galvanic isolation and bidirectional power flow are required. In addition to its modularity and simplicity, Zero Voltage Switching (ZVS) operation allows a higher switching frequency, which in turn enables a higher power density. Compared to the single-phase DAB, the three-phase topology variant, depicted in Fig. 1, allows reducing the input and output filters and transformer size [1–6].

Moreover, since several vector groups (Yy, ΔΔ and even YΔ) can be chosen for the three-phase transformer, a new degree of freedom can be used to optimize the design. Some papers have already studied the three-phase DAB, but the use of a single three-phase medium frequency transformer (instead of three single-phase transformers) is not well documented. Moreover, in this paper, the authors propose a global comparison between two Medium Frequency Transformers (MFT) with two vector groups (Yy and ΔΔ).

After a short introduction of the topology and the DC-DC converter prototype, this paper compares the performance of the transformers and the voltage source inverters (VSI). Then, the performances are summarized in order to identify the configuration which presents the best compromise for this application.

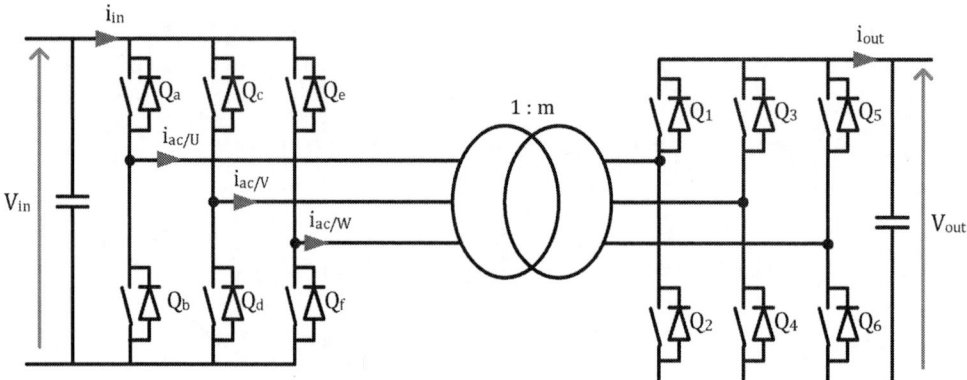

Fig. 1. Three-phase dual active bridge converter

Three phase dual active bridge converter

The three-phase DAB is composed of two VSIs and a MFT. The theoretical analysis of the circuit [2] allows the calculation of the expressions for the transmitted power and the ZVS operating range as a function of the input voltage (V_{in}), the reactance of leakage inductance seen from the primary side of the transformer (X_l) and the ratio between the input and output voltages, calculated by (1)

$$k = \frac{V_{out}}{mV_{in}} \qquad (1)$$

where m is the transformer's turns ratio.

Fig. 2 compares the theoretical RMS currents in the switches and the ZVS operating limits for Yy and ΔΔ vector groups.

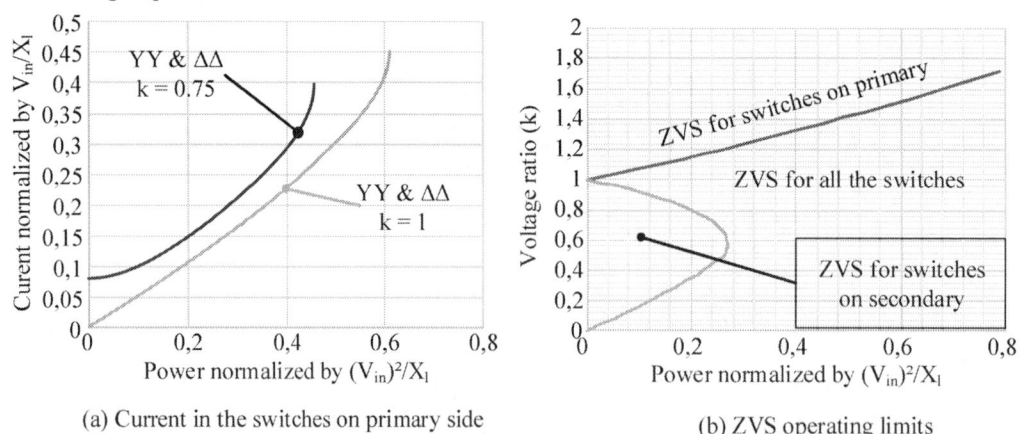

(a) Current in the switches on primary side

(b) ZVS operating limits

Fig. 2. Theoretical current in the switches and ZVS operating ranges for Yy and ΔΔ vector groups calculated using the equations of [2].

As it can be seen, there is no difference between the two solutions. Consequently, the performances and losses of the power switches are expected to be similar.

Experimental validation

Two three-phase MFT prototypes have been developed in order to perform a comparative analysis. For the first prototype T1, the vector group can be changed between Yy and ΔΔ while the Yy vector group has been selected for the second prototype T2. Both transformer prototypes were built using MnZn ferrite 3C90 and Litz wire. The design details are presented in [7], [8]. The main transformer specifications for $k = 1$ are presented in Table 1. The core power loss calculation is based on [9], [10] and the winding power loss is based on [11], [12].

Table 1. Three-phase medium frequency transformer specifications at k = 1: MFT T1 (Yy and ΔΔ) and MFT T2 (Yy only)

	T1 ΔΔ	T1 Yy	T2 Yy
Winding voltage (V)	980	566	566
Winding current (A)	36	65	65
Core flux density (T)	0.22	0.15	0.27
Winding current density (A/mm²)	1.2	2.1	2.1
Core power loss at 100°C (W)	432	122	230
Winding power loss at 80°C (W)	155	229	214
Equivalent series resistance (mΩ)	13.7	18.1	17.1
Equivalent magnetizing inductance (mH)	0.2 ... 0.7	0.5 ... 2	1 ... 2.5
Equivalent leakage inductance (μH)	11.3	34	15.8
Dimensions (cm)	67 x 20 x 35		45 x 20 x 30
Weight (kg)	57		36

Even if the theoretical studies have shown similar performances for the two configurations, experimental validations were necessary to validate and compare the behavior of the converter. In order to investigate this, a Silicon Carbide (SiC) based 1.2 kV – 100 kW – 20 kHz DC-DC converter prototype was developed and presented in [1]. The three-phase DAB test bench implementation is presented in Fig. 3.

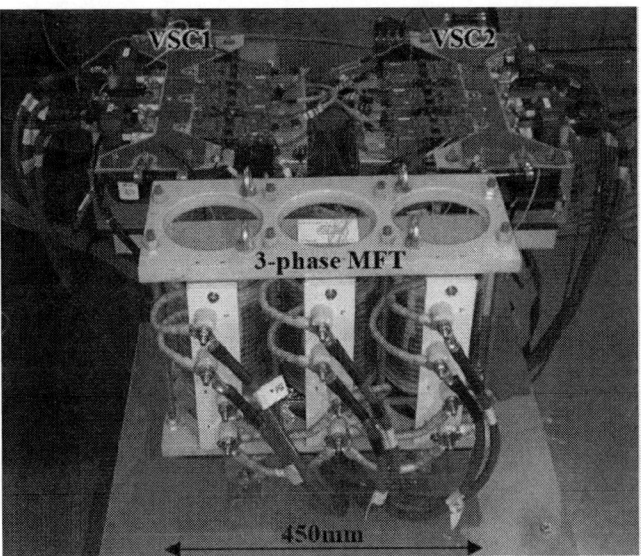

Fig. 3. Three-phase DAB test bench implementation (T2 MFT)

The power circuit is arranged in back-to-back fashion in order to test the DC-DC converter at full power with a minimum energy consumption. The converter output is connected to its input and the whole is supplied from the DC power supply which sets the voltage reference and compensates for the power losses in the converter.

Fig. 4 presents the experimental waveforms for Yy and $\Delta\Delta$ vector groups, obtained with the MFT T1 prototype.

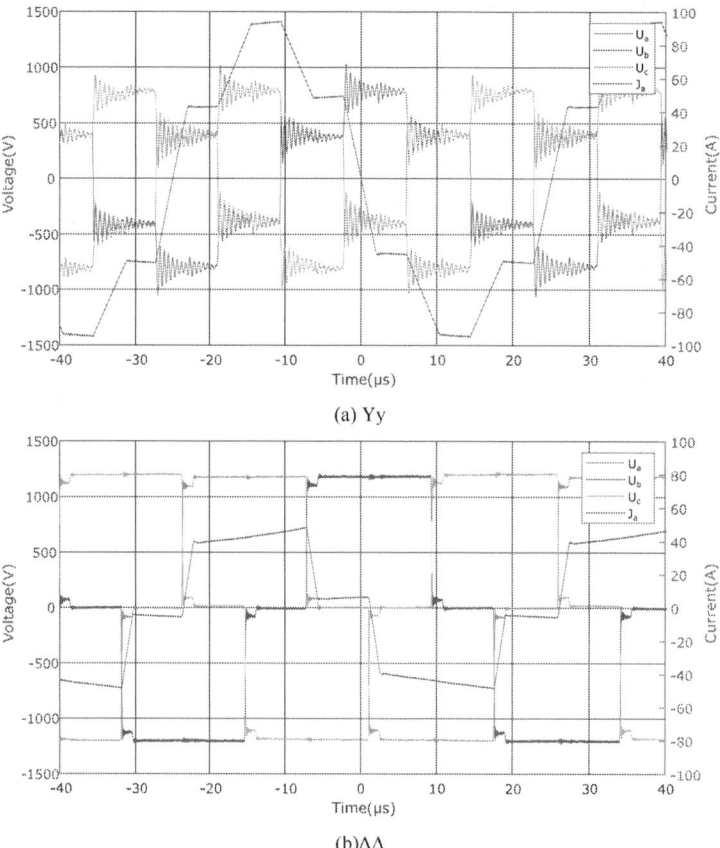

(a) Yy

(b)$\Delta\Delta$

Fig. 4. Experimental waveforms showing the voltages and currents across the windings of the transformer ($V_{in} = 1$ kV and $P_{dc} = 80$ kW)

Fig. 5 presents the measured efficiency for all the configurations.

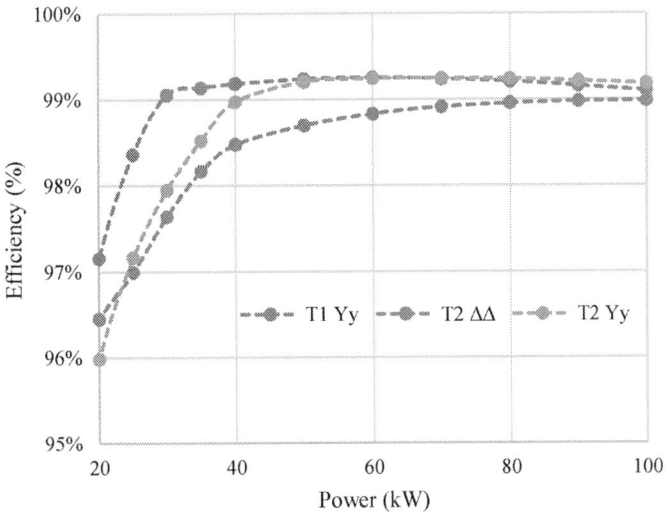

Fig. 5. Measured efficiency for the three configurations at $V_{in} = 1.2$ kV and $k = 1$.

As it can be seen, the Yy vector groups present the best efficiency over the entire power range. The following sections explain this difference.

Study of the power losses in the transformer

Fig. 6 presents the RMS current values in the windings on the primary side of the transformer. The currents are similar on the secondary side.

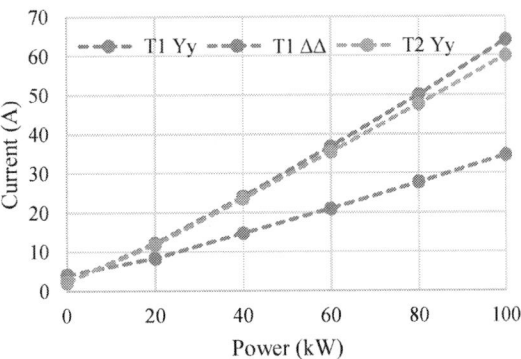

Fig. 6. Comparison of the RMS currents in the transformer primary winding

As expected, the currents in the windings are lower for the ΔΔ configuration, except at low power, where the contribution of the transformer magnetizing current is higher. This causes higher copper losses with ΔΔ configuration for a power lower than 7 kW.

At high power, the ΔΔ currents are lower by a factor of about $\sqrt{3}$. Consequently, at nominal power, the copper losses (RI_{ac}^2) are expected to be smaller, by a ratio of 3 $\left(\left(\sqrt{3}\right)^2\right)$.

Regarding the core losses, the Improved General Steinmetz equation (2-4) [10] can be used to understand their evolution. These losses depend on the voltage across the i-th winding (u_i where $i = 1\ldots6$), the maximum flux density (B_{max}), the core volume (V_c), the core cross-section (A_c), the frequency (f_{ac}) and the temperature (T). The coefficients used in the equations are depicted in Table 2. With the ΔΔ vector group, the voltage applied on the windings is $\sqrt{3}$ higher than with Yy. However, the maximum flux density is less than $\sqrt{3}$ higher (Table 1). Considering the same transformer design for ΔΔ and Yy, the core losses are obviously higher with ΔΔ. However, if there were two independent designs, one for ΔΔ and one for Yy, then core losses would be lower with ΔΔ.

Table 2. Ferrite 3C90 Steinmetz coefficients [13]

Parameter	Value
k	3.2
α	1.46
β	2.75
c_0	2.45
c_1	3.1e-2
c_2	1.65e-4

$$P_c = k_s k_T V_c (2B_{max})^{\beta-\alpha} f_{ac} \int_0^T \left| \frac{u_i}{N_1 A_c} \right|^\alpha dt \tag{2}$$

$$k_s = \frac{k}{2^{\beta+1}\pi^{\alpha-1}\left(0.2761 + \dfrac{1.7061}{\alpha + 1.354}\right)} \tag{3}$$

$$k_T = c_0 - c_1 T + c_2 T^2 \tag{4}$$

With the ΔΔ vector group, the lower copper losses obtained with the lower currents are not sufficient to compensate the increase in core losses caused by the higher voltage. This is the reason why losses are more important in ΔΔ. Moreover, as the core losses depend only on the voltage, they are expected to be constant with the transmitted power. This would explain why the difference in efficiency reduces when the transmitted power increases.

Study of the power losses in the VSI

Fig. 7 presents the RMS current value in the switches of the primary VSI. The currents are similar on the secondary side.

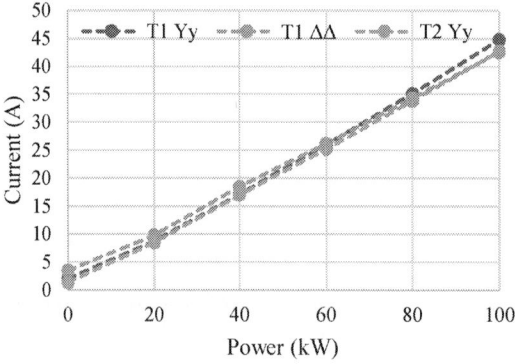

Fig. 7. Comparison of the RMS currents in the switches for the primary VSI (V_{in} = 1.2 kV, k = 1).

It can be seen that the currents are approximately the same for all the configurations. However, for low power (< 60 kW), the higher magnetizing current and higher core losses cause a small increase in the switches current for the ΔΔ vector group. However, the corresponding effect on the conduction losses of the switches is not significant enough to explain the difference in the efficiency.

Regarding the switching losses, a previous study [14] has shown that the soft switching operation depends on the parameters and the configuration of the circuit (leakage inductance, current, voltage). For some conditions, proper soft switching operation might not be obtained, resulting in higher switching losses. In [14], three cases have been identified as shown in Fig. 8, for the case of our prototype:

1. Case 1: Spontaneous turn-on failure. In this case, the energy stored in the leakage inductance is not important enough to fully discharge the Drain-Source Capacitance (C_{ds}) of the MOSFETs
2. Case 2: Dead time too long. In this case, the energy stored in the leakage inductance is sufficient to discharge the C_{ds} capacitance but because of the freewheeling sequence and the voltage applied by the secondary inverter, the current in the MOSFET cancels before the MOSFET turns-off.
3. Case 3: Soft switching operation. In this case the spontaneous commutation is obtained allowing the reduction of the switching losses.

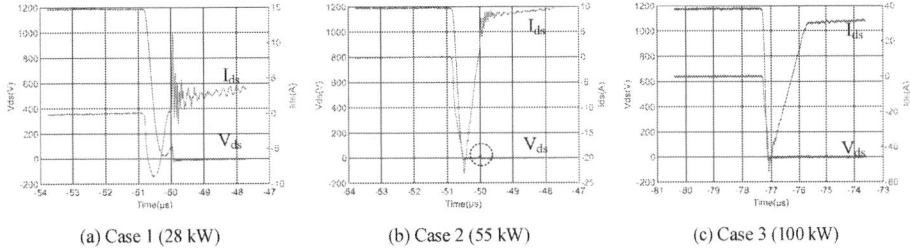

(a) Case 1 (28 kW)   (b) Case 2 (55 kW)   (c) Case 3 (100 kW)

Fig. 8. Switching waveforms for the three cases with the ΔΔ vector group (Vin = 1.2 kV, k = 1). The measured soft switching operating ranges obtained with all the configurations are summarized in Table 3.

Table 3. Soft switching operating range for all configurations (dead time = 500 ns, Vin = 1.2 kV, k = 1)

	T1 ΔΔ	T1 Yy	T2 Yy
Leakage inductance (μH)	11.3	34	15.8
Case 1: Spontaneous turn-on failure	0 → 28 kW	0 → 25 kW	0 → 40 kW
Case 2: Dead time too long	0 → 55 kW	N.A.	N.A.
Case 3: Soft switching operation	55→ 100 kW	25 → 100 kW	40 → 100 kW

As it can be seen, the Yy vector group allows the soft switching operation at lower power.

Fig. 9 depicts the equivalent scheme seen from the inverter's leg on the primary side for the two vector groups before the turn-off of the upper switch. The operation is similar for the other switches.

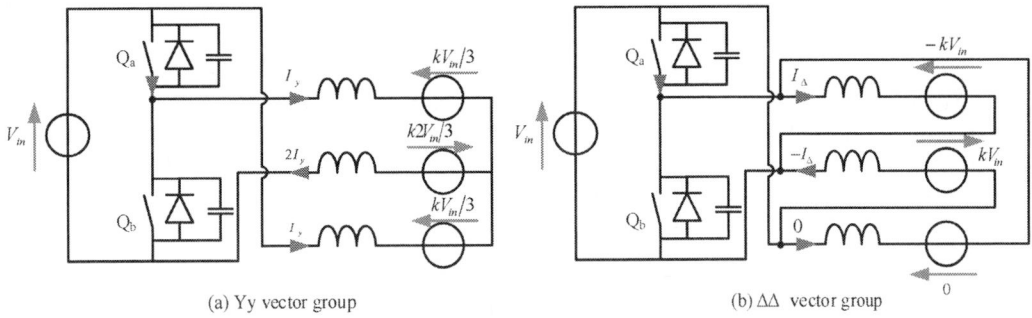

(a) Yy vector group   (b) ΔΔ vector group

Fig. 9. Equivalent scheme of the primary inverter before the turn-off of Qa (phase A) for the two vector groups

As it can be seen, the configuration, voltage and current before the turn-off are different. This explains the differences in the soft switching operating range. Moreover, simplified simulations have confirmed the experimental observations. A complete study, out of the scope of this paper, is nevertheless required to give more details. However, it clearly appears that the Yy vector group is more favorable to the ZVS operation because of the following reasons:

1. Higher leakage inductance with a higher current in the windings, allowing to get more energy stored in the leakage inductance.
2. Lower voltage applied by the VSI which causes a lower di/dt after the spontaneous commutation of the diodes and a shorter switching duration [14].

Summary

Thanks to the measurement performed and presented in this paper, Table 4 summarizes the observations made in this paper.

Table 4. General comparison of the different vector groups with a single MFT design

	Losses	The most favorable solution
Transformer	Core	Yy
	Windings	ΔΔ
VSI	Switching	Yy
	Conduction	Equivalent

Because of the high switching frequency of the prototype, switching losses have the biggest impact on the efficiency of the converter. It explains why Yy vector group shows the best efficiency.

Conclusion

The three-phase Dual Active Bridge is a promising topology when galvanic insulation and bidirectional power flow are required. Among all the degrees of freedom available for the design of the converter, the transformer vector group can be used to optimize the converter. At first glance, Yy and ΔΔ vector groups can be expected to show similar performance. However, technical aspects experimentally differentiated the two solutions and the Yy configuration showed the best performance.

References

[1] T. Lagier, L. Chédot, F. W. L. Ghossein, B. Lefebvre, P. Dworakowski, M. Mermet-Guyennet, and C. Buttay, "A 100 kW 1.2 kV 20 kHz DC-DC converter prototype based on the Dual Active Bridge topology," in *2018 IEEE International Conference on Industrial Technology (ICIT)*, 2018, pp. 559–564.

[2] R. W. A. A. D. Doncker, D. M. Divan, and M. H. Kheraluwala, "A three-phase soft-switched high-power-density DC/DC converter for high-power applications," *IEEE Transactions on Industry Applications*, vol. 27, no. 1, pp. 63–73, Jan. 1991.

[3] N. Soltau, H. Stagge, R. W. D. Doncker, and O. Apeldoorn, "Development and demonstration of a medium-voltage high-power DC-DC converter for DC distribution systems," in *2014 IEEE 5th International Symposium on Power Electronics for Distributed Generation Systems (PEDG)*, 2014, pp. 1–8.

[4] Y. Lee, G. Vakil, A. J. Watson, and P. W. Wheeler, "Geometry optimization and characterization of three-phase medium frequency transformer for 10kVA Isolated DC-DC converter," in *2017 IEEE Energy Conversion Congress and Exposition (ECCE)*, 2017, pp. 511–518.

[5] J. Xue, F. Wang, D. Boroyevich, and Z. Shen, "Single-phase vs. three-phase high density power transformers," in *2010 IEEE Energy Conversion Congress and Exposition*, 2010, pp. 4368–4375.

[6] A. Garcia-Bediaga, I. Villar, A. Rujas, I. Etxeberria-Otadui, and A. Rufer, "Analytical Models of Multiphase Isolated Medium-Frequency DC–DC Converters," *IEEE Transactions on Power Electronics*, vol. 32, no. 4, pp. 2508–2520, 2017.

[7] P. Dworakowski, A. Wilk, M. Michna, B. Lefebvre, and T. Lagier, "3-phase medium frequency transformer for a 100kW 1.2kV 20kHz Dual Active Bridge converter," in *2019 45th Annual Conference of the IEEE Industrial Electronics Society (IES)*, 2019.

[8] P. Dworakowski, A. Wilk, M. Michna, B. Lefebvre, F. Sixdenier, and M. Mermet-Guyennet, "Effective Permeability of Multi Air Gap Ferrite Core 3-Phase Medium Frequency Transformer in Isolated DC-DC Converters," *Energies*, 2020.

[9] C. P. Steinmetz, "On the Law of Hysteresis," *Transactions of the American Institute of Electrical Engineers*, vol. IX, no. 1, pp. 1–64, Jan. 1892.

[10] K. Venkatachalam, C. R. Sullivan, T. Abdallah, and H. Tacca, "Accurate prediction of ferrite core loss with nonsinusoidal waveforms using only Steinmetz parameters," in *2002 IEEE Workshop on Computers in Power Electronics, 2002. Proceedings.*, 2002, pp. 36–41.

[11] P. L. Dowell, "Effects of eddy currents in transformer windings," *Proceedings of the Institution of Electrical Engineers*, vol. 113, no. 8, pp. 1387–1394, Aug. 1966.

[12] F. Tourkhani and P. Viarouge, "Accurate analytical model of winding losses in round Litz wire windings," *IEEE Transactions on Magnetics*, vol. 37, no. 1, pp. 538–543, Jan. 2001.

[13] Ferroxcube, "Design of planar power transformers, application note," http://ferroxcube.home.pl/appl/info/plandesi.pdf." .

[14] T. Lagier and P. Ladoux, "Theoretical and experimental analysis of the soft switching process for SiC MOSFETs based Dual Active Bridge converters," in *2018 International Symposium on Power Electronics, Electrical Drives, Automation and Motion (SPEEDAM)*, 2018, pp. 262–267.

Impedance Analysis of an Automotive DC Bus

Michael Schlüter; Marius Gentejohann; Sibylle Dieckerhoff
Technische Universität Berlin
Institute of Energy and Automation Technology
Department of Power Electronics
Einsteinufer 19; Sec. E2
10587 Berlin, Germany
Tel.: +49 / (030) – 314 29217
E-Mail: m.schlueter@tu-berlin.de
URL: https://www.pe.tu-berlin.de/

Acknowledgements

This work was supported by the German Federal Ministry of Education and Research (BMBF) under contract number 16EMO0262 (SiCWell).

Keywords

«Impedance measurement», «Automotive application», «Power converters for EV», « Frequency-Domain Analysis», «Electric vehicle»

Abstract

This paper analyses the impact of the fundamental frequency, the switching frequency and the deadtime on the DC current and DC voltage spectrum of a traction inverter connected to the high voltage (HV) DC bus of an electric vehicle (EV). A method is presented to calculate the DC bus impedance using the perturbations generated by the inverter. This method is verified by simulations and measurements based on a test setup and then used to calculate the Thevenin's impedance at the terminals of the traction inverter in an EV.

Introduction

The ageing of the traction battery is one of the major concerns for the operation of EVs. The influence of the current ripple on the battery lifetime is discussed e.g. in [1], where "a detrimental impact of periodic pulses on the cell performance compared to profiles with constant current" is found. In [2], a sinusoidal AC ripple in the frequency range of 0 – 14,8 kHz is superimposed in cycling tests of a three-cells stack, and the AC ripple leads to a higher divergence of the cell properties like power density and 1C-capacity. In [3], Lithium-Ion batteries are cycled with a superimposed AC frequency up to 30 kHz, and no accelerated degradation occurs at higher AC frequencies. This shows that an accurate knowledge of the harmonic spectrum of the battery current is important to analyze the impact of its ripple on battery ageing. To determine the current ripple, a detailed knowledge of the DC bus impedance is necessary. This impedance depends on several parameters: the length of the cables, the DC-link capacitors, the filters of the high voltage components connected to the DC bus, and the battery, consisting of many individual cells.

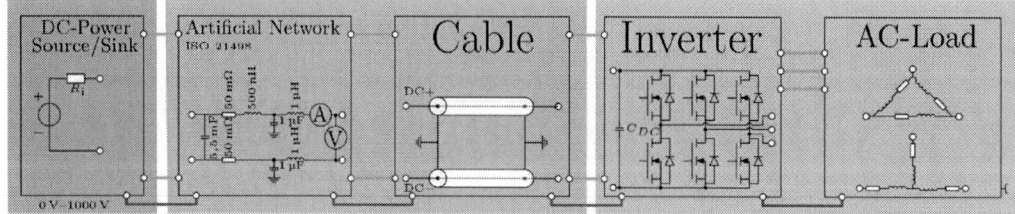

Fig. 1: Test setup

The analysis of the DC current ripple generated by a 3-phase voltage source inverter (VSI) is principally covered in literature. However, most references investigate the current ripple in the DC-link capacitor [4] or consider ideal conditions like a DC voltage source [5]. Reference [6] also focuses on the current ripple in electric vehicles. However, only the battery and the DC-link, but not the DC bus impedance are considered.

This paper presents a measurement-based method to calculate the impedance of the DC bus under rated conditions by analyzing the current and voltage ripples generated by the inverter. The standard ISO 21498 [7] suggests a generic artificial network, which is the basis for the simulation and the experimental test setup in this work. The proposed method is used to determine the impedance of a real automotive HV DC bus, including all high voltage (HV) loads as well as the main battery without any additional measurement signals.

Test Setup

The test setup (Fig. 1) consists of a DC-power source, the artificial network according to standard ISO 21498 [7], a shielded cable of 2 meters length, a Silicon Carbide (SiC) MOSFET two-level VSI and a three-phase passive load. The VSI represents the main inverter of an EV. A generic automotive DC bus is represented by the artificial network and the cable.

The artificial network includes an internal battery resistance of 100 mΩ. The internal resistance (R_i) of the DC-power-source can be neglected. The impedance of the artificial network is defined in the frequency range from 10 Hz up to 150 kHz in ISO 21498. To compare the impedance of the artificial network with the standard, the impedance is measured with open input terminals. The corresponding results are shown in Fig. 2. For current ripple measurements, a capacitance of more than 5 mF is required. A 5.5 mF capacitance is used in the setup and the measured impedance complies with the given tolerances of the standard ([7] annex E.2). Since the tolerances of the standard are only defined for a 10 mF capacitor, this impedance is also shown (red dotted line). The main parameters of the test setup are listed in the annex in Table II.

Simulation

Parasitic components may influence the system characteristics especially in the high frequency range. Therefore, the impedances of all components are measured individually and included in the simulation model. The inverter consists of ideally switched MOSFETs with anti-parallel diodes and is controlled by a space vector modulation scheme. Since the modulation scheme has a significant impact on the high frequency components of currents und voltages [8], the modulator used in the experimental setup also

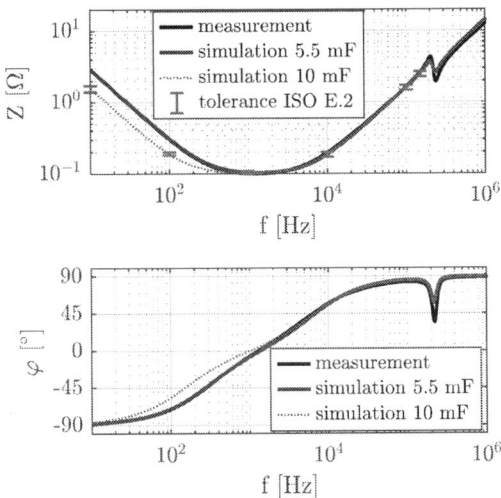

Fig. 2: DC bus impedance according to ISO 21498

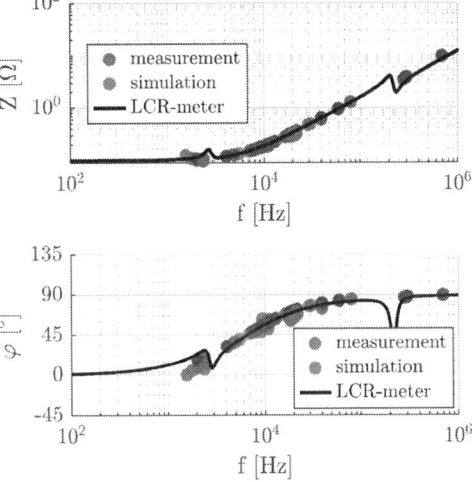

Fig. 3: Calculated impedance from measured and simulated data vs. LCR-measurement

runs as software-in-the-loop in the time domain simulation, in this guaranteeing the same pulse patterns for measurement and simulation. To simulate the FPGA implementation of the control, Xilinx System Generator is used, while the hardware is simulated with PLECS. Because of the FPGA clock, a fixed step size of 10 ns or less must be chosen.

The cable is modeled by two single-phase π-section transmission lines with lumped parameters. The parameters of the AC-load are chosen to match the real power resistor in the laboratory so that each simulated operation point can be compared to measurements in the test setup.

Fig. 4 shows the DC-bus current and voltage spectra exemplarily for one operation point. Only the major harmonics are plotted. The frequency of each harmonic can be calculated depending on multiples of the fundamental frequency f_g and the switching frequency f_s [9]. The amplitudes of the inverter current harmonics are derived based on the AC current and the modulation scheme [9], while the amplitudes of the DC bus current harmonics depend on the impedance of the DC bus, the DC link capacitor, the load, and further parameters like the deadtime or the modulation scheme [10]. In the following, the modulation index is set to $m = 0.97$, and the fundamental frequency f_g, the switching frequency f_s and the deadtime t_{dead} are varied, analyzing the respective impact on the harmonics. To permit a simple comparison between the operation points, the hardware parameters of the test setup are not changed within this paper (see table II).

Fundamental frequency

The fundamental frequency of traction inverters changes over a wide range during motor operation. For a battery electric vehicle with single-speed transmission, which is state-of-the-art, the fundamental frequency is 0 Hz if the vehicle stands still, and reaches more than 1000 Hz at top speed.

The simulation parameters are set to $t_{dead} = 400$ ns, $m = 0.97$ and $f_s = 10$ kHz. For an entire frequency sweep up to 1 kHz, the two highest harmonics in the current spectrum occur in the first carrier band at $f_s \pm 3f_g$. An exemplary current und voltage spectrum for one operation point is shown in Fig 4.

Assuming an acceleration of an electric vehicle from stand still to top speed, the impedance of the DC bus is therefore estimated in a frequency range from 7 kHz to 13 kHz, only considering these two highest peaks in the current and voltage spectrum. If more harmonics are to be considered, the analyzed frequency range is getting a lot wider and e.g. covers 1 kHz to 19 kHz, if the next sidebands at $f_s \pm 9f_g$ are included.

Switching frequency

The switching frequency defines the main location of the dominant harmonics. A sweep of the switching frequency therefore provides information on the DC bus characteristics in a wider frequency range. However, in a real system only a small variation of the switching frequency can be implemented without changes to the hardware. Larger frequency changes are limited e.g. by the control hardware or the gate

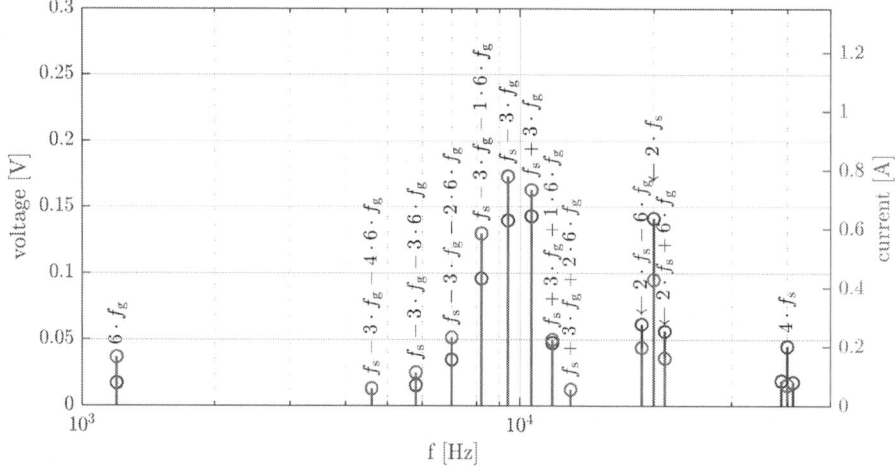

Fig. 4: Current und voltage spectra for major harmonic peaks ($f_g = 200$ Hz, $f_s = 10$ kHz)

driver. Generally, a higher switching frequency leads to smaller harmonic peaks, if the capacitance of the DC-link is not changed, while a lower switching frequency will increase these peaks. For instance, in the simulation model, the highest harmonic peaks observed are 1.3 A/ 218 mV, setting the switching frequency to 8 kHz, and 27 mA/ 17 mV for a switching frequency of 40 kHz, both simulated at 200 Hz fundamental frequency.

Deadtime

In state-of-the-art inverters, the deadtime is a control parameter and can be adjusted by software. Usually, it is set to be as small as possible considering safe switching. For MOSFET-based inverters, longer deadtimes lead to higher body diode losses and therefore to a worse inverter efficiency, due to the inferior conduction behavior of the diodes. Applying a longer deadtime, lower harmonic peaks in the first carrier band but higher peaks in the second one can be observed. Furthermore, the harmonic amplitude at six times the fundamental frequency rises if the deadtime is reduced. In our investigation, the deadtime is adjusted from 400 ns to 10 µs, and the major harmonic peaks obtained by simulation and measurement are summarized in table I. It should be mentioned that such a large change of the deadtime also changes the rms output current which has a significant impact on the amplitudes. In the example, the rms value of the line current changes from 10.39 A for 400 ns deadtime to 8.6 A for 10 µs deadtime. However, only the deadtime is adjusted and all other parameters are kept constant, because the load resistance in the measurements cannot be changed, and a change of the modulation index would also change the spectrum.

Table I: Voltage and current amplitudes for major harmonics with 400 ns and 10 µs deadtime

($f_g = 50$ Hz, $f_s = 10$ kHz, $m = 0,97$ and $V_{DC} = 400$ V)

| | t_{dead} = 400 ns | | | | t_{dead} = 10 µs | | | |
| | Simulation | | Measurement | | Simulation | | Measurement | |
f	I	V	I	V	I	V	I	V
300 Hz ($6 \cdot f_g$)	1.040 A	0.108 V	0.933 A	0.117 V	0.430 A	0.045 V	0.437 A	0.057 V
9850 Hz ($f_s - 3 \cdot f_g$)	0.649 A	0.128 V	0.664 A	0.121 V	0.520 A	0.102 V	0.456 A	0.080 V
10150 Hz ($f_s + 3 \cdot f_g$)	0.631 A	0.127 V	0.648 A	0.121 V	0.514 A	0.103 V	0.454 A	0.082 V
19700 Hz ($2 \cdot f_s - 6 \cdot f_g$)	0.313 A	0.110 V	0.297 A	0.099 V	0.175 A	0.061 V	0.155 A	0.051 V
20000 Hz ($2 \cdot f_s$)	0.579 A	0.205 V	0.559 A	0.185 V	1.431 A	0.508 V	1.344 A	0.436 V
20300 Hz ($2 + 6 \cdot f_g$)	0.293 A	0.105 V	0.277 A	0.092 V	0.162 A	0.058 V	0.139 A	0.044 V

Simulation Results

The further analysis of the test setup impedance is based on 21 different operation points with switching frequencies f_s from 10 kHz to 30 kHz, fundamental frequencies f_g in the range of 220 Hz to 400 Hz, a DC voltage V_{DC} between 300 V and 450 V, a deadtime t_{dead} of 400 ns and a modulation index m set to 0.97. Three 16 Ω resistors in either delta or star connection form the load. The current and the voltage at the output of the artificial network (see Fig. 1) are simulated in the time domain, followed by a Discrete-Fourier-Transformation (DFT) to obtain the harmonic spectrum. Only the major peaks are used to calculate the impedance, which is then represented by amplitude and phase. As discussed before, amplitude and frequency of the major peaks depend on the switching frequency, fundamental frequency and deadtime. By changing fundamental and switching frequency, the major peaks are distributed over a wide frequency range, and the impedance can be calculated for a larger number of different frequencies. In Fig. 3, the simulation results are compared to the LCR-meter measurement of the passive circuit. As expected, the impedance measured with the LCR-meter and the one calculated from the simulation are in quite good agreement.

EPE'20 ECCE Europe

Assigned jointly to the European Power Electronics and Drives Association & the Institute of Electrical and Electronics Engineers (IEEE)

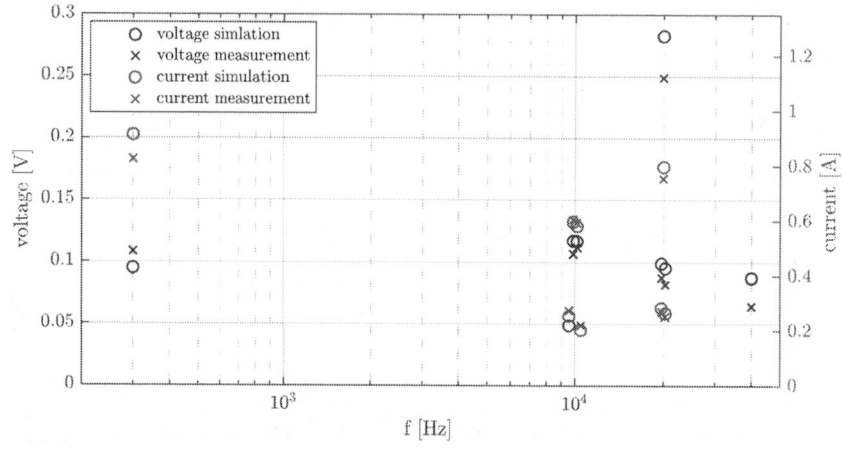

Fig. 5: Voltage and current spectra: measurement and simulation
$t_{\text{dead}} = 2\,\mu s\; f_{\text{g}} = 50\,\text{Hz}\,,\, f_{\text{s}} = 10\,\text{kHz},\, m = 0{,}97$ and $U_{\text{DC}} = 400\,\text{V}$

Measurement Results

This section consists of two parts. The first paragraph analyzes measurement results of the laboratory test setup. This setup is the basis for the simulation and used to verify the proposed impedance calculation method. In the second paragraph, measurements are conducted in an EV, and the results are presented and discussed.

Test Setup

During this research project, the SiC inverter, the control software and the artificial network have been designed and built. Therefore, all components in the simulation model are known, and all software parameters and some hardware parameters can be easily changed during the investigation (see Table II). Fig. 1 illustrates the schematic of the test setup. A small signal analysis of the assembled artificial network is done with an LCR-meter (Sourcetronic ST2827c), and this impedance is compared to the impedance calculated from the voltage and current measurements. The voltage is acquired with a PMK BumbleBee voltage probe, and the current is measured with a Pearson CM 1025 current sensor, both connected to a 12-bit 12,5 Mpts/Ch oscilloscope. For the same 21 different operation points as used in the simulation, the DFT of the voltage and current is performed. In table I as well as in Fig. 5, the measurements are compared to simulation results. The maximum deviation of the harmonics between simulation and measurements are 72 mV/ 86 mA as shown in table I. To compare the LCR-meter measurements to the calculated impedance (Fig. 3), the input of the network is short-circuited. The results in Fig. 3 show a good agreement between simulation and measurements, the maximum deviation is 42,3 mΩ at 2 kHz. This proves that the proposed impedance calculation method based on dominant harmonics of the current and voltage spectra is feasible to determine the impedance of a real HV DC bus.

Measurement Accuracy

Due to measurement uncertainty and limited resolution of the measurement equipment, only the major peaks of the voltage und current spectra can be used to evaluate the impedance. The 11th highest peak of the current spectrum has an amplitude of less than 1/10th compared to the highest peak. Depending on the type of current probe, also the DC component, e.g. in Fig. 4 $I_{\text{DC}} = 11{,}6\,\text{A}$, must be considered to calculate the vertical resolution due to the quantization of the oscilloscope. Since only few measured frequency components can be used to estimate the impedance, several operation points are analyzed in order to change the current and voltage spectra and cover a wide frequency range with only the respective major harmonic peaks.

Electric Vehicle

Measurements in real EVs are a challenge due to various reasons. An EV has a complex DC Bus with numerous HV components, and details are often only known by the manufacturer. Road safety and electrical safety must always be considered.

Setup

The EVs voltage and current are measured directly at the terminals of the traction inverter. To access DC+ und DC- and to measure the current without the shielding of the cables, the housing of the main inverter must be opened. To measure the voltage, two very short wires are connected to DC+ and DC- inside the inverter. The voltage is measured with a differential probe (Testek TT-SI9191(V_1)). To measure the current, a Rogowski coil (PEM CWT6B (I_1)) is placed inside the inverter housing. Due to an interlock sensor on the lid and to ensure electrical safety, the housing had to be completely closed while driving the EV. In order to sense the DC-bus current including its DC component, an additional current sensor (Tektronix A6304xl (I_3)) is placed in the engine compartment around the DC+ cable. As all HV cables in the EV are completely shielded, the high frequency components of this sensor are damped. A 12-bit 12,5 Mpts/Ch oscilloscope (LeCroy HDO4054A) is used to acquire the data of all sensors. Fig. 6 shows the engine compartment with the location of the sensors, and Fig. 8 illustrates a basic schematic of the HV bus including the measurement points.

Current and Voltage Measurements in Time Domain and Frequency Domain

An exemplary measurement (OP 11) of voltage V_1 and current I_1 is plotted in Fig. 7. The voltage ripple has an amplitude up to 10 V, and at maximum torque, the current ripple exceeds an amplitude of almost 60 A for a 100 ms window. As explained in Fig. 4, the sinusoidal frequency visible in the current and voltage measurement is supposed to be six times the fundamental frequency. Therefore, the fundamental frequency in the exemplary operation point is approximately 160 Hz. This frequency can also be observed in the current and voltage spectra in Fig. 9 (first and second carrier band: $f_s \pm 3f_g$ and $2f_s \pm 6f_g$). From this plot, also the switching frequency $f_s = 10$ kHz can be estimated.

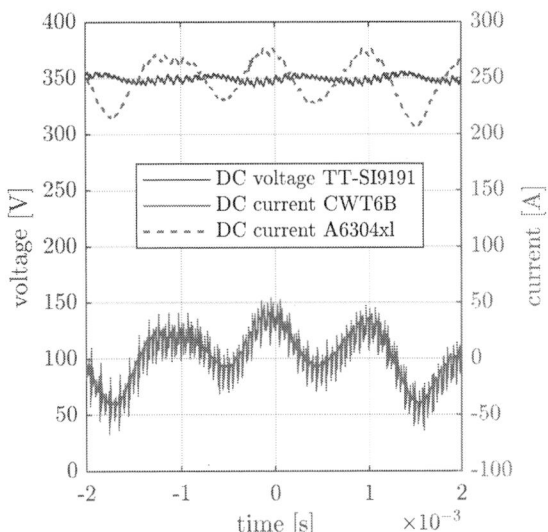

Fig. 6: Engine compartment of the EV
1 - main inverter
2 - differential voltage probe (V_1)
3 - current probe inverter DC+ (I_1)
4 - current probe inverter DC- (I_2)
5 - current probe cable DC+ (I_3)

Fig. 7: Exemplary current and voltage measurement at 85.92 kW electric power (OP 11)

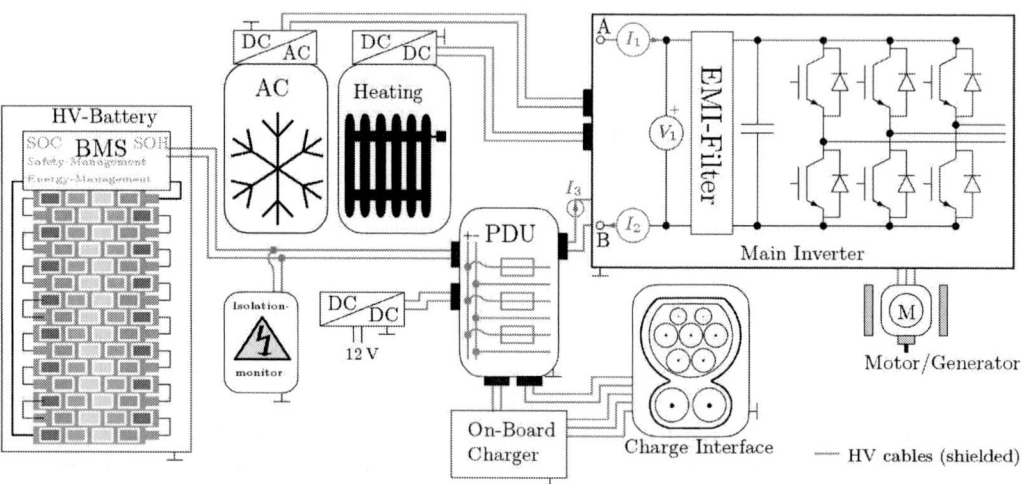

Fig. 8: Basic schematic of the HV Bus of the Electric Vehicle

The bottom plot in Fig. 9 depicts the calculated impedance amplitude and phase for the exemplary operation point. Analogous to Fig. 3, this impedance represents the overall impedance at the terminals of the main inverter where the voltage and current probes are connected (cf. Fig. 8).

The exact configuration of the DC bus of the EV is unknown. It is assumed that the topology is similar to the schematic in Fig. 8, i.e. the HV heater and air conditioner (AC) are connected to the main inverter terminals, the main inverter is connected to a Power Distribution Unit (PDU), and the PDU is connected to the HV battery. The charger and the 12 V DC/DC converter are connected to the HV bus via the PDU. The lengths and impedances of the HV cables are unknown, as are the X- and Y-capacitors of the HV components.

Compared to the impedance of the test setup according ISO 21498 (cf. Fig. 3), the frequency characteristic of the EV differs substantially. This is could be expected due to the complex configuration of the HV DC bus, but shows the importance of a measurement-based analysis of the impedance.

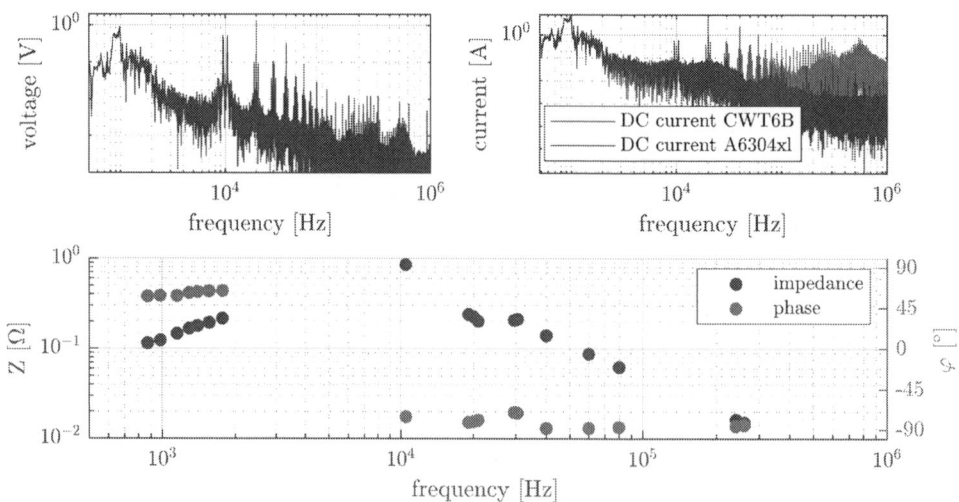

Fig. 9: Harmonic spectrum of voltage, current and calculated impedance (OP 11)

Operation Points

To analyze the impact of different parameters, measurements are conducted in more than 300 different operation points, e.g. at full torque, with constant speed, during recuperative breaking, at different SoC and so on. Especially the impedance of the battery cells is not linear and varies over time, since it highly depends on the SoC, the current and the temperature. Furthermore, the operating condition of other HV components can influence the impedance measurement at the main inverter terminals.

Thevenin's Impedance

The shape of the impedance curve (Fig. 9) suggests a resistive-inductive behavior between 1 kHz-2 kHz and a capacitive behavior from 10 kHz to 200 kHz. The impedance between 1 kHz – 2 kHz noticeably depends on the operation point, especially if high and low power operation points are compared, while the calculated impedance between 10 kHz up to 200 kHz varies only slightly with the operation point. To analyze the DC bus independent of the main inverter, Thévenin's and Norton's theorem are used to calculate the Thévenin's impedance Z_{th} of the DC bus. Thévenins's theorem for AC circuits can be used to reduce a complex network to an equivalent voltage source with frequency-dependent amplitude and phase in series with an impedance Z_{th}. The theorem is applied to substitute the complex HV bus at terminals A and B of Fig. 8 with a voltage source and a single impedance for further analysis e.g. of the current ripple.

Generally, the open circuit voltage and the short circuit current can be used to calculate Thévenin's impedance according to equation (1). Here, the calculation is performed for distinct frequencies based on the voltage and current spectra. The impedance is only evaluated for frequencies with voltage and current amplitudes above a defined threshold. As the short circuit current and the open circuit voltage are unknown, two different operation points are used according to equation (2).

$$Z_{th}(\omega) = \frac{u_0(\omega)}{i_k(\omega)} \tag{1}$$

$$Z_{th}(\omega) = \frac{\Delta u(\omega)}{\Delta i(\omega)} = \frac{u_{op_i}(\omega) - u_{op_n}(\omega)}{i_{op_n}(\omega) - i_{op_i}(\omega)} \; for \; n \neq i \tag{2}$$

A nested "for-loop" to compare all operation points with each other (eq. (2)) is implemented, and the calculated impedance in every cycle of the loop is plotted in Fig. 10 using low opacity in order to visualize the deviance and incidence of the calculated impedance values for amplitude and phase.

In the EV setup, neither the switching frequency f_s nor the deadtime t_{dead} can be changed. By accelerating and decelerating with different torque at various speeds, the fundamental frequency f_g, the current amplitude I_{DC} and the modulation index m vary with the operation point. Based on the measured voltage and current, the impedance at the inverter terminals can be calculated for distinct frequencies (c.f. Fig. 9). Applying Thévenin's theorem an equivalent circuit for the HV DC bus is calculated, again for distinct frequencies only.

With the knowledge of the equivalent circuit of the HV DC bus at the terminals of the main inverter, a current ripple, originally measured with a known inverter and the test setup according to ISO 21498, can be transformed into the ripple appearing on the real HV DC bus, if the same inverter is used.

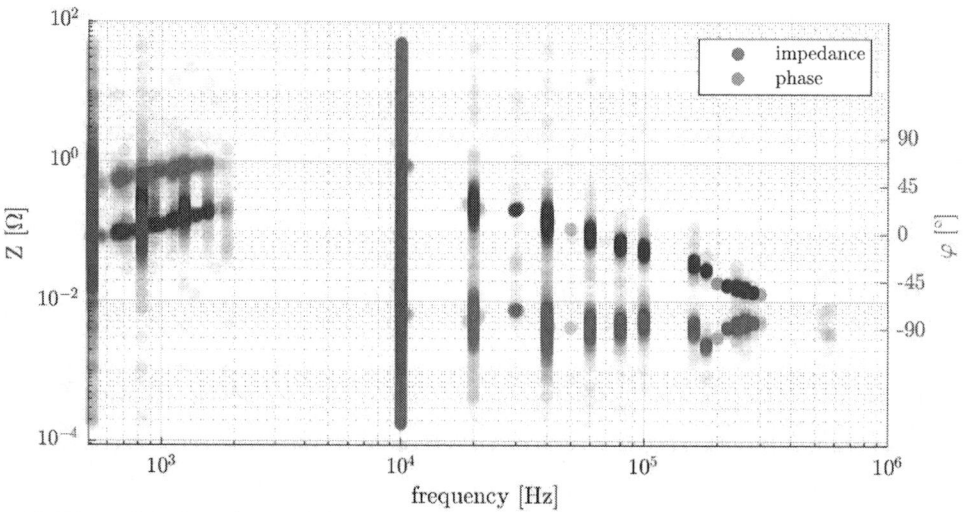

Fig. 10: Thevenin's impedance of the HV DC bus at the main inverter terminals

Conclusion

With the artificial network suggested in standard ISO 21498, automotive HV components and their impact on the DC bus can be studied under standardized test conditions. In order to calculate the current ripple in a real automotive HV DC bus from measurements based on the artificial network, the DC bus impedance must be known. The analysis in this paper verifies that the DC bus impedance at the terminals of the main inverter can be calculated only using the perturbations generated by the inverter itself. The limits of this method are defined by the measurement accuracy and bandwidth of the current and voltage sensors.

The calculation of the Thevenin's impedance is a powerful tool to analyze the impedance seen by the main inverter in real operating conditions. The presented measurement results of an EV show that the real HV DC bus impedance differs noticeably from the one defined in the standard. This leads to the conclusion that the actual impedance is an important parameter for the development of an EV and should be considered to analyze the current ripple of the inverter. The proposed method presents a feasible way to identify the DC bus impedance for real use conditions. The approach is limited to distinct frequencies defined by the switching frequency and the measuring accuracy.

References

[1] F. Savoye, P. Venet, M. Millet, and J. Groot, "Impact of Periodic Current Pulses on Li-Ion Battery Performance," IEEE Trans. Ind. Electron., vol. 59, no. 9, pp. 3481–3488, 2012.

[2] K. Uddin, A. D. Moore, A. Barai, and J. Marco, "The effects of high frequency current ripple on electric vehicle battery performance," Applied Energy, vol. 178, pp. 142–154, 2016.

[3] M. J. Brand, M. H. Hofmann, S. S. Schuster, P. Keil, and A. Jossen, "The Influence of Current Ripples on the Lifetime of Lithium-Ion Batteries," IEEE Trans. Veh. Technol., vol. 67, no. 11, pp. 10438–10445, 2018.

[4] A. Mariscotti, "Analysis of the DC-link current spectrum in voltage source inverters," IEEE Trans. Circuits Syst. I, vol. 49, no. 4, pp. 484–491, 2002.

[5] U. Ayhan and A. M. Hava, "Analysis and characterization of DC Bus ripple current of two-level inverters using the equivalent centered harmonic approach," in 2011 IEEE Energy Conversion Congress and Exposition, Phoenix, AZ, USA, Sep. 2011 - Sep. 2011, pp. 3830–3837.

[6] O. Satilmis and E. Mese, "Investigating DC link current ripple and PWM modulation methods in Electric Vehicles," in 2013 3rd International Conference on Electric Power and Energy Conversion Systems, Istanbul, Turkey, Oct. 2013 - Oct. 2013, pp. 1–6.

[7] Electrically propelled road vehicles - Electrical tests for voltage class B components, ISO/DIS 21498:2018(E), 2018.

[8] F. Yao, L. Geng, A. Janabi, and B. Wang, "Impact of modulation schemes on DC-link capacitor of VSI in HEV applications," in 2017 IEEE International Electric Machines and Drives Conference (IEMDC), Miami, FL, USA, May. 2017 - May. 2017, pp. 1–7.

[9] B. P. McGrath and D. G. Holmes, "A General Analytical Method for Calculating Inverter DC-Link Current Harmonics," IEEE Trans. on Ind. Applicat., vol. 45, no. 5, pp. 1851–1859, 2009.

[10] M. Gentejohann, M. Schlueter and S. Dieckerhoff, "Investigation of the Current Ripple caused by the Main Inverter in the High Voltage DC Bus of Electric Vehicles", CIPS 2020; 11th International Conference on Integrated Power Electronics Systems, Berlin, Germany, 2020.

Annex

Table II: List of main parameters for simulation and test setup

Symbol	Parameter	Value	Comment
V_{DC}	Battery voltage	300 – 400 V	
C_{DC}	DC Link Capacitor	115 µF	Additional Snubber of 1 µF
$R_{ds,on}$	On Resistance MOSFET	23 mΩ	Datasheet Semikron 26ACM112V17
$R_{d,on}$	On Resistance Diode	9 mΩ	Datasheet Semikron 26ACM112V17
V_{F0}	Threshold voltage Diode	0.95 V	Datasheet Semikron 26ACM112V17
$C_{S,iso}$	Blocking capacitor	5.5 mF	ISO 21498
$R_{i,iso}$	Internal resistance	100 mΩ	ISO 21498
$L_{C,iso}$	Power supply inductance	1.1 µH	ISO 21498
$C_{y,iso}$	Y capacitors	1 µF	ISO 21498
R_{Load}	Load resistance	16.3 Ω	Measurements from Test setup
L_{Load}	Load inductance	110 µH	Measurements from Test setup
$L_{c,PU}$	Cable inductance per unit length	105 nH	Measurements / Datasheet
$R_{c,PU}$	Cable resistance per unit length	368 mΩ	Measurements / Datasheet
$C_{c,PU}$	Cable capacitance per unit length	670 pF	Measurements / Datasheet
$G_{c,PU}$	Cable conductance per unit length	1 MΩ	Measurements / Datasheet
l_c	Cable length	2 m	ISO 21498
f_s	Switching frequency	6 – 50 kHz	
f_g	Fundamental frequency	1 – 1000 Hz	
t_{dead}	Dead time / turn on delay	0.2 – 10 µs	
t_s	Sample Time	10 ns	For Simulation step size and FPGA clock

A New Dual-Mode MPPT Algorithm Applied to a Quadratic Converter in a Solar Energy System

Ahmad GHAMRAWI, Jean-Paul GAUBERT, Driss MEHDI
UNIVERSITY OF POITIERS
Laboratoire d'Informatique et d'Automatique pour les Systèmes (LIAS)
2 Rue Pierre Brousse – TSA 41105
86073 POITIERS CEDEX
Poitiers, France
Tel.: +33 / (0) – 6 78 24 68 36.
Email: ahmad.ghamrawi@univ-poitiers.fr; ahmad_ghamrawi@hotmail.com
jean.paul.gaubert@univ-poitiers.fr; driss.mehdi@univ-poitiers.fr
URL: http://www.lias-lab.fr/

Keywords

«Photovoltaic», «Maximum power point tracking», «Solar cell system», «Converter control», «Renewable energy systems», «Converter circuit», «Efficiency».

Abstract

The work on this paper aims to improve the efficiency of a photovoltaic energy system. Reducing the losses in the adaptation stage allows to improve this efficiency. For that, in this paper we will work on the adaptation stage between the solar panels and the loads to reduce the losses. As a first step, we will present a comparison between a conventional DC-DC boost converter usually used in solar system and a quadratic boost with a single switch. This comparison aims to show the advantage of using quadratic converters with high gain in solar systems. The results show that the quadratic converter has a higher efficiency compared to the conventional one, it allows also to reach higher values of gain for a practically acceptable values of duty cycle. This comparison justifies the choice of the quadratic converter. Then, a new dual-mode variable step-size maximum power point tracking (MPPT) algorithm is proposed. The proposed algorithm is used to control the quadratic converter. The proposed algorithm is based on the perturb and observe (P&O) algorithm. The P&O is one of the most known and used MPPT methods, its drawbacks appear during fast solar irradiance variation. The proposed control method is a double mode algorithm. The first mode is active when the operating point is near of the MPP. This mode allows to stabilize the operating point to reduce the steady-state oscillations that occur using the P&O. The second mode is activated when the operating point moves away from the MPP. This mode is a P&O with a large step size that allows tracking the MPP quickly which also reduces the losses. The proposed algorithm is tested through simulations using Simulink/MATLAB. Simulations results prove the efficiency of the proposed algorithm and its advantage compared to the P&O.

Introduction

During the last years, the need to find new energy sources became crucial. The depletion of fossil fuels on the one hand, and the increasing needs due to the increase in the world's population on the other, made the exploitation of renewable energies a necessity. According to the Enerdata's Global Energy Statistical Yearbook, the world energy consumption is equal to 14000 Mtoe in 2018 comparing to 8500 Mtoe in 1990. More than 80% of these energies is providing by fossils like coal, oil and gas. This augmentation has led to a growth in the CO_2 emission from 20000 Mt in 1990 to 33000 Mt in 2018 [1]. The solar energy is one of many renewable sources that can be used to reduce our dependence on the fossils. The main drawback of using solar energy is the high initial installation cost of a photovoltaic panel [2]. To make a solar system more beneficial, we should increase its efficiency, which can be achieved by using new topologies of power converters [3-11] and by applying new control algorithms that allow a maximum power extraction from the panels [12-15]. Using these algorithms is called

Maximum Power Point Tracking (MPPT). Many MPPT algorithms exist in the literature such the Perturb and Observe (P&O), the Hill Climbing, the Incremental Conductance and other algorithms based on the artificial neural network or the fuzzy logic. One of the most known algorithms is the P&O. The problem of this algorithm arises while choosing its incrementation's step. A big step causes permanent oscillations of the operating point around the maximum power point while a small one slow the response of the system and increase the losses, it may also causes the divergence of the operating point, especially during fast solar irradiance changes.

In this paper, we will modify the P&O algorithm to present a new dual mode variable step-size MPPT algorithm. The proposed algorithm is a hybrid of the P&O algorithm and a fixed voltage operation. The dual-mode operation allows choosing two step-sizes, which reduces the inconvenience of the P&O algorithm, without making the implementation of the algorithm complicated. The proposed algorithm is applied to a quadratic DC/DC boost converter. The choice of the converter is justified in the paper. Simulations and experimental results are presented to compare the performance of the new algorithm to the classic P&O in order to show its advantage.

The quadratic boost converter

DC-DC converters are usually used during the adaptation stage between the solar panel and the load (Fig. 1). The importance of this stage is not just adapting the impedance to allow the connection between the source and the load, but it allows also to maximize the efficiency of the system and the adjustment of the output voltage.

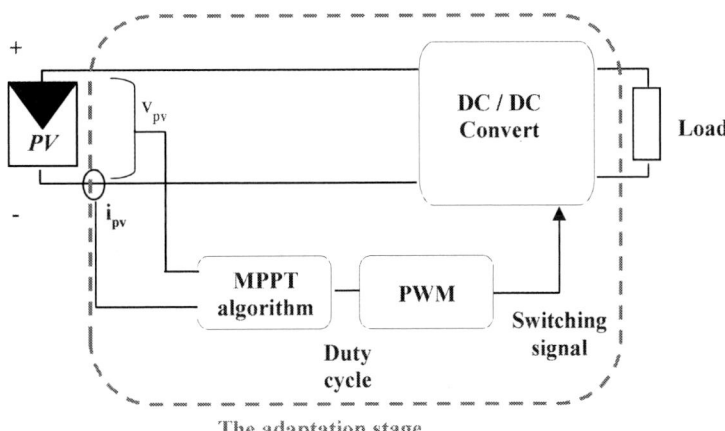

Fig. 1: The adaptation stage in a solar energy system

In this section, we will compare the classic boost used usually in solar system to the quadratic one used in our system to prove the advantage of quadratic converters in solar systems. The quadratic converter used in our system contains a single controllable switch, which makes its driving easier. The quadratic dependence between the duty cycle and the gain of these converters allows to reach high gains while operating at duty cycles far from one, which reduces the constraints on the switch. It allows also to broaden the range of adjustment of the duty cycle, in fact while using a quadratic converter, an increment of the duty cycle, generated by the MPPT controller, leads to smaller variation of the operating point, which reduce losses [3-8].

Fig. 2 and Fig. 3 show the scheme of the boost and the quadratic boost converters respectively. To compare the two topologies, we will consider the losses in the inductances, the diodes and the controllable switches (MOSFET). r_L, r_{L1} and r_{L2} are the internal resistances of the inductances L, L_1 and L_2 respectively. r_D and e_D are the small-signal diodes resistance and their forward voltage, while r_T is the internal resistance of the MOSFET.

The values of the components used in our system are:

- V_{pv} = 48V (considered constant for this comparative study) and R = 50Ω
- L = L_2 = 10mH and L_1 = 5mH; r_L = r_{L2} = 134mΩ and r_{L1} = 53mΩ

- $C = C_2 = 220\mu F$; $C_1 = 470\mu F$
- The diodes are DSEI 30-06A with $e_D = 1.01V$ and $r_D = 7.1m\Omega$
- The MOSFET is STW45NM50 with $r_T = 82m\Omega$.

The theoretical expressions of the gain G of the conventional and the quadratic boost converters depending on the duty cycle μ are given respectively by these equations:

$$G_{simple} = \frac{(1-\mu)[1-\frac{(1-\mu)e_D}{V_{pv}}]}{(1-\mu)^2 + \frac{r_L}{R} + \frac{\mu * r_T}{R} + \frac{(1-\mu)r_D}{R}} \tag{1}$$

$$G_{quadratic} = \frac{(1-\mu)^2[1-\frac{(1+(1-\mu)^2)e_D}{V_{pv}}]}{(1-\mu)^4 + \frac{r_{L1}}{R} + \frac{(1-\mu)^2 r_{L2}}{R} + \frac{\mu(2-\mu)^2 r_T}{R} + \frac{(1+(1-\mu)^3)r_D}{R}} \tag{2}$$

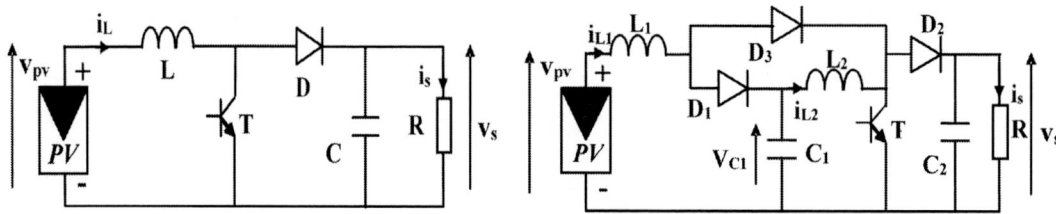

Fig. 2: The scheme of the simple boost converter   Fig. 3 : The scheme of the quadratic boost converter with one controllable switch

Using these two relations, we will plot the variation of the gain depending on the duty cycle for the two converters (Fig. 4(a)).

The losses in each converter is the sum of the losses in each of its components. The expression of the losses for the conventional and the quadratic boost converters are given respectively by these equations:

$$p_{simple} = [\frac{r_L + \mu * r_T}{(1-\mu)^2 R} + \frac{r_D}{(1-\mu)R} + \frac{e_D}{V_S}] * P_s \tag{3}$$

$$p_{quadratic} = [\frac{r_{L1} + r_{L2} + r_{D3} + \mu * r_T}{(1-\mu)^2 R} + \frac{r_{D1} + r_{D2}}{(1-\mu)R} + \frac{e_D}{(1-\mu)V_{C1}} + \frac{e_D}{V_S}] * P_s \tag{4}$$

While P_S is the output power of the converter.

The converter's efficiency ρ is given by:

$$\rho = \frac{P_S}{P_S + p} \tag{5}$$

Using these relations, we will plot the variation of the efficiency depending on the gain (Fig. 4(b)).

Table 1 presents the results of the comparison of the converters. We can clearly see that the maximum gain in voltage for the quadratic boost is higher (8.6 comparing to 7.7), and it is reached for a lower value of duty cycle (0.77 comparing to 0.93). In fact, a duty cycle equal to 0.93 cannot be reached practically, while the duty cycle equal to 0.77 can easily be reached.

The results show also that the efficiency of the quadratic boost is higher in comparison with the conventional one for high gain values (76% comparing to 70% at a gain equal to 7). However, for the low gain values, the efficiency of the simple boost is a little bit higher (92.6% comparing to 91.2% at a gain equal to 4). We can also see that to reach the same gain, the duty cycle needed by the quadratic converter is always lower, which reduces the constraints on the switches.

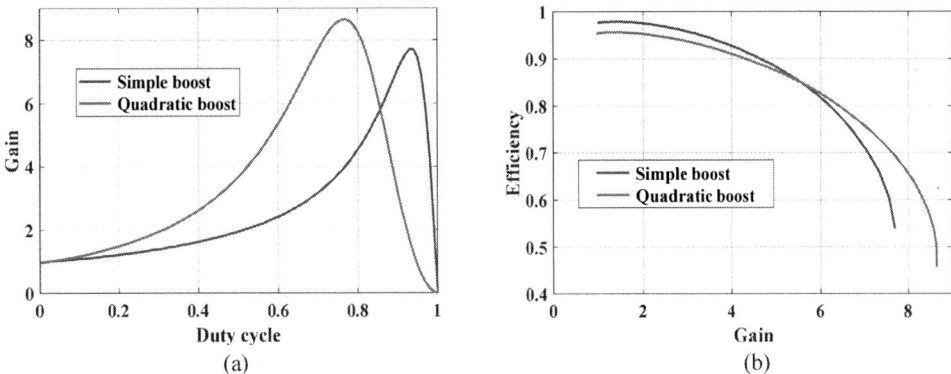

(a)　　　　　　　　　　　　　　　　　　(b)

Fig. 4: The comparison between the simple and quadratic boost

Table I: The comparison between the two converters

Topologies		Simple boost	Quadratic boost
Maximum gain in voltage	Duty cycle	0,93	0,77
	Maximum Gain	7,7	8,6
	Efficiency	53%	46%
Gain = 7	Duty cycle	0,9	0,67
	Efficiency	70%	76%
Gain = 5,6 (same efficiency)	Duty cycle	0,85	0,61
	Efficiency	84%	84%
Gain = 4	Duty cycle	0,77	0,52
	Efficiency	92,6%	91,2%

The solar panel

In this section, we will present the P-V characteristic of the photovoltaic panel Conergy PowerPlus 214P used in our system. In fact, this characteristic depends on many parameters, especially on the solar irradiance and the temperature [16-17].
We can see that the variation of the power depending on the voltage is not linear. The maximum power point depends on many parameters especially the solar irradiance and the temperature. It also depends on the shadow, the dust on the solar panel and the aging of the photovoltaic cells. All these factors make tracking and reaching the MPP harder [18-20].

The maximum power point tracking (MPPT)

Despite its simplicity, the P&O algorithm has so many drawbacks. The main problems arise during fast solar irradiance changes. The performances of this method depend on the choice of the step size. In fact, a large step size that allows reaching the MPP rapidly, increases the losses during the steady state because of the oscillations of the operating point around the MPP, while a small step size that reduces the oscillations during the steady state, increases the response time and produces more losses during the climatic conditions changes [12-14,18].

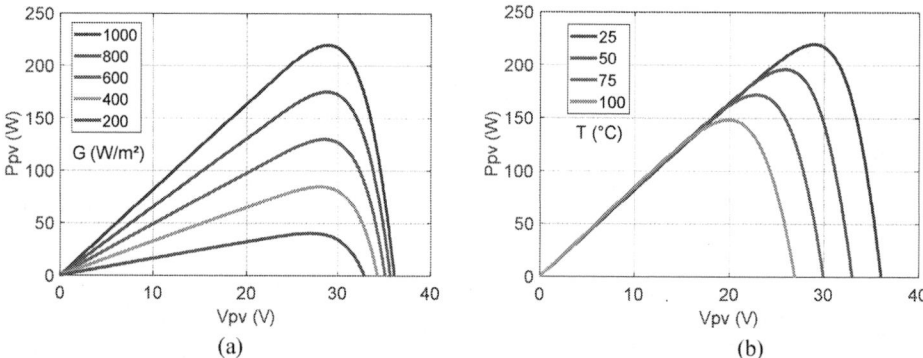

(a)   (b)

Fig. 5: The influence of the solar irradiance (a) and the temperature (b) on the P-V curve of the solar panel

In this section, we will propose an algorithm based on the P&O that is able to cure its two biggest problems: the oscillations of the operating point around the maximum power point MPP and the choice of the step size. The proposed algorithm detects the value of the derivative of the power with respect to the voltage at the output of the solar panel P'_{PV}. This derivative must be equal to zero at the MPP, near of this point, its value is lower than a threshold value P'_{th}. While the value of P'_{PV} is bigger than P'_{th} the first mode will stay active, which is a variable step-size P&O with a step given by:

$$\delta\mu = -\beta * P'_{PV} \tag{6}$$

β is a constant that can have a big value without causing permanent oscillations of the operating point. When the value of P'_{PV} became smaller than P'_{th} the second mode is activated to fix the voltage of the operating point at a constant value V_{opt} that depends on the voltage at the moment of the switching between the two modes V_i. In fact, V_{MPP} varies in a small interval and the variation of P'_{PV} is almost linear in the vicinity of the MPP as shown in figure 6. These two facts allow us to estimate the value of V_{MPP} using the equation (7).

$$V_{opt} = V_i + \gamma * sign(P'_{PV}) \tag{7}$$

Fixing the voltage at the output of the photovoltaic panel allows to reduce the oscillations of the operating point during the steady state, which reduce the permanent losses.

The value of γ cannot be constant because of the non-linear variation of P'_{PV} depending on the irradiance. An expression of γ that allows to reach the MPP for any value of irradiance is given by:

$$\gamma = \Gamma \left| \frac{d^2 P_{PV}}{dV_{PV}^2} \right| = \Gamma \left| P''_{PV} \right| \tag{8}$$

Using the almost linear variation of P''_{PV} depending on the irradiance near of the MPP, With Γ is a constant currently chosen manually by looking at the characteristic of the solar panel. The step is given by:

$$\delta\mu = \alpha * (V_{PV} - V_{opt}) \tag{9}$$

For example, using a conergy PowerPlus 214P, for Γ=0.5 and and P'_{th}=0.5, the results are given in the table 2.

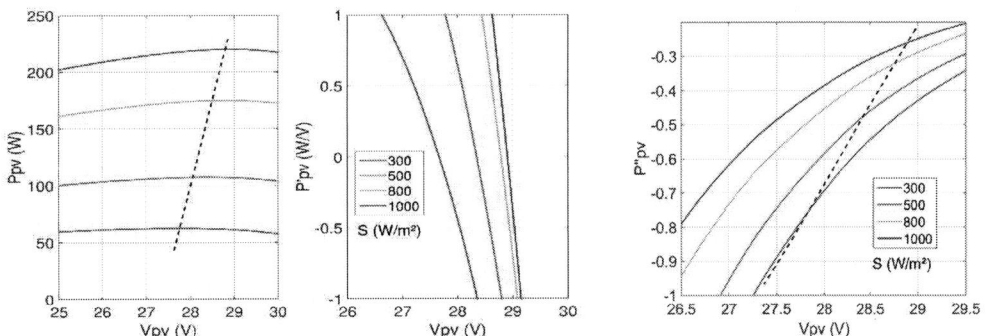

Fig. 6: The variation of P_{pv} and its first and second derivative with respect to the voltage at the output of the solar panel in the vicinity of the MPP

Table II: The values of V_{opt} in the increasing and decreasing direction comparing to V_{MPP}

Irradiance (W/m²)	300	500	800	1000
Vmpp (V)	27.62	28.35	28.79	28.90
Vopt+ (V)	27.72	28.37	28.79	29.00
Vopt- (V)	27.70	28.36	28.79	29.10

The flowchart of the proposed algorithm is given in Fig. 7.

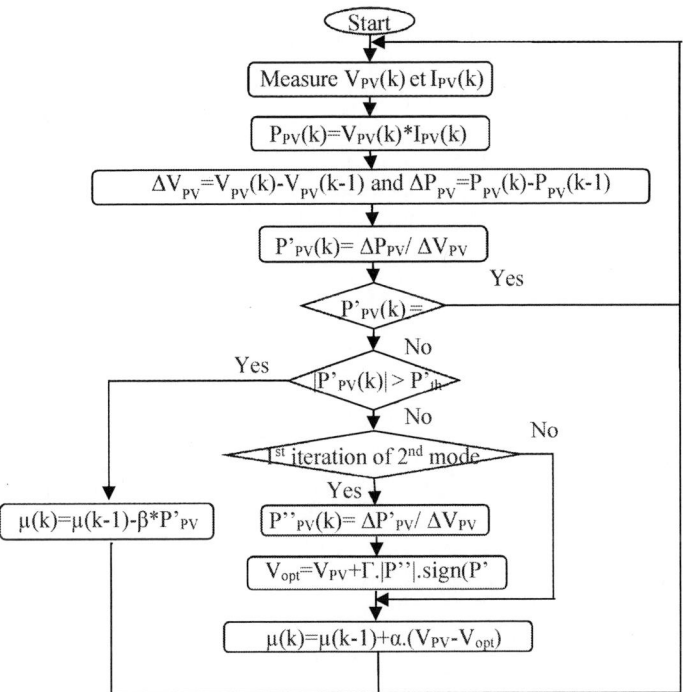

Fig. 7: The flowchart of the proposed algorithm

Simulations through Matlab/Simulink are exposed to validate the proposed algorithm. First, the proposed algorithm is tested with a trapezoidal variation of the solar irradiance, to verify the switching between the two modes. The result is given in Fig. 8. This result shows validate the switching between the two modes. The mode 2 stays active as long as the irradiance doesn't change, which means that the MPP stays the same. When the solar irradiance changes, the switching between two modes assures a the tracking of the new MPP.

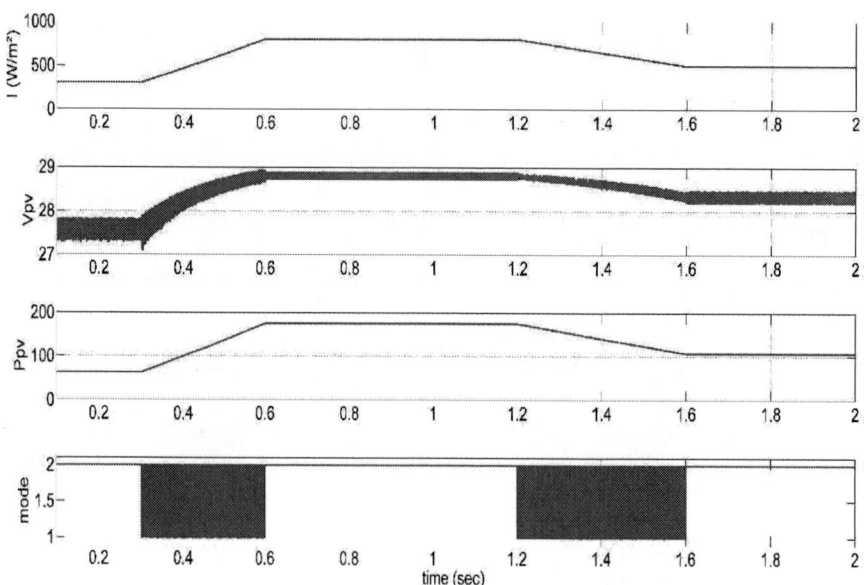

Fig. 8: The response of the proposed algorithm to a trapezoidal irradiance variation

The performances of the proposed algorithm are compared to those given by a conventional P&O algorithm.

Fig. 9 shows the step response of the two algorithms to a constant solar irradiation equal to 800 W/m². We can clearly see that the establishment of the steady state using the dual mode proposed algorithm is faster. The proposed algorithm has a good response time comparing to the P&O algorithm.

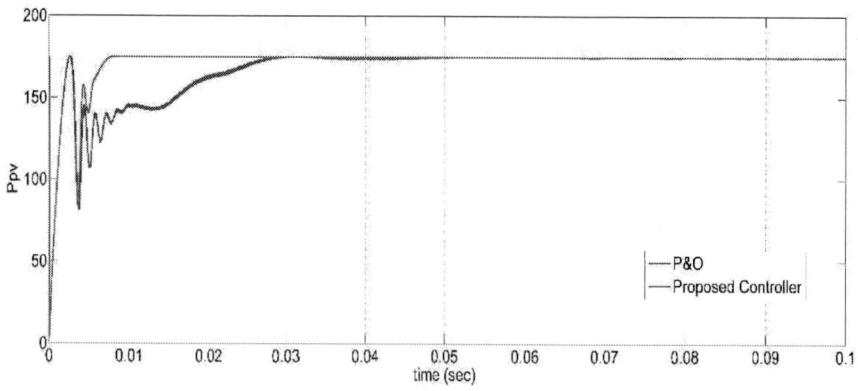

Fig. 9: Comparison between the response time of the proposed algorithm and a conventional P&O

A zoom on the steady state response is given in Fig. 10. We can see that the proposed algorithm reduce significantly the oscillations of the operating point during the steady-state, the permanent losses are smaller using the proposed algorithm which allows a higher power extraction and increases the global efficiency of the system.

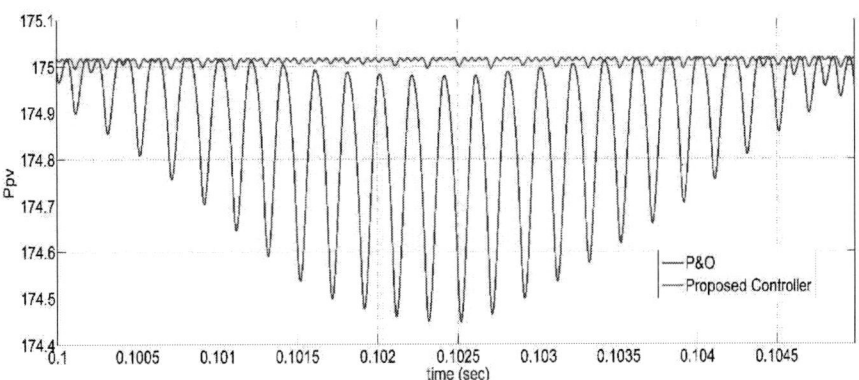

Fig. 10: Zoom on the steady-state response

Simulations results show that the proposed algorithm reduce the oscillation of the operating point after reaching the MPP, which reduce the losses. It also has smaller response time, which allows to reach the MPP quickly.

Conclusion

In this paper, firstly, we have presented a comparison between a conventional boost converter and a quadratic one to prove the advantage of using quadratic converters in solar systems. The quadratic converter allows to reach higher gain with duty cycle far from one.

In a second time we have proposed a new dual-mode variable step-size MPPT algorithm. The new algorithm is based on the P&O and aims to cure its problems. Simulations results proved that the proposed algorithm reduce the oscillations of the operation point around the MPP. It also allows to reach the MPP faster.

References

[1] https://yearbook.enerdata.net/

[2] Farah Al-Chaaban, Ahmad Ghamrawi, Chaiban Haykal and Nazih Moubayed "Comparative Study on Photovoltaic and Thermal Solar Energy Concentrators" The International Conference on Electrical and Electronics Engineering, Clean Energy and Green Computing (EEECEGC 2013), 2013.

[3] J.-P. Gaubert, G. Chanedeau, "Evaluation of DC-to-DC converters topologies with quadratic conversion ratios for photovoltaic power systems", European Conference on Power Electronics and Applications-EPE, Barcelona, Spain, September 2009.

[4] H. Kanoun, J.-P. Gaubert, "Analysis and Modeling of a New Quadratic Boost DC-DC Converter", International Conference on Sciences and Techniques of Automatic Control, STA, Hammamet, Tunisie, December 2009.

[5] Oswaldo López-Santos, Luis Martínez-Salamero, Germain García, Hugo Valderrama-Blavi, and Daniel O Mercuri, "Efficiency analysis of a sliding-mode controlled quadratic boost converter", IET Power Electronics, 6(2), pp. 364-373, 2013.

[6] Gorji, Saman A., et al. "Non-isolated buck–boost dc–dc converter with quadratic voltage gain ratio." IET Power Electronics 12.6 (2019): 1425-1433.

[7] Mirzaee, Afshin, Sajad Arab Ansari, and Javad Shokrollahi Moghani. "Single switch quadratic boost converter with continuous input current for high voltage applications." International Journal of Circuit Theory and Applications 48.4 (2020): 587-602.

[8] Lopez-Santos, Oswaldo, et al. "Quadratic boost converter with low-output-voltage ripple." IET Power Electronics 13.8 (2020): 1605-1612.

[9] Palanisamy, R., et al. "Simulation of various DC-DC converters for photovoltaic system." International Journal of Electrical & Computer Engineering (2088-8708) 9.2 (2019).

[10] Amir, Asim, et al. "Comparative analysis of high voltage gain DC-DC converter topologies for photovoltaic systems." Renewable energy 136 (2019): 1147-1163.

[11] Mojtaba Forouzesh, Yam P Siwakoti, Saman A Gorji, Frede Blaabjerg, and Brad Lehman, "Step-up DC-DC converters: a comprehensive review of voltage-boosting techniques, topologies, and applications", IEEE Transactions on Power Electronics, 32(12), pp. 9143-9178, 2017.

[12] Ahmed, Jubaer, and Zainal Salam. "An improved perturb and observe (P&O) maximum power point tracking (MPPT) algorithm for higher efficiency." Applied Energy 150 (2015): 97-108.

[13] Pilakkat, Deepthi, and S. Kanthalakshmi. "An improved P&O algorithm integrated with artificial bee colony for photovoltaic systems under partial shading conditions." Solar Energy 178 (2019): 37-47.

[14] Harrag, Abdelghani, and Sabir Messalti. "PSO□based SMC variable step size P&O MPPT controller for PV systems under fast changing atmospheric conditions." International Journal of Numerical Modelling: Electronic Networks, Devices and Fields 32.5 (2019).

[15] Yilmaz, Unal, Omer Turksoy, and Ahmet Teke. "Improved MPPT method to increase accuracy and speed in photovoltaic systems under variable atmospheric conditions." International Journal of Electrical Power & Energy Systems 113 (2019): 634-651.

[16] Marcelo Gradella Villalva, Jonas Rafael Gazoli, and Ernesto Ruppert Filho, "Comprehensive approach to modeling and simulation of photovoltaic arrays", IEEE Transactions on Power Electronics, 24(5), pp. 1198-1208, 2009.

[17] Ram, J. Prasanth, N. Rajasekar, and Masafumi Miyatake. "Design and overview of maximum power point tracking techniques in wind and solar photovoltaic systems: A review." Renewable and Sustainable Energy Reviews 73 (2017): 1138-1159.

[18] S. K. Kollimalla, M. K. Mishra, "A novel adaptive P&O MPPT algorithm considering sudden changes in the irradiance", IEEE Trans. on Energy Conversion, vol. 29, no. 3, pp. 602-610, September 2014.

[19] Ramli, Mohd Zulkifli, and Zainal Salam. "Analysis and experimental validation of partial shading mitigation in photovoltaic system using integrated dc–dc converter with maximum power point tracker." IET Renewable Power Generation 13.13 (2019): 2356-2366.

[20] Javed, Kashif, Haroon Ashfaq, and Rajveer Singh. "A new simple MPPT algorithm to track MPP under partial shading for solar photovoltaic systems." International Journal of Green Energy 17.1 (2020): 48-61.

Thermal Model Development for SiC MOSFETs Robustness Analysis under Repetitive Short Circuit Tests

M. Pulvirenti[*], D. Cavallaro[*], N. Bentivegna[*], S. Cascino[*], E. Zanetti[*], M. Saggio[*]

[*]STMicroelectronics
Str. Primosole 50
Catania, Italy
Tel.: +39/ (095) – 7404083

mario.pulvirenti@st.com, daniela.cavallaro@st.com, nella.bentivegna@st.com,
salvatore.cascino@st.com , edoardo.zanetti@st.com, mario.saggio@st.com

Keywords

«Silicon Carbide (SiC)»,«Semiconductor device», « Short Circuit Test (SCT)»,«Finite Element Method (FEM) »

Abstract

The aim of this paper is to analyze the SiC MOSFETs behavior under repetitive short circuit tests. In particular, the activity is focused on a deep evaluation of short circuit dynamic by dedicated laboratory measurements conducted at different conditions supported and compared by means of a robust physical model developed by Finite Element Approach (FEA) and Failure Analysis (FA).

Introduction

Until recently, silicon IGBTs have been the most used devices in power electronics converter, however, developments in the field of semiconductors have allowed the realization of Silicon Carbide (SiC) MOSFETs power devices with superior characteristics, especially in terms of: power density, breakdown voltage capability, switching frequency and operating temperature. [1]. Thanks to the previous properties, it is possible to increase significantly the efficiency in different topology of AC/DC and DC/DC power converters. Together with efficiency improvements, in some applications fields, as the automotive one for example, it is necessary to guarantee a specific level of reliability and robustness: the power converters must be capable to sustain operating conditions different than the rated ones during undesired faults events. Faults that can catastrophically compromise the converter are mainly short circuits [2], among them, for example in a three-phase inverter, the most severe can be due to: failure of the DC link capacitor bank, failure in one leg of the converter in which all the devices are in conduction at the same instant and phase to phase failure, where the devices of two legs of the converter are involved. Many efforts have been recently address to study SiC MOSFETs failure mechanisms under short circuit tests [3]-[11], [14]-[16].

In this paper, repetitive short circuit tests have been implemented to study the reliability of SiC MOSFETs and to explain the failure mode experimentally observed. To achieve this goal, starting from the model presented in [8], suitable dedicated electro-thermal model has been used to estimate the temperature and to understand the phenomenology occurring inside the MOSFET structure under repetitive short circuit stress. Moreover, the temperature estimation and its correlation with known critical values allow to explain failure modes observed experimentally.

In this paper, it is provided a general description of experimental repetitive Short Circuit tests performed on 1200V, 70mΩ SiC Power MOSFET samples, together with relevant failure analysis and modeling strategy.

Repetitive Short Circuit Tests and Setup Description

To deal with severe short circuit tests, the dedicated laboratory setup is realized connecting the drain-source terminals of the device under test (DUT) directly to the DC bus voltage, while a command pulse of few μs is applied at the gate-source terminals by means of a gate driver board. The gate driver board allows to set the gate-source voltage amplitude, that in turns acts on the current capability of the DUT. The equivalent electrical scheme is reported in Fig. 1a and the circuit layout has been optimized in order to ensure very low equivalent circuit impedance in the power loop.

Fig.1b and Fig.1c show respectively the block diagram of the experimental setup and the real test bench used for these tests. As can be noted, from Fig. 1b, a control unit, with a Matlab® code, coordinate the function generator, the DC bus voltage and the digital scope. The code sets the number of pulses, the period and the delay time among one pulse and the next one; during the delay time all relevant waveforms, such as V_{GS}, V_{DS}, I_D and short circuit power P_{SC} are stored. In case of device failure, a protection turns off the DC bus voltage, if the current exceeds a properly set threshold value.

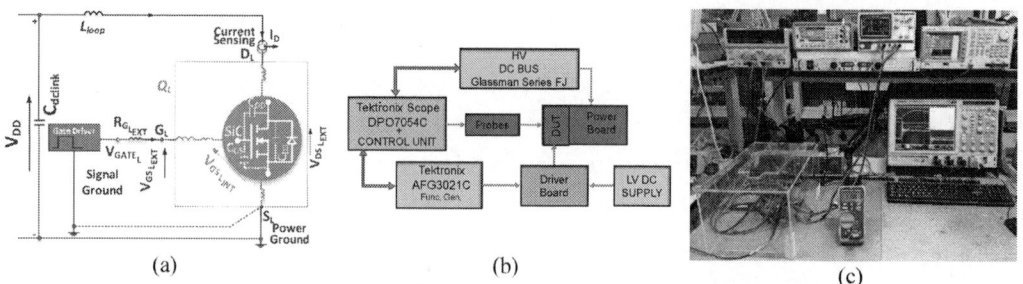

(a) (b) (c)

Fig. 1 Short circuit test: a) Equivalent electric scheme, b) block diagram of the experimental setup, c) picture of the test bench.

After this brief setup description, preliminary discussion about experimental tests is introduced here.

First of all, it is worth to distinguish two type of short circuit tests, destructive and non-destructive, both tests are typically used to quantify the device robustness, identifying parameters, such as: short circuit energy and withstanding time. The short circuit energy, E_{SC}, is obtained by integrating, for the withstand time, the short circuit power, P_{SC}, given by the product between V_{DS} and I_D.

Destructive short circuit tests are, usually, conducted by switching on/off the DUT, with a pulse command with incremental period, in this way the energy dissipated by the device will increase till failure will take place, as shown in Fig. 2(b). This kind of failure can be related to the thermal runway phenomena that lead to device explosion.

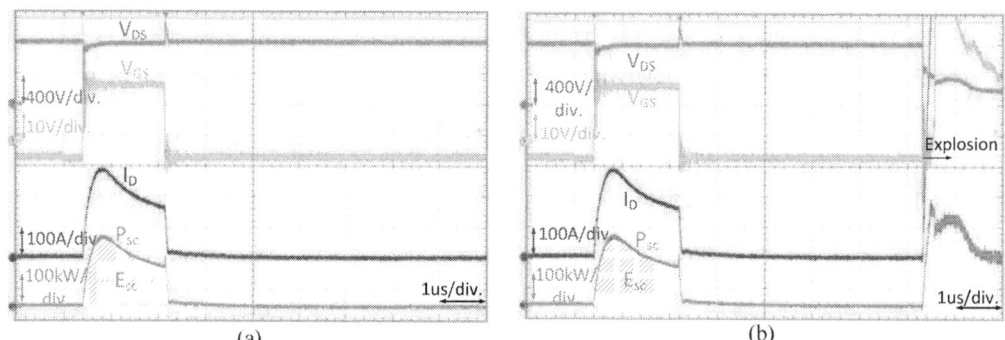

(a) (b)

Fig. 2 Destructive short circuit test: a) last pulse before the failure b) final pulse with device explosion due to thermal runaway.

Another type of failure can occur, associated to gate-source damaging which does not cause uncontrolled explosion and can be considered non-destructive. In this last case, despite the gate degradation, the device is still capable to turn on/off but with reduced performance capability[8]. The critical short circuit time corresponds to the pulse period that has caused the failure: in relation to this period, the critical short circuit energy will be calculated.

In the presented study device robustness has been investigated by means of repetitive short circuit tests, which have been performed, applying a gate command with suitable constant pulse period that define a short circuit energy that can be responsible for effects different than the aforementioned ones and will be better illustrated later.

For the purpose of this paper robustness of 20 samples of 1200V 70mΩ SiC MOSFETs under repetitive short circuit tests is presented at the following conditions V_{DD}=800V, V_{GS}=-5/18V, R_G=4,7Ω, case temperature T_c=175°C.

In order to define T_{sc} value for repetitive short circuit experiments, preliminary assessment of the technology under evaluation has been conducted, performing destructive short circuit tests on 10 samples. Among these samples, the minimum value recorded, T_{cmin}, has been considered to define T_{sc}, according to relation (1)

$$T_{sc}=0,7 \cdot T_{cmin} \qquad (1)$$

After this assessment, T_{sc} for repetitive short circuit tests has been set at 1.1μs with delay time of 5s from one pulse and the next one.

Before starting short circuit tests, all devices have been characterized at the Curve Tracer, checking principal static parameters such as: V_{th}, R_{on}, V_{BD}, I_{DSS} and I_{GSS}, typical values are reported in Table I

Table I SiC MOSFET under test nameplate.

V_{th} [V] @ 1mA	$R_{(on)}$ [mΩ] @18V, 20A	V_{BD} [V] @ 1mA	V_{GS} [V] abs. max. rating	I_{DSS} [μA] @ V_{DS}=1200V, V_{GS}=0	I_{GSS} [nA] @ V_{DS}=0V, V_{GS}=-10 to +22V
3	70	1200	-10/+22	50	±100

During repetitive short circuit tests, two devices fail over the considered sample of 20 SiC MOSFETs, one device fails after 950 pulses while another one after 3392 pulses: they will be indicated in the following respectively as "Sample 1" and "Sample 2".

Fig. 3 shows overlaid waveforms of the latest 6 pulses, from 945 to 950 of repetitive short circuit tests relative to V_{GS}, V_{DS}, I_D and P_{SC} of Sample 1, sustaining a short circuit energy, E_{SC}=232mJ.

When the pulse command T_{sc} is applied, the gate-source voltage switches from -5 to 18V, therefore the SiC MOSFET under test starts to conduct very high current. Due to this very high drain current, the temperature inside the "die" suddenly rises therefore R_{on} increases and after I_D reaches its peak value, starts to decrease.

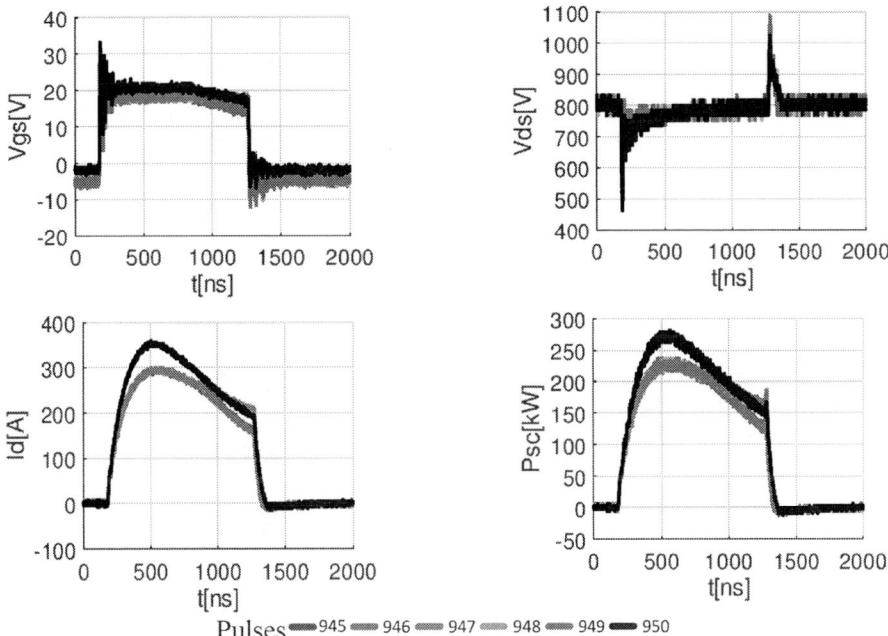

Fig. 3 Short circuit tests typical waveforms of V_{GS}, V_{DS} I_D and P_{sc} before , during and after the failure.

As can be noted, the device failure is principally identified from V_{GS} waveforms and can be related to gate-source damage, in fact, during the last pulse (black curve on graph), V_{GS} is no more fixed at -5V during the off state but it is ≈-2V, this means that the capacitive behaviour of the gate is compromised. This is also confirmed from I_{GS} measurement performed at Curve Tracer. Fig. 4 shows I_{GS} tests performed before starting repetitive short circuit test and after the failure, the other static parameters of this damaged device, do not show significant deviations.

Similar waveforms have been recorded for Sample 2.

Major explanations about the recorded failure can be given performing the failure analysis, which is included in the next paragraph.

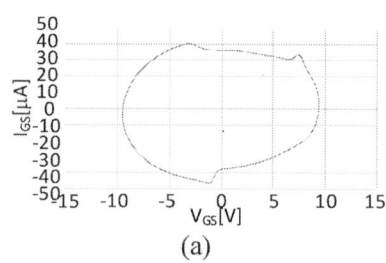
(a)

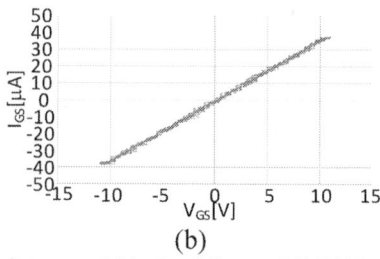
(b)

Fig. 4 I_{GS} vs V_{GS} measurement (a) before repetitive short circuit tests and (b) after failure at 950/1000 pulse.

Physical analysis of failed device

After repetitive Short-Circuit Test failure analysis has been performed on Sample 1 and Sample 2. The photo captured during X-ray is shown in Fig. 5a, and SAM (Scanning Acoustic Microscopy) in Fig. 5b. As expected, the almost 1000 repetitive short circuit does not induce any delamination neither in molding compound or in die-attach.

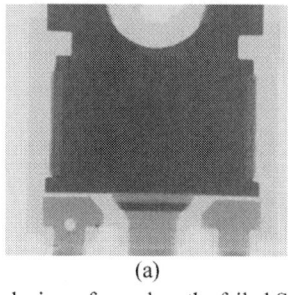
(a)

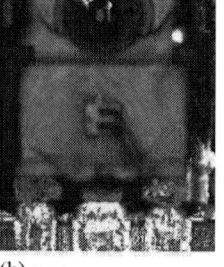
(b)

Fig. 5 Analysis performed on the failed Sample1. picture captured at X-Ray (a) frontal section from back-side, (b) SAM picture

After this non-destructive FA, resin has been removed and photos have been taken to check the top integrity (die and wires).

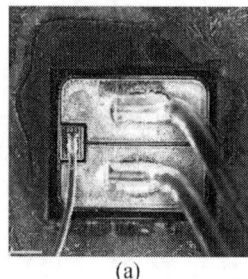
(a)

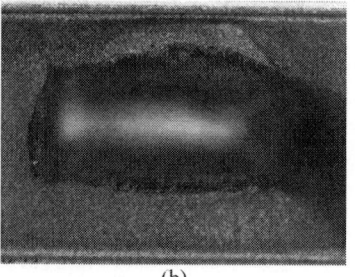
(b)

Fig.6 Visual inspection of Sample1 after resin removal: general view(a) and source bond wire(b)

The visual inspection of failed samples after resin removal is indicating aging effects occurring at top metallization level, Fig.6(a) which show high reflective metal layer while Fig 6(b) shows an halo along the contour of source bonding wire, suggesting repetitive cycles of melting/solidification due to high temperature swing which involves thermomechanical stress on wires.

The visual analysis shows a metal fatigue on both devices, with respect to a virgin sample, used for comparison, as illustrated in Fig 7.

Fig. 7 Optical Microscope comparison: (a) Sample 2, (b) Sample 1, (c) "virgin" Sample.

The optical microscope analysis has highlighted a more pronounced aging effect on Sample 2, in accordance of sustained number of pulses and additional FA (Themos and FIB) has been performed on Sample2.

Thermal Emission Microscopy (Themos) has been used for fault localization. As shown in Fig. 8 an emission zone is located near the device corner.

Fig. 8: Themos Analysis on Sample 2.

Then we have analyzed Sample2 on Scanning Electron Microscopy (SEM) equipment and we have performed a Focused Ion Beam (FIB) cut in the emission area.

Top view SEM analysis, Fig.9(a) and Fig.9(b), confirms metal aging already seen by optical microscopy Fig 7(a). Moreover FIB cut in the emission area show visible cracks near the contact area Fig 9(c) suggesting thermo-mechanical stress occurring during repetitive short circuit which could be responsible of intermediate dielectric cracks and short between gate and source as already observed by Reigosa et al. [17].

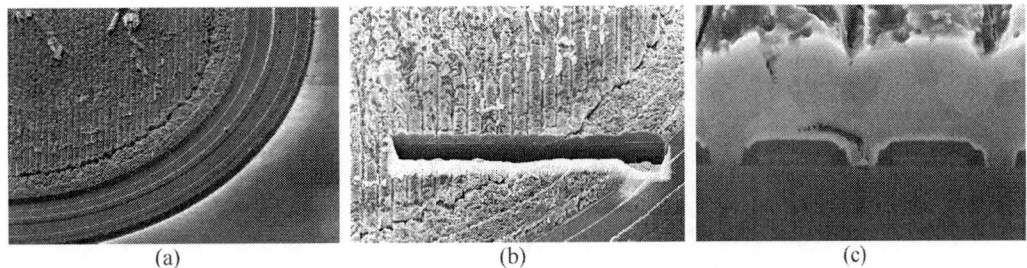

Fig. 9 SEM analysis: (a) corner top view, (b) section in "crack" region, (c) zoomed view of crack region.

TCAD simulations: electrothermal response of a SiC Power MOSFET during short circuit phenomena

Once the failure mode due to repetitive short circuit is investigated and understood, we developed two different models based on Finite Element Numerical Method to explain what experimentally observed. The first one based on TCAD simulations has been used to investigate the phenomena occurring inside the SiC structure by using an electro-thermal model. The second one is a physical structural model which has been used to investigate the temperature distribution on top metal where the failure is occurring.

Purpose of TCAD simulations was investigating the failure mechanisms related to short-circuits phenomena and the thermal impact on the top metal after the gate voltage is switched-off at time ~1.1 μs that is considerably lower than the single pulse critical short circuit time (~1.62 μs). One half of a symmetric SiC power MOSFET active cell was reproduced with Athena simulator [12] of SILVACO. The transient behavior of the device was simulated by a high voltage short-circuit performed at a DC-link voltage of 800 V, sweeping the gate voltage from − 5 V to + 18 V in a 175 °C fixed temperature environment.

Since electrical parameters of power device are related to the temperature evolution, obviously, in the electrical model, self-heating effects and their dependence on temperature were considered. Isothermal condition (T=175°C) on the back of the device and thermal exchange with air were considered on the top of device. Mixed-mode simulations [13] were performed using a simple circuital configuration shown in Fig. 10, in which the terminals of physical device were appropriately connected to simulate short circuit. Drain inductance L was fixed to 40 nH, V_{dd}=800 V, R_g=4.7 V and V_{gs} was switched-on from -5 V to 18 V and switched-off at two different times (1.1μs, 1.62μs).

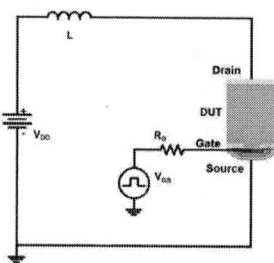

Fig. 10 Mixed-mode schematic used to simulate short-circuit

As showed in Fig.11(a), when device is switched-off at 1.62μs, a hole current density is observed and drain current increases during time: the hole generation, although the gate has been turned off, is able to sustain itself, causing increasing of temperature, Fig.11(b), and device failure. When device is switched-off at t=1.1μs, hole generation and temperature are lower than t=1.62μs switch-off case and phenomena does not occur.

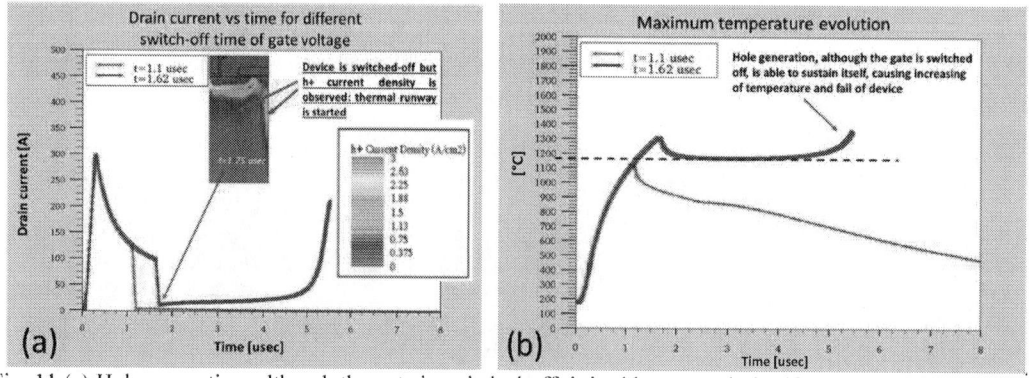

Fig. 11 (a) Hole generation, although the gate is switched-off, it is able to sustain itself, causing device failure. Hole current density is represented in log scale. (b) Maximum temperature evolution is showed for two different switching-off time.

However, the temperatures evolution inside the MOSFET structure after the 1.1 µs switching time can cause a slight degradation of the metal present on the top of the device and the effect could be evident after repetitive short circuit tests cycles.

So, to investigate Aluminum temperature evolution on the top of device after single pulse switched-off at 1.1 µs, snapshots of MOSFET elementary cell has been extracted at different times during Short circuit event and results are showed in fig.12. Simulations have been performed not considering phase changing of Aluminum, metal heat capacity was modelled as reported in [14] and it was considered constant above 620°C. As can be observed, simulation results show that a short time range (~ 4 µs) exists in which temperature can reach values above the melting point of Aluminum. Experimental evidence after repetitive short circuit cycles could be explained considering that although for brief instants, Aluminum temperature could approach critical values.

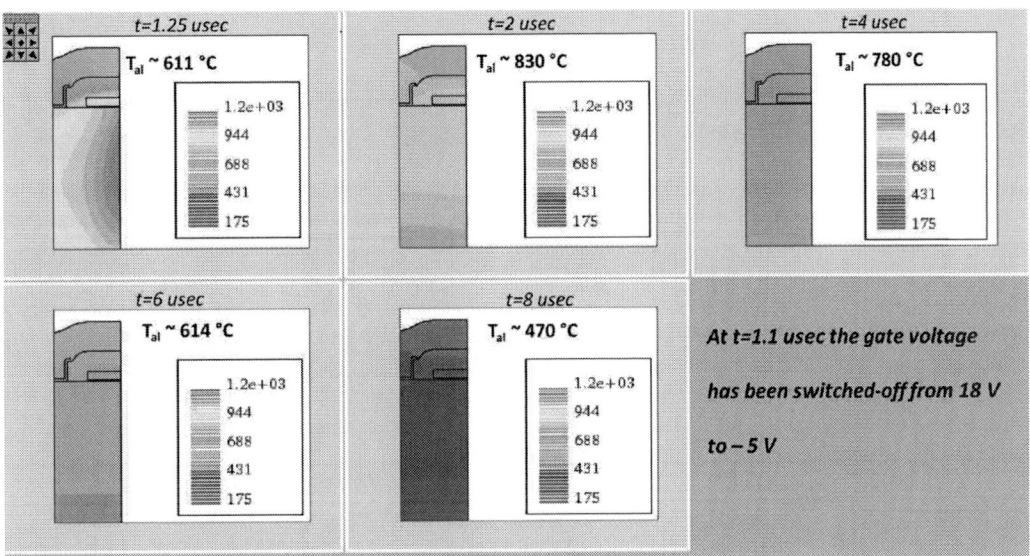

Fig.12 Temperature evolution after gate is switched-off at 1.1 µs.

Temperature estimation by FEA

The physical analysis of the failed samples has shown that repetitive short-circuit pulses, involving high temperature variations, induce thermomechanical stress at source wires and aging of Aluminum top metallization. Starting from this statement the last session of the paper has been focused on top metal thermal behavior. To get this goal a physical model able to describe the latest generation of ST SiC Power MOSFET and to explain the phenomena occurring in few microseconds, such as short circuit, has been implemented.

The proposed methodology, already described in [8] consists in providing to the physical model the experimental time-dependent power profiles, P_{SC}, (see Fig. 3) according to the distribution coming from structural TCAD model described above. The thermal model can simulate the Aluminum melting in terms of latent heat of fusion considered at 670°C where the heat capacity is considered almost doubling during phase transition.

The temperature profiles of device active area inside semiconductor, at the interface between top metal and SiC and at top metal surface have been evaluated for the last measured pulse before failure of Sample 2 and shown in Fig.13.

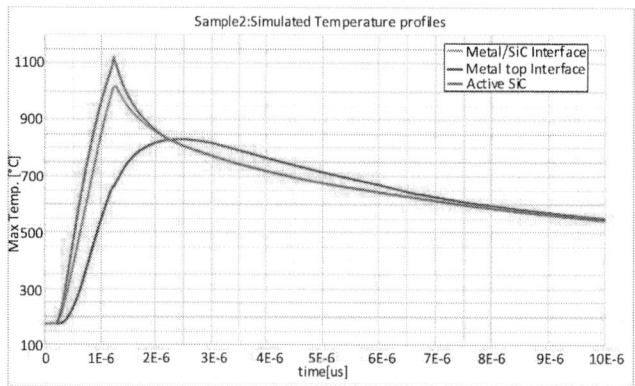

Fig.13 Simulated temperature profiles of Sample 2 at last measured pulse before failure

Looking at Fig. 13 we can observe that the temperature at Aluminum/SiC interface (orange curve) evolves rapidly following the temperature profile of SiC (green curve). On the other hand, the temperature monitored on the upper surface of the metal (blue curve) takes about 2us to reach its peak value (830°C) and the Aluminum remains on average at a temperature above 670°C (melting point) for about 4.5us at each pulse which explains what has been experimentally observed and through failure analysis.

To complete the analysis done by FEA, the thermal maps of Aluminum top metal have been extracted and reported in Fig. 14, at different time instants. They allow to monitor where the maximum temperature is located at each time instant and how the heat is propagating across the metal.

Looking at the Fig. 14 (a), (b), (c), we can observe that till 2us the maximum temperature stands at the interface between active SiC and front metal which behaves as heatsink for the power generated inside the structure. At about 2.5us (Fig.14 (d)) the metal is almost saturated. The temperature gradient inside the Aluminum (top-to-bottom) begins decreasing after 5us.

The temperature reached under these test conditions do not involve failure in single pulse but the repetition of such pulses leaves a mark which cycle after cycle undergoes thermomechanical stresses and induces aging in the metal, as observed in physical analysis.

Conclusions

In this paper, evaluations of repetitive short circuit tests performed on SiC MOSFETs have been introduced and discussed. Experimental results show a fault on gate isolation on two over 20 samples subjected to 950 and 3392 short circuit pulses respectively. The fault has been correlated with Aluminum metal aging revealed by failure analysis on failed parts, this in turn can be explained by Aluminum temperature profile extracted by simulation models.

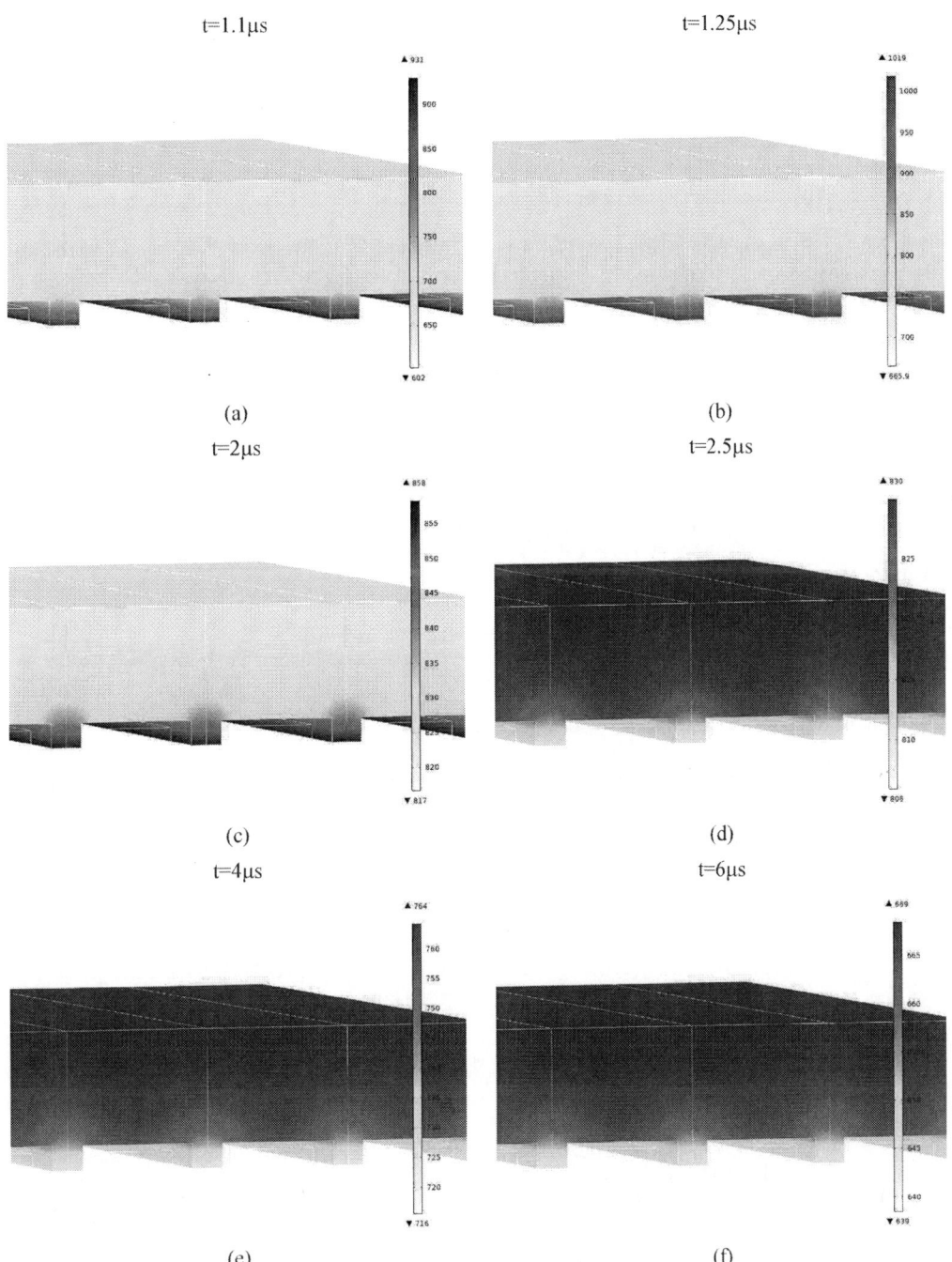

Fig. 14: 3D Thermal maps of Aluminum at different time instants: (a)1.1μs, (b)1.25μs, (c)2μs, (d)4μs, (e)6μs.

References

[1] F. Wang and Z. Zhang "Overview of Silicon Carbide Technology: Device, Converter, System, and Application," Power Electr. and Appl. Trans on. CPSS, vol. 1, no. 1, pp. 13-32, Dec. 2016.

[2] B. Mirafzal "Survey of Fault-Tolerance Techniques for Three-Phase Voltage Source Inverters," IEEE Trans. on Ind. Elec. Vol.61, no.10, pp. 5192-5202, Oct.2014.

[3] L. Ceccarelli, P.D. Reigosa, F. Iannuzzo, F. Blaabjerg, "A survey of SiC power MOSFETs short-circuit robustness and failure mode analysis," Microelec. Rel., vol. 76–77, 2017, pp.: 272-276.

[4] G. Romano, at al., "A Comprehensive Study of Short-Circuit Ruggedness of Silicon Carbide Power MOSFETs," IEEE J. Em. and Sel. Top. In Power Elec., vol. 4, no. 3, 2016, pp.978-986.

[5] V. Mulpuri, S. Choi "Degradation of SiC MOSFETs with Gate Oxide Breakdown under Short Circuit and High Temperature Operation," 2017 IEEE En. Conv. Cong. and Expo., pp.:2527-2532.

[6] Z.Wang at al., "Temperature-Dependent Short-Circuit Capability of Silicon Carbide Power MOSFETs," IEEE Trans. On Power Electr., Vol. 31, no. 2, Feb. 2016.

[7] X. Zhou, H. Su, Y. Wang, R. Yue, G. Dai, and J. Li, "Investigations on the Degradation of 1.2-kV 4H-SiC MOSFETs Under Repetitive Short-Circuit Tests," IEEE Trans. on Elec. Dev., vol. 63, no. 11, Nov. 2016, pp. 4346-4351.

[8] D. Cavallaro, M. Pulvirenti, E. Zanetti, M. G. Saggio "Capability of SiC MOSFETs under Short-Circuit tests and development of a Thermal Model by Finite Element Analysis," Materials Science and Engineering Journal, pp. 788-791, 2018.

[9] Gianpaolo Romano; Asad Fayyaz; Michele Riccio; Luca Maresca; Giovanni Breglio; Alberto Castellazzi, A Comprehensive Study of Short-Circuit Ruggedness of Silicon Carbide Power MOSFETs, IEEE Journal of Emerging and Selected Topics in Power Electronics (Volume: 4 , Issue: 3 , Sept. 2016)

[10] Alexander Tsibizov; Ivana Kovačević-Badstübner; Bhagyalakshmi Kakarla; Ulrike Grossner, Accurate Temperature Estimation of SiC Power MOSFETs under Extreme Operating Conditions, IEEE Transactions on Power Electronics (Volume: 35, Issue: 2, Feb. 2020).

[11] Paul Paret; Gilberto Moreno; Bidzina Kekelia; Ramchandra Kotecha; Xuhui Feng; Kevin Bennion; Barry Mather; Andriy Zakutayev; Sreekant Narumanchi; Samuel Kim; Samuel Graham, Thermal and Thermomechanical Modeling to Design a Gallium Oxide Power Electronics Package, 2018 IEEE 6th Workshop on Wide Bandgap Power Devices and Applications (WiPDA)

[12] ATHENA User's Manual, SILVACO, Inc., CA, USA, 2016.

[13] ATLAS User's Manual, SILVACO, Inc., CA, USA, 2016.

[14] Panagis Foteinopoulos, Alexios Papacharalampopoulos, Panagiotis Stavropoulos, On thermal modeling of Additive Manufacturing processes, CIRP Journal of Manufacturing Science and Technology Volume 20, January 2018, Pages 66-83.

[15] F. Richardeau, F. Boige, Circuit-type modelling of SiC power Mosfet in short-circuit operation including selective fail-to-open and fail-to-short modes competition, Microelectronics Reliability Volumes 100–101, September 2019, 113501.

[16] Huan Li,1 Jue Wang;Na Ren; Hongyi Xu; Kuang Sheng, Investigation of 1200 V SiC MOSFETs' Surge Reliability, Micromachines (Basel). 2019 Jul; 10(7): 485.

[17] P. D. Reigosa, F. Iannuzzo and L. Ceccarelli, "Failure Analysis of a Degraded 1.2 kV SiC MOSFET after Short Circuit at High Temperature," 2018 IEEE International Symposium on the Physical and Failure Analysis of Integrated Circuits (IPFA), Singapore, 2018, pp. 1-5.

Compensation of the Radial and Circumferential Mode 0 Vibration of a Permanent Magnet Electric Machine based on an Experimental Characterisation

Jan Andresen, Stephan Vip, Axel Mertens
Institute for Drive Systems and Power Electronics
Leibniz Universität Hannover
Welfengarten 1
30167 Hannover, Germany
Phone: +49 (0) 511-762-14569
Email: Jan.Andresen@ial.uni-hannover.de
URL: https://www.ial.uni-hannover.de

Sebastian Paulus
Robert Bosch GmbH
Frankfurter Str. 10
71732 Tamm, Germany

Keywords

≪Force Control≫, ≪Active damping≫, ≪Permanent magnet motor≫, ≪Electric vehicle≫, ≪Acoustic noise≫

Abstract

This paper presents an approach how to compensate radial and circumferential mode 0 vibrations based on harmonic currents. First, the influence of harmonic currents and the mechanic transfer function are identified. Using these results, harmonic currents are derived which compensate the pre-existing vibrations.

Introduction

In an electric power train design, it is a challenging task to achieve an acceptable acoustic behaviour. By their very nature, electric machines typically emit few, but distinct frequencies which lead to a very unpleasant sound. Furthermore, mechanical resonance frequencies of the housing and the power train can amplify the noise emissions at certain speeds. Two main sources of noise emissions have to be considered: noises that are emitted by radial forces acting on the housing of the machines and torque pulsations. For the first category, mode 0 vibration has been identified as the main source by [1–3]. Mode 0 causes the whole housing to shrink and expand in radial direction at the same time, as shown in Fig. 1. Naturally, a profound electric machine design forms the basis of a good overall power train behaviour. However, if the noise requirements are not achievable owing to design constraints or disadvantages with respect to other parameters, e.g. efficiency, additional measures are necessary. In the paper, a noise compensation approach through modifying the supply currents is analysed as a promising additional measure.

a) Radial (R)
mode 0 vibration

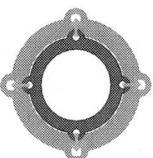

b) Circumferential (θ')
mode 0 vibration

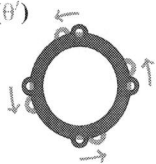

Fig. 1: Schematic view of the radial mode 0 vibration (left) (cf. [4]) and the circumferential mode 0 vibration (right).

Successful current-based approaches can be found in literature. By suppressing current harmonics and thus supplying a more sinusoidal current, the acoustic behaviour can be significantly improved [5]. Further improvements can be achieved by injecting specific harmonic currents. The radial accelerations can be improved as reported, e.g., in [6–12]. In a similar manner, torque pulsations can be compensated, e.g. [13, 14]. That both directions can be improved at the same time is shown by using numeric optimisation [15], dynamic programming [16] and for switched reluctance motors [17].

In the experimental approach presented, torque pulsations are not directly considered. Instead, the acceleration in circumferential direction of the machine housing is used. In analogy to the radial mode 0 vibration, the mode 0 of the circumferential direction is the considered shape. For the circumferential mode 0, the whole stator moves in the same tangential direction. The force exciting this mode to some extent is the reaction force to torque pulsation. This paper shows that both mode 0 vibrations can independently be influenced using harmonic currents. This enables the novel approach to compensate both modes independently. In order to achieve this compensation, an unintentional cross-coupling between the radial and circumferential direction, as reported in [6], has to be overcome. When achieving a compensation of both modes, two different phenomena can be compensated at the same time, thus enabling a comprehensive compensation of acoustic phenomena.

Influencing the mechanical vibrations of an electric machine is a multiphysics problem. Fig. 2 gives an overview of physical domains involved. The current harmonics represent the excitation. These generate magnetic forces exciting the mechanical system. The transfer characteristic of the mechanical system determines which vibrations occur. These vibrations are measured using acceleration sensors. The airborne noise is measured by means of a microphone. In order to reduce vibrations, a three-step approach is presented: First, an injection direction for the harmonic current is chosen. Then, the transfer function is estimated. The injection direction and the transfer function are then used to calculate the compensation currents. The effectiveness is experimentally validated.

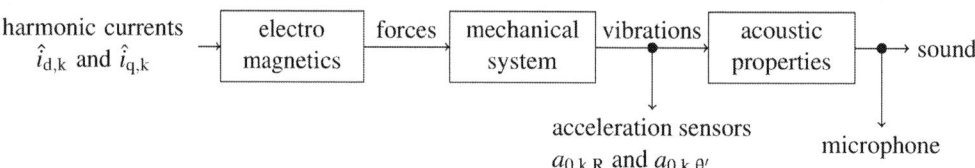

Fig. 2: Overview of the coupling between the involved physical domains.

Experimental Setup

The experimental setup consists of a permanent magnet electric traction motor, connected to a load machine via a jaw-type coupling. Fig. 3 illustrates the experimental setup.

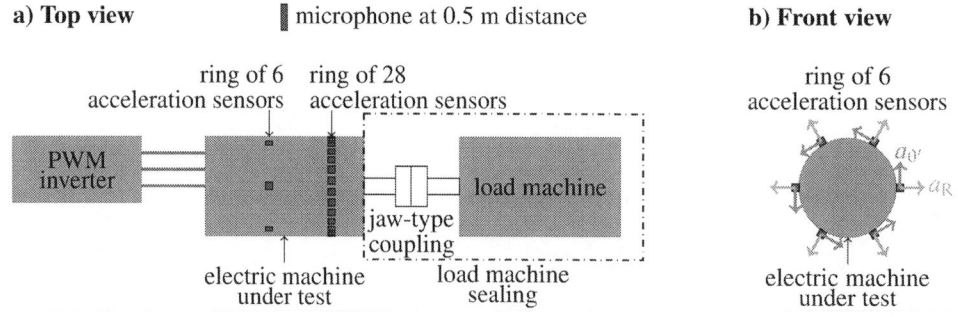

Fig. 3: Experimental test setup.

The machine is powered by a current-controlled PWM-type inverter. Using this inverter, currents in the form

$$
\begin{array}{ll}
\overbrace{i_{\mathrm{d}}}^{\text{d-current}} = \overbrace{\hat{i}_{\mathrm{d},0}}^{\text{fundamental d-component}} + \overbrace{\hat{i}_{\mathrm{d},\mathrm{k}}\cos\left(k\delta' - \varphi_{\mathrm{d},\mathrm{k}}\right)}^{\text{harmonic d-component}} \\
\underbrace{i_{\mathrm{q}}}_{\text{q-current}} = \underbrace{\hat{i}_{\mathrm{q},0}}_{\text{fundamental q-component}} + \underbrace{\hat{i}_{\mathrm{q},\mathrm{k}}\cos\left(k\delta' - \varphi_{\mathrm{q},\mathrm{k}}\right)}_{\text{harmonic q-component}}
\end{array}
\tag{1}
$$

can be injected. In the equation, δ' is the mechanical angle between d-axis and the fixed coordinate system, k is the rotational order, $\hat{i}_{\mathrm{d},0}/\hat{i}_{\mathrm{q},0}$ the fundamental components, $\hat{i}_{\mathrm{d},\mathrm{k}}/\hat{i}_{\mathrm{q},\mathrm{k}}$ the amplitudes of the kth harmonic components, and $\varphi_{\mathrm{d},\mathrm{k}}/\varphi_{\mathrm{q},\mathrm{k}}$ are the rotational phase shifts. The operating points considered in the paper are in the field weakening range ($\hat{i}_{\mathrm{d},0} < 0$). Field weakening is necessary in order to have the required voltage margin to operate in the whole considered speed range. Furthermore, acceleration sensors are placed on the electric motor. One ring consists of 6 acceleration sensors and the second ring is equipped with 28 acceleration sensors. All sensors are glued on the housing of the motor. For each sensor, both the acceleration in normal direction and the circumferential direction is recorded. The ring consisting of 6 sensors is used for online measurement and compensation, while the 28-sensor ring is used for an independent confirmatory measurement. Throughout the measurement, the average value of all sensor values of the same direction is extensively used. For the radial direction, the average value represents the radial mode 0 vibration and for the circumferential direction the circumferential mode 0 vibration. The amplitudes ($\hat{a}_{0,\mathrm{k},\mathrm{R}}$ and $\hat{a}_{0,\mathrm{k},\theta'}$) and phase angles ($\varphi_{\mathrm{a},\mathrm{k},\mathrm{R}}$ and $\varphi_{\mathrm{a},\mathrm{k},\theta'}$) of the kth order mode 0 vibrations were identified online as

$$
a_{0,\mathrm{k},\mathrm{R}} = \hat{a}_{0,\mathrm{k},\mathrm{R}}\cos\left(k\delta' - \varphi_{\mathrm{a},\mathrm{k},\mathrm{R}}\right) \text{ and } a_{0,\mathrm{k},\theta'} = \hat{a}_{0,\mathrm{k},\theta'}\cos\left(k\delta' - \varphi_{\mathrm{a},\mathrm{k},\theta'}\right).
\tag{2}
$$

Additionally, the sound is measured using a microphone placed at a distance of $0.5\,\mathrm{m}$.

Identification of the Influence of Harmonic Currents

The first step is to quantify the influence of the d- and q-current of the radial and circumferential acceleration. This analysis is performed at a fixed reference speed. In order to be able to match a direction to each current, only alternating currents in the dq-reference frame are used

$$
\begin{bmatrix} i_{\mathrm{d},\mathrm{k}} \\ i_{\mathrm{q},\mathrm{k}} \end{bmatrix} = \begin{bmatrix} \hat{i}_{\mathrm{d},\mathrm{k}} \\ \hat{i}_{\mathrm{q},\mathrm{k}} \end{bmatrix} \cos\left(k\delta' - \varphi_{\mathrm{i},\mathrm{k}}\right) = \hat{i}_{\mathrm{k}} \begin{bmatrix} \cos(\gamma) \\ \sin(\gamma) \end{bmatrix} \cos\left(k\delta' - \varphi_{\mathrm{i},\mathrm{k}}\right).
\tag{3}
$$

Here, the angle γ determines the excitation direction, $\hat{i}_{\mathrm{k}}$ is the amplitude of the respective current and φ_{k} is the phase angle of the current. The aim of this section is to describe the influence of a harmonic current on the acceleration in a compact form. This is achieved by representing the ratio between the acceleration caused by the harmonic current ($\Delta\hat{a}$) and the current ($\hat{i}_{\mathrm{k}}$) generated by a harmonic function of the excitation direction (γ)

$$
\frac{\Delta\hat{a}}{\hat{i}_{\mathrm{k}}} \approx \eta_{\mathrm{k}}\cos(\gamma - \alpha).
\tag{4}
$$

An offset angle (α) is introduced. Equation (4) shows that the current has the maximum influence if the excitation direction is chosen equal to the offset angle ($\alpha = \gamma$). η_{k} is called the influence factor and α the influence angle. This analysis is performed for both the radial and circumferential direction. Note that the representation (4) does not include any phase shift between the acceleration and the current which will be introduced in (6) in order to represent complex-valued η. The identification of the influence of the excitation direction starts with the recording of the acceleration with $\hat{i}_{\mathrm{k}} = 0\,\mathrm{A}$, referred to as $a_{\mathrm{k},\hat{i}_{\mathrm{k}}=0\,\mathrm{A}}$. Afterwards, multiple measurements with a constant amplitude $\hat{i}_{\mathrm{k}}$ are performed where γ and $\varphi_{\mathrm{i},\mathrm{k}}$ are varied. From each acceleration, $a_{\mathrm{k},\hat{i}_{\mathrm{k}}=0\,\mathrm{A}}$ is subtracted to calculate the acceleration caused by the respective harmonic current. Next, the acceleration and the current are represented as complex values using their amplitudes and phase shifts ($\underline{\Delta\hat{a}}_{k,n} = \Delta\hat{a}_{k,n}e^{j\varphi_{a,k,n}}$ and $\hat{\underline{i}}_k = \hat{i}_k e^{j\varphi_{i,k}}$). Using these values, a relationship between the acceleration and the current can be derived with two coefficients $\underline{\eta}_c$ and $\underline{\eta}_s$ using a least square approximation

$$\min_{\underline{\eta}_c,\underline{\eta}_s} \sum_{n=1}^{N} \left(\Delta\hat{a}_{k,n} - \left(\underline{\eta}_c \cos(\gamma_n) + \underline{\eta}_s \sin(\gamma_n) \right) \hat{i}_{k,n} \right)^2 . \tag{5}$$

In order to transform these complex factors into the real coefficients of (4), an approximation is used

$$\begin{bmatrix} \underline{\eta}_c \\ \underline{\eta}_s \end{bmatrix} \approx \eta_k \begin{bmatrix} \cos(\alpha) \\ \sin(\alpha) \end{bmatrix} e^{j\varphi_n} . \tag{6}$$

To find the best approximation, the phase shift between the acceleration and the current (φ_n) is determined first. Since the other coefficients are real, a multiplication by $e^{-j\varphi_n}$ makes the approximation of a real vector. Based on this observation, the angle is chosen that leads to the largest real part when multiplied with the complex factor

$$\max_{\varphi_n} \left(\left(\Re\{\underline{\eta}_c e^{-j\varphi_n}\} \right)^2 + \left(\Re\{\underline{\eta}_s e^{-j\varphi_n}\} \right)^2 \right) . \tag{7}$$

This equation is solved analytically. The remaining factors can be determined as

$$\eta_k^2 = \left(\Re\{\underline{\eta}_c e^{-j\varphi_n}\} \right)^2 + \left(\Re\{\underline{\eta}_s e^{-j\varphi_n}\} \right)^2 \text{ and } \alpha = \mathrm{atan2}\left\{ \Re\{\underline{\eta}_s e^{-j\varphi_n}\} ; \Re\{\underline{\eta}_c e^{-j\varphi_n}\} \right\} . \tag{8}$$

Since φ_n and $\varphi_n + \pi$ result in the same value in (7), different α can result. This is used to limit $-\pi/2 \le \alpha_R \le \pi/2$ and $0 \le \alpha_{\theta'} \le \pi$. These values are chosen to centre around the known excitation direction for a machine without armature reaction of $\alpha_R = 0$ and $\alpha_{\theta'} = \pi/2$ [18].

Fig. 4 illustrates the results for operation under load ($\hat{i}_{q,0} > 0$). The figure proves that a sinusoidal approximation of the influence between the angle γ and the acceleration closely matches the measurements. Furthermore, it can be observed that the error when transforming a complex influence factor into a real influence factor is small, too. Fig. 5 shows the same results in idle mode ($\hat{i}_{q,0} = 0$). It can be observed that both influence angles are operating point-dependent. Here, the radial influence angle has the most significant change. The influence factors in idle mode are close to the expected values without armature reaction. The radial excitation factor is significantly reduced in idle mode. The reduced gain makes measurement noise more dominant in the plot.

The results of each operating point can be represented in a matrix equation

$$\begin{bmatrix} \underline{\Delta\hat{a}}_{0,k,R} \cdot e^{-j\varphi_{n,R}} \\ \underline{\Delta\hat{a}}_{0,k,\theta'} \cdot e^{-j\varphi_{n,\theta'}} \end{bmatrix} = \begin{bmatrix} \eta_{k,R} \cos(\alpha_R) & \eta_{k,R} \sin(\alpha_R) \\ \eta_{k,\theta'} \cos(\alpha_{\theta'}) & \eta_{k,\theta'} \sin(\alpha_{\theta'}) \end{bmatrix} \begin{bmatrix} \hat{i}_{d,k} \\ \hat{i}_{q,k} \end{bmatrix} . \tag{9}$$

Using a matrix inversion, the current required for a certain acceleration can be derived

$$\begin{bmatrix} \hat{i}_{d,k} \\ \hat{i}_{q,k} \end{bmatrix} = \frac{1}{\sin(\alpha_{\theta'} - \alpha_R)} \cdot \begin{bmatrix} \sin(\alpha_{\theta'}) & -\sin(\alpha_R) \\ -\cos(\alpha_{\theta'}) & \cos(\alpha_R) \end{bmatrix} \begin{bmatrix} \frac{1}{\eta_{k,R}} \underline{\Delta\hat{a}}_{0,k,R} \cdot e^{-j\varphi_{n,R}} \\ \frac{1}{\eta_{k,\theta'}} \underline{\Delta\hat{a}}_{0,k,0'} \cdot e^{-j\varphi_{n,\theta'}} \end{bmatrix} . \tag{10}$$

The inversion is possible if $\alpha_{\theta'} \ne \alpha_R$. The invertibility proves that any amplitude can be generated. Moreover, only the radial direction is excited if $\gamma = \beta_R = \alpha_{\theta'} - \pi/2$ is chosen and the circumferential direction is excited if $\gamma = \beta_{\theta'} = \alpha_R + \pi/2$ is chosen. β_R and $\beta_{\theta'}$ are hereinafter called injection angles. They avoid an excitation of the other (unwanted) direction.

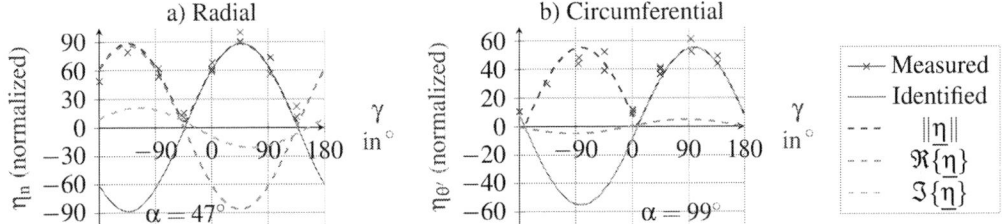

Fig. 4: Result of the excitation direction identification under load ($\hat{i}_{q,0} > 0$) for the normal direction (left) and for the circumferential direction (right).

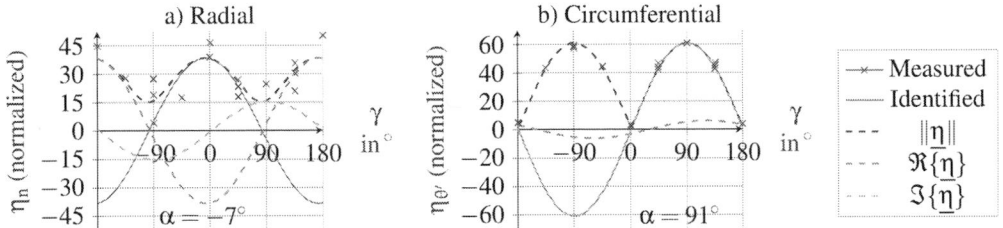

Fig. 5: Result of the excitation direction identification in idle mode ($\hat{i}_{q,0} = 0$) the normal direction (left) and for the circumferential direction (right).

Identification of the Transfer Function

The transfer function identified in this section describes the phase shift (φ_η) as well as the changing in gain if the speed is changed. The measurement is performed similar to the excitation direction identification. For each speed n_m, a measurement with $\hat{i}_k = 0\,\text{A}$ is performed. Then, alternating currents are first injected at the injection angle β_R in order to identify the transfer function for the radial direction. Next, a current at $\beta_{\theta'}$ is injected to identify the circumferential behaviour. For each injection angle, multiple rotational angles of the currents ($\varphi_{i,k}$) are used. Finally, the transfer ratio for the frequency of the current harmonic $f_m = k \cdot n_m$ can be calculated as

$$\underline{G}(f_m) = \frac{1}{\eta_k \sin(\alpha_{\theta'} - \alpha_R)} \frac{1}{N} \sum_{n=1}^{N} \frac{\Delta \hat{a}_{k,n} e^{j\varphi_{a,k,n}}}{\hat{i}_{k,n} e^{j\varphi_{i,k}}}. \tag{11}$$

By determining the transfer ratio at multiple speeds, the transfer functions displayed in Fig. 6 can be derived. The transfer ratios indicate two clear resonances: The first occurs in the radial direction around $f_{R,res} \approx 6021\,\text{Hz}$ and the second occurs in the circumferential direction at $f_{R,res} \approx 780\,\text{Hz}$. In the proximity of these frequencies, the transfer function can be approximated by a second-order transfer function

$$\underline{G}_{approx} = \frac{k s^2}{s^2 + 2D\omega_{res}s + \omega_{res}^2} e^{j\varphi_{offset}} \tag{12}$$

with $\omega_{res} = 2\pi 6021\,\text{Hz}$, $D = 0.03$ and $\varphi_{offset} = 90°$ for the radial resonance ($\underline{G}_R$) and $\omega_{res} = 2\pi 780\,\text{Hz}$, $D = 0.03$ and $\varphi_{offset} = 12°$ for the circumferential resonance ($\underline{G}_{\theta'}$).

Calculation of the Compensation Currents

Using both the identified injection angles and the known transfer ratios (using local approximations of (11)), the compensation currents can be calculated. An iterative process is proposed: First, an accelera-

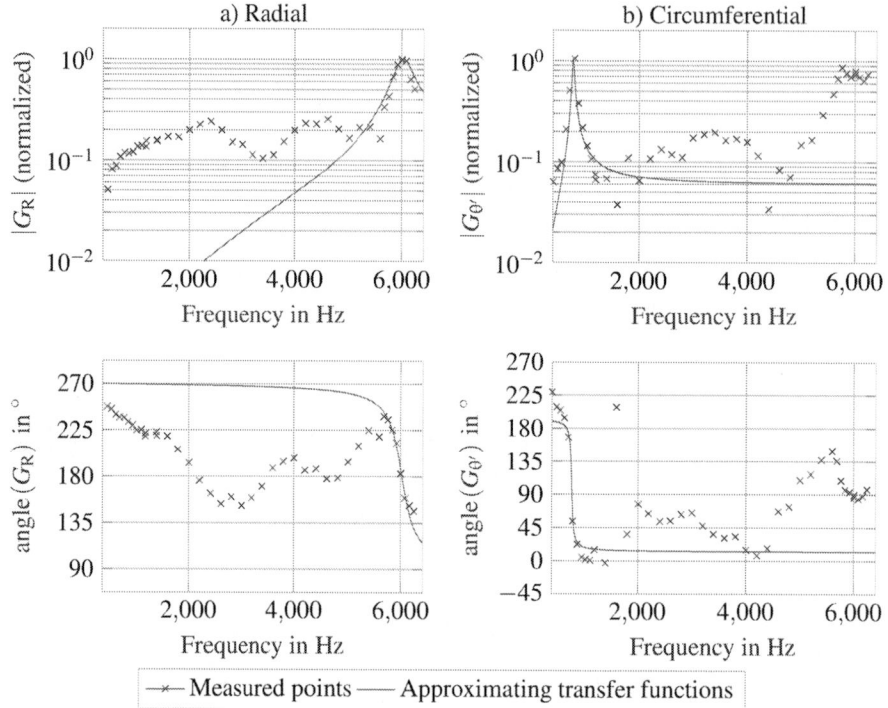

Fig. 6: Result of the frequency response identification. (Outlier $|G_{\theta'}| = 374.5$ at $1400\,\text{Hz}$ outside of scope)

tion with $\hat{i}_k = 0\,\text{A}$ is measured. This can then be used to calculate a compensation current as

$$
\begin{bmatrix} \hat{\underline{i}}_{d,k} \\ \hat{\underline{i}}_{q,k} \end{bmatrix} = - \begin{bmatrix} \cos(\beta_R) \\ \sin(\beta_R) \end{bmatrix} \frac{1}{\underline{G}_R(f)} \frac{1}{\eta_{k,R}\sin(\alpha_{\theta'} - \alpha_R)} \hat{\underline{a}}_{0,k,R}(\hat{i}_k = 0\,\text{A})
$$
$$
\quad - \begin{bmatrix} \cos(\beta_{\theta'}) \\ \sin(\beta_{\theta'}) \end{bmatrix} \frac{1}{\underline{G}_{\theta'}(f)} \frac{1}{\eta_{k,\theta'}\sin(\alpha_{\theta'} - \alpha_R)} \hat{\underline{a}}_{0,k,\theta'}(\hat{i}_k = 0\,\text{A}).
$$

(13)

As the equation reveals, compensation currents for both the radial and the circumferential direction can be calculated. However, by neglecting one part, only the radial or circumferential part can be compensated independently. Furthermore, the compensation is performed iteratively by using the residual acceleration after the first compensation to calculate an additional compensation current. The sum of the pre-existing and the additional current is used to achieve a better compensation.

Results of the Compensation Currents

In order to prove the effectiveness of the compensation, the 6-sensor ring is used for the compensation, while the 28-sensor ring is used to prove the effectiveness of the approach. By using independent sensors to confirm the effectiveness, it can be ensured that actual accelerations and not measurement errors are compensated. In the following, the compensation results are presented for two speeds under load. During idle operation, the acceleration of the example machine, especially for the radial direction, is already small without compensation. For each speed, the iterative compensation is performed for both directions and only for the radial and circumferential direction, respectively. The whole experiment is performed twice.

Fig. 7 illustrates the compensation at a speed of $1000\,1/\text{min}$. At this speed, the rotational order $k = 12$ excites the circumferential mode 0 vibration resonance frequency. This leads to high accelerations in the circumferential direction. Since no radial direction resonance frequency is excited, the accelerations are rather moderate. The graph illustrates that for the circumferential acceleration compensation, a reduction of about 1/5th is achieved. Furthermore, the figure proves that the case of a radial compensation does practically not affect the circumferential acceleration. The radial acceleration measurements unfortunately do not display the same level of improvement as the circumferential compensation. Especially, the case of only circumferential compensation does not display any improvement. In order to explain the lack of compensation effect, the sensor measurements are analysed in more detail in Fig. 8. The figure indicates the amplitude of each individual sensor, the amplitude of the mode 0 vibration and the residual. For the radial case, it can be seen that the radial mode 0 vibration is not the dominant mode excited in the machine. Thus, slight deviations in the measurement of the 6-sensor ring and the 28-sensor ring lead to different amplitudes of the mode 0 vibration. Therefore, an improvement in the measured mode 0 vibration of the 6-sensor ring does not necessarily lead to an improvement in the 28-sensor ring. For the radial case, it is the other way round: Here, the mode 0 vibration is dominant. Thus, small deviations between the two sensor rings have a minor effect leading to an overall good compensation.

Fig. 9 illustrates the compensation at 7500 1/min. Compared to the case of 1000 1/min, the radial and circumferential direction switch roles: the rotational order $k = 12$ excites the radial mode 0 vibration resonance frequency. This leads to high accelerations in the radial direction. Since no circumferential direction resonance frequency is excited, the circumferential accelerations are rather moderate. As a result, the radial acceleration can be significantly compensated, while the circumferential one does not perform that well for this case. This is again a consequence of the non-dominant mode 0 vibration, as proven in Fig. 10.

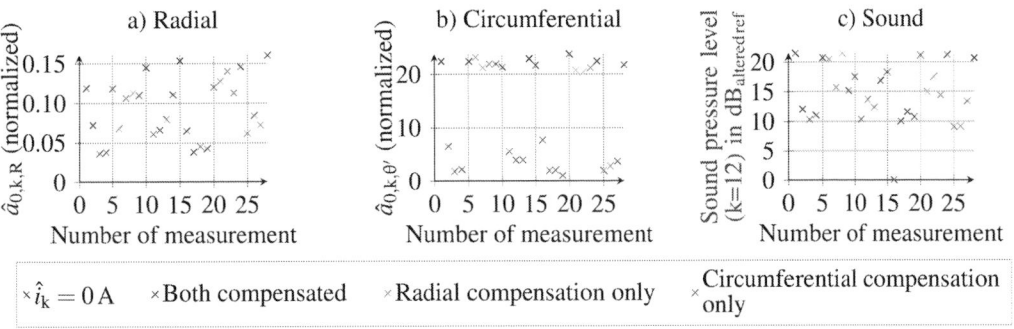

Fig. 7: Radial and circumferential compensation results at a speed of 1000 1/min.

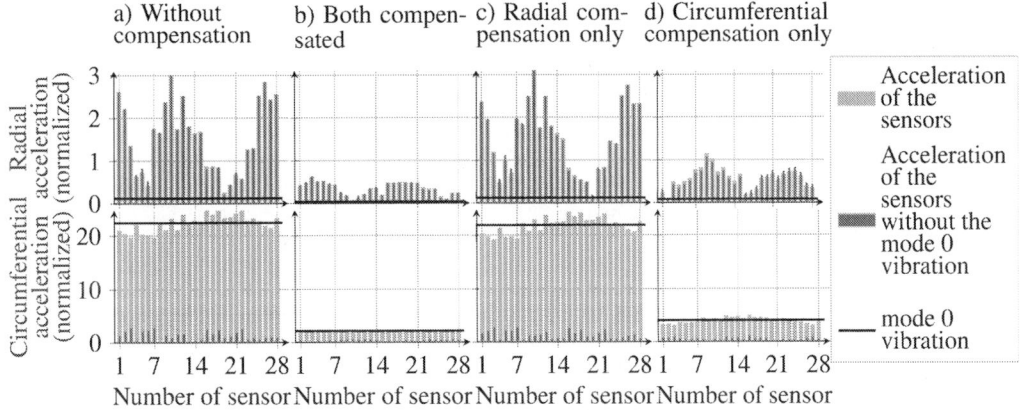

Fig. 8: Details of measurement 1 (without compensation), 4 (both compensated), 8 (radial compensation only), and 13 (circumferential compensation only) measured at 1000 1/min.

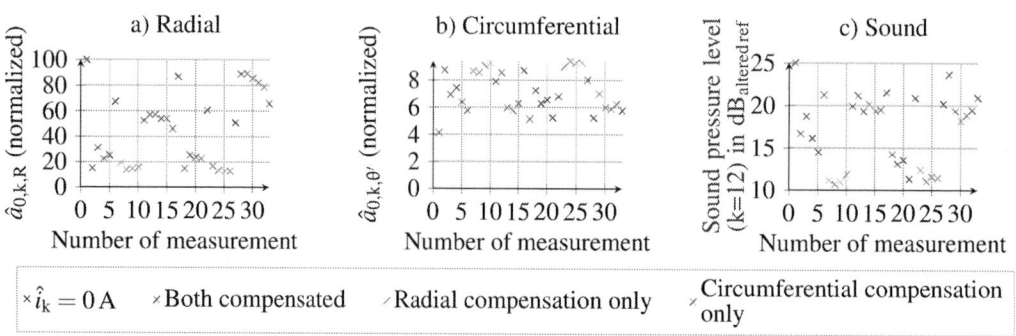

Fig. 9: Radial and circumferential compensation at 7500 1/min.

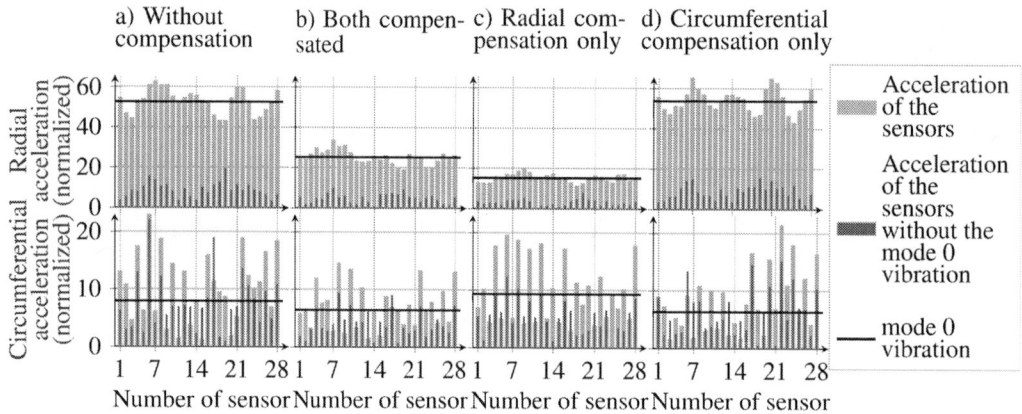

Fig. 10: Details of measurement 11 a) without compensation, 5 b) both compensated, 10 c) radial compensation only, and 15 d) circumferential compensation only measured at 7500 1/min.

As a first indication of the acoustic behaviour, Fig. 7 and 9 show the influence on the sound pressure level of the kth order. The figures demonstrate that the compensation of the dominant mode 0 vibration leads to a reduction in acoustic sound.

Conclusion

A novel approach to compensate radial and circumferential mode 0 vibrations independently has been introduced in this paper. The identified influence of the current and the identified transfer characteristic of the system have successfully been used to calculate compensation currents. The paper has additionally revealed that a compensation only improves the system's behaviour considerably if the respective direction is significantly excited. For the present machine, this is the case for the radial and circumferential mode 0 vibration resonance frequency. The current approach relies on the usage of acceleration sensors. Future work will analyse whether it is possible to calculate the compensation currents using numerical or analytical approaches.

References

[1] Hofmann A., Qi F., Lange T., and de Doncker R. W.: The breathing mode-shape 0: Is it the main acoustic issue in the PMSMs of today's electric vehicles? in Proc. ICEMS, Hangzhou, China, 2014, pp. 3067- 3073.

[2] Klarin B., Knaus K., Schneider J., Diwoky F., Resch T., and Brandl S.: PMSM noise - simulation measurement comparison, SAE Technical Paper Series 2018-01-1552, 2018.

[3] Wang S., Jouvray J.-L., and Kalos T.: NVH technologies and challenges on electric powertrain, SAE Technical Paper Series 2018-01-1551, 2018.

[4] Braunisch D.: Kombinierte analytisch-numerische Berechnung der Magnetgeräusche elektrischer Maschinen, Ph.D. dissertation, Fakultät für Elektrotechnik und Informatik, Leibniz Universität Hannover, Germany, 2015.

[5] Pellerey P., Favennec G., Lanfranchi V., and Friedrich G.: Active reduction of electrical machines magnetic noise by the control of low frequency current harmonics, in Proc. IECON, Montreal, QC, Canada, 2012, pp. 1654- 1659.

[6] Harries M., Hensgens M., and de Doncker R. W.: Noise reduction via harmonic current injection for concentrated-winding permanent magnet synchronous machines, in Proc. ICEMS, Jeju, South Korea, 2018, pp. 1157- 1162.

[7] Franck D., van der Giet M., and Hameyer K.: Active reduction of audible noise exciting radial force-density waves in induction motors, in Proc. IEMDC, Niagara Falls, ON, Canada, 2011, pp.1213- 1218.

[8] Cassoret B., Corton R., Roger D., and Brudny J.-F.: Magnetic noise reduction of induction machines, IEEE Trans. Power Electron., vol. 18, no. 2, pp. 570- 579, Mar. 2003.

[9] Belkhayat D., Roger D., and Brudny J. F.: Active reduction of magnetic noise in asynchronous machine controlled by stator current harmonics, in Proc. 8th Int. Conf. on Electrical Machines and Drives (Conf. Publ. No. 444), Cambridge, UK, 1997, pp. 400- 405.

[10] Kanematsu M., Miyajima T., Fujimoto H., Hori Y., Enomoto T., Kondou M., Komiya H., Yoshimoto K., and Miyakawa T.: Proposal of 6th radial force control based on flux linkage, in Proc. IPEC ECCE-Asia, Hiroshima, Japan, 2014, pp. 2421- 2426.

[11] Liu Y. and Zhang X.: Noise and vibration reduction of direct drive permanent magnet synchronous wind power generator, in Proc. ICEMS, Pattaya, Thailand, 2015, pp. 1257- 1260.

[12] Le Besnerais J.: Réduction du bruit audible d'origine magnétique dans les machines asynchrones alimentées par MLI, règles de conception silencieuse et optimisation multi-objectif, Ph.D. dissertation, l'École Centrale de Lille, France, 2008.

[13] Piippo A. and Luomi J.: Torque ripple reduction in sensorless PMSM drives, in Proc. IECON, Paris, France, 2006, pp. 920- 925.

[14] Mattavelli P., Tubiana L., and Zigliotto M.: Torque-ripple reduction in PM synchronous motor drives using repetitive current control, IEEE Trans. Power Electron., vol. 20, no. 6, pp. 1423- 1431, Nov. 2005.

[15] Mao Y., Liu G., Chen Q., and Zhou H.: Mitigation of acoustic noise by minimize torque and radial force fluctuation in fault tolerant permanent magnet machines, in Proc. ICEMS, Hangzhou, China, 2014, pp. 60- 64.

[16] Nägelkrämer J., Heitmann A., and Parspour N.: Application of dynamic programming for active noise reduction of PMSM by reducing torque ripple and radial force harmonics, in Proc. AEIT, Bari, Italy, 2018, pp. 1- 6.

[17] Klein-Hessling A., Hofmann A., and de Doncker R. W.: Direct instantaneous torque and force control: a control approach for switched reluctance machines, IET Electric Power Appl., vol. 11, no. 5, pp. 935- 943, 2017.

[18] Zhu W., Pekarek S., and Fahimi B.: On the effect of stator excitation on radial and tangential flux and force densities in a permanent magnet synchronous machine, in Proc. IEMDC, San Antonio, TX, USA, 2005, pp. 346- 353.

Measurement Based Model for the Calculation of Current Distributions Between Paralleled Power Semiconductors during High Current Operation

M.Sc. Julian da Cunha
University of Rostock
Albert-Einstein Strae 2
Rostock, Germany
Phone: +49(0) 381 498-7117
Email: julian.cunha@uni-rostock.de
URL: http://www.uni-rostock.de

Keywords

≪Thermal stress≫, ≪Modelling≫, ≪Conduction losses≫, ≪Test bench≫, ≪Power semiconductor device≫.

Abstract

To simulate current sharing between parallel power semiconductors and their temperatures during surge current under the circumstance of intensity and starting temperature variation, a new method is proposed. With a set of surge current measurements with varying starting temperatures, a high temperature thermal model and the temperature and current dependent on-state voltage are extracted.

Introduction

Simulating the current sharing between paralleled semiconductors in surge current operation requires the simulation of the temperature of each semiconductor with their temperature dependent conduction characteristics. An example for this operation might be the freewheeling of the resonant absorption circuit as a result of a short circuit. The resulting current will commutate into the freewheeling diode of a breaking unit on the one hand, which further is much colder and not sufficiently dimensioned in comparison to the parallel diodes of the inverter phases. This method though is not limited to diodes. IGBTs can be described as well which can be beneficial when simulating the balancing of a short circuit as shown in [1] and [4], where IGBTs will conduct the surge current in parallel. The temperatures which are reached during surge current operation clearly exceed the limit of what can be applied statically for conventional output characteristic measurements as shown in [3].

Multiple approaches have been published recently with the goal to monitor a semiconductors surge current temperature behavior. In [3] a method was demonstrated, calculating a Cauer-model from semiconductor parameters and dimensions. The drawback of this approach is the required knowledge of detailed parameters which have to be provided by the manufacturer as described by the author.

In [2] the device is heated in a surge current up to a certain point and the temperature is thereafter estimated by the help of a measurement pulse, evaluating a temperature sensitive semiconductor parameter. The downsides of this method are that such a parameter has to be monitored additionally with a more complex test setup and therefore its temperature characteristic has to be known. Though this approach can be used to estimate the quality of the proposed fitting.

The two previously mentioned methods either have the disadvantage to rely on the knowledge of certain semiconductor parameters, only available from the device's manufacturer, or need a complex test setup besides the surge current generator.

To bypass those additional demands, the method presented in this paper can be applied. For this method, only a number of surge current measurements with varying starting temperatures and shapes need to be conducted.

Under the assumption that the conduction voltage for a given current depends only on the temperature, a thermo-electrical conduction characteristic can be extracted. For this, a thermal model needs to be fitted to the measured conduction losses of different surge currents to meet the given assumption above.

Fundamental idea

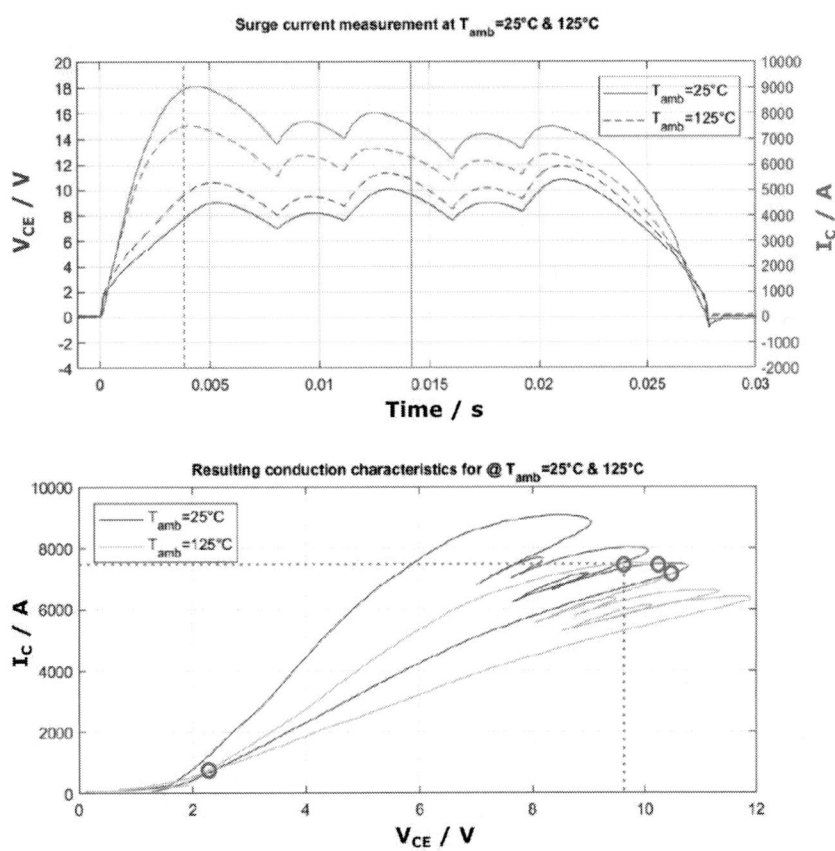

Fig. 1: Example for intersections between different surge current measurements of an IGBT

In Fig. 1 two measurements, that differ only with respect to initial ambient temperature and have been conducted in the surge current test bench, are displayed. Due to the more complex thermal dependency of the on-state voltage due to the desaturation behaviour, the proposed fitting method was developed and tested with diode and IGBT measurements as used for Fig. 1. The test-bench setup itself will be shown in the following section. In the upper plot of Fig. 1, the collector-current as well as the collector-emitter-voltage for both measurements are displayed. The lower plot is illustrates the same measurements, with collector current plotted in dependence to V_{CE}. Inside such a conduction characteristic, lower junction-temperatures lead to a steeper trajectory and a higher ambient temperature further causes an increased V_{CE} for same collector-currents.

The fundamental assumption to form the basis of the presented method of computing the thermal model from measurements says that for a given current, every value of V_{CE} can be assigned to a specific and rising temperature. Following this, every intersection in the V_{CE}-I_C plot of Fig. 1 can be assumed as the same chip temperature. Due to the different ambient temperatures and dissipated energies of the two intersecting measurements, the time passed just to this operating point within one curve differs. In Fig. 1

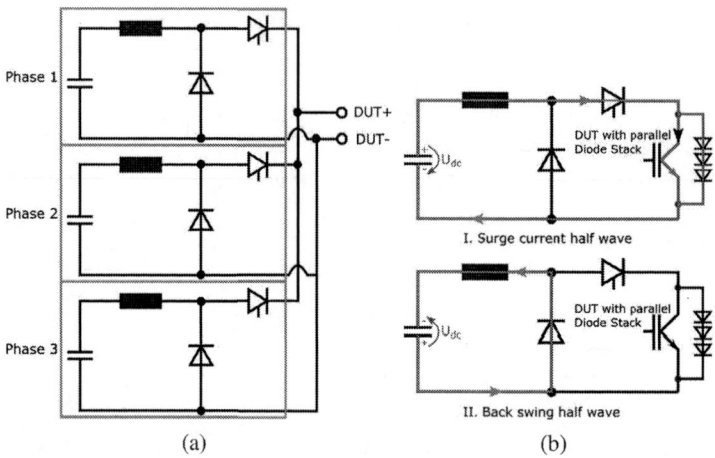

Fig. 2: (a)Setup of the used surge current test-bench (b)generation of the surge current inside of phase of the test-bench and back swing of the current

the intersections are exemplary displayed with red circles. Such a simplification only applies, where the influence of a slow plasma generation inside the semiconductor on V_{CE} on the one hand and the inductive voltage drop across parasitic inductances due to a steep current slope on the other hand can be neglected due to a high surge current pulse length. The example in Fig. 1 displays in the upper plot in green those two points in time where the temperature during the 25°C measurement after 14ms is the same as the temperature during the 125°C measurement after 4ms.

Test Setup

The test-bench, used to generate the surge currents as shown in Fig. 1 is displayed in Fig. 2(a). The basic surge current setup consists of a dc-link capacitor, an inductor, a thyristor and a diode. When the thyristor is ignited, the energy stored inside the resonant circuits capacitor leads to a current through the inductor, the thyristor and the DUT as depicted in Fig. 2(b). After all the energy has been transfered to the coil, the inductor current charges the capacitor inversely until the current is zero again and the thyristor is cleared. A diode stack is further installed in parallel to the DUT to clamp the maximum forward voltage by bypassing excess current prior to the desaturation of the IGBT and thus avoid destruction during surge current. Afterwards a negative current recharges the capacitor with help of the diode and the inductor into the initial condition as further shown in Fig. 2(b).

The test-bench in total features 3 equal surge current phases which can be ignited independently. In Fig. 1 all three phases have been used, while triggered with a 8ms time delay. This leads to the characteristic 5 pulse shape of the current, where the second and fourth pulse are summations of the previous and the following phase currents with a single pulse length of 11ms.

Temperature model fitting and 3D-V_{CE} curve

$$M(t) \circ\!\!-\!\!\bullet M(s) = \frac{K}{T^2 \cdot s^2 + 2 \cdot D \cdot T \cdot s + 1} \tag{1}$$

$$Pv(s) \cdot M(s) \bullet\!\!-\!\!\circ Pv(t) * M(t) = \int_{-\infty}^{+\infty} Pv(t-\tau) \cdot M(\tau) d\tau \tag{2}$$

After the surge current measurements have been conducted and all the necessary intersection points are found, it needs to be defined which thermal model should be assumed. This can be a cauer or foster

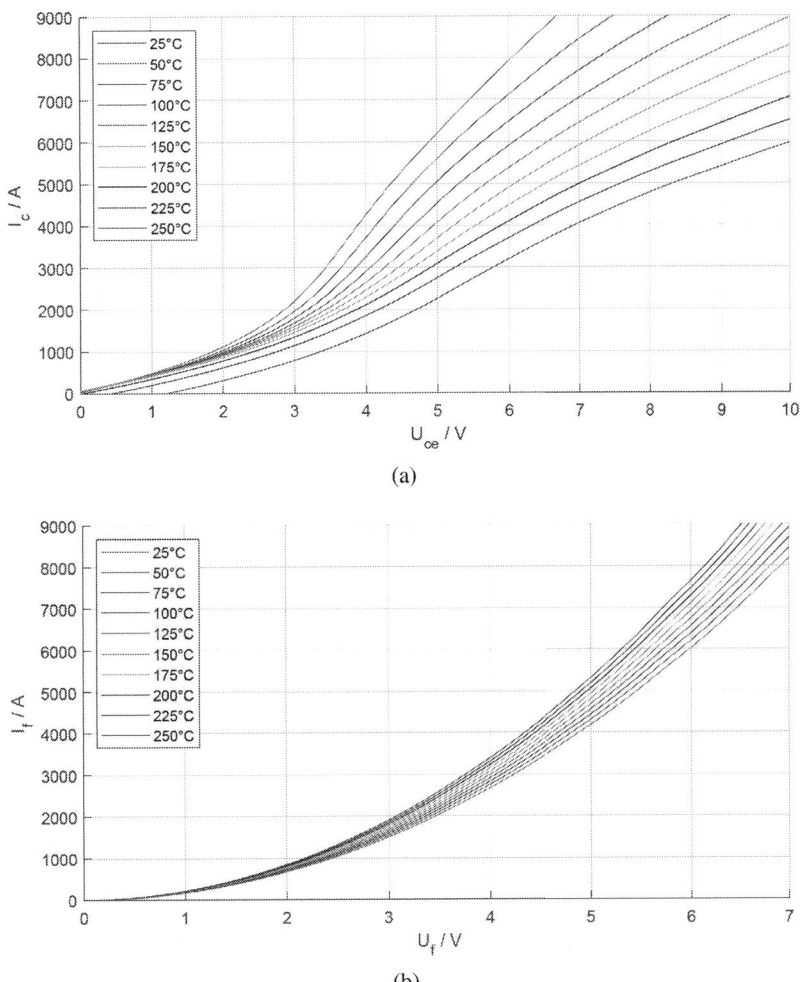

(a)

(b)

Fig. 3: 3D Conduction Characteristic Simplified into constant temperature lines (a) of the measured IGBT (b) of the measured diode

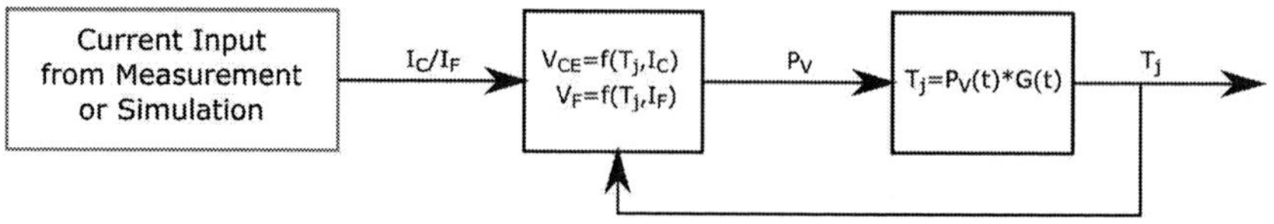

Current Input from Measurement or Simulation — I_C/I_F → $V_{CE}=f(T_j,I_C)$ $V_F=f(T_j,I_F)$ — P_V → $T_j=P_V(t)*G(t)$ — T_j →

Fig. 4: Schematic presentation of the process of simulation with the thermal model and the 3D-characteristic

model but also a simple function such as a PT2 characteristic, with its transfer function displayed in Eq.1, for instance. A more detailed model with more different time constants might increase the resolution of the model for lower currents, which is not needed for surge current simulations. Such a simple thermal model has the upside of accelerating the fitting process and showed to be sufficient for the measured diodes and IGBTs. By convolving the PT2 impulse response in time domain with the curve of the dissipated electrical power as shown in Eq. 2, the temperature development can be obtained. By doing this for both measurement of the same intersection point, an error value can be generated from the difference between both temperatures at this point. Fitting the PT2 until the smallest possible total error in all intersection points is reached will give the thermal model of the semiconductor.

After having found the best fitting thermal model, the temperature curve can be simulated for every single measurement and added as a third dimension to the current and ON-state voltage curves to create a three-dimensional-conduction characteristic. Those further called 3D-conduction characteristic are now being interpolated to generate the collector emitter voltage as a function of the device current and junction temperature inside the semiconductor. Extracting the lines of constant temperature from this interpolated functions can give a set of curves as shown in Fig. 3(a) and 3(b) which are based on the 3D-conduction-characteristic displayed as the interpolated functions $V_{CE}=f(I_C,T_j)$ and $V_F=f(I_F,T_j)$ for the two measured devices.

With both, the thermal transfer function as well as the 3D conduction characteristic, a simulation process for temperature dependent ON-state voltage as shown in Fig. 4 can be build which is taking self heating into account. With a current I_C defined by the simulation and a starting temperature, the voltage V_{CE} can be estimated with the help of the 3D-conduction-characteristic depicted in Fig. 3(a). The resulting conduction-losses inside the semiconductor will hence lead to a shift in the junction temperature which can be calculated by convolving with the generated thermal model of the semiconductor at surge current operation.

With the updating of the momentary junction temperature one simulation cycle ends. Beginning with the following simulation cycle, the updated temperature will lead to a changed ON-state voltage calculated by the 3D-characteristic. The change of the forward-voltage will influence conduction losses and consequently the development of the junction temperature during this second cycle. Now the just explained process will start again with the corresponding device current and the newly calculated junction temperature.

To evaluate, whether the device would survive the simulated surge current it is further necessary to obtain a number of destructive measurements. With such measurements the limitations of device operations can be found. In simulations of complete converters with configurations of paralleled devices and different starting conditions, utilizing such temperature estimations and ON-state characteristics, devices with critical temperatures can be identified.

First measured modules and simulative verification

To verify the simulations, the two 3,3kV, 1500A single switch IGBT modules have been tested together in the test-bench shown in Fig. 2. One operated as an IGBT and the other module in forward direction of the anti-parallel diode. The resulting on-state characteristics are shown in Fig. 3. After those individual characterization surge currents, further measurements for verification purposes have been conducted where both characterized devices have been operated in parallel during surge current, for V_{CE} to equal V_F. To test the generated models, a verification based on those parallel measurement has been conducted.

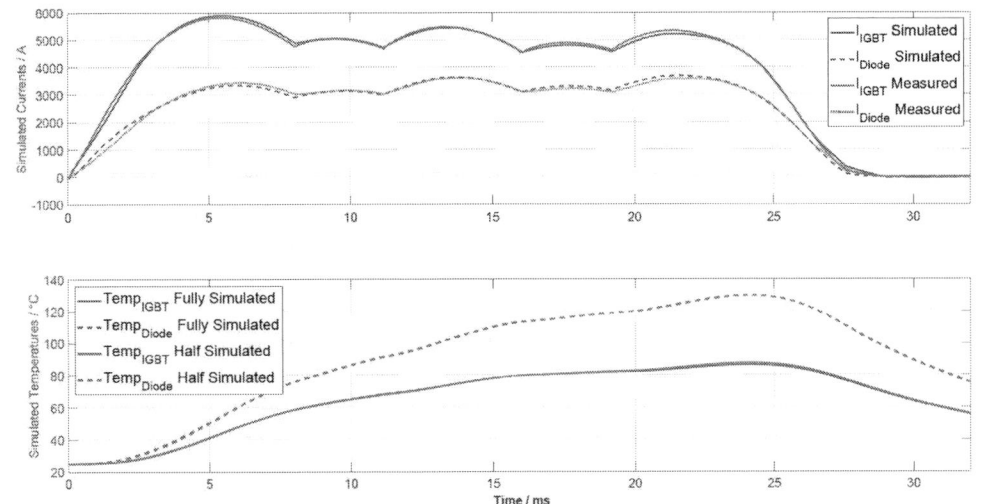

Fig. 5: Top: Measured and simulated current sharing between two parallel devices @T_j=25°C/ Bottom: Parallel temperature behaviour calculated from measured and simulated current sharing @T_j=25°C

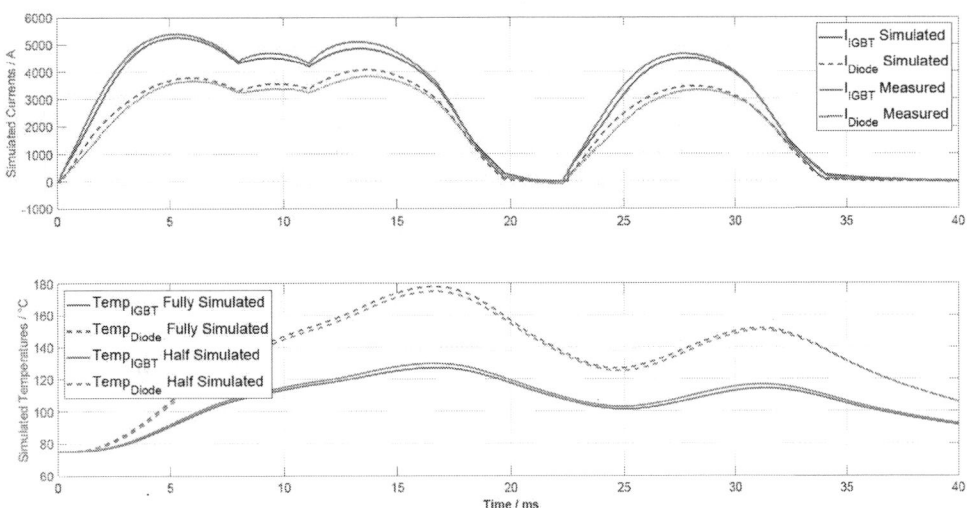

Fig. 6: Top: Measured and simulated current sharing between two parallel devices @T_j=75°C/ Bottom: Parallel temperature behaviour calculated from measured and simulated current sharing @T_j=75°C

Therefore the measured current sharing between the devices will be compared with a current sharing that origins in the boundary condition mentioned before, $V_{CE}=f(I_C,T_j)$ equals $V_F=f(I_F,T_j)$. Only the starting temperatures of the devices and the sum of the device currents will be provided as input. The procedure by which those results where generated will be described closely in the following section.

In the upper plot of Fig. 5 the measured and simulated current sharing is displayed for a starting temperature of 25°C. It shows decreasing IGBT currents and increasing diode currents for increasing junction temperatures. The deviations of ΔI_C and ΔI_F between measurement and simulation account for less than 4%.

In the lower plot of Fig. 5 the temperature curves of both devices are displayed. Here the temperature simulation based on the measured device currents (half simulated) is compared to the temperature simulation based on the simulated device currents (fully simulated) as they have all four been displayed in the upper plot of Fig. 5. The procedure to receive those simulations is described in the following section as well. A very small error of less than 1% confirms the quality of the simulation and hence of the modelling method further.

Figure 6 displays a second parallel measurement in comparison with its simulated curves similar to the procedure shown in the previous figure. The initial ambient temperature hereby has been increased to 75°C. Additionally it can be seen that the current in the parallel connection goes to zero for a short moment, before a second surge current is generated. This pulse pattern allows to check whether the transfer function used for the thermal modelling is able to reproduce thermal cool down phases. In comparison to Figure 5 an increased error can be seen between the measured and simulated current curves as well as the different temperature curves.

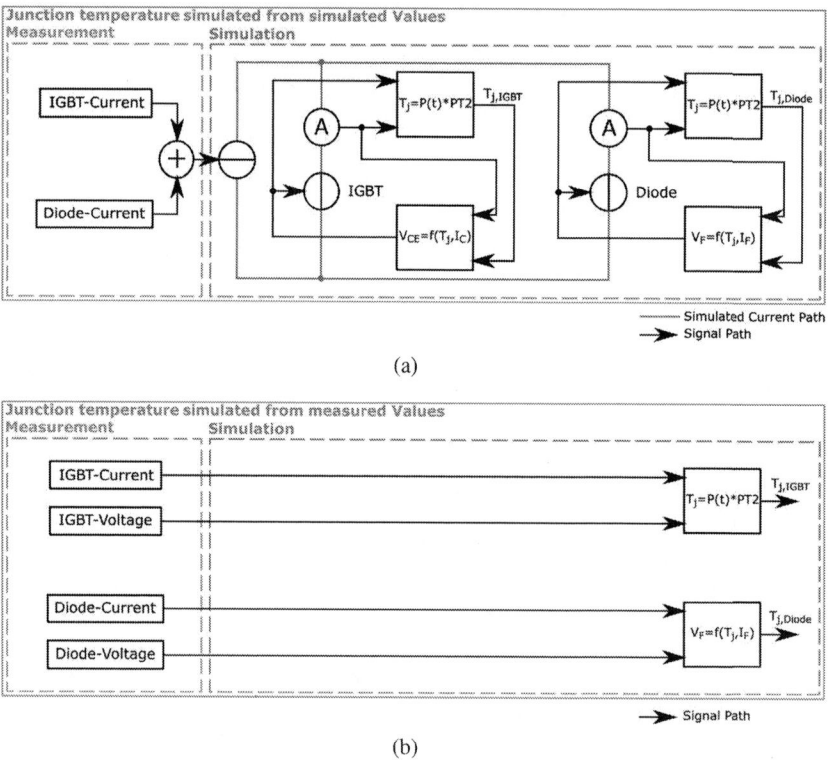

(a)

(b)

Fig. 7: Schematic setup (a) ..of the process of the parallel simulation (b) ..of the reference simulation which is based on the measured current sharing and voltage (*half simulated temperature* is calculated from measured electrical losses convoluted with thermal model)

Simulation procedure for parallel devices

Fig. 7(a) shows the principle of the closed loop simulation for the paralleled modules. The measured parallel currents are summed up and a current source is used to generate a current pulse combined current pulse to be inserted in the simulative electrical circuit drawn in red. The current in each path is measured and used together with the individual device temperature to calculate the device voltage equations $V_{CE}=f(I_C,T_j)$ and $V_F=f(I_F,T_j)$. With the help of a controlled voltage source, those voltages are fed back into the circuit and thereby force the current to split up until the voltages equal each other. The currents which have been displayed in the upper plot of Fig. 5 therefore are the measured currents in the left of Fig. 7(a) and the two parallel currents simulated in the right of Fig. 7(a).

To qualify the temperature estimation of the simulation based on a simulated current sharing and simulated voltages, the temperature for both devices was estimated a second time, this time using measured current and voltage values with the estimated thermal models according to $V_{CE}=f(I_C,T_j)$ and $V_F=f(I_F,T_j)$ as described in Fig. 7(b). Those results have beed displayed in the lower plot of Fig. 5.

Summary

A method to extract the thermo-electric model as well as the temperature dependent on state voltage and current characteristic of a semiconductor for high current and high temperature situations was presented. The main upside of this method is that only surge current measurements are needed and no additional observation of any measurable semiconductor parameter nor knowledge about the internal device structure is needed. Solely a computational fitting process has to be conducted. Further it was demonstrated how a surge current behavior can be simulated for a given current pulse even in a parallel connection.

References

[1] J.-E. Colasse , A. Delporte, L. Masselus. SYMETRISATION OF ASYMETRICAL SHORT-CIRCUITS IN IGBT DRIVES. *EPE 2001*.

[2] Y. Du. A Measurement Method to Extract the Transient Junction Temperature Profile of Power Semiconductors at Surge Conditions. *ECCE 2017*.

[3] P. Hofstetter and M.-m. Bakran. Predicting Failure of SiC MOSFETs under Short Circuit and Surge Current Conditions with a single Thermal Model. *EPE 2018*.

[4] J. Kowalsky and M. Geske. Surge Current Behaviour of Different IGBT Designs Behaviour of IGBTs at Surge Current Events. *PCIM 2015*.

Dual-loop Control Scheme with Optimized Type-III Controller based on Genetic Algorithm for 6-phase Interleaved Converter in Electric Vehicle Drivetrains

Dai-Duong Tran[a,b], Sajib Chakraborty[a,b], Thomas Geury[a,b], Joeri Van Mierlo[a,b],

Mohamed El Baghdadi [a,b], Omar Hegazy[a,b*]

[a] Vrije Universiteit Brussel, Pleinlaan 2, 1050 Brussels, Belgium
[b] Flanders Make, 3001 Heverlee, Belgium

* Corresponding author: omar.hegazy@vub.be (O.Hegazy)

Acknowledgments

We acknowledge Flanders Make for the support to our research group.

Keywords

«Non-isolated interleaved DC/DC converter», «Optimization algorithm», «Small-signal average model», «Genetic algorithm», «Field-programmable gate array».

Abstract

This paper presents an optimization procedure using a genetic algorithm (GA) for the dual-loop control scheme based on type-III controllers in a 6-phase interleaved converter. Four different objective functions (i.e. integration of squared and absolute errors and their time-weighted variants) are selected as the fitness functions of the GA. Based on the optimization procedure, step and dynamic responses of current and voltage controllers are investigated for those objective functions in terms of overshoot and settling time. In this study, another approach called classical 'k-factor' is utilized as a benchmark for the proposed type-III controller. Furthermore, two experimental tests are performed on a 60kW SiC-based prototype to verify the controller performances. The obtained results have demonstrated that the optimized controller design exhibits better dynamic response compared to the conventional counterpart.

Introduction

Multiphase DC/DC converters (e.g. 3-phase [1], 4-phase [2], 8–phase [3], from 16-phase [4] up to 36-phase [5]) have been used in electric vehicle (EV) and plug-in hybrid electric vehicle (PHEV) drivetrains. In such high-power applications, non-isolated (bidirectional) interleaved topologies are gaining strong attention thanks to their simplicity, high reliability, and economical total cost of ownership (TCO). By applying different approaches (such as coupled inductors [1], interleaving control [6], or multi-device [7]), the total input current ripples and power losses can be reduced, resulting in high power density for these converters. Specifically, Figure 1 shows the configuration of a hybrid energy storage system (HESS). In this case, a multiphase DC/DC converter is used to manage power flow between a Li-ion battery (LiB) and a Li-ion capacitor (LiC) [8].

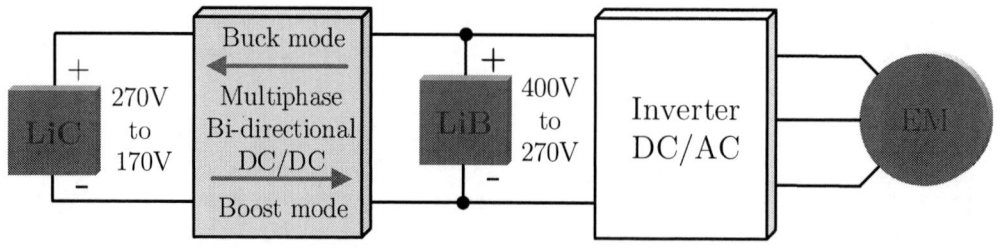

Figure 1. Multiphase DC/DC converter in semi-active configuration for HESS in EV drivertrain.

Using this multiphase concept, this paper focuses on a 6-phase converter topology as shown in Figure 2. It is highlighted that the 6-phase converter is reconfigured from two three-phase interleaved converters, which could work as two separate ports in a multiport converter (MPC) developed in [9]. The specifications of the 6-phase converter are provided as follows: rated power P_O =60kW, switching frequency f_{sw} =60kHz, output voltage V_O =400V, input voltage V_{in} =170V-270V, $L_1 = L_2 = L_3$ =175µH, $L_4 = L_5 = L_6$ =145µH, $C_{in} = C_{out}$ =160µF. This reconfigurable approach, also known as automated and multifunctional converters [7], can improve the modularity and scalability, which results in a significant reduction in cost, volume, and weight for the EV drivetrains. The modified 6-phase prototype can work bidirectionally in buck and boost modes.

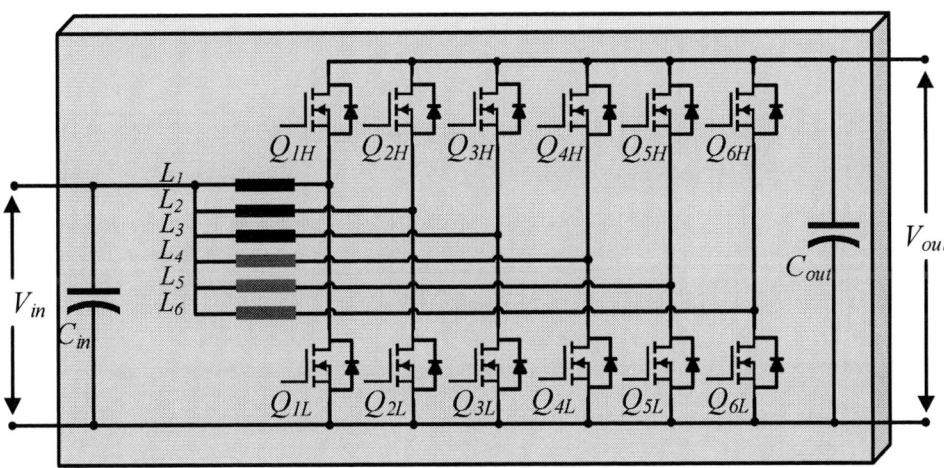

Figure 2. Non-isolated 6-phase interleaved converter reconfigured from a multiport converter in [9].

Regarding control design as the main focus of this paper, the design of the boost mode controller has more challenges compared to the buck mode counterpart. This is due to a non-minimum phase problem caused by a right-half plane zero (RHPZ) [10]. Therefore, the boost converter suffers an undesirable undershoot when the output load increases. In the literature, there exist various types of controllers for boost based converters [11]. It can be seen that the proportional-integral (PI) and lead-lag controllers have provided slow dynamic responses when output load changes. Compared to those conventional controllers, a type-III controller can improve the dynamic performance of the boost converter under the line and load change.

Traditionally, the 'k-factor' approach [11] has been used to determine double poles, double zeros, and a control gain in the transfer function of the type-III controller. Those control parameters can be optimized further by optimization techniques such as particle swarm optimization (PSO), gravitational search algorithm (GSA) [10], and queen-bee genetic algorithm (QBGA) [12]. However, in [12], only voltage-mode control was considered to compensate for the output voltage, which may reduce the robustness and overall stability of the converter. Recently, limited reports have addressed the applicability of a genetic algorithm (GA) for both type-III current and voltage controllers in a dual-loop control strategy related to the interleaved boost converters (IBCs).

Thus, the main contributions of this paper are as follows: (*i*) an optimization procedure based on the GA is proposed to optimize the control parameters for both current and voltage controllers nested in a dual-loop control scheme; (*ii*) four different objective functions, known as optimization criteria, are selected for the minimization of control cost, and their dynamic performances are analyzed using step responses of closed-loop current and voltage controllers; (*iii*) two experimental test campaigns including static testing (40kW, V_{in} =250V, V_O =395V) and dynamic testing with output load disturbance are performed to validate the proposed control design.

Dual-loop Control Scheme and Control-Oriented Modeling

In this paper, the 6-phase converter is reconstructed from two three-phase interleaved converters by reconfiguring the connections of the liquid-cooling MPC which has been developed in [9]. It is noticed that three out of six phases have ideally the same inductance value $L_1 = L_2 = L_3 = 175\mu H$, whereas the other three phase-legs connect to three inductors $L_4 = L_5 = L_6 = 145\mu H$. Due to the inductance differences, a dual-loop control strategy as shown in Figure 3 is implemented to regulate the output voltage and balance the inductor currents between phases. Voltage controllers (i.e. C_{v_P1} and C_{v_P2}) and current controllers (i.e. C_{i_P1} and C_{i_P2}) are designed for phase legs connecting to $L_{1,2,3}$ and $L_{4,5,6}$, respectively. The phase shift is equally 60 degrees ($360 / N_{ph}$), which is realized by delay blocks between 60kHz carriers in the PWM blocks.

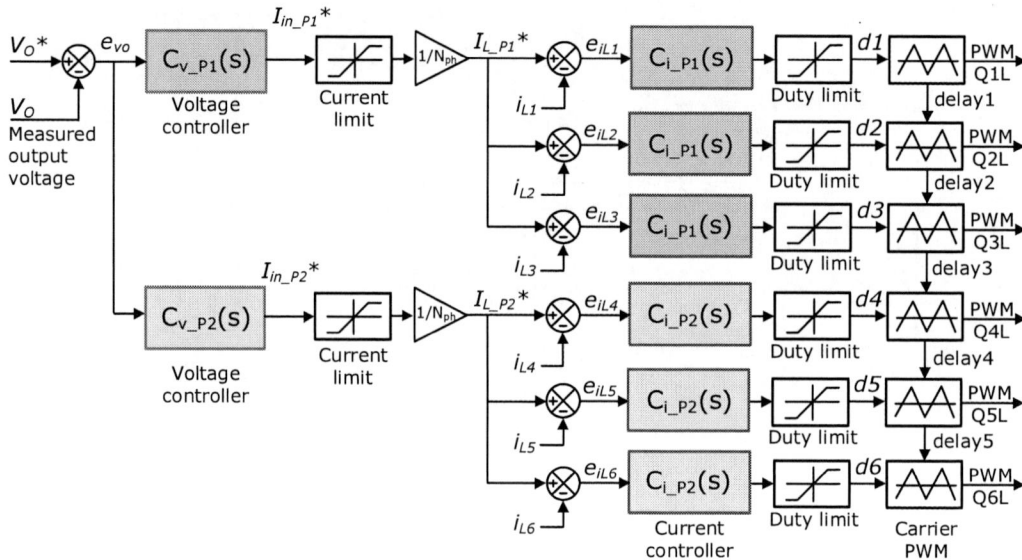

Figure 3. Dual-loop control strategy for 6-phase interleaved boost converter.

In this paper, both C_i and C_v adopt the type-III controller that is a lead-lead controller with one pole at the origin and two other poles and two zeros. The transfer function of type-III controllers are described as in (1) and (2). The main purpose is to optimize the position of the poles and the zeros as well as the control gains by using the GA.

$$C_i(s) = K_{Ci} \cdot \frac{\left(s + z_{1(Ci)}\right)\left(s + z_{2(Ci)}\right)}{s\left(s + p_{1(Ci)}\right)\left(s + p_{2(Ci)}\right)} \tag{1}$$

$$C_v(s) = K_{Cv} \cdot \frac{\left(s + z_{1(Cv)}\right)\left(s + z_{2(Cv)}\right)}{s\left(s + p_{1(Cv)}\right)\left(s + p_{2(Cv)}\right)} \tag{2}$$

The control-oriented plant models are developed using small-signal model averaging techniques [2]. The transfer functions of duty cycle-to-inductor current $G_{id}(s)$ and duty cycle-to-output voltage $G_{vd}(s)$ are derived in continuous conduction mode (CCM) and summarized in Table I. Where C is output capacitance, R_C is capacitor internal resistance, L is boost inductance, R_L is inductor internal resistance, N_{ph} is a number of phases, D is nominal duty ratio, and R_O is load resistance.

Table I. Transfer functions of $G_{id}(s)$ and $G_{vd}(s)$.

	Duty cycle-to-inductor current		Duty cycle-to-output voltage	
Transfer function	$G_{id}(s) = \dfrac{\tilde{i}_L(s)}{\tilde{d}(s)} = G_{di}\dfrac{1 + \dfrac{s}{\omega_{zi}}}{\Delta}$	(3)	$G_{vd}(s) = \dfrac{\tilde{v}_o(s)}{\tilde{d}(s)}$ $= G_{dv}\dfrac{\left(1 + \dfrac{s}{\omega_{zv\_ESR}}\right)\left(1 - \dfrac{s}{\omega_{zv\_RHP}}\right)}{\Delta}$	(4)
Open-loop gain	$G_{di} = \dfrac{2V_O}{R_L + N_{ph}(1-D)^2 R_O}$	(5)	$G_{dv} = \dfrac{V_O}{1-D}$	(6)
Zero	$\omega_{zi} = \dfrac{1}{C\left(R_C + \dfrac{R_O}{2}\right)}$	(7)	$\omega_{zv\_ESR} = \dfrac{1}{C \cdot R_C}$ $\omega_{zv\_RHP} = \dfrac{-R_L + N_{ph}(1-D)^2 R_O}{L}$	(8)
The denominator in (3) and (4)			$\Delta = \dfrac{s^2}{\omega_0^2} + \dfrac{s}{Q \cdot \omega_0} + 1$	(9)
Nature frequency ω_0			$\omega_0 = \sqrt{\dfrac{R_L + N_{ph}(1-D)^2 R_O}{L \cdot C(R_O + R_C)}}$	(10)
Quality factor Q			$Q = \dfrac{1}{2\zeta}$	(11)
System damped ratio ζ			$\zeta = \dfrac{L + C\left[R_L(R_O + R_C) + N_{ph}(1-D)^2 R_O \cdot R_C\right]}{2\sqrt{L \cdot C(R_O + R_C)\left[R_L + N_{ph}(1-D)^2 R_O\right]}}$	(12)

Objective Functions for Control Optimization

Generally, control cost is known as the effort of the controller to drive the actual signal towards the reference control signal. The dynamic error signal $e(t)$ between actual and reference control signals along the transient time should be minimized to obtain a desired dynamic response under load disturbance. In this study, a step response is used to evaluate the dynamic performance of the closed-loop of current and voltage controllers. The optimization problem is formulated as in (13), where the variable vector X involves two poles, two zeros, and control gain of a type-III controller.

$$\min_{X \in \Omega} Objfnc \tag{13}$$

In this study, four integral performance indices can be considered for $Objfnc$ including Integral of Squared Error (ISE) as in (14), Integral of Absolute Error (IAE) as in (15), Integral of Time-weighted Squared Error (ITSE) as in (16) and Integral of Time-weighted Absolute Error (ITAE) as in (17). The upper limit τ is chosen as the steady-state value.

$$ISE = \int_0^\tau e^2(t)dt \tag{14}$$

$$IAE = \int_0^\tau |e(t)| dt \tag{15}$$

$$ITSE = \int_0^\tau t \cdot e^2(t)dt \tag{16}$$

$$ITAE = \int_0^\tau t |e(t)| dt \tag{17}$$

The optimized current and voltage controllers according to the aforementioned objective functions are determined using an optimization flow chart as described in the next section.

Optimization Flow Chart and Simulation Results

The GA is a population-based algorithm that is biologically inspired by natural evolution operators such as selection, crossover, and mutation. The well-established principle of the GA can be found in [13]. Figure 4 illustrates an optimization procedure based on the GA to obtain a set of 5 control parameters (i.e. control gain K_C, two zeros z_1, z_2, and two poles p_1, p_2) for current and voltage type-III controllers, respectively.

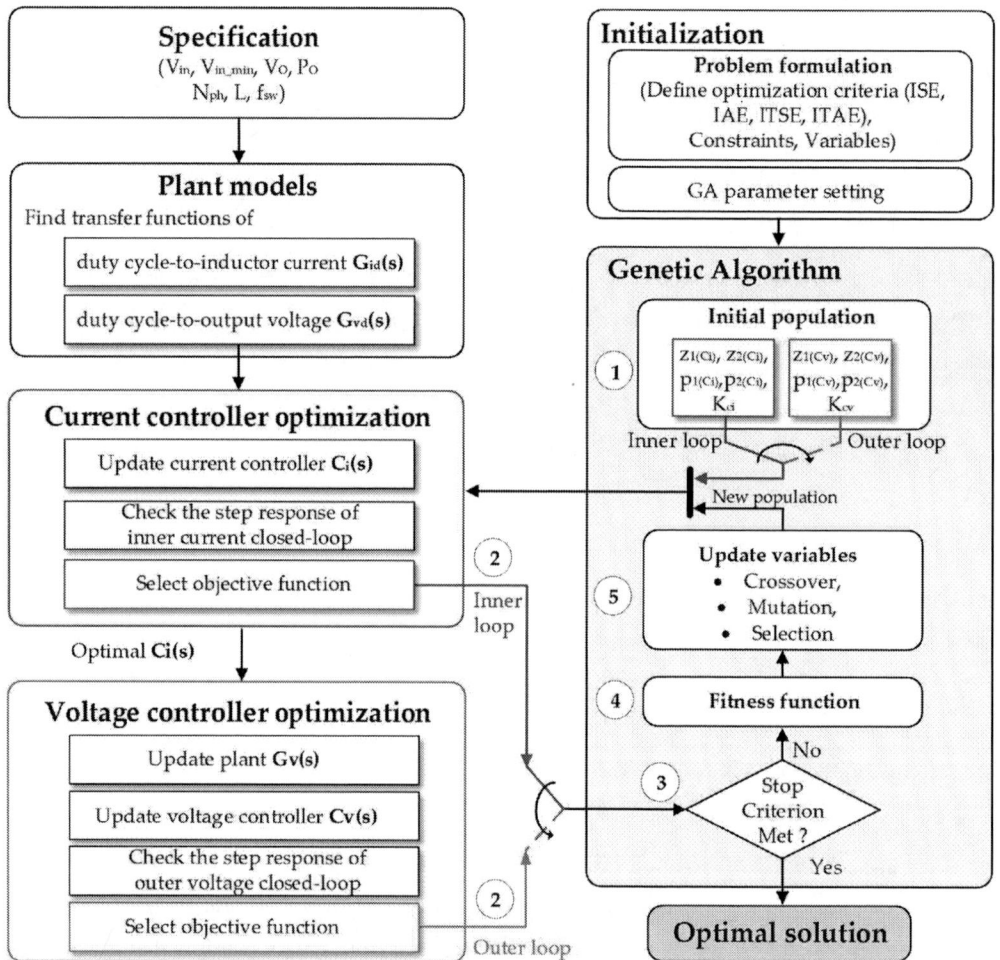

Figure 4. Optimization flow chart based on GA.

In principle, the optimization procedure based on GA includes five steps. In the first step, an initial population with each chromosome representing the design variables is initiated to feed into either current or voltage controller optimization blocks. In the second step, by assigning the design variables into the type-III transfer function and checking the step response, objective function (i.e. fitness value) for the entire population can be calculated. In the third step, the maximum generation as the stop criteria are checked. If the stop criteria are not met, a new population will be generated by GA operators (i.e. crossover, mutation and selection). The output of the current controller optimization is an optimal current controller $C_i(s)$ that is fed into the optimization loop of voltage controller. It is noted that in the dual-loop control architecture, the plant transfer function $G_v(s)$ for the outer voltage loop is determined as (18).

$$G_v(s) = \frac{C_i(s) \cdot G_{vd}(s)}{1 + C_i(s) \cdot G_{id}(s)} \tag{18}$$

Figure 5(a) and Figure 5(b) demonstrate the GA optimization process for current and voltage controllers, respectively, using the ITAE-based objective function. The optimization process is terminated when its generation reaches the predefined maximum number of 20 generations. It can be observed that ITAE-based fitness value can quickly converge to minimum value after 5 generations. The transfer functions, maximum overshoot and settling time of controllers using four different objective functions in (14)-(17) and a benchmark design using the 'k-factor' approach are tabulated in Table II.

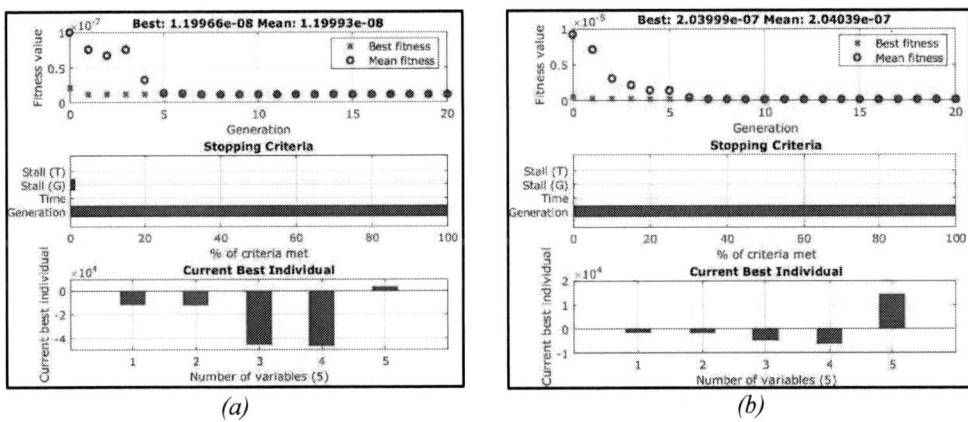

| (a) | (b) |

Figure 5. GA optimization evolution with ITAE criteria for (a) current and (b) voltage controllers.

Table II. Comparative performance of current and voltage controllers.

Controller	Objective functions	Controller transfer function	Maximum overshoot	Settling time
Current controller (inner loop)	Benchmark design	$\dfrac{725.7 \cdot (s + 5717.2)^2}{s(s + 43157)^2}$	1.01	1ms
	ISE	$\dfrac{988.67 \cdot (s + 7039)(s + 8396)}{s(s + 43490)(s + 37940)}$	1.1	0.2ms
	IAE	$\dfrac{1027.5 \cdot (s + 8085)(s + 7711)}{s(s + 26870)(s + 22400)}$	1.25	0.2ms
	ITSE	$\dfrac{905.68 \cdot (s + 8143)(s + 7825)}{s(s + 26660)(s + 35140)}$	1.05	0.2ms
	ITAE	$\dfrac{768.73 \cdot (s + 6696)(s + 8162)}{s(s + 29040)(s + 32200)}$	1.15	0.2ms
Voltage controller (outer loop)	Benchmark design	$\dfrac{3763.3 \cdot (s + 571.7)^2}{s(s + 4315.7)^2}$	0	6ms
	ISE	$\dfrac{3175.2 \cdot (s + 695.7)(s + 779.3)}{s(s + 2287)(s + 3305)}$	1.3	2ms
	IAE	$\dfrac{3036.2 \cdot (s + 832.1)(s + 823.6)}{s(s + 3741)(s + 2343)}$	1.28	2ms
	ITSE	$\dfrac{2665.8 \cdot (s + 533.3)(s + 845.7)}{s(s + 2470)(s + 2661)}$	1.27	2ms
	ITAE	$\dfrac{2665.8 \cdot (s + 749)(s + 841.5)}{s(s + 2189)(s + 3403)}$	1.2	2ms

Figure 6 shows comparative step responses of the controllers based on four different objective functions. As can be seen from Figure 6(b), the ITAE penalizes the error voltage signal along with a transient duration. As a result, the error signal is forced to reduce to zero as soon as possible. Hence, the ITAE has the smallest overshoot and fastest settling time in the step response for the voltage controller.

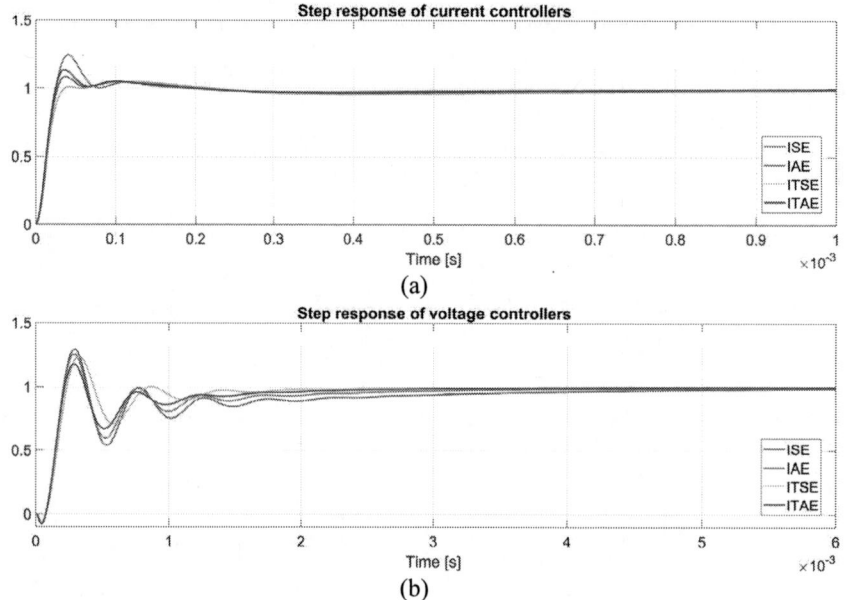

(a)

(b)

Figure 6. Comparative step response between 4 objective functions for (a) current and (b) voltage controllers.

To compare the solutions generated by the algorithms, the controller transfer functions in Table II are placed into a high-fidelity Matlab Simulink model developed in [14] using the SimPowerSystems toolbox. The 6-phase interleaved model is simulated over a time of 0.15s considering V_{in} =200V, V_O =400V, and other converter specifications. When the converter operates in a steady-state condition at P_O =30kW, the output load increases double at t=0.05s, and decreases back to the initial state at t=0.1s.

Figure 7 plots the comparative dynamic responses of output voltage according to four different objective functions. As can be seen in this figure, ISE, IAE, ITSE, and ITAE show a slight difference in the dynamic response of output voltage.

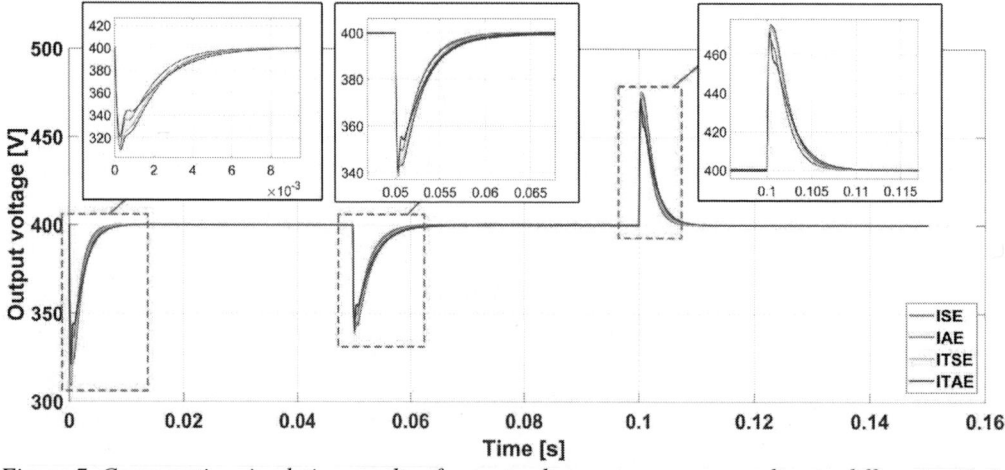

Figure 7. Comparative simulation results of output voltage responses according to different optimized controllers under output load disturbance.

Figure 8 shows the comparison of step response between the optimized type-III controller using ITAE and the conventional type-III controller using the 'k-factor' approach. It is noted that in both step responses of current and voltage controllers, the optimized design has a smaller settling time, which is 5 times faster compared to the classical 'k-factor' design.

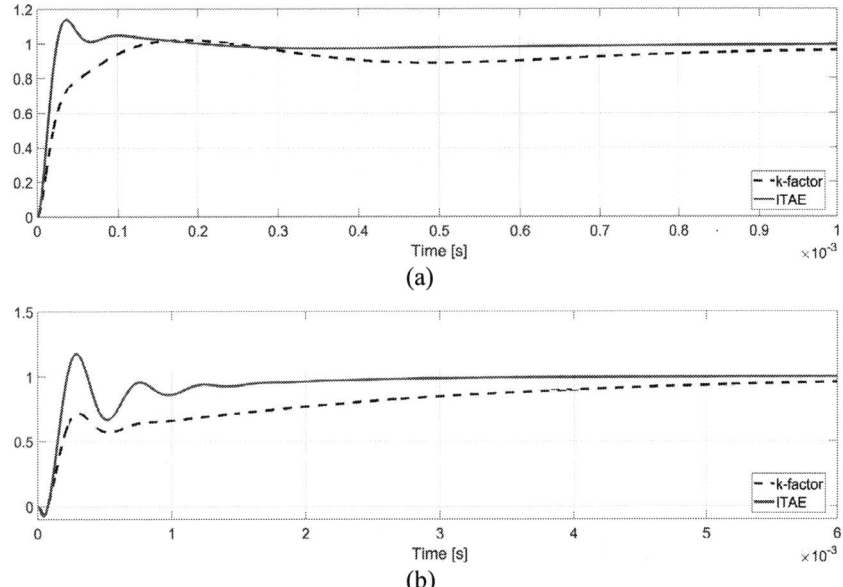

(a)

(b)

Figure 8. Comparative step response between ITAE-based optimal design and conventional 'k-factor' approach for (a) current and (b) voltage controllers.

Experiment Results

The experimental test bench has been set up in our laboratory as shown in Figure 9. A Xilinx Kintex-7 XC7K325T FPGA board (10ns period clock) embedded in the dSPACE MicroLabBox is used to implement the digital controllers based on the redesign approach. An interface circuit has been designed to measure the feedback signals obtained from LEM current and voltage sensors (<1% error). The converter efficiency is acquired from power analyzer YOKOGAWA W1804E (0.02% error). Two testing scenarios are considered. First, static testing is conducted to measure the converter efficiency and verify the proposed dual-loop control scheme. Second, dynamic testing is carried out to compare the controller performance between the optimized design and a conventional design based on the 'k-factor' approach.

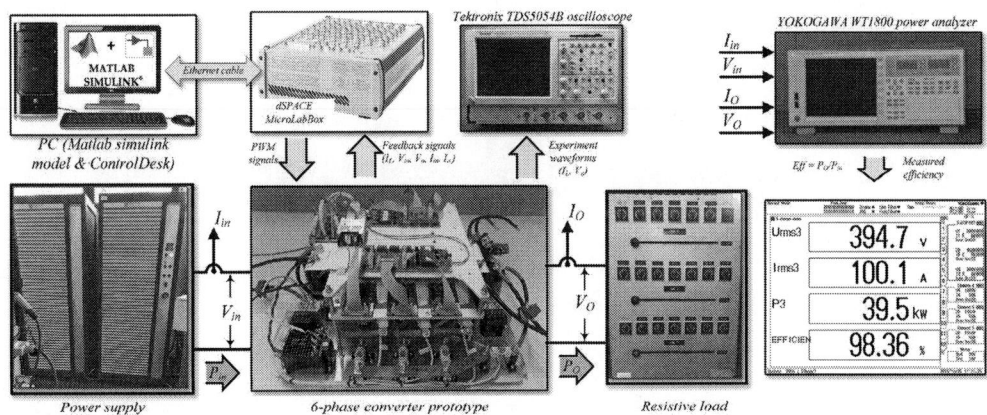

Figure 9. Experimental setup for 6-phase interleaved boost converter.

As shown in Figure 9, the 6-phase converter can achieve high efficiency of 98.4% at the static testing condition V_{in}=250V, V_O=395V, P_O=39.5kW. In this testing, the inductor currents are balanced and shifted 60 degrees between each phase (see Figure 10), which confirms the proposed dual-loop control scheme. The current controllers (i.e. C_{i_P1} and C_{i_P2}) in the inner loop can regulate equally the inductor currents to reduce the total input current ripple. In the dynamic testing, positive and negative step loads are made varying the output power from 6kW to 12kW and vice versa, respectively, with V_{in} =250V, V_O=385V. As shown in Figure 11, the output voltage controllers (i.e. C_{v_P1} and C_{v_P2}) in the outer loop can regulate a constant V_O=385V regardless of output load disturbance. Small undershoot and overshoot in the output voltage V_O can be observed during transient periods when the output current increases instantaneously to 31A and it decreases to 15.5A, respectively.

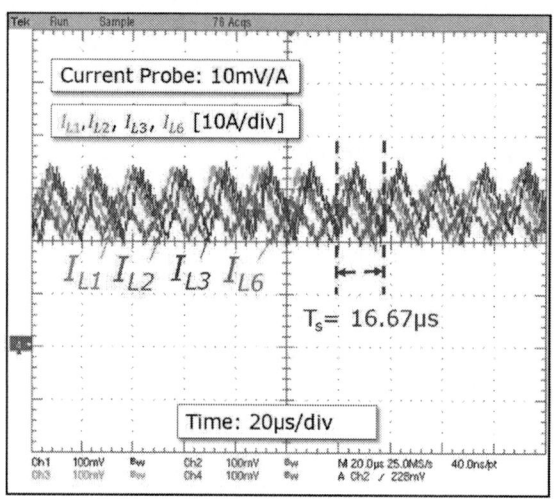

Figure 10. Static experimental waveforms of interleaving inductor currents ($V_{in} = 250V$, $V_O = 395V$, $P_O = 40kW$, $f_{sw} = 60kHz$).

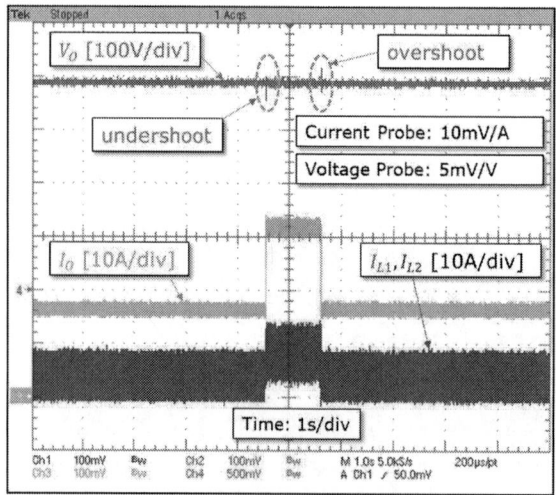

Figure 11. Dynamic experimental waveforms of inductor currents, output current, and output voltage ($V_{in} = 250V$, $V_O = 385V$, $P_O = 6kW \rightarrow 12kW \rightarrow 6kW$).

Figure 12 and Figure 13 show the zoom-in experimental waveforms for the classical 'k-factor'-based controllers and the optimized controllers, respectively. As can be seen, the optimized controller offers better performance in dynamic transient responses for the output voltage. The optimized controller design has less undershoot/overshoot (with 30% reduction) and faster settling time compared to the 'k-factor' counterpart. Under a negative load change as illustrated in Figure 12(b), the 'k-factor' design approach shows an oscillation (during 35ms) in the output voltage before the voltage error reduces to zero. In contrast, the optimized controller can regulate the output voltage within 20ms as shown in Figure 13(b). This means the optimized controller requires less control energy to shape the converter to be stable under the load disturbance.

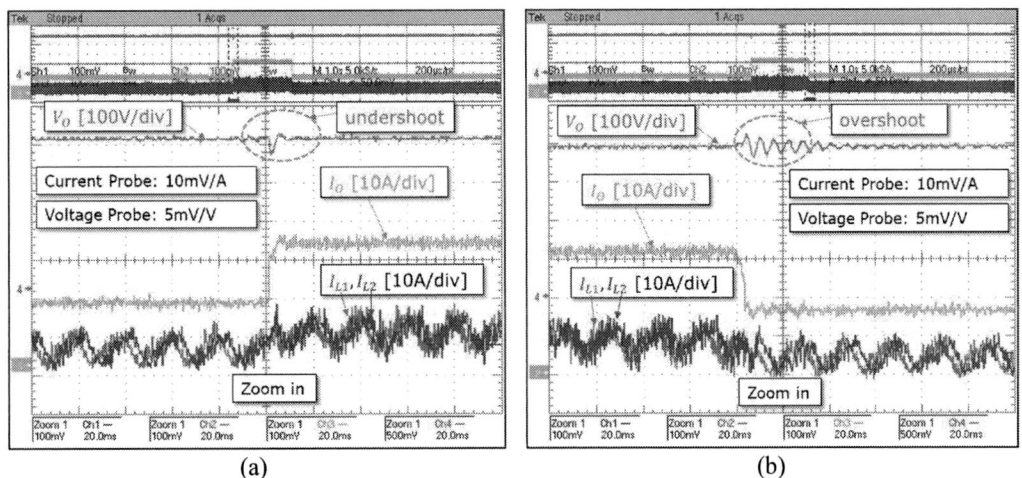

Figure 12. Zoom-in experiment waveforms for conventional 'k-factor' based controller ($V_{in} = 250V$, $V_O = 385V$) when (a) positive and (b) negative load change.

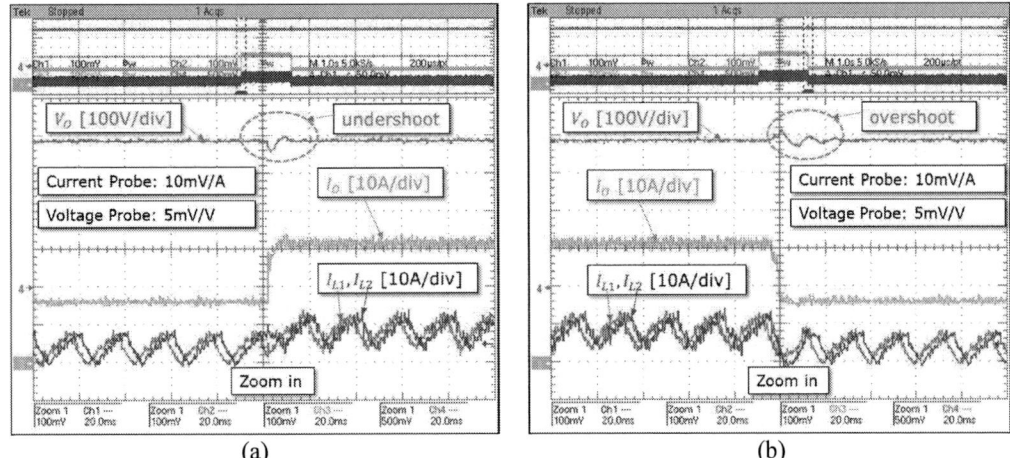

Figure 13. Zoom-in experiment waveforms for optimized type-III controller ($V_{in} = 250V$, $V_O = 385V$) when (a) positive and (b) negative load change.

Conclusion

This paper has presented the optimization procedure based on the GA for type-III controllers, which are used in a dual-loop control scheme for a 6-phase interleaved converter. To this end, four different objective functions called Integral of Squared Error (ISE), Integral of Absolute Error (IAE), Integral of Time-weighted Squared Error (ITSE) and Integral of Time-weighted Absolute Error (ITAE) have been selected as the fitness function of the GA. The converter and its controller have been fully modeled and designed in the Matlab environment. Simulation results of the dynamic response in the output voltage have demonstrated that only minor differences can be observed between the four objective functions. To validate the feasibility of the proposed optimization procedure, the ITAE has been selected to be compared with the classical 'k-factor' approach. Experimental results have shown that the optimal controller has a better dynamic response in the output voltage, resulting in reductions of 30% and 43% in undershoot and settling time, respectively, compared to the conventional 'k-factor' based controller.

References

[1] J. Zhang, J.-S. Lai, R.-Y. Kim, and W. Yu, "High-Power Density Design of a Soft-Switching High-Power Bidirectional dc–dc Converter," *IEEE Trans. Power Electron.*, vol. 22, no. 4, pp. 1145–1153, Jul. 2007.

[2] O. Hegazy, J. Van Mierlo, and P. Lataire, "Analysis, Modeling, and Implementation of a Multidevice Interleaved DC/DC Converter for Fuel Cell Hybrid Electric Vehicles," *IEEE Trans. Power Electron.*, vol. 27, no. 11, pp. 4445–4458, Nov. 2012.

[3] A. Rujas, V. M. Lopez, A. Garcia-Bediaga, A. Berasategi, and T. Nieva, "Influence of SiC technology in a railway traction DC-DC converter design evolution," in *2017 IEEE Energy Conversion Congress and Exposition (ECCE)*, 2017, vol. 2017-Janua, pp. 931–938.

[4] L. Ni, D. J. Patterson, and J. L. Hudgins, "High Power Current Sensorless Bidirectional 16-Phase Interleaved DC-DC Converter for Hybrid Vehicle Application," *IEEE Trans. Power Electron.*, vol. 27, no. 3, pp. 1141–1151, Mar. 2012.

[5] O. Garcia, P. Zumel, A. de Castro, and A. Cobos, "Automotive DC-DC bidirectional converter made with many interleaved buck stages," *IEEE Trans. Power Electron.*, vol. 21, no. 3, pp. 578–586, May 2006.

[6] F. Sobrino-Manzanares and A. Garrigos, "Bidirectional, Interleaved, Multiphase, Multidevice, Soft-Switching, FPGA-Controlled, Buck–Boost Converter With PWM Real-Time Reconfiguration," *IEEE Trans. Power Electron.*, vol. 33, no. 11, pp. 9710–9721, Nov. 2018.

[7] O. Hegazy, R. Barrero, J. Van Mierlo, P. Lataire, N. Omar, and T. Coosemans, "An Advanced Power Electronics Interface for Electric Vehicles Applications," *IEEE Trans. Power Electron.*, vol. 28, no. 12, pp. 5508–5521, Dec. 2013.

[8] P. J. Kollmeyer *et al.*, "Real-time control of a full scale Li-Ion battery and Li-Ion capacitor hybrid energy storage system for a plug-in hybrid vehicle," *IEEE Trans. Ind. Appl.*, vol. 55, no. 4, pp. 4204–4214, 2019.

[9] D. Tran, S. Chakraborty, Y. Lan, J. Van Mierlo, and O. Hegazy, "Optimized Multiport DC/DC Converter for Vehicle Drivetrains: Topology and Design Optimization," *Appl. Sci.*, vol. 8, no. 8, p. 1351, Aug. 2018.

[10] M. Veerachary and A. R. Saxena, "Optimized Power Stage Design of Low Source Current Ripple Fourth-Order Boost DC–DC Converter: A PSO Approach," *IEEE Trans. Ind. Electron.*, vol. 62, no. 3, pp. 1491–1502, Mar. 2015.

[11] A. Ghosh, S. Banerjee, M. K. Sarkar, and P. Dutta, "Design and implementation of type-II and type-III controller for DC–DC switched-mode boost converter by using K-factor approach and optimisation techniques," *IET Power Electron.*, vol. 9, no. 5, pp. 938–950, Apr. 2016.

[12] K. Sundareswaran and V. T. Sreedevi, "Boost Converter Controller Design Using Queen-Bee-Assisted GA," *IEEE Trans. Ind. Electron.*, vol. 56, no. 3, pp. 778–783, Mar. 2009.

[13] K. F. Man, K. S. Tang, and S. Kwong, "Genetic algorithms: concepts and applications," *IEEE Trans. Ind. Electron.*, vol. 43, no. 5, pp. 519–534, 1996.

[14] S. Chakraborty *et al.*, "Scalable Modeling Approach and Robust Hardware-in-the-Loop Testing of an Optimized Interleaved Bidirectional HV DC/DC Converter for Electric Vehicle Drivetrains," *IEEE Access*, vol. 8, pp. 115515–115536, 2020.

High Sensitivity Current Transformer with low Settling Time, for Magnified AC Current Measurements in Pulsed Applications

Georgios Tsolaridis, Pascal Seiler, and Juergen Biela
Laboratory for High Power Electronics, ETH Zurich
E-Mail: tsolaridis@hpe.ee.ethz.ch

Keywords

"Current Measurement", "Current Transformer", "Pulsed Power", "High Sensitivity", "High Precision"

Abstract

In solid-state pulsed power systems based on switching converters often a low amplitude, high frequency ripple current is superimposed on the high flat-top pulse current. Despite its low amplitude (usually measured in ppm) the ripple current could have a major impact on the performance of the system and therefore needs to be quantified/reduced. However, measuring such currents is challenging, due to the limited resolution of the current probe, especially when the DC offset current of the pulse is in the kA range. Therefore a suitable AC current probe able to perform magnified ripple current measurements with high fidelity is proposed in this paper. The probe is based on the principle of current transformers and filters the DC offset of the pulsed current, enabling to measure the ripple with a very high resolution. The design trade-offs of the conventional CT for such an application are explained, and an adaptive burden resistance topology is proposed in order to achieve simultaneously a low settling time, which is essential for measuring short pulses, and the needed high sensitivity, while maintaining a relatively high bandwidth.

1 Introduction

Solid-state pulsed power supplies with strict requirements regarding their dynamic performance and accuracy are required nowadays in a wide range of applications such as particle accelerator systems for fundamental physics research [1], [2], medical applications like cancer treatment [3], and magnetic resonance imaging (MRI), as well as power hardware-in-the-loop systems (P-HiL) [4]. These applications often require a highly precise, high amplitude DC current, and ideally an ultra-low current ripple as shown in Fig. 1b. In many of these applications, interleaved converters, like the one given in Fig. 1a, are an attractive solution, since they can increase the maximum output current while providing a superior output ripple performance.

In the interleaved converter systems, a load current measurement (i_load in Fig. 1a) is typically not necessary for control purposes, and only the module currents $i_\mathrm{L,i}$ are often measured [5]. The total converter current (i_con in Fig. 1a) is then the sum of all module currents, while i_load can be predicted/observed, provided that the parameters of the filter and load are known. The module current sensors are only measuring a fraction of the total converter current i_con, which is typically in the kA range. By using current sensors with a lower amplitude range typically a higher sensitivity and a higher bandwidth current measurement could be achieved, without the need for a special current probe. In this way, i_con can be reconstructed with accuracy and relatively low cost and complexity.

Nevertheless, the ripple of i_load is a crucial parameter for many applications and therefore needs to be accurately defined. Additionally, if it is measured with high precision and low phase-delay, the ripple can be reduced by using an active filter connected in parallel to the load, as shown in Fig. 1a. The importance of such filters in fulfilling high attenuation requirements has been presented in [6], [7]. However, the precise measurement of i_load is very challenging

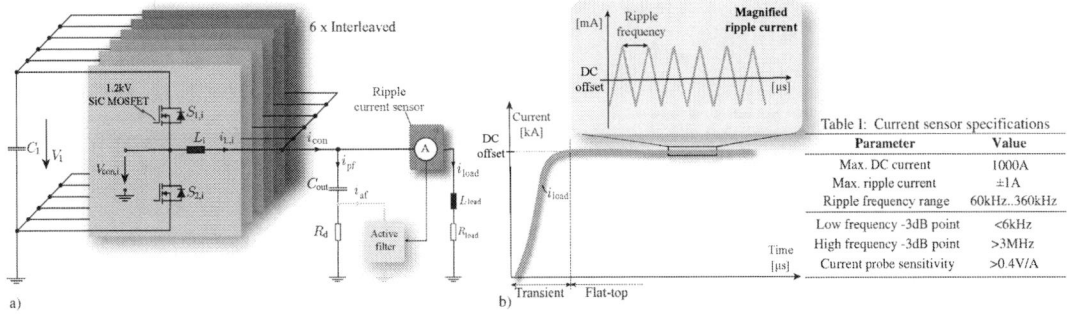

Figure 1: a) Schematic of a high-end current source converter. The converter generates the current waveform shown in b). The load current is comprised of a high DC offset current in the kA range and a high frequency ripple current part in the mA range.

and usually expensive current probes are required [8]. These probes need to deliver high accuracy and wide bandwidth that often ranges from DC up to hundreds of kHz, at high current magnitudes.

Direct-current current transducers (DCCTs) are the preferred technology for high-end applications since they combine excellent accuracy in the ppm range with high bandwidth and current ranges up to tens of kA [9]. Their use, however, is associated with a high cost. On the other hand, commercially available current sensing devices, like Hall sensors, Rogowsky coils, shunt resistors and current transformers (CTs) are usually more affordable but present a trade-off between bandwidth, accuracy and maximum current amplitude range [10]. For instance measuring the current ripple of the exemplary current given in Fig. 1b would be difficult with a conventional sensor due to their limited sensitivity.

In order to measure the ripple of i_{load} with high accuracy, a CT with a special design and an active burden resistance is proposed in this paper. The CT filters the DC offset of the current, which is measured and reconstructed by the sensors of the individual modules (sum of $i_{\text{L,i}}$), and allows the precise measurement of the high frequency current ripple, during flat top only shortly after the transient. In general, the use of a relatively high value for the burden resistance of the CT, increases the sensitivity and reduces the settling time after a pulse, as is revealed in section 2. However, a fast settling time leads to a higher minimum ripple frequency that can be measured without amplitude or phase distortion and therefore reduced bandwidth. In order to avoid this trade-off and achieve a faster settling time, which enables the sensing of the current ripple even in pulses with short flat-tops (e.g. few tens of μs), this paper proposes an adaptive topology that varies the frequency response of the CT during transient and facilitates the measurement of the ripple current with high sensitivity, already shortly after the flat-top of the pulse is reached.

This paper is structured as follows: In section 2, a CT model is presented, the performance trade-offs in terms of frequency response and sensitivity are shown, and the parasitic components that govern the performance are identified. Moreover, an exemplary design procedure is shown. In section 3, a topology with adaptive burden resistance is proposed for improving the transient behavior of the conventional CT. In section 4 the proposed circuit is verified by experimental results based on a designed prototype system that aims to fulfill the specifications listed on Table I. Section 5 shows full scale simulation results of the proposed system. Finally, section 6 summarizes the main outcomes of the paper.

2 Current Transformer Modeling

Detailed models of the CT have been presented in the literature along with experimental verification [11], [12]. In this section, the modeling of the CT is shortly revised and the main equations and design trade-offs are explained. Moreover, the special design considerations that need to be taken into account to fulfill the requirements of the considered application are described, and a parameter selection procedure is shown. A detailed model of the CT is shown in Fig. 2a. Its response is essentially similar to a band-pass filter. The derivation of the complete transfer function $\underline{Z}(s) = v_{\text{b}}(s)/i_{\text{p}}(s)$ does not allow for a simple analytical approach for the design procedure due to its complexity, so a simplified approach is typically followed, by using a low and a high frequency equivalent model.

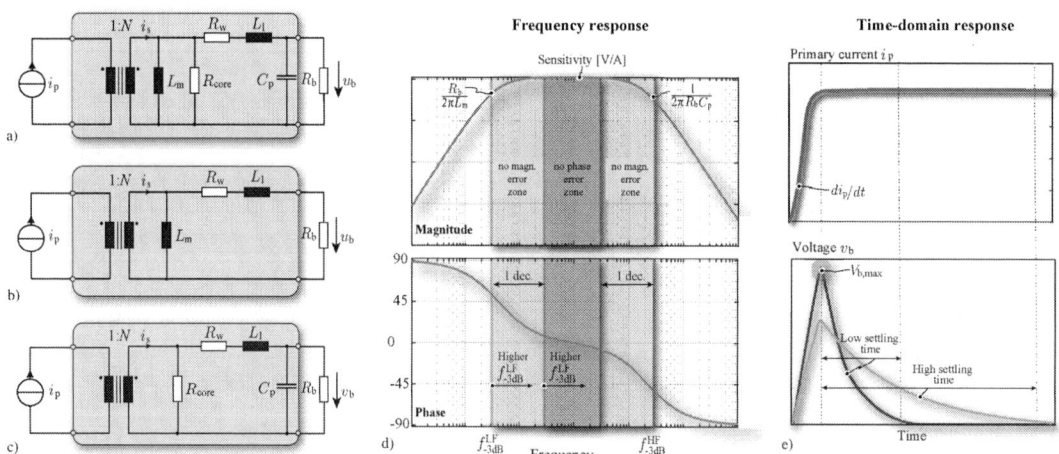

Figure 2: a) Conventional CT model. b) Low frequency equivalent model. c) High frequency equivalent model. d) Frequency response of a CT. e) Time-domain response of a CT to a fast transient (**light gray:** CT with a low $f_{\text{-3dB}}^{\text{LF}}$ and high settling time, **dark gray:** CT with high $f_{\text{-3dB}}^{\text{LF}}$) and low settling time.

2.1 Low Frequency Model

Fig. 2b shows the low frequency model of the CT. In the low frequency range, the core losses and the parasitic capacitance C_p (usually in the pF range), are negligible. The leakage inductance of the secondary winding L_l is usually low compared to the magnetizing inductance and therefore its effect can be neglected. If $R_b \gg R_w$ (which is the case when high sensitivity is required), the winding resistance can also be neglected, in a first approximation. The low frequency response is then given by the transfer function in (1) and the dominant pole is approximately given by (2).

$$Z_{LF}(s) = \frac{v_b(s)}{i_p(s)} = \frac{R_b}{N} \cdot \frac{L_m \cdot s}{(L_m + L_l) \cdot s + (R_w + R_b)} \approx \frac{R_b}{N} \cdot \frac{L_m \cdot s}{L_m \cdot s + R_b} \quad (1) \qquad f_{-3dB}^{LF} \approx \frac{R_b}{2\pi \cdot L_m} \quad (2)$$

Fig. 2e, shows the response of a CT to a pulse current transient. When i_p reaches steady state, the output voltage starts to decrease with a time constant that is determined by f_{-3dB}^{LF}. The voltage $v_{b,rise}$ across the burden resistor of the CT during a rising current input in the time domain is given by (3). The voltage $v_{b,ss}$ during the flat top of the pulsed current is given by (4), where $V_{b,max}$ is the voltage across the burden at the end of the rising edge of the current pulse.

$$v_{b,rise}(t) = \frac{L_m}{N} \cdot \frac{di_p}{dt} \cdot \left(1 - e^{-t \cdot 2\pi \cdot f_{-3dB}^{LF}}\right) \quad (3) \qquad v_{b,ss}(t) = V_{b,max} \cdot e^{-t \cdot 2\pi \cdot f_{-3dB}^{LF}} \quad (4)$$

In many applications, the settling time of the CT is maximized and the low frequency bandwidth is extended. However, in order to characterize and measure the high frequency ripple current with precision for short pulses (often in the μs), the settling time needs to be minimized and the CT is operated as a high pass filter, filtering the DC offset of the current and allowing the measurement of only the high frequency ripple current. In order to do so, f_{-3dB}^{LF} needs to be increased, as shown in Fig. 2e. Nevertheless, increasing f_{-3dB}^{LF} causes a decrease in the bandwidth of the CT, since it increases the minimum frequency that can be measured without error. Fig. 2d shows the frequency range that can be measured without a magnitude error (combination of light blue and dark blue region) and the frequency range that can be measured without a phase error (only dark blue region). Clearly, a higher f_{-3dB}^{LF} would shrink these regions and reduce the frequencies that can be measured with fidelity by the CT.

2.2 High Frequency Model

Fig. 2c shows the simplified high frequency model of the CT, where the magnetization inductance is considered to be an open circuit. Furthermore, neglecting the core losses simplifies the analysis and gives a better insight into the operation of the CT. The core losses can be neglected because in CTs they are usually very low due to the low loss material being used, and especially in the chosen application due to the low amplitude AC current that needs to be measured. Furthermore, the response in the high frequency range is dominated by the parallel combination of R_b-C_p, as reported in [11]. Equation (5) gives the transfer function of the equivalent circuit of Fig. 2c, and (6) gives the dominant pole in the high frequency range.

$$Z_{HF}(s) = \frac{v_b(s)}{i_p(s)} = \frac{R_b}{N} \cdot \frac{1}{R_b \cdot C_p \cdot s + 1} \quad (5) \qquad f_{-3dB}^{HF} = \frac{1}{2\pi \cdot R_b \cdot C_p} \quad (6)$$

Since the burden resistor is chosen based on the needed sensitivity $S = \frac{R_b}{N}$, it becomes evident that the parasitic capacitance of the secondary winding C_p needs to be minimized in order to increase the bandwidth of the CT. In practice, the parasitic capacitance is increasing for an increasing number of turns, due to the inter-winding capacitance. It should be noted that in the high frequency range, the distribution of the secondary winding capacitance and leakage inductance results in a resonant behavior, as discussed shortly in section 4.

2.3 Parameter Selection

The selection procedure is based on the equations (7)-(12), that describe the magnetization inductance L_m, the maximum DC bias of the magnetic flux B_{max}, the permeability μ_r including the drop due to the DC bias, the parasitic leakage inductance L_l, the winding resistance R_w and the core losses P_c. In the following, μ_0 is the permeability of air, μ_r is the relative permeability of the core material, μ_{init} is the initial permeability without DC bias, A_c is the core cross section area, l_e is the magnetic path length, D_w is the winding diameter and l_w is the winding length. A_L is the permeance in [nH/T^2].

$$L_m = \mu_0 \cdot \mu_r \cdot \frac{N^2 \cdot A_c}{l_e} = A_L \cdot N^2 \quad (7) \qquad B_{max} = \mu_0 \cdot \mu_r \cdot \frac{N_p \cdot i_p}{l_e} \quad (8) \qquad \mu_r = \mu_{init} \cdot \frac{1}{(\alpha_p + \beta_p \cdot H^{c_p})} \quad (9)$$

$$L_l = \frac{292 \cdot N^{1.065} \cdot A_c}{l_e} [\mu H] \quad (10) \qquad P_c = \alpha_s \cdot B_{pk}^{\beta_s} \cdot f^{c_s} \quad (11) \qquad R_w = \frac{4 \cdot N \cdot l_w}{\sigma \cdot \pi \cdot D_w^2} \quad (12)$$

The leakage inductance L_l is estimated based on the empirical formula (10) given by the manufacturer of the iron powder cores that are used in this work [13]. Similarly, the core losses which are modeled with an equivalent resistance R_c in Fig. 2, are calculated based on the basic Steinmetz equation (11). Additionally, for the effect of the DC offset, the permeability drop at the maximum considered current offset is taken into consideration, as in (9). Finally, the AC winding losses (i.e. skin and proximity losses) are not taken into consideration, since they are negligible compared to the DC winding losses. The fitting coefficients (α_p, β_p, c_p, α_s, β_s, c_s) for the above models are extracted from the core manufacturer's datasheet.

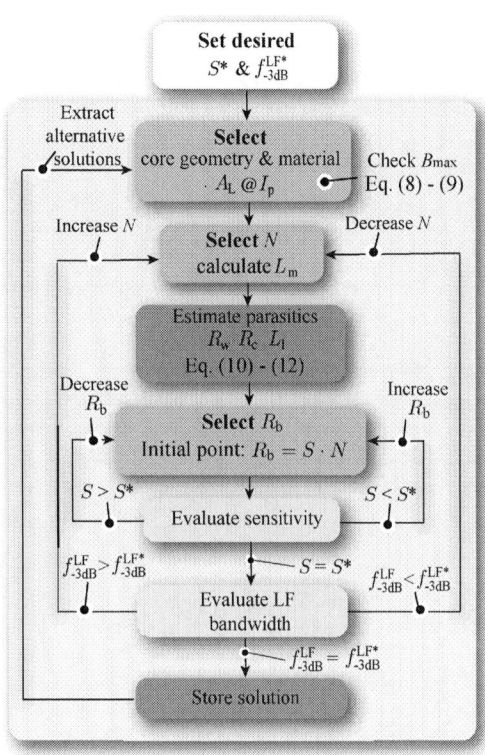

Figure 3: Parameter selection procedure.

The selection procedure for the core geometry, the burden resistance R_b, and the number of turns N, is shown on a flow-chart in Fig. 3. First, the desired sensitivity S^* is selected along with the desired f_{-3dB}^{LF*}. This frequency is selected based on the considerations shown in section 2.1 and the trade-off between the minimum frequency that can be tracked without a phase delay and the allowed settling time for the application. After selecting the core and the material, the procedure checks the maximum DC flux B_{max} based on (8), in order to avoid saturation of the core at the maximum operating DC offset. Then the algorithm iterates N, starting from a low value. For every N, it calculates the magnetization inductance L_m based on (7) and estimates the parasitics R_w, R_c, and L_l based on (10)-(12). Then, it chooses an initial point for R_b based on the set S^* and the simplified LF model. Finally, the complete model's response is derived and the real sensitivity S is compared with S^*. If S is lower than S^* then R_b is increased accordingly and the LF model is re-evaluated. On the contrary, if S is higher than S^* then R_b is decreased. When the needed sensitivity S^* is reached, f_{-3dB}^{LF} is calculated and compared with the set reference f_{-3dB}^{LF*}. If f_{-3dB}^{LF} is higher than f_{-3dB}^{LF*} then N is increased and the procedure is repeated. If f_{-3dB}^{LF} is lower than f_{-3dB}^{LF*} then N is decreased. The final solution is stored when the desired f_{-3dB}^{LF} is achieved. Thereafter, the algorithm selects a different core from a core library (i.e. different size, material, A_L) and repeats the described procedure. It should be noted that multiple alternative solutions exist, depending on the available core permeance.

3 CT with Adaptive Response

In the previous section, the trade-off between minimum measurable frequency and settling time after a transient, is highlighted. In order to improve the performance of the CT and allow it to measure the current ripple shortly after the primary current pulse has settled (during flat-top), an adaptive topology is presented in the following. The topology simply allows to use a high f_{-3dB}^{LF} during the pulse transient, enabling the CT to settle faster, and to use a lower f_{-3dB}^{LF} during the pulse flat-top, in order to extend the bandwidth of the current measurement.

The proposed topology is shown in Fig. 4a and its operational principle in Fig. 4d. The resistance value R_t should be at least an order of magnitude higher than the burden resistance R_b. As a result, R_b dominates and the circuit (Fig. 4a) is equivalent to the circuit of the conventional CT in Fig. 2a, when the bi-directional switch T is on.

When a transient in the primary current is detected, the bi-directional switch T is turned off and only the transient resistor R_t acts as burden resistance. Due to its high value, f_{-3dB}^{LF} is shifted to higher frequencies and therefore the settling time of the CT is reduced, as demonstrated graphically in Fig. 2e. Shortly after the pulse has reached its flat top, T is turned on again, allowing the measurement at the full frequency bandwidth of the current ripple, as shown in Fig. 4e. Additionally, voltage V_{dt} shown in Fig. 4a is measured, to detect the beginning and the end of the transient. The detection circuit is critical for the operation of the topology and its implementation is discussed in section 3.1.

Fig. 4e shows the frequency response of the proposed CT in steady state (light blue) and in transient state (darker blue). It can be seen that the use of a higher resistance R_t significantly increases f_{-3dB}^{LF} and reduces the measurable

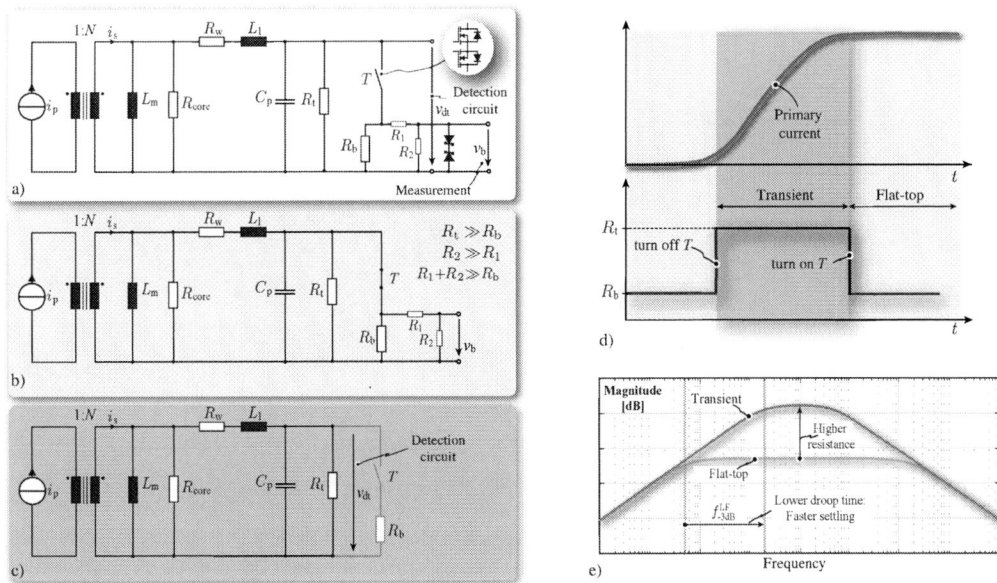

Figure 4: a) Proposed topology for decreased settling time. b) Flat-top equivalent model. c) Transient equivalent model. d) Operational principle of the proposed CT. e) Frequency response during flat-top and during transient.

bandwidth. Moreover, the magnitude gain of the voltage (i.e. the sensitivity) is increased, resulting in a relatively high voltage. To protect the analog components (connected in parallel to R_b), zener clamping diodes are used. However, these components and their parasitics have an influence on the resulting bandwidth of the CT. The complete circuit of the proposed topology is discussed in more detail in section 3.2.

3.1 Transient Detection

Ideally, the current probe should be able to control the bi-directional switch T without the need for external communication interfaces (e.g. master control signal), that would signify for example the start and the end of the transient state and the start of the flat-top. A transient detection circuit is therefore crucial for an autonomous operation of the proposed topology.

To make the operation autonomous, the detection circuit shown in Fig. 5a is implemented. The voltage across the transient burden resistor R_t is constantly measured, through a compensated RC voltage divider. The analog signal is then buffered, filtered with a passive filter and converted to a digital signal with a fast ADC. The output of the ADC is then sent to an FPGA. Based on the measured v_{dt}, the FPGA controls the bi-directional switch T, according to the state machine shown in Fig. 5b.

Starting from steady state, T is turned on. The digital value of the measured voltage v_{dt} is averaged by the use of a moving average, with an average frequency equal to the lowest frequency expected in the system (in the studied system 60kHz) and compared to a threshold value $V_{thr,on}$. If $\bar{v}_{dt} < V_{thr,on}$, the system remains at steady state.

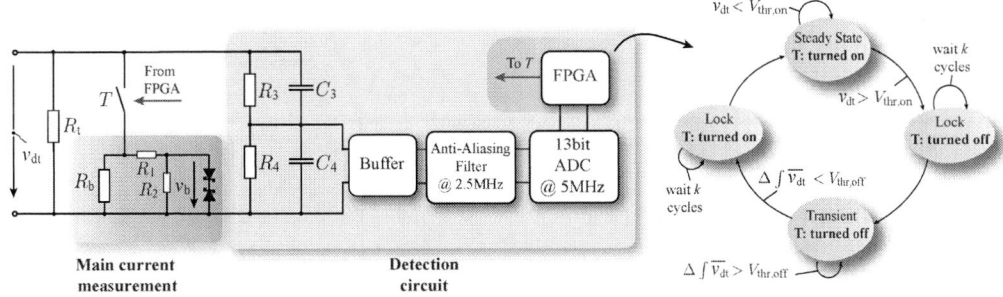

Figure 5: a) Overview of the transient detection circuit. b) State machine of the transient detection algorithm, implemented on the board's FPGA.

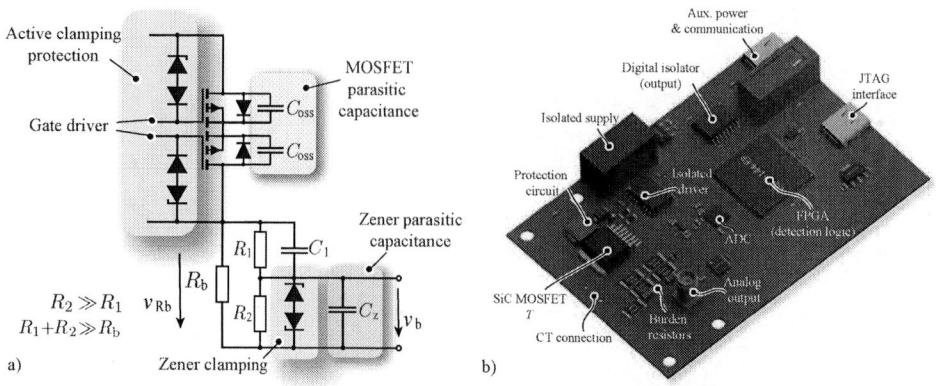

Figure 6: a) Detailed view of the circuit for the main voltage measurement and its parasitics. b) Illustration of the designed PCB.

When a transient occurs, voltage v_{dt} follows (3) and eventually $\bar{v}_{dt} > V_{thr,on}$. Therefore, switch T is turned off and the algorithm waits for k clock cycles (e.g. in this work $k = 100$ cycles with a clock frequency of 3.5MHz), to prevent it from bouncing. Then, the algorithm enters the transient state and checks the difference of the integral of the averaged voltage between consecutive samples. When the rate of rise/fall of the integral is high $\Delta \int \bar{v}_{dt} > V_{thr,off}$, the system remains in the transient state and T remains turned off. When $\Delta \int \bar{v}_{dt} < V_{thr,off}$, the system turns T on, waits again for k cycles and returns to the steady state.

3.2 Circuit Design and Component Choice

The detailed circuit of the main measurement path is shown in Fig. 6a, along with the parasitic components. At first, when T is turned off, a significant over-voltage may arise, the magnitude of which depends on the detection time. In order to protect the switch, an active clamping circuit is used, with a zener diode connected in series to a diode as shown in Fig. 6a.

The parasitic capacitance of the MOSFET is depicted in Fig. 6a. As long as the output capacitance C_{oss} is sufficiently low, this part does not have an influence on the performance of the circuit. Furthermore, during transients the voltage across the burden resistor is high and the measured voltage needs to be limited. In the circuit in Fig. 6a, the bi-directional zener diodes are used to clamp the measured voltage to the zener breakdown voltage and protect the analog circuitry that is typically in parallel to the burden resistance. However, if zener clamping is used directly in parallel to the burden resistance R_b, the current of the diodes will not be limited, and inevitably they will be destroyed. For this reason, another divider R_1 and R_2 is inserted, with $R_2 \gg R_1$. In this case, R_1 is chosen appropriately to limit the current of the zener diode during transients and the majority of the voltage across R_b is still sensed across R_2 ($v_{Rb} \approx v_b$).

A disadvantage of the circuit is that the zener diodes have also a parasitic output capacitance (C_z) and devices with low output capacitance need to be chosen, in order to not limit further the high frequency performance of the topology. More specifically, due to the parasitic C_z, shown in Fig. 6a, another pole is inserted in the voltage measurement, due to the protection circuit. The transfer function assuming $R_2 \gg R_1$ is (13). In order to cancel the effect of R_1, a capacitor C_1 is inserted, effectively comprising an RC divider. C_1 can be selected according to (14), in order to minimize the effect of the additional circuit. Nevertheless, the parasitic capacitance C_z needs to be minimized, since the series combination of C_1 and C_z, which is approximately equal to C_z, appears to be in parallel to R_b and therefore the total impedance of the main measurement circuit $\underline{Z}_{main}$ is given by (15).

$$\frac{v_b(s)}{v_{Rb}(s)} = \frac{1}{1 + s \cdot R_1 \cdot C_z} \quad (13) \qquad C_1 = C_z \cdot \frac{R_2}{R_1} \quad (14) \qquad \underline{Z}_{main}(s) = \frac{1}{1 + s \cdot R_b \cdot C_z} \quad (15)$$

4 Experimental Validation

In order to validate the models and the design procedure described in section 2 as well as the proposed circuit described in section 3, a current probe fulfilling the specifications listed in Table I is presented in this section together with measurement results.

Regarding the design of the CT, Fig. 7 can be calculated based on the LF model of the CT, showing the f_{-3dB}^{LF} as a function of N for different cores and a given $S = 0.5$V/A. It can be observed that in general the higher the permeance A_L of the core, the lower the number of turns needed for a given sensitivity. Furthermore, Table II shows two alternative CT designs that fulfill the specifications given in Table I. Both designs achieve the required S and f_{-3dB}^{LF}. However,

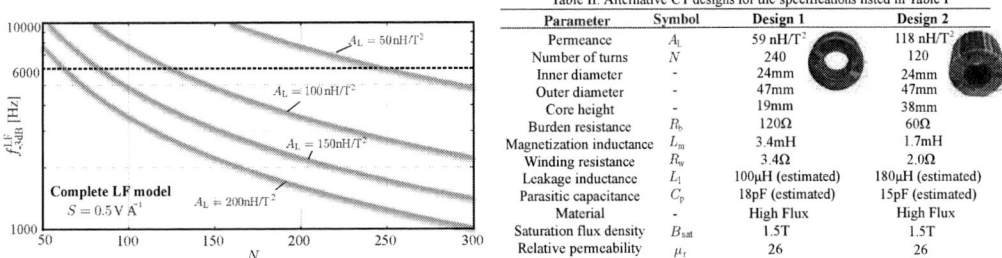

Table II: Alternative CT designs for the specifications listed in Table I

Parameter	Symbol	Design 1	Design 2
Permeance	A_l	59 nH/T²	118 nH/T²
Number of turns	N	240	120
Inner diameter	-	24mm	24mm
Outer diameter	-	47mm	47mm
Core height	-	19mm	38mm
Burden resistance	R_b	120Ω	60Ω
Magnetization inductance	L_m	3.4mH	1.7mH
Winding resistance	R_w	3.4Ω	2.0Ω
Leakage inductance	L_l	100μH (estimated)	180μH (estimated)
Parasitic capacitance	C_p	18pF (estimated)	15pF (estimated)
Material	-	High Flux	High Flux
Saturation flux density	B_{sat}	1.5T	1.5T
Relative permeability	μ_r	26	26

Figure 7: Selection of number of turns N for a given resolution of 0.5V/A. Table II shows the parameters of two CTs that fulfill the requirements shown in Table I.

design 2 has a higher f_{-3dB}^{HF} due to its lower number of turns and is therefore used in the following.

Regarding the proposed circuit topology, the hardware prototype of the designed current probe is shown in Fig. 6b. In the studied circuit, switch T is supplied by an isolated gate driver which is supplied by an isolated power supply. The main measurement v_b is directly connected to an analog output (SMA connector). The ADC is also shown along with the FPGA. In this work, a 900V SiC MOSFET (C3M0065090J) switch is chosen for the bidirectional switch T due to its high switching performance. The clamping voltage of the active clamping protection circuit (see Fig. 6) is set to 500V to allow for sufficient margin. The device has a low C_{oss} (less than 1nF at $V_{ds} = 10$V) and its influence on the circuit can be neglected. Moreover, a 13-bit 5MSps SAR ADC (LTC2311-12) is chosen along with an Intel MAX10 FPGA with 8k logical elements.

4.1 Steady State

At first, the impedance of the CT seen from the secondary side, with the primary side open circuited is measured with an impedance analyzer (Fig. 8a). Based on the measurements, it can be seen that the CT behaves inductively for frequencies up to its self-resonance frequency, which is given by the parasitic capacitance and the total inductance ($L_m + L_l$). However, for frequencies higher than the self-resonance, the model of Fig. 2b, with a lumped secondary winding needs to be redefined. In order to improve the high frequency performance of the model, the secondary winding needs to be distributed, as discussed in [14].

Fig. 8b shows the frequency response of the designed CT with a burden resistance of 60Ω. It can be seen that for frequencies below 3MHz the measurement behaves similar to the model and a sensitivity of 0.45V/A is achieved between 6kHz and approximately 3MHz. It can also be observed that the measurement does not introduce any phase between 60kHz and 3MHz.

Additionally, Fig. 9a&b show experimental results of the design CT in the time-domain. It can be seen that the CT can follow triangular currents of 60kHz and 300kHz (max. frequency of the setup), with precision and low phase delay. The achieved sensitivity is approximately 0.45V/A (constant in the relevant frequency range). It is also worth noting,

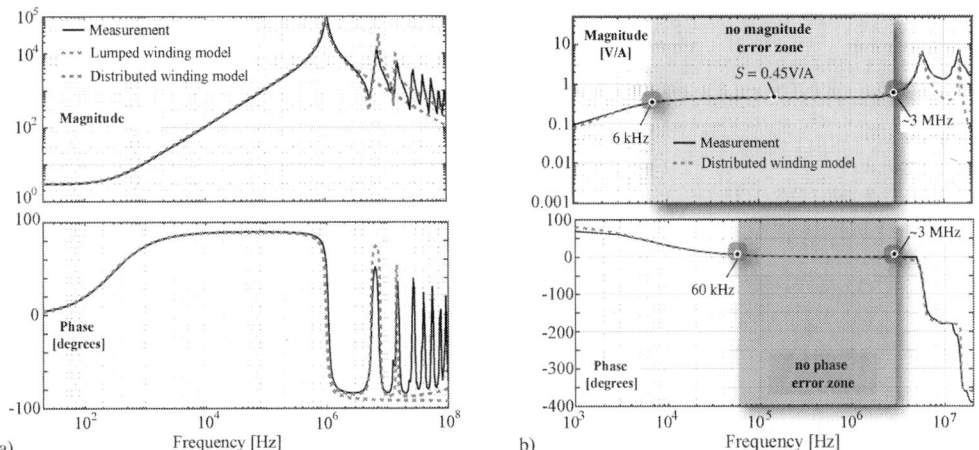

Figure 8: Frequency domain measurements. a) Output impedance of the CT, measured from the secondary side with the primary side open. b) Frequency response of the CT with a burden resistance (60Ω). A sensitivity of 0.45V/A is achieved.

EPE'20 ECCE Europe

Assigned jointly to the European Power Electronics and Drives Association & the Institute of Electrical and Electronics Engineers (IEEE)

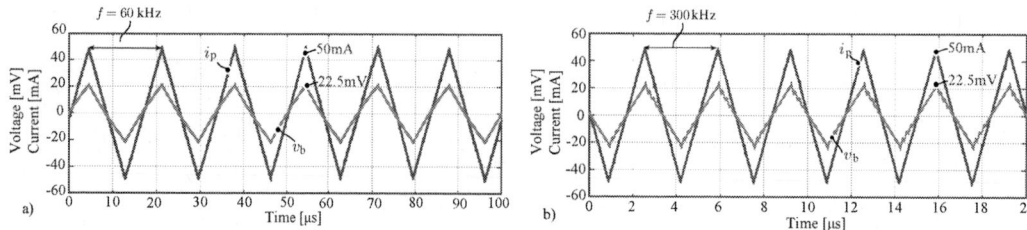

Figure 9: Time-domain measurements. a) Measurements with a primary current of 60kHz. b) Measurements with a primary current of 300kHz. Primary current i_p (blue) and sensed voltage v_b (light blue).

that in the 60kHz case, a slight distortion is noted, indicating that f_{-3dB}^{LF*} needs to be set to a lower value, for a better measurement of the 60kHz component. To achieve this, R_b can be selected to be slightly smaller, reducing sensitivity S. As shown in Fig. 3b, in order to reduce f_{-3dB}^{LF} without changing S, either N should be increased along with increasing R_b, or the core permeance A_L should be increased (e.g. bigger core size). Finally, in the 300kHz case, a ringing can be seen in the measured voltage waveform, due to the high frequency resonances that were noted with the frequency response measurements in Fig. 8b.

4.2 Transient State

In the following, the proposed CT with adaptive burden resistance is experimentally validated with a scaled down version of the current source shown in Fig. 1. The experimental setup used to generate the pulsed-shaped current through the designed CT is shown in Fig. 10a, and consists of a single high-power module of the interleaved converter system shown in Fig. 1. The main parameters of the setup are listed in Table III. The module is operating with open loop control with a higher duty ratio during the ramp up and a lower duty ratio during flat-top, in order to achieve a constant DC offset current of approximately 110A, as shown in Fig. 10b.

As a benchmark, a high-end current probe (Tektronix CP500 [15]) is used for sensing the load current. The probe is an AC/DC current measurement device which can measure up to 500A, with a bandwidth from DC up to 2MHz and a sensitivity of 10mV/A. The performance of the proposed current probe for a pulsed-shaped primary current can also be depicted in Fig. 10b, where the inferred measured current based on the v_b measurement is shown.

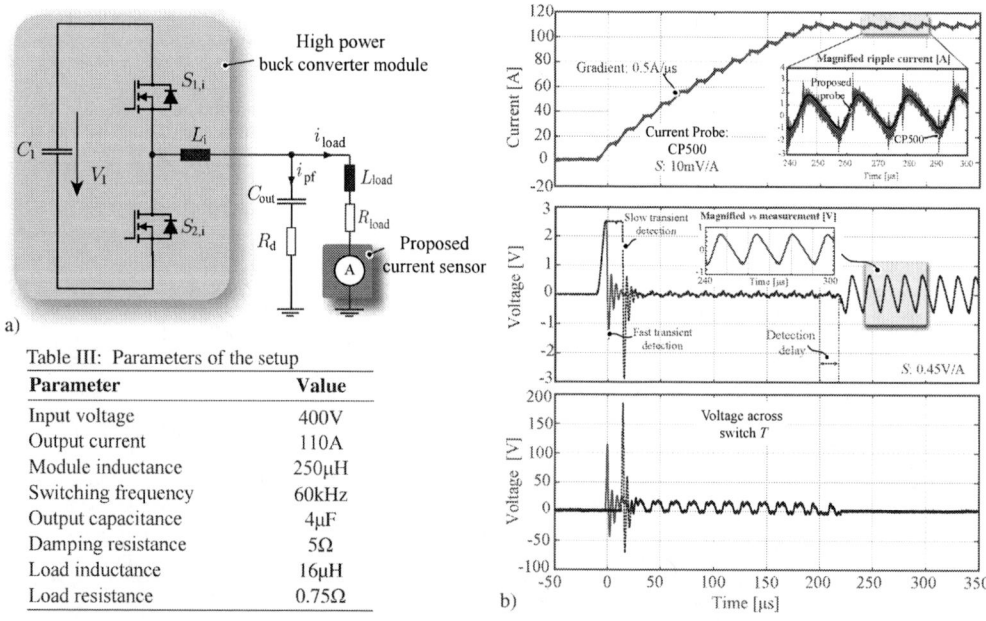

Table III: Parameters of the setup

Parameter	Value
Input voltage	400V
Output current	110A
Module inductance	250µH
Switching frequency	60kHz
Output capacitance	4µF
Damping resistance	5Ω
Load inductance	16µH
Load resistance	0.75Ω

Figure 10: a) Schematic of the experimental setup with a high power buck module used to generate the current.. b) Experimental measurements with the proposed CT. The load current is also sensed with the Tektronix CP500 current probe, that acts as a benchmark [15]. The main voltage measurement across the burden resistance v_b and the voltage across switch T are shown, too.

Furthermore, the effect of the detection time is examined, by simply varying the threshold limit $V_{thr,on}$, which has been described in section 3.1. In both cases the voltage across the burden resistor remains below 3V, keeping the analog circuit safe. Furthermore, in case of the slow detection, it is evident that the clamping protection circuit, limits the measured voltage to the breakdown voltage of the zener diode. Additionally, Fig. 10b, shows the voltage across switch T. It can be seen, that when the detection is fast, the over-voltage during the turn off of the switch is significantly lower than when the detection acts with a delay. Nevertheless, as shown in 6, the design includes an active clamping circuit that will clamp the voltage and protect the switches, if the voltage across T exceeds 500V.

When the primary current across the CT reaches it's flat-top, after approximately 200μs the condition for turning on switch T is fulfilled, and the expected voltage with the sensitivity of 0.45V/A is seen across the burden resistance. The detection delay time is noted on Fig. 10b. The detection delay occurs due to the inevitable delay of the moving average filter that is used and is described in section 3.1. The delay is approximately one switching period in this case, which is approximately 17μs. During flat-top, the proposed current probe measures the ripple with high fidelity as shown in the magnified ripple current comparison with the benchmark probe.

5 Simulation Results with Full Scale System

Based on the analytical models shown in section 2 and the performed measurements, a detailed model of the proposed current probe can be established for frequencies up to the resonant frequency of the transformer ($\approx$3MHz). In this section, simulation results of the proposed CT with adaptive burden resistance are presented and the performance is compared with the same CT, a single burden resistance and the same sensitivity S. All the identified parasitic components are included (e.g. parasitic capacitances of MOSFET and zener diodes) along with the protection circuit connected in parallel to R_b, to limit v_b as shown in Fig. 4a. The simulated transformer parameters are given in Table II (Design 2). The transformer is simulated with the magnetics package of PLECS [16]. Communication delays and finite sampling times are also taken into consideration for the control of the detection circuit.

Fig. 11a shows the current that excites the primary and Fig. 11b shows the corresponding v_b with the conventional design (i.e. Fig. 2a) and with the proposed topology (i.e. Fig. 4a). The primary current is pulsed shaped, with an

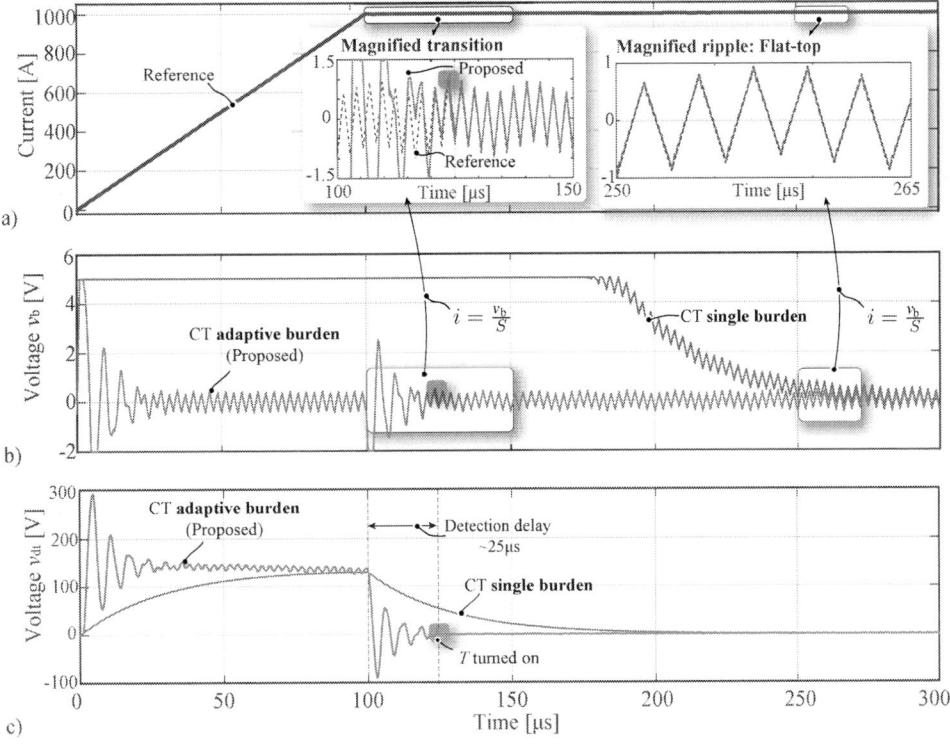

Figure 11: a) Primary current simulation and magnified current ripple with 360kHz and 60kHz triangular currents. b) Comparison of the sensed voltage v_b with the conventional CT with a single burden resistance value (green line) and with the proposed adaptive burden resistance using the same CT (light blue line). The parameters of the simulated CT are shown in Table II (Design 2). The proposed topology is using an $R_t = 10k\Omega$. Both circuits are simulated with ideal zener clamping diodes of 4V and $R_1 = 10k\Omega$ and $R_2 = 100k\Omega$. c) Voltage v_{dt} across the output of the CT.

amplitude of 1kA and a high current gradient during transient. At flat-top, the simulated current ripple has a magnitude of approximately ± 1A, two main frequency components at 360kHz and 60kHz and a triangular shape. This current ripple is a typical ripple at the output of an interleaved converter like the one in Fig. 1, with each module switching at 60kHz and slightly mismatched module inductances L_i [17]. Fig. 11a shows also a comparison of the reference current ripple that needs to be measured and the inferred current with the proposed probe. The inferred current of the proposed probe is essentially the measured voltage v_b scaled by a factor of $1/S$. It can be observed, that the CT manages to follow with precision the reference. However, the resonant behavior at high frequencies causes some oscillations on the measured current, too.

Fig. 11b, shows the measured voltage v_b in the proposed design with the adaptive burden resistance and compares it to the conventional design with a single burden and the same CT. The high voltage during the transient is limited in both cases by the protection circuit (i.e. clamping diodes). It can be seen that with the proposed topology, the settling time can be significantly decreased, and the current ripple can be measured with high precision shortly after the flat-top of i_p is reached (in the simulated case the settling time is less than 25μs). The delay is a result of the moving average that is used in the measurement of v_{dt}, as well as additional delays due to the non-ideal turn on times of switch T.

6 Conclusion

In this paper, the conventional CT design procedure is reconsidered, in order to fit the requirements of a variety of high-end applications. The aim of the proposed CT is to measure high frequency ripple currents in the mA range, which are superimposed on pulsed currents with high DC offset in the kA range. The designed CT acts as a high pass filter, blocking the DC offset of the current and allows measuring the ripple with high resolution. The paper describes the trade-offs that need to be considered and presents a parameter selection procedure to fulfill a given set of specifications. An adaptive topology that detects the transient and varies the CT's frequency response is also presented. The proposed topology reduces significantly the settling time of the CT after the transient and enables a faster measurement of the current ripple during the pulse flat-top. Simulation results for the dynamic response of the developed probe are shown and experimental results of its steady state and transient performance verify the design considerations.

Acknowledgment

This research is part of the activities of the Swiss Center for Competence in Energy Research on the Future Swiss Electrical Infrastructure (SCCER-FURIES), which is financially supported by the Swiss Innovation Agency.

References

[1] F. Cabaleiro Magallanes, D. Aguglia, C. de Almeida Martins, and P. Viarouge, "Review of design solutions for high performance pulsed power converters," *15th Int. Power Electronics and Motion Control Conf. (EPE/PEMC)*, Sep. 2012.

[2] R. Visintini, "Power converters for accelerators," *Proc. of the CAS-CERN Accelerator School*.

[3] M. Incurvati, C. Sanelli, and I. De Cesaris, "Feasibility study of high-precision power supply for ramping dipoles of a carbon/proton medical synchrotron," *IEEE Trans. on Applied Superconductivity*, vol. 16, June 2006.

[4] C. Carstensen and J. Biela, "10kV/30kA unipolar arbitrary voltage source for hardware-in-the-loop simulation systems for HVDC circuit breakers," in *Proc. of the European Conf. on Power Electronics and Applications (EPE)*, Aug 2011.

[5] G. Tsolaridis and J. Biela, "Flexible, highly dynamic, and precise 30-kA arbitrary current source," *IEEE Trans. on Plasma Science*, vol. 46, no. 10, Oct 2018.

[6] M. R. Pavan Kumar and J. M. S. Kim, "Capacitor current feedback for output filter damping in switched-mode magnet power supplies," *IEEE Trans. on Magnetics*, July 1994.

[7] F. Cabaleiro Magallanes, D. Aguglia, C. de Almeida Martins, and P. Viarouge, "Active damping filter for high bandwidth - Low ripple pulsed converters," in *15th Inter. Power Electronics and Motion Control Conf. (EPE/PEMC)*, Sep. 2012.

[8] M. C. Bastos, "High precision current measurement for power converters," *CERN*, May. 2014.

[9] L. Callegaro, C. Cassiago, and E. Gasparotto, "On the calibration of direct-current current transformers (DCCT)," *IEEE Trans. on Instr. and Measurement*, March 2015.

[10] S. Ziegler, R. C. Woodward, H. H. Iu, and L. J. Borle, "Current sensing techniques: A review," *IEEE Sensors Journal*, April 2009.

[11] N. Kondrath and M. K. Kazimierczuk, "Bandwidth of current transformers," *IEEE Trans. on Instr. and Measurement*, vol. 58, no. 6, June 2009.

[12] G. Laimer and J. W. Kolar, "Wide bandwidth low complexity isolated current sensor to be employed in a 10 kW/500 khz three-phase unity power factor PWM rectifier system," in *33rd Annual IEEE Power Electronics Specialists Conf.*, vol. 3, 2002.

[13] Magnetics, "Powder cores," 2017.

[14] F. Costa, E. Laboure, F. Forest, and C. Gautier, "Wide bandwidth, large AC current probe for power electronics and EMI measurements," *IEEE Trans. on Industrial Electronics*, vol. 44, no. 4, 1997.

[15] Teledyne, "CP500 current probe: Instruction manual," Jan. 2013.

[16] J. Allmeling, W. Hammer, and J. Schönberger, "Transient simulation of magnetic circuits using the permeance-capacitance analogy," in *13th Workshop on Control and Modeling for Power Electronics (COMPEL)*, June 2012.

[17] G. Tsolaridis, N. Patzelt, and J. Biela, "Output filter with adaptive damping for interleaved converters with low ripple and high dynamics," in *21st European Conf. on Power Electronics and Applications (EPE)*, 2019.

Loss Separation in Hard- and Soft-Switching GaN HEMTs operated in a 10 kW Isolated DC/DC Converter

Jan Böcker, Sören Heucke, Sibylle Dieckerhoff
TECHNICAL UNIVERSITY BERLIN
Einsteinufer 19, 10587 Berlin
Berlin, Germany
Tel.: +49 / (0) – 30 314 29218
E-Mail: jan.boecker@alumni.tu-berlin.de
URL: https://www.pe.tu-berlin.de

Keywords

«Gallium Nitride (GaN)», «Power semiconductor device», «ZCZVS converters», «Power converters for EV», «Conduction losses», «Switching losses», «Measurement», «Cooling»

Abstract

In this paper, GaN power transistors are operated in a 10 kW isolated DC/DC converter, investigating their loss distribution. Based on a combination of calorimetric loss-, and clamped drain source voltage measurements, the losses are separated in conduction and switching losses. Furthermore, the influence of dynamic R_{on} effects is identified during continuous operation. The chosen ZCZVS topology allows for comparison of hard- and soft-switching operation of the GaN devices. The loss separation reveals almost twice of additional dynamic R_{on} losses in the hard switching leg. In this case, the dynamic losses account for about one third of the conduction losses.

Introduction

Modern power electronic applications like on-board vehicle battery chargers have a rising demand for high power density. Furthermore, efficiency over a wide operation range is decisive, since many applications often operate in partial load conditions. Therefore, fast switching GaN transistors gain more and more attention in such applications. In order to investigate the benefits but also possible issues of current GaN HEMTs in a realistic application, an isolated DC/DC converter designed for on-board battery charging is built and analyzed.

In particular, normally-off GaN HEMTs with low gate leakage currents can still show notable dynamic R_{on} effects which depend on many parameters such as blocking voltage, current, switching speed or temperature [1], and charge trapping can even accumulate during continuous operation [2]. Therefore, state-of-the-art double-pulse tests are not sufficient to predict the device losses in application if a dynamic R_{on} increase is present. Consequently, they should be investigated in the final converter system to reveal all interactions. However, different loss mechanisms and especially the thermal rise of the on-state resistance and trapping effects are difficult to distinguish. Hence, systematic measurement methods for the DC/DC converter are developed in order to separate transistor losses over a wide operation range and finally evaluate additional dynamic R_{on} losses. This separation is based on a calorimetric loss measurement and a fast clamped measurement of the on-state voltages. The benefits of a calorimetric loss measurement system to reliably determine transistor losses are discussed in [3, 4] based on two different implementations. In [4], a strategy to separate switching- from conduction losses in continuous operation, using an additional transistor is presented. In our setup, a measurement system based on a water cooling circuit is used. Methods to separate trapping and heating effects on the dynamic R_{on} have been presented in [2, 5]. While these methods rely on a half-bridge topology operating in double-pulse tests, startup and continuous tests, the methods to separate losses and identify dynamic R_{on} effects presented in this paper are applicable independent from the converter topology.

To fulfill the demand for high power density, a prototype is designed towards high frequency switching, exploiting GaN HEMTs potential. Conventional LLC resonant converter suffer from narrow operation range in terms of output voltage, switching frequency and power, especially with minimized resonance inductors [6]. In contrast, the phase shifted full-bridge (PSFB) has a wide operation range and achieves soft switching from a certain load current depending on the selected semiconductors and their output capacitance. Hence, the PSFB is a promising topology for on-board vehicle battery charging systems [7] and is well suited for investigation of GaN device losses depending on the switching frequency. However, the usually applied inductive output filter causes overvoltage ringing across the diode rectifier stage when turned-off. This ringing is affected by the leakage inductance, the winding capacitances and the rectifier diode capacitance, since two of the rectifier diodes are reverse biased. Depending on the setup parameters, the ringing can reach more than twice the nominal output voltage [8, 9]. To overcome this effect, an additional auxiliary circuit is suggested in [7]. In order to avoid additional passive elements or oversizing the rectifier diodes, a capacitive output filter is proposed in [3, 10-12], which clamps the overvoltage and maintains zero voltage (ZVS) and zero current (ZCS) switching. In this work, a similar topology is chosen because it allows a strong minimization of the magnetic components. This benefit comes with higher turn-on losses (ZCS), where GaN transistors appear very promising. Nevertheless, the uneven transistor loss distribution is a disadvantage compared to resonant topologies which can achieve ZVS for all transistors in each operation point, but it enables an interesting comparison of GaN transistors in soft and hard switching conditions.

DC/DC converter setup

An isolated DC/DC converter is designed consisting of a controlled GaN full-bridge converter at the input, a HF-transformer providing isolation and voltage level conversion, and a SiC diode-bridge rectifier as output stage (see Fig. 1, top). The full-bridge is operated in phase-shift- and discontinuous current mode. In Fig. 2, the measured drain source voltages v_{ds} of the bottom transistors and the transformer current i_{Tr} are shown exemplarily for two points of operation. The phase shift angle is adjusted to achieve either low output power with a small phase shift (Fig. 2, left) or high output power with a large phase shift (Fig. 2, right). Both half-bridge legs are operated with a duty cycle of 50 %. Hence, both act identically for a phase shift angle Φ of zero, which results in zero power transmission from the converter input to the output. With increasing Φ the half-bridge voltages are shifted against each other, and the transformer primary winding is connected to either positive or negative input voltage. In between, a freewheeling period takes place where both top or both bottom transistors are conducting. During this period, the transformer current quickly falls to zero due to the relatively small stray inductance of the transformer (see Fig. 2). To clamp the resonant diode overvoltage during the power transmission stage, a capacitive filter (C2) is applied. In such a configuration, only one converter leg is

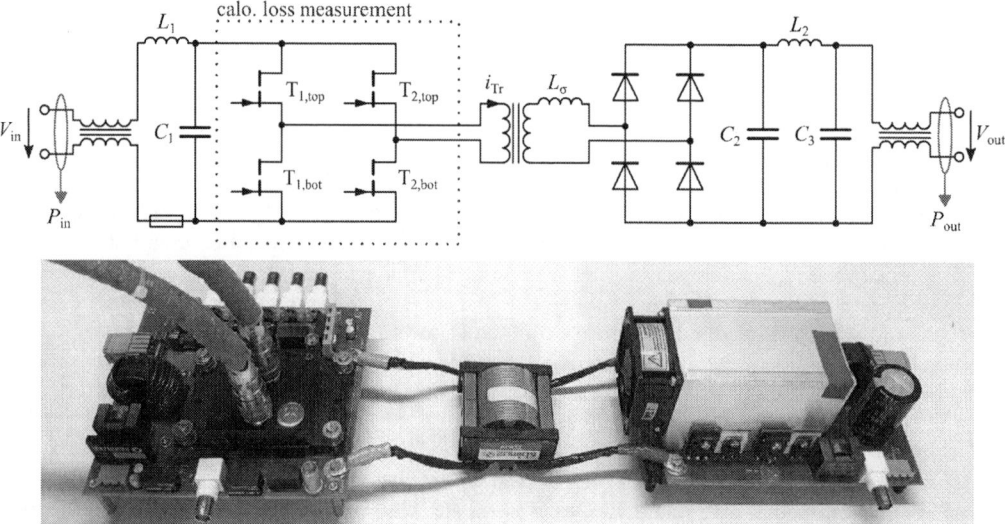

Fig. 1: Equivalent circuit of the DC/DC converter (top), picture of the experimental setup with connected calorimetric loss measurement system (bottom)

operated with ZVS whereas low current turn-on (~ ZCS) is achieved for the other leg. The transformer's leakage inductance supports ZVS as soon as a minimum primary current amplitude is reached, and its stored energy fully charges the transistor output capacitance during the applied deadtime. This ZVS is achieved for the transition from the power transmission period to the freewheeling stage. ZCS occurs for the transition from freewheeling to power transmission stage when the primary current is close to zero. However, a resonance of leakage inductance and diode capacitance appears during freewheeling stage as the rectifier diodes turn off. This can be seen in high frequency oscillations of the transformer current in Fig. 2. These oscillations can be damped efficiently with R-C-snubber circuits at the diode-bridge, but this reduces efficiency in all operation points. Besides, this current oscillation also affects ZCS, since the current can be either positive or negative during the switching events. This is part of the following loss analysis.

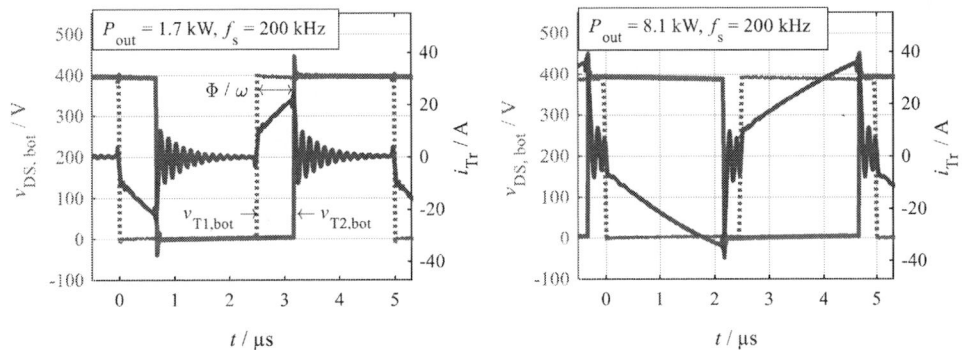

Fig. 2: Measurement of the drain source voltage transients of the bottom GaN-HEMTs and the respective transformer current for two different phase shifts and thus different output powers

Eventually, the transistor's hard (low current but full voltage) and soft switching (ZVS) operation can be investigated in the same test setup. The nominal input and output voltages during the following investigations are 400 V and 550 V to 575 V. Tab. 1 contains further converter design and relevant system parameters.

Table I: DC/DC converter parameters

Component	Type	Parameter	Value
GaN-HEMT	GS66516T	$R_{on,25°}$	25 mΩ
Gate Driver	UCC27511	$R_{G,on}$, $R_{G,off}$, t_{dead}	4.7 Ω, 1 Ω, 60 ns
SiC-Diode	SCS230AE2	V_R	650 V
Transformer	ETD59 (1:1.5)	L_σ, L_H	1.7 µH, 1.5 mH
	Litz wire	$R_{1,DC}$, $R_{2,DC}$	7.8 mΩ, 11 mΩ
		Core	3F3
		Transformer weight	450 g

Calorimetric measurement setup

The calorimetric measurement is based on the transistor losses which are dissipated in form of heat. The thermal equivalent circuit from transistor chip junction to heatsink contains resistive and capacitive elements leading to notable time constants and limiting the calorimetric measurement setup's bandwidth. Therefore, this method is suited for average loss measurements in continuous steady-state converter operation. In this case, the measurement enables very accurate overall transistor loss acquisition if other cooling paths can be suppressed.

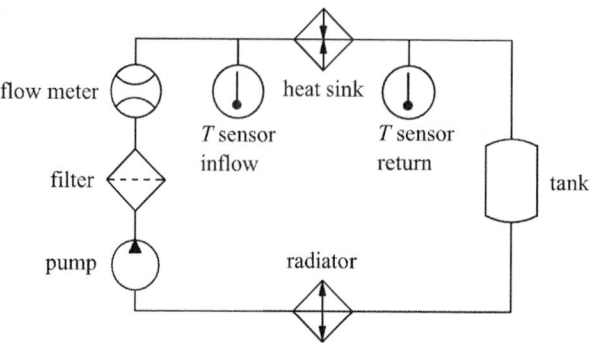

Fig. 3: Process diagram of the calorimetric measuring system using a water cooling circuit (cf. [13])

The system's process diagram is shown in Fig. 3. The main functionality is relying on a water cooling circuit, two thermal sensors and a flow sensor. Knowing the difference between inlet and outlet temperature of the heatsink ΔT, as well as its volume flow $\dot{V}$, the specific heat capacity c and its density ρ, the overall transistor losses in steady state can be calculated with

$$P_{\text{loss}} = c \cdot \rho \cdot \dot{V} \cdot \Delta T . \tag{1}$$

A good thermal interface between device and heatsink obviously improves the measurement accuracy since it reduces the share of heat flow that does not reach the water. Therefore, a thin AlN ceramic with high thermal conductivity (180 W/m K) is used as isolator. In total, a very small thermal resistance from junction to water in the range 1 K/W is achieved. A calibration with a DC current in the final setup with simultaneous electric ($v_{\text{ds}} \cdot i_{\text{d}}$) and calorimetric loss measurement further improves the accuracy to ± 0.5 W (Fig. 4 left). This is much more accurate than a loss calculation based on transient voltage and current measurements of the transistors. The accuracy was improved mainly by two adaptions: first, by using a large water tank to minimize temperature fluctuations in the water and second, by controlling the volume flow in relation to the current transistor losses. This ensures a minimization of the total measurement uncertainty that is mainly caused by the flow sensor and the temperature difference ($\dot{V} \cdot \Delta T$).

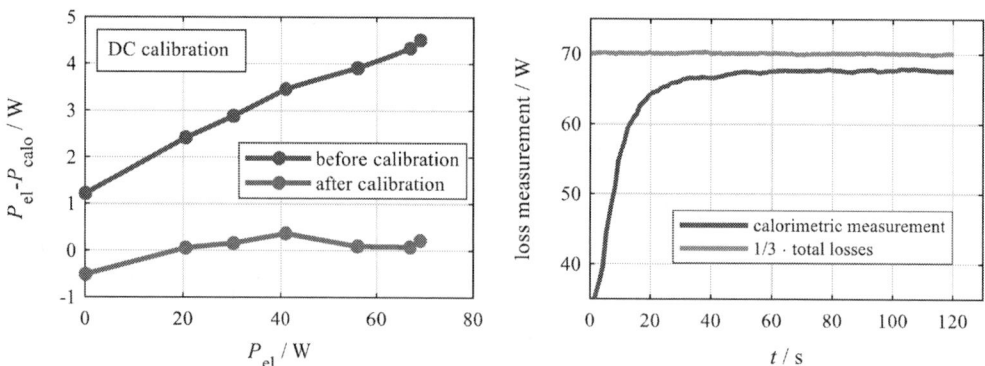

Fig. 4: Deviation of the calorimetric measurement to the electrical loss measurement under DC condition before and after calibration (left), step response of the calorimetric measurement system connected to the DC/DC converter ($f_{\text{s}} = 200$ kHz, $P_{\text{out}} = 9$ kW) (right),

The time constant of the measurement system is presented in Fig. 4 (right). The measurement after the DC/DC converter starts operating with 200 kHz and 9 kW reveals a steady and accurate measurement after ~60 s.

Loss analysis

For detailed loss analysis, input and output power are measured using a power analyzer (Zimmer LMG670). The transistor losses are additionally monitored by the calorimetric system. Thus, they can be separated from other converter losses (dominated by transformer and diode, additional filter, DC link etc.). Auxiliary circuits like drivers or fans are not included in the loss measurement. The entire test setup is fully automated, enabling a precise setting of specific operation points and therefore leading to a good comparability of the results in the following parameter variation studies.

Giving an overview of the loss analysis, Fig. 5 presents the influence of the switching frequency f_s and output voltage V_{out} on converter efficiency η and loss distribution for a wide power range. Increasing the switching frequency f_s at constant V_{out}, the efficiency decreases in the partial load range, as expected (top, left). This results from a rise of the transistor losses as well as of the remaining other losses (top, right). The transistor losses obviously increase due to the switching losses, whereas the other losses are probably dominated by increased diode and transformer losses. For higher output power P_{out}, the efficiencies for different switching frequencies approach each other and reach a high maximum value of 98 %. For higher f_s, the phase shift increases to maintain constant V_{out} and P_{out}, resulting in a smaller current amplitude in the transformer. That is in favor of the conduction losses of transformer and transistors. The chosen frequency range from 150 kHz to 350 kHz results from the transformer, which was designed for ~300 kHz. Therefore, frequencies below 150 kHz would lead to notable saturation of the core material. Switching frequencies above 350 kHz are possible, but they would limit either the output voltage or the output power due to the chosen stray inductance of the transformer.

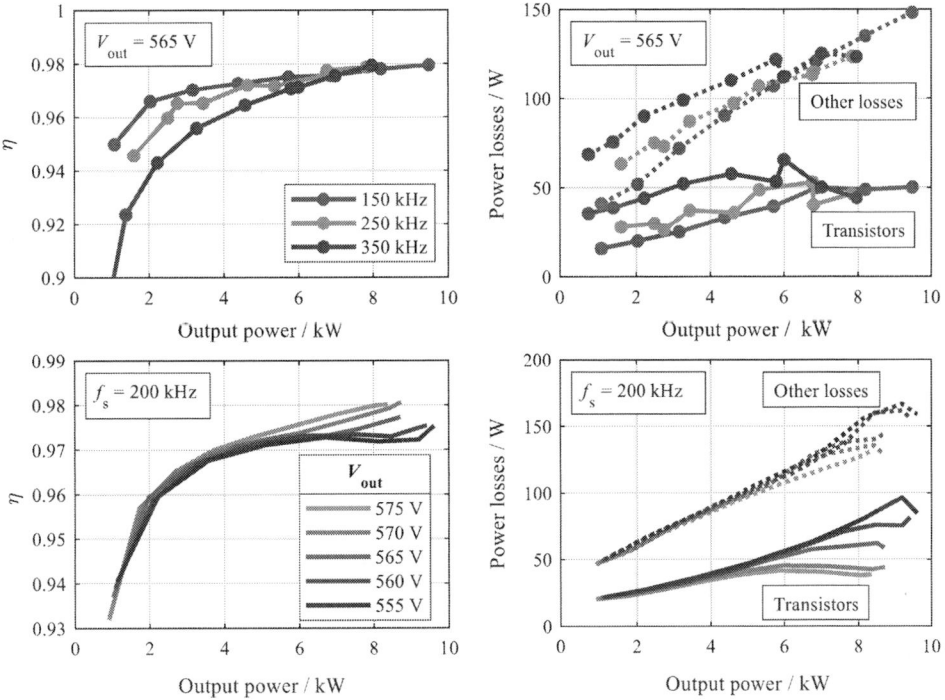

Fig. 5: Efficiency (η) (left) and loss distribution (right) for different switching frequencies at constant output voltage (top) and different output voltages at constant switching frequency (bottom)

Analyzing the impact of V_{out} on the efficiency, a similar effect can be observed (see Fig. 5 bottom, left). The transistor and other losses increase with decreasing V_{out}, because the current peak value rises according to

$$\frac{di_{Tr}}{dt} = \frac{v_{in} - v_{out}/n}{L_\sigma} \qquad (2)$$

where n denotes the transformer's winding ratio.

Unexpectedly, for some points of operation, the transistor losses even decrease for higher P_{out} (Fig. 5 bottom, right). This effect is not caused by measurement distortions but by small variations in the phase shift angle. Due to the high frequency ringing between diode capacitance and transformer leakage inductance, the turn-on after freewheeling occurs with either small positive or negative current. This leads to significant variations in the hard turn-on losses which is shown for a phase shift variation in Fig. 6, left. The right graph depicts the transients of the "ZCS" turn-on process for two indicated operation points 1 and 2. In operation point 1, a negative transformer current almost enables ZVS. However, in operation point 2, a positive transformer current and full blocking voltage V_{ds} lead to a notable increase of switching losses. In addition to these fluctuating switching losses, the dynamic on-state resistance usually depends on the magnitude of turn-on current as well. Hard switching can lead to hot electron trapping [14, 15] that increases with the load current [1]. Thus, the transistor losses vary even more in very narrow converter operation points. Consequently, the topology would benefit from a driver with adaptive dead time control to influence the precise turn-on instant and reduce losses.

The DC/DC converter reaches a peak efficiency of 98 % for high output power and 350 kHz switching frequency. A further increase of f_s does not seem appropriate regarding the partial load efficiency. The main cause are the transistor turn-on losses and the dynamic R_{on} losses which will be separated in the next section.

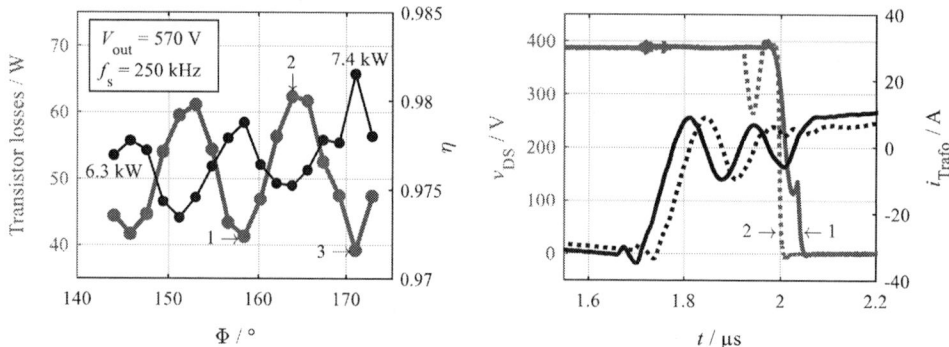

Fig. 6: Dependency of transistor losses and converter efficiency on small variations of the phase shift (left), drain source voltage and transformer current during a favorable and an unfavorable turn-on moment (right)

Dynamic R_{on} separation

In order to analyze the impact of the dynamic R_{on} increase, the drain source voltage is measured using a passive clamping circuit (see [2]) in addition to the calorimetric loss measurement. The passive clamping can be easily used even in complex topologies and can be applied to multiple transistors, since no further clamping control signal is necessary.

In continuous operation, a separation of thermal effects from trapped charges causing the dynamic R_{on} increase is difficult. However, knowing the overall transistor losses allows for a comparison with the on-state resistance under DC conditions, where only minor trapping is to be expected. By generating the same amount of losses with a DC current and in switched operation, the thermal increase of R_{on} can be determined based on the comparison. The remaining difference of the on-state resistance to switched operation can then be related to the dynamic R_{on} effect. This method assumes an even loss distribution. Unfortunately, an uneven loss distribution is expected in the ZCZVS topology. As all four transistors are attached to the same heatsink, it is not possible to directly determine the individual loss distribution. Instead, a best and worst case scenario is analyzed. In the best case, all four transistors share the losses equally, as ideally assumed. The estimated thermal R_{on}-rise in continuous operation is then smaller than in case of a realistic loss distribution. In the worst case, all losses are generated only in one leg of the full bridge, which leads to an overestimation of the thermal increase.

Fig. 7 (top) shows a comparison of R_{on} for the hard and the soft switching leg, as well as the corresponding DC equivalent for 7.5 kW output power. Only the DC worst case, where the thermal rise of R_{on} should be overestimated, is shown (dashed line). Still, in continuous operation every measured R_{on} is significantly higher than the DC resistance. The hard switched leg (blue) shows a notably higher R_{on} increase than the soft switched one (black). This difference is larger than only thermal effects due to more switching losses would explain. Furthermore, a partially soft switched operation with only slightly higher phase shift (green) is shown (see case 1 vs. 2 in section above). This represents a case of turn-on with reduced voltage and emphasizes the benefits of an adaptive controller. This qualitative comparison confirms that hard switching influences the conduction losses. However, the soft switched leg shows a significant increase in R_{on} as well.

For a quantitative analysis, the overall losses are first divided into transistor losses and others based on the calorimetric measurement. The transistor losses can be further separated into switching and conduction losses, using the clamped drain-source voltage measurement. This allows to calculate their conduction losses considering the measured transformer current. Finally, these conduction losses can be divided into thermal and dynamic R_{on} losses for hard and soft switched operation, using the thermal DC reference as explained before. Obviously, this method suffers from different measurement uncertainties where the clamped voltage measurement is the dominating part. The clamping circuit has a clamping delay of ~200 ns. Therefore, the low switching frequency of 150 kHz is presented, to minimize this influence. The uncertainty of the loss separation in conduction and switching losses can be in the range of 10 % of the total transistor losses. However, a comparison between both half-bridges should indicate differences with higher accuracy, since they share some systematic errors like the clamping delay.

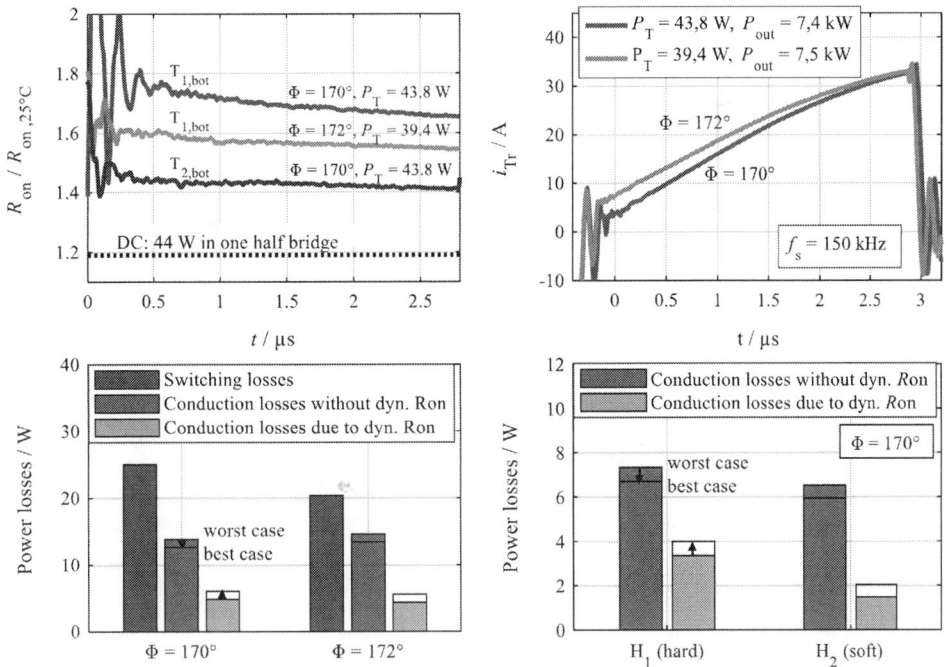

Fig. 7: Dynamic R_{on} at 150 kHz switching frequency for the hard switching ($T_{1,bot}$) and soft switching ($T_{2,bot}$) half-bridge in comparison with a DC resistance measurement (top, left); respective transformer current for two different phase shifts (top, right); loss distribution for two different phase shifts in the hard switching half-bridge H1 (bottom, left); loss distribution for both half-bridges (H1 and H2) at a phase shift of 170° (bottom, right)

Eventually, the loss distribution for the GaN full-bridge for the favorable ($\Phi = 172°$) and unfavorable ($\Phi = 170°$) switching moment is compared in Fig. 7 (bottom, left). Lower switching losses are obtained for a phase shift $\Phi = 172°$, which was expected due to softer switching conditions in the ZCS leg. The dynamic R_{on} losses are slightly reduced for the same reason.

In Fig 7 (bottom, right), the ZVS leg and the hard switching (ZCS) leg are compared at $\Phi = 170°$. For the hard-switched half-bridge, the conduction losses increase by 31-37 % due to dynamic R_{on} effects. In comparison, the increase is 19-25 % for the soft-switched leg. These additional losses therefore especially affect the hard switching leg, enhancing the unbalanced loss distribution within the converter and limit P_{out} or f_s.

In summary, the presented measurements prove the influence of the dynamic R_{on} effect on the conduction losses. This influence can be reduced but not fully avoided using soft switching techniques.

Conclusion

A method to separate switching losses, conduction losses and in particular additional dynamic R_{on} losses of GaN transistors is successfully demonstrated in a 10 kW DC/DC converter application. The dynamic R_{on} losses are still in a notable range for the investigated devices, especially in hard turn-on conditions. In this case, soft switching not only reduces switching losses but also decreases dynamic conduction losses. However, the dynamic R_{on} losses contribute only with approximately 3.5 % to the total system losses. Therefore, a compact and efficient DC/DC converter system can be realized applying state-of-the-art GaN HEMTs.

References

[1] Badawi N., O. Hilt, E. Bahat-Treidel, J. Böcker, J. Würfl, and S. Dieckerhoff, "Investigation of the Dynamic On-State Resistance of 600 V Normally-Off and Normally-On GaN HEMTs", IEEE Transactions on Industry Applications, vol. 52, no. 6, pp. 4955–4964, 2016. doi: 10.1109/TIA.2016.2585564..

[2] J. Böcker, C. Kuring, M. Tannhäuser and S. Dieckerhoff, "R_{on} Increase in GaN HEMTs — Temperature or Trapping Effects", 2017 IEEE Energy Conversion Congress and Exposition (ECCE), Cincinnati, OH, 2017, pp. 1975-1981. doi: 10.1109/ECCE.2017.8096398

[3] A. E. Awwad, N. Badawi and S. Dieckerhoff, "Efficiency analysis of a high frequency PS-ZVS isolated unidirectional full-bridge DC-DC converter based on SiC MOSFETs", 18th European Conference on Power Electronics and Applications (EPE'16 ECCE Europe), Karlsruhe, 2016, pp. 1-10. doi: 10.1109/EPE.2016.7695601

[4] D. Rothmund, D. Bortis and J. W. Kolar, "Accurate Transient Calorimetric Measurement of Soft-Switching Losses of 10-kV SiC MOSFETs and Diodes", in IEEE Transactions on Power Electronics, vol. 33, no. 6, pp. 5240-5250, June 2018. doi: 10.1109/TPEL.2017.2729892

[5] C. Kuring, M. Tannhaeuser and S. Dieckerhoff, "Improvements on Dynamic On-State Resistance in Normally-off GaN HEMTs", PCIM Europe 2019; International Exhibition and Conference for Power Electronics, Intelligent Motion, Renewable Energy and Energy Management, Nuremberg, Germany, 2019

[6] Fairchild, "AN-4151 Half-Bridge LLC Resonant Converter Design Using FSFR-Series Fairchild Power Switch", Fairchild Semiconductor Corporation, Application Note, Rev. 1.0.2, 2014

[7] M. Pahlevaninezhad, P. Das, J. Drobnik, P. K. Jain and A. Bakhshai, "A Novel ZVZCS Full-Bridge DC/DC Converter Used for Electric Vehicles", in IEEE Transactions on Power Electronics, vol. 27, no. 6, pp. 2752-2769, June 2012. doi: 10.1109/TPEL.2011.2178103

[8] X. Ruan, "Soft-Switching PWM Full-Bridge Converters Topologies, Control, and Design", Nanjing University of Aeronautics and Astronautics, China, John Wiley and Sons Singapore Pte. Ltd. Science Press, 2014, doi: 978-1-118-70220-8

[9] J. A. Sabate, V. Vlatkovic, R. B. Ridley, F. C. Lee and B. H. Cho, "Design considerations for high-voltage high-power full-bridge zero-voltage-switched PWM converter", Fifth Annual Proceedings on Applied Power Electronics Conference and Exposition, Los Angeles, CA, USA, 1990, pp. 275-284, doi: 10.1109/APEC.1990.66420.

[10] M. Aghaei, Y. Karimi and S. Kaboli, "Analysis of Phase-Shifted Full-Bridge Based dc-dc Converter considering Transformer Parasitic Elements in Discontinuous Current Mode", The 5th Annual International Power Electronics, Drive Systems and Technologies Conference (PEDSTC 2014), Tehran, 2014, pp. 366-372. doi: 10.1109/PEDSTC.2014.6799402

[11] A. E. Awwad, "On the Perspectives of SiC MOSFETs in High-Frequency and High-Power Isolated DC/DC Converters", Doctoral Thesis, Deposit Once, Technische Universität Berlin, 2018. doi: 10.14279/depositonce-7362.

[12] J. Morren, S. W. H. de Haan and J. A. Ferreira, "Design study and scaled experiments for high-power DC-DC conversion for HVDC-systems", 2001 IEEE 32nd Annual Power Electronics Specialists Conference (IEEE Cat. No.01CH37230), Vancouver, BC, 2001, pp. 1529-1534 vol. 3, doi: 10.1109/PESC.2001.954336.

[13] J. Böcker, "Analyse und Optimierung von AlGaN/GaN-HEMTs in der leistungselektronischen Anwendung", Doctoral Thesis, Deposit Once, Technische Universität Berlin, 2020 doi: 10.14279/depositonce-9678 (*in German*)

[14] I. Rossetto, M. Meneghini, A. Tajalli, S. Dalcanale, C. de Santi, P. Moens, A.Banerjee, E. Zanoni, and G. Meneghesso, „Evidence of Hot-Electron Effects During Hard Switching of AlGaN/GaN HEMTs", IEEE Transactions on Electron Devices, vol. 64, no. 9, pp. 3734–3739, 2017, issn: 0018-9383. doi: 10.1109/TED.2017.2728785.

[15] E. Fabris, M. Meneghini, C. de Santi, M. Borga, Y. Kinoshita, K. Tanaka, H. Ishida, T. Ueda, G. Meneghesso, and E. Zanoni, „Hot-Electron Trapping and Hole-Induced Detrapping in GaN-Based GITs and HD-GITs", IEEE Transactions on Electron Devices, vol. 66, no. 1, pp. 337–342, 2019. doi: 10.1109/TED.2018.2877905.

A Switched-Mode Power Amplifier for Ion Energy Control In Plasma Etching

Qihao Yu, Erik Lemmen, Korneel Wijnands, Bas Vermulst
Department of Electrical Engineering, Eindhoven University of Technology
P.O. Box 513, 5600MB
Eindhoven, The Netherlands
Email: q.yu@tue.nl
URL: http://www.tue.nl/epe

Acknowledgments

This research was supported by Prodrive Technologies B.V.. Assistance provided by Javier Escandon-Lopez and Erik Heijdra was greatly appreciated.

Keywords

≪Physics research≫, ≪Industrial application≫, ≪Amplifiers≫, ≪Plasma≫.

Abstract

Plasma etching is an important process in the semiconductor manufacturing process. In order to precisely control the ion energy for better process quality, a tailored pulse-shape voltage waveform is applied to the plasma reactor table. Traditionally, a linear amplifier is used to generate this waveform, which results in poor efficiency. This paper proposes a switched-mode power amplifier as a substitute to the traditional linear amplifier. The electric equivalent circuit of the plasma reactor is introduced and a basic topology for the switched-mode power amplifier is derived. The basic topology is able to generate the required waveform but it has a low efficiency of charging the capacitive load in practice. Therefore, an efficiency-improved topology is proposed by adopting resonant charging. A prototype is built in order to validate the research. The experiments show that the presented solution yields a significantly reduced input power compared to the normally used linear amplifier in this application.

Introduction

Plasma consists of positive ions, negative electrons and neutral particles with an approximately neutral net charge [1]. It is widely used in semiconductor manufacturing, such as plasma etching and deposition process. Fig. 1 (a) shows a schematic diagram of a typical setup of an inductively coupled plasma etching chamber. Gas, such as argon or hydrogen, is infused from the top of the chamber. Plasma is ignited and sustained in the chamber by applying an external radio frequency (RF) power supply which is coupled with the gas through a matching network and coils. The pressure in the chamber is kept low by a vacuum pump in order to increase the distance between particles thereby reducing particle collisions. A substrate wafer is placed on the table and the material on the substrate surface is removed by chemical reaction or physical sputtering in a certain pattern, which is called etching. A chemical reactant gas can also be injected into the chamber to react with the surface of the substrate wafer.

An ion sheath is formed where plasma is near the substrate, which is a region where the ion density is larger than the electron density. The fast-moving electrons in plasma easily get lost when confronted with a surface like the substrate, thus leaving excess ions near the surface. As a result, the surface is negatively charged, forming an electric field which attracts ions and repels electrons [2]. By applying a

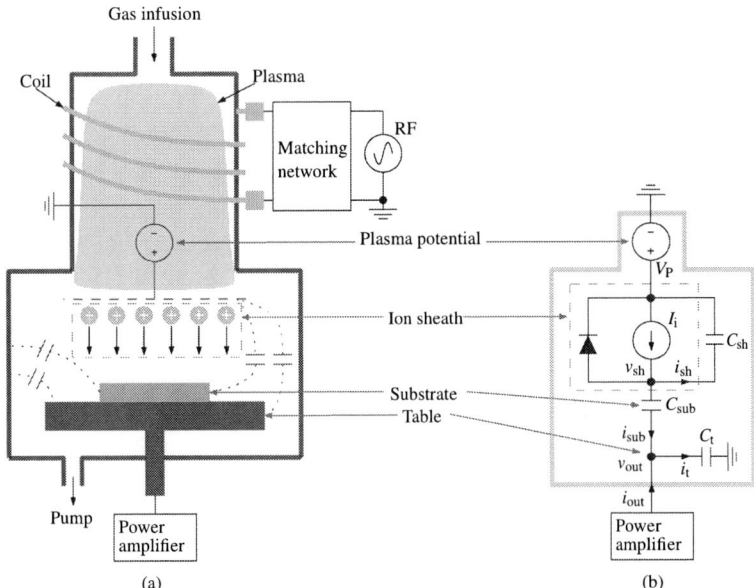

Fig. 1: (a) A typical setup of an inductively coupled plasma chamber and (b) its equivalent electric model.

negative voltage to the substrate with a power amplifier connected to the conductive table, the voltage potential of substrate can be controlled. As a result, ions are accelerated by the electric field and bombard the substrate surface. The bombarding ions supply additional energy to the surface and accelerate the chemical reactions. Additionally, the normal direction of ion bombardment to the material surface also enhances the anisotropy of the etching.

For a high-selectivity etching recipe, the ion energy must be within a specific limited range. Ions with too low energy cause a slow reaction rate while ions with too high energy causes sputtering, which decreases the selectivity. With the increasing demands on smaller semiconductor feature size, precise control of ion energy becomes critical in both plasma deposition and etching. In general, a narrow ion energy distribution (IED) falling into a specific energy range is desired.

A narrow IED requires a constant negative voltage potential on the substrate material surface. For a conductive substrate material, a negative dc bias voltage can be directly connected to the table in the plasma chamber. By varying the dc value of the voltage, the energy of the ions can be controlled correspondingly. However, for a dielectric substrate material, there is an equivalent substrate capacitance. An equivalent electric current caused by the bombarding ions is charging the substrate capacitance during etching, resulting in an increasing voltage potential on the substrate surface, thus making the ion energy variant during the process if left uncompensated.

To control the ion energy with a dielectric substrate, different kinds of voltage waveforms have been studied, including the radio-frequency (RF) sinusoidal waveform [3], square waveform [4] and tailored pulse-shape waveform [5], among which the tailored pulse-shape waveform is the most precise one. All of the tailored pulse-shape waveforms in previous researches were generated by class-A linear amplifiers [5, 6, 7], which are typically expensive and inefficient. In our literature research, no paper or combination of papers was found to consider the power amplifier from the perspective of the power consumption in the physics research. Therefore, in this paper, a power electronics concept based on a switched-mode power amplifier is introduced. It is able to generate the required waveform in order to obtain a narrow IED and at the same time reduce the input power by about 90% compared to the linear amplifier by improved amplifier efficiency. A prototype was built based on this concept and experiments were conducted on a plasma etching machine for validating the concept.

Power Electronics Concept

In order to illustrate the concept, a simplified equivalent electric model is built as depicted in Fig. 1 (b). The bulk plasma is modelled as a voltage source with a positive potential V_P with respect to ground, the magnitude of which is dependent on the temperature and pressure of the chamber and the power of the RF source. V_P is assumed to be constant during etching and is normally tens of volts.

During etching when the substrate potential v_{sh} is negatively biased, the positive ions are attracted to the substrate. As a result, the bombardment of ions generates an equivalent current I_i, which is assumed to be constant in this paper [8]. A sheath capacitance is present between the plasma potential and substrate surface. Besides, there is also a parasitic capacitor between the substrate and the chamber wall [3]. Since the wall is grounded and the plasma potential is constant during the etching phase, they are modelled in total as C_{sh}. C_{sh} is variant during etching phase unless the substrate potential is constant.

Depending on the material and composition of the substrate, the substrate can be either dielectric or conductive. For a dielectric substrate, it is modelled as a capacitor C_{sub}. In addition, there are also parasitic capacitances between the table and the chamber wall and between the table and the sheath respectively, which are modelled in total as C_t. In practice, both C_{sub} and C_t are much larger than C_{sh}. As a result, during etching, a series of equations can be used to describe the electric model and given by

$$i_t = C_t \frac{dv_{out}}{dt}, \tag{1}$$

$$i_{sub} = C_{sub} \left(\frac{dv_{sh}}{dt} - \frac{dv_{out}}{dt} \right), \tag{2}$$

$$i_{sh} = C_{sh} \frac{dv_{sh}}{dt}, \tag{3}$$

$$I_i = i_{sh} + i_{sub}, \tag{4}$$

$$i_{out} = i_t - i_{sub}. \tag{5}$$

In order to obtain a narrow ion energy distribution, a constant v_{sh} is desired thus $\frac{dv_{sh}}{dt} = 0$. By further manipulating the above equations, the required characteristics of the power amplifier can be derived and are governed by

$$\frac{dv_{out}}{dt} = -\frac{I_i}{C_{sub}}, \tag{6}$$

$$i_{out} = -I_i - \frac{I_i C_t}{C_{sub}}. \tag{7}$$

Therefore, the output voltage v_{out} of the amplifier is a negative voltage ramp and the output current i_{out} of the amplifier is a negative constant current in etching process.

However, since C_{sub} is continuously charging during etching, it is required to discharge it periodically to avoid over-voltage on substrate. A short positive voltage pulse can be used to discharge the substrate. During discharge, the voltage potential of the substrate surface is reset to a positive value by the discharge voltage. The fast-moving electrons are then attracted to the substrate surface rapidly so as to compensate the excess ions accumulated on the substrate surface, after which a negative voltage pulse should be applied to the table to restart the etching process. It should be noted that the equivalent electric model is only applicable to the etching phase. The discharge phase is rather complex and lasts for only a very short moment, moreover it is not the main focus of this paper.

The typical waveforms of v_{out}, v_{sh}, $v_{C_{sub}}$ and i_{out} are shown in Fig. 2, among which the waveform of v_{out} is the so-called tailored pulse-shape voltage waveform. The waveform is divided into a discharge phase and an etching phase with a duration of T_{dis} and T_{pro} respectively. V_d, V_s and V_e are the discharge voltage, start and end voltage of the etching respectively. The ion energy is controlled by varying V_s and it is approximately equal to $e(V_P + V_s)$, where e is the element charge.

EPE'20 ECCE Europe

Assigned jointly to the European Power Electronics and Drives Association & the Institute of Electrical and Electronics Engineers (IEEE)

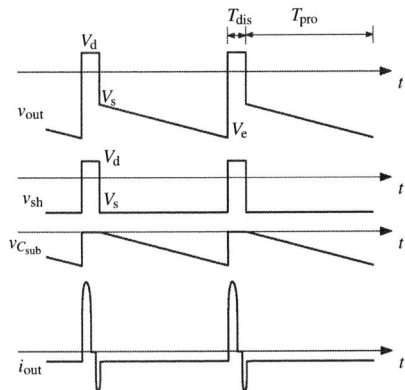

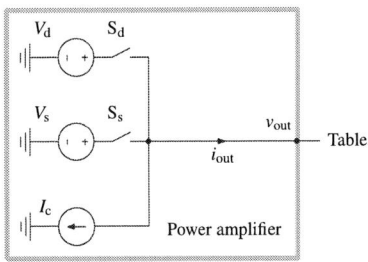

Fig. 3: The basic topology switched-mode power amplifier concept.

Fig. 2: The typical waveforms of v_{out}, v_{sh}, $v_{C_{sub}}$ and i_{out}.

At the beginning of the discharge and etching phase, the voltage of C_t should be charged promptly to V_d and V_s respectively. During the etching phase, there are two possibilities for generating a narrow IED: using a voltage source with a negative voltage ramp output or using a dc current source with a constant negative current output. The dc current source has the benefit of lower bandwidth requirement. Therefore, a basic switched-mode power amplifier concept is derived, which consists of two dc voltage sources and one dc current source, as shown in Fig. 3. By turning on S_d, V_d is connected to the table and the discharge phase starts. Similarly, by turning off S_d and turning on S_s, it turns to the etching phase. By turning off both S_d and S_s, etching starts properly.

Topology

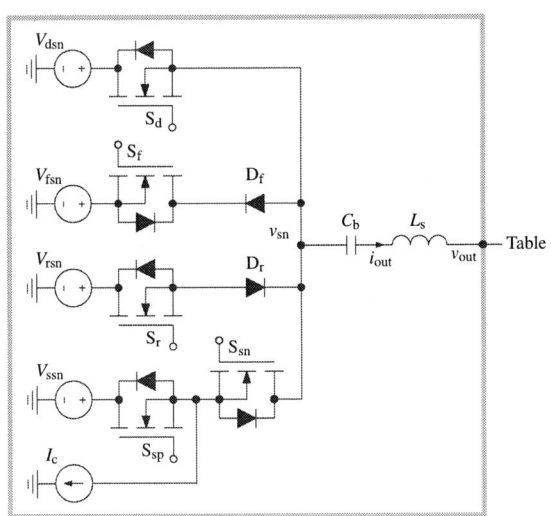

Fig. 4: The topology of the switched-mode power amplifier.

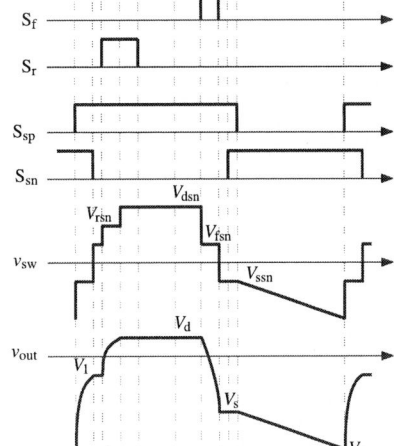

Fig. 5: The operation of the switched-mode power amplifier.

The basic topology shown in Fig. 3 illustrates a concept of generating the bias waveform by power electronics solution. As shown in Fig. 1, the load of the amplifier is capacitive and the equivalent capacitance C_{eq} is given by

$$C_{eq} = C_t + \frac{C_{sub} + C_{sh}}{C_{sub}C_{sh}}. \tag{8}$$

However, charging and discharging a capacitor by directly connecting it to a voltage source leads to only a maximum 50% power efficiency. In order to reduce the power loss, resonant charging can be adopted by adding an inductor between the voltage source and the load [9, 10].

The power amplifier depicted in Fig. 4 shows one feasible topology, which is able to charge and discharge the capacitive load softly thus significantly reducing the power loss. The amplifier consists of four controllable dc voltage levels V_{dsn}, V_{fsn}, V_{rsn} and V_{ssn} and one controllable dc current source I_c. V_{dsn} and V_{ssn} are utilized to generate V_d and V_s at the output respectively. V_{fsn} and V_{rsn} are two intermediate voltage levels used to reduce the resonance. The dc current source can be realized by an inductor in series with a voltage source. L_s is the resonant inductor and it can be the stray inductance of the chamber and connection. A blocking capacitor C_b is used at the switch-node in order to produce an equal current flowing into both side of the amplifier [11]. A self-biasing dc offset voltage V_b is then formed over C_b, the value of which is determined by the plasma condition and the output waveform. Normally, the value of C_b should be much larger than C_t and C_{sub} so that V_b can be assumed constant during steady state operation. In this case, V_{dsn}, V_r and V_{ssn} are governed by

$$V_{dsn} = V_d + V_b, \tag{9}$$

$$V_r = V_r + V_b, \tag{10}$$

$$V_{ssn} = V_s + V_b. \tag{11}$$

The operation of the power amplifier is depicted in Fig. 5. The time of a complete cycle is $T_{10} - T_0$. The details of the operation is explained as follows.

- T_0: at the end of the etching phase, S_d is turned on. The current I_c is connected in series with the voltage source V_{ssn}. At the same time, the switch-node voltage $v_{sw} = V_{ssn}$. The resonance between L_s and C_{eq} is then triggered. Since the resistance in the loop is small and can be neglected, the output voltage can be described by

$$v_{out}(t) = V_e + (V_s - V_e)\left[1 - \cos\left(\omega_r\left(t - T_0\right)\right)\right], \tag{12}$$

where ω_r is the resonant frequency and determined by $\omega_r = \frac{1}{\sqrt{L_s C_{eq}}}$. The output current is given by

$$i_{out}(t) = \frac{V_s - V_e}{\omega_o L}\sin\left(\omega_r\left(t - T_0\right)\right) - I_c. \tag{13}$$

I_c is typically much smaller than the magnitude of the current resonance thus it can be neglected.

- T_1: S_{sn} is turned off and the current source I_c is disconnected from the switch-node. In order to minimize the resonance, the energy stored in L_s should be fully discharged, which leads to

$$i_{out}(T_1) \simeq \frac{V_s - V_e}{\omega_o L}\sin\left(\omega_r\left(T_1 - T_0\right)\right) = 0. \tag{14}$$

Therefore, $T_1 = T_0 + \frac{1}{2}T_r$, where $T_r = \frac{2\pi}{\omega_r}$. In this time interval, $v_{out} = 2V_s - V_e$ and $i_L \simeq 0$.

- T_2: S_r is turned on after S_{sn} is turned off. The switch-node voltage v_{sw} is equal to the intermediate voltage V_{rsn}. In order to prevent short circuit, $T_2 - T_1$ should be larger than the dead-time of the MOSFET. Similar to T_0, the resonance is triggered again and the output voltage is given by

$$v_{out}(t) = 2V_s - V_e + (V_r - 2V_s + V_e)\left[1 - \cos\left(\omega_r\left(t - T_2\right)\right)\right]. \tag{15}$$

- T_3: S_d is turned on. Similar to T_1, in order to minimize the resonance, T_3 should be equal to $T_3 = T_2 + \frac{1}{2}T_r$. In this time interval, $v_{out} = 2V_r - 2V_s + V_e$ and $i_L \simeq 0$. Since v_{out} should be equal to V_d, V_r is then given by

$$V_r = V_s + \frac{1}{2}(V_d - V_e). \tag{16}$$

- T_4: S_r is turned off after S_d is turned on. The output voltage remains V_d.

- T_5: S_d is turned off. The output voltage remains V_d.
- T_6: S_f is turned on after S_d is turned off. The switch-node voltage v_{sw} is equal to the intermediate voltage V_{fsn}. In order to prevent short circuit, $T_6 - T_5$ should be larger than the dead-time of the MOSFET. The resonance is triggered and the output voltage is given by

$$v_{out}(t) = V_d + (V_f - V_d)\left[1 - \cos\left(\omega_r\left(t - T_6\right)\right)\right]. \tag{17}$$

- T_7: S_f is turned off and the etching phase is started. The resonance is minimized if $T_7 = T_6 + \frac{1}{2}T_r$. The output voltage in this interval is given by $v_{out} = 2V_f - V_d$. Since v_{out} should be equal to V_s, V_f is then governed by

$$V_f = \frac{1}{2}\left(V_s + V_d\right). \tag{18}$$

- T_8: S_{sn} is turned on after S_f is turned off. The current source I_c is connected to the switch-node and ready to sink current from the load. In order to prevent short circuit, $T_8 - T_7$ should be larger than the dead-time of the MOSFET. The output voltage remains V_s.
- T_9: S_{sp} is turned off after S_{sn} is turned on. All the voltage source are disconnected. The output voltage is decreasing linearly due to the the current source I_c. The slope rate of the output voltage is determined by

$$\frac{dv_{out}}{dt} = \frac{I_i C_{sub} - I_c\left(C_{sub} + C_{sh}\right)}{C_{sh}C_{sub} + C_t C_{sh} + C_t C_{sub}} \tag{19}$$

- T_{10}: the etching phase is ended. The end voltage V_e is approximated by

$$V_e = \frac{I_i C_{sub} - I_c\left(C_{sub} + C_{sh}\right)}{C_{sh}C_{sub} + C_t C_{sh} + C_t C_{sub}}T_{pro} + V_s, \tag{20}$$

where $T_{pro} \simeq T_{10} - T_7$.

Experimental validation

A prototype was built based on the introduced topology to validate the analysis. Experiments were conducted by testing the prototype in an Oxford Instruments FlexAL 2 plasma reactor, which is an atomic layer etching /deposition (ALE/D) machine, as shown in Fig. 6. The original integrated linear amplifier is uninstalled and the designed power amplifier is connected to the table instead. Argon plasma is excited and sustained with an RF power of 600 W and the constant pressure of the chamber is of 2.2 mTorr. A dielectric wafer is used as a substrate. It is hard to measure v_{sh} during the process inside the chamber but the ion energy distribution is measured by a retarding-field energy analyser (RFEA), as shown in Fig. 7.

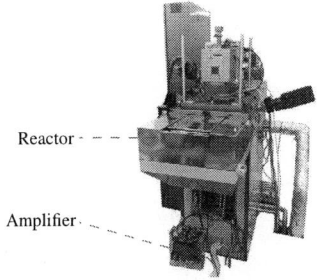

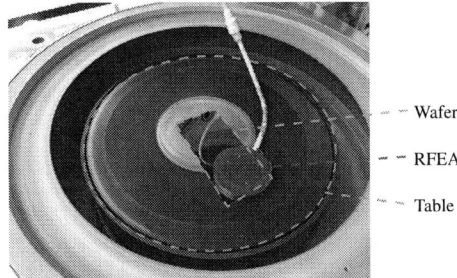

Fig. 6: The FlexAL 2 ALE/D machine.   Fig. 7: The reactor table, wafer and RFEA.

The waveforms of v_{out} and i_{out} are measured with high-bandwidth differential voltage probe and current probe respectively and shown in Fig. 8. V_d is about 30 V and V_s is about -190 V. Constant I_c is measured to be about 15 mA. The repetition rate is 100 kHz and duty cycle of etching is 90%, resulting in $V_e = -228$ V. The measured normalized ion energy distribution is shown in Fig. 9. It can be seen that the ion energy is mainly distributed around 232 eV, which is basically consistent with V_s. The plasma potential during etching can then be calculated by $V_P = 232 - 190 = 42$ V. The input power supplying to the power amplifier is only about 20 W. Compared to the linear amplifier used under the same condition [3] with a 200 W input power, a reduction of the input power of more than 90% is achieved while a narrow and single-peaked IED is obtained.

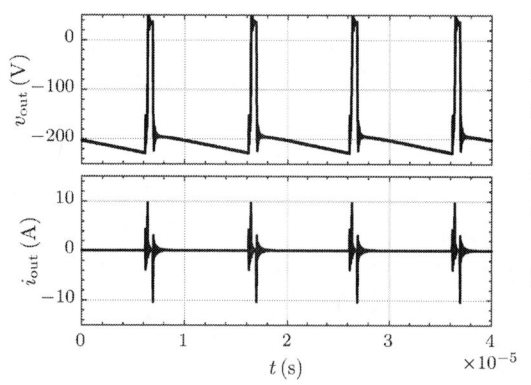

Fig. 8: The measured waveforms of v_{out} and i_{out}.

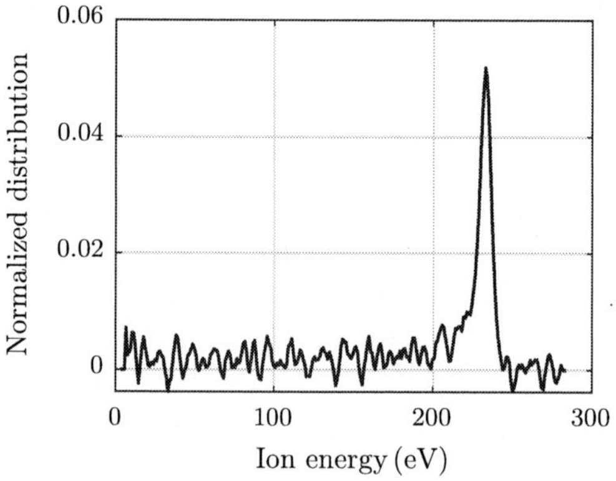

Fig. 9: The measured ion energy distribution.

Conclusion

Plasma etching requires a narrow ion energy distribution to improve the etching quality. In order to obtain precise ion energy control, a tailored pulse-shape voltage waveform generator is used for biasing the substrate. Typically a class-A linear amplifier is used, which is inefficient and expensive. In this paper, an electric equivalent model of the plasma chamber during etching is introduced and described analytically, based on which a switched-mode power amplifier concept is derived. Furthermore, an efficiency-improved topology is proposed by adopting resonant charging. The topology consists of four switched dc voltage sources and a dc current source. Two of the intermediate voltage sources are utilized during rise and fall transient of the voltage pulse in order to minimized the resonance.

A prototype amplifier is built and tested on a plasma etching machine. The measured ion energy yields a very narrow distribution while the input power of the prototype is reduced by more than 90% compared to that of a linear amplifier used in previous research, underlining the effectiveness of the concept. The paper shows a significant potential and prospect of bringing power electronics to the plasma physics application.

References

[1] M. A. Lieberman and A. J. Lichtenberg, *Principle of plasma discharges and materials processing*, 2nd ed. John Wiley & Sons, Inc., 2005, no. 1.

[2] D. J. Economou, "Tailored ion energy distributions on plasma electrodes," *J. Vac. Sci. Technol. A Vacuum, Surfaces, Film.*, vol. 31, no. 5, p. 050823, 2013.

[3] H. B. Profijt, M. C. M. van de Sanden, and W. M. M. Kessels, "Substrate-biasing during plasma-assisted atomic layer deposition to tailor metal-oxide thin film growth," *J. Vac. Sci. Technol. A Vacuum, Surfaces, Film.*, vol. 31, no. 1, p. 01A106, jan 2013.

[4] S. Rauf, "Effect of bias voltage waveform on ion energy distribution," *J. Appl. Phys.*, vol. 87, no. 11, pp. 7647–7651, 2000.

[5] P. Kudlacek, R. F. Rumphorst, and M. C. M. van de Sanden, "Accurate control of ion bombardment in remote plasmas using pulse-shaped biasing," *J. Appl. Phys.*, vol. 106, no. 7, p. 073303, oct 2009.

[6] S.-B. Wang and A. E. Wendt, "Control of ion energy distribution at substrates during plasma processing," *J. Appl. Phys.*, vol. 88, no. 2, pp. 643–646, jul 2000.

[7] I. T. Martin, M. A. Wank, M. A. Blauw, R. A. C. M. M. van Swaaij, W. M. M. Kessels, and M. C. M. van de Sanden, "The effect of low frequency pulse-shaped substrate bias on the remote plasma deposition of a-Si : H thin films," *Plasma Sources Sci. Technol.*, vol. 19, no. 1, p. 015012, jan 2010.

[8] F. F. Chen and J. P. Chang, *Lecture Notes on Principles of Plasma Processing*. Boston, MA: Springer US, 2003.

[9] Y. P. B. Yeung, K. W. E. Cheng, S. L. Ho, K. K. Law, and D. Sutanto, "Unified analysis of switched-capacitor resonant converters," *IEEE Trans. Ind. Electron.*, vol. 51, no. 4, pp. 864–873, 2004.

[10] A. Cervera, M. Evzelman, M. M. Peretz, and S. S. Ben-Yaakov, "A high-efficiency resonant switched capacitor converter with continuous conversion ratio," *IEEE Trans. Power Electron.*, vol. 30, no. 3, pp. 1373–1382, 2015.

[11] S.-H. Song and M. J. Kushner, "Role of the blocking capacitor in control of ion energy distributions in pulsed capacitively coupled plasmas sustained in Ar/CF4/O2," *J. Vac. Sci. Technol. A Vacuum, Surfaces, Film.*, vol. 32, no. 2, p. 021306, mar 2014.

Exploring the Boundaries and Effects of the Discontinuous Conduction Mode in H-Bridge Inverter with Dead-time

Qihao Yu, Erik Lemmen, Korneel Wijnands, Bas Vermulst
Department of Electrical Engineering, Eindhoven University of Technology
P.O. Box 513, 5600MB
Eindhoven, The Netherlands
Email: q.yu@tue.nl
URL: http://www.tue.nl/epe

Keywords

≪Time-Domain Analysis≫, ≪Pulse Width Modulation (PWM)≫, ≪Dead-time≫.

Abstract

Dead-time of an H-bridge inverter can cause nonlinear error on the inverter output. In different switching cycles during a fundamental period, the effect of dead-time might be different. The H-bridge inverter in different switching cycles can operate in three kinds of modes, including soft-switching continuous conduction mode, discontinuous conduction mode and hard-switching continuous conduction mode. In addition, the discontinuous conduction mode can be further classified into four different types, which have not been fully studied in previous research. In this paper, four different kinds of switching cycle in the discontinuous conduction mode are investigated. The effect of the dead-time on the voltage error is elaborated and the boundaries of each kind of switching cycle are determined by a series of constraint functions. Based on the analysis, a complete mathematical expression of the voltage error in a fundamental period is given and it yields a better accuracy compared to previous publications.

Introduction

The sine-pulse-width-modulated (SPWM) H-bridge inverter is widely used in industry, such as lithography and magnetic resonance imaging (MRI). Those applications demand high precision and low distortion on the inverter output [1, 2]. In practice, a certain dead-time should be applied to the H-bridge inverter to prevent short circuit. This dead-time introduces nonideal effect to the inverter output [3, 4, 5], which is typically the most dominant source of the distortion [6, 7]. In order to further reduce the distortion for high-precision applications, the effects of the dead-time should be investigated accurately.

During the dead-time, switches of both the upper and lower legs in the H-bridge inverter are turned-off. The switch-node voltage is then determined by the direction of the filter inductor current and the conductivity of the anti-parallel diodes, such as the body diodes of the MOSFETs [8]. Most of the previous researches have only considered the case of a distinctly positive or negative inductor current by neglecting the current ripple during the dead-time [9, 10, 11]. Neglecting the current ripple results in a fixed voltage error to the switch-node voltage. However, as illustrated in [12], soft-switching is achieved near the zero-crossing region of the inductor current, where there is no voltage error. Furthermore, discontinuous conduction mode (DCM) of the inductor current might be present in some switching cycles, which causes a variable voltage error to the switch-node voltage [13]. In [14], a mathematical expression was given to describe the values of the switch-node voltage under different conditions of the inductor current in a fundamental period. The expression was derived based on the assumption that the moving average value of the inductor current is equal to the reference output current. However, due to the voltage error in some switching cycles, current error is introduced to the output. Therefore, there is a deviation between the filter inductor current and the ideal output current. This deviation leads to an inaccurate estimation of the voltage error and the boundaries of DCM, especially in case of a high inductor current ripple.

The aim of this paper is to investigate the dead-time effect on the H-bridge inverter output in DCM. The effect of different DCM are analysed in detail. The clamping time during which inductor current is clamped to zero during the time is calculated. Moreover, an accuracy-improved model of the voltage error in DCM compared to [14] is proposed by taking the deviation of inductor current from the ideal current into consideration. The boundaries of the DCM are determined with a numerical method. Simulations are conducted to verify the model, underlining the proposed model have significantly increased the accuracy.

Switching cycle analysis

The topology of an H-bridge inverter is shown in Fig. 1. In this research, digital natural sampling PWM and bipolar modulation are used for illustration, as shown in Fig. 2. The output capacitances of the switches are neglected thus the rise and fall time of the switching transient are assumed to be zero.

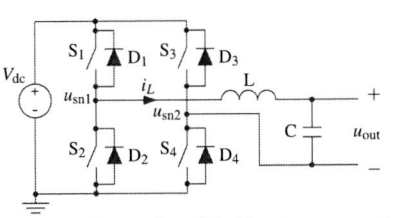

Fig. 1: The topology of an H-bridge inverter with output filter.

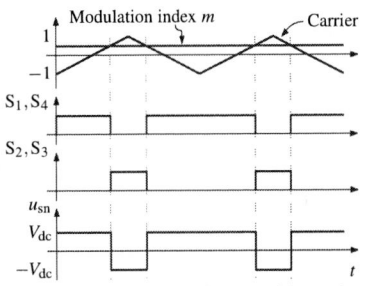

Fig. 2: The ideal PWM waveforms without dead-time. m is the modulation index. S_1,S_4 and S_2,S_3 represent the on-and-off state of each switch. $u_{sn} = u_{sn1} - u_{sn2}$.

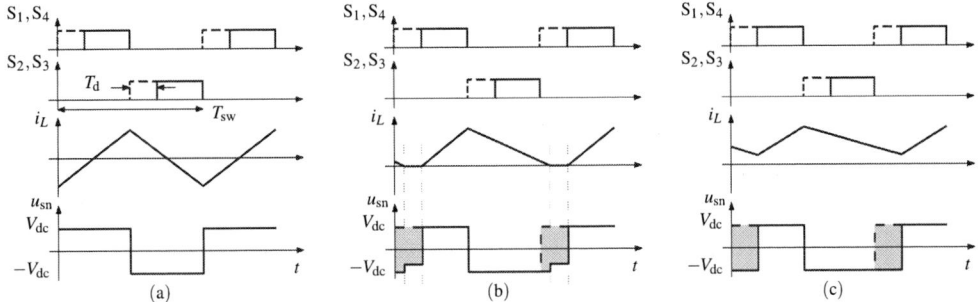

Fig. 3: Waveforms during a switching cycle. (a): soft-switching continuous conduction mode; (b): discontinuous conduction mode; (c): hard-switching continuous conduction mode. The dashed and solid line represent the ideal case without dead-time and practical case with dead-time respectively.

During a switching cycle $[nT_{sw}, (n+1)T_{sw}]$ with T_{sw} representing for the switching period and n a non-negative integer, the modulation index for a digital PWM is defined by

$$m(n) = M \sin\left(\frac{2\pi n}{N_{sw}}\right), \tag{1}$$

where M is the modulation depth and N_{sw} is the ratio of the switching frequency f_{sw} to the output frequency f_o. The moving average value of the switch-node voltage and the output voltage are defined by $\langle u_{sn}(n)\rangle$ and $\langle u_{out}(n)\rangle$ respectively and the operator $\langle \cdot \rangle$ is the average value in a switching cycle. Ideally $\langle u_{sn}(n)\rangle = \langle u_{out}(n)\rangle = V_{dc}m(n)$. However, a certain dead-time T_d is required to prevent short circuit in practice, which results in a voltage error and is given by

$$\langle u_{\mathrm{e}}(n)\rangle = \langle u_{\mathrm{out}}(n)\rangle^* - \langle u_{\mathrm{out}}(n)\rangle, \tag{2}$$

where $\langle u_{\mathrm{out}}(n)\rangle^*$ is the moving average value of the reference voltage. As shown in Fig. 3, the switching cycle can be divided into three modes depending on the inductor current waveform: soft-switching continuous conduction mode (SSCCM), discontinuous conduction mode (DCM) and hard-switching continuous conduction mode (HSCCM). SSCCM and HSCCM have been investigated into in [14] and they are not the focus in this paper. While for DCM, it can be further classified into four different switching modes, as shown in Fig. 4.

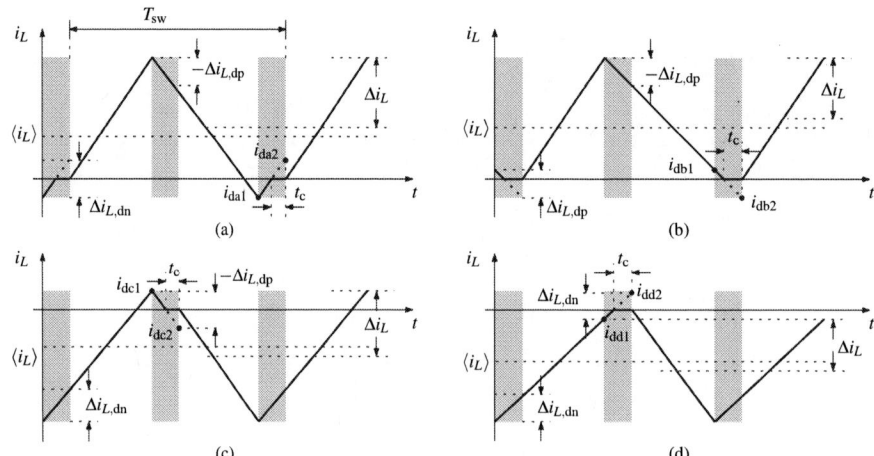

Fig. 4: The inductor current waveforms for DCM. The inductor current (a) rises from negative to zero and is clamped during dead-time with a positive average value in a switching cycle; (b) falls from positive to zero and is clamped during dead-time with a positive average value in a switching cycle; (c) falls from positive to zero and is clamped during dead-time with a negative average value in a switching cycle; (d) rises from negative to zero and is clamped during dead-time with a negative average value in a switching cycle.

In order to describe each DCM switching cycle, several parameters should be defined as follows. The ideal moving average value of filter inductor current without dead-time during a switching cycle is

$$\langle i_L(n)\rangle^* = \frac{MV_{\mathrm{dc}}}{Z}\sin\left(\frac{2\pi n}{N_{\mathrm{sw}}} - \varphi_o\right), \tag{3}$$

where Z and φ_o are the impedance and phase angle of the load. If dead-time is neglected, the peak-to-average inductor current ripple $\Delta i_L(n)$ during a switching cycle is given by

$$\Delta i_L(n) = \frac{T_{\mathrm{sw}}V_{\mathrm{dc}}}{4L}\left[1 - m(n)^2\right]. \tag{4}$$

The inductor current change during dead-time when the inductor current is distinctly positive and negative, which is achieved in SSCCM and HSCCM, is defined by

$$\Delta i_{L,\mathrm{dp}}(n) = -\frac{V_{\mathrm{dc}}T_{\mathrm{d}}}{L}\left[1 + m(n)\right] \tag{5}$$

and

$$\Delta i_{L,\mathrm{dn}}(n) = \frac{V_{\mathrm{dc}}T_{\mathrm{d}}}{L}\left[1 - m(n)\right] \tag{6}$$

respectively. Taking the switching cycle analysis of DCM shown in Fig. 4 (a) for an example, the actual

Table I: Clamping time $t_c(n)$ and moving average voltage error $\langle u_e(n) \rangle$ of each discontinuous conduction mode

Mode	$t_c(n)$	$\langle u_e(n) \rangle$
(a)	$\dfrac{\langle i_L(n) \rangle^* - \Delta i_L(n) + \Delta i_{L,\mathrm{dn}}(n)}{\frac{\Delta i_{L,\mathrm{dn}}(n)}{T_\mathrm{d}} - \frac{\Delta i_L(n)}{T_\mathrm{sw}} + \frac{[1-m(n)]}{ZT_\mathrm{sw}}}$	$\dfrac{V_\mathrm{dc}\left[1 - m(n)\right] t_c(n)}{T_\mathrm{sw}}$
(b)	$\dfrac{\langle i_L(n) \rangle^* - 2\frac{V_\mathrm{dc}T_\mathrm{d}}{ZT_\mathrm{sw}} - \Delta i_L(n) + \frac{1}{2}\Delta i_{L,\mathrm{dp}}(n)}{\frac{\Delta i_{L,\mathrm{dp}}(n)}{2T_\mathrm{d}} - \frac{\Delta i_L(n)}{T_\mathrm{sw}} - \frac{[1+m(n)]}{ZT_\mathrm{sw}}}$	$\dfrac{2V_\mathrm{dc}T_\mathrm{d} - (1 + m(n)) V_\mathrm{dc} t_c(n)}{T_\mathrm{sw}}$
(c)	$\dfrac{\langle i_L(n) \rangle^* + \Delta i_L(n) + \Delta i_{L,\mathrm{dp}}(n)}{\frac{\Delta i_{L,\mathrm{dp}}(n)}{T_\mathrm{d}} + \frac{\Delta i_L(n)}{T_\mathrm{sw}} - \frac{[1+m(n)]}{ZT_\mathrm{sw}}}$	$-\dfrac{V_\mathrm{dc}\left[1 + m(n)\right] t_c(n)}{T_\mathrm{sw}}$
(d)	$\dfrac{\langle i_L(n) \rangle^* + 2\frac{V_\mathrm{dc}T_\mathrm{d}}{ZT_\mathrm{sw}} + \Delta i_L(n) + \frac{1}{2}\Delta i_{L,\mathrm{dn}}(n)}{\frac{\Delta i_{L,\mathrm{dn}}(n)}{2T_\mathrm{d}} + \frac{\Delta i_L(n)}{T_\mathrm{sw}} + \frac{[1-m(n)]}{ZT_\mathrm{sw}}}$	$\dfrac{-2V_\mathrm{dc}T_\mathrm{d} + (1 - m(n)) V_\mathrm{dc} t_c(n)}{T_\mathrm{sw}}$

moving average inductor current can be written as

$$\langle i_L(n) \rangle = \frac{T_\mathrm{sw} - t_c(n)}{2T_\mathrm{sw}} \cdot 2\Delta i_L(n) - \frac{2\left[T_\mathrm{d} - t_c(n)\right]}{\left[1 + m(n)\right] T_\mathrm{sw}} \cdot 2\Delta i_L(n), \tag{7}$$

where $t_c(n)$ is the clamping time during which the inductor current is clamped to zero. In previous research [14], $\langle i_L(n) \rangle$ is assumed to be $\langle i_L(n) \rangle^*$. However, a current error introduced by the voltage error during the dead-time is introduced. For a inductive load with dominant resistance, the actual moving average inductor current can be simplified by

$$\langle i_L(n) \rangle = \langle i_L(n) \rangle^* - \frac{\langle u_e(n) \rangle}{Z}. \tag{8}$$

The moving average voltage error $\langle u_e(n) \rangle$ in Fig. 3 (a) is governed by

$$\langle u_e(n) \rangle = \frac{\left[1 - m(n)\right] V_\mathrm{dc} t_c(n)}{T_\mathrm{sw}}. \tag{9}$$

Solving (7),(8) and (9) for $t_c(n)$ leads to

$$t_c(n) = \frac{\langle i_L(n) \rangle^* - \Delta i_L(n) + \Delta i_{L,\mathrm{dn}}(n)}{\frac{\Delta i_{L,\mathrm{dn}}(n)}{T_\mathrm{d}} - \frac{\Delta i_L(n)}{T_\mathrm{sw}} + \frac{[1-m(n)]}{ZT_\mathrm{sw}}}. \tag{10}$$

Therefore, the moving average voltage error $\langle u_e(n) \rangle$ in Fig. 3 (a) can be solved by substituting (10) into (9). Similarly, the clamping time and the moving average voltage error $\langle u_e(n) \rangle$ of DCM described in Fig. 3 (b), (c) and (d) can be analytically solved, which are listed in Table I.

In order to obtain a complete expression of the moving average voltage error $\langle u_e(n) \rangle$ in a fundamental period, the boundaries of each DCM should be found. As shown in Fig. 4, the boundaries of the switching cycle depicted in Fig. 4 (a) can be described by

$$\begin{cases} \langle i_L(n) \rangle - \Delta i_L(n) + \Delta i_{L,\mathrm{dn}}(n) \geq 0, \\ \langle i_L(n) \rangle - \frac{V_\mathrm{dc}T_\mathrm{d}(1-m(n))}{ZT_\mathrm{sw}} - \Delta i_L(n) + \frac{T_\mathrm{d}}{T_\mathrm{sw}}\Delta i_L(n) < 0. \end{cases} \tag{11}$$

Based on the above analysis, two constraint functions $y_\mathrm{sn}(n)$ and $y_\mathrm{dn}(n)$ can be defined to simplify the expressions as

$$\begin{cases} y_\mathrm{sn}(n) = \langle i_L(n) \rangle - \Delta i_L(n) + \Delta i_{L,\mathrm{dn}}(n), \\ y_\mathrm{dn}(n) = \langle i_L(n) \rangle - \frac{V_\mathrm{dc}T_\mathrm{d}(1-m(n))}{ZT_\mathrm{sw}} - \Delta i_L(n) + \frac{T_\mathrm{d}}{T_\mathrm{sw}}. \end{cases} \tag{12}$$

Table II: The boundaries and time intervals of each switching modes

Mode	Boundary	Time interval
(a)	$\begin{cases} y_{\mathrm{sn}}(n) = \langle i_L(n) \rangle - \Delta i_L(n) + \Delta i_{L,\mathrm{dn}}(n) \geq 0, \\ y_{\mathrm{dn}}(n) = \langle i_L(n) \rangle - \frac{V_{\mathrm{dc}} T_{\mathrm{d}}(1-m(n))}{Z T_{\mathrm{sw}}} - \Delta i_L(n) + \frac{T_{\mathrm{d}}}{T_{\mathrm{sw}}} \Delta i_L(n) < 0. \end{cases}$	$[N_{\mathrm{sn1}}, N_{\mathrm{dn1}}) \cap (N_{\mathrm{dn2}}, N_{\mathrm{sn2}}]$
(b)	$\begin{cases} y_{\mathrm{dn}}(n) = \langle i_L(n) \rangle - \frac{V_{\mathrm{dc}} T_{\mathrm{d}}(1-m(n))}{Z T_{\mathrm{sw}}} - \Delta i_L(n) + \frac{T_{\mathrm{d}}}{T_{\mathrm{sw}}} \Delta i_L(n) \geq 0, \\ y_{\mathrm{hn}}(n) = \langle i_L(n) \rangle - \frac{2 V_{\mathrm{dc}} T_{\mathrm{d}}}{Z T_{\mathrm{sw}}} - \Delta i_L(n) + \frac{1}{2} \Delta i_{L,\mathrm{dp}}(n) < 0. \end{cases}$	$[N_{\mathrm{dn1}}, N_{\mathrm{hn1}}) \cap (N_{\mathrm{hn2}}, N_{\mathrm{dn2}}]$
(c)	$\begin{cases} y_{\mathrm{sp}}(n) = \langle i_L(n) \rangle + \Delta i_L(n) + \Delta i_{L,\mathrm{dp}}(n) \leq 0, \\ y_{\mathrm{dp}}(n) = \langle i_L(n) \rangle + \frac{V_{\mathrm{dc}} T_{\mathrm{d}}(1+m(n))}{Z T_{\mathrm{sw}}} + \Delta i_L(n) - \frac{T_{\mathrm{d}}}{T_{\mathrm{sw}}} \Delta i_L(n) > 0. \end{cases}$	$[N_{\mathrm{sp1}}, N_{\mathrm{dp1}}) \cap (N_{\mathrm{dp2}}, N_{\mathrm{sp2}}]$
(d)	$\begin{cases} y_{\mathrm{dp}}(n) = \langle i_L(n) \rangle + \frac{V_{\mathrm{dc}} T_{\mathrm{d}}(1+m(n))}{Z T_{\mathrm{sw}}} + \Delta i_L(n) - \frac{T_{\mathrm{d}}}{T_{\mathrm{sw}}} \Delta i_L(n) \leq 0, \\ y_{\mathrm{hp}}(n) = \langle i_L(n) \rangle + \frac{2 V_{\mathrm{dc}} T_{\mathrm{d}}}{Z T_{\mathrm{sw}}} + \Delta i_L(n) + \frac{1}{2} \Delta i_{L,\mathrm{dn}}(n) > 0. \end{cases}$	$[N_{\mathrm{dp1}}, N_{\mathrm{hp1}}) \cap (N_{\mathrm{hp2}}, N_{\mathrm{dp2}}]$

Solving $y_{\mathrm{sn}}(n)$ and $y_{\mathrm{dn}}(n)$ in the period of $[0, N_{\mathrm{sw}}]$ gives zero-crossing $N_{\mathrm{sn1}}, N_{\mathrm{sn2}}$ ($N_{\mathrm{sn2}} > N_{\mathrm{sn1}}$) and N_{dp1}, N_{dp2} ($N_{\mathrm{dp2}} > N_{\mathrm{dp1}}$) respectively. Therefore, when $n \in [N_{\mathrm{sn1}}, N_{\mathrm{dn1}}) \cap (N_{\mathrm{dn2}}, N_{\mathrm{sn2}}]$ and $n \in [N_{\mathrm{dp2}}, N_{\mathrm{sn2}}]$, the voltage error is given by (9). The similar analysis also applies to other DCM switching cycles illustrated in Fig. 4 (b), (c) and (d) in order to obtain the voltage error. A complete expression of the boundaries of DCM is given in Table II. A typical plot of the constraint functions is depicted in Fig. 5.

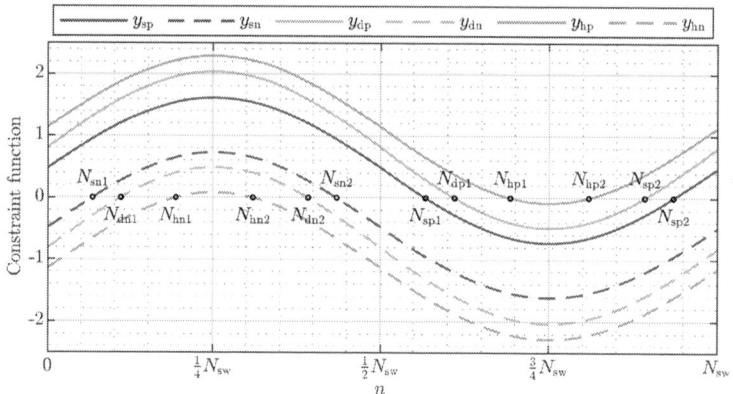

Fig. 5: A typical plot of constraint functions.

With the boundaries of DCM clear, it is easy to derive that SSCCM is achieved in $[0, N_{\mathrm{sn1}}] \cup [N_{\mathrm{sn2}}, N_{\mathrm{sp1}}] \cup [N_{\mathrm{sp2}}, N_{\mathrm{sw}} - 1]$ and the moving average voltage error $\langle u_{\mathrm{e}}(n) \rangle$ is zero. HSCCM happens in $[N_{\mathrm{hn1}}, N_{\mathrm{hn2}}]$ and $[N_{\mathrm{hp1}}, N_{\mathrm{hp2}}]$. The moving average voltage error $\langle u_{\mathrm{e}}(n) \rangle$ is $\frac{2}{T_{\mathrm{sw}}} V_{\mathrm{dc}} T_{\mathrm{d}}$ in $[N_{\mathrm{hn1}}, N_{\mathrm{hn2}}]$ and $-\frac{2}{T_{\mathrm{sw}}} V_{\mathrm{dc}} T_{\mathrm{d}}$ in $[N_{\mathrm{hp1}}, N_{\mathrm{hp2}}]$. As a result, a complete expression of the voltage error in a fundamental period can be easily derived.

It should be noted that the constraint functions have not necessarily two zero-crossings. For example, if $y_{\mathrm{sp}}(n)$ and $y_{\mathrm{sn}}(n)$ have no zero-crossing point in a fundamental period, then there is no DCM or HSCCM switching cycle and thus soft-switching is achieved all the time. In this case, the dead-time causes no voltage error. Similarly, if $y_{\mathrm{hp}}(n)$ and $y_{\mathrm{hn}}(n)$ have no zero-crossing point in a fundamental period, then there are only SSCCM and DCM but no HSCCM switching cycle. The HSCCM cycle exists if and only if all the constraint functions have two zero-crossing points. Fig. 6 depicts three typical combinations of the constraint functions and the calculated voltage error for an H-bridge inverter working under different modulation depths. As shown in the figure, when solely changing the modulation depth, the time intervals of the switching modes in a fundamental period change accordingly. When the modulation depth is low, the output current is low making the inductor current ripple relatively high and SSCCM can be achieved all the time. When increasing the modulation index, DCM and HSCCM appear one after another.

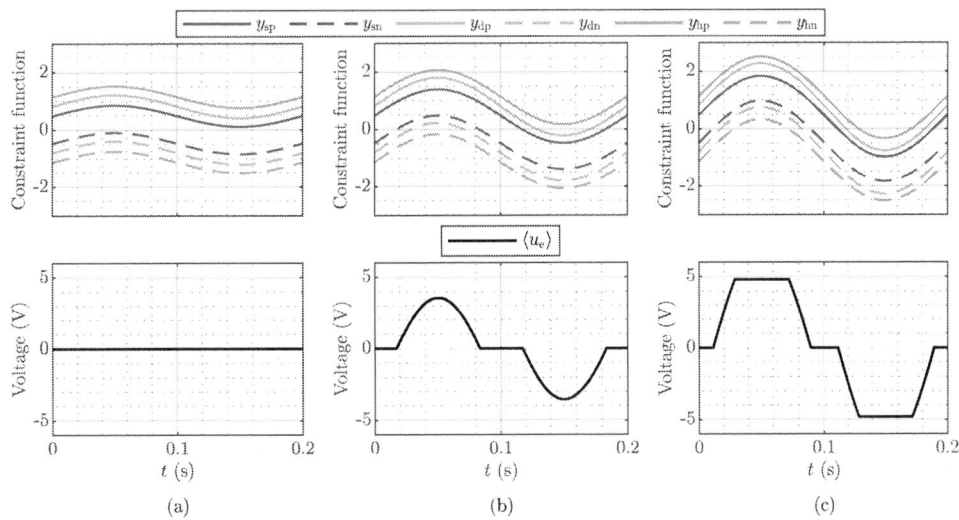

(a)   (b)   (c)

Fig. 6: Waveforms of the calculated constraint functions y_{sp}, y_{sn}, y_{dp}, y_{dn}, y_{hp} and y_{hn} and the calculated moving average voltage error $\langle u_e \rangle$ of an H-bridge inverter working at $V_{dc} = 48$ V, $T_d = 5$ μs, $f_{sw} = 10$ kHz, $f_o = 5$ Hz, $L = 2$ mH, $C = 30$ μF, $R = 10$ Ω and (a) $M = 0.08$, (b) $M = 0.2$, (c) $M = 0.3$.

Experimental validation

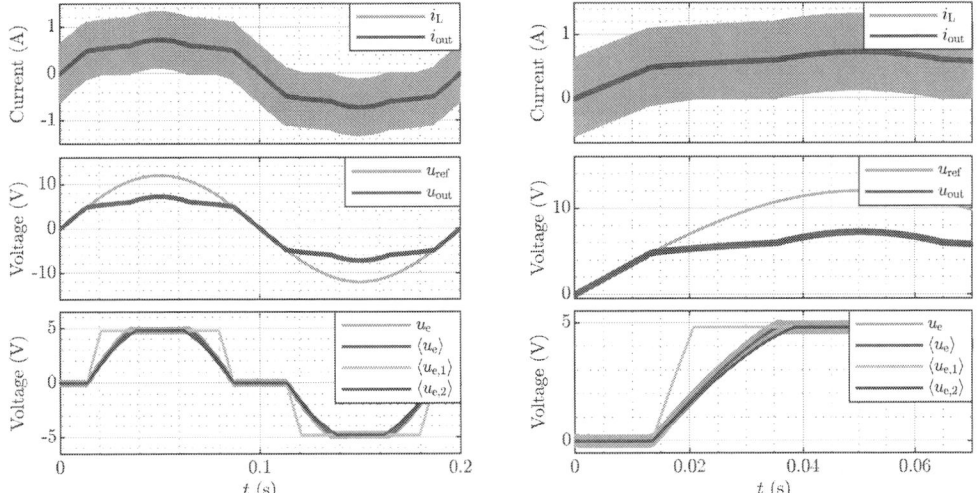

Fig. 7: The simulated waveforms of the H-bridge inverter.

Fig. 8: A zoom-in of the simulated waveforms in DCM.

Simulations were done using MATLAB/Simulink by adopting ideal MOSFETs with no parasitic capacitance to verify the presented model. The simulation is conducted under the condition that $V_{dc} = 48$ V, $M = 0.25$, $f_o = 5$ Hz, $f_{sw} = 10$ kHz, $T_d = 5$ μs, $L_l = 0$, $R = 10$ Ω, $L = 2$ mH and $C = 30$ μF. The simulated waveforms in a fundamental period are depicted in Fig. 7 and a zoom-in of the waveforms in DCM is shown in Fig. 8, where u_e, $\langle u_e \rangle$, $\langle u_{e,1} \rangle$ and $\langle u_{e,2} \rangle$ are the simulated voltage error, the moving average value of the simulated voltage error, calculated moving average value of voltage error based on the method in [14] and in this paper respectively. As can be seen, both the voltage error magnitude and the boundaries of DCM estimated by this paper shows an improved accuracy compared to the previous one.

Euclidean distance is a mathematical indicator of the similarities of two different sequence [15]. By using the simulation results as a benchmark, the accuracy of the calculated results by the different methods can

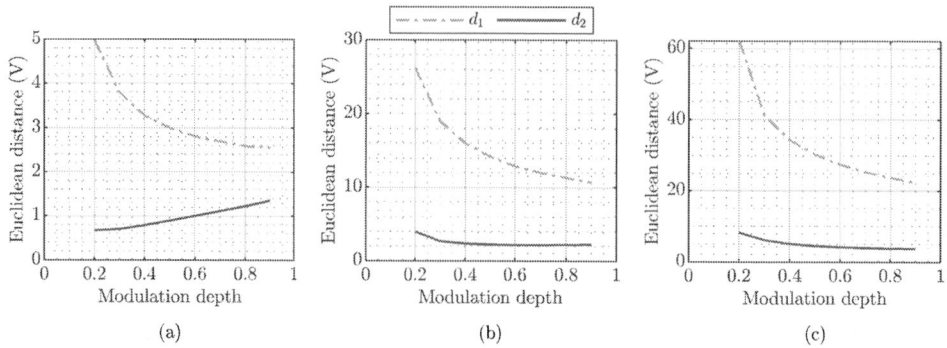

Fig. 9: The Euclidean distances for the different methods under (a) $T_\mathrm{d} = 1\ \mu s$; (b) $T_\mathrm{d} = 3\ \mu s$; (c) $T_\mathrm{d} = 5\ \mu s$.

be compared. The Euclidean distance is defined by

$$d_x = \sqrt{\sum_{n=0}^{N_\mathrm{sw}-1} \left(\langle u_{\mathrm{e},x}(n)\rangle - \langle u_\mathrm{e}(n)\rangle\right)^2},\ x = 1, 2. \tag{13}$$

As an example of the waveforms shown in Fig. 7 and Fig. 8, the differences of the calculated results by the method from [14] and the method presented in this paper yield to $d_1 = 47.4$ V and $d_2 = 7.59$ V. In general, a smaller Euclidean distance indicates a better similarity of two waveforms.

In order to see the accuracy comparison under different conditions, simulations are done by varying the modulation depth and the dead-time value while other properties remaining unchanged and the results are shown in Fig. 9. As can be seen, the presented model in this paper has a better accuracy than the previous one under different dead-time and modulation depth. Normally, a greater dead-time results in a larger Euclidean distance for both methods since a larger dead-time causes a larger portion of DCM and the calculation error in DCM accumulates as regards Euclidean distance. Not only the value of Euclidean distance increases with an increasing dead-time, but also the difference of the value is exaggerated, underlining the accuracy and the effectiveness of the presented model.

In general, the Euclidean distance is smaller with a higher modulation depth due to the fact that the portion of HSCCM is larger while that of DCM is smaller thus the calculation error in DCM becomes less significant. However, as shown in Fig. 9 (a), Euclidean distance calculated by this method increases with an increasing modulation depth. The reason is that with such small dead-time, the portion of DCM is small thus the calculation error in DCM is not dominant in the final result of Euclidean distance. On the other hand, the voltage error introduced by the SPWM becomes dominant and with an increasing modulation depth this error becomes governing on Euclidean distance.

Conclusion

Previous researches have not fully explored the dead-time effect in DCM of an H-bridge inverter. In this paper, four different DCMs are elaborated and the effect of dead-time to the output voltage error in different DCM is analysed and described by mathematical models. A series of constraint functions are defined such that the boundaries and the time intervals of each DCM can be determined. The combination of the voltage error models and constraint functions enables a complete expression of the voltage error in a fundamental period. Compared to previous researches, by taking the current error of the inductor current compared to the ideal current into consideration, the calculation done in this paper yields a better accuracy which is validated by simulations. The method and the results of the calculation introduced in this paper can be used for output harmonics or distortions modelling and feed-forward dead-time compensation, especially for high-precision applications.

References

[1] S. J. Settels, J. Duarte, J. van Duivenbode, and E. Lomonova, "A 2kV Charge-based ZVS Three-level Inverter," *IEEE Trans. Power Electron.*, vol. 8993, no. c, pp. 1–1, 2019.

[2] Q. Yu, R. Baeten, E. Lemmen, B. Vermulst, and K. Wijnands, "A 1 MHz Wide Bandgap Power Amplifier for High-Precision Applications," in *2019 21st Eur. Conf. Power Electron. Appl. (EPE '19 ECCE Eur.* IEEE, sep 2019, pp. P.1–P.7.

[3] F. Chierchie and E. Paolini, "Analytical and numerical analysis of dead-time distortion in power inverters," *Argentine Sch. Micro-Nanoelectronics Technol. Appl. (EAMTA), 2010*, pp. 6–11, 2010.

[4] N. Urasaki, T. Senjyu, T. Kinjo, T. Funabashi, and H. Sekine, "Dead-time compensation strategy for permanent magnet synchronous motor drive taking zero-current clamp and parasitic capacitance effects into account," *IEE Proc. - Electr. Power Appl.*, vol. 152, no. 4, p. 845, 2005.

[5] Z. Zhang and L. Xu, "Dead-time compensation of inverters considering snubber and parasitic capacitance," *IEEE Trans. Power Electron.*, vol. 29, no. 6, pp. 3179–3187, 2014.

[6] R. J. Kerkman, D. Leggate, D. W. Schlegel, and C. Winterhalter, "Effects of parasitics on the control of voltage source inverters," *IEEE Trans. Power Electron.*, vol. 18, no. 1 I, pp. 140–150, 2003.

[7] S. Y. Kim, W. Lee, M. S. Rho, and S. Y. Park, "Effective dead-time compensation using a simple vectorial disturbance estimator in PMSM drives," *IEEE Trans. Ind. Electron.*, vol. 57, no. 5, pp. 1609–1614, 2010.

[8] A. Munoz and T. Lipo, "On-line dead-time compensation technique for open-loop PWM-VSI drives," *IEEE Trans. Power Electron.*, vol. 14, no. 4, pp. 683–689, jul 1999.

[9] C. Wu, W. H. Lau, and H. S. H. Chung, "Analytical technique for calculating the output harmonics of an H-bridge inverter with dead time," *IEEE Trans. Circuits Syst. I Fundam. Theory Appl.*, vol. 46, no. 5, pp. 617–627, 1999.

[10] C. Attaianese, V. Nardi, and G. Tomasso, "A novel SVM strategy for VSI dead-time-effect reduction," *IEEE Trans. Ind. Appl.*, vol. 41, no. 6, pp. 1667–1674, 2005.

[11] Yong-Kai Lin and Yen-Shin Lai, "Dead-Time Elimination of PWM-Controlled Inverter/Converter Without Separate Power Sources for Current Polarity Detection Circuit," *IEEE Trans. Ind. Electron.*, vol. 56, no. 6, pp. 2121–2127, 2009.

[12] J. M. Schellekens, R. A. M. Bierbooms, and J. L. Duarte, "Dead-time compensation for PWM amplifiers using simple feed-forward techniques," *19th Int. Conf. Electr. Mach. ICEM 2010*, pp. 1–6, 2010.

[13] M. A. Herrán, J. R. Fischer, S. A. González, M. G. Judewicz, and D. O. Carrica, "Adaptive dead-time compensation for grid-connected PWM inverters of single-stage PV systems," *IEEE Trans. Power Electron.*, vol. 28, no. 6, pp. 2816–2825, 2013.

[14] Q. Yu, E. Lemmen, and B. Vermulst, "A Numerical Method for Calculating the Output Spectrum of an H-Bridge Inverter with Dead-time Based on Switching Mode Analysis," in *2019 IEEE Energy Convers. Congr. Expo.* IEEE, sep 2019, pp. 2245–2251.

[15] C. Faloutsos, M. Ranganathan, and Y. Manolopoulos, "Fast Subsequence Matching in Time-Series Databases," *ACM SIGMOD Rec.*, vol. 23, no. 2, pp. 419–429, 1994.

Figures-of-Merit and current metric for the comparison of IGCTs and IGBTs in Modular Multilevel Converters

Arthur Boutry[12], Cyril Buttay[2], Dong Dong[3], Rolando Burgos[3], Bruno Lefebvre[1],
Florent Morel[1], and Colin Davidson[4]

[1] SuperGrid Institute
23 rue Cyprian
F-69100 Villeurbanne, France

[2] Univ Lyon
INSA Lyon, CNRS
Laboratoire Ampère
F-69621 Villeurbanne, France

[3] Center for Power Electronics Systems
The Bradley Dept. of Electrical and Computer Eng.
Virginia Polytechnic Institute and State University
Blacksburg, VA 24061, USA

[4] GE Grid Solutions
Redhill Business Park,
Harry Kerr Dr,
Stafford, United Kingdom

E-mail: arthur.boutry@supergrid-institute.com

Acknowledgements

This work was supported by a grant overseen by the French National Research Agency (ANR) as part of the "Investissements d'Avenir" Program ANE-ITE-002-01.

Keywords

≪Multilevel converters≫, ≪IGCT≫, ≪IGBT≫, ≪Switching losses≫, ≪Conduction losses≫.

Abstract

IGCTs and IGBTs are compared in the case of a HVDC MMC. Specific figures of merit, and a current metric providing simple means to compare them, are introduced and discussed. Simulation results of a MMC model and figures of merit are shown to provide consistent results, proving that the proposed figures of merit are a very simple and fast way to select the best semiconductor switch. Furthermore, our analysis supports the growing interest in IGCTs for MMCs, as they are found to produce the lowest level of losses.

1 Introduction

The Modular Multilevel Converter (MMC) is a Voltage Source Converter (VSC) developed and used for Medium or High Voltage Direct Current (MVDC or HVDC) applications. It is a reversible, ac/dc Converter. The MMC (fig. 1a) is based on submodules (SMs), its elementary building blocks. Mainly composed of switches and a capacitor, a submodule can be seen as a small voltage source that can be inserted or not (depending on the switching sequence and the current sign along the submodules). The main type of submodule has a half-bridge topology, consisting in two switches and their associated anti-parallel diodes, one capacitor and auxiliary systems.

Most submodule designs rely on IGBTs as the semiconductor switch. This is the case, for example in HVDC-MMCs produced by General Electric, ABB, Siemens [3] or RXPE [4]. Indeed, IGBT modules

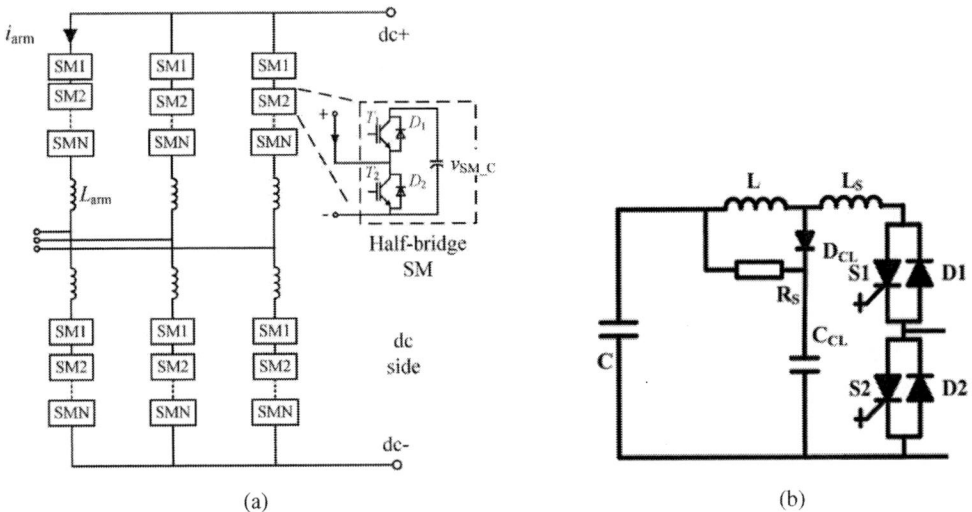

Fig. 1: (a) circuit diagram of the modular multilevel converter [1] and of one of its submodules (SM); (b) Snubber circuit for an IGCT-based submodule, extracted from [2].

offer high voltage ratings up to 6.5 kV, high current ratings, are fully controllable with little power, and can be sourced from many suppliers. On the contrary, thyristors, which are often used in other HVDC converters would require a complex circuitry to be turned-off. In theory, high voltage SiC MOSFETs would be ideal components for such application due to their low on-state voltage and low switching losses, but their cost and current ratings are currently prohibitive. Several authors [5, 6, 7] have investigated the possibility of using IGCTs (a type of gate-controlled thyristor that can be turned off without parallel-type snubber) in a MMC. They demonstrated that doing so would increase the conversion efficiency of a MMC as compared to using IGBTs. Note that IGCTs require a series-type snubber (as shown on fig. 1b) to limit the switching speed and protect the associated freewheeling diode when the IGCT turns on [2, 8]. Even considering the additional power dissipation caused by this snubber circuit, IGCTs were found to be more efficient than IGBTs in the case of a MMC [5]. In [9], the cost of the IGCT-based submodules is compared to IGBT-based submodules and is found to offer lower capital cost per megawatt and lower life-cycle cost per megawatt.

However, because of the very different principles IGBTs and IGCTs operate on, they cannot be compared directly from the figures quoted in their datasheets. Therefore, there is a concern that the advantage of IGCTs over IGBT could result from a biased comparison, using devices with very different ratings. In this paper, we introduce new Figures Of Merit (FOMs), as a means for easy and accurate selection of semiconductor switches for a MMC application. The validity of these FOMs is then assessed against a dynamic MMC model: unlike an average arm model, the current waveforms, the on-state voltages and the losses are calculated at each instant. Such model provides realistic waveforms and losses. In particular, the actual switching frequency and switching instants of the semiconductor devices are an outcome of the simulation (MMCs do not operate at a constant frequency) and not an input parameter.

2 Comparison between IGCTs and IGBTs

2.1 General considerations and features

Table I presents an overview of the main features and parameters of of IGCTs and IGBTs. The IGCT is the result of the evolution of the thyristor technology. The main type of thyristor is the SCR (Silicon Controlled Rectifier). It is a semiconductor device made of four layers ($P^+N^-PN^+$ structure) which is turned-on by applying a current pulse on its gate, and turns off naturally when current stops flowing through its anode. The GTO (Gate Turn-Off) thyristor was introduced in 1980 [10]. Compared to the

SCR, it offers a controlled turn-off capability by applying a negative voltage to the gate. The GCT is an evolution of the GTO. Two improvements are the base of the GCT: a low-inductive housing-design allowing a quick, more homogeneous turn-off without filamentation (non-uniform current distribution), and a "transparent emitter buffer layer". The buffer-layer and transparent emitter technologies consist in a weakly doped n-layer located between the n-base and the p-emitter. This results in reductions in on-state voltage and turn-off energies. The low-inductive structure of the GCT allows to rapidly redirect all the anode current from the cathode to the gate at turn-off. This prevents filamentation [11] and reduces turn-off times (typical values: 10 μs for GCTs compared to 100 μs for GTOs). Another consequence is a much better immunity to dV/dt, removing the need for a turn-off snubber. The IGCT (Integrated GCT) corresponds to a GCT switch attached to its gate drive circuit. The IGCT switching frequency is limited by its driver [11], and at constant turning-off current, the higher the switching frequency the higher the power consumption of the gate driver. Furthermore, the lifetime of the gate drive is limited by that of its electrolytic capacitors, which is affected by the operating temperature.

The IGBT is a voltage-driven semiconductor developed in the early 1980s using a combination of a MOSFET and a N-P-N transistor structure [12]. Its relatively low losses and its ability to operate at high frequencies (several kilohertz) make it widely used in many fields. It requires a simple gate unit with a low power consumption. Compared to the GCT or other thyristors, which are made on a whole semiconductor wafer, IGBT modules are formed by paralleling a number of smaller semiconductor chips, making it scalable to different power ratings.

The technologies are still being improved with new structures for the IGBT (Enhanced Trench IGBTs – TSPT+ – for ABB [13], or the Injection Enhancement Gate Transistor – IEGT – for Toshiba [14]) and the IGCT (Reverse Conducting IGCT, Reverse Blocking IGCT... [15]), pushing their limits with higher ratings [13, 15, 16, 17], lower on-state voltage and switching losses [18, 15].

Table I: Global comparison of IGCTs and IGBTs (Press-Pack)

Semiconductor	IGCT	IGBT (Press-Pack)
Snubber	Needed (series, to limit turn-on dI/dt)	No
On-state Voltage	Low (around 2V at high current)	High (more than 3V at high current)
Turn-on Energy loss	Around 2 J (2.8 kV, 2 kA)	Around 10 J (2.8 kV, 2 kA)
Turn-off Energy loss	Around 10 J (2.8 kV, 2 kA)	Around 10 J (2.8 kV, 2 kA)
Gate circuit	Large, and high power consumption [19]	Small, low power consumption [19]
Switching Frequency	Lower (up to 1 kHz, typical max. value) [11]	Higher (up to tens of kHz) [20]

2.2 Current ratings of semiconductor switches

Table II: Description of the current-related variables mentioned in the paper.

Current	Switch	Source	Description	Symbol
dc-current	IGBT	Datasheet	Max. dc-current that the IGBT can conduct	$I_{dc-igbt}$
Peak Current	IGBT	Datasheet	Max. peak (1ms) current that the IGBT can switch	$I_{pk-igbt}$
Max. average on-state current	IGCT	Datasheet	Based on half-sine, no real practical meaning according to [21]	$I_{av-igct}$
Max. RMS on-state current	IGCT	Datasheet	Based on half-sine, no real practical meaning according to [21]	$I_{rms-igct}$
Max. controllable turn-off current	IGCT	Datasheet	Max. anode current that can be turned-off	$I_{mto-igct}$
Switching Current	Both	Datasheet	Condition test current for the switching energy in the datasheet	I_{swi}
Max. av. current MMC	Both	This paper	Av. current going through the switches (worst case)	$I_{av-semi}$
Max. instantaneous current MMC	Both	This paper	Max. instantaneous current seen by the switches	$I_{max-semi}$
Equivalent current IGCT	IGCT	This paper	Built with the datasheet currents, equivalent of $I_{dc-igbt}$ for the IGCT	$I_{eq-igct}$

Many values are quoted in the datasheet of a semiconductor switch regarding the on-state current it can manage. These current values and those introduced in this paper are described in table II. For the IGCT, none of them can directly and simply be linked to the MMC operation, unlike for the IGBT (for which the dc-current rating on the IGBT datasheet, $I_{dc-igbt}$, corresponds to the maximum current going through the semiconductor during MMC operation, $I_{max-semi}$). As a consequence, the suitability of an IGCT for a given MMC cannot be assessed directly, and no direct comparison can be done between IGBTs and IGCTs.

Thus, a new current metric ($I_{eq-IGCT}$) has to be defined to compare efficiently IGCTs and IGBTs current capabilities. This new current metric should correspond to a generic MMC case (it is not specific to one MMC implementation in particular) and has of course to be based on data available in the IGCT datasheet. The goal is to build a current metric comparable to the dc-current quoted in IGBT datasheets $I_{dc-igbt}$, considering the actual current waveforms in a MMC.

The worst case for the IGCT – i.e. when the IGCT is subject to the maximum possible current – is when the submodule is inserted for an entire ac period. Considering the asymmetric operation of the MMC (because arm currents have DC and AC components), one of the IGCTs will be subject to more current than the other. These IGCTs currents have their waveforms drawn in figure 2, assuming the current of the ac output of the MMC is perfectly sinusoidal. It can be seen that this waveform is close to a 50 Hz half-sine. The adopted approach is to calculate the average current of that waveform in the general case and link it to the maximum current value of the same waveform, i.e. finding a relation between $I_{av-semi}$ and $I_{max-semi}$. By establishing this connection between $I_{av-semi}$ and $I_{max-semi}$, and based on $I_{av-IGCT}$ we can introduce $I_{eq-igct}$, which is the equivalent for the IGCT of the dc current for an IGBT ($I_{dc-igbt}$, quoted in the device datasheet). The waveforms of the MMC are described in eqs. (1a) and (1b), with i_a, the current on the ac-side, I_{arm} the current through the considered arm and I_d the current on the dc-side. Here, it will be considered that the arm is the one described in equation 1b, but the same reasoning can be used with any arm and any power factor.

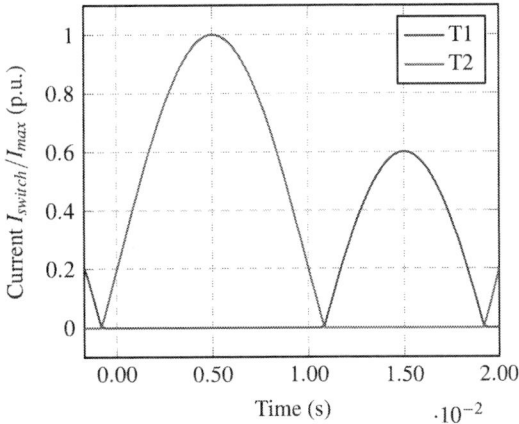

Fig. 2: Currents in the power switches, if the submodule is kept inserted (worst case).

$$i_a = I_a * sin(\omega t) \tag{1a}$$

$$I_{arm}(t) = \frac{I_d}{3} + \frac{I_a}{2} \sin \omega t \tag{1b}$$

A MMC can be described with the following design-related ratios (eqs. (2a) and (2b)), k being the ratio between the ac and dc currents, and m being the ratio between the ac and dc voltages. The relation between these ratios and the power factor is the eq. (2c), resulting from the hypothesis that the powers from each side (ac and dc) are equal. Typical values for m is around 0.8 ± 0.1 and for $\cos(\phi)$ (power factor) is around 0.85 ± 0.1 too. So the typical value of k is around 3 ± 0.8.

$$k = \frac{3I_a}{2I_d} \tag{2a}$$

$$m = \frac{2U_a}{U_d} \tag{2b}$$

$$k * m * \cos(\phi) = 2 \tag{2c}$$

The calculation of the average current through the IGCT in the worst case is described in eqs.(3a) to (3d), with the integration of the positive component of the arm current, corresponding to the current in the most loaded IGCT. The arm current is positive between t_1 and t_2, described in (3a). The average current of the IGCT is obtained in eq. (3d).

$$t_1 = \frac{\arcsin\left(-\frac{1}{k}\right)}{\omega}; t_2 = \frac{\pi - \arcsin\left(-\frac{1}{k}\right)}{\omega} \tag{3a}$$

$$I_{av-semi} = \frac{1}{T} \int_{t_1}^{t_2} I_{arm}(t)dt = \frac{1}{T} \int_{t_1}^{t_2} \left[\frac{I_d}{3} + \frac{I_a}{2} \sin \omega t\right]dt \tag{3b}$$

$$I_a = \frac{2k}{3}I_d \tag{3c}$$

$$I_{av-semi} = \frac{I_d}{6\pi} * \left(\pi - 2\arcsin\left(-\frac{1}{k}\right)\right) + \frac{kI_d}{3\pi}\left[\cos\left(\arcsin\left(-\frac{1}{k}\right)\right)\right] \tag{3d}$$

This expression of $I_{av-semi}$ can be simplified using a linear regression, shown in eq. (4a). This regression is done on a limited range of values of k, between 2 (minimum of k in half-bridge based MMCs according to according eq. 2c) and 10 (corresponding to normal half-bridge based MMCs, 10 being chosen high for a HVDC MMC). The regression gives a R^2 value of 0.99992.

$$I_{av-semi} = I_d * [\alpha * k + \beta] \quad ; \quad \alpha = 0.103398 \quad ; \quad \beta = 0.19019 \tag{4a}$$

$I_{max-semi}$ is obtained from equation (1b) and equation (2a) by replacing the sinus by maximal value. The function f is then introduced in equation 5b, as the ratio between $I_{max-semi}$ and $I_{av-semi}$.

$$I_{max-semi} = I_d * \left(\frac{1+k}{3}\right) \tag{5a}$$

$$f(k) = \frac{I_{max-semi}}{I_{av-semi}} = \frac{1+k}{3 \times (\alpha \times k + \beta)} \tag{5b}$$

An equivalent current $I_{eq-IGCT}$, that can be directly compared to the $I_{dc-igbt}$ of an IGBT, is defined in equation 6. $I_{eq-IGCT}$ is equal to the minimum of $I_{mto-IGCT}$ (maximum current at turn-off) and $I_{av-IGCT} \times f(k)$ to ensure it does not exceed the current that can safely be interrupted, in a normal operation. In this paper, only the normal operation is taken into account, not the behaviour in case of fault.

$$I_{eq-IGCT} = \min(I_{mto-IGCT}, I_{av-IGCT} \times f(k)) \tag{6}$$

f is a function that is increasing with k. Here, we consider the case in which $k = 2$, i.e. $f(k) = 2.5116$, to stay in the most general case. In this case, for many IGCTs, $I_{av-IGCT} \times f(k)$ has a higher value than $I_{mto-igct}$. Two things can be concluded from that result. First, $I_{mto-igct}$ can be most of the times compared to $I_{dc-igbt}$ (even though it needs to be checked afterwards with this method). Then, most available IGCTs have very high $I_{mto-igct}$ (3 to 8 kA), which means they can handle much larger currents than available IGBTs ($I_{dc-igbt} = 1$ to 3 kA for >3.3 kV IGBTs) in a MMC application. This is illustrated in Fig. 3 where $I_{dc-igbt}$ and $I_{eq-IGCT}$ are plotted. Finally, $I_{mto-igct}$ is a limit of the IGCT related to the gate circuit (how much current the gate can divert to open the IGCT) and $I_{av-IGCT} \times f(k)$ is a thermal related limit of the semiconductor. If $I_{eq-IGCT}$ is equal to $I_{mto-igct}$, the limit is not related to the semiconductor itself: the semiconductor could withstand more current.

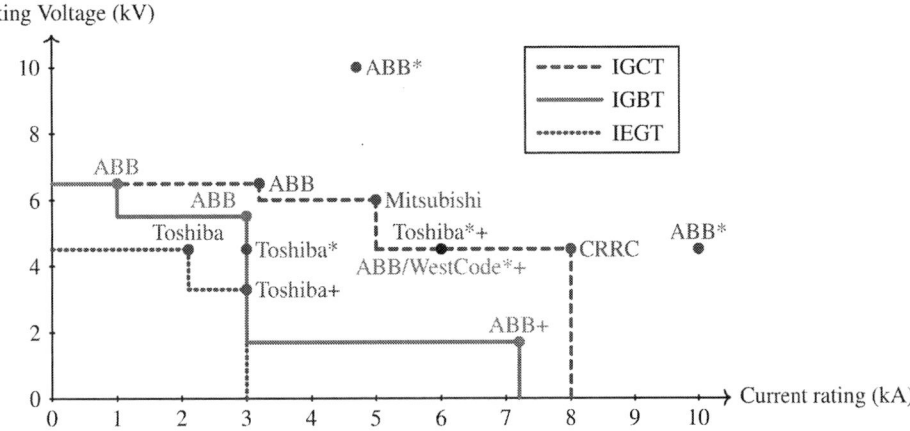

Fig. 3: Ratings of semiconductors (IGBTs, IEGTs and IGCTs), with the blocking voltage and the current calculated in the paper: $I_{dc-igbt}$ for IEGTs and IGBTs and $I_{eq-igct}$ for the IGCTs. '+' means 'not found in catalog but in literature', '*' means 'prototype'. Sources: ABB, Mitsubishi, CRRC, Dynex, Toshiba online catalogs (2019) and [19, 22, 23].

2.3 Figures of merit for conduction and switching losses

After we established a consistent current rating for IGBT and IGCTs in the previous section, we can now use it to compare the merits of different switches. These Figures of Merit (FOMs) are related to the conduction and switching losses.

For the conduction losses, a FOM that can be used is the on-state voltage. We consider in this paper the on-state voltage model as described in eq (7a) for the following reasons: it is very accurate for IGCTs over their entire SOA and acceptable for IGBTs, even though an exponential fit would be more accurate for these devices. As this on-state voltage is dependent on the current flowing through the component, we consider the MMC waveforms to calculate an average on-state voltage value. It is equal to the on-state voltage for the average current, as it is demonstrated in eqs (7a) to (7b). Thus, this approach only uses values that are available in the datasheets, which makes it interesting for at the early stages of design, for the pre-selection of devices.

$$V_{on}(I) = V_0 + R.I \tag{7a}$$

$$\text{average}(V_{on}) = V_0 + R.\text{average}(I) = V_{on}(I_{av-semi}) \tag{7b}$$

The average on-state voltage is then normalised with respect to the blocking voltage of the switch (considered with a de-rating to reach a reliability of 100 Failure-In-Time – FIT – defined in particular in [24]), to allow the comparison between semiconductors with different voltage ratings. The final figure of merit is then:

$$FM_{cond-loss}(IGCT) = \frac{V_{on}(I_{av-igct})}{V_{block-100FIT}} \tag{8a}$$

$$FM_{cond-loss}(IGBT) = \frac{V_{on}(\frac{I_{dc-igbt}}{2.5116})}{V_{block-100FIT}} \tag{8b}$$

For the switching losses, turn-on and turn-off energies are used to compare semiconductors. In datasheets, the losses figures are always quoted at maximum rated current (I_{swi}), which differs from one semiconductor to another and may differ from the actual current in the application. The IGBT and IGCT switching losses can be approximated by a linear function depending on the current. For the IGBT this is only true up to the dc-current rating, but this is sufficient in the case of the MMC. To compare the switching losses, the chosen figure of merit is the following:

$$FM_{swi-loss} = \frac{E_{datasheet}}{V_{block-100FIT} \times I_{swi}} \tag{9}$$

Table III: Loss studies results

	Total losses IGBT (%)	Total losses IGCT (%)	Inverter or rectifier	Loss red. (%)
[7]	0.48	0.41	Average rectifier and inverter	14.6
[5]	0.64	0.44	Rectifier	31
[5]	0.32	0.27	Inverter	15.6
[6]	0.76	0.74	Rectifier	2.6
[6]	0.83	0.64	Inverter	22.9

3 FOM validation, simulation and loss study

3.1 Model Description

As it can be seen in the literature (see Tab. III), IGCTs are always found to cause less losses than IGBTs. However, the estimated amount actually varies noticeably from one publication to another because it depends on many factors: MMC ratings, control methods, operating points, not all of which are disclosed in the corresponding publications. As a consequence, it is difficult to use published results to validate our FOMs. That is why a simple MMC model has been developed, focusing on the individual behaviour of each submodule (capacitor and semiconductors). This model is used to validate the FOM approach.

The model is written as a Matlab script. The modelling assumptions are as follows: perfect sinusoidal current in the arm and perfect sinusoidal voltage across the arm for the ac components of current and voltage; arm inductor, line resistors and energy exchange between arms not modelled; only one arm considered.

The model calculates the switching instants of the semiconductors switches with the nearest level modulation control method [25] and the reduced switching algorithm for capacitor voltage balancing [26] (a balancing capacitor algorithm based on a voltage tolerance band). Although simple, this model produces realistic waveforms for the operating voltage and current of each submodule, which allows for a more accurate estimation of the losses than average arm models. Examples of current and voltage waveforms for the upper arm, and the capacitor voltages for some (randomly chosen) submodules are displayed in figure 4. The waveforms in figure 4b show that, as it is the case in actual MMCs, the different submodules have different switching instants, different voltages at a given time, and they respect the set voltage ripple (10% in that case) around the average value of the capacitors voltages. The waveforms in figure 4a show that the semiconductor conduction losses are calculated with realistic current waveforms. This model has been used to simulate MMC operation over different length of time, and it was found that simulating operation over 5 ac-periods was sufficient to get consistent results.

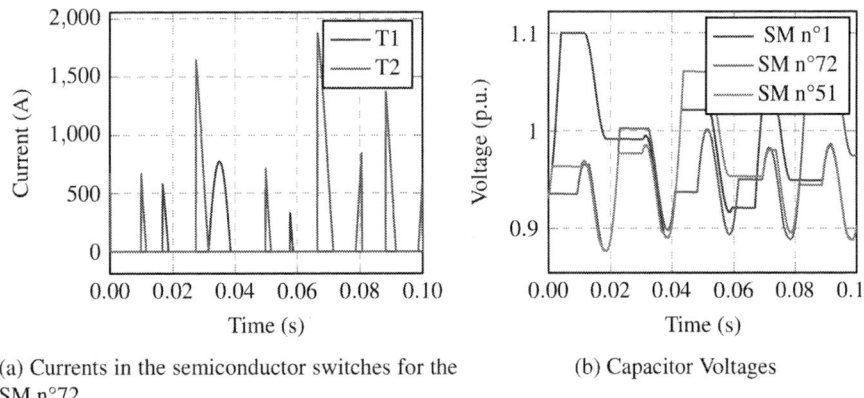

(a) Currents in the semiconductor switches for the SM n°72

(b) Capacitor Voltages

Fig. 4: Waveforms of the arm and some submodules, after simulation of the same MMC as [6]. (p.u. means per unit and is the instantaneous voltage of the submodule divided by the rated voltage of the submodule)

3.2 Loss Study

A simulation was realised considering the following MMC: 1 GVA, 400 submodules per arm, a mean submodule voltage of 1600 V, a dc voltage of 640 kV (pole to pole). The considered devices are: IGCTs (5SHY 35L4521) and associated diodes (5SDF 20L4520), or press-pack IGBTs (5SNA 2000K450300) which include their own diodes. Using the model, losses of 5.2 MW (0.52% of the transferred power) in the case of the IGCT and 6.72 MW (0.672%) in the case of the IGBT have been calculated. This corresponds to a loss reduction of 23.6% for the IGCT case compared to the IGBT case, which is consistent with the results presented in Tab. III.

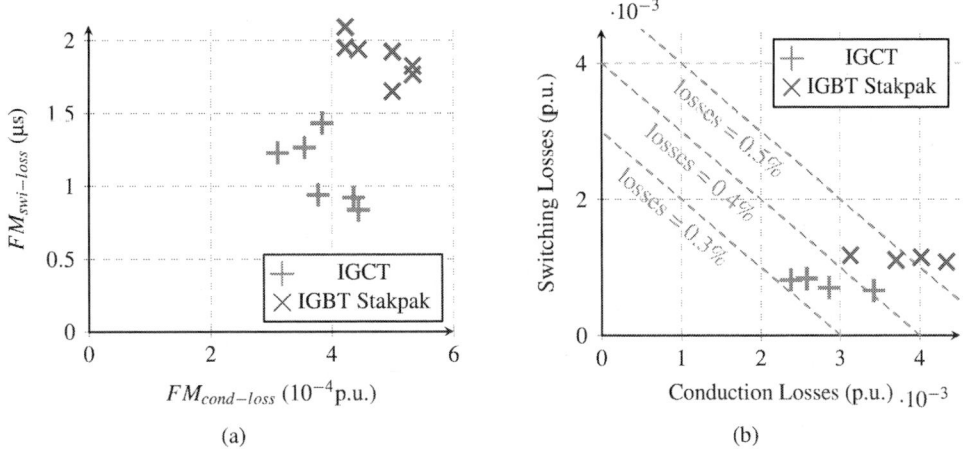

Fig. 5: (a) Figures of merit applied to ABB online catalog for IGCTs and Stakpak IGBTs (2019); (b) corresponding Switching and conduction losses for 4.5 kV ABB IGCTs and Stakpak IGBTs, simulated for the same MMC as [6] (inverter case), with iso-loss curves drawn in orange.

3.3 Analysis of the figures of merit

The figures of merit for the ABB IGCTs and Stakpak IGBTs for different voltage ratings are displayed in Fig. 5a. They are in good agreement with the corresponding switching and conduction losses calculated using the MMC model (for 4.5 kV devices only, Fig. 5b). Both approaches confirm the superiority of IGCTs over IGBTs regarding losses in general. Fig. 5a provides a comparison which is independent from the actual characteristics of the target MMC, while Fig. 5b is specific to one particular MMC. This demonstrates the interest of these figures of merit: they provide a quick, general and efficient way to compare devices even before modelling and simulating a specific MMC.

As expected, the IGCT has lower conduction losses on average, although in some cases, with the worst IGCT and the best IGBT in terms of conduction losses, the IGBT is better. This appears both in the figure of merit and in the simulated conduction losses (see figure 5). It can be explained by the use of average on-state voltage in the figure of merit – calculated as the on-state voltage at the average current of each semiconductor – and by the use of the same MMC for the simulations – the current of this MMC is the same for all the semiconductors. This implies an under-utilisation of the IGCTs, whose on-state voltage grows slowly with the current. Indeed, as it can be seen in figure 6, the on-state voltage of an IGCT grows slower than the on-state voltage of an IGBT. In fact, the IGCT is more suitable for MMCs with higher current ratings. But it has to be kept in mind that figure 6 is a particular case of one IGCT and one IGBT.

Furthermore, the IGCT is often described as having more switching losses than the IGBT: with the figures of merit (and confirmed by the simulation), it can be seen that any considered IGCT actually has lower switching losses than any IGBT. This is due to the fact that the turn-on losses are very small for the IGCT. The biggest conduction losses difference (the best IGCT and the worst IGBT in terms of conduction losses) is 0.2% while the biggest switching losses difference is 0.055%. This is due to the low switching frequency of the devices in the case of HVDC MMCs.

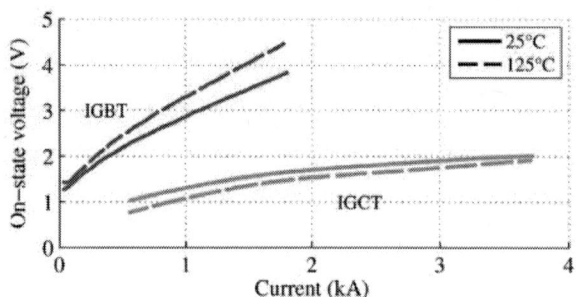

Fig. 6: On-state voltage of one IGCT and one IGBT [19]

4 Conclusion

In this paper, new metrics have been developed to compare efficiently and quickly IGCTs and IGBTs in the particular case of HVDC MMCs. Figures of merit have been proposed to allow for a fair comparison in the current ratings and losses between devices. The current rating has shown that the available IGCTs have higher usable current ratings than the available IGBTs. The proposed FOMs offer a simple, analytical way to compare given IGBTs and IGCTs for a HVDC MMC application, and to easily select the best device. The different simulations have validated the FOMs and confirmed their necessity. Indeed, not all IGCTs are superior (in terms of losses) than IGBTs, and the selection must be performed on a device-per-device basis. But the FOMs confirm an advantage – in general – of IGCTs over IGBTs regarding converter efficiency, and are consistent with more complex converter-level simulations.

References

[1] Y. Tang, *Modular Multilevel Converter: Submodule Dimensioning, Testing Method, and Topology Innovation*. PhD thesis, University of Warwick, 2015.

[2] T. Wei, Q. Song, J. Li, B. Zhao, Z. Chen, and R. Zeng, "Experimental evaluation of IGCT converters with reduced di/dt limiting inductance," in *2018 IEEE Applied Power Electronics Conference and Exposition (APEC)*, IEEE, mar 2018.

[3] K. Sharifabadi, L. Harnefors, H.-P. Nee, S. Norrga, and R. Teodorescu, *Design, Control, and Application of Modular Multilevel Converters for HVDC Transmission Systems*. John Wiley & Sons Inc, 2016.

[4] H. Chen, W. Cao, P. Bordignon, R. Yi, H. Zhang, and W. Shi, "Design and testing of the world's first single-level press-pack IGBT based submodule for MMC VSC HVDC applications," in *2015 IEEE Energy Conversion Congress and Exposition (ECCE)*, IEEE, sep 2015.

[5] D. Guédon, P. Ladoux, M. Kanoun, and S. Sanchez, "IGCTs in HVDC systems: Analysis and assessment of losses," in *PCIM Europe 2019*, 2019.

[6] B. Zhao, R. Zeng, J. Li, T. Wei, Z. Chen, Q. Song, and Z. Yu, "Practical analytical model and comprehensive comparison of power loss performance for various MMCs based on IGCT in HVDC application," *IEEE Journal of Emerging and Selected Topics in Power Electronics*, vol. 7, pp. 1071–1083, jun 2019.

[7] T. Modeer, H.-P. Nee, and S. Norrga, "Loss comparison of different sub-module implementations for modular multilevel converters in HVDC applications," *EPE Journal*, vol. 22, pp. 32–38, sep 2012.

[8] R. Zeng, B. Zhao, T. Wei, C. Xu, Z. Chen, J. Liu, W. Zhou, Q. Song, and Z. Yu, "Integrated gate commutated thyristor-based modular multilevel converters: A promising solution for high-voltage dc applications," *IEEE Industrial Electronics Magazine*, vol. 13, pp. 4–16, jun 2019.

[9] P. Ladoux, N. Serbia, and E. I. Carroll, "On the potential of igcts in hvdc," *IEEE Journal of Emerging and Selected Topics in Power Electronics*, vol. 3, no. 3, pp. 780–793, 2015.

[10] J. Lutz, H. Schlangenotto, U. Scheuermann, and R. D. Doncker, *Semiconductor Power Devices*. Springer Berlin Heidelberg, 2011.

[11] S. Alvarez-Hidalgo, *Characterisation of 3.3kV IGCTs for Medium Power Applications*. PhD thesis, ENSEEIHT, 2005.

[12] B. J. Baliga, *Fundamentals of Power Semiconductor Devices*. Springer, 2008.

[13] C. Corvasce, M. Andenna, S. Matthias, L. Storasta, A. Kopta, M. Rahimo, L. De-Michielis, S. Geissmann, and R. Schnell, "3300v hipak2 modules with enhanced trench (tspt+) igbts and field charge extraction diodes rated up to 1800a," in *PCIM Europe 2016; International Exhibition and Conference for Power Electronics, Intelligent Motion, Renewable Energy and Energy Management*, pp. 1–8, May 2016.

[14] Toshiba, "High-power electric solutions - system catalog," tech. rep., Toshiba, 2015.

[15] U. Vemulapati, M. Rahimo, M. Arnold, T. Wikstrom, J. Vobecky, B. Backlund, and T. Stiasny, "Recent advancements in IGCT technologies for high power electronics applications," in *2015 17th European Conference on Power Electronics and Applications (EPE'15 ECCE-Europe)*, IEEE, sep 2015.

[16] U. R. Vemulapati, T. Wikström, and M. Lüscher, "An rc-igct for application at up to 5.3kv," in *ICPE 2019 - ECCE Asia*, 2019.

[17] T. Wikstrom, M. Arnold, T. Stiasny, C. Waltisberg, H. Ravener, and M. Rahimo, "The 150 mm RC-IGCT: A device for the highest power requirements," in *2014 IEEE 26th International Symposium on Power Semiconductor Devices & IC's (ISPSD)*, IEEE, jun 2014.

[18] M. Rahimo, "Future trends in high power mos controlled power semiconductors," 2012.

[19] F. Filsecker, R. Alvarez, and S. Bernet, "Comparison of 4.5-kV press-pack IGBTs and IGCTs for medium-voltage converters," *IEEE Transactions on Industrial Electronics*, vol. 60, pp. 440–449, Feb 2013.

[20] Toshiba, "Trends in and future outlook for semiconductor devices with enhanced energy efficiency," tech. rep., Toshiba, 2018.

[21] ABB, "Applying IGCTs," tech. rep., ABB, 2016.

[22] Z. Chen, Z. Yu, X. Liu, J. Liu, and R. Zeng, "Stray impedance measurement and improvement of high-power IGCT gate driver units," *IEEE Transactions on Power Electronics*, vol. 34, no. 7, pp. 6639–6647, 2019.

[23] I. Omura, T. Domon, E. Miyake, Y. Sakiyama, T. Ogura, M. Hiyoshi, N. Yamano, and H. Ohashi, "Electrical and mechanical package design for 4.5kV ultra high power IEGT with 6kA turn-off capability," in *ISPSD '03. 2003 IEEE 15th International Symposium on Power Semiconductor Devices and ICs, 2003. Proceedings.*, pp. 114–117, April 2003.

[24] ABB, "Failure rates of igbt modules due to cosmic rays," tech. rep., ABB, 2017.

[25] M. Perez, J. Rodriguez, J. Pontt, and S. Kouro, "Power distribution in hybrid multi-cell converter with nearest level modulation," in *2007 IEEE International Symposium on Industrial Electronics*, pp. 736–741, June 2007.

[26] Q. Tu, Z. Xu, and L. Xu, "Reduced switching-frequency modulation and circulating current suppression for modular multilevel converters," *IEEE Transactions on Power Delivery*, vol. 26, pp. 2009–2017, jul 2011.

Zero-current switching with LC resonant tank circuit and capacitor isolation DC-DC converter.

Hideki Jonokuchi, Osamu Nakashima, Daichi Hiwatari, Hiroshi Hirayama.
Nagoya Institute of Technology
Gokiso-Cho, Showa-Ku,
Nagoya, 4668555, Japan
Tel.: +81 / (52) – 732-2111, ext. 5639.
Fax: +81 / (52) -735-5518.
E-Mail: h.jonokuchi.894@nitech.jp
URL: https://www.nitech.ac.jp/

Keywords

«Capacitor», «Coupling», «Converter», «Isolation», «Zero-cross», «Switch», «Resonant», «EMI noise», «Spice», «PFM: Pulse Frequency Modulation», «PDM: Pulse Density Modulation».

Abstract

Insulated DC-DC switching power supplies have been using a magnetic coupling method using a transformer for several decades. Transformers have excellent properties as a material that can change the voltage while insulating. On the other hand, it can be considered that there is a limit to the reduction in size and weight with increasing switching frequency. Therefore, we are interested in the insulation type converter by the capacitor coupling. The principle of the capacitor insulation method was introduced in 1994, but it was pointed out that the problem of noise that occurs when the power device is turned on and off. After that, a case where noise was suppressed by using a series resonant circuit of a capacitor and an inductor was reported. However, it had no means to control the output voltage, and had the problem that fluctuations in the input voltage would directly cause fluctuations in the output. Therefore, we maintain "resonance state" by providing "two sets of LC resonance tank circuits corresponding to the resonance frequency" and changing the operating frequency for ON width pulse of fixed width (pulse frequency modulation or pulse density modulation). We propose that the output voltage can be controlled while. The voltage control characteristics are analyzed using Spice Simulation, and the Co-relation with the experimental value is confirmed. By using frequency/density modulation, the inverter is always switched at the current zero crossing point, enabling Noiseless switching. Regarding the possibility, we will analyze the switching characteristics due to the deviation from the resonance point by using ADS; Electronic Design Automation tool, for the noise that may occur in the actual PCB (circuit pattern), and show the results.

Introduction

With the advancement of power devices and the realization of higher switching frequencies, it has been pointed out that the magnetic coupling method (transformer) has a limit to further improvement in performance. Capacitor coupling method may be one solution that enables downsizing and weight saving. The principle of the insulation method using capacitors was introduced in 1994 [1], but it was pointed out that the problem of noise that occurs when the power device is turned on and off. After that, a case was reported where noise was suppressed by using a series resonant circuit of a capacitor and an inductor [2]. The circuit reminiscent of the circuit proposed this time is shown in the paper [3], [4], [5], [6]. However, none of them has a means to control the output voltage, and there is a problem that the fluctuation of the input voltage directly causes the fluctuation of the output. Therefore, we maintain "resonance state" by providing "two sets of LC resonance tank circuits corresponding to the resonance frequency" and changing the operating frequency for ON width pulse of fixed width (pulse frequency modulation or pulse density modulation). We propose that the output voltage can be controlled while.

Capacitor coupled DC-DC converter with LC tank.

The proposed topology consists of 2 "switching device with wheeling diode or an inverter arms", 2 "capacitor; LC tank circuit" and a "full bridge diode rectifier with filter capacitor".

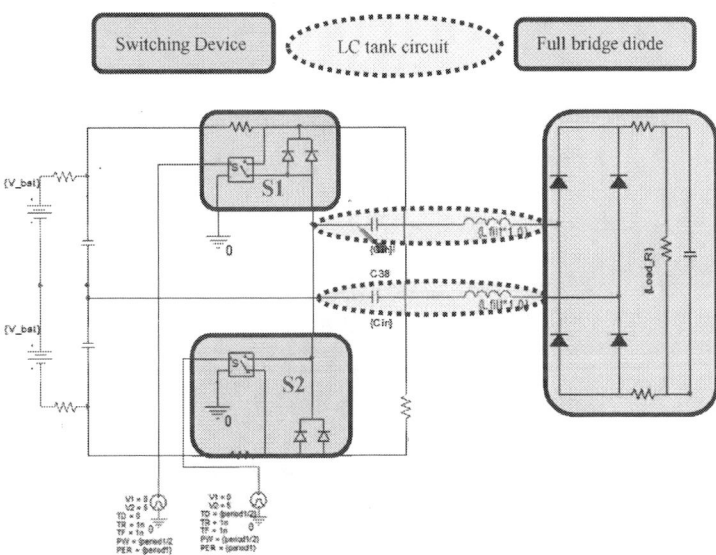

Fig. 1: Configuration of capacitor Coupled DCDC converter with LC tank.

The inverter operation sequence is shown in Fig.2.

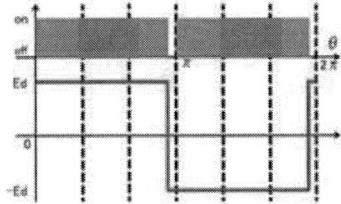

Fig. 2: a half bridge inverter operation timing chart.

Zero-cross switching frequency and LC tank circuit.

The "Zero-cross (ZC) switching" can be established by the resonant frequency switching shown in (1). And the simulation has been done with the "simple switching elements and diodes" based on the conditions in the table below.

$$fo = \frac{1}{2\pi\sqrt{Lf*Cin}} \quad ---(1)$$

Table I: Zero-cross switching and LC tank constant.

Cin [uF]	2.2	1	1	4
Lf [uH]	2.2	1	4	4
[kHz]	72.3	159.1549	79.57747	39.78874
Simulation Results	Fig.4	Fig.5	Fig.6	Fig.7

 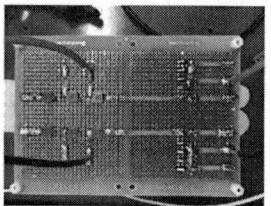

Fig. 3: (a) Test setup.   (b)LC tank circuit and rectifier.

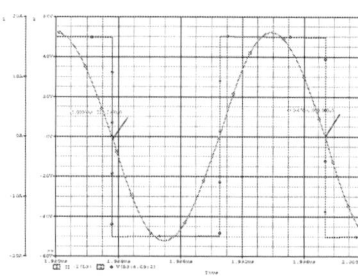 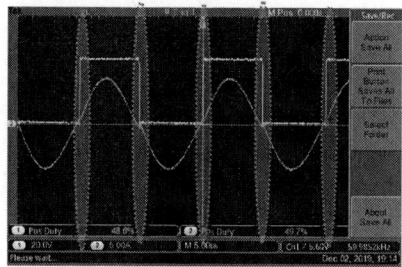

Fig. 4. (a)ZC switching simulation (Left), (b)Test result(Right) with 4uH & 1uF LC tank.

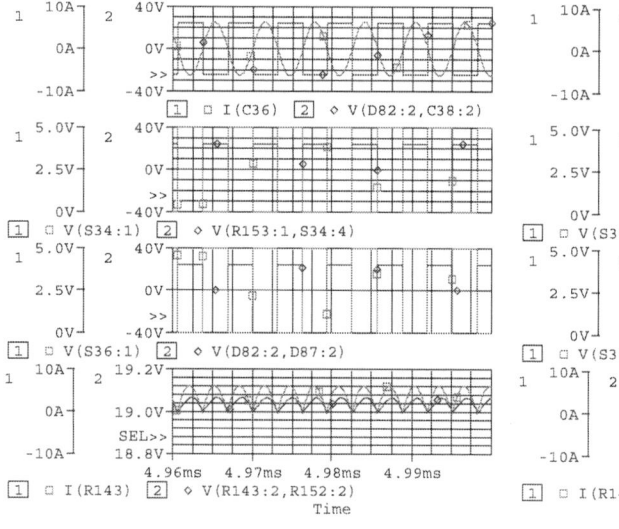 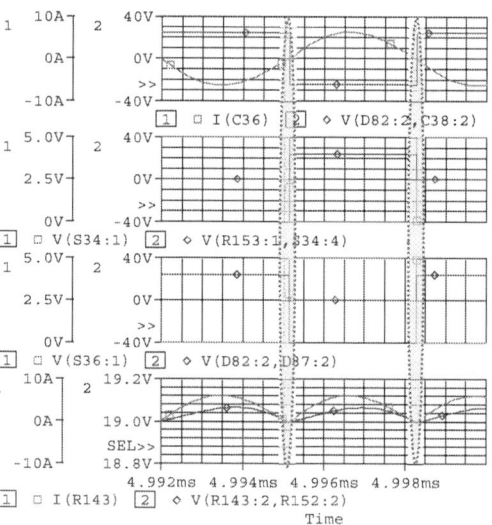

Fig. 5.: Zero-cross 159.15kHz switching frequency with LC tank; 1uH and 1uF.
Left one; long span, right one; short span.
Top; Inverter output voltage and current.
Second; Gate on/off signal for S1 and S2.
Third; Gate on/off signal for S3 and S4.

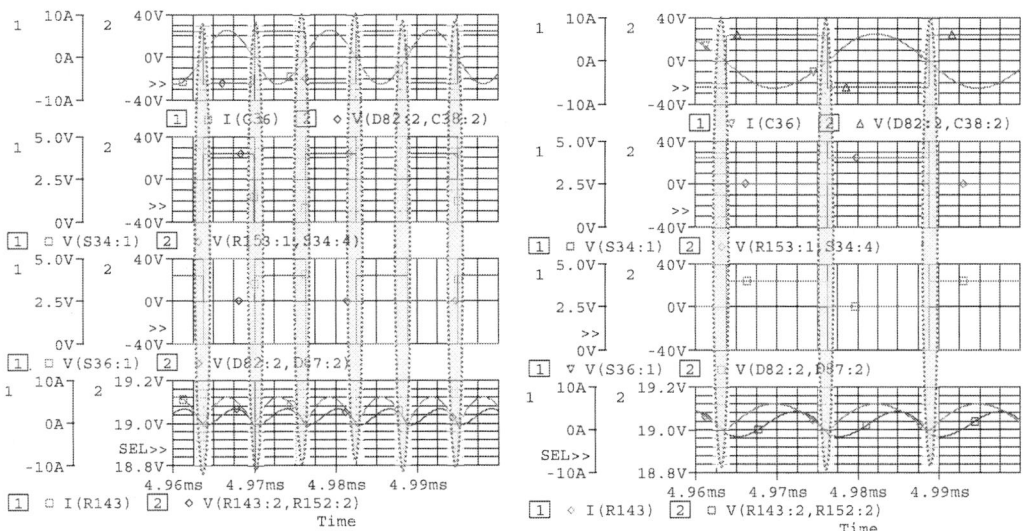

Fig.6: Zero-cross 79.58kHz switching frequency with LC tank; 1uH and 4uF.

Fig. 7: Zero-cross 39.79kHz switching frequency with LC tank; 4uH and 4uF.

And it shows a good correlation between simulation results and measured waveforms.

Voltage controllability and its coverage.

Two methods are proposed to control the output voltage.

(a) The output voltage is controlled by setting the capacitance of the capacitor (C_p) that generates the neutral point voltage in the half-bridge inverter configuration to be smaller than that of the resonance capacitor (Cin).

(b) The voltage is controlled by switching with an ON time width corresponding to the resonance frequency and changing the period of the ON pulse (frequency density modulation). Here, the frequency density modulation defines a frequency Index as "1.0" for a frequency having a positive side ON pulse and a negative side ON pulse as one cycle, and means that the cycle becomes longer according to the frequency Index.

An example of voltage control will be shown below.

Fig.8 shows the operation reference, and the output voltage (about 50V for 100Vdc input) that is rectified half bridge inverter output is output. Fig.9 shows the characteristics when the neutral point generation capacitor (C_p) is 0.1uF, and the output voltage is about 36V. Fig.10(a) shows the characteristics when the frequency density is modulated and FrIn=4.0 (output voltage is about 9V). Fig. 10(b) shows the operating waveforms in the actual circuit.

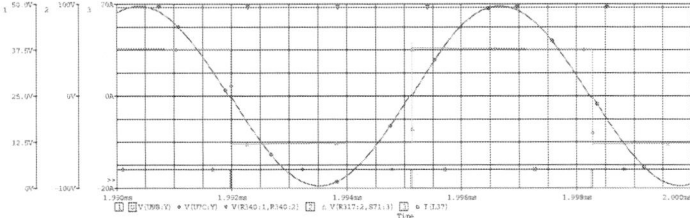

Fig.8: Basic operation waveform with
C_p=10uF, Cin=1uF, Lf=1uF, FrIn=1.0, Load_R=4-→ Vout=48V

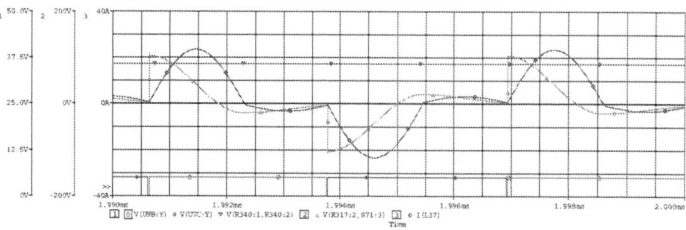

Fig.9: Less C_p value operation waveform with
C_p=0.1uF, Cin=1uF, Lf=1uF, FrIn=1.0, Load_R=4 -→ Vout=36V

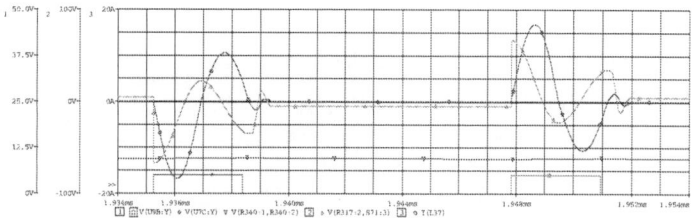

Fig.10(a): Bigger FrIn value operation waveform with
C_p=0.1uF, Cin=1uF, Lf=1uF, FrIn=4.0, Load_R=4 -→ Vout=9V

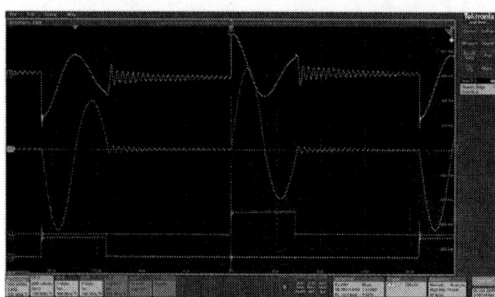

Fig.10(b): Bigger FrIn value operation waveform with FrIn=4.0, Load_R=4

Next, the voltage controllable range will be considered.
At an input voltage of 100 Vdc, Cin=1uF, Lf=1uH, the frequency density modulation index (FrIn) was defined, and the relationship with the output voltage was simulated according to Table 2. The results are shown in the Fig.11., 12. and 13.

Table II: Simulation matrix for output voltage versus FrIn.

C_p [uF]	0.05	0.1	0.2					
Load_R	2	4	6	8	10	12	14	16

FrIn is Frequency density modulation Index, and expressed by "k1.0 – k4.0"in figures.

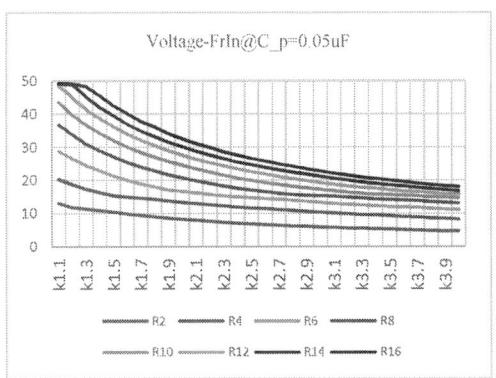
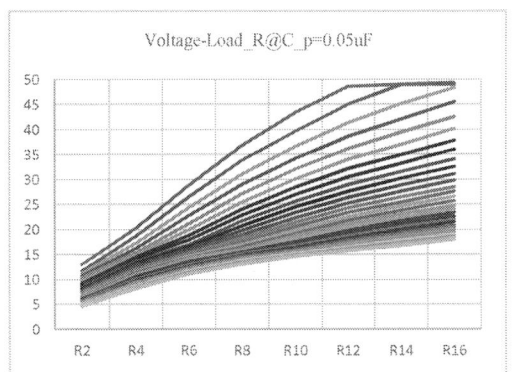

Fig.11(a): Output Voltage-FrIn with C_p=0.05uF.
Parameter is Load_R.

Fig.11(b): Output Voltage-Load_R @ C_p=0.05uF
Parameter is FrIn: k1.0 – k4.0

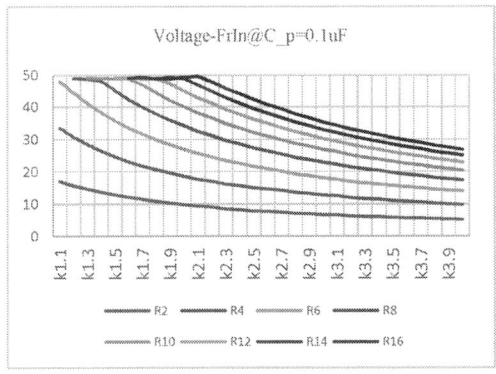
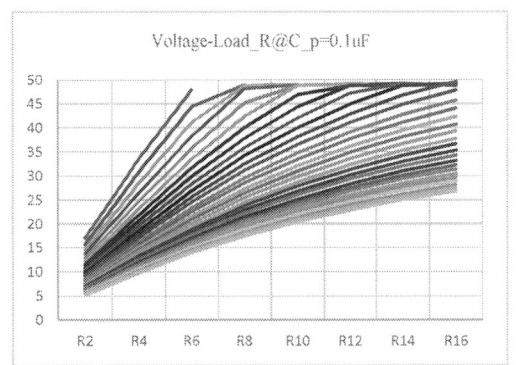

Fig.12(a): Output Voltage-FrIn with C_p=0.1uF.
Parameter is Load_R.

Fig.12(b): Output Voltage-Load_R @ C_p=0.1uF
Parameter is FrIn: k1.0 – k4.0

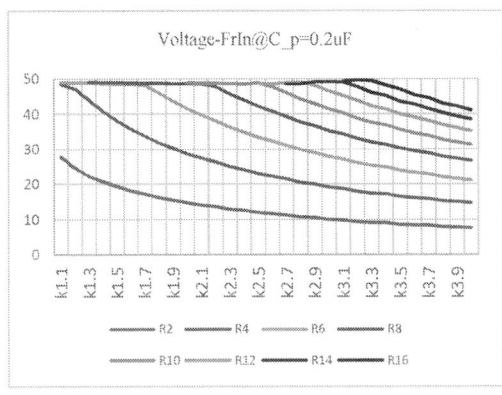
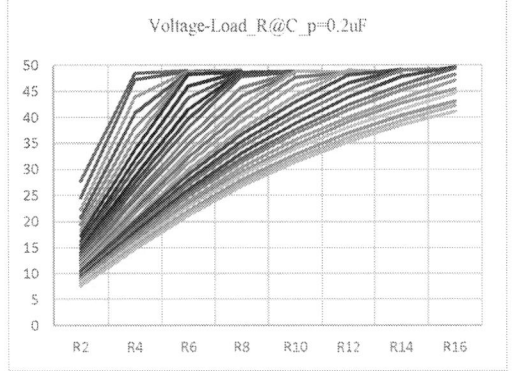

Fig.13(a): Output Voltage-FrIn with C_p=0.2uF.
Parameter is Load_R.

Fig.13(b): Output Voltage-Load_R @ C_p=0.2uF
Parameter is FrIn: k1.0 – k4.0

These results show that the output voltage can be controlled by changing FrIn. However, its control range is not as wide as that of PWM control. For example, in Fig.12 (a), the output voltage is 20V at FrIn=1.8, the output voltage is 24V at FrIn=1.5, and the output voltage is 16V at FrIn=2.4 with respect to the load resistance 4Ω. These controllability means that a constant voltage can be output with respect to an input voltage fluctuation. The following figures are simulation and tested data with the condition; Outputvoltage:12V, C_p=0.1uF, Cin=1.2uF, Lf=5.6uH and Load_R=4

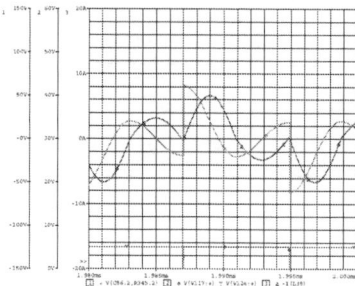

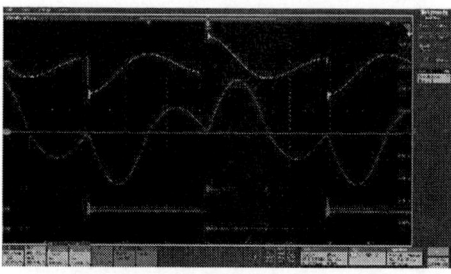

Fig.14(a): Simulation, @ input V=80V & FrIn=1.0.   Fig.14(b): Tested result.

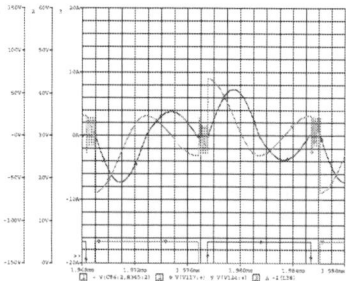

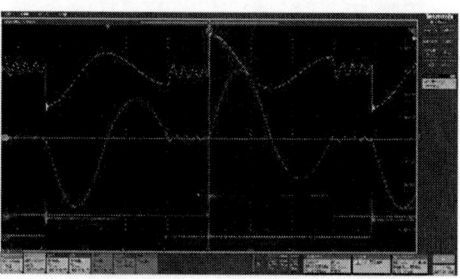

Fig.15(a): Simulation, @ input V=90V & FrIn=1.1.   Fig.15(b): Tested result.

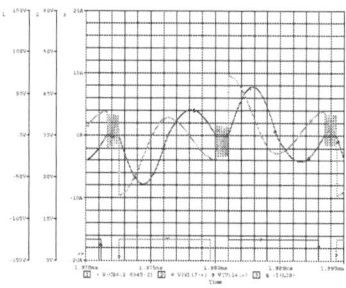

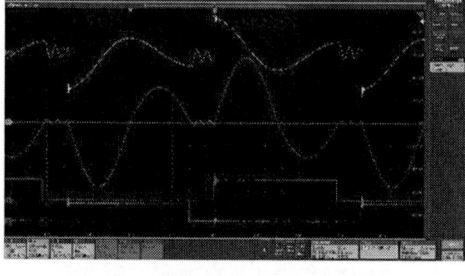

Fig.16(a): Simulation, @ input V=100V & FrIn=1.2.   Fig.16(b): Tested result.

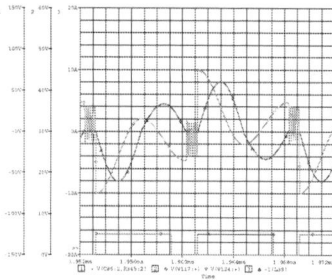

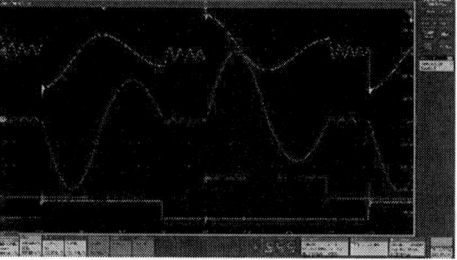

Fig.17(a): Simulation, @ input V=110V & FrIn=1.35.   Fig.17(b): Tested result.

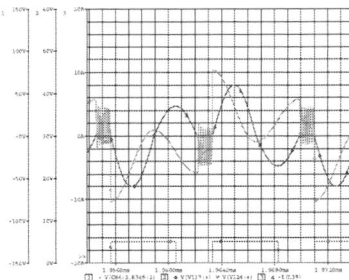

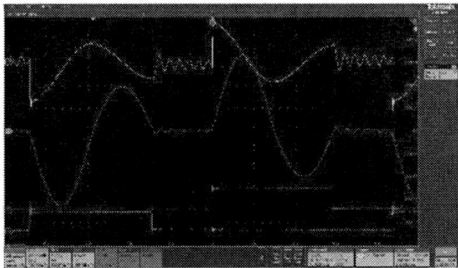

Fig.18(a): Simulation, @ input V=110V & FrIn=1.5.   Fig.18(b): Tested result.

The operating frequency in the actual experiment is slightly lower. This is because the capacitance of the capacitor changes depending on the applied voltage[10].

DC-DC converter efficiency is measured versus Input Voltage, shown in Fig. 19. Conversion efficiency depends on the input voltage: FrIn.

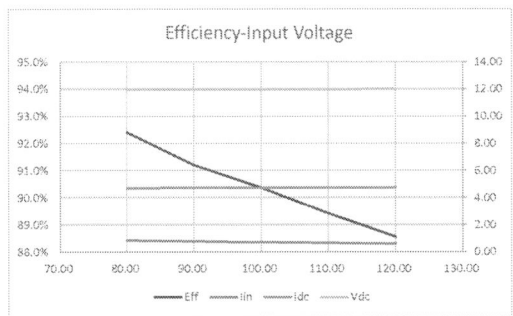

Fig.19: Tested DC-DC Converter Efficiency.

Noise simulation of zero-current switching on PCB.

We have evaluated the noise on the PCB shown in Fig.20 by ADS; simulator [11]. Measurement points are P17 and P18 for voltage and its current for P18 to P19 through capacitor.

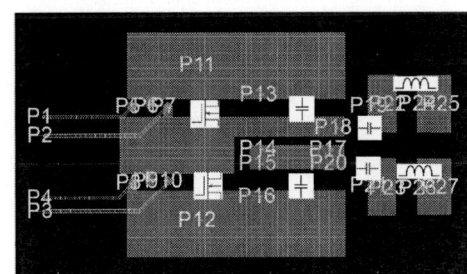

Fig.20: Noise simulation on PCB by ADS.

The results are shown in Fig.21 and Fig.22.
Fig.21 is the simulation result of Zero-cross (ZC) switching for 0.7-0.8MHz operation with 0.1uF and 0.5uH LC tank. Fig.22 is the simulation result of Non-ZC switching with "current peak cut off".

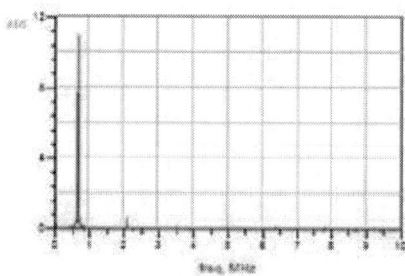 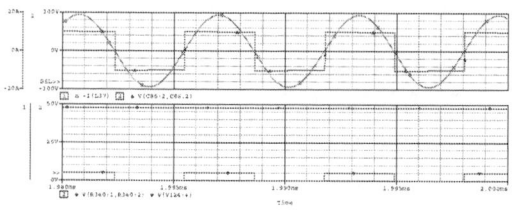

(b) Time domain @ZC Switching.

Fig.21: (a) Frequency Domain @ZC Switching current waveform: Upper Green.

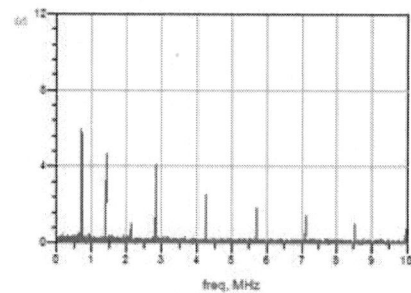 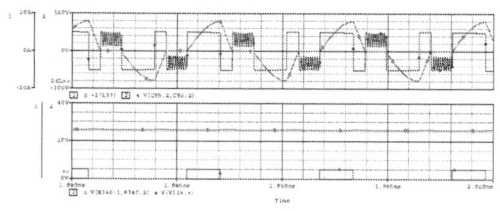

(b) Time domain @Non-ZC Switching.

Fig.22: (a) Frequency Domain @Non-ZC Switching current waveform: Upper Green.

Conclusion

In this paper, we proposed to use two sets of LC tank circuits to switch at zero current or zero voltage for the capacitor coupling method proposed as a transformer-less insulation method. In particular, it has been shown that the output voltage can be adjusted while maintaining the zero current switching state by switching at the resonance frequency (ON pulse of a certain width) of the LC tank circuit and changing the density of the resonance frequency. For the design of variable voltage range, we proposed the Spice model, and verified it with the experimental data, and found a very good correlation. In addition, an example of noiseless operation due to zero current switching on an actual PC board has been analyzed by ADS and the possibility was shown. In the future, we will proceed with actual measurement confirmation on the EMC site.

Understanding that the control range of pulse frequency/density modulation is not so wide, but I think that it can be applied to data centers with relatively small load fluctuation range and power supplies for servers.

References

[1] Yuuji Sayama, Hirotami Nakano, Akira Nabae, etc.: Switching Power Supply Using Capacitive Isolation, T. IEE Japan , Vol. 114-D, No. 9, '94.

[2] Yusuke Hayashi, Yoshikatsu Matsugaki, Tamotsu Ninomiya.: Capacitively Isolated Multicell Dc-Dc Transformer for Future Dc Distribution System, IEEJ Journal of Industry Application. Vol.6, 2017, No.4.

[3] Mitchell KlineIgor IzyuminIgor IzyuminBernhard E. BoserBernhard E. BoserSeth Sanders: Capacitive power transfer for contactless charging, Applied Power Electronics Conference and Exposition (APEC), 2011 Twenty-Sixth Annual IEEE.

[4] M. Kline, I. Izyumin, B. Boser, and S. Sanders: "Capacitive Power Transfer for Contactless Charging", Proceedings of Twenty-Sixth Annual IEEE Applied Power Electronics Conference and Exposition (APEC), Fort Worth, U.S., pp. 1398-1404 (2011)

[5] Mitchell Kline: Capacitive Power Transfer, Electrical Engineering and Computer Sciences University of California at Berkeley, Technical Report No. UCB/EECS-2010-155, December 15, 2010

[6] Xuan Zhang: Switched Capacitor Circuit Based Isolated Power Converter, Graduate Program in Electrical and Computer Science, The Ohio State University, 2016.

[7] Zichao Ye | Yutian Lei | Robert C. N. Pilawa-Podgurski,: A 48-to-12 V Cascaded Resonant Switched-Capacitor Converter for Data Centers with 99% Peak Efficiency and 2500 W/in 3 Power Density, 2019 IEEE Applied Power Electronics Conference and Exposition (APEC), PP. 13-18.

[8] OOTA Ichirou, HARADA Ikko, INOUE Takahiro: High Efficiency Control Method of Switched-Capacitor AC-DC Converter, The transactions of the Institute of Electronics, Information and Communication Engineers 76(6), 422-431, 1993-06-25.

[9] SANO Kenichiro, FUJITA Hideaki: Dynamic Control and Performance of a Resonant Switched-Capacitor Converter Based on Phase-Shift Control, The transactions of the Institute of Electrical Engineers of Japan. D, A publication of Industry Applications Society 128(10), 1190-1197, 2008-10-01.

[10] https://www.murata.com/en-us/support/faqs/products/capacitor/mlcc/char/0005?intcid5=com_xxx_xxx_cmn_hd_xxx

[11] https://www.keysight.com/ja/pd-1385381-pn-W2200BP/advanced-design-system-ads-core?pm=PL&nid=-34333.804581&cc=JP&lc=jpn

[12] https://www.ema-eda.com/products/cadence-orcad/why-orcad#

A Full State-Variable Predictive Control of Bi-directional Boost Converters with Guaranteed Stability

Yu Li[1], Zhenbin Zhang[1] and Ralph Kennel[2,1]

[1]School of Electrical Engineering
Shandong University
Jishi Road 17923, 250061, Jinan, China.
Email: yu.li@mail.sdu.edu.cn and zbz@sdu.edu.cn

[2]Institute for Electrical Drive Systems and Power Electronics
Technical University of Munich (TUM)
Munich, Germany.
Email: ralph.kennel@tum.de

Keywords

≪Bi-directional power conversion≫, ≪Boost converter≫, ≪model predictive control (MPC)≫, ≪stability≫.

Abstract

This work proposes a new direct predictive control with full state-variable for bi-directional DC-DC converters. With the proposed controller, both inductor current and capacitor voltage are incorporated into a unified cost function. The proposed controller allows all state variables to converge to a predefined region resulting in a good steady-state performance. Additionally, the weighting factor matrix based on the Lyapunov theory is analytically designed. Moreover, the steady-state error of the output voltage caused by model mismatch is addressed with a simple integral compensator. To the end, the proposed method is validated on a lab-constructed test bench using experimental results.

Introduction

In recent years, bi-directional power flow is becoming a preferred and even mandatory requirement in various applications of power conversion, such as DC microgrids, EV charger, and energy storage systems. The promising features of the bi-directional power conversion capability enable energy storage equipment to actively participate in the energy management process [1]. A variety of power topologies can be used to realize dc to dc power conversion, among which, the Boost converter is a simple and pragmatic topology for voltage step-up applications.

The control of Boost converters is inherently difficult due to its switching non-linear behavior. Moreover, the presence of right-half-plane (RHP) zero in the small-signal control-to-output transfer function makes it a more formidable task to have a high crossover frequency operating in continuous conduction mode (CCM) [2]. Specifically, as the duty cycle increases, the inductor current increases accordingly. However, the average current of the diode decreases since its conduction duration is reduced at the initial stage, leading to an output voltage dip temporarily. An enormous amount of research efforts have been devoted to improve the dynamic response of Boost power converters. Linear control schemes based on feedback theory are commonly employed, and is considered as an industry-standard method. The control parameters are selected by deciding the placement of compensation poles and zeros. However, the controller is usually implemented by analog operational amplifiers which limits its flexibility and

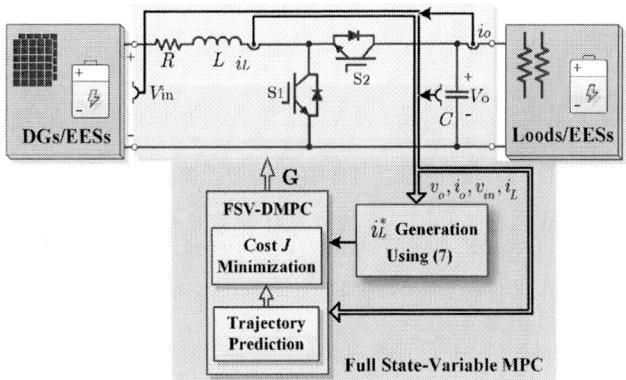

Fig. 1: Power circuit topology of Boost converter and proposed full state-variable MPC.

compatibility. Other control techniques, such as sliding mode, fuzzy logic, and model predictive control, have also been reported in previous literature [3, 4].

The existing control methods achieve considerable progress in terms of steady and transient performance. Nevertheless, some challenges need to be further addressed. For instance, cascaded loops are typically adopted in the linear control law. Consequently, dynamic performance is degraded. Other challenges include complex tuning processes and poor robustness to system parameter mismatches.

In the past two decades, model predictive control (MPC), especially finite control set MPC (FCS-MPC), has been considered as an advanced control method for power converters and electrical drives [5, 6]. Successful implementations of MPC for Boost converters are reported in recent years as well [3, 7, 8]. In [3], the authors employed a long prediction horizon to overcome the non-minimum phase character of the output voltage. Move blocking techniques are used to reduce the computational burden. In [7], suitable references for the inductor current are generated using the dynamic reference method. Thus, only a single-step prediction horizon is required to achieve satisfactory performance. In [8], double control loops are employed, with MPC current control as its inner loop. However, the generation of the current reference is relatively complicated.

This work mainly focuses on the Boost mode of the bi-directional DC-DC converter. To overcome the drawback of the non-minimum phase system character caused by the RHP zero, we propose a novel model predictive control that incorporates the full state variables into a unified cost function. In the procedure of reference step up, the proposed controller drives both control objectives to their references. This is an effective way that can overcome the voltage dip during the reference or load step-up process. The contribution of this work is summarized as follows.

- Removing nested control loops, we design a cost function with full state-variable in a unified fashion. This simplifies the controller design significantly.
- The weighting factors are obtained using Lyapunov stability theory. Therefore, system stability is guaranteed. Meanwhile, the trial and error tuning process is avoided.

Model of the bi-directional Boost converter

A bi-directional Boost converter and proposed control scheme are depicted in Fig. 1. Using the active power switch to replace the freewheeling diode allows a bi-directional current flow through S_2, resulting in a continuous inductor current under any load conditions.

A. System dynamics

S_1 and S_2 are operating in a complementary manner if the dead-time effect is ignored. The dynamics inductor current and capacitor voltage are derived as follows

$$L\frac{di_L}{dt} = V_{in} - Ri_L - (1-u)v_o \qquad\qquad C\frac{dv_o}{dt} = (1-u)i_L - i_o, \tag{1}$$

where V_{in} is the input voltage which might be the EV batteries or energy storage equipment. L and R denote the inductor and its serious resistor. i_L is the inductor current, accordingly. C corresponds to the output filter capacitor. v_o and i_o are the output voltage and output current, respectively. $u \in \mathcal{U} \triangleq \{0,1\}$ indicts the switching position of S_1 and S_2, i.e, $u=0$ if S_1 is turned off and S_2 is turned on, while $u=1$ when S_1 is turned on and S_2 is turned off. Obviously, the switching behavior exhibits non-linearity.

B. Prediction equations

To implement the control algorithm on the digital controller, the discrete prediction equations are obtained by using the forward Euler method as

$$i_L(k+1) = (1-\frac{RT_s}{L})i_L(k) + \frac{T_s}{L}\big(v_{in}(k) - \big(1-u(k)\big)v_o(k)\big), \tag{2}$$

$$v_o(k+1) = v_o(k) + \frac{T_s}{C}\big(\big(1-u(k)\big)i_{Lavg} - i_o(k)\big), \tag{3}$$

where T_s is the sampling time. In practical applications, the current ripple on the inductor is usually large (as a design rule of thumb, 20%-50% of peak value). Therefore, we use an average value of inductor current for capacitor voltage prediction, i.e.,

$$i_{Lavg} = \big(i_L(k) + i_L(k+1)\big)/2. \tag{4}$$

Proposed full sate-variable MPC

As well investigated, the output voltage exhibits a non-minimum phase behavior regarding the switching action. For instance, the increase of the output voltage leads to a larger duty cycle of S_1. However, the output voltage does not always change in phase with the duty cycle. In fact, the presence of RHP zeros causes the output to change in the opposite direction when the duty cycle changes too fast, which limits the bandwidth of the control loop. To overcome this problem, we propose a full-state variable MPC in this section.

A. Inductor current reference

Apparently, the output voltage reference can be obtained by a higher level control system. The inductor current reference needs to be derived according to its dynamic equation. If the conversion power loss is ignored, the following relation (5) holds in a specific sampling interval

$$v_{in}(k)\cdot i_{Lavg} = v_o(k)\cdot i_o(k) \qquad\Rightarrow\qquad i_{Lavg} = \frac{v_o(k)}{v_{in}(k)}\cdot i_o(k). \tag{5}$$

Considering equation $i_{Lavg} = \big(i_L(k) + i_L(k+1)\big)/2$,

$$i_L^*(k+1) = 2\cdot i_{Lavg} - i_L(k) = \frac{2v_o(k)}{v_{in}(k)}\cdot i_o(k) - i_L(k). \tag{6}$$

The above equation is derived from the assumption of zero power loss. Nevertheless, the power loss always exists due to the presence of an equivalent series resistor of the inductor, conduction and switching losses of power devices. Simulation results show that a steady-state error of output voltage arises if equation (6) is applied. A compensator using an integral controller is introduced, resulting in the complete inductor current reference as

$$i_L^\star(k+1) = \frac{2v_o(k)}{v_{in}(k)} \cdot i_o(k) - i_L(k) + K_i T_s \sum \left(v_o^\star(k) - v_o(k)\right), \tag{7}$$

where K_i is integral coefficients of the compensator, respectively. Note the first two items of (7) are dominating during the transient procedure as a feed-forward term, while the compensator only copes with the steady-state error caused by conversion losses.

B. MPC problem formulation

A two step of prediction horizon is considered in this work to enhance the system performance and stability [6]. Define $\mathbf{U}(k) = [u(k), u(k+1)]^\top \in \mathbb{U}$ as the control sequence, where $\mathbb{U} = \mathcal{U}^2$, and $\mathbf{x} = [i_L, v_o]^\top$ is the state variable. Based on the above observations, the cost function is defined as

$$J = \|\mathbf{x}(k+1) - \mathbf{x}^*(k+1)\|_Q^2 + \|\mathbf{x}(k+2) - \mathbf{x}^*(k+2)\|_P^2 + \sum_{\ell=k}^{k+1} \lambda_u \|\Delta u(\ell)\|_2^2, \tag{8}$$

where Q is the weighting matrix to balance the importance of control objectives, and $\|\xi\|_Q^2 = \xi^\top Q \xi$ denotes the squared weighted Euclidean norm. The final state penalty has a different weighting matrix P to ensure local stability. Note Q and P are positive semi-definite, and $\lambda_u > 0$. A more detailed discussion will be conducted in the sequel. Thus far, the MPC problem is formulated as

$$\mathbf{U}(k) = \arg \min_{\mathbf{U}(k) \in \mathbb{U}} J, \tag{9a}$$

$$\text{subject to}: (2), (3), \text{and } u \in \mathcal{U} \triangleq \{0, 1\}. \tag{9b}$$

C. Weighting matrix design

Choosing appropriate values of Q and P is usually a difficult task. In [10], a Lyapunov based stabilizing cost function design is proposed to provide sufficient conditions for local stability. However, this approach is based on representing power converters as a linear system which is not the case for a Boost converter. Motivated by the previous work of [11, 12, 13], we linearized the Boost converter around a specific operating point. We define the nominal operating point as the equilibrium. Therefore, the equilibrium of state variables can be obtained as $i_L^{eq} = I_L = V_o^2/(R_L \cdot V_{in})$, $v_o^{eq} = V_o$, $u^{eq} = D = (V_o - V_{in})/V_o$, where R_L, V_o and V_{in} are nominal values. According to (1), the dynamic equations are written in the state-space format as

$$f_1(\mathbf{x}, u) = \dot{x}_1 = -\frac{R}{L}x_1 + \frac{u-1}{L}x_2 + \frac{1}{L}v_{in}, \tag{10a}$$

$$f_2(\mathbf{x}, u) = \dot{x}_2 = \frac{1-u}{C}x_1 - \frac{1}{R_L \cdot C}x_2. \tag{10b}$$

Adopting the first-order Taylor approximation, the continuous linearized system model is obtained as

$$\dot{\mathbf{x}} = A_c \mathbf{x} + B_c u + B_{cd} d, \tag{11}$$

where

$$A_c = \begin{bmatrix} \frac{\partial f_1}{\partial x_1} & \frac{\partial f_1}{\partial x_2} \\ \frac{\partial f_2}{\partial x_1} & \frac{\partial f_2}{\partial x_2} \end{bmatrix}\Bigg|_{\mathbf{x}^{eq}, u^{eq}} = \begin{bmatrix} \frac{-R}{L} & \frac{D-1}{L} \\ \frac{1-D}{L} & -\frac{1}{R_L \cdot C} \end{bmatrix}, \ B_c = \begin{bmatrix} \frac{\partial f_1}{\partial u} \\ \frac{\partial f_2}{\partial u} \end{bmatrix}\Bigg|_{\mathbf{x}^{eq}, u^{eq}} = \begin{bmatrix} \frac{V_o}{L} \\ \frac{-I_L}{C} \end{bmatrix}, \ B_{cd} = \begin{bmatrix} \frac{\partial f_1}{\partial d} \\ \frac{\partial f_2}{\partial d} \end{bmatrix}\Bigg|_{\mathbf{x}^{eq}, u^{eq}} = \begin{bmatrix} \frac{1}{L} \\ 0 \end{bmatrix}. \tag{12}$$

Thereafter, exact Euler method can be used to obtain the linearized discrete model $A = e^{A_c T_s}$, $B = -A_c^{-1}(I-A)B_c$. Thus far, we are ready to design the weighting matrix Q and P. Assuming i_L and v_o has the same importance, then $Q = I$ holds as the state variables are transformed into the per unit value.

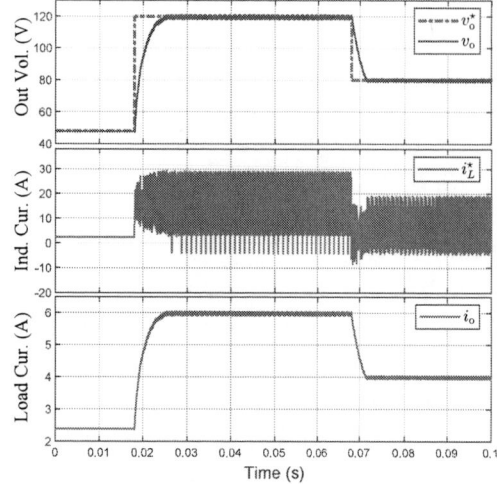

Fig. 2: Start-up and reference change performance

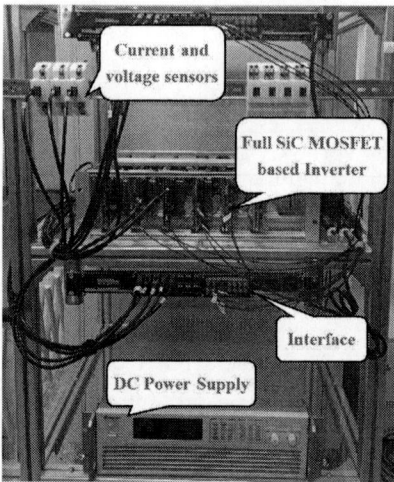

Fig. 3: Lab-constructed test bench

The terminal state penalty weighting matrix P is obtained by solving the algebraic Riccati equation. The definition of algebraic Riccati equation is $A_K^\top P A_K - P + Q + K^\top R K = 0$, where $K = -W^{-1} B^\top P A$, $W = B^\top P B + R$, $A_K = A + BK$, and $R = \lambda_u > 0$. Designing the weighting matrix Q and P using the presented method obtains the desired performance while guarantees practical stability of the Boost converter. The interested readers refer to [7] for a more detailed explanation.

Algorithm synthesis and implementation

Sampling and computational delay in digital control systems render a deteriorated steady-state performance. A simple and effective method is to employ a one-step prediction on the measurements to compensate for the delay effect [14]. The proposed control algorithm is described as follows.

1. Obtain the new measurements at time step k.
2. Estimate state variables at time step $k+1$.
3. Calculate the inductor current reference $i_L^\star(k+2)$ and $i_L^\star(k+3)$ using (7).
4. Compute the cost function (8) for each control sequence.
5. Select the control sequence which minimizes (8), and employ the first element to the converter.

The above procedure is repeated in the next sampling interval.

Simulation and experimental results

The proposed full state-variable MPC is validated through simulation and experimental measurements in this section. A dc power supply is used to emulate the behavior of distributed generation. A full SiC Mosfet based Boost converter is used for this test, which is shown in Fig. 3. The nominal values of input voltage, output voltage, and output power are 48V, 80V, and 160W, respectively. The filter inductor is 300uH, and the output capacitor is 260uF. Besides, the sampling frequency is set as 50kHz.

A. Simulation validation

The proposed control strategy is validated by the off-line simulation first. The start-up process and reference step-down is investigated under rated load condition. One can observe that as the reference steps up to 120V, the output voltage tracks the reference rapidly. The inductor current jumps up accordingly. Then, the output voltage steps down back to the rated value, i.e., 80V. The output voltage converges to 80V with a fast dynamic response. In summary, the proposed control strategy achieves good dynamic response and satisfactory steady-state performance.

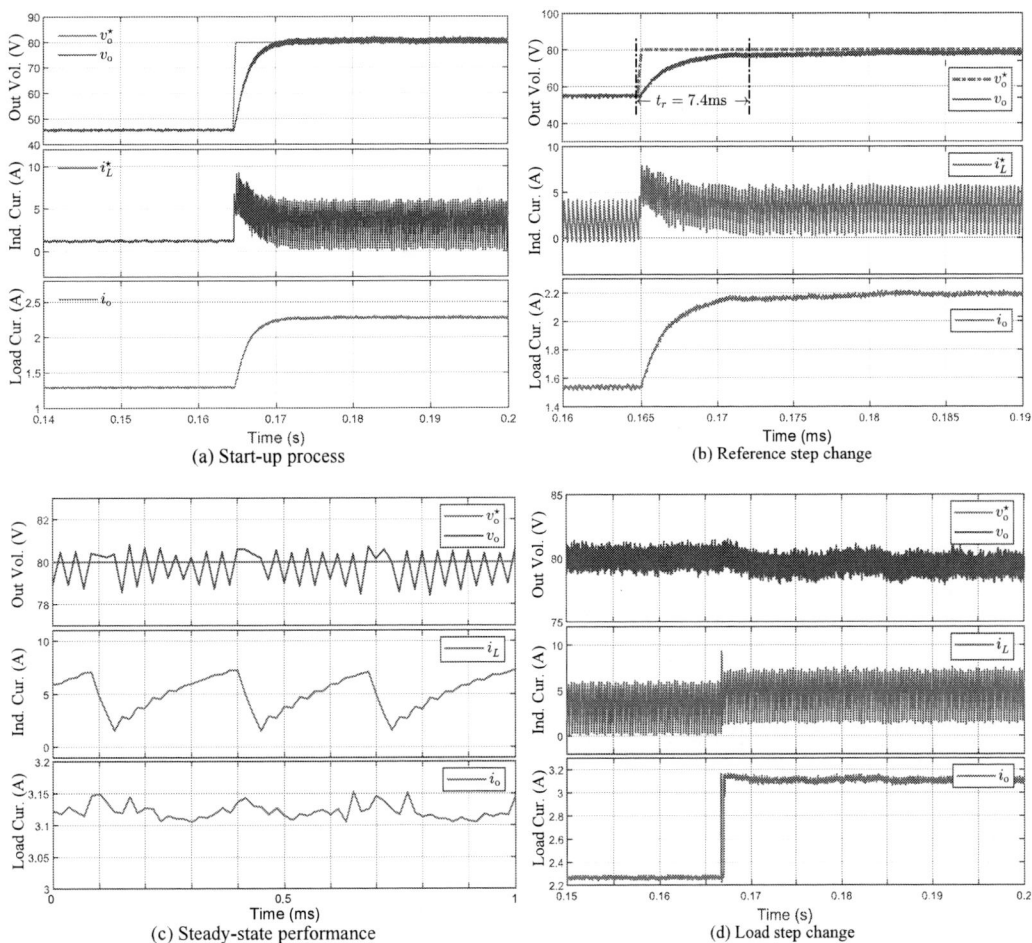

Fig. 4: Experimental measurements of bi-directional Boost converters under different operation modes.

B. Experimental validation

To further validate the control strategy, experiments on a lab-constructed test bench are conducted. The start up behavior under rated load is investigated. As observed in Fig. 4(a), the output voltage ramped up to the reference value rapidly without obvious overshoot. Accordingly, the inductor current responds quickly as well thanks to the full state-variable predictive controller. As the converter reaches the steady-state, both capacitor voltage and inductor current are regulated well around their reference value.

The transient procedure, i.e., a reference step change from 55V to 80V, is investigated then. The performance is shown in (b) of Fig. 4. As observed clearly, the inductor current ramps up rapidly, and the output voltage converges to 90% of its new reference value with a quite short rising time of 7.4ms. The proposed full state-variable predictive controller successfully achieves the fast dynamic response with respect to the output reference step-up. The tinny steady-state error will be eliminated by the integral compensator.

The steady-state performance is shown in (c) of Fig. 4. The peak-to-peak output voltage ripple is around 1.9V, and this metric can be further optimized with higher sampling frequency and higher output capacitance. The inductor current presents periodic quasi sawtooth waveform, indicting a stable operation.

A load step change test is carried out in this part, using a DC circuit breaker. A very sharp load step with a 30% nominal load change is shown in (d) of Fig. 4. Note that the load current shows rapid jumps and drops, which is caused by the mechanical contact impedance character of the DC breaker in the transient process. The output voltage maintains its rated value very well without obvious voltage dip. The inductor current achieves ultra fast response with respect to the load change. In fact, the inductor current reference contains a feed-forward term of the load current, which is clearly depicted by (7). Therefore, the load current change can be directly responded in a feed-through manner.

Conclusion

Bi-directional DC-DC converters have been widely used in DC microgrid and energy storage systems in recent days. The motivation of this work is to address the difficulty of the control problems for bi-directional Boost DC-DC converters. The presence of the RHP zeros entails a low control bandwidth for conventional control strategy. To overcome this issue, a full state-variable model predictive control approach is proposed. Both inductor current and capacitor voltage are taken into account to constitute a unified cost function. This allows the output voltage to converge to its new reference with very fast dynamic response. The tuning efforts are avoided by designing weighting matrices based on the Lyapunov theory, ensuring the local stability of the system. To the end, experimental results show the effectiveness of the proposed scheme.

References

[1] T. Dragievi, X. Lu, J. C. Vasquez and J. M. Guerrero, "DC MicrogridsPart I: A Review of Control Strategies and Stabilization Techniques," in IEEE Transactions on Power Electronics, vol. 31, no. 7, pp. 4876-4891, July 2016.

[2] K. Kittipeerachon and C. Bunlaksananusorn, "Feedback compensation design for switched mode power supplies with a right-half plane (RHP) zero," Second International Conference on Power Electronics, Machines and Drives (PEMD 2004)., Edinburgh, UK, 2004, pp. 236-241 Vol.1.

[3] P. Karamanakos, T. Geyer and S. Manias, "Direct Voltage Control of DCDC Boost Converters Using Enumeration-Based Model Predictive Control," in IEEE Transactions on Power Electronics, vol. 29, no. 2, pp. 968-978, Feb. 2014.

[4] P. Mattavelli, L. Rossetto and G. Spiazzi, "Small-signal analysis of DC-DC converters with sliding mode control," in IEEE Transactions on Power Electronics, vol. 12, no. 1, pp. 96-102, Jan. 1997.

[5] Z. Zhang, F. Wang, T. Sun, J. Rodrguez and R. Kennel, "FPGA-Based Experimental Investigation of a Quasi-Centralized Model Predictive Control for Back-to-Back Converters," in IEEE Transactions on Power Electronics, vol. 31, no. 1, pp. 662-674, Jan. 2016.

[6] J. Rodriguez et al., "State of the Art of Finite Control Set Model Predictive Control in Power Electronics," in IEEE Transactions on Industrial Informatics, vol. 9, no. 2, pp. 1003-1016, May 2013.

[7] L. Cheng et al., "Model Predictive Control for DCDC Boost Converters With Reduced-Prediction Horizon and Constant Switching Frequency," in IEEE Transactions on Power Electronics, vol. 33, no. 10, pp. 9064-9075, Oct. 2018.

[8] M. Hejri and H. Mokhtari, "Hybrid modeling and control of a DC-DC boost converter via Extended Mixed Logical Dynamical systems (EMLDs)," The 5th Annual International Power Electronics, Drive Systems and Technologies Conference (PEDSTC 2014), Tehran, 2014, pp. 373-378.

[9] P. Karamanakos and T. Geyer, "Guidelines for the Design of Finite Control Set Model Predictive Controllers," in IEEE Transactions on Power Electronics.

[10] R. P. Aguilera and D. E. Quevedo, "Predictive Control of Power Converters: Designs With Guaranteed Performance," in IEEE Transactions on Industrial Informatics, vol. 11, no. 1, pp. 53-63, Feb. 2015.

[11] F. Grimm, Z. Zhang, M. Abdelrahem and R. Kennel, "Computationally efficient predictive control of three-level NPC converters with DC-link voltage balancing: A priori state selection approach," 2017 IEEE International Symposium on Predictive Control of Electrical Drives and Power Electronics (PRECEDE), Pilsen, 2017, pp. 72-77.

[12] G. Darivianakis, T. Geyer, and W. van der Merwe, Model predictive current control of modular multilevel converters, in Proc. IEEE Energy Convers. Congr. Expo., Pittsburgh, PA, Sep. 2014, pp. 50165023.

[13] E. Liegmann, P. Karamanakos, T. Geyer, T. Mouton, and R. Kennel, Long-horizon direct model predictive control with active balancing of the neutral point potential, in Proc. IEEE Int. Symp. Pred. Control of Elect. Drives and Power Electron., Pilsen, Czech Republic, Sep. 2017, pp. 8994.

[14] P. Cortes, J. Rodriguez, C. Silva and A. Flores, "Delay Compensation in Model Predictive Current Control of a Three-Phase Inverter," in IEEE Transactions on Industrial Electronics, vol. 59, no. 2, pp. 1323-1325, Feb. 2012.

System-Level Reliability Analysis of a Repairable Power Electronic-Based Power System Considering Non-Constant Failure Rates

Amirali Davoodi[1], Yongheng Yang[1], Tomislav Dragičević[2], and Frede Blaabjerg[1]

[1] Department of Energy Technology, Aalborg University, Aalborg, Denmark

[2] Dept. of Electrical Engineering, Technical University of Denmark, Copenhagen, Denmark

E-Mail: amirali.davoodi@et.aau.dk

Keywords

«Reliability», «Microgrid», «Photovoltaic», «Wind energy»

Abstract

Conventionally, for reliability studies in power systems, a constant failure rate is assumed for power generation units. If these units consist of power electronic converters, this assumption might not be valid due to the aging of power components, and it will lead to an unrealistic prediction of reliability. On the other hand, at the system-level, commonly-used reliability calculation tools, such as Monte Carlo Simulation (MCS) and Continuous Markov Process (CMP), are either time-consuming or not possible to be applied to the systems with non-constant failure rates. Therefore, in this paper, a methodology is proposed to calculate system-level reliability for a Power-Electronic-based Power System (PEPS), consisting of several converters with non-constant failure rates. By doing so, not only is the effect of mission profiles integrated into the system-level reliability model, but also the wear-out failures and corrective maintenance are considered. Finally, for a case study PEPS, the system-level indices are calculated using the proposed method. It is shown that assuming constant failure rates for PEPS units is inaccurate and misleading. Moreover, the impacts of various factors, e.g., mission profile, repair rate, topology, and rating of converters, on the system-level reliability are investigated and analyzed.

I. Introduction

A power system directly supplies a wide variety of industrial, commercial, and household loads. As a result, any interruptions in its normal operation will have severe consequences. Such interruptions, even for a short period, will directly affect a large number of people and may lead to huge economic losses. Therefore, reliability and security are two substantial principles in the power system operation and planning phases.

Moreover, with the ever-increasing use of power electronic converters in the grid, the modern power system will be a Power-Electronic-based Power System (PEPS). On the other hand, there are several concerns regarding the reliability of Power Electronic (PE) converters. The relatively high failure rate of certain components [1], direct dependency on operating conditions, complex topologies and control structures, and wear-out of components are of importance, which can cast doubt on the viability of a PEPS in terms of reliability. To address these concerns, the system-level reliability must be analyzed for a PEPS to benchmark with relevant requirements. As a result, a quantitative system-level reliability model of the reliability of the PEPS is essential.

Typically, in power system reliability studies, a constant failure rate is assigned to each generation unit, based on the historical data of failure [2]. However, in PE reliability, a non-constant failure rate is calculated for the converter when using the Stress-Strain Analysis (SSA). The details on how the SSA is performed can be found in [3]. In practice, mission profiles and the physics of failure are contributing to the stress level and lifetime of a converter, both of which are considered in the SSA approach [1]. In addition, while the wear-out of the components is modeled in the SSA, the constant failure rate is unable to take this fact into account. Therefore, considering a non-constant failure rate will be a more accurate reliability model for the converters, and will, in return, lead to a more realistic estimation of system-level indices in a PEPS.

In PE literature, quite a few works are dedicated to study and model the reliability of power converters. A considerable part of these works concentrates on the component and converter-level reliability in power electronic systems. For example, in [4], a failure Probability Density Function (PDF) was predicted for power semiconductors under a certain mission profile. In [5], the reliability of a power electronic system, including five DC-DC converters, was analyzed, where the reliability model of the converters was built based on the SSA, and the system-level reliability is analyzed through Reliability Block Diagram (RBD) approach. Notably, one major drawback of RBD is its inability to take repair and maintenance into account. Similarly, in [6], a new methodology is adopted to assess the reliability of a power electronic system more accurately. In that paper, the impact of component degradation on the operating conditions is considered by using a multistate degraded system approach. Notably, having no repair or maintenance is the preassumption of using that approach. However, repair and maintenance are particularly important when considering that a PEPS is not a mission-oriented system. In other words, it does not have a specific end-of-life target and is always expected to operate correctly and have high availability. Hence, repair and maintenance play a key role in its availability, and modeling a PEPS without considering the repair is not a practical assumption. Nevertheless, further attempts have been made to model the system-level reliability of a PEPS based on the SSA, an overview of which can be found in [7].

In the power system reliability context, Monte Carlo Simulation (MCS) [8] and Continuous Markov Process (CMP) [9] are used extensively to calculate the system level indices [10]. The MCS can model both constant and non-constant failures. However, due to the high computational burden, the reliability evaluation using the MCS can be a time-consuming task [11]. The CMP, on the other hand, can be a good solution for systems with a few number of states. Nonetheless, due to the Markovian property, it can only model the systems with constant failure rates [12]. Therefore, one of the challenges in the literature is the lack of an appropriate tool for system-level modeling of a PEPS with non-constant failure rates.

With the above concerns, a Time-Inhomogeneous Markov Chain (TIMC) [13] approach is used to tackle this challenge. In this paper, the proposed method aims at analyzing the system-level reliability in the transient and steady-state modes with relatively high accuracy, while it maintains the simplicity and a low computational burden. Moreover, corrective maintenance is considered in the proposed approach using a constant repair rate. Additionally, converter-level failure rates are acquired through the SSA. In other words, the impact of mission profiles, physics of failure [1], and wear-out of components can be observed on the system-level indices. Finally, the method is applied to a case study PEPS. With this case study, the impact of the repair rate, mission profile, converter topology, comparison between constant and non-constant failure rate, and converter ratings, on the index of EENS (Expected Energy Not Supplied) [10] is investigated.

In Section II, the proposed method to calculate the system level indices for transient and steady-state modes is explained. In Section III, the proposed methodology is used to calculate the EENS for a case study PEPS. Various scenarios are considered and compared to explore the impact of different parameters on the EENS. Finally, the conclusions are drawn in Section IV.

II. The proposed method

A power system with several generation units can be in various states (in terms of generation capacity) depending on the failure of its units. To calculate system-level reliability indices in such a system, it is essential to know the probability of being in each state. A single-unit system is exemplified in this section, which can be either in State 1 (Healthy) or State 2 (Faulty). The reliability function of the unit, $R(t)$, can be expressed in terms of its failure PDF, $f(t)$, as a function of time, t:

$$R(t) = 1 - \int_0^t f(\tau)d\tau \qquad (1)$$

Moreover, the failure rate function, $\lambda(t)$ is defined as:

$$\lambda(t) = f(t) / R(t) \qquad (2)$$

As can be seen in Fig. 1, this unit can make the transition from State 1 to State 2 with the failure rate of $\lambda_{eq}(t)$ and from State 2 to State 1 with the constant repair rate of μ. In other words, the converter can fail with the rate of $\lambda_{eq}(t)$, and once it fails, it is restored to the initial condition with the rate of μ. If $\lambda_{eq}(t)$ is a constant value (i.e., time-independent), the Markov property is satisfied, and the probability of residing in each state can be calculated by using the CMP approach.

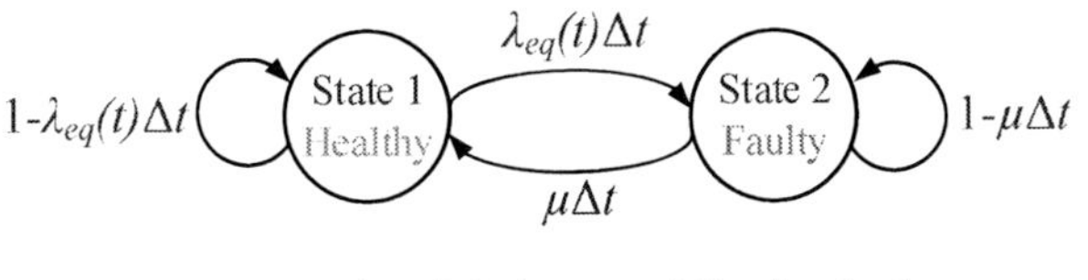

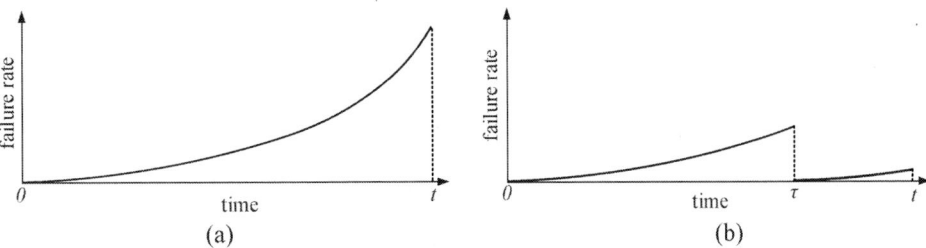

Fig. 1: State-space representation (Markov model) of a single-converter system.

On the one hand, in power system reliability studies, traditionally, an exponential distribution is assumed to model the failures of the units. The failure PDF of this distribution can be written as:

$$f_e(t) = he^{-ht} \tag{3}$$

In this case, by using (2), the failure rate function will be a constant value equal to h, i.e., independent of time. As a result, a system comprised of several units can be modeled using the CMP approach.

In PE reliability literature, on the other hand, the failure rates of power converters are typically modeled through a Weibull distribution. The failure PDF of the Weibull distribution, as a function of time, t, is shown as:

$$f_W(t) = \frac{\beta}{\eta^\beta} t^{\beta-1} e^{-(t/\eta)^\beta} \tag{4}$$

where β and η are shape and scale parameters, respectively. $R_W(t)$, $\lambda_W(t)$ can be calculated using (1) and (2), respectively. It should be emphasized that, in this case, since the failure rate is non-constant over time, the CMP approach cannot be employed to calculate the state probabilities.

In this paper, a TIMC-based approach is then introduced to tackle this challenge. In the proposed method, a time-dependent equivalent failure rate, $\lambda_{eq}(t)$, is calculated for each converter. Afterward, time is discretized into several instants, and for each moment of time, t_i, a probability transition matrix, T_i, is calculated. Finally, the probability of residing in each state is obtained with simple matrix multiplications, and subsequently, the system-level indices can be obtained accordingly.

A. Analysis of probabilities in the transient mode

Calculating the equivalent failure rate:

Provided that no failure happens before the instant t, the failure rate of the Weibull distribution is shown in Fig. 2 (a). Assuming that this is the only possible scenario, $\lambda_{eq}(t)$ can be written as:

$$\lambda_{eq}(t) = \lambda_W(t) \tag{5}$$

where $\lambda_W(t)$ is the Weibull failure rate. However, in theory, an infinite number of failures could have happened before the instant t, and (5) is only valid in the case that the converter has been failure-free up to the moment t. The probability of the converter being without failure before t can be calculated as:

$$P_0(t) = R_W(t) \tag{6}$$

where $R_W(t)$ is the Weibull reliability function. As long as the value of (6) is close to 1, the previous assumption can be justified. Given that C is a desired coefficient close to 1, the range within which the previous assumption is valid (validity range) can be extracted according to:

$$0 \le t \le \eta \sqrt[\beta]{\ln(1/C)} \tag{7}$$

If smaller values are chosen for C, the validity range increases, but the accuracy of approximation in (5) decreases. It should be mentioned that a similar assumption was made in [14] for maintenance scheduling in overhead lines in the power distribution network. However, in that paper, the validity range of this assumption was not discussed.

Fig. 2: Failure rate over time: (a) no failure before t and (b) one failure before t at instant τ.

Now, according to Fig. 2 (b), assuming that only one failure happens before t at the moment τ, and the converter is restored to the initial condition shortly and continues working until t. In this case, the effective failure rate at the moment t would be equal to $\lambda_W(t-\tau)$. The probability of only one failure at the moment τ (and no failure, before and after that) is equal to $R_W(\tau)\lambda_W(\tau)R_W(t-\tau)d\tau$. Since τ can hold any values within [0 - t], to find the probability of only one failure before t, it should be integrated over [0 - t]. Also, in the previous case, where no failure could happen before t, the effective failure rate was $\lambda_W(t)$, and the probability of being failure-free was $R_W(t)$.

Assuming that there is either no failure before t or only one failure, $\lambda_{eq}(t)$ can be calculated according to:

$$\lambda_{eq}(t) = f_W(t) + (f_W * f_W)(t) \tag{8}$$

where $(f_W * f_W)(t)$ is the convolution of the PDF with itself, defined as:

$$(f_W * f_W)(t) = \int_0^t f_W(\tau) f_W(t-\tau) d\tau \tag{9}$$

Equation (8) is valid as long as the following condition is satisfied:

$$R_W(\tau) + \int_0^t f_W(\tau) R_W(t-\tau) d\tau \geq C \quad , \quad t \geq 0 \tag{10}$$

in which C is a desired coefficient close to 1. If smaller values are chosen for C, the validity range increases, but the accuracy of equation (8) decreases.

In reality, an infinite number of failures can occur before t. If cases with more failures are assumed, new terms would be added to $\lambda_{eq}(t)$, and the validity range increases. However, taking all possible scenarios into consideration will inevitably increase the computational burden and complexity. As a result, in this paper, only two approximations of $\lambda_{eq}(t)$ are proposed in (5) and (8). These approximations are valid within a certain time range, which is also modeled in (7) and (10), respectively. Thus, depending on the validity range, this method can be used for short-term or mid-term system-level reliability studies.

Calculating the probability transition matrix and state probabilities:

Once $\lambda_{eq}(t)$ is calculated, time is quantized into several discrete instants, t_i as:

$$t_i = i\Delta t, \quad i \in \mathbb{N} \tag{11}$$

in which Δt is a very small time increment. It is assumed that Δt is small enough so that the probability of the occurrence of more than one failure or repair during one increment of time is negligible. Moreover, it can be adjusted with respect to the compromise between the accuracy and the computational burden of the proposed method.

For each instant of time, t_i, a probability transition matrix, T_i, is formed as:

$$T_i = \begin{bmatrix} 1 - \lambda_{eq}(t_i)\Delta t & \lambda_{eq}(t_i)\Delta t \\ \mu\Delta t & 1 - \mu\Delta t \end{bmatrix} \begin{matrix} \text{State 1} \\ \text{State 2} \end{matrix} \tag{12}$$

Considering that the system starts in the healthy condition at $t = 0$, S_0, the initial probability vector can be written as $S_0 = [1 \ 0]$. Accordingly, S_i, the probability vector of residing in each state at the instant t_i can be calculated according to:

$$S_i = S_0 \prod_{n=1}^{i} T_n \tag{13}$$

This approach can be applied to a PEPS with several converters to assess the system-level indices in the transient state. First, the equivalent failure rate must be calculated, and afterward, the probability transition matrix must be formed for each instant of time. Finally, the probability of being in each state and system-level indices are calculated.

Validation:

To evaluate this method, it is applied to a single-converter system, similar to the system shown in Fig. 1, considering three different Weibull distributions. It is assumed that the converter starts in the healthy state, $\mu = 20$, and $C = 0.9$. The probability of finding the converter in the healthy state ($P_{healthy}$) at $t = 10$ years is calculated using the proposed method. Moreover, the same system is simulated using the MCS approach, and the same probability is calculated. In Table I, the results from the proposed method are compared to the results from the MCS. The details of simulating a system using the MCS can be found in [8]. As can be concluded from Table I, the error of the proposed method (within its validity range) is negligible. In addition, as shown in Fig. 3, for case #2 of Table I, the result of the proposed method exactly follows the results obtained from the MCS.

Table I: Probability of finding the converter in the healthy state, comparison of the results obtained using the proposed method with the Monte Carlo Simulation (MCS) results for three different Weibull cases.

	(β, η)	Validity range of the proposed method (years)	$P_{healthy}$ @t = 10 years from MCS	$P_{healthy}$ @t = 10 years from the proposed method	Error (%)
1	(3,10)	$0 \le t \le 11.98$	0.99346186	0.99348199	0.002
2	(5,10)	$0 \le t \le 14.46$	0.99068133	0.99076100	0.008
3	(3,20)	$0 \le t \le 23.95$	0.99833962	0.99829800	0.004

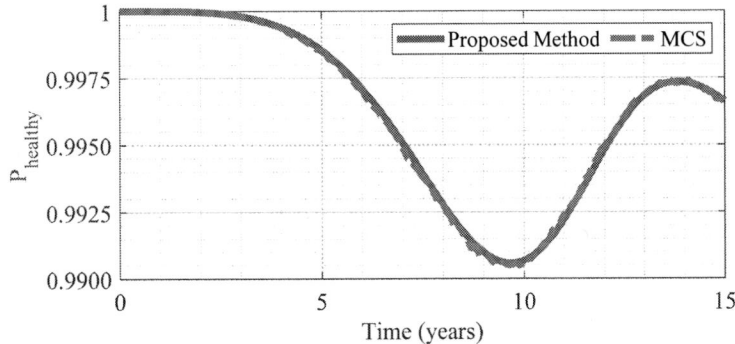

Fig. 3: Probability of finding the converter with $(\beta, \eta) = (5, 10)$ in the healthy state, comparison of the results obtained using the proposed method with the Monte Carlo Simulation (MCS) results.

B. Analysis of steady-state probabilities

The previous procedure can be applied to a PEPS with several converters to assess the system-level indices in the transient state. However, in the long run, the probabilities of residing in each state will reach a steady-state value. For a single converter system, the probability of being in the healthy state is called availability, A. Similarly, U is the unavailability—the probability of finding the converter in the faulty state. When $t \to \infty$, the average model, shown in Fig. 4, can be used, where MTTF (Mean Time To Failure) is the expected value of the failure PDF, and MTTR (Mean Time To Repair) is the expected value of the repair PDF.

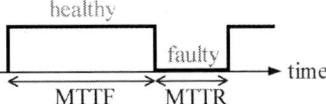

Fig. 4: Average model of reliability of the converter, valid when $t \to \infty$.

Accordingly, when $t \to \infty$, A and U can be calculated as:

$$A = \frac{MTTF}{MTTF + MTTR}, \quad U = 1 - A \quad (14)$$

Considering a Weibull distribution for failures, and exponential distribution for repair with the rate of μ, the steady-state availability and unavailability can be calculated according to:

$$A = \frac{\eta\Gamma(1+1/\beta)}{\eta\Gamma(1+1/\beta)+1/\mu}, \quad U = \frac{1/\mu}{\eta\Gamma(1+1/\beta)+1/\mu} \quad (15)$$

For a system with N converters, the probability of residing in each state can be calculated according to:

$$P_{Sj} = \prod_{i=1}^{N} P_i \quad (16)$$

where P_{Sj}, the probability of the system being in state j, is calculated by simply multiplying the availability or unavailability of the converters. If, in state j, the converter i is considered healthy, $P_i = A$.

If, in state j, the converter i is considered faulty, $P_i=U$. Hence, the probability of being in all states can be calculated, and thereby calculating the system-level indices.

III. Case study and results

In this section, the reliability of a case study PEPS is assessed using the proposed method. As shown in Fig. 5 (a), the case study PEPS is a four-converter system, including two wind converters and two PV systems. The structure of the wind systems is shown in Fig. 6, where a 2-level back-to-back converter is used in a Doubly-Fed Induction Generator (DFIG) wind turbine system. More details of such wind systems and their thermal modeling are provided in [15]. The rated power of the back-to-back converter is 5 kW, and the converter-level reliability metrics are calculated based on the component-level data provided in [7] and reported in Table II. Similarly, the structure of the PV system is illustrated in Fig. 7, where PV strings are connected to the grid through a boost converter, 2-Level Voltage-Source Inverter (2-L VSI), and the output filter. The maximum power of the PV string is 4 kW. The PV system is simulated, and its reliability metrics are extracted by performing the SSA. In this regard, the ratings of converters, as well as the extracted Weibull parameters, are presented in Table II.

As for the load model, if the hourly peak loads are sorted in descending order, an hourly peak Load Duration Curve (LDC) can be obtained. In [10], the procedure for acquiring the LDC from the hourly load profile is explained in detail. In this paper, an arbitrary hourly peak LDC is chosen to model the load in the case study, which is shown in Fig. 5 (b).

Concerning the system-level index, EENS (Expected Energy Not Supplied) is chosen. It indicates the number of kWh (or MWh) of energy, which is expected not to be supplied by the generation system in one year. The EENS is one of the most commonly-used severity indices in power system reliability assessment, and the procedure for calculating it is explained in [16].

Table II: Rated power, Weibull parameters, and repair rate of the converters used in the case study PEPS.

	Rated Power (kW)	Weibull		Repair Rate
		β	η	
Converter 1	5	3	8.5	150
Converter 2	5	3	8.5	150
Converter 3	4	3.3	10.9	100
Converter 4	4	3.3	10.9	100

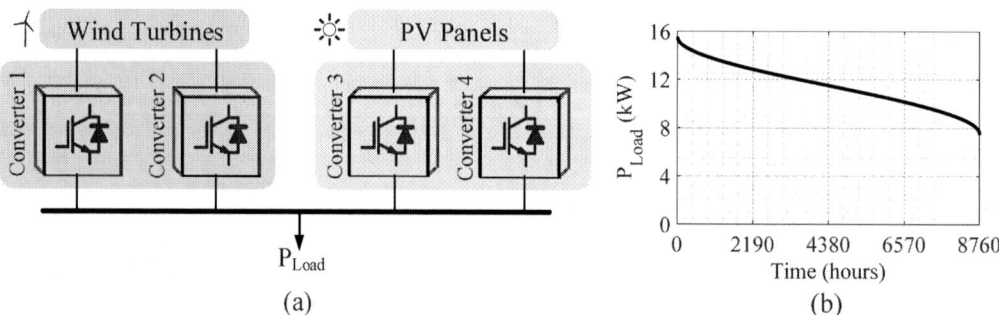

(a)   (b)

Fig. 5: (a) Schematic of the case study PEPS and (b) the Load Duration Curve (LDC) model.

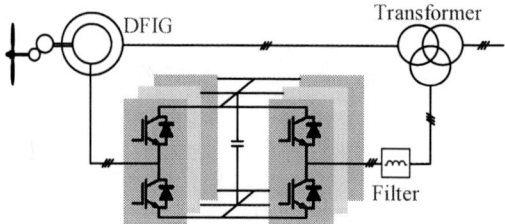

Fig. 6: Structure of the wind system used for modeling Converters 1 and 2 of the case study [15].

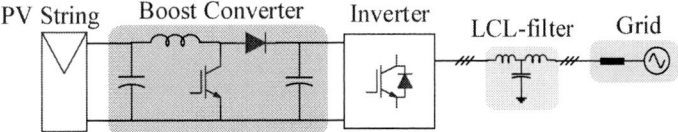

Fig. 7: Structure of the PV system used for modeling Converters 3 and 4 of the case study.

A. Comparison of cases with a constant and non-constant failure rate

As mentioned before, in power system reliability studies, a constant failure rate is usually assigned to the generation units. Interestingly, in a special case of the Weibull distribution, if $\beta = 1$, the failure rate would be a constant value equal to $1/\eta$. In this section, the EENS is calculated and compared for two different cases: first, the Weibull case with $\beta > 1$ (non-constant failure rate), and second, $\beta = 1$ (constant failure rate). As shown in Fig. 8, the EENS index, in the latter case, reaches a limiting value shortly and remains constant over time. In contrast, with the non-constant failure rate, the EENS value is negligible in the beginning and gradually increases over time. From Fig. 8, it can be understood that, in the short term, using the constant failure rates for the converters leads to an overestimation of the EENS. This, in turn, leads to the over-scheduling of the system, which is not cost-effective. Furthermore, in the long-term, using the constant failure rates provides an underestimated value of the EENS. Consequently, the under-scheduling would be inevitable, which leads to the unreliability.

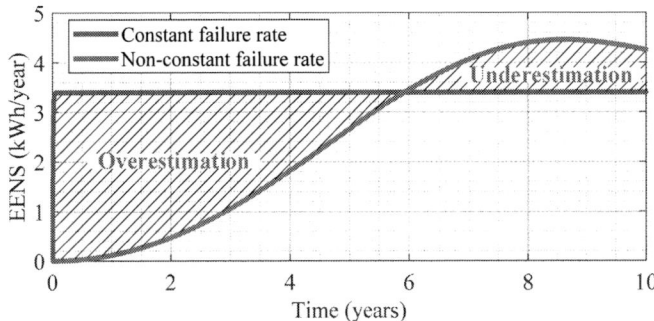

Fig. 8: Comparison of the EENS in a PEPS with constant and non-constant failure rates.

B. Sensitivity of EENS to the repair rate

In mission-oriented systems, it is important that the device survives until the end of the mission time. In such systems, the index of reliability, $R(t)$, is of importance. Generally, power electronic converters are treated as such systems in the PE reliability literature. Therefore, repair and maintenance are not of interest in the PE reliability context and usually are neglected. However, a power system is not a mission-oriented system; in other words, it is expected to operate all the time. In such systems, other indices such as availability, EENS, and LOLE are more commonly used [16], rather than $R(t)$. Corrective maintenance is vital to keep such a system up all the time, which can be modeled with a constant repair rate, μ. In corrective maintenance, once a converter fails, it is restored to the initial condition, as soon as possible.

For the case study PEPS, it is assumed for simplicity that all the converters have the same repair rate, μ. This repair rate is then varied from 5 to 150, and the EENS index is calculated in steady-state accordingly. It can be seen from Fig. 9 that, for small values of μ, the EENS can be extremely large. In the worst case, when no repair is considered, the steady-state EENS will be infinite, which is impractical for a real power system. From this analysis, it can be concluded that, if a PE converter is to be studied in the PS reliability context, considering repair is essential for having a realistic and practical analysis.

It is worth mentioning that increasing the repair rate can be realized by taking various measures, such as having more maintenance personnel and spare parts. As a result, as the repair rate increases, the cost of maintenance increases. As shown in Fig. 9, at the beginning of the curve, the increase of μ reduces the EENS drastically. However, after some point, a significant increase of μ is required to reduce the

EENS. Therefore, this feature can be used to find an optimal value for the repair rate to obtain a balance between the EENS and the maintenance cost.

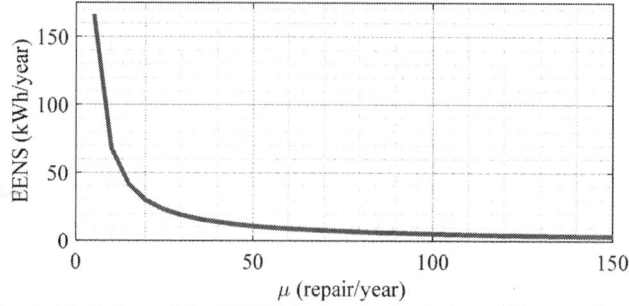

Fig. 9: Variation of the EENS due to the variation of the repair rate μ.

C. Impact of the mission profile

In a real PEPS, each converter can be subjected to different mission profiles. In return, different mission profiles can induce different stress levels on the converters, and therefore, may result in different reliability performances of converters. To illustrate the impact of the mission profile on the system-level indices in the case study PEPS, it is assumed that Converters 1 and 2 are located in a different location with a more severe wind speed profile. Wind speed profiles for two different locations, as well as the component reliability data (using the SSA approach) under these profiles, are presented in [7]. According to this data, the failure distribution of the converters is extracted under the new mission profile. (β, η) of the Weibull distribution for Converters 1 and 2 changed from $(3, 8.5)$ to $(2.8, 5.8)$. A decrease of η can generally be interpreted as an increase in the failure rate. Furthermore, the EENS of the system is recalculated and compared with the previous EENS, as shown in Fig. 10. As can be seen in Fig. 10, in a more severe wind profile, where the failure rate of converters 1 and 2 increases, the EENS also increases. Notably, the peak EENS changes from 4.4 (kWh/year) for the wind profile A to 5.5 (kWh/year) for the wind profile B.

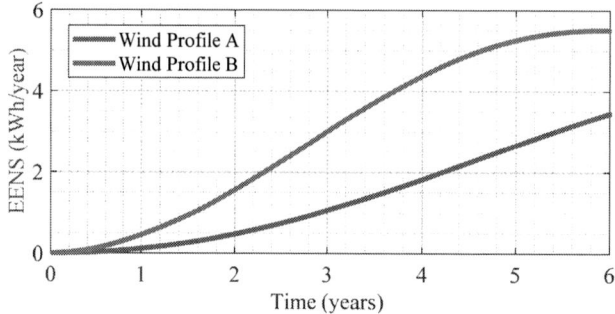

Fig. 10: Comparison of the EENS when wind profile of converters 1 and 2 changes.

D. Impact of converter-level design

Changing the topology of inverters: To illustrate the impact of the converter-level design on the system-level indices, another topology for the PV inverter is taken into account. In this case, in Converters 3 and 4, the topology of inverters is changed from 2-L VSI to 3-L NPC, and the failure distribution of the converters is recalculated through the SSA approach. More information regarding the topology, modulation, and control strategies can be found in [17]. Since more switches and voltage levels are employed in the 3-level inverter, the stress is distributed among the components. Therefore, it has a better reliability performance, and (β, η) of the PV systems changed from $(3.3, 10.9)$ to $(3.4, 24.8)$. Afterward, the EENS is calculated considering two different topologies and shown in Fig. 11. As can be seen in Fig. 11, the EENS decreases from 4.4 (kWh/year) to 3.5 (kWh/year) when a 3-L NPC is used rather than a 2-L VSI. Such system-level indices can further be used for decision-making in the system-level to justify such changes at the converter level and compare various options with each other.

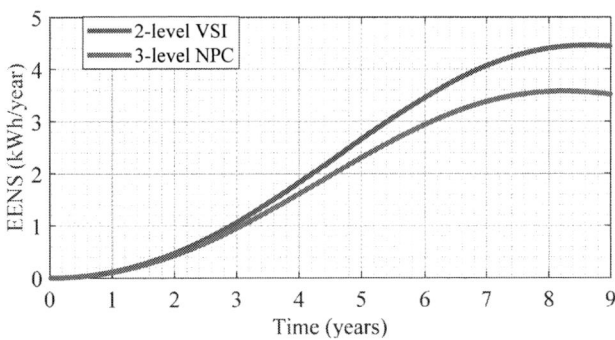

Fig. 11: Comparison of the EENS when the topology of Converters 3 and 4 is changed from 2-level voltage-source inverter (VSI) to 3-level Neutral-Point Clamped (NPC) inverter

Derating the converters: As mentioned before, Converters 3 and 4 are designed for supplying a 4-kW load. In this case, it is assumed that the converters remain the same (with 4-kW design), but connected to a 3-kW PV string. In this case, the failure distribution of Converters 3 and 4 was recalculated, and it turned out that (β, η) of the converters changed from $(3.3, 10.9)$ to $(3.3, 27)$. Subsequently, the EENS is calculated as shown in Fig. 12, assuming the load has also decreased by 2 kW. As shown in Fig. 12, the EENS of the system has decreased in the new condition, since the stress on the components of the converters has decreased (and their η has increased).

It is worth mentioning that the derating of converters might be justifiable as an option in some specific cases where certain system-level reliability is required. For instance, such an idea is demonstrated and elaborated in [18] to find the optimal derating level for wind converters. However, in [18], system-level reliability indices are not considered for optimization. Analysis of the system-level indices, namely the EENS, can help the decision-makers to have a quantitative tool for assessing all the possible options such as derating of converters.

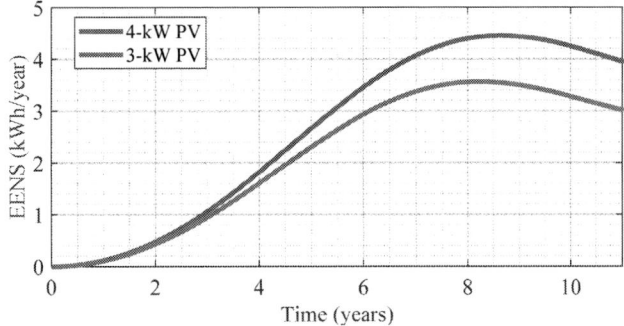

Fig. 12: Change in the EENS when converters 3 and 4 are derated from 4 kW to 3 kW.

IV. Conclusions

In this paper, a new method was proposed to calculate the system-level reliability indices in a Power Electronic based Power System (PEPS) with non-constant failure rates. By using this method, in addition to modeling the corrective maintenance, the impact of mission profiles and the physics of failure can be incorporated into the system-level reliability analysis. It has been revealed that the accuracy of the proposed method is comparable to the Monte Carlo Simulation (MCS), while it does not require a large number of simulations. Subsequently, for a case study PEPS, the Expected Energy Not Supplied (EENS) index was calculated and compared in different cases to study the impact of various factors on the system reliability. Through the analysis, it was shown that considering a constant failure rate for converters was not a realistic and accurate assumption. In other words, it will result in an overestimation of the system-level indices in the short term and underestimation in the long run. Furthermore, it was shown that the impact of the mission profile was considerable on the reliability of the PEPS. For instance, the peak EENS increased from 4.4 (kWh/year) to 5.5 (kWh/year) when two wind converters were exposed to a more severe wind profile. Moreover, the impact of corrective maintenance on the EENS was investigated by performing a sensitivity analysis, where the repair rate varied from 5 to 150. As a result, the EENS varied from 166.2 (kWh/year) to 3.5 (kWh/year), and it was concluded that not modeling the repair is an impractical assumption for system-level reliability studies. Afterward, the impact of converter-level design parameters on the EENS was studied by two examples. First, the topology of PV inverters was changed from 2-Level Voltage-Source Inverter (2-L VSI) to 3-Level Neutral-Point Clamped (3-L NPC), where the EENS decreased from 4.4 (kWh/year) to 3.5 (kWh/year), following the decrease of stress at the converter level. Second, when 4-kW inverters were used for converting 3-kW power, the EENS decreased due to the reduction of stress at the converter level.

References

[1]   H. Wang, M. Liserre, F. Blaabjerg *et al.*, "Transitioning to physics-of-failure as a reliability driver in power electronics," *IEEE J. Emerg. Sel. Top. Power Electron.*, vol. 2, no. 1, pp. 97–114, Mar. 2014.

[2]   S. Peyghami, P. Palensky, and F. Blaabjerg, "An overview on the reliability of modern power electronic based power systems," *IEEE Open J. Power Electron.*, vol. 1, pp. 34–50, Feb. 2020.

[3]   S. Peyghami, Z. Wang, and F. Blaabjerg, "A guideline for reliability prediction in power electronic converters," *IEEE Trans. Power Electron.*, pp. 1–1, Mar. 2020.

[4]   K. Ma, U.-M. Choi, and F. Blaabjerg, "Prediction and validation of wear-out reliability metrics for power semiconductor devices with mission profiles in motor drive application," *IEEE Trans. Power Electron.*, vol. 33, no. 11, pp. 9843–9853, Nov. 2018.

[5]   D. Zhou, H. Wang, and F. Blaabjerg, "Mission profile based system-level reliability analysis of DC/DC converters for a backup power application," *IEEE Trans. Power Electron.*, vol. 33, no. 9, pp. 8030–8039, Sep. 2018.

[6]   V. Samavatian, H. Iman-Eini, and Y. Avenas, "Reliability assessment of multistate degraded systems: an application to power electronic systems," *IEEE Trans. Power Electron.*, vol. 35, no. 4, pp. 4024–4032, Apr. 2020.

[7]   S. Peyghami, F. Blaabjerg, and P. Palensky, "Incorporating power electronic converters reliability into modern power system reliability analysis," *IEEE J. Emerg. Sel. Top. Power Electron.*, pp. 1–1, Jan. 2020.

[8]   "IEEE guide for selecting and using reliability predictions based on IEEE 1413," *IEEE Std 1413.1-2002*, pp. 1–106, Feb. 2003.

[9]   X. Jia, J. Shen, and R. Xing, "Reliability analysis for repairable multistate two-unit series systems when repair time can be neglected," *IEEE Trans. Reliab.*, vol. 65, no. 1, pp. 208–216, Mar. 2016.

[10]   R. Billinton and R. N. Allan, *Reliability Evaluation of Power Systems*. Boston, MA: Springer US, 1996.

[11]   "IEEE guide for selecting and using reliability predictions based on IEEE 1413," *IEEE Std 1413.1-2002*, pp. 1–97, 2003.

[12]   Y. Liu and C. Singh, "Reliability evaluation of composite power systems using Markov cut-set method," *IEEE Trans. Power Syst.*, vol. 25, no. 2, pp. 777–785, May 2010.

[13]   E. B. Iversen, J. K. Moller, J. M. Morales, and H. Madsen, "Inhomogeneous Markov models for describing driving patterns," *IEEE Trans. Smart Grid*, vol. 8, no. 2, pp. 581–588, Mar. 2017.

[14]   A. Abiri-Jahromi, M. Fotuhi-Firuzabad, and E. Abbasi, "An efficient mixed-integer linear formulation for long-term overhead lines maintenance scheduling in power distribution systems," *IEEE Trans. Power Deliv.*, vol. 24, no. 4, pp. 2043–2053, Oct. 2009.

[15]   D. Zhou, F. Blaabjerg, M. Lau, and M. Tonnes, "Thermal cycling overview of multi-megawatt two-level wind power converter at full grid code operation," *IEEE J. Ind. Appl.*, vol. 2, no. 4, pp. 173–182, Jul. 2013.

[16]   A. K. Verma, S. Ajit, and D. R. Karanki, *Reliability and Safety Engineering*. London: Springer, 2016.

[17]   K. Chen, W. Jiang, and P. Wang, "An extended DPWM strategy with unconditional balanced neutral point voltage for neutral point clamped three-level converter," *IEEE Trans. Ind. Electron.*, vol. 66, no. 11, pp. 8402–8413, Nov. 2019.

[18]   I. Vernica, K. Ma, and F. Blaabjerg, "Optimal derating strategy of power electronics converter for maximum wind energy production with lifetime information of power devices," *IEEE J. Emerg. Sel. Top. Power Electron.*, vol. 6, no. 1, pp. 267–276, Mar. 2018.

An Efficiency Analysis of a Ferrite Magnet assisted Synchronous Reluctance Machine for Low Power Drives including Flux Weakening

Matthias Hofer, Mario Nikowitz, Thomas Kirowitz, Manfred Schrödl
Technische Universität Wien
Institute of Energy Systems and Electrical Drives
Gußhausstraße 25-27
A-1040 Vienna, Austria
Phone: +43 1 58801-370230
Email: matthias.hofer@tuwien.ac.at
URL: http://www.tuwien.ac.at

Keywords

≪Synchronous Motor≫, ≪Electrical Machine≫, ≪Electrical Drive≫, ≪Permanent Magnet assisted Synchronous Reluctance Machine≫, ≪Ferrite Magnet assisted Synchronous Reluctance Machine≫, ≪Efficiency≫.

Abstract

In this paper a ferrite magnet assisted synchronous reluctance machine for a low power industrial spooling drive with a constant power characteristic at high rotational speeds is discussed. To combine an efficient machine topology at rather low product cost the permanent magnet assisted synchronous reluctance machine is proposed. Different rotor topologies are investigated and compared by simulation. Finally a three layer magnet topology was chosen, constructed and experimentally tested. The experimental results confirm a high torque and power capability even in the flux weakening operation above the nominal speed. Further, this machine type shows a significant high efficiency in a wide operation range.

Introduction

Today, in industrial drive applications the implementation of highly efficient electrical drives is still growing. The demand of a sustainable global energy usage as well as the severe efficiency standards for electrical machines and drives (e.g. IEC60034-30-2) require the replacement of electrical machines and variable speed drives in a broad field of applications. The synchronous reluctance machines (SynRM) represent the simplest electrical machine topology without any permanent magnets (PM) or rotor windings and are recommended by electric machine manufacturers because SynRMs can reach a higher efficiency than induction machines (IM) [1] at low costs. For variable speed drives including flux weakening operation with speeds much higher than the nominal speeds, SynRM's restricted power capability is a clear limitation criteria because SynRMs have a small constant power speed range (CPSR) and a decreasing power characteristic at increasing speed [2],[3],[4]. Thus, the implementation of SynRMs is critical or unsuitable for applications with high power demands in flux weakening operation.

With interior permanent magnet synchronous machines (IPMSM) highest efficiencies and highest torque and power densities can be reached, but due to unsustainable and expensive rare earth magnet materials the machine costs are high. Therefore, other electrical drive concepts optimized in a way to just fulfill the requirements without any power margins at very low cost get more and more in the focus of industrial drives. This is exactly the chance for permanent magnet assisted synchronous machines (PMaSynRMs) and also focus of this work. They can easily compete with power and torque of SynRMs because of a high saliency and offer the possibility to adjust the CPSR by the right choice of the permanent magnet flux level [4],[5]. In case of utilization with ferrite magnets, the machine cost can be still rather low at a respectable power in flux weakening range.

The PMaSynRM model

The PMaSynRM can be described by the following mathematical equations. The stator flux linkage $\underline{\Psi}_{S,dq}$ in the complex rotor oriented dq-frame is noted as

$$\underline{\Psi}_{S,dq} = \Psi_{S,d} + j\,\Psi_{S,q} = L_d\,I_d + j\{L_q\,I_q - \Psi_M\}. \tag{1}$$

The inductances $L_d(I_d, I_q)$ and $L_q(I_d, I_q)$ strongly depend on the machine's operation point by the current components I_d and I_q. The electromagnetic machine torque T_e is given as

$$T_e = \frac{3}{2}\,p\,\left[(L_d - L_q)\,I_d\,I_q + \Psi_M\,I_d\right] \tag{2}$$

where p is the number of pole pairs and Ψ_M is the flux linkage caused the permanent magnets. Usually, at PMaSynRMs the saliency $(L_d - L_q)$ is the dominant torque component and an additional torque is generated by the permanent magnet flux linkage Ψ_M. The stationary phasor diagram is shown in Fig. 5. Due to the flux linkage Ψ_M caused by the permanent magnet the PMaSynRM's stator voltage is affected and results in a smaller phase angle φ between stator voltage $\underline{U}_S$ and stator current $\underline{I}_S$ compared to SynRMs. Further, the flux weakening characteristic shows an enlarged CPSR by higher values of Ψ_M as also shown in [5]. The design criteria for the optimal choice of the PM flux can be selected in different ways, e.g optimization for a high torque, for a wide CPSR or for a high efficiency [6],[7],[8].

The PMaSynRM design

In this work the PMaSynRM design is focused on an industrial spooling application with the target to enhance the machine output power and machine efficiency for a given machine package. Currently a modular and standardized IM is implemented in this application. For modularity even the stator design is not changed from an existing IM (4 poles, size 63) to achieve low production cost and also the electrical quantities from the inverter remain the same ($280V_{RMS}$ line to line voltage, $0.86A_{RMS}$ phase current). The application requires a minimum torque of 0.6Nm and a constant power of 50W in flux weakening operation up to 6000rpm (see also Fig. 6).

PMaSynRM rotor topologies

PMaSynRM rotors are often realized by the application of permanent magnets inside the rotor, more or less inserted in the flux barriers of flux barrier type SynRMs. Thus, a high machine saliency ratio l_d/l_q is caused by the flux barriers and a certain permanent magnet flux linkage Ψ_M is gained by the PMs. Due to cost reasons PMaSynRMs are mainly designed without rare earth PMs and low cost and low energy magnets like ferrites are used. In this work various rotor topologies are investigated and compared as shown in Fig. 1 - Fig. 4, but the magnet material Y30 is not changed. For a high reluctance three and

 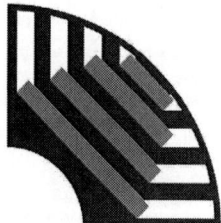 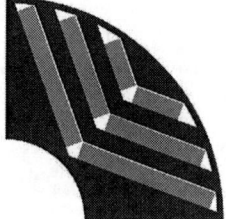 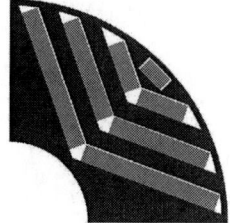

Fig. 1: Design U3PM   Fig. 2: Design U4PM   Fig. 3: Design V3PM   Fig. 4: Design V4PM

four layer topologies in U and V-shape are analyzed by a 2D finite element simulation. The main characteristics are presented in Table I. The V-shape designs V3PM and V4PM have a higher amount of PM material and provide a higher peak torque compared to the U-shape designs, but they show a lower peak power at maximum speed and a higher torque ripple at maximum speed. This high torque ripple would require additional design measures to reach acceptable values for the target application. Both U-shape topologies have similar performance although the PM content and manufacturing effort is lower at the design U3PM. Obviously, the implementation of a 4th magnet layer does not really improve the machine performance and a three layer topology is already sufficient. Considering the overall performance it can be stated, that by the V-shape topology the power is shifted from the high speed to the low speed range similar to pure SynRMs, which reduces the CPSR and power capability in flux weakening operation.

Table I: Design comparison of the electromagnetic performance (unskewed)

	Flux linkage Ψ_M	Magnet mass	Peak torque	Peak power @6000rpm	Torque ripple @6000rpm
U3PM	0.097mVs	185.4g	1.085Nm	370W	20.3%
U4PM	0.115mVs	203.5g	1.084Nm	370W	20.5%
V3PM	0.176mVs	234.7g	1.298Nm	312W	70.5%
V4PM	0.210mVs	241.4g	1.296Nm	312W	70.5%

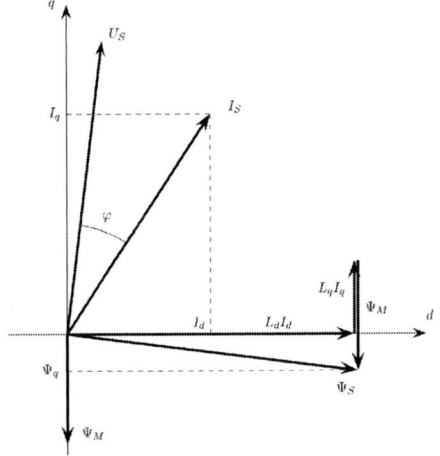

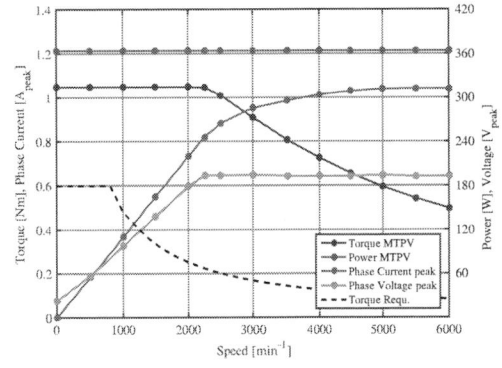

Fig. 5: Phasor diagram of the PMaSynRM

Fig. 6: Simulation result of design U3PM, skewed rotor (reduced voltage $U_{PH} = 138V_{RMS}$)

The prototype design

Due to the simplicity of the geometry and a low total number of magnets the design U3PM (Fig. 1) was chosen for realization of a prototype machine according to Fig. 7. For a further improvement of the torque ripple and machine harmonics a discrete rotor skewing with 3 segments was implemented. Thus, a small reduction in the peak torque from 1.085Nm to 1.05Nm has to be considered, but the torque ripple reduces significantly from 0.238Nm (20.5%) to 0.134Nm (12.7%). The simulated peak performance characteristic is depicted in Fig. 6. An electromagnetic torque of 1.05Nm is reached at maximum current up to a corner speed 2250rpm. In the flux weakening range above 2250rpm the peak current 1.21A remains constant, whereas the current angle increases from 38.5° to 74.5° in the maximum torque per voltage (MTPV) operation. Finally, at 6000rpm an electromagnetic power of 312W is reached, even at a lower phase voltage 138Vrms. This power is several times the shaft power requirement of 50W as also shown in Fig. 6.

For SynRMs and PMaSynRMs the bridges and ribs within the rotor have to be designed for a compromise between an acceptable mechanical stress and a low q-axis flux. In this work, all four designs consider a circumferential iron bridge at the rotor surface of 0.8mm. A mechanical stress analysis of the design U3PM has shown, that the von-Mises stress is within the linear material properties even at nearly four times of maximum speed. Thus, a further potential for improvement of the machine performance is identified with an optimized rib width with respect to a sustainable and high quality industrial manufacturing process.

Fig. 7: Prototype Machine U3PM

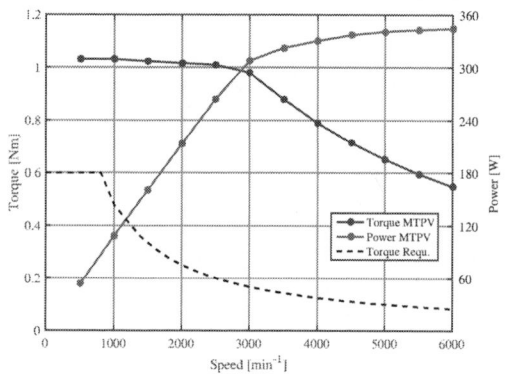

Fig. 8: Measured torque/power vs. speed characteristic (electromagnetic) of the prototype U3PM (reduced voltage $U_{PH} = 138V_{RMS}$)

Experimental results

In this section the experimental test results of the prototype U3PM are presented. First, the machine parameters are identified and discussed. Second, the machine torque and power characteristic including the flux weakening operation is shown. Finally the machine's efficiency and loss maps are presented.

Machine parameters

For characterization of the machine itself and an optimal field oriented control the knowledge of the machine parameters are essential. In Fig. 9 the measured flux linkage characteristic $\psi_d(I_d)$ and $\psi_q(I_q)$ according to Equ.(1) is shown. At the d-axis curve the iron path saturation can be seen (black) because it is differing from a linear characteristic. From the q-axis behavior (red) the rated permanent magnet flux ψ_M=0.39 is identified at the value $\psi_q(0)$. The application of the nominal current directly in q-axis leads to a vanished total q-axis flux linkage ψ_q (red). In Fig. 10 the inductance characteristic is

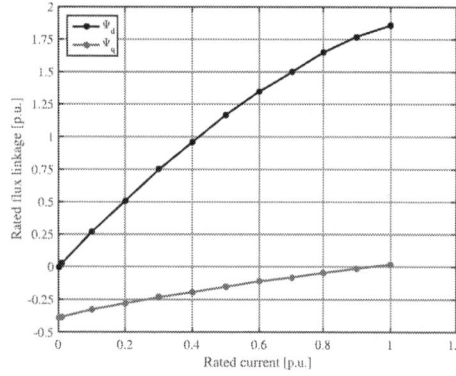

Fig. 9: Measured flux linkages of prototype U3PM

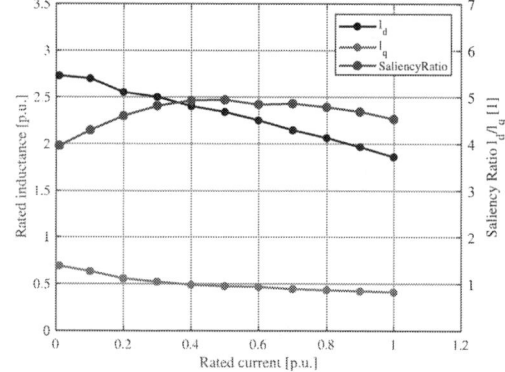

Fig. 10: Measured inductances of prototype U3PM

shown accordingly. The permanent magnet flux already ensures sufficient rib saturation of the rated inductance l_q (red), because even at a low stator current i_q the rated inductance is already $l_q < 0.7$. Due to saturation of the iron path the rated inductance l_d decreases from $l_d \approx 2.7$ by higher rated currents i_d to approximately 1.8. Nevertheless, a high saliency ratio $l_d/l_q \approx 4$ - 4.9 is given in the complete operation range due to a sufficient pre-saturation of the ribs caused by the ferrite magnets.

Machine power and torque

The peak performance measurement in maximum torque per ampere mode (up to 2500rpm) and maximum torque per voltage operation for higher speeds above 2500rpm is depicted in Fig. 8. The machine provides an electromagnetic peak torque of 1.04Nm and an electromagnetic peak power of 342W. These values differ slightly from the FE simulation results, although the PM flux linkage of the prototype is 14 % higher and the saliency ratio is slightly lower, because the circumferential rib of the prototype is measured as 0.7mm instead of 0.8mm. However, the expected wide CPSR and high torque and power capability are confirmed by these experiments.

Efficiency

The investigation of the PMaSynRM's efficiency is based on a normalized machine drag torque. Previous investigations have shown high mechanical friction loss at existing SynRM prototypes for this application [3]. Although this PMaSynRM setup with ball bearings of another supplier shows significantly lower friction loss compared to the salient pole and flux barrier type SynRM, the higher normalized drag losses are used (Fig. 13) for efficiency comparison. The PMaSynRM prototype shows only one third of the drag torque from the SynRMs. This is a mechanical loss difference up to approx. 40W, which is quite high for an application with only 50W shaft power. Furthermore, this measurement shows the potential efficiency improvement for low power drives considering suitable ball bearing parameters [3]. The efficiency map for normalized drag torque is shown in Fig. 11. The highest efficiency of 83.0% is reached at 246W shaft power, at 0.94Nm torque and 2500rpm shaft speed. The loss map is depicted in Fig. 12. At the same normalized mechanical drag losses for all machines investigated, the zero torque losses of 68.6W at 6000rpm are lower compared to 74.8W for the salient pole type SynRM [3] due to the higher torque/current ratio of the PMaSynRM. An efficiency comparison with the salient pole type SynRM is presented in Fig. 14. At 3000rpm the PMaSynRM's efficiency is significantly higher than the salient pole SynRM in a wide range with values up to 7.3% and at 6000rpm a difference of up to even 7.5% is given. The highest efficiency difference is gained at 500rpm with values up to 15.4%. The required power of 50W is far in the part load range of the PMaSynRM's performance (Fig. 8, Fig. 11) and compared in Table II. But even in the part load region a significant efficiency improvement up to 6.8% is gained by the PMaSynRM approach.

Table II: Efficiency comparison of SP-SynRM and PMaSynRM with normalized drag torque

	Salient Pole SynRM	PMaSynRM
50W @ 1000rpm	61.4%	73.3%
50W @ 3000rpm	53.8%	61.1%
50W @ 6000rpm	38.0%	41.2%

Conclusion

In this paper an analysis of a ferrite magnet assisted SynRM for low power drives with regard to power and efficiency is presented. Within a common stator the PMaSynRM rotor provides a much higher maximum torque and due to a wide constant power speed range the maximum shaft output power is 5 times higher than the application's requirement. These results show the high potentials for PMaSynRM usage, because two possible strategies with regard to the power can be applied. Either downsizing the electric machine by utilization of a smaller power class (size), or this machine can be applied to applications for higher power requirements. Finally, the PMaSynRM reduces the electric drive costs per power at an increased efficiency.

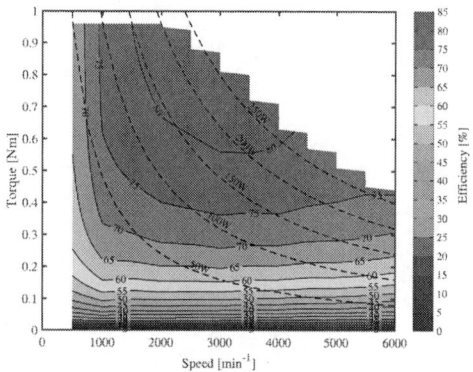

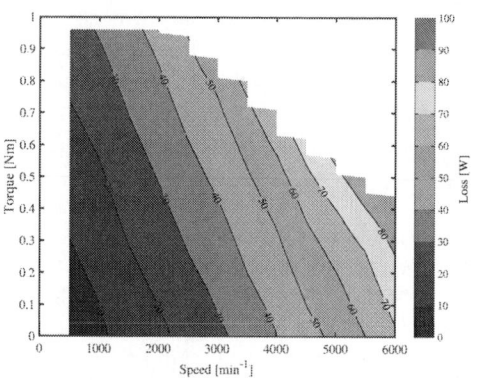

Fig. 11: Measured efficiency of the U3PM prototype at normalized drag torque

Fig. 12: Measured loss map of the U3PM prototype at normalized drag torque

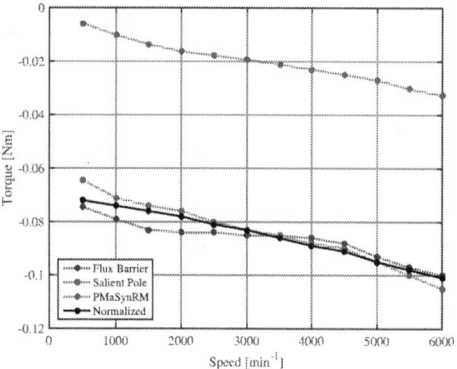

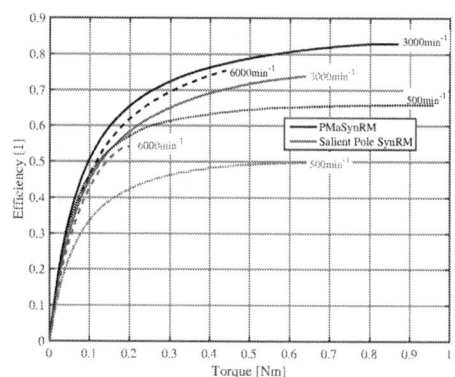

Fig. 13: Drag torque comparison and normalized drag torque (at zero current)

Fig. 14: Efficiency comparison of SynRM and PMaSynRM at normalized drag torque

References

[1] Almeida A. de, Ferreira F., Baoming G.: Beyond induction motors technology, trends to move up efficiency, IEEE Transactions on Industry Applications, vol. 50, no. 3 May/June 2014.

[2] Hofer M., Nikowitz M., Schrödl M.: Power and Efficiency of Salient Pole and Flux Barrier Type Synchronous Reluctance Machines in Low Power Drives, The 20th European Conference on Power Electronics and Applications (EPE'18 ECCE Europe), Riga, Latvia, Sept. 2018

[3] Hofer M., Nikowitz M., Schrödl M.: An Efficiency Analysis of a Salient Pole and a Flux Barrier Synchronous Reluctance Machine including Flux Weakening, IEEE Workshop on Electrical Machines Design, Control and Diagnosis (WEMDCD), Athens, Greece, April 2019

[4] Pellegrino G., Jahns T.M., Bianchi N., Soong W., Cupertino F.: The rediscovery of synchronous reluctance and ferrite permanent magnet motors, Springer Briefs in Electrical and Computer Engineering, 2016.

[5] Hofer M., Schrödl M.: An Analysis of Ferrite Magnet Assisted Synchronous Reluctance Machines for Low Power Drives Including Flux Weakening, IEEE International Electric Machines & Drives Conference (IEMDC), San Diego, USA, May 2019

[6] Prakht V., Dmitrievskii V., Kazakbaev V., Mathematical modeling ultra premium efficiency (ie5 class) PM assisted synchronous reluctance motor with ferrite magnets, in 25th International Workshop on Electric Drives: Optimization in Control of Electric Drives (IWED), 2018.

[7] Barcaro M., Bianchi N., Magnussen F.: Permanent-Magnet Optimization in Permanent-Magnet-Assisted Synchronous Reluctance Motor for a Wide Constant-Power Speed Range, IEEE Transactions on Industrial Electronics, Volume: 59, Issue: 6, June 2012

[8] Bianchi N., Fornasiero E., Soong W.: Optimal selection of PM flux linkage in a PM assisted synchronous reluctance machine, The International Conference on Electrical Machines (ICEM), Berlin, Germany, Sept. 2014

High Performance LQR Control of Modular Multilevel Converters with Simple Control Structure and Implementation

Min Jeong, Simon Fuchs, and Jürgen Biela
Laboratory for High Power Electronic Systems (HPE)
ETH Zürich, Switzerland
Email: jeong@hpe.ee.ethz.com, jbiela@ethz.ch

Keywords

≪Optimal control≫, ≪Multilevel converters≫, ≪MPC (Model-based Predictive Control)≫, ≪Modelling≫

Abstract

In this paper, a novel feedback controller concept for grid-connected Modular Multilevel Converters (MMC) is proposed. The controller is based on the Linear Quadratic Regulator (LQR) method and shows outstanding control performance since it is a multi-input multi-output (MIMO) controller. The proposed controller achieves efficient energy balancing control with high bandwidth, such that the required margin for the module capacitance value for dynamic control is reduced and an excellent transient behavior can be accomplished. Alternatively, compact MMC designs can be realized with a reduced capacitance value by sacrificing some transient performance. The implementation requires a relatively low computational effort, and the simple control structure enables a straightforward tuning and time-delay compensation.

1 Introduction

Over the past decade, the MMC has been applied in many high and medium voltage applications, including high-voltage dc transmission (HVDC) and static synchronous compensators (STATCOM), due to its numerous merits [1]-[2]. However, controlling the MMC is a challenging task because multiple control objectives have to be satisfied simultaneously: control of input and output power, sub-module capacitor voltages averaging and balancing, and circulating currents suppression. In literatures, different control schemes of the MMC has been presented to improve the control performance or to increase the system efficiency [3].

Based on the well-known standard cascaded PI-controller [4], many attempts have been made with resonant controllers or nonlinear controllers (e.g. [5]-[6]) to increase the robustness and the performance of the control system. Nonetheless, the multi-input multi-output (MIMO) characteristic of the MMC results in a complex structure for such cascaded controllers. Moreover, since the MMC is a nonlinear system due to arm energy dynamics, tuning of the controller parameters is challenging. When it comes to the controller implementation, delays due to the sensing and the communication aggravate the situation.

MIMO controllers have recently gained a lot of attention to overcome limitations such as the complicated control system design, the ambiguous tuning process, and the limited bandwidth of the control system due to the cascaded structure [7]. In particular, many proposals are based on model predictive control (MPC) methods because MPC methods show superior control performance when subject to constraints [8]-[9]. In [9], a linear MPC method is proposed which linearizes the energy related equations of the MMC to translate a complex nonlinear optimization problem into a constrained linear-quadratic regulator (LQR) problem. Even though the linear MPC method dramatically reduces the computational burden, which typically limits implementing the MPC, it still requires complicated implementation steps using an FPGA or SoC FPGA to solve optimization problems in real-time [10].

In this paper, a simple state-feedback controller, which avoids the complicated controller implementation step, is proposed as an alternative MIMO control scheme on the basis of the linear MPC method introduced in [9]. The state-feedback law is computed by solving an unconstrained LQR problem, which omits the constraints from the constrained LQR problem. The simple feedback law can be implemented in most embedded systems without complex implementation steps while keeping all outstanding features of a MIMO controller. However, with the proposed feedback controller, some trade-offs are necessary to indirectly keep the constraints that are omitted in the

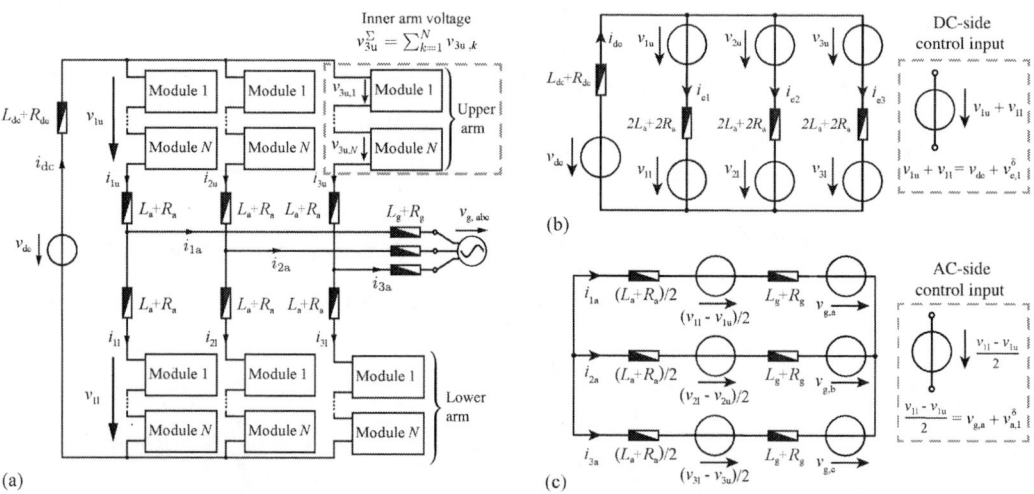

Fig. 1: (a) Grid-connected three phase MMC. (b) DC-side MMC equivalent circuit, where the summation of the arm voltage $(v_{1u} + v_{1l})$ behaves as a control input to drive the DC current ($i_{e1} = i_{1u}/2 + i_{1l}/2$). (c) AC-side MMC equivalent circuit, where the difference between the arm voltage ($v_{1l}/2 - v_{1u}/2$) behaves as a control input to drive the AC current.

control design process: either increasing the module capacitance value or sacrificing some transient performance. The details of these trade-offs are also analyzed and evaluated in the paper.

The paper is organized as follows: Section 2 provides a review of linear time-varying models of the MMC and thereafter section 3 introduces a condensed formulation of the unconstrained LQR problem. The resulting state-feedback law is presented in section 4 with the overall control structure. Simulation results are shown and discussed in section 5, followed by the conclusion in section 6.

2 Linear Time-varying Model of the MMC

A typical grid-connected MMC is shown in Fig. 1(a). If each arm is modeled as an independent voltage source, the overall MMC system can be represented in a more general way with six independent voltage sources regardless of the number of submodules [11]. The so-called averaged model of the MMC is enabled by using a modulator, which translates a given reference voltage for each arm into switching signals of the submodules. Well-designed modulators can implement an arbitrary voltage level with small errors and low computational burden while balancing all capacitor voltages around their mean value [12]. Consequently, the MMC can be modeled at a higher level, which reduces the complexity of the modeling process and considers only currents and arm energies.

Based on the averaged model, detailed equations are presented in [9] to derive a time-varying model of the MMC with three parts: DC side currents, AC side currents, and arm energies. Fig. 1(b)-(c) illustrate the equivalent circuits of the DC and AC currents when each arm is represented as a controllable voltage source. The current dynamics are represented in the $\alpha\beta0$-frame with the Clarke-transformation and described by a linear time-invariant form as

$$\frac{d}{dt}\begin{bmatrix} i_{e,\alpha} \\ i_{e,\beta} \\ i_{e,0} \end{bmatrix} = \underbrace{\begin{bmatrix} \frac{-R_a}{L_a} & 0 & 0 \\ 0 & \frac{-R_a}{L_a} & 0 \\ 0 & 0 & \frac{-2R_a - 3R_{dc}}{2L_a + 3L_{dc}} \end{bmatrix}}_{\mathbf{A}_e} \cdot \underbrace{\begin{bmatrix} i_{e,\alpha} \\ i_{e,\beta} \\ i_{e,0} \end{bmatrix}}_{\mathbf{i}_{e,\alpha\beta0}} + \underbrace{\begin{bmatrix} \frac{-1}{2L_a} & 0 & 0 \\ 0 & \frac{-1}{2L_a} & 0 \\ 0 & 0 & \frac{-1}{2L_a + 3L_{dc}} \end{bmatrix}}_{\mathbf{B}_e} \cdot \underbrace{\begin{bmatrix} v_{e,\alpha}^{\delta} \\ v_{e,\beta}^{\delta} \\ v_{e,0}^{\delta} \end{bmatrix}}_{\mathbf{v}_{e,\alpha\beta0}^{\delta}}, \tag{1}$$

$$\frac{d}{dt}\begin{bmatrix} i_{a,\alpha} \\ i_{a,\beta} \end{bmatrix} = \underbrace{\begin{bmatrix} \frac{-R_a/2 - R_g}{L_a/2 + L_g} & 0 \\ 0 & \frac{-R_a/2 - R_g}{L_a/2 + L_g} \end{bmatrix}}_{\mathbf{A}_a} \cdot \underbrace{\begin{bmatrix} i_{a,\alpha} \\ i_{a,\beta} \end{bmatrix}}_{\mathbf{i}_{a,\alpha\beta}} + \underbrace{\begin{bmatrix} \frac{1}{L_a/2 + L_g} & 0 & 0 \\ 0 & \frac{1}{L_a/2 + L_g} & 0 \end{bmatrix}}_{\mathbf{B}_a} \cdot \underbrace{\begin{bmatrix} v_{a,\alpha}^{\delta} \\ v_{a,\beta}^{\delta} \\ v_{a,0}^{\delta} \end{bmatrix}}_{\mathbf{v}_{a,\alpha\beta0}^{\delta}}. \tag{2}$$

Note that the inputs of the system are the driving voltages, given as

$$
\mathbf{v}^{\delta}_{e,\alpha\beta0} = \mathbf{K}_{\alpha\beta0} \cdot \underbrace{\left(\begin{bmatrix} v_{1u} \\ v_{2u} \\ v_{3u} \end{bmatrix} + \begin{bmatrix} v_{1l} \\ v_{2l} \\ v_{3l} \end{bmatrix} \right)}_{ve,\alpha\beta0} - \underbrace{\begin{bmatrix} 0 \\ 0 \\ v_{dc} \end{bmatrix}}_{vdc,\alpha\beta0} \quad \text{and} \quad \mathbf{v}^{\delta}_{a,\alpha\beta0} = \frac{\mathbf{K}_{\alpha\beta0}}{2} \cdot \underbrace{\left(-\begin{bmatrix} v_{1u} \\ v_{2u} \\ v_{3u} \end{bmatrix} + \begin{bmatrix} v_{1l} \\ v_{2l} \\ v_{3l} \end{bmatrix} \right)}_{va,\alpha\beta0} - \underbrace{\begin{bmatrix} v_{g,\alpha} \\ v_{g,\beta} \\ v_{g,0} \end{bmatrix}}_{vg,\alpha\beta0} \quad (3)
$$

where $\mathbf{K}_{\alpha\beta0}$ is the Clarke transformation matrix. All bold and capital letters represent matrices, and all bold and lower-case letters represent vectors.

The arm energy dynamics are described in the abc-frame and linearized by neglecting the driving voltages ($\mathbf{v}^{\delta}_{e,\alpha\beta0}$ and $\mathbf{v}^{\delta}_{a,\alpha\beta0}$) since the driving voltages are small compared to the grid and DC voltage values. This can be written as

$$
\frac{d\mathbf{w}}{dt} \approx \underbrace{\begin{bmatrix} \mathbf{K}^{-1}_{\alpha\beta0} \cdot (-\mathbf{v}_{g,\alpha\beta0}(\varphi_g) + 1/2\, \mathbf{v}_{dc,\alpha\beta0}) \\ \mathbf{K}^{-1}_{\alpha\beta0} \cdot (+\mathbf{v}_{g,\alpha\beta0}(\varphi_g) + 1/2\, \mathbf{v}_{dc,\alpha\beta0}) \end{bmatrix}}_{\text{Approximated arm inserted voltages}} \cdot \underbrace{\begin{bmatrix} \mathbf{K}^{-1}_{\alpha\beta0} \cdot (+1/2\, \mathbf{i}_{a,\alpha\beta0} + \mathbf{i}_{e,\alpha\beta0}) \\ \mathbf{K}^{-1}_{\alpha\beta0} \cdot (-1/2\, \mathbf{i}_{a,\alpha\beta0} + \mathbf{i}_{e,\alpha\beta0}) \end{bmatrix}}_{\text{Arm currents}} = \mathbf{A}_w(\varphi_g) \cdot \begin{bmatrix} \mathbf{i}_{e,\alpha\beta0} \\ \mathbf{i}_{a,\alpha\beta0} \end{bmatrix}, \quad (4)
$$

where $\mathbf{w} = [w_{1u}, w_{2u}, w_{3u}, w_{1l}, w_{2l}, w_{3l}]^{T}$ is the arm energy vector, $\mathbf{i}_{a,\alpha\beta0} = [\mathbf{i}^{T}_{a,\alpha\beta}, 0]^{T}$, and φ_g is the grid angle.

Since the linearized arm energy dynamic matrix $\mathbf{A}_w(\varphi_g)$ is known a priori as a function of the grid angle, the MMC can be discretized and represented as a discrete linear model as

$$
\mathbf{x}_{k+1} = \mathbf{A}_d(\varphi_{g,k}) \cdot \mathbf{x}_k + \mathbf{B}_d(\varphi_{g,k}) \cdot \mathbf{u}_k, \quad (5)
$$

where $\mathbf{x}_k = [\mathbf{i}^{T}_{e,\alpha\beta0,k}, \mathbf{i}^{T}_{a,\alpha\beta,k}, \mathbf{w}^{T}_k]^{T} \in \mathbb{R}^{11}$ is the state vector, $\mathbf{u}_k = [\mathbf{v}^{\delta T}_{e,\alpha\beta0,k}, \mathbf{v}^{\delta T}_{a,\alpha\beta0,k}]^{T} \in \mathbb{R}^{6}$ is the control input vector, and $\mathbf{A}_d(\varphi_{g,k})$ and $\mathbf{B}_d(\varphi_{g,k})$ are the periodic time-varying discrete model matrices depending on the grid angle $\varphi_{g,k}$.

3   Condensed Formulation of Unconstrained LQR Problem

With the MMC being represented as a periodic discrete linear model, an unconstrained LQR can be formulated to control the MMC with

$$
\min_{\mathbf{U}_K} \sum_{l=0}^{N_p-1} \left\| \mathbf{x}_{k+l+1} - \mathbf{x}_{\text{ref},k+l+1} \right\|^{2}_{\mathbf{Q}} + \left\| \mathbf{u}_{k+l} - \mathbf{u}_{\text{ref},k+l} \right\|^{2}_{\mathbf{R}} \quad (6a)
$$

$$
s.t. \quad \mathbf{x}_{k+l+1} = \mathbf{A}_d(\varphi_{g,k+l}) \mathbf{x}_{k+l} + \mathbf{B}_d(\varphi_{g,k+l}) \mathbf{u}_{k+l}, \qquad \forall l \in I \quad (6b)
$$

where $\mathbf{U}_k = [\mathbf{u}^{T}_k, \cdots, \mathbf{u}^{T}_{k+N_p-1}]^{T} \in \mathbb{R}^{6 \cdot N_p}$ is the complete control input vector, N_p is the prediction horizon, $\mathbf{Q} \geq 0$ and $\mathbf{R} \geq 0$ are weighting matrices, $I = \{0, 1, \cdots, N_p - 1\}$, and $\|\mathbf{z}\|^{2}_{\mathbf{P}}$ denotes a 2-norm with the weighting matrix $\mathbf{P}$. $\mathbf{x}_{\text{ref},k+l+1}$ and $\mathbf{u}_{\text{ref},k+l}$ are the l-th step reference values for the states and inputs - refer to [9] for further details. Note that $\mathbf{A}_d(\varphi_g)$ and $\mathbf{B}_d(\varphi_g)$ are periodic matrices and satisfy $\mathbf{A}_d(\varphi_{g,k}) = \mathbf{A}_d(\varphi_{g,k+N_s})$ and $\mathbf{B}_d(\varphi_{g,k}) = \mathbf{B}_d(\varphi_{g,k+N_s})$, where N_s is the number of samples in one grid-period.

When the control inputs are chosen as decision variables, the future states can be expressed as a linear function of the initial state ($\mathbf{x}_k$) and the control inputs ($\mathbf{U}_k$) as

$$
\underbrace{\begin{bmatrix} \mathbf{x}_{k+1} \\ \mathbf{x}_{k+2} \\ \vdots \\ \mathbf{x}_{k+N_p} \end{bmatrix}}_{\mathbf{X}_k} = \underbrace{\begin{bmatrix} \mathbf{B}_d(\varphi_{g,k}) & \cdots & 0 \\ \mathbf{A}_d(\varphi_{g,k+1}) \cdot \mathbf{B}_d(\varphi_{g,k}) & \cdots & 0 \\ \vdots & \vdots & \vdots \\ \mathbf{A}_d(\varphi_{g,k+N_p})... \mathbf{A}_d(\varphi_{g,k}) \mathbf{B}_d(\varphi_{g,k}) & \cdots & \mathbf{B}_k \end{bmatrix}}_{\mathbf{S}_k} \cdot \underbrace{\begin{bmatrix} \mathbf{u}_k \\ \mathbf{u}_{k+1} \\ \vdots \\ \mathbf{u}_{k+N_p-1} \end{bmatrix}}_{\mathbf{U}_k} + \underbrace{\begin{bmatrix} \mathbf{A}_d(\varphi_{g,k}) \\ \mathbf{A}_d(\varphi_{g,k+1}) \cdot \mathbf{A}_d(\varphi_{g,k}) \\ \vdots \\ \mathbf{A}_d(\varphi_{g,k+N_p})... \mathbf{A}_d(\varphi_{g,k+1}) \end{bmatrix}}_{\mathbf{T}_k} \cdot \mathbf{x}_k \quad (7)
$$

where $\mathbf{X}_k \in \mathbb{R}^{11 \cdot N_p}$ is the future state vector, and $\mathbf{S}_k \in \mathbb{R}^{11 \cdot N_p \times 6 \cdot N_p}$ and $\mathbf{T}_k \in \mathbb{R}^{11 \cdot N_p \times 11}$ are the relation matrices. Note that even though the future states and the control inputs are not matrices, they are denoted as bold and capital letters to emphasize that they are stacked vectors. Then, the cost function of the constrained LQR (6) can be reformulated as a condensed quadratic function of the complete control input vector ($\mathbf{U}_k$) by substituting the future

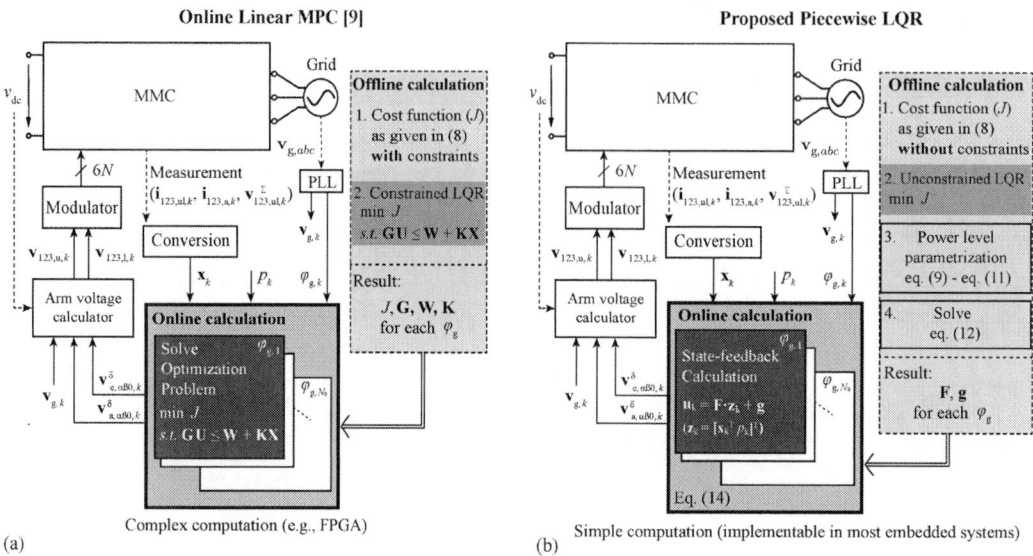

Fig. 2: (a) Control system diagram for the linear MPC. (b) Control system diagram for the piecewise LQR. Same color indicates the counterpart step between two methods.

states with (7) as

$$J = \sum_{l=0}^{N_{\mathrm{p}}-1} \left\| \mathbf{x}_{k+l+1} - \mathbf{x}_{\mathrm{ref},k+l+1} \right\|_{\mathbf{Q}}^2 + \left\| \mathbf{u}_{k+l} - \mathbf{u}_{\mathrm{ref},k+l} \right\|_{\mathbf{R}}^2 \tag{8a}$$

$$= (\mathbf{X}_k - \mathbf{X}_{\mathrm{ref},k})^{\mathrm{T}} \mathbf{Q}_{\mathrm{L}} (\mathbf{X}_k - \mathbf{X}_{\mathrm{ref},k}) + (\mathbf{U}_k - \mathbf{U}_{\mathrm{ref},k})^{\mathrm{T}} \mathbf{R}_{\mathrm{L}} (\mathbf{U}_k - \mathbf{U}_{\mathrm{ref},k}) \tag{8b}$$

$$= \frac{1}{2} \mathbf{U}_k^{\mathrm{T}} \underbrace{(2\mathbf{S}_k \mathbf{Q}_{\mathrm{L}} \mathbf{S}_k + 2\mathbf{R}_{\mathrm{L}})}_{\mathbf{H}_k} \mathbf{U}_k + (\mathbf{x}_k^{\mathrm{T}} \underbrace{(2\mathbf{T}_k^{\mathrm{T}} \mathbf{Q}_{\mathrm{L}} \mathbf{S}_k)}_{\mathbf{F}_{1,k}} + \underbrace{(-2\mathbf{U}_{\mathrm{ref},k}^{\mathrm{T}} \mathbf{R}_{\mathrm{L}} - 2\mathbf{X}_{\mathrm{ref},k}^{\mathrm{T}} \mathbf{Q}_{\mathrm{L}} \mathbf{S}_k)}_{\mathbf{f}_{2,k}}) \mathbf{U}_k + \mathit{Const.} \tag{8c}$$

$$= \frac{1}{2} \mathbf{U}_k^{\mathrm{T}} \mathbf{H}_k \mathbf{U}_k + (\mathbf{x}_k^{\mathrm{T}} \mathbf{F}_{1,k} + \mathbf{f}_{2,k}) \mathbf{U}_k + \mathit{Const.} \tag{8d}$$

where $\mathbf{Q}_{\mathrm{L}} = \mathrm{diag}(\mathbf{Q}, \cdots, \mathbf{Q}) \in \mathbb{R}^{11 \cdot N_{\mathrm{P}} \times 11 \cdot N_{\mathrm{P}}}$ and $\mathbf{R}_{\mathrm{L}} = \mathrm{diag}(\mathbf{R}, \cdots, \mathbf{R}) \in \mathbb{R}^{6 \cdot N_{\mathrm{P}} \times 6 \cdot N_{\mathrm{P}}}$ are the large weighting matrices, and $\mathbf{X}_{\mathrm{ref},k} = [\mathbf{x}_{\mathrm{ref},k+1}^{\mathrm{T}}, \cdots, \mathbf{x}_{\mathrm{ref},k+N_{N_p}}^{\mathrm{T}}]^{\mathrm{T}} \in \mathbb{R}^{11 \cdot N_{\mathrm{P}}}$ and $\mathbf{U}_{\mathrm{ref},k} = [\mathbf{u}_{\mathrm{ref},k}^{\mathrm{T}}, \cdots, \mathbf{u}_{\mathrm{ref},k+N_{N_p}-1}^{\mathrm{T}}]^{\mathrm{T}} \in \mathbb{R}^{6 \cdot N_{\mathrm{P}}}$ are the state reference vector and the input reference vector, respectively. All the terms which do not include the input vector ($\mathbf{U}_k$) are grouped into the constant term, since these terms do not influence the value of the cost function. In the next section, an optimal controller is derived by finding the value of the complete control input vector that minimizes the cost function.

4 Piecewise LQR (P-LQR)

In this section, a state-feedback controller is proposed which solves the unconstrained LQR problem in the condensed formulation, such that the control inputs can be calculated as a simple feedback law based on the measured system states $\mathbf{x}_k$. A control system diagram for the new feedback law is illustrated in Fig. 2(b). The new feedback law can mitigate the complexity of the control system dramatically, because the control inputs can be calculated through a simple multiplication and an addition without any iteration steps. In Fig. 2(a), a control system diagram of the linear MPC method from [9] is presented to give a comparison between the two methods. Even though they have similar structures, implementing the linear MPC is more challenging since large computational power is required to solve an optimization problem at each sampling period, which is typically very short (≤ 1 ms) in the field of power electronics.

The control laws from the unconstrained LQR problem, however, sometimes cannot be directly applied to the MMC since the constraints are not considered. Therefore, first, the process to derive the control laws is investigated in this section. Second, the relation between the constraints and the unconstrained LQR is analyzed so that the proposed control laws can account for the constraints indirectly. Lastly, the implementation of the overall control structure is explained, together with a method to compensate time-delays.

4.1 Linear State-feedback Law for Arbitrary Power Level

As an optimal controller, the unconstrained LQR problem determines a control law that minimizes the cost function. This can be expressed by $\nabla_{U_k} J = 0$, with the cost function from (8), and the analytical solution can be described as

$$\mathbf{U}_k = -\mathbf{H}_k^{-1} \cdot \left(\mathbf{F}_{1,k}^{\mathrm{T}} \cdot \mathbf{x}_k + \mathbf{f}_{2,k}^{\mathrm{T}} \right),$$

where $-\mathbf{H}_k^{-1} \cdot \mathbf{F}_{1,k}^{\mathrm{T}}$ is the state-feedback law for the current state ($\mathbf{x}_k$) and $-\mathbf{H}_k^{-1} \cdot \mathbf{f}_{2,k}^{\mathrm{T}}$ is the trajectory tracking term to track varying reference values ($\mathbf{x}_{\mathrm{ref}}$, $\mathbf{u}_{\mathrm{ref}}$). Unless the references values are updated on-line, the reference values are determined based on the steady-state behavior of the MMC as given in [9]. They can be described as a function of the grid angle and the power level of the MMC. Therefore, the trajectory tracking term of $-\mathbf{H}_k^{-1} \cdot \mathbf{f}_{2,k}^{\mathrm{T}}$ can be represented with a smaller number of parameters instead of taking all states and input references to compute the values.

Since the MMC dynamics are already described as a function of the grid angle, the reference values need to be parameterized only with the power level. This can be represented with the relative power level, $p_k \in [-1, 1]$, describing the required power level with respect to the rated power level. Then, the reference values can be written as

$$\mathbf{X}_{\mathrm{ref},k} = p_k \cdot \mathbf{X}_{\mathrm{ref,rated}} + \mathbf{X}_0 \qquad \text{and} \qquad \mathbf{U}_{\mathrm{ref},k} = p_k \cdot \mathbf{U}_{\mathrm{ref,rated}} \qquad (9)$$

where $\mathbf{X}_0$ is the mean energy value of the arm. Note that the other states and input references have a mean value of zero and are expressed proportional to the rated power level. Then, the cost function of (8d) changes into

$$J = \frac{1}{2} \mathbf{U}_k^{\mathrm{T}} \mathbf{H}_k \mathbf{U}_k + (\mathbf{x}_k^{\mathrm{T}} \mathbf{F}_{1,k} + p_k \mathbf{f}_{2a,k} + \mathbf{f}_{2b,k}) \mathbf{U}_k, \qquad (10)$$

where

$$\mathbf{f}_{2a,k} = -2\mathbf{U}_{\mathrm{ref,rated}}^{\mathrm{T}} \mathbf{R}_{\mathrm{L}} - 2\mathbf{X}_{\mathrm{ref,rated}}^{\mathrm{T}} \mathbf{Q}_{\mathrm{L}} \mathbf{S}_k \qquad \text{and} \qquad \mathbf{f}_{2b,k} = -2\mathbf{X}_0^{\mathrm{T}} \mathbf{Q}_{\mathrm{L}} \mathbf{S}_k. \qquad (11)$$

The control inputs over the complete prediction horizon can be achieved by $\nabla_{U_k} J = 0$ as

$$\mathbf{U}_k = -\mathbf{H}_k^{-1} \cdot \left(\begin{bmatrix} \mathbf{F}_{1,k}^{\mathrm{T}} & \mathbf{f}_{2a,k}^{\mathrm{T}} \end{bmatrix} \cdot \underbrace{\begin{bmatrix} \mathbf{x}_k \\ p_k \end{bmatrix}}_{\mathbf{z}_k} + \mathbf{f}_{2b,k}^{\mathrm{T}} \right) = \mathbf{F}_{\mathrm{all},k} \cdot \mathbf{z}_k + \mathbf{g}_{\mathrm{all},k}, \qquad (12)$$

where $\mathbf{F}_{\mathrm{all},k} = -\mathbf{H}_k^{-1} \cdot \begin{bmatrix} \mathbf{F}_{1,k}^{\mathrm{T}} & \mathbf{f}_{2a,k}^{\mathrm{T}} \end{bmatrix}$, $\mathbf{g}_{\mathrm{all},k} = -\mathbf{H}_k^{-1} \cdot \mathbf{f}_{2b,k}^{\mathrm{T}}$ and $\mathbf{z}_k$ is the new initial state combined with the actual initial state ($\mathbf{x}_k$) and the relative power level (p_k). The control law can be obtained by taking the control inputs of the next time step, i.e., the first 6 entries of the complete control input vector, in a receding horizon manner as

$$\mathbf{u}_k = \mathbf{F}_k \cdot \mathbf{z}_k + \mathbf{g}_k, \qquad (13)$$

where $\mathbf{F}_k$ and $\mathbf{g}_k$ are the linear state-feedback matrix achieved by taking the first 6 rows of $\mathbf{F}_{\mathrm{all},k}$ and $\mathbf{g}_{\mathrm{all},k}$, respectively. The time-varying characteristic of the MMC model results in varying state-feedback laws depending on the grid angle ($\varphi_{\mathrm{g},k}$), yet the number of the feedback laws is limited to a finite number as they are periodic. The set of all feedback laws can be expressed in a piecewise function of the grid angle ($\varphi_{\mathrm{g},k}$) as

$$\mathbf{u}_k^*(\mathbf{z}_k, \varphi_{\mathrm{g},k}) = \begin{cases} \mathbf{F}_1 \cdot \mathbf{z}_k + \mathbf{g}_1 & \text{if} \quad \varphi_{\mathrm{g},k} \in \varphi_{\mathrm{g},1} \\ \quad \vdots \\ \mathbf{F}_{N_{\mathrm{s}}} \cdot \mathbf{z}_k + \mathbf{g}_{N_{\mathrm{s}}} & \text{if} \quad \varphi_{\mathrm{g},k} \in \varphi_{\mathrm{g},N_{\mathrm{s}}} \end{cases} \qquad (14)$$

where N_{s} is the number of samples in one grid-period. The piecewise control law is denoted as piecewise LQR (P-LQR) in the following.

4.2 Relations between constraints and piecewise LQR

A distinctive characteristic of the MMC is the distributed capacitors in the submodules. Although the distributed capacitors allow the MMC to operate as a multi-level converter, this also induces difficulties of controlling the MMC. The submodule capacitors of each arm need to have a proper arm energy level to buffer energy variations

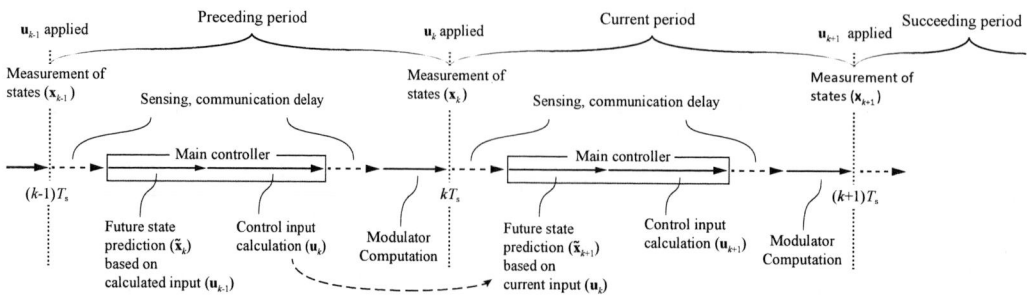

Fig. 3: Timing diagram of the unit-delay method to compensate time-delays in a control system.

without exceeding the maximum voltage for the protection of the submodule capacitors. At the same time, the arm voltage level must be sufficiently high to generate the arm inserted voltage such that the submodule capacitors can operate as a voltage source. The control is particularly challenging during transients because the MMC and the capacitance values are usually designed based on the steady-state behavior. In the constrained LQR problem, four kinds of constraints are considered for the safe operation of the MMC: arm currents, grid currents, arm energy (module capacitor voltages), and arm output voltages. Therefore, the MMC can exploit the full dynamic potential, e.g., operating the submodules near the maximum allowed voltage level, and achieve the fastest physically possible transient behavior even with system parameters which are designed based on the steady-state operation.

However, for the P-LQR, the fastest possible transient behavior cannot be guaranteed since the P-LQR does not consider such constraints. Therefore, similar to conventional cascaded methods, some margins need to be added to the module capacitance values, which are designed based on the steady-state, for the dynamic control to avoid saturation effects and overcurrents/-voltages during transients. In other words, the constraints can be compensated indirectly for example by increasing the module capacitance values. A faster transient behavior can be achieved with higher module capacitance values. Nevertheless, the scale of margins required for the same transient performance is significantly different between the P-LQR and cascaded control methods (PI, PR, etc.). With the cascaded control methods, a large margin is required because the arm energy balancing is located at the outer loop of the cascaded control loops and therefore the bandwidth of the arm energy balancing is limited. Compared to that, the P-LQR has a fast and efficient arm energy balancing since it is a MIMO controller, and considerably smaller margins are needed. If a fast transient performance can be completely sacrificed at the cost of a compact design, the required module capacitance values can be further reduced by including harmonics in the circulating currents and/or a common-mode voltage as proposed in [13].

4.3 Simple Control Structure with Delay Compensation

In standard cascaded controllers, the control structure is complicated because there are multiple layers to handle multiple control objectives. In such a structure, the tuning of the controller parameters is challenging as different layers are coupled to each other. The complexity further increases drastically if the MMC is designed to have harmonics in circulating currents and/or a common-mode voltage as mentioned in the previous section [13]. Compared to that, MIMO controllers have a relatively simple control structure as seen in Fig. 2, because all control objectives are considered simultaneously. Therefore, the tuning can be done intuitively by adjusting the weighting matrices of $\mathbf{Q}$ and $\mathbf{R}$ in the cost function of (8), and the harmonics can be included easily by modifying the state and input references.

Since the MMC is a complex system, several time-delays are present in the actual implementation of the control system. For example, sensing delays for all measurements, communication delays to send measurement values and control input values, and computation delays of the main controller and the modulator. A major benefit of the model-based MIMO controllers is that such time-delays can be compensated easily using the unit-delay method described in [14] as shown in Fig. 3.

The core idea of the unit-delay is that control inputs of the current period are determined in the preceding period by predicting the current states. Therefore, time-delays do not aggravate the tuning process and all delays can be modeled in the simulations. The only condition is that the sum of all delays has to be shorter than the sampling interval. New measurements at each sampling instant bring a feedback loop and the time-delays can be compensated.

5 Simulation

In this section, the control performance of the proposed piecewise linear-quadratic regulator (P-LQR) is evaluated with simulations and compared to the well-known cascaded PI controller [4] and the linear MPC [9]. The MMC parameters for the simulations are given in Table I. To highlight the efficient energy balancing control of the P-LQR, the results of two simulation scenarios are illustrated with different module capacitance values. The simulations are performed for worst-case operation conditions, i.e. a power flow reverse at rated power, to demonstrate the benefits of the proposed control method. For the cascaded PI-controller and the P-LQR, the power reference changes are given as a ramp function to slow down the response and avoid saturation and voltage restrictions, i.e. the internal arm voltages become too low or too high during transients.

The simulation results of the first scenario are shown in Fig. 4. There, a module capacitance of $105\,\mu F$ is used based on the design procedure given e.g. in [15]. In this case, a large margin is added, and the control performance of the P-LQR is equivalent to that of the MPC method because no constraints are necessary during the transients. This results in an extremely fast transient behavior to track reference changes. Compared to the cascaded PI controller, the P-LQR outperforms in all domains: settling time is ninety times faster and the good energy balancing results in smaller circulating currents to achieve smooth and balanced energy transitions.

The simulation results of the second scenario are given in Fig. 5, where a module capacitance of $62\,\mu F$ is used. The module capacitance is determined by adding a small margin (10 %) to the minimum required module capacitance value ($56\,\mu F$). The minimum value is calculated with an extended version of the procedure presented in [16], where the design is performed based on the steady-state behavior assuming no circulating currents and no common-mode voltage injections. Compared to the first scenario, where a large margin (87.5 % or $49\,\mu F$) is added for the dynamic

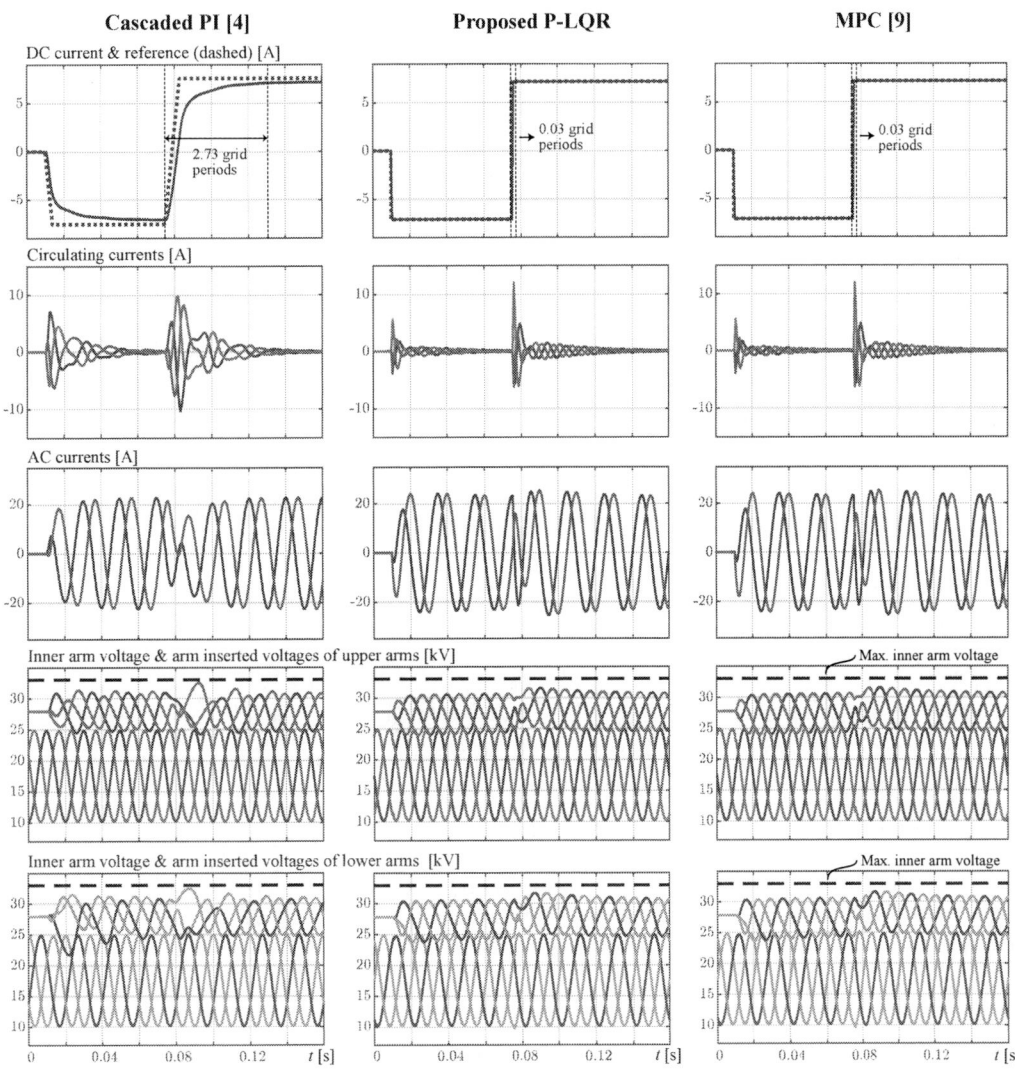

Fig. 4: Simulation results of scenario 1 with a module capacitance of $105\,\mu F$ for the cascaded PI-controller, the P-LQR, and the MPC . The DC current settling time in the first row indicate the different dynamic performance between the control methods.

Table I: MMC SYSTEM PARAMETERS

Symbol		SI value	PU value	Symbol		SI value	PU value
V_{g}	Grid voltage	9 kV	$\sqrt{3/2}$	$I_{\mathrm{g,r}}$	Rated grid current	22.681 A	1
S_{r}	Rated power	250 kVA	1	$\omega_{\mathrm{g,r}}$	Rated grid frequency	$2\pi\,50$ Hz	1
$V_{\mathrm{dc,r}}$	Rated DC voltage	35 kV	4.763	N	Module number / arm	15	
L_{g}	Grid inductance	5 mH	0.005	R_{g}	Grid resistance	$0.5\,\Omega$	0.0015
L_{a}	Arm inductance	26.8 mH	0.026	R_{a}	Arm resistance	$1\,\Omega$	0.0031
L_{dc}	DC inductance	$1.4\,\mu$H	$1.4\cdot10^{-6}$	R_{dc}	DC resistance	$20.6\,$mΩ	$63\cdot10^{-6}$

control, the margin for dynamic control in the second scenario is reduced so that the control performance of the P-LQR cannot be equivalent to that of the MPC. Note that the MPC can keep the same transient behavior by utilizing larger circulating currents to balance energies in the module capacitors up to their physical limit. Even though the P-LQR cannot achieve the same ideal control actions like the MPC, it shows a symmetric control behavior and still achieves very fast transients with accurate steady-state tracking performance. Especially, compared to the cascaded PI controller, the P-LQR shows a ten times faster transient behavior regarding the settling time with much smaller circulating currents. The amplitude of the circulating currents is reduced by two thirds.

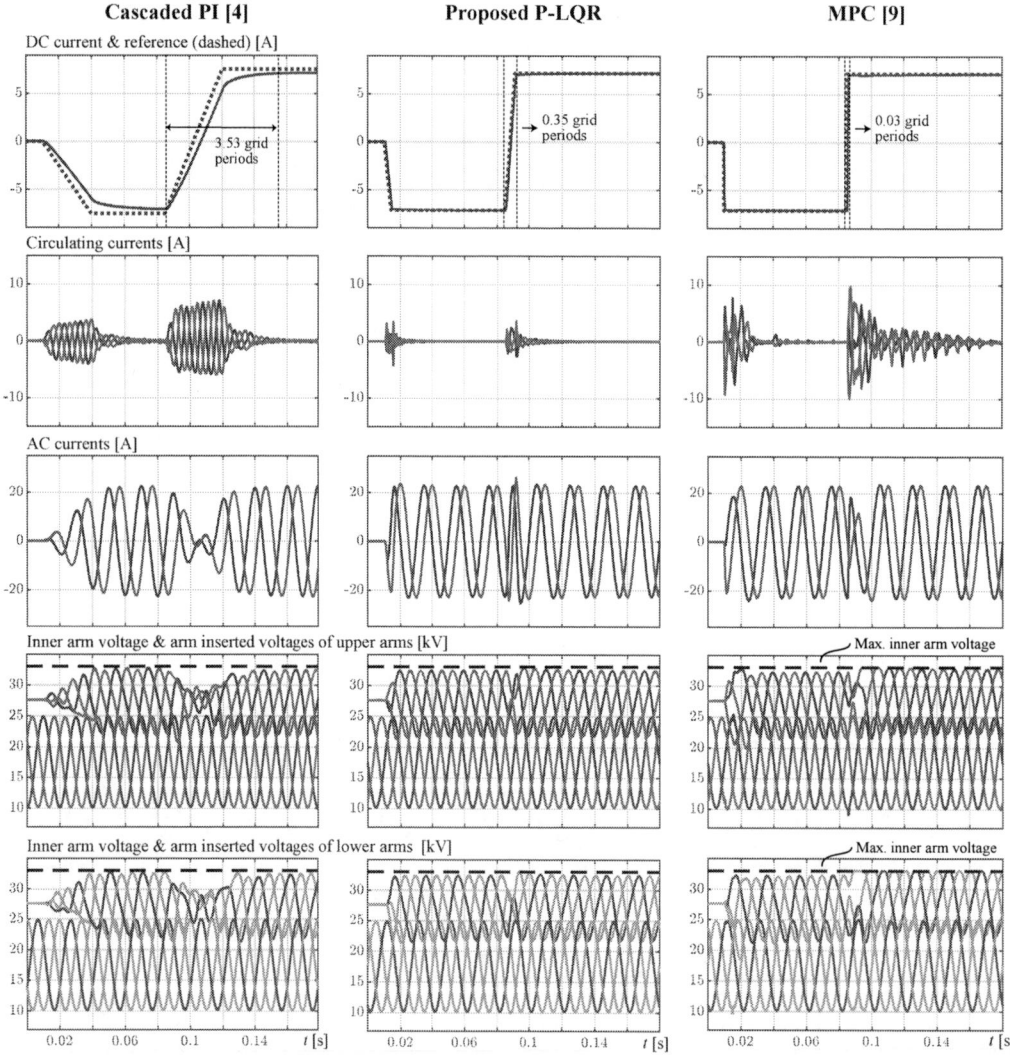

Fig. 5: Simulation results of scenario 2 with a module capacitance of $62\,\mu$F for the cascaded PI-controller, the P-LQR, and the MPC. The DC current settling time in the first row indicate the different dynamic performance between the control methods.

Table II: Controller Parameters

Symbol	Parameter	Scenario 1		Scenario 2	
		P-LQR	MPC	P-LQR	MPC
$Q_{e,\alpha\beta}$	Circulating currents	0	0	0	0
$Q_{e,0}$	DC side zero sequence current	2.5×10^4	2.5×10^4	10^4	2.5×10^4
$Q_{a,\alpha\beta}$	AC side currents	0	0	0	0
Q_w	Arm energies	5	5	10	5
$R_{e,\alpha\beta}$	Circulating currents control effort	10^2	10^2	1	10^2
$R_{e,0}$	DC side zero sequence control effort	5×10^2	5×10^2	10^3	5×10^2
$R_{a,\alpha\beta0}$	AC side control effort	10^4	10^4	10^3	10^4
f_s	Sampling frequency	1.5 kHz*			

* : The number of samples in one grid-period, $N_s = 30$.

6 Conclusion

This paper presents a novel state-feedback controller for grid-connected MMCs. Based on a linear MPC method, a condensed formulation of the unconstrained LQR problem for arbitrary power levels is derived, and a piecewise feedback law is proposed as a function of the current states, the reference power level, and the grid angle.

The P-LQR is a MIMO controller and has many merits compared to common cascaded control systems. First, the P-LQR shows extremely faster behaviors regarding transient performance. The simulation results prove that the settling time is reduced by more than a factor of ten. Second, less margin is required for the module capacitance values to achieve dynamic control since it has the efficient energy balancing control with high bandwidth. This results in a compact MMC realization. Lastly, for the implementation of the control system, all time-delays due to sensing, communication, and computation can be efficiently compensated with the unit-delay method.

In addition, the P-LQR can be implemented relatively easily in most embedded systems, from a microcontroller to an FPGA, due to its simple control laws. This is the biggest advantage of the P-LQR compared to other MIMO controllers. Though the linear MPC outperforms the P-LQR regarding the control performance and also has all mentioned merits as a MIMO controller, it can only be achieved at the cost of a complicated implementation to solve an optimization problem at each sampling instant. Therefore, unless the application has strict specifications regarding the transient performance and/or the dimension of the MMC system, the P-LQR can be a superb option to achieve improved control performance with a simple real-time implementation process.

Acknowledgements

This research is part of the activities of the Swiss Centre for Competence in Energy Research on the Future Swiss Electrical Infrastructure (SCCER-FURIES), which is financially supported by the Swiss Innovation Agency (Innosuisse - SCCER program).

References

[1] M. A. Perez, S. Bernet, J. Rodriguez, S. Kouro, and R. Lizana, "Circuit topologies, modeling, control schemes, and applications of modular multilevel converters," *IEEE Trans. on Power Electronics*, 2015.

[2] S. Debnath, J. Qin, B. Bahrani, M. Saeedifard, and P. Barbosa, "Operation, control and applications of the modular multilevel converter: A review," *IEEE Trans. on Power Electronics*, 2015.

[3] J. I. Leon, S. Vazquez, and L. G. Franquelo, "Multilevel converters: Control and modulation techniques for their operation and industrial applications," *Proc. IEEE*, vol. 105, no. 11, Nov 2017.

[4] J. Kolb, F. Kammerer, M. Gommeringer, and M. Braun, "Cascaded control system of the modular multilevel converter for feeding variable-speed drives," *IEEE Trans. on Power Electronics*, vol. 30, no. 1, pp. 349–357, 2015.

[5] S. Li, X. Wang, Z. Yao, T. Li, and Z. Peng, "Circulating current suppressing strategy for MMC-HVDC based on nonideal proportional resonant controllers under unbalanced grid conditions," *IEEE Trans. on Power Electronics*, vol. 30, no. 1, pp. 387–397, Jan 2015.

[6] S. Yang, P. Wang, and Y. Tang, "Feedback linearization-based current control strategy for modular multilevel converters," *IEEE Trans. on Power Electronics*, vol. 33, no. 1, pp. 161–174, Jan 2018.

[7] P. Münch, D. Görges, M. Izák, and S. Liu, "Integrated current control, energy control and energy balancing of modular multilevel converters," in *Conf. of IEEE Industrial Electronics Society (IECON)*, Nov 2010.

[8] B. S. Riar, T. Geyer, and U. K. Madawala, "Model predictive direct current control of modular multilevel converters: Modeling, analysis, and experimental evaluation," *IEEE Trans. on Power Electronics*, vol. 30, no. 1, pp. 431–439, Jan 2015.

[9] S. Fuchs, M. Jeong, and J. Biela, "Long horizon, quadratic programming based model predictive control (MPC) for grid connected modular multilevel converters (MMC)," in *Conf. of the IEEE Industrial Electronics Society (IECON)*, Oct 2019.

[10] I. McInerney, G. A. Constantinides, and E. C. Kerrigan, "A survey of the implementation of linear model predictive control on fpgas," *IFAC-PapersOnLine*, vol. 51, no. 20, pp. 381 – 387, 2018, 6th IFAC Conference on Nonlinear Model Predictive Control NMPC.

[11] H. Bärnklau, A. Gensior, and S. Bernet, "Derivation of an equivalent submodule per arm for modular multilevel converters," in *Power Electronics and Motion Control Conference (EPE/PEMC)*, Sep. 2012.

[12] S. Fuchs, S. Beck, and J. Biela, "Analysis and reduction of the output voltage error of PWM for modular multilevel converters," *IEEE Trans. on Industrial Electronics*, vol. 66, no. 3, March 2019.

[13] S. Fuchs, M. Jeong, and J. Biela, "Reducing the Energy Storage Requirements of Modular Multilevel Converters with Optimal Capacitor Voltage Trajectory Shaping," in *22nd Europ. Conf. on Power Electronics and Applications (EPE ECCE Europe)*, 2020, in press.

[14] P. Cortes, J. Rodriguez, C. Silva, and A. Flores, "Delay compensation in model predictive current control of a three-phase inverter," *IEEE Trans. on Industrial Electronics*, vol. 59, no. 2, pp. 1323–1325, Feb 2012.

[15] A. Hillers and J. Biela, "Optimal design of the modular multilevel converter for an energy storage system based on split batteries," in *15th Europ. Conf. on Power Electronics and Applications (EPE)*, Sep. 2013.

[16] K. Ilves, S. Norrga, L. Harnefors, and H. Nee, "On energy storage requirements in modular multilevel converters," *IEEE Trans. on Power Electronics*, vol. 29, no. 1, pp. 77–88, Jan 2014.

Fault detection and classification based on deep learning in LVDC off-grid system

Iurii Demidov, Antti Pinomaa, Andrey Lana, Olli Pyrhönen
LUT University
Yliopistonkatu 34, P.O.Box 20, 53850
Lappeenranta, Finland
Phone: +358 413688395
Email: Iurii.Demidov@lut.fi

July 1, 2020

Keywords

≪Deep learning≫, ≪Distributed generation (DG)≫, ≪Fault-detection≫, ≪Long short-term memory (LSTM)≫,≪Low-voltage direct current (LVDC)≫, ≪Off-grid≫

Abstract

The integration of information and communication technologies (ICT) into energy power systems provides new applications and possibilities for grid control, operation, and protection. In this paper, a smart self-sustained off-grid concept based on photovoltaics- and battery energy storage system and low-voltage direct current (LVDC) power distribution network is studied. Due to small electrification and poor internet coverage, Sub-Saharan Africa is the target area of the concept realization. Focusing on the application of deep learning, this research presents new approaches to the power system's protection, as fault detection and classification are one of the most essential LVDC electricity distribution issues.

Introduction

There are over 1 billion people in the world without access to electricity. Less than 50% of the population in SSA are electrified. Considering rural areas in Sub-Saharan Africa (SSA), only 16% of locals have access to electricity [1]. In addition, there are over 3.5 billion people without access to internet globally. The overall percentage of internet connectivity in the SSA is approximately 26% [2]. There is no specific statistics about how many people in rural areas have internet access but based on the electrification, the number is rather low. Therefore, rural electrification and digitalization in developing countries has become a hot topic. To tackle both of these, we propose a smart grid off-grid concept, which provides electricity, connectivity with internet access, and technology base for digital services that are run on the system, all these in an integrated package. This platform solution is aimed to and focused on remote locations in SSA. The main idea behind the concept is to improve the livelihood of communities in remote locations through electrification and digitalization. The concept platform components comprise an off-grid power system with a build-in Long-term Evolution (LTE) base transceiver station (BTS), which provides mobile network coverage and Internet access. The backhaul connectivity for the BTS can be arranged for instance with satellite connection. The off-grid power system consists of photovoltaic (PV) panels, for supplying LTE BTS, that is, the primary load as well as the other customer loads in day time, battery energy storage system (BESS), for storing the energy and providing the power throughout night and while sun absence, and the required power electronics equipment; maximum power point tracker (MPPT), inverter/converter, and control devices for the smart system operation. The system is designed to be easily scalable and modular, if/when the power and energy demands are increased, and for replicating the system to new locations.

As we are proposing the off-grid concept for the rural areas of SSA, there is a need to take into account several specific conditions typical for these environment and surroundings. Small experience of electricity usage, low levels of maintenance and safety knowledge can negatively effect on power system stability, and operational capability. Most of the existing off-grid systems in developing countries have failed before estimated operational lifespan [3]. In order to provide safe and reliable power system operation, off-grid should have high-efficiency control and protection systems. Fault detection and localization are one of the most essential issues in LVDC microgrids; an example of protection approaches applied in those are described in [4]. Integration of power system with ICT systems within the proposed smart off-grid concept provides possibilities for the new protection approaches implementation. Deep learning networks have a high potential and utilized in many industries. Applying different network configurations, it is possible to define non-obvious dependencies between input and output data. Long short-term memory (LSTM) model of a deep learning network used for memorizing and analysing long sequences of input data consequently can be applied for the fault detection issue within the time domain [5].

The paper focuses on the application of LSTM model to the fault detection issues in LVDC off-grids. The second and the third Sections overview questions of LVDC microgrids design and protection issues in such kind of grids. In the fourth Section, the proposed smart off-grid concept is described. The fifth Section of the paper presents fault detection and classification methodology based on LSTM model of a deep learning network. In the results and conclusion Sections, the outcome of proposed fault detection and classification methodology is discussed and the next steps in the research are proposed.

Low-voltage DC microgrid concept design

Economically, it is not feasible to build traditional utility-scale alternating current (AC) transmission and distribution power grid infrastructures, including high-, medium- and low-voltage (HV/MV/LV) stages to sparsely populated rural areas, but alternative solutions for electrifying such locations needs to be applied. An off-grid system based on photovoltaics (PV) is an efficient technical solution for rural areas, which are isolated from the traditional utility grids in SSA region, as this geographical locations are one of the optimal ones in regards of solar irradiance.

Infiltration of distributed generation (DG) based on renewable energy resources (RES), in this case PV and BESS, both operating with DC, makes approach utilizing LVDC for power distribution attractive. Following the previous statement, most of the consumer appliances, such as mobile phones and devices, computers, and TVs work with DC power. Those come with standard 230 VAC plug, but inside the device the 230 VAC is rectified to DC. An LVDC approach in power distribution allows to avoid DC/AC and AC/DC voltage transformation, which take place in inverting and rectifying stages in standard 230 VAC power distribution solution. Accordingly, the power losses in power conversions, which can be significant especially in with low loads in low-loaded grids, are decreased with DC distribution approach [6]. Moreover, PV-based systems, delivering already DC to output, are optimal for the establishment LVDC system. The energy efficiency is even more highlighted in off-grid systems, where the available power and energy are limited. Thus, this can be considered as effective and smart usage of the system, as one needs to remember that affordability is one of the key things in rural electrification and those investments. Expediency of LVDC power distribution implementation in rural conditions has been discussed in [7]. In addition to consumer appliances, which operate with DC, there is also a need to consider unitization AC appliances such as fridges, pumps, electric drives, etc. There are consumer appliances with extra low-voltage power supply (e.g. 24 VDC), but due to low production volumes the price levels are higher compared to standard AC ones, at least at the moment. And again in the context of developing countries the affordability is a prerequisite and has to keep in mind. Therefore, inverters with standard one phase 230 VAC output should also be optionally included in the system design.

Protection and grounding

LVDC distribution utilization in PV and BESS based off-grid system increases power efficiency, reduces transmission losses, eliminates skin effect and excludes source synchronization [8]. According to simulation and test comparison presented at [9], the DC grid is 2–20% more efficient than the AC one. However, the implementation of DC distribution leads to operational and control challenges; most essential are protection and safety issues [10]. There is no nominal current zero crossing in the DC grid and fault current rate is higher than in AC one, which is challenging from the fault isolation point of view [10]. The first step in the fault clearance is the fault detection. Existing protection schemes are less effective in the detection of high resistive fault current [10].

Line to line (LL), double line to ground (LLG) and line to ground (LG) are possible kinds of faults in the DC system. Accurate and fast fault detection operation allows for minimizing fault damage. The methods of fault detection strongly depend on the grounding system of the off-grid system. There are several ways of grounding. Figure 1 presents two configurations TN-S (T – direct connection to the earth; N - the neutral connection is supplied by the electricity supply network; S- protective earth (PE) and neutral (N) are separate conductors) and IT (I – no or high impedance connection to the earth; T - earth connection is by a local direct connection to earth) groundings. In the system with TN-S grounding, converter and battery connected to the ground and two separate wires are used for neutral and PE. Implementation of TN-S grounding leads to large current and high DC-link voltage transient in case of low-resistance ground faults. Moreover, TN-S grounding is rather selective in fault detection, but large voltage transient may damage the load devices connected to the fault pole. IT grounding system has a connection of positive pole of the converter near the power source, with an impedance in a grounding wiring. Such kind of grounding system has a small current and voltage transient in case of a ground fault, which ensures stable operation while a single ground fault. However, the small ground-fault current makes the fault detection challenging [11, 12].

System description

The smart grid concept proposed in this study is a de-centralized off-grid system, which consists of self-sustained power modules that can be interconnected or operated as independent power cells. Each module is a subsystem which includes PV-panels, a BESS, a DC/DC converter or a DC/AC inverter, depending on consumer needs. The PV array and BESS powers are dimensioned based on the expected loads to provide an uninterrupted power supply for the customer load 24/7. The modules can be interconnected with power cables and the DC/DC transformation from subsystem's bus voltage to line one is applied with a bidirectional DC/DC power converter. For power flows management, each power line is equipped with additional current and voltage measurement sensors on both line ends. The installation of sensors to each line end protects from energy theft and provides data required for fault detection

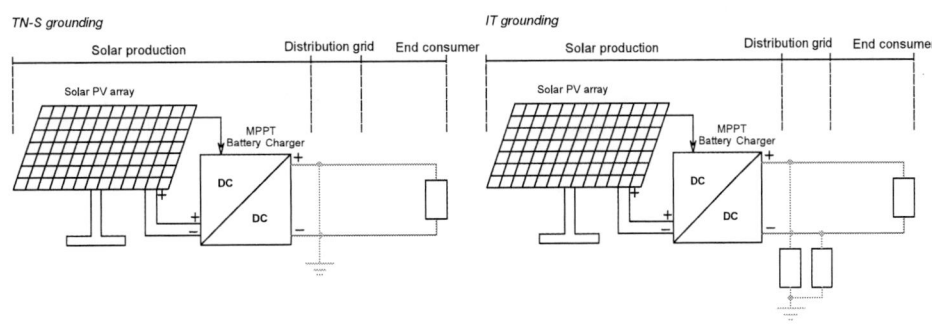

Fig. 1: Grounding systems for DC unipolar system configuration.

system. The concept's built-in ICT system enables the possibility to establish a power grid local-level supervision system, which allows data gathering, exchange and analysis from equipment within different system modules simultaneously. This significantly improves the control and protection opportunities.

Based on the literature review, most of the studies concerning de-centralized generation cover some basic topologies like radial or ring [13]. However, most of the existing grids have meshed topology which is a combination of radial and ring one. Differences in theoretical studies and real existing grid's topologies create challenges with the implementation of many studies. To bring this study as close as possible to the real conditions and cover more operation issues in the DC grid, meshed topology consisting of four PV-based subsystems is considered as a case study. A load of a subsystem consists of DC and AC consumer appliances. Figure 2a illustrates the scheme of the de-centralized off-grid topology applied as a case study. Three of the subsystems are connected into the ring and one has a radial connection, concerning the ring. Based on Figure 2b and estimating the average distances between subsystems up to 1 km and cable cross-section 10 mm^2, the line DC voltage applied is set to unipolar 120 VDC. The red curve illustrates cable power limitation for the 10 mm^2 cable and unipolar line voltage of 120 VDC. The blue curve shows minimal line voltage for transmitting 5 kW through the cable of the same cross section.

A detailed model of the proposed system designed in Simulink is illustrated in Figure 3. In addition to electrical equipment and its connection, it includes control of different system components:

- Consumer-end DC/AC Inverter is a device that has (proportional integral derivative) PID based control, which focuses on supplying consumers with a reference voltage of 230 VAC, and having input voltage and AC load variation as a disturbance. It is used for supply traditionally AC devices, such as fridges, pumps, etc [14].
- Consumer-end DC/DC converter is a device which has PID based control, which focuses on supplying consumers with reference voltage of 5/12 VDC, coupling with input voltage and DC load variation disturbance. It is used for supply devices that can be powered directly with DC, such as phones, laptops, TVs, LED lighting, etc [15].
- BESS charger provides power and energy balance in the system. It consists of bidirectional converter with current reference PID controller [16].
- MPPT adjust output voltage from PV array, providing maximum output power to the system and BESS charging [17].

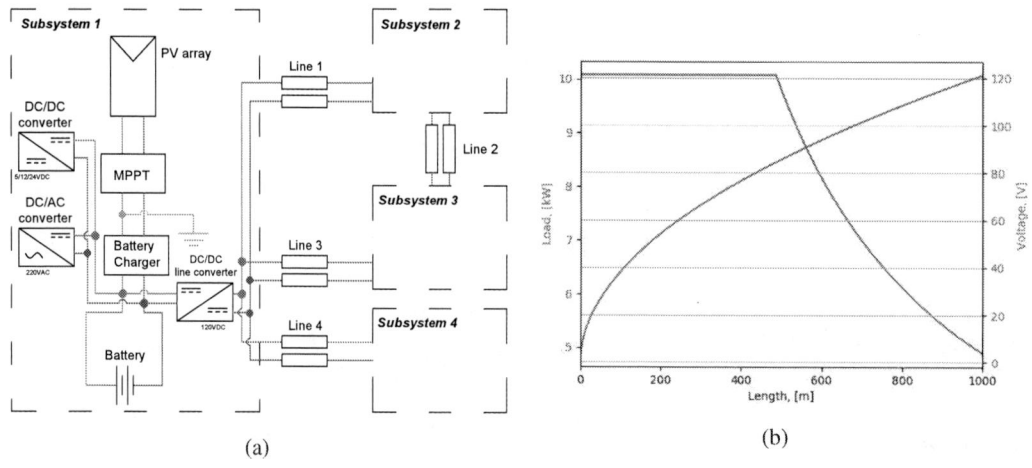

Fig. 2: a) Schematic of the case study grid; b) Power and voltage curves and limitations depending on the line/cable length.

- Bidirectional DC/DC line converter has grid forming and grid following mode of operation. In the grid forming mode the device settles 120 VDC line reference voltage. Grid following is a current reference base mode, insuring presetted power flow. There is a control interconnection between converters on the opposite ends of the line. Only one converter per line can be operates in a grid forming mode [18]

Fault detection methodology

The main focus in this paper is on line-fault detection and classification in the studied off-grid system with meshed topology and IT grounding, as LG fault detection is one essential issue in such off-grid systems. As mentioned above, the proposed concept's ICT system collects data from all measurement devices with minimal latency. Figure 4 illustrates the scheme of the centralized fault detection system, which consists of two stages. Data from all devices is collected into a local edge-cloud-based main controller. The first stage of the fault detection method focuses on fault detection, while the second one is responsible for fault classification. Therefore, such a system allows not only to identify and locate a place of a fault but also to propose the fault type. LSTM networks applied in deep learning are implemented

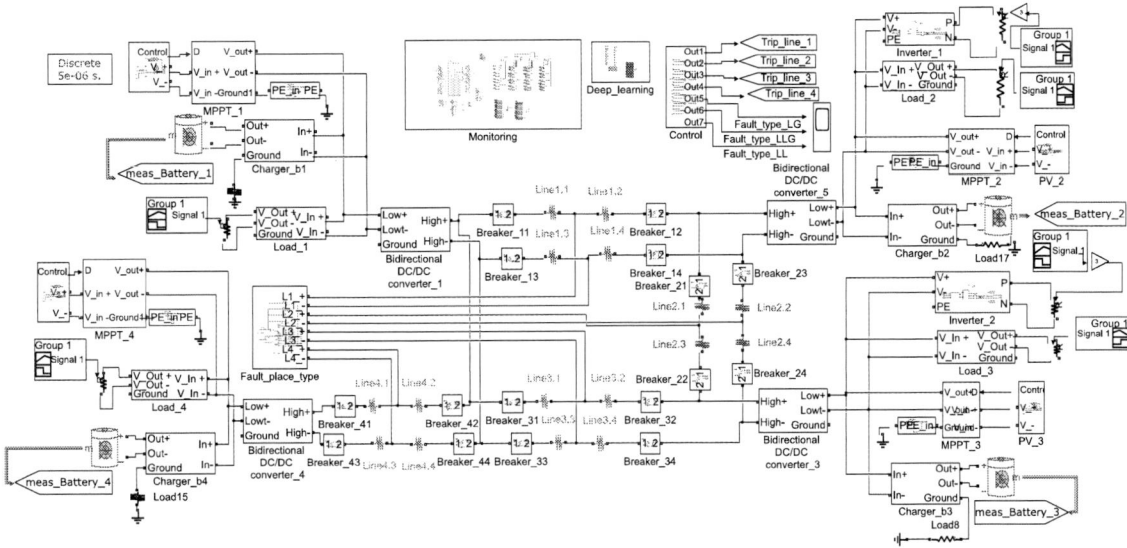

Fig. 3: Simulink model of the proposed off-grid system and the case study power network.

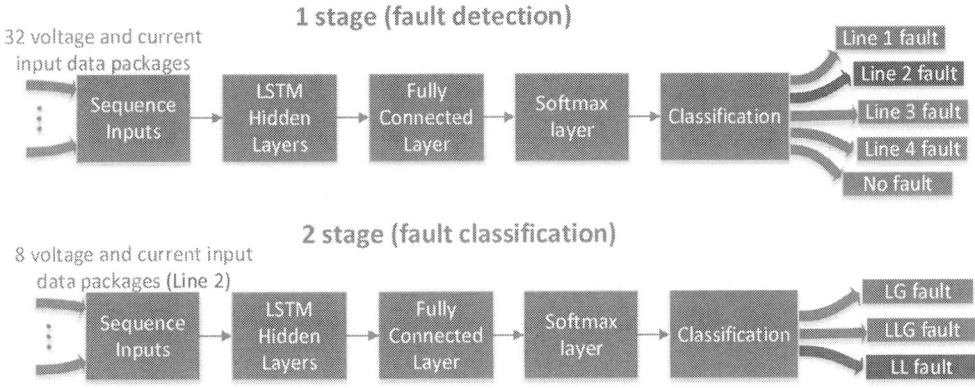

Fig. 4: Fault detection and classification structure (example of LL fault in the Line 2).

for fault detection and classification (time-series classification issues). There are several main settings of a deep learning network, such as configuration and number of layers, amount of epoch, and size of a bunch, described in [19, 20]. Proposed LSTM network architecture consists of sequence input with a bidirectional LSTM layer with 60 hidden units, fully connected layer, followed by softmax and classification layer. Implementation of the LSTM method to time-series classification is described in [20].

The LSTM network of the first stage of the fault detection system is trained on fault, non-fault, and external fault simulation data. Current and voltage measurements from both ends of each line in the system are inputs for this stage (8 per line, 32 in total). There are five outputs of the LSTM network: one per line location and one additional for the normal operation mode. After the detection of the fault in one of the lines, the main controller extracts only the data related to the fault line. These current and voltage measurements are the inputs for the second stage of the fault detection system, which is responsible for fault classification. Therefore, the second stage consists of the LSTM network having eight inputs and three outputs for LG, LLG, and LL faults.

The training set consists of data from 1600 simulations, generated by the Simulink model. The duration of the simulation is 0.23 seconds. The fault location within a line, fault line, and time of a fault is chosen randomly within the preset boundary for each simulation. The fault location line limits are 5–95% of line length and fault time varies from 0.1 to 0.2 seconds. The starting 0.1 seconds isn't considered in-network training and testing to exclude transient state during the system startup. The proportion of fault and non-fault mode is defined as 75/25%, respectively. The model frequency of discretization is 10 kHz, which provides 2300 data entries of each parameter. In order to provide fast and stable operation of the fault detection system, there is a need to decrease the length of entry data packages for the LSTM network. Reduction of entry data package length allows to increase the time of the system response as less data need to be collected for analysis. Therefore, data within $t_{\text{fault}} \pm 1$ ms, $t_{\text{fault}} \pm 2$ ms and $t_{\text{fault}} \pm 3$ ms forms 21, 31 and 41 values length input packages, respectively.

There are Two different approach applied in the testing of the fault detection and classification system. The first one is the simulation of the testing data sets, using the same approach as for the training data. Such method allows to generate a high amount of testing data sets and define the accuracy of the LSTM networks with different settings. The testing data set consists of 600 simulations. The second approach for the system testing is the implementation of fault detection and classification system into the Simulink model and testing the system in online mode. This method provides a visual illustration of the fault detection system operation and brings testing closer to real-world conditions. In addition, the second approach allows to evaluate the time of fault detection and classification with different input package lengths for the LSTM networks.

Results

Table I presents results of the first testing approach. 54 fault detection and classification systems based on LSTM networks with different settings were tested. There are three groups of tested LSTM networks with different lengths of an input data package. As it is seen from the results, the accuracy of fault detection rise with package length increment. It happens because a bigger input package provides greater short-circuit time coverage which allows to define all fault features in a more specific way. However, the main disadvantage of the long input package is a delay in fault detection which is presented in Table II.

The accuracy of the LSTM network is a ratio of matches between predicted and actual output data to the total number of tested data. As it is seen from the results, higher amount of layers or number of an Epoch does not presuppose higher accuracy, because LSTM network can be over-trained on training data and accordingly, does not detect real features of faults. LSTM network, with 60 layers, 20 Epochs, and 40 Bunch size settings, has the highest 91.5%accuracy among networks with 21 lengths of an input package.The highest accuracies among LSTM networks with 31 and 41 lengths of an input package

Table I: First testing approach results.

simulation	Input Package size	Layer	Max Epoch	Bunch size	Accuracy, %
1-3	21	40	10	15 – 27 – 40	89 – 89.33 – 89.33
4-6	21	40	20	15 – 27 – 40	89.33 – 90.5 – 90.17
7-9	21	60	10	15–27–40	89.33 – 89.33 – 89.33
10-12	21	60	20	15 – 27 – 40	89.33 – 86.33 – 91.5
13-15	21	80	10	15 – 27 – 40	80.17 – 89.33 – 84.67
16-18	21	80	20	15 – 27 – 40	88.83 – 89.83 – 90.67
19-21	31	40	10	15 – 27 – 40	89.33 – 89.33 – 93.5
22-24	31	40	20	15 – 27 – 40	93.67 – 95.17 – 96.67
25-27	31	60	10	15–27–40	89.33 – 89.33 – 89.67
28-30	31	60	20	15 – 27 – 40	93.5 – 96.83 – 95.67
31-33	31	80	10	15 – 27 – 40	89.33 – 90.83 – 89.33
34-36	31	80	20	15 – 27 – 40	92.83 – 95.83 – 97.5
37-39	41	40	10	15 – 27 – 40	91.67 – 94.83 – 94
40-42	41	40	20	15 – 27 – 40	96.83 – 99.83 – 97.5
43-45	41	60	10	15–27–40	90.83 – 95.17 – 94
46-48	41	60	20	15 – 27 – 40	99.67 – 99.83 – 99.83
49-51	41	80	10	15 – 27 – 40	90.33 – 89.33 – 94
52-54	41	80	20	15 – 27 – 40	96.5 – 99.33 – 99.83

are 97.5% and 99.83%, respectively. In a case when different settings provide equal accuracy, LSTM network with less number of layers is more preferable because it has a simpler structure that decreases computational power.

Table II shows outcomes of the second approach testing, which is done with the implementation of fault detection and classification system directly in the model. After analysis of the first approach results, three LSTM networks with the highest accuracy for the different length of input package are chosen for the second approach. In this approach, time of fault detection is the main characteristic for comparison. Frequency of discretization in the model is 10 kHz, meaning a data entry per microsecond (μs). Data entries form the input packages for the LSTM networks, on basis of which fault detection and classification is to be carried out. From the results presented in the Table II, it is seen that delay in fault

Table II: Second testing approach results.

N simulation	Fault Line	Input Packet size, Layer, Epoch, Bunch size	Line distance proportion	Fault type	t_{fault}, ms	$t_{detection}$, ms
1	1	21 – 60 – 20 – 40	0.2	LG	300	300.3
2	2	21 – 60 – 20 – 40	0.3	LLG	300	300.5
3	3	21 – 60 – 20 – 40	0.6	LL	300	300.5
4	4	21 – 60 – 20 – 40	0.9	LG	300	300.2
5	1	31 – 80 – 20 – 40	0.9	LG	300	301.0
6	2	31 – 80 – 20 – 40	0.6	LLG	300	301.1
7	3	31 – 80 – 20 – 40	0.3	LL	300	301.3
8	4	31 – 80 – 20 – 40	0.2	LG	300	300.8
9	1	41 – 60 – 20 – 27	0.6	LG	300	301.1
10	2	41 – 60 – 20 – 27	0.2	LLG	300	301.4
11	3	41 – 60 – 20 – 27	0.9	LL	300	301.3
12	4	41 – 60 – 20 – 27	0.3	LG	300	301.1

detection varies from 3–5 μs, 8–13 μs and 11–14μs for the systems with the input data package length of 21, 31, and 41, respectively. Comparing the times of the fault detection between systems with different input package lengths, within the model with low fault current rate and high discretization frequency, the factual difference in fault detection delay (8–9 μs) does not seem significant. However, considering the lower frequency and higher fault current, which can damage power systems components, the balance in accuracy and input package length needs to be achieved.

Figure 5 illustrates the results of the fault detection and classification system implemented into the Simulink model. Three cases marked in Table II are presented. Plots represent current, voltage, and fault classification time serial signals. The moment of simulation when line current equals zero indicates switching on corresponding line breakers. Current and voltage pairs I_line1:1 and I_line1:3 (V_line1:1 and V_line1:3), and I_line1:1 and I_line1:3 (V_line1:1 and V_line1:3) are measurements located in the opposite ends of a line. Fault detection and classification system successfully switch off a fault line, providing and remaining the system stability. Voltage drops up to 90 V after a fault but then restores to the nominal 120 V within 0.1 seconds. The highest peak current during the LLG fault in the line 4 reaches 6 A, while the current in the nominal operation mode is 0.2 A.

Presented fault detection and classification system has a few advantages compared to traditional over-current and differential protection schemes. In the low powered lines, it is hard to define the tripping threshold because of the possible high variation of nominal currents and low fault current. Ground faults can also have some challenges in fault detection for the differential protection scheme when the difference between in and out currents can be lower than a threshold. In most cases, protection against voltage drop helps to deal with the above-mentioned issues. However, the voltage drop in the low-powered LVDC grid can be insignificant and has no effect on fault detection. Applied LSTM networks based system allows to analyze not only amplitude or value of the parameter change but the character of this change in the time domain. Moreover, it allows not only to define the location but also to identify the type of a fault that helps to define the reason for the damaged line.

Conclusion

The paper discovers the application of deep learning to the protection issues of LVDC lines in the off-grid. The proposed power system is design to be modular and has a mesh topology, which brings studied case closer to real-world conditions. Implementation of IT grounding decrease fault currents but makes fault detection more challenging.
According to the acquired results, the LSTM network has high application potential in the field of fault protection. Acquired results show that the proposed methodology allows to achieve high accuracy of fault detection. The fault classification stage provides information about the type of the fault, that can be used for possible fault identification or definition, consequently possible approaches of solution. This possibility is especially essential or practical for the rural SSA areas with poor maintenance conditions, as fault clearance can be done automatically, and implemented locally on the ICT system, or by providing support remotely.

There are several steps of development for proposed methodology. First of all it should be applied for the protection of another power system's components such as end-user loads, batteries, PV-arrays and DC busses. The next step of the current research will be the verification of the Simulink model based on real data and testing the system within the verified model. Discovering more sophisticated fault features will allow to broaden protection schemes. There are plans to extend and scale the utilization of the proposed approach for the different grid typologies and as a final step is to implement and test the proposed fault detection and classification system in the existing laboratory and field installations.

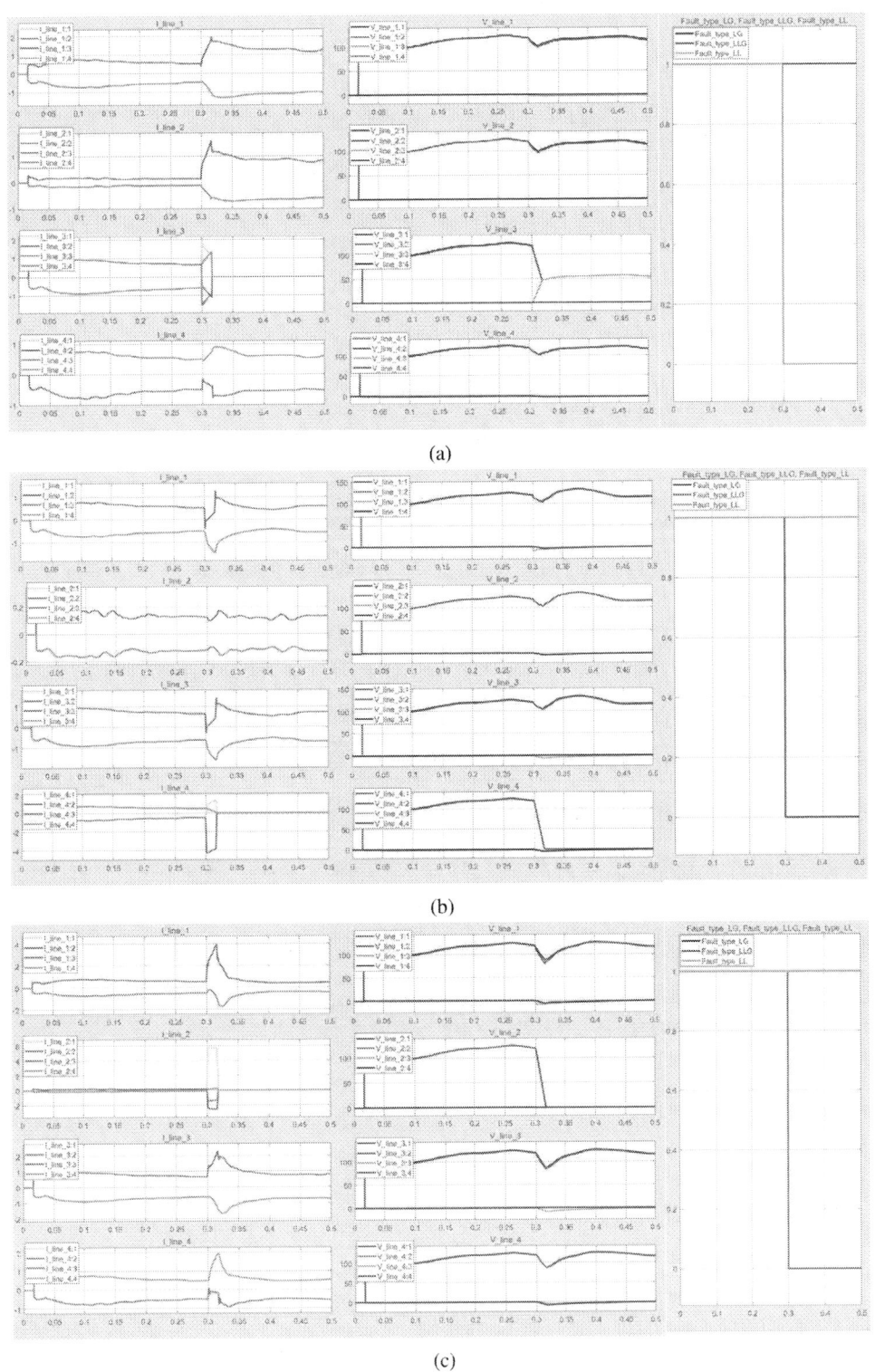

(a)

(b)

(c)

Fig. 5: Simulation model results: (a) Line 3 LL fault (21 input package), (b) Line 4 LG fault (31 input package), (c) Line 2 LLG fault (41 input package).

References

[1] Tagliapietra S.:Electrification in sub-Saharan Africa: The role of international institutions,Oxford Energy forum 2018, issue 115

[2] Mahler D. G.: Internet Access in Sub-Saharan Africa, World Bank Group 2019, paper 13

[3] Akinyele D.: Challenges of Microgrids in Remote Communities: A STEEP Model Application, Energies, 2018

[4] Salomonsson D.: Protection of Low-Voltage DC Microgrids, IEEE Transactions on Power Delivery, pp. 1045-1053, July 2009

[5] Siami-Namini S.:The Performance of LSTM and BiLSTM in Forecasting Time Series, 2019 IEEE International Conference on Big Data (Big Data), Los Angeles, CA, USA, 2019

[6] Mohanty R.: An Accurate Non-Iterative Fault Location Technique for Low Voltage DC Microgrid, IEEE Transactions on Power Delivery, Vol.31, Issue.2

[7] Techno-economic analysis of network configuration of PV based off-grid distribution system, CIRED 2019, paper 1874

[8] Pugliese H.: Discovering DC: A Primer on DC Circuit Breaker, Their Advantages, and Design, IEEE Industry Applications Magazine 2013, Vol. 19, No. 5

[9] Savage P.:From Silos to Systems:Issues in Clean Energy and Climate Change: DC microgrids: benefitsand barriers, Yale School of Forestry Environmental Sciences 2010

[10] Wang D.: Fault analysis of an active LVDC distribution network for utility applications,51st International Universities Power Engineering Conference (UPEC) 2016, pp. 1-6

[11] Salomonsson D.: Protection of Low-Voltage DC Microgrids, IEEE Transactions on Power Delivery 2009, vol. 24, no. 3, pp. 1045-1053

[12] ChewS. H.: A-line to ground fault detection technique for ELVDC distribution system in the built environment, IEEE 7th International Symposium on Power Electronics for Distributed Generation Systems (PEDG) 2016, pp. 1-7

[13] Guerrero J. M.: Advanced Control Architectures for Intelligent Microgrids—Part I: Decentralized and Hierarchical Control,in IEEE Transactions on Industrial Electronics 2013, vol.60, no. 4, pp. 1254-1262

[14] Samerchur S.: Power control of single-phase voltage source inverter for grid-connected photovoltaic systems, 2011 IEEE/PES Power Systems Conference and Exposition, Phoenix, AZ, 2011, pp. 1-6

[15] Kapat S.: Formulation of PID control for dc-dc converters based on capacitor current: A geometric context, 2010 IEEE 12th Workshop on Control and Modeling for Power Electronics (COMPEL), Boulder, CO, 2010, pp. 1-6

[16] Banguero E.: A review on battery charging and discharging control 3 strategies: Application to renewable energy systems, Energies 2018

[17] Ahmed A. S.: MPPT algorithms: Performance and evaluation, 2016 11th International Conference on Computer Engineering Systems (ICCES), Cairo, 2016, pp. 461-467

[18] Umaid A.: Fuzzy-PI Control of Bidirectional DC-DC Converter for 250KW Distributed Solar-Solar Off-Grid System, International Journal of Electrical Energy, 2019 pp. 26-29

[19] Brownlee J.: How to Tune LSTM Hyperparameters with Keras for Time Series Forecasting, Machine Learning Mastery, 2019

[20] Adam K.: Memristive LSTM network hardware architecture for time-series predictive modeling problems, IEEE Asia Pacific Conference on Circuits and Systems (APCCAS)2018, pp. 459-462

An Input-Series Output-Independent Full-Bridge Dual Active Bridge Converter with Soft-Switching Characteristics for Charging and Balancing Electric Vehicle Battery Stacks

Alex V. Mirtchev, Emmanuel C. Tatakis
UNIVERSITY OF PATRAS
Laboratory of Electromechanical Energy Conversion
Electrical and Computer Engineering Department
Department of Electrical and Computer Engineering
Rion-Patras, Greece 26504
Tel.: +30.2610.996414
E-Mail: a.mirtchev@upatras.gr, e.c.tatakis@ece.upatras.gr
URL: http://lemec.ece.upatras.gr

Keywords

«Battery charger», «Energy system management», «Soft switching», «Electric Vehicle», «Resonant converter».

Abstract

In this paper a multifunctional configuration for charging and balancing a battery stack, using modular converter design approach is proposed. Battery stacks consist of multiple cells to reach the desire voltage level and are conventionally charged altogether. The author proposes that the battery stack is divided into Battery Groups (BGs) that can be independently charged, controllably discharged under load and also balanced. The equations that describe the converter reference current for any state of operation and also permit minimization of the required sensors in the system, are presented. Mathematical modeling and detailed characteristic curves are derived based on the charging profile of the selected battery. The proposed control method relies on Pulse Frequency Modulation (PFM) and Phase Shift Modulation (PSM) to maintain Zero Voltage Switching (ZVS) and accomplish the desired output. The main advantages are the soft-switching characteristics, the flexibility in variable voltage and power levels due to modular design, and the online equalization of the batteries during charging and discharging.

Introduction

The overindulgence of fossil fuels for energy production along with the exceeding emissions of CO_2 indicate that, the total transition from internal combustion engines to electric motors seems inevitable, since not only CO_2 emissions from Electric Vehicles (EVs) are zero, but also the efficiency of electric motors is much higher (>80%) compared to fuel powered engines (<30%) [1]. Fast and efficient charging methods along with optimal battery management systems can increase the driving range and prolong the Energy Storage System (ESS) End-Of-Life (EOL). The off-board chargers reduce the vehicle weight and offer high power transfer compared to on-board chargers, however, the latest are much more convenient in terms of power supply, since even a typical power outlet would be sufficient [2], making it suitable for small EVs and urban transportation. In addition, on-board chargers can be tailored to specific ESS, meeting its detailed requirements and can detect possible malfunctions of a battery stack. Charging topologies require isolation, which can be implemented at the AC/DC stage and as a result non-isolated DC/DC converter can be used [3]. However, due to the low frequency transformer at the AC/DC stage (along with the rectifier bridge), the overall efficiency is decreased, and the volume and the weight are increased. In order to provide isolation via the DC/DC converter a Full-Bridge (FB) topology can be adopted with high frequency transformer [4], [5], but the soft-switching region, due to the presence of leakage inductance and parasitic capacitance, is relatively small [6], [7] and depends on the load. In the interest of expanding the soft-switching region, resonant topologies can be implemented [8], [9], by compensating the leakage inductance with series or parallel capacitors and even by adding extra inductors to accomplish the required resonance. However, the analysis and the control in these cases is more challenging since, the complexity is higher and there is still dependence from the output load. Nevertheless, due to the great reduction of the switching losses, high power density can be achieved with this type of converter. The CLLC-type resonant FB offers wide soft-switching range with Zero-Voltage-Switching (ZVS) at the inverting side and Zero-Current-Switching (ZCS) at the rectifying bridge [6]. For bi-directional power flow and synchronous rectification, the rectifying bridge at the output can be substituted with another active H-Bridge, resulting in a Full-Bridge Dual-Active-Bridge (FBDAB) converter [6]. Even greater power rating is possible when multiple converters are synthesized in modular manner or via interleaving, as proposed in [10], [11].

A typical charger configuration and a charging profile for Li-Ion batteries are shown in Fig. 1 and Fig. 2, respectively, where ESR_{wire} is the resistance of the wires connecting the converter to the battery. In high capacity charging applications, the power dissipated on this resistance is not negligible. The purpose of Constant Current (CC) and Constant Voltage (CV) stages is to maximize the capacity at the end of the cycle and avoid any damage from overcurrent or overvoltage, that could lead to destruction or degradation of the cells [12], [13].

In the present paper, an on-board charger with Input-Series Output-Independent (ISOI) configuration is proposed, where multiple converter modules are interconnected. By connecting the input of each converter in series, the voltage stress of the components is improved. The battery stack is divided into Battery Groups (BGs) with each one having its own module dedicated to it. As a result, every BG can be independently charged and controllably balanced, with the prospect to maintain their voltages at same level. With the suggested topology configuration, balancing between the BGs is succeeded with the most charged groups transferring energy to the least ones. Therefore, inevitable differences of the State-Of-Charge (SOC) between cells, that in the short or long term can damage them, are decreased and are limited within each BG. As an advantage, a Battery Management System (BMS) of lower count components is needed and the equalizing process is speeded up. For that, a CLLC-type FBDAB converter is designed for charging high capacity batteries of a small EV and maintaining soft-switching characteristics for the full battery charging profile. Finally, a closed loop control system is used to ensure CC and CV modes by applying both Phase Shift Modulation (PSM) and Pulse Frequency Modulation (PFM). The behavior of the system is analyzed with mathematical equations and evaluated in charging and discharging states. The overall system is tested in simulation environment using real component characteristics.

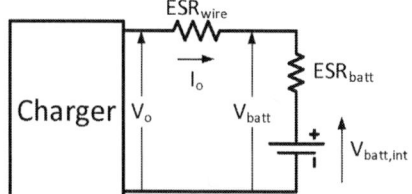

Fig. 1: Typical charger and battery connection with parasitic ohmic resistances.

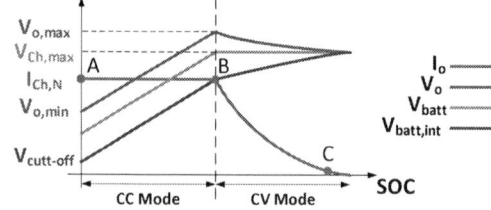

Fig. 2: Charging profile of a Li-ion battery with constant current and saturation charge modes.

ISOI Configuration for equal charging and balancing

In this section the proposed topology configuration is analyzed for different operating states and the equations that unify the converter currents during balancing mode are also presented with ostensive examples.

Topology Description

Each converter module is dedicated to a BG that consists of X cells and altogether form a battery stack of Y BGs. In Fig. 3 the schematic diagram of the proposed system is presented and SWx serve as switches to change between charging and discharging modes. Adjusting the numbers X and Y, the system can be adjusted to various power and voltage ratings depending on the battery stack or the input supply. Each mode of operation is described as follows.

Charging [SW1→On, SW1N→Off, SW2→Off]: In Fig. 2, the charging profile of each BG is depicted. Each module takes measures for its corresponding BG and if its voltage is below $V_{ch,max}$, then the converter operates under CC mode. When the voltage reaches the maximum value, then the module switches to CV mode. Usually, the CC mode is short in period and the batteries are charged until >60% of SOC [13], while CV is slow, it is an optimum and safe way to fully charge Li-ion batteries. Therefore, a much lower or much higher SOC of a BG wouldn't affect the rest of the stack, due to the independency. On top of that, the batteries are charged faster and more reliably than conventional methods.

Discharging [SW1→Off, SW1N→On, SW2→On]: The LOAD at the output in Fig. 3 represents the EV's loads (electronic circuits, drivetrain, etc.). During operation, it is common for some cells to discharge faster than others. By connecting the output of the converter to the input, part of the energy from the whole stack can be transferred to the most discharged BG. The estimation of the lowest SOC can be done by measuring the BG voltage. Although, this method is inaccurate due to the small slope of the cell voltage with respect to SOC, it can detect when the voltage tends to reach the lowest permittable value. Since, its purpose is not monitoring but charging, it is considered as a quite sufficient method. This balancing method during discharging has the advantage of keeping all BGs to similar voltage level and protecting them from over-discharge.

Battery Group Balancing [SW1→Off, SW1N→Off, SW2→On]: Even after charging or discharging and without any load, any deviations in voltages that may remain can be compensated. Once again, the battery stack acts as power source for the converter and the most charged BGs transfer energy directly to those with lower voltage. The equations that determine the priorities and the amount of current supplied from each module will be given in the next subsection.

Capacitor Balancing: Since, uneven power distribution is required and there is in series connected capacitors at the input to supply each module, a necessary active balancing circuit is implemented like the one depicted in Fig. 4. For example, when a converter module is OFF while other are ON, energy from its capacitor is transferred to the rest, in order to maintain their voltage level at the same value. The neighboring MOSFETs are driven with complementary pulses of 50% duty cycle. Towards minimizing the losses with soft-switching commutations, the pulses are driven at the resonance frequency $\omega_r = 1/\sqrt{L_b C_b}$. Similar topology can be used for balancing the cells within every BG [14].

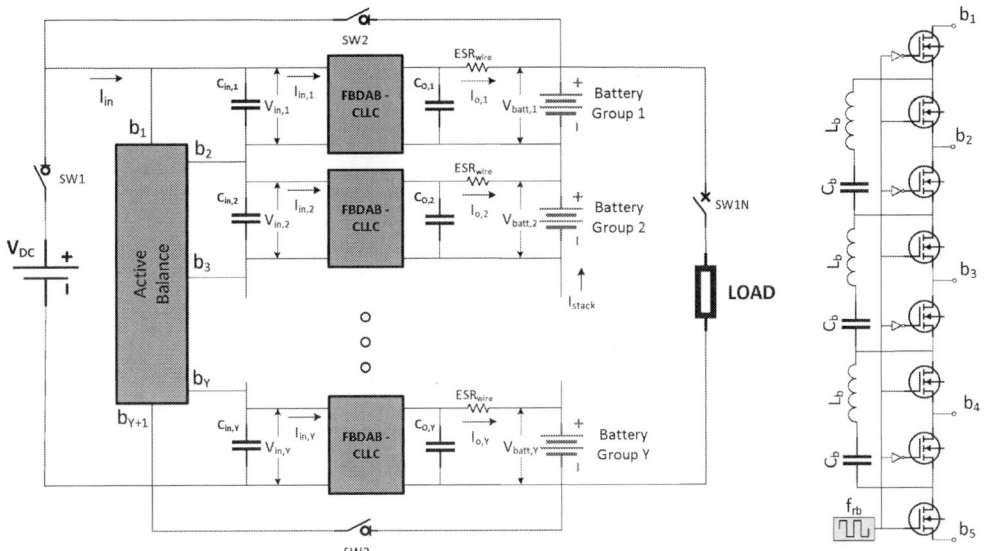

Fig. 3: Proposed ISOI configuration for charging and balancing of battery stack with Y BGs.

Fig. 4: Input capacitor balance circuit (Y=4).

Balancing Equations

The balancing method should be expressed in mathematical functions, so that the reference signal for each control unit of the modules can be calculated. Moreover, the circuit currents should be examined and be ensured that they don 't exceed the maximum permittable values. By using only, the converter output current sensors and the BGs voltage, all the calculations can be executed online, with the additional computational effort that comes with it. For that reason, the derived equations must describe the converter output current in relation to the total battery stack and each BG currents. In order for that to happen, the specifications of the system must be defined, especially the battery characteristics.

By measuring the voltage level of a BG, a decent estimation of its SOC can be made. First, the number of different balancing levels N_{levels} is defined, as well as the maximum output current I_{max} of the module in the state of balancing. Also, the boundary voltage levels of the BG, V_{max} and V_{min}, based on the battery specifications from the datasheet for discharging mode are selected. In Fig. 5, a graphical visualization of three different balancing levels is depicted and the general eq. (1) describes the amount of current that a BG should be balanced with, as a function of the measured battery voltage $V_{m,i}$. As already mentioned, this power converter configuration can support two states of balancing and for each one separate analysis is required.

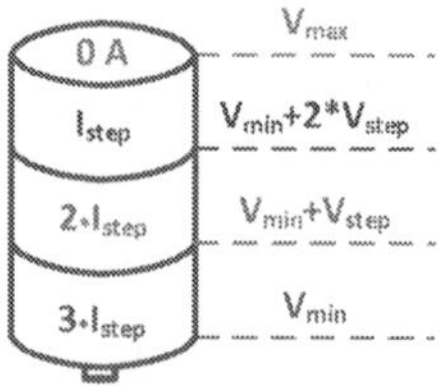

$$I_{balance,i} = \begin{cases} N_{levels}I_{step} & , & V_{m,i} = V_{min} \\ I_{step}\left(\left.\left|\dfrac{V_{max}-V_{m,i}}{V_{step}}\right|_{DIV}\right.+1\right) & , & V_{min} < V_{m,i} < V_{max} \\ 0 & , & V_{m,i} = V_{max} \end{cases} \quad (1)$$

Fig. 5: Different levels of balancing.

Where, $V_{step} = \dfrac{V_{max}-V_{min}}{N_{levels}}$, $I_{step} = \dfrac{I_{max}}{N_{levels}}$

Balancing during discharging

For the case of having a load connected to the output, the battery current is reduced from the stack current by the amount of the current injected from the corresponding module that the BG is connected to. For calculating the reference control signal $I_{o,i}$ for each converter, the eq. (1) is used, therefore $I_{o,i}=I_{balance,i}=I_{o,Inject,i}$. Given that the load current I_{load} is known, since it is the main supply current for several circuits (e.g. power inverter, electronic circuits, etc.), the actual rate of discharge of each BG can be determined, via eq. (2). Instead of, inserting an additional voltage sensor for measuring each converter output voltage, it can be estimated from eq. (3). The converter efficiency n_j can be withdrawn from a table that is constructed beforehand, with respect to output power. Alternatively, extra sensor for the total input current can be used.

$$I_{batt,i} = I_{load} + I_{in} - I_{o,i} \Rightarrow I_{batt,i} = I_{load} + \frac{1}{V_{stack}}\sum_{j=1}^{Y}\left(\frac{V_{o,j}I_{o,j}}{n_j} - I_{o,j}^{2}ESR_{wire}\right) - I_{o,i} \quad (2)$$

$$V_{o,j} = V_{batt,j} + I_{o,j}\left(ESR_{wire} + ESR_{batt}\right) \quad (3)$$

The battery internal resistance (ESR_{batt}) as well as the resistance of the copper/wire connection (ESR_{wire}) are also measured in advance. Since, the purpose of this circuit is not monitoring accurately the SOC of each cell, charge and temperature dependencies of the resistances can be neglected, without great impact on the circuit.

For visualization of the balancing logic, let for example have Y=4 BGs with X=6 cells each, loaded with I_{load}=15A, two of them being fully charged with maximum voltage $V_{batt,Inject,3}=V_{batt,Inject,4}=V_{max}$, and one being a little below V_{max}=21.6V, specifically $V_{batt,Inject,2}$=20.916V. In Fig. 6 – Fig. 9, input and output currents are depicted with respect to BG_1 voltage for the whole SOC spectrum, according to eq. (1), while the other BGs remain at the same voltage level.

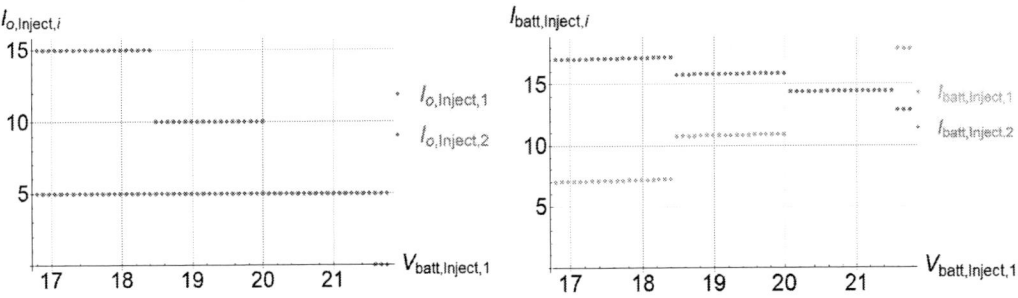

Fig. 6: Output injected currents for each FBDAB converter module of BG_1 & BG_2.

Fig. 7: Battery currents that discharge each BG_1 & BG_2 with different ratios.

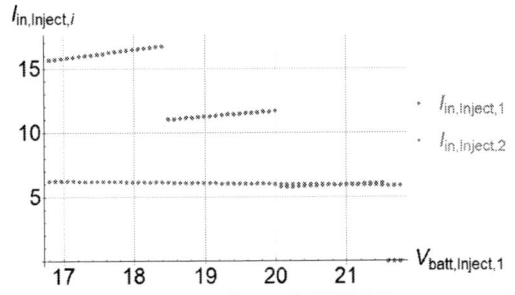
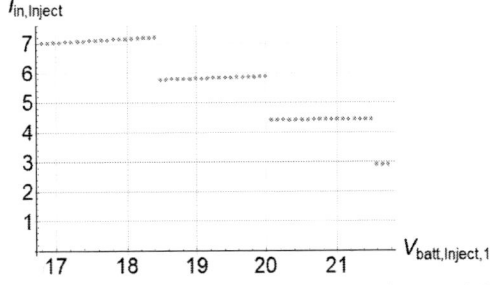

Fig. 8: Input currents for each FBDAB converter module of BG_1 & BG_2.

Fig. 9: Input current of the whole modular converter, drawn from the battery stack.

As it can be seen from Fig. 6, as the BG charges and its voltage rises, the reference current gradually decreases until it becomes zero at the end of the balance. Respectively, in Fig. 7 can be observed that the amount of current supplied from BG_1 increases as the voltage (and hence the SOC) gets higher, preventing it this way from over-discharge. Moreover, in combination with Fig. 9, for the lowest BG_1 voltage, where the injected current matches the load current, BG_1 supplies only the total converter current. If the load current was low enough then the battery current would become negative and it would be charged instead of discharged.

Balancing between BGs

When there is no load connected to the battery stack, pure balancing can be accomplished. In this case, some BGs transfer power to the least charged ones. The determination of the converter reference output current signal is more complicated since the stack current (I_{stack}) passes through the converter module and adds up to the output current, when the corresponding BG is being charged. It should be noted, that once again eq. (1) is implemented, but for this situation $I_{batt,i}=I_{balance,i}$. The function that describes the converter output current is given by eq. (4) and to solve the Y equation system initial conditions are assumed as $I_{o,i}=I_{batt,i}$.

$$I_{o,i} = I_{batt,i} + I_{in} \Rightarrow$$

$$\Rightarrow I_{o,i} = \frac{1}{1 - \dfrac{V_{o,i}}{V_{stack}}\dfrac{1}{n_i}} \left[I_{batt,i} + \frac{1}{V_{stack}} \sum_{\substack{j=1 \\ j \neq i}}^{Y} \left(\frac{V_{o,j} I_{o,j}}{n_j} - I_{o,j}{}^2 ESR_{wire} \right) - I_{o,i}{}^2 ESR_{wire} \right] \tag{4}$$

For the same example of voltages for each BG, as previously described, in the following Fig. 10 - -Fig. 13 the current flow in the system is depicted. While, for the previous case the positive battery current indicates discharging, for this mode of operation signifies charging of them. It is clear that, the converter is loaded with the total amount of currents, that is why care should be taken in the number of BGs charged as well as the maximum output current, to prevent the converter exceeding its maximum ratings. This is the reason why this analysis is mandatory.

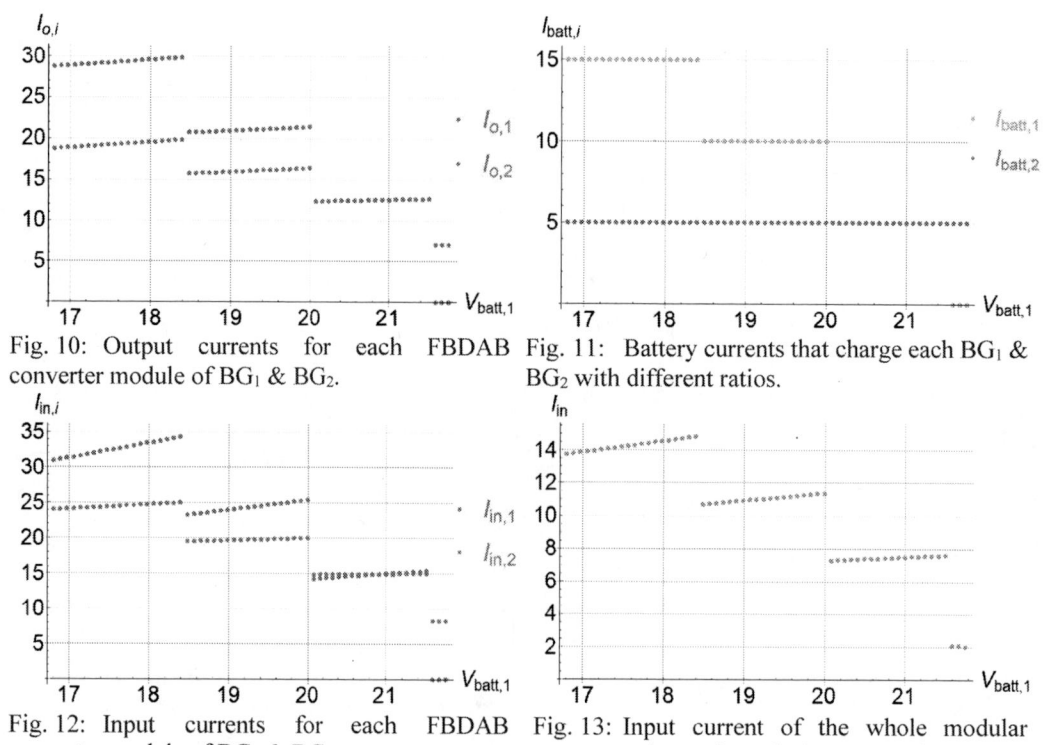

Fig. 10: Output currents for each FBDAB converter module of BG_1 & BG_2.

Fig. 11: Battery currents that charge each BG_1 & BG_2 with different ratios.

Fig. 12: Input currents for each FBDAB converter module of BG_1 & BG_2.

Fig. 13: Input current of the whole modular converter, drawn from the battery stack.

Converter Analysis and Control Method

The circuit diagram of the converter is presented in Fig. 14, where r_1, r_2 represent the equivalent ohmic resistance of two MOSFETs, the transformer, the inductor and the resonant capacitor, on the primary and secondary, respectively. The DC/DC stage of the charger consists of a CLLC-type FBDAB converters in ISOI configuration. Each output is connected to a BG of equal number of cells and is dedicated to charging and monitoring that stack. As already mentioned, due to the resonant circuit, the output of the converter is load dependent and for that reason, detailed analysis is necessary. Moreover, due to reactive power circulating in the resonant tank, care should be taken in the voltage stress of the components during operation. For the circuit analysis, Fundamental Harmonic Approximation (FHA) is implemented [9], [15] meaning that only the first harmonic of the AC signals is taken into consideration. This method is relatively simple, but it produces errors as the operating frequency deviates from the resonance frequency. However, its simplicity makes it ideal for circuit design and output estimation under various loads. Considering only the conduction losses, since soft-switching operation is considered and the turn-off losses of the MOSFETs at the primary can be neglected, the results of the analysis produce the eq. (5), (6) for the voltage gain and the output current, where δ_1, δ_2 are the duty cycles for each H-Bridge.

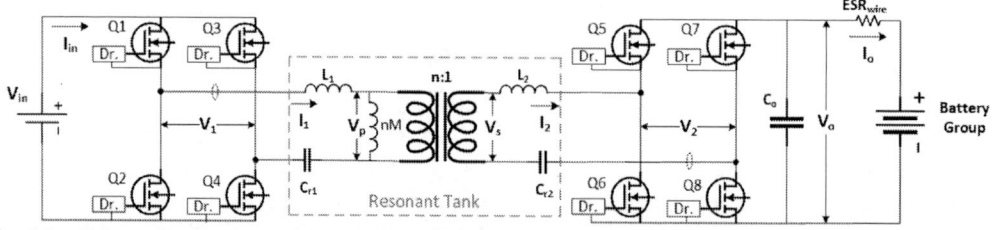

Fig. 14: Schematic diagram of CLLC-type FBDAB converter.

$$M_v = \frac{nR_{oe}Z_{r1}\sqrt{k_1}\csc(\delta_2\pi)\sin(\delta_1\pi)f_{or1n}^{\;3}}{(1+k_1)^{3/2}\pi^2\sqrt{A+B^2}} \qquad \text{where,}$$

$$A = \frac{k_1 Z_{r1}^2\left((1+k_1)n^2\left(k_c^2 r_1 + r_2 + R_{oe}\right) - f_{or1n}^2\left((1+k_2)r_1 + (1+k_1)n^2(r_2+R_{oe})\right)\right)^2 f_{or1n}^2}{(1+k_1)^3} \qquad (5)$$

$$B = \frac{Z_{r1}^2(k_1+k_2+k_1k_2)f_{or1n}^4}{(1+k_1)^2} + k_c^2 n^2 Z_{r1}^2 - \frac{\left(\left(1+k_2+k_c^2 n^2\right)Z_{r1}^2 + k_1 n^2\left(r_1(r_2+R_{oe})+k_c^2 Z_{r1}^2\right)\right)f_{or1n}^2}{1+k_1}$$

$$I_o = \frac{8nV_{in,avg}Z_{r1}\sqrt{k_1}\sin(\delta_1\pi)f_{or1n}^{\;3}}{(1+k_1)^{3/2}\pi^2\sqrt{A+B^2}} \qquad (6)$$

The definitions of important parameters are given in eq. (7) - (16).

$$k_1 = L_1/nM \qquad (7) \qquad k_2 = L_2/(M/n) \qquad (8) \qquad k_c = \sqrt{C_{r1}/C_{r2}} \qquad (9) \qquad n = \sqrt{L_p/L_s} \qquad (10)$$

$$f_{or1n} = f_{sw}/f_{or1} \qquad (11) \qquad f_{or1} = 1/2\pi\sqrt{(L_1+nM)C_{r1}} \qquad (12) \qquad f_{sr1} = 1/2\pi\sqrt{L_1 C_{r1}} \qquad (13)$$

$$M = k\sqrt{L_p L_s} \qquad (14) \qquad Z_{r1} = 2\pi f_{sr1}L_1 \qquad (15) \qquad R_{oe} = \frac{8}{\pi^2}R_{load} \qquad (16)$$

The parameter R_{load} is calculated by dividing the output voltage V_o with the output current I_o and varies according to the charging conditions of the batteries. In Fig. 15 and Fig. 16 the functions of the voltage gain and the output current are plotted with respect to the normalized operating frequency, with zero phase-shift (δ_1=0.5, δ_2=0.5), in order to verify that the converter can support the maximum charging load. The red graph (CC) corresponds to the point **A** of Fig. 2, the orange one (CC/CV) to the point **B** and the green one (CV) to the point **C**. For the converter design, it is assumed that k_1=k_2, k_c=1/n, meaning that the resonance frequencies of primary and secondary are equal, therefore f_{sr1}=f_{sr2}=f_{sr}. and f_{or1}=f_{or2}=f_{or}.

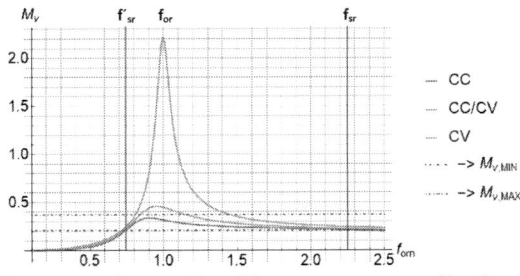

Fig. 15: Voltage gain with respect to normalized switching frequency for different charging states.

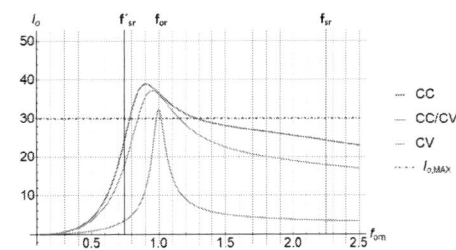

Fig. 16: Output current with respect to normalized switching frequency for different charging states.

In an ideal topology, when the operating frequency is equal to the (open circuit) resonance frequency f_{or}, the converter acts like a Current Source (CS), therefore, the output current is independent of the load R_{load}. This is true, only if the ohmic losses in the primary are neglected, otherwise r_1 decreases the equal Thevenin impedance and the current source is dependent on the load, as shown in Fig. 16 and Fig. 17. Also, around f_{or} the maximum power transfer is achieved. When the frequency is equal to the series resonance frequency f_{sr} or f_{sr}' (Fig. 15), the converter acts like a Voltage Source (VS), meaning that the resonant tank is consisted only of the transformer and the voltage gain is equal to the winding ratio 1/n. However, due to voltage drop on r_1 and r_2 and the variable current at f_{sr}, the output is affected by the load variations. So, in order to decouple the output current and voltage from the load, the ratios Q_{r1}=r_1/R_{load} and Q_{r2}=r_2/R_{load} should be minimized.

The ZCS at the rectifying bridge is achieved naturally [9], due to the resonant tank that makes the output of the secondary to act as current-fed source. Meaning that, the current flowing through the second H-Bridge determines the polarity of voltage V_2, which is equal to Vo in absolute value. To accomplish

ZVS, the phase angle of the voltage and the current at the input φ_{VI} must be greater than zero, as shown in Fig. 18. This means that, when the transistor is turned on, the current already flows through its antiparallel diode. Since, both OFC and PSC are implemented, the ZVS range is a function of both frequency and phase-shift, and their relation is expressed with eq. (17). For simplicity, it is assumed that the real current and its first harmonic component cross the x-axis at the same point.

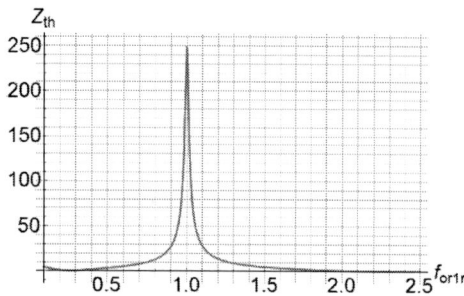

Fig. 17: Equivalent Thevenin impedance of the converter.

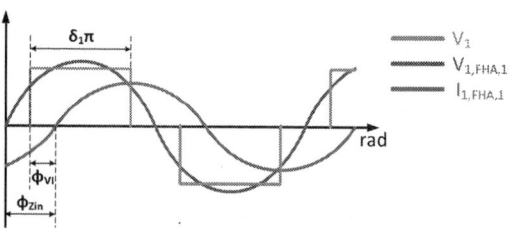

Fig. 18: Real voltage V_1, first harmonic component of the voltage $V_{1,FHA,1}$ and current $I_{1,FHA,1}$.

$$\varphi_{VI} = \varphi_{Zin} - \varphi_{shift} = \varphi_{Zin} - \frac{\pi(1-2\delta_1)}{2} \tag{17}$$

Naturally, if no bifurcation phenomenon occurs, for $f_{sw} < f_{or}$ the input impedance is capacitive and for $f_{sw} > f_{or}$ is inductive, otherwise there are three points, instead of one, where the input impedance angle φ_{Zin} becomes zero, making it much harder to apply close loop control, due to non-linearity. An example of non-careful designed topology is shown in Fig. 19, where the green dot indicates the critical point. For that reason, the circuit is designed such as that no bifurcation occurs for the required operation spectrum.

In order to achieve ZVS at the inverting bridge, φ_{VI} must be greater or equal to zero, therefore the input impedance has to be inductive, if not resistive. In reality though, a dead-time between complementary pulses is necessary, hence a minimum phase-shift φ_{shift} exists and as a result a minimum phase angle φ_{Zin} is required. The latest is adjusted via the operating frequency f_{sw} and the phase-shift can be expressed as duty cycle δ_1 (eq. 17). In Fig. 20 is shown the region for ZVS as frequency and phase-shift vary. At the start of CC mode, the range is limited (Fig.20), however, in combination with Fig.15 and Fig. 16, it can be seen that since the output current is high with low equivalent load resistance, it requires large duty cycle. As R_{oe} increases so is the ZVS region, since the orange graph is elevated with respect to the blue surface indicating the zero angle. For the converter control two closed loops are used, one for the output current for CC mode or the battery voltage for CV mode and the other loop for the phase angle φ_{VI}. The converter control variables are the phase-shift angle and the switching frequency, respectively.

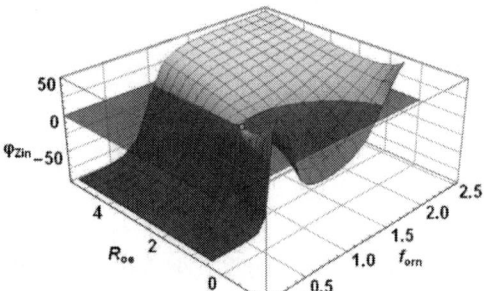

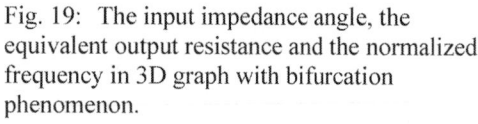

Fig. 19: The input impedance angle, the equivalent output resistance and the normalized frequency in 3D graph with bifurcation phenomenon.

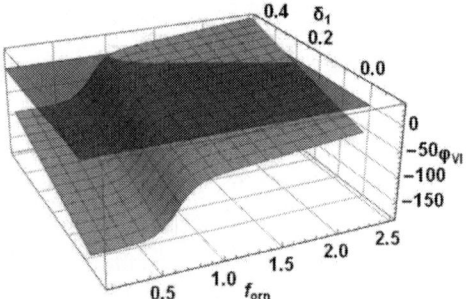

Fig. 20: The relation of the ZVS angle, the duty cycle and the normalized frequency in 3D graph at the start of the CC charging mode.

Simulation Results

An on-board charger for small EV is designed and simulated for different conditions. A rectified AC supply voltage of 230 V is assumed and f_{or} is selected to be 40 kHz. At the output, there is a battery stack of total 24 cells divided in Y=4 BGs of X=6 cells, with each battery having 90Ah capacity C, 4V maximum charge voltage and is charged with C/3. Indicative results are presented in Fig. 21 – Fig. 26 to verify the stability of the converter control, for various cases with unequal BG voltages.

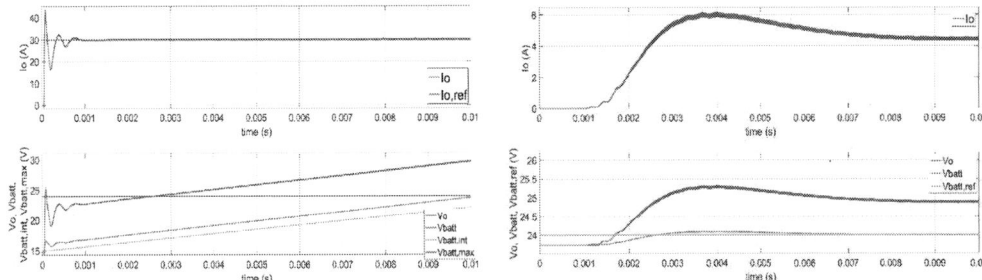

Fig. 21: Dynamic CC charging of BG, modeled with a capacitor to speed up the process.

Fig. 22: Static CV charging of BG at the end of the charge.

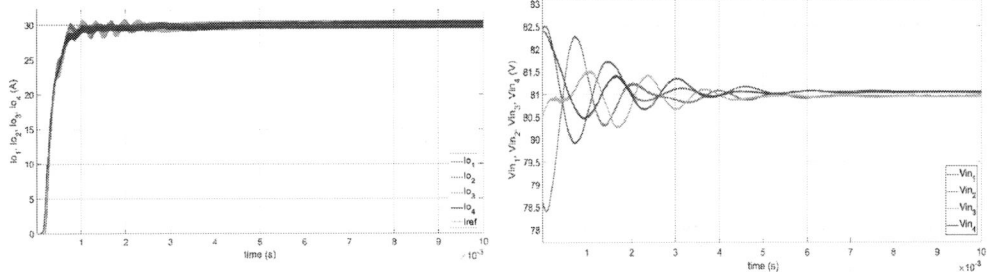

Fig. 23: Output currents in static CC charge when $V_{batt,int,1}$=**15V**, $V_{batt,int,2}$=**21.6V**, $V_{batt,int,3}$=**22.2V**, $V_{batt,int,4}$=**15V**.

Fig. 24: Voltage of input capacitors for the case in Fig. 23.

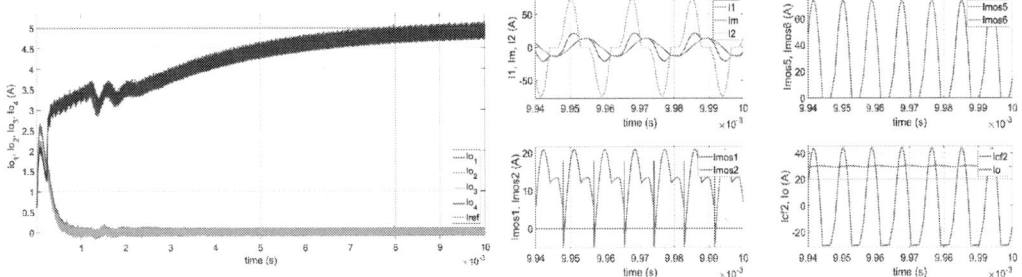

Fig. 25: Output currents while balancing when $V_{batt,int,1}$=**16.8V**, $V_{batt,int,2}$=**21.6V**, $V_{batt,int,3}$=**18V**, $V_{batt,int,4}$=**16.8V** with external load.

Fig. 26: Primary and secondary currents with ZVS on Q1 & Q2 and ZCS on Q5 & Q6.

In Fig. 21, a simulation of a BG is presented, in CC mode. Due to simulation time constructions, a capacitor with a series DC voltage is used instead of a real battery model. As it can be seen, the output current follows the reference control signal through the whole process, while the battery/capacitor voltage rises. When the BG voltage reaches the maximum value (X*4 V), the control method switches automatically to CV mode and in Fig. 22 it is shown a point of operation during that interval. In Fig. 23 and Fig. 24, the independent charging of each BG is verified as well as the equalization of the input capacitors. In Fig. 25, the case of injecting currents to the batteries with external load is shown. Finally, in Fig. 26 the primary and secondary currents are shown, proving that ZVS and ZCS are accomplished.

Conclusion

The proposed configuration is well suited for charging an energy storage system for its whole battery profile and equalizing BGs during charging and discharging, as the topology analysis and the simulation results prove. Each output can be independently adjusted, while maintaining equal voltage at the input capacitors. The total stack voltage is lower than the total input voltage during balancing mode and as a result the converter output current is limited to lower value compared to charging mode. Nevertheless, the extensive analysis and the mathematical equations presented in this paper lead to exact determination of the operating range. The presented closed loop dual control method permits ZVS and ZCS for the total load region and the converter efficiency is almost entirely dependent on the conduction losses.

References

[1] S. Mehar, S. Zeadally, G. Rémy and S. M. Senouci, "Sustainable Transportation Management System for a Fleet of Electric Vehicles," in IEEE Transactions on Intelligent Transportation Systems, vol. 16, no. 3, pp. 1401-1414, June 2015.

[2] S. Habib, M. M. Khan, F. Abbas, and H. Tang, "Assessment of electric vehicles concerning impacts, charging infrastructure with unidirectional and bidirectional chargers, and power flow comparisons," in Int. J. Energy Res., vol. 42, no. 11, pp. 3416–3441, 2018.

[3] M. Jung, G. Lempidis, D. Hölsch and J. Steffen, "Control and optimization strategies for interleaved dc-dc converters for EV battery charging applications," 2015 IEEE Energy Conversion Congress and Exposition (ECCE), Montreal,

[4] M. Ryu, D. Jung, J. Baek and H. Kim, "An optimized design of bi-directional dual active bridge converter for low voltage battery charger," 2014 16th International Power Electronics and Motion Control Conference and Exposition, Antalya, 2014, pp. 177-183.

[5] J. Lai, L. Zhang, Z. Zahid, N. Tseng, C. Lee and C. Lin, "A high-efficiency 3.3-kW bidirectional on-board charger," 2015 IEEE 2nd International Future Energy Electronics Conference (IFEEC), Taipei, 2015, pp. 1-5.

[6] B. Zhao, Q. Song, W. Liu and Y. Sun, "Overview of Dual-Active-Bridge Isolated Bidirectional DC–DC Converter for High-Frequency-Link Power-Conversion System," in IEEE Transactions on Power Electronics, vol. 29, no. 8, pp. 4091-4106, Aug. 2014.

[7] K. V. Ravi Kishore, N. Brahmendra, B. Sivaneasan, P. L. So and C. C. Chan, "An APWM soft switched DC-DC converter for PV and EV," 2014 IEEE 40th Photovoltaic Specialist Conference (PVSC), Denver, CO, 2014, pp. 3707-3712.

[8] Z. Fang, T. Cai, S. Duan and C. Chen, "Optimal Design Methodology for LLC Resonant Converter in Battery Charging Applications Based on Time-Weighted Average Efficiency," in IEEE Transactions on Power Electronics, vol. 30, no. 10, pp. 5469-5483, Oct. 2015.

[9] P. He and A. Khaligh, "Comprehensive Analyses and Comparison of 1 kW Isolated DC–DC Converters for Bidirectional EV Charging Systems," in IEEE Transactions on Transportation Electrification, vol. 3, no. 1, pp. 147-156, March 2017.

[10] Yue Zhang, Zheng Wang and Ming Cheng, "An interleaved current-fed bidirectional full-bridge DC/DC converter for on-board charger," IECON 2016 - 42nd Annual Conference of the IEEE Industrial Electronics Society, Florence, 2016, pp. 4376-4381.

[11] Q. Tian et al., "A novel energy balanced variable frequency control for input-series-output-parallel modular EV fast charging stations," 2016 IEEE Energy Conversion Congress and Exposition (ECCE), Milwaukee, WI, 2016, pp. 1-6.

[12] E. Chemali, M. Preindl, P. Malysz and A. Emadi, "Electrochemical and Electrostatic Energy Storage and Management Systems for Electric Drive Vehicles: State-of-the-Art Review and Future Trends," in IEEE Journal of Emerging and Selected Topics in Power Electronics, vol. 4, no. 3, pp. 1117-1134, Sept. 2016.

[13] "BU-409: Charging Lithium-ion," 2018. [Online]. Available: [Accessed: 02-Nov-2019] https://batteryuniversity.com/learn/article/charging_lithium_ion_batteries.

[14] Y. Ye and K. W. E. Cheng, "Analysis and Design of Zero-Current Switching Switched-Capacitor Cell Balancing Circuit for Series-Connected Battery/Supercapacitor," in IEEE Transactions on Vehicular Technology, vol. 67, no. 2, pp. 948-955, Feb. 2018.

[15] Y. Jiang, L. Wang, Y. Wang, J. Liu, M. Wu and G. Ning, "Analysis, Design, and Implementation of WPT System for EV's Battery Charging Based on Optimal Operation Frequency Range," in IEEE Transactions on Power Electronics, vol. 34, no. 7, pp. 6890-6905, July 2019.

A Method to search Global Maxima by Permanent Monitoring of Voltage and Current of each PV Panel

Shailendra Rajput, Moshe Averbukh
Ariel University, Kiryat Hamada, Ariel, Israel-40700
Tel.: +972 / (3) – 9765726, Fax: +972 / (3) – 9765726.
E-Mail: mosheav@ariel.ac.il
URL: http://www.ariel.ac.il

Abstract

The maximum power point tracking (MPPT) systems relate to photovoltaic (PV) power plants for maximum energy production. An efficient MPPT systems is needed for large solar plants including several serially and parallel-connected PV panels especially functioning in partial shading conditions. A proportional decrease in electrical power will be a result of inhomogeneous irradiation (partial shading), as well as multiple local maximums may appear. The presence of multiple local maximums is the most difficult obstacle for traditional MPPT which is based on a sequential search of an optimum working point. The present article submits a new algorithm that can be corresponded to the mathematical modeling with elements of Artificial Intelligence (AI). It combines the advantages of AI with those are typical to conventional methods applicable for homogeneous irradiation. A proposed method is to use permanent monitoring of a voltage, a current and temperature of each PV panel placed in the string. An MPPT algorithm determines the position of a global maximum (GM) based on this information and in accordance with the previously obtained math model of individual PV panels. Owing principles of AI, math models should be periodically précised during the service life of the PV plant. Since none of the presented math algorithms can provide localization of GM with the accuracy, required for the modern MPPT, the proposed method desire to be complemented by a conventional approach, let say perturbation and observation or incremental conductance techniques. For example, an algorithm finding zero roots of a power derivative versus current change was used in our work. Proposed algorithm can achieve GM with relatively high speed that is only restricted by digital control ability. Currently, this task would take no more than 50-100 ms maximum. Therefore, the global maximum can be found for any rapidly changing solar irradiation.

Keywords: PV solar plant, MPPT, partial shading, permanent monitoring, global maximum.

Introduction

The photovoltaic (PV) solar power plants include hundreds or even thousands of individual panels linked together in serial and parallel connections. Entire PV arrangement to be connected to a DC/AC inverter and must be controlled by a maximum power point tracker (MPPT). The primary aim of this device is to maintain optimal parameters of PV panels (i.e. an optimal combination of a voltage and a current), that providing maximum achievable power of a plant. Alterations of environmental conditions (i.e., solar irradiation magnitude, panel temperature, etc.) cause variation of the optimal point, requiring real-time parameters adjustment. The additional complexity of this task provides inhomogeneous irradiation or partial shading of PV panels causes the appearance of numerous local maximums that make difficult a finding of a global maxima (GM) between local ones.

A large spectrum of different MPPT techniques has been developed during a time. Three big groups can classify them: a) based on a periodical search of PV characteristics causing an interruption of energy producing, b) stochastic probing including simulating plants or animal's behavior, c) methods of artificial intelligence (AI) and math modeling. Methods from the first two groups are characterized either by prolonged searching time or by low accuracy. A representative overview of the variety of conventional algorithms for locating the MPP can be found in [1-15]. Included therein are traditional algorithms of perturbation and observation (P&O) [1-3], fuzzy logic control [4, 5] and sliding mode

MPPT control [6, 7]. There multiple are methods based on a simulation of wildlife nature. Such as particle swarm optimization [8, 9] or a grey wolf-assisted perturb & observe MPPT algorithm [10]. Comprehensive analysis of existing MPPT techniques is done in [11] and for those applicable during partial shading in [12]. Some methods are based on techniques distinctive for AI, including different predictive and computational methods, which offer improved prognosis of MPP location and thereby diminishing tracking time [13-15]. The stability issue of MPPT electronic evaluation is discussed in [16], for example. An MPPT approach is typical for AI and based on mathematical modeling of PV modules using scanning monitoring of PV array by the optical camera was proposed in [17]. The images of individual panels captured by the camera provide an estimation of actual solar irradiation with the following prediction of GM by a mathematical model.

Our work uses AI components such as math modelling with permanent monitoring of PV panel's parameters. An MPPT algorithm using this method provides fast control response with high accuracy in finding GM. An article consists of three main sections: development of the proposed method based on permanent monitoring of each panel, development of the electronic device simulation for finding GM during partial shading of PV panels and verification of a developed approach.

Proposed method of GM localization

Explanation about the presence of multiple maxima

Simplified equivalent circuit based on the entire equivalent circuit (Fig.1) can do a description of multiple maximums.

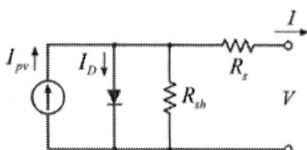

Fig. 1: Equivalent circuit of PV panel. I_{pv} is photocurrent, I_D is diode current, R_{sh} is shunt resistance, I_{sh} is shunt current, R_s is series resistance, I is output current, and V is output voltage.

The electrical behavior of panel parameters is described by an algebraic equation:

$$ I = I_{pv} - I_0 \left[\exp\left(\frac{V + IR_s}{aV_T} \right) - 1 \right] - \frac{V + IR_s}{R_{sh}}, \tag{1} $$

Where I, V, I_0, R_s, R_{sh}, a are output panel current, output voltage, diode reverse saturation current, series resistance, shunt resistance, and panel quality factor respectively. The thermal voltage ($V_T = k_B T / q$) is equal to ~0.025V for absolute temperature T=300°K (27°C). The q is electron charge (1.602·10⁻¹⁹ C), and k_B is Boltzmann constant (=1.3806×10⁻²³ m²·kg·s⁻²·K⁻¹). Resistance magnitudes (R_s, and R_{sh}), I_0, and a may be estimated by different approximating procedures [18] based on manufacturer data or a real panel set of V-I curves measured under various irradiation levels and different operating temperatures. Above mentioned parameters R_s, R_{sh} and a remain constant for silicon crystal and polycrystalline panels [19] under a wide range of temperatures and solar radiation intensities. In contrast to constant values, three variable parameters (I_{pv}, I_0, and V_T) are strongly dependent on solar irradiation and temperature. Their analytical expressions due to [19] are shown below:

$$ I_0 = \left(I_0^* \right) \left(\frac{T}{T_{ref}} \right)^3 \exp\left\{ \frac{q_e E_g}{a k_B} \left(\frac{1}{T_{ref}} - \frac{1}{T} \right) \right\}, \tag{2} $$

where I_0^* is a reverse diode current for the reference temperature T_{ref}=289 °K, T is current temperature, E_g is the band gap of the semiconductor material. For silicon panels including n-serially connected solar cells, the E_g value (in eV) can be calculated as [19]:

$$E_g = n \cdot \left(1.16 - 7.02 \times 10^{-4} \frac{T^2}{T-1108}\right). \tag{3}$$

The current source, or photocurrent I_{pv}, supplies primary electrical panel energy, and depends on solar irradiation level S and absolute temperature T. Its analytical approximation due to [19] yields:

$$I_{PV} = (I_{SC})_0 \left(\frac{S}{1000}\right) + J_0 \left(T - T_{ref}\right), \tag{4}$$

where $(I_{SC})_0$ is a short circuit current at reference state and J_0 is temperature coefficient for short circuit current. In general, J_0 has a relatively small value, thus the photocurrent remains closely proportional to the solar irradiation. Once all parameters of the equivalent circuit are determined, they may serve different mathematical manipulations. In a real PV panel, R_{sh} is relatively big and R_s is relatively small that allow neglecting them both. In addition, I-V shunt diode curve could be simplified to that is represented in Fig.2.

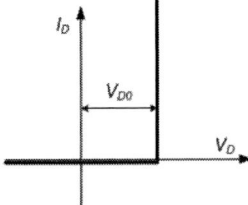

Fig. 2: Simplified diode I-V characteristic.

In accordance with the mentioned above, simplifications solar panel V-I curve looks likes in Fig.3. The short circuit current is equal to the I_{pv} of the equivalent scheme Fig. 3.

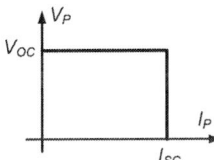

Fig. 3: Simplified V-I characteristic of solar panel. The V_O is an open-circuit voltage and I_{sc} is short circuit current.

The solar array comprised of n-serially connected panels and represented by simplified equivalent circuits is shown in Fig. 4. Parallel bypass diodes are connected to each panel.

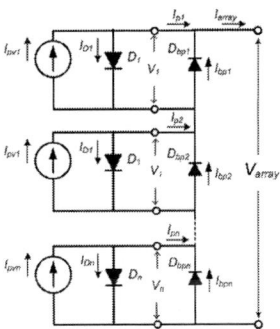

Fig. 4: Equivalent simplified circuit of solar array including n-individual solar panels with connected in parallel bypass diodes.

The uniform solar irradiation produces the output characteristic of this array like that in Fig.3. Here and in the sequel, the constancy of panel temperatures is assumed. The open-circuit voltage for array $(V_{OC})_n$ is equal to V_{OC} of a single panel multiplied by the number of serially connected panels-n. The short circuit current I_{SC} I_{SC} will be like the photovoltaic current that is the same as the short circuit current of each panel (because of neglecting R_{sh} and R_s). A significantly different V-I curve occurs when panels are under diverse light intensities due to inequality of panel short circuit currents in this case. Due to the presence of bypass diodes, the V-I curve of the entire array will be a composition of different sized rectangles. To illustrate this situation, let us assume that only one between n panels is being lighted differently from $(n-1)$ other panels that are illuminated equally. The graphs of voltage array and array power versus array current ($V_P = \Phi(I_P)$; $P_P = \Psi(I_P)$) are represented in Fig.7(a, b). Fig.7a demonstrates a situation when one panel is illuminated more strongly than $(n-1)$ others; therefore, its I_{SC} is larger than others. Fig.7b represents the opposite, when one panel is less illuminated than the remaining ones. Graph $V_P = \Phi(I_P)$ begins as a straight-line segment parallel to the current axes from the open-circuit voltage $(V_{OC})_n = nV_{OC}$ until the array current will achieve the smallest I_{SC} equal in our denomination to $(I_{pv})_1$.

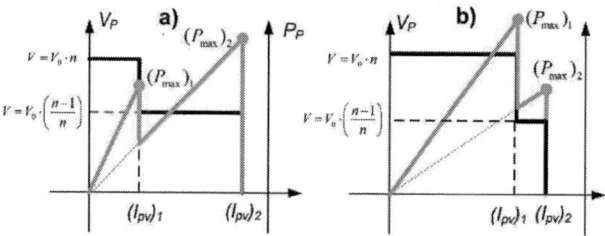

Fig. 5: The V-I and P-I curves for solar array for one panel is irradiated differently than (n-1) equally lighted panels **(a)** the second maximum is GM, **(b)** the first maximum is GM.

This time all bypass diodes are in the reverse bias with zero currents. The situation changes meaningfully after the array current rises above $(I_{pv})_1$. Increasing the current will cause a forward bias in the diode (diodes) of the weakest panel (panels) (in Fig.5a, diode $(D_{bp})_l$; in Fig.5b diodes $(D_{bp})_j$ - $(D_{bp})_k$). An additional portion $(I_p - (I_{pv})_n)$ of current will flow through open diodes and their voltages will be much lower than the panel voltages and could be neglected. Thus, magnitude of output voltage after $(I_{pv})_1$ will drop down either on V_{OC} value (Fig.5a) or on $(n-1)\,V_{OC}$ (Fig.5b). Henceforth, the array output voltage remains constant up to the current that will achieve maximum short circuit current value $(I_{pv})_2$. The array voltage becomes equal to zero after the current increase above $(I_{pv})_2$.

Array power is a multiplication of current-voltage magnitudes. Hence, power dependence versus current for homogeneous array irradiation represents an inclined straight segment beginning from axes commencement and finishing at the point where current is equal to short circuit current value. Maximum power is achieved in the current equal to I_{pv}. The power curve slope is proportional to the voltage value. Therefore, for the solar array with panels dissimilarly lighted, there are a set of straight-line segments with dissimilar slope angles finishing in different current points. Consequently, each segment achieves its own maximum and the entire array has several local maximums. However, only one of them is global. Its location and value are dependent on magnitudes of short circuit currents and on voltage values of differently illuminated panel groups.

The combination of two power straight-lines are shown in Fig. 5a, b. As a result, two power graphs have (in our example) two local maximums. The global maximum (Fig. 5a) is near $(I_{pv})_2$, although in the vicinity of $(I_{pv})_1$ for Fig. 5b.

Methodology for determination of global maximum

The methodology of permanent monitoring is based on our previous considerations. Let us introduce two main assumptions. First, all parameters included in equations (1-4) of the equivalent circuit for solar panels are determined. Second, permanent measurement provides in real-time the accurate knowledge of all panel voltages and they are current, which is equal for all of them since the panels are linked in a series. As a result, a monitoring algorithm was developed and is represented in Fig. 8.

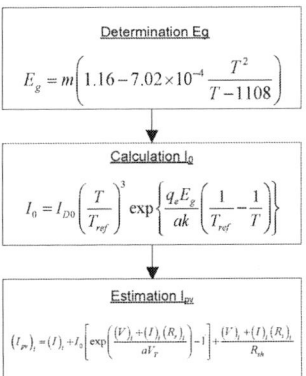

Fig. 6: Algorithm estimation of short circuit currents for solar panels serially connected in solar array.

The V-I and P-I curves are drawn after determination of short circuit currents and all critical voltages for all panels in the solar array. The example is represented in Fig.7.

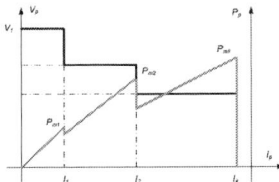

Fig. 7: V-I and P-I curves for solar array with various irradiation levels on different solar panels.

After voltages $V_1...V_K$ and $I_1...I_K$ were found, maximum powers are calculated as:

$$\left(P_m\right)_i = V_i \cdot I_i \qquad (5)$$

The global maximum is chosen as, and its location is close to the Ii that provides maximum power.

Verification of proposed MPPT algorithm

It was done by PSIM soft. Graph below shows the results of MPPT functionality: the algorithm during finding procedure is ignoring intermittent local maximums and move the current search to the GM where it is locked around. The total search time was less than 800ms.

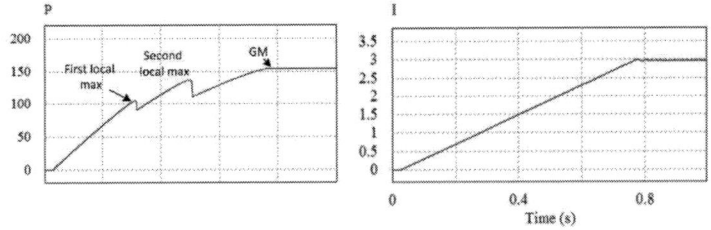

Fig. 8: Results of MPPT algorithm verification.

This sentence will be repeated to represent the body of the text. This is where you continue to write the extended paper, including subparagraphs, sub-sub paragraphs, figures, formulae, tables and eventual images.

Conclusions

The present article describes the original algorithm and based on it an MPPT system for finding GM in partial shading conditions. The essence of the method can be determined as a combination of PV mathematical modeling based on real measurements of panels parameters. During a sequential search of an optimal PV current toward the first maximum on P-I characteristic curve, the control system determines with some precision lower and upper boundaries of a GM. MPPT control shifts the current straightforward to one of these boundaries (depends on a situation) ignoring all local maximums, which can be located on this path, and applies one of the conventional algorithms (P&O or incremental conductance methods) immediately after achievement of one of these points. Further MPPT continue the monitoring of PV parameters constantly determining boundaries of a GM current. If the system will detect that its present location went beyond them, it will shift a current forthwith to one of them.

The functionality of such an MPPT system was proven by a simulation approach on PSIM software. The finding of a GM is very fast and takes time is approximately equal to that of conventional methods. Therefore, the duration of GM search is much shorter than that for most of the methods were designated to work in partial shading.

The exactness and efficiency in the GM search may be like of the conventional methods. In addition, these parameters remain constant when time of GM finding is being decreased. This disadvantage is typical for many methods are suggested for MPPT working in partial shading. This system could be recommended for the implementation in real MPPT appliances require efficiently to operate during partial shading conditions which are escorted by the presence of multiple maximums.

References

[1]   C. Manickam, G. R. Raman, G. P. Raman, S. I. Ganesan and C. Nagamani.:A hybrid algorithm for tracking of GMPP based on P&O and PSO with reduced power oscillation in string inverters, IEEE Transactions on Industrial Electronics, vol. 63, no. 10, pp. 6097-6106 2016.

[2]   J. Ahmed and Z. Salam.: A modified P&O maximum power point tracking method with reduced steady-state oscillation and improved tracking efficiency, IEEE Transactions on Sustainable Energy, vol. 7, no. 4, pp. 1506-1515 2016.

[3]   M. Averbukh, Y. Ben-Galim and A. Uhananov.: Development of a quick dynamic response maximum power point tracking algorithm for off-grid system with adaptive switching (On–Off) control of dc/dc converter, Journal of Solar Energy Engineering, vol. 135, no. 2, pp. 021003 2013.

[4]   A. Al Nabulsi and R. Dhaouadi.: Efficiency optimization of a DSP-based standalone PV system using fuzzy logic and dual-MPPT control, IEEE Transactions on Industrial Informatics, vol. 8, no. 3, pp. 573-584 2012.

[5]   A. El Khateb, N. A. Rahim, J. Selvaraj and M. N. Uddin.: Fuzzy-logic-controller-based SEPIC converter for maximum power point tracking, IEEE Transactions on Industry Applications, vol. 50, no. 4, pp. 2349-2358

[6]   R. Pradhan and B. Subudhi.: Double integral sliding mode MPPT control of a photovoltaic system, IEEE Transactions on Control Systems Technology, vol. 24, no. 1, pp. 285-292 2016.

[7]   A. Costabeber, M. Carraro and M. Zigliotto.: Convergence analysis and tuning of a sliding-mode ripple-correlation MPPT, IEEE Transactions on Energy Conversion, vol. 30, no. 2, pp. 696-706 2015.

[8]   R. B. A. Koad, A. F. Zobaa and A. El-Shahat.: A novel MPPT algorithm based on particle swarm optimization for photovoltaic systems, IEEE Transactions on Sustainable Energy, vol. 8, no. 2, pp. 468-476 2017.

[9]   H. Renaudineau, F. Donatantonio, J. Fontchastagner, G. Petrone, G. Spagnuolo, J. P. Martin and S. Pierfederici.: A PSO-based global MPPT technique for distributed PV power generation, IEEE Transactions on Industrial Electronics, vol. 62, no. 2, pp. 1047-1058 2015.

[10]   S. Mohanty, B. Subudhi and P. K. Ray.: A grey wolf-assisted perturb & observe MPPT algorithm for a PV system, IEEE Transactions on Energy Conversion, vol. 32, no. 1, pp. 340-347 2017.

[11]   N. Karami, N. Moubayed and R. Outbib.: General review and classification of different MPPT techniques, Renewable and Sustainable Energy Reviews, vol. 68, pp. 1-18 2017.

[12]   A. Mohapatra, B. Nayak, P. Das and K.B. Mohanty.: A review on MPPT techniques of PV system under partial shading condition, Renewable and Sustainable Energy Reviews, vol. 80, pp. 854-867 2017.

[13]   M. Seyedmahmoudian, B. Horan, T.K. Soon, R. Rahmani, A.M.T. Oo, S. Mekhilef and A. Stojcevski.: State of the art artificial intelligence-based MPPT techniques for mitigating partial shading effects on PV systems–A review, Renewable and Sustainable Energy Reviews, vol. 64, pp. 435-455 2016.

[14]   M. Metry, M. B. Shadmand, R. S. Balog and H. Abu-Rub.: MPPT of photovoltaic systems using sensorless current-based model predictive control, IEEE Transactions on Industry Applications, vol. 53, no. 2, pp. 1157-1167 2017.

[15]   M. Uoya and H. Koizumi.: A calculation method of photovoltaic array's operating point for MPPT evaluation based on one-dimensional Newton–Raphson method, IEEE Transactions on Industry Applications, vol. 51, no. 1, pp. 567-575 2015.

[16]   A. Kuperman, M. Sitbon, S. Gadelovits, M. Averbukh and T. Suntio, "Single-source multi-battery solar charger: Analysis and stability issues," Energies, vol. 8, no. 7, pp. 6427-6450 2015.

[17] Y. Mahmoud and E. F. El-Saadany.: A novel MPPT technique based on an image of PV modules, IEEE Transactions on Energy Conversion, vol. 32, no. 1, pp. 213-221 2017.

[18] V. Franzitta, A. Orioli and A. Di Gangi, "Assessment of the usability and accuracy of the simplified one-diode models for photovoltaic modules," Energies, vol. 9, no. 12, pp. 1019 2016.

[19] S. Lineykin, M. Averbukh and A. Kuperman.: Issues in modeling amorphous silicon photovoltaic modules by single-diode equivalent circuit, IEEE Transactions on Industrial Electronics, vol. 61, no. 12, pp. 6785-6793 2014.

Survey and Comparison of
1D/2D analytical Models of HF Losses in Litz Wire

Qingchao Meng and Jürgen Biela
ETH Zurich, Laboratory for High Power Electronic Systems (HPE)
Email: meng@hpe.ee.ethz.ch
URL: http://www.hpe.ee.ethz.ch

Keywords

≪Transformer≫, ≪Modelling≫, ≪Magnetic device≫, ≪Design≫, ≪Litz wire≫

Abstract

Litz wire (LW) is essential for reducing HF winding losses in medium frequency transformers and inductors. In order to predict the losses and optimise the parameters of LW, several analytical loss models have been presented. However, a detailed comparison of the models and the accuracy is missing. Therefore, this paper presents a comprehensive comparison of four loss models with respect to general mathematical approach, computational effort, as well as the impact of different parameters on the accuracy of the different models. Guidelines and insights for choosing the most suitable model are also provided.

1 Introduction

Medium frequency transformers and inductors are key components in many power electronic systems. By using a high operation frequency, the size of the transformer can be reduced, thus helping to increase the power density and reduce the costs of power electronic systems. However, the winding losses increase significantly with frequency due to the skin and the proximity effect. Accordingly, it is crucial to calculate and minimize the winding losses during the design of transformers and inductors.

In the medium frequency range, Litz wire (LW) is usually applied to reduce winding losses. To predict the losses in LW, different calculation techniques can be used, including numerical methods, analytical methods, and semi-numerical methods. Commonly used numerical methods such as FEM [1]-[4] and PEEC [5]-[8] are very time consuming and require much computing power due to the complex structure of LW. Therefore, numerical methods are too computationally expensive to be used in the optimization process of magnetics, in which the losses must be calculated numerous times. By contrast, analytical methods allow the losses to be calculated in a few milliseconds and require much less computing power. Furthermore, they provide a physical insight into the generation of eddy-current losses which is useful in the optimization process. Nevertheless, the accuracy of analytical methods is constrained by the winding geometry. By combining numerical and analytical methods results in semi-numerical methods [9]-[12] , which usually use numerical methods to calculate the magnetic field, and then import this magnetic field in the analytical models to calculate the losses. Semi-numerical methods are more computationally efficient than numerical methods and more accurate than analytical methods. However, the calculation time of semi-numerical methods is still several orders of magnitude higher than that of analytical methods and typically too high for a comprehensive converter optimization procedure.

In the past, many analytical models for calculating eddy-current losses have been presented. For example, in [13], [14] the losses of foil winding and other shaped windings are modeled based on a 1D field assumption. In [15], the 2D exact solution is derived for round conductors. These models serve as the theoretical basis for modeling losses in LW. In [16]-[22] three different LW models, which are called M2, M3, and M4 in the following, are derived based on the 2D solution of round conductors. In [23], [24] a LW model is proposed which is based on the Dowell's equations, and which is labelled as M1. All the models (M1-M4) are based on the assumption of perfect twisting, i.e. they neglect any 3D effects. Besides, several 2.5D/3D analytical models [25]-[28] have been proposed, which involve the impact of twisting pattern on losses. Since this paper aims to compare 1D/2D models, the 2.5/3D models are not further considered here. As all 4 selected 1D/2D LW loss models are derived for transformer windings, this paper also considers transformers as application examples. However, the 4 models can also be used for calculating winding losses in inductors. There, the magnetic field calculation must be adapted.

Since a comprehensive evaluation and comparison of the models is missing, it is difficult to choose the most accurate and flexible model for the loss calculation. Furthermore, the impact of different parameters on the model accuracy has not been analysed yet, which makes the scope of application of different models unclear. Therefore, this paper compares different models and investigates how LW parameters affect the model accuracy. The model description and comparison are presented in section 2. The accuracy and the impact of different parameters on the accuracy are calculated by using 2D FEM simulations and given in section 3.

2 Model description and comparison

The model comparison begins by describing the different assumptions, which are the basis of each LW model. In order to compare the models in detail, the expressions of different loss types (skin, internal/external proximity effect) are extracted from the original equations and are reformulated in a comparable form. Furthermore, calculation results of each model for a reference case with the parameters given in Table IV are compared to the results calculated by 2D FEM simulation, and a comparison of the computational effort is presented.

2.1 Basic assumptions

As shown in Fig. 2, the four models can be categorised into two groups: hyperbolic-function (Dowell [14]) based (M1) and Bessel-function (Ferreira [16]-[19]) based (M2-M4). All methods are derived based on the three following assumptions:

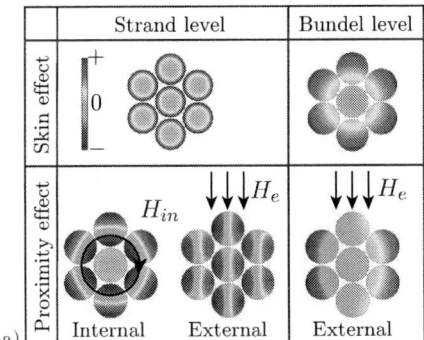 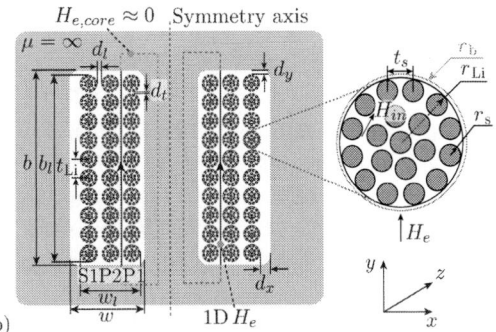

Fig. 1: a) High frequency effects in strand and bundle level. b) Parameter definition. Besides the shown parameters, also packing factor $\beta = n_s \pi r_s^2 / \pi r_{Li}^2$, skin depth $\delta = 1/\sqrt{\pi f \sigma \mu}$, f: Frequency, σ: Conductivity, μ: Permeability, N_l: Number of turns in one layer , d_s: Diameter of a single strand.

- **A1.1: 2D approximation/perfect twisting.**

With the assumption of perfect twisting, an equal current sharing of all strands is given and the flow of eddy currents among different strands is neglected, so that the high frequency effect at the bundle level as shown in Fig. 1a is eliminated (3D → 2D). With the 2D approximation, it is assumed that the winding is a straight, infinite long conductor. A cross-section of such a winding is illustrated in Fig. 1b. Based on this assumption, the loss density changes only in x and y direction. Hence, the total losses are equal to the integral of the loss density over the xy-plane multiplied by the length of turn.

- **A1.2: 1D magnetic field.**

The magnetic field generated by a coil flows through the core and the coil, and forms a closed loop as shown in Fig. 1b. However, the core material usually has a high permeability μ and in models M1-M4 the permeability of the core is assumed to be infinite, which results in zero magnetic field in the core. Hence, only the magnetic field H_e, which is assumed to be 1D, remains in the window.

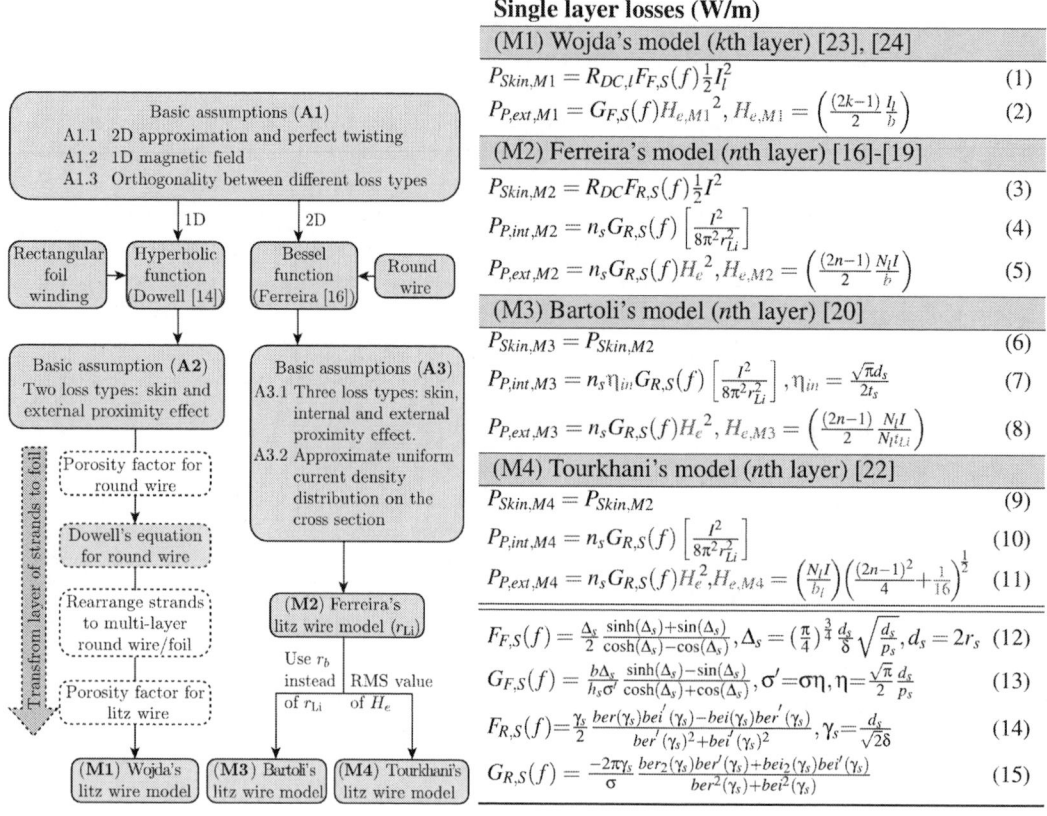

Single layer losses (W/m)

(M1) Wojda's model (*k*th layer) [23], [24]	
$P_{Skin,M1} = R_{DC,l}F_{F,S}(f)\frac{1}{2}I_l^2$	(1)
$P_{P,ext,M1} = G_{F,S}(f)H_{e,M1}^2, H_{e,M1} = \left(\frac{(2k-1)}{2}\frac{I_l}{b}\right)$	(2)
(M2) Ferreira's model (*n*th layer) [16]-[19]	
$P_{Skin,M2} = R_{DC}F_{R,S}(f)\frac{1}{2}I^2$	(3)
$P_{P,int,M2} = n_s G_{R,S}(f)\left[\frac{I^2}{8\pi^2 r_{Li}^2}\right]$	(4)
$P_{P,ext,M2} = n_s G_{R,S}(f)H_e^2, H_{e,M2} = \left(\frac{(2n-1)}{2}\frac{N_l I}{b}\right)$	(5)
(M3) Bartoli's model (*n*th layer) [20]	
$P_{Skin,M3} = P_{Skin,M2}$	(6)
$P_{P,int,M3} = n_s \eta_{in} G_{R,S}(f)\left[\frac{I^2}{8\pi^2 r_{Li}^2}\right], \eta_{in} = \frac{\sqrt{\pi}d_s}{2t_s}$	(7)
$P_{P,ext,M3} = n_s G_{R,S}(f)H_e^2, H_{e,M3} = \left(\frac{(2n-1)}{2}\frac{N_l I}{N_l t_{Li}}\right)$	(8)
(M4) Tourkhani's model (*n*th layer) [22]	
$P_{Skin,M4} = P_{Skin,M2}$	(9)
$P_{P,int,M4} = n_s G_{R,S}(f)\left[\frac{I^2}{8\pi^2 r_{Li}^2}\right]$	(10)
$P_{P,ext,M4} = n_s G_{R,S}(f)H_e^2, H_{e,M4} = \left(\frac{N_l I}{b_l}\right)\left(\frac{(2n-1)^2}{4}+\frac{1}{16}\right)^{\frac{1}{2}}$	(11)
$F_{F,S}(f) = \frac{\Delta_s}{2}\frac{\sinh(\Delta_s)+\sin(\Delta_s)}{\cosh(\Delta_s)-\cos(\Delta_s)}, \Delta_s = (\frac{\pi}{4})^{\frac{3}{4}}\frac{d_s}{\delta}\sqrt{\frac{d_s}{p_s}}, d_s = 2r_s$	(12)
$G_{F,S}(f) = \frac{b\Delta_s}{h_s\sigma'}\frac{\sinh(\Delta_s)-\sin(\Delta_s)}{\cosh(\Delta_s)+\cos(\Delta_s)}, \sigma'=\sigma\eta, \eta=\frac{\sqrt{\pi}}{2}\frac{d_s}{p_s}$	(13)
$F_{R,S}(f) = \frac{\gamma_s}{2}\frac{ber(\gamma_s)bei'(\gamma_s)-bei(\gamma_s)ber'(\gamma_s)}{ber'(\gamma_s)^2+bei'(\gamma_s)^2}, \gamma_s = \frac{d_s}{\sqrt{2}\delta}$	(14)
$G_{R,S}(f) = \frac{-2\pi\gamma_s}{\sigma}\frac{ber_2(\gamma_s)ber'(\gamma_s)+bei_2(\gamma_s)bei'(\gamma_s)}{ber^2(\gamma_s)+bei^2(\gamma_s)}$	(15)

Fig. 2: Model assumptions, model M1 is based on assumptions A1 and A2 and models M2-M4 are based on assumptions A1 and A3.

Table I: Formulas of different loss types. The equations are reformulated to be written in the same form. The differences between the equations are highlighted.

- **A1.3: Orthogonality between skin and proximity effect.**

This assumption has been well interpreted in [16]. As the current density J_s reflecting the skin effect is an even function and the current density J_p describing the proximity effect is an odd function, the integral of the product of both current density functions over the cross section of a conductor is zero. Consequently, the skin effect is orthogonal to the proximity effect, and both loss types can be calculated separately.

2.2 Ferreira's Litz wire model (M2) [16]-[19]

Model M2 is based on the 2D model of round conductors presented in [15]. In a perfect twisted LW, each strand conducts the same alternating current, which causes skin effect losses in the strand. Furthermore, each strand is exposed to the internal magnetic field H_{in} as shown in Fig. 1a, which is caused by the current in the strands of the same bundle. For LW with a high packing factor β at a frequency where $r_s < \delta$, the current density can be assumed to be uniformly distributed in the cross section of the LW (A3.2). Hence, the internal magnetic field is equal to $\vec{H}_{in}(r) = \frac{I}{2\pi r_{Li}^2}r\vec{e}_\varphi$.

Besides $\vec{H}_{in}$, each strand is also exposed to the magnetic field caused by other litz bundles, which is called external magnetic field $\vec{H}_e$. Based on assumption A1.2 $\vec{H}_e$ is 1D. As

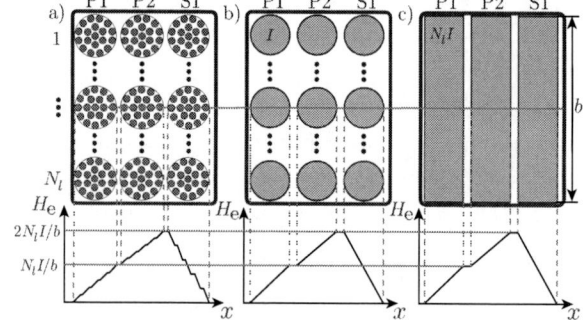

Fig. 3: H_e in LW, solid, and foil windings. In a) and b) each turn carries the same current I, in c) each foil layer carries the current $N_l I$. The winding height is equal to the window height.

shown in Fig. 3b and 3c, the 1D magnetic field in a solid wire winding is equal to their equivalent foil winding with the same winding height, and the increment of the magnetic field in a single layer is $\Delta H_e = N_l I/b$. In LW winding, the magnetic field increases in the strands and remains constant in the insulation between strands as shown in Fig. 3a. However, the increment of the magnetic field in a single layer LW winding is equal to the solid wire winding, as each turn of LW conducts the same current as the solid wire. Therefore, the total field, including $\vec{H}_{in}$ and $\vec{H}_e$, is equal to $\vec{H} = H_{in}\cos(\varphi)\vec{e}_x + (H_{in}\sin(\varphi) + H_e)\vec{e}_y$. The loss density caused by the proximity effect can be calculated with $P'_p = \frac{1}{\pi r_{Li}^2}G(f)H(r,\varphi)^2$. The losses caused by the internal and the external proximity effect P_p is equal to the integral of the loss density over the cross-section of LW:

$$P_p = \frac{n_s G_{R,S}(f)}{\pi r_{Li}^2}\int_0^{2\pi}\int_0^{r_{Li}}\left(H_{in}^2(r) + H_e^2 + 2H_{in}(r)H_e\sin(\varphi)\right)r\,dr\,d\varphi \tag{16}$$

$$= \underbrace{n_s G_{R,S}(f)H_e^2}_{P_{P,ext}} + \underbrace{\frac{n_s G_{R,S}(f)}{\pi r_{Li}^2}\int_0^{2\pi}\int_0^{r_{Li}}H_{in}^2(r)r\,dr\,d\varphi}_{P_{P,int}} \tag{17}$$

From (17) it can be concluded that $P_{P,ext}$ and $P_{P,int}$ are orthogonal, since the integral over the cross-section of $2H_{in}H_e\sin(\varphi)$ is equal to zero. There, the average H_e for the nth layer $H_{e,M2} = \frac{H_{left}+H_{right}}{2} = \frac{(2n-1)}{2}\frac{N_l I}{b}$ is used for calculating the losses caused by external proximity effect. The equations for each loss type are given in Table I (3)-(5).

2.3 Bartoli's Litz wire model (M3) [20], [21]

Model M3 is based on assumptions A1/A3. The losses caused by the skin effect are the same as for model M2. Furthermore, model M3 adopts the result for the losses caused by the internal proximity effect from model M2 and modifies this result with a self-defined internal porosity factor $\eta_{in} = \frac{\sqrt{\pi}d_s}{2t_s}$. This internal porosity factor takes the influence of distance from strand to strand into account, which is equivalent to extending the LW radius from r_{Li} to r_b (highlighted in (18)). The equation for $P_{P,int,M3}$ is given by:

$$P_{P,int,M3} = n_s\eta_{in}^2 G_{R,S}(f)\frac{I^2}{8\pi^2 r_{Li}^2} = n_s G_{R,S}(f)\frac{I^2\beta}{8\pi n_s t_s^2} = n_s G_{R,S}(f)\left[\frac{I^2}{8\pi^2 r_b^2}\right], \quad n_s t_s^2 = \beta\pi r_b^2 \tag{18}$$

For the losses caused by the external proximity effect, the AC resistance $R_{AC,ext}$ is derived in [20]. To obtain the loss expression in the form of $G(f)H_e^2$, the amplitude of the external field $\vec{H}_{e,M3}$ for the nth layer must be calculated and is given in (19), where t_{Li} is defined in Fig. 1. The expression for each loss type is given in Table I (6)-(8).

$$H_{e,M3} = \sqrt{\frac{\frac{1}{2}I^2 R_{AC,ext}}{G_{R,S}(f)n_s}} = \frac{N_l I}{2N_l t_{Li}} \text{ (for the 1st layer)}, \quad H_{e,M3} = \frac{(2n-1)}{2}\frac{N_l I}{N_l t_{Li}} \text{ (for the nth layer)} \tag{19}$$

$$R_{AC,ext} = R_{DC,s}\frac{\gamma_s}{2}\left(-2\pi n_s\eta_{ex}^2\frac{ber_2(\gamma_s)ber'(\gamma_s) + bei_2(\gamma_s)bei'(\gamma_s)}{ber^2(\gamma_s) + bei^2(\gamma_s)}\right), \quad \eta_{ex} = \frac{2r_s}{t_{Li}}\sqrt{\frac{\pi}{4}} \tag{20}$$

2.4 Tourkhani's Litz wire model (M4) [22]

Model M4 is based on Ferreira's model [16]-[19]. The only difference between model M4 and M2 is the calculation of the 1D magnetic field. As shown in Fig. 4, a linearly increasing magnetic field is considered in model M4 which is similar to the magnetic field in model M2. However, instead of using the average amplitude of H_e for each layer, the RMS value of H_e is utilized. As shown in Fig. 4, the amplitude of $\vec{H}_e$ of a arbitrary strand in a LW of the nth layer depends on its position in x-direction, which is given by (21). As in model M4 the current per winding length is used for calculating H_e, the winding height b_l which is highlighted in (21), is used instead of the window height b.

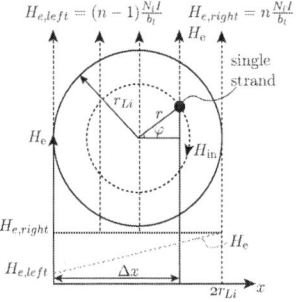

Fig. 4: 1D magnetic field H_e for a LW turn in the n-th layer.

$$H_e(n,r,\varphi) = H_{e,left} + \frac{\Delta x}{2r_{Li}}\Delta H_e = \left(n - \frac{1}{2} + \frac{r\cos\varphi}{2r_{Li}}\right)\frac{N_l I}{b_l} \tag{21}$$

Moreover, model M4 uses the same expressions for the losses caused by the skin and the internal proximity effect as model M2. Substituting $H_e(n,r,\varphi)$ in (16), it can be concluded that the losses caused by the internal and the external proximity effect are orthogonal, as $\int_0^{2\pi}\int_0^{r_{Li}} 2H_{in}H_e \sin(\varphi)rdrd\varphi = 0$. Furthermore, the factor $G_{R,S}(f)$ is a function of frequency for a given LW, thus only $H_e(n,r,\varphi)^2$ need to be integrated over the cross-section of LW. Finally, the integral of losses per unit area is converted to an integral of H_e^2 over the cross-section of LW, and hence, the RMS value of $\vec{H}_e$, as given in (22), is used in model M4 for calculating $P_{P,ext}$ instead of the average value of H_e.

$$H_{e,RMS} = \frac{N_l I}{b_l}\sqrt{\frac{1}{\pi r_0^2}\left(\int_0^{2\pi}\int_0^{r_0}(H_e(n,r,\varphi))^2 rdrd\theta\right)} = \frac{N_l I}{b_l}\sqrt{\left(\frac{2n-1}{2}\right)^2 + \frac{1}{16}} = H_{e,M4} \tag{22}$$

The equations for each loss type of model M4 are given in Table I (9)-(11). Although the equations in [22] are given in a different form, the same numerical results can be obtained.

2.5 Wojda's Litz wire model (M1) [23], [24]

Model M1 is based on the 1D foil model presented in [13] and [14]. Wojda extends Dowell's concept of transforming a round wire to a foil to transforming LW to a foil winding. As shown in Fig. 5, the strands are first rearranged into multi-layer round wires, which can be further transformed into the resistance equivalent square wire and finally a foil. Since the DC resistance must be kept constant during the transformation, the porosity factor has been proposed for each transformed winding. The rearrangement of LW leads to $\sqrt{n_s}$ times more layers and $\sqrt{n_s}$ times more turns than the LW winding. With the transformed foil also the internal proximity effect is eliminated (see assumption A2).

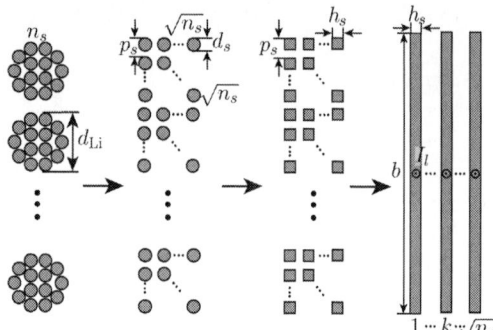

Fig. 5: Transformation of LW into a multi-layer foil winding.

Winding	η	Modified Δ
Foil	1	$\Delta = \frac{h}{\delta}$
Square wire	$\frac{h}{p}$	$\Delta_{sq} = \frac{h}{\delta}\sqrt{\frac{h}{p}}$
Round wire	$\frac{\sqrt{\pi}}{2}\frac{d}{p}$	$\Delta_r = \left(\frac{\pi}{4}\right)^{\frac{3}{4}}\frac{d}{\delta}\sqrt{\frac{d}{p}}$
Litz wire	$\frac{\sqrt{\pi}}{2}\frac{d_s}{p_s}$	$\Delta_s = \left(\frac{\pi}{4}\right)^{\frac{3}{4}}\frac{d_s}{\delta}\sqrt{\frac{d_s}{p_s}}$

Table II: Porosity factor η and modified penetration ratio Δ for different winding types. h, d, and p represent the thickness of square wire, diameter of round wire and center point distance of 2 adjacent turns respectively.

In order to obtain the equations for the losses caused by the skin P_{Skin} and the proximity effect $P_{P,ext}$ separately. The equations for calculating the losses of LW given in [23] [24] are reformulated based on the equations derived in [29] for calculating the losses of the foil winding. As in [29], the losses P_{Skin} and $P_{P,ext}$ are derived based on the boundary condition for the skin and the proximity effect respectively. The equations are rewritten in the same form as the Bessel-function based models and are given by:

$$P_{Skin} = R_{DC}\left[\frac{\Delta}{2}\frac{\sinh(\Delta)+\sin(\Delta)}{\cosh(\Delta)-\cos(\Delta)}\right]\frac{1}{2}I^2 = R_{DC}F_F(f)\frac{1}{2}I^2 \tag{23}$$

$$P_{P,ext} = \frac{b\Delta}{h\sigma}\frac{\sinh(\Delta)-\sin(\Delta)}{\cosh(\Delta)+\cos(\Delta)}H_p^2 = G_F(f)H_e^2, H_e = \frac{1}{2}\left(H_{e,left}+H_{e,right}\right) \tag{24}$$

In this paper, the number of turns in each layer, the current in a single LW, and the resistance per unit length of a single strand are defined as N_l, I and $R_{DC,s}$ respectively. Therefore, the current in each transformed foil is $I_l = IN_l/\sqrt{n_s}$. The DC resistance of each transformed foil is $R_{DC,l} = R_{DC,s}/(N_l\sqrt{n_s})$ and the average amplitude of the magnetic field of the kth transformed foil layer is $H_{e,M1} = \frac{(2k-1)}{2}\frac{I_l}{b}$. Substituting I_l, $R_{DC,l}$, $H_{e,M1}$, the porosity factor η for LW, and the modified penetration ratio Δ_s from Table II, in (23) and (24), the losses of LW caused by the skin and the proximity effect given in Table I (1) and (2) can be derived. Furthermore, the overall losses of a n layer LW winding is equal to the losses of an $m = n\sqrt{n_s}$ layer transformed foil winding as given in (25).

$$P_{total,M1} = mP_{Skin,M1} + \sum_{k=1}^{k=m} P_{P,ext,M1,k} = mR_{DC,l}F_{F,S}(f)\frac{1}{2}I_l^2 + G_{F,S}(f)\left(m + \frac{4m(m^2-1)}{3}\right)\left(\frac{I_l}{2b}\right)^2 \quad (25)$$

2.6 Comparison of 4 Litz wire models

In the following, the general mathematical differences and the computational effort of the 4 considered models are compared.

• General mathematical differences

In Fig. 6, the losses for the 2 layer primary winding with the parameters given in Table IV (reference case) is shown as an example. In general, it is only reasonable to use LW in a frequency range where the losses caused by the LW are lower than the losses of an equivalent round wire. In [30], a frequency range is given by $0.394 < r_s/\delta < 0.881$ for $7 \leq n_s \leq 175$ (excluding the external proximity effect). However, the frequency range becomes smaller if the external proximity effect is also considered. In Fig. 6b the loss ratios $\eta_p = P_{Litz}/P_{round}$ are given for different number of strands and their equivalent round wire. The range of penetration ratio $\Delta = d_s/\delta$, where the loss ratio $\eta_p < 1$, decreases with the increasing number of strands. In order to utilize the maximal advantage of LW, LW is usually applied around the optimum Δ_{op} at which the power loss ratio η_p is the smallest. Therefore, in this paper the considered frequency range for the loss calculation is set to $0.1 \leq \Delta \leq 1.2$, which covers all relevant Δ_{op} for the LW with $n_s \geq 19$.

As shown in Fig. 6a, the losses caused by the skin effect P_{Skin} are almost constant and equal to the DC losses of LW, since Δ is very low and the skin effect is not significant. In the zoomed figure of P_{Skin}, model M1 results in slightly higher skin effect losses than models M2-M4, but this difference is neglectable. However, the losses caused by the proximity effect are different for the four models, which can be explained by their mathematical expressions. In model M1, $P_{P,int}$ does not exist, because of the transformation of the strands to a foil. Model M3 uses r_b instead of r_{Li} to calculate $P_{P,int}$ and $r_b > r_{Li}$, so that $P_{P,int,M2} = P_{P,int,M4} > P_{P,int,M3}$. Furthermore, $P_{P,ext}$ of model M1 is not comparable to models M2-M4 based on mathematics, since model M1 calculates $P_{P,ext}$ by transforming LW into a multi-layer foil winding. The differences between models M2-M3 are the increment of the external magnetic field pro layer ΔH_e and the value of the external magnetic field amplitude, as shown in Table III. Because $H_{e,M2} \leq H_{e,M3} < H_{e,M4}$, if the distance between two adjacent turns d_t is zero, the losses caused by external proximity effect can be sorted as $P_{P,ext,M2} \leq P_{P,ext,M3} < P_{P,ext,M4}$. The major differences of the total losses of models M1-M4 are dominated by the losses caused by external proximity effect, as P_{Skin} is almost the same for all four models, and $P_{P,int}$ is relatively small compare to the other loss types.

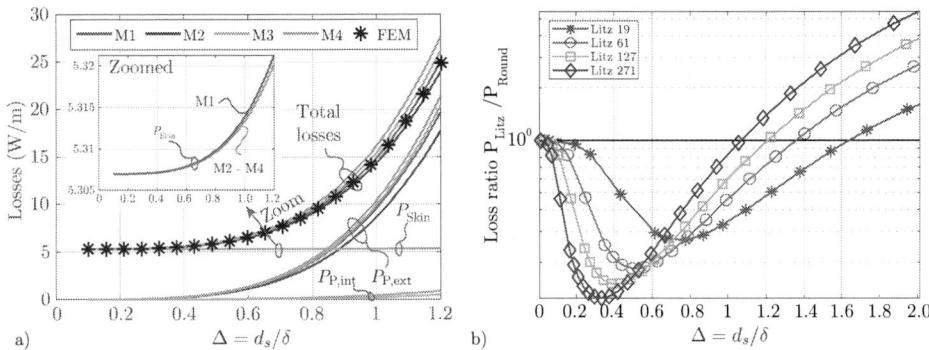

Fig. 6: a) Losses calculated for the reference case given in Table IV (Reference case). b) Power losses ratio for LW with different n_s (19-271) and their equivalent round wire with the same DC resistance. The power loss ratio is calculated by 2D FEM simulations based on the parameters given in Table IV (Reference case). As the DC resistance of the equivalent round wire is fixed, the strands become thinner when n_s increases.

• Computational effort

The calculation time t_{cal} is measured by using the function 'timeit' from Matlab on the desktop with the CPU i7-7700 and the 16 GB RAM, as given in Table III. For each model, 100 frequency points are calculated, and the final calculation time is the mean value of t_{cal} for the 100 results. As shown in Table III, t_{cal} of model M1 is one order of magnitude smaller than the Bessel-function based models M2-M4, because the Bessel function is an open form function with higher calculation efforts than the hyperbolic function. As the differences in the mathematical expressions of models M2-M4 are only the coefficients of the Bessel-function, t_{cal} of models M2-M4 is the same.

Table III: Comparison table for different LW models, the parameters used in models M1, M3 and M4 which are different from model M2 are highlighted in red.

Differences	M1	M2	M3	M4
Basic function	Hyperbolic	Bessel	Bessel	Bessel
Loss types	$P_{Skin}+P_{P,ext}$	$P_{Skin}+P_{P,int}+P_{P,ext}$	$P_{Skin}+P_{P,int}+P_{P,ext}$	$P_{Skin}+P_{P,int}+P_{P,ext}$
Effective layers	$n\sqrt{n_s}$	n	n	n
$H_{in}(r)$	—	$H_{in}(r)=\frac{Ir}{2\pi r_{L_i}^2}$	$H_{in}(r)=\frac{Ir}{2\pi r_{l_i}^2}$	$H_{in}(r)=\frac{Ir}{2\pi r_{L_i}^2}$
Increase in H_e pro layer (ΔH_e)	$\frac{I}{b}$	$\frac{N_l I}{b}$	$\frac{N_l I}{N_l l_{l_i}}$	$\frac{N_l I}{b_{l_i}}$
Value type of H_e	Average	Average	Average	RMS
$t_{cal}/\mu s$ (i7-7700/4 cores/4.2GHz)	12 ± 1	175 ± 5	175 ± 5	175 ± 5
Comparison of P_{Skin}		$P_{Skin,M1}\approx P_{Skin,M2}=P_{Skin,M3}=P_{Skin,M4}$		
Comparison of $P_{P,int}$		$P_{P,int,M2}=P_{P,int,M4}>P_{P,int,M3}$		
Comparison of $P_{P,ext}$		$P_{P,ext,M2}\leq P_{P,ext,M3}<P_{P,ext,M4}$, (if $d_t=0$) $P_{P,ext,M4}-P_{P,ext,M3}=G_{R,S}(f)\frac{1}{16}\Delta H_e^2$, (if $d_t=0$)		

3 Accuracy and scope of application

In this section, the accuracy of the four models is investigated based on 2D FEM simulations, since all considered models are 2D. For obtaining reliable results from FEM, the meshing process plays an important role. Because of the skin effect, the current concentrates on the surface of the conductor. Hence, the mesh near the conductor surface must be very fine. In this paper, the edge length of mesh elements near the surface is maximal 1/5 of the smallest skin depth in the considered frequency range.

Table IV: Parameters of the considered cases and for scaling. n_P: Number of layers in primary winding, n_S : Number of layers in secondary winding, k_{cu} : Copper fill factor. The selected diameters of a single strand $2r_s$ are based on the standard EN 60317-11. Turn distance $d_t=0$ for transformer cases 1-13.

Core size	N_p	N_s	n_P	n_S	$2r_s$(mm)	n_s	k_{cu}	b/b_l	d_y/d_x	w/w_l
① EE16-Z	6	6	1	1	0.2	19	0.1824	1.2074	1.69	1.36
② EE25/19-Z	10	5	2	1	0.25	19	0.1789	1.5015	4.38	1.19
③ EE30/30/7-Z	14	7	2	1	0.3	19	0.2259	1.4189	22.24	1.04
④ EE35-Z	10	5	2	1	0.355	19	0.1874	1.6698	10.09	1.11
⑤ EE40-Z	18	9	2	1	0.3	19	0.2118	1.1368	1.05	1.39
⑥ EE42/42/15-Z	12	6	2	1	0.4	19	0.1616	1.9514	14.61	1.13
⑦ EE50-Z	30	20	3	2	0.25	19	0.1866	1.4933	6.67	1.15
⑧ EE55/55/21-Z	20	10	2	1	0.5	19	0.2897	1.1881	7.62	1.08
⑨ EE70/91/19	30	15	2	1	0.5	37	0.3176	1.0761	3.85	1.10
⑩ EE90/56/16	28	14	4	2	0.4	19	0.1736	1.6781	3.54	1.22
⑪ EE70/108/31N	40	20	2	1	0.4	37	0.2705	1.1918	10.79	1.12
⑫ EE80/76/20	33	22	3	2	0.4	37	0.2415	1.3923	11.97	1.07
⑬ UU120/160/20	36	54	2	3	0.5	37	0.1108	1.2630	0.56	2.68
Scaling of number of layers n ($d_t=0$)										
EE90/56/16	$7n_P$	$7n_S$	1:1:6	1	0.355	19	0.1736	1.6781		
Scaling of number of strands n_s ($d_t=0$)										
E40-Z	18	9	2	1	$1/\sqrt{n_s}$	19-397	0.1238			
Reference case ($d_t=0$)										
EE55/55/21-Z	20	10	2	1	0.46	19	0.2462	1.29	7.62	1.17
Frequency	Copper fill factor: k_{cu}		Aspect ratio of the window: b/w			Packing factor for LW: β				
$\Delta=0.1:0.1:1.2$	0.1 - 0.4		1.6 - 7.93			0.5 - 0.6				

EPE'20 ECCE Europe

Assigned jointly to the European Power Electronics and Drives Association & the Institute of Electrical and Electronics Engineers (IEEE)

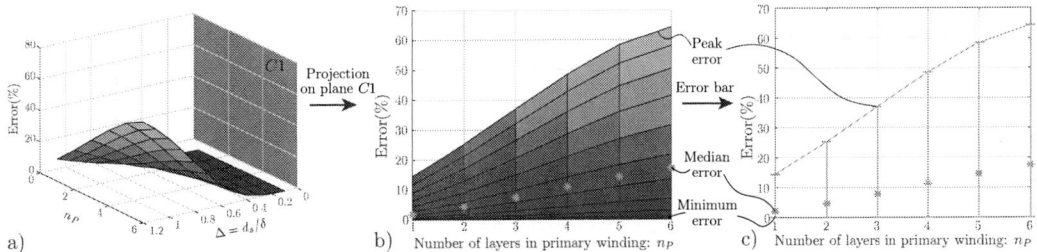

a)   b) Number of layers in primary winding: n_P c) Number of layers in primary winding: n_P

Fig. 7: Transformation of the 3D figure (Error vs. Δ vs. n_P) to a 2D error bar figure for model M4 (Tourkhani's model). The 3D figure a) is projected on the plane $C1$, the minimum , the median and the peak error are extracted from the projection b), which results in the error bar figure c).

In order to compare the accuracy of the 4 models more comprehensively, many different design cases are considered. First, 13 different cores from EE16-Z to UU120/160/20 are selected which cover a large range of core sizes. For each core, the number of turns and number of layers in the primary and the secondary winding are defined as given in Table IV. The diameter and the number of strands are selected based on the standard EN 60317-11. The chosen packing factor for the selected LW is between $0.5 - 0.6$, the chosen copper fill factor of the window k_{cu} is between $0.1 - 0.4$, and the aspect ratio b/w of the selected core windows is between $1.6 - 7.93$. Furthermore, the parameters of the cases for investigating the impact of the number of layers n and the number of strands n_s on the accuracy, are given in Table IV. Besides, for isolation purpose, the insulation thickness of strands t_{iso} and the distance between the primary and the secondary winding d_{PS} are also modified. According to the standard EN 60317-11, there are three typical thickness values for t_{iso}. However, the difference between these three kinds of thickness is so small, that the impact of changing t_{iso} on the accuracy of the loss calculation is neglectable. Furthermore, increasing d_{PS} does not change the magnetic field in the winding, so that changing d_{PS} does not affect the accuracy neither.

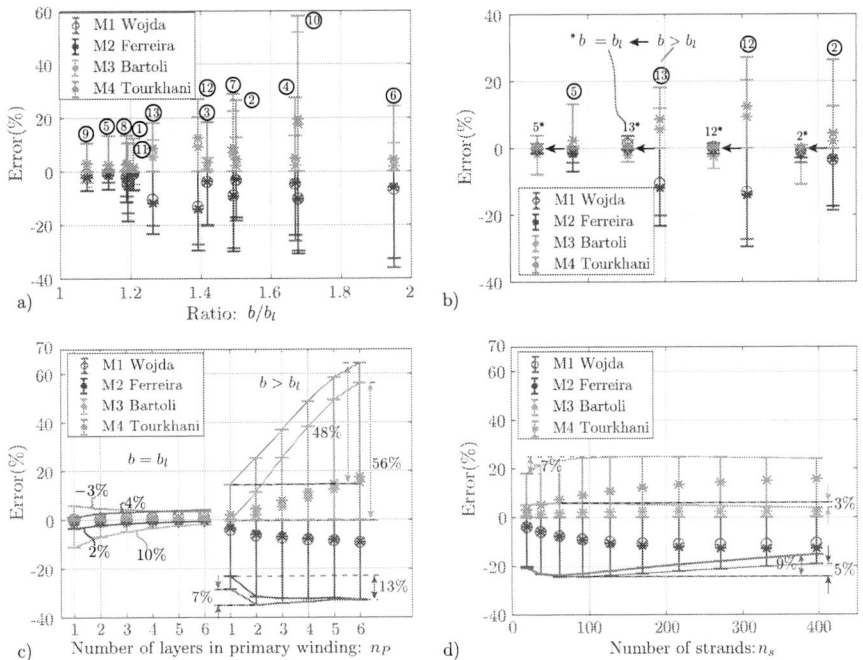

Fig. 8: a) Error bar for the transformer cases 1 -13 with $d_t = 0$ and $b > b_l$ for all cases (Transformer case 10 has in total 6 layers). b) Comparison of the error bar for selected transformers cases, where the cases highlighted by $*$ have the same core size and winding configuration, but with $d_t > 0$, $b = b_l$ and $w = w_l$. c) Error bar for scaling the number of layers for the transformer case 10 with $b = b_l$ as well as with $b > b_l$. d) Error bar for scaling the number of strands based on the transformer case 5.

Fig. 7 shows the error of scaling the number of layers, where the error is calculated by $(P_{M4} - P_{FEM})/P_{FEM}$ and the losses P_{M4} and P_{FEM} are calculated for the primary and the secondary winding. To show the error of the four models in the same figure comprehensively, the 3D results in Fig.7a are transformed to a 2D error bar figure as shown in Fig.7c. There, the calculation starts at frequency $\Delta = 0.1$, which is so small, that the losses cause by high frequency effect are almost zero and all error bars start from 0 and increase with frequency. The considered error bar in Fig. 7 which is shown as an example presents, how the error changes with the number of layers n at different frequencies.

In Fig. 8a, the error as function of the ratio b/b_l for the 13 considered transformer cases is shown. The absolute error of the 4 models increases with the ratio b/b_l. Since the winding height b_l is smaller than window height b, the magnetic field H_e becomes more 2D and also the amplitude of the magnetic field H_e decreases as shown in Fig. 9b and 9d. As models M1-M4 are all based on a 1D field assumption, they can not predict the losses $P_{P,ext}$ accurately when H_e becomes more 2D, what results in a higher error. The error of some transformer cases with a similar ratio b/b_l, such as transformer cases 4 and 10 in Fig. 8a, is different, because other winding parameters of these transformer cases, e.g. the number of layers n, are different.

Fig. 8b shows the comparison of selected transformer cases with $b > b_l$ and with $b = b_l$. The transformer cases highlighted by $*$ are transformed by extending the turn and the layer distance until $b = b_l$ and $w = w_l$ as shown in Fig. 9a. The error typically becomes smaller with such a transformation. Since the magnetic field for the windings which reach up to the yoke, is almost 1D and the length of the field path in the window is approximately equal to the window height b. Therefore, the field amplitude H_e is close to the 2D FEM simulation results, what results in a smaller error. However, by increasing the turn distance d_t, also the calculated amplitude of the magnetic field is reduced in case of model M3 ($\Delta H_{e,M3} = {}^{N_l l}/{}_{N_l t_{Li}}$). With the reduced magnetic field also the calculated losses for the external proximity effect are reduced, so that the error of model M3 typically increases with an increasing turn distance d_t. This effect can outweigh the reduced error caused by the smaller distance between the yoke and the winding, so that the total error increases as this is true for the transformer case 5 in Fig. 8b.

In Fig. 8c, the error for scaling the number of layers for transformer case 10 with $b = b_l$ and with $b > b_l$ is shown. The error for transformer cases with $b > b_l$ increases and is significant bigger than for the transformer case with $b = b_l$. Because $b > b_l$, the magnetic field is more 2D as can be seen in Fig. 9b and 9d. The loss share caused by the external proximity effect $P_{P,ext}$ increases, and the error is dominated by the error caused by the calculation of $P_{P,ext}$. For the transformer cases with $b = b_l$ the error is generally much smaller. Since $b = b_l$, the magnetic field is more 1D as shown in Fig. 9a and 9c, and the 4 models predict the losses more accurately what results in a smaller error.

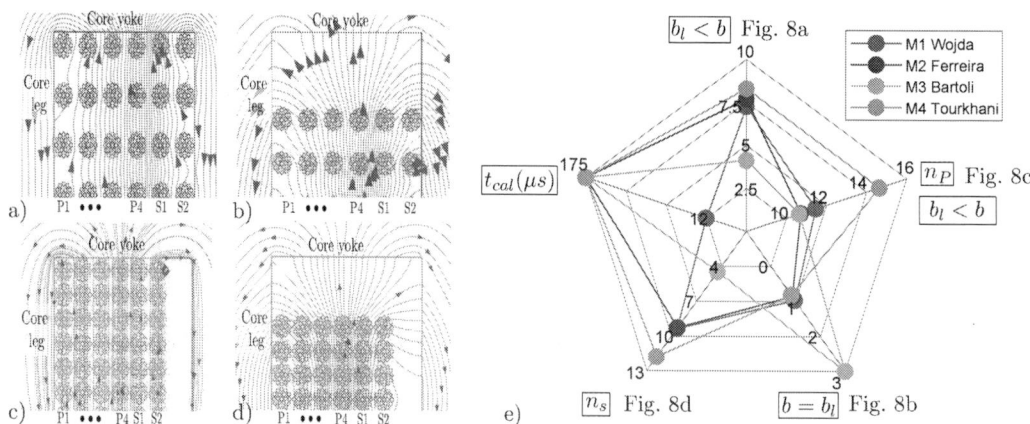

Fig. 9: a) Magnetic field lines for the transformer case 10 with $d_t > 0$, $b = b_l$ and $w = w_l$. b) Magnetic field lines for the transformer case 10 with $d_t > 0$, $b > b_l$ and $w = w_l$. c) Magnetic field lines for the transformer case 10 with $d_t = 0$, $b = b_l$. d) Magnetic field lines for the transformer case 10 with $d_t = 0$, $b > b_l$. The frequency for $a - d$ is $\Delta = 1.2$. e) Radar diagram for the error performance of the 4 models. The values of 5 different parameters except for t_{cal} are given in percentage and indicate the average error over the whole frequency and scaling range.

In Fig. 8d, the error for scaling the number of strands is shown. Since the DC resistance per unit length of the LW is kept constant during scaling, the radius of the strands becomes thinner, when the number of strands increases. The error of models M1-M4 only slowly increases with the number of strands, because increasing the number of strands does not affect the magnetic field H_e significantly.

3.2 Comparison of model performance

Fig. 9e shows the calculation time and the average error over the frequency and the considered scaling cases given in Fig. 8. Models M1-M4 predict the losses more accurately for transformers with $b_l = b$ than for transformers with $b_l < b$, since the magnetic field H_e is more 1D in the transformers with $b_l = b$. The average error of all models as shown in Fig. 9e ($b = b_l$) is below 3%. Furthermore, model M1 has the lowest calculation time and is 10 times faster than the Bessel-function based models M2-M4.

Conclusion

This paper compares four different 1D/2D analytical HF loss models of litz wire. The differences of the models are clearly indicated in terms of general mathematical approach, computational effort, and the impact of different parameters on the accuracy. Based on the comparison results, it can be concluded that models M1-M4 can predict the losses relative accurately, if the winding height b_l is close to the window height b. However, if $b_l < b$, the magnetic field becomes more 2D and all four considered models do not predict the losses accurately. Furthermore, the hyperbolic-function based model M1 is almost 10 times faster than the Bessel-function based models. In case that computing time is critical, model M1 is recommended, since its accuracy is comparable to the one of the other models.

References

[1] P. Reddy, T. Jahns, and T. Bohn, "Transposition effects on bundle proximity losses in high-speed PM machines," *2009 IEEE Energy Conv. Cong. and Expo.*, 2009.

[2] E. Plumed, J. Acero, I. Lope, and C. Carretero, "3D finite element simulation of litz wires with multilevel bundle structure," in *Proc. Conf. of the IEEE Ind. Electron. Society IECON*, pp. 3479–3484, Oct. 2018.

[3] A. Rosskopf, E. Bar, and C. Joffe, "Influence of inner skin- and proximity effects on conduction in litz wires," *IEEE Trans. on Power Electron.*, vol. 29, no. 10, p. 5454–5461, 2014.

[4] X. Nan and C. R. Sullivan, "An equivalent complex permeability model for litz-wire windings," in *IEEE Ind. Appl. Society Annual Meeting*, vol. 3, pp. 2229–2235 Vol. 3, Oct. 2005.

[5] R. Y. Zhang, Jacob K.White and C. R.Sullivan, "Real litz wire characterization using fast numerical simulations," *IEEE Applied Power Electro. Conf. and Expo. - APEC*, 2014.

[6] R. Y. Zhang, J. K. White, and J. G. Kassakian, "Fast simulation of complicated 3-D structures above lossy magnetic media," *IEEE Trans. on Magnetics*, vol. 50, no. 10, pp. 1–16, Oct. 2014.

[7] A. Roskopf, E. Bar, C. Joffe, and C. Bonse, "Calculation of power losses in litz wire systems by coupling FEM and PEEC method," *IEEE Trans. on Power Electron.*, vol. 31, no. 9, p. 6442–6449, 2016.

[8] T. Guillod, J. Huber, F. Krismer, and J. W. Kolar, "Litz wire losses: Effects of twisting imperfections," *Workshop on Control and Modeling for Power Electronics (COMPEL)*, 2017.

[9] C. R. Sullivan, "Computationally efficient winding loss calculation with multiple windings, arbitrary waveforms, and two-dimensional or three-dimensional field geometry," *IEEE Trans. on Power Electron.*, vol. 16, no. 1, pp. 142–150, Jan. 2001.

[10] A. D. Podoltsev, I. N. Kucheryavaya, and B. B. Lebedev, "Analysis of effective resistance and eddy-current losses in multiturn winding of high-frequency magnetic components," *IEEE Trans. on Magnetics*, vol. 39, no. 1, pp. 539–548, Jan. 2003.

[11] A. Roßkopf, C. Joffe, and E. Bär, "Calculation of ohmic losses in litz wires by coupling analytical and numerical methods," in *Proc. Int. Electric Drives Production Conf. (EDPC)*, Sep. 2014.

[12] J. Liu, Q. Deng, D. Czarkowski, M. K. Kazimierczuk, H. Zhou, and W. Hu, "Frequency optimization for inductive power transfer based on ac resistance evaluation in litz-wire coil," *IEEE Trans. on Power Electron.*, vol. 34, no. 3, p. 2355–2363, 2019.

[13] E. Bennett and S. C. Larson, "Effective resistance to alternating currents of multilayer windings," *Electrical Engineering*, vol. 59, no. 12, p. 1010–1016, 1940.

[14] P.Dowell, "Effects of eddy current in transformer windings," *Proc. of the IEEE*, vol. 113, no. 8, pp. 1387–1394, 1966.

[15] M. J.Lammeraner, *Eddy Current.*   ILIFFE BOOKS LTD.LONDON, 1966.

[16] J. A. Ferreira, "Power conditioning in electronic circuits," *Electromagnetic Modelling of Power Electronic Converters*, p. 83–85, 1989.

[17] J. Ferreira, "Improved analytical modeling of conductive losses in magnetic components," *IEEE Trans. on Power Electron.*, vol. 9, no. 1, pp. 127–131, 1994.

[18] J. Ferreira, "Analytical computation of ac resistance of round and rectangular litz wire windings," *IEEE Proceedings-B*, vol. 139, no. 1, pp. 21–25, 1992.

[19] J. Ferreira, "Appropriate modelling of conductive losses in the design of magnetic components," *IEEE Conf. on Power Electron. Spec. (PESC)*, 1990.

[20] M. Bartoli, N. Noferi, A. Reatti, and M. Kazimierczuk, "Modeling litz-wire winding losses in high-frequency power inductors," *IEEE Power Electron. Spec. Conf. (PESC)*,1996.

[21] M. Bartoli, N. Noferi, A. Reatti, and M. Kazimierczuk, "Modeling winding losses in high-frequency power inductors," *Journal of Circuits, Systems, and Computers*, 1995.

[22] F. Tourkhani and P. Viarouge, "Accurate analytical model of winding losses in round litz wire windings," *IEEE Trans. on Magnetics*, vol. 37, no. 1, p. 538–543, 2001.

[23] R. Wojda and M. Kazimierczuk, "Winding resistance of litz-wire and multi-strand inductors," *IET Power Electron.*, vol. 5, no. 2, p. 257, 2012.

[24] M. K. Kazimierczuk, *High-frequency magnetic components.*   Wiley, 2014.

[25] H. Rossmanith, M. Doebroenti, M. Albach, and D. Exner, "Measurement and characterization of high frequency losses in nonideal litz wires," *IEEE Trans. on Power Electron.*, vol. 26, no. 11, p. 3386–3394, 2011.

[26] R. Y. Charles R.Sullivan, "Analytical model for effects of twisting on litz wire losses," *Workshop on Control and Modeling for Power Electronics (COMPEL)*, 2014.

[27] Z. Liu, J. Zhu, and L. Zhu, "Accurate calculation of eddy current loss in litz-wired high-frequency transformer windings," *IEEE Trans. on Magnetics*, vol. 54, no. 11, pp. 1–5, Nov. 2018.

[28] H. Li, N. Zhang, S. Wang, and J. Zhu, "An analytical loss model of litz-wire windings for transformers excited by converters with winding configurations considered," *IEEE Trans. on Magnetics*, vol. 55, no. 9, 2019.

[29] J. Biela, "Optimierung des elektromagnetisch integrierten Serien-Parallel-Resonanzkonverters mit eingeprägtem Ausgangsstrom," *Ph.D. Dissertation, ETH, Zürich*, 2005.

[30] A. Lotfi and F. Lee, "A high frequency model for litz wire for switch-mode magnetics," *IEEE Industry Appl. Conf.*, 1993.

High-Frequency SiC-Based Medium Voltage Quasi-2-Level Flying Capacitor DC/DC Converter With Zero Voltage Switching

Rafał Kopacz, Przemysław Trochimiuk, Grzegorz Wrona, Jacek Rąbkowski
WARSAW UNIVERSITY OF TECHNOLOGY
Koszykowa 75
Warsaw, Poland
Tel.: +48 / (22) – 234.76.15.
Fax: +48 / (22) – 234.60.23.
E-Mail: rafal.kopacz@ee.pw.edu.pl
URL: http://www.ee.pw.edu.pl

Acknowledgements

Work described in this paper has been financed from National Science Center, Poland project no. 2017/27/B/ST7/00970.

Keywords

Medium Voltage, High frequency power converter, Multilevel converters, Silicon Carbide (SiC), Modulation strategy.

Abstract

This paper presents a medium voltage (1.5 kV) flying capacitor DC/DC converter rated at 30 kW. The fast switching 1.2 kV/11 mΩ SiC MOSFETs converter operates in quasi-2-level (Q2L) mode at high frequency (100 kHz) to obtain a low size of the flying capacitor (330 nF). Moreover, a novel switching pattern is introduced and ZVS at turn-on is also enabled. Such control method leads to lower switching losses, higher switching frequency and minimized inductor volume compared to more conventional topologies. Presented experimental results prove correct operation of the proposed converter and control method.

Introduction

The constant demand to reduce the greenhouse gas emission and carbonization levels combined with never-ending increase in energy consumption and finite amounts of fossil fuels have strengthened the need to maximize the efficiency of electrical energy conversion systems and globally employ renewable energy sources (RES), such as photovoltaic (PV) systems. Moreover, employing medium voltage (MV) into these systems is considered beneficial in terms of, amongst many, efficiency [1, 2] and has already been introduced into the industry. Recently, a large part of the PV market have started to adopt the 1.5 kV DC voltage rating for their PV strings instead of standard levels (maximally 1 kV), as it provides superior performance, mainly lower copper and switchgear cost, as well as greater possible operational area for energy capture under wider temperature and irradiance ranges for the system [3]. However, this is associated with the design and construction of highly performant MV DC/DC converters, as currently two-stage energy conversion is considered advantageous in PV systems [4]. Moreover, such converters in MV range are predicted to be applicable in other systems utilizing these voltage levels, for example in future distribution grids and EV charging stations [5], where bidirectional operation is expected.

The standard approach to construct such converters would be to employ commonly-known Si IGBTs capable of blocking voltages as high as 6.5 kV. On the other hand, the recent advancements in SiC power semiconductor technology, especially regarding SiC MOSFETs [6], have introduced superior power devices capable of high switching speeds and low power losses into kV range [7, 8]. However, even if devices reaching voltages as high as 10 kV are being constructed [9] they are still not commercially available and used, as they are limited by reliability issues, more complicated driving, as well as very high cost. Thus, nowadays SiC MOSFETs with blocking voltage only up to 1.7 kV are widely available at the market. Therefore, in order to employ SiC devices and its many advantages in MV converters supplementary means have to be utilized. This can be achieved by either using multilevel topologies or series connection. Even though the former are well-known and provide high blocking voltage and exemplary parameters even with low voltage devices series connection is generally characterized by lower cost and higher efficiency. However, it is not easily applicable to a system, as due to the voltage imbalance between the devices in stack auxiliary voltage balancing circuits, either passive [10, 11] or active [12], have to be added, increasing the cost and complexity of the circuit [13].

Therefore, in this paper, a novel converter based on a the convergence of the advantages of these two methods is presented. Namely, the three-level flying capacitor converter (3L-FCC) DC/DC topology with quasi-2-level (Q2L) control [14-17] and zero voltage switching (ZVS) at turn-on for all of the power devices referred as a modified triangular current mode (TCM) [18, 19], which combined provides a worthy alternative for other solutions in MV range (1.5 kV DC in this case). More specifically, when compared to standard two-level boost converters with single power devices higher possible voltages and lower stresses are achievable [20] as 1.2 kV SiC MOSFETs may be applied. Comparing with the same topology, but with series-connected devices the requirement of employing auxiliary voltage compensation methods is omitted, since the balancing is assured due to the use of the flying capacitor. When similarities with standard three-level boost topologies are considered the proposed system is characterized by lower volume of the flying capacitor thanks to the Q2L modulation and higher possible frequencies through the employment of the ZVS algorithm. Moreover, when resonant switched capacitor converters are considered [21] remarkable efficiency is only achievable for constant input-to-output voltage ratios, whereas the proposed converter is capable of operation in a wide range.

In this paper the quasi-2-level flying capacitor DC/DC converter with ZVS (Q2L-ZVS-FCC) is proposed. At first, operations principles of the converter are described, including both Q2L and ZVS modes. Moreover, simulation study have been conducted, including bidirectional operation. Finally, a prototype of the converter rated at 1.5 kV and 30 kW have been designed and constructed and then has validated experimentally confirming its noteworthy characteristics of lower passive component density in comparison to more conventional solutions in this power/voltage range.

Operation principles

The topology for the proposed Q2L-ZVS-FCC is based on the standard and well-known three-level flying capacitor converter first introduced in [22] and applied in a bidirectional buck-boost DC/DC converter as shown in Fig. 1. In this topology, energy can be transferred either from high level (V_H) to low level (V_L) using the buck part of the converter or in the opposite direction employing the boost part. In regard to basic buck operation considered in this paper states B1 and B2 are used (see Table I). Moreover, since voltage equal to the half of the high voltage V_H is required at the flying capacitor for proper operating conditions supplementary charging/discharging states C1 and C2 have to be introduced into the control of the converter in a closed-loop system based on the voltage of the capacitor C_F. However, the essence of this paper are the two more sophisticated operation modes, the Q2L and ZVS. These will be elaborated on in the subsequent paragraphs.

Table I: Main switching states of the transistors of the Q2L-ZVS-FCC in buck mode

State	T_{H1}	T_{H2}	T_{L1}	T_{L2}	v_M	Status
B1	1	1	0	0	V_H	Energy transferred from V_H
B2	0	0	1	1	0	Energy supplied from the inductor to the output
C1	0	1	0	1	$V_H/2$	Discharging of the flying capacitor
C2	1	0	1	0	$V_H/2$	Charging of the flying capacitor

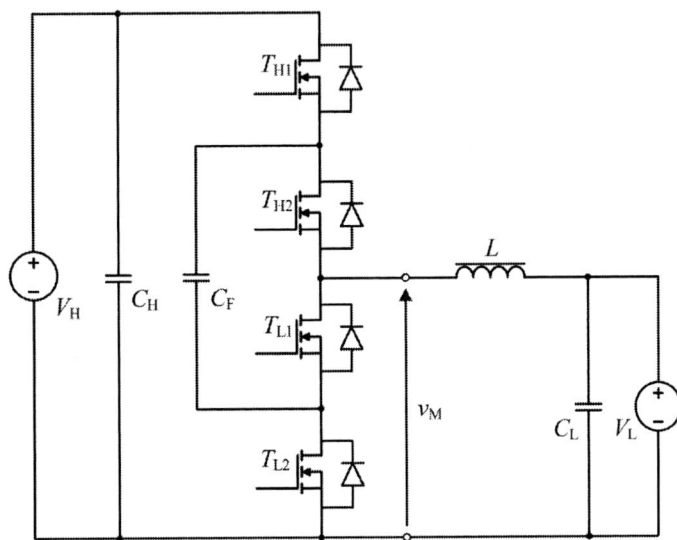

Fig. 1: Topology of the proposed bidirectional Q2L-ZVS-FCC

Quasi-2-level (Q2L) mode

In a conventional FCC auxiliary states C1 and C2 are used for an extended amount of time to deliver third level of the v_M voltage and reduce dv/dt stress of the inductor. Therefore, the required capacitance and total volume of the flying capacitors have to be relatively high. In Q2L operating mode these states are only employed to balance the voltages across the transistor pairs (T_{H1} and T_{H2} or T_{L1} and T_{L2}) and thus they are much shorter in comparison to the classic multilevel approach. Therefore, the capacitance C_F may be also significantly reduced. However, in order to have the system operate properly flying capacitor voltage equal to the half of the V_H level is still required. Thus, the main issue here is to balance these voltages.

This can be easily achieved by adding a closed-loop control based on the voltage measurement of the flying capacitor (V_{CF}) and the high level (V_H) and a simple PI controller. In this way the length of the states responsible for charging and discharging of the flying capacitor (C1 and C2) can be adjusted, which in result will stabilize the voltage at a desired level. Moreover, using state sequences shown in (1) and (2) alternately, which correspond to states shown in Tab. I and supplementary dead time transition states DT1-4 required for protecting the circuit from short-circuits and also affecting the ZVS conditions, will further balance the flying capacitor voltage. This can be justified when we consider a situation when only one sequence is used. Then, state C1 always occurs alongside high inductor current (I_M) whereas state C2 takes place when the inductor current is lower and equal to I_R. Thus, different amount of energy is being delivered to or from the flying capacitor and in result this will lead to an unbalanced voltage. Applying the suggested sequence order assures that the same amount of energy is transferred from and into the flying capacitor and therefore leads to natural balancing.

$$Seq1 = \boldsymbol{B1} \rightarrow DT2 \rightarrow \boldsymbol{C1} \rightarrow DT4 \rightarrow \boldsymbol{B2} \rightarrow DT3 \rightarrow \boldsymbol{C2} \rightarrow DT1 \tag{1}$$

$$Seq2 = \boldsymbol{B1} \rightarrow DT1 \rightarrow \boldsymbol{C2} \rightarrow DT3 \rightarrow \boldsymbol{B2} \rightarrow DT4 \rightarrow \boldsymbol{C1} \rightarrow DT2 \tag{2}$$

Zero Voltage Switching (ZVS) mode

In a simplest approach to Q2L modulation only states shown in Tab. I can be employed. However, this comes with the danger of short-circuiting the flying capacitor when transistor switching delays are considered. Thus, employing auxiliary transition states DT when only one transistor is turned-on at a time in specific dead time intervals t_{dt} presented in Tab. II is a necessity. Furthermore, the supplementary states can be also utilized for achieving ZVS at the turn-on for all of the devices. To describe the ZVS process with the proposed modulation pattern shown in Fig. 2 on an example the transition from state B2 to C2 will be considered. At the start of the state DT3 transistor T_{L2} is being turned-off and the current starts to flow through the flying capacitor C_F and the body diode of T_{H1}. Thus, if a sufficient reverse current I_R is assured within a specific dead time interval (t_{dt}) the output capacitance of the transistor T_{H1} will discharge. Therefore, when T_{H1} is switched on during state C2 zero voltage switching condition for the turn-on is met eliminating the switching power losses completely. When we consider transitions in other states, as in, for example, B1 to C1 the process is nearly identical, however as I_M is large the ZVS condition is always met.

Furthermore, it is worth noting that the transition states DT affect neither the voltage v_M nor the output of the converter since the order of switchings and the duration of the states between the devices is kept unchanged, the difference comes only in a matter whether the commutation occurs between the diode or the transistor from a certain switch.

Table II: Supplementary transition switching states of the transistors in a FCC employing ZVS modulation scheme

State	T_{H1}	T_{H2}	T_{L1}	T_{L2}	Status
DT1	1	0	0	0	Transition between state B1 and C2
DT2	0	1	0	0	Transition between state B1 and C1
DT3	0	0	1	0	Transition between state B2 and C2
DT4	0	0	0	1	Transition between state B2 and C1

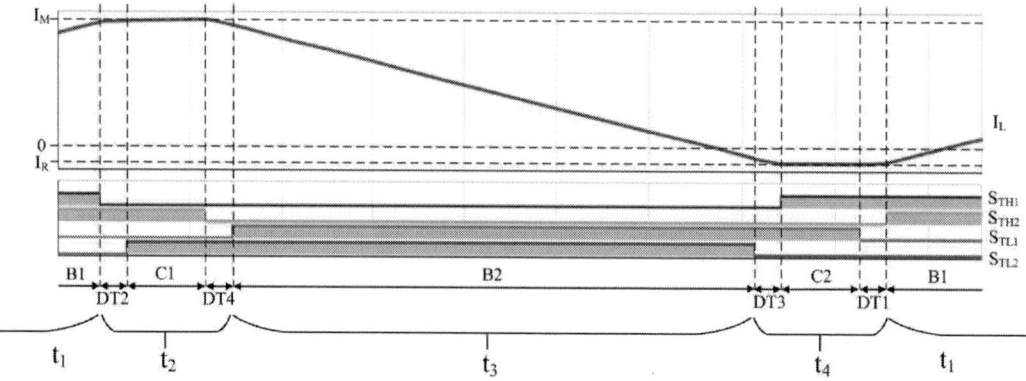

Fig. 2: Proposed modulation pattern for the Q2L-FCC with ZVS – basic sequence no. 1 (Seq1) shown for the converter operating in buck mode

Utilizing this ZVS approach requires to carefully adjust the reverse current I_R and the dead time intervals t_{dt}, so that the transistor output capacitance will always discharge, which is affected by the V_H and V_L voltage levels, the applied inductance L, as well as the output capacitance of the MOSFET C_{oss}. Thus, when the load or the output voltage is changed, the reverse current I_R will be also affected, which may result in non-ZVS switching conditions. Therefore, adjusting the frequency, which in consequence alters the time t_{dt} accordingly, is a necessity in order to ensure proper operation of the converter in a wide operating range. Moreover, it is worth noting that in this ZVS mode the switching losses are lowered, but the conduction losses rise slightly. Thus, this method is only noteworthy in high frequency applications. Furthermore, as body diode conducts during transition states adding Schottky diodes in parallel to transistors should be considered efficiency-wise.

Conversion ratio

In order to define the conversion ratio first we have to define switching period t_s and times t_1, t_2, t_3, t_4 as in Fig. 2, where t_1 and t_3 are described as the length of states B1 and B2 respectively and t_2 and t_4 are bound to states C1 and C2 and its surrounding states DT accordingly to the current sequence. Then, the output voltage for the converter in buck mode may be described as:

$$V_L = V_H * D_{buck},$$
(3)

where:

$$D_{buck} = \frac{2t_1 + t_2 + t_4}{2t_s}$$
(4)

Simulation study

The discussed Q2L-FCC with ZVS have been validated in simulations performed in Synopsys Saber and using the Power MOSFET modelling tool so that the results would be as accurate as possible. 1.2 kV SiC MOSFETs (FF11MR12W1M1_B11) from Infineon have chosen as the transistors based on prior modelling, simulation and selection among the group of state-of-the-art devices. The high level voltage (input voltage for buck mode) has been set at 1.5 kV. The converter rated at 30 kW have been simulated in both buck and boost modes. The main parameters are shown in Tab. III.

Table III: Converter simulation and prototype parameters

Element	Parameters/Model
Rated power	30 kW
Operating frequency	100 kHz
SiC power transistors	1.2 kV and 100 A rms/FF11MR12W1M1_B11
Inductor	30 µH/100 A rms peak
Flying capacitor	330 nF/1000 V
Output/input capacitors	3 µH/1600 V

The results for the converter operating in step-down and Q2L modes are shown in Fig. 3, on which we can observe the characteristic shape of the v_M voltage, where the mid-level states are only employed for a small amount of time (5% of the period) . Hence, the amount of energy drawn from the flying capacitor is low and much smaller capacitance values are possible in comparison to standard three-level converters – according to performed simulations just 330 nF capacitance is enough to maintain the voltage ripple roughly at 7% for the presented converter.

The next Fig. 4 validates the ZVS operation at the turn-on process for the top transistor T_{H1} during the transition between states B2 – DT3 – C2 – DT1 – B1. As shown, at first transistor T_{H1} is turned off and sustains half of the V_H voltage. Then, transistor T_{H4} is turned off and since the current is negative (flowing from V_L to V_H) it starts to pass through diode D_{H1} and thus the voltage V_{DS} drops to 0. After a certain amount of time sufficient to discharge the transistor output capacitance under specific current the transistor is turned on (S_{TH1}) under ZVS conditions thus without any switching loss and the MOSFET current starts to slowly rise and conducts alongside the diode.

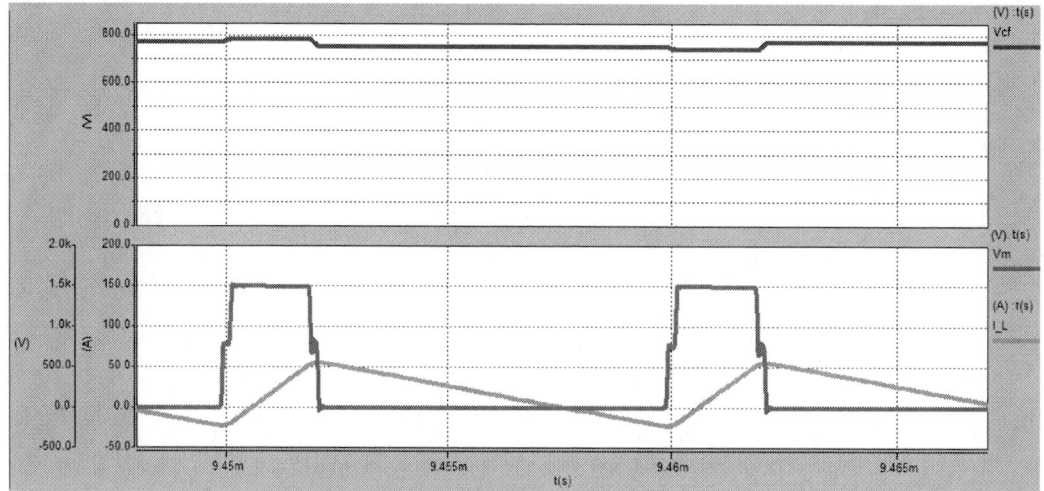

Fig. 3: Results from the simulation study for the converter with $D_{buck} = 0.2$ and $C_F = 330$ nF at 1.5 kV. From the top: balanced flying capacitor voltage (V_{CF}), Q2L phase voltage of the converter (v_M) and inductor current (I_L).

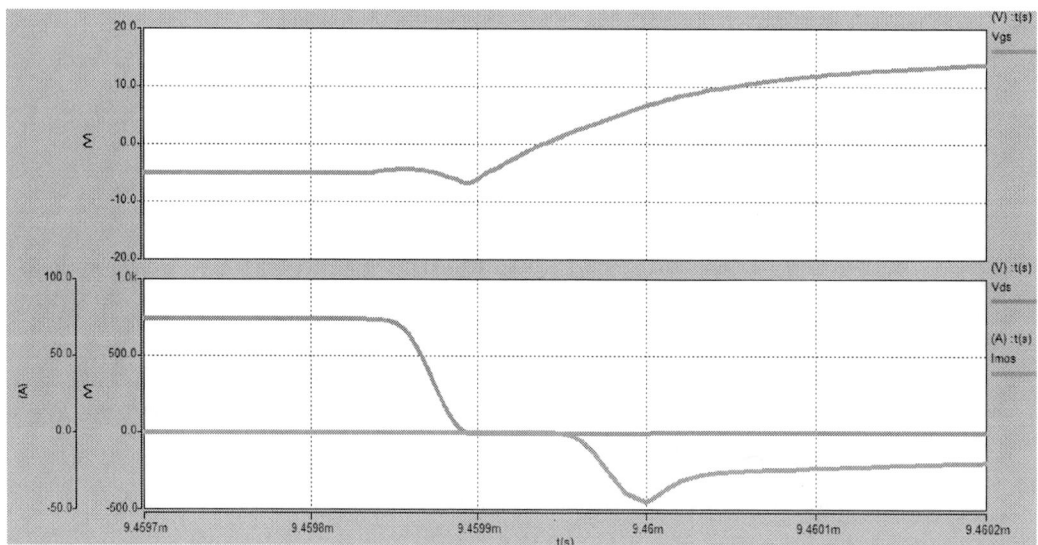

Fig. 4: Results from the simulation study: ZVS validation at turn-on for transistor T_{H1}. From the top: V_{GS} voltage, MOSFET current (I_{MOS}) and drain-source voltage of the transistor (V_{DS}).

Converter prototype

In order to validate the proposed Q2L-FCC with ZVS, a prototype, preliminarily verified in simulations, has been designed on the basis of electro-thermal calculations from the acquired data and then constructed. The parameters are the same as in the simulations and are depicted in Table III. Since the converter operates at high frequency (100 kHz) 3 µH/1600 V capacitors have been applied to the V_H and V_L sides. Thanks to the employment of the Q2L mode flying capacitance settled at only 330 nF, whereas due to the use of the ZVS modulation the choke inductance has been set as roughly 30 µH, which is a significantly lower value in comparison to more common approaches for MV DC/DC converters in with similar efficiency in the same voltage range.

Furthermore, special care has been given to the design of the powerboard, minimizing the path length in order to reduce parasitic inductances between the switching elements, which is especially important since the converter operates at high dv/dt rates. Due to the substantial power LA V 8 with forced air convection have been applied as the heatsink. Moreover, the necessary V_{CF} and V_H measurements have been implemented via fully analogue circuits based on resistive dividers and isolated voltage followers. The system has been controlled via a floating-point DSP system that have been mounted on the side of the converter shielded from interferences by grounded cover attached to the heatsink. A photo of the constructed converter prototype is shown in Fig. 5.

Fig. 5: Photo of the constructed converter prototype.

Experimental study

So that the converter concept could be verified a series of experimental tests on the prototype have been performed. The input voltage (V_H) was set to 1.3 kV, a 18 Ω resistor was used as a load and the converter operated in the buck mode at 100 kHz with a 30 µH choke. D_{buck} factor was established as 0.2.

Fig. 6 showcases the operation of the Q2L-FCC, which very closely correspond to the results from the simulation study. As can be seen the flying capacitor voltage (V_{CF}) is properly balanced with ripples as low as 5% of the nominal value ($V_H/2$) even though only 330 nF capacitance has been applied. This is a benefit from Q2L operation of the converter. Moreover, very characteristic shape of the v_M voltage can be observed in the same figure, where the third, middle voltage level is only applied for a very short time. Furthermore, Fig. 7 shows that due to balanced capacitor voltage the drain-source voltage across two transistors T_{H1} and T_{H2} is divided almost equally. During one of the sequences the difference is nearly imperceptible, whereas in the second sequence there is a difference of roughly 5% of the V_{CF} voltage. This experimentally validates the Q2L method at high-switching frequency as a viable alternative to connect MOSFETs in series, which is as effective as active voltage balancing approaches even if it requires much less effort.

High-Frequency SiC-Based Medium Voltage Quasi-2-Level Flying Capacitor DC/DC Converter With Zero Voltage Switching KOPACZ Rafal

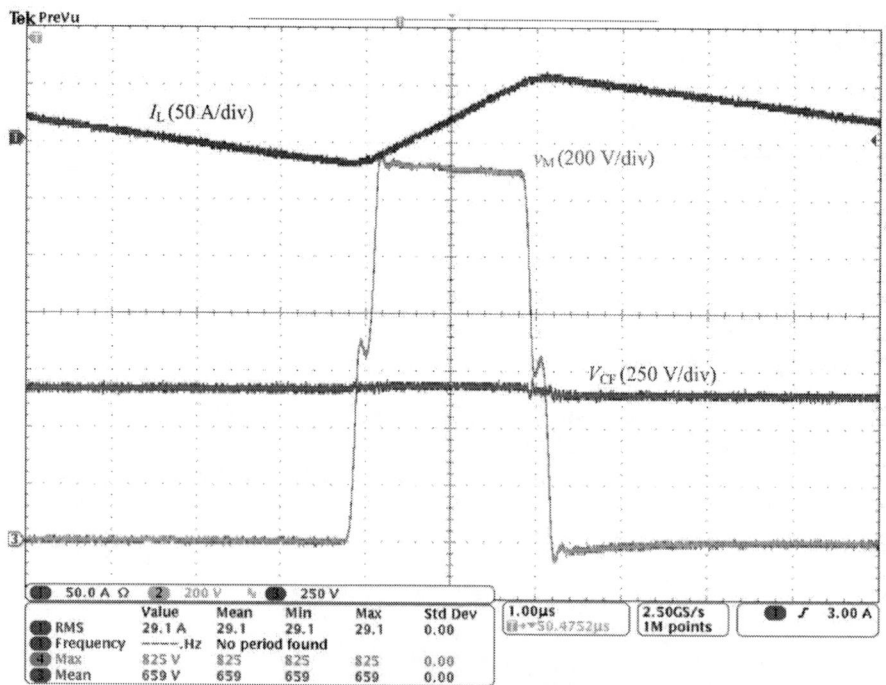

Fig. 6: Q2L operation – experimental results at $D_{buck} = 0.2$ and $V_H = 1300$ V. From the top: inductor current (I_L), Q2L phase voltage of the converter (v_M), balanced flying capacitor voltage (V_{CF}).

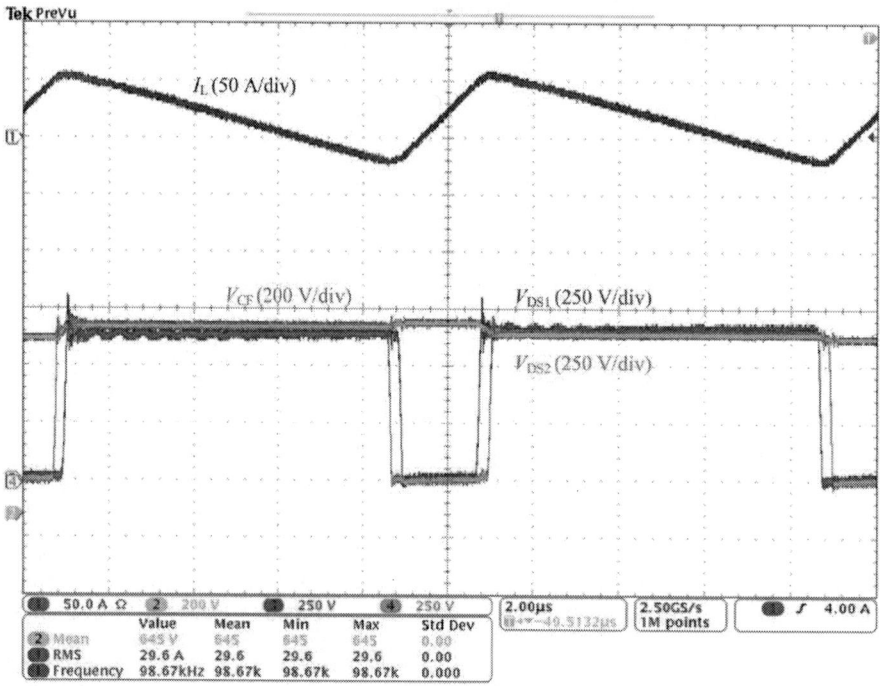

Fig. 7 Validation of proper balancing of the flying capacitor voltage ($C_F = 330$ nF) at $D_{buck} = 0.2$ and $V_H = 1300$ V. From the top: inductor current (I_L), balanced voltages across transistors T_{H1} and T_{H2} (V_{DS1} and V_{DS2}) and the voltage of the flying capacitor (V_{CF}).

EPE'20 ECCE Europe

Assigned jointly to the European Power Electronics and Drives Association & the Institute of Electrical and Electronics Engineers (IEEE)

Finally, the last Fig. 8 presents exemplary turn-on switching process for transistor T_{H1}. It can be noticed that the V_{GS} voltage reaches MOSFET $V_{GS(th)}$ value (4.5 V according to the datasheet) when the V_{DS1} voltage is already at zero. Thus, ZVS condition at turn-on is confirmed. The turn-on process for all the other transistors in the converter is analogous, thus all turn-on switching losses are vastly minimized and higher efficiency levels of the converter can be achieved.

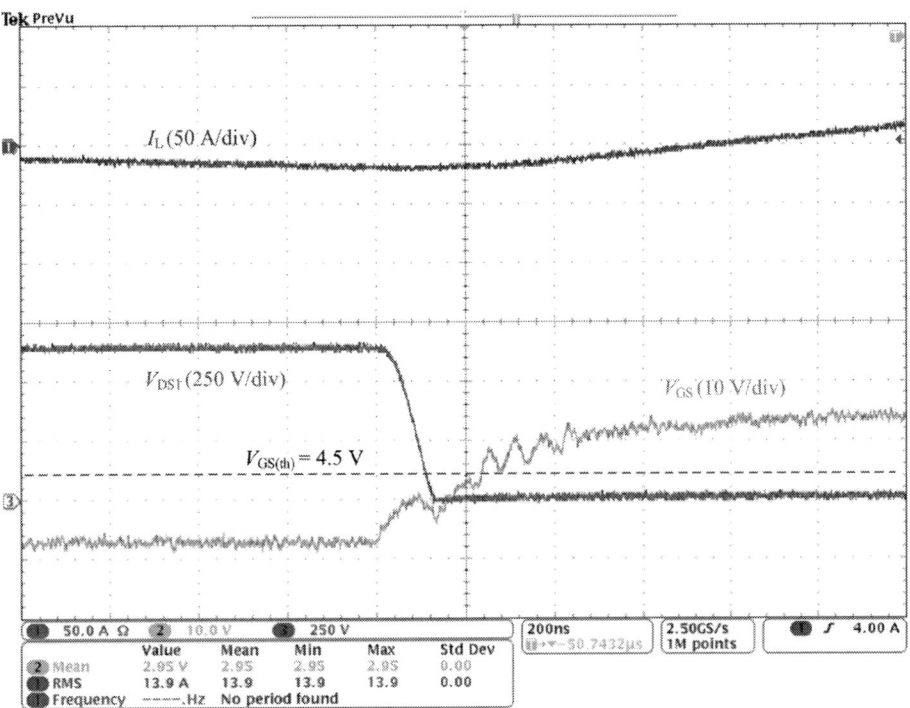

Fig. 8: ZVS validation at turn-on for transistor T_{H1} at $D_{buck} = 0.2$ and $V_H = 1300$ V. From the top: inductor current (I_L), drain-source voltage of the transistor (V_{DS1}) and the gate-source voltage (V_{GS}).

Conclusion

The proposed and validated in this paper novel control method converging Q2L and ZVS methods applied in a flying capacitor DC/DC converter may be a viable solution for highly efficient energy conversion in MV applications. Q2L operation enables use of fast and affordable 1200 V SiC MOSFETs with flying capacitor value reduced to as low as 330 nF. The proposed modulation pattern introducing ZVS conditions results in higher frequency such as discussed 100 kHz without significant efficiency drop. Another advantage of the discussed DC/DC converter is low volume and size of the applied inductor. Performed simulations and experimental tests confirm the noteworthy characteristics of the FCC DC/DC with proposed control method. After experimental validation of basic characteristics further measurements of efficiency in different operation modes are planned.

References

[1] De Doncker R. W., Meyer C., Lenke R.U and Mura F., "Power Electronics for Future Utility Applications", 2007 7th International Conference on Power Electronics and Drive Systems, , pp. K-1-K-8, Bangkok, 2007.

[2] Reed G. F., Grainger B. M., Sparacino A. R., Taylor E. J., Korytowski M. J., and Mao Z.-H., "Medium Voltage DC Technology Developments, Applications, and Trends", CIGRE, 2012 Grid of the Future Symposium, 2012

[3] Serban E., Ordonez M. and Pondiche C., "DC-Bus Voltage Range Extension in 1500 V Photovoltaic Inverters," in *IEEE Journal of Emerging and Selected Topics in Power Electronics*, vol. 3, no. 4, pp. 901-917, Dec. 2015.

[4] Kim J., Kwon J. and Kwon B., "High-Efficiency Two-Stage Three-Level Grid-Connected Photovoltaic Inverter," in *IEEE Transactions on Industrial Electronics*, vol. 65, no. 3, pp. 2368-2377, March 2018.

[5] Rivera S. and Wu B., "Electric Vehicle Charging Station With an Energy Storage Stage for Split-DC Bus Voltage Balancing," in *IEEE Transactions on Power Electronics*, vol. 32, no. 3, pp. 2376-2386, March 2017.

[6] Rabkowski J., Peftitsis D. and Nee H. P., "Silicon Carbide Power Transistors: A New Era in Power Electronics Is Initiated", IEEE Industrial Electronics Magazine, vol. 6, no. 2, pp. 17-26, June 2012.

[7] Bhatnagar M., Baliga B. J., "Comparison of 6H-SiC, 3C-SiC, and Si for Power Devices", IEEE Transactions on Electron Devices, vol. 40, pp. 645 - 655, March 1993

[8] Biela J., Schweizer M., Waffler S., Kolar J. W., "SiC versus Si -Evaluation of Potentials for Performance improvement of Inverter and DC-DC Converter Systems by SiC Power Semiconductors", IEEE Transactions on

[9] Palmour J. W. *et al.*, "Silicon carbide power MOSFETs: Breakthrough performance from 900 V up to 15 kV," *2014 IEEE 26th International Symposium on Power Semiconductor Devices & IC's (ISPSD)*, Waikoloa, HI, 2014, pp. 79-82.

[10] Vechalapu K. , Hazra S., Raheja U., Negi A. and Bhattacharya S., "High-speed medium voltage (MV) drive applications enabled by series connection of 1.7 kV SiC MOSFET devices", 2017 IEEE Energy Conversion Congress and Exposition (ECCE), pp. 808-815, October 2017

[11] Kopacz R., Peftitsis D. and Rabkowski J., "Experimental study on fast- switching series-connected SiC MOSFETs", 2017 19th European Conference on Power Electronics and Applications (EPE'17 ECCE Europe), September 2017.

[12] Marzoughi A., Burgos R. and Boroyevich D., "Active Gate-Driver with dv/dt Controller for Dynamic Voltage Balancing in Series-Connected SiC MOSFETs", IEEE Transactions on Industrial Electronics, 2019

[13] Trochimiuk P., Kopacz R., Wrona G. and Rabkowski J., "Medium voltage power switch based on 1.7 kV SiC MOSFETs connected in series inside power modules", 2019 21th European Conference on Power Electronics and Applications (EPE'19 ECCE Europe), pp. P.1- P.10, September 2019.

[14] Adam G. P., Finney S. J., Massoud A. M. and Williams B. W., "Capacitor Balance Issues of the Diode-Clamped Multilevel Inverter Operated in a Quasi Two-State Mode," in *IEEE Transactions on Industrial Electronics*, vol. 55, no. 8, pp. 3088-3099, Aug. 2008.

[15] Gowaid I. A., Adam G. P., Massoud A. M., Ahmed S., Holliday D. and Williams B. W., "Quasi Two-Level Operation of Modular Multilevel Converter for Use in a High-Power DC Transformer With DC Fault Isolation Capability," in *IEEE Transactions on Power Electronics*, vol. 30, no. 1, pp. 108-123, Jan. 2015.

[16] Schweizer M. and Soeiro T. B., "Heatsink-less Quasi 3-level flying capacitor inverter based on low voltage SMD MOSFETs," *2017 19th European Conference on Power Electronics and Applications (EPE'17 ECCE Europe)*, Warsaw, 2017, pp. P1-P.10.

[17] Czyz P., Papamanolis P., Guillod T., Krismer F. and Kolar J. W., "New 40kV / 300kVA Quasi-2-Level Operated 5-Level Flying Capacitor SiC "Super-Switch" IPM," *2019 10th International Conference on Power Electronics and ECCE Asia (ICPE 2019 - ECCE Asia)*, Busan, Korea (South), 2019, pp. 813-820.

[18] Wang B., Yuan Y., Zhou Y. and Sun X., "Buck/boost bidirectional converter TCM control without zero-crossing detection," *2016 IEEE 8th International Power Electronics and Motion Control Conference (IPEMC-ECCE Asia)*, Hefei, 2016, pp. 3073-3078.

[19] Wang Y. *et al.*, "TCM Controller Design for Three-Level Bidirectional Soft-Switching DC-DC Converter," *2019 IEEE 28th International Symposium on Industrial Electronics (ISIE)*, Vancouver, BC, Canada, 2019, pp. 996-1001.

[20] Sayed S., Elmenshawy M., Ben-Brahim L. and Massoud A., "Design and analysis of high-gain medium-voltage DC-DC converters for high-power PV applications," *2018 IEEE 12th International Conference on Compatibility, Power Electronics and Power Engineering (CPE-POWERENG 2018)*, Doha, 2018, pp. 1-5.

[21] Stevanović B., Serrano D., Vasić M., Alou P., Oliver J. A. and Cobos J. A., "900V SiC Based, Hybrid, Multilevel DC/DC Topology for 1500VDC PV Application," *2019 IEEE Applied Power Electronics Conference and Exposition (APEC)*, Anaheim, CA, USA, 2019, pp. 6-12.

[22] Meynard T. A. and Foch H., "Multi-level conversion: high voltage choppers and voltage-source inverters," *PESC '92 Record. 23rd Annual IEEE Power Electronics Specialists Conference*, Toledo, Spain, 1992, pp. 397-403 vol.1.

Smart fuel cell module (6.5 kW) for a range extender application

Pascal BAZIN, Bruno BERANGER, Jacques ECRABEY,
Laurent GARNIER, Sylvain MERCIER
Univ. Grenoble Alpes, CEA, LITEN, DEHT, LAEH,
38000 Grenoble, France
Tel.: +33 / (0) – 438.78.24.47.
E-Mail: sylvain.mercier@cea.fr
URL: http://www-liten.cea.fr

Keywords

«Fuel Cell Electric Vehicle (FCEV)», «Silicon Carbide (SiC)», «Interleaved converters», «Thermal design», «System integration»

Abstract

To extend the autonomy of electrical vehicles, fuel cells are a possible complementary power source. This source must be electrically adapted through a DC-DC converter and mechanically easily integrated. This paper shows the design of a 6-phase interleaved boost converter (6.5 kW max.) for this application. Due to integration constraints, the converter should be mounted on the terminal plates of the fuel cell and cooled by the same cooling circuit. To dissipate the losses of the semiconductors, the terminal plate embeds 30 cooling channels under the printed circuit board. The converter efficiency has been measured up to 96.5% for a minimum output power (1.5 kW) and a minimum output voltage (240 V). To move forward the design of the smart fuel cell module (i.e. with power electronics), a new development is ongoing for an aircraft application by designing a specific and optimized cooling system.

Introduction

One of the major barrier for the spreading of electrical vehicles (EVs) is their limited autonomy. A smart fuel cell module (i.e. with power electronics) of 6.5 kW has been developed to extend the EVs autonomy (European research project 3Ccar). The 6.5 kW module has the following particular properties: a facilitated parallelization to increase the available power and a smart connecting cooling plates to easily integrate the power electronics. This paper focuses on the DC-DC converter design according to the integration and industrial (components supply) constraints.

First, the fuel cell stack is briefly described and then some explanations are given about the DC-DC converter design. Secondly, the DC-DC mechanical integration and manufacturing are illustrated. Finally, the converter performances experimentally obtained are analyzed and compared to theoretical data.

1. Smart module design

1.1 Fuel cell stack

1.1.1 Mechanical features

The fuel cell stack is composed of 100 cells of 100 cm² active surface. These features allow optimizing the bipolar plate cost and the DC/DC converter efficiency thanks to a relatively high fuel cell voltage. The gas supply is located under the fuel cell stack while the power electronics are set on the top. Fig. 1 shows the main parts of the module.

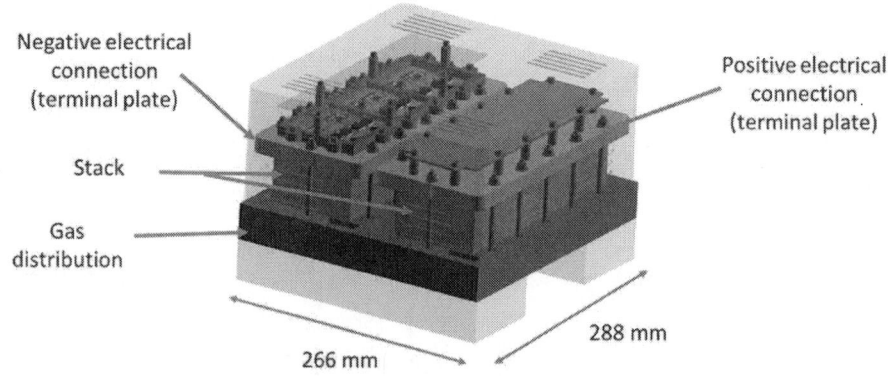

Fig. 1: Fuel cell design (SolidWorks 3D CAD)

In order to integrate the power electronics, bipolar plates are dissociated in two parts (one for the positive electrode and one for the negative): the converter electrical connections to the fuel cell are located on those plates (of same dimensions). This leads to limit the converter printed circuit boards to a maximum dimension of 110 mm x 250 mm.

1.1.2 Electrical features

The fuel cell voltage range is [60 - 102.5] V with at the end of life a maximum current of 113 A (at 60 V, i.e. maximum power of 6.78 kW). Assuming that the high frequency current ripples do not affect the lifetime of a fuel cell [1], the input filtering of the converter is only assured with the phase inductors of the DC/DC converter. The phase inductors are directly connected to the positive electrical connection and the fuel cell behaves as a resistance connected in series with an inductor for the ripple frequency (few hundred of kHz) [2] [3] [4].

1.1.3 Thermal features

The semiconductors are soldered on an Insulated Metal Substrate (IMS). Their losses should be evacuated through the fuel cell terminal plates which embed 30 cooling channels of 0.5 x 1 mm² spaced 1 mm apart (Fig. 2). Aluminum alloy (ref. 7075) having a high thermal conductivity (130 W/(mK)), it is the chosen material for the channels.

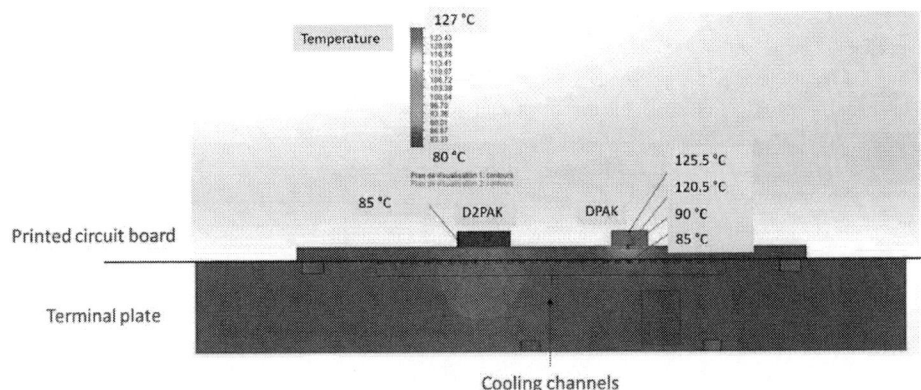

Fig. 2: Terminal plate and printed circuit board (cross section)
(SolidWorks Flow Simulation)

Since the converter shares the cooling system with the fuel cell, the channel input liquid temperature is 80 °C (allowing a high efficiency for the fuel cell). Some simulations have been done to evaluate the thermal performances of the thermal system. Two different semiconductors packages are simulated: D2PAK and DPAK. DPAK heats at 9 W and D2PAK at 11 W. For a D2PAK package, the thermal resistance between the case and the liquid is estimated to about 1.8 °C/W. For a DPAK package (smaller than D2PAK), the thermal resistance is about two times higher: about 4.0 °C/W. The insulation thickness of the IMS is set to 0.125 mm.

The liquid temperature rise between the channels input and output is estimated to about 5 °C with a flow rate of 0.9 L/min.

1.2 Step-up converter

1.2.1 State-of-the art

The selection of the proper converter topology for this application has been made among non-insulated DC-DC converters (since insulation was not required). Fig. 3 shows 7 candidate topologies. The usual topology of the boost converter is illustrated ref. A. on Fig. 3. The other topologies are designed to achieve a higher voltage gain at the expense of more devices and more complex control.

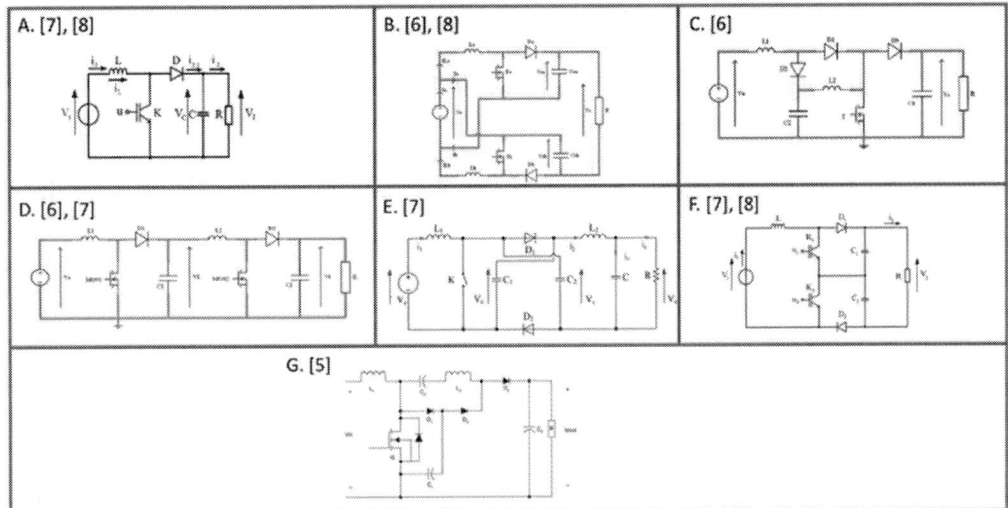

Fig. 3: Step-up converter topologies (non-insulated)

Besides, the converter should be easily integrated on the fuel cell terminals. Firstly, the number of required devices should be therefore minimized. Secondly, the topology should naturally fit the fuel cell configuration. These are the two main reasons why the usual boost converter topology (ref. A., Fig. 3) has been chosen. It embeds the lowest number of components and the phase inductors can be connected directly on the top of the positive plate and, the switching devices on the top of the negative plate (Fig. 4). Besides, the design and control are facilitated. Finally, the efficiency should be sufficiently high, about 95% for the maximum voltage gain (6.3), compared to the fuel cell efficiency. With the same philosophy, the DC/DC converter conduction mode is chosen to be continuous. Indeed, the ripples are reduced and the minimum output power is limited to 1.5 kW because the fuel cell operation for low currents is not relevant (Fig. 4).

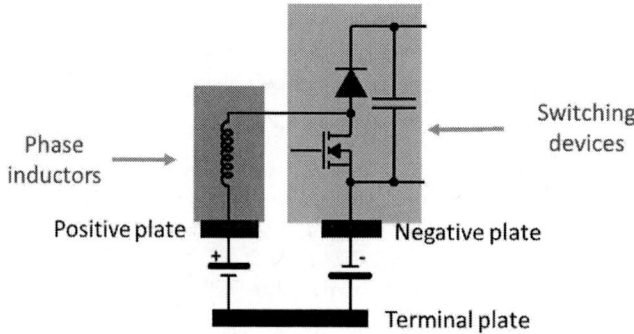

Fig. 4: Mechanical integration of the circuit schematic

The topology is interleaved in the aim to reduce the low frequency input current ripple (to rise fuel cell durability) and to reduce the average and root mean square currents through the components. Higher is the number of phases, lower will be the ripples and currents. But in the same time, there are some devices with an incompressible package area. To fit with the size of the terminal plate, a 6-phase interleaved boost converter is the best tradeoff between the mechanical and electrical constraints of the fuel cell.

1.2.2 Main power devices choice

The power devices have been chosen regarding input and output ripples, device losses and thermal behaviors (use of PTC Mathcad and PSIM softwares). These properties are linked to the switching frequency: 100 kHz has been chosen (then ripples frequency is 600 kHz).

In the aim to limit the electromagnetic interferences (EMI) produced by the transistors, their gate resistance (C3M0065090J Silicon Carbide (SiC) MOSFET from Wolfspeed) is as high as the thermal constraints allowed. For helping on this constraint, two devices are connected in parallel per phase to divide per two the thermal resistance and to decrease their switching and conduction losses. The total power dissipated per phase is estimated to about 35 W (LTspice) for a 113 A fuel cell current.

One SiC diode from Wolfspeed, ref. C3D10065E, is used per phase. The total power dissipated per phase is equal to about 5 W (LTspice) for a 113 A fuel cell current.

Commercial devices from Coilcraft, ref. VER2923-103KL, are used for the phase inductors. These devices are efficient and easy to purchase. Moreover, the value of the inductances should be very close between the devices and should facilitate the interleaving. To reach the expected inductance on each phase and keep continuous conduction mode, ten inductors are connected in series per phase. The power dissipated per phase is estimated to 14 W (Coilcraft tool) for a 113 A fuel cell current.

A PI-filter is used at the output of the converter. Four ceramic capacitors from TDK, ref. B58031U5105M062, are connected in parallel per phase and placed on the negative plate (Fig. 4 and

Fig. 6). The PI-filter integrates also four inductors set in parallel (ref. MSS1246-153MLC from Coilcraft), and two capacitors in parallel (ref. MKP1848S55050JK2A from TDK).

2. Converter manufacturing

Fig. 5 shows the mechanical integration of the converter inside the module. The phase inductors and the current sensors (for current control) are located on the top of the positive plate. The switching devices are put on the IMS directly screw on the negative plate.

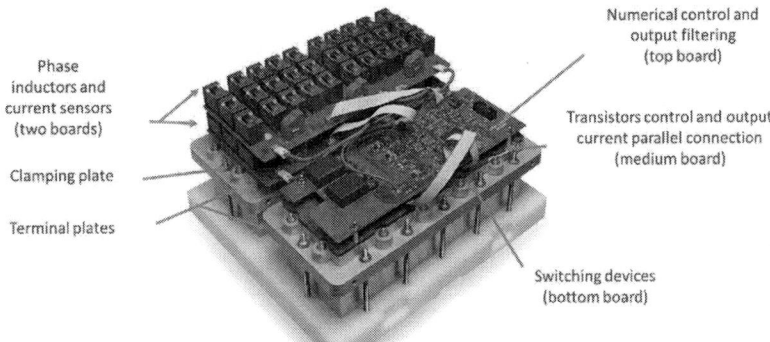

Fig. 5: Converter assembling

To connect the electrical potentials between the boards and the plates and between the two inductors boards, some spacers are used. The board with the switching devices (bottom board) and the board with the drivers (medium board) are electrically connected together with several two pins connectors, as illustrated on Fig. 6. Four connectors are used for each phase, to compile with the desired current intensity.

Fig. 6: Switching devices board assembling

The thickness of the IMS insulation layer is 0.1 mm: this thickness must be as low as possible to improve the cooling efficiency by respecting the insulation strength. The aluminum thickness is 0.9 mm.

3. Converter tests

3.1 Mechanical performances

The weight of the converter is 3.5 kg and its volume is 3.8 L. About 66% of the weight is involved by the phase inductors (2.3 kg) which also represents 75% of the total volume. For a 6.5 kW maximum output power, the converter power density is 1.7 kW/L and 1.8 kW/kg. To increase the converter power density, either the input current ripple could be increased to a higher value (if the fuel cell is able to withstand it), either the phase inductors could be the results of a dedicated design instead of off the shelf components (if the components supply is not critical).

3.2 Thermal performances

Junction to case thermal resistance of semi-conductors are in general well indicated in manufacturer datasheets. The case to liquid thermal resistance is also a necessary data to design the converter, meanwhile its acquisition is not straight forward since require data (insulation thickness, flow rate, channels design, etc.) are not fixed yet.

In this study case, the simulation has been performed with a flow rate of 0.9 L/min and an IMS insulation layer thickness of 0.125 mm and gave a thermal resistance between the case and the liquid of 1.8 °C/W for D2PAK package. During testing (Fig. 7), the thermal resistance is estimated thanks to the characterization of the on-state resistance according to the junction temperature. Indeed, the efficiency of the cooling prevents the measurement of device temperatures with a thermal camera or thermocouple sensor. The experiments gave a range between 1.7 °C/W and 2.0 °C/W with a flow rate of 1.5 L/min and an insulation thickness of 0.1 mm. It is of the same order of magnitude, which validates the thermal management of the SiC semiconductors.

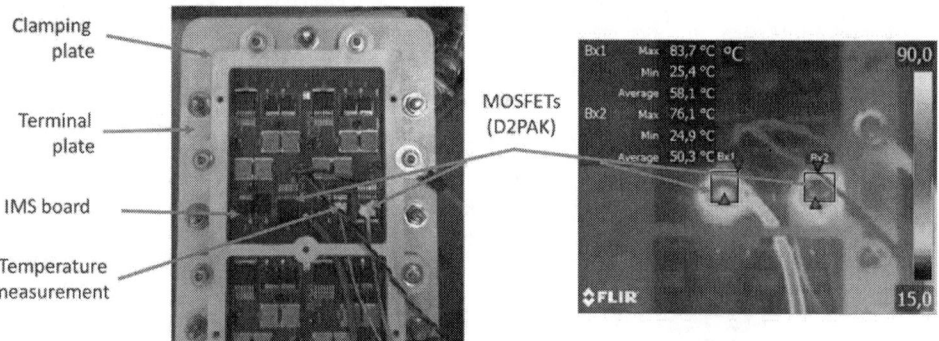

Fig. 7: MOSFETs thermal resistance measurement test bench

3.3 Electrical performances

The maximum input current ripple occurs at the lowest input current (20 A) and 330 V output voltage (P = 1.5 kW). The measured ripple for a close operating point is 1.2 A for an estimated 1.35 A (12% error, Fig. 8). The difference is mainly due to the operation in discontinuous mode of some phases at the tested operation point. The maximum output voltage ripple is measured before the L-C output voltage filter. The measured value is 0.8 V (7% error with simulation).

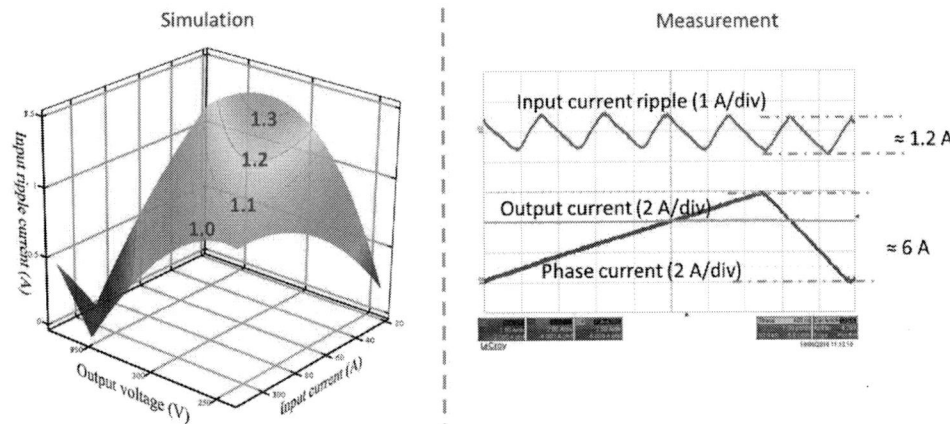

Fig. 8: Input current ripple estimated (PTC Mathcad) and measured

Fig. 9 shows the measured efficiency of the power converter for different output power and the two extrema output voltages (240 V and 380 V).

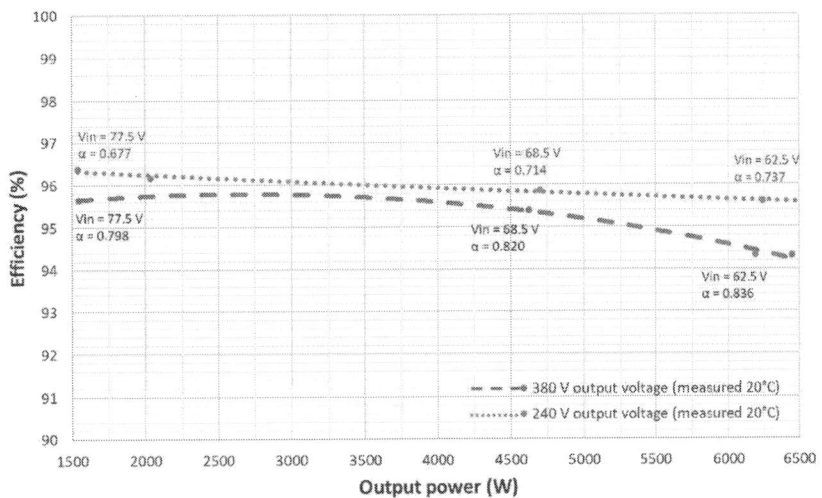

Fig. 9: Efficiency versus output power and voltage

The maximum converter efficiency (96.5%) is obtained at minimum output power and voltage, while the minimum efficiency (94%) is obtained for a maximum output power and voltage, as estimated theoretically. The temperature of the liquid cooling has no impact on the converter efficiency (20 °C up to 75 °C). The dissipated power by the transistors and the phase inductors represent respectively 57% and 23% of the total dissipated power. To increase the converter efficiency, the switching times could be reduced but with probably an impact on the EMI filter. Because fuel cell efficiency is about 50%, increasing the converter efficiency above 94% at the expense of the converter power density is not relevant.

Besides, this efficiency is quite good regarding the required gain of the converter (6.3 from a minimum input voltage of 60 V up to a maximum output voltage of 380 V).

Conclusion

A smart fuel cell module of 6.5 kW, based on a dedicated fuel cell design and an integrated step-up converter, has been developed. The fuel cell stack is composed of 100 cells with an active surface of each cell around 100 cm². The step-up converter is integrated on the terminal plates of the fuel cell and is cooled by the same cooling circuit. To dissipate the semiconductors losses of the 6-phase interleaved boost converter, the terminal plate embeds 30 cooling channels under the printed circuit board. The thermal resistance of the cooling system has been measured between 1.7 °C/W and 2.0 °C/W for a D2PAK device. The converter efficiency has been measured up to 96.5% for a minimum output power (1.5 kW) and a minimum output voltage (240 V).

To move forward the design of the smart fuel cell module, a new development is on-going as part of a European research project (FLHYSAFE). This new development integrates power inductors custom designs and an additional cooling solution (immersion).

References

[1] C. Restrepo, T. Konjedic, J. Calvente, R. Giral: A review of the main power electronics' advances in order to ensure efficient operation and durability of PEMFCs, journal Automatika 53 (2012), pp. 184-198

[2] I. Saldi : Modélisation par impédance d'une pile à combustible PEM pour utilisation en électronique de puissance, Ph. D. thesis, INPL, Nancy (France), December 2006

[3] G. Fontes, C. Turpin, S. Astier, T. A. Meynard: Interactions between fuel cells and power converters: influence of current harmonics on a fuel cell stack, IEEE Transactions on Power Electronics, Vol. 22, issue 2, 2007

[4] G. Fontes : Modélisation et caractérisation de la pile PEM pour l'étude des interactions avec les convertisseurs statiques, Ph. D. thesis, ENSEEIHT, Toulouse (France), September 20 05

[5] X. Yu, M. R. Starke, L. M. Tolbert, B. Ozpineci: Fuel cell power conditioning for electric power applications: a summary, IET Electr. Power Appl., Vol. 1, issue 5, 2007

[6] S. Vighetti : Systèmes photovoltaïques raccordés au réseau : choix et dimensionnement des étages de conversion, Ph. D. thesis, INPG, Grenoble (France), September 2010

[7] B. Huang : Convertisseur continu-continu à rapport de transformation élevé pour applications pile à combustible, Ph. D. thesis, INPL, Nancy (France), May 2009

[8] B. V. Dang : Conception d'une interface d'électronique de puissance pour Pile à Combustible, Ph. D. thesis, UJF, Grenoble (France), December 2006

Impact of the Initial Transient Interruption Voltage (ITIV) on the Design and Operation of Hybrid Current-Injection DC Circuit Breakers

Andreas Jehle and Jürgen Biela
Laboratory for high power electronic systems, ETH Zürich
Email:jehle@hpe.ee.ethz.ch, URL:http://www.hpe.ee.ethz.ch

Keywords

≪HVDC≫, ≪Multiterminal HVDC≫, ≪Fault handling strategy≫, ≪Power transmission≫

Abstract

In hybrid DC circuit breakers with current-injection, the arc in the mechanical switch is extinguished with a current pulse. After the arc extinction, the voltage increase across the mechanical switch must be limited in order to avoid an arc reignition. The voltage across the MS is usually assumed to be equal to the voltage across the capacitor C_{CI} in the current-injection circuit, which increases slowly due to slow charging of the capacitor by the fault current. However, in reality the voltage across capacitor C_{CI} is usually not zero at the point in time of the arc extinction. This results in high transient voltages across the mechanical switch, the so called so-called initial transient interruption voltage (ITIV) after the arc extinguishes. However, a fast increasing ITIV can lead to a thermal reignition of the arc. Therefore, it is crucial for the design of a DC-CB to understand, which effects contribute to a high ITIV. In this paper, these causes and the influence of the ITIV are investigated. Furthermore, the beneficial influence of grading capacitors parallel to the MS is investigated. Finally, an optimization of a DC-CB including the ITIV and its results are presented for a typical implementation of a current-injection DC-CB.

1 Introduction

In recent years, interest in bulk HVDC transmission has significantly increased due to the need for offshore and long distance energy transmission with low losses. In addition, voltage source converters (VSC) enable a power reversal without the need for a voltage reversal, which is a first step to a meshed multi-terminal DC grid with low transmission losses [1]. One of the major remaining problems of HVDC transmission is to turn lines rapidly off, especially in case of a fault in a meshed DC grid. Besides turning off the complete DC grid [2], DC circuit breaker (DC-CB) are an attractive solution.

Such DC-CB must interrupt a (fault) current in the line, must block an increasing transient interruption voltage (TIV) across the DC-CB, and must dissipate the remaining energy in the line inductances for deenergizing the lines[1]. To interrupt a current and block the TIV, DC-CB can use either mechanical switches (MS),

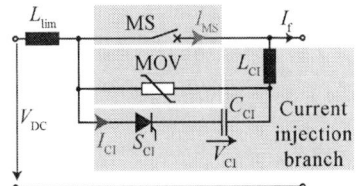

Fig. 1: Current-injection DC-CB with precharged capacitor C_{CI} ($V_{CI}(t_0) = -V_{C0}$): A current pulse I_{CI} is triggered with thyristor S_{CI}, which generates a ZCC in the MS and extinguishes the arc. After the arc extinction, the voltage slope across the MS is limited with capacitance C_{CI} in order to prohibit a reignition of the MS.

semiconductor switches or in so called hybrid circuit breaker (HCB) - a combination of both, which combines the fast interruption of semiconductors with the low on-state losses of MS.

One possible implementation of such a HCB is the current-injection DC-CB presented in [3]. In a current-injection DC-CB, the MS is opened under current with an arc, which must be extinguished by generating a zero current crossing (ZCC) in the MS. This is performed by injecting a single current pulse into the MS in opposite direction of the fault current, so that the current in the MS crosses zero. A common implementation of this topology is shown in Fig.1.

EPE'20 ECCE Europe

Assigned jointly to the European Power Electronics and Drives Association & the Institute of Electrical and Electronics Engineers (IEEE)

After the arc extinction caused by the ZCC, the voltage across the MS must be limited to avoid a reignition of the arc. In current-injection DC-CB, the voltage slope is limited by the capacitor C_{CI} (Fig.1), which is charged by the fault current. However, the capacitor voltage V_{CI} is usually not zero at the point in time of the ZCC/arc extinction. This causes (along other causes) a fast charging of the parasitic capacitances of the MS after the arc extinction [3, 4]. The resulting fast transient voltage across the MS is called Initial Transient Interruption Voltage (ITIV) [4] and it can lead to a reignition of the arc. Although the ITIV is generally recognized as possible problem, the different causes of the ITIV and its effect on the HCB design have not been investigated yet. Therefore, this paper presents a comprehensive investigation of the ITIV and its impact on the HCB design considering the HCB-topology in Fig.1. First, the general turn-off procedure of the current-injection DC-CB is shortly summarized in section 2. Thereafter, parameters, which influence the ITIV, are determined and the effect of the parasitic capacitance of the MS and of grading capacitors on the ITIV are investigated in section 3. Finally, the influence of the ITIV on an optimized design of a current-injection DC-CB are presented in section 4.

Remark: For the investigations, a current-injection DC-CB for a DC voltage of $V_{DC} = 200kV$ and a fault current of $I_{f,t_{open},max} = 10.6kA$ at the start of the current-injection is used as example. Furthermore, a 245kV-SF6 CB with a black box model for the arc [5] is used as example for the MS.

2 Turn-off procedure of current-injection DC-CB

In this section, the structure and the turn-off procedure of the current-injection DC-CB in Fig.1 is shortly explained [3, 6]. The current-injection DC-CB can basically be divided into three parallel branches:

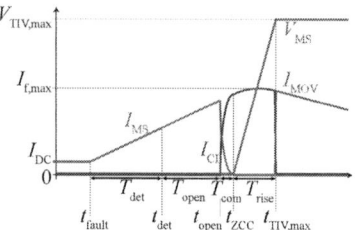

Fig. 2: Timeline of a turn-off of the current-injection DC-CB in Fig.1

1. The main current branch consisting of a MS, which conducts the current during normal operation at low conduction losses.

2. The current-injection branch consisting of a precharged capacitor C_{CI} with initial capacitor voltage $V_{CI}(t_0) = V_{C0}$ [3, 7, 8], a switch S_{CI} (e.g. a thyristor) and an inductance L_{CI}.

3. The energy dissipation branch consisting of an energy absorbing MOV to dissipate the stored energy of the line inductances and the current limiting inductance L_{lim} after the interruption of the fault current.

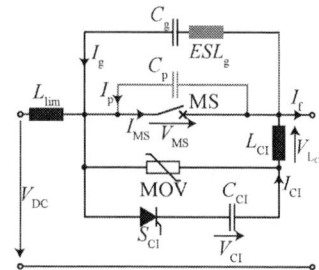

Fig. 3: Current-injection DC-CB with parasitic capacitor C_p of the MS. In addition, a capacitor C_g across the MS could be used as grading capacitor. (Parasitic elements are shown in grey)

The timeline of a turn-off process of the current-injection DC-CB is shown in Fig.2. During normal operation, the MS conducts the current at low conduction losses. In case of a fault at $t = t_{fault}$, the MS starts to open at t_{det} after a detection time T_{det}. The MS is opened within T_{open} under current. This causes an arc, which must be extinguished by the current-injection branch. The arc extinction is performed by injecting the current I_{CI} into the MS in opposite direction to the fault current I_f. The injected current results in a ZCC at $t = t_{ZCC}$ and the arc is extinguished. After the arc extinction, the voltage across the DC-CB increases to the maximum transient interruption voltage (TIV) $V_{TIV,max}$ before the current is commutated to the energy absorber MOV and then decreases to zero. During the rise time T_{rise} of V_{MS}, the voltage slope dv_{MS}/dt is limited by the capacitor C_{CI} so that a reignition of the arc is usually prohibited. However, an additional fast increasing voltage transient can be observed across the MS, which can reignite the arc. This transient voltage, the resulting post-arc current and the measures to avoid a reignition are investigated in the following section 3.

3 ITIV in current-injection DC-CB

As mentioned above, the voltage V_{MS} across the MS must be limited after the arc extinction in order to avoid a reignition of the arc. Reignitions can be either thermally or dielectrically triggered [9]. For DC-CB, dielectric reignitions typically are no problem since the ZCC is generated when the contact distance of the MS is sufficiently large. However, thermal reignitions can occur relatively easy. After the ZCC, the temperature of the arc channel/plasma and therefore the conductivity is still high. If the voltage across the MS increases too fast, the post-arc current can reheat the arc channel/plasma and lead to a reignition of the arc. Therefore, it is especially important to prohibit a fast increase of the voltage across the MS directly after the ZCC. This is ideally achieved

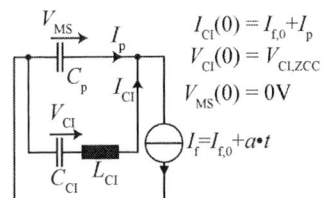

Fig. 4: Equivalent LC resonant circuit of the current-injection DC-CB for the calculation of the TIV including the ITIV under the assumption of a fault current with a constant slope of $a = \mathrm{d}I_f/\mathrm{d}t$ and $t_{ZCC} = 0$.

with the capacitor C_{CI}, which should be completely discharged when the arc extinguishes at $t = t_{ZCC}$, and has a voltage slope of $\mathrm{d}v_{CI}/\mathrm{d}t = i_f(t)/C_{CI}$. Ideally, this voltage slope is also the voltage slope of the TIV across the MS. However, a fast increasing transient voltage, the so called Initial Transient Interruption Voltage (ITIV), can be observed in addition to the ideal TIV due to the parasitic capacitance C_p and the inductor L_{CI} in the current-injection DC-CB (Fig.3), which represent together with capacitor C_{CI} a LC resonant circuit. The ITIV (with and without grading capacitors) and its effect on the arc extinction are investigated in the following sequence: The ITIV without grading capacitors C_g is inspected in section 3.1. The resulting post-arc current, which can lead to a reignition but also influences the ITIV, is investigated in section 3.2. Finally, the (required) improvement of the arc extinction with grading capacitors is shown in section 3.3.

3.1 ITIV

In this section, the causes of the ITIV are investigated for a DC-CB without grading capacitor C_g. An equivalent LC-resonant circuit (Fig.4) and a fault current $I_f(t) = I_{f,0} + a \cdot t$ with slope $a = \mathrm{d}I_f/\mathrm{d}t$ are assumed in the following. There, three effects contribute to the ITIV:

E1 $V_{CI,ZCC} \neq 0$: The first contribution results from the fact that the capacitor voltage $V_{CI,ZCC}$ is not zero at the point of time of the arc extinction $t = t_{ZCC}$. Depending on the fault current $I_{f,0}$ at t_{ZCC}, the capacitor voltage is $v_{CI,ZCC} = -\sqrt{V_{C0}^2 - L_{CI}/C_{CI} I_{f,0}^2}$ [4], if the arc voltage and the losses in the current-injection circuit are neglected. Since $V_{L_{CI}} = V_{arc} + V_{CI}$ during the current-injection, the arc voltage supports the current-injection and the capacitor voltage $V_{CI,ZCC}$ is actually even higher. For example, the capacitor voltage in Fig.5 is $V_{CI,ZCC} = -27.5kV$, while without the arc voltage, the capacitor voltage $V_{CI,ZCC}$ for the same current would be $-26.43kV$. In addition, the slope of the injected current I_{CI} is at least equal

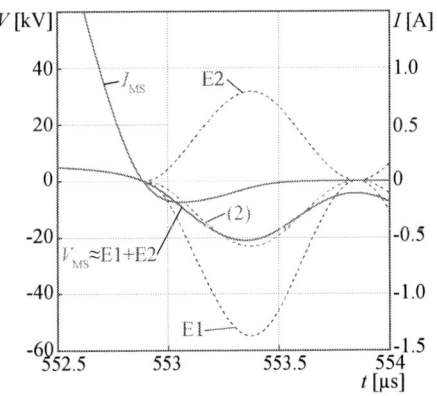

Fig. 5: The two main contributions to the ITIV without grading capacitor C_g are the capacitor voltage $V_{CI,ZCC}$ ($-27.5kV$) at t_{ZCC} (E1) and the voltage ($15.8kV$) across inductor L_{CI} due to the fault current slope a (E2). E1 and E2 add basically up to the MS voltage V_{MS}, since the third source of the ITIV and the ideal TIV add only a few volt. The approximation (2) of the MS voltage is also shown in the figure. Parameters are shown in Tab .I.

to the slope a of the fault current I_f at t_{ZCC}, which means that $V_{CI,ZCC} \geq L_{CI} \cdot a$ after t_{ZCC}. The capacitor voltage $V_{CI,ZCC}$ charges the parasitic capacitance C_p of the MS, which results in an additional oscillating voltage of approximately $V_{CI,ZCC}(\cos(\omega t) - 1)$ across the MS.

E2 $\mathrm{d}i/\mathrm{d}t$: A second contribution to the ITIV is the slope a of the fault current I_f. Since typically $C_{CI} \gg C_p$, most of the fault current I_f flows through inductor L_{CI} so that $\mathrm{d}I_{L_{CI}}/\mathrm{d}t \approx a$. The average voltage $a \cdot L_{CI}$ across inductor L_{CI} results in an oscillating voltage of approximately $a \cdot L_{CI}(1 - \cos(\omega t))$

across the MS. Since the fault current is in general increasing ($a > 0$), this effect can partly compensate the aforementioned effect E1 and decrease the amplitude of the ITIV. However, the fault current slope a also can be negative for short time periods, e.g. due to oscillations in the grid, and therefore increase the amplitude of the ITIV.

E3 Current sharing: A third contribution is that the current sharing between capacitors C_p and C_{CI} is not optimal at the ZCC ($C_{CI}I_p \neq C_pI_{CI}$). The current I_{CI} is equal to the fault current $I_{f,0}$ at the arc extinction at t_{ZCC}, when the current through the parasitic capacitance $I_p = I_{MS}$ is zero. As a result, the current $I_{f,0}$ leads to an oscillating voltage of approximately $I_{f,0}/(C_{CI}\omega)\sin(\omega t)$ across the MS. However, this third contribution is relatively small as long as no grading capacitors are present.

For illustration purposes, a typical ZCC is shown in Fig.5. The TIV including the ITIV for a fault current $I_f = I_{f,0} + a \cdot t$ with constant slope $a = \mathrm{d}I_f/\mathrm{d}t$ can be summarized with:

$$
V_{MS}(t) = \underbrace{\frac{\frac{1}{2}at^2 + I_{f,0}t}{C_{CI} + C_p}}_{\text{Ideal TIV}} + \underbrace{\frac{V_{CI,ZCC}C_{CI} - C_{CI}V_{CI,ZCC}\cos(\omega t)}{C_{CI} + C_p}}_{\text{E1: ITIV due to } V_{CI,ZCC}} - \underbrace{\frac{\frac{aC_{CI}}{\omega^2 C_p}(\cos(\omega t) - 1)}{C_{CI} + C_p}}_{\text{E2: ITIV due to slope } a \text{ of } I_f} - \underbrace{\frac{\frac{(I_pC_{CI} + (I_{f,0} + I_p)C_p)}{C_p\omega}\sin(\omega t)}{C_{CI} + C_p}}_{\text{E3: ITIV due to } I_{f,0} \text{ and } I_p}
$$

$$(1)$$

$$
\approx \frac{1}{C_{CI}}\left(\frac{1}{2}at^2 + I_{f,0}t\right) + V_{CI,ZCC}(1 - \cos(\omega t)) - aL_{CI}(\cos(\omega t) - 1) - \frac{I_{f,0}}{C_{CI}\omega}\sin(\omega t) \tag{2}
$$

$$
\omega = \sqrt{\frac{C_{CI} + C_p}{L_{CI}C_{CI}C_p}} \approx \sqrt{\frac{1}{L_{CI}C_p}} \tag{3}
$$

The approximation (2) (for $C_p \ll C_{CI}$ and $I_p = 0$) shows that the capacitor voltage $V_{CI,ZCC}$ at the ZCC, the inductance of the current-injection path L_{CI}, and the parasitic capacitance C_p (within ω) are the main parameters, which influence the ITIV. Changing these parameters, respectively V_{C0}, C_{CI} and L_{CI} for $V_{CI,ZCC} = -\sqrt{V_{C0}^2 - L_{CI}/C_{CI}I_{f,0}^2}$, allows to decrease the ITIV. Alternatively C_p could be increased artificially (section

Table I: Parameters of the simulated curves in Fig.5, Fig.8 and Fig.9.

		Fig.5	Fig.5 & Fig.9
Current-injection capacitance	C_{CI}	$240\mu F$	$35\mu F$
Current-injection inductance	L_{CI}	$2000\mu H$	$400\mu H$
Initial capacitor voltage of C_{CI}	V_{C0}	$40kV$	$40kV$
Capacitor voltage at t_{ZCC}	$V_{CI,ZCC}$	$-27.5kV$	$-23.5kV$
Fault current at t_{ZCC}	$I_{f,0}$	$10.4kA$	$9.818kA$
Current injection duration	$t_{ZCC} - t_{open}$	$552.88\mu s$	$109.67\mu s$
Fault current after t_{ZCC}	$I_f = I_{f,0} + t \cdot a = I_{f,0} + t \cdot 7.94A/\mu s$		
Parasitic capacitance of MS	C_p	$12pF$	$12pF$
Grading capacitor	C_g	-	$300 - 2500pF$
Parasitic inductance of C_g	ESL_g	-	$840nH$

3.3) to slow down the increase of the ITIV until the MS can block a high voltage.

The ITIV is not directly responsible the arc reignition, but by causing a post-arc current, which reheats the arc. Therefore, the post-arc current and the requirements for a reigniton are inspected in section 3.2. Additionally, the post-arc current after the ZCC is also missing in the equations and results in a deviation, which is also inspected in section 3.2.

3.2 Post-arc current & reignition

Shortly after the arc extinction, the plasma has still a relatively high conductivity, which also depends on the current slope $\mathrm{d}i/\mathrm{d}t$ before the ZCC [10]. Depending on this conductivity and the voltage V_{MS} across the MS after the ZCC, a so called post-arc current can flow through the plasma in the MS after t_{ZCC}. If V_{MS} is low, the arc channel cools down further after the ZCC and the conductivity decreases so that also the post-arc current decreases as a result of the decreasing conductivity. As soon as the conductivity is sufficiently low, even relatively high voltages across the MS only lead to small post-arc currents and do not cause a reignition. However, high voltages across the MS relatively shortly after the ZCC result in high post-arc currents and an increasing arc temperature, what finally leads to a thermal reignition.

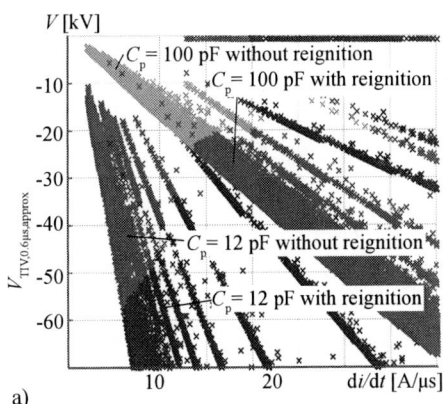
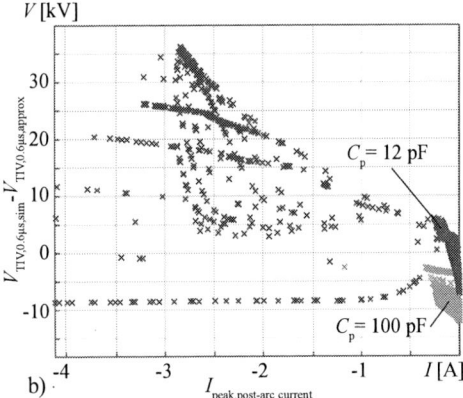

Fig. 7: a) At $t = t_{ZCC} + 0.6\mu s$, the voltage across the MS is according to approximation (2) already in the range of several kilovolts. For such a fast voltage increase, only a very low di/dt before the ZCC results in a successful arc extinction. Remark: In approximation (2), it is assumed that the interruption is successful. However, no real ITIV exists for reignitions. b) The error between simulated values of the ITIV at $t = t_{ZCC} + 0.6\mu s$ and the ITIV calculated according to approximation (2) is due to the post-arc current in the order of several kilovolts and must be taken into account during the optimization.

Therefore, relatively high post-arc currents are an indicator that the arc is close to a reignition [11]. For example post-arc currents in a well designed SF6 CB must be below a few hundred milliampere [10]. Consequently, a well-designed DC-CB must prohibit a fast increase of the ITIV while the conductivity is still high.

The duration during which the increase of the ITIV must be low is mainly defined by the slowest time constant of the arc, respectively of the arc model [12]. However, the duration depends also on the ITIV itself, since the conductivity decreases faster if the post-arc current is low, respectively if the ITIV is low. An indication until which point in time the ITIV must be limited is the amplitude of the post-arc current. As soon as the post-arc current starts to decrease, the energy for heating the arc decreases and the interruption will eventually be successful. A further increase of the ITIV after the peak-post-arc current is not relevant for a successful interruption (conductivity decreases faster than the voltage increases). Since the time of the peak-post-arc current cannot be predicted without simulations and also differs slightly depending on the conditions of the ZCC, an estimated value for the time of

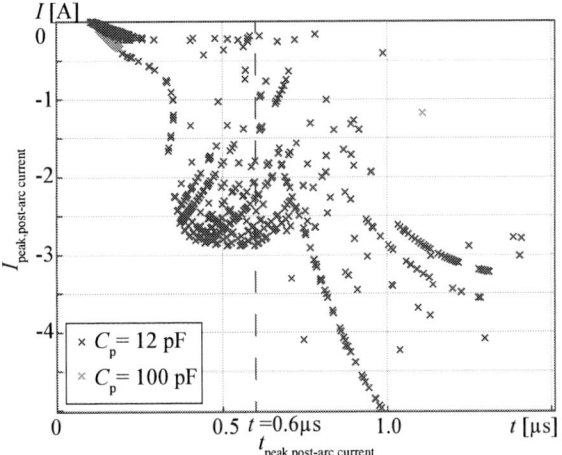

Fig. 6: The low post-arc currents, which signal a successful interruption without high probability of a reignition, have the peak-post-arc current below $1\mu s$, usually between $0.1\mu s - 0.3\mu s$. As soon as the post-arc current starts to decrease, a thermal reignition is unlikely. Therefore, the increasing ITIV is important up to this point in time. For comparing the increase of the ITIV with parameters in Tab.IV, $0.6\mu s$ are used.

the peak-post-arc current is used for evaluating the increase of the ITIV. Since the ITIV is continuously increasing, basically any point in time close to a typical peak-post-arc current can be used to evaluate the increase of the ITIV. In the following, $t = t_{ZCC} + 0.6\mu s$ is used, which is also used for transients in AC-CB [13].

By relating the ITIV at $t = t_{ZCC} + 0.6\mu s$ to the conductivity of the arc at the ZCC, respectively to the current slope di/dt before the ZCC on which the conductivity depends [10], a limit for successful arc extinctions is given. This relation is used in the optimization of the DC-CB (in Fig.13) to determine if a successful ZCC is possible for a given set of parameters. Furthermore, equation (2) with $v_{CI,ZCC} = -\sqrt{v_{C0}^2 - L_{CI}/C_{CI} I_{f,0}^2}$ is used for approximating the ITIV. The current slope at the ZCC is estimated with $di/dt \approx v_{CI,ZCC}/L_{CI}$. This allows to speed up the optimization of a current-injection DC-CB (section 4). However, if exact values are required a black box model of the arc is required.

For justifying the choice of $t = t_{ZCC} + 0.6\mu s$ for the peak post-arc current, the most relevant results for $C_p = 12pF$ and $C_p = 100pF$ from the optimization presented in section 4 are shown in the following. As can be seen in Fig. 6, the peak-post-arc current occurs usually within $1\mu s$ after the ZCC, so that $t = t_{ZCC} + 0.6\mu s$ is a good choice. In Fig.7, the relation between the ITIV at $t_{ZCC} + 0.6\mu s$ and the current slope di/dt before the ZCC is shown, which is used for estimating if a set of parameters will result in a successful arc extinction for all possible fault currents. It can be observed, that small values of the parasitic capacitance C_p result in a relatively fast increasing ITIV. Therefore, the di/dt before the ZCC must be relatively low for a successful interruption. However, for negative fault current slopes at the ZCC, e.g. due to oscillations, the di/dt is relatively high. Accordingly, the increase of the ITIV must be low to avoid a reignition, which is, however, hard to achieve for low fault currents since in this case the

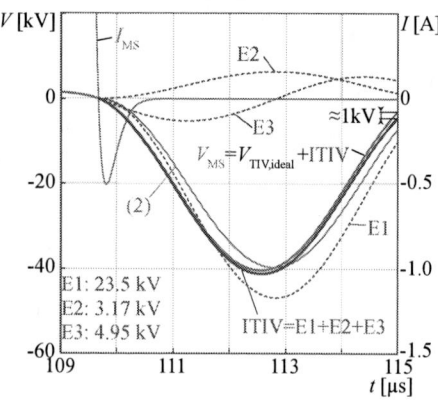

Fig. 8: The use of the grading capacitor C_g results in a lower contribution E2 (lower inductance value L_{CI} due to higher allowed di/dt), a higher contribution E3 due to the capacitor discharge current I_g and a higher frequency of the ITIV. Additionally, the smaller inductance L_{CI} allows to decrease capacitance C_{CI}, which results in an faster increase of the ideal TIV. The approximation (2) of the MS voltage is also shown in the figure. Parameters are shown in Tab .I.

capacitor voltage $V_{CI,ZCC}$ is still high at the ZCC. As a result, the increase of the ITIV must be limited, which is possible by increasing the capacitance C_p. This is achieved by adding a grading capacitor in parallel to the MS as explained in section 3.3.

Remark: Transients in AC-CB

In AC-grids, a so called initial transient recovery voltage (ITRV) exists, which is not related to a current-injection circuit but to traveling waves on the busbar and the reflections from the first discontinuity along the busbar [13]. Nevertheless, the ITRV is as the ITIV a fast transient voltage, which can lead to a reignition shortly after the ZCC. The authors of [13] provide an insight into the ITRV, which an AC-CB should be able to handle. For the standard values of ITRV, a multiplying factor f_i is used to determine the first peak u_i as a function of the RMS value of the short-circuit breaking current. This results for example for a 245kV/40kA MS in a maximum allowed ITRV of 2.76kV after $0.6\mu s$ (for $di/dt = 21.33A/\mu s$). In comparison, the ITIV in the DC-CB is in the range of several tens of kilovolt.

3.3 ITIV with grading capacitors

Grading capacitors are typically used for equal voltage sharing of series connected MS or to limit transient voltages. A grading capacitor is connected in parallel to the MS, respectively to the parasitic capacitance C_p to slow down the increase of the ITIV. The parasitic capacitance C_p is usually relatively small ([14] uses $12pF$ for SF6 CB and vacuum CB have a stray capacitance below $22pF$ [15]) compared to grading capacitors ($C_g = 300pF \ldots 2500pF$ with $L_g = 840nH$ [14]). The influence of the parasitic capacitor is despite the large grading capacitor still detectable, which is not the case for the effect of the parasitic inductance L_g. Therefore, equations (1) & (2) can be reused in this case by replacing $C_p = C_g + C_p$ and $L_{CI} = L_{CI} + L_g$. As a result of adding the grading capacitor C_g, the frequency ω (3) of the ITIV decreases. Therefore, the voltage increase after the ZCC is slower, so that the post-arc current is lower, which leads to a faster decrease of the conductivity.

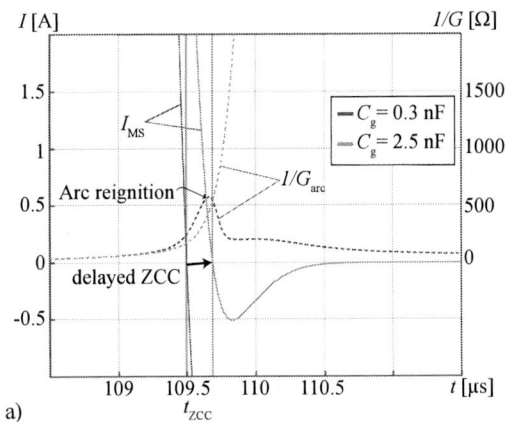
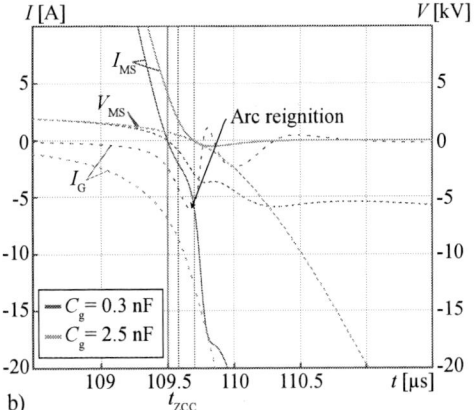

a) b)

Fig. 9: Influence of the grading capacitor on the ZCC. a) Higher values for the grading capacitor C_g result in a lower conductivity G_{arc} of the arc plasma in the MS at t_{ZCC} and therefore improve the arc extinction. b) Higher values for the grading capacitor C_g decrease the slope of the MS voltage V_{MS} after the ZCC. However, higher values of the grading capacitor C_g also increase the discharge current I_g at the ZCC, which increases the third contribution (E3) to the ITIV.

In addition to decreasing the voltage increase after the ZCC, grading capacitors also improve the interruption capability by decreasing the conductivity of the arc at the ZCC. While the current in the MS decreases due to the injected current in the opposite direction of the fault current, the conductivity starts to decrease. This leads to an increasing arc voltage. If a grading capacitor C_g is in parallel to the MS, the increasing arc voltage forces a part of the fault current into the parallel capacitor C_g.

Close to the ZCC, the arc voltage decreases again and the grading capacitance C_g is discharged across the MS. Due to this discharge current of capacitor C_g, the injected current must be slightly higher at the ZCC and the ZCC is delayed. As a result, the duration with low MS current I_{MS} before the ZCC is longer and the average di/dt is lower, so that the plasma in the arc has more time to cool down before the ZCC. The resulting lower conductivity of the arc is beneficial for the interruption [16, 17].

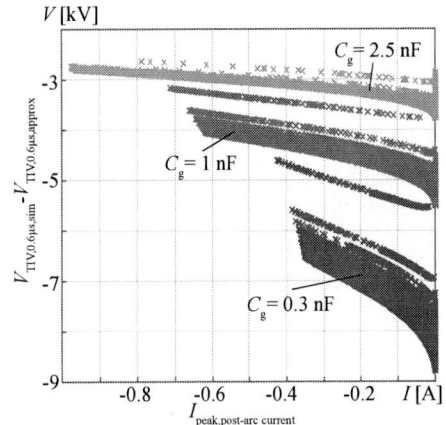

Fig. 10: Error between simulated TIV at $t = t_{ZCC} + 0.6\mu s$ and the TIV according to approximation (2). The error of the approximation is still in the range of a few kilovolts. However, the main error results due to neglecting the discharge current I_g of the grading capacitor at the ZCC. The error due to the post-arc current is smaller due to the decreased post-arc current.

A side effect of the discharge current of C_g is that the contribution to the ITIV due to non-ideal current sharing between capacitors $C_g + C_p$ and C_{CI} (E3) increases considerably (Fig.8). The discharge current of capacitor C_g at the ZCC can be considered in equation (1) with $I_p = I_g$. However, in the approximation (2), the discharge current of C_g is neglected since I_g depends on the arc behavior and, therefore, cannot be determined without a simulation of the arc. This results in an additional deviation/error of the values of the approximation (2) from the simulated ITIV values.

Table II: Discharge currents of the grading capacitor C_g at t_{ZCC} for the ZCC depicted in Fig.9b

C_g	0.3 nF	1.0 nF	2.5 nF
$I_{g,ZCC}$	2.4 A	6.3 A	12.4 A

The two effects of using grading capacitors (lower conductivity at the ZCC and slower increase of the TIV) result in lower post-arc currents and, therefore, in an improved interruption of the arc. For illustration, the waveforms for two ZCCs in the MS are depicted in Fig.9a) for the parameters given in Tab.I for different grading capacitance values. As mentioned, the ZCC is delayed for increasing values of C_g and therefore also the conductivity at the ZCC is lower. In Fig.9b), the lower increase of the ITIV can be observed. Both effects combined result in a successful interruption for higher values of grading capacitor C_g.

In Fig.9b), also the grading capacitor discharge current I_g can be observed, which increases for higher grading capacitance values (Tab.II). As a result, the error between the simulated TIV after $0.6\mu s$ and the ap-

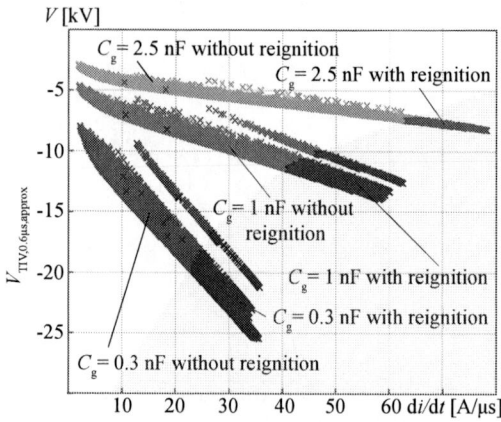

Fig. 11: Current slope di/dt before the ZCC versus TIV at $t = t_{ZCC} + 0.6\mu s$. Due to the beneficial effect of the grading capacitor, which decreases the di/dt shortly before the ZCC and slows down the increase of the ITIV, the interruption can be successful for higher di/dt. The area for an unsuccessful interruption is highlighted in yellow.

proximation (2) (Fig.10b) is still in the range of several kilovolt. The reason for this is the neglected discharge current in approximation (2) (despite the fact that the post-arc current is lower, which results in a lower error caused by this post-arc current). Fig. 11 shows the overall advantage of grading capacitors. Since grading capacitors slow down the increase of the TIV across the MS, the conductivity at the ZCC can be relatively high. Therefore, the interruption is successful even for high current slopes di/dt.

3.4 Summary

In the following, the main findings of section 3 are summarized:

- A fast transient voltage (ITIV) across the MS occurs after the arc extinction due to:

  1. the capacitor voltage $V_{CI,ZCC}$ at the ZCC,
  2. the voltage across the inductor L_{CI} from the fault current slope a,
  3. and the nonideal current sharing between the capacitances C_{CI} and $C_p + C_g$.

- By approximating the ITIV shortly after the ZCC (e.g. at $t = t_{ZCC} + 0.6\mu s$) and the current slope di/dt before the ZCC, a prediction is possible if the arc extinction is successful. However, compared to AC-CB, the transient voltage is much higher and with a low parasitic capacitance, a successful arc extinction is harder to achieve.

- Grading capacitors in parallel to the MS decrease the di/dt before the ZCC and delay the ITIV, which allows DC-CB designs with higher current slopes of the injected current.

$$I_{MS} \rightarrow \boxed{\frac{dg_1}{dt} = \frac{1}{P_1 \tau_1} g_1^{-0.6} I_{MS}^2 - \frac{1}{\tau_1} g_1} \boxed{\frac{dg_2}{dt} = \frac{1}{P_2 \tau_2} g_2^{-0.1} I_{MS}^2 - \frac{1}{\tau_2} g_2} \boxed{\frac{dg_3}{dt} = \frac{1}{P_3 \tau_3} I_{MS}^2 - \frac{1}{\tau_3} g_3}$$

V_{MS}

Fig. 12: Composite black box model of an arc, which combines the classical Cassie-model, the Mayr-model and a hybrid model [18–20].

Table III: Assumed arc parameters of the black box model in Fig.12 from [19].

P_1	P_2	P_3	τ_1	τ_2	τ_3
$3178.64 \cdot 10^3$	$139.61 \cdot 10^3$	$6345W$	$2.541\mu s$	$0.5082\mu s$	$0.1016\mu s$

4 Influence of the ITIV on the DC-CB design

In this section, the optimization of the current-injection DC-CB given in Fig.3 is presented and the influence of the parasitic capacitance C_p, the grading capacitance C_g and the parasitic inductance L_g on the design are discussed. For the optimization, the procedure in Fig.13 is used. In the optimization, the parameters L_{CI}, C_{CI} and V_{C0} of the current-injection circuit are varied in order to identify the best design. For a given set of parameters, it is determined in the first step S1 if a ZCC can be generated for the maximum fault current value.

In the second step S2, the di/dt and the ITIV at $t = t_{ZCC} + 0.6\mu s$ are calculated and compared to the di/dt and the ITIV at $t = t_{ZCC} + 0.6\mu s$ of previous simulations to asses if a successful arc extinction is possible, and, if not, to skip the the time consuming third step S3. This step requires results of the next step S3, so that it is not executed in the first iterations of the optimization. In the third step S3, the current-injection is simulated for different current slopes and current amplitudes. There, a 245kV-SF6 CB with the black box model of the arc [5] shown in Fig.12 and with the parameters in Tab.III is used for the MS. The results of this simulation are also stored for the comparison performed in the previous step S2 of the optimization. Since the simulation requires a relatively long time due to the short time constants of the arc model and its non-linearity, unnecessary simulations are disregarded in steps S1 and S2. Finally, a Pareto-front is plotted based on all successful combinations of the parameters L_{CI}, C_{CI}, and V_{C0} for the operating parameters in Tab.IV.

The optimization shows that no successful DC-CB design is possible without a grading capacitor. The results in Fig.14 show that increasing values for the grading capacitances C_g result in a decrease of the required values of the inductor L_{CI} and the capacitor C_{CI} of the current-injection circuit. The decrease of the required value of inductance L_{CI} results from the higher allowed di/dt values due to the reduced ITIV. As a consequence, the capacitance C_{CI} can be decreased since less energy is required for generating the required current pulse. A higher initial capacitor voltage V_{C0} requires a higher inductance L_{CI} for generating the same low di/dt. However, the

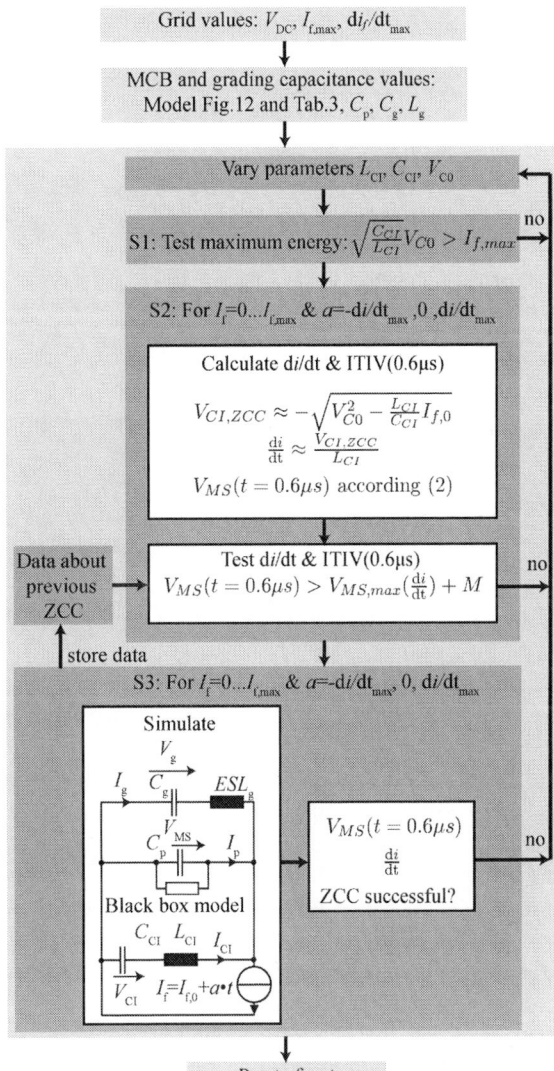

Fig. 13: Optimization procedure of DC-CB with ITIV: The interruption process is simulated with the black box model for different fault scenarios during the optimization. Due to the small time step, the simulation requires a relatively long time. By storing the ITIV at $t = t_{ZCC} + 0.6\mu s$ and the di/dt of previous simulations, the success of other combinations L_{CI}, C_{CI} and V_{C0} can be estimated and similar to the case that the stored energy is not sufficient for generating a ZCC for all fault currents, unnecessary simulations are avoided.

required capacitance value of C_{CI} decreases for higher initial capacitor voltages V_{C0}. Therefore, a trade-off between the capacitance C_{CI} and the inductance L_{CI} must be made. A change of the parasitic inductance L_g has no impact on the results.

5 Conclusions

The paper investigates the processes in the Current-Injection DC-CB after the arc extinction and its influence on the design of the DC-CB. The arc in the mechanical switch (MS) in the DC-CB is extinguished by generating a zero current crossing (ZCC) in the MS with a current pulse in opposite direction of the fault current. However, the conductivity of the arc channel is still high for the first microseconds after the ZCC. A high voltage across the MS must be prohibited during this time period, since the voltage results in high post-arc currents, which heat up the arc channel and possibly reignite the arc. However, the transient interruption voltage (TIV), which is ideally slowly increasing, includes a high fast transient voltage, the so called initial transient interruption voltage (ITIV).

Table IV: Specifications used for the DC-CB optimization

Nominal DC voltage	$V_{DC} = 200kV$
Maximum fault current at t_{open}	$I_{f,t_{open},max} = 10.6kA$
Maximum fault current slope	$a = \pm 7.94 A/\mu s$
Range of current-injection capacitance	$C_{CI} = 2.5 - 200\mu F$
Range of current-injection inductance	$L_{CI} = 0.15 - 10mH$
Range of initial capacitor voltage	$V_{C0} = 20 - 200kV$

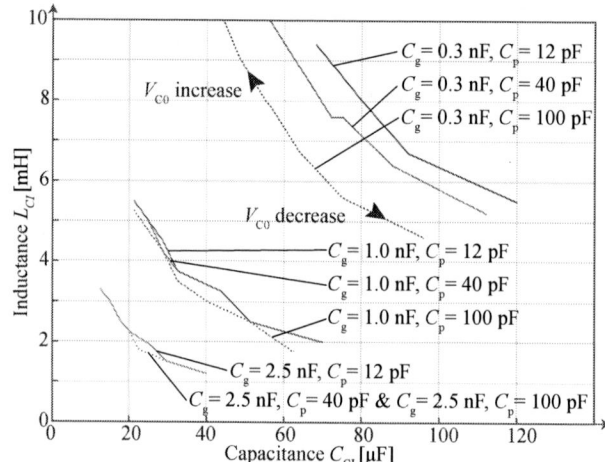

Fig. 14: Result of the optimization: Higher values for the parasitic capacitance C_p and for the grading capacitance C_g allow to decrease the current-injection inductance L_{CI} since higher current slopes di/dt are possible. As a result, the capacitance C_{CI} can also be decreased. The influence of the parasitic capacitance C_p is at least for small grading capacitance values still high.

This ITIV is mainly determined by the capacitor voltage $V_{CI,ZCC}$ at the ZCC, the di/dt of the increasing fault current and the unequal current sharing between the parallel capacitors C_p, C_{CI} and C_g in the DC-CB. Simulating the current-injection DC-CB with a black box arc model shows that the ITIV rises within a relatively short time to high values, what could reignite the arc. Only with relatively small current slopes before the ZCC a successful arc extinction is possible. However, these small current slopes cannot be generated for all fault current values and fault current slopes. Accordingly, the rise of the ITIV must be delayed by additional grading capacitors in parallel to the MS in order to always guarantee a successful interruption. The optimization shows that grading capacitors are a successful means against this ITIV and enable a successful arc extinction. Higher grading capacitor values allow in addition a significant decrease of the component values in the injection circuit.

Acknowledgment

This project is carried out in the frame of the Swiss Centre for Competence in Energy Research on the Future Swiss Electrical Infrastructure (SCCER-FURIES) with the financial support of the Swiss Commission for Technology and Innovation (CTI - SCCER program).

References

[1] C. Franck, "HVDC Circuit Breakers: A Review Identifying Future Research Needs," *IEEE Trans. Power Del.*, vol. 26, no. 2, pp. 998–1007, April 2011.

[2] D. Schmitt, Y. Wang, T. Weyh, and R. Marquardt, "DC-side fault current management in extended multiterminal-HVDC-grids," in *9th Int. Multi-Conf. on Systems, Signals and Devices (SSD)*, March 2012.

[3] A. N. Greenwood and T. H. Lee, "Theory and application of the commutation principle for HVDC circuit breakers," *IEEE Trans. Power App. Syst.*, vol. PAS-91, no. 4, pp. 1570–1574, July 1972.

[4] *Technical requirements and specifications of state-of-the-art HVDC switching equipment.* CIGRE, 2017.

[5] A. Ahmethodzic, M. Kapetanovic, K. Sokolija, R. P. P. Smeets, and V. Kertesz, "Linking a physical arc model with a black box arc model and verification," *IEEE Trans. Dielectr. Electr. Insul.*, vol. 18, no. 4, pp. 1029–1037, August 2011.

[6] Y. Zhang, Z. Shi, Q. Wang, Z. Gao, S. Jia, and L. Wang, "Experimental investigation on HVDC vacuum circuit breaker based on artificial current zero," in *27th Int. Symp. on Discharges and Electrical Insulation in Vacuum (ISDEIV)*, vol. 2, Sept 2016.

[7] T. Heinz, V. Hinrichsen, L. R. J´anicke, E. D. Taylor, and J. Teichmann, "Direct current interruption with commercially available vacuum interrupters," in *Int. Symp. on Discharges and Electrical Insulation in Vacuum (ISDEIV)*, Sept 2014, pp. 425–428.

[8] A. Jehle and J. Biela, "Design procedure and control of a hybrid circuit breaker with adaptable pulse current injection," in *Int. Power Electronics Conf. (IPEC-Niigata -ECCE Asia)*, May 2018, pp. 1509–1516.

[9] K. Niayesh and M. Runde, *Power Switching Components.* Springer International Publishing, 2017.

[10] R. P. P. Smeets, V. Kertesz, S. Nishiwaki, and K. Suzuki, "Performance evaluation of high-voltage circuit breakers by means of current zero analysis," in *IEEE/PES Transmission and Distribution Conference and Exhibition*, vol. 1, 2002, pp. 424–429 vol.1.

[11] R. P. P. Smeets and V. Kertesz, "Evaluation of high-voltage circuit breaker performance with a validated arc model," *IEE Proc. - Generation, Transmission and Distribution*, vol. 147, Mar 2000.

[12] R. SMEETS1 and V. KERTÉSZ, "A new arc parameter database for characterisation of short-line fault interruption capability of high-voltage circuit breakers," *CIGRE*, 2006.

[13] *Guide for application of IEC 62271-100 and IEC 62271-1 - Part 2 Making and breaking tests.* CIGRE, 2006.

[14] *Operating environment of voltage grading capacitors applied to high voltage circuit-breakers.* CIGRE, 2009.

[15] T. Betz and D. Koenig, "Influence of grading capacitors on the breaking capability of two vacuum circuit-breakers in series," in *Proc. Int. Symp. on Discharges and Electrical Insulation in Vacuum (ISDEIV)*, vol. 2, Aug 1998, pp. 679–683 vol.2.

[16] R. P. P. Smeets, V. Kertesz, D. Dufournet, D. Penache, and M. Schlaug, "Interaction of a vacuum arc with an SF6 arc in a hybrid circuit breaker during high-current interruption," in *Int. Symp. on Discharges and Electrical Insulation in Vacuum*, vol. 1, 2006, pp. 220–223.

[17] A. Ahmethodzic, R. P. P. Smeets, V. Kertesz, M. Kapetanovic, and K. Sokolija, "Design improvement of a 245-kV sf_6 circuit breaker with double-speed mechanism through current zero analysis," *IEEE Transactions on Power Delivery*, vol. 25, no. 4, pp. 2496–2503, 2010.

[18] H. Urai, Y. Ooshita, M. Koizumi, N. Yaginuma, M. Tsukushi, and R. P. P. Smeets, "Estimation of 80kA short-line fault interrupting capability in an SF6 gas circuit breaker based on arc model calculation," in *17. Int. Conf. on Gas Discharges and Their Appl.*, Sep. 2008, pp. 129–132.

[19] A. Ahmethodzic, M. Kapetanovic, K. Sokolija, R. P. P. Smeets, and V. Kertesz, "Linking a physical arc model with a black box arc model and verification," *IEEE Trans. Dielectr. Electr. Insul.*, vol. 18, no. 4, pp. 1029–1037, August 2011.

[20] H. A. Darwish and N. I. Elkalashy, "Universal arc representation using EMTP," *IEEE Trans. Power Del.*, vol. 20, no. 2, pp. 772–779, April 2005.

Four Quadrant Bus-Tie Switch for Protection of Shipboard Power Systems

Gabriele Ulissi[1], Seong-Yong Lee[2] and Drazen Dujic[1]

[1]Power Electronics Laboratory
École polytechnique fédérale de Lausanne (EPFL) 1015 Lausanne, Swtzerland
[2]Hyundai Electrical & Energy Systems (HE) 16891 Yongin, Republic of Korea
Email: gabriele.ulissi@epfl.ch, lee.seongyong@hyundai-electric.com, drazen.dujic@epfl.ch
URL: http://pel.epfl.ch

Acknowledgments

This work has been supported by Hyundai Electric & Energy Systems Co., LTD., Republic of Korea.

Keywords

≪Protection Device≫, ≪Marine≫, ≪DC Grid≫, ≪Solid-State≫.

Abstract

The desire to increase the voltage of DC shipboard power distribution networks to the medium voltage level derives from the pressure to reduce the operating costs of such systems by increasing their efficiency. Solid state bus-tie switches are accepted to be an essential component of such installations at the low voltage level, as they allow system reconfiguration and prevent fault propagation through ultrafast fault current identification and interruption. Nevertheless, the lack of standardisation in medium voltage DC shipboard power systems hinders the development of such technologies as custom, *ad hoc* solutions must be found according to the selected voltage level. This paper presents a solid state bus-tie switch topology that is scalable in both power and voltage rating and relies exclusively on existing, commercially available technologies. This provides a simple, readily employable solution with the flexibility needed to bridge the technological gap in the time required for medium voltage system operating voltages to become standardised. This paper presents the prototype of the bus-tie switch and validates its scalability through extensive experimental tests.

Introduction

DC power distribution networks (PDNs) have been widely reported to provide increased flexibility of operation and efficiency in existing commercial applications [1, 2]. A voltage increase in such networks from the low voltage (LV) level to the medium voltage (MV) level would further increase the system efficiency, with the corresponding advantages in terms of operating costs, and enabling an increase of maximum installed power over the 20 MW - 30 MW that are generally accepted to be achievable in an LV system [3, 4, 5, 6]. Due to their safety critical role, shipboard PDNs at all voltage levels employ redundancy to avoid system-wide failures in the event of faults [7, 8]. To achieve this, the PDN is separated into multiple switchboards each clustering various loads and power supplies, that are then interconnected through the use of solid state bus tie switches (SSBTSs). In addition to the reconfigurability offered by the solutions, these devices prevent the propagation of a fault by quickly isolating the malfunctioning sector in the event of a fault, effectively acting as a first line of defence and providing selectivity as a part of the protection coordination scheme in which they operate [8, 9, 10]. There are several characteristics that an SSBTS must have in order to fulfil this role, differentiating it from a circuit breaker in the traditional sense:

- Provide current interruption in the range of a few μs.
- Have limited conduction losses, subordinated to its protection ability.
- Allow four-quadrant operation.
- Include on-board fault detection logic, to minimise reaction time.

Due to the lack of standardisation of MVDC shipboard PDNs, that include voltage ranging from 3 kV up to 25 kV, a standard SSBTS solution for bus interface is not available [11, 12]. In this context, a scalable SSBTS that can operate at increasing voltage and power ratings through series and parallel connection of standardised building blocks can provide significant advantages in terms of flexibility, providing a solution based on available technology that can accommodate the different needs of MVDC systems as the industry moves towards standardisation. This paper presents such a scalable SSBTS topology and it's implementation in two identical prototype units, evaluating their ability to achieve increased current and voltage ratings in parallel and series connection, respectively.

Scalable SSBTS Topology

Fig. 1a displays the proposed SSBTS topology, based on a well known four quadrant switch and including the addition of protective circuitry to enable SSBTS operation. This additional circuitry includes metal oxide varistors (MOVs) to limit the voltage on the device terminals, an *RC* snubber to limit voltage peaks on the active semiconductor device upon current interruption. Additionally, a current rate limiting inductor L_{didt} is inserted in the current path to limit the current rise rate in the event of a fault. This allows for controllable (depending on the selected inductor value) addition of reaction time for the control to detect the fault and turn off the device. A general rule for the sizing of this inductor, based on the system in which it operates, is provided. The antiparallel diode of the inductor, D_L, provides a self contained freewheeling path that allows for the dissipation of the stored energy of L_{didt} through the diode and inductor's own internal resistance upon the device opening. Note that the device only needs to be connected between the positive terminals of the DC bus, as shown in Fig. 2. Fig. 3 displays the operation of the topology during a fault, time instant by time instant. Before instant t_0, the device is in conduction and no fault condition is present. The current path is through two of the rectifier diodes (depending on the current direction), the IGBT and current rate limiting inductor. Then, at time t_0, a fault happens and a voltage appears across the SSBTS terminals as a consequence. This results in a progressive increase of the current in the device at a rate determined by the applied voltage and L_{didt}. At time t_1 the device has detected the fault and turned off the IGBT. This initiates the interruption process and the path of the current goes from being that represented in Fig. 1b, to that in Fig. 1c. The current that was conducted by the IGBT now flows through the parallel RC snubber, gradually increasing the voltage and slowing the current rise. At time t_2 the voltage on the snubber is sufficient to stop the current increase by reaching the same value as the DC voltage applied at the SSBTS terminals. This forward biases diode D_L that enters conduction allowing the current level in L_{didt} to remain almost constant for the duration for the rest of the interruption process. After t_2 the voltage on the snubber keeps increasing until at t_3 it reaches

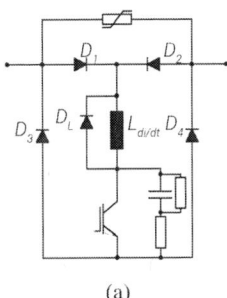
(a)

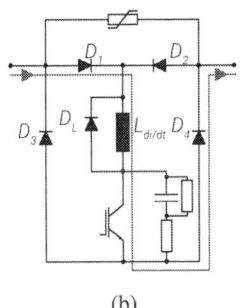
(b)

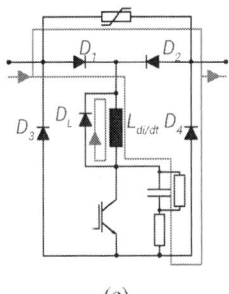
(c)

Fig. 1: (a) Proposed scalable SSBTS topology; (b) SSBTS current path during conduction; (c) SSBTS current path during breaking.

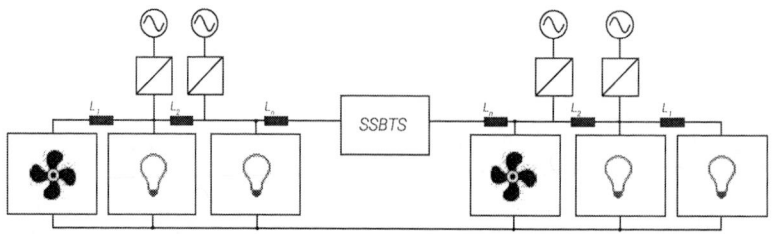

Fig. 2: The SSBTS separates switchboard clustering power courses and loads. The device is connected only to the positive terminals of the DC bus, and includes a current rate limiting inductor to limit current rise rate in the event of a fault.

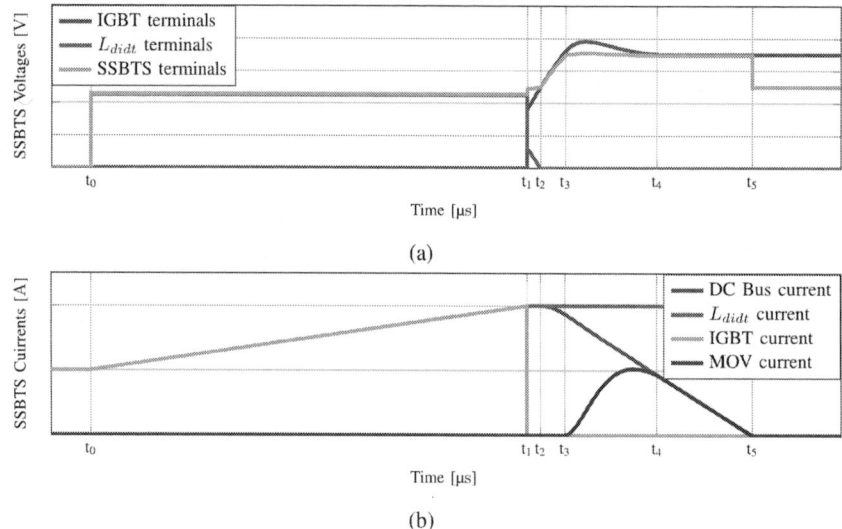

Fig. 3: (a) View from above of the SSBTS prototype, without current rate limiting inductor; (b) Side view of the prototype with active cooling fans.

the clamping voltage of the MOV. The MOV therefore enters conduction maintaining the voltage at its terminals (and therefore at the SSBTS terminals) at a constant value. Between t_3 and t_4 the voltage on the snubber is higher than that at the MOV terminals. This is due to the internal stray inductance of the device. The DC bus current in this interval drops at a rate determined by the difference between the MOV clamping voltage and the voltage applied to the SSBTS terminals. From t_4 onwards the whole DC bus current is conducted through the MOV and none through the snubber, until at t_5 the current reaches 0 A and the interruption is complete.

The analysis of the operation of the topology highlights how its ability to be series connected in a simple and effective way makes it particularly suitable for operation at increased voltage ratings. This is mainly due to three characteristics:

- Connecting the terminals of the switch is a MOV able to clamp the voltage at a predetermined value. This guarantees correct voltage sharing of series connected SSBTS units during breaking.
- The current in the L_{didt} current rate limiting inductor is allowed to freewheel internally, without requiring access to the negative DC bus bars, as is the case in other topologies [8].
- The current rise rate in series or parallel connected devices at the time when the fault takes place

is determined by L_{didt}, and will ensure all SSBTS units are conducting the same current level. On the other hand, the diode bridge structure of the topology forces current flow through three semiconductor devices in the *ON* state, causing an increase of losses compared to other existing topologies. Nevertheless, as the SSBTS is first and foremost a protection device, an increase in losses is outweighed by an increased protection capability that allows an extended range of employment for the device. Additionally, the conducted current in the SSBTS is rarely equal to it's nominal current, as a balanced operation of the PDN with generated power similar to the load power in each switchboard is generally preferred [9]. Overall, the advantages offered by the simplicity of connection, single active semiconductor device and most importantly scalability result in the topology being a very suitable for the application. Note also that even in series or parallel connection the efficiency of the device remains the same, as the power conduction ability increases at the same rate as device losses.

SSBTS Prototype Units

Two SSBTS prototype units as in Fig. 4a and 4b are assembled with the goal of validating the scalability of current and voltage rating through parallel and series connection respectively. The prototypes result from the downscaling of an SSBTS operating in a 1 kV, 8 MW system. This full-scale device has a nominal current of 8 kA and a maximum interruption current of 16 kA, and this ratio of 2 : 1 is maintained in the presented prototype. The ratings for the device assembled in the laboratory are:

- A nominal voltage of the PDN of 500 V.
- A nominal device current I_{nom} of 100 A.
- A maximum breaking current I_{max} of 200 A.
- An interruption time $t_{reaction}$ of 10 μs.

It is worth noting that the relatively low value of I_{max} with respect to I_{nom} can be achieved thanks to the fast interruption time. Upon a fault taking place, the current in the SSBTS increases gradually limited by the inductance of L_{didt}. If the inductance is appropriately sized, the current in the device will still be below I_{max} once the SSBTS identifies the fault and interrupts. The inductor L_{didt} is sized according to:

$$L_{didt} = \frac{V_{DC}}{I_{max} - I_{nom}} t_{reaction} = \frac{500\,V}{100\,A} * 10\,\mu s = 50\,\mu H \tag{1}$$

In the prototype the values is adapted to 48 μH due to availability of components. The fact that an air core inductors is selected removes possible issues due to saturation. With the chosen L_{didt} value, the controller has 10 μs to interrupt after the fault takes place, under the hypothesis of a worst case scenario where the value of the current in the DC bus is already equal to the full nominal current I_{nom}. The reaction time $t_{reaction}$ is a parameter to be chosen by the system designer based on the specific needs of the application, and is selected here to be equal to 10 μs to demonstrate ultrafast interruption ability of the device. Upon interruption, the energy stored in the stray inductance of the DC bus and the internal stray inductance of the SSBTS needs to be dissipated. This task is shared between *RC* and MOV, where the former is charged with storing the energy in the stray inductance of the device itself, while the latter dissipates the energy in the DC bus inductance, by conducting at a clamped terminal voltage. As a 10 m DC bus has a stray inductance of approximately 10 μH according to [13], even at 200 A the energy stored will not exceed

$$E_{bus} = \frac{1}{2} L_{bus} I_{max}^2 = \frac{1}{2} \times 10\,\mu H \times 200\,A^2 = 0.2\,J \tag{2}$$

This is a relatively modest amount of stored energy. Therefore, the MOVs are selected not based on their energy dissipation ability, but rather on their voltage to current characteristic to ensure the desired clamping voltage. The selected device is a *Littlefuse V421HG34*, of which three are paralleled to limit the clamping voltage below 1 kV at 200 A, as the semiconductors are rated for 1.2 kV. The sizing of the *RC* is more challenging as the internal stray inductance of the SSBTS is not well known. The capacitor is selected to have a value of 1 μF, which allows, with a voltage increase of 500 V over the DC bus voltage,

the storage of the energy contained in

$$L_{stray,max} = \frac{C_{snub}V_{DC}}{I_{max}^2} = \frac{1\,\mu F \times 500\,V}{200\,A^2} \approx 6\,\mu H, \tag{3}$$

which is the energy stored in more than half of the DC bus. In testing, the size of the capacitor proved to be comfortably sufficient and a smaller value could be chosen if needed. The value of the resistor is selected so that a current value of I_{max} through the snubber immediately results in the application of 500 V to the snubber terminals, which results in a resistor value of $2.5\,\Omega$, but due to component availability, a value of $1.8\,\Omega$ is finally selected. Before series and parallel connection of the units is evaluated, the prototype is individually tested in its ability to conduct and interrupt current. A block schematic of the switching test setup for the characterisation of the prototype is shown in Fig. 5a and its results are in Fig. 5b. In this test, the level of current flowing through the SSBTS is sensed and sampled by the controller, which turns off the device if threshold of 100 A is exceeded. The setup is such that an artificial short circuit current is generated by charging capacitor $C = 230\,\mu F$ to 500 V, and then closing the SSBTS effectively short circuiting the capacitor. The current increases linearly through the device, until the controller detects the current value exceeding the set threshold and switches off. $L_{external}$ is added in the current path to reproduce the effect of DC bus stray inductance and has a value of $5\,\mu H$. In the results in Fig. 5b, sensed through oscilloscope connected voltage and current probes, one can see the linear current increase starting once the SSBTS turns on, and the interruption of the current after turn off. Note that during interruption the voltage on the device terminals and on the active switch is kept below 800 V. The voltage on the IGBT terminals remains at this level for an extended amount of time, as the snubber capacitor slowly discharges over a time significantly longer time interval (up to 10 ms). Note that in spite of the presence of the *RC* snubber in parallel with the IGBT position, there is a voltage spike at the moment of turn off before the charging of the snubber takes place. This is due to the stray inductance present in the snubber path, that reacts with an overvoltage as the snubber is forced to take over the IGBT current as they turn off. This overvoltage can likely be reduced through redesign of the snubber board, as the version used to obtain the results presented in this paper was designed for the adaptability of resistor and capacitor values, and not optimised for the reduction of stray inductance. Nevertheless, the voltage peak at the time of switching remains always smaller than the maximum voltage at the IGBT terminals reached through snubber capacitor charging, and therefore does not constitute an impediment to the operation of the device.

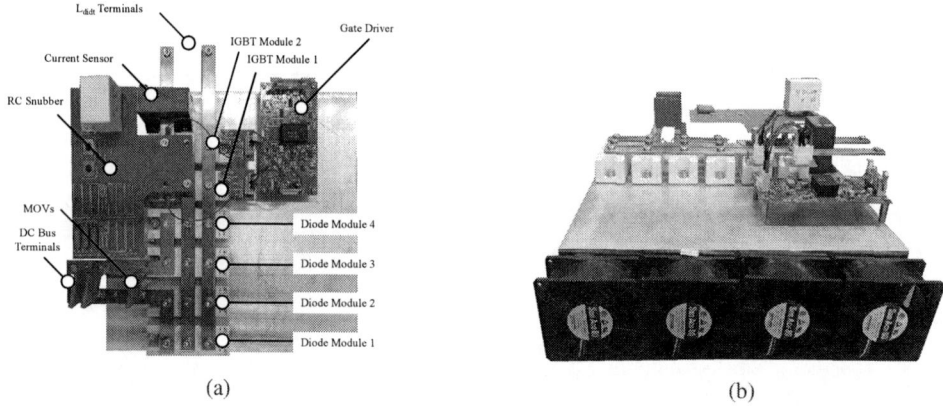

Fig. 4: (a) View from above of the SSBTS prototype, without current rate limiting inductor; (b) Side view of the prototype with active cooling fans.

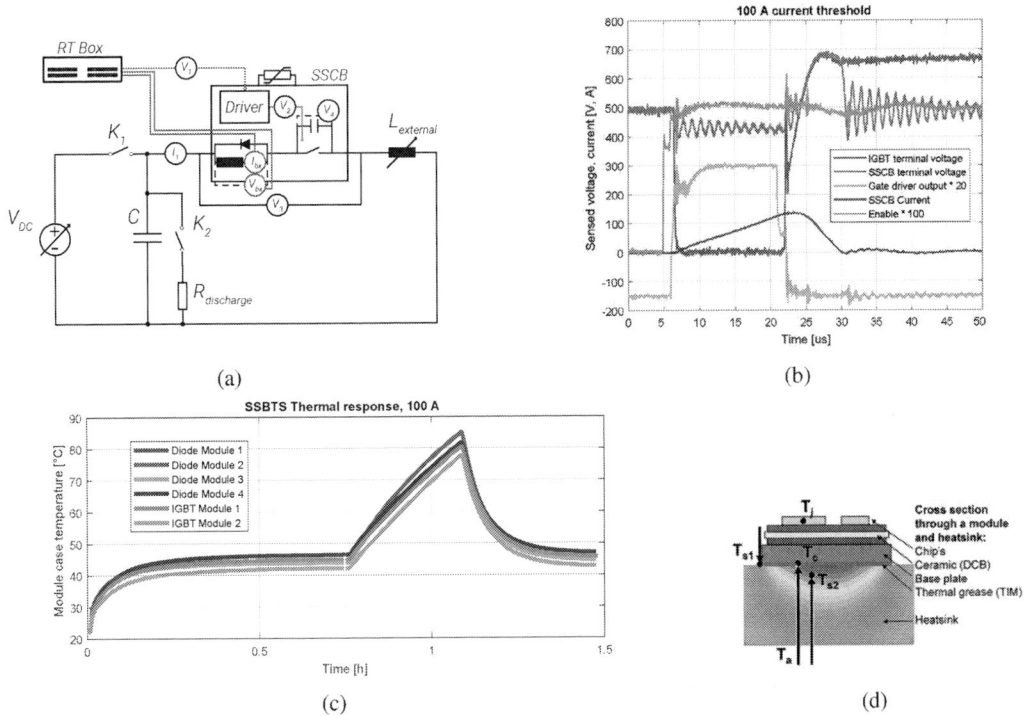

Fig. 5: (a) Block schematic of of the test setup for the characterisation of prototype switching; (b) Switching test results; (c) Thermal conduction test results; (d) Case temperature sensing is performed by inserting thermocouples in a channel in the heatsink reaching under the modules [14].

Fig. 5c displays the results of the thermal testing of the device, in which the nominal current of 100 A is circulated through the SSBTS and the case temperature of the semiconductor devices is sensed as in Fig. 5d. This is relevant as it was chosen that all positions in the device should be constituted by two paralleled semiconductor devices. The reason for this choice is that this is necessary in most full scale SSBTSs for marine applications. Interruption currents in the range of tens of kA can easily be reached, which are beyond the safe operating area (SOA) of individual commercially available semiconductor modules. Therefore, an SSBTS prototype employing paralleled semiconductors offers a more realistic solution than using individual devices. Fig. 5c shows that the temperature sharing between the semiconductor modules is satisfactory. The shape of the curve results from the way the test is performed. Initially, the devices starts cold and is allowed to reach steady state with cooling fans turned on. The fans are turned off and the case temperature is allowed to increase up to 85 °C, showing the temperature sharing of the modules is effective also at increased case temperature. Then, cooling fans are once again turned on bringing the temperatures back to the steady state initially achieved.

Parallel Operation

Having determined that the SSBTS prototype individually performs as desired, the two devices are tested in parallel to evaluate whether they can provide equivalent performance at twice the current ratings. Therefore, in parallel configuration the nominal current is considered to be $I_{nom,p} = 200$ A, and the maximum breaking current $I_{max,p} = 400$ A. The current interruption ability of the device is evaluated first. Fig. 6a shows how the test setup is adapted to accommodate for two parallel connected SSBTS units. The paralleling of the units is done through symmetric bus bars to avoid issues with current sharing linked to different resistance of the conductors. From the point of view of control, the controller is coded such that if the current in either of the devices exceeds the value of 100 A, then both units are tripped.

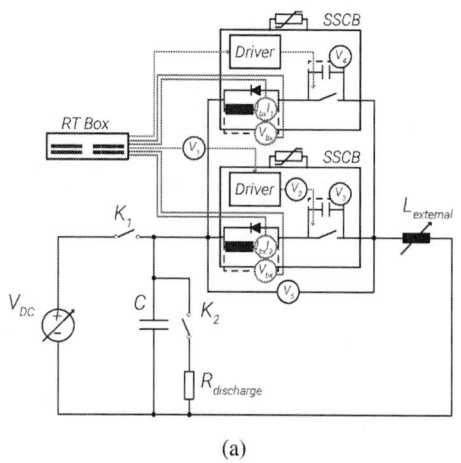
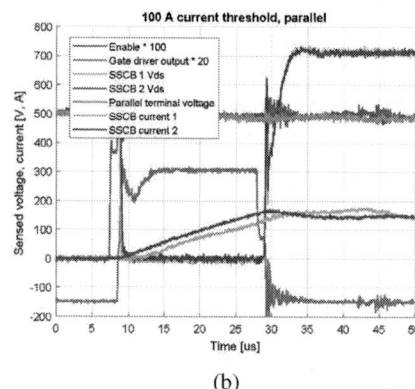

(a)   (b)

Fig. 6: (a) The interruption test setup is expanded to accomodate two parallel connected SSBTSs units; (b) The results of the switching test show adequate current sharing.

This prevent units from tripping individually resulting in unwanted current peaks. Fig. 8b displays the results of this test. Here, one can see how, after the devices are turned on, the increase of current happens at the same rate in each SSBTS. This is determined by the equal value of L_{didt} in each unit, which are both equal to 48 µH. The equal rise rate of current in the devices results in correct fault current sharing, and tripping happens due to the current in the device 2, which is slightly higher than than in the device 1. Operation with less effective current sharing than that shown in Fig. 8b, while undesirable, is still possible as long as the maximum current in both devices is lower than their $I_{max} = 200$ A.

Fig. 7 displays the results of thermal conduction tests for the two paralleled units. The goal of the test is to show that the temperature reached by the cases of the semiconductors is similar is the two paralleled units. This in turn shows that the current conducting capability of each unit is not decreased by paralleling. The results do not display the temperature of each individual semiconductor module, as was the case in Fig. 5c, but of the averaged temperature of the cases of IGBTs and diodes of each unit. The figure shows clearly that the temperature of both these modules are almost identical in the two units, and that therefore the conducted and interrupted current of the device can be linearly increased through paralleling.

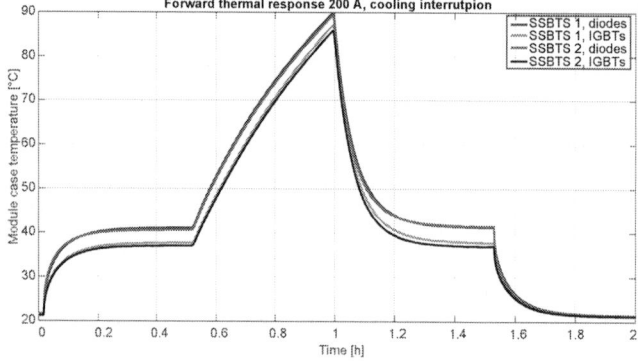

Fig. 7: The nominal current two parallel devices are able to conduct is 200 A, thanks to excellent temperature distribution of the modules in the two units.

EPE'20 ECCE Europe

Assigned jointly to the European Power Electronics and Drives Association & the Institute of Electrical and Electronics Engineers (IEEE)

Series Operation

Finally, series operation of two units is evaluated. The test carried out in this section aims to determine whether the operating voltage of the SSBTS can be increased by connecting multiple prototype units in series. For this test, the series connected devices perform an interruption at the voltage level of 1 kV and the voltage sharing between the two is evaluated. Fig. 8a displays the series connection of the two units in the test setup. Unlike in the previous two tests, in this test capacitor C is loaded up to 1 kV instead of 500 V. Since each SSBTS unit contains a current rate limiting inductance of 48 µH, this results in a current rate of increase of:

$$\frac{di}{dt} = \frac{V_{DC}}{2L_{didt}} = \frac{1000\,V}{96\,\mu H} = \frac{500\,V}{48\,\mu H} \tag{4}$$

Therefore, in spite of the increase of DC voltage, the current rate of increase remains the same in this test as it was by using a single unit at 500 V. As in the previous parallel connected test, the two devices are both tripped if the current in either one exceeds $I_{nom} = 100\,A$. As the devices are series connected, the current in both is exactly the same. Nevertheless, due to sensing delays and noise one of the two devices will inevitably cause tripping before the other. In this case, both units need to be tripped to prevent a single one from blocking the full DC voltage of 1 kV.

Fig. 8b displays the results of the switching tests. The figure shows how the voltage of both units is kept below 800 V and that the voltage and current stresses in series connected operation are equivalent to those of a single device operating at 500 V, demonstrating that series connected operation is an effective way of scaling up the SSBTS voltage. Nevertheless, it can be seen that before the devices are turned on, the voltage sharing between the two is not perfect, with unit 1 blocking around 550 V and unit 2 only blocking around 450 V. The trend toward this disparity in blocking voltage is visible also after the devices have turned off, and the voltage gradually settles back to this level. This behaviour is due to the MOVs in parallel with each device. In particular, to the fact that the current to voltage characteristic of MOVs is very variable for low values of conducted current. Therefore, a small variation of conducted current results in a large terminal voltage variation. Due to manufacturing tolerances, it is unlikely for two MOVs to exhibit exactly the same characteristic. For series connected SSBTS units, this means that since the same leakage current in the present in the MOVs of the two devices, any variation in current to voltage characteristic is reflected on the terminal voltage.

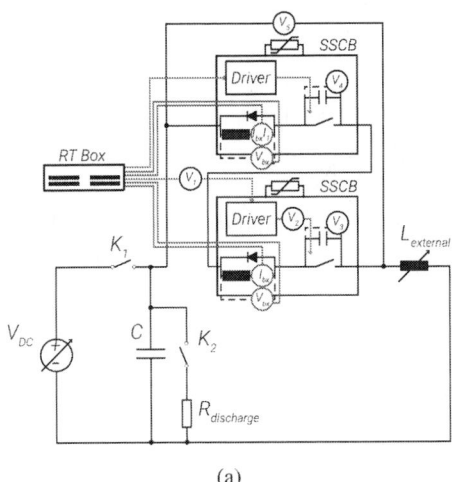

(a)

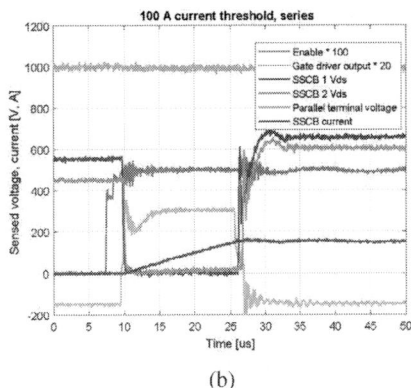

(b)

Fig. 8: (a) Block schematic of of the test setup for the characterisation of series connected prototype switching; (b) Switching test results.

Nevertheless, it should be noted that if the leakage current in the series connected devices were to increase because of this, then the relative difference of the terminal voltage of the SSBTSs would drop, due to the MOV characteristic flattening out at increased levels of current. Therefore, the voltage sharing between the units one can see in the figure is a stable configuration, and there is no risk of it increasing for varying values of leakage current.

Conclusion

This paper has presented a novel SSBTS topology for marine applications intended to operate in parallel and series connection to allow the increase of current and voltage rating of the shipboard PDNs in which it operates. The operating principles of the topology have been illustrated and the ability of the constructed prototype to conduct and interrupt its rated current at its rated voltage has been demonstrated experimentally. Additionally, the operation of the topology in series and parallel connection is also experimentally validated, demonstrating the desired ability to increase device's ratings through the connection of multiple units.

References

[1] J. F. Hansen, J. O. Lindtjørn, and K. Vanska. "Onboard DC Grid for enhanced DP operation in ships". In: *Dynamic Positioning Conference, Houston, 2011*.

[2] U. Javaid, D. Dujić, and W. van der Merwe. "MVDC marine electrical distribution: Are we ready?" In: *IECON 2015 - 41st Annual Conference of the IEEE Industrial Electronics Society*. 2015, pp. 000823–000828.

[3] U. Javaid et al. "MVDC supply technologies for marine electrical distribution systems". In: *CPSS Transactions on Power Electronics and Applications* 3.1 (Mar. 2018), pp. 65–76. ISSN: 2475-742X.

[4] P. Cheetham et al. "High Temperature Superconducting Power Cables for MVDC Power Systems of Navy Ships". In: *2019 IEEE Electric Ship Technologies Symposium (ESTS)*. Aug. 2019, pp. 548–555.

[5] J. T. Leman, E. J. William, and B. K. Johnson. "DC fault dynamics in a VSC based MVDC shipboard distribution". In: *2015 IEEE Power Energy Society General Meeting*. July 2015, pp. 1–5.

[6] T. J. McCoy. "Trends in ship electric propulsion". In: *IEEE Power Engineering Society Summer Meeting*, vol. 1. July 2002, 343–346 vol.1.

[7] M. M. Islam. "Electrical system redundancy requirementshipboard". In: *Handbook to IEEE Standard 45: A Guide to Electrical Installations on Shipboard*. IEEE, 2004. ISBN: null.

[8] E. Haugan et al. "Discrimination in offshore and marine dc distribution systems". In: *2016 IEEE 17th Workshop on Control and Modeling for Power Electronics (COMPEL)*. June 2016, pp. 1–7.

[9] S. Kim, S. Kim, and D. Dujic. "Extending Protection Selectivity in DC Shipboard Power Systems by Means of Additional Bus Capacitance". In: *IEEE Transactions on Industrial Electronics* 67.5 (2020), pp. 3673–3683.

[10] S. Kim, S. Kim, and D. Dujic. "Impact of Synchronous Generator De-excitation Dynamics on the Protection in Marine DC Power Distribution Networks". In: *IEEE Transactions on Transportation Electrification* (2020), pp. 1–1.

[11] N. Doerry and J. Amy. "Electric ship power and energy system architectures". In: *2017 IEEE Electric Ship Technologies Symposium (ESTS)*. Aug. 2017, pp. 1–64.

[12] "IEEE Recommended Practice for 1 kV to 35 kV Medium-Voltage DC Power Systems on Ships". In: *IEEE Std 1709-2010* (Nov. 2010), pp. 1–54. ISSN: null.

[13] Eaton Corporation. "Aluminium Busbar rated values 800 – 4000A". In: ().

[14] SEMIKRON International Gmbh. "Application Manual Power Semiconductors". In: 2015, p. 298.

Estimation of an Unbalanced Grid Impedance Using a Three-Phase Power Converter

Jarno Kukkola, Ville Pirsto, Mikko Routimo and Marko Hinkkanen
Aalto University
Department of Electrical Engineering and Automation
Espoo, Finland
Email: jarno.kukkola@aalto.fi

Keywords

≪Converter control≫, ≪Estimation technique≫, ≪Impedance measurement≫, ≪Three-phase system≫

Abstract

This paper proposes a real-time method for estimating an unbalanced grid impedance using a three-phase converter. In the method, a periodic single-frequency or multi-frequency excitation signal is added to the converter voltage reference. The converter measures currents and voltages at the point of common coupling. The impedance estimate is obtained from the measurements using sliding discrete Fourier transform (SDFT). The method is experimentally validated.

1. Introduction

Grid converters provide an interface between renewable energy sources and the grid. For control of these converters, the grid is often modeled as a three-phase voltage source and a balanced series impedance. The balanced impedance Z_g can be expressed with a transfer function such as $Z_g(s) = sL_g + R_g$, where L_g is the inductance and R_g is the resistance of the grid. Several methods for estimating the balanced grid impedance using a converter have been proposed, see [1–10] and the papers cited therein. Generally, the impedance can be estimated either in the frequency domain, e.g., using the discrete Fourier transform (DFT) [1–4] or band-pass filters [5], or in the time domain applying model-based methods [6–10]. The estimation result is typically the frequency response $Z_g(j\omega)$ or the values of L_g and R_g. The impedance estimate can be used, e.g., to detect islanding conditions [1,5] or to optimize converter control tuning [6].

When the per-phase impedances are unequal, the grid impedance is unbalanced (asymmetric). Then, the space-vector model of the impedance becomes an asymmetric 2×2 transfer-function matrix in stationary $\alpha\beta$ coordinates [11]. Real-time estimation of the unbalanced impedance has been addressed in [12–14]. Compared to estimation of the balanced impedance, estimation of the unbalanced impedance is more complicated due to increased amount of parameters. In [12] and [13], the estimation relies on the assumption of the inductive-resistive grid. In [12], a model-based recursive least squares algorithm is used. In [13], the estimation is based on dual-frequency injection and DFTs of the currents and voltages. If the assumption of the inductive-resistive grid is violated, such as in the case of capacitive elements in the grid, these methods provide biased results. In a wavelet-based estimation method [14], this assumption is not needed. However, the method requires measurement of three line-to-neutral voltages. Therefore, contrary to the methods in [12, 13], it cannot be applied if the neutral-point is not available or only line-to-line voltages are measured.

Estimation of a 2×2 impedance matrix in synchronous (dq) coordinates has been studied in [3, 15] assuming time-invariant impedance during the estimation. If an $\alpha\beta$-asymmetric impedance is transformed to synchronous coordinates, the resulting 2×2 transfer-function matrix becomes time variant,

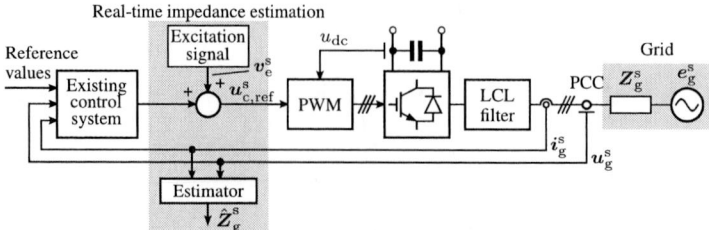

Fig. 1: Real-time grid impedance estimation as a part of the converter control system.

i.e., it depends on the angle of the coordinate system [11]. Therefore, instead of $\alpha\beta$-asymmetric systems, the methods in [3, 15] are suitable only for time-invariant dq-asymmetric grid impedances. The output impedance of a three-phase grid converter is typically asymmetric in dq coordinates [16, 17]. Hence, the grid impedance may also be time-invariant dq-asymmetric when the converters have notable effect on it.

This paper presents an SDFT-based method for real-time frequency-domain estimation of an unbalanced grid impedance in $\alpha\beta$ or dq coordinates. The method estimates the elements of the impedance matrix from two successive tests at a single frequency or multiple frequencies at the same time. To provide two linearly independent tests, an excitation signal is added to converter voltage reference in two different directions periodically, first in one direction and then in its orthogonal direction. Depending on the frequencies of interest, the excitation signal is a pulsating single-frequency or a multi-frequency space vector. Line-to-line voltages can be used in the estimator, and no a-priori assumptions on the grid type are needed. If the inductive-resistive grid is assumed, the obtained estimation result can be translated to per-phase inductances and resistances. The method is validated through simulations and experiments.

2. Model

Fig. 1 shows a control system for a three-phase grid converter. The system is augmented with a real-time grid impedance estimator. The estimator is considered as a module that can be plugged in to the existing control system. The converter is connected to the grid at the point of common coupling (PCC), and the grid is modeled as a voltage source $e_{\mathrm{g}}^{\mathrm{s}}$ with a series impedance $\boldsymbol{Z}_{\mathrm{g}}^{\mathrm{s}}$. Three-wire system is assumed, i.e., the sum of the grid phase currents is zero $i_{\mathrm{ga}} + i_{\mathrm{gb}} + i_{\mathrm{gc}} = 0$, where a, b, and c mark the phases. The system is modeled with real-valued space vectors. The vector of the PCC voltage is $\boldsymbol{u}_{\mathrm{g}}^{\mathrm{s}} = [u_{\mathrm{g}\alpha}, u_{\mathrm{g}\beta}]^{\mathrm{T}}$, where $u_{\mathrm{g}\alpha}$ and $u_{\mathrm{g}\beta}$ are the space-vector components. The superscript s denotes stationary coordinates. The PCC voltage can be further expressed as

$$\boldsymbol{u}_{\mathrm{g}}^{\mathrm{s}}(s) = \boldsymbol{Z}_{\mathrm{g}}^{\mathrm{s}}(s)\boldsymbol{i}_{\mathrm{g}}^{\mathrm{s}}(s) + \boldsymbol{e}_{\mathrm{g}}^{\mathrm{s}}(s) \tag{1}$$

where $\boldsymbol{i}_{\mathrm{g}}^{\mathrm{s}} = [i_{\mathrm{g}\alpha}, i_{\mathrm{g}\beta}]^{\mathrm{T}}$ is the grid current vector and $\boldsymbol{Z}_{\mathrm{g}}^{\mathrm{s}}$ is the grid impedance

$$\boldsymbol{Z}_{\mathrm{g}}^{\mathrm{s}}(s) = \begin{bmatrix} Z_{\alpha\alpha}(s) & Z_{\alpha\beta}(s) \\ Z_{\beta\alpha}(s) & Z_{\beta\beta}(s) \end{bmatrix} \tag{2}$$

The model can be transformed to synchronous coordinates resulting in $\boldsymbol{u}_{\mathrm{g}}(s) = \boldsymbol{Z}_{\mathrm{g}}(s)\boldsymbol{i}_{\mathrm{g}}(s) + \boldsymbol{e}_{\mathrm{g}}(s)$, where $\boldsymbol{u}_{\mathrm{g}} = \exp\left\{-\mathbf{J}\vartheta_{\mathrm{g}}(t)\right\}\boldsymbol{u}_{\mathrm{g}}^{\mathrm{s}}$, $\boldsymbol{e}_{\mathrm{g}} = \exp\left\{-\mathbf{J}\vartheta_{\mathrm{g}}(t)\right\}\boldsymbol{e}_{\mathrm{g}}^{\mathrm{s}}$, $\boldsymbol{i}_{\mathrm{g}} = \exp\left\{-\mathbf{J}\vartheta_{\mathrm{g}}(t)\right\}\boldsymbol{i}_{\mathrm{g}}^{\mathrm{s}}$, and $\boldsymbol{Z}_{\mathrm{g}}$ is the grid impedance with the elements Z_{dd}, Z_{dq}, Z_{qd}, and Z_{qq}. In the transformation, ϑ_{g} is the angle of the coordinate system, and $\mathbf{J} = \begin{bmatrix} 0 & -1 \\ 1 & 0 \end{bmatrix}$.

The impedance $\boldsymbol{Z}_{\mathrm{g}}^{\mathrm{s}}(s)$ can represent a balanced or unbalanced system. To provide an example, let us consider a three-phase inductive-resistive grid with the per-phase inductances $L_{\mathrm{ga}}, L_{\mathrm{gb}}$, and L_{gc}, and resistances $R_{\mathrm{ga}}, R_{\mathrm{gb}}$, and R_{gc}. The per-phase impedances are $Z_{\mathrm{a}}(s) = sL_{\mathrm{ga}} + R_{\mathrm{ga}}$, $Z_{\mathrm{b}}(s) = sL_{\mathrm{gb}} + R_{\mathrm{gb}}$, and $Z_{\mathrm{c}}(s) = sL_{\mathrm{gc}} + R_{\mathrm{gc}}$. If the coupling between the phases is neglected, the phase voltages at the PCC can be written

$$[u_{\mathrm{ga}}, u_{\mathrm{gb}}, u_{\mathrm{gc}}]^{\mathrm{T}} = \boldsymbol{Z}_{\mathrm{g}}^{\mathrm{abc}}[i_{\mathrm{ga}}, i_{\mathrm{gb}}, i_{\mathrm{gc}}]^{\mathrm{T}} + [e_{\mathrm{ga}}, e_{\mathrm{gb}}, e_{\mathrm{gc}}]^{\mathrm{T}} \tag{3}$$

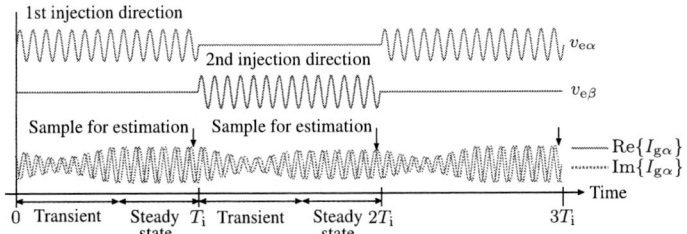

Fig. 2: Timing of the injection and estimation.

where $\boldsymbol{Z}_\mathrm{g}^\mathrm{abc}$ is a diagonal matrix $\boldsymbol{Z}_\mathrm{g}^\mathrm{abc} = \mathrm{diag}(Z_\mathrm{a}, Z_\mathrm{b}, Z_\mathrm{c})$. Transforming the phase quantities to space vectors as presented in [11], the impedance of this system in stationary coordinates becomes

$$\boldsymbol{Z}_\mathrm{g}^\mathrm{s}(s) = \boldsymbol{T}_{32}\boldsymbol{Z}_\mathrm{g}^\mathrm{abc}(s)\boldsymbol{T}_{23} = \begin{bmatrix} \frac{4Z_\mathrm{a}(s)+Z_\mathrm{b}(s)+Z_\mathrm{c}(s)}{6} & \frac{\sqrt{3}[Z_\mathrm{c}(s)-Z_\mathrm{b}(s)]}{6} \\ \frac{\sqrt{3}[Z_\mathrm{c}(s)-Z_\mathrm{b}(s)]}{6} & \frac{Z_\mathrm{b}(s)+Z_\mathrm{c}(s)}{2} \end{bmatrix}, \quad \boldsymbol{T}_{32} = \begin{bmatrix} \frac{2}{3} & -\frac{1}{3} & -\frac{1}{3} \\ 0 & \frac{1}{\sqrt{3}} & -\frac{1}{\sqrt{3}} \end{bmatrix} \tag{4}$$

where $\boldsymbol{T}_{23} = (3/2)\boldsymbol{T}_{32}^\mathrm{T}$. From (4), it can seen that if the per-phase impedances are equal $Z_\mathrm{a} = Z_\mathrm{b} = Z_\mathrm{c} = Z_\mathrm{g}$, the impedance $\boldsymbol{Z}_\mathrm{g}^\mathrm{s}$ reduces to the diagonal matrix $\boldsymbol{Z}_\mathrm{g}^\mathrm{s} = \mathrm{diag}(Z_\mathrm{g}, Z_\mathrm{g})$, but in the case of an unbalanced impedance, all elements of $\boldsymbol{Z}_\mathrm{g}^\mathrm{s}$ are generally nonzero.

3. Impedance Estimator

The frequency response of the grid impedance is estimated at a single frequency ω_e or multiple frequencies $\omega_{\mathrm{e},1}, \dots, \omega_{\mathrm{e},n}$ at a time. For the estimation, a pulsating excitation signal $\boldsymbol{v}_\mathrm{e}^\mathrm{s}$ is added to the converter voltage reference $\boldsymbol{u}_{\mathrm{c,ref}}^\mathrm{s}$ as shown in Fig. 1.

3.1. Single-Frequency Estimation

For the single-frequency estimation, the excitation signal is

$$\boldsymbol{v}_\mathrm{e}^\mathrm{s}(t) = \begin{bmatrix} v_{\mathrm{e}\alpha}(t) \\ v_{\mathrm{e}\beta}(t) \end{bmatrix} = \mathrm{e}^{\mathbf{J}\vartheta_\mathrm{e}(t)} \begin{bmatrix} v_\mathrm{e}\sin(\omega_\mathrm{e}t) \\ 0 \end{bmatrix}, \qquad \mathbf{J} = \begin{bmatrix} 0 & -1 \\ 1 & 0 \end{bmatrix} \tag{5}$$

where v_e is the magnitude and ϑ_e determines the angle of the pulsating vector in stationary coordinates. In this work, the angle is alternated between $\vartheta_\mathrm{e} = 0$ and $\vartheta_\mathrm{e} = \pi/2$ to excite the system in α-axis and β-axis directions periodically

$$\begin{aligned} \vartheta_\mathrm{e}(t) &= 0, \qquad \text{when} \quad \sin(\pi t/T_\mathrm{i}) > 0 \\ \vartheta_\mathrm{e}(t) &= \pi/2, \quad \text{otherwise} \end{aligned} \tag{6}$$

where T_i is the sampling interval of the impedance estimation. Two different injection directions are needed to provide two linearly independent tests for the impedance matrix identification. It is assumed that the injection-frequency component is not present in the grid voltage or current spectrum before the injection. Therefore, the frequency ω_e of the signal is a grid-frequency interharmonic. An excitation signal corresponding to (5) and (6) is shown in Fig. 2.

Fig. 3(a) shows the structure of the proposed impedance estimator in the case of single-frequency estimation. The impedance estimator utilizes the PCC voltage and current samples measured by the converter. Sampled line-to-line voltages and phase currents are turned into corresponding space-vector components $u_{\mathrm{g}\alpha}$, $u_{\mathrm{g}\beta}$, $i_{\mathrm{g}\alpha}$, and $i_{\mathrm{g}\beta}$. The complex phasors $I_{\mathrm{g}\alpha}(\omega_\mathrm{e})$ and $I_{\mathrm{g}\beta}(\omega_\mathrm{e})$ of the current components and the complex phasors $U_{\mathrm{g}\alpha}(\omega_\mathrm{e})$ and $U_{\mathrm{g}\beta}(\omega_\mathrm{e})$ of the voltage components are continuously calculated in real-time applying the modulated SDFT algorithm [18]. The internal structure of the SDFT module used in the proposed method is shown in Fig. 3(b). It is to be noted that due to the sliding window of the SDFT

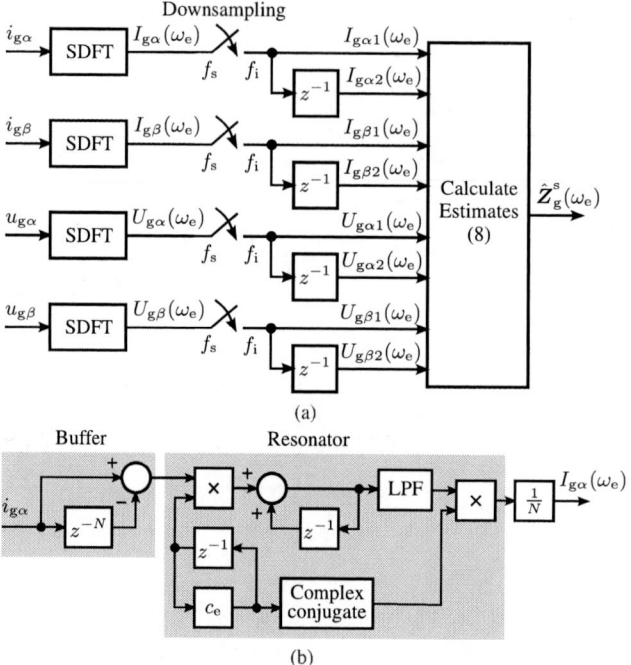

(a)

(b)

Fig. 3: (a) Proposed impedance estimator. (b) Internal structure of the SDFT module. The resonator coefficient $c_e = \exp(-j\omega_e T_s)$ is set corresponding the angular frequency ω_e of the excitation.

algorithm, the calculated phasors are rotating as shown in Fig. 2.

Since the injection-frequency component of e_g^s is assumed to be zero, the phasors of the PCC voltage at ω_e can be expressed as

$$\underbrace{\begin{bmatrix} U_{g\alpha1}(\omega_e) & U_{g\alpha2}(\omega_e) \\ U_{g\beta1}(\omega_e) & U_{g\beta2}(\omega_e) \end{bmatrix}}_{\mathbf{U}_m(\omega_e)} = \mathbf{Z}_g^s(j\omega_e) \underbrace{\begin{bmatrix} I_{g\alpha1}(\omega_e) & I_{g\alpha2}(\omega_e) \\ I_{g\beta1}(\omega_e) & I_{g\beta2}(\omega_e) \end{bmatrix}}_{\mathbf{I}_m(\omega_e)} \tag{7}$$

where $U_{g\alpha1}, U_{g\beta1}, I_{g\alpha1}, I_{g\beta1}$ are obtained during the first injection direction (first test) and $U_{g\alpha2}, U_{g\beta2},$ $I_{g\alpha2}, I_{g\beta2}$ during the second injection direction (second test). Two different tests provide enough information for the impedance matrix estimation, and the estimate is directly calculated as

$$\hat{\mathbf{Z}}_g^s(j\omega_e) = \hat{\mathbf{R}}_g^s(\omega_e) + j\hat{\mathbf{X}}_g^s(\omega_e) = \mathbf{U}_m(\omega_e)\mathbf{I}_m^{-1}(\omega_e) \tag{8}$$

where $\hat{\mathbf{R}}_g^s$ and $\hat{\mathbf{X}}_g^s$ are the resistive and reactive parts of the impedance matrix.

The phasors of the voltages and currents are applicable to impedance estimation only when the system is in steady state. Since the angle ϑ_e of the pulsating injection (6) is periodically changed, the steady state has to be reached before the next change in ϑ_e. Therefore, the impedance estimation is synchronized with the excitation signal such that the rotating phasors of the currents and voltages for (8) are sampled just before the change in ϑ_e with the sampling frequency of $f_i = 1/T_i$ as illustrated in Fig. 2. The steady-state requirement sets the minimum value for the impedance estimation interval T_i. Moreover, the time required to reach the steady state depends on the settling time of the existing control system and phasor calculation. The settling time of the phasor calculation is related to the buffer length N of the SDFT modules, and N depends on the desired frequency resolution and sampling frequency f_s of the DFT. Therefore, the settling time of the phasor calculation is approximately one fundamental period of the DFT frequency resolution. The frequency resolution has to be a common divisor of the fundamental frequency of the grid and the injection frequency. To give an example, if the frequency resolution is 10

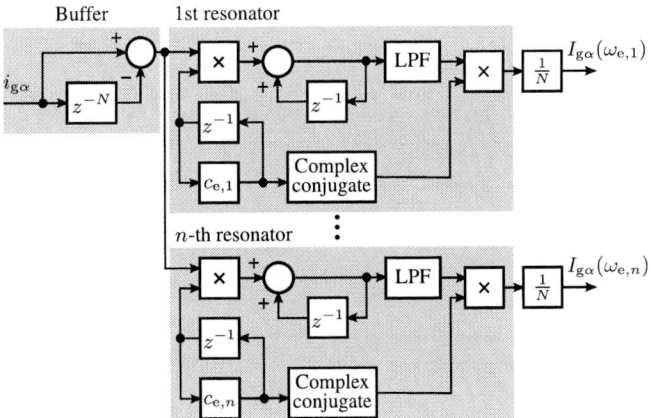

Fig. 4: SDFT module for multi-frequency phasor calculation. The resonator coefficients are set corresponding the frequencies of interest, i.e., $c_{e,1} = \exp(-j\omega_{e,1}T_s), \ldots, c_{e,n} = \exp(-j\omega_{e,n}T_s)$.

Hz and the sampling frequency is $f_s = 10$ kHz, the buffer size is $N = 1000$ and the settling time of the SDFT modules and the phasor calculation is approximately $1/(10 \text{ Hz}) = 100$ ms.

When the system is in steady state, the calculated phasors can be averaged to reduce effect of measurement noise and to increase accuracy of the impedance estimation. For averaging the phasors, the SDFT structure presented in [18] is augmented with a low-pass filter (LPF). In this paper, a first-order LPF with the transfer function of $\alpha_f/(s+\alpha_f)$ and bandwidth of α_f is applied due to its simplicity. It is transformed to discrete time via the mapping $s \to (z-1)/T_s$. Alternatively, the LPF can be, e.g., a moving-average filter which is naturally averaging the phasors over its buffer length. The LPF is inserted inside the resonator, as shown in Fig. 3(b), where the signals are dc-valued in steady state.

In the SDFT module, the resonator coefficient $c_e = \exp(-j\omega_e T_s)$ selects the frequency of the single DFT bin for the phasor calculation as explained in [19]. It is to be noted that, ω_e has to be integer multiple of the angular frequency resolution of the SDFT module as in the case of the conventional DFT. Finally, as Fig. 3(b) shows, only a few multiplications and additions are needed in one sampling period $T_s = 1/f_s$ per SDFT module meaning low computation burden that is an advantage in a real-time control system.

3.2. Multi-Frequency Estimation

The presented single-frequency estimator can be easily extended for multi-frequency impedance estimation with the following modifications: 1) n pulsating sinusoidal components are added at the frequencies of $\omega_{e,1}, \ldots, \omega_{e,n}$ to the converter voltage reference. 2) The SDFT modules are extended for multi-frequency phasor calculation adding extra resonators, one per frequency, in parallel with the first resonator as demonstrated in Fig. 4. 3) At every frequency component $\omega_{e,1}, \ldots, \omega_{e,n}$, the calculated phasors are sampled and impedance estimate (8) is calculated as explained in the previous section for a single frequency component.

The multi-frequency excitation signal is

$$\boldsymbol{v}_e^s(t) = \sum_{k=1}^{n} e^{\mathbf{J}\vartheta_{e,k}(t)} \begin{bmatrix} v_{e,k}\sin(\omega_{e,k}t + \phi_{e,k}) \\ 0 \end{bmatrix} \tag{9}$$

where $v_{e,k}$ is the magnitude and $\vartheta_{e,k}$ is the angle of the k:th pulsating vector. In addition, every pulsating component can have different initial phase $\phi_{e,k}$ at $t = 0$. The angles $\vartheta_{e,k}$, $k = 1 \ldots n$, are alternated periodically in order to excite the system in different directions and to provide linearly independent tests for the impedance matrix identification. If the angles $\vartheta_{e,k}$ are set as in (6) for all components, the components will excite the system in the same direction, i.e., in α-axis and β-axis directions periodically. On the contrary, if the angle $\vartheta_{e,k}$ is not the same for every component, the magnitude $|\boldsymbol{v}_e^s(t)|$ of the signal can be reduced [20]. For example, in the case of dual-frequency excitation, selecting the angles to provide orthogonal components would minimize the magnitude. The initial phases $\phi_{\mathrm{e},k}$, $k = 1 \ldots n$, provide an additional degree of freedom to shape $\boldsymbol{v}_{\mathrm{e}}^{\mathrm{s}}(t)$. For multi-tone signals, several approaches to have been presented to select these initial phases to minimize the signal magnitude or the crest factor, e.g., [20, 21].

3.3. Estimation in Synchronous Coordinates

In Sections 3.1 and 3.2, the proposed estimator is introduced in stationary coordinates for $\alpha\beta$-asymmetric systems. In order to estimate dq-asymmetric impedances, the estimator can be transformed to synchronous (dq) coordinates. Only two modifications are needed. Firstly, the excitation signal $\boldsymbol{v}_{\mathrm{e}}$ is generated in synchronous coordinates similarly to (5) or (9) and transformed to stationary coordinates as $\boldsymbol{v}_{\mathrm{e}}^{\mathrm{s}} = \exp\left\{\mathbf{J}\vartheta_{\mathrm{g}}(t)\right\}\boldsymbol{v}_{\mathrm{e}}$, where ϑ_{g} is the angle of the synchronous coordinate system. Secondly, the PCC voltage and current vectors are transformed to synchronous coordinates before the SDFT-based phasor calculation and impedance estimation as $\boldsymbol{u}_{\mathrm{g}} = \exp\left\{-\mathbf{J}\vartheta_{\mathrm{g}}(t)\right\}\boldsymbol{u}_{\mathrm{g}}^{\mathrm{s}}$, and $\boldsymbol{i}_{\mathrm{g}} = \exp\left\{-\mathbf{J}\vartheta_{\mathrm{g}}(t)\right\}\boldsymbol{i}_{\mathrm{g}}^{\mathrm{s}}$, respectively. The angle ϑ_{g} for the transformations can be obtained, e.g., from a phase-locked loop (PLL) tracking the PCC voltage. The PLL has to include a band-stop filter at the injection frequency to provide injection-frequency free angle. If ϑ_{g} oscillates at the injection frequency, the estimated impedance can be significantly biased. With these modifications, the proposed estimator operates in synchronous coordinates and produces an impedance estimate in these coordinates instead of stationary coordinates. The estimation of a dq-asymmetric impedance may be of interest if other converters connected to the same grid have a notable impact on the impedance seen from the PCC.

3.4. Comparison

Properties of the proposed estimator are compared with those of the state-of-the-art converter-based estimators capable for unbalanced grid-impedance estimation. Whereas the methods in [13, 14] estimate the impedance in the natural (abc) reference frame, the method in [12] in the $\alpha\beta$ reference frame, and the methods in [3, 15] in the dq reference frame, the proposed estimator can be configured to estimate either an $\alpha\beta$- or dq-asymmetric impedance. Access to the neutral-point potential and phase-to-neutral voltage measurements are required in [14]. On the contrary, when the space vectors are applied in the estimation, line-to-line voltage measurements can be used and neutral point potential is not needed.

The estimation methods in [12, 13] assume inductive-resistive grid model whereas the impedance model is not fixed in [3, 14] and in the proposed method. In [15], a parametric differential-equation model is iteratively fitted to measured data. Although a parametric grid model is obtained as an estimation result in [15], the model order and structure has to be determined which complicates the estimation.

The DFT is used in the frequency-domain impedance estimation in [3, 13] which requires collecting the measurement data in a buffer before the impedance estimate can be calculated. If the number of data points is large, the calculation has to be run as a background process of the converter delaying the estimation. The method in [15] also requires buffers and background processing the measured data due to its iterative nature. On the contrary, the impedance estimate is updated recursively on a sample-by-sample basis in [12]. The advantage of the recursive calculation is that the computational load is spread over the excitation period. In the proposed method, data buffers are needed for the SDFT but the phasors are recursively calculated on a sample-by-sample basis. This reduces the computational burden compared to the conventional DFT in the case of a few frequency components in the estimation.

As presented in Sections 3.1 and 3.2, the proposed estimator can be configured for selective single- or multi-frequency estimation in a flexible manner thanks to its modular structure. In [13] two frequency components are required in the excitation signal to obtain the estimated inductance and resistance values. In [3, 12, 15] the excitation signal has wide frequency band since it is either required in the selected parametric estimation method [12, 15] or a wide-band frequency response is of interest [3]. Even though in the proposed method, the number of frequency components can be increased in the excitation signal and SDFT modules, the approach [3] with the pseudo-random binary signal excitation and the conventional DFT becomes more attractive if tens or hundreds of frequency components are of interest.

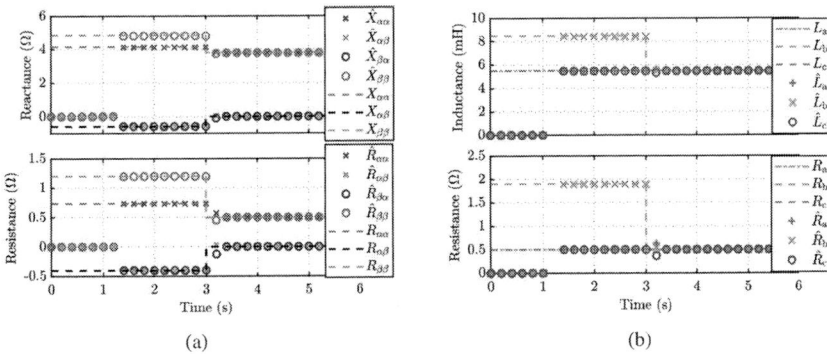

(a)   (b)

Fig. 5: Simulation results: (a) estimated elements of the grid impedance matrix $\hat{\boldsymbol{Z}}_{\mathrm{g}}^{\mathrm{s}}(\omega_{\mathrm{e}}) = \hat{\boldsymbol{R}}_{\mathrm{g}}^{\mathrm{s}}(\omega_{\mathrm{e}}) + \mathrm{j}\hat{\boldsymbol{X}}_{\mathrm{g}}^{\mathrm{s}}(\omega_{\mathrm{e}})$; (b) estimated per-phase inductances and resistances. The dashed lines mark the actual values.

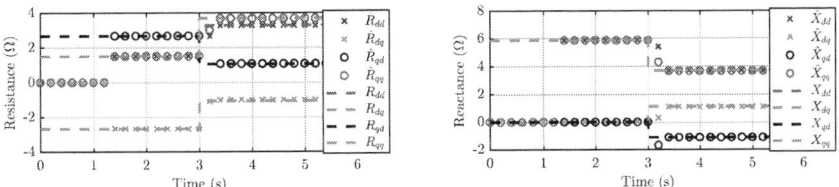

Fig. 6: Simulation results: estimated elements of the grid impedance matrix $\hat{\boldsymbol{Z}}_{\mathrm{g}}(\omega_{\mathrm{e}}) = \hat{\boldsymbol{R}}_{\mathrm{g}}(\omega_{\mathrm{e}}) + \mathrm{j}\hat{\boldsymbol{X}}_{\mathrm{g}}(\omega_{\mathrm{e}})$. The actual values are marked with dashed lines.

4. Results

4.1. Simulations

The proposed estimator is first verified with simulations considering a 400-V 12.5-kVA grid converter system. The existing control system [22] comprises a current-control loop and a PLL. The switching frequency of the converter is 5 kHz. The sampling frequency of the control system and the SDFT is $f_{\mathrm{s}} = 10$ kHz. The buffer size for the SDFT is $N = 1000$ samples per measured current or voltage component. The impedance estimation interval is set to $T_{\mathrm{i}} = 200$ ms to ensure that steady state is reached for the phasor calculation. A single-frequency excitation signal, given in (5) and (6), is used. Its frequency is an inter-harmonic of $\omega_{\mathrm{e}} = 2\pi \cdot 110$ rad/s, and its magnitude is $v_{\mathrm{e}} = 0.02$ p.u. During the verification test, the converter injects the power of $p_{\mathrm{g}} = 0.5$ p.u. to the 50-Hz grid.

Fig. 5(a) shows the estimated elements of the grid impedance matrix when the proposed method is started at $t = 1$ s to estimate an unbalanced inductive-resistive grid impedance. Per-phase inductances are $L_{\mathrm{ga}} = L_{\mathrm{gc}} = 5.5$ mH and $L_{\mathrm{gb}} = 8.5$ mH and resistances are $R_{\mathrm{ga}} = R_{\mathrm{gc}} = 0.5\ \Omega$ and $R_{\mathrm{gb}} = 1.9\ \Omega$. At $t = 3$ s the b-phase inductance L_{gb} is reduced from 8.5 mH to 5.5 mH and resistance R_{gb} is reduced from 1.9 Ω to 0.5 Ω, i.e., the impedance becomes balanced. If the inductive-resistive grid is assumed in the estimator, the estimated impedance matrix elements can be translated to per-phase inductance and resistance estimates based on (4) as $Z_{\mathrm{a}} = (3Z_{\alpha\alpha} - Z_{\beta\beta})/2$, $Z_{\mathrm{b}} = Z_{\beta\beta} - \sqrt{3}/2 \cdot (Z_{\alpha\beta} + Z_{\beta\alpha})$, and $Z_{\mathrm{c}} = Z_{\beta\beta} + \sqrt{3}/2 \cdot (Z_{\alpha\beta} + Z_{\beta\alpha})$. The per-phase estimates are demonstrated in Fig. 5(b). As the figure shows, the method correctly estimates the unbalanced and balanced grid impedance.

The capability to estimate a dq-asymmetric impedance is also verified with simulations. The estimator is configured as in the first simulation test and transformed to synchronous coordinates as described in Section 3.3. Fig. 6 shows the estimated elements of the impedance matrix when the estimator is started at $t = 1$ s to estimate a balanced grid impedance $R_{\mathrm{g}} = 1.5\ \Omega$ and $L_{\mathrm{g}} = 8.5$ mH. At $t = 3$ s, another 12.5-kVA three-phase converter with an LCL filter ($L_1 = 3.3$ mH, $C = 8.8\ \mu$F, $L_2 = 3.0$ mH) is connected to the PCC and it starts to inject active power of $p_{\mathrm{g}} = 0.5$ p.u. to the grid. Its switching frequency is 4 kHz, and its control system comprises proportional-integral (PI) grid current controller in dq coordinates

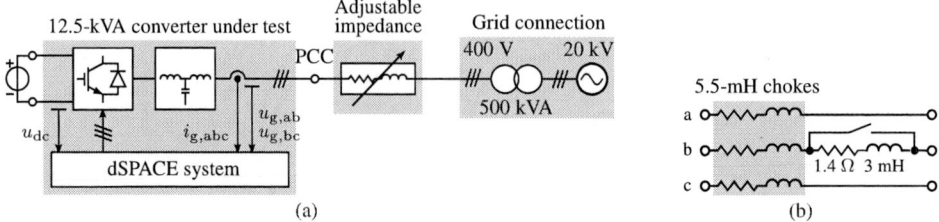

(a)

(b)

Fig. 7: (a) Experimental test setup. The converter control software and the impedance estimator are running on the dSPACE DS1006 processor board where the sampling of the measured signals is synchronized with the pulse-width modulation (PWM). (b) Adjusted impedance under test.

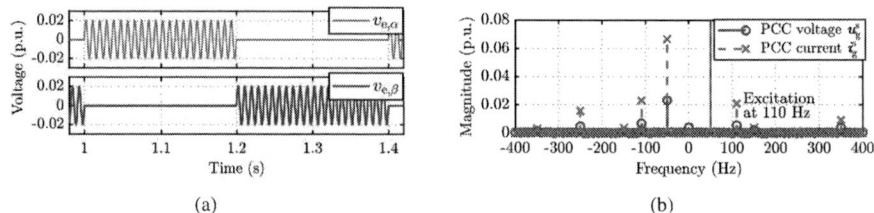

(a)   (b)

Fig. 8: Measurement results: (a) excitation voltage; (b) spectra of the PCC voltage and current during the estimation.

with the cross-coupling compensation and a synchronous-reference-frame PLL (SFR-PLL). The gains of the PI controller correspond the design presented in [23] with the phase margin of $\pi/4$. The gains of the SRF-PLL are $k_{\mathrm{p,pll}} = 0.272\ (1/\mathrm{Vs})$ and $k_{\mathrm{i,pll}} = 12.1\ (1/\mathrm{Vs}^2)$.

As Fig. 6 shows, the other converter at the PCC significantly changes the overall grid impedance seen by the estimating converter (dashed-lines show the theoretical overall impedance). Nevertheless, the proposed estimator accurately estimates the impedance in dq coordinates with and without the other converter in the system. Furthermore, it was observed that the switching frequency components originating from the other converter can cause significant bias in the estimates. To alleviate this bias, a simple 3-kHz first-order filter was employed as an anti-aliasing filter for the PCC voltage and current measurements.

4.2. Experiments

The verification test was experimentally repeated. A diagram of the setup used in the experiments is shown in Fig. 7(a). The unbalanced impedance is emulated with 5.5-mH chokes in all three phases and an extra 3-mH choke and resistance of 1.4 Ω in the b phase as demonstrated in Fig. 7(b). The chokes are connected between the PCC and the 50-Hz electric power system. Fig. 8(a) shows the excitation signal during the experiment, and Fig. 8(b) shows the PCC current and voltage spectra when the excitation signal is enabled. The spectra are calculated over five grid-voltage periods (0.1 s) during one injection direction when the grid impedance is unbalanced. It can be seen that the magnitudes of the injection-frequency components are at the same level as typical grid current harmonics. The -50 Hz negative sequence component originates from the unbalanced impedance.

Fig. 9(a) shows the estimated elements of the grid impedance matrix and Fig. 9(b) the estimated per-phase inductances and resistances. The estimated per-phase impedances are obtained from the elements of the impedance matrix as in Section 4.1. The proposed method is started at $t = 1$ s, and the extra 3-mH choke and resistance in the b-phase is bypassed around $t = 3.5$ s [cf. Fig 7(b)] to change the impedance from unbalanced to balanced. As the results demonstrate, the proposed method can identify the balanced and unbalanced grid impedance and detect changes well in real time. Compared to the simulation results, the estimated resistive components are influenced by the noise in the measured signals causing some variance. The effect of noise can be reduced with averaging the calculated phasors as explained in Section 3.1. Here, the LPF bandwidth for this purpose was set to $\alpha_{\mathrm{f}} = 2\pi \cdot 10$ Hz.

Finally, the same test was repeated using a multi-frequency excitation signal (9) to demonstrate the

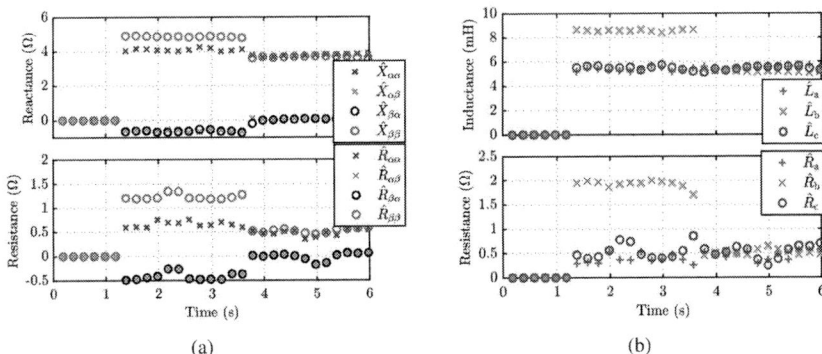

(a)   (b)

Fig. 9: Measurement results with single-frequency excitation: (a) estimated elements of the grid impedance matrix $\hat{\boldsymbol{Z}}_{\mathrm{g}}^{\mathrm{s}}(\omega_{\mathrm{e}}) = \hat{\boldsymbol{R}}_{\mathrm{g}}^{\mathrm{s}}(\omega_{\mathrm{e}}) + \mathrm{j}\hat{\boldsymbol{X}}_{\mathrm{g}}^{\mathrm{s}}(\omega_{\mathrm{e}})$; (b) estimated per-phase inductances and resistances.

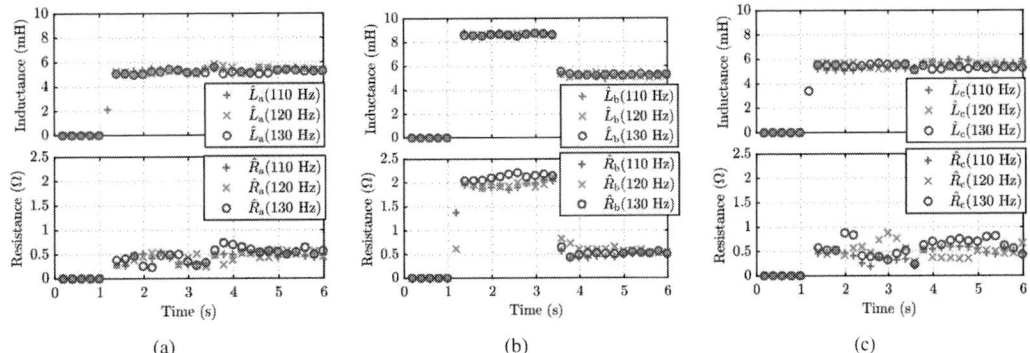

(a)   (b)   (c)

Fig. 10: Measurement results with multi-frequency excitation: (a) estimated a-phase inductances and resistances; (b) estimated b-phase inductances and resistances; (c) estimated c-phase inductances and resistances.

multi-frequency estimation. The injection frequencies are $\omega_{\mathrm{e},1} = 2\pi \cdot 110$ Hz, $\omega_{\mathrm{e},2} = 2\pi \cdot 120$ Hz, and $\omega_{\mathrm{e},3} = 2\pi \cdot 130$ Hz, and the amplitudes of the signals are $v_{\mathrm{e},1} = v_{\mathrm{e},2} = v_{\mathrm{e},3} = 0.02$ p.u. For simplicity, the injection angles $\vartheta_{\mathrm{e},k}$ are selected according to (6) and the phase shifts are $\phi_{\mathrm{e},k} = 0$ for all $k = 1 \ldots 3$. Fig. 10 shows the estimated per-phase inductances and resistances at all injection frequencies. Again, the impedance change is well detected and the estimation results at these frequencies agree with each other. A drawback of the multi-frequency estimation is increased distortion of grid currents due to multiple injection components. However, simultaneous estimation of multiple frequency components can provide more information of the grid impedance within same estimation interval. In addition, calculating the average of the estimated resistances obtained at different frequencies, e.g., $[R_{\mathrm{a}}(110\,\mathrm{Hz}) + R_{\mathrm{a}}(120\,\mathrm{Hz}) + R_{\mathrm{a}}(130\,\mathrm{Hz})]/3$, helps to reduce variance of the estimated resistances.

5. Conclusions

This paper presented a real-time grid impedance estimation method for three-phase power converters. While a periodic excitation signal is added to the converter voltage reference, the proposed method can continuously estimate either unbalanced or balanced impedance at the injection frequency. Due to lightweight computation of the SDFT algorithm used in the estimation, the proposed method can be easily implemented in a converter control system. The estimated impedance can be internally applied in the control system, e.g., to improve control performance. In remote monitoring of converters, the estimated impedance may provide added value when analyzing a converter or power system status or possible grid–converter interactions. In addition to grid converter systems, the presented estimation technique can be applied in other converter systems with similar interfaces, such as in motor drives.

References

[1] L. Asiminoaei, R. Teodorescu, F. Blaabjerg, and U. Borup, "Implementation and test of an online embedded grid impedance estimation technique for PV inverters," *IEEE Trans. Ind. Electron.*, vol. 52, no. 4, pp. 1136–1144, Aug. 2005.

[2] M. Céspedes and J. Sun, "Online grid impedance identification for adaptive control of grid-connected inverters," in *Proc. IEEE ECCE 2012*, Raleigh, NC, Sep. 2012, pp. 914–921.

[3] T. Roinila, T. Messo, and E. Santi, "MIMO-identification techniques for rapid impedance-based stability assessment of three-phase systems in DQ domain," *IEEE Trans. Power Electron.*, vol. 33, no. 5, pp. 4015–4022, May 2018.

[4] A. Riccobono, M. Mirz, and A. Monti, "Noninvasive online parametric identification of three-phase AC power impedances to assess the stability of grid-tied power electronic inverters in LV networks," *IEEE Trans. Emerg. Sel. Topics Power Electron.*, vol. 6, no. 2, pp. 629–647, Jun. 2018.

[5] D. Reigosa, F. Briz, C. B. Charro, P. García, and J. M. Guerrero, "Active islanding detection using high-frequency signal injection," *IEEE Trans. Ind. Appl.*, vol. 48, no. 5, pp. 1588–1597, Sep. 2012.

[6] A. Vidal, A. G. Yepes, F. D. Freijedo, Ó. López, J. Malvar, F. Baneira, and J. Doval-Gandoy, "A method for identification of the equivalent inductance and resistance in the plant model of current-controlled grid-tied converters," *IEEE Trans. Power Electron.*, vol. 30, no. 12, pp. 7245–7261, Dec. 2015.

[7] S. Cobreces, E. J. Bueno, D. Pizarro, F. J. Rodriguez, and F. Huerta, "Grid impedance monitoring system for distributed power generation electronic interfaces," *IEEE Trans. Instrum. Meas.*, vol. 58, no. 9, pp. 3112–3121, Sep. 2009.

[8] A. K. Broen, M. Amin, E. Skjong, and M. Molinas, "Instantaneous frequency tracking of harmonic distortions for grid impedance identification based on Kalman filtering," in *Proc. IEEE COMPEL*, Trondheim, Norway, Jun. 2016.

[9] N. Hoffmann and F. W. Fuchs, "Minimal invasive equivalent grid impedance estimation in inductive-resistive power networks using extended Kalman filter," *IEEE Trans. Power Electron.*, vol. 29, no. 2, pp. 631–641, Feb. 2014.

[10] M. Ciobotaru, V. Agelidis, and R. Teodorescu, "Line impedance estimation using model based identification technique," in *Proc. EPE*, Birmingham, UK, Aug. 2011.

[11] L. Harnefors, "Modeling of three-phase dynamic systems using complex transfer functions and transfer matrices," *IEEE Trans. Ind. Electron.*, vol. 54, no. 4, pp. 2239–2248, Aug. 2007.

[12] P. García, M. Sumner, Á. Navarro-Rodríguez, J. M. Guerrero, and J. García, "Observer-based pulsed signal injection for grid impedance estimation in three-phase systems," *IEEE Trans. Ind. Electron.*, vol. 65, no. 10, pp. 7888–7899, Oct. 2018.

[13] W. Cai, B. Liu, S. Duan, and C. Zou, "An islanding detection method based on dual-frequency harmonic current injection under grid impedance unbalanced condition," *IEEE Trans. Ind. Informat.*, vol. 9, no. 2, pp. 1178–1187, May 2013.

[14] D. K. Alves, R. Ribeiro, F. B. Costa, T. O. A. Rocha, and J. M. Guerrero, "Wavelet-based monitor for grid impedance estimation of three-phase networks," *IEEE Trans. Ind. Electron.*, 2020, Early access.

[15] M. A. Azzouz and E. F. El-Saadany, "Multivariable grid admittance identification for impedance stabilization of active distribution networks," *IEEE Trans. Smart Grid*, vol. 8, no. 3, pp. 1116–1128, May 2017.

[16] B. Wen, D. Boroyevich, R. Burgos, P. Mattavelli, and Z. Shen, "Analysis of D-Q small-signal impedance of grid-tied inverters," *IEEE Trans. Power Electron.*, vol. 31, no. 1, pp. 675–687, Jan. 2016.

[17] X. Wang, L. Harnefors, and F. Blaabjerg, "Unified impedance model of grid-connected voltage-source converters," *IEEE Trans. Power Electron.*, vol. 33, no. 2, pp. 1775–1787, Feb. 2018.

[18] K. Duda, "Accurate, guaranteed stable, sliding discrete Fourier transform [DSP tips & tricks]," *IEEE Signal Process. Mag.*, vol. 27, no. 6, pp. 124–127, Nov. 2010.

[19] J. Kukkola, M. Routimo, and M. Hinkkanen, "Real-time grid impedance estimation using a converter," in *Proc. IEEE ECCE 2019*, Baltimore, MD, Sep. 2019, pp. 6005–6012.

[20] M. Friese, "Multitone signals with low crest factor," *IEEE Trans. Commun.*, vol. 45, no. 10, pp. 1338–1344, Oct. 1997.

[21] M. Schroeder, "Synthesis of low-peak-factor signals and binary sequences with low autocorrelation," *IEEE Trans. Inf. Theory*, vol. 16, no. 1, pp. 85–89, Jan. 1970.

[22] F. M. M. Rahman, U. Riaz, J. Kukkola, M. Routimo, and M. Hinkkanen, "Observer-based current control for converters with an LCL filter: Robust design for weak grids," in *Proc. IEEE SLED*, Helsinki, Finland, Sep. 2018, pp. 36–41.

[23] S. G. Parker, B. P. McGrath, and D. G. Holmes, "Regions of active damping control for LCL filters," *IEEE Trans. Ind. Appl.*, vol. 50, no. 1, pp. 424–432, Jan./Feb. 2014.

Fault Diagnosis of HVDC Transmission System Using Wavelet Energy Entropy and the Wavelet Neural Network

Cuicui Liu[1], Feng Wang[1], Fang Zhuo[1], Ziqian Zhang[2]
[1]State Key Laboratory of Electrical Insulation and Power Equipment
[1]School of Electrical Engineering, Xi'an Jiaotong University
[1]Xi'an, China
[2]Institute of Electrical Power Systems, Graz University of Technology
[2]Inffeldgasse 18/I, 8010 Graz, Austria
Tel.:+86 18710809616
E-Mail: liuc16@stu.xjtu.edu.cn
URL:http://ee.xjtu.edu.cn

Acknowledgements

This project has been partially supported by the fundamental research funds for the central universities "Research on Key Technologies of Active Thermal Balance Control and Reliability Improvement of Energy Router for DC Distribution Grid".

Keywords

«Neural network », «Faults», «HVDC», «Reliability»

Abstract

The failure of the HVDC transmission system is the main factor affecting its reliability. There are many types of faults in the actual project. When a fault occurs, timely and effective identification of the fault type to determine the specific cause of the failure has important research value for improving the reliability of the system. Therefore, this paper focuses on the fault diagnosis method of HVDC transmission system. In this paper, a new fault diagnosis method combining wavelet energy spectrum entropy and wavelet neural network is proposed. In this method, the inverter-side converter bus voltage signal is analyzed as an electrical quantity, and the energy spectrum entropy value of the signal is used to distinguish the normal operating state from each fault state. First, the db10 wavelet is used to decompose and reconstruct the inverter-side converter bus voltage signal collected during the system operation into 10 layers and to obtain the detailed signal of wavelet reconstruction at various scales, and then calculate the wavelet energy spectrum information entropy value of each layer. Use the extracted feature energy spectrum entropy as the input feature vector of wavelet neural network, so as to realize the diagnosis of each fault type of HVDC transmission. The results show that the diagnosis method can accurately diagnose the diagnosis cause of the reduced reliability of the converter valve system.

1 Introduction

HVDC transmission system has been widely used due to its large capacity and long distance transmission characteristics[1-3]. However, with the continuous use of HVDC projects, more and more problems emerge in the actual project[4-5]. There are many types of faults in HVDC system, mainly including faults on the inverter side, rectifier side, and DC transmission line. When one fault occurs in the HVDC transmission system, the inverter bus voltage of the inverter side will change suddenly, and the current signal of the inverter bus will also change suddenly[6-8]. When the fault occurs in the inverter side and the DC transmission line, the inverter side bus voltage will drop. Instantaneously, which will cause commutation failure. However, the failure of the rectifier side will not cause commutation failure, but it will also cause system failure and reduce the reliability of system operation. Therefore, how to quickly identify the fault type when a fault occurs is of great significance for timely and effective maintenance of the HVDC system. In addition, the power output of the system can be improved by stopping the loss in time, and finally the reliability of the system can be improved.

There have some references researching the problem of the fault diagnosis[9-12]. In HVDC transmission system, the most common problem is commutation failure, scholars have studied a variety of schemes from control system to main circuit system to monitor and effectively suppress commutation failure.However, in the actual engineering, there also has other failure problems. Reference [9] calculated the wavelet energy skewness of the voltage signal of the inverter-side commutation bus when the fault occurred through wavelet decomposition, and then analyzed the fault type that caused the commutation failure by using the wavelet energy skewness. However, in the HVDC system, in addition to the failure of the commutation failure caused by the fault in the inverter side and the DC transmission line, the fault in the rectifier side also occurs frequently; Reference [11] proposes a fault diagnosis method for the flexible DC transmission converter that extracts fault features with optimized wavelet packets. This method mainly optimizes the wavelet basis by calculating the angle cosine between the feature vectors of the fault state and the normal state. Compared with directly extracting the feature vectors by performing wavelet decomposition on the fault signal, the calculation amount of this method is too large, and will cause an increase in diagnosis time; Reference [12] uses the wavelet packet energy after the wavelet packet decomposition of the fault current component as the characteristic of the fault signal, and combined with the generalized neural network to identify several types of faults that cause commutation failure, but this method has a relatively low failure recognition rate and is prone to recognition errors.

The work done in this paper is to propose a fault diagnosis method based on the combination of wavelet decomposition and BP neural network. By performing wavelet decomposition on different types of fault voltage signals, the wavelet energy entropy value of each layer is calculated, and then the difference is extracted. Large layers of values construct the input feature vector of the wavelet neural network(WNN), and the extracted values under various types are training samples, and finally realize the diagnosis and identification of the system fault types on the rectifier side, DC transmission line, and inverter side of the HVDC transmission system.

2 HVDC system fault signal

In this paper, the standard simulation model of HVDC 12-pulse transmission system 1000MW 50/60 Hz provided by CIGRE in MATLAB is used to simulate the fault operation states of the inverter side, rectifier side and DC link transmission line. At the same time, the normal state is also be considered, and the causes of every fault are shown in Table I.

Table I: Fault causes

Fault type	Fault causes	Fault type
N	Normal state	Normal
F_1	DC link transmission line fault	DClink
F_2	Single-phase ground fault on the inverter side	I--ASLG, I--BSLG, I--CSLG
F_3	Two-phase ground fault on the inverter side	I--ABDLG, I--ACDLG, I--BCDLG
F_4	Short-circuit fault between two phases on the inverter side	I--ABLL, I--ACLL, I--BCLL
F_5	Three-phase ground fault on the inverter side	I--ABC
F_6	Single-phase ground fault on the rectifier side	R--ASLG, R--BSLG, R--CSLG
F_7	Two-phase ground fault on the rectifier side	R--ABDLG, R--ACDLG,R--BCDLG
F_8	Short-circuit fault between two phases on the rectifier side	R--ABLL, R--ACLL, R--BCLL
F_9	Three-phase ground fault on the rectifier side	R--ABC

In different fault operating states, the inverter side converter bus voltages V_{dc} is collected as the signal to be nanlyzed. Figures 1, 2 and 3 respectively show the V_{dc} under inverter side fault, rectifier side fault, and DC link transmission line fault.

In the simulation model, it is set that different types of faults occur in 1.5s and the fault duration is 0.1s. It can be seen from the figure that the fault can recover itself after a period of time. Although the recovery time of different types of faults is inconsistent, recovery can be completed before 2.4s. When the fault occurs, the voltage of the converter bus rises or falls instantaneously, and the depth of change is different. Therefore, the signal between 1.4s and 2.4s is selected for wavelet decomposition and analysis.

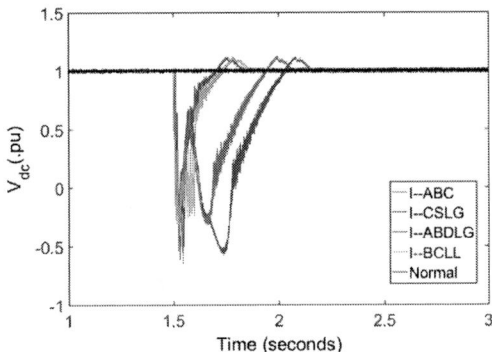

Fig. 1: V_{dc} waveform in inverter-side fault state

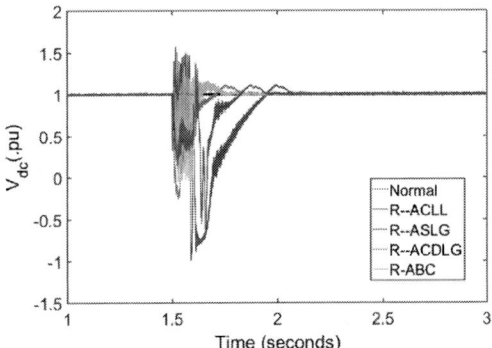

Fig. 2: V_{dc} waveform in rectifier side fault state

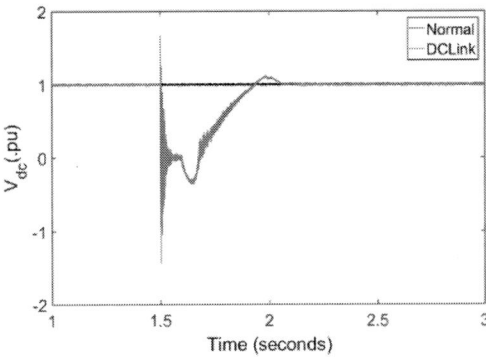

Fig. 3: V_{dc} waveform in DC link fault state

3 Fault feature extraction with wavelet transform

Wavelet transform is a multi-scale signal analysis method, and the inverter side converter bus voltages signal of a HVDC system usually contains fault information, so wo choose to perform wavelet analysis over such voltage signals via MATLAB program.

3.1 Wavelet transform

Wavelet transform refers to the expansion and translation of the base wavelet function $\psi(t)$ to obtain a standard orthogonal wavelet subfunction (1), and then do inner product of the timedomain signal to be analyzed and the wavelet subfunction (2), in the formula (1) and (2), a is the expansion coefficient, b is the translation coefficient, $WT_x(a,b)$ is the wavelet transform coefficient, and the $x(t)$ is the timedomain signal.

$$\psi_{a,b}(t) = \frac{1}{\sqrt{a}} \psi(\frac{t-b}{a}) \tag{1}$$

$$WT_x(a,b) = \frac{1}{\sqrt{a}} \int_{-\infty}^{+\infty} x(t)\psi(\frac{t-b}{a})dt \tag{2}$$

The wavelet decomposition process is to decompose the original signal into an average signal a_1 and a detail signal d_1, and then decompose the obtained average signal into another average signal and detail signal again. The specific decomposition process is shown in Figure 4.

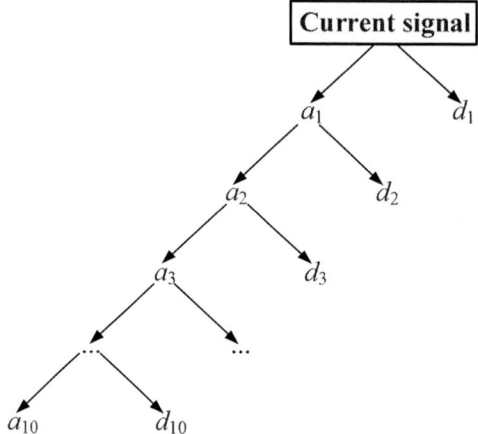

Fig. 4: 10-layer wavelet decomposition schematic

The main principle of wavelet decomposition for fault diagnosis is to decompose the wavelet signal in the fault state and reconstruct the detail signals of each layer, and then extract the fault feature such as wavelet energy entropy that can intuitively reflect the operating state of the HVDC transmission system, and the energy entropy(EN) be used as the input value of the neural network for fault diagnosis.

Taking the C-phase ground fault on the inverter side as an example, the 10-layer wavelet decomposition is performed on the inverter side converter bus voltage signal when the fault occurs, and the detailed signals of the each layer after decomposition are reconstructed, and the reconstructed detail signals d_1-d_{10} are shown in Figure 5.

3.2 Feature extraction

Information entropy is an information measure of system positioning under a certain state, which represents a measure of the degree of sequence unknown, and can be used to estimate the complexity of random signals. The wavelet transform has good time-frequency localization ability. The wavelet transform is combined with information entropy to obtain the wavelet energy entropy of the signal.

Perform j-level wavelet decomposition and reconstruction on the signal $f(x)$ to obtain the energy E_j and wavelet energy entropy EN_j of the signal at different scales. Details are given as follows:

$$E_j = \sum_{k=1}^{N} E_j(k) = \sum_{k=1}^{N} \left| d_j(k) \right|^2 \tag{3}$$

$$E_{\text{total}} = \sum_{j=1}^{L} E_j, \quad p_j = \frac{E_j}{E_{\text{total}}} \tag{4}$$

$$EN_j = -\sum_{j=1}^{L} p_j * \log p_j \tag{5}$$

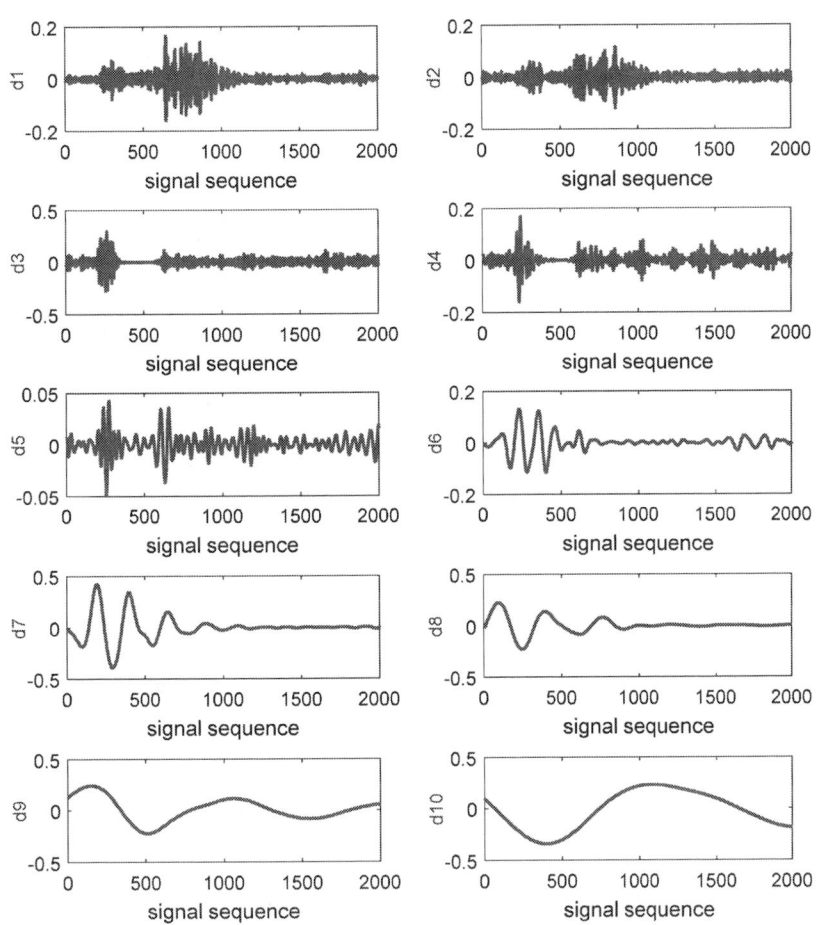

Fig. 5: Detail signals of wavelet reconstruction of the converter bus voltage on inverter side

Table II shows the every layer energy entropy values under N, F_2, F_3 and F_7 fault types. It can be seen from Table II that the EN value of each layer of different types of fault signals has a big difference, and the EN value of each layer of the same type of fault signals is not much different. Therefore, according to the difference characteristics of each fault types combined with the fault diagnosis method, each fault can be effectively diagnosed. In a nutshell, the EN is effective indicator for the fault diagnosis of the HVDC system.

Table II: Test samples of the WNN in different fault states

Fault type	EN_1	EN_2	EN_3	EN_4	EN_5	EN_6	EN_7	EN_8	EN_9	EN_{10}
N	8.1310	3.9924	1.4742	1.4161	1.6762	2.3664	1.7039	4.4082	4.0309	4.0860
F_1	6.6546	4.4825	3.0580	2.4474	2.1543	1.0317	3.6212	2.8866	2.1816	1.6812
F_2	11.8393	7.4841	4.0575	3.7422	3.2184	4.6757	2.8184	1.5705	1.5546	0.8435
	12.1397	7.8259	4.3318	3.8870	3.4089	5.2778	2.9366	1.4368	2.5232	0.5855
	11.8726	7.5516	4.1682	3.5652	2.8620	4.9179	2.5397	1.8328	1.5417	0.8247
F_7	13.7146	9.6494	6.9334	5.3803	4.0093	2.6988	0.1497	4.9321	5.5433	3.3006
	13.4654	9.4334	6.7790	5.1588	3.8109	2.8129	0.1618	5.1881	4.5367	3.1246
	13.6138	9.5141	6.9914	5.1193	3.4907	2.7393	0.1651	5.4001	5.6245	3.1699

4 Fault diagnosis with wavelet neural network

4.1 Construction of WNN

Wavelet neural network(WNN) is a neural network based on the BP（Back Propagation） neural network topology structure, using the wavelet basis function as the transfer function of the hidden layer, and the signal propagates forward while the error propagates backward. Figure 6 shows the structure of wavelet neural network.

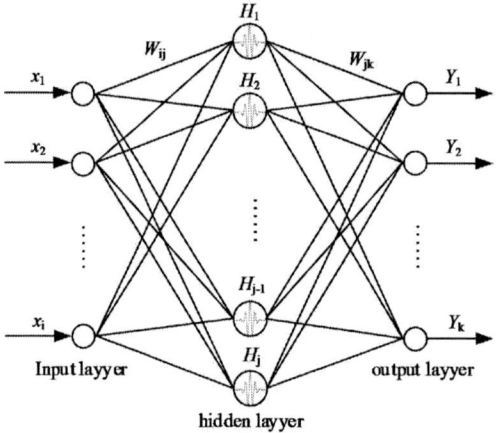

Fig. 6: structure of WNN

Furthermore, the main parameters, such as the weights w_{ij} and w_{jk}, the scaling factor a_j, and the translation factor b_j, are computed iteratively by the following update laws:

$$Error = \frac{1}{2}\sum_{K=1}^{9}(Y_k - d_k)^2 \tag{6}$$

$$w_{jk} = -\eta_1 \frac{\delta Error}{\delta w_{jk}} + w_{jk}, w_{ij} = -\eta_1 \frac{\delta Error}{\delta w_{ij}} + w_{ij} \tag{7}$$

$$a_j = -\eta_2 \frac{\delta Error}{\delta a_j} + a_j, b_j = -\eta_2 \frac{\delta Error}{\delta b_j} + b_j \tag{8}$$

where Error is the object function to minimize, d_k is the desired output of the kth node of the output layer, and η_1, η_2 are the learning rates. In this paper, the network parameters w_{ij}, w_{jk}, a_j, and b_j are randomly initialized in the interval (0, 1), and the two learning rates are set as 0.05.

According to the above description, db10 wavelet is used to decompose and reconstruct 10-layer wavelet on the inverter side converter bus voltage signal in HVDC system fault states, and calculate the wavelet energy entropy value of the signal at different scales, as $EN=[EN_1\ EN_2\ EN_3\ EN_4\ EN_5\ EN_6\ EN_7\ EN_8\ EN_9\ EN_{10}]$, the input samples of the wavelet neural network are established, so the input of the wavelet neural network has 10 nodes. The wavelet transfer function of the hidden layer is morlet wavelet, and the 20 nodes are established into by empirical design. For the 10 types of faults shown in Table 1, 6 digits are used to characterize the fault status, as $R=[R_1\ R_2\ R_3\ R_4\ R_5\ R_6]$, so the number of neurons in the output layer of the wavelet neural network is 6, and the target output corresponding to the network is shown in Table III.

Table III: The target output of WNN

I	R	S	D	L-L	ABC	Fault type
0	0	0	0	0	0	N
1	0	1	0	0	0	F_1
1	0	0	1	0	0	F_2
1	0	0	0	1	0	F_3
1	0	0	0	0	1	F_4
1	1	1	1	1	1	F_5
0	1	1	0	0	0	F_6
0	1	0	1	0	0	F_7
0	1	0	0	1	0	F_8
0	1	0	0	0	1	F_9

In the table III, "I" indicates the inverter side fault, "R" indicates the rectifier side fault, "S" indicates the single-phase ground fault, "D" indicates the two-phase ground fault, " L-L " indicates the Short-circuit fault, "ABC "Indicates a three-phase ground fault.Each bit indicates failure with 1 and 0 means no failure. In addition, the normal operating state is represented by all 0, and the DC transmission line fault is represented by all 1.

4.2 Results of fault diagnosis

The energy spectrum entropy values under various fault types are used as input of wavelet neural network. The value of R shown in table III is the output of the wavelet neural network. With input and output samples, the network training is performed first. After the training, use the test samples shown in table II to verify the feasibility and accuracy of the proposed method. Table IV shows the test results of wavelet neural network. Take an integer for the output result in table IV, and obtain a diagnosis output result that represents each bit of data with 0 or 1. From the diagnosis result, it can be seen that the method can accurately determine the type of failure that occurs in the system.

Table IV: The test result of WNN

R_1	R_2	R_3	R_4	R_5	R_6	Test result
-0.0055	0.9745	1.0547	1.0063	0.9880	-0.0093	N
0.0100	0.9915	-0.0590	0.0269	0.0185	0.9528	F1
0.0090	1.0112	0.9167	0.8246	0.9722	0.0351	
-0.0075	0.9808	0.0449	-0.0277	0.0044	0.9523	F2
0.0008	0.9987	0.0209	-0.0645	-0.0706	-0.0134	
0.0024	0.9680	0.0465	0.0052	0.0499	-0.0194	
-0.0055	0.9745	1.0547	1.0063	0.9880	-0.0093	F7
0.0100	0.9915	-0.0590	0.0269	0.0185	0.9528	

5 conclusion

In the HVDC system, being able to quickly and accurately diagnose the location of a fault has specific guiding significance for the maintenance of the system, thereby reducing the loss caused by the unstable operation of the system, which can effectively promote the steady improvement of system reliability. In this paper, wavelet transform are used to extract the wavelet energy entropy of the inverter side converter bus voltage signal of HVDC system under different fault types, and combined with wavelet neural network for fault diagnosis. The diagnosis results show that the wavelet neural network has extremely strong convergence, the diagnosis results are completely consistent with the expected output results, and the error is within the allowable range, that is, this method can effectively diagnose the fault in the operation of the converter valve system, and then develop detailed realtime maintenance program to achieve the purpose of improving the HVDC system reliability.

References

[1] Zhao Wanjun . High voltage direct current project technology[M]. 2nded. Beijing: China Electric Power Press, 2011: 124.

[2] B. Hu, K. Xie and H. Tai, "Optimal Reliability Allocation of ±800 kV Ultra HVDC Transmission Systems," in IEEE Transactions on Power Delivery, vol. 33, no. 3, pp. 1174-1184, June 2018.

[3] B. Hu, K. Xie and H. Tai, "Reliability Evaluation and Weak Component Identification of ±500-kV HVDC Transmission Systems With Double-Circuit Lines on the Same Tower," in IEEE Transactions on Power Delivery, vol. 33, no. 4, pp. 1716-1726, Aug. 2018.

[4] K. Xie, B. Hu and C. Singh, "Reliability Evaluation of Double 12-Pulse Ultra HVDC Transmission Systems," in IEEE Transactions on Power Delivery, vol. 31, no. 1, pp. 210-218, Feb. 2016.

[5] S. R. Mohammed, J. Teh and M. Kamarol, "Upgrading of the Existing Bi-Pole to the New Four-Pole Back-to-Back HVDC Converter for Greater Reliability and Power Quality," in IEEE Access, vol. 7, pp. 145532-145545, 2019.

[6] J. Yuan and Y. Tian, "A Multiscale Feature Learning Scheme Based on Deep Learning for Industrial Process Monitoring and Fault Diagnosis," in IEEE Access, vol. 7, pp. 151189-151202, 2019.

[7] G. Zhang, L. Jing, M. Liu, B. Wang and X. Dong, "An Improved Continuous Commutation Failure Mitigation Method in High Voltage Direct Current Transmission System," 2018 China International Conference on Electricity Distribution (CICED), Tianjin, 2018, pp. 1132-1136.

[8] V. Vanitha, D. Kavitha, R. Resmi and M. S. Pranav, "Fault classification and location in Three Phase Transmission Lines using Discrete Wavelet Transform," 2019 IEEE International Conference on Electrical, Computer and Communication Technologies (ICECCT), Coimbatore, India, 2019, pp. 1-4

[9] Zhu Yan, Wang Yuhong, Ding Zhilin, Song Liang, Li Xingyuan, Qiu Daqiang. Fault diagnosis of HVDC commutation failure based on wavelet energy skew neural network [J] .Journal of Electric Power System and Automation, 2017,29 (02): 39-44.

[10] Zheng Xiaoxia, Peng Peng. Fault diagnosis of flexible HVDC converter based on optimized wavelet packet and AdaBoost-SVM [J]. Journal of Electric Power System and Automation, 2019, 31 (03): 42-49.

[11] A. Dileep Kumar S. Raghunath Sagar "Discrimination of Faults and Their Location Identification on a High Voltage Transmission Lines Using the Discrete Wavelet Transform" International Journal of Education and Applied Research vol. 4 no. 1 pp. 107-111 2014.

[12] Liu Fei. Analysis and application of HVDC fault signal based on wavelet packet and generalized regression neural network [D] .South China University of Technology, 2017.

Reducing the Energy Storage Requirements of Modular Multilevel Converters with Optimal Capacitor Voltage Trajectory Shaping

Simon Fuchs, Min Jeong, Jürgen Biela
Laboratory for High Power Electronic Systems (HPE)
Email: fuchssim@ethz.ch, jbiela@ethz.ch
ETH Zürich, Switzerland

Keywords

≪HVDC≫, ≪Optimal control≫, ≪High power density systems≫, ≪High voltage power converters≫, ≪Harmonics≫

Abstract

The required module capacitance value is a driving factor of the volume and cost of modular multilevel converters (MMC). In order to minimize the required capacitance value, an optimal circulating current and common mode (CM) voltage injection strategy which keeps the conduction losses and die CM voltage as low as possible is proposed in this paper. Unlike previous methods from literature that only focus on minimizing the amplitude of energy fluctuation in the arm capacitors, the proposed optimization procedure is based on the optimal shaping of both the arm voltages and energies. As a result, the proposed optimization scheme achieves a further reduction in the required capacitance value of up to 42 % compared to existing methods. The procedure is validated with closed loop simulation results covering the full operating range of an exemplary MMC.

1 Introduction

The Modular Multilevel Converter (MMC, [1]) represents one of the standard topologies for converters operating at medium to high voltages [2, 3]. However, a major drawback of the MMC is its relatively high energy storage requirements resulting in large module capacitors that significantly contribute to the MMC's volume, weight, and cost.

The large module capacitors are required because of the single phase characteristics of the six MMC arms (cf. Fig. 1). In each arm, the instantaneous AC and DC power are not equal, which causes a fluctuation in the energy stored in the module capacitors. This leads to a fluctuation/ripple in the module capacitor voltages and therefore also in the available arm voltage v^Σ (sum of all capacitor voltages) as shown in Fig. 2a. The smaller the module capacitance value, the higher the amplitude of this voltage fluctuation. In [4], the minimum required capacitance value for the standard operation of an MMC is derived: The available arm voltage v_{1u}^Σ is always higher or just equal to the required arm output voltage v_{1u} and always less or just equal to the maximum allowed available arm voltage $(N \cdot V_{C,max})$ as shown in Fig. 2b.

In order to decrease the amplitude of the arm energy/capacitor voltage fluctuation, many examples that inject circulating currents are proposed in literature. The amplitude and phase of the injected harmonics are either

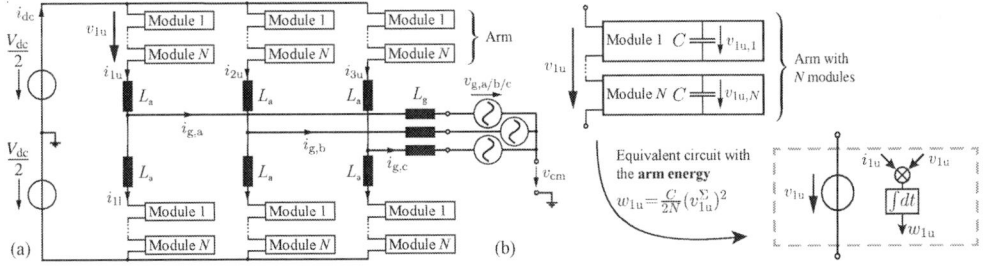

Fig. 1: (a) Three phase MMC. N modules form one out of six arms. (b) One arm can be represented with an energy storage, that integrates the arm power $\dot{w}_{1u} = i_{1u} \cdot v_{1u}$. The indices represent the upper arm of the first phase.

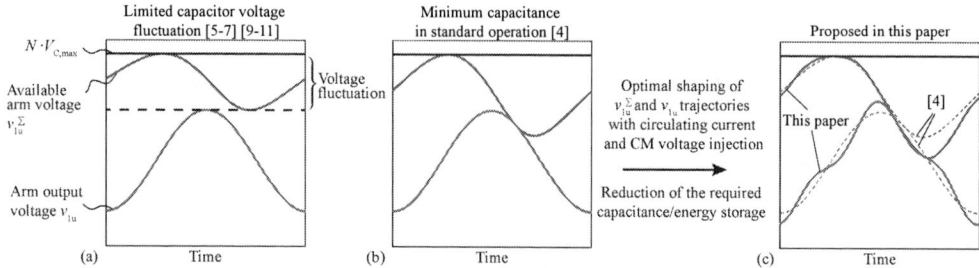

Fig. 2: Different (available) arm voltage trajectories for active power transfer: (a) For MMC design with limited capacitor voltage ripple, (b) minimum required capacitance/energy storage with standard operation (no circulating currents, no CM voltage injection) as shown in [4], (c) Optimal shaping of the available arm voltage (v_{1u}^{Σ}) and arm output voltage (v_{1u}) by circulating current and common mode (CM) voltage injection. The indices represent the upper arm of the first phase.

determined analytically [5, 6, 7] or using offline optimizations [8, 9, 10]. In some cases, the circulating current injection is also combined with additional harmonics in the CM voltage v_{cm} at the grid side of the MMC [11]. If there is no star-point connection at the grid side (see Fig. 1a), this has no influence on the grid currents but decreases the maximum required arm output voltage [4, 12].

The reduced energy fluctuation in the MMC arms can be used to reduce the required module capacitance value because the same voltage fluctuation can be achieved with a lower capacitance value. However, reducing only the arm energy fluctuation does not necessarily minimize the required capacitance value because it does not consider the relation between the trajectories of the required arm output voltage and the available arm voltage as described in [4]. All listed references aim to decrease the fluctuation of the capacitor voltage/energy and do not consider the actually required capacitance value or the trajectories of the arm output voltage and the available arm voltage. An exception is the procedure described in [8], which aligns the peak of the available arm voltage with the peak of the required arm output voltage by using a second harmonic circulating current. The analysis is however limited to a second harmonic in the circulating current.

To complete the analysis and further reduce the required capacitance value, this paper proposes an optimization procedure that identifies the optimal trade-off between the reduction in the required capacitance, the (semiconductor conduction) losses, and the CM voltage amplitude. The optimization procedure searches for optimal combinations of the second and fourth harmonics in the circulating current as well as the third and ninth harmonic in the CM voltage. The required capacitance value is determined with an extended version of the procedure presented in [4], in order to consider the actual trajectories of the available arm voltage and the arm output voltage (Fig. 2c). As a result, a three-dimensional Pareto front/surface is calculated, which allows to analyse the influence of combining a CM voltage and a circulating current injection on the reduction of the required capacitance. Variable output frequencies as required in drive applications are not explicitly considered.

The paper is organized as follows: In section 2, all important modelling equations required to determine the steady state trajectories of the MMC are reviewed. The proposed optimization scheme is presented in section 3. In section 4, the optimization results for an exemplary MMC parameter set are shown, analysed, and compared to some benchmark schemes given in literature. Section 5 presents simulation results and proposes a possible controller implementation to realize the harmonics injection within a closed loop control system.

2 Steady State Trajectories of the MMC

In the following, analytical expressions for the steady state trajectories of the MMC as shown in Fig. 1 are introduced. For the sake of simplicity, the analysis is limited to transferring purely active power P only. The expressions are given for the upper arm of the first phase (index '1u').

Arm Currents

As introduced e.g. in [4], the arm currents consist of a DC component $i_{1u,dc}$ and an AC component $i_{1u,ac}$ that has the same frequency as the grid. These two components are necessary for the operation of the MMC because they exchange the power with the DC side and AC side respectively. They can be expressed as

$$ i_{1u,dc} = \frac{i_{dc}}{3} = \frac{P}{3 \cdot V_{dc}}, \qquad i_{1u,ac} = \frac{i_g}{2} \cdot \cos(\omega t) = \frac{P}{2\sqrt{3} \cdot V_g} \cdot \cos(\omega t). \tag{1} $$

In this paper, a second and fourth harmonic circulating current are added as suggested in [9]. Note, that only even harmonics that are not divisible by three are allowed for circulating currents, because all other harmonics have an effect on either the DC or the AC side currents. With the circulating current harmonics, the total arm current is

$$i_{1u} = i_{1u,dc} + i_{1u,ac} + \hat{i}_{c2} \cdot \cos(2\omega t + \varphi_{c2}) + \hat{i}_{c4} \cdot \cos(4\omega t + \varphi_{c4}), \tag{2}$$

where $\hat{i}_{c2}$ as well as $\hat{i}_{c4}$ are the amplitudes and φ_{c2} as well as φ_{c4} are the phase angles of the circulating current harmonics.

Arm Output Voltages

To simplify the analysis, the MMC arms are regarded as continuously controllable voltage sources [13]. Therefore, the arm output voltage is composed of a DC and an AC component given as

$$v_{1u,dc} = \frac{V_{dc}}{2}, \qquad v_{1u,ac} = \sqrt{\frac{2}{3}} \cdot V_g \cdot \cos(\omega t). \tag{3}$$

In this paper, a third and a ninth harmonic Common Mode (CM) voltage are added to the AC component. CM harmonics divisible by three do not drive a current on the DC side nor on the AC side, if the AC side star point is not grounded. However, the 6th harmonic is omitted because even CM harmonics lead to a non-symmetric energy fluctuation in the upper and lower MMC arms, which is not desirable within the presented optimization routine, because one would have to consider the upper and lower arms separately. With the CM voltage harmonics, the total arm voltage is given as

$$v_{1u} = v_{1u,dc} - v_{1u,ac} - \hat{v}_{cm3} \cdot \cos(3\omega t + \varphi_{cm3}) - \hat{v}_{cm9} \cdot \cos(9\omega t + \varphi_{cm9}), \tag{4}$$

where $\hat{v}_{cm3}$ as well as $\hat{v}_{cm9}$ are the amplitudes and φ_{cm3} as well as φ_{cm9} are the phase angles of the CM voltage harmonics. The voltages necessary to drive the arm currents through the arm inductors and resistors are neglected here, as they are typically very small compared to the other parts of the arm output voltage. However, the procedure presented in this paper does not change in case this voltage drop is included.

Arm Energy and Available Arm Voltage

The available arm voltage v_{1u}^{Σ} can be expressed with the energy w_{1u} stored in all capacitors of the arm. As depicted in Fig. 1(b) his energy can be derived with the arm power p_{1u} given as

$$w_{1u}(t) = \int_0^t p_{1u}(\tau)d\tau + w_{dc}, \qquad \text{with} \qquad p_{1u} = v_{1u} \cdot i_{1u}, \tag{5}$$

where $w_{dc} = C/N \cdot (v_{dc}^{\Sigma})^2/2$ is the average arm energy during a grid period defined by the DC offset v_{dc}^{Σ} of the available arm voltage. It will be shown in the first part of section 3 how the optimal average arm energy/DC offset voltage is determined. Inserting (1), (2), (3) and (4) into (5) results in a long expression that is omitted here for space reasons.

3 Optimization Procedure

Based on the equations presented in section 2, an optimization procedure determining the optimal harmonics injection is presented in the following. First, the determination of the minimum capacitance value is described based on fixed trajectories with given circulating currents and CM voltages. Then, the trajectory shaping optimization procedure is explained to find the optimal combination(s) of circulating current and CM voltage harmonics injection.

Minimum Capacitance Determination

[4] provides a procedure to determine the minimum module capacitor value for a given set of MMC ratings (AC/DC voltage, power, number of modules, maximum module voltage) when no harmonics in the arm currents/CM voltage are assumed. In the following, the relations derived in [4] are shortly revised and adopted to be used with given harmonics in the arm currents/CM voltage.

There are two important factors k_{max} (relation between the max. allowed available arm voltage and the DC voltage) and k_{dc} (relation between the DC offset voltage v_{dc}^{Σ} and the DC voltage) given as

$$k_{max} = \frac{N \cdot V_{C,max}}{V_{dc}}, \qquad k_{dc} = \frac{v_{dc}^{\Sigma}}{V_{dc}}, \tag{6}$$

where $V_{C,\max}$ is the maximum allowed capacitor voltage of the MMC's modules. It can be shown (cf. [4]) that the minimum k_{dc} fulfilling

$$k_{dc} \geq \sqrt{\frac{\frac{v_{1u}^2(t)}{V_{dc}^2} - k_{\max}^2 \cdot \frac{w_{1u}(t) - w_{dc}}{w_{1u,\max} - w_{dc}}}{1 - \frac{w_{1u}(t) - w_{dc}}{w_{1u,\max} - w_{dc}}}}, \quad \forall t \in \left[0, \frac{2\pi}{\omega}\right] \qquad \text{with} \qquad w_{1u,\max} = \max_{t \in \left[0, \frac{2\pi}{\omega}\right]} w_{1u}(t) \tag{7}$$

for the complete grid period yields the minimum required module capacitance

$$C_{\min} = \frac{2N}{V_{dc}^2} \cdot \frac{w_{1u,\max} - w_{dc}}{k_{\max}^2 - k_{dc}^2}. \tag{8}$$

If no harmonics in the circulating current and the CM voltage are injected, the trajectories given in Fig. 2b result, where the available arm voltage is always just bigger or equal to the required arm voltage and always just smaller or equal to $N \cdot V_{C,\max}$.

Note that w_{dc} does not have to be known to solve this. The difference $w_{1u} - w_{dc}$ is just used to express the integral part of (5). Therefore, w_{dc} and v_{dc}^Σ are rather the second result of the capacitance value minimization, because they are given with the chosen k_{dc} as

$$v_{dc}^\Sigma = k_{dc} \cdot \frac{V_{dc}}{N} \qquad \text{and} \qquad w_{dc} = N \cdot C \cdot \frac{(v_{dc}^\Sigma)^2}{2}. \tag{9}$$

Optimal Trajectory Shaping

The vector of optimization variables $\mathbf{x}$ contains the amplitude and phase angles of the 2nd and 4th harmonics in the circulating currents as well as the 3rd and 9th harmonics in the CM voltage, such that

$$\mathbf{x} = \begin{bmatrix} \hat{i}_{c2} & \varphi_{c2} & \hat{i}_{c4} & \varphi_{c4} & \hat{v}_{cm3} & \varphi_{cm3} & \hat{v}_{cm9} & \varphi_{cm9} \end{bmatrix}. \tag{10}$$

Modifying the shape of the energy/available arm voltage trajectory comes at a cost of a CM voltage at the MMC's AC side and/or changed losses in the MMC's semiconductors. There is no general optimally shaped trajectory but rather a Pareto front representing the optimal trade-off between required module capacitance value, semiconductor losses and CM voltage amplitude. Therefore, three fitness functions are used to perform a multi-objective optimization:

1. The reduction in the required capacitance value for circulating current and CM voltage harmonics given in $\mathbf{x}$ can be stated as

$$J_{cap} = C_{\min}(\mathbf{x}) / C_{\min}(\mathbf{x} = 0). \tag{11}$$

    The value of the required capacitance is determined by the procedure described in the previous subsection 'Minimum Capacitance Determination' (7) - (9).

2. By injecting circulating current harmonics, the arm currents that cause losses in the semiconductors and arm inductors are changed. Here, the change in the arm current mean rectified value is considered. Using (1) and (2), this change can be expressed with

$$J_{current} = I_{av,rect}(\mathbf{x}) / I_{av,rect}(\mathbf{x} = \mathbf{0}), \qquad \text{where} \qquad I_{av,rect}(\mathbf{x}) = \frac{\omega}{2\pi} \int_{t=0}^{\omega/2\pi} |i_{1u}(t, \mathbf{x})| dt. \tag{12}$$

    This fitness function can also be replaced by the increase in the RMS value of the arm current or any combination J_{loss} of RMS and mean rectified current to represent the increase of semiconductor and arm inductor losses due to the injected circulating currents ($J_{loss} = a \cdot I_{av,rect}(\mathbf{x}) + b \cdot I_{RMS}(\mathbf{x})$, where a and b depend on the candidate semiconductor switches and the specific inductor implementation). This does not cause any major change in the proposed algorithm.

3. The third important parameter is the maximum amplitude of the CM voltage. This is the voltage of the star-point of a transformer which defines the isolation requirements of the grid transformer as denoted in Fig. 1a. Therefore,

$$J_{CM} = \max_{t \in [0, 2\pi/\omega]} |\hat{v}_{cm3} \cdot \cos(3\omega t + \varphi_{cm3}) + \hat{v}_{cm9} \cdot \cos(9\omega t + \varphi_{cm9})| \tag{13}$$

can be used to formulate the cost associated with the injected common mode voltage harmonics.

There are also constraints that must be considered:

1. All entries of $\mathbf{x}$ must be greater than zero and the phase angles are constrained to the interval $[0\ldots2\pi]$.

2. In order to consider also the maximum current capabilities of the used semiconductors and/or the saturation current of the used inductors, the arm currents must be constraint to a maximum absolute value i_{max}, such that

$$|i_{1u}(t,\mathbf{x})| \leq i_{max} \qquad \forall t \in [0, 2\pi/\omega].\tag{14}$$

3. In case of half-bridge based MMC modules, the arm output voltage has to be constrained to be greater than zero. This results in

$$v_{1u,dc} - v_{1u,ac} - \hat{v}_{cm3} \cdot \cos(3\omega t + \varphi_{cm3}) - \hat{v}_{cm9} \cdot \cos(9\omega t + \varphi_{cm9}) \geq 0 \qquad \forall t \in [0, 2\pi/\omega].\tag{15}$$

In case of full-bridge MMC modules this constraint is not needed.

To perform a multi-criteria optimization, one could use a brute force approach, but with $\mathbf{x}$ having eight entries this takes a very long time if accurate results should be achieved. Therefore, the genetic algorithm [14] contained in the Matlab Global Optimization Toolbox is used to generate the results presented in the next section.

4 Optimization Results and Discussion

In this section the results of the optimization procedure described in section 3 for the exemplary MMC parameters given in Table I (MV case, see Fig. 3) and Table II (HVDC case, see Fig. 5) are presented and discussed. The resulting three-dimensional Pareto front is shown in part (c) of both figures. Part (a), (b), and (d) represent the top and side view of this Pareto front. Every black dot represents a Pareto optimal optimization result, which means, that it is better in two out of three of the considered cost functions than all other solutions. There are also solid lines with three different colours drawn within the Pareto fronts:

- The red lines — represent the Pareto front, when the CM voltage amplitude is not restricted. In part (c), one of the red lines is the projection of the other red line onto the xy-plane. If the CM voltage amplitude is not considered, the points on the red line provide the best trade-off between capacitance value reduction and arm current increase with the given maximum arm current. Note that the red line also represents the maximum meaningful CM voltage that helps to reduce the required capacitance value. Higher values would not result in a lower required capacitance value as shown in part (d) of the figures.

- The green lines — represent the Pareto front if no CM voltage injection is allowed.

- The blue lines — represent the Pareto front, when the CM voltage amplitude is limited to $\hat{V}_g/6$.

There is a point marked with a green star * in both figures. This point features the combination of CM voltage and circulating harmonic injection that results in the lowest possible required capacitance value with the given parameters and maximum arm current. Besides this, there is also a point marked with a pink star * representing the lowest possible required capacitance value when avoiding CM voltage injection. The time domain trajectories of the voltages and currents of the upper arm of the first phase for these design points are shown in Fig. 4 for one grid period using exactly the minimum required capacitance value.

Comparison with Existing Methods

Fig. 3 and 5 also contain the results for different methods given in the literature, if the fitness functions proposed in this paper are applied to the given circulating current and CM voltage trajectories proposed in the respective publication. It can be seen that the optimization proposed in this paper leads to a lower required capacitance value with a lower mean rectified current in any case. Furthermore, the CM voltage injection proposed in the literature references does always lead to an increase in the required capacitance value (compare [5] with and without CM as well as [7] Methods B and C in Fig. 3c).

The procedure presented in [9, 10, 11] minimizes the energy fluctuation Δw_{1u} in the module capacitors given by

$$\Delta w_{1u} = \max_{t \in [0, \frac{2\pi}{\omega}]} w_{1u}(t) - \min_{t \in [0, \frac{2\pi}{\omega}]} w_{1u}(t).\tag{16}$$

Table I: MMC PARAMETERS (MV CASE)

Symbol		Value	Symbol		Value
V_g	Rated grid voltage	9 kV	I_g	Rated grid current	22.681 A
P	Rated power	250 kW	ω	Rated grid frequency	$2\pi\,50$ Hz
V_dc	Rated DC voltage	35 kV	N	Module number / arm	15
C	Capacitance (std. operation [4])	$56\,\mu$F	v_dc^Σ	DC offset voltage (std. op. [4])	27.765 kV
$v_\text{C,max}$	Maximum module voltage	2.2 kV	i_max	Maximum arm current	34 A
L_g	Grid inductance	0 mH	R_g	Grid resistance	$0\,\Omega$
L_a	Arm inductance	26.8 mH	R_a	Arm resistance	$0.1\,\Omega$

Table II: MMC PARAMETERS (HVDC CASE, based on [15])

Symbol		Value	Symbol		Value
V_g	Rated grid voltage	220 kV	I_g	Rated grid current	4.454 kA
P	Rated power	1200 MW	ω	Rated grid frequency	$2\pi\,50$ Hz
V_dc	Rated DC voltage	400 kV	N	Module number / arm	130
C	Capacitance (std. operation [4])	35 mF	v_dc^Σ	DC offset voltage (std. op. [4])	380.22 kV
$v_\text{C,max}$	Maximum module voltage	3 kV	i_max	Maximum arm current	5.7 kA (+5 %)

To compare the results of [9, 10, 11] with those of the procedure presented in this paper, the first fitness function of the optimization scheme given in Sec. 3 is modified to represent the change in energy fluctuation $\Delta w_{1\text{u}}(\mathbf{x})/\Delta w_{1\text{u}}(\mathbf{x} = \mathbf{0})$. Note that $\Delta w_{1\text{u}}(\mathbf{x})$ is independent of w_dc, such that it is not necessary to find to minimum k_dc as done in (7) - (9) and the fitness function boils down to evaluating (5) and finding its maximum and minimum. With this change, the optimization is performed again and the resulting solutions are mapped on the originally proposed fitness function by applying (11). The resulting Pareto fronts are also plotted in Fig. 3 and 5 using dashed lines in the same colour coding as described before. Besides the overview in Fig. 3 and 5, the numerical values for the comparison are also given in Table III.

The presented optimization procedure outperforms all methods from literature in all cases. In the MV case, capacitance value savings of up to 35 % are possible without increasing the mean rectified arm current. However, this comes at the price of a comparably high CM voltage. If CM voltage injection must be avoided, the required capacitance value can be decreased by up to 26.6 % while the mean rectified arm current increases by 2.8 % only. When comparing to the methods that optimize the energy fluctuation only, the procedure proposed in this paper results in a 13.3 % smaller required capacitance value if CM voltage injection is allowed.

In the HVDC case, capacitance value savings of up to 82.7 % are possible without increasing the mean rectified arm current and with an increase in the maximum arm current of less than 5 %. The injected CM voltage is always less than the usual $\hat{v}_\text{g}/6$ in this case. If CM voltage injection must be avoided completely, the required capacitance value can be decreased by up to 51 % while the mean rectified arm current is still not increased. When comparing to the methods that optimize the energy fluctuation only, the procedure proposed in this paper results in a 42.5 % smaller required capacitance value if CM voltage injection is allowed.

5 Closed Loop Simulation

Generally speaking, any cascaded PI control system for MMCs can be adopted to successfully implement the harmonics injection presented in this paper. However, the controllers must be tuned for a comparably low bandwidth in order to not exceed the maximum allowed available arm voltage and to keep the available arm voltage high enough to avoid saturation. For better control performance, the control algorithms presented in [16, 17] can also be used.

The simulation results presented in Fig. 6 have been obtained with a slightly adopted version of the algorithm presented in [17]: The reference trajectories of the arm energies, the circulating currents and the controller output voltages are matched with the optimization results scaled by the reference power rated with the nominal power of the converter. This means that e.g. no CM voltage and no circulating current are injected at zero reference power. The simulation model is realized with an average arm model similar to the one proposed in [13]. Because of the voltage drop across the arm inductors which is not considered within the optimization procedure, the capacitance value obtained by the optimization procedure is increased by 5 % as well as the average arm voltage DC offset

Reducing the Energy Storage Requirements of Modular Multilevel Converters with
Optimal Capacitor Voltage Trajectory Shaping

FUCHS Simon

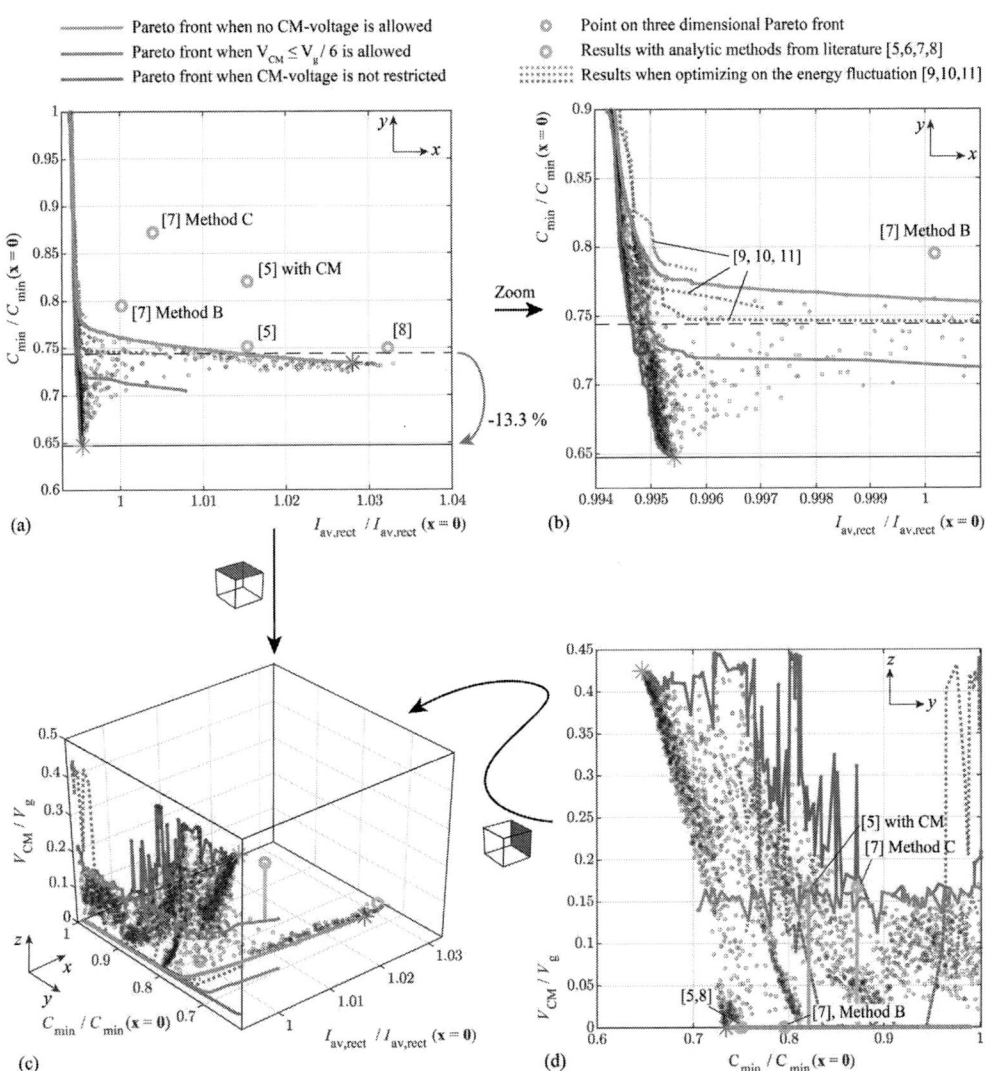

Fig. 3: Optimization results for the MV application MMC parameters in Tab. I. The dashed lines represent the results achieved with the methods from [9, 10, 11]. (a) Top view of c. Black horizontal lines show the difference in the minimum required capacitance value with the presented method vs. [9, 10, 11]. (b) Detail view of a. (c) Three dimensional Pareto front. (d) Side view of c.

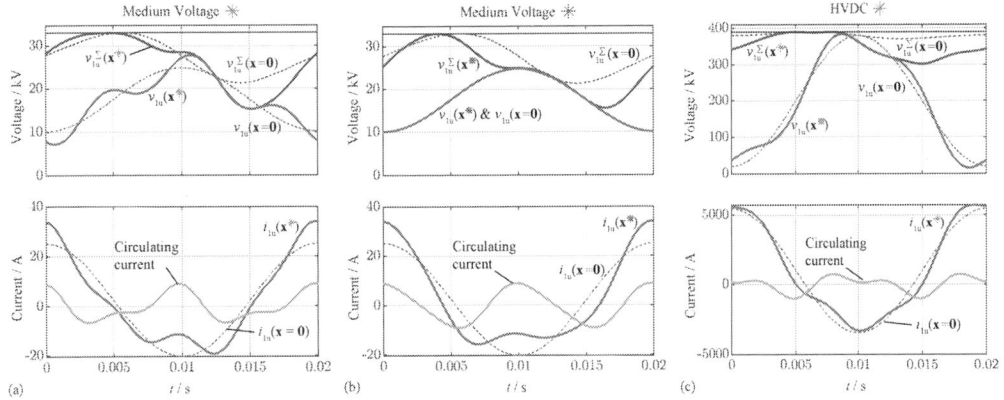

Fig. 4: Time domain trajectories for the points marked in Fig. 3 and 5. The dashed lines represent the trajectories with the minimum required capacitance value without any circulating current or CM voltage injection ($\mathbf{x} = \mathbf{0}$).

EPE'20 ECCE Europe

Assigned jointly to the European Power Electronics and Drives Association & the Institute of Electrical and Electronics Engineers (IEEE)

Reducing the Energy Storage Requirements of Modular Multilevel Converters with Optimal Capacitor Voltage Trajectory Shaping — FUCHS Simon

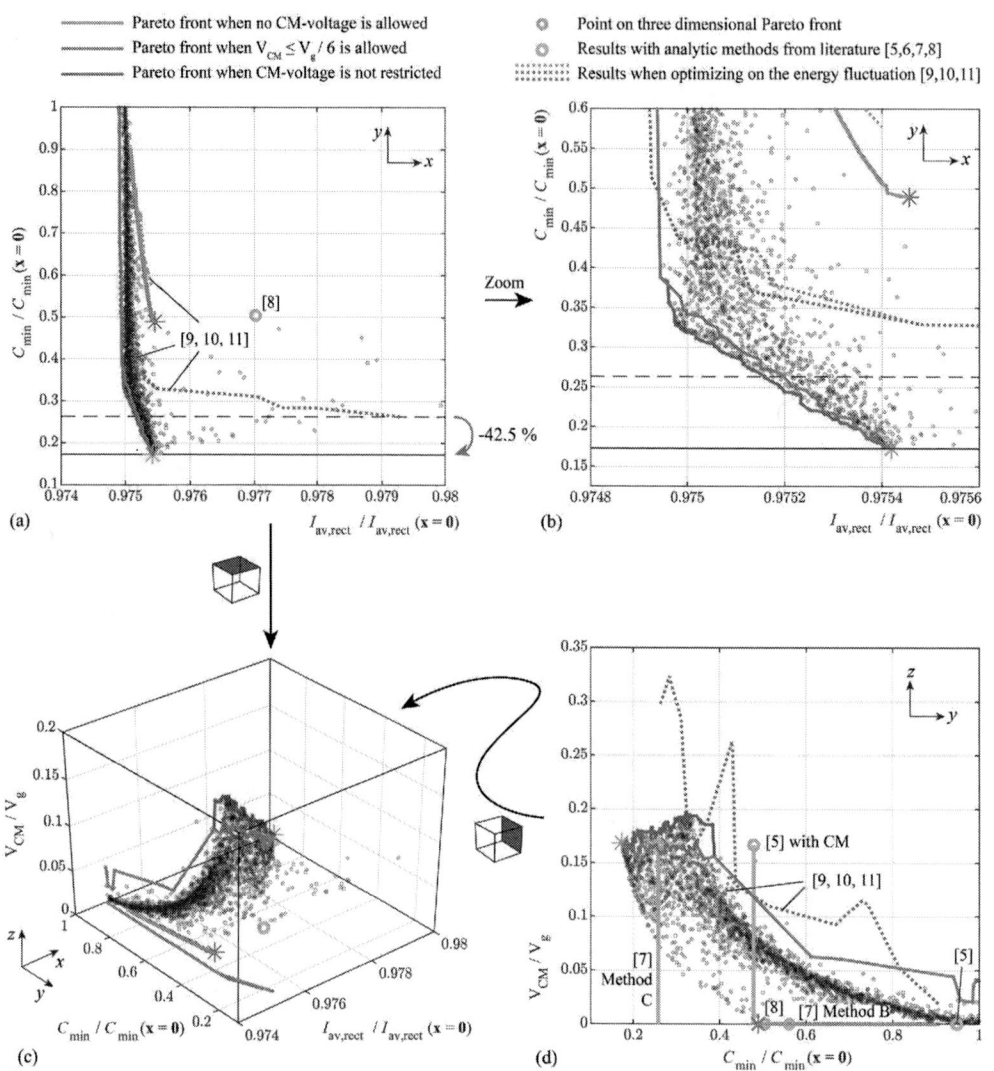

Fig. 5: Optimization results for the HVDC application MMC parameters in Tab. II. The dashed lines represent the results achieved with the methods from [9, 10, 11]. (a) Top view of c. Black horizontal lines show the difference in the minimum required capacitance value with the presented method vs. [9, 10, 11]. (b) Detail view of a. (c) Three dimensional Pareto front. (d) Side view of c.

Table III: COMPARISON WITH METHODS FROM LITERATURE

Reference	MV (Tab. I, Fig. 3)			HVDC (Tab. II, Fig. 5)		
	J_{cap}	$J_{current}$	J_{CM}	J_{cap}	$J_{current}$	J_{CM}
Point *	0.734	1.028	0	0.489	0.976	0
Point *	0.647	0.995	0.425	0.173	0.975	0.1684
[8]	0.750	1.032	0	0.505	0.977	0
[5]	0.751	1.015	0	0.948	1.053	0
[5] with CM	0.820	1.015	0.166	0.479	1.053	0.166
[7] method B	0.795	1.000	0	0.562	0.998	0
[7] method C	0.872	1.004	0.166	0.258	1.012	0.166
[9, 10, 11] without CM	0.783	0.996	0	0.576	0.975	0
[9, 10, 11] with CM	0.744	1.004	0.310	0.263	0.979	0.293

EPE'20 ECCE Europe

Assigned jointly to the European Power Electronics and Drives Association & the Institute of Electrical and Electronics Engineers (IEEE)

Table IV: CONTROLLER PARAMETERS

Parameter	Value	Parameter	Value
Sampling frequency	7.5 kHz	DC current weighting factor	100
Circ. current weighting factor	0	AC current weighting factor	0
Arm energy weighting factor	100	Control effort weighting factors (e)	1000
Control effort weighting factors (a)	10^5		

v_{dc}^{Σ} by 1 %. Therefore, the arm voltage trajectories do not look exactly the same as in Fig. 4a. Due to the very low margin between the required and the available arm voltage, the power reference changes are implemented as ramp commands and not as step commands to avoid saturation effects. Moreover, the increase in the amplitude of the injected circulating currents and CM voltage during the increase of the power reference can be shown nicely during the power reference changes ($t = 10\ldots50$ ms and $t = 150\ldots240$ ms). All controller parameters are given in Tab. IV.

6 Conclusion

In this paper, an optimization scheme minimizing the energy storage requirements and therefore the necessary module capacitance value of MMCs is proposed. The scheme identifies the optimal combinations of an AC common mode voltage and a circulating current injection in order to shape the trajectories of the available arm voltages and the required arm output voltages such that the required capacitance value is minimized rather than the capacitor voltage fluctuation/ripple. The result of the optimization scheme is presented as a Pareto front to visualize the trade-off between required module capacitance value, semiconductor conduction losses, and the maximum injected AC common mode voltage.

For an exemplary MV MMC, a capacitance value reduction of almost 35 % is achieved without increasing the mean rectified arm current compared to the standard operation without harmonics injection. Methods from literature achieve a reduction in the required capacitance value of only 25 % compared to the standard operation. In case CM voltage injection is not allowed (e.g. due to a grounded star point at the AC side), the proposed optimization scheme achieves a reduction of the required capacitance value by up to 27 %. Here, the increase of the mean rectified arm current is lower than 3 %. For the considered HVDC application, a capacitance value reduction of almost 83 % is achieved while the maximum arm current is constrained to be only 5 % higher than in the standard operation without harmonics injection. The mean rectified arm current is not increased. State-of-the-art methods achieve a reduction in the required capacitance value of 52 % only. Therefore, the capacitance value with the proposed optimization scheme is 42.5 % lower compared to the state-of-the-art. Without CM voltage injection, the proposed optimization scheme decreases the required capacitance value by up to 51 % without increasing the mean rectified arm current. Therefore, the presented optimal trajectory shaping scheme can be used to dramatically decrease the energy storage requirements of MMCs while keeping the conduction losses as low as possible.

Acknowledgements

This research is part of the activities of the Swiss Centre for Competence in Energy Research on the Future Swiss Electrical Infrastructure (SCCER-FURIES), which is financially supported by the Swiss Innovation Agency (Innosuisse - SCCER program).

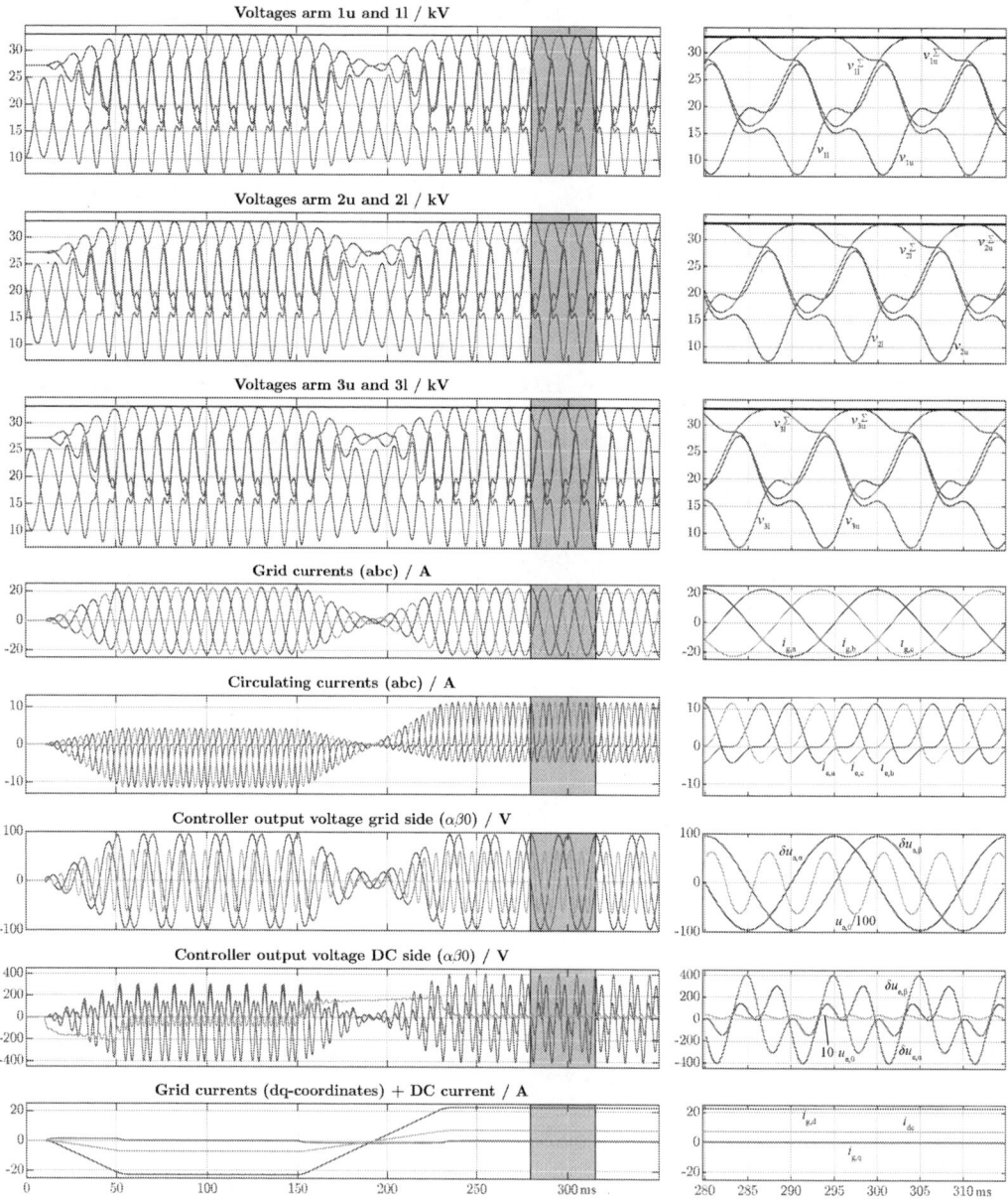

Fig. 6: Simulation results for the MMC design for the parameters given in Tab. I except for a reduced module capacitance value of $38\,\mu$F and the optimized trajectories marked with $*$ in Fig. 3. A slightly adopted version of the controller presented in [17] is used. The controller parameters are given in Tab. IV.

The left part presents an overview, while the right part represents a zoom-in of the time range marked in grey in the left part.

References

[1] A. Lesnicar and R. Marquardt, "An innovative modular multilevel converter topology suitable for a wide power range," in *Power Tech Conf.*, vol. 3, 2003.

[2] M. A. Perez, S. Bernet, J. Rodriguez, S. Kouro, and R. Lizana, "Circuit topologies, modeling, control schemes, and applications of modular multilevel converters," *IEEE Trans. on Power Electron.*, vol. 30, no. 1, Jan. 2015.

[3] S. Debnath, J. Qin, B. Bahrani, M. Saeedifard, and P. Barbosa, "Operation, Control and Appl. of the Modular Multilevel Converter: A Review," *IEEE Trans. on Power Electron.*, vol. 30, no. 1, 1 2015.

[4] K. Ilves, S. Norrga, L. Harnefors, and H.-P. Nee, "On Energy Storage Requirements in Modular Multilevel Converters," *IEEE Trans. on Power Electron.*, vol. 29, no. 1, 2014.

[5] J. Pou, S. Ceballos, G. Konstantinou, V. G. Agelidis, R. Picas, and J. Zaragoza, "Circulating current injection methods based on instantaneous information for the modular multilevel converter," *IEEE Trans. on Ind. Electron.*, vol. 62, no. 2, Feb 2015.

[6] J. Wang, X. Han, H. Ma, and Z. Bai, "Analysis and injection control of circulating current for modular multilevel converters," *IEEE Trans. on Ind. Electron.*, vol. 66, no. 3, March 2019.

[7] H. Fehr and A. Gensior, "Model-based circulating current references for mmc cell votlge ripple reduction and loss-equivalent arm current assessment," in *21st Europ. Conf. on Power Electron. and Appl. (EPE, ECCE Europe)*, 2019.

[8] K. Ilves, A. Antonopoulos, L. Harnefors, S. Norrga, L. Ängquist, and H. Nee, "Capacitor voltage ripple shaping in modular multilevel converters allowing for operating region extension," in *37th An. Conf. of the IEEE Ind. Electron. Soc. (IECON)*, Nov 2011.

[9] S. P. Engel and R. W. De Doncker, "Control of the Modular Multi-Level Converter for minimized cell capacitance," in *14th Europ. Conf. on Power Electron. and Appl. (EPE)*, 2011.

[10] R. Picas, J. Pou, S. Ceballos, J. Zaragoza, G. Konstantinou, and V. G. Agelidis, "Optimal injection of harmonics in circulating currents of modular multilevel converters for capacitor voltage ripple minimization," in *IEEE ECCE Asia Downunder*, June 2013.

[11] D. Townsend, G. Mirzaeva, and G. C. Goodwin, "Capacitance minimization in modular multilevel converters: A reliable and computationally efficient algorithm to identify optimal circulating currents and zero-sequence voltages," in *IEEE 12th Int. Conf. on Power Electron. and Drive Sys. (PEDS)*, Dec 2017.

[12] J. A. Houldsworth and D. A. Grant, "The use of harmonic distortion to increase the output voltage of a three-phase pwm inverter," *IEEE Trans. on Ind. Appl.*, vol. IA-20, no. 5, Sep. 1984.

[13] H. Bärnklau, A. Gensior, and S. Bernet, "Derivation of an equivalent submodule per arm for modular multilevel converters," in *15th Int. Power Electron. and Motion Control Conf. (EPE/PEMC)*, 2012.

[14] MathWorks. (2019) gamultiobj - find pareto front of multiple fitness functions using genetic algorithm. [Online]. Available: https://www.mathworks.com/help/gads/gamultiobj.html

[15] CIGRE, "Guide for the development of models for HVDC converters in a HVDC grid," 2014.

[16] S. Fuchs, M. Jeong, and J. Biela, "Long horizon, quadratic programming based model predictive control (MPC) for grid connected modular multilevel converters (MMC)," in *45th Ann. Conf. of the IEEE Ind. Electron. Soc. (IECON)*, Oct 2019.

[17] M. Jeong, S. Fuchs, and J. Biela, "High performance LQR control of modular multilevel converters with simple control structure and implementation," in *22nd Europ. Conf. on Power Electron. and Appl. (EPE'20, ECCE Europe)*, Sept 2020.

Leakage Inductance Modelling of Transformers: Accurate and Fast Models to Scale the Leakage Inductance Per Unit Length

Richard Schlesinger, Jürgen Biela
ETH Zurich, Laboratory for High Power Electronic Systems (HPE)
Email: schlesinger@hpe.ee.ethz.ch, jbiela@ethz.ch

Keywords

≪Transformer≫, ≪Modelling≫, ≪Design≫, ≪Magnetic device≫, ≪Passive Component≫

Abstract

Fast and accurate transformer leakage inductance models are crucial for optimisation-based design of galvanically isolated converters. Analytical models are rapidly executable and therefore specially suitable for such optimisations. These analytical leakage inductance models typically consist of two steps: First, acquire the leakage inductance per unit length and second, scale this value with a suitable length. In this paper, the term leakage length is introduced for the scaling length. It is shown that the leakage length depends on the magnetic energy distribution and the most influential factors are determined. Furthermore, two accurate and fast leakage length models for E-core and U-core transformers with concentric windings are proposed: The Empirically Corrected Axial Flux (ECAF) model is based on a compact modification of the known axial flux formula. The cut line (CL) model pursues a semi-analytical approach and achieves high accuracy at the cost of higher computational effort. The models are verified with more than 6000 FEM simulations and the error of both models is significantly lower than the error of the known axial flux formula.

1 Introduction

Leakage inductance is an important property of transformers in galvanically isolated power electronic converters as it significantly influences the operation of the converter. Galvanically isolated converters are key components in several applications that enable a more sustainable energy system such as photovoltaic inverters [1], electric traction systems [2], and power quality control of the electric power grid [3].

In the design stage of such converters, the operating and design parameters are usually determined in an optimisation procedure, before the components are physically built [2, 4]. During this optimisation procedure, the parameters are recalculated several thousand times (see e.g. [4]). Consequently, the employed models have to be accurate and fast to deliver accurate results within a reasonable amount of time. Analytical models are very well suited for this purpose as they are more compact and therefore faster than reluctance network modelling and numerical methods such as the finite element method (FEM) [5–7]. Analytical modelling of the transformer leakage inductance typically consists of two steps [8–13]:

1. Calculate the leakage inductance per unit length L'_σ from a 2D cross section (Fig 1a).
2. Scale the leakage inductance per unit length L'_σ to the actual leakage inductance L_σ (Fig. 1b).

In previous work [5,7], we compared leakage inductance per unit length models with respect to accuracy and computational effort. The present work examines the appropriate scaling length. In [14–17], the mean length of turns MLT is used to scale the leakage inductance per unit length. However, using the MLT as scaling length is not exactly accurate as the leakage inductance is deduced from the stored magnetic energy. In case of curved winding sections such as around a circular center leg and at the edges of a rectangular center leg, the magnetic field is curved. Hence, an energy weighted mean length is required to take the curvature of the magnetic energy distribution into account. This energy weighted mean length is referred to as "leakage length" l_σ in this paper to avoid confusion with other terms. Hence, the leakage inductance assuming an axisymmetric arrangement as shown in Figs. 1a,b is analytically

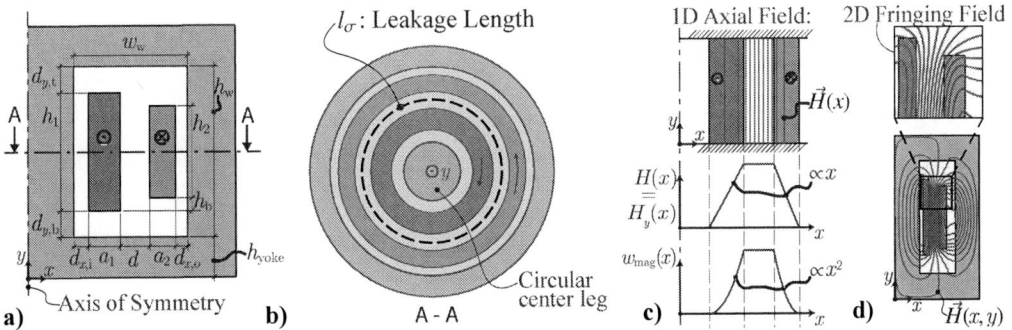

Fig. 1: **a)** Fundamental 2D cross section of transformer including geometric parameter definitions, **b)** Leakage length l_σ required to obtain leakage inductance L_σ. The leakage length needs to take the curved field due to curved windings into account. **c)** Axial leakage flux $H_y(x)$ (1D-field) and magnetic energy density distribution $w_{\mathrm{mag}}(x)$ in case of windings reaching up to the transformer yokes. **d)** The bend of the flux lines towards x-direction at the edges of the windings is referred to as flux fringing.

calculated according to

$$\underbrace{L'_\sigma}_{\substack{\text{1. Leakage inductance per unit length (2D)} \\ \rightarrow \text{Fig. 1a}}} \quad * \quad \underbrace{l_\sigma}_{\substack{\text{2. Leakage length ("1D")} \\ \rightarrow \text{Fig. 1b}}} \quad = \quad \underbrace{L_\sigma}_{\text{Leakage inductance (3D)}}$$

The leakage inductance models proposed in [8–10, 13] do not give a specific equation to calculate the leakage length at all. Margueron [11, 12] uses the local maximum of the magnetic energy density to calculate the leakage length. This is disadvantageous because the magnetic field needs to be calculated over the whole 2D cross section to obtain the maximum value, which leads to considerable computational effort. In [18], a known formula for the leakage length is given, however, without derivation. A derivation of the known formula can be found in [19]. However, this formula assumes purely axial leakage flux (1D-field) in the transformer window (see Fig. 1c) and is therefore not generally applicable. Hence, a comprehensive treatise and general model of the leakage length has not been published yet.

Therefore, this paper derives the leakage length for curved winding sections based on axisymmetric 2D transformer cross sections. Furthermore, two models of the leakage length are proposed: The first proposed model is called empirically corrected axial flux (ECAF) model (sec. 4.1) and is based on a modification of the known axial flux formula. The second proposed model is called cut line (CL) model (sec. 4.2) and pursues a semi-analytical approach. The proposed models are applicable to E-core and U-core transformers with concentric windings.

The rest of this paper is organised as follows: First, fundamental assumptions are briefly summarised and the typical leakage inductance modelling steps are elucidated in sec. 2. Sec. 3 describes the general concept of the leakage length, the derivation of the axial flux formula, and identifies the main geometric influences on the leakage length. Sec. 4 presents the proposed ECAF and CL models. The models are validated by more than 6000 2D-FEM simulations in sec. 5.

2 Analytical Leakage Inductance Modelling

2.1 Fundamentals

In analytical leakage inductance modelling, the windings are unified to rectangular winding blocks as shown in Figs. 2a&b. The exact winding type such as Litz, round, and foil is usually neglected [5, 8–12]. The magnetomotive forces MMF of primary and secondary side of the transformer are balanced according to (1) to obtain the leakage field distribution.

$$MMF_1 = N_1 \cdot I_1 = -MMF_2 = -N_2 \cdot I_2 \tag{1}$$

where N and I are the number of turns and the current through the winding. Indices 1 and 2 indicate primary and secondary side of the transformer. Further details on basic principles of leakage inductance

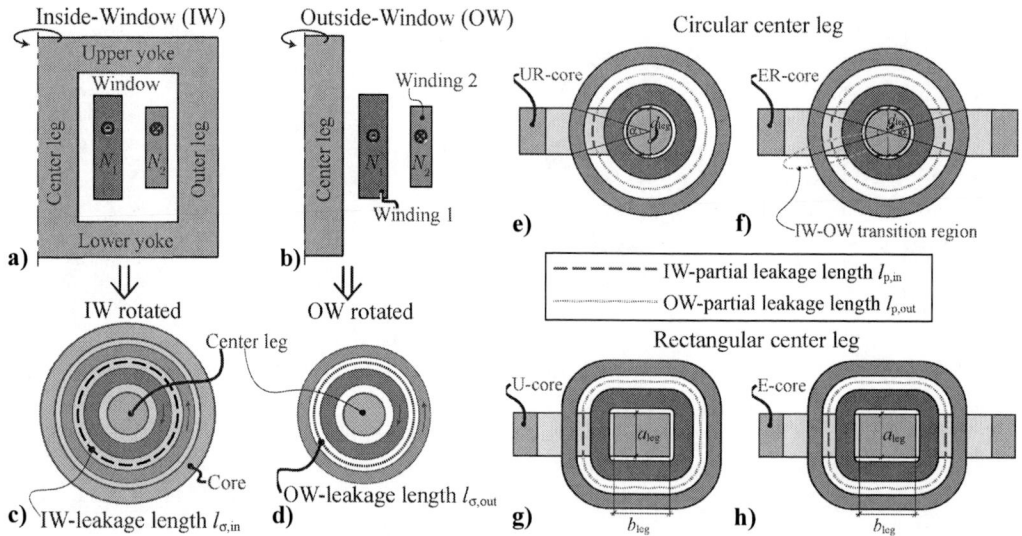

Fig. 2: **a&b)** Basic 2D cross sections of Double-2D approach. **c&d)** leakage lengths resulting from the rotated cross sections, **e)–h)** IW and OW partial leakage lengths of UR-, ER, U-, and E-core.

calculations can be found in standard literature (e.g. [20, 21]).

2.2 Modelling Steps

Recent modelling approaches pursue a "Double-2D" concept [5, 6, 22–25]. This means that the calculation is based on two cross sections of the transformer: inside the core window (IW) and outside the core window (OW) as depicted in Figs. 2a&b. This approach is more accurate than "Single-2D" modelling [5] and is therefore also used in this paper. Double-2D modelling consists of two steps:

1. The leakage inductance per unit length is calculated for both, IW and OW cross section with an appropriate analytical model, resulting in $L'_{\sigma,\text{in}}$ and $L'_{\sigma,\text{out}}$. See [5] for a comparison of different leakage inductance per unit length models with respect to accuracy and computational effort. The assumed geometries imply straight windings of infinite length.

2. The IW and OW leakage inductance per unit lengths are scaled with their respective partial leakage length $l_{\text{p,in}}$ and $l_{\text{p,out}}$ as shown in Figs. 2e–h. The total leakage inductance according to the Double-2D approach is then obtained with (2).

$$L_\sigma = L'_{\sigma,\text{in}} \cdot l_{\text{p,in}} + L'_{\sigma,\text{out}} \cdot l_{\text{p,out}} \tag{2}$$

The partial leakage lengths $l_{\text{p,in}}$ and $l_{\text{p,out}}$ depend on the core shape and can be derived from the IW and OW leakage lengths $l_{\sigma,\text{in}}$ and $l_{\sigma,\text{out}}$. The IW and OW leakage lengths $l_{\sigma,\text{in}}$ and $l_{\sigma,\text{out}}$ refer to the rotation of the particular cross section and are visualised in Figs. 2c&d.

2.2.1 Circular Center Leg (UR- and ER-core)

In case of UR- and ER-cores with circular center leg, the windings are curved along the whole circumference as depicted in Figs. 2e&f. Here, the partial leakage lengths can be derived from the IW and OW leakage length $l_{\sigma,\text{in}}$ and $l_{\sigma,\text{out}}$ with the equations (3) and (4).

$$l_{\text{p,in,circ}} = s_{\text{core}} \cdot \frac{\alpha[\text{rad}]}{2\pi} \cdot l_{\sigma,\text{in}} \tag{3} \qquad l_{\text{p,out,circ}} = \frac{2\pi - s_{\text{core}}\,\alpha[\text{rad}]}{2\pi} \cdot l_{\sigma,\text{out}} \tag{4}$$

The factor s_{core} (5) takes the proportion of IW and OW cross section of the core type into account.

$$s_{\text{core}} = 1 \ldots \text{ for UR- and U-core}; \qquad s_{\text{core}} = 2 \ldots \text{ for ER- and E-core} \tag{5}$$

The magnetic energy distributions of IW and OW cross section are typically different. Hence, $L'_{\sigma,\text{in}} \neq L'_{\sigma,\text{out}}$ and $l_{\sigma,\text{in}} \neq l_{\sigma,\text{out}}$. Between the IW and the OW cross section, there is a continuous transition region as indicated in Fig. 2f. However, the calculation is based only on two 2D cross sections (IW and OW) for simplicity reasons. Therefore, an abrupt transition from IW to OW cross section has to be chosen. In this paper, the IW-OW transition is assumed to take place at the angle $\alpha = 2 \arcsin\left(\frac{d_c}{}\right)$.

EPE'20 ECCE Europe

Assigned jointly to the European Power Electronics and Drives Association & the Institute of Electrical and Electronics Engineers (IEEE)

2.2.2 Rectangular Center Leg (U- and E-core)

For U- and E-cores with rectangular center legs as shown in Figs. 2g&h, the partial leakage lengths consist of straight and curved parts. The straight parts of the partial leakage lengths can simply be derived from width and depth of the rectangular center leg. For the curved parts around the center leg edges, the OW-leakage length $l_{\sigma,\text{out}}$ with $d_{\text{leg}} \to 0$ is used.

$$l_{\text{p,in,rect}} = s_{\text{core}} \cdot a_{\text{leg}} \qquad (6) \qquad l_{\text{p,out,rect}} = (2 - s_{\text{core}}) a_{\text{leg}} + 2 b_{\text{leg}} + l_{\sigma,\text{out},d_{\text{leg}} \to 0} \qquad (7)$$

3 Modelling the Leakage Length

For calculating the partial leakage lengths, the curved winding sections are most crucial for accuracy. In the curved winding sections, a more sophisticated model is required than for non-curved winding sections. Hence, this paper focuses on the calculation of the inside-window (IW) and outside-window (OW) leakage lengths resulting from the rotated IW and OW cross sections.

3.1 Fundamental Concept

As mentioned in the introduction, the leakage length is derived from the magnetic energy distribution as leakage inductance depends on the stored magnetic energy. In curved winding sections, the curvature of the magnetic field needs to be taken into account. Hence, an energy weighted mean length is required to correctly scale the leakage inductance per unit length. In this paper, the term "leakage length l_σ" is introduced for this energy weighted mean length.

This leakage length refers to the rotation of the transformer cross section and is therefore derived from an axisymmetric arrangement with circular center leg as shown in Figs. 2a–d. Here, the windings are curved over the whole rotational axis. The contribution of the magnetic energy density w_{mag} to the total magnetic energy W_{mag} increases with the distance to the rotational axis (i.e. increasing values of x). Since inductance and energy are proportional, the leakage length l_σ can be calculated according to (8) for the considered axisymmetric arrangement.

$$l_\sigma = \frac{W_{\text{mag}}}{W'_{\text{mag}}} \quad \overset{\text{rotational}}{\underset{\text{symmetry}}{=}} \quad \frac{2\pi \iint_{\mathcal{A}} x \cdot w_{\text{mag}}(x,y) \, \mathrm{d}A}{\iint_{\mathcal{A}} w_{\text{mag}}(x,y) \, \mathrm{d}A} \qquad (8)$$

where W'_{mag} is the stored magnetic energy per unit length.

3.2 Leakage Length for Axial Field (1D-Approximation)

In case of purely axial leakage flux inside the transformer window as shown in Fig. 1c, the magnetic field $\vec{H}$ only has a y-component $H_y(x)$ depending on the x-coordinate. Hence, this case is often referred to as 1D-field. In this case, the leakage length (8) simplifies to the known formula (9) mentioned in the introduction. The formula can be found in [18].

$$l_{\sigma,\text{axial}} = \pi \cdot \left(d_{\text{leg}} + 2 d_{x,\text{i}} + a_1 + d + a_2 - \frac{a_2 - a_1}{2} \frac{a_1 + a_2 + 4d}{a_1 + a_2 + 3d} \right) \qquad (9)$$

A derivation of the total rotated magnetic energy can be found in [19]. Dividing the total rotated magnetic energy by the energy per unit length according to (8) results in (9).

However, the leakage field is only purely axial when 1.) the windings are modelled with a rectangular block and 2.) reach up to the transformer yokes as displayed in Fig. 1c. The former is a common assumption in leakage inductance models. The latter is technically not feasible due to required isolation distances.

3.3 Leakage Length of Fringing Field (2D-Field)

In real transformers, an isolation distance between windings and yokes is required. Fig. 1d shows the magnetic field of a feasible geometry featuring isolation distances with the only assumption of a rectangular winding block. Here, the field contains axial and radial components (fringing field) and the leakage length cannot be calculated with (9).

The presence of a fringing field does not lead to big deviations from the axial flux formula (9) a priori. In fact, with aligned windings of similar dimensions, the deviation is negligible. Significant deviations from the axial flux assumption arise if windings are not aligned and/or have different dimensions.

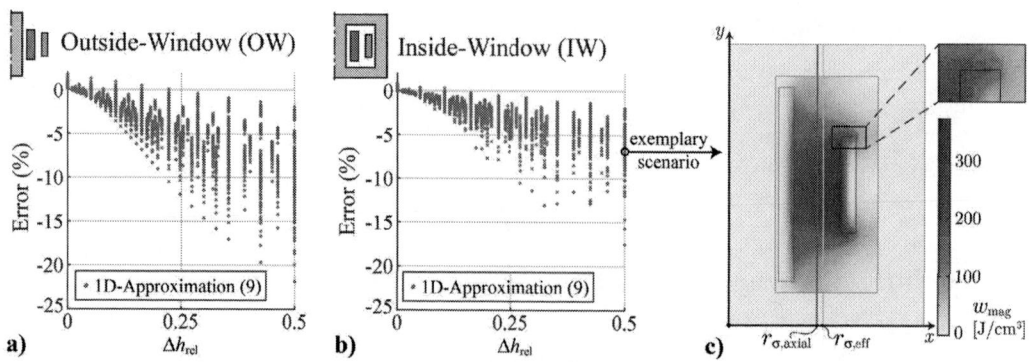

Fig. 3: Error of the axial flux approximation (9) for the conducted 2D FEM simulations: **a)** outside-window (OW), **b)** inside-window (IW). See Tab. II and III for simulation parameters. **c)** Effect of flux fringing on the leakage radius (11) and leakage length explained based on an exemplary scenario: The effective leakage radius $r_{\sigma,\text{eff}}$ is pushed towards the smaller winding due to the distribution of the stored magnetic energy. The shift in magnetic energy is especially pronounced at the edges of the smaller winding (see inset). A considerable amount of magnetic energy is stored in this area.

Especially the relative height difference of the windings Δh_{rel} (10) leads to a deviation from the axial flux approximation.

$$\Delta h_{\text{rel}} = \frac{\Delta h}{h_{\text{avg}}} \overset{2 \text{ windings}}{=} \frac{h_1 - h_2}{\frac{1}{2}(h_1 + h_2)} \tag{10}$$

The relative winding height difference Δh_{rel} leads to a deviation between axial flux approximation (9) and effective leakage length (8) as shown in Figs. 3a&b. These figures show the error of the axial flux approximation for the conducted FEM simulations of outside-window (OW) and inside-window (IW) cross section. The simulation parameters are listed in Tabs. II and III in the Appendix.

From a physical point of view, the fringing flux (Fig. 1d) causes the deviation between effective leakage length and the leakage length assuming an axial field. Flux fringing leads to an altered magnetic energy density distribution compared to a purely axial flux.

The deviation between effective leakage length and leakage length assuming axial flux is best explained introducing the leakage radius r_σ. As the leakage length l_σ is effectively a perimeter due to the rotation of the arrangement, a corresponding leakage radius r_σ can be defined according to (11).

$$r_\sigma = \frac{l_\sigma}{2\pi} \tag{11}$$

The leakage radius represents the horizontal center of the magnetic energy. The flux fringing effect usually pushes the effective leakage radius $r_{\sigma,\text{eff}}$ towards the smaller winding compared to the leakage radius assuming axial flux $r_{\sigma,\text{axial}}$ as shown in Fig. 3c. Here, the effective leakage radius $r_{\sigma,\text{eff}}$ deduced from (8) is closer to the smaller winding compared to the leakage radius resulting from an axial field $r_{\sigma,\text{axial}}$. Consequently, the resulting effective leakage length l_σ is higher in the case of a geometry such as the displayed geometry in Fig. 3c.

This deviation between the effective leakage length and the leakage length resulting from the axial flux approximation (9) is barely discussed in literature. Therefore, two accurate and computationally efficient leakage length models are proposed in the following section.

4 Proposed Leakage Length Models

4.1 Empirically Corrected Axial Flux (ECAF) Model

The ECAF model is based on a compact empirical correction to the axial leakage flux approximation (9) and thus very simple to apply. Particularly, the difference Δr between effective leakage radius $r_{\sigma\text{eff}}$ and 1D-leakage radius $r_{\sigma,\text{axial}}$ shown in Fig. 4a is empirically modelled. This difference is added to the 1D-approximation to obtain the effective leakage length resulting in (12)

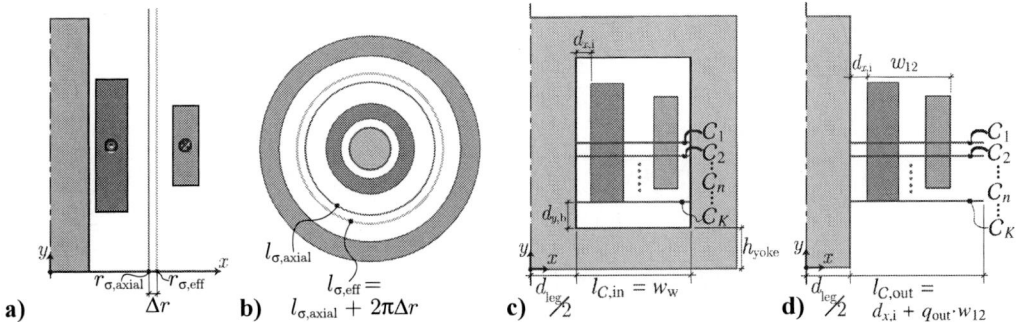

Fig. 4: **a)** Difference between effective leakage radius and 1D-leakage radius $\Delta r = r_{\sigma,\text{eff}} - r_{\sigma,\text{axial}}$. **b)** Concept of ECAF model. **c&d)** Concept of cut line model: A set of representative cut lines $\mathbf{U} = \{C_1, C_2, ..., C_K\}$ is defined as integrational domain of (16) for IW and OW cross section.

$$l_{\sigma,\text{ECAF}} = l_{\sigma,\text{axial}} + 2\pi\, \Delta r_{\text{ECAF}} \tag{12}$$

The concept of the ECAF model is visualised in Fig. 4b. The correction radius according to the ECAF model Δr_{ECAF} is calculated according to (13)

$$\Delta r_{\text{ECAF}} = m\,(h_1 - h_2)\left(\sqrt{\frac{h_1 - h_2}{d + \frac{a_1 + a_2}{3}}} + \sqrt{\frac{h_1 - h_2}{h_1 + h_2}}\right) \approx \Delta r \tag{13}$$

The geometry parameters are defined in Fig. 1a and m is a prefactor that depends on whether the equation is applied to inside-window (IW) or outside-window (OW) cross section. Eqs. (14) and (15) show the proper prefactors for both cross sections.

$$\text{Inside-Window (IW): } m = \tfrac{1}{42} \tag{14} \qquad \text{Outside-Window (OW): } m = \tfrac{1}{26} \tag{15}$$

The correction radius Δr_{ECAF} is deduced from the performed FEM simulations listed in Tabs. II and III. Here, a wide range of technically sound transformer geometries is covered. These simulations show that the difference between the effective leakage radius and the 1D-leakage radius $\Delta r = r_{\sigma,\text{eff}} - r_{\sigma,\text{axial}}$ correlates almost linearly with the height difference of the windings $h_1 - h_2$ if the rest of the geometrical parameters are constant. Hence, the other geometrical parameters determine the slope of the almost linear correlation between Δr and $h_1 - h_2$. The prefactor m and the terms in brackets in (13) are empirically determined multipliers that match Δr_{ECAF} to Δr.

With the chosen wide range of winding/core parameters, the presence of the transformer yokes and legs do not significantly influence the leakage length. Therefore, these winding-core distances are neglected in the ECAF model. This approach is chosen to keep the model compact and to keep the focus on the major geometric influence, i.e. the height difference of the windings.

4.2 Cut Line (CL) Model

The cut line (CL) model pursues a semi-analytical approach with the target of achieving high accuracy without calculating the complete field distribution. To achieve this, a set of horizontal cut lines $\mathbf{U}$ is defined, along which the field is integrated. This avoids integrating the field over the total 2D cross section. The concept is visualised in Figs. 4c&d. With this approach, l_σ (8) is approximated by (16).

$$l_{\sigma,\text{Cut Line}} = 2\pi \sum_{n=1}^{K} \frac{\int_{C_n} x \cdot w_{\text{mag}}(x, y_n)\,dx}{\int_{C_n} w_{\text{mag}}(x, y_n)\,dx} \tag{16}$$

$\mathbf{U} = \{C_1, C_2, ..., C_K\}$ is the set of cut lines and K is the total amount of cut lines. This set of cut lines $\mathbf{U}$ needs to be chosen such that the leakage length is well approximated.

To achieve high accuracy independent of the geometry, different sets of cut lines $\mathbf{U}$ are defined depending on the considered geometry. Figs. 5a&b illustrate the choice of the cut line set depending on geometric characteristic numbers. The cut line sets are specified below (see Fig. 1a for definition of geometry parameters):

- **Center Line Model**: The center line model uses only one cut line at center winding height

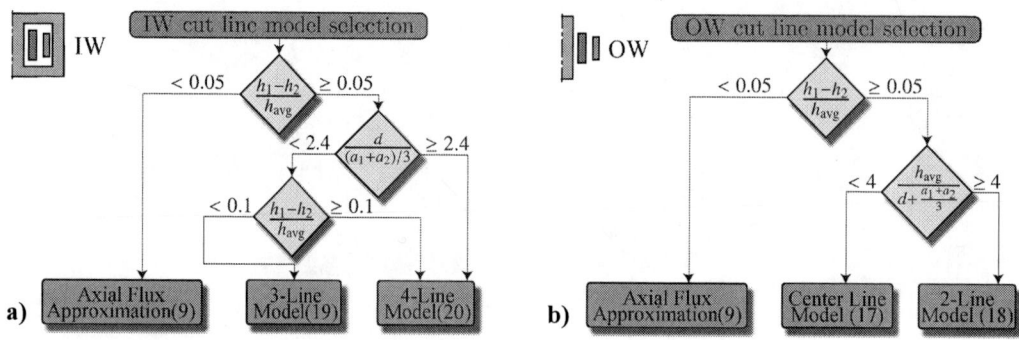

Fig. 5: Model selection flowcharts of cut line model depending on geometric characteristic numbers. **a)** inside-window (IW), **b)** outside-window(OW)

$$y_1 = h_{\text{yoke}} + d_{y,b} + \tfrac{h_1}{2} \tag{17}$$

- **2-Line Model**: The 2-line model uses two cut lines at

$$y_1 = h_{\text{yoke}} + d_{y,b} + \tfrac{h_1}{2}, \quad y_2 = h_{\text{yoke}} + d_{y,b} + h_b \tag{18}$$

- **3-Line Model**: The 3-line model uses three cut lines at

$$y_1 = h_{\text{yoke}} + d_{y,b} + \tfrac{h_1}{2}, \quad y_2 = h_{\text{yoke}} + d_{y,b} + \tfrac{3}{8}h_1, \quad y_3 = h_{\text{yoke}} + d_{y,b} + \tfrac{h_b}{2} \tag{19}$$

- **4-Line Model**: The 4-line model uses four cut lines at

$$y_1 = h_{\text{yoke}} + d_{y,b} + \tfrac{h_1}{2}, \quad y_2 = h_{\text{yoke}} + d_{y,b} + \tfrac{3}{8}h_1, \quad y_3 = h_{\text{yoke}} + d_{y,b} + h_b, \quad y_4 = h_{\text{yoke}} + d_{y,b} \tag{20}$$

The inside-window cut line length $l_{C,\text{in}}$ shown in Fig. 4c is set to the window width according to (21) as the amount of magnetic energy in the core is negligible due to the high relative permeability of the core.

$$l_{C,\text{in}} = w_w \tag{21}$$

The outside-window cut line length $l_{C,\text{out}}$ shown in Fig. 4d needs to be adjusted as there is no transformer window that confines the magnetic energy to a certain domain. FEM simulations show that the length of cut lines needs to be adjusted depending on the geometry. The cut line length $l_{C,\text{out}}$ according to (22) yields satisfactory results

$$l_{C,\text{out}} = d_{x,i} + q_{\text{out}} \cdot w_{12} \quad \text{with} \quad q_{\text{out}} = 3 + 2\frac{h_1 - h_2}{d + (a_1 + a_2)/3} \quad \text{and} \quad w_{12} = d + a_1 + a_2 \tag{22}$$

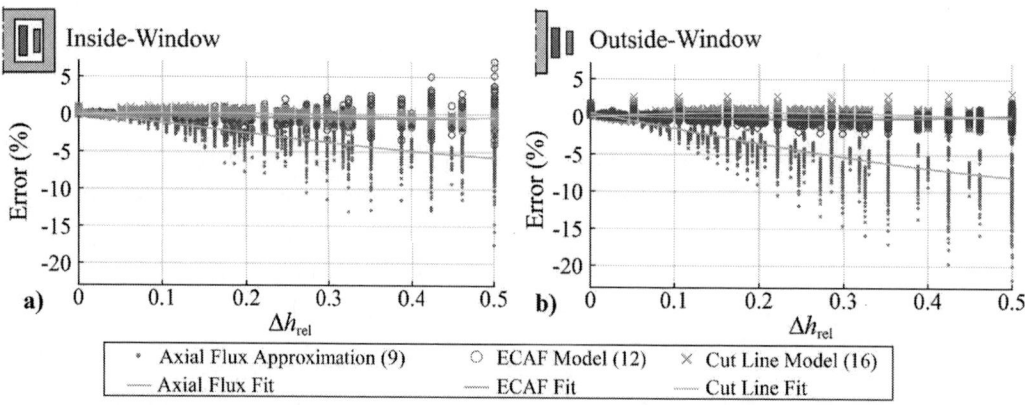

Fig. 6: Error of the axial flux approximation, the ECAF model, and the cut line model for the simulated scenarios from Tabs. II and III. The fitted curves are given because some data points are obscured by others. Fits are 3rd order polynomials and determined with the least squares method.

Table I: Performance indicators of the proposed models. The normalised Root Mean Square Error (nRMSE) gives the approximate scale of error to expect from a model.

Cross section		**Axial Flux** (9)	**ECAF** (12)	**Cut Line** (16)
IW (Fig. 2a)	**nRMSE**	5.89%	1.36%	0.65%
	Worst Case Error	17.54%	7.11%	2.14%
OW (Fig. 2b)	**nRMSE**	8.15%	0.90%	1.17%
	Worst Case Error	21.84%	2.19%	3.10%

5   Model Validation

Both, ECAF and CL model give a good approximation of the effective leakage length. This can be seen in Figs. 6a&b which show the error of each model as a function of the relative winding height difference Δh_{rel} (10) for the performed simulations listed in Tab. II and III. The figure shows that both, ECAF (12) and cut line model (16) result in a significantly lower error compared to the axial flux approximation (9). Tab. I shows the overall performance of the proposed models by listing the normalised root mean square error (nRMSE) and the worst case error. The nRMSE indicates the approximate scale of error to expect from a model and can be calculated according to (23).

$$\text{nRMSE} = \frac{\sqrt{\frac{1}{n}\sum_{i=1}^{n}\left(l_{\sigma,\text{model},i} - l_{\sigma,\text{FEM},i}\right)^2}}{\max(l_{\sigma,\text{FEM}}) - \min(l_{\sigma,\text{FEM}})} \tag{23}$$

Regarding the IW cross section, the cut line model performs especially well with an nRMSE as low as 0.65 %. This is because the magnetic energy is confined to the transformer window and therefore, the cut lines give a very good approximation of the total energy distribution. The ECAF model is also relatively accurate within a very big range, which is indicated by the low nRMSE of 1.36 %. Only for individual scenarios with very high Δh_{rel}, the error reaches up to 7.11 % for the set of simulated scenarios.

Regarding the OW cross section, the ECAF model performs especially well with an nRMSE of only 0.90 %. This is because the winding-yoke distances $d_{y,b}$ and $d_{y,t}$ and the outer winding-leg distance $d_{x,o}$ as defined in Fig. 1a do not exist in the OW cross section. In the ECAF model, these distances practically represent disturbance quantitites and therefore the model is very accurate when these parameters are absent. The cut line model also performs quite well with an nRMSE of 1.17 %. The accuracy is slightly worse for the OW cross section because the magnetic energy is not confined to the transformer window in the OW cross section.

The previously assessed performance of the models is reflected in detail by the histograms presented in Fig. 7. The histograms show the error density distribution of the simulated scenarios.

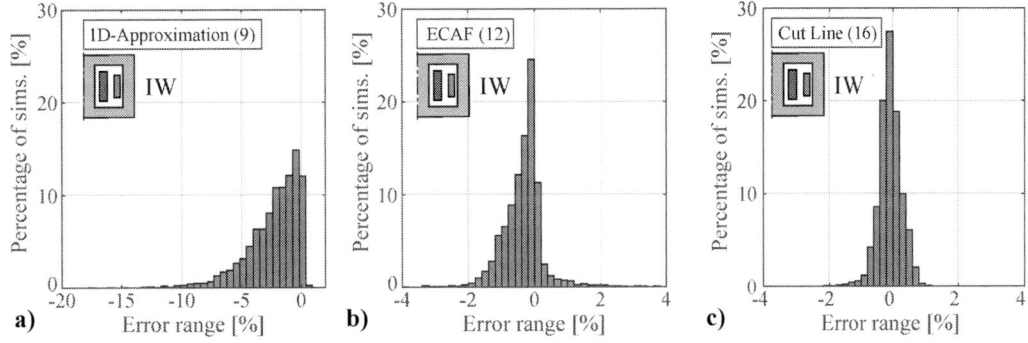

Fig. 7: Error range histogram for IW cross section. **a)** 1D-Approximation, **b)** ECAF model, **c)** CL model

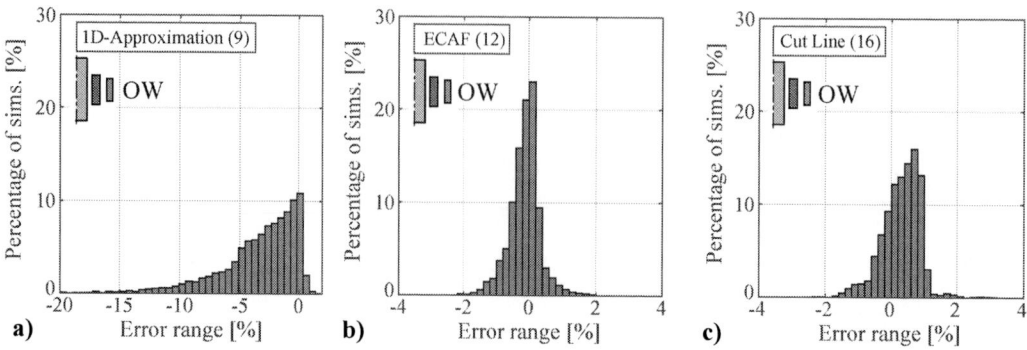

Fig. 8: Error range histogram for OW cross section. **a)** 1D-Approximation, **b)** ECAF model, **c)** CL model

6 Conclusion

The leakage length which is required for analytical leakage inductance modelling of curved winding sections has been investigated. More than 6000 performed FEM simulations showed that a height difference of the windings leads to error if the leakage length is calculated with the known formula that assumes purely axial leakage flux, i.e. a 1D-field. The error is caused by the altered magnetic energy density distribution of a fringing field, i.e. 2D-field.

The proposed ECAF model is based on a compact empirical modification of the axial flux approximation. With this approach, the error is significantly decreased manifesting in an nRMSE as low as 1.36 % and 0.90 % for the simulated inside-window (IW) and outside-window (OW) scenarios, respectively. The ECAF model is compact, easy to implement, and rapidly computable. Considering that the outside-window winding section typically exceeds the inside-window winding section, this model becomes even more relevant.

The proposed cut line model is even more accurate for the inside-window scenario. The cut line model is a semi-analytical model yielding an nRMSE of 0.65 % for the IW scenarios and 1.17 % for the OW scenarios. The increased accuracy comes at the cost of higher complexity and higher computational effort.

Both proposed models can be implemented in an optimisation procedure. The model choice depends on the desired trade-off between accuracy and computational effort. Using the ECAF model for IW and OW cross section will lead to high accuracy and very low computational effort. If accuracy is the most important criterion, the cut line model should be used for the inside-window cross section and the ECAF model for the outside-window cross section.

Appendix

More than 6000 FEM simulations are performed to derive the proposed models. Both, inside-window (IW) and outside-window (OW) cross sections are evaluated. The geometrical parameters in Tabs. II and III are chosen such that a wide variety of technically sound geometrical winding/core parameters are covered. Only extreme aspect ratios are avoided.

The OW cross section is approximated with a center core leg of infinite height as FEM simulations have shown that the difference in leakage length resulting from a center leg of finite height and a center leg of infinite height is negligibly small.

Table II: Geometrical parameter range of inside-window (IW) and outside-window (OW) simulations. A wide range of technically sound transformers is covered, only extreme aspect ratios are avoided. See Fig. 1a for definition of geometry parameters.

Cross section	Δh_{rel} (10)	$\frac{h_1-h_2}{2(d+(a_1+a_2)/3)}$	$\frac{d}{(a_1+a_2)/3}$	$\frac{\max(a_1,a_2)}{\min(a_1,a_2)}$	$\frac{(h_1+h_2)/2}{d+(a_1+a_2)/3}$	Simulated scenarios
IW (Fig. 2a)	0–0.5	0–4	0.6–5	1–5	1.06–20	3030
OW (Fig. 2b)	0–0.5	0–4	0.6–5	1–5	1.06–20	3030

Table III: Winding/core parameter range of IW and OW simulations. Parameters have been set randomly within the given range. See Fig. 1a for definition of geometry parameters.

Cross section	$\dfrac{d_{x,i}}{d+(a_1+a_2)/3}$	$\dfrac{d_{x,o}}{d+(a_1+a_2)/3}$	$\dfrac{d_{y,b}=d_{y,t}}{d+(a_1+a_2)/3}$
IW (Fig. 2a)	0–0.2	0.2–3.6	0.01–0.44
OW (Fig. 2b)	0	-	-

References

[1] D. Leuenberger and J. Biela, "PV-Module-Integrated AC Inverters (AC Modules) with Subpanel MPP Tracking," *IEEE Trans. on Power Electronics*, vol. 32, no. 8, pp. 6105–6118, 2017.

[2] M. Stojadinovic and J. Biela, "Modelling and Design of a Medium Frequency Transformer for High Power DC-DC Converters," *IEEJ Journal of Industry Applications*, vol. 8, no. 4, pp. 685–693, 2019.

[3] J. W. Kolar and G. Ortiz, "Solid-State-Transformers : Key Components of Future Traction and Smart Grid Systems," in *Int. Power Electronics Conf.*, 2014.

[4] J. Biela and J. W. Kolar, "Pareto-optimal design and performance mapping of telecom rectifier concepts," in *Proc. of the Power Conversion and Intelligent Motion (PCIM) Conf.*, Shanghai, 2010.

[5] R. Schlesinger and J. Biela, "Comparison of Analytical Models of Transformer Leakage Inductance: Accuracy vs. Computational Effort," *IEEE Trans. on Power Electronics (Early Access)*, 2020.

[6] A. Fouineau, M. A. Raulet, B. Lefebvre, N. Burais, and F. Sixdenier, "Semi-Analytical Methods for Calculation of Leakage Inductance and Frequency-Dependent Resistance of Windings in Transformers," *IEEE Trans. on Magnetics*, vol. 54, no. 10, 2018.

[7] R. Schlesinger and J. Biela, "Comparison of Analytical Transformer Leakage Inductance Models: Accuracy vs. Computational Effort," in *Proc. of European Conf. on Power Electronics and Applications (EPE ECCE Europe)*, 2019.

[8] W. Rogowski, "Ueber das Streufeld und den Streuinduktionskoeffizienten eines Transformators mit Scheibenwicklung und geteilten Endspulen," Ph.D. dissertation, Technische Hochschule zu Danzig, 1908.

[9] E. Roth, "Étude analytique du champ de fuites des transformateurs et des efforts mécaniques exercés sur les enroulements," *Revue générale de l'électricité*, vol. 23, pp. 773–787, 1928.

[10] G. Petrov, "Weitere Entwicklung der allgemeinen Methode zur Berechnung der Streuung von Transformatoren," *Elektrotechnik und Maschinenbau, Zeitschrift des Elektrotechnischen Vereines in Wien*, vol. 52, no. 34, pp. 396–400, 1934.

[11] X. Margueron, J. P. Keradec, and D. Magot, "Analytical calculation of static leakage inductances of HF transformers using PEEC formulas," *IEEE Trans. on Industry Applications*, vol. 43, no. 4, pp. 884–892, 2007.

[12] X. Margueron, A. Besri, P. O. Jeannin, J. P. Keradec, and G. Parent, "Complete analytical calculation of static leakage parameters: A step toward HF transformer optimization," *IEEE Trans. on Industry Applications*, vol. 46, no. 3, pp. 1055–1063, 2010.

[13] P. Dowell, "Effects of eddy currents in transformer windings," *Proc. of the Institution of Electrical Engineers*, vol. 113, no. 8, p. 1387, 1966.

[14] A. Dauhajre and R. Middlebrook, "Modelling and estimation of leakage phenomena in magnetic circuits," in *IEEE Power Electronics Specialists Conf.*, 1986.

[15] W. T. McLyman, *Transformer and Inductor Design Handbook, Fourth Edition*. CRC Press Taylor & Francis Group, 2011.

[16] M. K. Kazimierczuk, *High-Frequency Magnetic Components*. Wiley, 2014.

[17] W. G. Hurley and W. H. Wölfle, *Transformers and Inductors for Power Electronics*. Wiley, 2013.

[18] A. Amromin, L. Kubarev, and V. Pavlov, "Average Length of Turn in Calculation of Transformer Leakage Inductance," *Elektrotekhnika*, vol. 46, no. 4, 1975.

[19] M. Albach, *Induktivitäten in der Leistungselektronik*. Springer Vieweg, 2017.

[20] A. Van den Bossche and V. C. Valchev, *Inductors and Transformers for Power Electronics*. CRC Press Taylor & Francis Group, 2005.

[21] S. V. Kulkarni and S. A. Kharpade, *Transformer Engineering - Design, Technology and Diagnostics*. CRC Press Taylor & Francis Group, 2013.

[22] R. Prieto, J. A. Cobos, O. García, P. Alou, and J. Uceda, "Study of 3-D magnetic components by means of "Double 2-D" methodology," *IEEE Trans. on Industrial Electronics*, vol. 50, no. 1, pp. 183–192, 2003.

[23] A. Hoke and C. Sullivan, "An improved two-dimensional numerical modeling method for E-core transformers," in *Annual IEEE Applied Power Electronics Conference and Exposition*, 2002.

[24] M. Eslamian, M. Kharezy, and T. Thiringer, "Calculation of the Leakage Inductance of Medium Frequency Transformers with Rectangular-Shaped Windings using an Accurate Analytical Method," in *Proc. of European Conf. on Power Electronics and Applications (EPE ECCE Europe)*, 2019.

[25] M. Mogorovic and D. Dujic, "100 kW, 10 kHz Medium-Frequency Transformer Design Optimization and Experimental Verification," *IEEE Trans. on Power Electronics*, vol. 34, no. 2, pp. 1696–1708, 2019.

A GaN-based DC/DC converter for e-vehicles applications

Eduardo F. de Oliveira[1], Sebastian Sprunck[2], Jonas Pfeiffer[2] and Peter Zacharias[2]

[1] HUAWEI TECHNOLOGIES	[2] UNIVERSITY OF KASSEL
Nuremberg Research Center	KDEE-EVS
Südwestpark 48	Wilhelmshöher Allee 71
90449 Nürnberg, Germany	34121 Kassel, Germany
Tel.: +49 / (0)1590 - 4450835	Tel.: +49 / (0)561 - 804 6344
Fax: +49 / (0)911 - 255223090	Fax: +49 / (0)561 - 804 6521
E-Mail: eduardof.oliveira@huawei.com	E-Mail: peter.zacharias@uni-kassel.de
URL: http://www.huawei.com	URL: http://www.uni-kassel.de/eecs/evs

Acknowledgements

The authors acknowledge the financial support by the German Federal Ministry of Education and Research (BMBF) in the framework of project HELENE number 16EMO0234. Responsibility for the contents of this publications lies with the authors.

Keywords

«Power converters for EV», «Wide bandgap devices», «Gallium Nitride (GaN)», «Thermal design», «On-board auxiliary power supply system»

Abstract

This paper presents a 4 kW GaN-based galvanically isolated bidirectional DC/DC converter suitable for on-board auxiliary power supply systems, interconnecting the HV battery with the LV power system. To enable the trend towards higher levels of electrification, a second LV level of 48 V is introduced to supply the high-power consumers, relieving the 12 V grid on the one hand and/or avoiding using additional costly isolated converters connected to the HV batteries on the other. The built converter prototypes aim to demonstrate higher efficiency, compactness and lower filter requirements of the proposed circuit.

Introduction

In the course of the debate on the reduction of fossil fuel emissions, electromobility became an important topic in politics and society. In conjunction with the expansion of renewable energies, the electrification of transport can be a useful approach. Especially in urban areas, electric vehicles can be a solution to reduce fossil fuel emissions drastically. The development of electric vehicles has increased substantially in recent years. However, automotive manufactures still have to deal with numerous challenges, especially relating to range extension and load management. Wide band gap semiconductors have the potential to improve the efficiency of the used power electronic devices and can therefore be a part of the required solutions.

The common structure of a BEV is composed by two on-board power systems. The HV power system comprises the traction batteries with voltage levels up to 500 V and is responsible for feeding the drive train of the vehicle. In turn, the LV power system, with a rated voltage of 12 V, feeds most of the electric consumers. Its power demand easily reaches around 3 kW in contemporary vehicles [1]. The increase of the on-board power with the addition of new electric components motivates the introduction of a third voltage level of 48 V, which enables the supply of high-power consumers, especially in high-end model cars [2], [3]. Compared to the 12 V battery, the 48 V network can provide 4 times more power at the same current level, while remaining below the automotive HV

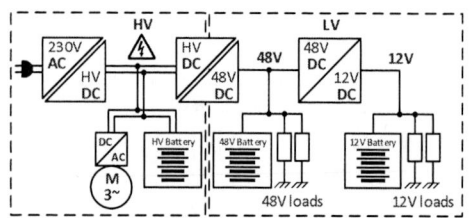

Fig. 1: State-of-the-art on-board power supply system block diagram [1]

threshold of 60 V, where the standards require the use of costly shielding, conduits and connectors that would increase the cost rapidly [3]. The 48 V system reliefs the 12 V grid on the one hand and/or avoids using additional costly isolated converters connected to the HV batteries on the other [2][4]. Moreover, 12 V and 48 V electrical grids can co-exist for the foreseeable future. A joint effort from GM, HELLA and COSMX has developed a way to provide both voltage levels from a single system [5]. Called MODACS (Multiple Output Dynamically Adjustable Capacity System), the invention eliminates not only the need for separate 12 V and 48 V batteries, but also the losses coming from the DC/DC converter between both grids [6].

In turn, the connection between the 48 V supply and the HV battery must be executed by using a galvanically isolated DC/DC converter. Using a highly dynamic converter would allow the reduction or even the elimination of the 48 V battery. The converter proposed in this paper works as an on-board power supply system connecting the HV side to the 48 V batteries. The focus of this paper lies on the optimal electrical implementation of gallium nitride semiconductors and the inherent challenge of removing its dissipated heat. Further focuses are the thermal management of the magnetics in combination with their arrangement to achieve a high power density. By iterative optimization, a setup with minimum losses and volume is identified, constructed and validated experimentally.

Investigated topology

One of the most attractive soft-switching isolated DC-DC converter topologies for high power applications, where bidirectional power flow, robust operation and step-up/step-down functionality are required, is the three-phase DAB presented in [7]. In recent years, some new topologies based on the 3ph-DAB have been developed, for instance the Y-YY 3ph-DAB proposed in [8]. A second low voltage bridge is added to the original circuit, so that the topology now works with six transformers, each phase having two transformers. This configuration presents two transformers per phase connected in series on the high voltage side and paralleled three-phase bridges on the low voltage side. The new turn ratio of each transformer is now divided by two. Therefore, the voltage across the transformer's HV winding as well as the currents through the transformer's LV winding and through the LV switches are also divided by two. A similar approach is proposed in [1]. However, in contrast to the previous circuit, one of the secondaries presents a delta connection, originating the wye wye-delta (Y-YΔ) 3ph-DAB (see Fig. 2). The Y-YΔ connection results in a 12-pulse converter, reducing therefore the output filter requirements due to the higher frequency seen by the output filter. Other than for the conventional 3ph-DAB, power flow is controlled by applying a phase-shift angle of Φ and

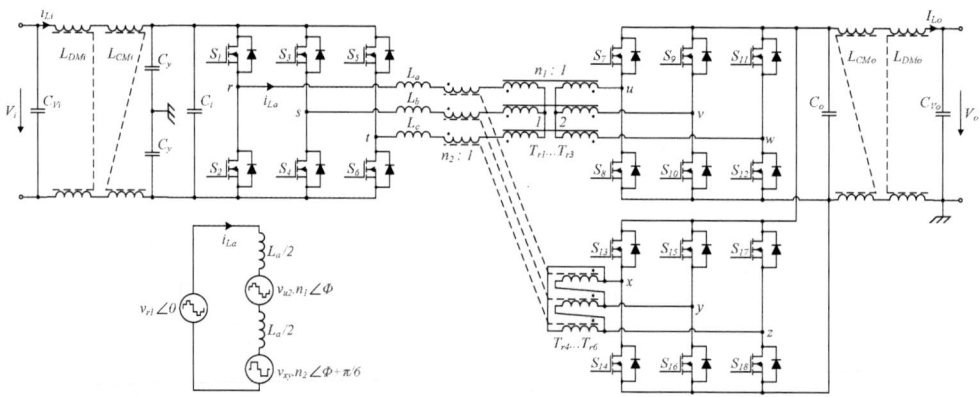

Fig. 2: Simplified topology of the 12-pulse Y-YΔ 3ph-DAB

$\Phi+\pi/6$ between the gating signals of the input and the Y and Δoutput bridges, respectively. After an extensive topology benchmark, the Y-YΔ 3ph-DAB is chosen as the more advantageous alternative regarding losses, volume and output filter efforts for this application. A detailed description of the operation modes and an extensive steady-state analysis of the Y-YΔ 3ph-DAB can be found in [1].

Integrated Magnetics

The storage chokes are essential for the operation of DAB converters. They can be designed as discrete magnetic components or integrated within the transformer's leakage inductance. Regarding the integrated approach, three different setups were investigated. In the first design, primary and secondary windings are wound on the entire length of the center leg of the used PQ-core (helical windings). The leakage inductance is adjusted by varying the horizontal distance between both windings using spacers. Due to the high coupling factor between the coils, the achievable value of the leakage inductance is quite limited. In the second design, the primary winding is wound on the upper part of the center leg, whereas the secondary winding is placed on the lower part (disc windings). Vertical spacers or winding chambers are used to adjust the required leakage inductance. In comparison to the first method, a higher leakage inductance can be reached, whilst degrading the efficiency. The third design extends the second one by introducing ferrite plates between the coils. The ferrite plates concentrate the flux leakage, which results in a higher leakage inductance. However, the design and construction processes of the magnetic device become more difficult, especially if a concrete leakage inductance value is required. Furthermore, it is important to prevent a saturation of the ferrite plates what would lead to a significant drop of the transformer's flux leakage. For these reasons, standard disc windings are used in the final converter. Due to geometric constrains, the delivered transformers do by far not reach the desired leakage inductance of 10.8 µH (being for each transformer the half), presenting an average total value of ca. 7.8 µH. Nonetheless, no additional inductors are added to the final prototype.

Characterization of GaN E-HEMT switching losses

The characterization of GaN E-HEMTs is performed using a standard double pulse test (DPT). The test environment has to be improved regarding its dv/dt and di/dt immunity, which is realized using suitable common mode filters on every external connection. The measured switching losses, though subject to uncertainties stemming from the used current shunt [9], are used as a benchmark for several gate driver and switching cell setups. Using an automated measurement setup, the GaN semiconductor's switching losses are examined regarding their dependency on DC voltage, switching current, gate turn-on and turn-off voltages, gate resistances, dead time and junction temperature. These investigations are repeated for several combinations of different gate drivers and gate networks, resulting in a broad database for application-specific loss minimization. An excerpt of the measurements' results at room temperature is displayed in Fig. 3.

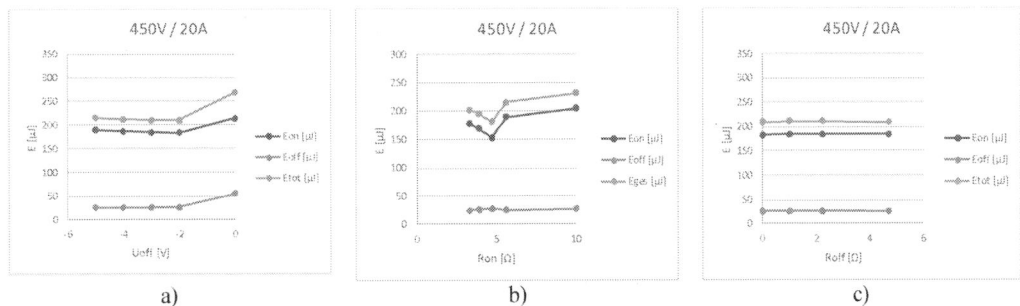

Fig. 3: Influence of different setup parameters on the measured switching energies. a) E_{sw} vs. U_{off} for R_{on} = 5.6 Ω, R_{off} = 0 Ω and U_{on} = 5 V. b) E_{sw} vs. R_{on} for R_{off} = 0 Ω, U_{on} = 5 V and U_{off} = -2 V. c) E_{sw} vs. R_{off} for R_{on} = 10 Ω, U_{on} = 6 V and U_{off} = -2 V

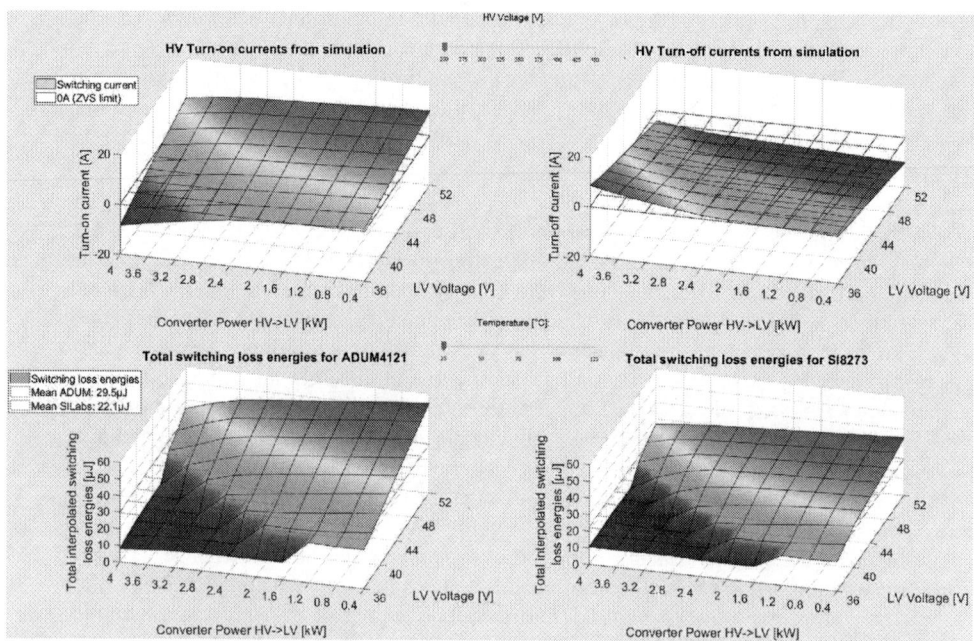

Fig. 4: GUI of the candidate comparison tool. Top row: Turn-on and turn-off currents from the simulation according to the selected operation conditions. Bottom row: Switching loss results for two different drivers. The HV voltage and chip temperature can be adjusted using the sliders

It is evident that the turn-off losses are an order of magnitude below the turn-on losses and can hardly be influenced at all, save for turn-off voltages between 0 V and -2 V. Turn-on losses, on the other hand, dominate the overall switching losses and can be influenced by the gate resistor network as well as the turn-on and turn-off voltages. The latter effect is rather surprising and was therefore investigated more closely in [10]. The data recorded by this test bench are then integrated into an evaluation tool that uses the simulated operation points of the selected topology to calculate the total semiconductor switching losses. These losses are calculated for the most promising driver and setup candidates and consecutively compared regarding certain criteria, such as operating voltage and chip temperature. An excerpt of the tool's GUI is shown in Fig. 4.

Physical setup and thermal design

Although GaN can operate at very high junction temperatures, it still needs to be packaged in a case. This case limits the maximum allowable operating temperature, making an effective cooling system a necessity. Liquid cooling systems, present in most cars, can be used for this purpose but it has to be ensured that the cooling terminal of the GaN semiconductor package is thermally well connected to the heat sink. The required creepage distance for the HV switches as well as the amount of LV switches pose challenges regarding the required PCB and cooler surface areas. If the switching cells were arranged side-by-side, a very large area would be required, adding a lot of volume through the vertical size of the transformers alone and thus reducing the achievable power density. Through iterative testing, a 3D setup is developed to minimize the converter's volume. This allows the transformers, switches as well as filters to be connected closely to the cooling channel, while keeping the base area of the converter in an acceptable range (see Fig. 5).

The top-side cooled GaN semiconductors are mounted on the bottom of the PCB. Their substrates are internally connected to the thermal pads on top, which in turn are coupled to the heatsink by means of a flexible thermal interface material (TIM). To avoid device failure due to thermomechanical bending stress, the heat sink is designed with cavities to fit the semiconductors and the TIMs, as shown in Fig. 6a. The PCB area opposite to the switches is lightly pressed against the heatsink using 3D printed

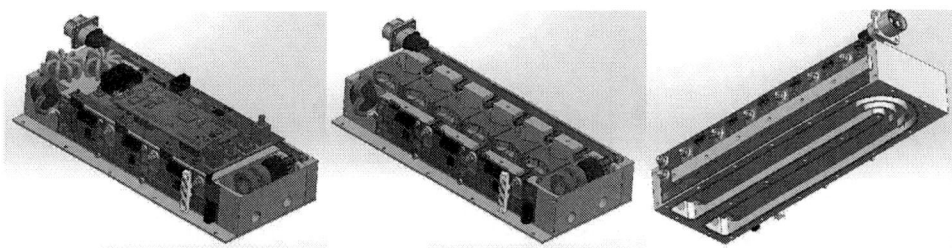

Fig. 5: CAD model of the 3D-PCB-concept and the cooling system (casing not shown)

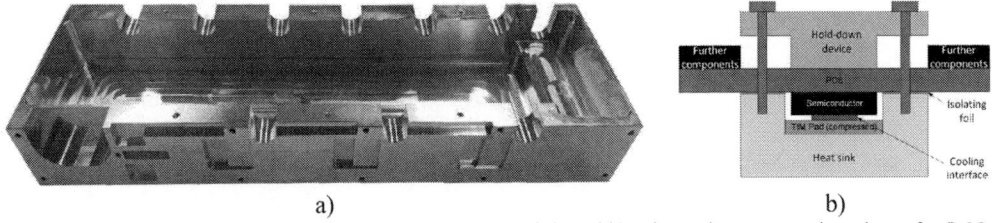

a)   b)

Fig. 6: a) Details of the cavities on the water cooling heatsink and b) schematic cross-section view of a GaN semiconductor mounted in a cavity (not to scale)

hold-down devices (Fig. 6b) to ensure constant thermal contact between semiconductors, TIMs and the heat sink. In this manner, the GaN devices are protected from excessive bending stresses due to the screw mounting and thus are almost solely subjected to thermal expansion [11]. This approach provides a compact and thermally stable connection, but great care has to be taken regarding device tolerances, TIM compression and manufacturing precision, especially when using very thin TIMs.

As the transformers have to be placed in-between the HV and LV switching cells, they are embedded into a cavity below the control and measurement PCB which is placed on top. This board is protected against the transformers' shifting magnetic fields by inserting a grounded metal sheet between them, acting as an EMI shield. As the LV filters are subjected to high currents, they too are embedded into the cavity for cooling purposes. The HV filters on the other hand are loaded with lower currents, requiring no tight coupling to the cooling system. As the cooling fluid should pass all the semiconductors and the transformers to avoid overheating, a 180° turn of the flow was designed into the channel. Several simulative iterations were performed, accounting for chip orientation, cooling channel design and chip heating. With this approach, an optimal overall setup could be found, allowing a total volume of 298 x 127 x 77 mm (≈ 2.9 liters), resulting in a power density of approx. 1.38 kW/L.

Optimization and 3ph-DAB circuits' benchmark

Reconfiguring the transformer's arrangement, the implemented prototypes are not only able to operate as a 12-pulse Y-YΔ 3ph-DAB, but also as both conventional 3ph-DAB [7] and Y-YY 3ph-DAB [8]. An optimization procedure, based on comprehensive analytical models of these three topologies, has been developed and leads to the optimal parameters to achieve the best compromise between compactness and efficiency. The best possible combination between losses, volume and required dc-link capacitance can be found after millions of iterations (see Fig. 7).

Comparing the best obtained designs of the 12-pulse Y-YΔ 3ph-DAB and of the conventional 3ph-DAB, the calculations indicate a reduction of approximately 22 %, 9 % and 70 % concerning overall losses, transformer volume and required output capacitance, respectively. In turn, comparing the 12-pulse Y-YΔ 3ph-DAB to the Y-YY 3-phase DAB, the total transformer volume remains the same for both, however the overall losses and the output capacitance are reduced by 6 % and 70 %, respectively. Besides exhibiting lower average total losses and lower total magnetic core volume, the biggest advantage of the Y-YΔ 3ph-DAB compared to the conventional 3ph-DAB and to the Y-YY 3ph-DAB is the much lower current ripple at the secondary side, thereby lower VA rating of filter capacitors owing to the interleaving effect of the Y-YΔ connection [1].

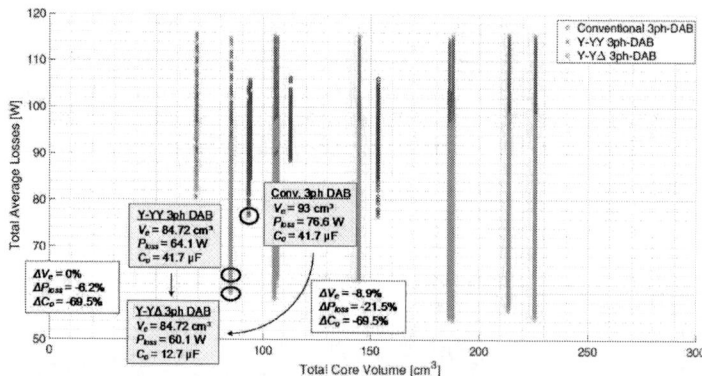

Fig. 7: Selection of the best configuration results regarding total losses, total magnetic core volume and output capacitance. Marked are the chosen designs

Constructed prototypes

In order to verify the functionality of the circuits, a total of three 4 kW 3ph-DAB prototypes with two secondary bridges (six independent legs) have been built. The first two prototypes are air-cooled, whereas the final demonstrator is water-cooled. The HV and LV sides are rated from 250 V to 450 V and from 36 V to 52 V, respectively. Moreover, the use of GaN power switches and a maximum volume of 4 liters are required for the final construction.

The following lines describe the chronological path of the executed constructions, from the first prototype, used to check the functionality of both power and control circuits, until reaching the final demonstrator, fulfilling the requirement of maximum volume (see Fig. 8).

The first prototype (Fig. 8a), aimed to verify the basic functionality of both power and control circuits of the also proposed Y-YY 3ph-DAB, uses six Rohm 650 V SiC MOSFETs SCT3030AL, located below the main PCB and screwed onto the heat sink, that are driven with six Analog Devices drivers ADuM4121CRIZ. Twelve Infineon 80 V Si MOSFETs IAUT300N08S5N012 and six TI drivers LM25101A form the low side bridges. The switches are mounted on the PCB top, thermally connected to the heat sink through hundreds of filled & capped vias positioned under the packages. They are isolated using a flexible thermal interface material between PCB and heat sink and lightly pressed using a hold-down device made of aluminium. The six transformers are built with N97 PQ 50/40 cores, placed on an acrylic support directly above the main PCB.

The second prototype (Fig. 8b), introduces for the first time not only the Y-YΔ 3ph-DAB, but also the use of GaN power switches. Moreover, the constructed EMI filters were added to the circuit. The second prototype serves as an air-cooled preliminary version for the final water-cooled construction. Two large air-cooled heatsinks are chosen in order to compare the performance of four different power circuit boards. Two of them comprise the power circuit of the high voltage side, whereas the other two, the one of the low voltage side. The transformers are located between the heatsinks. Both PCBAs of the HV side utilise six GaN Systems 650 V top-side cooled GS66516T GaN HEMTs mounted on the bottom of the PCB. The difference between both HV PCBAs lies in the gate drivers' design. In one PCBA, the power switches are driven with three TI drivers UCC21520ADW, providing isolated bipolar supplies for high- and low-side switches. In the other setup, six Analog Devices drivers ADuM4121ARIZ-1 with active Miller clamping are adopted, providing a unipolar supply and using bootstrap diodes to supply the high side switches.

In turn, one of the LV PCBAs consists of twelve Infineon 100 V IAUT300N10S5N015 Si MOSFETs and six TI drivers UCC27212A-Q1 with unipolar supplies. The high side drivers are operated through internal bootstrap diodes. The cooling method is similar to the one described for the LV Si MOSFETs in the first prototype. GaN Systems 100 V GS61008T GaN HEMTs are adopted for benchmarking in the other LV PCBA. The power switches are driven with a unipolar supply through six TI drivers LM5113-Q1 with internal bootstrap diodes. The cooling method is similar to the one described for the HV GaN switches.

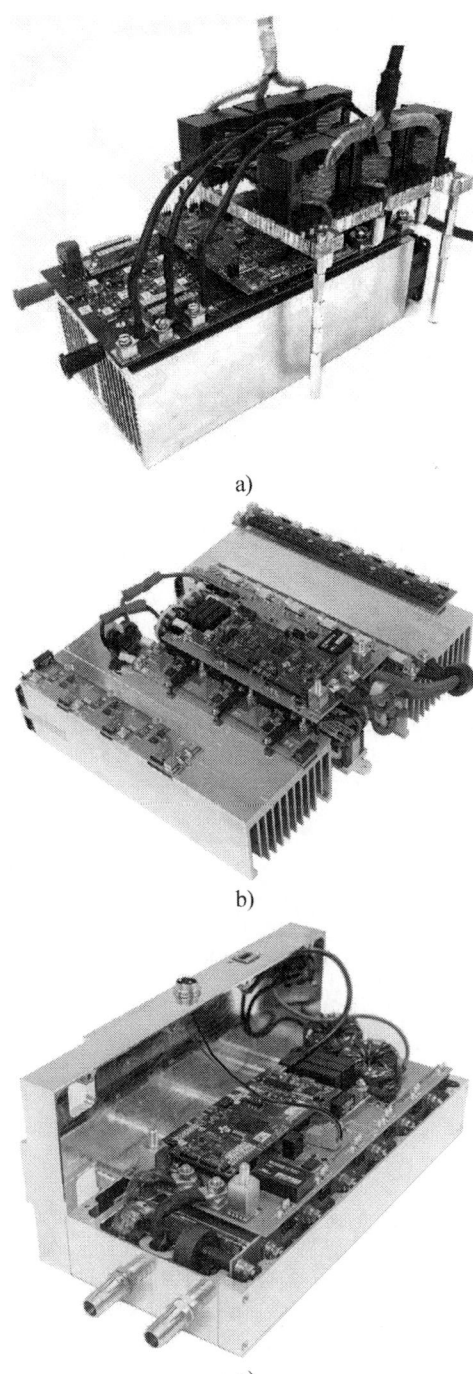

a)

b)

c)

Fig. 8: Chronological evolution of the executed constructions: a) first, b) second and c) third prototypes

The hardware of the third prototype is very similar to the one of the second construction. The main differences between both constructions are the introduction of a water cooled heatsink and the "3D" arrangement of PCBs and transformers, as shown in Fig. 8 c). The six transformers are built with N97 PQ 40/30 cores and are encapsulated inside a cavity in the water-cooled heat sink, directly below the controller PCB. The complete enclosed casing serves as an electromagnetic shield. The litz wire conductors are terminated using self-developed wire-to-board connectors, which are screwed to SMT bolts from the PCB backside through cut-outs in the heat sink, as can be seen in Fig. 6a. The reduction in the leg length has a direct effect on the value of the leakage inductance. This value has been further reduced from 9.1 μH (second prototype) to 7.8 μH on average, getting even further away from the specified value of 10.8 μH and, consequently, from the calculated optimal design. Finally, the second and third prototypes use the same control board. It is composed by, among other things, input and output power connectors, measurement circuitry, auxiliary power supplies, digital isolators and a piggyback TI C2000™ LAUNCHXL-F28379D LaunchPad™. This LaunchPad development kit uses a dual core 32-bit MCU TMS320F28379D, which in turn is responsible for the control loop, PWM generation and communication. The final demonstrator has a volume of 2.9 L and weighs 4.5 kg. Further measures to reduce both weight and volume are already identified.

As already mentioned, due to geometric constrains, the delivered transformers did by far not reach the desired leakage inductance. The obtained performance regarding power losses is far away from the designed one. The calculations indicate an increase concerning the overall average losses of approximately 19.8 %, as shown in Fig. 9a. In turn, Fig. 9b illustrates the deviation of the operating area from the constructed second prototype and the specified system at nominal power.

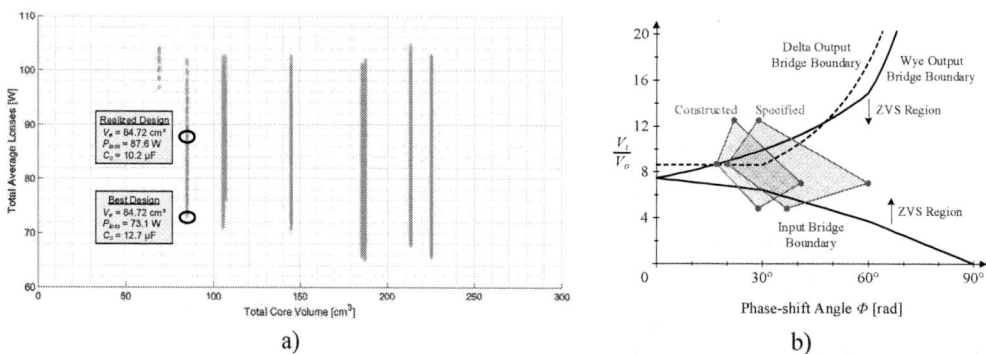

a)

b)

Fig. 9: a) Selection of the best configuration results for the Y-YΔ 3ph-DAB using GaN switches. Marked are the best configuration and the realized construction and b) divergent operating areas at nominal power of the specified system and the constructed second prototype

Experimental results

The oscilloscope waveforms in Fig. 10 present the current through the transformer coils and voltage across the power switches over the time during operation in steady state for P_o = 4 kW, V_i = 400 V and V_o = 48 V. In addition, a curve obtained by simulation under identical condition is also provided to verify the correct functionality of the circuit. The HV drain-to-source voltage across the switch s_1 (U_{ds_S1}) is represented by the magenta waveform and the LV drain-to-source voltage across s_7 (U_{ds_S7}) by the yellow one. The transformer primary side current i_{La_p} is represented by the green waveform and the secondary side i_{La_s} by the blue one. The vertical scales of each channel are given in the picture and the time scale is 2 μs/div. The current through the primary side power switch s_1 can be extracted directly from i_{La_s} during the on-time, when U_{ds_S1} is zero. On the other hand, the current through s_7 is the inverse of i_{La_s} during the on-time. These currents are represented through dashed lines. As the currents are completely symmetrical, the waveforms of the other switches are identical except for their phase delays. The juxtaposition of the experimental and simulation results clearly shows that the constructed circuit behaves as expected.

Furthermore, numerous efficiency curves versus transferred power, taken from the third prototype working in direct and reverse power flow directions by keeping the high voltage level constant at 400 V and varying the low voltage side from 36 V to 52 V in five steps, are presented in Fig. 11. A conservative dead time of 200 ns is adopted as a safety precaution against damage in the initial testing phase. There is an efficiency penalty due to increased diode's forward / HEMT's reverse voltage drop, but it can be minimized by optimizing the system dead time [12]. The obtained efficiency curves are very similar, while the values for direct power transfer are slightly higher than for the reverse operation. For nominal power, all curves are above 94 %, reaching a peak of approximately 97 %.

Tests are performed under the worst-case voltage combinations at nominal power with ambient temperature of 25 °C and cooling fluid temperature of 20 °C. Thermal images (Fig. 12) show the HV and LV power switches and the whole system in thermal stationary state. The temperatures of all components are considerably higher than the ambient temperature, however in an acceptable range since the system has to work under higher ambient temperatures of up to 85 °C and with a cooling fluid temperature of up to 55 °C. On the other hand, the temperature increase on the LV DM chokes is very critical. The measured temperature exceeds 92 °C under operation with output voltage equal to 36 V, an increase of approximately 67 °C considering an ambient temperature of 25 °C. It is noteworthy that the LV EMI chokes are not water-cooled in this setup. For operation under higher ambient temperatures, it is necessary to improve their cooling, e.g. through encapsulation. Temperatures on the encapsulated transformers are also observed through five type J thermocouples located between two cores, approximately at the height of their gaps. With the temperature of the cooling medium regulated at 20 °C, a maximum temperature variation of approx. 50 K is achieved, which is consistent with the value adopted in the design. It is noteworthy that the adopted ferrite material N97 reaches its lowest core losses at approximately 100 °C [13], which it is expected to reach with the adopted design using a cooling medium temperature value of 55 °C.

A full-order small-signal model has been derived based on phasor transformation technique [14]. Adopting this model, an output current control loop is implemented using a simple digital PI

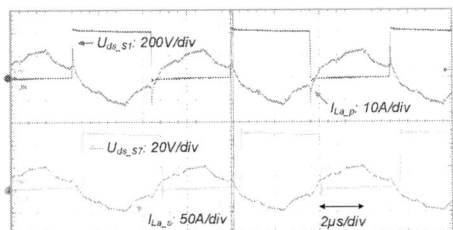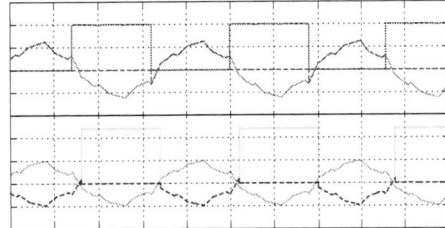

Fig. 10: Juxtaposition of experimental (left) and simulation (right) results for direct power flow direction at rated voltage and power (P_o = 4 kW, V_i = 400 V and V_o = 48 V)

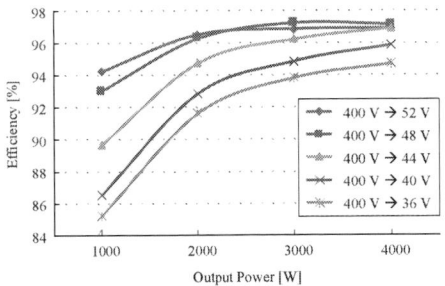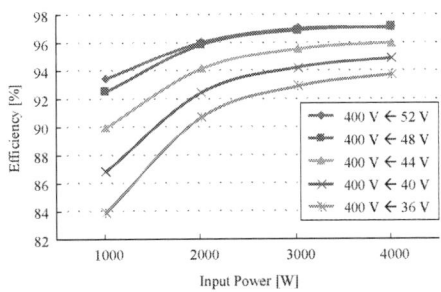

Fig. 11: Efficiency versus power for V_i = 400 V and V_o = 36...52 V for direct (left) and reverse (right) power flow directions

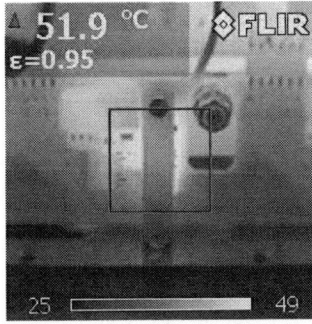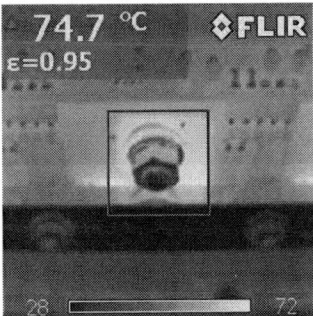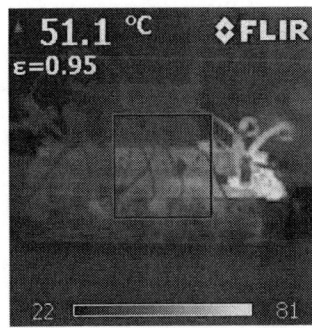

Fig. 12: Worst-case temperatures on the HV GaN switch located on the PCB bottom and measured from the PCB top (left), on the LV switch located under an aluminum hold-down device on the PCB top (middle) and on the complete system viewed from the HV lateral side (right)

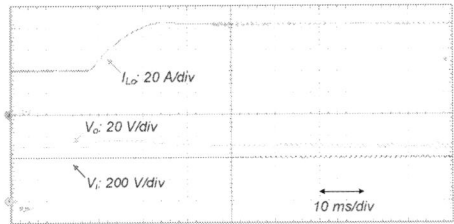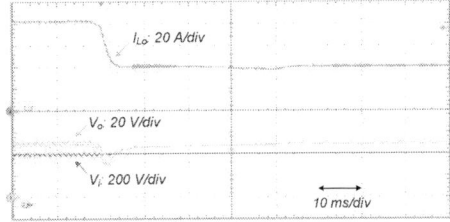

Fig. 13: Output current steps from 42 A to 84 A (left) and vice-versa (right) for (400 V → 48 V)

controller. Output current steps from 50 % to 100 % (nominal value) and vice versa performed in direct power flow direction (HV → LV) under nominal voltage (400 V, 48 V) are shown in Fig. 13. HV input is represented by the magenta waveform and the controlled DC output current by the blue one. Horizontal and vertical scales are shown individually. The LV output, represented in yellow, is fully controlled by an electronic load. Unfortunately, the used e-load is not able to keep its voltage constant during the step event due to the large time constant of its control loop. If the control loop is realized in a faster way, the circuit can be damaged due to overvoltage.

Conclusion

This paper introduced a 4 kW GaN-based isolated bidirectional DC/DC converter intended to interconnect the HV batteries with the 48 V on-board power supply system in electrical vehicles. An innovative, highly compact 3D setup has been developed, allowing not only the power switches, but also the magnetic components to be actively cooled. The heat sink also acts as an EMI shield, protecting the boards against the transformers' magnetic fields. Finally, three 4 kW prototypes with integrated magnetics were built in order to validate the feasibility of the topology and of the electrical and thermal behaviours of the GaN switches.

References

[1] E. F. de Oliveira and P. Zacharias, "A modified 12-Pulse Three-Phase Bidirectional Dual Active Bridge DC/DC Converter for E-Vehicles Applications," *PCIM Europe 2019*, Nuremberg, Germany, 2019, pp. 1-8.

[2] A. Körner, C. Buhlheller, L. Keuck, E. F. de Oliveira. "Spannung im Wandel – Leistungselektronische Wandler für Hybrid- und Elektrofahrzeuge", *7. ETG-Fachtagung*, Bad Nauheim, Germany, 2017.

[3] M. G. Txapartegi, "Power electronics for electric & hybrid vehicles," Yole Développement, Lyon, France, Rep. YD18014, May 2018.

[4] E. F. de Oliveira and P. Zacharias, "Comprehensive Mode Analysis and Optimal Design Methodology of a Bidirectional CLLC Resonant Converter for E-Vehicles Applications," *EPE '19 ECCE Europe*, Genova, Italy, 2019, pp. P.1-P.10.

[5] D. Rich. (2020). What are the latest updates on dynamically adjustable, dual voltage batteries DC/DC for future electrification?. Innovations in 48V Technology Online. [online]. Available: https://www.automotive-iq.com/events-innovations-in-48v-technology-online/

[6] D. Rich, "Dynamically Adjustable, Dual Voltage Batteries Without a DC/DC for Future Electrification," *The Battery Show and Electric & Hybrid Vehicle Technology Expo 2019*, Novi, MI, USA, 2019.

[7] R. W. A. A. De Doncker, D. M. Divan and M. H. Kheraluwala, "A three-phase soft-switched high-power-density DC/DC converter for high-power applications", in *IEEE Transactions on Industry Applications*, vol. 27, no. 1, pp. 63-73, Jan/Feb 1991.

[8] F. B. da Silva, E. F. de Oliveira, T. R. F. Neto and P. Zacharias, "Design Optimization of a Three-Phase Bidirectional Dual Active Bridge DC/DC Converter," *PCIM Europe 2018*, Nuremberg, Germany, 2018, pp. 1-8.

[9] S. Sprunck, M. Muench and P. Zacharias, "Transient Current Sensors for Wide Band Gap Semiconductor Switching Loss Measurements," *PCIM Europe 2019*, Nuremberg, Germany, 2019, pp. 1-8.

[10] L. Will, S. Sprunck and P. Zacharias, "Impact of Negative Turn-Off Voltage On Turn-On Losses in GaN EHEMTs," *PCIM Europe 2020*, Nuremberg, Germany, 2020

[11] GaN Systems, Ontario, Canada. GN002 Application Note - Thermal Design for GaN Systems' Top-side cooled GaNPX®-T packaged devices. Accessed: Jun. 28, 2020. [Online]. Available: https://gansystems.com /wp-content/uploads/2018/10/GN002_Thermal-Design-Guide-for-Top-Side-Cooled-GaNpx-T-Devices_Rev-181030.pdf

[12] S. Havanur, "Beware of Zero-Voltage Switching," *How2Power Today*, 80th issue, Apr. 2016, pp. 1-8. Accessed on Apr. 02, 2020. [Online]. Available: http://www.how2power.com/pdf_view.php?url=/ newsletters/1604/articles/H2PToday1604_design_VishaySiliconix.pdf

[13] TDK Electronics, Munich, Germany. Ferrites and accessories - SIFERRIT material N97 (2017). Accessed: Apr. 22, 2020. [Online]. Available: https://www.tdk-electronics.tdk.com/download/187242/ cf0d9784d3d2438b52ade72fbb86ecf6/pdf-n97.pdf

[14] W. Han and L. Corradini, "Analytical Small-Signal Transfer Functions for Phase Shift Modulated Dual Active Bridge Converters Using Phasor Transformation," *ECCE 2018*, Portland, OR, 2018, pp. 1442-1448.

Theory of Influencing the Breathing Mode and Torque Pulsations of Permanent Magnet Electric Machines with Harmonic Currents

Jan Andresen, Stephan Vip, Axel Mertens
Institute for Drive Systems and Power Electronics
Leibniz Universität Hannover
Welfengarten 1
30167 Hannover, Germany
Phone: +49 (0) 511-762-14569
Email: Jan.Andresen@ial.uni-hannover.de
URL: https://www.ial.uni-hannover.de

Sebastian Paulus
Robert Bosch GmbH
Frankfurter Str. 10
71732 Tamm, Germany

Keywords

≪Force Control≫, ≪Permanent magnet motor≫, ≪Harmonics≫, ≪Electric vehicle≫, ≪Acoustic noise≫

Abstract

This paper analytically demonstrates how both breathing mode and torque pulsations can be independently influenced in permanent magnet synchronous machines. This is achieved through superimposing an alternating magnetomotive force component on the pre-existing magnetomotive force in the dq-coordinate system. An FEM simulation is used to validate the results.

Introduction

Acoustic noise emissions are an important criterion for benchmarking electric machines. This is especially the case for traction applications where strict noise requirements have a long history as well as for wind turbines where the public acceptance increasingly depends on the acoustic performance. For an electric machine, there are two electromagnetic phenomena that can cause noise emissions: radial deflections of the stator and torque pulsations. While torque itself is a scalar quantity, radial deflections can have multiple shapes potentially emitting noise. However, the breathing mode 0 is typically the main cause among the radial modes [1–3]. By only considering the mode 0 radial deflection, the whole radial deflection can be represented as a scalar value, too. Breathing mode 0 vibrations are excited by a pulsation of the mean radial force density.

Acoustic noise improvements can be achieved through mitigating the radial force pulsations using current harmonics [4–10] or through mitigating torque pulsations [11, 12]. An improvement achieved by reducing harmonic current components is confirmed in [13]. Reference [14] analyses the generation of both torque and radial forces considering the fundamental current. Here, it is reported that both the d-current and q-current influence the radial force. However, the analysed machine does not exhibit a reluctance torque, thus making the torque independent of the d-current. Reluctance torque can, however, have a significant influence, as demonstrated in this paper.

Using the approach presented in this paper, torque pulsations and mean radial force density pulsations can independently be influenced while taking into account the reluctance torque. This has the advantage that both pulsations can be reduced. In existing approaches, a trade-off between radial force improvement and torque pulsation improvement had to be found [4] or a full numeric optimisation was performed [15, 16]. An approach for switched reluctance motors is described in [17].

This paper is particularly motivated by buried permanent magnet machines, as sketched in Fig. 1, which is one common permanent machine type especially among traction applications [18]. Typical for this kind of machines is that the dominant spacial harmonic is the fundamental spacial harmonic centred around the d-axis. Additionally, rotors with buried magnets have a significant saliency.

In order to analyse the dominant effects analytically, a simplified and idealised configuration is considered in this paper (Fig. 1). The permanent magnet excitation is assumed to be sinusoidal. However, the height of the permanent magnet is modulated in order to achieve saliency. Furthermore, the excitation by the stator currents is assumed to be sinusoidal.

In the following, the magnetic flux density is described analytically for the fundamental field component and the injected harmonics. The flux density is then used to calculate the forces which unveils the direction-dependent influence of the current. Finally, the results are interpreted and compared to FEM.

a) Rotor with buried magnets

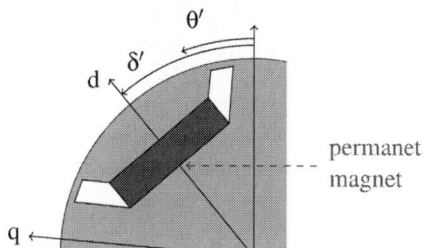

b) Simplified configuration

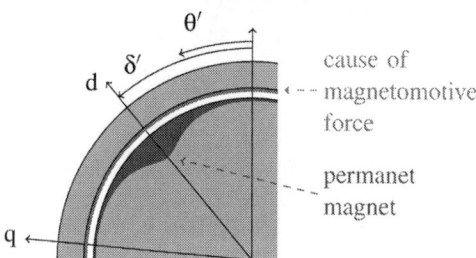

Fig. 1: Rotor of a four-pole permanent magnet machine with buried magnets (a) and exaggerated representation of the simplified four-pole configuration including smooth stator (b). The prime symbol indicates mechanical angles.

Analytical Magnetic Field Calculation

The analytical description of the magnetic field forms the basis of the following force calculation. The field description inside the air-gap is based on the one-dimensional field analysis [19–21]. The fields are represented in the dq-coordinate system [22] where the d-axis is aligned with the direction of magnetisation of the permanent magnet. In order to achieve a compact representation, the flux density B and the magnetomotive force (MMF) $\mathcal{F}$ are represented as spatial harmonics. For instance, the radial field can be written as

$$B_{\mathrm{R}}\left(\theta', \delta'\right) = \sum_{n'} \sum_{k'} \hat{B}_{\mathrm{d},n',k',\mathrm{R}} \cos(n'\theta' - k'\delta') + \hat{B}_{\mathrm{q},n',k',\mathrm{R}} \cos(n'\theta' - k'\delta' - \frac{\pi}{2}), \qquad (1)$$

where θ' is the mechanical circumferential angle, n' is the circumferential order, δ' is the mechanical angle between d-axis and the fixed coordinate system and k' is the rotational order. In the following, only the fundamental and third harmonic field are considered. Using these assumptions, the radial field strength can be calculated as

$$
\begin{aligned}
\hat{B}_{\mathrm{d},p,p,\mathrm{R}} &= \Lambda_{\mathrm{d},p\to p} \hat{\mathcal{F}}_{\mathrm{d},p,p} + \hat{B}_{\mathrm{PM}} && \text{with} \Lambda_{\mathrm{d},p\to p} = \Lambda_0 - \Lambda_{2p}/2 \\
\hat{B}_{\mathrm{q},p,p,\mathrm{R}} &= \Lambda_{\mathrm{q},p\to p} \hat{\mathcal{F}}_{\mathrm{q},p,p} && \text{with} \Lambda_{\mathrm{q},p\to p} = \Lambda_0 + \Lambda_{2p}/2 \\
\hat{B}_{\mathrm{d},3p,3p,\mathrm{R}} &= \Lambda_{\mathrm{d},p\to 3p} \hat{\mathcal{F}}_{\mathrm{d},p,p} && \text{with} \Lambda_{\mathrm{d},p\to 3p} = -\Lambda_{2p}/2 \\
\hat{B}_{\mathrm{q},3p,3p,\mathrm{R}} &= \Lambda_{\mathrm{q},p\to 3p} \hat{\mathcal{F}}_{\mathrm{q},p,p} && \text{with} \Lambda_{\mathrm{q},p\to 3p} = -\Lambda_{2p}/2.
\end{aligned}
\qquad (2)
$$

Λ_0 and Λ_{2p} can be derived based on the one-dimensional field analysis [19–21]. According to the one-dimensional field analysis, the radial air-gap field can be calculated as

$$B_R\left(\theta',\delta'\right) = \frac{\mu_0}{l_{tot}}\mathcal{F}_{p,p}\left(\theta',\delta'\right) = \frac{\mu_0}{1\,\text{mm}}\left(\frac{3}{5} + \frac{2}{5}\cos(2p\theta' - 2p\delta')\right)\mathcal{F}_{p,p}\left(\theta',\delta'\right)$$
$$= \left(\Lambda_0 + \Lambda_{2p}\cos(2p\theta' - 2p\delta')\right)\mathcal{F}_{p,p}\left(\theta',\delta'\right) \tag{3}$$

$$\text{with}\quad \Lambda_0 = \frac{3\mu_0}{5\,\text{mm}}\quad\text{and}\quad \Lambda_{2p} = \frac{2\mu_0}{5\,\text{mm}}. \tag{4}$$

In the equation, B_R is the radial flux density, μ_0 is the vacuum permeability, l_{tot} is the length between stator and rotor iron, r_o the outer radius, p the number of pole pairs, $\mathcal{F}$ the MMF, θ' is the mechanical circumferential angle, δ' is the mechanical angle between d-axis and the fixed coordinate system and Λ the factor between MMF and flux density. The geometry parameters are given in Table I.

Table I: Parameters of the simplified machine.

Air-Gap	$0.8\,\text{mm}$
Length between stator and rotor iron	$l_{tot} = \frac{1\,\text{mm}}{\frac{3}{5}+\frac{2}{5}\cos(2p(\theta'-\delta'))}$
Outer radius	$r_o = 0.5\,\text{m}$
Field strength	$B_{PM} = 1\,\text{T}\cos(p(\theta'-\delta'))$
Number of pole pairs	$p = 2$

Equation (2) contains two MMFs: $\mathcal{F}_{d,p,p}$ is exciting the d-axis component and $\mathcal{F}_{q,p,p}$ is exciting the q-axis component, respectively. Fig. 2 shows an example for radial spatial harmonics and the total radial flux density. In a real machine, the MMFs $\mathcal{F}_{d,p,p}$ and $\mathcal{F}_{q,p,p}$ would be caused by the fundamental spacial

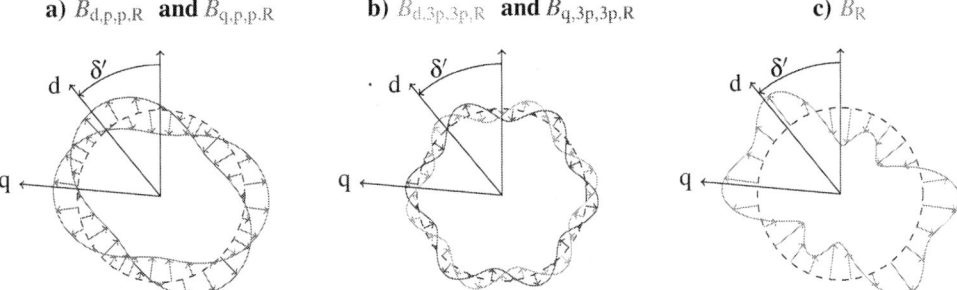

a) $B_{d,p,p,R}$ **and** $B_{q,p,p,R}$ b) $B_{d,3p,3p,R}$ **and** $B_{q,3p,3p,R}$ c) B_R

Fig. 2: Example for spacial harmonics of the radial air-gap flux density (a and b) and total radial air-gap flux density (c) (cf. [19]).

harmonic of a constant d- and q-current. The electric loading causing the MMF is adjacent to the stator (Fig. 1). The stator is modelled as a smooth cylinder of infinite permeability. The circumferential flux density at the air-gap side of the electric loading is [22] (with μ_0 vacuum permeability)

$$\hat{B}_{d,p,p,0'} = \frac{p}{r_o}\mu_0\hat{\mathcal{F}}_{q,p,p}\quad\text{and}\quad \hat{B}_{q,p,p,0'} = -\frac{p}{r_o}\mu_0\hat{\mathcal{F}}_{d,p,p}. \tag{5}$$

Analytical Calculation of Force Density Pulsations

The fields described in (2) and (5) are assumed to be the pre-existing magnetic field. An additional alternating field in dq-coordinates is introduced to cause force density pulsations. This alternating field can be described as

$$B_{\mathrm{R}}\left(\theta',\delta'\right)=\hat{B}_{\mathrm{d,p,p,R,kc}}\cos(p\theta'-p\delta')\cos(k\delta')+\hat{B}_{\mathrm{d,p,p,R,ks}}\cos(p\theta'-p\delta')\sin(k\delta')+\dots \tag{6}$$

For simplicity, only a cosinusoidal dependency on $k\delta$ is used during the derivation. Alternating fields of this form can, for instance, be generated by injecting an alternating MMF in the d-axis which could be generated by an alternating d-current in a real machine. These additional alternating MMFs result in additional fields, both in radial

$$\hat{B}_{\mathrm{d,p,p,R,kc}}\cos(k\delta')=\Lambda_{\mathrm{d,p\to p}}\hat{\mathcal{F}}_{\mathrm{p,p,kc}}\cos(\gamma)\cos(k\delta')$$
$$\hat{B}_{\mathrm{q,p,p,R,kc}}\cos(k\delta')=\Lambda_{\mathrm{q,p\to p}}\hat{\mathcal{F}}_{\mathrm{p,p,kc}}\sin(\gamma)\cos(k\delta')$$
$$\hat{B}_{\mathrm{d,3p,3p,R,kc}}\cos(k\delta')=\Lambda_{\mathrm{d,p\to 3p}}\hat{\mathcal{F}}_{\mathrm{p,p,kc}}\cos(\gamma)\cos(k\delta') \tag{7}$$
$$\hat{B}_{\mathrm{q,3p,3p,R,kc}}\cos(k\delta')=\Lambda_{\mathrm{q,p\to 3p}}\hat{\mathcal{F}}_{\mathrm{p,p,kc}}\sin(\gamma)\cos(k\delta')$$

and circumferential direction

$$\hat{B}_{\mathrm{d,p,p,\theta',kc}}\cos(k\delta')=\ \frac{p}{r_{\mathrm{o}}}\mu_0\hat{\mathcal{F}}_{\mathrm{p,p,kc}}\sin(\gamma)\cos(k\delta')$$
$$\hat{B}_{\mathrm{q,p,p,\theta',kc}}\cos(k\delta')=-\frac{p}{r_{\mathrm{o}}}\mu_0\hat{\mathcal{F}}_{\mathrm{p,p,kc}}\cos(\gamma)\cos(k\delta'). \tag{8}$$

The angle γ describes where the alternating MMF is located. Using an angle instead of two amplitudes to describe the location of the alternating MMF is a key feature of the presented field description. The advantage will be explained during the force density calculation. For $\gamma = 0°$, the d-axis is excited whereas $\gamma = 90°$ leads to an excitation of the q-axis. Using the fields described in (2), (5), (7), and (8), the force densities (σ) in radial (R) and circumferential direction (θ) can be calculated [23]

$$\sigma_{\mathrm{R}}\left(\theta',\delta'\right)=\frac{1}{2\mu_0}\left(B_{\mathrm{R}}^2\left(\theta',\delta'\right)-B_{\theta}^2\left(\theta',\delta'\right)\right)\quad\text{and}\quad\sigma_{\theta'}\left(\theta',\delta'\right)=\frac{1}{\mu_0}\left(B_{\mathrm{R}}\left(\theta',\delta'\right)\cdot B_{\theta'}\left(\theta',\delta'\right)\right). \tag{9}$$

In the following, force waves that result in a mean radial force density pulsation of the order k are of interest. These force densities are the result of a combination of the fundamental field and the alternating field

$$\hat{B}_{\mathrm{d,p,p,R}}\cos(p\theta'-p\delta')\hat{B}_{\mathrm{d,p,p,R,kc}}\cos(p\theta'-p\delta')\cos(k\delta')=$$
$$\frac{\hat{B}_{\mathrm{d,p,p,R}}\hat{B}_{\mathrm{d,p,p,R,kc}}}{2}\left(1+\cos(2p\theta'-2p\delta')\right)\cos(k\delta'), \tag{10}$$

where the circumferential order $2p$ is excited as well [7]. The resulting mean radial force density pulsation of the order k is

$$\hat{\sigma}_{0,0,\mathrm{R,kc}}=\hat{\mathcal{F}}_{\mathrm{p,p,kc}}\left[\cos(\gamma)\underbrace{\frac{\hat{B}_{\mathrm{PM}}\Lambda_{\mathrm{p,d}}+\left(\Lambda_{\mathrm{p,d}}^2+\Lambda_{\mathrm{3p,d}}^2-\frac{p^2\mu_0^2}{r_{\mathrm{o}}^2}\right)\hat{\mathcal{F}}_{\mathrm{d,p,p}}}{2\mu_0}}_{\eta_{\mathrm{c,R}}}+\sin(\gamma)\underbrace{\frac{\left(\Lambda_{\mathrm{p,q}}^2+\Lambda_{\mathrm{3p,q}}^2-\frac{p^2\mu_0^2}{r_{\mathrm{o}}^2}\right)\hat{\mathcal{F}}_{\mathrm{q,p,p}}}{2\mu_0}}_{\eta_{\mathrm{s,R}}}\right]. \tag{11}$$

$\eta_{\mathrm{c,R}}$ represents the influence of the alternating MMF located in the d-axis on the mean radial force density pulsation. $\eta_{\mathrm{s,R}}$ analogously represents the influence of an alternating MMF located in the q-axis. With $\eta_{\mathrm{R}}=\sqrt{\eta_{\mathrm{c,R}}^2+\eta_{\mathrm{s,R}}^2}$ and $\alpha_{\mathrm{R}}=\mathrm{atan2}\{\eta_{\mathrm{s,R}},\eta_{\mathrm{c,R}}\}$, the equation can be simplified to

$$\hat{\sigma}_{0,0,\mathrm{R,kc}}=\hat{\mathcal{F}}_{\mathrm{p,p,kc}}\eta_{\mathrm{R}}\cos\left(\gamma-\alpha_{\mathrm{R}}\right). \tag{12}$$

The resulting mean radial force density pulsation is proportional to the amplitude of the alternating MMF.

Furthermore, it depends on the angle γ in a harmonious manner. If the alternating MMF is located at $\gamma = \alpha_R$, the result is maximum possible force density per alternating MMF. This highlights that the introduction of the angle γ helps to understand the impact of an alternating MMF. A graphical exemplification is shown in the appendix. Analogously, this leads to

$$\hat{\sigma}_{0,0,\theta',kc} = \hat{\mathcal{F}}_{p,p,kc}\frac{p}{r_o}\left[\sin(\gamma)\frac{\hat{B}_{PM} + \left(\Lambda_{p,d} - \Lambda_{p,q}\right)\hat{\mathcal{F}}_{d,p,p}}{2} + \cos(\gamma)\frac{\left(\Lambda_{p,d} - \Lambda_{p,q}\right)\hat{\mathcal{F}}_{q,p,p}}{2}\right]$$

$$= \hat{\mathcal{F}}_{p,p,kc}\eta_{\theta'}\cos\left(\gamma - \alpha_{\theta'}\right) \tag{13}$$

for the circumferential direction. It represents the torque pulsation since the torque is proportional to the mean circumferential force density. The results of the radial and circumferential force density can be rewritten in a matrix representation

$$\begin{bmatrix}\hat{\sigma}_{0,0,R,kc}\\\hat{\sigma}_{0,0,\theta',kc}\end{bmatrix} = \begin{bmatrix}\eta_R\cos(\alpha_R) & \eta_R\sin(\alpha_R)\\\eta_{\theta'}\cos(\alpha_{\theta'}) & \eta_{\theta'}\sin(\alpha_{\theta'})\end{bmatrix}\begin{bmatrix}\hat{\mathcal{F}}_{d,p,p,kc}\\\hat{\mathcal{F}}_{q,p,p,kc}\end{bmatrix}. \tag{14}$$

If $\alpha_{\theta'} \neq \alpha_R$, this matrix can be inverted

$$\begin{bmatrix}\hat{\mathcal{F}}_{d,p,p,kc}\\\hat{\mathcal{F}}_{q,p,p,kc}\end{bmatrix} = \frac{1}{\sin(\alpha_{\theta'} - \alpha_R)}\begin{bmatrix}\sin(\alpha_{\theta'}) & -\sin(\alpha_R)\\-\cos(\alpha_{\theta'}) & \cos(\alpha_R)\end{bmatrix}\begin{bmatrix}\frac{1}{\eta_R}\hat{\sigma}_{0,0,R,kc}\\\frac{1}{\eta_{\theta'}}\hat{\sigma}_{0,0,\theta',kc}\end{bmatrix}. \tag{15}$$

The invertibility of (15) demonstrates that there is only one alternating MMF that can influence the cosine fraction of the mean radial and circumferential force density in the desired manner. The same considerations apply to the sinusoidal fraction. The whole alternating MMF results in

$$\begin{bmatrix}\hat{\mathcal{F}}_{d,p,p,k}\\\hat{\mathcal{F}}_{q,p,p,k}\end{bmatrix} = \begin{bmatrix}\hat{\mathcal{F}}_{d,p,p,kc}\\\hat{\mathcal{F}}_{q,p,p,kc}\end{bmatrix} + \begin{bmatrix}\hat{\mathcal{F}}_{d,p,p,ks}\\\hat{\mathcal{F}}_{q,p,p,ks}\end{bmatrix} =$$

$$\begin{bmatrix}\cos(\beta_R)\\\sin(\beta_R)\end{bmatrix}\frac{\hat{\sigma}_{0,0,R,kc}\cos(k\delta') + \hat{\sigma}_{0,0,R,ks}\sin(k\delta')}{\sin(\alpha_{\theta'} - \alpha_R)\eta_R} + \begin{bmatrix}\cos(\beta_{\theta'})\\\sin(\beta_{\theta'})\end{bmatrix}\frac{\hat{\sigma}_{0,0,\theta',kc}\cos(k\delta') + \hat{\sigma}_{0,0,\theta',ks}\sin(k\delta')}{\sin(\alpha_{\theta'} - \alpha_R)\eta_{\theta'}} \tag{16}$$

$$\text{with}\quad \beta_R = \alpha_{\theta'} - \pi/2 \quad\text{and}\quad \beta_{\theta'} = \alpha_R + \pi/2. \tag{17}$$

β_R and $\beta_{\theta'}$ represent the angle γ at which alternating MMFs can be injected with the aim of only influencing the mean radial or the circumferential force density, respectively. This is a noteworthy result, since it also allows to influence the mean radial force density pulsation and torque pulsations independently. The core characteristics that β_R does not influence the circumferential direction is also reflected by (17). Here, β_R is chosen to be $\alpha_{\theta'}$ with an offset of $-90°$. This ensures that the influence on the circumferential direction is zero. The angle with maximum influence on the radial direction, α_R, does not affect the choice of β_R (except for the invertibility of the matrix in (14)). In an analogous manner, $\beta_{\theta'}$ could be analysed.

Fig. 3 presents the calculation results. At zero fundamental MMF ($\hat{\mathcal{F}}_{d,p,p} = \hat{\mathcal{F}}_{q,p,p} = 0$), only an alternating MMF in the d-axis influences the mean radial force density pulsation ($\alpha_R = 0°$) and only an alternating q-axis MMF influences the torque pulsations ($\alpha_{\theta'} = 90°$). Since the angles are perpendicular, the influence directions behave in the same manner ($\beta_R = 0°$ and $\beta_{\theta'} = 90°$). This is a well-known fact often used for the compensation of torque pulsations. Both α_R and $\alpha_{\theta'}$ increase or decrease with greater or smaller fundamental q-axis MMF, respectively. As a consequence, alternating MMFs that are not purely in one axis have the maximum effect. It can also be observed that α_R and $\alpha_{\theta'}$ are no longer perpendicular. As a result, $\alpha_R \neq \beta_R$. Therefore, the alternating MMF causing the maximum mean radial force density pulsation is not used. Instead, the alternating MMF that does not influence the other quantity is used. Negative fundamental MMFs $\hat{\mathcal{F}}_{d,p,p}$ have a minor but observable influence on the angles. The gains of both influence factors (η_R, $\eta_{\theta'}$) increase with the absolute value of $\hat{\mathcal{F}}_{q,p,p}$. Negative fundamental

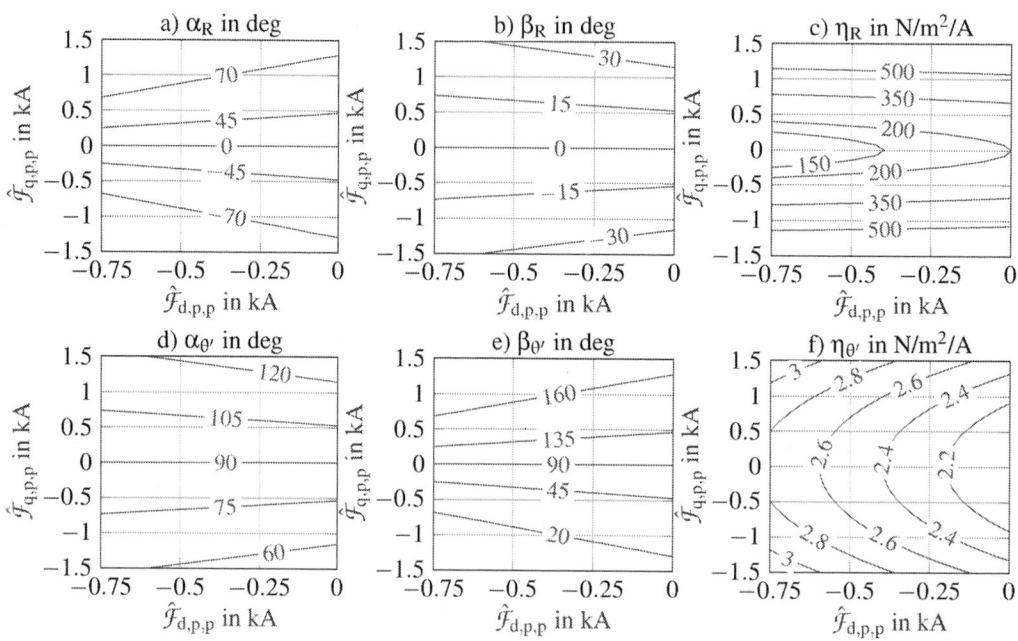

Fig. 3: Analytical calculation results where α is the angle at which the maximum force density is caused, β is the angle at which only one direction is influenced, and η the gain for the respective direction.

MMFs $\hat{\mathcal{F}}_{d,p,p}$ decrease the radial influence factor while increasing the circumferential influence factor. Both behaviours are plausible. The change in radial force density generation stems from an increased or decreased fundamental field component. The change in circumferential force density generation stems from reluctance torques and thus from the interactions between $\mathcal{F}_d$ and $\mathcal{F}_q$.

FEM Validation

Using a commercial FEM software, the analytical findings have been validated. Fig. 4 shows the FEM model. Unlike the exaggerated representation in Fig. 1, the radius of the rotor is considerably bigger than the thickness of the permanent magnet. Apart from the permeability of the iron which was chosen

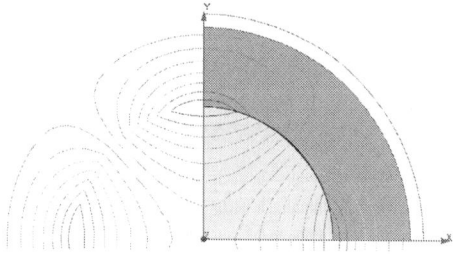

Fig. 4: FEM model including flux lines at zero current excited by the permanent magnet.

as $\mu_R = 10^5$ rather than infinite, the model matches the values from Table I. Table II gives a comparison. The numerical results (especially the angles) agree with the analytical results (maximum error $< 13\%$), the analytical results can thus be considered valid. In the simulation, MMFs at different angles γ are injected and resulting force densities inside of the air-gap are calculated. Through combining multiple simulations, the results shown in Table II are interpolated.

Table II: Comparison of analytical and FEM results.

$\hat{\mathcal{F}}_{d,p,p}$ in A	$\hat{\mathcal{F}}_{q,p,p}$ in A	$\alpha_{R,\text{analytical}}$ in °	$\eta_{R,\text{analytical}}$ in N/A/m²	$\alpha_{\theta,\text{analytical}}$ in °	$\eta_{\theta,\text{analytical}}$ in N/A/m²	$\alpha_{R,\text{FEM}}$ in °	$\eta_{R,\text{FEM}}$ in N/A/m²	$\alpha_{\theta,\text{FEM}}$ in °	$\eta_{\theta,\text{FEM}}$ in N/A/m²
0	0	0.00	200	90.00	2.00	0.01	199	89.72	2.256
-125	250	30.10	213	96.74	2.14	30.16	212	96.72	2.360
-250	500	51.72	272	102.59	2.31	51.79	275	101.71	2.532

Conclusion

This paper analytically describes that alternating MMFs can independently influence the mean radial force density pulsation and the torque pulsation. The results found facilitate an improvement of existing torque pulsation compensations by mitigating the impact on the mean radial force density pulsation. They are also the basis for novel approaches that compensate both pulsations at the same time. Since the mean radial force density pulsation excites the breathing mode, the presented approach is an appropriate means for influencing the dominant source of noise in permanent magnet synchronous machines.

Appendix: Example of the Influence of the Radial Flux Density on the Radial Force Density

Fig. 5 shows a simplified example. Here, the influence of an additional alternating radial flux density on the radial force density is shown. Each subfigure exhibits the fundamental radial flux density $B_{p,p,R}$ and the additional alternating radial flux density $B_{p,p,R,kc}$ as well as the sum $B_R = B_{p,p,R} + B_{p,p,R,kc}$. Further flux densities are neglected for reasons of simplicity. Additionally, each subfigure shows the contributions of the radial flux density to the radial force density (9). B_R^2 is plotted for this purpose. In the example, the alternating flux density is injected at three different angles: $\gamma = 0°$ (a, b, c), $\gamma = 45°$ (d, e, f) and $\gamma = 90°$ (g, h, i). For each γ, three positions are shown: $\delta' = 0°$ (a, d, g), $\delta' = 7.5°$ (b, e, h) and $\delta' = 15°$ (c, f, i). The change in δ' causes a forward rotation of the fundamental field and shows the alternating nature of $B_{p,p,R,kc}$, since $\cos(k\delta') = \cos(2 \cdot 6 \cdot 0°) = 1$, $\cos(2 \cdot 6 \cdot 7.5°) = 0$, and $\cos(2 \cdot 6 \cdot 15°) = -1$. The figure shows that $\gamma = 0°$ leads to the maximum mean value variation (dashed line) of the order k. In the case of $\gamma = 90°$, no variation of the order $k = 6p$ is visible, since the variation of the order $2 \cdot 6p$ is more dominant. However, the total contribution to the mean force density pulsation is significantly smaller. This suggests that $\alpha_R \approx 0°$ in the example.

References

[1] Hofmann A., Qi F., Lange T., and de Doncker R. W.: The breathing mode-shape 0: Is it the main acoustic issue in the PMSMs of today's electric vehicles?, in Proc. ICEMS, Hangzhou, China, 2014, pp. 3067- 3073.

[2] Klarin B., Knaus K., Schneider J., Diwoky F., Resch T., and Brandl S.: PMSM noise - simulation measurement comparison, SAE Technical Paper Series 2018-01-1552, 2018.

[3] Wang S., Jouvray J.-L., and Kalos T.: NVH technologies and challenges on electric powertrain, SAE Technical Paper Series 2018-01-1551, 2018.

[4] Harries M., Hensgens M., and de Doncker R. W.: Noise reduction via harmonic current injection for concentrated-winding permanent magnet synchronous machines, in Proc. ICEMS, Jeju, South Korea, 2018, pp. 1157- 1162.

[5] Franck D., van der Giet M., and Hameyer K.: Active reduction of audible noise exciting radial force-density waves in induction motors, in Proc. IEMDC, Niagara Falls, ON, Canada, 2011, pp. 1213- 1218.

[6] Le Besnerais J.: Réduction du bruit audible d'origine magnétique dans les machines asynchrones alimentées par MLI, règles de conception silencieuse et optimisation multi-objectif, Ph.D. dissertation, l'École Centrale de Lille, France, 2008.

[7] Cassoret B., Corton R., Roger D., and Brudny J.-F.: Magnetic noise reduction of induction machines, IEEE Trans. Power Electron., vol. 18, no. 2, pp. 570- 579, Mar. 2003.

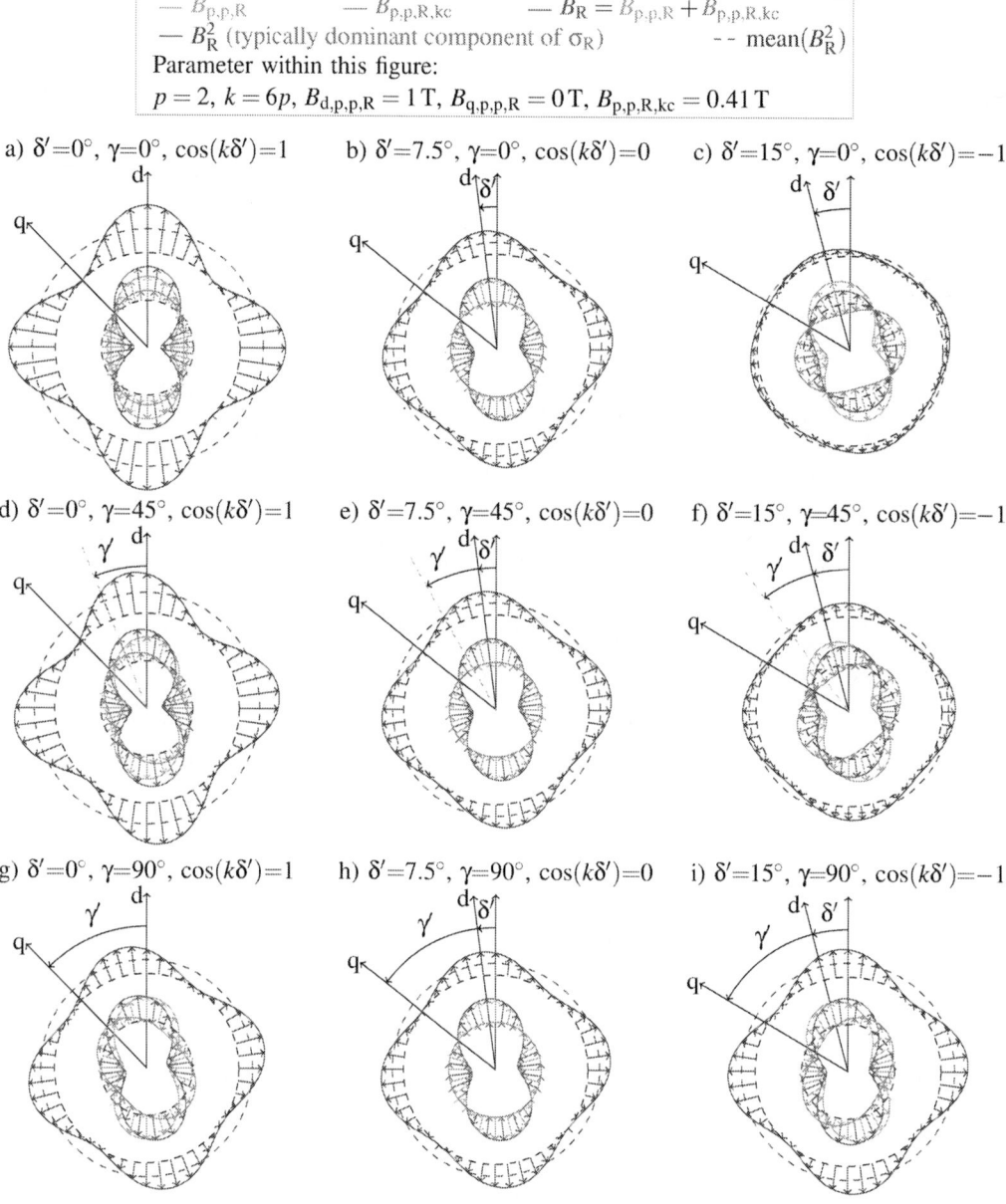

Fig. 5: Example of the influence of the radial flux density on the radial force density (cf. [19]).

[8] Belkhayat D., Roger D., and Brudny J. F.: Active reduction of magnetic noise in asynchronous machine controlled by stator current harmonics, in Proc. 8th Int. Conf. on Electrical Machines and Drives (Conf. Publ. No. 444), Cambridge, UK, 1997, pp. 400- 405.

[9] Kanematsu M., Miyajima T., Fujimoto H., Hori Y., Enomoto T., Kondou M., Komiya H., Yoshimoto K., and Miyakawa T.: Proposal of 6th radial force control based on flux linkage, in Proc. IPEC ECCE-Asia, Hiroshima, Japan, 2014, pp. 2421- 2426.

[10] Liu Y. and Zhang X.: Noise and vibration reduction of direct drive permanent magnet synchronous wind power generator, in Proc. ICEMS, Pattaya, Thailand, 2015, pp. 1257- 1260.

[11] Piippo A. and Luomi J.: Torque ripple reduction in sensorless PMSM drives, in Proc. IECON, Paris, France, 2006, pp. 920- 925.

[12] Mattavelli P., Tubiana L., and Zigliotto M.: Torque-ripple reduction in PM synchronous motor drives using repetitive current control, IEEE Trans. Power Electron., vol. 20, no. 6, pp. 1423- 1431, Nov. 2005.

[13] Pellerey P., Favennec G., Lanfranchi V., and Friedrich G.: Active reduction of electrical machines magnetic noise by the control of low frequency current harmonics, in Proc. IECON, Montreal, QC, Canada, 2012, pp. 1654- 1659.

[14] Zhu W., Pekarek S., and Fahimi B.: On the effect of stator excitation on radial and tangential flux and force densities in a permanent magnet synchronous machine, in Proc. IEMDC, San Antonio, TX, USA, 2005, pp. 346- 353.

[15] Mao Y., Liu G., Chen Q., and Zhou H.: Mitigation of acoustic noise by minimize torque and radial force fluctuation in fault tolerant permanent magnet machines, in Proc. ICEMS, Hangzhou, China, 2014, pp. 60- 64.

[16] Nägelkrämer J., Heitmann A., and Parspour N.: Application of dynamic programming for active noise reduction of PMSM by reducing torque ripple and radial force harmonics, in Proc. AEIT, Bari, Italy, 2018, pp. 1- 6.

[17] Klein-Hessling A., Hofmann A., and de Doncker R. W.: Direct instantaneous torque and force control: a control approach for switched reluctance machines, IET Electric Power Appl., vol. 11, no. 5, pp. 935- 943, 2017.

[18] Sarlioglu B., Morris C. T., Han D., and Li S.: Benchmarking of electric and hybrid vehicle electric machines, power electronics, and batteries, in Proc. ACEMP OPTIM ELECTROMOTION, Side, Turkey, 2015, pp. 519- 526.

[19] Seinsch H. O., Oberfelderscheinungen in Drehfeldmaschinen: Grundlagen zur analytischen und numerischen Berechnung. Stuttgart, Germany: Teubner, 1992.

[20] Gieras J. F., Wang C., and Lai J. C.: Noise of polyphase electric motors, Ser. Electrical and computer engineering. Boca Raton, FL, USA: CRC/Taylor & Francis, 2006, vol. 129.

[21] Lipo T. A.: Introduction to AC machine design. Hoboken, NJ, USA: Wiley IEEE Press, 2017, vol. 63.

[22] Andresen J., Ponick B., and Mertens A.: Direct radial and circumferential analytical air-gap field calculation for electrical machines, in Proc. ACEMP OPTIM, Istanbul, Turkey, 2019.

[23] Braunisch D.: Kombinierte analytisch-numerische Berechnung der Magnetgeräusche elektrischer Maschinen, Ph.D. dissertation, Fakultät für Elektrotechnik und Informatik, Leibniz Universität Hannover, Germany, 2015.

Power Hardware in the Loop System based on Interleaved Converter and FPGA - Application to DC and AC side Emulation for Photovoltaic Inverter Testing

R. KADRI[*], R. BAKRI[*], A. OMRANE[*], F. COLAS[*], F. DELPECH[§]

[*]Univ. Lille, Arts et Metiers ParisTech, Centrale Lille, HEI, EA 2697-L2EP-Laboratoire
d'Electrotechnique et d'Electronique de Puissance,
59000 Lille, France

[§] Puissance + Groupe Spherea, 500 Avenue du Danemark, Z.I. Albasud
82000 Montauban, France

E-Mail: riad.kadri@ensam.eu, frederic.colas@ensam.eu

Keywords

«Real-time simulation», «Power Hardware-in-the-Loop», «FPGA», «photovoltaic», «Coupled Inductors», «Multilevel interleaved converter».

Abstract

In this paper, the development of a high bandwidth power amplifier for power hardware-in-the-loop system is proposed. This amplifier consists of a multilevel interleaved converter, controlled by a real time model of a studied system that is implemented in a field programmable gate array. The accuracy of the proposed topology is illustrated with the design of a test bench for PV inverter testing. This topology is able to emulate both sides (AC and DC) of PV inverter. With this scheme it is possible to analyze the behavior of the complete photovoltaic system under special conditions like partial shading and disturbed grid voltage. The simulation and experimental results verify that the proposed topology exhibits good performance compared with classical topology.

Introduction

With the evolution of digital technologies, Real Time Simulator (RTS) [1] provides to system designers a helpful approach in prototyping and testing process. One field related to RTS is the concept of Power Hardware in the Loop (PHIL) simulation [2] where parts of an actual system is emulated by a RTS driving a power interface. It is thus an extension of Hardware In the Loop (HIL) simulation where a Device Under Test (DUT) can exchange not only information signals but power signals also. An illustration of such application is presented in Fig. 1.

In this example, the mechanical parts and the generator are replaced by virtual ones. The power interface is used to interconnect all the components. It can be useful if parts of the system are not available or to test and tune the controller in the same conditions (here with the same simulated wind conditions). However, stability issues can appear if the power interface is too slow [3]. One way to tackle these issues is to decrease the simulation time step and the communication delay between the RTS and the power interface [4] by introducing an FPGA used as an RTS [5]. In recent years, PHIL gave promising results on several research projects dealing for example with simulation of large-scale power system [6], emulation of permanent magnet motor [7], reduced scale MTDC grid [8]. However, the existing works use a classical RTS and linear amplifier structure or cannot be used for different kind of application (like simulation DC or AC or both).

One interesting field for PHIL application is linked to renewable energy as testing such energy production could be complicated: wind turbine testing require wind, Photovoltaic (PV) system require sun, etc. As an alternative, PHIL based emulator that is consistent, time and weather independent and free from space constraints is nearly mandatory to undertake precise simulation analysis and experimental validations.

In the specific topic of PV converter system testing, a PV emulator can be a useful tool for testing the performance of power converters during the design and developmental phase. This emulator is composed of: on the DC side, an emulation of PV system dynamic behavior and, on the AC side, an AC grid voltage waveforms generator.

A lot of such emulators exist [9] however they are either focused on the DC side or on the AC side simulation. The objective of this paper is to present an original laboratory test-bench which can be used to simulate the DC side or the AC side and both of them. This PHIL system is composed of two main parts: a power stage constituted of several bridges interconnected with coupling chokes and a local controller and an RTS embedded in a FPGA component which is used to simulate the PV behavior in this paper.

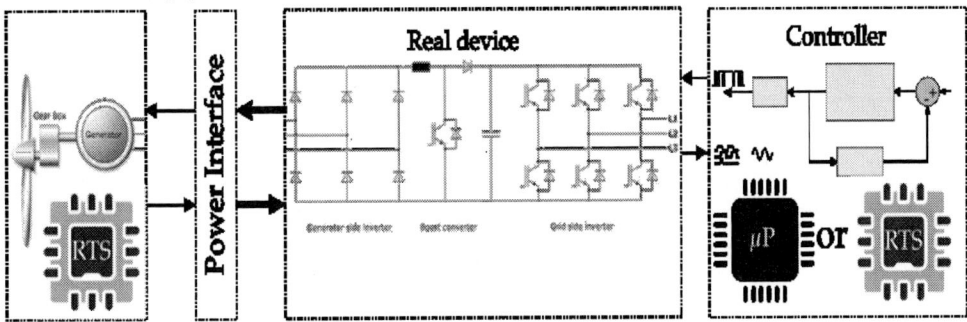

Fig. 1: PHIL simulation illustration

The paper is organized as follow: The section I present the converter topology and design, the section II is focused on the application of proposed emulator structure on PV inverter testing and the controller structure and the last section presents the real-time implementation of PV system in FPGA and analysis of the experimental results obtained from the implemented prototype.

I. Power amplifier presentation

I.1. Topology presentation

The most important requirements for a power amplifier used in power hardware-in-the-loop system are a high bandwidth, high dynamics, low harmonic distortion, high accuracy, and low internal resistance. In this paper, the proposed power amplifier is a multilevel interleaved converter based on the paralleling of several standard half-bridges coupling between them by a coupled inductor connected in a cyclic cascade [10]- [11]. The scheme of this topology is presented in Fig. 2. Phase parallelizing presents two capital advantages. In one hand, it increases the output current without increasing the current rate of the semiconductors switches, so a high power can be handled. In the other hand, it allows to reduce magnetic components' sizes thanks to the effective switching frequency in the converter output, which is equal to the switching frequency multiplied by the total number of legs. A such topology can be used for DC-DC and DC-AC. The last point makes this topology adapted for many PHIL applications.

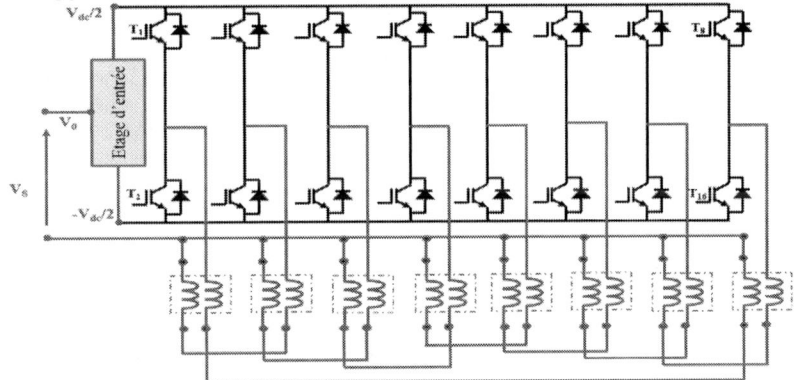

Fig. 2: converter topology with cyclic cascade coupled inductors

I.2. Coupling chokes design

To couple the N legs of the converter, separated chokes can be used. In this paper coupled inductors are used to optimize the use of the core, then the fluxes generated by the currents of the two windings are compensating each other, and only magnetizing and leakage fluxes are taken into account on the choice of the core. A simplified equivalent circuit of a coupling choke is shown in Fig. 3 (a). The design of the coupled inductor is a key point in order to achieve specifications in terms of current ripple and dynamic performances. Thanks to the cyclic cascade scheme, output inductance of the converter is equal to the total leakage inductance of one choke L_{lk} divided by phases number N as presented in the equivalent circuit of the converter shown in Fig. 3 (b). This low output inductance associated with an increased effective frequency allows good dynamic performances, which make this topology suitable for PHIL applications. Moreover, this low inductance will induce a very small voltage change with current charge variation. The design of such chokes is different from a simple transformer design, because the values of the leakage and magnetizing inductance must be tuned in order to achieve the converter specifications. The leakage inductance must be high enough to reduce the output current ripple and magnetizing inductance must be as high as possible to reduce magnetizing current (differential current).

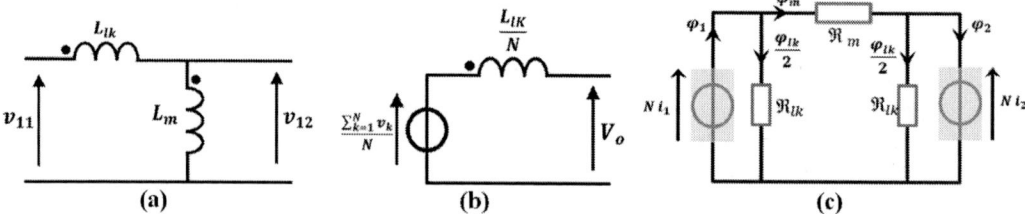

Fig. 3: (a) coupled inductor equivalent circuit, (b) converter equivalent circuit, and (c) equivalent magnetic circuit of the coupler

In order to choose a core for the chokes, the expression of the area product must be expressed. The magnetic equivalent circuit of the choke represented in Fig. 3 (c) is considered. From the equivalent magnetic circuit, the total flux could be expressed also by introducing windings current, magnetizing and leakage inductance in one hand $N\,\varphi_1 = L_m(i_1 - i_2) + \frac{L_{lk}}{2}i_1$. In the other hand it can be expressed by definition using the magnetic induction $BN\,\varphi_1 = N \cdot B \cdot A_c$. From the two expressions the section of the core is given by:

$$A_c = \frac{L_m(i_1 - i_2) + \frac{L_{lk}}{2}i_1}{N \cdot B_{max}} \qquad (1)$$

Where is B_{max} is the maximal allowed induction of the core.

According to this formula, it is important to choose a core with enough section in order to handle the differential current. To compute the area product, the window area is needed. It can be expressed from the RMS value of the winding current, the winding filling factor k_f, and the current density J.

$$A_w = 2N \, \frac{I_{rms}}{k_f J} \qquad (2)$$

Then, the area product could be expressed by:

$$AP = A_w \cdot A_c = 2\,k_b \frac{I_{eff}}{J\,B_{max}} \left(\frac{L_{lk}}{2}i_1 + L_m(i_1 - i_2) \right) \qquad (3)$$

In this application, coupling chokes are designed with a toroid core, the value of the leakage and magnetizing inductances are $L_{lk} = 10\mu H$ and $L_m = 20mH$ repectively.

I.3. Power amplifier output waveforms

The multilevel interleaved converter with coupled inductor is presented in Fig. 2. The mentioned converter has 16 controllable switches and their respective freewheeling diodes. Each set of switches that constitutes one of the leg is connected to the load through coupled inductors with cyclic cascade configuration. The interleaving technique is applied to the system shown in Fig. 2 in order to procure apparent switching frequency N times higher than the individual switching frequency of each leg. When operating with a carrier-based modulation strategy, this is achieved by using an N number of shifted carriers. In this configuration, the number of voltage levels will be N + 1, N being the number of legs interleaved. Another characteristic of this structure is the increasing of the load voltage levels, thus contributing to reduce the common mode voltage when compared with standard structures. Moreover, the use of coupled inductors with cyclic cascade configuration allows the division of the load phase current through the switches, in a balanced way so that the current in each arm of the converter is equal to 1/N of the load current, reducing the current values for the switches.

In this section, the proposed converter was simulated with output voltage with a frequency of 500Hz, switching frequency of 20 kHz and symmetrical triangular carriers shifted by 45°. The simulation results are presented in Fig. 4. From the results of simulation shown in this figure, it is possible to verify that the output voltage has 9 levels; this favors the harmonic content reduction and minimizes the size of the output filter. On the other hand, it is verified from this simulation that the load current is divided in a balanced way in the eight coupled inductor.

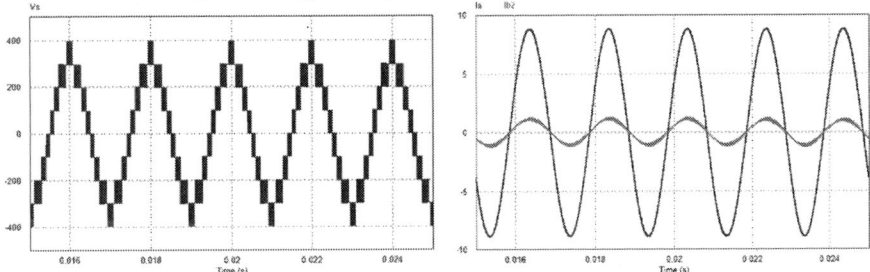

Fig. 3: Output current and voltage waveforms

II. Application for PV inverter testing

The PV systems are modular; hence the major advantage of these systems is that they can be simply adopted in existing buildings and can be installed anywhere. In addition, manufacturers have designed various panels types, which can be placed at a variety of different types of houses or buildings to achieve better performance. However, in this new trend of integrated PV arrays, it is difficult to avoid partial shading of array due to neighboring buildings throughout the day in all the seasons. This makes the performance study of PV inverters without real PV energy source under partial shading conditions and grid disturbances an important issue.

Field testing of PV inverters is conducted to ensure the quality of installation and the performance of the product, however, is costly, time consuming and strongly dependent on the actual grid voltage and climatic conditions. Thus, a solution for developing experimentations without need of real PV panels is very important. This has increased the interest on the development of laboratory test benches for carrying out measurements and analysis, with no need to perform field tests.

A wide range of photovoltaic array emulators have been proposed and developed during the past few years. Some of them present limitations in the output power dynamics and also in the number of points of the I-V panel curve reproduction. Other works use power converters with PWM principle to increase the accuracy of the I-V curve, but the influence of temperature and inhomogeneous irradiance is not taken into account. In addition of these drawbacks, we can note that: these emulation systems, does not allows performance analysis of photovoltaic inverters under disturbed grid voltage and the dynamics limitation appears when the MPPT efficiency is needed to test.

II.1. Proposed Power Hardware-in-the-Loop topology

In this section, the development of Power Hardware-in-the-Loop testing system for PV inverter is presented. This topology is able to emulate both sides of PV inverter. With this scheme it is possible to analyze at the laboratory the behavior of the complete photovoltaic system under special conditions like partial shading and disturbed grid voltage. The block diagram of the proposed PV inverter test bench topology is shown in Fig. . The photovoltaic array emulation is based on closed-loop reference model structure. The power amplifier used in the proposed work is based on multilevel interleaved converter. The control is achieved by PI regulators with feedback of outputs voltage and current to regulate the operating point. The measurements of currant and voltage, controls strategies and modulations are implanted into TEXAS TMS320F28379D and the PV model is embedded into a field programmable gate array (FPGA). On the other hand, for the grid side emulation, a second power amplifier based on a multilevel interleaved converter to generate grid faults and disturbed grid voltage conditions. The test bench is applied to generate different reproducible types of disturbed grid voltages like overvoltage, voltage sag, frequency variations and harmonic disturbances.

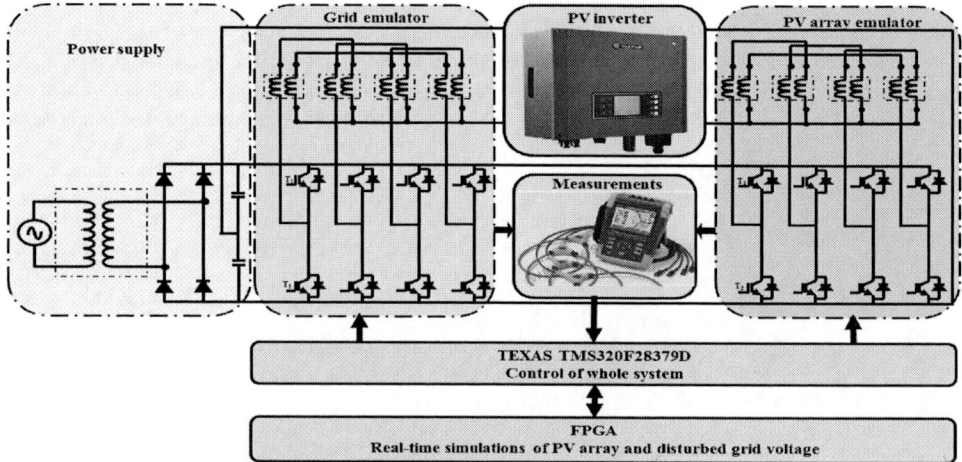

Fig. 5: PHIL test bench for PV inverter testing.

II.2. Photovoltaic array emulation model

A photovoltaic panel is the whole assembly of solar cells, connections, protective parts, supports etc. In the literature, several equivalent circuits have been proposed in order to review the behavior of the PV cell [12]–[14]. In the present modeling, the focus is only on PV cells. PV cells consist of a silicon p-n junction that when exposed to light release electrons around a closed electrical circuit. From this physical interpretation of the photoelectric effect the equivalent circuit of a solar panel can be modeled through the circuit shown in Fig. (a). The output of the current source is directly proportional to the light falling on the cell (photocurrent). The output current I of solar panel is given by (3).

$$I = [I_{sc} + k_1(T - 298)]\frac{G}{1000} - I_0\left(exp(\frac{q.V_{do}}{n.k.T}) - 1\right) - \frac{V_{do}}{R_{sh}} \tag{4}$$

Where K_1 is cell's short-circuit current temperature coefficient, I_{sc} is cell short-circuit current at 25 °C, T is the cell's temperature and G is the solar irradiance in W/m², I_0 is the reverse saturation current, n is the diode factor, q is the electron charge, k is Boltzmann's constant. R_s is the intrinsic series resistance of the solar cell, this value is normally very small. R_{sh} is the equivalent shunt resistance of the solar array and its value is very large. In general, the resistances R_s and R_{sh} can be neglected.

On the other hand, the PV system, consisting of more panels in series. The series connection forces all panels to operate at the same current, the shaded panel within a module becomes reverse biased which leads to power dissipation and thus to heating effects.

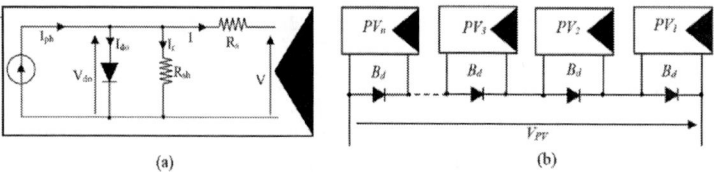

Fig. 6: Equivalent circuit model of: (a) PV panel, (b) PV array bridged by bypass diodes.

To avoid thermal overload, PV panel are bridged by bypass diodes Fig. 6(b). This measure limits the bias voltage at the shaded panel and thus the dissipated power. Another reason to use bypass diodes is to preserve more of the power output of the module in case of partial shading. This phenomenon can move the maximum power point to unexpected places. Hence, in order to obtain the I–V characteristics of the series-connected modules, the voltages across these panels are added to determine the resultant output voltage.

II.3. Power amplifier controllers design

For both sides of the proposed topology, the current controller of each amplifier, which is a part of the inner loop, is designed followed by the voltage controller of the outer loop. In the controllers design, the equivalents output inductance of the coupling inductors connected in a cyclic cascade is represented by simple inductor L with a parasitic resistor R. Additionally a very small shunt capacitor C, is designed to remove the high-frequency switching components originated by the pulse-width modulation.

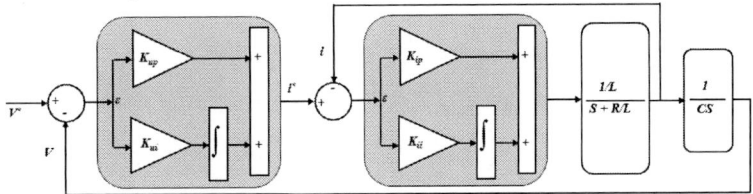

Fig.7: Control structure

II.3.1. Current control mode

By using PI-type regulators, a fast-dynamic response and zero steady-state errors can be achieved in the current controller [15]. The diagram of the current regulator (inner current control loop) is shown in Fig. 7. Since the switching frequency is much higher than the line frequency, the sampling and hold delay is neglected. In the diagram, kip and kii are the proportional and integral parameters, respectively; i^* is the reference current signal, and i is the feedback current. The closed-loop transfer function of the current loops is given by (1):

$$\frac{i(s)}{i^*(s)} = \frac{k_{ii}}{L} * \frac{S+1}{S^2 + \frac{(k_{ip}+R)}{L}S + \frac{k_{ii}}{L}} \tag{5}$$

The damping ratio $\zeta = \frac{(k_{ip}+R)}{2L\sqrt{k_{ii}/L}}$, and $\omega_{ni}^2 = \frac{k_{ii}}{L}$. Thus, the parameters of the current regulator can be designed as follows: $k_{ip} = 2\zeta\omega_{ni}L - R$ and $k_{ii} = L\omega_{ni}^2$

II.3.2. Voltage control mode

In order to regulate the dc voltage at a fixed value, the error ε is passed through a PI-type compensator, as shown in Fig. 7. In the diagram, the voltage loop is an outer loop, while the current loop is an inner loop. The internal loop has been designed to achieve short settling times in order to achieve a fast correction of the error. The outer loop can be designed to be slower. Thus, the inner and outer loops can be considered decoupled, and they can be linearized [16]. Consequently, the current loop transfer function is approximately considered as Gc = 1. The closed-loop transfer function of dc voltage regulation, obtained from Fig. 4, has the following form:

$$\frac{V(s)}{V^*(s)} = \frac{k_{up}}{C} \frac{\frac{k_{ui}}{k_{up}} + S}{S^2 + \frac{k_{up}}{C}S + \frac{k_{ui}}{C}} \tag{6}$$

In the same way as the design process of the current loop, the voltage regulator parameters can be given as follows: $k_{up} = 2\zeta C\omega_{nu}$ and $K_{ui} = C\omega_{nu}^2$

III Experimental results

III.1. Code generation

The photovoltaic panel was implemented using Simulink coder combined with Vivado High Level Synthesis tool. First, the Simulink Coder generates standalone C code from the PV model. Then, after adding the appropriate HLS constraints, the Vivado HLS tool translates the C Code into a hardware description language (VHDL in this case), to get the register transfer level description. The design process workflow is summarized in Fig. 8 HLS constraints include setting the target hardware, the clock frequency and the user directives used to optimize the design.

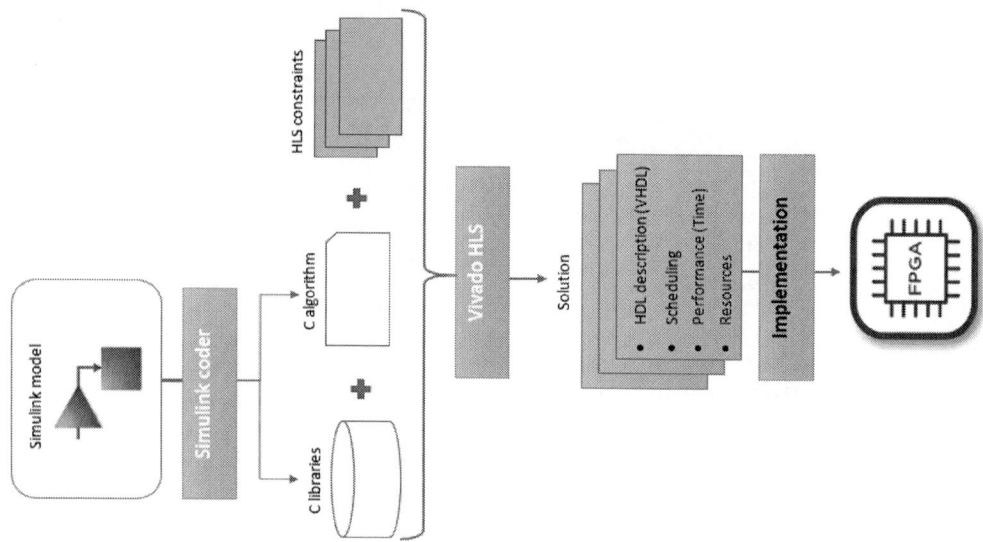

Fig.8: Code generation for FPGA

Before Implementation, the system has been simulated using the Vivado HLS tools, and the results were verified with the ones obtained in Matlab Simulink. While operating in real time, a calculation step of less than 1 µs was achieved (680 ns).In order to interface the system running on FPGA with the control board, physical connections were used. Fig. 9 shows a picture with the complete system implemented in the laboratory.

Fig.9: PHIL Experimental setup based on interleaved converter and FPGA

III.2. Experimental results

A single phase prototype was developed to evaluate the performance of the proposed PHIL system based on an interleaved topology and controlled by a TEXAS TMS320F28379D and FPGA. The whole developed Power Hardware-in-the-Loop system is presented in Fig.9. In this section, all experimentations are based on a configuration of a PV array with four modules MSX60 60-W in series-parallels configuration. Each PV module considered in this paper is comprised of 36 PV cells connected in series providing an open circuit voltage equal to 21.05 V and a short-circuit current equal to 3.87 A.

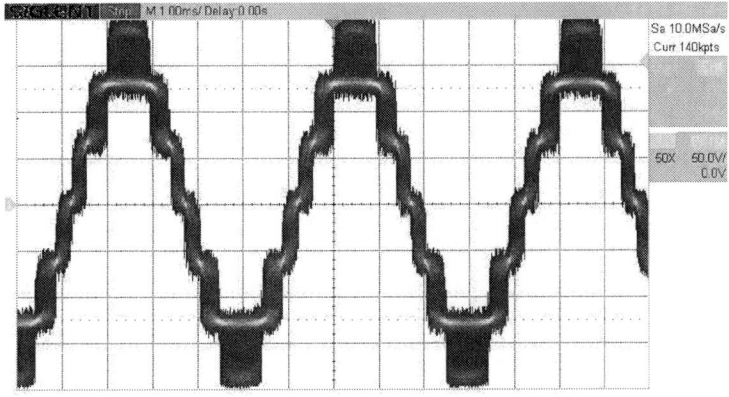

Fig.10: Open-loop voltage waveform

Fig.10 illustrates experimental result obtained from the multilevel interleaved converter prototype operating as inverter in open-loop. The switching frequency was 20 kHz, the DC link voltage was 500V, and modulation index was 0.6 and the output voltage with a frequency of 200Hz. The switch gate signals were generated by the digital signal processor of TEXAS TMS320F28379D. The presence of the seven voltage levels can be clearly observed in Fig.10, thus proving the theoretical analysis and the simulation previously presented.

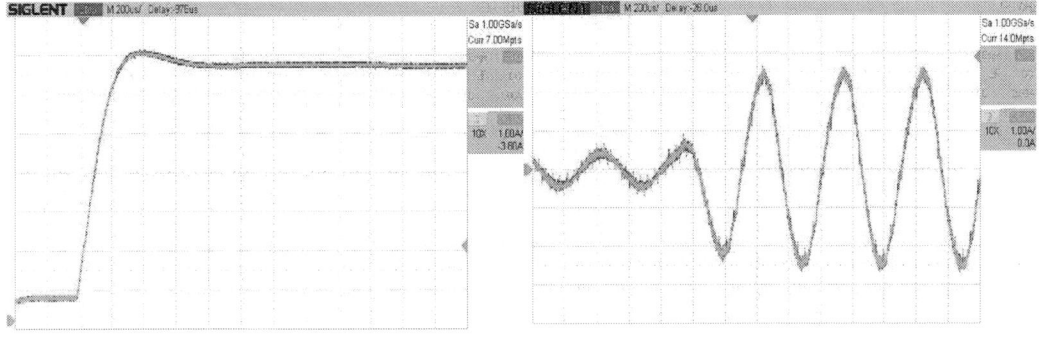

Fig.11: Closed-loop response of current control: (a) for DC current and (b) for AC current

Afterwards, the designed controllers performances is evaluated by means of a set of experimental tests to check the power amplifier during transient and steady-state conditions, as well as the system dynamics for DC or AC working configurations. Fig. 11 shows the step response of the current loop of the system: for DC configuration in Fig. 11(a) and AC configuration in in Fig. 11(b). Fig. 12 shows the response of the outer voltage loop of the system under undisturbed and disturbed grid voltage. It is clear from the voltage and, current waveforms in both configurations presented in Fig. 11 and Fig. 12 that the developed controller is powerful in steady state and transient performance.

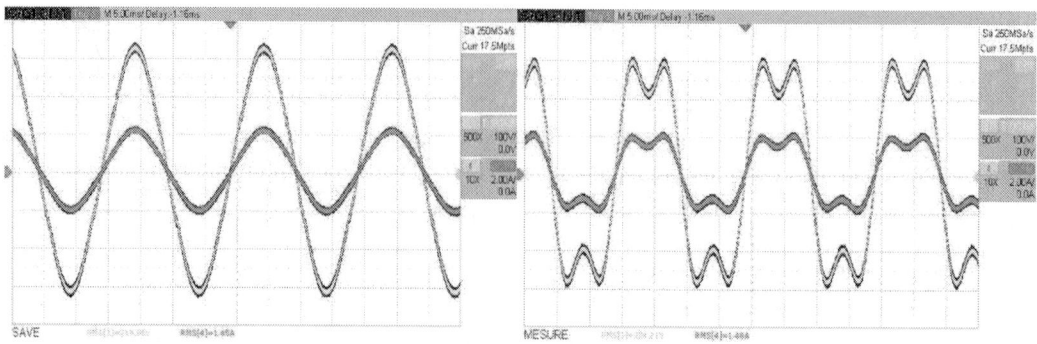

Fig.12: Closed-loop response of voltage control for undisturbed and disturbed grid voltage

On the other hand, for observing the performance of the PV emulator, the four PV modules are subjected to the same temperature of 25 °C and same irradiance level of 500 W/m². Fig.13 shows the obtained *I-V* and *P-V* characteristics obtained with dynamic load. Red line corresponds to the reference model results and blue are experimental results. These results show that, the real-time emulator output follows perfectly the trajectories of the statement PV arrays.

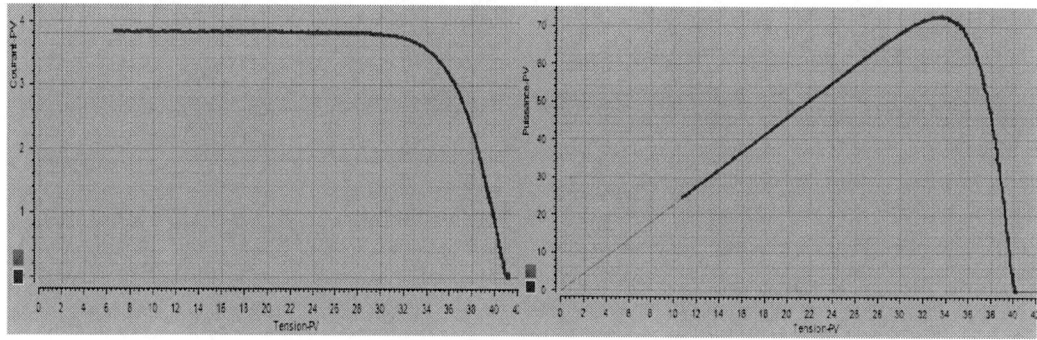

Fig.13: I-V and P-V characteristics of PV array: theoretical model (red line); experimental (blue line).

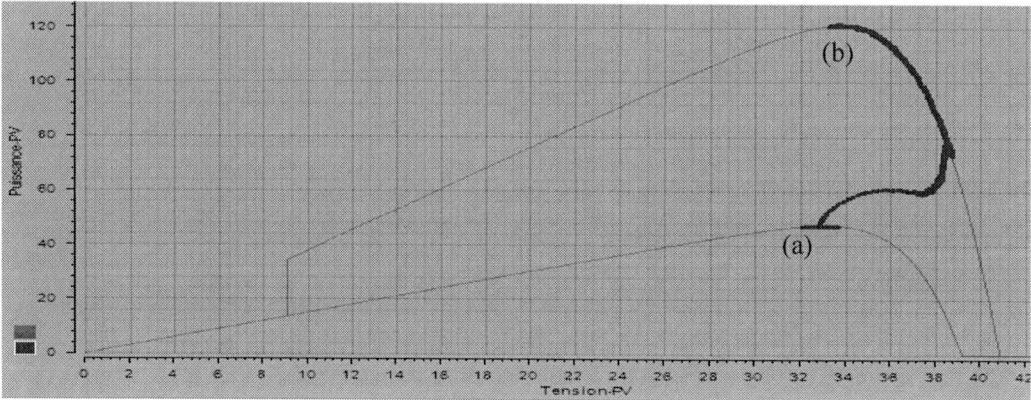

Fig.14: Comparison of P-V curves obtained by the PV emulator (blue line) and theoretical model (red line) at various irradiation levels.

In order to evaluate commercial PV inverter efficiency, the MPPT algorithm must be evaluated. The efficiency of this algorithm and it proper operations are essentials to inject the maximum amount of energy produced by the PV panels. The commercial PV inverter that has been studied is the APS YC250-EU micro-inverter which can be used for one PV panel composed by 60 to 72 cell modules. The recommended PV module power range goes from 180 to 310W. Its rated output power is equal to 250W. In Fig.15 firstly, at point (a) in we can see output power of emulated PV panel when the MPP is achieved by the commercial inverter. According to the theoretical model (red line) of the PV panel, the MPP static efficiency is validate. In order to analyze the dynamic MPPT efficiency to reach the MPP in variable conditions of irradiance, a step in the irradiance from 200 W/m2 to 500 W/m2 has been done. The response of the MPP algorithm is depicted in Fig.14 with blue line.

Conclusion

This paper proposes an original PHIL system based on an interleaved topology and controlled by an FPGA. An application for PV inverter testing has been proposed to validate the PHIL concept. The experimental results presented on this paper validate the control strategy of proposed power amplifier. Hence, the PV emulator properly reproduces the characteristic curve of the PV array which is embedded into a FPGA. It device is suitable for testing commercial PV inverter at low power level. Also, it is possible to program any irradiance and temperature profiles which allow evaluating MPP tracking algorithm efficiency and power energy conversion performances.

References

[1] J. Belanger, P. Venne, and J.-N. Paquin, "The What, Where and Why of Real-Time Simulation," *PES IEEE Gen. Meet.*, pp. 37–49, 2010.

[2] X. Guillaud *et al.*, "Applications of Real-Time Simulation Technologies in Power and Energy Systems," *IEEE Power Energy Technol. Syst. J.*, vol. 2, no. 3, pp. 103–115, 2015.

[3] A. Viehweider, G. Lauss, and F. Lehfuss, "Interface and Stability Issues for SISO and MIMO Power Hardware in the Loop Simulation of Distribution Networks with Photovoltaic Generation," vol. 2, no. 4, 2012.

[4] A. Viehweider, G. Lauss, and L. Felix, "Stabilization of Power Hardware-in-the-Loop simulations of electric energy systems," *Simul. Model. Pract. Theory*, vol. 19, no. 7, pp. 1699–1708, 2011.

[5] Y. Chen and V. Dinavahi, "FPGA-based real-time EMTP," *IEEE Trans. Power Deliv.*, vol. 24, no. 2, 2009.

[6] C. Mao *et al.*, "A 400-V/50-kVA Digital-Physical Hybrid Real-Time Simulation Platform for Power Systems," *IEEE Trans. Ind. Electron.*, vol. 65, no. 5, pp. 3666–3676, 2018.

[7] A. Schmitt, J. Richter, M. Gommeringer, T. Wersal, and M. Braun, "A novel 100 kW power hardware-in-the-loop emulation test bench for permanent magnet synchronous machines with nonlinear magnetics," *IET Conf. Publ.*, vol. 2016, no. CP684, pp. 1–6, 2016.

[8] S.-A. Amamra, F. Colas, X. Guillaud, P. Rault, and S. Nguefeu, "Laboratory Demonstration of a Multiterminal VSC-HVDC Power Grid," *IEEE Trans. Power Deliv.*, vol. 32, no. 5, 2017.

[9] J. P. Ram, H. Manghani, D. S. Pillai, T. S. Babu, M. Miyatake, and N. Rajasekar, "Analysis on solar PV emulators: A review," *Renew. Sustain. Energy Rev.*, vol. 81, no. July 2017, pp. 149–160, 2018.

[10] A. Schmitt, M. Gommeringer, J. Kolb, and M. Braun, "A high current, high frequency modular multiphase multilevel converter for power hardware-in-the-loop emulation," *PCIM Eur. Conf. Proc.*, pp. 1537–1544, 2014.

[11] I. G. Park and S. I. Kim, "Modeling and analysis of multi-interphase transformers for connecting power converters in parallel," *PESC Rec. - IEEE Annu. Power Electron. Spec. Conf.*, vol. 2, pp. 1164–1170, 1997.

[12] M. G. Jaboori, M. M. Saied, and A. A. R. Hanafy, "A contribution to the simulation and design optimization of photovoltaic systems," *IEEE Trans. Energy Convers.*, vol. 6, no. 3, pp. 401–406, 1991.

[13] J. T. Bialasiewicz, "Renewable energy systems with photovoltaic power generators: Operation and modeling," *IEEE Trans. Ind. Electron.*, vol. 55, no. 7, pp. 2752–2758, 2008.

[14] F. Spertino and J. S. Akilimali, "Are manufacturing I-V mismatch and reverse currents key factors in large photovoltaic arrays," *IEEE Trans. Ind. Electron.*, vol. 56, no. 11, pp. 4520–4531, 2009.

[15] T. qoria, F. gruson, F. colas, X. guillaud, "Tuning of cascaded controllers for robust grid-forming Voltage Source Converter," Power Systems Computation Conference, 06/2018.

[16] T. qoria, Li chuanyue, O. ko, F. gruson, F. colas, X. guillaud, "Direct ac voltage control for grid-forming inverters," journal of power electronics, 12/2019.

Implementation of TAPIR switching cells with integrated direct air-cooling for SiC power devices

Wendpanga Fadel BIKINGA, Kouceila ALKAMA, Bachir MEZRAG, Jean Michel Guichon,
Yvan Avenas

UNIV.GRENOBLE ALPES, CNRS, GRENOBLE INP G2ELAB

21 Avenue des Martyrs,

38000 Grenoble, FRANCE

E-Mail: wendpanga-abdoul.bikinga@grenoble-inp.fr, yvan.avenas@g2elab.grenoble-inp.fr

URL: ww.g2elab.grenoble-inp.fr

Acknowledgements

The authors would like to thank SATT LINKSIUM for funding the work.

Keywords

«Power module», «3D packaging», «SiC devices», «Thermal management», «Switching cell».

Abstract

This paper presents the design of an SiC switching cell made with the TAPIR technology. Electromagnetic and thermal simulations are carried out in order to analyze the performance of this switching cell. This cell is then compared to commercial power modules and discrete components and exhibits excellent junction to ambient thermal resistance and stray inductance.

Introduction

The energy transition involves a deep change in energy production and consumption patterns and involves an increase in the electrification of systems. According to the U.S. Advanced Research Projects Agency-Energy (ARPA-E)' more than 80% of electrical energy in the United States will use static converters by 2030 [1]. Modern power electronics is thus facing a substantial demand on improving efficiency while reducing size and cost of systems for a broad range of areas, including electric vehicles, renewable energies, industrial motors and generators, and distribution grid applications [2]. Size reduction and efficiency of converters have been improved in recent years due to the growth of wide-bandgap semiconductors use [3]. The increase in switching frequency allows reducing the size of passive components and thus obtaining converters with higher power density [4] [5]. However, the packaging of these semiconductors requires special attention due to their high switching speeds [6].

Therefore, the challenges of tomorrow's power electronics designers include improving semiconductor packaging to reduce electrical and electromagnetic disturbances, and optimizing cooling to lighten converters to meet the needs of electric mobility where weight is a significant constraint. Indeed, the weight of the cooling system is generally non-negligible. For example, Delhommais et al. [7] propose a global optimization of a SiC switching cell and show that about 25% of the global weight is due to the heatsink and the fan.

The TAPIR (compacT and modulAr Power modules with IntegRated cooling) technology [8] seems to be able to reduce the mass of power modules and cooling systems by a ratio of 3 to 5 by distributing the heat sources (chips) in the volume of the heatsink. Contrarily to most researches on double side cooling [1], the heat extraction is made by air-cooling and the heatsinks are used as electrodes as it is the case with press-pack power converters [9] or Power Chip on Bus (PCoB) modules [10]. TAPIR technology consists in sandwiching vertical semiconductor components between two heat sinks. Then two sandwiches and a decoupling capacitor are assembled to make a low voltage-switching cell as shown in

Fig. 1. Finally, the global power module is made by assembling several switching cells together. An example of a 4-phase converter is given in [8].

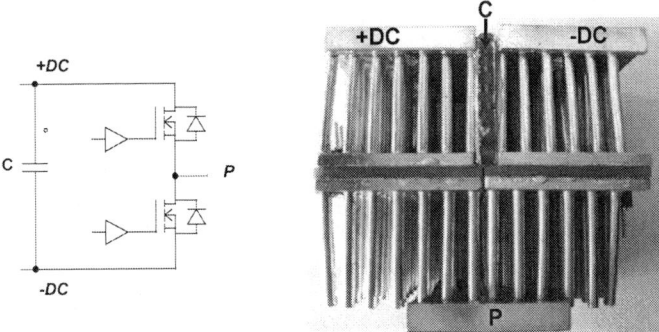

Fig. 1: TAPIR low voltage Switching cell

In [8], two packaged 60V-100A power MOSFETs are implemented in a switching cell that is characterized using the double pulse method. The parasitic inductance of the switching loop is less than 2 nH. Also the thermal behavior is estimated using thermo-sensitive electrical parameters. The measured junction to ambient thermal resistance is close to 1.30 K/W. The coupling thermal resistance is relatively low and limited to 0.2 K/W. However, the junction-to-ambient thermal resistance can be further reduced by using bare chips and improving the heat sink as proposed by Xu et al. [10]. In this reference, the authors propose a sandwich similar to the one of the TAPIR method and including SiC devices with a junction-to-ambient thermal resistance of only 0.5 °C/W.

The aim of the present paper is to propose an implementation of the TAPIR technology on a high voltage SiC switching cell. The goal is to analyze the electromagnetic and thermal performances of the cell and compare it to commercially available power modules and discrete components. The implementation of the SiC switching cell is presented, followed by electromagnetic and thermal simulations. Before concluding, a comparison will be made with commercially available components.

Implementation of an SiC switching cell

SiC MOSFETs CPM2_1200_0080b (1200V-36A, R_{DSon}=80mΩ) from CREE company were used to build a switching cell. A 4-layer printed circuit board (PCB) has been designed to connect the MOSFET to the gate drive circuit but also to ensure a connection of the die to the heatsinks with few thermal interfaces (Fig. 2 and 3).

In Fig. 3, it can be seen that the source and gate pads of the chip are soldered on PCB layer 2. For this purpose, a nickel-gold coating has been made on the gate and source metallizations. Beneath the two source electrodes of the chip, a copper inlay is inserted between layer 2 and layer 4. Layer 1 is cut so that the Drain is accessible from the PCB surface through a hole (Fig. 2 and 3).

This implementation has several advantages in terms of thermal management:
- one heatsink can be directly connected to the drain of the chip,
- the number of thermal interfaces between the source metallization and the other heatsink is relatively low.

The gate drive circuit (buffer) is located on layer 4 of the PCB and vias provide electrical connections between the control signals and the chip. Layer 1 is thin (35µm) to minimize the distance between the chip and the top of the PCB in order to have the thinnest soldering thickness and ensure good electrical and thermal conductions. Layer 3 is used to connect the gate to the driver circuit avoiding the guard rings. Vias connecting the source are located on either side of the gate track. This minimizes the inductance of the gate circuit and creates electromagnetic shielding of the gate. Layer 4 thickness is 120µm for allowing heat spreading of the heat flux and thus reducing the thermal resistance of the solder between the PCB and the heatsink.

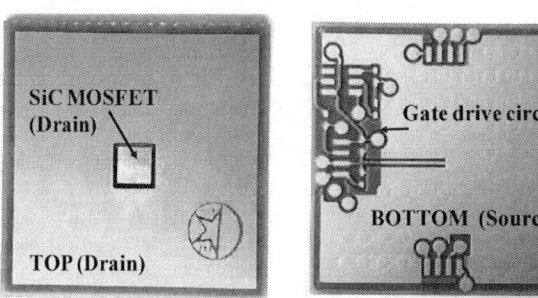

Fig. 2: PCB for SiC MOSFETs

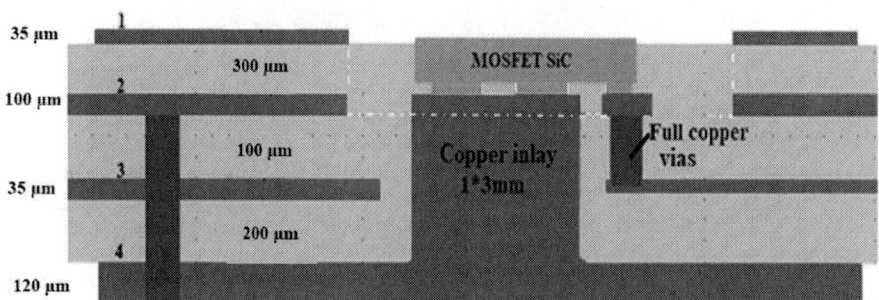

Fig. 3: Architecture of the PCB

To make the switching cell, the heatsinks are cut (Fig. 4) to avoid any electrical contact with the components of the gate control circuit on the PCB board. The thickest component is the gate driver which has a height lower than 2 mm. This choice is made in order not to degrade too much the thermal performances by improving the electromagnetic one. Note that a gate driver chip could be directly integrated inside the PCB to simplify the assembly but the technological process is not done already.

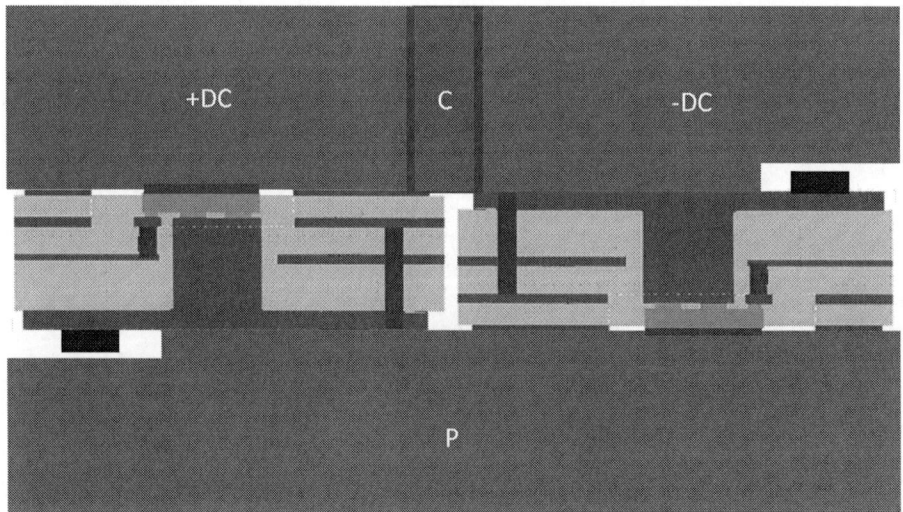

Fig. 4: SiC Switching cell

Once the switching cell is presented, electromagnetic simulations will be proposed to estimate the parasitic inductance of the switching loop, then thermal simulations will be carried out to analyse the thermal performances of this assembly.

Stray inductance of the SiC switching cell

Electromagnetic simulations are carried out with Flux3D software based on the finite element method. Indeed, the PEEC method that is usually used for evaluating power electronics packages is not adapted to the TAPIR geometry that does not include any thin electrical conductors like it is the case in classical power modules. The MIPSE software still under development at G2Elab which uses integral methods [11] is also used here to visualize the distribution of the current density in order to refine the mesh on Flux3D, but also to confirm the result obtained with Flux3D.

In Fux3D, a magneto-harmonic application at a frequency of 100 MHz is used. The simulation is done at this frequency because the equivalent frequency measured in [8] was about 100MHz. The decoupling capacitor is replaced by a current source of 1A injected on the surface over a thickness of 2mm (see Fig. 5). This consideration was made after having noted that during switching time only the lower part of the capacitor injects current in the structure (see Fig. 4). The voltage is measured across the current injection areas and the stray inductance is deducted. The capacitive effects are not included in this simulation.

The simulated geometry is shown in Fig. 5. The chip (3mm*3mm*0.35 mm) is assumed to be directly attached to the heatsinks, as well as the copper inserts (3mm *1mm*0.4mm). Only one internal layer is considered in this simulation because adding the second one does not change much the result. This inner layer is not connected to the copper inlay as can be seen on the PCB architecture in Fig. 3. The micro vias that link the inner layers to the source potential are represented as two bars.

All components of the switching cell are assumed to be made of aluminum of resistivity 2.68E-8 Ω.m. This approximation is valid since the size of the chips and inserts is small compared to that of the heatsinks.

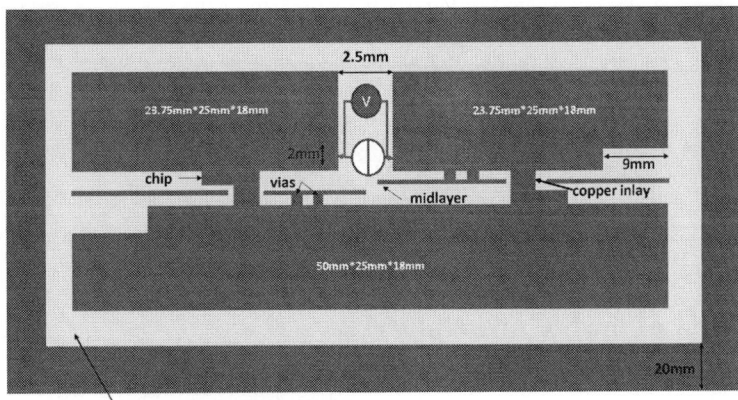

Fig. 5: Simulated switching cell geometry

For applying boundary conditions, an infinite box is used. Its dimension is twice the size of the device to be tested as recommended in the software documentation. The thickness of the infinite box is 20mm. Simulations have been performed to analyze the impact of increasing the size of the infinite box. This has almost no influence on the results.

As in any finite element resolution, the quality of the mesh has an impact on the veracity of the results. For this purpose, it is necessary to know which is the main current path and thus to adapt the mesh by decreasing the elements size in critical areas. Several simulations have been done with Flux3D and MIPSE sofwares to determine the current path at the proposed frequency. Fig. 6 shows the distribution of the current density in logarithmic form obtained after simulation on MIPSE. It can be seen that the current flows mainly on the edges (skin effect) of the heatsinks and in the semiconductor chips and copper inlay. The mesh is densified at these places by adding 100µm plates shown in red in Fig. 5, so as to have at least two meshes in the skin thickness which is about 60μm. The mesh creates more than 500,000 1st order volume elements.

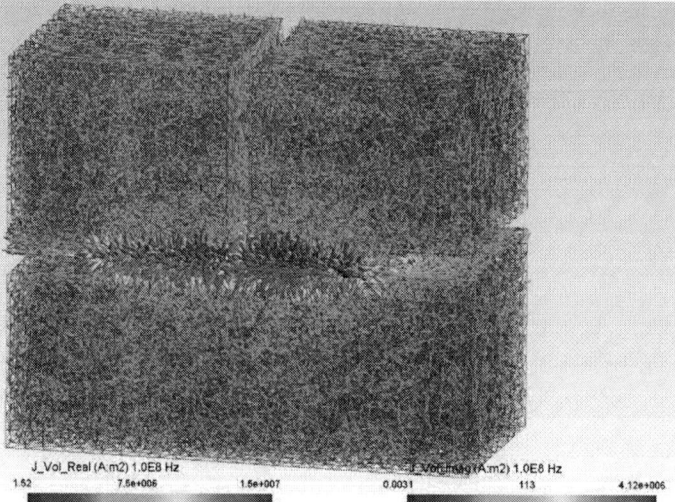

Fig. 6: Current path in the switching cell

The parasitic inductance obtained after simulation with Flux3D is 1.41nH. To carry out the simulation with MIPSE, the appropriate mesh is exported from Flux3D. The application is always magneto-harmonic and the injected current is 1A. The elements here are assumed to be in copper simply because that is the only available conductive material. This time the parasitic inductance obtained is 1.38nH that confirms the value obtained with Flux3D. Note that the simulation time is largely decreased using MIPSE compared with Flux3D, it will allow optimizing the geometry in the future.

The parasitic inductance of the simulated SiC cell is thus less than 1.5 nH while the parasitic inductance measured in [8] is 2nH. This is explained by the fact that in [8] the MOSFETs are packaged and thus the PCB is thicker that increases the stray inductance of the assembly.

The SiC switching cell shows the advantages that were expected from an electromagnetic point of view. The thermal behavior will now be studied.

Thermal behavior of the SiC switching cell

Thermal simulations are performed to estimate the junction to ambient thermal resistance of a MOSFET and the thermal coupling that would exist between the two MOSFETs of the switching cell. They are carried out with the Flotherm software. The simulated architecture is shown in Fig. 7 and Fig. 8. This assembly is fairly representative of the actual switching cell. The solder thickness of 200μm between the heatsink and the drain of the chip (fig. 7) is present in practice due to a too large margin taken by the subcontractor (CIBEL) in order to avoid the breakdown of the chip during the PCB fabrication. This thickness will be considerably reduced in the next phases of production. Contact resistances are not considered in these simulations. The thermal conductivity of the different materials used for the simulation are shown in Table I. The heat transfer coefficient varies between 500 and 1500 W.m-². K-1, these values are estimated from heat sink datasheets [8]. The dissipated power is 50W in MOS1 and 0W in MOS2. The average temperature of both MOSFETs is measured to determine the junction to ambient thermal resistance of MOS1 and the coupling thermal resistance between MOS1 and MOS2. The results are shown in Fig. 9.

Then MOS2 dissipates 50W and the other 0W to check if the heat is extracted in the same way. The thermal resistance value is slightly lower than that of the MOS1 and is 1.38K/W for a heat transfer coefficient of 1500 W.m⁻². K-1 instead of 1.46 K/W for MOS1. The coupling remains the same 0.074K/W.

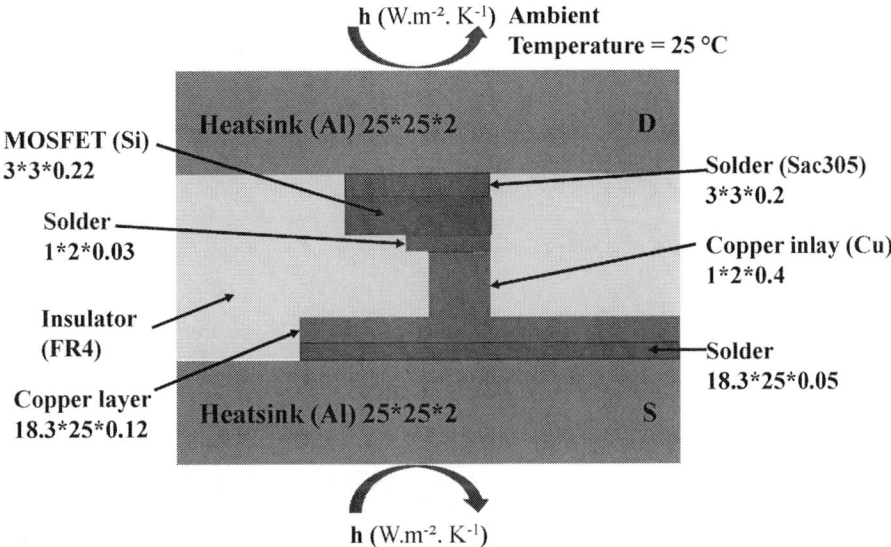

Fig. 7: Cross-section simulated for one MOSFET (dimensions are given in mm)

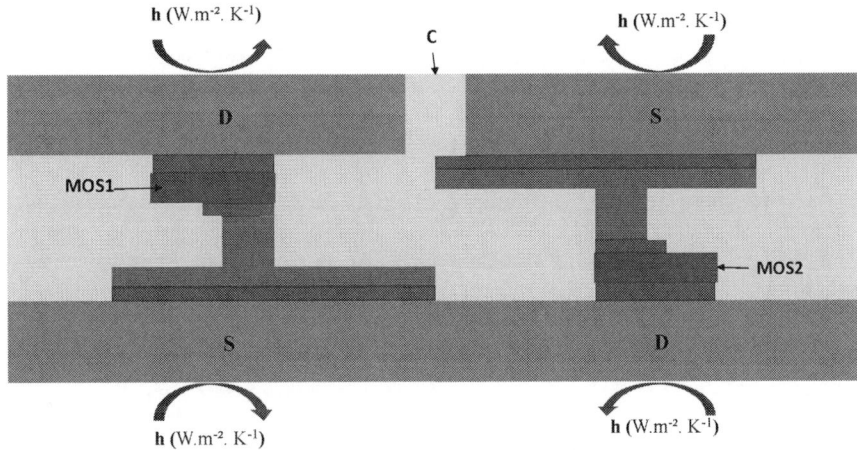

Fig. 8: Cross-section of the simulated switching cell

Table I: Thermal conductivity of materials

Material	Thermal conductivity W.m⁻¹. K⁻¹
Aluminum	201
SiC	350
Copper	385
Solder (SAC305)	58.7
FR4	0.5

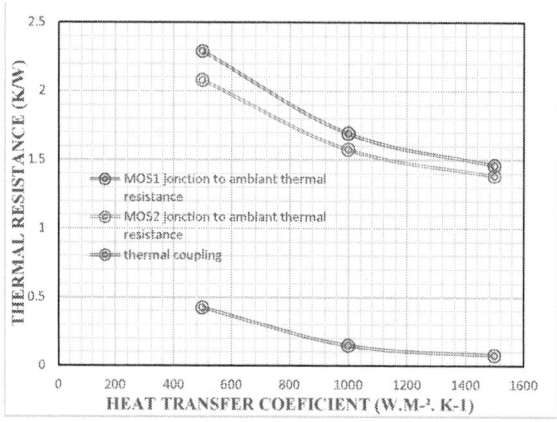

Fig. 9: Junction to ambient thermal resistance and thermal coupling as a function of the heat transfer coefficient

It is then interesting to analyze the temperature map inside the assembly to understand what can be done to improve its thermal behavior. For that purpose, Fig. 10 presents the evolution of the temperature along two paths: the first is perpendicular to the chip surface and crosses its center, the second is also perpendicular to the chip but crosses the center of the copper inlay. First it can be observed that the temperature is not uniform in the chip, the temperature being largely higher at the chip center than at the inlay center. It is something usual but that is accentuated here by the non-uniform dissipation area on the source side of the device. It is also observed that the temperature drop in the inlay is relatively high. It is due to its small cross section compared to the chip surface area. This problem induces a non-symmetrical heat flux diffusion inside the assembly: using Flotherm, it is estimated that about 2/3 of the total dissipated heat flows through the top heat sink (ΦD in the figure) and thus only 1/3 of the remaining heat goes through the bottom heat sink (ΦS in the figure). Therefore, a good way to improve the global thermal behavior should be to propose a solution for increasing the average cross section of the copper inlay to obtain a better heat repartition between the two faces of the chip.

Also, it is observed that there are large temperature drops in the heat sink base plates and between these base plates and the ambient temperature. This is the reason why the temperature evolution inside the assembly using a copper base plate was also shown. It can be clearly seen that the better thermal conductivity induces a lower temperature gradient and a better heat spreading (the temperature at the base plate surface is decreased). The last way to improve the thermal behavior would be to use a more performant heat sink to increase the global heat transfer coefficient.

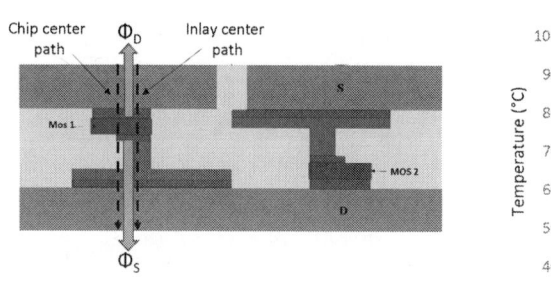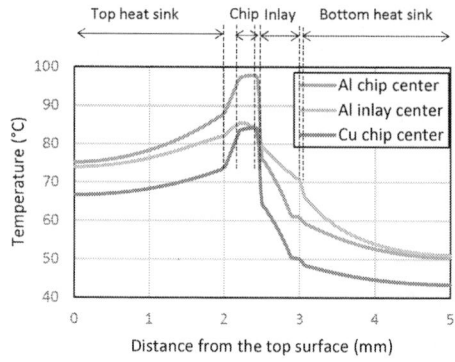

Fig. 10: Temperature evolution inside the assembly (dissipated power in MOS1 50W, h=1500 W.m^{-2}. K^{-1} and ambient temperature 25°C)

The thermal performances obtained are those expected. However, it would be possible to optimize the heatsink in order to have higher heat transfer coefficients and therefore a lower junction-to-ambient thermal resistance. The TAPIR SiC switching cell will then be compared with commercially available discrete SiC power modules and discrete components.

Benchmark of TAPIR SiC switching cell with commercially available power modules and discrete components

For this comparison power modules and discrete components using 1200V-80mΩ MOSFET SiC devices are chosen. It is obvious that the thermal resistance depends on the size of the devices and on the number of paralleled chips but it can even give fist ideas on the performance of the proposed package. Data are given in Table II.

Considering the stray inductance parameter, the TAPIR switching cell exhibits very interesting performances (1.5 nH) compared with classical power modules [12] and circuits using discrete devices with global stray inductances that are higher than 10 nH. Today commercial power modules are improved, by including decoupling capacitors for example [13], but such a low stray inductance is not reported. Note that if N devices are put in parallel in the TAPIR technology, the stray inductance is divided by N which is hardly obtained when implementing 2D packages.

Considering thermal aspects, it can be seen that the junction to ambient thermal resistance of the TAPIR switching cell is close to the junction to sink thermal resistance of power modules. That proves its very good thermal behavior. The use of better ceramic materials in power modules could reduce their thermal resistances but it has also to be noted that the thermal resistance of the TAPIR switching cell could be largely reduced by implementing better heat sinks. It was proved by Xu et al. [10] that obtained a junction to ambient thermal resistance of 0.5 K.W-1 with a geometry close to the TAPIR one. Finally, it is important to mention that the coupling thermal resistance of the TAPIR technology is very low especially when the heat transfer coefficient is high because the cooling is made at chip level. In the case of power modules, a thermal coupling exists through the power module substrate but also through the heat sink that induces more important values. For example, in [15], the thermal coupling resistance between devices was until 25 % of the junction to ambient thermal resistance using water cooling.

Table II: Electromagnetic and thermal performances benchmark

Components	Package	Stray inductance	Thermal resistance
TAPIR SiC Switching cell		Ls = 1.5 nH	Rth(j-a) = 1.4 K/W @ h=1500 W.m^{-2}. K Rth(j-a) = 2.3 K/W @ h=500 W.m^{-2}. K
SK25MH120SCTp, SEMIKRON [12]	Power module	Ls= 18 nH	Rth(j-s) = 1.52 K/W
10-PZ126PA080ME-M909F18Y Vincotech [13]	Power module	Not provided but includes a decoupling capacitor	Rth(j-s) = 1.79 K/W
C2M0080120D , CREE ([14]	TO247	Depends on the PCB layout (> 10 nH)	Rth(j-c) = 0.65 K/W

Conclusion

An SiC switching cell was designed and the structure of the related PCB was presented. The results obtained from the electromagnetic simulations show the quality of the switching loop and therefore makes this technology suitable for wideband gap components. The TAPIR SiC switching cell has also been evaluated from a thermal point of view. It has be seen that this technology is more efficient than those currently on the market. The TAPIR solution has a junction to ambient thermal resistance close to the junction to sink thermal resistance of some modules. The next step will therefore be to electrically and thermally characterize the SiC switching cell. The breakdown voltage, the common mode perturbations and the assembly of several switching cells to make a more complex topology will also be specific points to be addressed in the future.

References

[1] J. Broughton et al., "Review of thermal packaging technologies for automotive power electronics for traction purpose", Journal of Electronic Packaging; Vol. 140, n°4, pp. 1-11, 2018.

[2] H. Lee, V. Smet and R. Tummala, "A Review of SiC Power Module Packaging Technologies: Challenges, Advances, and Emerging Issues," in IEEE Journal of Emerging and Selected Topics in Power Electronics, vol. 8, no. 1, pp. 239-255, March 2020.

[3] M. O¨ stling, R. Ghandi, and C.-M. Zetterling, "SiC power devices present status, applications and future perspective," in Power Semiconductor Devices and ICs (ISPSD), 2011 IEEE 23rd International Symposium on. IEEE, 2011, pp. 10–15.

[4] M. Adamowicz, S. Giziewski, J. Pietryka, and Z. Krzeminski, "Performance comparison of sic schottky diodes and silicon ultra-fast recovery diodes," in Compatibility and Power Electronics (CPE), 2011 7th International Conference-Workshop. IEEE, 2011, pp. 144–149.

[5] P. Ranstad and H.-P. Nee, "On dynamic effects influencing IGBT losses in soft-switching converters," IEEE Transactions on Power Electronics, vol. 26, no. 1, pp. 260–271, 2011.

[6] G. Regnat et al., "Optimized Power Modules for Silicon Carbide mosfet," in IEEE Transactions on Industry Applications, vol. 54, no. 2, pp. 1634-1644, 2018.

[7] M. Delhommais, G. Dadanema, Y. Avenas, J. L. Schanen, F. Costa and C. Vollaire, "Using design by optimization for reducing the weight of a SiC switching cell," 2016 IEEE Energy Conversion Congress and Exposition (ECCE), Milwaukee, WI, 2016, pp. 1-8.

[8] W. F. Bikinga et al., "Low voltage switching cell for high density and modular 3D power module with integrated air-cooling," CIPS 2020; 11th International Conference on Integrated Power Electronics Systems, Berlin, Germany, 2020, pp. 1-6.

[9] O. S. Senturk et al., "Converter Structure-Based Power Loss and Static Thermal Modeling of The Press-Pack IGBT Three-Level ANPC VSC Applied to Multi-MW Wind Turbines", IEEE Transactions on Industry Applications, vol. 47, no. 6, pp. 2505-2515, 2011.

[10] Y. Xu, I. Husain, H. West, W. Yu and D. Hopkins, "Development of an ultra-high-density Power Chip on Bus (PCoB) module," 2016 IEEE Energy Conversion Congress and Exposition (ECCE), Milwaukee, WI, 2016, pp. 1-7.

[11] T. Le-Duc, G. Meunier, O. Chadebec and J. -. Guichon, "A New Integral Formulation for Eddy Current Computation in Thin Conductive Shells," in IEEE Transactions on Magnetics, vol. 48, no. 2, pp. 427-430, Feb. 2012

[12] www.semikron.com

[13] www.vincotech.com

[14] www.wolfspeed.com

[15] M. Dbeiss, Y. Avenas, "Power semiconductor ageing test bench dedicated to photovoltaic applications", IEEE Transactions on Industry Applications, vol.55, no. 3, pp. 3003-3010, 2019.

Effect of Unipolar and Bipolar SPWM on the Lifetime of DC-link Capacitors in Single-Phase Voltage Source Inverters

Silpa Baburajan, Saeed Peyghami
Aalborg University,
Pontoppidanstraede 101, Aalborg
East 9220, Aalborg, Denmark
Tel.: +45-22205664/+45-93562431
Email: sbb@et.aau.dk,
sap@et.aau.dk

Dinesh Kumar
Danfoss Drives A/S
Global Research and
Development Centre,6300
Gråsten, Denmark,
Email:
dineshr30@ieee.org

Frede Blaabjerg, Pooya Davari
Aalborg University,
Pontoppidanstraede 101, Aalborg
East 9220, Aalborg, Denmark
Tel.: +45-21292454/+45-31478845
Email: fbl@et.aau.dk,
pda@et.aau.dk

Keywords

<<DC power supply>>, << Pulse width modulation (PWM)>>,<< Frequency-Domain Analysis >>, <<Reliability>>,<< Voltage Source Inverters (VSI)>>

Abstract

This paper explores the impact of different modulation techniques on the lifetime of DC-link capacitors for different modulation indices and varying operating power of voltage source inverter system. The conducted analysis show that the capacitor lifetime improves when using unipolar Sinusoidal pulse-width modulation (SPWM), which introduces the highest lifetime for DC-link capacitors when a single unit voltage source inverter is used. Likewise, for two paralleled voltage source inverter unit' topology, both units using unipolar SPWM, having the same switching frequency produces the least thermal stress on the capacitor and consequently giving the highest lifetime.

Introduction

DC-link capacitor, also known as an intermediate circuit capacitor is one of the most crucial power conversion stages for several industrial applications such as adjustable speed motor drives, wind power systems, photovoltaic systems, and electric vehicles [1], [2]. The main functions of DC-link capacitors are to compensate the instantaneous power difference between the source (usually a rectifier bridge) and output load (usually an inverter whose mean value is constant in steady-state operation), absorb the high-frequency currents generated by inverter modulation technique, thereby preventing the flow of switching frequency current harmonics into the source and to smoothen voltage ripple superimposed on the DC-link voltage [3], [4]. Aluminum electrolytic capacitors (El-caps) are one of the most commonly used DC-link capacitors, due to their advantage of providing a high capacitance per unit volume at low costs compared to other types of capacitors. However, recent studies show that capacitors, in general, contribute to 30% of the total failure root causes in the power electronic systems [5] and they are considered as the most fragile components in a power electronic system [3]–[6].

Many studies [3], [7] have shown that for different pulse width modulation (PWM) techniques, the root means square (RMS) value of DC-link capacitor current only depends on the fundamental output current, modulation amplitude index and load power factor. This might be true from the analytical point of view. However, different modulation techniques affect the switching characteristics of semiconductor devices, resulting in a change of the harmonic spectrum of the DC-link capacitor [8]. Similarly, connecting multiple converters onto one common DC-bus affects the DC-link capacitor's harmonic spectrum. These research gaps are not well discussed in the existing literature. The capacitor's full current spectrum, which affects its lifetime is of paramount importance. Firstly, it affects the hotspot temperature (core temperature increases with increase in capacitor current) of the capacitor, which contributes to self-heating and thereby, decreasing the lifetime of capacitors [9]. Secondly, the power loss of the capacitor, which is a function of the current harmonic spectrum and its Equivalent Series Resistance (ESR) increases as the capacitor current increases. To have a more accurate representation of the actual current flowing through the capacitor, the harmonic spectrum of capacitor, which includes the DC component, carrier frequency, and all other harmonic currents are used in this paper to accurately size the DC-link capacitors of a power converter operating with a specific modulation method. Furthermore, the lifetime of DC-link capacitors is also evaluated using the capacitor current value calculated based on the entire capacitor current harmonic spectrum.

In this paper, the lifetime of DC-link capacitors is analysed with different modulation schemes such as unipolar and bipolar SPWM for a single inverter and then for two inverter systems sharing a common DC-bus, in order to find which of the modulation techniques produce the least capacitor ripple current at different output power levels and different values of the modulation amplitude index. The key feature of this paper is the lifetime estimation of the DC-link capacitor considering the DC-link capacitor ripple term, which includes the entire range of harmonic ripple currents. Based on the findings, this paper proposes modulation schemes that can improve the lifetime of the DC-link capacitor in multi converter systems.

System Description

To analyse the effect of using unipolar and bipolar SPWM technique on the lifetime of DC-link capacitor, a single-phase full-bridge, commonly known as H-bridge voltage source inverter (VSI) shown in Fig.1 is considered in this paper. The four switches $Q1, Q2, Q3, Q4$ of the VSI are controlled using the SPWM technique (explained in detail in [10]).

The specifications of the VSI are listed in Table I and output filter information is provided in Table II and the DC-link capacitor used in this study is obtained from the capacitor datasheet [11]. Fig. 1(a) shows the first topology, where a single VSI is connected to a DC-bus and Fig. 1(b) presents the second topology containing two VSI units connected to a common DC-bus. Both the topologies have a large DC-link bank ($C_{dc\ bank}$) with El-caps to limit the DC-link voltage fluctuation. In Fig 1(a), two parallel electrolytic capacitors are implemented in the $C_{dc\ bank}$, and in Fig 1(b), it contains four parallel capacitors. In this way, the RMS value of the high-frequency ripple current (I_{cap}) flowing through the DC-link bank is distributed within the capacitors, which reduces the ripple current stress on the individual DC-link capacitor. Hence, the heating and power losses within the DC-link capacitor reduces, which improve the lifetime of C_{dc} (single DC-link capacitor). In this power converter topology, a Proportional – Integral (PI) controller is employed to regulate the DC-link voltage (V_{dc}) at a rated value of 400V as shown in Fig. 1(c). A Proportional-Resonant (PR) controller in Fig. 1(d) is used to ensure that the output voltage is the rated $400V_o$. The specifications of PI and PR controller are given in Table III and Table IV.

Table I: Specifications of the output load

Parameter	Symbol	Value
Rated Power (kW)	P_o	2.5
Load (Ω)	R_L	20
Rated RMS Load Voltage (V)	$V_{o\text{-}rated}$	230
Load Frequency (Hz)	f_o	50
Rated DC-link Volatge (V)	$V_{dc\text{-}rated}$	400

Table II: Specifications of the output filter

Parameter	Symbol	Value
Inductor (H)	L_f	$1*10^{-3}$
Capacitor (F)	C_f	$1*10^{-6}$
Resistor (Ω)	R_f	0.1

Table III: Specifications of the PI-controller in Fig. 1c

Parameter	Symbol	Value
Propotional Parameter	K_P	2
Integrator Parameter	K_I	100

Table IV: Specifications of the PR- controller in Fig. 1d.

Parameter	Symbol	Value
Propotional Parameter	K_P	1
Resonant frequency	ω	$2\pi(50)$
Resonant Coefficient	K_R	50

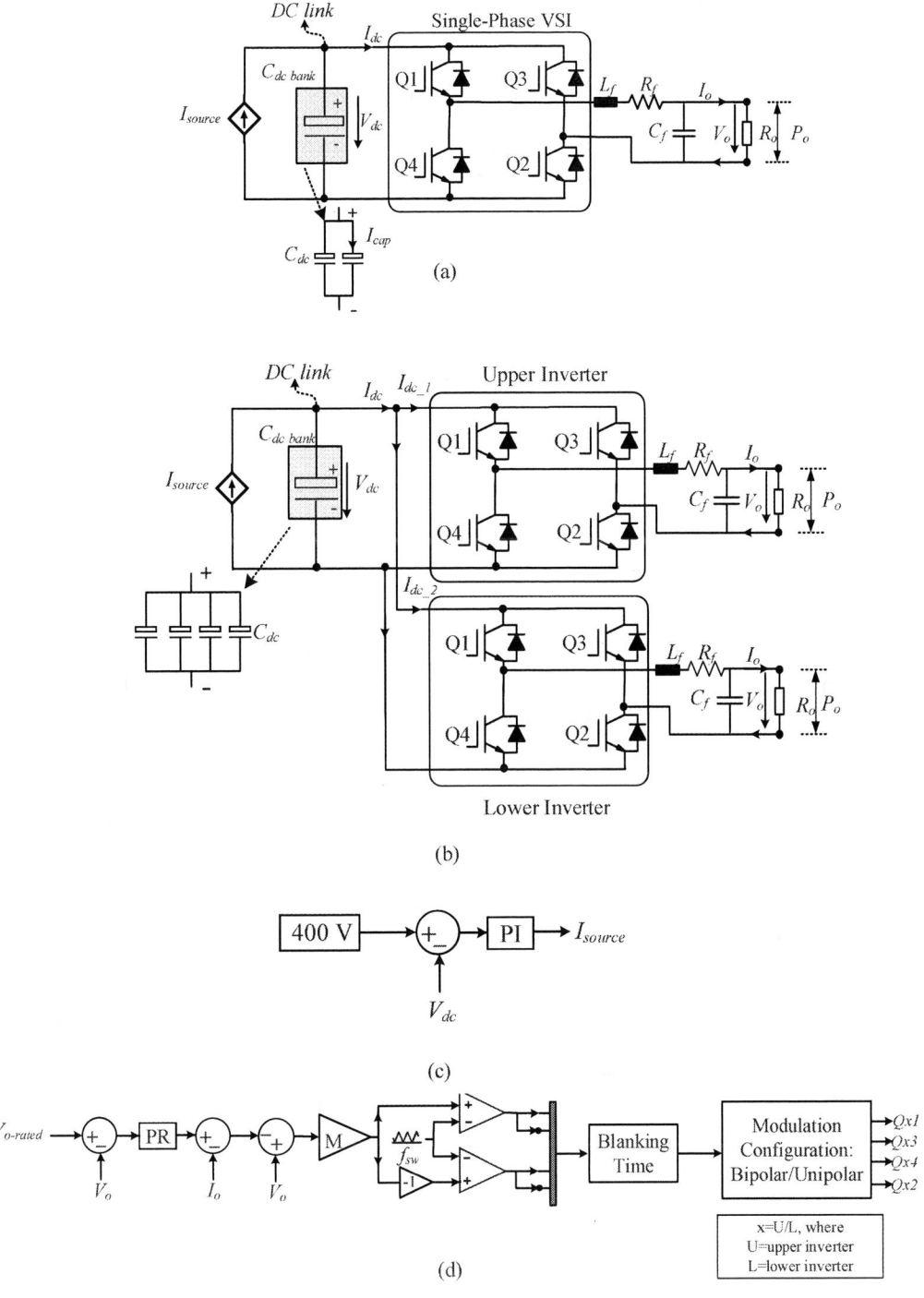

Fig. 1: Block diagram of (a) single phase VSI (b) two parallel VSI units connected to a common DC-bus (c) PI controller unit for input DC-link voltage (V_{dc}) control (d) PR controller unit for output voltage control using the different modulation techniques (Specifications: M=1/400).

Reliability Analysis of the DC-link Capacitor

The first step is to choose the C_{dc} based on the DC-link specifications (V_{dc}) and maximum allowed I_{cap}. The rated voltage and rated ripple current of chosen capacitor should be greater than the specified DC-link voltage and maximum I_{cap}. The next important step, while designing the DC-link capacitor bank is to determine if multiple capacitors are needed rather than connecting a single capacitor based on the required voltage rating and also to reduce the current stress on the C_{dc}. Finally, the electrical and thermal stress should be carefully evaluated for a precise capacitor lifetime estimation. In a VSI, high capacitance at the DC-link is required to suppress the high-frequency harmonic ripples from the switching of the inverters as well as the double-line frequency ripples. Electrical stress analysis and thermal stress analysis are given below [12].

Electrical Stress Analysis

The equivalent electrical circuit model and the thermal model of El-caps are shown in Fig.2(a). It consists of the ESR, ESL (Equivalent Series Inductance), and $R_{leakage}$ (connected in parallel with C_{dc} to provide a path for the flow of $I_{leakage}$). The ESL is not considered in this paper for calculations because the simulations are operated below the resonance frequency of the capacitor, where the inductive resistance of the winding and its terminals ($X_L = \omega ESL$) is considered to be very small and constant. Similarly, according to the datasheet [9], $I_{leakage}$ for the selected capacitor is very small (in values of μA).

$$ESR = R_o + R_d + R_e \qquad (1)$$

As explained in (1), the ESR of a capacitor, which is a sum of R_o (constant ohmic resistance of foil, connecting legs of capacitors, etc), R_d (frequency-dependent resistance of dielectric layer) and R_e (temperature-dependent resistance of electrolytic solution) varies with the frequency of ripple current (I_{cap}) as well the hot-spot (core temperature) of the capacitor. Therefore, it is necessary to consider the ESR to calculate the capacitor current ripple, which is a limiting factor of the capacitor lifetime as shown in (2)-(3).

$$I_{cap} = \sqrt{\left(I_{f1} \times K_{f1}\right)^2 + \left(I_{f2} \times K_{f2}\right)^2 + \ldots + \left(I_{fi} \times K_{fi}\right)^2} \qquad (2)$$

$$K_{fi} = \sqrt{\frac{ESR(f_i)}{ESR(f_{double})}} \qquad (3)$$

where I_{fi} is the RMS value of harmonic ripple current (calculated from the current harmonic spectrum) at frequency f_i. K_{fi} is the correction factor for I_{fi} at the respective frequency, which can be obtained from the datasheet of the specific capacitor used in this study. In (3), f_{double} represents the double-frequency or second harmonic. $ESR\ (f_i)$ and $ESR\ (f_{double})$ are the ESR of the ith frequency and at 100 Hz. For the capacitor chosen in this paper, the line fundamental frequency is 50 Hz. Hence, f_{double} =100Hz, from capacitor datasheet [11].

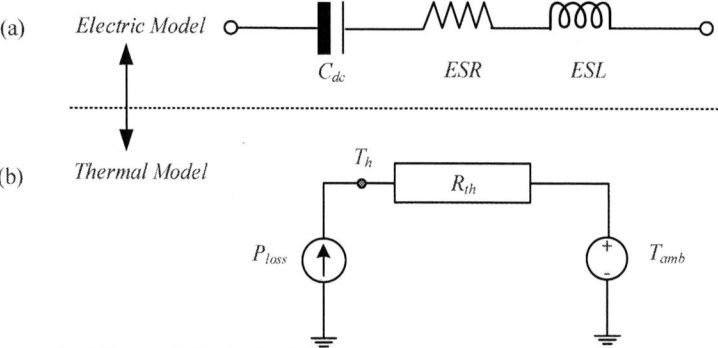

Fig. 2: Aluminium Electrolytic Capacitor (a) Electrical Model and (b)Thermal Model [13]

Thermal Stress Analysis

Besides the electrical stress, the thermal stress analysis is an equally critical stressor to capacitor wear out [14]. The thermal model of an electrolytic capacitor is shown in Fig. 2(b) and the thermal parameters, obtained from the datasheets are summarized in Table V. The core temperature is also known as the hot-spot temperature T_h which is determined by three factors [9] as shown in (4): T_a (operational temperature), P_{loss} (the power dissipation in the capacitor as shown in (5)), and the overall thermal resistance R_{th} between the capacitor core and the ambient air.

$$T_h = T_a - \left(P_{loss} \times R_{th}\right) \tag{4}$$

$$P_{loss} = \sum_{t=1}^{n} \left(I_{ft}\right)^2 \times ESR(f_t) \tag{5}$$

$$\Delta T_h = \left(P_{loss} \times R_{th}\right) \tag{6}$$

where ΔT_h is the rise in the core temperature of the capacitor. The above three equations show that an increase in the capacitor current increases both the power losses and capacitor core temperature within the capacitor. This will eventually degrade the capacitor reliability due to the loss of electrolyte, as described in previous sections.

Lifetime Estimation of the DC-link Capacitor

The lifetime of the El-caps, which includes the actual ripple current I_{cap} flowing through the capacitor can be estimated by using the (7) [15].

$$L_x = L_o \times \underbrace{2^{\left(\frac{T_r - T_a}{10}\right)}}_{K_T} \times \underbrace{\left(\frac{V_a}{V_r}\right)^{-p}}_{K_V} \times \underbrace{K_c^{\left[1 - \left(\frac{I_{cap}}{I_r}\right)^2\right] \times \frac{\Delta T_h}{10}}}_{K_{CR}} \tag{7}$$

where L_x and L_o are the estimated and rated lifetime of the capacitor. L_x depends upon the temperature factor K_T, voltage ripple factor K_V, and the current ripple factor K_{CR}. T_r is the rated upper category temperature, ΔT_h is the rated rise in the core temperature of the capacitor and T_a refers to the operating temperature. V_a and V_r are the actual operating voltage and rated voltage. Exponent p is a constant and depends on the ratio between V_a and V_r whereas parameter K_C is a temperature T_r dependent constant. In this paper, as discussed in [12], the values are taken as $p=5$ and $K_C=4$. According to (7), an increase in the capacitor current decreases the lifetime of the capacitor. The parameters of the chosen capacitor for this study, obtained from the capacitor's datasheet [11] are shown in Table V. I_r is the rated RMS value of capacitor current and I_{cap} is calculated using (2-3) [12].

Table V: Parameters of a Single DC-link Capacitor (BDK4345)

Parameter	Values
Single Capacitance	460 uF
ESR(f_{double})	0.18 Ω @ 100 Hz
ESL	20 nH
Thermal Resistance (R_{th})	5.74 °C /W
Rated Lifetime (L_o)	5000 hours @ T_r and I_r
Rated Upper Category Temperature (T_r)	105 °C
Rated Ripple Current (I_r)	2.54 A
Rated Voltage (V_r)	500 V
Operating Temperature (T_a)	55 °C
Rated core temperature increase (ΔT_h)	5 °C

Results and Discussion

To study the effect of the unipolar and bipolar SPWM technique on the lifetime of the DC-link capacitor, two scenarios are analysed in this paper. In the first scenario, the two topologies shown in Fig.1 (a) and Fig.1 (b) are studied under different output loading conditions, and in the second scenario, different modulation indices are considered. First, one single-phase VSI unit, as shown in Fig. 1(a) is simulated using different modulations for three cases (explained in Table VI). Next, two parallel single-phase VSI units connected to a common DC-bus as shown in Fig. 1(b) are simulated under four cases (as explained in Table VI). In the two-units system, both units are operated at the same loading conditions and are assumed to be synchronous. The modelling and simulations are carried out in *PLECS* to obtain the I_{cap} spectrum corresponding to each case. It is then exported to *MATLAB*, where the actual I_{cap} is calculated as explained in [6] from its spectrum. Finally, the L_x of the capacitor is estimated using (7) and the results are shown in Fig. 3 (a)-(d). It can be observed from Fig. 3 (a)-(b) that the lifetime of C_{dc} decreases with the increasing output power of the system. Furthermore, the lifetime of the C_{dc} increases with increasing M_a as it is shown in Fig.3 (c)-(d). The time-domain waveforms of the DC-link capacitor current are shown in Fig. 4, which illustrates the effect of different modulations on the I_{cap} ripples in a single unit and parallel-connected VSI system. The results from the simulations can be summarised as follows:

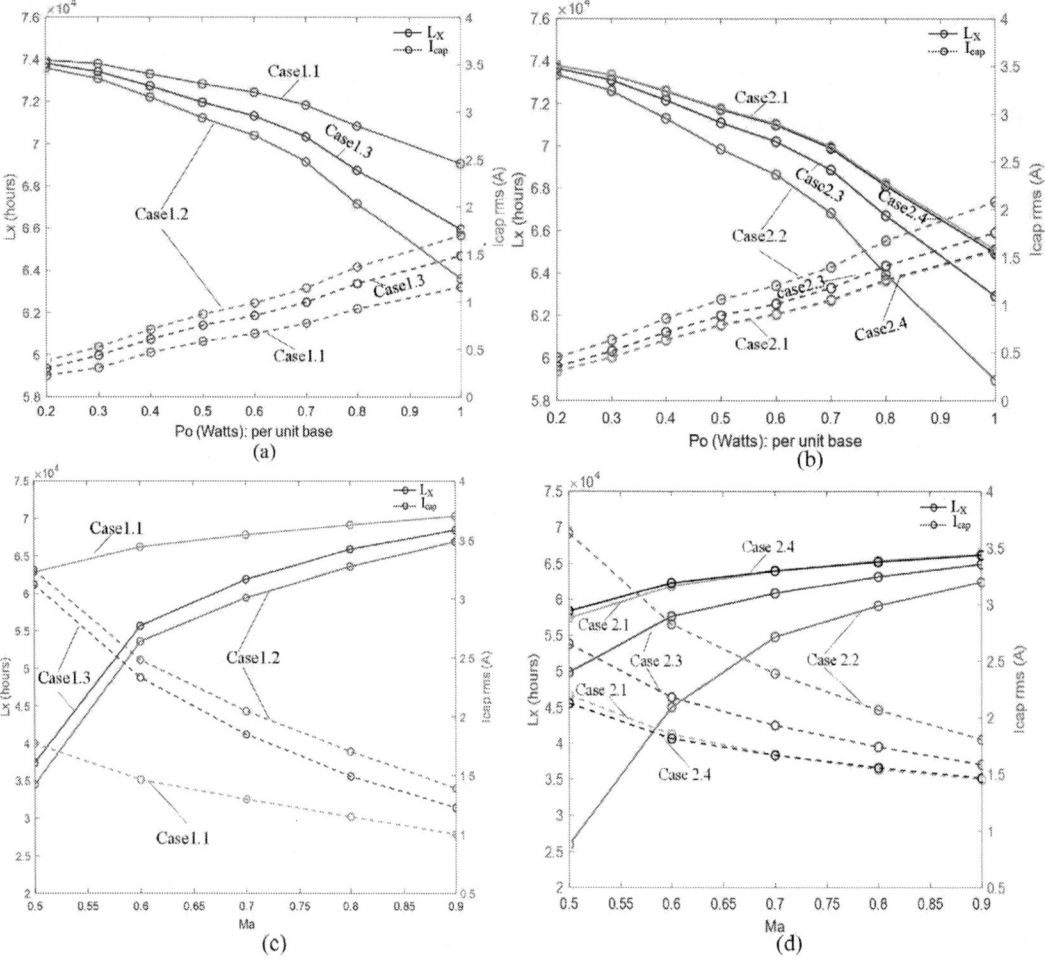

Fig. 3: Effect of modulation techniques on the lifetime of single DC-link El-cap, *under different output power conditions in*: (a) single VSI unit (Fig. 1(a)) (b) two parallel VSI units connected to a common DC-bus (Fig. 1(b)), and *under varying values of modulation index M_a in* : (c) single VSI unit (d) two parallel VSI units.

Table VI: Simulation case studies of systems in Fig. 1

Case	No. of Units	Modulation Type	
		Unipolar SPWM (unit #1)	Bipolar SPWM (unit #2)
1.1	1	✓ $f_{sw} = 20$ kHz	✗
1.2	1	✗	✓ $f_{sw} = 20$ kHz
1.3	1	✗	✓ $f_{sw} = 40$ kHz
2.1	2	✓ $f_{sw} = 20$ kHz	✗
2.2	2	✗	✓ $f_{sw} = 20$ kHz
2.3	2	✓ $f_{sw} = 20$ kHz	✓ $f_{sw} = 20$ kHz
2.4	2	✓ $f_{sw} = 20$ kHz	✓ $f_{sw} = 40$ kHz

Results of scenario 1-Different output power (P_o) conditions:

From the graphs in Fig. 3(a)-(b), it is observed that the current flowing through the DC-link capacitor increases with the increase in output power, thereby decreasing the lifetime of C_{dc}. Comparing the three cases shown in Fig. 3(a), the L_x of the DC-link capacitor is the highest in Case 1.1 when unipolar SPWM (with switching frequency $f_{sw} = 20$ kHz) is applied to a single unit VSI. The L_x of the capacitor in Case 1.2 and Case 1.3, which uses bipolar SPWM is less compared to when using unipolar modulation. This is because, as seen from the right y-axis of Fig. 3(a), I_{cap} flowing into the capacitor is less for unipolar SPWM, resulting in lower the electro-thermal stress on the capacitor than when using bipolar SPWM. Besides, it is visible from Fig. 3(a) that even though both Case 1.2 and Case 1.3 use bipolar SPWM, the L_x in case 1.3 (having $f_{sw} = 40$ kHz) is higher than in Case 1.2 (having $f_{sw} = 20$ kHz). The increase in the lifetime is because increasing the f_{sw} decreases the high-frequency current ripple components flowing into the capacitor, thus reducing the electro-thermal stress on C_{dc}.

Now, when two VSI units are implemented, it is expected from the results of single-unit topology that when both upper and lower units use unipolar SPWM modulation (at same f_{sw}), the L_x of C_{dc} will be the highest. Whereas, employing bipolar SPWM for both lower and upper VSI units will result in the lowest L_x. From the four cases shown in Fig. 3(b), it can be seen that the results match the expectation where Case 2.1 results in the highest L_x and Case 2.2 results in the lowest L_x of the DC-link capacitor. It is interesting to note that, Case 2.1 and case 2.4 results have almost the same lifetime for the C_{dc}. However, comparing Case 2.1 and Case 2.4, it is better to use Case 2.1 because of the lower switching frequency capacitor current ripple as shown in Fig. 4(c)-(d). This is due to the differences in the electro-thermal stresses on the capacitor which are a result of the current flowing into the capacitor.

Results of scenario 2-Different modulation amplitude indices (M_a):

For scenario-2 in Fig. 3(c)-(d), it is seen that the current flowing through the DC-link capacitor decreases with the increase in the modulation amplitude index. This is because the output voltage increases with the increase in modulation amplitude index, thereby decreasing the output current. Eventually, the I_{cap} decreases with the decrease in the output current, which increases the lifetime of C_{dc}. As the modulation amplitude index increases, for a single unit VSI (three cases implemented) it can be seen from the result in Fig. 3(c) that unipolar SPWM gives the highest lifetime for DC-link capacitors whereas bipolar SPWM produces the lowest L_x, as expected from the analysis of the scenario-1 results. Similarly, for the two VSI unit's topology, using Case 2.1 (both units having unipolar SPWM, having the same switching frequency) results in the highest lifetime for the C_{dc} as shown in Fig. 3(d). In contrast, the lowest L_x is observed when implementing Case 2.2.

Finally, as shown in Fig. 4(a)-(b), it can be concluded that unipolar SPWM gives the highest lifetime for DC-link capacitors when a single unit VSI is implemented. Likewise, for two paralleled VSI unit's topology, both units using unipolar SPWM, having the same switching frequency produces the least thermal stress on the capacitor and consequently giving the highest lifetime as shown in Fig. 4(c)-(d).

Furthermore, increasing the number of VSI units connected onto a common DC-bus increase the I_{cap}, thereby lowering the L_x of C_{dc}.

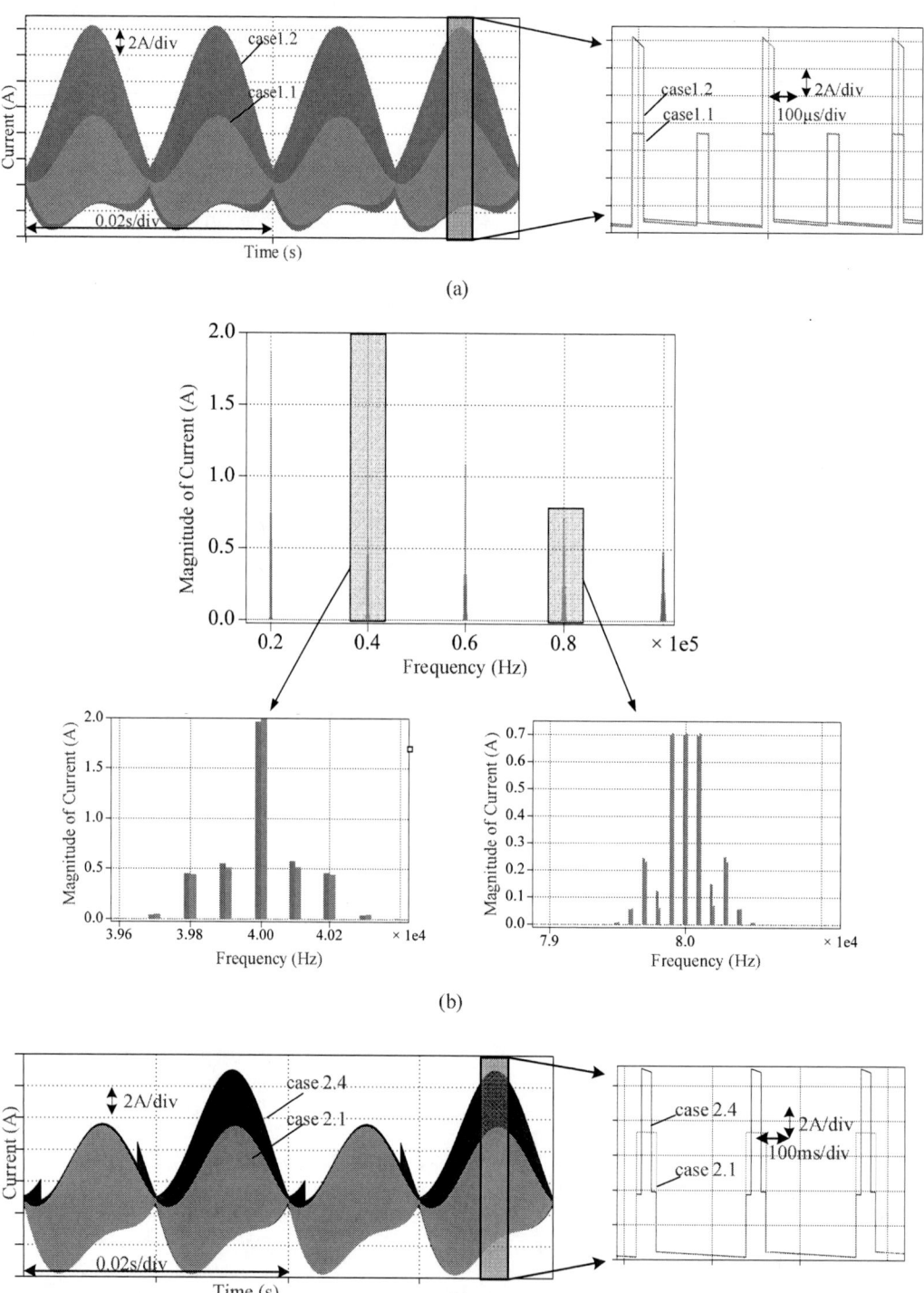

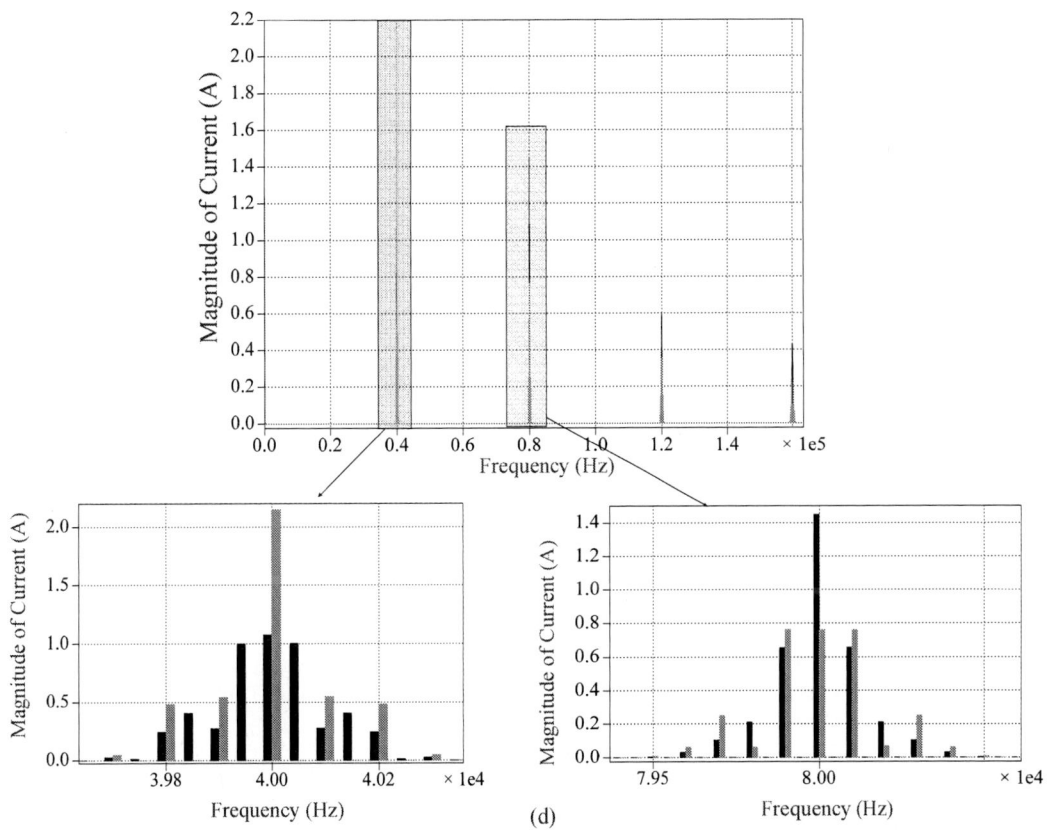

Fig. 4: At rated P_o = 2.5 kW and M_a=0.8 : (a) Time-domain waveforms of I_{cap} for Case 1.1 and Case 1.2 in a single unit VSI (b) FFT of I_{cap} for Case 1.1 and Case 1.2 in a single unit VSI (c) Time-domain waveforms of I_{cap} for Case 2.1 and Case 2.4 in parallel connected VSI units (d) FFT of I_{cap} for Case 2.1 and Case 2.4 in parallel connected VSI units.

Conclusion

This paper has explored the impact of different modulation techniques on the lifetime of the DC-link capacitor in a single voltage source inverter unit and two paralleled voltage source inverter unit topology. Two converter topologies having the same power ratings such as the same DC-link input voltage and rated output power were analysed for different cases. The obtained results indicate that the ripple/switching current components contained in the capacitor current depend on the number of VSI units implemented and the switching characteristics of the semiconductor switches in the inverter, which is controlled by the modulation technique and switching frequency. Hence, using different modulation techniques produces different harmonic currents in the I_{cap}, consequently having different impacts on the lifetime of the capacitors. The higher the value of I_{cap} is, the lower is the lifespan of the capacitor due to an increase in the core-temperature and power-losses within the capacitor. Therefore, in a power converter system, for an appropriate sizing and thermal design of the capacitor and thereby a more accurate estimation of the lifetime of a capacitor, the following factors are of equal paramount importance: the actual current (containing all harmonic ripple currents) flowing through the capacitor, the number of inverter units, switching frequency and the modulation technique which produces the lowest capacitor current ripple.

References

[1] S. Peyghami, P. Davari, H. Wang, and F. Blaabjerg, "The Impact of Topology and Mission Profile on the Reliability of Boost-type Converters in PV Applications," in *Proc. IEEE COMPEL*, pp. 1–8, 2018.

[2] H. Wang, P. Davari, D. Kumar, F. Zare, and F. Blaabjerg, "The impact of grid unbalances on the reliability of DC-link capacitors in a motor drive," in *Proc. EPE ECCE*, pp. 4345–4350, 2017.

[3] J. W. Kolar and S. D. Round, "Analytical calculation of the RMS current stress on the DC-link capacitor of voltage-PWM converter systems," *IEEE Trans. Ind. Electron.*, vol. 153, no. 4, pp. 535–543, 2006.

[4] H. Wang, "Capacitors in Power Electronics Applications-Reliability and Circuit Design," in *Proc. IEEE IECON Tutorial*, 2016.

[5] B. S. Gadalla, E. Schaltz, Y. Siwakoti, and F. Blaabjerg, "Analysis of loss distribution of Conventional Boost, Z-source and Y-source Converters for wide power and voltage range," in *Proc. IEEE EEEIC*, 2017.

[6] A. Tcai, I. M. Alsofyani, I.-Y. Seo, and K.-B. Lee, "DC-link Ripple Reduction in a DPWM-Based Two-Level VSI," *Energies*, vol. 11, no. 11, p. 3008, 2018.

[7] F. Renken, "The DC-link capacitor current in pulsed single-phase H-bridge inverters," in *Proc. EPE ECCE*, 2005.

[8] M. W. Bierhoff Friedrich Fuchs, "DC Link Harmonics of Three Phase Voltage Source Converters Influenced by the Pulse Width Modulation Strategy," *IEEE Trans. Power Electron.*, vol.55, no.5, pp.2085-2092, 2008.

[9] H. Wang, P. Davari, H. Wang, D. Kumar, F. Zare, and F. Blaabjerg, "Lifetime Estimation of DC-Link Capacitors in Adjustable Speed Drives under Grid Voltage Unbalances," in *IEEE Trans. Power Electron*, vol. 34, no. 5, pp. 4064–4078, 2019.

[10] S. A. Azmi, A. A. Shukor, and S. R. A. Rahim, "Performance Evaluation of Single-Phase H-Bridge Inverter Using Selective Harmonic Elimination and Sinusoidal PWM Techniques," in *Proc. IEEE PECON*, pp. 67–72, 2018.

[11] "Aluminum electrolytic capacitors Snap-in capacitors B43545-TDK," 2018. [Online, accessed 1 Sept. 2019]. Available: https://www.tdk-electronics.tdk.com › inf › aec › B43545.

[12] A. Albertsen, "DC-Link Capacitor Technology Comparison Aluminum Electrolytic vs. Film Capacitors," 2015.

[13] Y. Yang, K. Ma, H. Wang, and F. Blaabjerg, "Instantaneous thermal modeling of the DC-link capacitor in PhotoVoltaic systems,"*Proc. IEEE APEC*, 2015.

[14] H. Wang and F. Blaabjerg, "Reliability of Capacitors for DC-Link Applications in Power Electronic Converters—An Overview," *IEEE Trans. Ind. Appl.*, vol. 50, no. 5, pp. 3569–3578, 2014.

[15] Elna. Aluminium Electrolyte Capacitors Catalog- Technical Note. [Online, accessed 18 Sept. 2019]. Available: www.elna.co.jp › capacitor › alumi › catalog › pdf › tnot_al_e_p220-225

Transient thermal models of capacitors and inductors for system optimization

Vasilios Karaventzas[1,2], Juergen Biela[2], and Felix Rodriguez Mateos[1]

[1]Machine Protection and Electrical Integrity Group, TE Department, CERN Geneva
[2]Laboratory for High Power Electronic Systems, ETH Zurich
Email: vasilios.karaventzas@cern.ch

Keywords

≪Particle accelerator≫, ≪Superconductors≫, ≪Energy storage≫, ≪Modelling≫

Abstract

In accelerators using superconducting magnets, energy extraction systems (EES) are employed as means of protection in case of a quench, i.e. the spontaneous transition of the magnets from superconducting to resistive state. These EES operate only for short durations so that, for identifying an optimal design, the thermal transient properties/models of the system components are required. Therefore, a transient thermal model of cylindrical film capacitors based on physical properties is presented in detail in this paper. The model is based only on data sheet values and is validated experimentally. In addition, a transient thermal model of inductors is presented and validated by FEM simulations. Finally, it is shown that the power density of a converter for an energy extraction system could be more than doubled, if the presented transient thermal models of the passive components are used in the design optimization process.

1 Introduction

In proton accelerators, superconducting dipole magnets are used to force the proton particle trajectory to bend and follow the circular path of the accelerator. These magnets are capable of producing high magnetic fields and allow to reduce the operational losses. However, in the case of a quench, the stored magnetic energy can potentially cause severe damage due to excessive voltages and overheating. Currently, in such an event, resistors are inserted in series to the magnets for dissipating the energy.

For future accelerators using superconducting magnets an alternative approach for extracting the energy stored in the magnets in case of a quench is proposed in [1], which enables to recuperate the energy by an active energy extraction system (EES). Thus, a converter system which controls the voltage across the

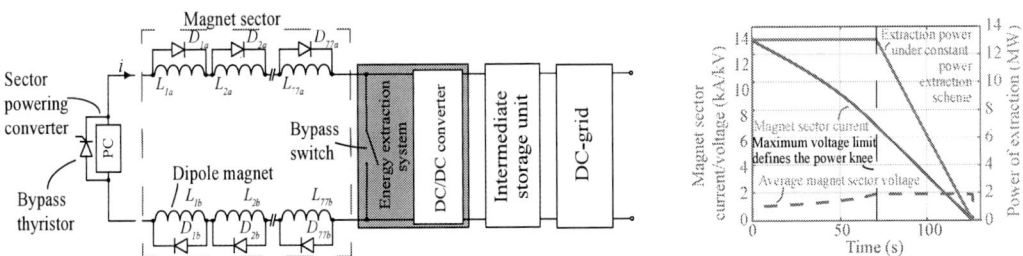

Fig. 1: Overview of main dipole magnet sector connected to an EES. Under normal operation, the bypass switch in EES is closed and in case of a quench, the bypass switch is opened and the EES transfers the energy to an intermediate storage unit [1].

magnets during the energy extraction process and which transfers the energy to a storage unit is proposed and the overview of such a system is given in Fig. 1. It allows to have full control of the extraction rate and the extraction process can be accelerated, compared to a purely passive extraction. With a faster current decay rate, the RMS current in the copper bus bars connecting the bypass diodes can be reduced by up to 15 % resulting in smaller bus bars. Additionally, the energy recuperation allows to reduce the operational losses and the environmental impact. Such a system operates in the exceptional case when a quench in the magnet occurs, i.e. a sudden transition from the superconducting to the normal state happens, so that the superconducting material becomes locally resistive.

For designing the active EES, an optimization method is developed for identifying the best converter topology and its design parameters for the best compromise between power density and efficiency. Usually, converter systems are designed to operate continuously or for a long period of time, so that the transient thermal characteristics of the components are not crucial for the design of the system and usually only the steady state thermal behavior is taken into account. In case of the energy extraction system, the system operates for a duration of less than 150 seconds as depicted in Fig. 1. Additionally, sufficient time passes for the temperature of the components of the converter to return to the ambient temperature before the next energy extraction happens, with the time required for recommencing the accelerator after a quench being relatively long. Thus, for identifying a cost effective system that occupies only a small volume, the thermal transient properties of the system components are crucial.

Regarding capacitors, the hot spot temperature is an important parameter for determining the maximum allowed RMS current in the capacitors and consecutively the number of capacitors needed to be connected in parallel. Therefore, many publications exist on the calculation of the hot spot to ambient thermal resistance (e.g. [2] - [4]), for steady state operation. Nevertheless, there are no publications investigating methods for calculating the transient response of the hot spot temperature of film capacitors based on physical properties of the capacitors. So far, there are only publications on modeling the transient thermal behavior of electrolytic capacitors. However, these models are fitted with experimental data and are not based on the physical properties/data sheet parameters and cannot be adapted for the case of film capacitors (e.g. [5], [6]). In the considered system the capacitors have to operate at voltages up to 2 kV and the RMS current is higher than 1 kA, so that only film capacitors are suitable for the active EES.

Regarding magnetic devices, there are many analytical thermal models of transformers considering the steady state behavior (e.g. [7], [8]) as well as the transient thermal response (e.g. [9]). Such models can be directly adapted to inductors. In this paper, such an adapted thermal model is presented, which is based on the model presented in [9] and which ignores any heat exchange to ambient. This model is also integrated in the inductor optimization for the pulsed operation.

The transient thermal models of film capacitors and inductors are presented in section 2. In section 2.1, the process for modeling the thermal behavior of cylindrical film capacitors is presented and in section 2.2, the process for thermal modeling of inductors is discussed. In section 3, the effect of the transient model of each component on the power density of the designed system is shown. Finally, section 4 summarizes the thermal models presented in the paper and their impact to the power density of the system's optimal solution.

2 Transient thermal models

The overview of the considered system is presented in Fig. 1 and this paper focuses only on the energy extraction system. The principle of operation of the EES can be divided in 4 stages: 1) In normal operation, the EES remains idle and the bypass switch is closed. 2) Once a fault is detected, the power converter (PC in Fig. 1) is turned-off and bypassed by the thyristor in parallel. Furthermore, the bypass switch is opened, so that the current commutes to the EES. 3) The EES controls its input voltage and transfers the energy of the magnet to the storage unit. 4) Finally, once the energy stored in the magnet sector is successfully extracted, the magnets are cooled down to enter once again the superconducting state and the stored energy can be reused by the accelerator.

The three main contributors to the volume of the EES are the heat sink for the semiconductors [1], the capacitors necessary for keeping the converter voltage ripple below a certain level and the inductors used as current filters at the output of the converter. The volume of these elements is mainly defined by the losses and the maximum allowed temperature. For the considered pulse system, the relatively short operating time allows the power density to reach high values that would not be possible for a continuous operation.

2.1 Capacitor transient thermal model

For film capacitors, it is important to keep the hot spot temperature below a certain level, in order to assure a long and reliable operation. For example, according to the capacitor data sheet of the considered TDK MKP type film capacitors [10], a FIT rate of <50 is possible when the hot spot temperature is kept below 70-75 °C at an operating voltage $<90\ \% \ V_{rated}$. For determining the hot spot temperature for pulse applications, the transient thermal model of the film capacitors used in the EES is derived. The process of thermal modeling is performed in 2 steps. First, a capacitor of the type B25620B1367K203 is cut in half and the thermal network is formulated by having access to the internal structure. In the second step, a thermal model for cylindrical MKP type capacitors is developed based only on the data provided on the data sheet.

For the development of the thermal model, it is assumed that the losses calculated are concentrated in the center of the film and not distributed along the film. Additionally, it is assumed that polyurethane is used as encapsulating material, as this is typically done in industry [10]. Due to its low thermal conductivity, various polyurethane composites are often used, which strive for a balance between increased thermal conductivity and adequate dielectric strength [11]. The exact properties of the encapsulating resin used in the studied capacitors is unknown, therefore the properties of pristine polyurethane (PUR) are used as these can be found in literature. Finally, the effect of the capacitor terminals to the thermal network is neglected due to the small area covered by the terminals.

Film capacitors consist of one film which is rolled and the film consists of 2 layers glued together as shown in Fig. 2(a). In the first step of thermal modeling, one layer with dimensions 66 mm x 157 mm was physically taken and its capacitance was measured at 34 nF, in order to identify the number of layers N_l in the B25620B1367K203 capacitor. From the total capacitance, the layer thickness has been determined: $d_l = \varepsilon_0 \varepsilon_{r,PP} \frac{A_l}{C_l} = 5.66\,\mu m$, where $\varepsilon_{r,PP} = 2.1$ is the relative permittivity of polypropylene [12], A_l is the area of the considered layer and C_l is the capacitance. The total number of parallel layers is $N_l = \frac{W_{Cap}}{d_l} = 7544$, where W_{Cap} is the total width of the rolled film as defined in Fig. 2.

Following this, the losses per layer $P_{l,n}$ are calculated and introduced in the thermal network. Assuming that the layers of the capacitor are concentric cylindrical shells with thickness d_l, the capacitance of each layer is calculated as a function of the radius of each layer (r_l in Fig. 2(a)), $C(r_l) = \varepsilon_0 \varepsilon_{r,PP} \frac{2\pi r_l h_l}{d_l}$. This means that the impedance of each layer $z_{l,n}$ is a function of the radius $z_{l,n} = R_{l,s} + \frac{1}{j2\pi f C(r_l)}$ and the fraction of the total current conducted by each layer is

$$\xi_{l,n} = \frac{I_{l,n}}{I_{IN}} = \left(|z_{l,n}| \sum_{n=1}^{N_l} |\frac{1}{z_{l,n}}| \right)^{-1} \quad n \in [1,\ N_l] \tag{1}$$

where h_l is the height of the film covered by aluminum coating, $R_{l,s}$ is the series resistance of each layer calculated as $R_{l,s} = N_l R_{s,total}$, f is the frequency of the current conducted by the capacitor, $I_{l,n}$ is the current conducted by each layer and I_{IN} is the total capacitor current. The equivalent series resistance of each layer is $ESR_l = R_{l,s} + \frac{tan\delta}{2\pi f C(r_l)}$ so that the losses per layer are $P_{l,n} = ESR_l I_{l,n}^2$.

The thermal resistance of a single layer can be divided into a radial R_{Rad} and an axial R_{Ax} component as shown in Fig. 2, since the metallic coating results in an anisotropic thermal conductivity in different directions [2], [3]. Assuming that the dielectric losses and the ohmic losses in the metallic coating are concentrated in the center of each layer, the radial and axial thermal resistances are R_{Rad} and R_{Ax} as shown in Fig. 2(c), with the axial thermal resistance being composed of the parallel connection between the polypropylene layer and the metallic coating thermal resistances.

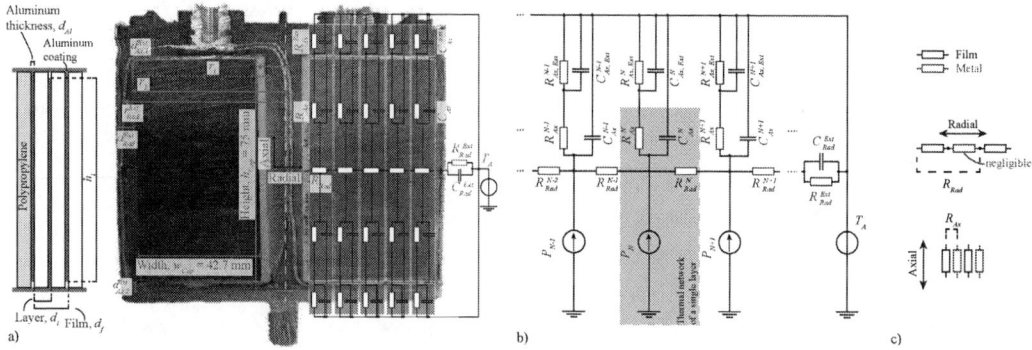

Fig. 2: a) Cross-section of a cylindrical film capacitor with the electrical equivalent thermal network. b) Thermal equivalent circuit of a cylindrical film capacitor. c) Anisotropic thermal resistances of a single layer in the radial and axial direction.

$$R_{Rad} = \frac{d_l}{\lambda_{PP} 2\pi r_l h_{Cap}} \quad (2) \qquad R_{Ax} = \left[\frac{4\pi r_l}{h_{Cap}} (\lambda_{PP} d_l + \lambda_{Al} d_{Al}) \right]^{-1} \quad (3)$$

Here, λ_{PP} is the thermal conductivity of polypropylene, h_{Cap} the total height of the film, λ_{Al} the thermal conductivity of the aluminum, and d_{Al} the thickness of the metalized coating, which is typically between 20 to 100 nm [13]. Since measuring the thickness of the aluminum coating is difficult, the thickness is determined by performing an optimization process for achieving a best fit of the model with the experimental data presented below. The best fit is achieved for a thickness of the metalized coating d_{Al} at 20 nm.

The resin encapsulation of the capacitor is hard polyurethane resin [10]. In the radial direction, a single thermal resistance R_{Rad}^{Ext} is assumed, connecting the film to the ambient. In the axial direction, the thermal resistance R_{Ax}^{Ext} connecting each layer to the ambient is assumed. Here, λ_{PUR} is the thermal conductivity of the polyurethane and the rest of the variables are defined in Fig. 2(a).

$$R_{Rad}^{Ext} = \frac{d_{Rad}^{Ext}}{\lambda_{PUR} 2\pi r_{Rad}^{Ext} h_{Cap}} \quad (4) \qquad R_{Ax}^{Ext} = \frac{(d_{Ax,1}^{Ext} + d_{Ax,2}^{Ext})}{2\lambda_{PUR} \pi [(r_l + d_l)^2 - r_l^2]} \quad (5)$$

With the aforementioned thermal resistances calculated, the circuit shown in Fig. 2(b) for N_l layers is formulated and numerically solved using PSpice. Initially, for evaluating the steady state performance of the model, only the resistive elements are considered and the hot spot temperature versus the radius is plotted in Fig. 3(a) for a nominal RMS current $I_{IN} = 70$ A and an ambient temperature $T_A = 25\,°C$ as calculated by the model as well as from FEM simulations. Thus, it can be observed that the model results in a higher steady state temperature than the one calculated by FEM simulations. In the model, the losses per layer are assumed to be concentrated at the center of the layer, so that for dissipation the complete length from the center of the layer to the ambient in the axial direction is considered. In contrast, in the FEM simulations the losses are homogeneously distributed in the complete volume of the film. The highest temperature is $T_{HS} = 51.5\,°C$, which is 18.6 % higher than the temperature calculated by the hot-spot thermal resistance given in the data sheet.

Afterwards, the thermal capacitance of the layers and the encapsulating resin are inserted in the thermal equivalent circuit as shown in Fig. 2(b). The thermal capacitance of each layer is calculated as $C_{Ax} = c_{PP} \rho_{PP} V_l$ where c_{PP} is the specific heat capacity of polypropylene, ρ_{PP} the polypropylene density and V_l the layer volume where $V_l = h_{Cap} \pi ((r_l + d_l)^2 - r_l^2)$. The thermal capacity of the metallic coating is 2 orders of magnitude smaller than that of polypropylene, due to the volume of the coating being 2 orders of magnitude smaller. Thus, its contribution to the total layer thermal capacitance is negligible and is omitted. The volume of the radial component of the polyurethane resin is $V_{Rad}^{Ext} = h_{cap} \pi \left[(r_2 + d_{Rad}^{Ext})^2 - r_2^2 \right]$. So, $C_{Rad}^{Ext} = \rho_{PUR} V_{Rad}^{Ext} c_{PUR}$ where c_{PUR} is the specific heat capacity

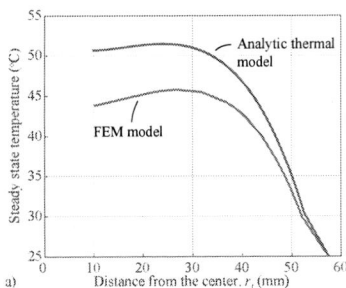 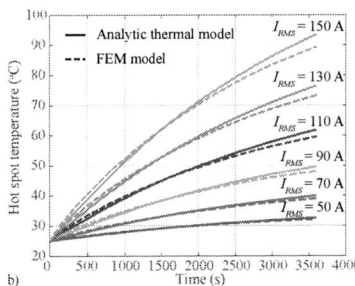

Fig. 3: a) Results obtained with the analytic thermal model and with FEM simulations of the steady state temperature along the middle of the capacitor as a function of the radius r_l. b) Transient response of the capacitor's hot spot temperature obtained with the analytic thermal model compared to values obtained with FEM simulations.

Table I: Material parameters [14], [15].

	Polypropylene (PP)	Polyurethane (PUR)	Aluminum (Al)
Thermal conductivity (λ)	Radial: $0.12\,\mathrm{W}/(\mathrm{m\,K})$ Axial: $\lambda_{PP} + \frac{d_{Al}}{d_l - d_{Al}}\lambda_{Al}$	$0.21\,\mathrm{W}/(\mathrm{m\,K})$	$205\,\mathrm{W}/(\mathrm{m\,K})$
Specific heat capacity (C_{th})	$2350\,\mathrm{J}/(\mathrm{kg\,K})$	$1800\,\mathrm{J}/(\mathrm{kg\,K})$	$915\,\mathrm{J}/(\mathrm{kg\,K})$
Density (ρ)	$946\,\mathrm{kg}/\mathrm{m}^3$	$1225\,\mathrm{kg}/\mathrm{m}^3$	$2698\,\mathrm{kg}/\mathrm{m}^3$

of polyurethane resin and ρ_{PUR} the polyurethane resin density. The volume of the axial component is $V_{Ax}^{Ext} = \frac{1}{2}\left(d_{Ax,1}^{Ext} + d_{Ax,2}^{Ext}\right)\pi\left[(r_l + d_l)^2 - r_l^2\right]$, so $C_{Ax}^{Ext} = \rho_{PUR}V_{Ax}^{Ext}c_{PUR}$. The material parameters used in the model are presented in Table I. The transient response of the model for the studied capacitor and for a range of RMS currents is presented in Fig. 3(b) and compared to transient FEM simulations.

In the second step of modeling, the thermal network of various capacitors is parameterized based only on the external dimensions found in the data sheets. There, a certain thickness for the polyurethane resin, i.e. d_{Rad}^{Ext} and d_{Ax}^{Ext}, is assumed so as no additional data to be required. It is expected that the resin thickness follows the graph shown in Fig. 4 for enabling both mechanical stability and dielectric strength. Nonetheless, as there is not sufficient data on the manufacturing process, it is assumed that the resin thickness is identical for all capacitors and equal to the capacitor studied during step 1. With these assumptions, the total film width w_{Cap} and height h_{Cap} can be determined by the external dimensions, thus the thickness of the layer d_l and the number of layers N_l are calculated and from there the process described above is followed. Consequently, a transient thermal model using only the data sheet values can be derived.

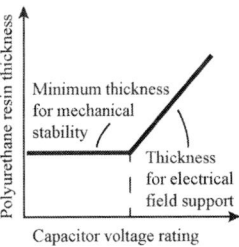

Fig. 4: Expected resin thickness in cylindrical capacitors.

For experimentally validating the model, a half bridge is used for generating a high ripple current in capacitors equipped with thermocouples as depicted in Fig. 5. The capacitors tested experimentally are listed in Table II. The capacitance values and voltage ratings have been chosen based on the first system optimizations and are the most suitable candidates. For generating the required RMS current, the switching frequency of the system was fixed at 3 kHz with a duty cycle of 50 % and the current was controlled by the input voltage. The current levels chosen for the thermal tests are 35 A RMS, 50 A RMS and 70 A RMS. The tests at 70 A were conducted only for 10 minutes, due to the high hot spot temperature of the inductor and therefore are not presented in the steady state comparison in Table II.

The fitness of the model to the experimental measurements is determined, initially, by the steady state temperature T_{SS}. In Table II, the temperature as calculated from the hot spot thermal resistance in the data sheet is presented together with the measured steady state temperature and the steady state temperature

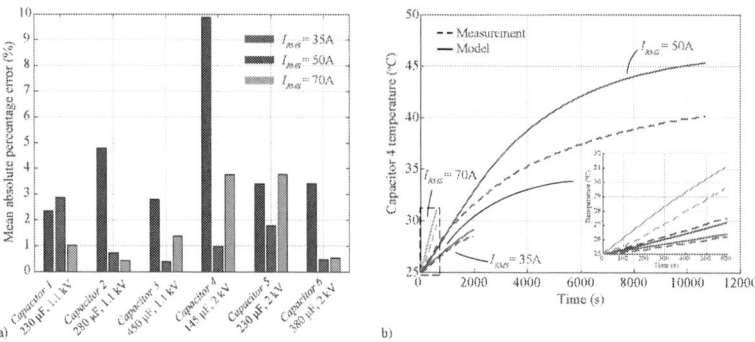

Fig. 5: Test set-up for experimentally testing the thermal transient characteristic of the capacitors.

Table II: Comparison of the steady state temperature obtained for I_{RMS} of 35 A and 50 A, from the data sheet, experimentally and by the discussed model.

	Epcos B25620*				
Type	Data sheet T_{SS}^{Data}	Experimental T_{SS}^{Exp}	$\frac{T_{SS}^{Exp}-T_{SS}^{Data}}{T_{SS}^{Data}}$	Model T_{SS}^{Model}	$\frac{T_{SS}^{Model}-T_{SS}^{Exp}}{T_{SS}^{Exp}}$
	Capacitor RMS current I_{RMS} = 35 A				
B25620B1237K101	31.9 °C	36.4 °C	14.1 %	36.0 °C	-1.1 %
B25620B1287K103	28.4 °C	28.6 °C	0.7 %	30.2 °C	5.6 %
B25620B1457K103	30.5 °C	32.2 °C	3.5 %	30.3 °C	-5.9 %
B25620B1147K981	31.1 °C	31.8 °C	2.3 %	35.5 °C	11.6 %
B25620B1237K983	28.5 °C	27.1 °C	-4.9 %	32.1 °C	18.5 %
B25620B1387K983	32.4 °C	35.2 °C	8.6 %	31.6 °C	-10.2 %
	Capacitor RMS current I_{RMS} = 50 A				
B25620B1237K101	40 °C	49.5 °C	23.8 %	45.4 °C	-8.3 %
B25620B1287K103	34 °C	34.5 °C	1.5 %	34.9 °C	1.2 %
B25620B1457K103	30.3 °C	34.9 °C	15.1 %	34.7 °C	-0.6 %
B25620B1147K981	36.9 °C	39.4 °C	6.8 %	45.4 °C	15.2 %
B25620B1237K983	33.7 °C	35.4 °C	5 %	38.7 °C	9.3 %
B25620B1387K983	32.5 °C	34.2 °C	5.2 %	37.1 °C	8.5 %

Fig. 6: a) Percentage of the fitness between the experimental data and the model. b) Behavior of the temperature in time as predicted by the model and measured experimentally for the worst case capacitor.

predicted by the model. It is observed that the deviation between the experimental results and the model is smaller than 20 %. A higher maximum deviation is observed between the experimental results and the values calculated from the hot-spot thermal resistance provided in the data sheet, indicating that a 20 % margin is well accepted also in the thermal models of the manufacturer.

Additionally, the time transient fitness of the model to the experimental data is evaluated. In Fig. 6, the experimental measurements are compared to the model predictions. A measure of how well the model fits the experimental data for different RMS current values is given on Fig. 6(a) based on the mean absolute percentage error calculated as

$$Fitness = \frac{1}{n} \sum_{i=1}^{n} \left| \frac{E_i - M_i}{E_i} \right| \tag{6}$$

where E_i is the ith experimental value, M_i the ith model value and n the number of measured/modeled points.

In Fig. 6(b) the case of Capacitor 4 (145 µF/2 kV) is shown in detail, as it was the one that showed the highest percentage of error. It is observed that the mean absolute error for all cases is no higher than 10 %. The higher temperature predicted by the model compared to the experimental can be explained by the assumption that the losses are concentrated in the center of the film (see Fig. 3). More importantly, for the initial few minutes, which are for the considered application the most crucial, the model closely follows the experimental results with an error of less than 5 %. It is therefore considered that the model can be used in the optimization algorithm [1] for determining the required capacitors for handling the short duration of high RMS ripple current.

2.2   Inductor transient thermal model

A relatively large core size is required for achieving a magnetic flux density of less than 80 % of B_{sat} for currents in the range of a few kA. Thus, the thermal capacitances of the core and the winding are expected to be large, resulting in a thermal time constant of the inductor far higher than the pulse duration. Therefore, no heat exchange with the ambient is considered in the transient thermal model of the inductor. This results in the thermal model shown in Fig. 7 which is based on the thermal model discussed in [7] and [9]. Initially, the winding losses are calculated taking the proximity effect, the skin effect and the DC losses [16] into account. The core losses are calculated with the improved generalized Steinmetz equation (iGSE) [17] by using the magnetic flux change rate in the core and the Steinmetz parameters for the core material.

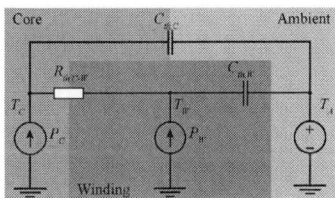

Fig. 7: Thermal equivalent circuit of an inductor.

The core and winding dimensions are parameters, which are optimized for minimizing the volume and the losses. Therefore, their exact dimensions are known and so, the elements shown in Fig. 7 can be calculated. $C_{th,W}$, the thermal capacity of the winding, is calculated as $C_{th,W} = c_{Cu}\rho_{Cu}V_W$, where c_{Cu} is the specific heat capacity of copper, ρ_{Cu} is the density of copper and V_W is the volume of the winding. $R_{th,C-W}$ represents the heat exchange through the gap between the winding and the core [1]. Finally, $C_{th,C}$, the thermal capacity of the core is calculated as $C_{th,C} = c_{core}\rho_{core}V_C$. The material properties of the considered core materials, i.e. Metglas 2605SA1, Vitroperm 500F and SiFe 3% can be found in [18].

For validating the presented thermal model, FEM simulations are performed. There, the iGSE method [17] is used in the simulations for calculating the core losses. In the model shown in Fig. 7, it is assumed that the relatively high thermal conductivity of the core results in an approximately homogeneous temperature within the core. In the FEM simulations, on the other hand, the finite thermal conductivity of the core is considered.

The optimal operating parameters in the design space are chosen based on the Pareto plots shown in Fig. 11. For reducing the computational effort for calculating the Pareto plots, the extraction/operation time of the EES is divided into $N_I = 15$ time intervals, in order to approximate the continuous operation. At the beginning of each time interval, the inductor current $i_L(t)$ is sampled as shown in Fig. 8 and for each time interval the losses are calculated for this sample current. Analogously, the piecewise linear inductor current for one switching period at the beginning of each time interval is used in the FEM simulation to calculate the losses for the complete time interval. This results in the loss step plot shown in Fig. 9. The algorithm for performing the FEM simulation is given in Fig. 8.

The analytically calculated core losses and the average losses of all the points in space calculated by the FEM simulation are depicted in Fig. 9(a). The resulting temperature rise of the analytical method compared to the FEM simulation is shown in Fig. 9(b), showing a good correlation both for the losses and the resulting temperature rise between the 2 methods. The higher temperature calculated by the FEM simulations can be attributed to edge effects resulting in localized points of higher magnetic flux density

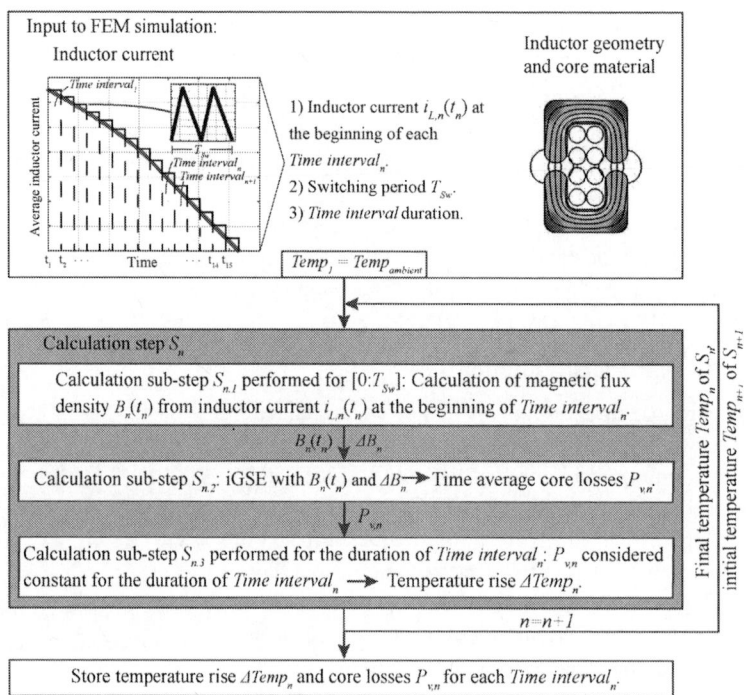

Fig. 8: FEM simulation algorithm of inductor core losses and temperature rise.

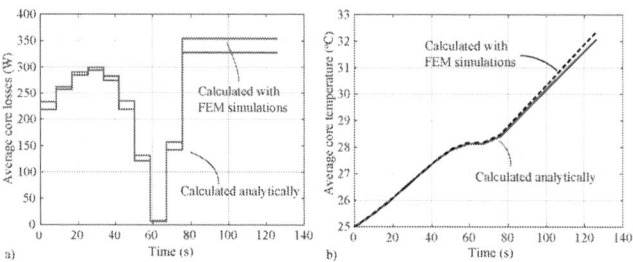

Fig. 9: a) Results obtained with the analytic thermal model and with FEM simulations of the average core losses of every point in space. b) Average core temperature of all the points in space obtained with FEM simulations compared to the values obtained with the analytic thermal model.

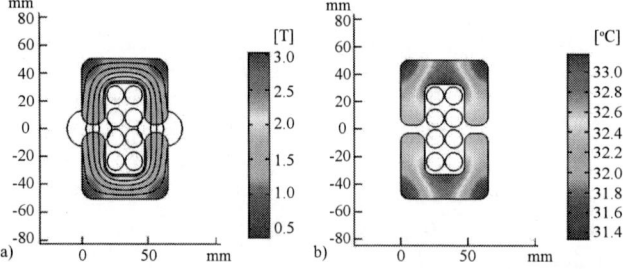

Fig. 10: a) Distribution of the magnetic flux density in the core for maximum winding current $I_w = 1845A$. b) Distribution of the temperature in the core in the final time step of the simulation, $t = 126.15s$.

and thus higher local losses. The contribution of the edge effect to the losses and the heat distribution in the core can be seen in Fig. 10. It is observed that the magnetic flux density and consecutively the temperature on the outer edges of the core is below the average value, whereas the maximum flux

Table III: Characteristics of the core material and its dimensions

Core characteristics			
Core material: SiFe 3%		Core dimensions	
Density (ρ)	$7530\,\mathrm{kg/m^3}$	Width	130 mm
Specific Heat Capacity (c_p)	$536\,\mathrm{J/(kg\,K)}$	Height	100 mm
Thermal conductivity (λ)	$19.6\,\mathrm{W/(m\,K)}$	Depth	130 mm
Relative permeability (μ_r)	10^4	Central leg width	35 mm

density and temperature is observed on the core edges closest to the winding. Such edge effects are not considered in the analytical method, where a homogeneous magnetic flux density is assumed for the complete core. Nevertheless, a good average matching is achieved and, moreover, the difference between maximum and minimum temperature throughout the complete core volume calculated by the FEM simulations is $\Delta T = 1.7\,^\circ\mathrm{C}$. The characteristics of the core material simulated and its dimensions are presented in Table III.

3 Design optimization

The design resulting from the optimization algorithm and the impact of the transient thermal models on the power density is presented in Fig. 11. With steady state thermal models, the power density of the optimal solution is $27.2\,\mathrm{kW\,dm^{-3}}$. By introducing a transient thermal model for the heat sinks, an increase of the power density to $28.5\,\mathrm{kW\,dm^{-3}}$ and a reduction of the total heat sink volume from $130.9\,\mathrm{dm^3}$ to $110\,\mathrm{dm^3}$, results. Furthermore, by introducing an inductor transient thermal model, the power density could be increased to $34.4\,\mathrm{kW\,dm^{-3}}$, and the volume occupied by the inductors decreases from $91.4\,\mathrm{dm^3}$ to $20.7\,\mathrm{dm^3}$. Finally, considering also the transient thermal model of the capacitors, the power density of the optimal solution increases to $61.2\,\mathrm{kW\,dm^{-3}}$ and the volume occupied by the capacitors drops from $246.4\,\mathrm{dm^3}$ to $67.1\,\mathrm{dm^3}$.

The complete optimization algorithm is described in [1]. In Fig. 12, a detailed view on the optimization procedure for determining the best designs for inductors and capacitors based on their transient thermal models is provided. Regarding the inductor design, initially the core geometries are determined and the the number of turns are calculated. The peak current is determined and the air gap is calculated for achieving a magnetic flux density of less than 80 % of B_{sat}. Then, the winding type (litz wire or foil) and its dimensions are iterated. With defined geometrical properties of the inductor, the average inductor current is divided into 15 intervals as shown in Fig. 12(a). There, the core and winding losses for each interval are calculated and with the thermal model the maximum temperature is determined. The solutions that result in a maximum core temperature below $110\,^\circ\mathrm{C}$ and a winding temperature below $120\,^\circ\mathrm{C}$ are stored.

For the case of the capacitor design, the operation time of the converter is divided again into 15 intervals. For each interval the instantaneous currents for each capacitor are deduced and their harmonic

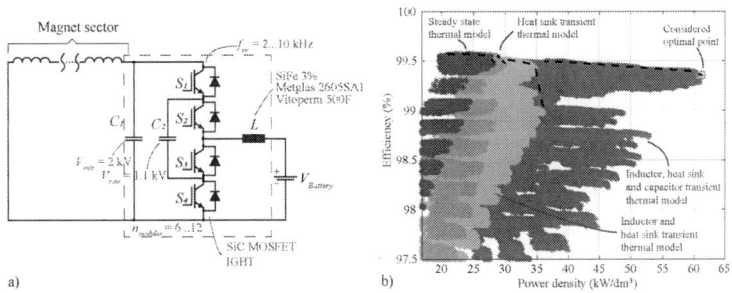

Fig. 11: a) Topology and parameters used in the optimization algorithm. b) Comparison of the impact of the transient thermal models on the power density of the designed systems. The power density is calculated as *(Max. extraction power × System efficiency)/Pure component volume.*

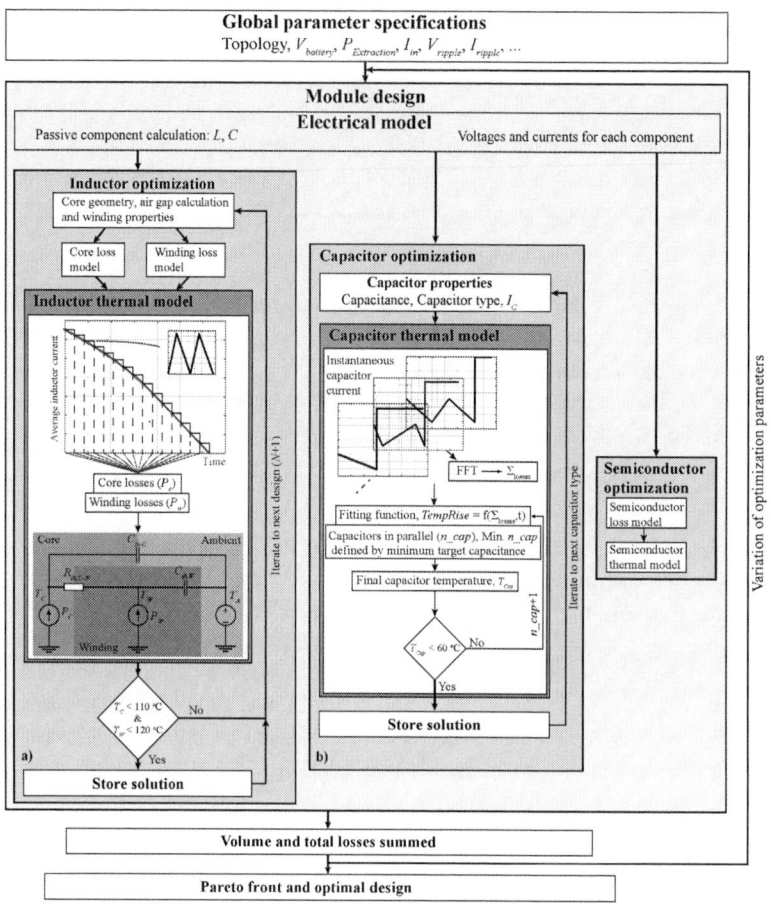

Fig. 12: The optimization algorithm: a) Algorithm for calculating the maximum temperature rise in the inductor. b) Algorithm for calculating the maximum temperature rise in the capacitor and the required number of capacitors placed in parallel.

components extracted. The losses are calculated for each harmonic component and the sum of losses per capacitor is determined. For accelerating the optimization procedure, a fitting function of the temperature rise versus time and losses, $TempRise = f(\sum_{losses}, t)$, is calculated for each capacitor type, from the transient thermal model discussed above. The minimum number of capacitors in parallel for each capacitor type is determined by the target capacitance, defined by the maximum allowed voltage ripple, i.e. 5 %. The maximum temperature of the capacitors is calculated and if the temperature is below 60 °C the solution is stored, else the number of capacitors placed in parallel is incremented. This procedure is followed for all the capacitor types considered in the optimization algorithm.

4 Conclusion

In this paper, transient thermal models of the passive elements of the EES converter are developed, in order to fully utilize the thermal capacity of the components and maximize the power density. The developed transient thermal model of cylindrical film capacitors is validated experimentally and shows an error of less than 20 % compared to the measured values. Also, the given model for an adiabatic heating of inductors is validated with FEM simulations and shows no significant deviation (less than 1 %). The transient thermal models are implemented in an optimization algorithm and the optimal solution shows a volume reduction of 56.3 % compared to the solution considering only steady state thermal models. With this, the power density of the converter increases from $27.2 \, \mathrm{kW \, dm^{-3}}$ to $61.2 \, \mathrm{kW \, dm^{-3}}$, without significant reduction of the system efficiency.

References

[1] V. Karaventzas, J. Biela, and F. Rodriguez Mateos, "Optimized design of a novel energy extraction system for superconducting magnets in future particle accelerators," in *Proc. of the European Conf. on Power Electronics and Applications (EPE)*, 2019.

[2] M. H. El-Husseini et al., "Thermal optimization of metalized polypropylene film capacitors," in *Proc. of the IEEE Industry Applications Conf.*, 2000.

[3] Zhiwei Li et al., "Temperature rise of metallized film capacitors in repetitive pulse applications," *IEEE Plasma Science*, vol. 43, no. 6, pp. 2038–2045, 2015.

[4] Se-Hee Lee et al., "Local heat source approximation technique for predicting temperature rise in power capacitors," *IEEE Trans. Magnetics*, vol. 45, no. 3, pp. 1250–1253, 2009.

[5] T. Furukawa, D. Senzai, and T. Yoshida, "Electrolytic capacitor thermal model and life study for forklift motor drive application," in *Proc. of the World Electric Vehicle Symposium and Exhibition*, 2013.

[6] P. Freiburger, "Transient thermal modeling of aluminum electrolytic capacitors under varying mounting boundary conditions," in *Proc. of the Int. Workshop on Thermal Investigations of ICs and Systems*, 2015.

[7] M. Jaritz, and J. Biela, "Analytical model for the thermal resistance of windings consisting of solid or litz wire," in *Proc. of the European Conf. on Power Electronics and Applications (EPE)*, 2013.

[8] M. Mogorovic, and D. Dujic, "100 kW, 10 kHz medium-frequency transformer design optimization and experimental verification," *IEEE Trans. on Power Electronics*, vol. 34, no. 2, pp. 1696–1708, 2019.

[9] I. Villar et al., "Transient thermal model of a medium frequency power transformer," in *Proc. of the Annu. Conf. of IEEE Industrial Electronics (IECON)*, 2008.

[10] TDK, "Power Electronic Capacitors," *MKP DC data sheet*, May 2018.

[11] Jin-Chao Zhao et al., "Thermal conductive and electrical properties of polyurethane/hyperbranched poly(urea-urethane)-grafted multi-walled carbon nanotube composites," *Composites Part B: Engineering*, vol. 42, no. 8, pp. 2111–2116, 2011.

[12] K. Kannus et al., "Electrical properties of polypropylene and polyaniline compounds," in *Proc. of the Int. Conf. on Solid Dielectrics (ICSD)*, 2004.

[13] C. W. Reed, and S. W. Cichanowskil, "The fundamentals of aging in HV polymer-film capacitors," *IEEE Trans. Dielectr. Electr. Insul.*, vol. 1, no. 5, pp. 904–922, 1994.

[14] E. V. Thompson, "Thermal Properties," in *Encyclopedia of Polymer Science and Technology*, American Cancer Society, 2010.

[15] E. V. Thompson, "Thermal conductivity of nonporous polyurethane," in *Proc. of the European Conf. on Thermophysical Properties*, 1999.

[16] J. Muhlethaler, J. W. Kolar, and A. Ecklebe, "Loss modeling of inductive components employed in power electronic systems," in *Proc. of the Int. Conf. on Power Electronics (ECCE Asia)*, 2011.

[17] K. Venkatachalam et al., "Accurate prediction of ferrite core loss with nonsinusoidal waveforms using only steinmetz parameters," in *Proc. of the IEEE Workshop on Computers in Power Electronics (COMPEL)*, 2002.

[18] M. S. Rylko et al., "Magnetic material comparisons for high-current gapped and gapless foil wound inductors in high frequency DC-DC converters," in *Proc. of the Int. Power Electronics and Motion Control Conf. (PCIM)*, 2008.

Energy Management for isolated renewable-powered microgrids using reinforcement learning and game theory

Rui Hu, Alexis Kwasinski
UNIVERSITY OF PITTSBURGH
3700 O'Hara St
Pittsburgh, USA
Tel.: +1 – 412.383.6744
E-Mail: ruh14@pitt.edu, akwasins@pitt.edu

Keywords

«Microgrid», «Energy system management», « Renewable energy systems», «Smart grids», «Machine learning».

Abstract

This paper presents a decentralized energy management system (EMS) solving for optimal load-response strategy applying reinforcement learning (RL) and game theory for islanded renewable-powered microgrids. The EMS enables the consumers in a microgrid to independently evaluate the tradeoff between satisfying load demand and maintaining sufficient stored energy to make load-response decisions correspondingly. The evaluation and decision-making process consists of two parts: an instant virtual two-player load-response game and a long-term linear-reward inaction (LR-I) learning process adjusting consumer power/load models. The virtual two-game solving process is an instantaneous decision-making system so that the consumers could make real-time decisions, while the LR-I process gradually improves the consumer payoff based on the system feedback during the operation. Simulation of a microgrid powered by PV cells and battery banks is conducted to evaluate the EMS performance. It is shown that the game-learning EMS has a better performance compared to both the direct virtual two-player game and the naïve LR-I approach. Additionally, compared to the naïve LR-I approach, the proposed game-learning algorithm has a faster converging-speed.

Introduction

As the concerns over the global environment and energy efficiency rise, alternative solutions to fossil fuel energy, such as renewable sources and related supporting microgrid systems, have been attracting considerable interest over the past two decades. Different from a conventional predominately centralized power grid with relatively few and large power plants, renewable power sources are often many and with relatively small power generation sites. Hence, microgrids may usually be better options to manage the distributed energy resources (DER), so their power is managed at a local level with relatively close consumers and likely energy storage devices. Additionally, microgrids are able to operate fully autonomously, both connected and disconnected from other grids [1]. Examples of existing microgrids include Ta'u island microgrid, Hartley Bay, and Isle of Eigg [2]–[4]. Microgrids present solutions to areas like these ones that are often isolated from the main power grid. Commonly, microgrids rely heavily on renewable power sources such as wind turbines and photovoltaic (PV) arrays. Additionally, because microgrids can operate autonomously, power-sensitive facilities such as hospitals, communication systems, and manufacturing factories are increasingly considering the use of microgrids to enhance their system resiliencies against natural disasters and other disruptive events causing extensive and long power outages. [5]–[7]. As a result, it can be expected that the penetration of renewable resources and microgrids will grow in the future with more consumers and industries are relying on microgrids for powering their facilities.

A main requirement in a microgrid is maintaining the power-load balance. This condition needs to be satisfied at all times. However, due to the varying power output of renewable power sources, the presence of energy storage devices is deemed to be indispensable. As a result, the energy management system (EMS) is required for the microgrids to control its stored energy and interface with the power network. A microgrid EMS is responsible for multiple tasks, including:

1. Monitoring power generation status, load demand profiles, energy storage devices status, and stored energy level;
2. Estimating system status and making power/load forecast;
3. Planning and conducting power/load-response to satisfy load needs.

Usually, these processes involve solving specific optimization problems. There have been various approaches focusing on computing the optimal energy management strategy for a microgrid in different scenarios. Analytical approaches, such as exhaustive search and linear programming, have been applied to microgrid design, battery scheduling, and power planning [8]–[10]. These approaches yield an optimal power budget given a well-defined environment. However, these algorithms face challenges when the system parameters are ill-defined/unknown or changing over time. An analogous example is an electricity market in which the bidders do not have the full knowledge of the market, and the behavior of each participant impacts other bidders' profits. To better address problems as such, multi-agent optimization approaches such as game theory and reinforcement learning (RL) algorithms have been introduced to tackle the optimal strategy problem in related research areas [11]–[15].

Game theory studies the behavior of rational decision-makers in a multi-agent system in which the agents' payoffs are at least partially dependent on the others' decisions. This theory proposes that the potential solution of a multi-agent system is found in the form of equilibriums, where no agent could gain higher payoff by single-sided deviating from it [16]. Equilibriums (in terms of the mixed strategies) have been proven to be existing in all finite games, which is deemed to be the optimal solution for all rational agents [17], [18]. However, the equilibrium(s) in a general n-player game has yet to be proven that it can be solved effectively, while all the present solving techniques empirically require a computation time/space complexity that is at least exponential to the number of variables in the game [19], [20]. Therefore, when applied to a system with a large number of agents, the optimal decision might take a long time to be computed.

The other approach, reinforcement learning, stemmed from the learning behavior of living creatures, has been an attractive research area as part of current machine learning focus. The mechanism of RL is similar to that of an adaptive controller because both of them utilize observed feedback from the environment (or the plant, in the control engineering realm) and adjust the agent's (controller) parameters/strategies to obtain specific objectives [21]–[23]. Typically, with limited or no prior knowledge of the system, an RL algorithm could obtain a strategy that yields the maximal long-term payoff defined by the reward or objective function. The optimality and convergence of the strategy searching process are often guaranteed in a single-agent environment [23]. However, these features are not always ensured in a multi-agent environment in which the agents could sometimes end up being locked in limited cycles or local optimums [24], [25]. There has been a significant effort in developing robust multi-agent RL algorithms. Some existing algorithm includes Nash-Q learning, JAL-Q-learning, distributed Q-learning, and Linear-Reward inaction (LR-I), most of which require communication between agents except LR-I [26]–[28].

This paper aims at integrating the two-player game and LR-I algorithm into a load-response system for a renewable-powered microgrid. In our past work, a naïve LR-I approach was applied to obtain a load-response strategy for the consumers, but this approach requires a relatively long training phase for the consumers' load-response strategies to converge, which typically takes 50-100 days [14]. The proposed combined load-response algorithm models the instantaneous load control decision as a two-player game for each consumer in the system and adjusts their power/load estimation using the LR-I algorithm. Therefore, the learning algorithm is freed from exploring the entire load-response decision space and is reduced to a simpler power/load model. The performance of the combined algorithm is compared to that of the virtual two-player game approach and the naïve LR-I algorithm in a simulated microgrid. As the simulation results show, the combined algorithm approach reduces the learning period and obtains a better strategy compared to other algorithms.

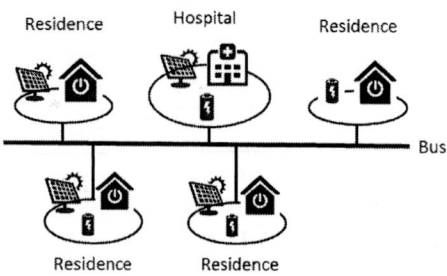

Figure 1: Residences and a hospital in a microgrid.

Microgrid Energy Management

The schematic of an isolated renewable-powered microgrid is shown in Fig.1. This microgrid consists of several residential consumers and a critical load consumer (e.g., a hospital), both of which are powered by local PV arrays and energy storage devices (batteries). The consumers are connected by a dc bus whose voltage is maintained by all the consumers in a de-centralized approach using drop control. It is assumed that all the consumers are capable of performing load-response to some extent. Possible methods include switching off non-critical loads and demand shedding on household appliances [15]. Based on these assumptions a consumer's load is expressed as

$$P_i(t) = rt_i(P_b + P_c\alpha(t)) \tag{1}$$

where rt_i is the ratio (ranging from 0 to 1.0) that the consumer's load takes in the total microgrid load consumption, P_b is the minimum load requirement of the microgrid, and P_c is the controllable load as a linear function of the load control index α. The load control index α indicates how much power consumption is reduced by conducting the load-response measure and is adjusted by the consumer load-response controller.

For the consumer load-response controllers, the objectives are to meet the load demand to the greatest extent while maintaining sufficient stored energy. The latter requirement comes from the fact that the battery units are the only fully-controllable power source capable of stabilizing the system bus. If the batteries were fully discharged during the nights when the PVs arrays do not harvest energy, then there is no possibility that the bus voltage could be maintained at the intended operating point, and the whole system needs to be turned off. Ideally, the load-response controller should lower the load control index in an insufficient-power scenario and could raise it when there is surplus power. Such an energy management approach is achieved by an objective function that evaluates the optimalities of applying different control index α in considering the present energy state. In this study, this objective function takes a piece-wise weighted summation form, as shown in (2)-(3).

$$\text{obj}(t) = \begin{cases} w_{load} \cdot \frac{\sum_{i=1}^n P_i(t)}{P_b + P_c} + w_{SoC} \cdot P(\text{SoC}(t_d) > \text{SoC}_{goal}), P(\text{SoC}(t_d) > \text{SoC}_{goal}) \geq 0.8 \\ (1 - w_{load} \cdot \frac{\sum_{i=1}^n P_i(t)}{P_b + P_c})/10, P(\text{SoC}(t_d) > \text{SoC}_{goal}) < 0.8 \end{cases} \tag{2}$$

$$w_{load} + w_{SoC} = 1, w_{SoC} \geq 0, w_{SoC} \geq 0 \tag{3}$$

where n is the number of consumer in this microgrid, $P_i(t)$ is the ith consumer's load consumption as shown in (1), state of charge (SoC) indicates the energy level in the battery units, w_{LSR} and w_{SoC} are weighting factors set by the operators indicating the importance levels of LSR and the SoC_{goal} satisfaction, SoC_{goal} is a preset SoC goal, and t_d is the SoC goal checking time. The first term $\frac{\sum_{i=1}^n P_i(t)}{P_b + P_c}$ calculates the proportion in which the load demand is met in the microgrid and is referred to as load-satisfaction-rate (LSR) in the following context. When the probability of the system SoC reaching its preset goal is lower than a preset value (0.8), the objective function is set to be negatively related to the LSR, as shown in the second equation in (2), so the load-response controllers would choose lower control indexes. This mechanism acts as an 'emergency brake' when the stored energy level is critical. $P(\text{SoC}(t_d) > \text{SoC}_{goal})$ is the probability of the system state of charge (SoC) reaching a level (SoC_{goal}) at t_d, also referred to as the system stored energy availability. [29]. In this study, t_d is set to be the last hour of every day ($t_d = 24$). The probability $P(\text{SoC}(t_d) > \text{SoC}_{goal})$ is calculated using the SoC distribution function, which depends on the load demand and renewable power distributions. Moreover, the load demand of the consumer and hospitals depends on a variety of variables such as temperature and weather (affecting airconditioner and heat usage), time (daily load curve), number of patients, and so on. Besides, renewable power from the PV array is partially stochastic. As a result, the SoC value SoC(t) is also stochastic at the given time t_d, and its distribution is estimated as a random variable following a distribution function, as shown in

$$\text{SoC}(t_d) \sim X(t, \theta_1, \theta_2, \dots, \theta_n, P_1(t), P_2(t), \dots, P_n(t)) \qquad (4)$$

where X is the distribution function of the battery SoC at time t_d, θ_i is the environment variables and $P_i(t)$ is the ith consumer's load from (1). The function X could be obtained using renewable power and load demand estimation/historical data with a discrete probability computing process. The details on how X is computed could be found in [29]. So, the objectives of the consumer load-response controllers are to search for a control index strategy $\alpha_i(t)$ that maximizes (2). In this microgrid, it is assumed that real-time communication between residential consumers is not available. As a result, the residential consumer load-response controllers need to take the other controllers' responses into consideration when maximizing (2).

Energy management using the two-player game and LR-I

For residential consumers in the microgrid, the optimal load-response problem could be represented as shown in

$$\max_{\alpha_i(t)} \text{obj}(t) = \max_{\alpha_i(t)} \left[w_{LSR} \cdot f_{load}(\alpha_1(t), \dots, \alpha_n(t)) + w_{SoC} \cdot f_{SoC}(\alpha_1(t), \dots, \alpha_n(t)) \right] \qquad (5)$$

$$f_{load}(\alpha_1(t), \dots, \alpha_n(t)) = \frac{\sum_{i=1}^{n} rt_i(P_b + P_c\alpha(t))}{P_b + P_c} \qquad (6)$$

$$f_{SoC}(\alpha_1(t), \dots, \alpha_n(t)) = P(\text{SoC}(t_d) > \text{SoC}_{goal} | \alpha_1(t), \dots, \alpha_n(t)) \qquad (7)$$

$$\sum_{i=1}^{n} rt_i(P_c\alpha(t)) = \sum_{i=1}^{n} rt_i(P_{c_i}\alpha_i(t)) \qquad (8)$$

where P_{c_i} is the ith consumer's controllable load. (5) could be fully simplified to

$$\max_{\alpha_i(t)} \text{obj}(t) = \max_{\alpha_i(t)} F_s(\alpha_1(t), \dots, \alpha_n(t), t, \theta_1, \theta_2, \dots, \theta_n) \qquad (9)$$

where the objective function for a residential consumer is a function F_s of all consumer control indexes in the system.

Assume every consumer load-response controller pursues the optimal load-response strategy in the form of (9), a multi-agent game is formed as shown in

$$\max_{\alpha_1(t), \dots, \alpha_n(t)} F_s(\alpha_1(t), \dots, \alpha_n(t), t, \theta_1, \theta_2, \dots, \theta_n) \qquad (10)$$

which is the optimization problem that each consumer controller needs to solve.

Ideally, (10) could be solved using the indifference principle, which yields the possible equilibriums in the game [16]. However, because the solving process involves solving for general high order equations, the analytical solutions are not guaranteed to be solvable according to the Abel–Ruffini theorem [30]. The existing numerical approaches could obtain a solution close to the equilibrium but usually require a computation time that is exponential to the number of agents/actions in the game, which is not a desired feature [31], [32]. The solving difficulty of a general n-player game is a part of the *P versus NP* problem

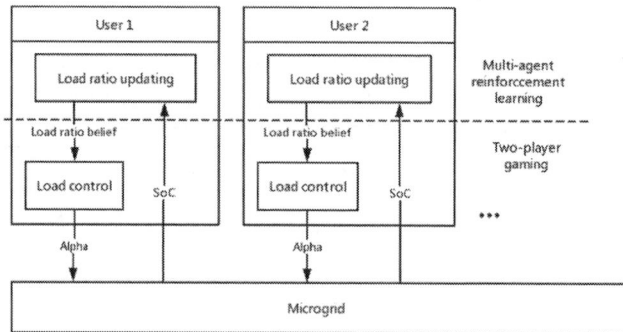

Figure 2: Learning-gaming algorithm information flow scheme.

that remains open in the computer science algorithm realm [33]. Therefore, instead of solving for the full multi-agent game, the consumer controllers are designed to solve for a virtual two-player game shown in

$$\max_{\alpha_i(t),\alpha_o(t)} F_s(\alpha_i(t),\alpha_o(t),t,\theta_1,\theta_2,\dots,\theta_n) \qquad (11)$$

where the actions of n-1 other consumers are attributed to one single virtual player, whose variables are marked with the subscript o. Correspondingly, the load ratio of the virtual player is the controllable load weighted summation of the other consumers as shown in

$$rt_o P_c \alpha_o(t) = \sum_{k\neq i}^{n} rt_k (P_{c_k}\alpha_k(t)) \qquad (12)$$

Due to the reduced size of the game variable and the connection between the two-player game and linear/non-linear programming problems, the computation cost of solving (11) is more manageable than solving for the original problem (10). This approach has been applied in our early work and showed a compatible performance compared to an exhaustive search [29]. The main drawback is that the algorithm performance slips as the number of consumers increases in the system, caused by the biased estimation of each consumer towards the action of the virtual player. This biased estimation is highly related to the load-ratio parameter because it directly affects the shape of the objective function curve. In this paper, the reinforcement learning algorithm is introduced to adjust the consumer load-ratios to compensate for the performance loss.

The proposed load-response algorithm consists of two parts: load-ratio multi-agent reinforcement learning and instantaneous load control two-player gaming. The functional architecture of this algorithm is shown in Fig. 2. The load-response controller of each consumer is referred to as an agent in the following context. At the two-payer gaming layer, each agent is given a load-ratio list

$$\boldsymbol{p} = [p_1, p_2, \dots, p_M] \qquad (13)$$

where p_i indicates the agent's confidence that its load takes i/M of the microgrid's total load demand. These values are initially obtained from the historical data and will be adjusted along with the operation. At the instant of t, the agents are set to pick their load-ratios according to the belief list:

$$rt_i(t) = \frac{i}{M} \ with \ probability \ p_i \qquad (14)$$

Also, the agents assume the virtual player load-ratio is

$$rt_{i_o}(t) = 1 - rt_i(t) \qquad (15)$$

Then, the probabilities of choosing control indexes $\alpha(t)$ are computed as a virtual two-player game, as shown in (11) and solved in two different ways:

1. For residential and non-critical loads, the control index is solved as a zero-sum game.

2. For critical load consumers, such as hospitals, the control index is solved as a common-interest game.

The main difference between the two solving technics is that the agents in a common-interest game assume other agents are cooperative, while the ones in the zero-sum game are self-interested. The motivation of the arrangement was made because the zero-sum assumption induces residential agents to apply more conservative strategies (lower control index), which potentially saves energy for the critical load consumers to maintain higher LSRs.

At the multi-agent reinforcement learning layer, the controllers measure or receive the battery SoC information from the microgrid sensors and SoC estimator, and update the confidence values in the load-ratio belief list L_b. The updating law is

$$\boldsymbol{p}(t+1) = \boldsymbol{p}(t) + b \cdot r(t)\big(\boldsymbol{e}_\delta - \boldsymbol{p}(t)\big), i = 1, \dots, N \tag{16}$$

where 0<b<1 is a learning rate parameter, $r(t)$ is the reward the agent receive after choosing the load-ratio, $\boldsymbol{e}_{\delta_i}$ is a unit vector with its δ_ith component unity, where the δth load-ratio was the last chosen load-ratio. This algorithm is known as Linear Reward-Inaction algorithm L_{R-I}, which could also be interpreted as a multiplayer load-ratio stochastic game where the agents search for equilibriums with mixed solutions [16]. The reward function $r(t)$ is designed to be

$$r_i(t) = wsoc \frac{1}{1+e^{-k \cdot SoC(t)/\alpha(t)}} + (1 - wsoc)\frac{1}{1+e^{-\alpha(t)SoC(t)}}, \tag{17}$$

which is an approximation of the objective function (2) using the consumer's local information. According to [16], the agents applying updating (16) converge to an equilibrium of the stochastic game given a sufficient small learning rate. The LR-I algorithm was applied directly to search for the optimal load-response strategies given different SoC levels for consumers in the microgrid in our previous work [14]. This naïve approach requires the consumer load-response controllers to explore a two-dimension SoC-control index space, which requires a relatively long training phase before the controllers' strategies converge to the equilibriums. The effect will be shown in the verification section below. However, when applied with the combined game-learning algorithm, the consumer load-response controllers only need to explore the one-dimension belief list (13), which potentially reduces the training time. The other benefits of the learning feature are preserved, such as adaptation capability to environment changes and robustness against system integrity damage.

Table I: Evaluation parameter values of one residential consumer

Symbol	PARAMETER	Value
w_{LSR}	Load satisfaction rate weight	0.5
w_{SoC}	Energy availability weight	0.5
$Ebat$	Battery fully charged energy	12 kWh
$\overline{E_{P_{solar}}}$	PV power generation expectation	1 kW
$\overline{E_{P_B}}$	Base load expectation	200 W
$\overline{E_c}$	Controllable load expectation	800 W
$\overline{V_{P_{solar}}}$	Solar power generation variance	400
$\overline{V_{P_c}}$	BS traffic depended load variance	400
SoC_{goal}	Desired battery SoC level	0.8
SoC_0	Initial Battery SoC level	0.7

Table II: Initial load-ratio list

Load-ratio value	Probability	
	Residential consumer	Critical consumer
0.2	60%	10%
0.4	10%	10%
0.6	10%	60%
0.8	10%	10%
1	10%	10%

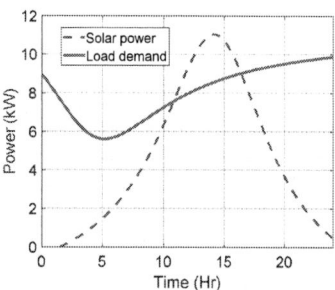

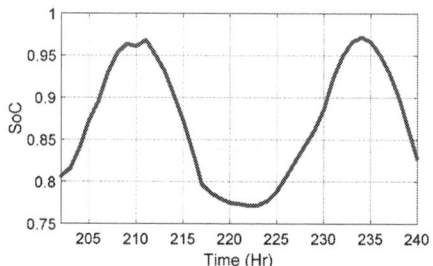

Figure 3: Typical power and solar curves of the microgrid.

Figure 4: Average battery unit SoC of the microgrid in two days.

Analysis Verification

A microgrid controlled by the proposed load-response algorithm was built and simulated in MATLAB. The microgrid consists of a 1:1000 proportionally scaled-down version of five groups of residential consumers and one hospital similar to the one shown in Fig. 1. The parameters of the residential consumers in the system are shown in table I. The hospital has a rated power of 5kW and a battery capacity of 60kWh. The simulated total microgrid PV array power output and load-demand curves are obtained from a PV power record in Cambridge, MA and the load demand from the U.S. Energy Information Administration, as shown in Fig. 3 [34], [35]. The available load-response control index for the residential consumers are [0.1,0.2, … ,1.0] and [0.5, 0.6, …, 1.0] for the hospital. The initial load-ratio lists provided to the consumers load-response controllers are shown in Table II. The virtual two-player game load-response decision-making and load-ratio adjusting using the LR-I algorithm are conducted by the consumer controllers at the beginning of every hour.

The system average battery SoC and LSRs after thirty days of training using the learning-gaming algorithm are shown in Fig. 4-5. As the figures show, the battery SoC in the microgrid is stabilized around the desired goal (80%) with some fluctuations. The LSRs of the consumers were adapted along the day, which were lower at the beginning of a day and graduated increased as time passed. As shown in Fig. 5, compared to the critical load consumer, the residential consumers experienced longer minimal-LSR periods due to the zero-sum game setting.

A simulation comparing the performance of the algorithms with the different number of consumers is also conducted. The results are shown in Fig. 6. The other two algorithms compared are virtual two-player zero-sum load-response game and the naïve linear reward-inaction approach. As the figure demonstrates, the proposed learning-gaming algorithm shows a better performance compared to both approaches, especially with a large number of consumers. Additionally, the learning curves shown in Fig. 7 demonstrates that the proposed algorithm has a converging speed that is faster than that of the

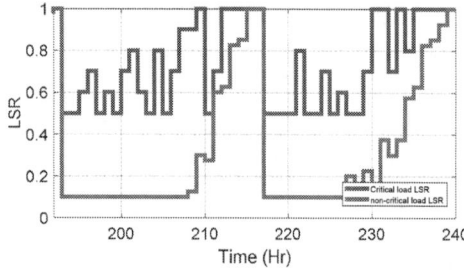

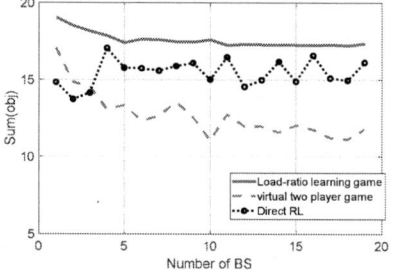

Figure 5: LSRs of different consumers after training.

Figure 6: Performance comparison of different algorithms.

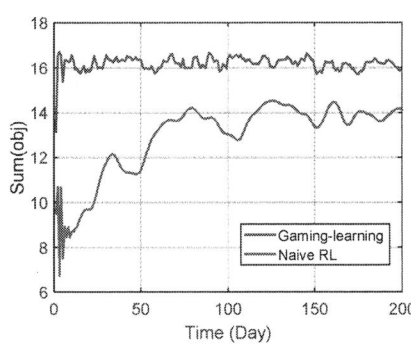

Figure 7: Converging speed comparison of RL and learning-gaming algorithm.

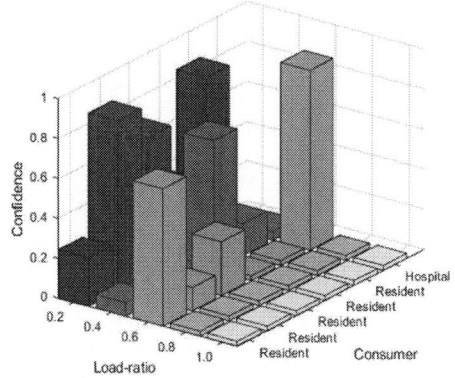

Figure 8: Obtained load-ratio lists.

naïve RL approach. The learning-gaming algorithm finds the optimal load-ratio strategies in less than twenty days, while the naïve RL approach takes up to around seventy days. Fig. 8 shows the trained consumer load-ratios lists. As the figure demonstrates, instead of following the preset load-ratio value (0.2), some resident consumer load-response controllers chose to solve the load-response game based on a larger load-ratio (0.4,0.6), which suggests the controllers are more "responsible" when evaluating the impacts of their choices of load control index, which potentially improved the system performance.

Conclusion

This paper presents a distributed energy management mechanism for an isolated renewable-powered microgrid applying game theory and LR-I algorithm. In this particular case, the analysis focuses on a system applicable to residences along with a hospital, but the same approach can be used in other applications with a partially controllable load. The simulation results show that the energy management strategy obtained by the learning-gaming algorithm has a better performance compared to naïve reinforcement learning and direct game theory methods. Also, benefiting from the reduced search space, the converging speed of the proposed algorithm is faster than the naïve RL algorithm.

References

[1] N. Hatziargyriou, *Microgrids: architectures and control*. John Wiley & Sons, 2014.

[2] Z. Chmiel and S. C. Bhattacharyya, "Analysis of off-grid electricity system at Isle of Eigg (Scotland): Lessons for developing countries," *Renew. Energy*, vol. 81, pp. 578–588, Sep. 2015, doi: 10.1016/j.renene.2015.03.061.

[3] N. R. Canada, "The First Canadian Smart Remote Microgrid: Hartley Bay, BC," Jan. 15, 2014. https://www.nrcan.gc.ca/maps-tools-publications/publications/energy-publications/technology-research-publications/first-canadian-smart-remote-microgrid-hartley-bay-bc/14421 (accessed May 18, 2020).

[4] "How a Pacific Island Changed From Diesel to 100% Solar Power," *National Geographic News*, Feb. 23, 2017. https://www.nationalgeographic.com/news/2017/02/tau-american-samoa-solar-power-microgrid-tesla-solarcity/ (accessed May 18, 2020).

[5] "Solar-Storage Microgrid Delivers Savings to Kaiser Permanente Hospital," *Microgrid Knowledge*, Jul. 16, 2018. https://microgridknowledge.com/solar-storage-microgrid-hospital/ (accessed May 18, 2020).

[6] M. Soshinskaya, W. H. J. Crijns-Graus, J. M. Guerrero, and J. C. Vasquez, "Microgrids: Experiences, barriers and success factors," *Renew. Sustain. Energy Rev.*, vol. 40, pp. 659–672, Dec. 2014, doi: 10.1016/j.rser.2014.07.198.

[7] E. Wood, "CHP Microgrids for Manufacturers Win $10M in DOE Funding," *Microgrid Knowledge*, Sep. 11, 2018. https://microgridknowledge.com/chp-microgrids-manufacturers-doe/ (accessed May 18, 2020).

[8] B. Yan, P. B. Luh, G. Warner, and P. Zhang, "Operation and Design Optimization of Microgrids With Renewables," *IEEE Trans. Autom. Sci. Eng.*, vol. 14, no. 2, pp. 573–585, Apr. 2017, doi: 10.1109/TASE.2016.2645761.

[9] G. Cardoso *et al.*, "Microgrid reliability modeling and battery scheduling using stochastic linear programming," *Electr. Power Syst. Res.*, vol. 103, pp. 61–69, Oct. 2013, doi: 10.1016/j.epsr.2013.05.005.

[10] A. Khodaei and M. Shahidehpour, "Microgrid-Based Co-Optimization of Generation and Transmission Planning in Power Systems," *IEEE Trans. Power Syst.*, vol. 28, no. 2, pp. 1582–1590, May 2013, doi: 10.1109/TPWRS.2012.2224676.

[11] P. Reddy and M. Veloso, *Strategy Learning for Autonomous Agents in Smart Grid Markets.* 2011, p. 1451.

[12] J. Wu, W. Zhou, W. Zhong, Y. Cheng, and J. Liu, "Dual Energy Scheduling for Microgrids in Energy Internet: A Non-Cooperative Game Approach," in *2017 IEEE International Conference on Energy Internet (ICEI)*, Apr. 2017, pp. 48–52, doi: 10.1109/ICEI.2017.16.

[13] C. Stevanoni, F. Vallee, Z. De Greve, and O. Deblecker, "On the use of game theory to study the planning and profitability of industrial microgrids connected to the distribution network," *CIRED - Open Access Proc. J.*, vol. 2017, no. 1, pp. 2444–2448, 2017, doi: 10.1049/oap-cired.2017.0394.

[14] R. Hu and A. Kwasinski, "Energy Management for Microgrids Using a Hierarchical Game-Machine Learning Algorithm," in *2019 1st International Conference on Control Systems, Mathematical Modelling, Automation and Energy Efficiency (SUMMA)*, Nov. 2019, pp. 546–551, doi: 10.1109/SUMMA48161.2019.8947473.

[15] R. Leo, R. S. Milton, and A. Kaviya, "Multi agent reinforcement learning based distributed optimization of solar microgrid," in *2014 IEEE International Conference on Computational Intelligence and Computing Research*, Dec. 2014, pp. 1–7, doi: 10.1109/ICCIC.2014.7238438.

[16] M. Maschler, E. Solan, and S. Zamir, *Game Theory*. Cambridge: Cambridge University Press, 2013.

[17] J. F. Nash, "Equilibrium points in n-person games," *Proc. Natl. Acad. Sci.*, vol. 36, no. 1, pp. 48–49, Jan. 1950, doi: 10.1073/pnas.36.1.48.

[18] J. Nash, "Non-Cooperative Games," *Ann. Math.*, vol. 54, no. 2, pp. 286–295, 1951, doi: 10.2307/1969529.

[19] K. Daskalakis and C. Papadimitriou, "Three-Player Games Are Hard," *Electron. Colloq. Comput. Complex. ECCC*, Jan. 2005.

[20] C. Daskalakis, P. W. Goldberg, and C. H. Papadimitriou, "The complexity of computing a Nash equilibrium," *SIAM J. Comput.*, vol. 39, no. 1, pp. 195–259, 2009.

[21] R. S. Sutton, "Sutton & Barto Book: Reinforcement Learning: An Introduction," in *A Bradford Book*, MIT Press Cambridge, MA, 1998.

[22] K. J. Åström and B. Wittenmark, *Adaptive control*. Courier Corporation, 2013.

[23] M. Wiering and M. Van Otterlo, "Reinforcement learning," *Adapt. Learn. Optim.*, vol. 12, p. 3, 2012.

[24] L. Panait and S. Luke, "Cooperative Multi-Agent Learning: The State of the Art," *Auton. Agents Multi-Agent Syst.*, vol. 11, no. 3, pp. 387–434, Nov. 2005, doi: 10.1007/s10458-005-2631-2.

[25] L. Buşoniu, R. Babuška, and B. De Schutter, "Multi-agent reinforcement learning: An overview," in *Innovations in multi-agent systems and applications-1*, Springer, 2010, pp. 183–221.

[26] J. Hu and M. P. Wellman, "Multiagent Reinforcement LeAarlgnoinrigt:hmTheoretical Framework and an," p. 9.

[27] C. Claus and C. Boutilier, "The Dynamics of Reinforcement Learning in Cooperative Multiagent Systems," p. 7.

[28] M. Lauer and M. Riedmiller, "An Algorithm for Distributed Reinforcement Learning in Cooperative Multi-Agent Systems," in *In Proceedings of the Seventeenth International Conference on Machine Learning*, 2000, pp. 535–542.

[29] R. Hu, A. Kwasinski, and A. Kwasinski, "Mixed strategy load management strategy for wireless communication network micro grid," presented at the 2016 IEEE International Telecommunications Energy Conference (INTELEC), 2016, pp. 1–8.

[30] J. Rotman, *Galois theory*. Springer Science & Business Media, 1998.

[31] S. Jorgensen, M. Quincampoix, and T. L. Vincent, *Advances in Dynamic Game Theory: Numerical Methods, Algorithms, and Applications to Ecology and Economics*, vol. 9. Springer Science & Business Media, 2007.

[32] S. Govindan and R. Wilson, "A global Newton method to compute Nash equilibria," *J. Econ. Theory*, vol. 110, no. 1, pp. 65–86, 2003.

[33] S. Cook, "The P versus NP problem," *Millenn. Prize Probl.*, pp. 87–104, 2006.

[34] "Solar Panels in Cambridge," *Cambridge Solar*. https://www.cambridge-solar.co.uk/solar-pv-cambridge (accessed May 29, 2020).

[35] "Demand for electricity changes through the day - Today in Energy - U.S. Energy Information Administration (EIA)." https://www.eia.gov/todayinenergy/detail.php?id=830 (accessed May 29, 2020).

All-GaN Bidirectional ANPC-based Resonant DC-DC Converter

Tino Kahl, Laurenz Wernicke,
Sibylle Dieckerhoff
Technische Universität Berlin
Chair of Power Electronics
Einsteinufer 19
10587 Berlin, Germany
Tel.: +49 (0)30 314-25513
E-Mail: t.kahl@tu-berlin.de
URL: https://www.pe.tu-berlin.de

Christopher Fromme, Marvin Tannhäuser
Siemens AG

Frauenauracher Str. 80,
91056 Erlangen, Germany

Keywords

«DC power supply», «Multilevel converters», «Switched-mode power supply»,
«Gallium Nitride (GaN)»,

Abstract

This paper presents an isolated all-GaN resonant DC-DC converter using a HV ANPC and a LV
full-bridge converter stage. The thermal characteristics and current limits as well as the switching
behavior of the commutation cells of both converter stages are discussed. The ANPC-stage is tested with
an output power of 1.8 kW operating as LLC converter, and the LV full-bridge is tested up to 27 A in a
step-down DC-DC configuration. Measurement results of the resonant all-GaN LLC-converter are
discussed and compared with simulation results.

Introduction

Power transmission between high and low voltage DC grids is a typical application for bidirectional
DC-DC converters [1]. Specifically, several power sources can be connected in parallel to control the
energy flow and enhance fail-safe operation of a grid. Using fast-switching Gallium Nitride (GaN)
transistors has the potential to noticeably increase the efficiency of such bidirectional DC-DC
converters. For this purpose, this paper presents an all-GaN Active Neutral Point Clamped (ANPC)
bidirectional resonant converter. It connects a 750 V high voltage (HV) bus to a 40 V low voltage (LV)
bus and is designed for a maximum rated power of 1.5 kW. The converter topology is depicted in Fig. 1,
consisting of the ANPC-stage (left) which requires six transistors, and the low voltage full-bridge stage
(right). An LLC resonant tank including the isolating transformer connects both converter stages.

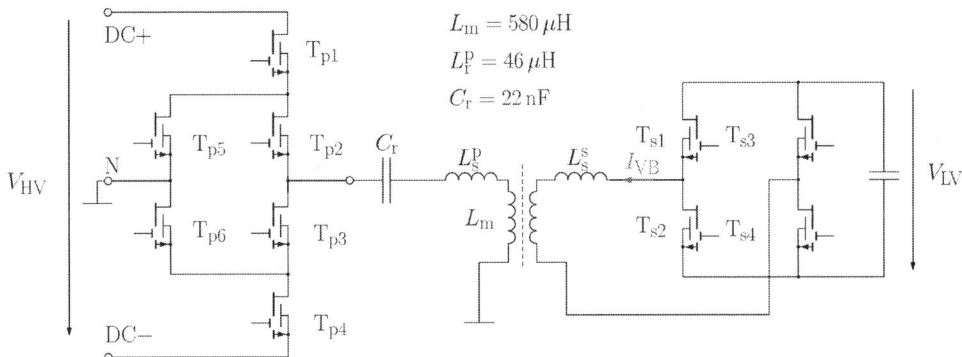

Fig. 1: ANPC- resonant DC-DC converter structure

The ANPC converter offers the advantage of operating at dc-link voltages above the voltage rating of the power semiconductors [2]. Because of widely available GaN-transistors with a maximum blocking voltage of 650 V, the ANPC-converter is a good choice for the HV-side. References [3] and [4] present a possible application for an ANPC-stage based on GaN transistors for a 400 V AC-grid and proposes an improved PCB layout with a low inductive connection between the transistors. The low stray inductance is achieved by mounting the transistors on the bottom and top layer of the PCB, at the cost of increased cooling effort. The sensitivity to the circuit design was also demonstrated in our previous work [5], where the ANPC converter was designed using three GaN half-bridge boards from Texas Instruments. A single half-bridge showed excellent switching behavior. However, the ANPC-stage produced high oscillations and turn-off overvoltage in hard-switching conditions. It should be noted that the bidirectional DC-DC converter does not operate in ZVS (Zero Voltage Switching) mode in all operation points. Therefore, the converter also needs to operate safely in hard-switching conditions.

The paper is structured as follows: First, the full-bridge converter is characterized. The DC voltage on the LV-side of the resonant tank is below 100 V. Accordingly, the full-bridge stage is designed with low-voltage, top-cooled 100 V / 7 mΩ normally-off GaN transistors (GS61008T) and mounted on a water-cooled heatsink. The thermal characterization and DC-DC operation with an output power up to 500 W are discussed. The second section introduces the ANPC-stage, applying 650 V / 50 mΩ normally-off GaN transistors (GS66508T). All transistors are placed on the same layer of the PCB and thermally connected to an air-forced cooled heatsink. The thermal properties are analyzed to verify the PCB design of the ANPC-stage. The switching behavior is characterized, compared to simulation results and cross-checked with a half-bridge design. To ensure the overload capability of the ANPC-stage, investigations on the resonant tank are conducted at an output power of 1.8 kW. Finally, the measurements of the all-GaN converter are presented and compared with simulation results.

Low Voltage Full-Bridge Stage

In this section, the electrical limits for one half-bridge leg of the full-bridge stage are derived. The PCB design is depicted in Fig. 2. Both GaN transistors (GS61008T) are closely mounted to each other on the top side with low ESL ceramic capacitors to minimize the commutation loop area, see Fig. 2a. To achieve a reduced stray inductance a four-layer PCB design is used.

Thermal coupling

The chosen transistors have a cooling pad located on the top side that is connected to the water-cooling system. Therefore, it is not necessary to include thermal aspects in the PCB design, which allows to improve the design according to electrical constraints.

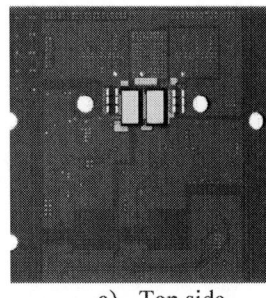
a) Top side

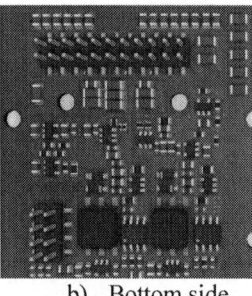
b) Bottom side

Fig. 2: Full-bridge PCB design

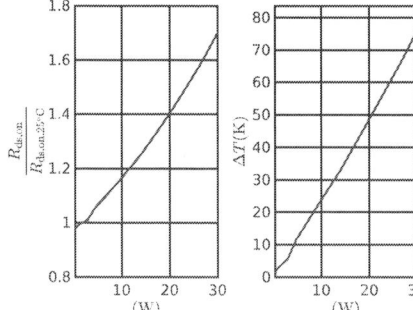

Fig. 3: Thermal characteristics of the GS61008T applied in the low-voltage full-bridge stage

Thermal tests based on self-heating of the transistors by a DC-current are conducted. The normalized on-state resistance (left), and the temperature change (right) as a function of power losses are shown in Fig. 3, representing the thermal coupling of the semiconductors to the heatsink. Since the characterization relies on a DC-measurement, no switching losses are included. Therefore, the junction temperature is calculated directly from the on-state resistance vs. temperature characteristics provided in the datasheet [6]. For example, at a measured ambient temperature of 27°C and a power dissipation of 20 W, an on-state resistance of 9.8 mΩ corresponds to a junction temperature of 80°C. From this measurement a thermal resistance of $R_{th,JA}$=2.7 K/W is calculated for one chip. Assuming an ambient temperature of 40°C and a maximum rated junction temperature of 110°C results in a maximum power loss of 25 W for each transistor.

Switching characterization

In order to determine the current limit of the half-bridge in switched operation, a simulation model is created and configured as buck converter (Fig. 5). To ensure that simulation results are close to reality, the parasitic inductance and capacitance are tuned using switching test results. The measurement results are depicted in Fig. 4a. In comparison with the simulation in Fig. 4b (turn-off), it can be noticed that the maximum voltage of 91 V at a load current of 57 A is almost identical. The time period of the oscillation has a deviation of 1 ns between simulation and measurement, and similar damping behavior. The turn-on shows approximately the same fall time and oscillation in the voltage. However, the final decay of the voltage during turn-on ("tail" at 1.15 µs, Fig. 4a right) cannot be modeled with linear passive components. Because of trapping effects, the channel resistance increases and then decays with a similar time constant [7],[8]. The dynamic $r_{ds,on}$ increase can be considered as additional resistance. The resistor value changes depending on the operation conditions, e.g. blocking voltage, blocking time and switched current of the transistor. Including an additional dynamic $\Delta r_{ds,on}$ in the simulation gives a more realistic turn-on behaviour (Fig. 4c). However, this model requires substantial simulation resources and thus is not suitable for circuit simulations. Therefore, for further investigations the model from Fig. 4b was used. It should be noted that in consequence the losses during the voltage "tail" are not included in the simulation results, and losses of real converter operation are underestimated.

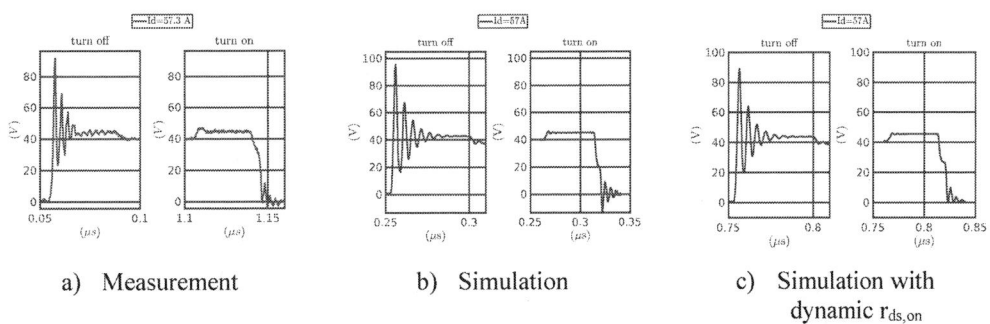

a) Measurement   b) Simulation   c) Simulation with dynamic $r_{ds,on}$

Fig. 4: GS61008T switching characteristic in a half-bridge topology

In the buck-converter investigation, the duty cycle is set to 50%, which ensures identical on-state times for both transistors. The switching and conduction losses are depending on the current direction. In case of positive current, according to the notation in Fig. 5, the bottom transistor is operated in hard-switching mode causing switching losses. The top transistor T_{s1} achieves ZVS, but reverse conduction during the dead times of the half-bridge contributes to the overall conduction losses. In the following, the dead time is set to 50 ns and the switching frequency f_s=100 kHz is applied. The simulation results show a junction temperature of approximately 55°C for both transistors, with 5.4 W losses including 1.3 W reverse conduction losses for the top transistor T_{s1}, and 4.3 W losses for the bottom transistor T_{s2} at a current of 27 A.

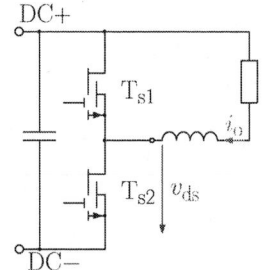

Fig. 5: Buck converter structure

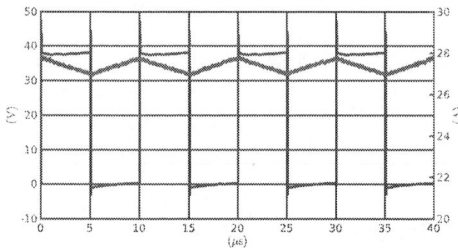

Fig. 6: Half-bridge measurement: output voltage and current

Hence, as explained above, the transistor T_{s1} operating in reverse conduction mode generates higher losses compared to the hard-switching transistor T_{s2}. Fig. 6 depicts the measured load current and voltage of the half-bridge operating under the same conditions with an average load current of 27 A. In this case, a calorimetric measurement results in losses of approximately 9 W for both transistors.

In order to compare simulated and measured on-state resistance of the bottom transistor T_{s2}, the on-state voltage is captured with a clamping circuit as described in [8]. In the simulation, a constant on-state resistance of 11 mΩ is calculated, as depicted in Fig. 7. In the in-circuit operation, the transistor has a higher on-state resistance of 12 mΩ at the beginning, and reaches 11 mΩ at the end of the on-state period (Fig. 8). An increase of the dynamic on-state resistance could be explained by the accumulation of trapping in the transistor channel [9]. On the other hand, the dynamic $r_{ds,on}$ variation is $\Delta r_{ds,on} = 1$ mΩ, corresponding to an on-state voltage variation of approximately $\Delta v_{ds,on} = 9$ mV. It should be considered that the clamping circuit has a measurement error, and the simulation does not include all parasitic elements. Furthermore, the clamping circuit has a low-pass filter characteristic with a time constant of $t_{3,db}$=4.5 ns, but the measured time constant of $r_{ds,on}$ is approximately $t_{3,db}$=1.7 µs and could therefore more probably be related to the dynamic characteristic of the transistor than to a filter effect of the clamping circuit. Overall, it is therefore not possible to identify the exact root cause of the measured $r_{ds,on}$ characteristic.

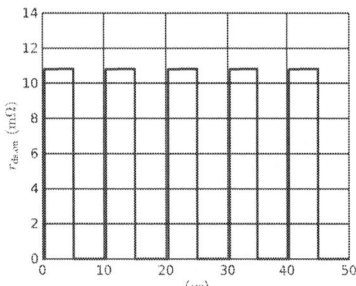

Fig. 7: Half-bridge: simulation of the $r_{ds,on}$

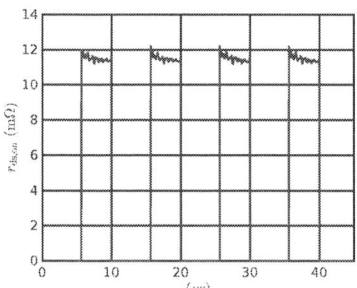

Fig. 8: Half-bridge: measurement of the $r_{ds,on}$

To avoid damage, the power loss limit of the entire full-bridge is set to 20 W, limiting the output power of the half-bridge to 520W. However, the maximum simulated output power of the half-bridge configuration is 900 W at a junction temperature of 121°C, at $T_{amb} = 40°C$ and semiconductor power losses of 36.5 W.

Active Neutral Point Clamped Converter

In this section, the ANPC-stage is presented, and it is shown that the thermal limit is not critical. To analyze the complex layout with six switches and the high voltage dc-link bus, the switching characteristic of the top and bottom switching cell, formed by transistors T_1, T_5 and T_6, T_4 respectively is discussed. The full functionality of the ANPC-stage is demonstrated for resonant operation at 1.8 kW output power.

Thermal coupling

The transistors GS66508T have a topside cooling pad to improve the thermal interface to the heatsink. The junction temperature dependency as a function of the losses is derived applying air-forced cooling. Under these conditions, losses of 7.5 W result in a junction temperature of approx. 116°C at an ambient temperature of $T_{amb} = 40°C$ (Fig. 9). The converter simulation (Fig. 17) shows that the maximum losses in the system are 3 W per transistor, so it can be concluded that the thermal interfacing is sufficient.

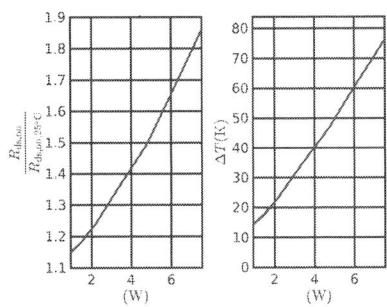

Fig. 9: ANPC-converter: Thermal characteristics

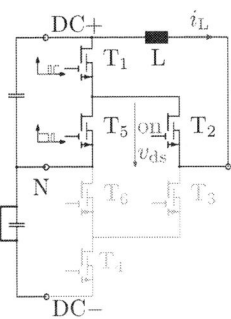

Fig. 10: ANPC: Test circuit for switching characterization

Switching characterization

The ANPC-stage is designed on a 4-layer PCB, illustrated in Fig. 11. The transistors form two groups, T_1,T_5,T_2 and T_6,T_4,T_3, and are placed closely together. Ceramic capacitors are mounted directly on the opposite side of the PCB to optimize the commutation area and minimize the stray inductance. In order to verify the layout of the PCB, the switching characteristics of the half-bridge cell formed by transistors T_1, T_5 is investigated (Fig. 10). The negative dc-link is shorted for this investigation to suppress the charge of this part of the dc-link. During normal operation of the ANPC-stage, when the output voltage changes from N to DC+ or from DC+ to N, the transistor T_2 must be turned on to connect the path with transistors T_1 and T_5 to the output of the ANPC-stage.

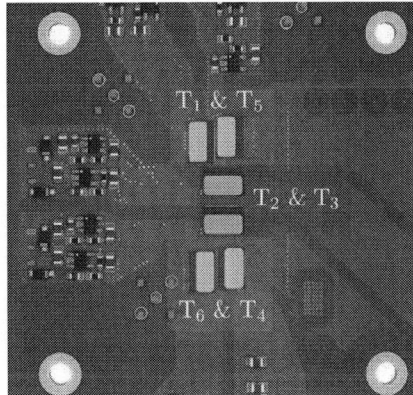

Fig. 11: PCB design of the ANPC-converter

The test circuit is shown in Fig. 10; turn-on and turn-off voltage transients for transistor T_5 are depicted in Fig. 12. The voltage transients for transistor T_4, which is part of the complementary half-bridge test circuit T_6, T_4 and T_3 with shorted positive dc-link, are shown in Fig. 13. Considering the complex design, the measured over-voltage of approximately 50 V at a drain current of 12 A and a dc-link voltage of 400 V is an acceptable result.

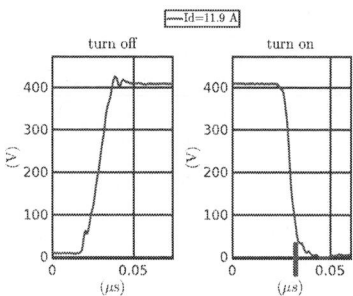

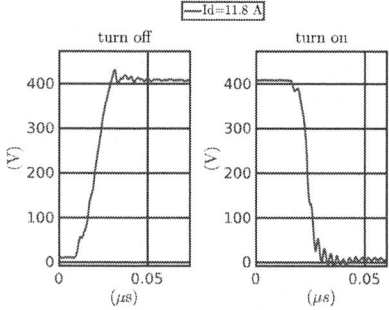

Fig. 12: ANPC: Switching test for cell T_1, T_5 Fig. 13: ANPC: Switching test for cell T_6, T_4

The results from the switching tests are used to create an LTSpice-model of the entire ANPC-stage based on level 3 transistor Spice models. These transistor models are available from the manufacturer (GaN Systems) and include the parasitic inductance of the transistor housing. Layout parasitics are included in the converter simulation model and adjusted according to the measurement results. The corresponding simulation of the voltage transients is depicted in Fig. 14. The simulated switching transients are slower, the voltage rise time differs by 7 ns from the measurement. It can be therefore be concluded that the output capacitance from transistor T_5 is higher compared to the real system.

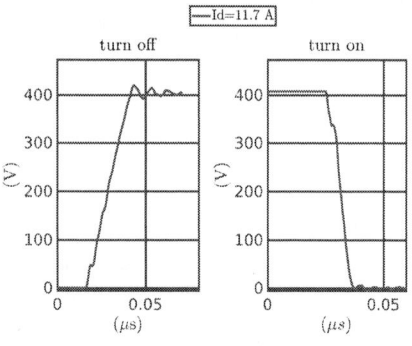

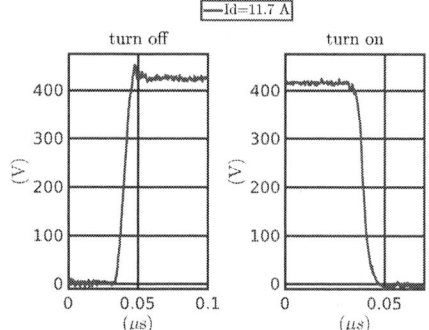

Fig. 14: ANPC Simulation: Switching test for cell T_1, T_5 Fig. 15: Optimized half-bridge design: Switching test

The last part of the turn-on process, where the drain-source voltage v_{ds} decays as a tail, cannot be modeled by linear parasitic components (marked red in Fig. 12). This effect could be explained by a dynamic $r_{ds,on}$, but it is too significant for that alone. Therefore, transistors with the same date code were investigated in an optimized half-bridge design. The corresponding switching test results for I_d=12 A are illustrated in Fig. 15. Since these measurements do not show a significant tail-effect, it can be concluded that the layout must be the primary cause of this effect. In order to model the effect, further investigations on PCB-level are necessary to identify the responsible parasitic components.

Resonant operation

The test circuit of the GaN ANPC-stage with an LLC resonant tank is shown in Fig. 16. The resonant tank is connected between the output and the neutral point of the converter. A diode rectifier is applied on the secondary side of the transformer, and the load is fixed to 1.8 kW output power. The higher output power compared to the rated power of 1.5 kW is investigated to ensure that the converter operates safely in short overload conditions. This topology allows to control the output power by changing the switching frequency or by changing the duty cycle of the ANPC-stage.

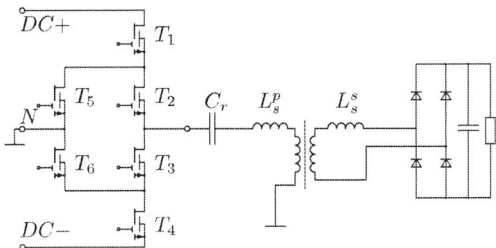

Fig. 16: ANPC: Resonant test circuit

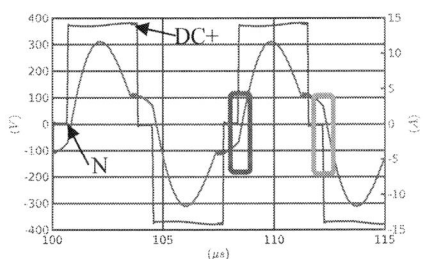

Fig. 17: ANPC: Simulation of resonant operation

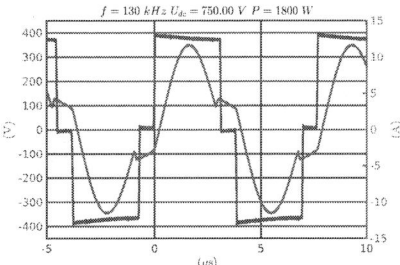

Fig. 18: ANPC: Measurement of resonant operation

The simulated current from the resonant tank and the output voltage from the ANPC-stage are illustrated in Fig. 17, the corresponding measurement results are given in Fig. 18. The current has an inductive phase-shift and enables ZVS of the ANPC transistors.

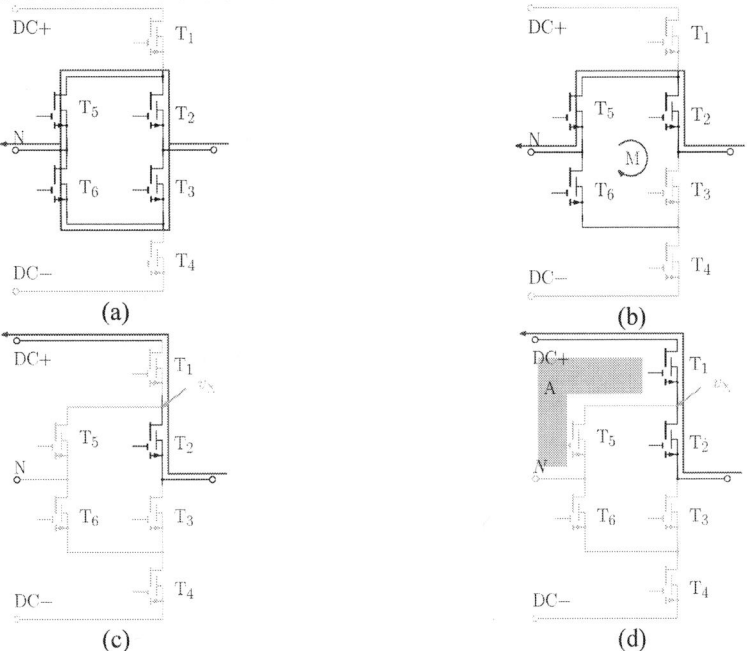

Fig. 19: ANPC: Commutation cycle switching the output from N to DC+

In order to clarify the ZVS behavior in the ANPC-stage, the commutation cycle switching the output potential from N to the DC+ is depicted in Fig. 19 and highlighted in red in Fig. 17. Starting with the N-state, the current flows through the inner transistors T_2; T_3; T_5; T_6 as shown in Fig. 19 (a). In the next step, the transistor T_3 turns off and forces the entire current through transistors T_5 and T_2. The transistor T_3 only has to block the forward voltage of T_5 and T_2 and consequently the di/dt as well as the dv/dt is minimal (Fig. 19 (b)). Then, transistors T_5 and T_6 switch off, and the current commutates to the reverse channel of transistor T_1. Due to the high reverse conducting voltage with the gate biased for off-state, this mode produces high power losses. The node v_x indicated in Fig. 19 (c) undergoes a potential change with the turn-off dv/dt. After the potential change is completed, transistors T_3 and T_4 block the entire dc-link, and T_5 blocks half the dc-link voltage (positive part). In this example the critical commutation cell T_1 and T_5 is identified (highlighted in Fig. 19 (d)). The last step is to activate transistor T_1, and the output voltage is fixed to DC+.

In the commutation cycle switching the output from N to DC-, the critical cell is formed by T_6 and T_4. This is the reason why the PCB layout groups transistors $T_{1,5,2}$ as well as $T_{6,4,3}$ as cells. Furthermore, the simulation results show that transistors T_5 and T_6 produce losses of only 110 mW, which is due to the fact that under the considered operation conditions, the current only flows for a short time through these transistors. Transistors T_1 and T_4 generate the major losses of 3.04 W.

Finally, the real system is tested under the same conditions. The measured current shown in Fig. 18. has an inductive phase-shift and is comparable to the simulated current. No relevant over-voltage can be detected in the output voltage of the ANPC-converter. In summary, stable operation of the ANPC-converter stage is verified at the required power level.

Evaluation of the all-GaN ANPC full-bridge converter

The simulation models obtained from the previous investigations are used to study the entire DC-DC converter. The structure is the same as in Fig. 1 but includes the identified parasitic elements. The power flow direction is chosen from the full-bridge stage to the ANPC-stage at a power rating of 500 W, which is chosen to investigate the function of the all-GaN converter without risking irreparable damage. An exemplary simulation result is given in Fig. 20 and shows the ANPC-stage output voltage and the full-bridge current measured in the direction depicted in Fig. 1. The output voltage of the ANPC-stage shows no relevant over-voltage peak. First measurements of the all-GaN converter (Fig. 21) verify the simulation results. Both, measurement and simulation show two oscillations in the current. The first fundamental oscillation, transferring the energy, is well known from the LLC converter (Fig. 18). The additional oscillation (red circle in Fig. 20) results from the used modulation technique.

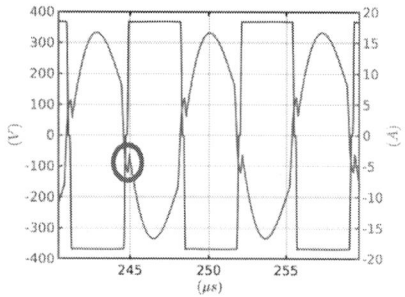

Fig. 20: Entire converter simulation: ANPC voltage and full-bridge current

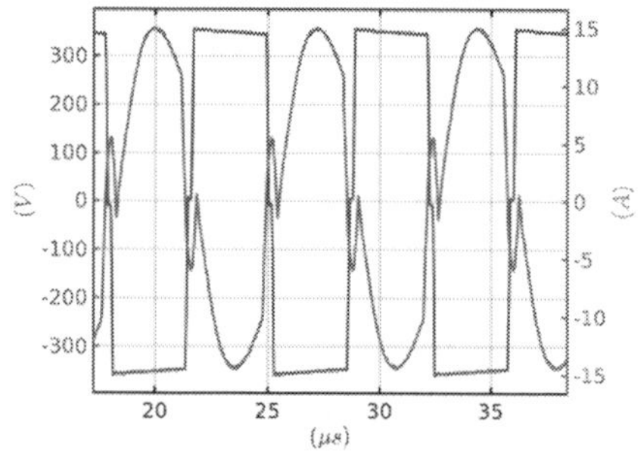

Fig. 21: Entire converter measurement: ANPC voltage and full-bridge current

In order to control the dc-link voltage of the ANPC-stage, the phase shift between the output voltages of the full-bridge and ANPC-stage is manipulated. Using this modulation technique, there exist states where one of the bridges short-circuits the resonant tank, while the other bridge still connects the output to DC+ or DC-.

Finally, simulations and measurements verify the all-GaN converter functionality and show an efficiency of approximately 94% at a power of 500W for the entire converter. The target for the rated converter power is 1.5 kW in resonant operation. For this operation point, Fig. 22 and Fig. 23 depict the simulated output current and voltage for the full-bridge stage and the ANPC-stage. The simulation of the controlled converter is based on the software PLECS and does not include the parasitic elements. The previously presented measurements are conducted without a closed-loop control, while a control of the ANPC dc-link voltage is implemented in the simulation. This control keeps the dc-link voltage constant and at the same time tries to keep the reactive power in the resonant tank as low as possible. This is important for converter operation, since a low reactive power is required to ensure ZVS of the transistors in the converter stages. In the 1.5 kW operating point, the RMS resonant tank current of I_{RMS}=46 A is supplied by the full-bridge converter with the transistors in the full-bridge having an RMS current of up to 32 A. Considering that the DC-DC measurements of a half-bridge generate a power loss of about 10 W at an RMS transistor current of I=19.5 A, it is assumed that the full-bridge stage can operate at 1.5 kW, if properly controlled. In the ANPC-stage flows an RMS output current of I_{rms}=5 A which is not critical compared to the experimentally achieved resonant ANPC-converter operation with P=1.8 kW.

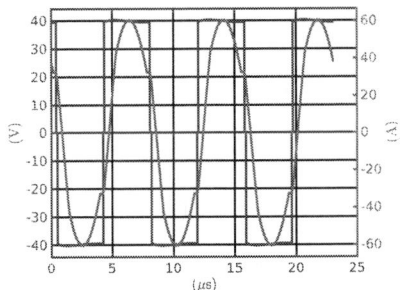

Fig. 22: Simulation: full-bridge stage: current and output voltage

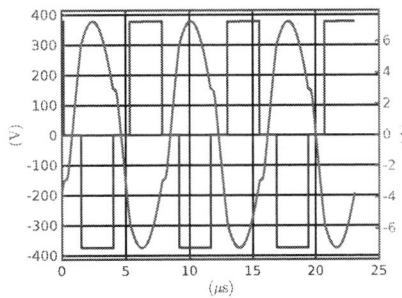

Fig. 23: Simulation: ANPC-stage: current and output voltage

Conclusion

A bidirectional DC-DC converter based on GaN semiconductors requiring a multi-level topology is realized as ANPC-stage for the HV bus, and as full-bridge stage with high current carrying capacity for the LV bus. The DC-characterization, verifying the thermal coupling to the cooling system, and the analysis of the switching transients show promising results for both converter stages. Continuous measurements of the individual converter stages show that the full-bridge stage operates up to a current of 27 A, resulting in a maximum RMS current of 19.5 A per transistor, while operation of the ANPC-stage was demonstrated up to 1.8 kW in an LLC configuration.

Simulation and measurement results of the bidirectional all-GaN ANPC resonant converter verify its functionality at 500 W with an efficiency of 94%. Taking into account that the power of the full-bridge stage is limited below its rated power in this proof-of-concept investigation, it can be expected that the all-GaN converter will have a higher efficiency at increased output power. A simulation including closed-loop control shows that the required power of 1.5 kW for the all-GaN converter could be achieved. Future work will focus on the implementation of a closed-loop control in the experimental setup in order to demonstrate the operation at full power.

References

[1] R. Ramachandran and M. Nymand, "A 98.8% efficient bidirectional full-bridge isolated dc-dc GaN converter," 2016 IEEE Applied Power Electronics Conference and Exposition (APEC), Long Beach, CA, 2016, pp. 609-614, doi: 10.1109/APEC.2016.7467934.

[2] T. Bruckner, S. Bernet and H. Guldner, "The active NPC converter and its loss-balancing control," in IEEE Transactions on Industrial Electronics, vol. 52, no. 3, pp. 855-868, June 2005, doi: 10.1109/TIE.2005.847586.

[3] E. Gurpinar, A. Castellazzi, F. Iannuzzo, Y. Yang and F. Blaabjerg, "Ultra-low inductance design for a GaN HEMT based 3L-ANPC inverter," *2016 IEEE Energy Conversion Congress and Exposition (ECCE)*, Milwaukee, WI, 2016, pp. 1-8.
doi: 10.1109/ECCE.2016.7855540

[4] M. Valente, F. Iannuzzo, Y. Yang and E. Gurpinar, "Performance Analysis of a Single-phase GaN-based 3L-ANPC Inverter for Photovoltaic Applications," *2018 IEEE 4th Southern Power Electronics Conference (SPEC)*, Singapore, Singapore, 2018, pp. 1-8.doi: 10.1109/SPEC.2018.8635942

[5] T. Kahl, C. Kuring, S. Dieckerhoff, C. Fromme and M. Tannhäuser, "Active Neutral Point Clamped Resonant DC/DC-Converter in Gallium Nitride Technology," 2018 20th European Conference on Power Electronics and Applications (EPE'18 ECCE Europe), Riga, 2018, pp. P.1-P.9.

[6] "Datasheet GS61008T" Accessed on: Feb. 22,2020. [Online]. Available: https://gansystems.com/wp-content/uploads/2018/04/GS61008T-DS-Rev-180420.pdf

[7] P. B. Klein et al., "Investigation of traps producing current collapse in AlGaN/GaN high electron mobility transistors," in Electronics Letters, vol. 37, no. 10, pp. 661-662, 10 May 2001, doi: 10.1049/el:20010434.

[8] C. Kuring, M. Tannhaeuser and S. Dieckerhoff, "Improvements on Dynamic On-State Resistance in Normally-off GaN HEMTs," in PCIM Europe 2019.

[9] J. Böcker, C. Kuring, M. Tannhäuser and S. Dieckerhoff, "Ron increase in GaN HEMTs - Temperature or Trapping Effects," 2017 IEEE Energy Conversion Congress and Exposition (ECCE), Cincinnati, OH, 2017, pp. 1975-1981. doi: 10.1109/ECCE.2017.8096398

Lifetime Estimation and Dimensioning of the Machine-Side Converter for Pumping-Cycle Airborne Wind Energy System

Bakr Bagaber[1], Patrick Junge[2] and Axel Mertens[1]
LEIBNIZ UNIVERSITY HANNOVER[1], SkySails Power[2]
Institute for Drive Systems and Power Electronics[1]
Hannover, Germany
Phone: +49 (0) 511-762-3766
Email: bakr.bagaber@ial.uni-hannover.de
URL: www.ial.uni-hannover.de

Acknowledgment

This work was supported by the German Ministry of Economics and Technology (BMWi) – 0324217D.

Keywords

≪Renewable energy systems≫, ≪Wind energy≫, ≪Windgenerator systems≫, ≪Mission profile≫, ≪Voltage Source Converter (VSC)≫, ≪Power cycling≫, ≪Thermal cycling≫, ≪Thermal stress≫

Abstract

Fostering of high altitude wind energy (HAWE) resources above 200 meters is a recent promising technology that seeks to capture the strong wind currents at high elevations. Among the many concepts of airborne wind energy (AWE) generators, the soft-kite pumping-cycle (PC) concept promises to provide a very lightweight, high power density, and cost-effective solution. In this study, the impact of the load-cycle on the lifetime of the machine-side converter (MSC) is examined. By employing a physics-of-failure estimation approach, the main pumping-cycles and the machine speed-reversal were identified as the primary adverse influencers of the IGBT and diode solder joints. Whereas, wind speeds around 12 m/s contribute the most to the predicted degradation. To fulfill the thermal limitations and the lifetime requirements of the application, an optimum converter dimension is found using linear scaling of the semiconductors chip-area and the heatsink thermal impedances. With the generation (reel-out) phase power defined as the base value, the results suggest that the converter needs to be scaled by at least 150 % to meet the thermal constraints, and by 350 % to approach the target lifetime of ten years.

Introduction

The wind energy sector is today one of the largest renewable energy sources in terms of installed capacity and annual growth. The estimated total installed capacity accounts for 651 GW worldwide as of 2019 with a steady +50 GW new annual installations in the last five years [1]. This ever-increasing penetration combined with the intermittent nature of wind energy raised network security and stability concerns. As a result, a tendency towards offshore wind farms and large-scale turbines was one of the implemented solutions to mitigate the problem. This move is, however, not clear from technological and economic challenges involving the construction and cost of such massive infrastructures [1, 2, 3].

A promising alternative solution would be the fostering of HAWE resources above 200 meters using AWE generators [2, 4]. This unique technology makes use of faster and more consistent wind currents at high altitudes to generate power [4]; this is illustrated with the help of Fig. 1. A typical wind distribution profile at various altitudes is depicted; the average wind speed increases at higher elevations as witnessed from the flattening and the broader distribution of the Weibull curves.

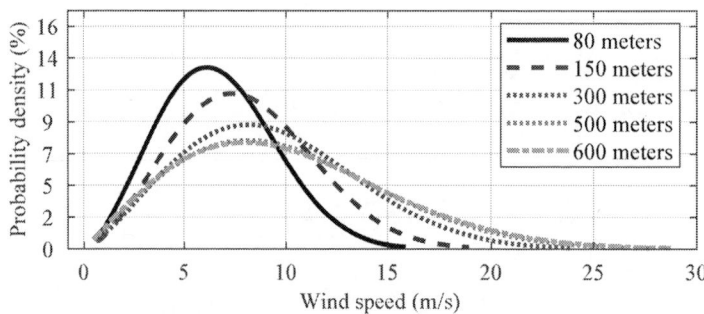

Fig. 1: Weibull distributions of wind speeds for an onshore location in northern Germany [5].

In a conventional wind turbine, the outer 30 % of the blades generate 50 % of the power. The rest of the turbine, namely: the tower and foundations are only needed to bring mechanical support [3], but they account for most of the material cost. Replacing the blades by a flying kite, and the turbine tower by a very long tether can cut down the material costs by a sizable margin [2, 4]. The result is a very lightweight, high power density and cost-effective system. Existing AWE prototypes have already achieved a power density 25-50 times that of equally rated conventional wind turbines. At the same time, cost estimates suggest a reduction of 50 % in investment costs as being very conceivable [4].

However, the replacement of stable structures with smart flying drones or kites requires sophisticated autopilot and launching/landing control systems [4]; besides newly imposed challenges concerning grid integration and power electronic system requirements. In-depth investigations addressing AWE system level and control challenges are already available in the literature [2, 4]. There is yet demand to address the challenges concerning the power electronics requirements for such systems.

There is a multitude of concepts of AWE systems. They can be classified according to the location of the electricity generation (fixed, moving, flying), the type of flight operation (crosswind, rotational, tether-aligned), the take-off and landing principle (propellers, lighter-than-air, wind-induced, mechanically induced), and the number of tethers or the structure of the flying system. In this work, we focus our attention on one specific type of AWE generators, namely the "pumping-cycle" or "Yo-Yo" system with ground-based generation unit and a soft-kite on a single tether.

This system has a unique load-cycle that could prove challenging for sizing the machine-side voltage source converter (VSC). Dimensioning of a battery-supported pumping-cycle AWE drivetrain based on the peak junction temperature requirements was studied in [6]. The authors found that the MSC is susceptible to 165 % overshoots in the IGBT/diode junction temperatures because of the reversal of the electrical machine speed. They estimated an over-dimensioning requirement of 250 % of the rated generation (reel-out) phase power under turbulent wind conditions. This was cut down to 155 % by adopting an active switching frequency control strategy [6].

In this paper, we carry out a physics-of-failure lifetime investigation procedure to evaluate the influence of the system load-cycle on the lifetime of the IGBTs and diodes. Using computer simulation, the system dynamics leading to the degradation of the modules are first analyzed. Afterwards, the dimensioning of a 2-level VSC to satisfy the lifetime requirements of 10 years is estimated applying a linear chip-area scaling process. This lifetime dimensioning approach can offer better insight into the true requirements of the power electronics. It shall also serve a comparative measure against the long-established solutions in the standard wind energy sector.

Principle of Operation

Pumping-cycle AWE systems consist mainly of a base station and a flying kite. Inside the ground station, the launch/release unit, electrical machine, rotating drum, power electronics, batteries, and control units

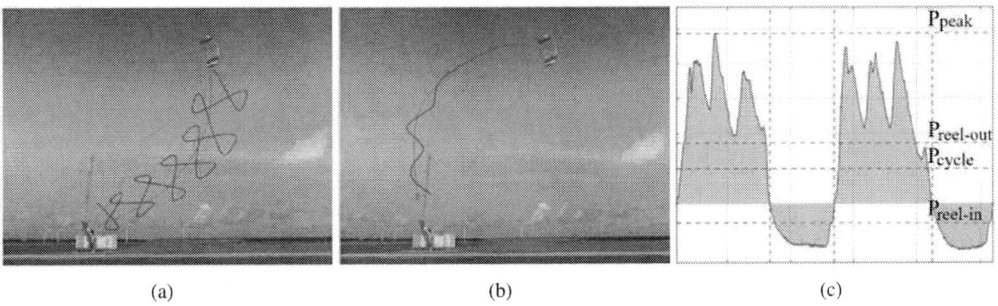

(a)   (b)   (c)

Fig. 2: Operation of a pumping-cycle AWE system, flight range 200-800 meters : a) Reel-out, b) Reel-in, c) mechanical power [2]. P_{peak} because of wind gusts; $P_{reel-out}$, $P_{reel-in}$, and P_{cycle} are average values.

are hosted. The kite is attached to the ground station through a strong tether; it is controlled wirelessly using a control pod attached to the end of the tether.

The load-cycle of this system is explained with the aid of Fig. 2. It involves a reel-out phase (energy generation), a reel-in phase (energy consumption) and the two transit phases in between. In the reel-out phase, the kite is released into the sky in an open-eight trajectory. This crosswind movement is essential to increase the relative velocity of the kite compared to wind speed, which ultimately intensifies the lift force applied to wings quadratically as in Equation 1. Loyd [7] estimated the generated mechanical power under ideal assumptions in Equation 2; where ρ is the air density, A is the airfoil area, v_a is the apparent wind speed at the wing, v_w is the wind speed, C_L and C_D are the lift and drag coefficients, respectively.

At the end of the reel-out phase, when the tether almost achieved its maximum length, the transit phase starts by steering the flying system out of the wind. This reduces the force on the tether and allows the kite to be reeled in with as little energy as possible. During the transit phase, the rotational speed of the generator is reduced. Once the speed reaches zero, the electrical machine operates as a motor to pull back the kite in a straight motion in the so-called "reel-in" phase. According to [2], to optimize the pumping-cycle power, the time without energy generation has to be minimized. Therefore, the reel-in speed is normally several times faster than the reel-out speed, and the transition between the two phases is performed quickly using high power.

$$F = \frac{1}{2} \rho A C_L v_a^2 \qquad (1) \qquad P = \frac{2}{27} \rho A v_w^3 C_L (\frac{C_L}{C_D})^2 \qquad (2)$$

After the tether has been pulled in and the kite has reached the desired height, the cycle restarts. The flight path is now considerably shorter because of the limited tether length; the motor also need not to apply prime torque to reverse the winch. Therefore, the transit from reel-in to reel-out is substantially faster than the transit from reel-out to reel-in. If the wind speed is too high to produce electricity, the system is flown to a safe position. Should the wind speed increase further, the system can be landed for safety. As the wind speed drops, the system can continue to produce electricity. If the wind speed is too low, the system is landed with little energy consumption. The wind speed required for take-off is usually higher than the wind speed required for electricity production.

Investigation Methodology

To examine the impact of pumping load cycles on the lifetime of the MSC, the algorithm in [8, 9, 10] is implemented, see Fig. 3. The primary goal is to find out the relation between the lifetime of the power electronic modules and the thermal cycles produced by the different dynamics of the system. Therefore, one-hour ideal load cycles (at constant wind speed without considering turbulence) are simulated using MATLAB Simulink for various wind speeds. A machine model is used to convert the mechanical loads

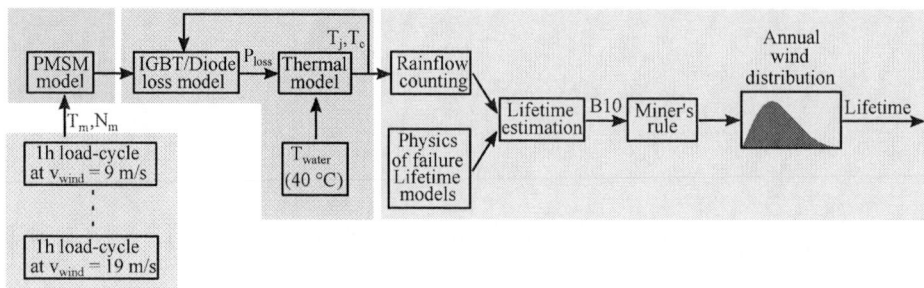

Fig. 3: Block diagram of the lifetime estimation algorithm.

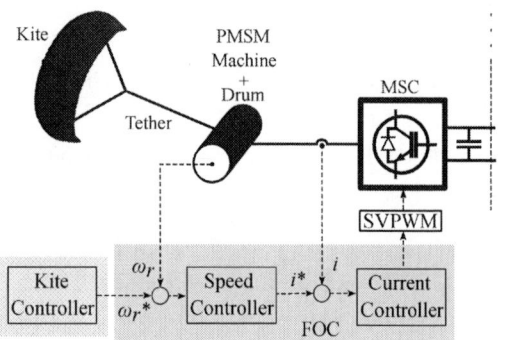

Fig. 4: Overview of the simulation model.

Table I: PMSM/VSC nominal parameters.

Apparent power	S_N	2.5 MVA
Line voltage	U_N	690 V
Phase current	I_N	2.1 kA
Power factor	$cos(\phi)$	0.8
Pair of poles	p	16
d-axis inductance	L_d	1.8 mH
q-axis inductance	L_q	3.4 mH
DC-Link voltage	V_{DC}	1.2 kV
Switching frequency	f_s	2 kHz

into voltage and current signals. Those signals are used by a thermal model to estimate the IGBT and diode temperatures. After that, the junction and case temperature cycles are counted using a Rainflow-counting algorithm implemented according to [11]. The results are compared against the lifetime models of the IGBT and the diode to predict the impact of individual thermal cycles. Finally, the linear accumulative impact of all thermal cycles is calculated using Miner's rule and extended for one year using an annual Weibull wind distribution.

System Modeling

An overview of the system model is illustrated in Fig. 4. The system under study is a direct drive AWE system where the mechanical system is coupled to the electrical system through a spinning drum. The flying kite is modeled as torque and speed profiles. The torque is defined as the electrical machine load, while the speed is applied as a reference signal for the speed controller.

Electrical model

The electrical system consists of an electrical machine, an MSC, and a DC bus, which could be an individual DC-link or a DC grid connecting many several kites. The MSC converter is realized as an average model with a sampling frequency equal to the switching frequency; Field oriented control (FOC) and space vector modulation (SVPWM) is the selected control structure.

The electrical machine is a ferrite permanent magnet synchronous machine (PMSM) with an outer rotor attached to the drum [12]. The machine is rated according to the average reel-out power $P_{reel-out}$, which is approximately two times the average pumping-cycle power P_{cycle}, it has a nominal voltage of 690 V. It is modeled as a voltage behind reactance using lookup-tables from a 2D FEM model running on maximum efficiency control. The electrical machine and the VSC parameters are listed in Table I.

Thermal model

The thermal model of the IGBT module comprises two parts; the temperature-dependent losses are first calculated using lookup-tables from manufacturers datasheets. In the second part, the losses are fed into the thermal impedance networks of the half-bridge. To reproduce the thermal cycles caused by the fundamental electrical frequency component, the instantaneous system variables are first discretized using a zero-order-hold sampled at the switching frequency. Afterwards, the losses within each switching period are calculated according to Equations 3 to 6. The presented loss models also depend on the chip-area scaling factor (SF) defined in Equation 7. A linear relation between the semiconductor chip-area and the associated losses has been assumed [13, 14]; this assumption could slightly reduce the accuracy of the model [15]. The electrical current is adapted by the scaling factor to find the equivalent current leading to the same current density.

$$P_{cond,S}(k,SF) = D(k) \cdot V_{CE}\left(\frac{I_{CE}(k)}{SF}, T_{j,S}(k)\right) \cdot I_{CE}(k) \tag{3}$$

$$P_{cond,D}(k,SF) = \left(1 - D(k)\right) \cdot V_F\left(\frac{I_F(k)}{SF}, T_{j,D}(k)\right) \cdot I_F(k) \tag{4}$$

$$P_{sw,S}(k,SF) = f_{sw} \cdot SF \cdot \left[E_{on,S}\left(U_{DC}(k), \frac{I_{CE}(k)}{SF}(k), T_{j,S}(k)\right) \right.$$
$$\left. + E_{off,S}\left(U_{DC}(k), \frac{I_{CE}(k)}{SF}, T_{j,S}(k)\right) \right] \tag{5}$$

$$P_{sw,D}(k,SF) = f_{sw} \cdot SF \cdot \left[E_{rec,D}\left(U_{DC}(k), \frac{I_F(k)}{SF}(k), T_{j,D}(k)\right) \right] \tag{6}$$

The thermal impedance is modeled as a 4-element Cauer network to filter any abrupt unrealistic changes in the case temperature level. For the initial analysis, a 1.7 kV, 2,4 kA HiPack IGBT module from ABB is used [16]. The liquid-cooled heatsink from Semikron [17] is selected for each half-bridge module. It is also modeled as a three-element Cauer network connected to the device models via the thermal interface resistance. Liquid cooling with a constant fluid temperature of 40 °C is regarded in this investigation. Both thermal impedances of the IGBT module and the heatsink are defined as a function of the scaling factor in Equations 8 and 9.

Based on the assumptions and compromises made in the thermal modeling of the converter, the results should only be regarded as reasonable estimates of the system performance and the underlying tendencies of the behavior.

$$SF = \frac{ChipArea^{new}}{ChipArea^{old}} \quad (7) \qquad R_{th}^{scaled}(SF) = \frac{R_{th}^{base}}{SF} \quad (8) \qquad \tau_{th}^{scaled}(SF) = \tau_{th}^{base} \quad (9)$$

Lifetime model

Several lifetime models are available in literature ranging from a simple Coffin-Manson model to more complicated physics-of failure models. In this work, the lifetime models from ABB are used [18], they are empirical models in which the number of cycles causing 10 % of the tested modules to fail (B_{10} lifetime) is estimated for specific joint types. The model differentiates between three categories of degradation, namely: chip solder joints, bond wire joints and substrate solder joints.

The model covers a specific range of thermal cycles in terms of temperature amplitude (T_{max}, T_{min}), variation range (dT), and cycle duration (t_{on}). In this investigation, certain necessary assumptions are considered. • In case the thermal cycle has a period longer than the maximum duration in the lifetime model, the maximum value in the model is applied based on the assumption that viscoplastic deformation is saturated [19]. • In case of short thermal cycles, the minimum cycle period in the lifetime model is used as a conservative approach. • Temperature amplitudes and variations beyond the lifetime model range are extrapolated. • Any temperature variation below 10 K is deemed insignificant for any impact on the lifetime [18, 20].

Discussion of Results

The complete system model is simulated for wind speeds ranging from 9 m/s to 19 m/s. The simulation results are discussed next.

Thermal simulation results

Fig. 5 depicts the IGBT/diode thermal cycles at the chip junction and module case levels. These cycles fall into two broad categories according to their dynamic origin. The first category contains cycles produced by the fundamental frequency of the current; their severity depends on the electrical frequency of the machine during the reel-out, reel-in, and transition phases. In the second category, we find thermal cycles generated by the crosswind open-eight kite trajectory and the primary pumping-cycles. Table II lists all cycles of interest according to their dynamic origin.

For extraction of maximum power, the electrical machine needs to supply high current to maintain constant torque during the transition from reel-out to reel-in phases while the speed is reversing. The decline of speed causes the absolute value of the power factor and the modulation index to reduce (the power factor sign is -negative during reel-out (generation) and +positive during reel-in (motor)). According to Equation 10, which describes the average conduction losses of the IGBT and the diode within one period of the fundamental frequency, the IGBT conduction losses will increase during this phase, whereas the diode conduction losses will fall. The change in the junction temperature produced by the conduction losses causes the switching losses to increase/decrease accordingly. Ultimately, the IGBT average junction and case temperatures rise during this transition phase, while the diode temperatures decline, this can be examined from Fig. 5(a).

$$P_{cond} = \left(V_{CEO,FO}(T_j).\frac{\hat{i}}{\pi} + r_{CE,F}(T_j).\frac{\hat{i}^2}{4} \right) \pm m.\cos(\phi).\left(V_{CEO,FO}(T_j).\frac{\hat{i}}{8} + \frac{1}{3\pi}.r_{CE,F}(T_j).\hat{i}^2 \right) \qquad (10)$$

The semiconductor devices could reach their highest temperatures during this period, as illustrated at the bottom of Fig. 5(b). The temperature overshoot is maximum when the machine speed reverses direction while the phase current approaches the top/bottom of the sine wave. Luckily enough, the coincidence of peak-current zero-speed combination does not strike the same semiconductor device in every pumping cycle. It instead circulates the three phases stochastically and depends among many factors on the phase shift and wind conditions. The transition from reel-in to reel-out is much faster; it is thus not as critical from the thermal point of view.

Table II: Categorization of thermal cycles according to their dynamic origin. T_j, T_c, T_s refer to junction, case, and heat sink temperatures respectively.

Domain	No.	Dynamic origin	Impact on	Cycle period
Electrical frequency	1	Reel-out speed	T_j	ms
	2	Reel-in speed	T_j	ms
	3	Transition phases speed	T_j, T_c	ms → s
	4	Speed zero-crossing Temperature overshoot (TO)	T_j, T_c	ms
Mechanical	5	Complete pumping-cycles	T_j, T_c, T_s	s → min.
	6	Open-eight trajectory	T_j	s

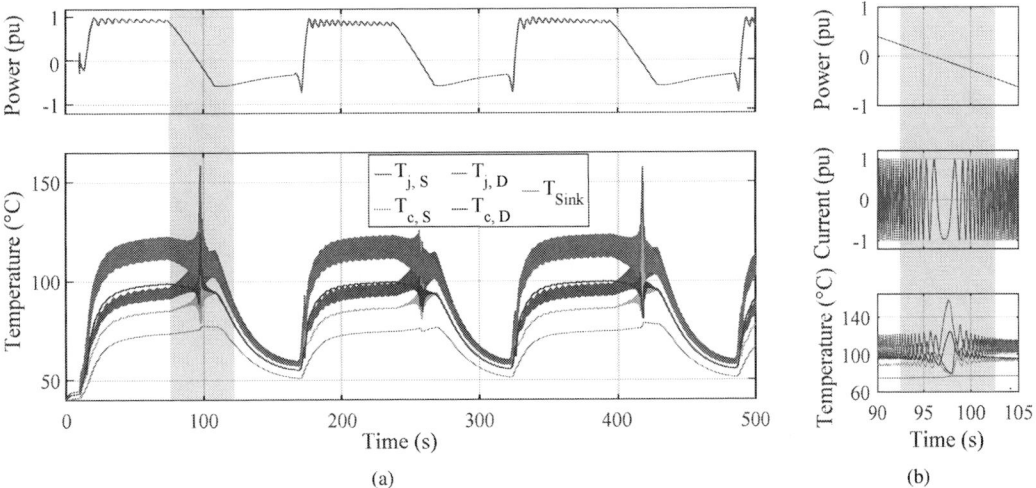

Fig. 5: Thermal simulation results at constant wind speed; power and current magnitudes are normalized to the reel-out phase values. (a) Complete thermal cycles due to reel-out/reel-in phases, (b) zoomed view at the instant nearby speed-reversal. $T_{x,S}$ and $T_{x,D}$ refer to IGBT and diode temperatures respectively.

Rainflow counting results

Based on the assumptions made in the lifetime model, the thermal cycles of the reel-in phase (No. 2) in Table II and crosswind movement (No. 6) are too small to develop plastic deformation in the solder joints; and therefore ruled out from the following discussion. The Rainflow results of the remaining active thermal cycles are depicted in Fig. 6 for the diode only; IGBT results are dropped due to analogy.

- Reel-out cycles (No. 1): they are very short cycles of small temperature variation range dT. There-fore, they reflect only on the junction temperature, and they do not propagate to the case level. Their count is very high due to their brief period and the long duration of the reel-out phase.

- Transit phase cycles (No. 3): they are slower cycles that contain more significant temperature variations dT; thus, they appear on both junction and case temperatures. In the case of the IGBT, they can also reach higher absolute temperatures compared to the reel-out phase because of the conduction losses. Their count, on the other hand, is far less than that of the reel-out cycles.

- The overlap between zero-speed temperature overshoots (No. 4) with the primary pumping-cycles (No. 5) generates the most significant stress on the semiconductor devices. They also have a fairly similar count to the transit phase cycles.

Lifetime estimation results

The accumulative impact of thermal cycles from all categories at individual wind speeds is estimated using Miner's rule for the wind distribution profile of Fig. 1. The results are depicted in Fig. 7 for the IGBT and the diode; they predict the following: • The substrate solder is the most affected joint, whereas the bond wire joints do not weaken so fast. • The degradation of the diode is quicker than that of the IGBT, mainly because of the higher temperatures during the reel-out phase. • Despite the higher transmitted power at faster wind speed, the impact of lower wind speeds around 12 m/s is the most critical; this suggests the more substantial impact of the annual Weibull wind distribution.

On the left side of Fig. 8, the lifetime results are broken down according to the dynamic origin of the thermal cycles. The pumping-cycles and temperature overshoot consume alone 100 % of the diode substrate lifetime within the first year. The exact stake of each of them is difficult to quantify due to the

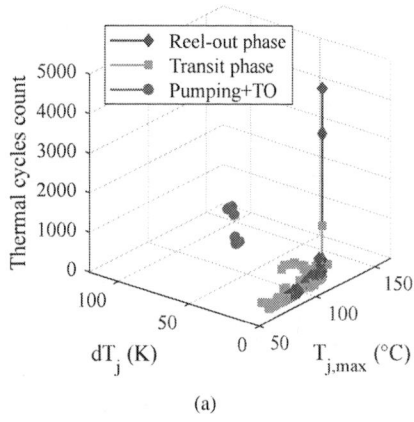
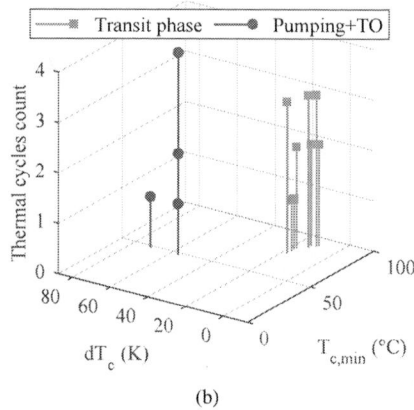

(a)   (b)

Fig. 6: Rainflow counting results for the diode effective thermal cycles at constant wind speed. (a) Thermal cycles of the junction temperature, (b) thermal cycles of the case temperature.

stochastic nature of the temperature overshoots. However, for a one-hour load profile at constant wind speed, they both contributed to approximately 50 % of the degradation; an equal proportion also applies to the chip solder joints albeit the lower overall consumption rate. On the other hand, the bond wire degradation depends equally on all classes of thermal cycles. The reason being that the bond wire lifetime models in use are independent of the period of the thermal cycle; therefore, the prolonged overheating effects of the primary pumping-cycles do not contribute to any further harm. The IGBT is also entirely dependent on the pumping-cycles and temperature overshoots. It is predicted that 70 % of its lifetime is consumed within the first year alone.

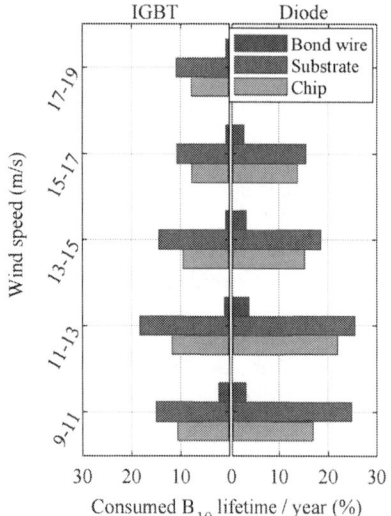
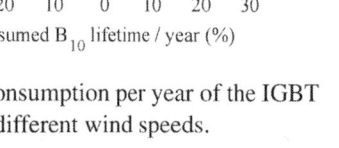

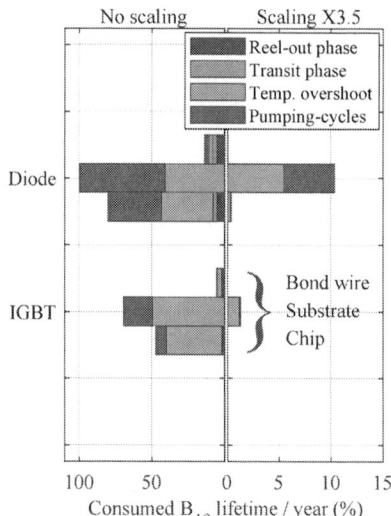

Fig. 7: Lifetime consumption per year of the IGBT and the diode for different wind speeds.

Fig. 8: Lifetime consumption per year of the IGBT and the diode according to the cycles category.

Converter Dimensioning

To determine the optimal converter size that can meet the lifetime target of 10 years, the IGBT, diode, and the heatsink are over-scaled. The thermal results are displayed in Fig. 9 for a constant-load variable-temperature scenario. Fig. 9(a) reveals that the maximum junction temperature of both devices reduces as the scaling factor increases. On the contrary, the overall losses in Fig. 9(b) decrease slightly for a

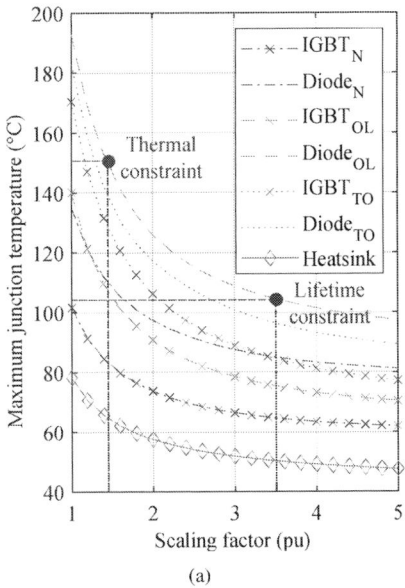 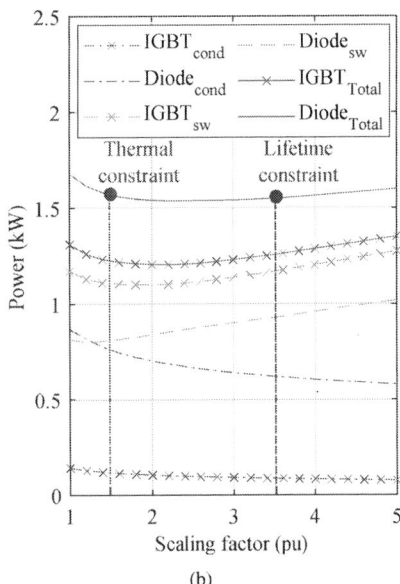

Fig. 9: Thermal simulation results for different converter sizes. (a) Maximum junction temperatures under nominal load (N), over-load because of wind guests (OL), and temperature overshoot (TO) scenarios, (b) conduction and switching losses for different converter sizes.

scaling factor between 1 and 2; it increases afterwards because of the switching losses. The reduction in temperatures can be accounted to the reduced junction to ambient thermal resistance. We conclude that selecting bigger converter systems can enhance the thermal performance of the system. However, once the build-up of switching losses overtakes the reduction in conduction losses and the maximum heat dissipation capacity of the heatsink, any further increase would only contribute to more losses; these findings agree with the results presented in [21]. For the selected IGBT module, which permits a maximum junction temperature of 175 °C, scaling the converter by at least 150 % would provide sufficient thermal capacity to limit the peak overload junction temperature of the diode to 150 °C. However, according to Fig. 10, the diode substrate can only last up to 1.5 years. From the same figure, we find that scaling the converter at least by 350 % is needed to reach the ten years lifetime target. Any further scaling beyond this value barely improves the lifetime because of the saturation of junction and case temperatures.

The breakdown of lifetime consumption for a scaling factor of 3.5 is depicted on the right side of Fig. 8. Notice that the reel-out and transit phase cycles contribute no more to the degradation process, whereas the total diode substrate consumption reduces to merely 10 % per year; the proportion between pumping-cycles and temperature overshoots remains unaffected.

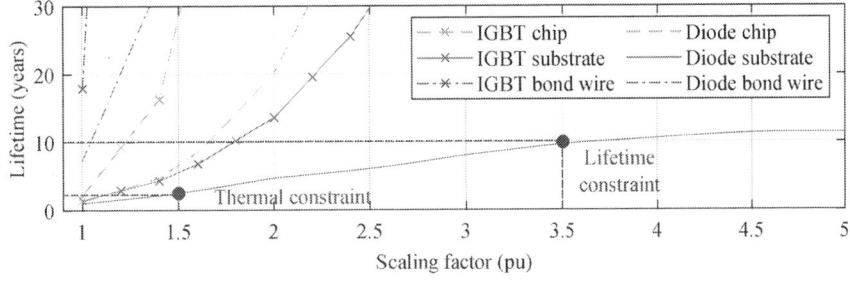

Fig. 10: Estimated lifetime of the module joints for different converter sizes.

Conclusion

The nature of the mission-profile of pumping-cycle AWE systems causes severe thermal cycles on the MSC. By using a physics-of-failure lifetime estimation approach, the primary pumping-cycles and machine speed-reversal were identified as the main contributors to the degradation of the power electronic modules. Whereas, wind speeds around 12 m/s have the most significant impact on the lifetime. Scaling of the silicon chip-area and heatsink thermal resistance is used to determine the optimum converter size that satisfies both the thermal and lifetime target constraints. It is found that there is a specific range within which larger semiconductor devices can reduce the overall losses of the system: whereas improving the performance of the heatsink has a better contribution to the overall thermal performance. An over-scaling of 350 % of the semiconductors chip-area and the heatsink is required to achieve a lifetime of ten years. To reduce the damage caused by the machine-speed reversal, it is possible to reduce the mechanical load in the critical speed range. The reversal could be delayed until the kite has flown to a low-force position. The associated small loss of efficiency in power generation would, by first estimation, be more economical than bearing the cost of an over-dimensioned electrical system.

From these results, we conclude that a standard converter solution might not be the best option for such an application. Therefore, alternative converter configurations, as well as active thermal control methods, will also be explored. Also keeping in mind that so far, only ideal wind conditions were assumed. In reality, the high volatility of the soft-kite in partnership with strong sudden wind gusts generates additional power spikes as in Fig. 2(c). Their influence on the MSC shall be investigated in future work.

References

[1] "Renewables 2019 Global Status Report," tech. rep., REN21, 2020.

[2] M. Diehl, U. Ahren, and R. Schmehl, *Airborne Wind Energy*. Springer, Berlin, Heidelberg, 2013.

[3] R. McKenna, P. Ostman v.d. Leye, and W. Fichtner, "Key challenges and prospects for large wind turbines," *Renewable and Sustainable Energy Reviews*, vol. 53, pp. 1212–1221, Jan. 2016.

[4] R. Schmehl, *Airborne Wind Energy*. Springer Berlin Heidelberg, 2018.

[5] K. D. Centre, "Dutch Offshore Wind Atlas." https://data.knmi.nl.

[6] D. von den Hoff, D. Haberschusz, N. Rotering, and R. W. De Doncker, "Design and Evaluation of a Battery-Supported Electric Drivetrain for Kite-Based High-Altitude Wind Energy Conversion," in *CWD*, 2019.

[7] M. L. Loyd, "Lawrence Livermore National Laboratory, Livermore, Calif.," *J. ENERGY*, p. 6, 1980.

[8] F. Fuchs and A. Mertens, "Steady state lifetime estimation of the power semiconductors in the rotor side converter of a 2 MW DFIG wind turbine via power cycling capability analysis," in *EPE*, 2011.

[9] K. Ma, M. Liserre, F. Blaabjerg, and T. Kerekes, "Thermal Loading and Lifetime Estimation for Power Device Considering Mission Profiles in Wind Power Converter," *IEEE Transactions on Power Electronics*, vol. 30, no. 2, pp. 590–602, 2015.

[10] M. Morisse, A. Bartschat, J. Wenske, and A. Mertens, "Dependency of the lifetime estimation of power modules in fully rated wind turbine converters on the modelling depth of the overall system," in *EPE*, 2016.

[11] E08 Committee, "Practices for Cycle Counting in Fatigue Analysis," tech. rep., ASTM International.

[12] S. Urbanek, D. Heide, B. Bagaber, M. Lohss, B. Specht, X. Paulig, A. Mertens, and B. Ponick, "Analysis of External Rotor Electric Drives for an All-Automatic Airborne Wind Energy System," in *IEMDC*, 2019.

[13] T. Friedli and J. W. Kolar, "A Semiconductor Area Based Assessment of AC Motor Drive Converter Topologies," in *Annual IEEE Applied Power Electronics Conference and Exposition*, pp. 336–342, 2009.

[14] A. Merkert, T. Krone, and A. Mertens, "Characterization and Scalable Modeling of Power Semiconductors for Optimized Design of Traction Inverters with Si- and SiC-Devices," *IEEE Transactions on Power Electronics*, vol. 29, no. 5, pp. 2238–2245, 2014.

[15] J. Kucka, *Quasi-Two-Level PWM Operation of Modular Multilevel Converters : Implementation, Analysis, and Application to Medium-Voltage Drives*. Doctoral Thesis, Leibniz Universität Hannover, 2019.

[16] ABB, "Data Sheet, 5SNA 2400E170305 HiPak IGBT Module," tech. rep., 2014.

[17] SEMIKRON, "DataSheet, SKiiP_2414_GB17E4_4DUW_V2," tech. rep., 2019.

[18] ABB, "Load-cycling capability of HiPak Modules," 2014.

[19] Infineon, "PC and TC Diagrams," p. 15.

[20] A. Wintrich, U. Nicolai, W. Tursky, and T. Reimann, *Application Manual Power Semiconductors*. Ilmenau: ISLE Verlag, 2nd revised edition ed., 2015.

[21] K. Ma, A. S. Bahman, S. Beczkowski, and F. Blaabjerg, "Complete Loss and Thermal Model of Power Semiconductors Including Device Rating Information," *IEEE Transactions on Power Electronics*, 2015.

A Design of High-Power Inverter Circuit Including GaN Power Devices

Takashi Sawada, Hiroshi Tadano, Koji Shiozaki
Institute of Materials and Systems for Sustainability
Nagoya University
Furo-cho, Chikusa-ku, Nagoya, Japan
Tel.: +81 / (0)-52-788-2620.
E-Mail: takashi.sawada@imass.nagoya-u.ac.jp,
htadano@imass.nagoya-u.ac.jp, kshiozaki@imass.nagoya-u.ac.jp

Acknowledgements

This research was supported by the Japanese national project of the Ministry of the Environment, "Technical innovation to create future ideal society and life style."

Keywords

Electric vehicle, Cooling, Gallium Nitride (GaN), Thermal design

Abstract

GaN power conversion circuits need to avoid overheating. This paper describes the design of inverter circuits including GaN power devices, focusing on dual cooling systems. and proposes DC-DC converter circuit which achieves the operation of up to 13kW.

Introduction

The ultra-high-speed switching power devices such as Gallium Nitride High Electron Mobility Transistor (GaN-HEMT) are expected to improve the performance in power conversion systems. Their switching characteristics [1] contribute to high efficiency and small size of circuits. When considering the high-power application of these high-speed switching devices, the electrical layout and cooling system should be carefully integrated.

The well-designed stray inductances in the power loop and the gate drive circuit contribute the safety operation and improve the efficiency of GaN based power electronics circuit [2]. Minimal stray inductance in the power loop keeps drain voltage surges low, and well-calibrated stray inductance in the gate loop suppresses the gate false turn-on phenomena [3]. That is also good approach for the automotive applications because minimizing stray inductance places electrical components close together, making the power conversion circuits smaller and lighter and confining the limited vehicle space.

The heat dissipation of power devices is also the essential issue to realize the best performance of power devices. The double-sided cooling structure is one of well-known solutions for reducing power devices' thermal resistance [4]. On the other hand, gate drive circuit components such as gate resistances, capacitors, buffers, and isolation ICs are difficult to be cooled due to the variety of their package structures. To protect these components from overheating, they must be isolated from the heat dissipation path of power devices.

The effective design of heat dissipation structure keeping the low stray inductance is one of the challenges in GaN based power electronics circuits.

In this paper, we propose a GaN inverter board for the automotive application. It achieves a double-sided cooling structure, and it is evaluated on the high-power continuous operation test.

The structure of half-bridge circuit and cooling unit

When one side of lateral GaN power device contacts a heat sink, the other side has to contact the circuit board. The proposed circuit board has bus-bars inside which transfer heat as well as drain current to the outside of the circuit board. These bus-bars provide an excellent thermal shield for gate drive components on the back side of this board. Therefore, the gate drive components are isolated from GaN devices' heat flow although they are located near the GaN devices.

Fig. 1: 4 paralleled GaN on the proposed half-bridge test circuit board

Fig. 1 shows the front of proposed inverter leg circuit board, which consists of 4 parallel GaN-HEMTs on each arm (8 devices total). In actual use of this board, two heat spreaders will contact with these GaN devices. Gate drive circuit components are located on the back side.

Fig 2. shows the half model of GaN half-bridge circuit board. This board wears several thermal interface materials. Heat spreaders contact to the top heat sinks through gap fillers, and bus-bars contact to the the side heat sinks through thermal sheets. The bus-bars with the thickness of 2.5 mm copper contained in circuit board are not only for large current conducting, but for heat dissipating. Therefore, this board provides "quasi-double-sided cooling." The bus bar also protects the gate drive circuit components on the opposite side of the board from high temperature.

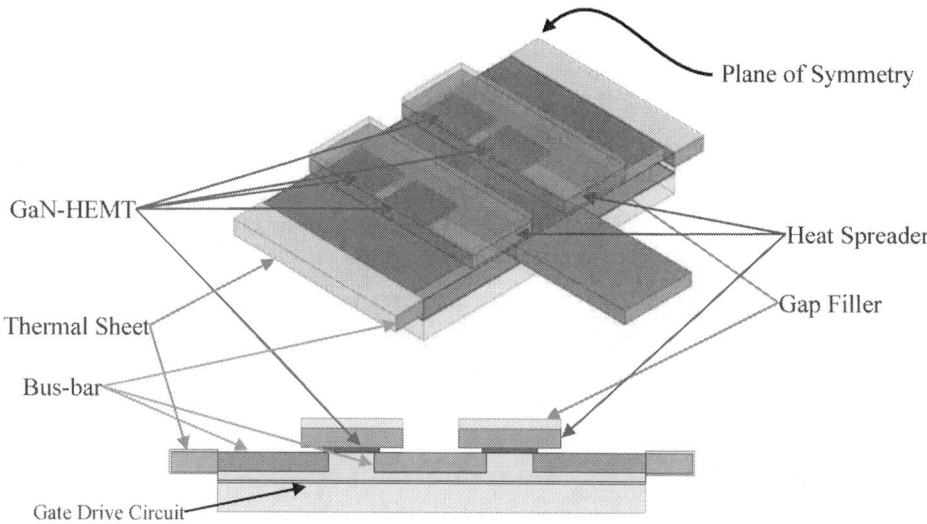

Fig. 2: Overall view and cross-section of the test circuit board half model

This half-bridge circuit board is intended to be finally installed on the electric vehicle (EV) inverter unit. Fig. 3 shows the shape of a combination of three GaN half-bridge circuit boards and heat sinks. This inverter unit installed to the three-phase inverter system including three half-bridge circuit boards for driving the in-wheel motors of EV. Fig. 4 shows exterior appearance of the proposed EV inverter.

Fig. 3: Three GaN half-bridge circuit boards and their heat sinks combination

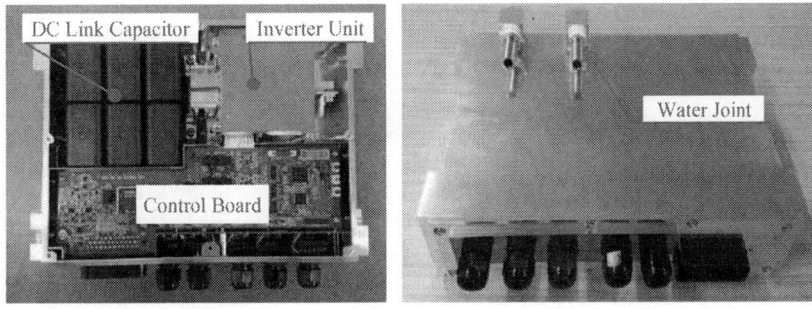

(a) top view　　　　　　　　　　(b) bottom view

Fig. 4: Exterior appearance of inverter unit including proposed GaN half-bridge circuit boards

Temperature distribution simulation of GaN half-bridge board

Before continuous current test of GaN half-bridge board, the effect of the heat shield of the bus-bars was confirmed by thermal simulation. Table I shows the conditions of the simulation, and table II shows the parameters of thermal interface materials.

Table I: Thermal simulation conditions

Simulator	ANSYS Mechanical Workbench 2019 R1
Heat Source and Density	GaN Devices 1.46 kW/cc
Boundary Condition	Surfaces in Contact with Heat Sink: Fixed Temperature 20 °C Other Surfaces: Insulated
Cooling Condition	Single-Sided Cooling: Top Surfaces are cooled Double-Sided Cooling: Top, Bottom and Side Surfaces are cooled

Table II: Parameters of the thermal interface materials

Components	Product Name	Thickness	Thermal Conductivity
Gap Filler	Henkel Bergquist GF3500LV	1.0 mm	3.5 W/mK
Thermal Sheet	FujiPoly 20GAR-AD	0.2 mm	3.0 W/mK

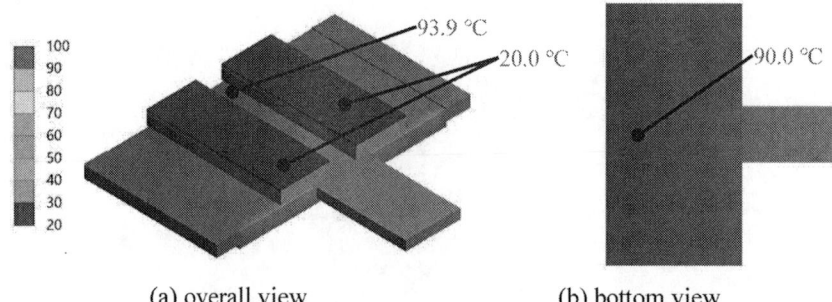

(a) overall view   (b) bottom view

Fig. 5: Thermal simulation results of single-sided cooling (half model)

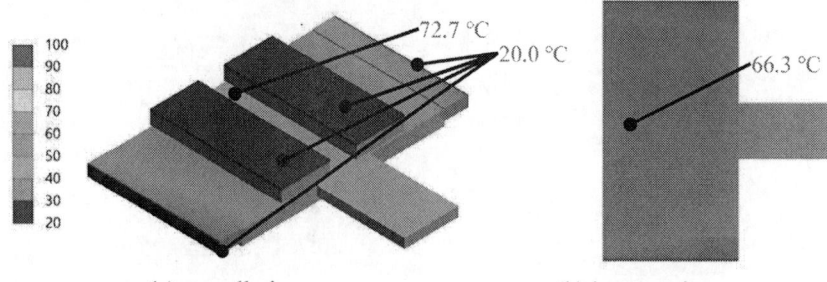

(a) overall view   (b) bottom view

Fig. 6: Thermal simulation results of quasi-double-sided cooling (half model)

Fig. 5 and Fig. 6 show a comparison of thermal simulation results for single-sided and quasi-double-sided cooling. The left sides are the overall views of the half-bridge circuit board half model, and the right side are their bottom views on which gate driver is mounted. The heat source is 4 paralleled GaN devices on each arm (8 devices total). The upper surface of the gap filler applied to the heat spreader is set to 20 degree C, which means they connect to infinite heat sink. Quasi-double-sided cooling uses the side heat sinks, so Fig. 6 shows that the surface temperatures of the both flanks are also set to 20 degrees C. Single-sided cooling does not use the side heat sink, so the both flanks are free to heat up as shown in Fig. 5.

These figures appear that the side heat sinks are effective in reducing the bottom side temperature rise by approximately 24 degrees C. This bus-bar cooling structure keeps the gate drive circuit components cool during high power inverter operation, which improve the reliability of the inverter.

Electrical experiment of GaN half-bridge board

Table III shows the major circuit components of proposed GaN half-bridge board. Because this board contains 4 parallel GaN power devices, the ferrite beads are placed in each miner loop of gate drive circuit in order to dampen the oscillation. The characteristics of the GaN half-bridge test board are evaluated on the buck converter continuous operation test as shown in the experimental setup and schematics Fig. 7.

Table III: Components of the GaN half-bridge circuit

Power Transistor	GaNSystems GS66516T
Gate ON / OFF Resistance	47 ohm / 4.7 ohm
Gate Ferrite Bead	33 ohm @ 100 MHz
Gate Drive IC	Rohm BM6104FV

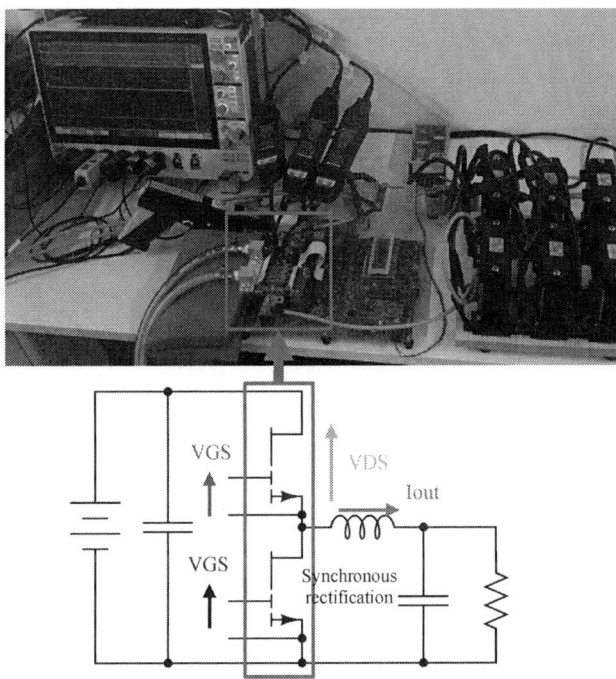

Fig. 7: Experimental setup and schematic of back converter continuous operation test

The GaN devices on high side are switching, and those on low side are synchronous rectifying. Table IV shows the test conditions. Since this test is high power continuous operation, the inverter's GaN half bridge board is water cooled. The GaN half-bridge board has test electrodes that measure the voltage waveforms of the gate and drain with Tektronix IsoVu [5], so the experiment runs safely without being disturbed by common mode noise.

Table IV: The evaluation conditions

DC Input Voltage	300 V
Output Power	2 kW - 13 kW
Duty	50 %
Switching Frequency	20 kHz
Dead Time	100 ns
Coolant Temperature	20 ℃

Experiment results of the GaN half-bridge board

Fig. 8 shows the examples of switching waveforms of buck converter continuous operation test at Vout=150V, Iout=90A. Blue waveform is GaN gate-source voltage (VGS) on high-side (switching side), black one is that of the low side (synchronous rectification side), and yellow one is high side drain-source voltage (VDS). Low-side VGS undershoot and rise occur simultaneously with changes in VDS. As shown in Fig. 8, a high-power continuous operation succeeded up to 13kW.

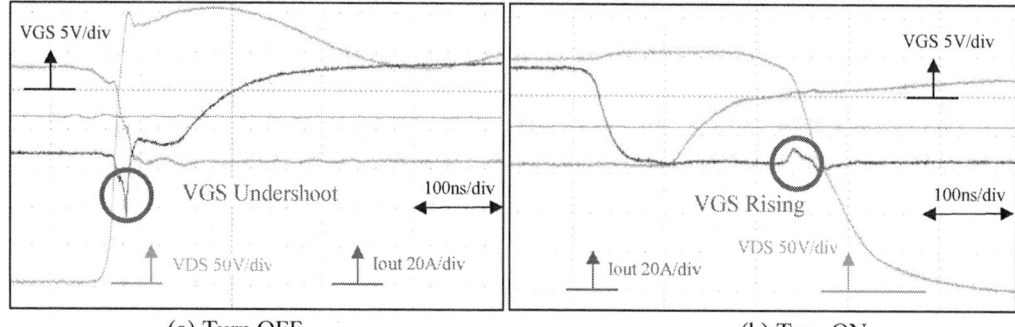

(a) Turn OFF (b) Turn ON

Fig. 8: Examples of switching waveforms of buck converter test (Vin=300V, Vout=150V, Iout=90A)

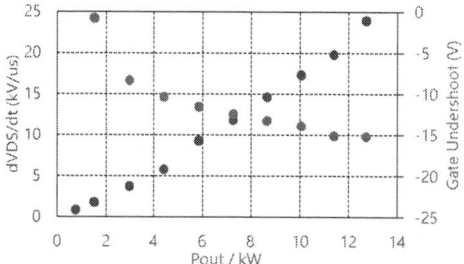

Fig. 9: Experimental results of the continuous operation of the buck converter.

Fig. 9 shows the rising ratio of high side VDS and the undershoot voltage of low side VGS vs. output power. This experiment performed up to 13kW to keep the undershoot below GaN device's negative gate voltage limit.

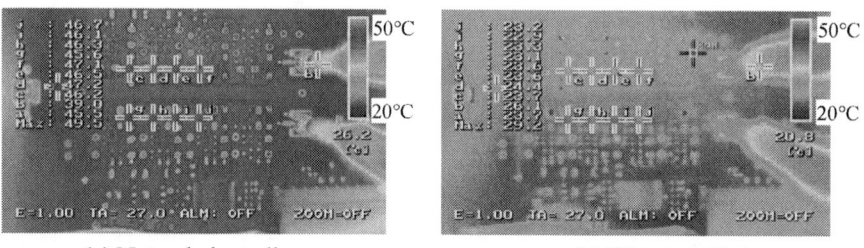

(a) Natural air cooling (b) Water cooling

Fig. 10: The comparison of back surface temperatures at the same output power 1.0kW

Fig. 10 is compering the surface temperatures of half-bridge board's gate driver side between natural air cooling and water cooling under the same output power conditions of 1.0kW. Output power is throttled to avoid device overheating even with natural air cooling. Looking at these figures, we can see that the bus-bar cooling is properly functioning as a thermal barrier and protecting the gate drive circuit components.

Conclusion

We proposed a GaN half-bridge circuit board structure with double-sided cooling to protect the gate drive circuit components from the heat of the GaN device. Successful 13kW continuous test. The surface temperature measurements of that test proved that the busbar was functioning properly as a heat shield. In order for this circuit to operate at higher power, gate drive conditions such as component values should be investigated to reduce gate undershoot.

References

[1] J. Millan, et. al., "A Survey of Wide Bandgap Power Semiconductor Devices," IEEE Trans. Power Electron., vol. 29, no. 5, pp. 2155-2163, May 2014.

[2] W. Kangping, et. al., "An Optimized Layout with Low Parasitic Inductances for GaN HEMTs Based DC-DC Converter," in 2015 IEEE APEC, Charlotte, NC, USA, 2015, pp. 948-951.

[3] A. Nishigaki, et. al., "An analysis of false turn-on mechanism on power devices," in 2014 IEEE Energy Conversion Congress and Exposition (ECCE), Pittsburgh, PA, USA, Sep. 2014, pp. 2988-2993

[4] Y. Xu, et. al., "Development of an Ultra-high Density Power Chip on Bus (PCoB) Module," in 2016 IEEE ECCE, Milwaukee, WI, USA, 2016, pp. 1-7.

[5] TIVM Series IsoVu Measurement System: Users Manual, Tektronix Inc.

Speed Sensorless Commissioning of Resonating Mechanical System in Electric Drives

A. Putkonen, N. Nevaranta, O. Liukkonen, M. Niemelä, and O. Pyrhönen,
LUT University, Department of Electrical Engineering
P.O. Box 20, 53851, Lappeenranta, Finland
Lappeenranta, Finland
Email: atte.putkonen@lut.fi

Keywords

≪Electrical Drive≫, ≪Identification≫, ≪Mechanical Resonance≫, ≪Sensorless≫.

Abstract

Artificial excitation signal based commissioning routines has been widely applied for the identification of the resonating mechanical system in electric drives. The commissioning tests encountered in the industry are mostly related to the detection of the first resonating mode, to be more specific, to the estimation of the two-mass-system model approximation needed for the velocity PI controller tuning. A standard identification procedure uses the torque reference as an input signal and the output is the measured velocity signal obtained from the encoder. However, most of the industrial drives are based on sensorless operation, thus there is no possibility to measure the velocity, but some information from the mechanical system is needed for the control design. This paper addresses to the issues related to the speed sensorless identification under closed loop control. A closed loop identification routine where a pseudo random binary sequence (PRBS) is superposed to the velocity control output is studied experimentally by considering a standard commercial frequency converter in sensorless mode driving an induction machine that is connected to resonating mechanical system.

Introduction

In many industrial applications, the performance of the electrical drive system is mostly limited by characteristics of the mechanical system, such as flexibilities. Depending on the application, the mechanical resonances can be often located in low frequencies range, hence limiting the dynamic performance of the closed loop control. In most cases, an approximation of the first resonating mode, namely, a two-mass-system model is needed for the control design purposes [1], [2]. As a result, several artificial excitation signal based commissioning routines has been proposed in the literature especially for the identification of the closed loop velocity controlled mechanical system [3]–[4]. The most typical practical identification test is based on the superposition of the excitation signal to the velocity controller output and after that the torque reference and measured velocity is used to estimate frequency response [5]. This approach is referred as direct identification routine, but lot of research has been in indirect identification where the influence of the closed loop controller is taken into account.

However in some industrial applications, like wind mills, the speed encoder is not typically present and the closed loop velocity control rely on the estimates. Depending on the application the dynamics of the mechanical system should be known based on the design or experiments to support the control commissioning. Naturally, the identification of the mechanical system can be based on the estimates produced by the electrical drives. In this case, the routine is speed sensorless identification that has been acknowledged in the covering literature. In [7] an encoderless identification based auto-tuning of PI controller is proposed based on the proposed speed estimator. It is noted, that the paper do not focus on the resonance identification and one-mass-system approximation is used for the tuning procedure. Another

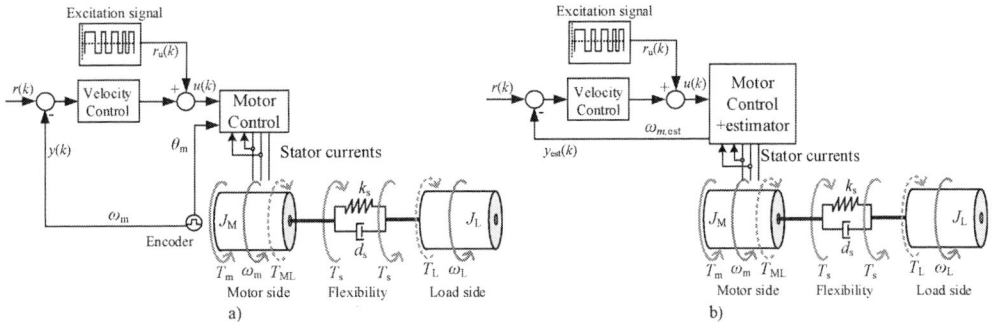

Fig. 1: The closed loop identification approach a) when speed encoder is applied as feedback and b) when the motor control includes a velocity estimator. The excitation signal is superposed to the velocity control output.

routine is studied in [8], [9], [10], [11] where a Luenberger adaptive observer structure is proposed to obtain the speed estimate needed for the identification routine. In [13] the diagnostics to detect torsional oscillations is studied and the results are based on tests obtained during sensorless operation. This paper addresses issues in the speed sensorless identification of a resonating mechanical system operating under closed loop velocity controller. The sensorless identification is studied with simulations and experimentally by considering a standard commercial frequency converter where the velocity estimation routine is based on the motor electrical parameters obtained from the identification run. The experimental results demonstrates the effectiveness of the sensorless identification especially as the particular focus is on the determination of the dominating resonance in the system.

Closed-Loop Identification

Due to safety constraints and practical limitations, in most industrial cases the identification must be carried out under closed loop control. A generalized principle of velocity controlled mechanical system is shown in Fig. 1 a) where the measured velocity $y(k) = \omega_\mathrm{m}(k)$ is used as feedback to the control law. Often with such a configuration, the artificially generated excitation signal $r_\mathrm{u}(k)$, e.g. a maximum length binary signal (MLBS), is superposed to the torque reference to obtain identification experiment data that is rich in frequencies. Usually the excitation signal must be designed accordingly for the system in consideration, otherwise, unwanted dynamics might be excited. Usually the limiting factor of the excitation signal design is the amplitude [12]. Depending on the closed loop identification routine, indirect [6] or direct [1], the estimated frequency response is based on the known excitation signal and/or measured input and output signals. In this paper, the direct identification routine is considered where the estimation is based on the input (torque) and output (angular velocity) signals. The well-known direct frequency response estimation by using the input and output signals is written as

$$\hat{G}(e^{j\omega}) = \frac{\hat{S}_\mathrm{uy}(e^{j\omega})}{\hat{S}_\mathrm{uu}(e^{j\omega})}, \tag{1}$$

where the $\hat{S}_\mathrm{uy}(e^{j\omega})$ is the cross spectral estimate between the input signal $u(k)$ and output signal $y(k)$ and the $\hat{S}_\mathrm{uu}(e^{j\omega})$ is the auto spectral estimate of the input signal, respectively. The spectral estimates are obtained using the Welch method where the input and output signals are divided to eight overlapping sections that are windowed with a Hamming window. The derivation of the Welch's frequency response estimate can be found e.g. in [14]. The closed loop identification experiments are carried out with an experiment corresponding to Fig. 1 b), where the closed loop feedback is based on the velocity estimate provided by the frequency converter. It is noted that the configuration in Fig. 1 a) is used to validate the obtained sensorless identification results.

Maximum Length Binary Signal

The utilization of maximum length binary signal (MLBS) give the possibility to design an excitation signal rich in frequencies. The length N of the MLBS period is defined by the amount of shift registers d

$$N = 2^d - 1. \tag{2}$$

A signal with eleven shift registers is considered and the signal has a generation frequency of 250 Hz. The length of the one MLBS period is 8.18 seconds when the switching time is $T_{sw} = 4$ ms. The parameters of the identification experiments are given in the Table I. The maximum sampling frequency f_s is limited by hardware to 1 kHz.

Table I: Key parameters of the identification conditions. The motor rated torque T_n is 49.5 Nm.

Parameter	Value
Proportional gain K_P	5
Integration time T_I	2.5 s
Feedback filter time T_f	0 s
MLBS amplitude A_{MLBS}	$0.2 \cdot T_n$
Shift register d	11
Control sampling time T_{sc}	500 μs

Simulations

The sensorless identification is studied with simulations by using modified Mathworks™simulation model (*ac3_sensorless*) [15] that is used with the values given in Tables I and II. It is worth remarking that the motor control approach do not correspond to the experimental frequency converter's one, but it is suitable to study the influence of the inner control functions to the identification result. The estimators used in the control uses the values given in Table II. In the simulation model a model reference adaptive system (MRAS)-type velocity estimator is used for the sensorless mode that forms the estimate from the terminal voltages and phase currents. This structure is depicted in Fig. 2, where there PI-control structure is used to form the estimate from the fluxes of the reference and adaptive models as follows

$$\hat{\omega}_r = \xi \cdot \left(K_{P,\,est} + \frac{K_{I,\,est}}{s} \right), \tag{3}$$

where the $K_{P,\,est}$ and $K_{I,\,est}$ are the parameters for the estimators PI-control. In the sensorless identification it is known that the estimator dynamics can have a major influence on the results [11]. Thus, the simulation model was simulated with different proportional gains of the estimator. The Fig. 3 shows the frequency response identification results for the sensorless operation as well as the Welch's power spectral density of the estimated velocity.

The result in Fig. 3 a) shows that by changing the estimator dynamics, namely the gain, the frequency response estimate is influenced by the changes. Noisier frequency response estimate would lead to poorer parametric estimate of the model. However, with only the mechanical resonance detection in mind the Fig. 3 (b) shows that the resonance can be seen in the spectrum of the velocity signal. The estimator's gain is highly affecting the width of the peaks and thus making them more difficult to notice. In [16] it is also shown the estimates of the motor's electrical parameters are highly affecting the MRAS-type velocity estimator's performance and hence would most likely also have a negative effect on the resonance detection.

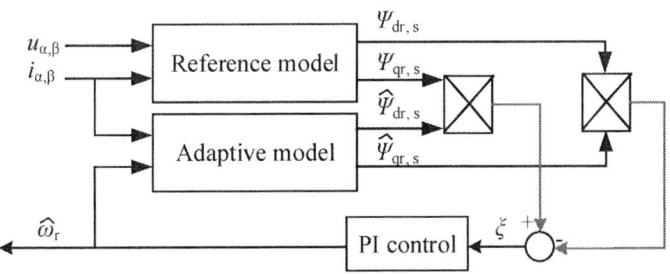

Fig. 2: The MRAS-type velocity estimator structure used in the simulations

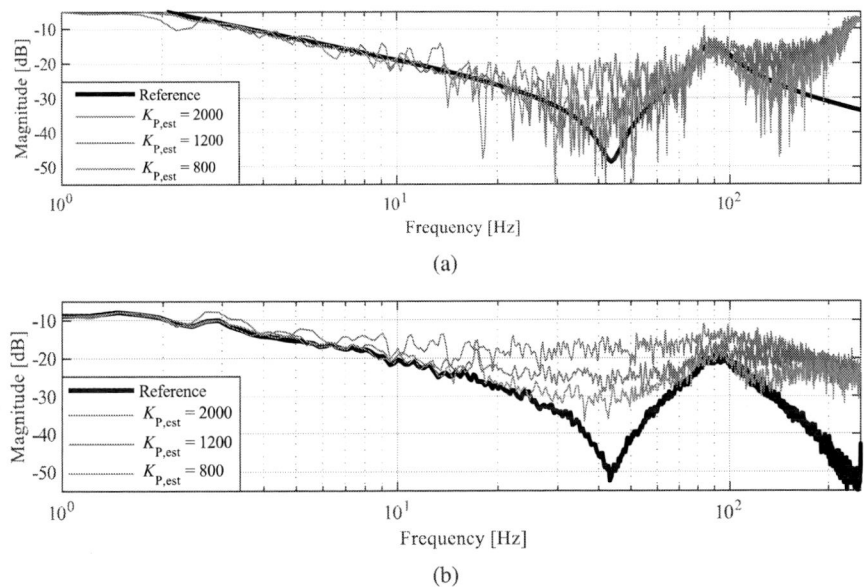

(a)

(b)

Fig. 3: Simulation results where the velocity reference was 500 rpm while the estimator's PI controller's proportional gain $K_{\text{P,est}}$ was varied. (a) shows the frequency response between torque and estimated speed and (b) shows the Welch's power spectral density estimate of the velocity signal. The reference of (b) is the velocity spectrum when estimator is not used.

Table II: The electrical parameters of a 7.5 kW 400 V 50 Hz 1440 rpm induction motor and the initial mechanical parameters used in the simulation. Note: the torsional stiffness and damping are estimated from the experimental results with 50 % of the nominal load and with KTR Rotex 92 Shore A coupling.

Electrical		Mechanical	
Parameter	Value	Parameter	Value
Stator resistance R_{s}	0.62276 Ω	Motor inertia J_{m}	0.034 kgm^2
Stator leakage inductance $L_{\text{s}\sigma}$	2.6493 mH	Motor friction b_{m}	0.005 Nm·s/rad
Rotor resistance R_{r}'	0.59088 Ω	Load inertia J_{l}	0.103 kgm^2
Rotor leakage inductance $L_{\text{s}\sigma}'$	5.2734 mH	Load friction b_{l}	0.005 Nm·s/rad
Magnetizing inductance L_{mag}	98.708 mH	Torsional stiffness k_{t}	$7.8 \cdot 10^3$ Nm/rad
Pole pairs p	2	Torsional damping c_{t}	3 Nm·s/rad

Experimental Tests

The identification tests are carried out on an experimental system depicted in Fig. 4, where a 7.5 kW induction machine is velocity controlled with an ACS880 frequency converter and an 11 kW induction motor is used as a loading machine. The shafts of the machines are connected with KTR Rotex coupling. The programmable logic controller (PLC) is used to control the frequency converters, data acquisition and to generate excitation signal, namely, a maximum length binary signal (MLBS) that is superposed to the velocity control output. The minimum task cycle of 1 ms of the used PLC limits the sampling frequency to only 1 kHz. However, the low sampling frequency is not a limiting factor in this case since the torsional resonances are located well below 1 kHz. The sensorless identification is evaluated using couplings with different hardness in the experimental system and compared with the results obtained using traditional closed loop identification with encoder. It is worth noting that the torsional stiffness of the couplings is a function of the total torque of the system and thus also the resonance will change for different torques.

Identification results

The sensorless identification is studied in different closed loop operation conditions; both velocities and loading. In Fig. 5 the frequency responses from the sensorless identification are shown with the ones obtained when the encoder is included in the feedback loop. Although there are slight difference between the estimated frequency responses, based on the results from sensorless identification the characteristics of the resonating mechanical system is evident and resonance frequency can be clearly observed from the result. To further analyse the resonance detection from the sensorless identification, in Fig. 6 the Welch's power spectral density estimates from the measured and estimated velocities are shown under different loading conditions, where the black dashed line represent the actual resonance of the mechanical system. Based on the velocity power spectral density estimates it can be seen that the identified resonances are close to each other, thus the velocity estimate gives a clear indication of the dominant dynamics. To further support this observation the identification has been also evaluated with different coupling configuration and the results can be found in Table III. By comparing the identification results with the actual resonance (reference) it can be noticed that the results are in good agreement with those of encoder tests.

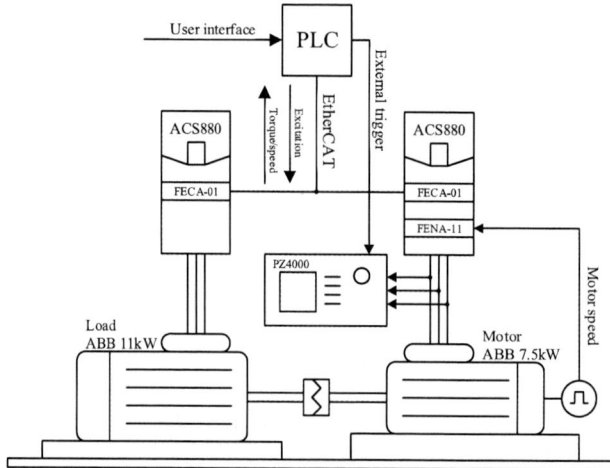

Fig. 4: The experimental system with a programmable logic controller (PLC), frequency converters and induction machines manufactured by ABB. [17] The shafts of the motors are connected with a shaft coupling clutch.

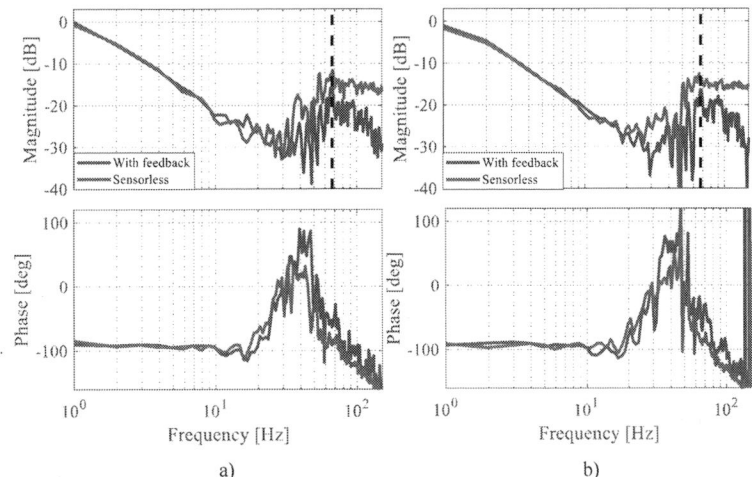

Fig. 5: The experimental identification results with 25 % of the nominal load a) when the velocity is 1000 rpm and b) 1447 rpm.

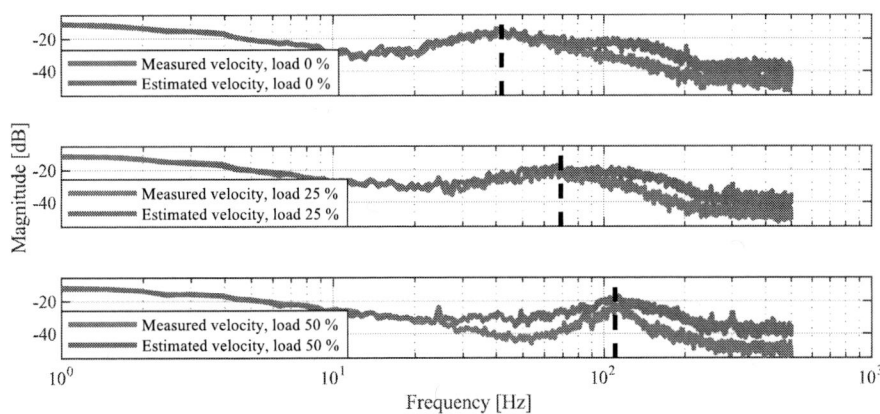

Fig. 6: The spectrum of the measured and the estimated velocity under different loading conditions. The dashed black line represent the actual resonance.

Table III: Identified resonances from velocity power spectral density estimates using different couplings in the system. [17]

Load	Coupling type	Reference resonance	Encoder	Sensorless
0 %	98 Shore A	42 Hz	41 Hz	41 Hz
25 %	98 Shore A	69 Hz	67 Hz	66 Hz
50 %	98 Shore A	110 Hz	111 Hz	110 Hz
0 %	92 Shore A	35 Hz	33 Hz	36 Hz
25 %	92 Shore A	64 Hz	63 Hz	64 Hz
50 %	92 Shore A	88 Hz	88 Hz	91 Hz

Conclusion

This paper studied the speed sensorless identification of resonating mechanical system in electric drives. Based on the experimental results it was shown that the resonance related to the mechanical system can be reasonably identified without velocity encoder in the control loop. One important remark is made based on the results; the spectrum of the estimated velocity gives useful information about the location of the resonance even though the estimated model is not sufficient enough. By using only the estimated velocity signal the resonance detection process will be faster and more straightforward. It's useful during commissioning since it can be used to quickly check and verify the dominating dynamics of the system onsite. By knowing the actual resonance it can be used e.g. for controller tuning if a maximum bandwidth is desired.

References

[1] Wahrburg A., Jelavic E., Klose S. and Listman K. D.: Robust semi-automatic identification of compliantly coupled two-mass system, *IFAC – PapersOnLine*, vol. 50, no. 1, pp. 14569–14574, July 2017.

[2] Wernholt E. and Gunnarsson S.: Nonlinear identification of a physically parameterized robot model, *14th IFAC Symp. on Syst. Ident.*, vol. 39, no. 1, pp. 143-148, 2006.

[3] Nevaranta N., Derammelaere S., Parkkinen J., Vervisch B., Lindh T., Niemelä M. and Pyrhönen O.: Online identification of a two-mass system in frequency domain using a Kalman filter, *Model. Ident. Control.*, vol. 37, no. 2, pp. 133–147, 2016.

[4] Pacas M. and Villwock S.: Methods for commissioning and identification in drives, *Int. J. Comput. Math. Elect. Electron. Eng.*, vol. 29, no. 1, pp. 53–71, Oct. 2010.

[5] Saarakkala S. E. and Hinkkanen M.: Identification of two-mass mechanical systems using torque excitation: Design and experimental evaluation, *Trans. on Ind. Appl.*, vol. 51, no. 5, pp. 4180–4189, Sep./Oct. 2015.

[6] Isaksson A.J., Lindkvist R., Zhang X., Nordin M.C. and Tallfors M.: Identification of Mechanical Parameters in Drive Train Systems, *13th IFAC Symp. on Syst. Ident.*, Rotterdam,Netherlands, 2003, pp. 1501–1506.

[7] Weber A.R., Weissbacher J., Steiner G. and Horn M.: An accurate auto-tuning procedure for encoderless AC motor drives in industrial environments, *Trans. on Electric. Eng.*, vol. 3, no. 1, pp. 1–7, 2014.

[8] Zoubek H. and Pacas M.: Encoderless Identification of Two-Mass-Systems Utilizing an Extended Speed Adaptive Observer Structure, IEEE Trans. on Ind. Electron., vol. 64, no. 1, pp. 595–604, Jan. 2017.

[9] Zoubek H. and Pacas M.: Two Steps towards Speed Estimation and Encoderless Identification of Two-Mass-Systems with Extended Speed Adaptive Observer Structure, IECON 2011 - 37th Annual Conf. of the IEEE Industrial Electronics Society, 2011, pp. 2072–2077.

[10] Zoubek H. and Pacas M.: An identification method for multi-mass-systems in speed-sensorless operation, 2011 IEEE International Symposium on Industrial Electronics, 2011, pp. 1895–1900.

[11] Zoubek H. and Pacas M.: A method for speed-sensorless identification of two-mass-systems, 2010 IEEE Energy Conversion Congress and Exposition, 2010, pp. 4461–4468.

[12] Pintelon R. and Schoukens J.: System Identification: A Frequency Domain Approach, IEEE Press, ISBN 0-7803-6000-1, 2008.

[13] Orkisz M. and Ottewill J.: Detecting Mechanical Problems by Examining Variable Speed Drive Signals, International Symposium on Power Electronics, Electrical Drives, Automation and Motion, 2012, pp. 1366–1371.

[14] Villwock S. and Pacas M.: Application of the Welch-Method for the Identification of Two- and Three-Mass-Systems, IEEE Trans. on Ind. Electron., vol. 55, no. 1, pp 457–466, Jan. 2008.

[15] Souleman N. M. and Dessaint Louis-A.: Mathworks Simulation model: https://www.mathworks.com/help/physmod/sps/examples/ac3-sensorless-field-oriented-control-induction-motor-drive.html

[16] Iqbal A. and Husain M. A.: MRAS based Sensorless Control of Induction Motor based on Rotor Flux, CCTES, 2018, pp. 152–155.

[17] Putkonen A.: Detection of Torsional Resonance Frequency With Variable Frequency Drive, Master's Thesis, LUT University, 2020.

Control of a two-stage, single-phase grid-tied, GaN based solar micro-inverter

Anthony Bier, Van Sang Nguyen, Stéphane Catellani, Jérémy Martin
CEA / Solar Technologies Department
50 avenue du Lac Léman
Le Bourget du Lac - 73375, France
Tel.: +33 / (0)4 79 79 21 54
E-Mail: anthony.bier@cea.fr
URL: www.cea.fr

Acknowledgements

This project was partially funded by an European project – No 792245 (project #SuperPV).

Keywords

«Converter control», «Digital control», «Voltage Source Inverters (VSI)», «Single phase system», «Photovoltaic», «Wide bandgap devices», «Gallium Nitride (GaN)», «Simulation», «Field Programmable Gate Array (FPGA)», «Pulse Width Modulation (PWM)», «Reactive power», «ZVS converters».

Abstract

This work presents the modelling and the implementation of the control of a single-phase grid-tied GaN micro-inverter dedicated to photovoltaic applications. First, the power hardware based wide-band-gap components is shown, and then, switches pulses control blocks and control loops of this specific grid-tied inverter are described.

Introduction

In a conventional photovoltaic (PV) power plant, PV modules are connected in series. Then, as the current is the same for each modules, unbalanced condition between several modules, in terms of irradiations, shading, orientation, aging, dirt or temperature, can significantly reduce the energy yield of the entire PV string, even with bypass-diode use [1]. However, using module-level converter like micro-inverter allows, by tracking individually the maximum power point (MPP), to extract the maximum available power of each PV modules. In case of converter failure, it can also ensure power production continuity compared to a centralized inverter. The trend to integrate the energy conversion part (DC/DC or DC/AC) into the PV module junction box and the emergence of the bifacial cell technology [2] brings new challenges for designing compact micro-inverter in which the high performances wide-band-gap (WBG) semiconductors can be implemented. The hardware of a WBG based micro-inverter is presented. Then the implementation of the required control blocks, designed to ensure good performances and stability are described.

PV micro-inverter power-stage prototype

The interleaved flyback topology is the most widespread of all the solar micro-inverter topologies in industrial products [3]. However, in our study, from the comparison of several micro-inverter topology, a two-stage converter (FED DC/DC and full bridge DC/AC) has been designed and prototyped [4]. The schematic of the power stages of the inverter is given Figure 1.

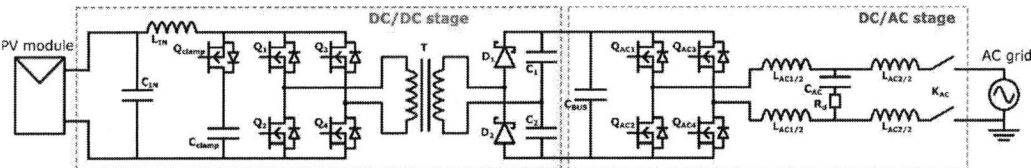

Figure 1: Power stage schematic of the micro-inverter

A picture of the prototype is shown Figure 2, it is based on 100V GaN HEMT components (GS61008T), 650V GaN FET components (TPH3206PSB). SiC diodes (C3D03065E) and Si82xx family gate drivers with bootstrap supplies. A full-custom planar transformer is built up in order to reach high performances and reduce the voltage stress on the primary GaN devices. The power density is about 0.85 kW per liter, including the control board along with sensors.

Values of the main passives components are listed in Table I.

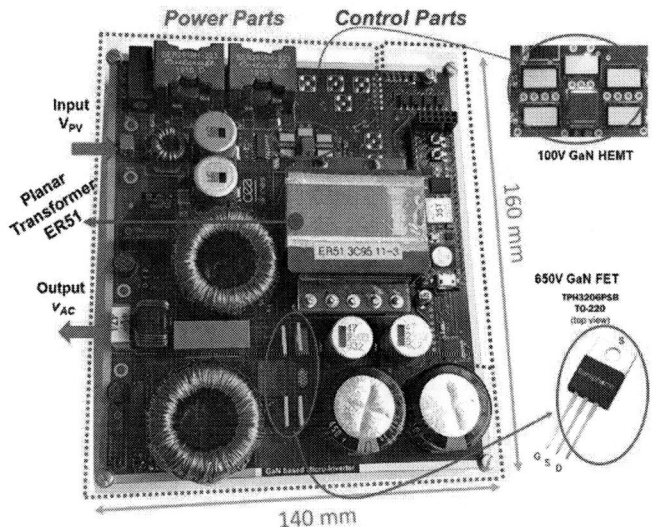

Figure 2: Prototype of the micro-inverter

Table I: List of the micro-inverter components and their values

C_{IN}	450 µF	C_1	47 µF	C_{AC}	1 µF
L_{IN}	44 µH	C_{BUS}	240 µF	R_D	1 Ω
T	Ratio 3:11	L_{AC1}	720 µH	L_{AC2}	5.3 µH

Inverter specification

The inverter has been designed to be associated with 400 W-peak, bi-facial, 72 cells PV modules. In standard condition, the module MPP voltage is 44 V, the short-circuit current is about 10 A and the open-circuit voltage is 55 V. The maximum operating input PV voltage is fixed to 65 V and the maximum input PV current to 15 A. The maximum power point tracker (MPPT) range is [25 V; 55 V]. This is illustrated in Figure 3.

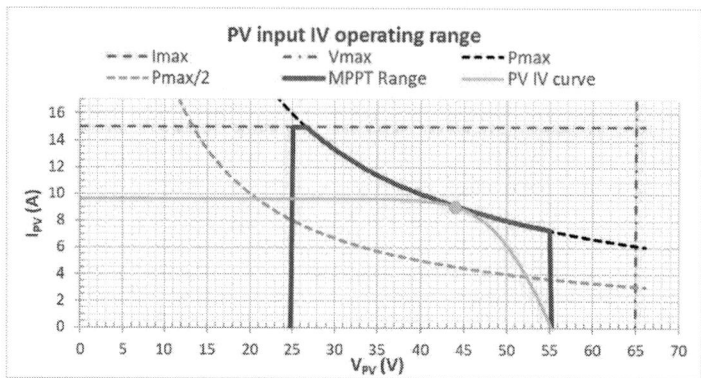

Figure 3: PV input voltage and current ranges

The micro-inverter is connected to the 50Hz, 230V$_{RMS}$ AC grid and is capable of reactive power injection. The size is 150 x 130 x 25 mm without the mechanical packaging. These characteristics can be summarized in the datasheet Table II.

Table II: Datasheet of the micro-inverter

Electrical data			
DC		**AC**	
PV/MPPT inputs	1	Rated power	400 W / 400 VA
Rated power	400 W	Rated voltage	230 V / 1 ph
Maximum voltage	65 V	Rated frequency	50 Hz
MPPT voltage range	25 V / 55 V	Power factor	[0.8; 1] (lag and lead)
Maximum current	15 A	Maximum current	2 A
General			
Dimension	150 x 130 x 25 mm		
Weight	0.7 kg (unpackaged)		

PV micro-inverter control stage

Previous works has shown the control loops for a three-phase current source inverter [5]. In this work, a global schematic including the control part for the single-phase PV micro-inverter is presented in Figure 4. The sensors acquire the measures of the micro-inverter PV input current I$_{PV}$ and voltage V$_{PV}$; the output grid AC current I$_{AC}$, and AC voltage V$_{AC}$. In addition, the DC-bus voltage V$_{DC}$ and the top and bottom side of the printed circuit board temperatures are measured.

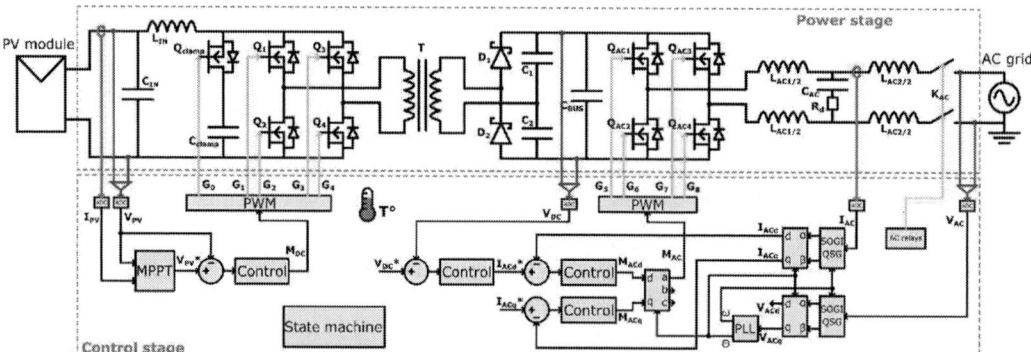

Figure 4: Global schematic of the single-phase grid-tied GaN based solar micro-inverter including the control stage blocks

The DC/AC stage control the AC grid-injected current with its inner loop, in order to maintain it at the same frequency of the grid AC voltage and at a given phase, by using a synchronization method. An outer loop control the DC-bus voltage at a fixed value. The DC/DC stage control the PV voltage, whose reference is given by the MPPT algorithm. For AC signals, it has been chosen to work in the rotating reference frame (RRF), using d and q components of AC voltage and current. Control methods of single-phase and three-phase PV inverter have been exposed in [6]

DC/AC control blocks

Pulse-width modulation

Figure 5 shows the four pulse-width modulation (PWM) signals driving the DC/AC stage control signal G5, G6, G7, G8. This is a unipolar PWM based on sinus-triangle comparison. A triangular carrier is compared to a sinusoidal modulating signal comprised between -1 and 1.

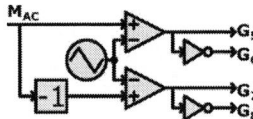

Figure 5: DC/AC full bridge single-phase inverter PWM

AC voltage and current transformation

The d and q component of AC voltage and current must be determined. As the RRF representation is based on three-phase system, they cannot be directly calculated from the single-phase measured signals. For this reason, a quadrature signal generator (QSG) based on a second-order generalized integrator (SOGI) is used to obtain the equivalent α and β components of the synchronous reference frame (SRF) for the single-phase signals. This generator, with the two integrations, is represented Figure 6.

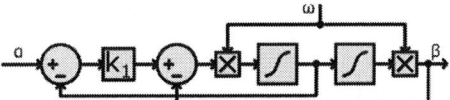

Figure 6: Second-order generalized integrator for quadrature signal generation

Then, from α and β components, V_{ACd} and V_{ACq}, d and q components of the AC voltage can be calculated using (1) with θ the estimated grid voltage angle. Same method is employed for the AC current.

$$\begin{bmatrix} V_{ACd} \\ V_{ACq} \end{bmatrix} = \begin{bmatrix} \cos\theta & \sin\theta \\ -\sin\theta & \cos\theta \end{bmatrix} \times \begin{bmatrix} V_{AC\alpha} \\ V_{AC\beta} \end{bmatrix} \tag{1}$$

Grid synchronization

Grid voltage frequency and angle can be easily evaluated because of working in the RRF. The AC voltage q component V_{ACq} is driven to zero by using a PI controller; it calculates the pulsation ω of the grid voltage, which is integrated to obtain its angle. A simplified representation of this method of phase locking loop (PLL) is given Figure 7.

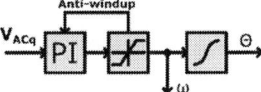

Figure 7: Phase locking loop for grid frequency and angle calculation

AC current control

The reference signal $I_{ACd}*$ is given by the DC-bus voltage control-loop and compared to the measured and calculated actual current I_{ACd}. Then, a PI controller followed by feed-forward blocks (compensation of I_{ACd} and I_{ACq} coupling and grid-voltage and DC-bus voltage fluctuations) generate the d component of the DC/AC modulation signal M_{ACd}. A limiter avoid M_{ACd} to flow over or under the right range [0; 1] and an anti-windup feedback allows to stop the integration of the PI. The q component of the AC current I_{ACq} is controlled by the same way and the reference $I_{ACq}*$ is calculated from the desired reactive power to be injected to the grid; M_{ACq} is limited in the range [-1; 1]. Figure 8 illustrate I_{ACd} and I_{ACq} control blocks.

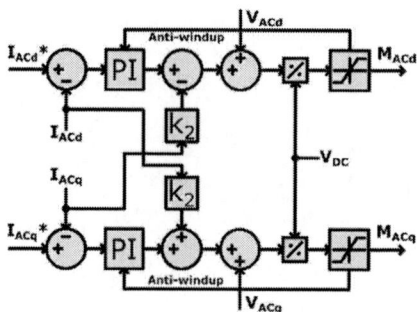

Figure 8: Grid-injected current control block

DC voltage control

The DC-bus voltage is controlled to a fixed value (390 V for this inverter) by comparing this reference $V_{DC}*$ to the measured voltage V_{DC}. Then, a PI controller estimate the DC capacitor instantaneous-average current, which is multiplied by V_{DC} to give the DC capacitor instantaneous-average power value. Subtracting this power to the measured PV power P_{PV}, evaluate the DC/AC instantaneous-average input power. Considering a converter efficiency of one, this power is considered as the inverter output active power $P_{AC}*$ reference. Reactive output power reference $Q_{AC}*$ can also be given to obtain a desired power factor (PF). From these power references and the d and q component of the grid voltage, AC current d and q component references $I_{ACd}*$ and $I_{ACq}*$ can be calculated using (2) and (3), considering that V_{ACq} is equal to zero due to the use of PLL.

$$I_{ACd}^* = 2 \times \frac{P_{ACd}^*}{V_{ACd}} \tag{2}$$

$$I_{ACq}^* = 2 \times \frac{Q_{ACd}^*}{V_{ACd}} \tag{3}$$

Moreover, the V_{DC} measured signal is band-stop filtered at 100 Hz to improve the loop dynamics without being influenced by the ripple caused by the 100 Hz fluctuating power of the single-phase inverters. V_{DC} control block giving AC current reference is shown Figure 9.

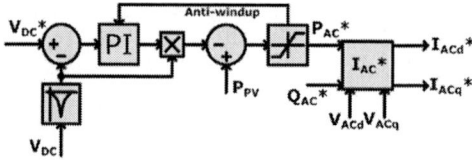

Figure 9: DC-bus voltage control block

DC/DC control blocks

Pulse-width modulation

Figure 10 presents the PWM block of the fed full-bridge with active clamp DC/DC stage. The five control signals G1, G2, G3, and G3 are used to drive the GaN devices, G0 is the control signal of the active clamp transistor placed in order to archive the ZVS operation.

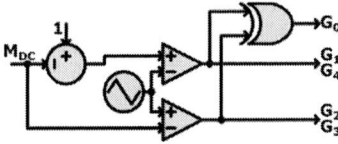

Figure 10: DC/DC FED converter PWM

PV voltage control

The voltage reference V_{PV}* given by the MPPT is compared to the measured PV voltage V_{PV}. Then a PI controller, followed by a feed-forwarded division by the DC bus voltage V_{DC} generate the modulation signal M_{DC}. A limiter avoid M_{DC} to flow over or under the right range [0; 0.5] and, if this occur, stop the integration of the PI thanks to an anti-windup feedback. See Figure 11.

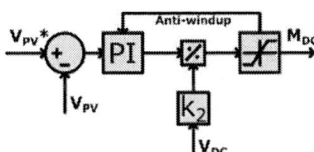

Figure 11: PV voltage control block

Maximum power point tracking

The MPPT must be implemented in order to determine the PV voltage that work the solar module at its maximum power point (MPP). The PV module maximum power voltage V_{MPP} must be reached, to work at the MPP as in the IV curve in the rigth of Figure 12.

A widely employed method [7] is to use the Perturb and Observe (P&O) algorithm in which the voltage reference V_{PV}* is perturbed (increased or decreased) and the resulting power P_{PV} is analyzed. The next voltage reference is determined accordingly, in order to work closer to the MPP. The right of Figure 12 shows a block diagram of an implemented P&O MPPT.

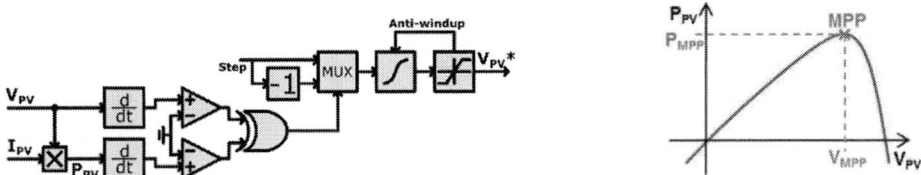

Figure 12: Maximum power point tracker control block and PV IV-curve

Controllers tuning

Proportional-integral controller are used in all the control loops. As in Figure 13, in each PI, the error between the reference value and the measured one is multiplied by a proportional gain K_P and an integral gain K_I. The output of the integral gain is integrated and the result added to the proportional gain output.

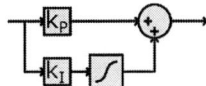

Figure 13: PI controller block

The two parameters K_P and K_I must be tuned for each control loop in order to obtain a sufficient rapidity and bandwidth but insuring the stability.

This is done following several steps:
1. Elaborate the average model in Laplace domain of the system to be controlled
2. For AC controlled signals, translate the model in the RRF
3. Create the open loop : PI controller followed by the system model followed by the sensor model
4. Tune the PI parameters
5. Generate the Bode plot of the open loop and check the stability margins: phase margin M_φ in degree, gain margin M_G in dB.

From the open loop transfer function Bode plot, M_φ and M_G can be calculated using (4) and (5) with H the transfer function of the open-loop, f_0 the frequency at which the magnitude of the function is zero and f_{180} the frequency at which the phase of the function reach -180°. To ensure stability, phase and gain margins must be positive.

$$M_\varphi = 180 - arg\big(H(f_0)\big) \tag{4}$$

$$M_G = -20 \times log\|H(f_{180})\| \tag{5}$$

The Bode plot with magnitude and phase, of the AC current inner control loop, is shown on the left of Figure 14. The system transfer function is based on the inverter average model followed the output AC-side LCL type damped filter. On the right of Figure 14, the Bode plot of the DC-bus outer control loop is given. The system transfer function is based here on the DC capacitor model, it include the transfer function of the AC current closed-loop.

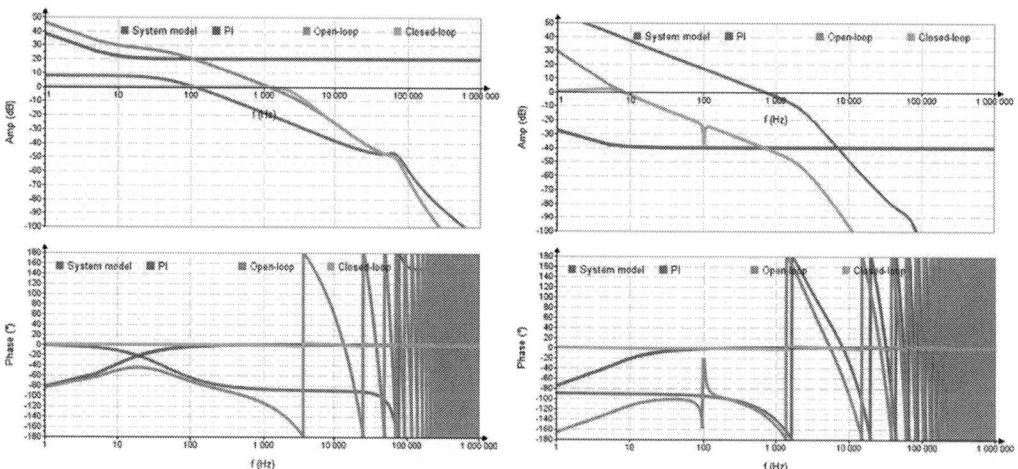

Figure 14: Bode plot of the AC current control loop and of the DC-bus control loop followed by the AC current control closed-loop

Similarly, Figure 15 represent the Bode plot of the PV voltage control loop.

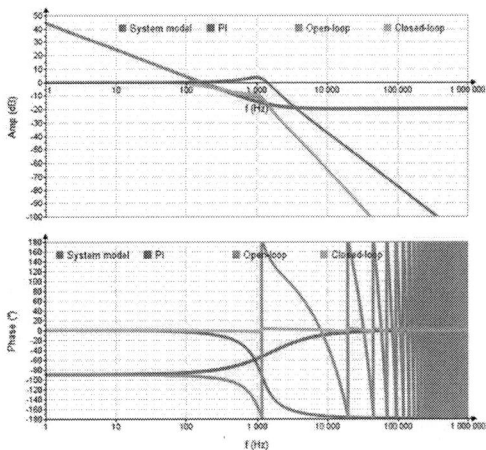

Figure 15: Bode plot of the PV voltage control loop

The Table III gives the list of the PI parameter and the margins of the three control loops.

Table III: PI controllers tunings and stability margins

I_{AC} control		V_{DC} control		V_{PV} control	
K_P	10	K_P	0.01	K_P	0.1
K_I	500	K_I	0.25	K_I	1000
M_φ	62°	M_φ	61°	M_φ	85°
M_G	12 dB	M_G	47 dB	M_G	13 dB

Simulation results

Figure 16 shows the simulated waveforms of the inverter electrical signals, considering the complete closed-loop control stage, in normal operation:
- V_{AC}, V_{ACd}, V_{ACq} the AC voltage and its d and q components
- I_{AC}, I_{ACd}, I_{ACq} the AC current and its d and q components
- DC-bus voltage V_{DC} with its 100 Hz perturbation
- PV voltage V_{PV} and its reference V_{PV}* given by the MPPT
- AC active, reactive and apparent power P_{AC}, Q_{AC}, S_{AC}
- PV power P_{PV} and the power factor PF.

From 0.16 s to 0.27 s, the inverter work at rated power (400 W) and the power factor PF = 1. From 0.27 s to 0.34 s, the power factor is fixed to PF = 0.8 (leading) and then from 0.34 s PF = 0.8 lagging.

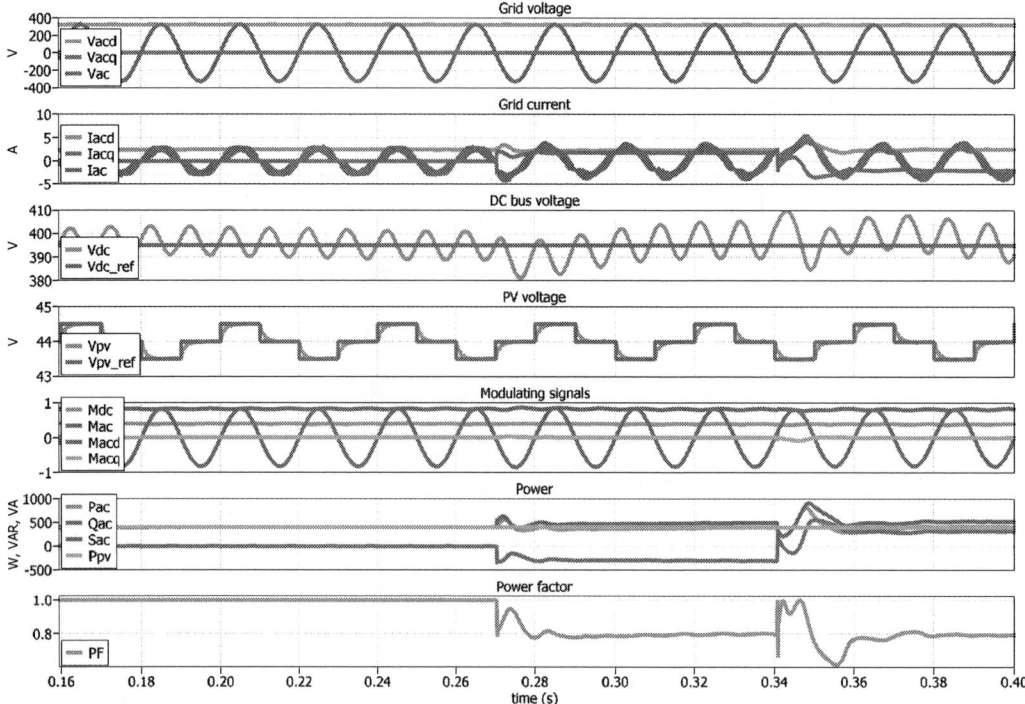

Figure 16: Simulated waveforms of the micro-inverter with PF = 1, then PF = 0.8 (leading), then PF = 0.8 (lagging)

Experimental results

The control blocks presented previously are fully implemented into a field-programmable-gate-array (FPGA) chip placed on an embedded board. Experiments have been realized by connecting the inverter to a PV generator simulator and a grid simulator.

An experimental scope view of the AC grid injected current and the grid voltage at rated power (400 W) is shown Figure 17. The current is sinusoidal, at the grid voltage frequency and in phase with it (PF = 1).

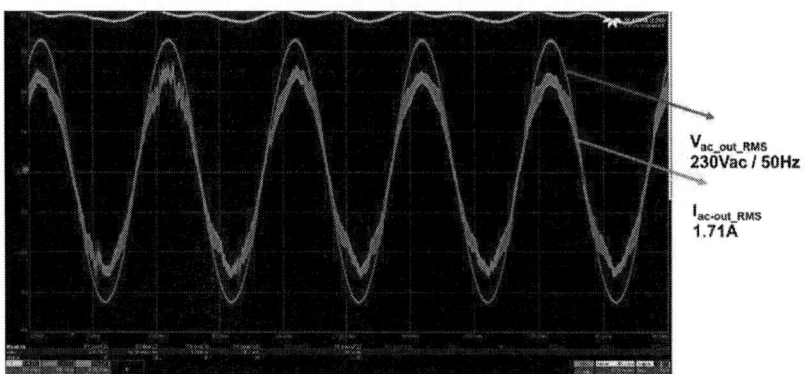

Figure 17: Experimental waveforms of grid voltage and current at rated power and PF = 1

Figure 18 shows the measurements and the graphic interface of the DC/DC stage and the DC/AC stage controls, implemented into a DSPACE rapid prototyping target.

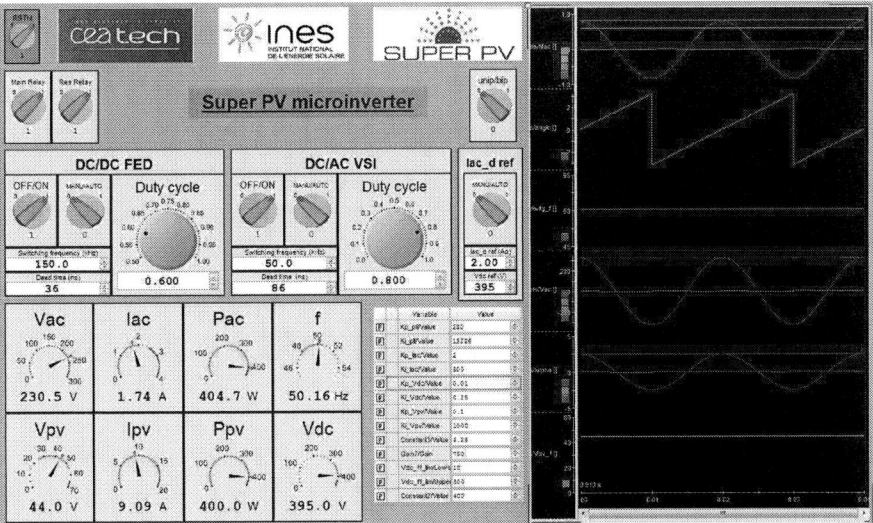

Figure 18: Experimental control bench interface using a DSPACE prototyping target

Future work

The future works will be focused on the standard characterization of the inverter in order to evaluate:
- The maximum and European conversion and MPPT efficiency
- The power quality and total harmonic distortion of the grid injected AC current
- The thermal behavior considering several hour of operation at full power
- The islanding detection capability of the inverter

Conclusion

The control principles of the grid-tied micro-inverter dedicated to photovoltaic application are discussed in this work. The implementation of the PWM blocks on the current fed full bridge DC/DC and the full-bridge DC/AC, the control loop using PI controllers in the (d,q) frame are proposed and some specificities of the grid-tied PV inverters are listed. Then, simulation results and partial experimental results are presented.

References

[1] J. C. Teoa, et al "Effects of bypass diode configurations to the maximum power of photovoltaic module", International Journal of Smart Grid and Clean Energy, Oct 2017

[2] VDMA "International technology roadmap for photovoltaic ITRPV", 10th edition, March 2019

[3] M. Johns et al "Grid-Connected Solar Electronics" EE-290N-3 – Contemporary Energy Issues, 2009

[4] V. S. Nguyen, S. Catellani, A. Bier, J. Martin, H. Zara, J. Aimé: A compact / high efficiency GaN based 400W solar micro inverter in ZVS operation, PCIM 2020

[5] A. Bier: Three-phase grid-tied current-source inverter sizing and control for photovoltaic application, SPEEDAM, June 2016

[6] R. Teodorescu, M. Liserre, P. Rodriguez, "Grid Converters for Photovoltaic and Wind Power Systems".

[7] T. Esram, P. L. Chapman, "Comparison of Photovoltaic Array Maximum Power Point Tracking Techniques" IEEE transactions on energy conversion, vol. 22, no. 2, june 2007.

A DC/DC buck-boost converter control using sliding surface mode controller and adaptive PID controller

Bassem Saleh, Ahmed Teirelbar, Amr Wasfi
KarmSolar, Freelance, KarmSolar
17 Bahgat Ali St.
Cairo, Egypt
Phone: +20 (111) 70-23498
Email: bassem.saleh@karmsolar.com, ateirelb@gmail.com, amr.wasfy@karmsolar.com
URL: http://www.karmsolar.com

Keywords

≪Converter Control≫, ≪Sliding mode control≫, ≪Modelling≫, ≪Non-linear control≫, ≪Simulation≫

Abstract

Integration of various intermittent renewable energy sources and diesel generators for microgrid and off-grid applications demands DC/DC converters to be more agnostic to input voltage and load changes. Novel combination of sliding surface mode control and adaptive PID is presented to control the output voltage of a buck-boost converter. The converter and controller are modeled using MATLAB/SIMULINK. Adaptive controller resulted in up to 89% drop in the RMS voltage error compared to typical single PID configuration. A zero steady state error was also achieved and the performance was consistent at different loads.

Introduction

The increased share of the renewable energy sources in the energy mix during the previous years has prompted the use of hybrid energy systems integrating renewable energy with various constant energy sources such as AC or DC generators, AC grid and energy storage sources including batteries and capacitors for a better control and smoother integration. The DC/DC converter is considered a fundamental component in hybrid systems; they can control the level of output DC voltage according to the input DC voltage by stepping it down (Buck), stepping it up (Boost) or both (Buck-boost). Couple of main issues, however, with DC/DC converters are, nonlinearity and unstable zero dynamic which makes their output control challenging [1]. Two major control techniques have been used in the literature to handle these challenges: closed loop PI (proportional Integral) controller and sliding surface mode controller. The PI controller was used in [2] [3] [4] and proved to have good behavior but in other studies [5] [6] it was shown that when combining another control technique with PI controller, better results can be obtained. The sliding surface mode controller was used to control DC/DC converters in [7] [8] [9]. One of its drawbacks mentioned in literature is: low robustness to variations of power source voltage and disturbance of load. In this paper, a sliding surface mode controller combined with adaptive PID is proposed to control the output voltage of a buck-boost converter to achieve an error-free steady state response. The buck-boost converter and the implementation of the control scheme are modeled in MATLAB/SIMULINK. Simulation results prove that the combined controllers provide stable and good dynamic response output, which is better than each controller alone and better than SSMC with a non-adaptive PID. An experimental setup is built consisted of a self-manufactured prototype buck-boost converter and an embedded microcontroller to implement the control algorithm. Its results are going to be published in a future publication.

Technical Background

Modeling of Buck-boost Converter

Figure 1 shows the topology of the buck-boost converter where:
E is the input DC source, V_{out} is the output voltage across the resistor R, i_L is the current passing through the inductor L.

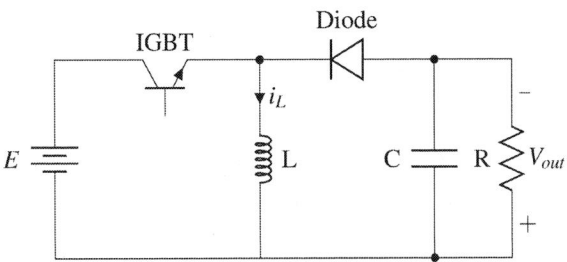

Fig. 1: Buck-boost Converter Topology

This can be modeled according to [7] by Equations (1,2):

$$L\frac{di}{dt} = (1-u)V_{out} + uE \tag{1}$$

$$C\frac{dv}{dt} = -(1-u)i_L - \frac{V_{out}}{R} \tag{2}$$

Where u is the control input to the switching device.
The normalized model of the Buck-boost converter is given by:

$$\frac{dx_1}{d\tau} = (1-u)x_2 + u \tag{3}$$

$$\frac{dx_2}{d\tau} = -(1-u)x_1 - \frac{x_2}{Q} \tag{4}$$

Where

$$x_1 = i\frac{1}{E}\sqrt{\frac{L}{C}}, \quad x_2 = \frac{V_{out}}{E}, \quad Q = R\sqrt{\frac{C}{L}}, \quad \tau = \frac{t}{\sqrt{LC}} \tag{5}$$

Sliding Surface Mode Control

Equation 6 represents the non-linear state space where:

$$\dot{x} = f(x) + g(x)u \quad and \quad y = h(x) \tag{6}$$

x is the system state, u is the control input, y is the output of the system. $f(x)$ is the drift vector field and $g(x)$ is the control input field. The sliding surface mode control proposed assumes a sliding surface S where:

$$S = \{x \in R^n \mid h(x) = 0\} \tag{7}$$

Assuming that there exists a feedback control $u(x)$ with respect to the condition $h(x) = 0$ that is satisfied by the state trajectory $x(t)$. By keeping the system state on the surface S results in a desired behavior of the system. The control signal should have a value satisfying: $u \in \{0,1\}$. The control design problem lies in specifying the function $h(x)$ which will depend on the system control objective.
Defining $u_{eq}(x)$ which is the equivalent control input, which preserves the state trajectory $x(t)$ on the

surface S, assuming that the initial state of the system is on S and when $h(x_0) = 0$. In this case, $h(x)$ has to follow the condition:

$$h'(x) = L_f h(x) + (L_g h(x))u_{eq}(x) = 0 \tag{8}$$

Where:

$$L_f h(x) = \frac{\partial h}{\partial x} f(x) \qquad L_g h(x) = \frac{\partial h}{\partial x} g(x) \tag{9}$$

The equivalent control input can be expressed as:

$$u_{eq}(x) = -\frac{L_f h(x)}{L_g h(x)} \tag{10}$$

From equations 3 and 4, $f(x)$ and $g(x)$ of the Buck-boost converter take the form:

$$f(x) = \begin{bmatrix} x_2 \\ -x_1 - \frac{1}{Q}x_2 \end{bmatrix} \quad g(x) = \begin{bmatrix} 1 - x_2 \\ x_1 \end{bmatrix} \tag{11}$$

The next step in the derivation is to specify the coordinate function $h(x)$. The literature suggests that the direct control, i.e. controlling the output voltage by means of the output voltage, results in unstable dynamics. Thus, the indirect control is adapted, which is controlling the output voltage by means of the current passing through the inductor L. In this case, $h(x)$ can be expressed as follows:

$$h(x) = \overline{x_1} - x_1 \tag{12}$$

Where $\overline{x_1}$ is the reference value of the first state of the system. The control input can be calculated using this equation:

$$u = \begin{cases} 1 & if \ \overline{x_1} - x_1 > 0 \\ 0 & if \ \overline{x_1} - x_1 < 0 \end{cases} \tag{13}$$

The equilibrium point of x_1 can be expressed as:

$$\overline{x_1} = -(1 - \overline{x_2})\frac{\overline{x_2}}{Q} \tag{14}$$

And the derivatives $L_f h(x)$ and $L_g h(x)$ are:

$$L_f h(x) = x_2 \qquad L_g h(x) = 1 - x_2 \tag{15}$$

According to Equation 10, the equivalent control input can be expressed as follows:

$$u_{eq}(x) = -\frac{x_2}{1 - x_2} \tag{16}$$

By substituting into equation 5, the base component of the reference inductor current is:

$$i_{ref-base} = -(1 + \frac{V_{ref}}{E})(\frac{-V_{ref}}{R}) \tag{17}$$

A measurement of the load current (i_{load}) is added to equation 17 to compensate the load consumption:

$$i_{ref} = i_{ref-base} + i_{load} \tag{18}$$

Figure 2 describes the control steps to control the buck-boost converter using sliding surface mode controller.

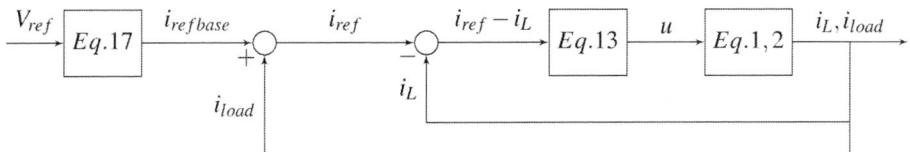

Fig. 2: Control Diagram with SSMC

Adaptive Proportional Integral Derivative Control

In the previous section, a control algorithm is introduced to control the output voltage of the buck-boost control by controlling the current passing through the inductor. In this section, an error compensation algorithm is introduced to add components to the reference current value in equation 18. A voltage measurement is used (V_{out}) and compared with the reference value desired (V_{ref}) where:

$$\Delta V = V_{ref} - V_{out} \tag{19}$$

Where ΔV is the error in the output voltage.

The PID controller dynamic response can be described by its aggressiveness. When it is more aggressive, this means it can reach its set point faster but also, it suffers more of oscillations and overshoot which can harm the converter components. When it is less aggressive, this means it can have a good dynamic response with low overshoot but it suffers from a steady state error. The approach introduced here to achieve a fast yet smooth and error-free response is the adaptive PID. The main idea is to have two PID controllers applied to the same error signal. Each controller has a penetration gain, which changes with time. The sum of both gains is equal to one and no gain shall exceed 1. The penetration gains reset at the beginning of the operation or in case of change of V_{ref}. Equation 20 and 21 demonstrate the adaptive PID idea:

$$PID_{add1}(t) = \left[K_{p1}\Delta V(t) + K_{i1}\int_0^t \Delta V(t)dt + K_{d1}\frac{d\Delta V(t)}{dt} \right] * \alpha^{n(t)} \tag{20}$$

$$PID_{add2}(t) = \left[K_{p2}\Delta V(t) + K_{i2}\int_0^t \Delta V(t)dt + K_{d2}\frac{d\Delta V(t)}{dt} \right] * (1 - \alpha^{n(t)}) \tag{21}$$

Where K_{p1}, K_{i1}, K_{d1} are the proportional, integral, derivative gains of the first PID controller respectively and K_{p2}, K_{i2}, K_{d2} are the proportional, integral, derivative gains of the second PID controller respectively, α is the penetration gain and $n(t)$ is a counter increasing each step forward in time.

With every step forward in time, $n(t)$ increases and then the term $\alpha^{n(t)}$ decreases, therefore the contribution of PID_{add1} to equation decreases. On the other hand, the term $1 - \alpha^{n(t)}$ is increasing and therefore the contribution of PID_{add2} to equation increases.

$$i_{ref} = i_{ref-base} + i_{load} + PID_{add1} + PID_{add2} \tag{22}$$

Figure 3 demonstrate the control methodology using combined SSMC with adaptive PID.

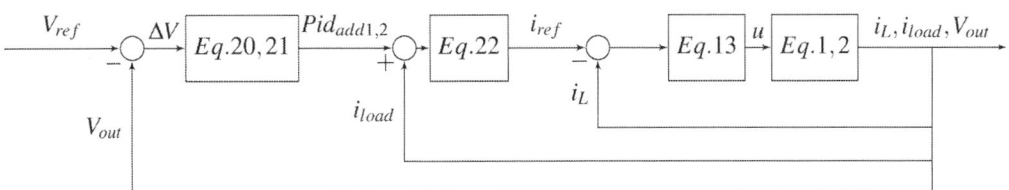

Fig. 3: Control Diagram with SSMC combined with adaptive PID

Modeling and Simulation Results

A model of the buck-boost converter and the proposed controller are implemented in Simulink. Figure 4 shows the overall model developed. The three blocks are: buck-boost controller, buck-boost converter and the load.

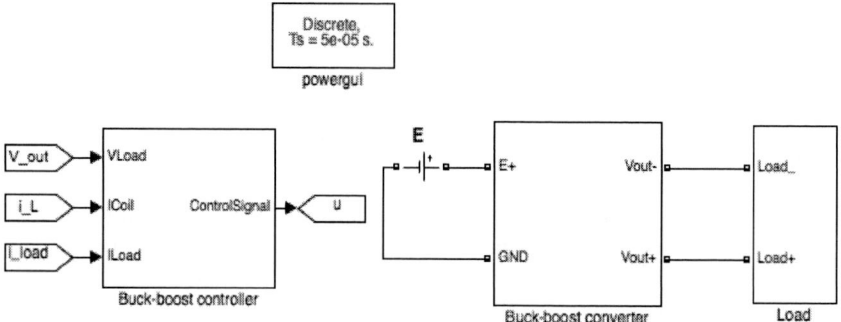

Fig. 4: Simulink Model of the Buck-boost converter

No load test

The first test scenario is testing the buck-boost converter without a load connected. Table I shows the test configuration parameters. Figures 5 and 6 show the voltage error when applying single PID and adaptive PID, respectively.

Parameter	Value	Parameter	Value
E	$48V$	K_{p1}	0.09
R	3000Ω	K_{i1}	0.09
L	$4mH$	K_{d1}	0.0001
C	$720\mu F$	K_{p2}	0.9
V_{ref}	$120V$	K_{i2}	0.95
f	$20KHz$	K_{d2}	0.001

Table I: Simulation Parameters - Test Scenario 1

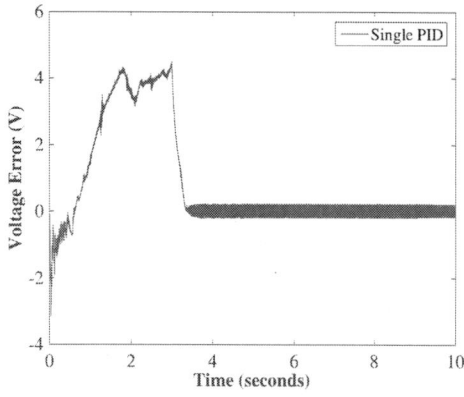

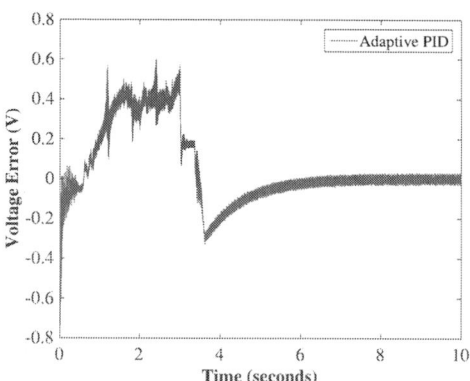

Fig. 5: Voltage error - Single PID applied   Fig. 6: Voltage error - Adaptive PID applied

In figure 7, results of the test are shown. e_{rms} is the root mean square of output voltage error, e_{max+} is the maximum positive value of the error, e_{max-} is the maximum negative value of the error and e_{ss}

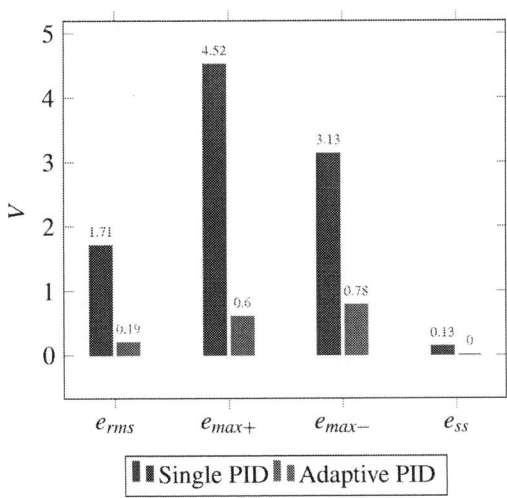

Fig. 7: No-load test results

is the steady state error value. It is demonstrated that when applying the adaptive PID, the root mean square error of the output voltage drops from $1.705V$ to $0.192V$, a drop of 88.7%. It also resulted in a 86.7% and 75% drop in the maximum positive and negative errors of the output voltage, respectively. In addition, the steady state error reached $0V$.

Load test

The second test scenario is testing the Buck-boost converter with a load connected, which is modeled as Permanent Magnet DC motor. The load current is varied by simulating the load at different torques: low, medium and high loads, as shown in table II with the parameters shown in table III. Figure 8 and 9 show the results when a single PID is applied and when adaptive PID is applied at the different load levels.

	T_L	i_{load}
Low	0.5	0.9A
Medium	1.5	2.68A
High	2.5	4.46A

Table II: Load Torque and Load current

Parameter	Value	Parameter	Value
E	$48V$	K_{p1}	0.1
R	3000Ω	K_{i1}	0.1
L	$4mH$	K_{d1}	0.0001
C	$720\mu F$	K_{p2}	0.95
V_{ref}	$120V$	K_{i2}	0.99
f	$20KHz$	K_{d2}	0.001

Table III: Simulation Parameters - Test Scenario 2

It is demonstrated that the proposed controller is able to achieve a good dynamic response with a steady state error less than 1% when applying single PID and almost 0% when the adaptive PID is applied. This is due to the i_{load} term added to equation 18. Results show that applying the adaptive PID results in an 88%, 86% and 45% lower RMS, maximum positive and maximum negative voltage errors, respectively.

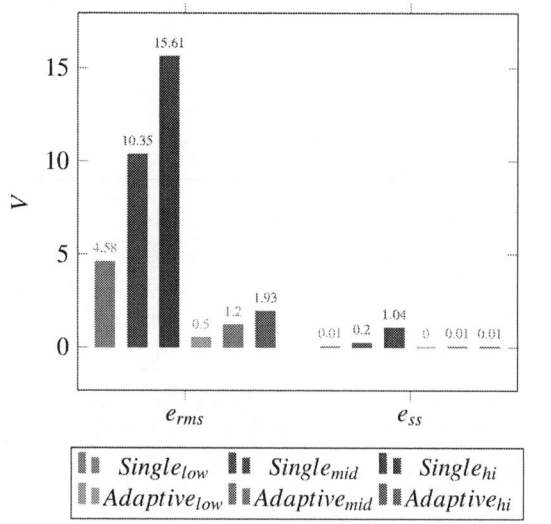

Fig. 8: Load test results (e_{rms} and e_{ss})

Fig. 9: Load test results (e_{max+} and e_{max-})

Figures 10 , 11 and 12 show the voltage error when a single PID is applied and when adaptive PID is applied at different load conditions.

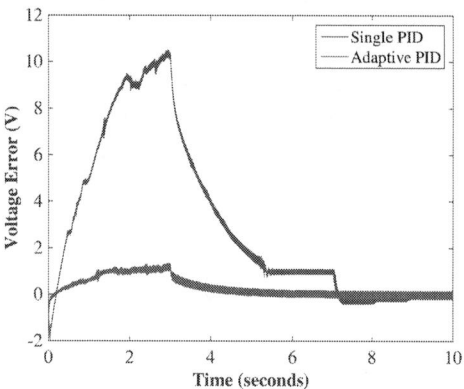

Fig. 10: Voltage error - low load

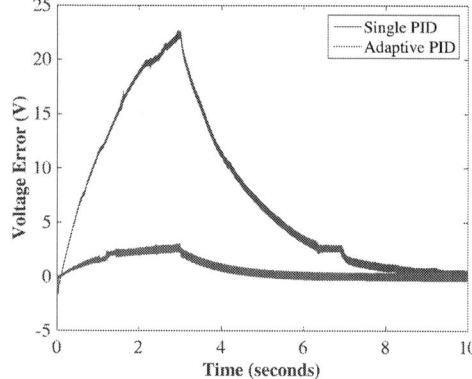

Fig. 11: Voltage error - medium load

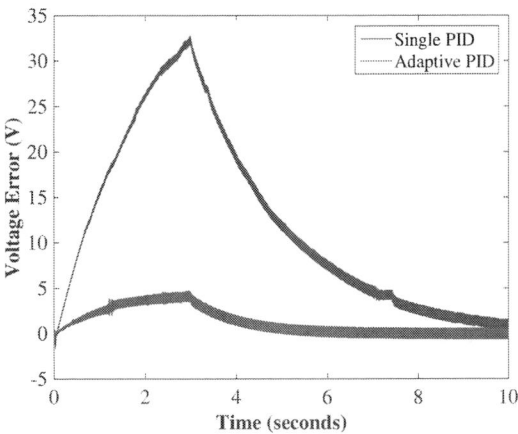

Fig. 12: Voltage error - high load

Conclusion

This paper presented a novel control technique for a DC/DC buck-boost converter, by combining sliding surface mode controller with adaptive PID. Simulation results show that the proposed adaptive PID can achieve better dynamic response compared to the conventional single PID by dropping the RMS of the output voltage around 88% and achieving zero steady state error. Also, the effectiveness of the controller against different load level was demonstrated.

Although, the controller was addressed to the buck-boost converter, it can be implemented to control different DC/DC converters such as buck or boost converters with some modifications. These converters are essential blocks in DC systems such as: DC chargers and hybrid DC controllers. The proposed technique will expedite the enhancement of these systems.

References

[1] J. Liu, W. Ming, and F. Gao: A new control strategy for improving performance of boost dc/dc converter based on input-output feedback linearization, 2010 8th World Congress on Intelligent Control and Automation, IEEE, 2010, pp. 2439-2444.

[2] Duong, Minh Quan, Gabriela Nicoleta Sava, Mircea Scripcariu, and Marco Mussetta: Design and simulation of PI-type control for the Buck Boost converter, 2017 International Conference on ENERGY and ENVIRONMENT (CIEM), pp. 79-82. IEEE, 2017.

[3] Hongwen, He, and Xiong Rui: DC/DC converters design and control for hybrid power system, 2010 International Conference on Intelligent Computation Technology and Automation, vol. 1, pp. 1089-1092. IEEE, 2010.

[4] Park, Hyun-Hee, and Gyu-Hyeong Cho: A DC-DC converter for a fully integrated PID compensator with a single capacitor, IEEE Transactions on Circuits and Systems II: Express Briefs 61, no. 8 (2014): 629-633.

[5] Guo, Liping, John Y. Hung, and R. Mark Nelms: Evaluation of DSP-based PID and fuzzy controllers for DC-DC converters, IEEE transactions on industrial electronics 56, no. 6 (2009): 2237-2248.

[6] Chang, Changyuan: FPGA Implementation of a Single-Input Fuzzy PID Controller for DC-DC Buck Converters, IET Power Electronics 9, no. 6 (2016).

[7] Sira-Ramirez, Hebertt J., and Ramn Silva-Ortigoza: Control design techniques in power electronics devices, Springer Science & Business Media, 2006.

[8] Seker, Murat, and Erkan Zergeroglu: A new sliding mode controller for the DC to DC flyback converter, 2011 IEEE International Conference on Automation Science and Engineering, pp. 720-724. IEEE, 2011.

[9] Yu, Zehui, Jun Zeng, Junfeng Liu, and Fei Luo: Terminal sliding mode control for dual active bridge dc-dc converter with structure of voltage and current double closed loop, 2018 Australian & New Zealand Control Conference (ANZCC), pp. 11-15. IEEE, 2018.

Sensorless Neutral Point Voltage Stabilization in Three-Phase Four-Wire Converters

Xinwei Xu, Gabriel Tibola, Jorge L. Duarte
Eindhoven University of Technology
Eindhoven, The Netherlands
x.xu@tue.nl

Acknowledgements

This work has been conducted within HiPERFORM project, which has received funding from the ECSEL Joint Undertaking (JU) under Grant Agreement No. 783174. The JU receives support from the European Union's Horizon 2020 research and innovation program and Austria, Spain, Belgium, Germany, Slovakia, Italy, Netherlands, Slovenia.

Keywords

«Sensorless control», «Converter circuit», «Interleaved converters», «Three-phase system», «Midpoint voltage balancer»

Abstract

This paper presents a midpoint voltage balancer (MVB) which provides neutral point voltage stabilization to three-phase four-wire converters. The MVB consists of dual switching legs, two neutral inductors, and two split capacitors. A sensorless approach with open-loop control is adopted. It removes current/voltage sensors and alleviates computational demand. It is a cost effective and robust alternative to the closed-loop MVB. Due to the zero-crossing of the neutral inductor current, all switches of the MVB operates in zero-voltage switching mode when neutral current is smaller than the nominal phase current of the three-phase four-wire converter. Interleaving operation of the MVB minimizes the high-frequency current circulation in the split capacitors. In addition, the two neutral inductors are magnetically coupled to decrease the inductor current ripple. As a result, the size of passive components of the MVB is reduced. The proposed sensorless approach is verified by a 20 kVA prototype.

Introduction

Three-phase four-wire (3P4W) converters, being standalone or grid-connected, provide a path for neutral current. While the standalone can support various single-phase loads in an isolated micro-grid [1], most of the grid-connected inject energy from distributed resources to utility grid. A 3P4W grid-connected converter can offer zero-sequential voltage imbalance compensation to the point of common connection (PCC) in a low-voltage distribution network [2]. Nowadays, the adoption of single-phase rooftop photovoltaic system and electric car are increasing. These may lead to degradation of voltage balance in utility grid. Therefore, research and development of 3P4W grid-connected converters that offers voltage imbalance compensation is necessary.

Popular configurations of 3P4W converters are the split dc bus (2-C) [3], the four-leg converter (4-Leg) [4], the actively control split dc bus (ACSB) [5] , and the midpoint voltage balancer (MVB) [6]. The 2-C configuration is easy to implement. However, it suffers from bulky size of electrolytic capacitors and incapability to process neutral current with dc component. The 4-Leg configuration adopts an extra switching leg to form a neutral point for the 3P4W GCC. Since the control of the fourth switching leg and the control of the three-phase inverter are not decoupled, it requires more development effort. The ACSB configuration combines the idea of 2-C and the 4-Leg, and it is easy to control. In addition, the split dc bus of an ACSB can be implemented using small capacitors.

The MVB configuration proposed in [6] doubles the neutral current handling capability of the 3P4W GCC by adopting dual switching legs, as shown in Fig. 1. Dual-loop control is implemented, and the neutral point voltage ripple is less than 1.5% under severe neutral current transient. It offers zero-voltage switching (ZVS) operation to the dual switching legs under a wide range of neutral current injection.

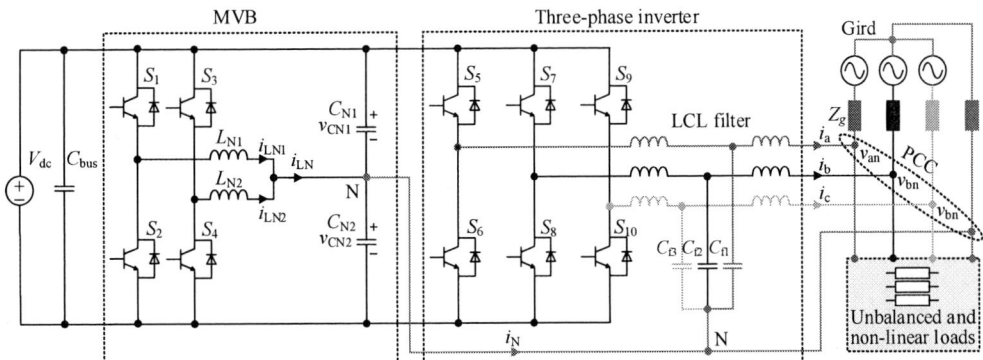

Fig. 1: Circuit diagram of the MVB in a three-phase four-wire grid-connected converter

The ZVS operation is achieved by adopting small neutral inductance so that inductor current cross zero at any switching cycle. Interleaved control is implemented to prevent high-frequency current from circulating into the split capacitor branch. Because of dual-loop control, usage of multiple sensors and high computational demand is inevitable.

In this paper, the control strategy of the MVB is redesigned to allow sensorless neutral point voltage stabilization in the 3P4W converter. This approach is considered as a cost-effective alternative to the MVB proposed in [6]. It provides majority of the advantages existed in its predecessor, while it requires no information on neutral inductor currents, split dc bus voltages, nor neutral current. The MVB operates in open loop with a fixed duty cycle and fixed switching frequency. The neutral inductance is half compared to that in [6]. Nevertheless, same inductor current ripple is maintained due to coupling. In addition, capacitance of the split dc bus is also reduced. Moreover, the neutral point voltage is tolerant to capacitance mismatch, which might happen due to parameter deviation and component degradation.

Description of MVB

The MVB is connected to a 3P4W grid-connected converter, as shown in Fig. 1. It consists of dual IGBT switching legs with S_1, S_2, S_3, S_4, two neutral inductors L_{N1}, L_{N2}, and two split capacitors C_{N1}, C_{N2}. Being connected to the common end of the neutral inductors, the midpoint of the split dc bus forms the neutral point N. Table 1 presents the specifications of the 20 kVA 3P4W converter, which adopts the MVB to stabilize the neutral point. This work focus on the MVB, while the design of the three-phase inverter and how it is controlled to offer voltage imbalance compensation is not covered. The MVB operates in open loop mode, requiring no current and voltage measurements. The PWM signals for each switching legs are 180° phase shifted so that the current ripple of the two neutral inductors are interleaved. As a result, the inductor output current i_{LN} is, theoretically, ripple free [7].

A three-phase voltage signal can be decomposed into positive-sequence components, negative-sequence components, and zero-sequence components. An unbalanced three-phase voltage measured at a PCC may contain negative-sequence components. During voltage imbalance compensation, the 3P4W converter injects unbalanced currents to the PCC. The injected currents contain suitable amount of zero sequential currents with reversed polarity to that of the voltage sequential components. The zero sequential currents, i.e. the neutral current directly flows into the midpoint of the dc bus. This current may contain components of fundamental frequency and certain harmonic frequencies based on the power quality compensation mode of the 3P4W converter.

Grid standard, EN 50160-2010, limits voltage total harmonic distortion (THD) at PCC up to 8%. To meet harmonic requirements of grid regulations, the adoption of 3P4W grid-connected converter is a good approach at mitigating harmonics, and compensating voltage imbalance. Since low-order voltage harmonics are dominant in utility grid, the 3P4W converter mentioned in this paper concerns harmonic order up to 13th. This means that the neutral current could contain harmonic components from 3rd up to 13th order with various amplitude.

Table 1: Specifications of the 20 kVA 3P4W converter with the midpoint voltage balancer

Parameter	Symbol	Value	Unit
Apparent power of the 3P4W grid-connected converter	P_a	20	kVA
Input dc voltage	V_{dc}	760	V
Phase-to neutral grid fundamental voltage	V_g	230	V_{rms}
Grid fundamental frequency	f_g	50	Hz
Nominal phase current	$I_{a,b,c}$	29	A_{rms}
Maximum neutral current at 50 Hz	I_{N_m}	58	A_{rms}
Desired maximum neutral-point voltage ripple	ΔV_{CN}	20	V
Harmonic orders under compensation	h	3,5…13	-
Inductance of the neutral inductors	L_{N1}, L_{N2}	110	μH
Capacitance of the split dc capacitors	C_{N1}, C_{N2}	10	μF
Duty cycle of the IGBTs in the MVB	D	50	%
Switching frequency	f_{sw}	20	kHz

With the proposed MVB, this harmonic-abundant neutral current is redirected to the neutral inductor path, keeping the split capacitor free from the neutral current. Therefore, the voltage of the neutral point is stabilized.

Operation principle

The MVB either operates in closed loop [6, 8] or open loop. In closed loop, current and voltage signals are fed back to the loop, which modifies duty cycle to alter the neutral point voltage to a desired condition. Fig. 2 (a) shows the control implementation of the MVB, where dual loop is implemented [8]. The voltages of the upper and bottom split capacitor v_{CN1}, v_{CN2}, currents of the neutral inductors i_{LN1}, i_{LN2}, and the neutral current i_N are measured. Due to the existence of LC resonance in the MVB circuit, active damping is typically adopted. Two control commands u_{k_LN1} and u_{k_LN2} are compared with the 180° phase shifted carrier signals, from which the gate signals G_1, G_2, G_3, and G_4 are derived.

The objective here is to keep the neutral point voltage as stable as possible. Since the neutral point voltage should be half of the total dc bus voltage, the duty cycle D of the switches is kept at 50%. Therefore, in open loop, control commands u_{k_LN1} and u_{k_LN2} are fixed values, which equal half of the carrier amplitude. As shown in Fig. 2 (b), same to the closed loop, the control commands are compared with the carrier signals, from which gate signals are derived.

According to Fig. 2 (b), at the midpoint N

$$i_{LN} + C_{N1}\frac{d}{dt}(V_{dc} - v_{CN2}) = C_{N2}\frac{d}{dt}v_{CN2} + i_N. \tag{1}$$

Assuming that the dc bus voltage is constant, so $dV_{dc}/dt=0$, (1) becomes

$$i_{LN} - i_N \triangleq C_N \frac{d}{dt}v_{CN}, \tag{2}$$

where $C_N = C_{N1} + C_{N2}$ and $v_{CN} = v_{CN2}$. Therefore, the MVB can be simplified into a buck converter with dual switching legs, as shown in Fig. 3 (a) [9]. This buck converter is connected to a capacitor C_N and a current source I_N.

As mentioned before, the duty cycle is 50% with switching pattern interleaved by 180° phase shift. Neglecting dead time effect, there are only two switching states. In state 1, S_1 and S_4 are switched on, S_2 and S_3 are switched off; in state 2, the switches work in the opposite. Both states last half switching period T_s. Fig. 3 (b) shows some key waveforms of the MVB under different switching states. Note that the doted lines in Fig. 3 (b) are meant for inductor currents of the MVB with coupled neutral inductor, which is covered in the next section.

In state 1, the inductor voltages are represented as

$$v_{LN1} = V_{dc} - V_{CN}, \tag{3}$$

$$v_{LN2} = V_{dc} - V_{CN}. \tag{4}$$

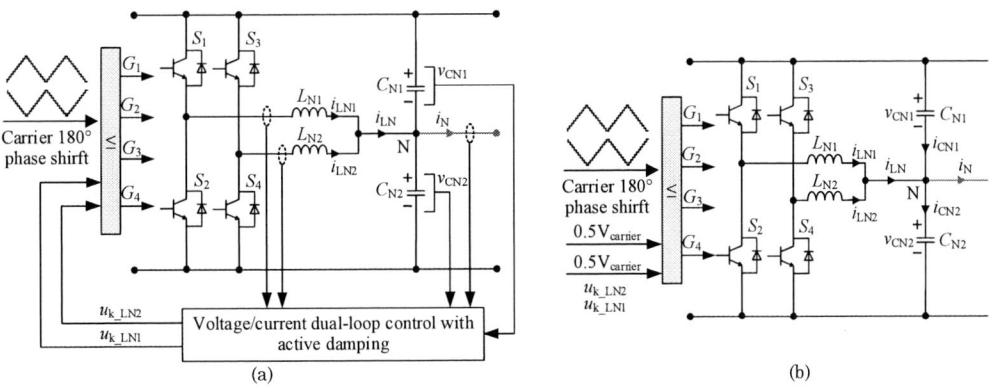

Fig. 2: Control implementation of the MVB with (a) dual loop with active damping and with (b) sensorless approach

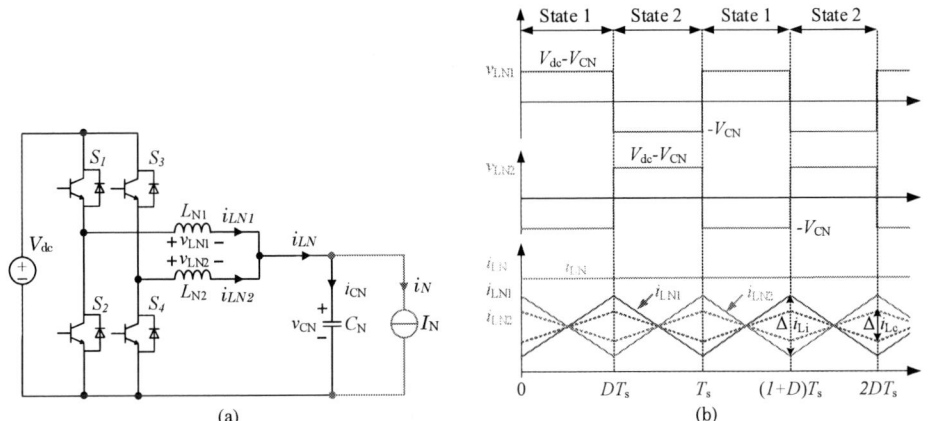

Fig. 3: (a) Simplified circuit of the MVB. (b) Key waveforms of the MVB under different switching states

The current through L_{N1} increases, while the current through L_{N2} decreases. Assume that the inductance of both inductors is equal, $L_{N1}=L_{N2}=L_N$. Since the duty cycle is 50%, $V_{CN}=V_{dc}/2$. The inductor current change during each switching cycle are expressed as

$$\left|\Delta i_{LN1}^{+}\right| = \left|\Delta i_{LN2}^{-}\right| = \frac{V_{CN}}{2L_N}T_s. \tag{5}$$

Similarly, in state 2, the inductor current change during each switching cycle are denoted as

$$\left|\Delta i_{LN1}^{-}\right| = \left|\Delta i_{LN2}^{+}\right| = \frac{V_{CN}}{2L_N}T_s. \tag{6}$$

Based on (5) and (6), the increment and decrement of neutral inductor current in each switching cycle are equal. Therefore, the current ripple on the two neutral inductor counteracts each other. Ideally, the total inductor current i_{LN} is a straight line and ripple-free, as shown in Fig. 3 (b)

Passive component design

Because of sensorless approach, design of the neutral inductors and dc split capacitors are particularly crucial. According to (2), if i_{LN} counteracts i_N, no current flows into C_N, which stabilizes the neutral voltage v_{CN}. This requires the impedance of the neutral inductor branch to be much smaller than that of the split capacitor branch at selected frequencies, which range from grid fundamental and its harmonic up to 13th order as indicated in Table 1. In closed loop, impedance reduction of the neutral inductor branch is achieved by paralleling a virtual impedance to the existing inductor [9, 10]. However, this requires multiple sensors and multiloop control implementation. The sensorless approach proposed in this paper removes the need for control loops and keeps the neutral point voltage stable by selecting appropriate neutral inductors and split capacitors.

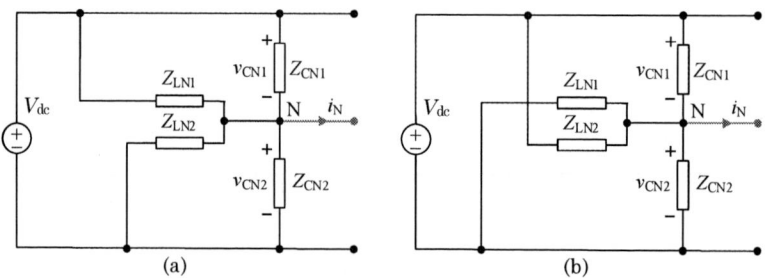

Fig. 4: MVB impedance model for (a) state 1 and (b) state 2

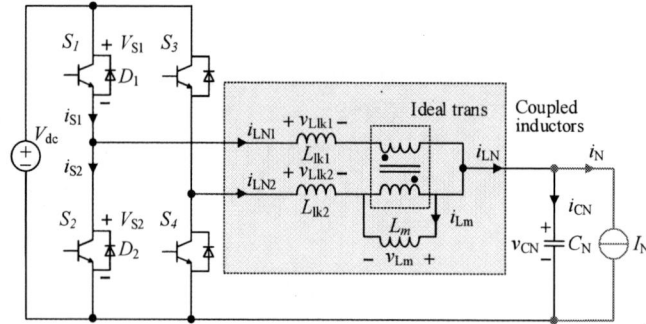

Fig. 5: Equivalent circuit of the MVB with coupled neutral inductors

Split capacitor

As mentioned before, the MVB only has two switching states. At any given time, one of the neutral inductors is connected to the positive potential of the dc bus, while the other is connected to the negative. The circuit diagram of an impedance model of the MVB is shown in Fig. 4, where the neutral inductors L_{N1}, L_{N2}, and the split capacitors C_{N1}, C_{N2} are represented by Z_{LN1}, Z_{LN2}, Z_{CN1}, Z_{CN2}, respectively. Based on the assumption that L_{N1} equals L_{N2}, thus $Z_{LN1}=Z_{LN2}=Z_{LN}$. Therefore, the voltage of the neutral point N is expressed as

$$V_{N} = \left[\frac{Z_{CN1}}{Z_{CN2}} \left(\frac{Z_{LN} + Z_{CN2}}{Z_{LN} + Z_{CN1}} \right) + 1 \right]^{-1} V_{dc}. \tag{7}$$

Given the premise that

$$\begin{cases} Z_{LN} \ll Z_{CN1} \\ Z_{LN} \ll Z_{CN2} \end{cases} \tag{8}$$

at frequencies of interest, equation (7) becomes

$$V_{N} = \frac{V_{dc}}{2}. \tag{9}$$

Therefore, as long as (8) is satisfied, the neutral point voltage V_N equals half of the dc bus voltage, even if the split capacitors does not match in capacitance.

In [6], the neutral inductance is designed to be $L_N=220$ µH, such that the IGBTs of the MVB operate in ZVS mode under a wide neutral current injection range ($I_N \leq 29$ A$_{rms}$). Under this circumstance, the neutral inductor currents i_{LN1}, i_{LN2} cross zero at any given switching cycle when the neutral current i_N is less than the nominal phase current I_{abc}. Re-arrangement of (8) yields

$$\left(C_{N1}, C_{N2} \right) \ll 1 / \omega^2 L_N, \ \omega \in \left\{ 50 \cdot 2\pi, \ 150 \cdot 2\pi, \ \cdots \ 650 \cdot 2\pi \right\}. \tag{10}$$

where $\omega = 2\pi f_N$ and f_N is the frequency of the neutral current, ranging from 50 Hz to 650 Hz.

Therefore, two split capacitors of 10 µF are selected to meet the requirement of (10). Note that C_{N1} and C_{N2} can be different without deteriorating the neutral point voltage stabilization of the 3P4W converter. In addition, the tolerance of the neutral point voltage stability to split capacitance mismatch is verified through experiments.

Neutral inductor

Interleaved control of two switching legs significantly reduces the output current ripple of i_{LN}. However, the inductor current ripple remains unchanged. On the one hand, inductor current ripple is deliberately increased to facilitate ZVS of the MVB. On the other hand, less inductor current ripple is desired to reduce copper loss. In this paper, neutral inductor coupling is employed to address the trade-off between current ripple and power loss.

By coupling the neutral inductors, as shown in Fig. 5, in a way that the derived ideal transformer is connected out of phase with the polarity dots on opposite ends, one obtains the MVB with coupled neutral inductors. The coupled inductor is represented as an ideal 1:1 transformer, two leakage inductors L_{lk1}, L_{lk2}, and a magnetizing inductor L_m [11]. Same to the MVB with independent inductors, there are only 2 operational states of the MVB with the coupled inductors.

In state 1, the inductor voltages are represented as (3) and (4). As elaborated in [11], the current changes through neutral inductors in this state are

$$\Delta i_{LN1}^+ = \frac{V_{CN}}{L_{lk}} \left(\frac{1-D-kD}{(1+k)D} \right) DT_s, \tag{11}$$

$$\Delta i_{LN2}^- = \frac{V_{CN}}{L_{lk}} \left(\frac{k-D-kD}{(1+k)D} \right) DT_s. \tag{12}$$

where k is the coupling coefficient of the coupled inductor, L_s is the self-inductance, and $k = L_m/L_s$, $L_s = L_{lk} + L_m$. Since the duty cycle D is always 50%, the inductor current changes are denoted

$$\left| \Delta i_{LN1}^+ \right| = \left| \Delta i_{LN2}^- \right| = \frac{V_{CN}}{2L_s} \left(\frac{1}{1+k} \right) T_s. \tag{13}$$

Similarly, in state 2, the current changes are expressed as

$$\left| \Delta i_{LN1}^- \right| = \left| \Delta i_{LN2}^+ \right| = \frac{V_{CN}}{2L_s} \left(\frac{1}{1+k} \right) T_s. \tag{14}$$

For a transformer with perfect coupling, the leakage inductance L_{lk} is zero, and the coupling coefficient k is then equal to 1. Compared to (5) and (6), it is beneficial to build a coupled inductor with $k \approx 1$. Therefore, the current ripple through the coupled neutral inductor Δi_{Lc} is half of that through the independent neutral inductors Δi_{Li}, as shown in Fig. 3 (b) and expressed as

$$\lim_{k \to 1} \Delta i_{Lc} = \frac{\lim_{k \to 1} \Delta i_{Li}}{2}. \tag{15}$$

However, as mentioned before, the neutral inductor current ripple is deliberately brought up to maintain ZVS operation under a wide neutral current injection range. To preserve the same amount of inductor current ripple, a 50% neutral inductance reduction becomes a good alternative, hence $L_{N1}=L_{N2}=110$ μH.

Experimental verification

In order to verify the effective of the proposed MVB, A 20 kVA 3P4W grid-connected converter prototype is built. Main components of the converter are shown in Fig. 6. Key specifications of the converter has already been presented in Table 1.The converter provides harmonic mitigation and voltage unbalance compensation to the PCC. Two three-phase IGBT modules are adopted in the prototype. A dSPACE MicroLabBox is employed to handle current and voltage signal acquisition and PWM signal generation.

Key waveforms of one switching leg during switching operation are shown in Fig. 7. These waveforms include gate signal of S_2, neutral inductor current i_{LN1}, current through S_2, and voltage across S_2. The conventions of these signals are referred in Fig. 5. The inductor current has a high current ripple, and it crosses zero at every switching cycle. During switching period T_1, gate signal G_2 is positive, however, $i_{S2} \leq 0$, diode D_2 conducts until the current reverses its polarity. During switching period T_2, G_2 is still positive, $i_{S2} > 0$, S_2 conducts until the G_2 goes to negative. After that, in T_3, D_1 conducts and in T_4, S_1 conducts. S_2 operates in hard turn-off at the transition between T_2 and T_3, as shown in Fig. 7. However,

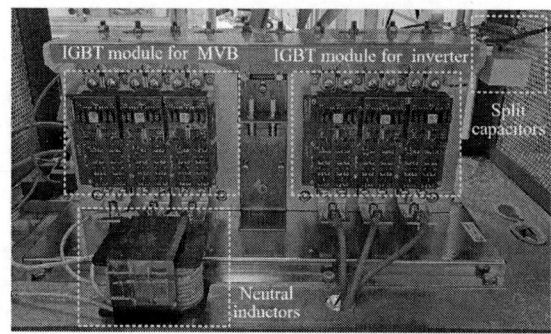

Fig. 6: Main components the 20-kVA 3P4W grid-connected converter with proposed MVB

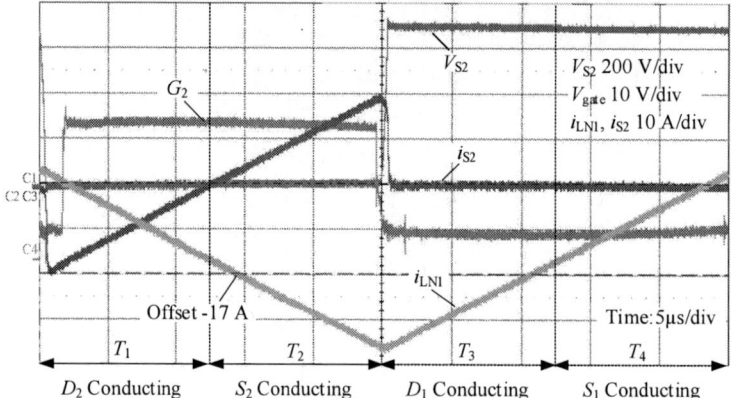

Fig. 7: Voltage and current waveforms of leg S_1, S_2 during switching operation

since the anti-paralleled diode D_2 conducts before S_2, there is only a diode forward voltage across the emitter and the collector during the turn-on of S_2. As a result, the turn-on loss of S_2 is almost zero. This operational principle also applies to S_1, S_3 and S_4. Therefore, the ZVS turn-on is great advantage of the MVB, since turn-on loss of IGBT is normally higher than turn-off losses due to diode reverse recovery.

The 3P4W converter injects unbalanced current i_a, i_b, i_c to the PCC, such that a certain amount of i_N is injected to the neutral point. Due to component availability, coupled inductor whose self-inductance is 85 µH is chosen instead of 110 µH. To verify the response of the proposed MVB, experimental tests under four different cases were performed, as described in the following.
- Case 1: Open loop based on independent inductors with $C_{N1}=C_{N2}=10$ µF, $L_{N1}=L_{N2}=220$ µH.
- Case 2: Closed loop based on independent inductors with $C_{N1}=C_{N2}=100$ µF, $L_{N1}=L_{N2}=220$ µH.
- Case 3: Open loop based on coupled inductor with $C_{N1}=C_{N2}=10$ µF, $L_{N1}=L_{N2}=85$ µH.
- Case 4: Open loop based on coupled inductor with $C_{N1}=5$ µF, $C_{N2}=10$ µF, and $L_{N1}=L_{N2}=85$ µH.

The test results of case 1 are shown in Fig. 8, which includes PCC voltage v_{an}, v_{bn}, v_{cn}, converter output currents i_a, i_b, i_c, neutral current i_N, neutral voltage v_{CN}, and inductor currents i_{LN1}, i_{LN2}, i_{LN}. A sudden output current change happens at 3.13 s, which results in a neutral current transient of 32 A_{peak}. The neutral current flows into the midpoint of the dc bus and it is re-directed to the inductor branch. Even with open-loop approach, the MVB reacts to the transient within half a grid cycle, maintaining the voltage ripple of the neutral point around 20 V. Due to interleaving operation, the current ripple across two inductors counteracts each other. As a result, the output current i_{LN} is almost ripple free, keeping the split capacitors from high-frequency current circulation.

Fig. 9 shows the test results of the MVB under case 2. It contains i_N, v_{CN}, i_{LN1}, i_{LN2}, and i_{LN}. Similarly, a neutral current transient of 60 A_{peak} is created at 2.44 s to verify the dynamic response of the MVB. It takes the MVB around two grid cycles to fully react to the neutral current transient, and to maintain the

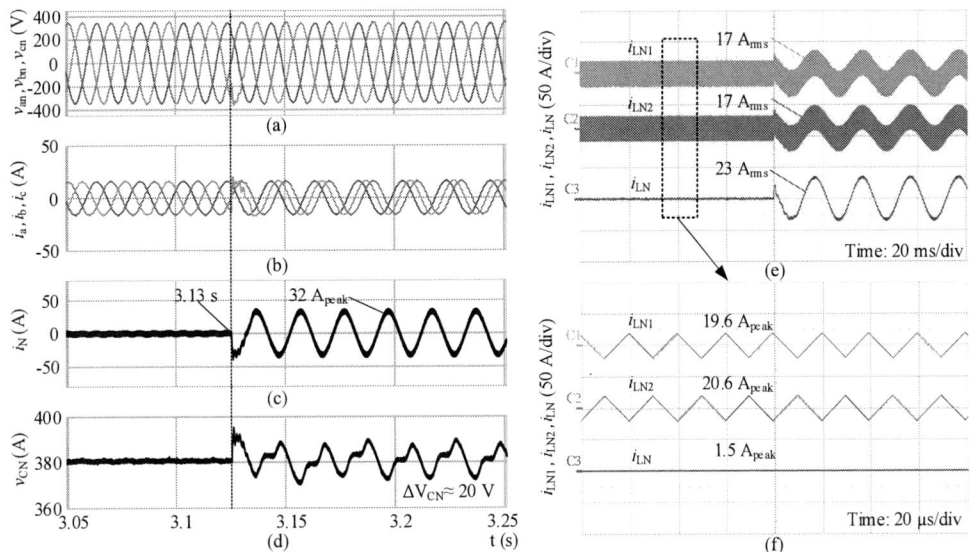

Fig. 8: Experimental results of case 1: (a) voltages v_{an}, v_{bn}, v_{cn}. (b) currents i_a, i_b, i_c. (c) i_N (d) voltage v_{CN}. (e) currents i_{LN1}, i_{LN2}, and i_{LN}. (f) currents i_{LN1}, i_{LN2}, and i_{LN} in high-frequency detail

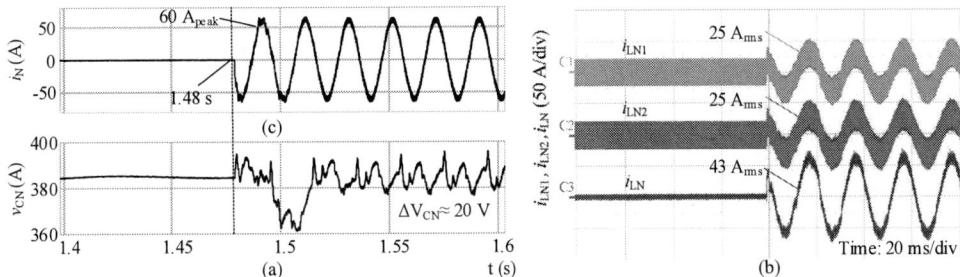

Fig. 9: Experimental results of case 2: (a) voltage v_{CN}. (b) currents i_{LN1}, i_{LN2}, and i_{LN}

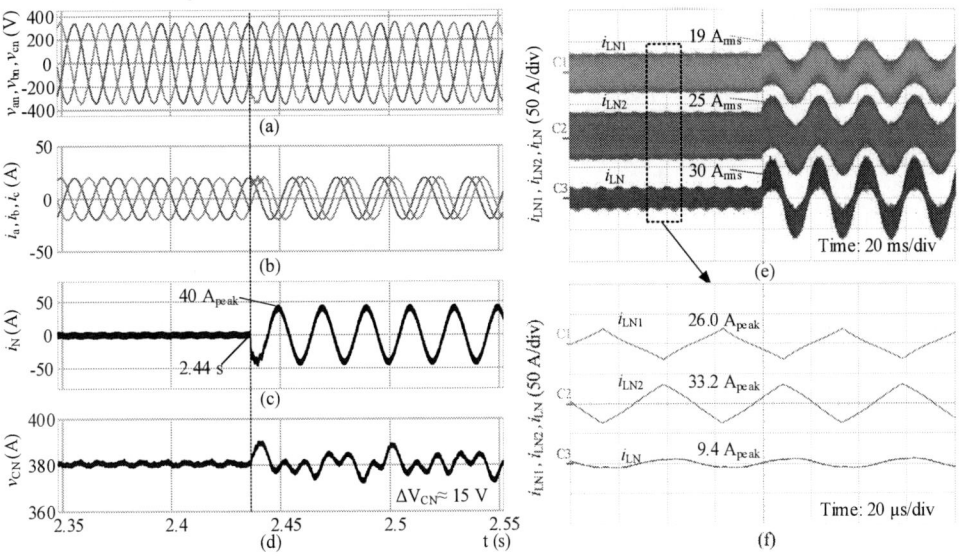

Fig. 10: Experimental results of case 3: (a) voltages v_{an}, v_{bn}, v_{cn}. (b) currents i_a, i_b, i_c. (c) i_N (d) voltage v_{CN}. (e) currents i_{LN1}, i_{LN2}, and i_{LN}. (f) currents i_{LN1}, i_{LN2}, and i_{LN} in high-frequency detail

neutral point voltage ripple around 20 V. Same to the test in case 1, the interleaving operation alleviates

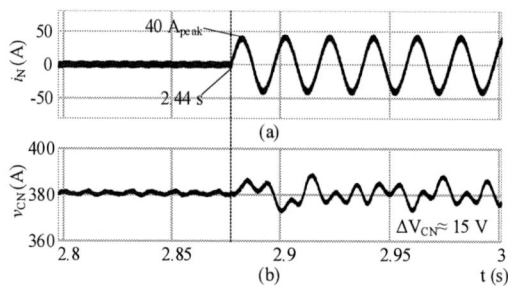

Fig. 11: Experimental results of case 4: (a) i_N. (b) v_{CN}

Table 2: Comparison of the open-loop MVB described in this paper and closed-loop MVB in [6]

Comparison	Parameter/Condition	Open-loop MVB	Close-loop MVB
Quantity of sensors	Voltage sensor	0	2
	Current sensor	0	2-3[1]
Control loop	Loop implementation	Open loop	Dual loop
Passive components	Neutral inductance	110 μH[2]	220 μH
	Split capacitance	10 μF	100 μF
Tolerance to	Split capacitance mismatch	High	Medium
	Neutral inductance mismatch	Medium	High
Steady-state performance	Max. neutral current at 50 Hz[3]	40 A_{peak}	60 A_{peak}
Dynamic performance	Neutral current transient	Fast	Relatively fast

1. Since the neutral current signal can be constructed by adding three-phase current measurements together, so the sensor necessary for the close-loop MVB ranges from 2 to 3.
2. In correspondence to the theoretical analysis, the neutral inductance should be 110 μH. However, due to component availably, a coupled inductor with self-inductance of 85 μH is used in the experiments.
3. The neutral current handling performance of the MVB is evaluated under a desired maximum neutral point voltage ripple.

the high-frequency current circulation in the split capacitors. This means that capacitors with low RMS rating are sufficient for the proposed MVB.

Test on case 3 is performed to the verify the effectiveness of the coupled neutral inductor, with results shown in Fig. 10. These results include v_{an}, v_{bn}, v_{cn}, i_a, i_b, i_c, i_N, v_{CN}, i_{LN1}, i_{LN2}, and i_{LN}. Neutral current transient is introduced at 2.44 s. The MVB reacts to the transient within half a grid cycle and keeps the neutral point voltage ripple around 15 V. Even though, the self-inductance of the coupled inductor is 39% of that in case 1, the average current ripple across the coupled inductor is just 47% more than that in case 1. Due to some piratical issues, the coupling coefficient of the inductor is never 1, and the leakage inductance L_{k1} and L_{k1} are not the same. As it shown in Fig. 10 (f), currents i_{LN1} and i_{LN2} does not cancel each other fully, which results a 9.4 A_{peak} current ripple in i_{LN}. Nevertheless, the coupled implementation of the neutral inductors is a way to reduce core and copper use of the proposed MVB.

In order to show voltage stability tolerance to split capacitance mismatch, C_{N1} is chosen to be 100% more than that of C_{N2}. The test results of case 4 is shown in Fig. 11, which includes waveforms of i_N and v_{CN}. The MVB regulates the neutral point voltage to half of the dc bus voltage with ripple around 15 V rapidly. It can be concluded that the dynamic performance of the MVB is not compromised under split capacitance mismatch.

The comparison between the open loop regulated MVB with sensorless approach and the closed loop controlled MVB is summarized in Table 2. The open loop regulated MVB is inferior in terms of neutral current handling capacity and tolerance to neutral inductance asymmetry. However, considering the fast-dynamic performance, absence of control loops and sensor, as well as small passive components, the proposed MVB is a cost-effective and robust alternative to the closed loop controlled MVB.

Conclusion

A voltage stable neutral point is essential to normal operation of 3P4W converters, especially those in voltage imbalance compensation applications. An unstable neutral point could lead to undesired output current distortion, which degrades the compensation performance of the converter at PCC. This paper proposes a cost-effective and robust MVB for neutral point voltage stabilization in 3P4W converters.

The MVB requires no current nor voltage measurements, and it operates in open loop mode. The passive components (110 μH and 10 μF respectively) are carefully selected so that most of the neutral current is re-directed into the inductor branch, keeping the neutral voltage stable. Coupled implementation of neutral inductor is adopted, which cuts the required inductance to half. In order to achieve the best result, high coupling coefficient and symmetrical inductor construction are desired. Due to interleaving operation, high-frequency current circulation inside the split capacitor branch is alleviated. All IGBT switches of the MVB operates under ZVS when the neutral current is less than the nominal phase current of the 3P4W converter.

Compared to the closed loop counterpart, the sensorless approach does reduce the neutral current handling capability of the MVB by about 33%. However, it stabilizes the neutral point voltage much faster by around 300%. In addition, the performance of the proposed MVB is not compromised even under a 100% split capacitance mismatch.

References

[1] M. Dai, M. N. Marwali, J.-W. Jung, and A. Keyhani, "A three-phase four-wire inverter control technique for a single distributed generation unit in island mode," *IEEE Transactions on power electronics,* vol. 23, pp. 322-331, 2008.

[2] R. S. R. Chilipi, N. A. Sayari, K. H. A. Hosani, and A. R. Beig, "Adaptive Notch Filter-Based Multipurpose Control Scheme for Grid-Interfaced Three-Phase Four-Wire DG Inverter," *IEEE Transactions on Industry Applications,* vol. 53, pp. 4015-4027, 2017.

[3] M. Aredes, J. Hafner, and K. Heumann, "Three-phase four-wire shunt active filter control strategies," *IEEE Transactions on Power Electronics,* vol. 12, pp. 311-318, 1997.

[4] R. Zhang, V. H. Prasad, D. Boroyevich, and F. C. Lee, "Three-dimensional space vector modulation for four-leg voltage-source converters," *IEEE Transactions on power electronics,* vol. 17, pp. 314-326, 2002.

[5] M. K. Mishra, A. Joshi, and A. Ghosh, "Control schemes for equalization of capacitor voltages in neutral clamped shunt compensator," *IEEE Transactions on Power Delivery,* vol. 18, pp. 538-544, 2003.

[6] X. Xu, G. Tibola, J. Duarte, and F. Wang, "Fast Voltage Stabilization Control of Split DC Bus Midpoint in 3P4W Shunt APFs," presented at the 21st European Conference on Power Electronics and Applications, Genova, 2019.

[7] W. Li and X. He, "Review of nonisolated high-step-up DC/DC converters in photovoltaic grid-connected applications," *IEEE Transactions on Industrial Electronics,* vol. 58, pp. 1239-1250, 2011.

[8] X. Xu, "Midpoint voltage stabilization of split DC bus in three-phase four-wire shunt active power filters: prototyping of a midpoint voltage balancer for the ECSEL JU funded project CONNECT," PDEng Thesis, Eindhoven University of Technology, Eindhoven, 2019.

[9] W. Zhao, X. Ruan, D. Yang, X. Chen, and L. Jia, "Neutral Point Voltage Ripple Suppression for a Three-Phase Four-Wire Inverter With an Independently Controlled Neutral Module," *IEEE Transactions on Industrial Electronics,* vol. 64, pp. 2608-2619, 2017.

[10] K. K. Patricio Cortes Estay, Armin Schmiegel, "Method for DC-Link Voltage Balancing for a Grid-Connected Inverter in Case of Asymmetric Operation," in *19th European Conference on Power Electronics and Applications,* Warsaw, Poland, 2017.

[11] J.-P. Lee, H. Cha, D. Shin, K.-J. Lee, D.-W. Yoo, and J.-Y. Yoo, "Analysis and design of coupled inductors for two-phase interleaved DC-DC converters," *Journal of power electronics,* vol. 13, pp. 339-348, 2013.

Bidirectional Isolated Ripple Cancel Triple Active Bridge DC-DC Converter

Takahiro Ohta, Pin-Yu Huang, Yuichi Kado
KYOTO INSTITUTE OF TECHNOLOGY
Matsugasaki, Sakyo-ku, Kyoto
Kyoto, Japan
Tel.: +81 / (075) –724–7451
Fax: +81 / (075) –724–7451
E-Mail: pyhuang@kit.ac.jp, kado@kit.ac.jp

Acknowledgements

This work was supported by JSPS KAKENHI Grant Number JP19K14963.

Keywords

«Converter circuit», «Reliability», «Emerging topology», «Switched-mode power supply », «High frequency power converter », «Microgrid», «Distributed power», «Soft switching», «ZVS converters»

Abstract

To achieve a highly reliable DC distribution microgrid network, a bidirectional isolated ripple cancel triple active bridge (TAB) converter is proposed in this paper. In a conventional full-bridge TAB converter, the DC-link capacitors suffer from high ripple current, which significantly reduces their lifetime. To solve this issue, the proposed converter can reduce the ripple current to nearly zero by adding the clamping capacitors as an internal ripple-cancellation circuit topology. In addition, the proposed converter inherits the advantages of the conventional full-bridge TAB converter such as a wide range of soft switching and bidirectional power conversion. This makes it easier to replace the conventional full-bridge TAB converters. Finally, a 1 kW prototype was built to demonstrate its feasibility. In the experiment, the ripple current reduced to nearly zero (0.16 A) under 400 V/400 V/400 V and rated power 1 kW operating conditions. The proposed converter could be used to make highly reliable and efficient DC distribution microgrid as a power router.

Introduction

A meshed DC distribution microgrid is desired to efficiently utilize distributed renewable energy resources and energy storages. With the increase in DC loads, the DC microgrid is considered as more efficient than conventional power system due to AC/DC conversion being unnecessary. In addition, the meshed distributed power system provides higher reliability against emergency blackouts caused by natural disasters such as typhoons and earthquakes.

Among different scales of a DC power grid, the bidirectional DC/DC converters are used to achieve an efficient energy management system as a power router. The required power routers feature high power density, high reliability, and continuously controlled bidirectional power flow. To achieve the meshed power network system, the multiport converter is receiving attention [1]-[3] because it can reduce the conversion steps and total number of power switches. Accordingly, the system efficiency can be further improved [4], and the power distribution system can be made more cost-effective.

Among various multiport converters, the bidirectional triple-active-bridge (TAB) converter is attracting attention for DC microgrid converters [5]. Since the TAB converter is galvanically isolated by a three-winding transformer, three different output voltages can be converted easily by adjusting the turns-ratio of the transformer. By using a single TAB converter as a power router, the TAB converter can be connected to the various systems such as grids, batteries, and electric vehicles (EVs) for integrating distributed power systems.

In a power system, the converters are usually one of the most critical parts in terms of failure rate, lifetime, and maintenance cost [6]. Especially, the DC-link capacitor has the highest failure ratio in power electronic systems as shown in Fig. 1. For capacitor reliability, the ripple current is the one of the critical causes reducing the lifetime [7]. The full-bridge configuration of the TAB converter causes the high ripple current on DC-link capacitor as shown in Fig. 2. The high ripple current in turn causes the increase in aluminum electrolytic capacitors internal heating and thus degrades the lifetime of DC-link capacitors. Proposed solutions to this problem were summarized in [8]. In [9] and [10], the active ripple reduction circuit is proposed. An additional circuit is connected to eliminate the ripple current of the DC-link capacitor. However, the additional circuit cannot transfer the energy and causes additional power losses, thus increasing the size of the converter.

In this paper, a novel bidirectional ripple cancel bidirectional TAB converter is proposed that can reduce ripple current flowing into DC-link capacitor to nearly zero. A 1 kW prototype with a 100 kHz switching frequency and 400 V/400 V/400 V input/output voltage is implemented, and its feasibility is demonstrated.

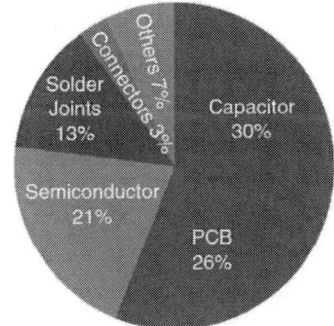

Fig. 1: Surveys on failure distribution among major components in power electronic system [6]

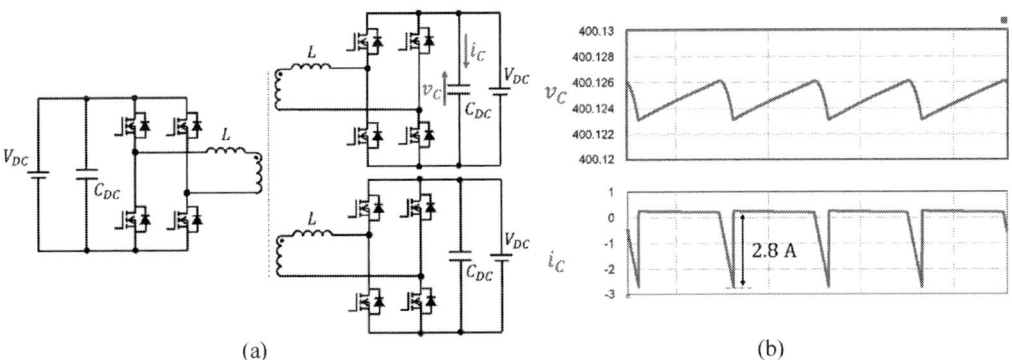

Fig. 2: (a) Circuit diagram and (b) voltage and current waveforms of DC-link capacitor of conventional full-bridge TAB converter. Simulated at V_{DC} = 400 V, 1 kW/0.5 kW/0.5 kW, 100 kHz, C_{DC} = 330 µF, L = 70 µH

Proposed circuit diagram

Fig. 3 shows the circuit diagram of the proposed bidirectional ripple cancel TAB converter. The proposed converter consists of six inductors (L_1, L_2, L_3, L_4, L_5, L_6), six wiring transformers (S_1 — S_2, S_3 — S_4, S_5 — S_6) with a turns-ratio of $1:1:n_2:n_2:n_3:n_3$, three complementarily operated switch pairs (Q_1 – Q_2, Q_3 – Q_4, Q_5 – Q_6), nine clamping capacitors (C_1, C_2, C_3, C_4, C_5, C_6, C_a, C_b, C_c), and three DC-link capacitors (C_p, C_s, C_t). The V_p, $V_s{}'$, and $V_t{}'$ are the DC bus voltages.

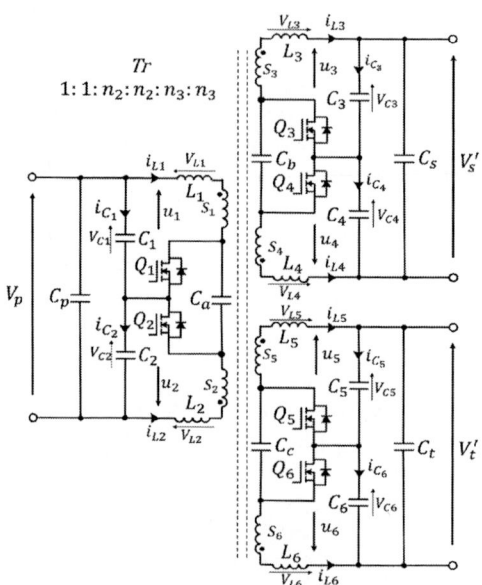

Fig. 3: Circuit diagram of proposed ripple cancel TAB converter

Circuit operation

To simplify the circuit analysis, several assumptions are made. The semiconductors, capacitors, inductors, and transformers are assumed to be ideal. The inductors in the same port are designed to have the same inductance, $L_1=L_2$, $L_3=L_4$, $L_5=L_6$. The turns-ratio of the transformer is designed to be $1:1:n_2:n_2:n_3:n_3$. Each switch pair is complementarily driven by a signal with a 50% duty cycle. To adjust the power transmission, the three switch pairs have a phase-shift angle δ_2 between the primary and secondary ports and δ_3 between the primary and tertiary ports.

To understand the circuit operation, the average voltage of the clamping capacitor is derived. On the primary port, the average voltage of clamping capacitors, V_{Ca}, is equal to the input voltage V_p, because the average voltage of inductors L_1 and L_2 and transformers S_1 and S_2 is zero. In the same way, the average voltages of V_{Cb} and V_{Cc} are equal to V_s' and V_t', respectively. The average voltages of clamping capacitors V_{C1} and V_{C2} evenly share the input voltage as $V_p/2$ and $-V_p/2$, while u_1 and u_2 repeat $V_p/2$ and $V_p/2$. The same goes here: V_{C3} and V_{C4} are $V_s'/2$ and $-V_s'/2$, and V_{C5} and V_{C6} are $V_t'/2$ and $-V_t'/2$ [11].

There are 12 operation stages within one switching period, T, during steady-state operation. Due to the analogous operation, only half a period is described. The key waveform of converter operation is shown in Fig.4.

A) Stage [t_0—t_1]

At t_0, Q_1 is turned on with zero voltage switching (ZVS). Voltage u_1 is clamped to $V_{C1} = V_p/2$ by capacitor C_1. Voltage u_2 is clamped to $V_{Ca} - V_{C2} = V_p/2$ by capacitors C_a and C_2. The voltage of transformer V_{tr} is represented by the sum of products of the inverter voltage as (1). The constant values of a, b, and c are calculated by the inductance L_1 and the turns-ratio of the transformer. Thus, the voltage of inductors L_1 and L_2 is $V_p/2 - V_{tr}$, and the inductor currents i_{L1} and i_{L2} increase linearly at the same amplitude.

$$V_{tr} = \frac{1}{A}\left(\frac{1}{L_1}u_1 + \frac{1}{L_3}u_3 + \frac{1}{L_5}u_5\right)$$
$$A = \frac{1}{L_1} + \frac{1}{L_3} + \frac{1}{L_5}$$

(1)

On the secondary port, voltages u_3 and u_4 are clamped to $V_{C3} - V_{Cb} = -V_s'/2$ and $-V_{C4} = -V_s'/2$, respectively, and voltages of inductors L_3 and L_4 are $-V_s'/2 - V_{tr}$, and the inductor currents i_{L3} and i_{L4} increase linearly in the same way as on the primary port.

B) Stage [$t_1 - t_2$]
    The phased shift δ_2 is controlled to adjust the power transmission between the primary and secondary ports. At t_1, Q_4 is turned off. The body diode of Q_3 is conducted by the inductor current. Q_3 can be turned on during this time interval to achieve soft switching. In addition, the voltage stress of Q_4 is well clamped by capacitor C_b. Voltages u_3 and u_4 are equal to $V_{C3} = V_S/2$ and $V_{Cb} - V_{C4} = V_s'/2$, respectively.

C) Stage [$t_2 - t_3$]
    At t_2, Q_3 is turned on with ZVS. The voltages u_3 and u_4 are equal to $V_{C3} = V_s'/2$ and $V_{Cb} - V_{C4} = V_s'/2$.

D) Stage [$t_3 - t_4$]
    The phase shift δ_3 is controlled to adjust the power transmission between the primary and tertiary ports. The same as stage [$t_1 - t_2$] on the secondary port, Q_6 is turned off at t_3, while the body diode of Q_5 is conducted and can achieve soft switching. The voltage stress of Q_6 is clamped.

E) Stage [$t_4 - t_5$]
    At t_4, Q_5 is turned on with ZVS. Voltages u_5 and u_6 are equal to $V_{C5} = V_t'/2$ and $V_{Cc} - V_{C6} = V_t'/2$, respectively. The energy is transferred from the primary port to the secondary and tertiary ports.

F) Stage [$t_5 - t_6$]
    At t_5, Q_1 is turned off. The body diode of Q_1 is conducted by the inductor current. Q_2 can be turned on during this time interval to achieve soft switching. In addition, the voltage stress of Q_1 is well clamped by capacitor C_a.

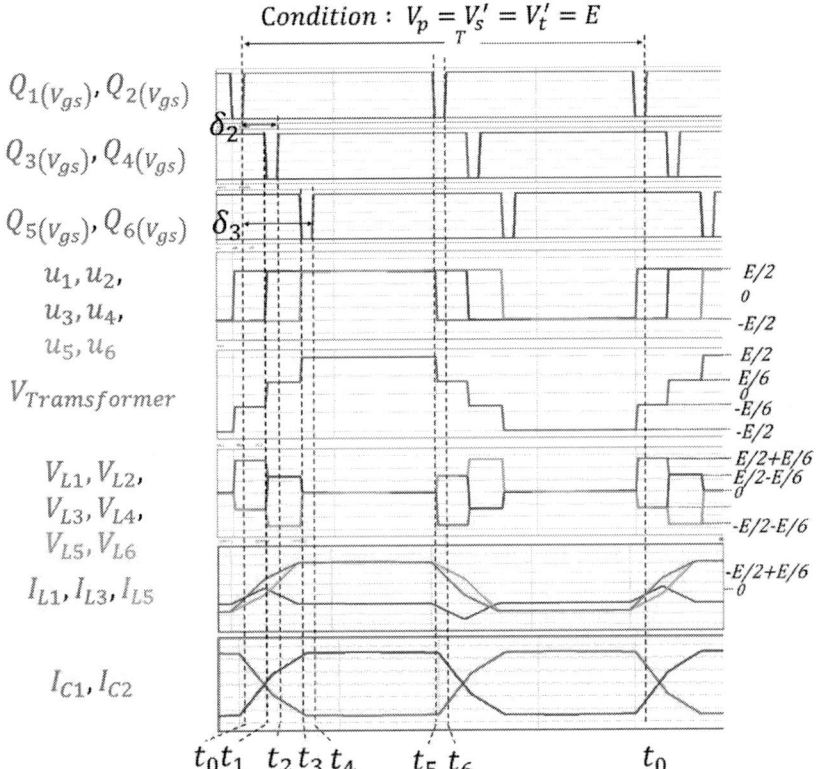

Fig. 4: The key waveforms of proposed converter

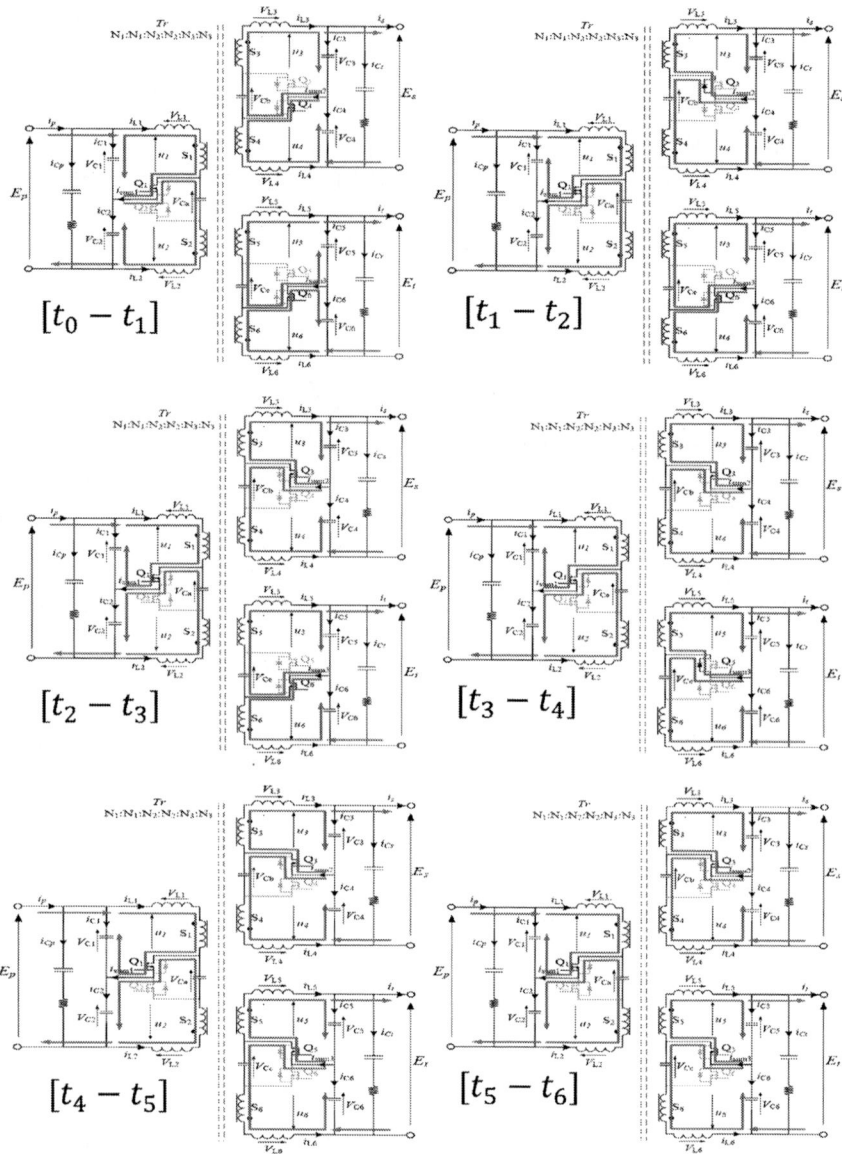

Fig. 5: The circuit operations for each half switching period

Ripple-canceling characteristic

The ripple currents of the DC-link capacitors, C_p, C_s, and C_t are canceled in the proposed circuit. Due to the analogous operation, only the secondary side of the transformer is described as an example. Fig. 6 shows the circuit operation and the key waveforms of ripple-canceling operation. Fig. 6 (a) shows the current flow when the upper-side switch is turned on, and thus, the upper-side clamping capacitor, C_3, is discharged. At the same time, the lower-side clamping capacitor, C_4, is charged. On the other hand, Fig. 6 (b) shows the current flow when the lower-side switch is turned on, the upper-side clamping capacitor is charged, and thus, the lower-side clamping capacitor is discharged, respectively [11].

By designing the symmetrical parameter of components such as the same turns-ratio of the transformer, the inductances, and the clamping capacitances in one side of the rectifier, the clamping capacitors, C_3 and C_4, can be charged and discharged with the same ripple current. Therefore, the ripple current of the output DC-link capacitor can be almost completely eliminated and both circuits can provide the path to transfer the power.

Accordingly, the design consideration of DC-link capacitor only needs to deal with the load transient response and has ripple-free characteristics at steady-state operation. In addition, the small capacitance film or ceramic capacitors can be applied to clamping capacitors, C_{1-6}, which have higher ripple current ratings. To make sure the proposed converter operates at a steady-state condition, the resonant frequency of the clamping capacitor and inductor is designed to be less than the half of the switching frequency. The capacitance of clamping capacitors, C_{1-6}, is expressed as (2)[11].

$$\pi\sqrt{L_1 C_1} \gg \frac{1}{2}T \qquad (2)$$

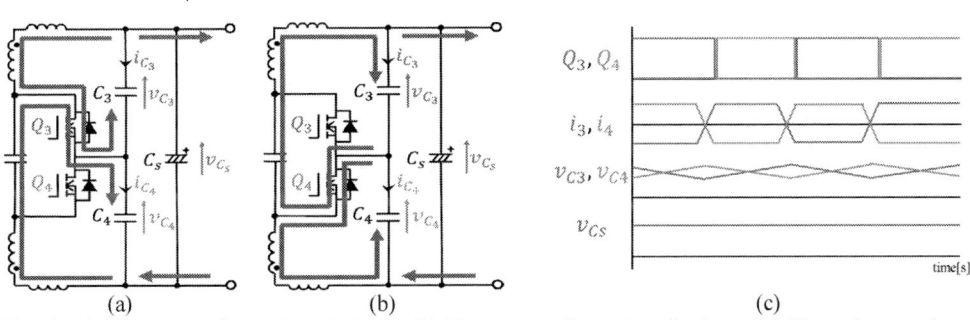

(a)　　　　　　　　(b)　　　　　　　　(c)

Fig. 6: (a) The current flow when Q_3 is on. (b) The current flow when Q_4 is on. (c) The voltage and current waveforms of clamping capacitor and DC-link capacitor

Power flow equation and equivalent circuit

To simplify the power flow analysis, several assumptions are made: the transformer is ideal, has large enough clamping capacitance, and has no power losses. First, the equivalent circuit of the proposed converter is illustrated as Fig. 7 (a). Second, all parameters are referred to the primary side of the transformer in Fig. 7 (b). Finally, the equivalent circuit becomes more simplified because of the identical inverter voltages, $u_1 = u_2$, $u_3 = u_4$, and $u_5 = u_6$, as shown in Fig. 7 (c).

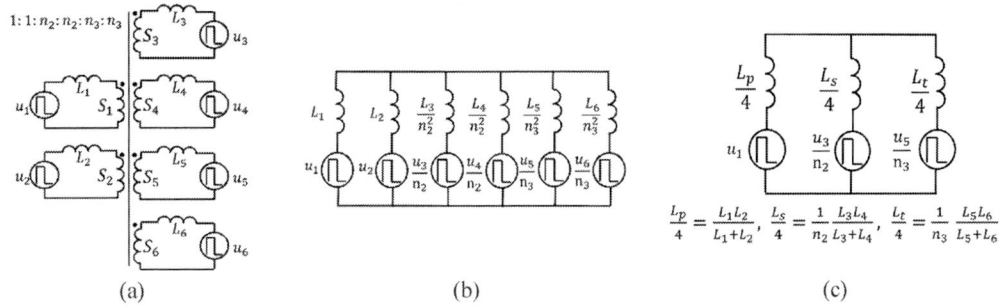

(a)　　　　　　　　(b)　　　　　　　　(c)

Fig. 7: (a) Equivalent circuit with six windings transformer, (b) primary-referred equivalent circuit, and (c) simplified primary-referred equivalent circuit

It can be seen that the equivalent circuit is the same as that of the conventional full-bridge TAB converter [12]. Thus, the power flow equation of proposed converter is expressed as (3), (4), and (5). Here, $\delta_2 > 0$, $\delta_3 > 0$, and $\delta_2 > \delta_3$ are assumed; the source voltages are referred to the primary side of the transformer and are represented by $V_s = V_s' / n_2$ and $V_t = V_t' / n_3$.

As shown in (3) to (5), the power flow of the proposed converter can be controlled by adjusting the phase-shift angles, δ_2 and δ_3. Furthermore, the power transmission characteristic is the same as that of the conventional full-bridge TAB converter [12].

$$P_p = \frac{\delta_2(\pi - \delta_2)V_pV_sL_t + \delta_3(\pi - \delta_3)V_pV_tL_s}{2\pi^2 f(L_pL_s + L_sL_t + L_tL_p)} \tag{3}$$

$$P_s = \frac{\delta_2(\delta_2 - \pi)V_pV_sL_t + (\delta_2 - \delta_3)(\delta_2 - \delta_3 - \pi)V_pV_tL_s}{2\pi^2 f(L_pL_s + L_sL_t + L_tL_p)} \tag{4}$$

$$P_t = \frac{\delta_3(\delta_3 - \pi)V_pV_tL_s + (\delta_2 - \delta_3)(\pi - \delta_2 + \delta_3)V_sV_tL_s}{2\pi^2 f(L_pL_s + L_sL_t + L_tL_p)} \tag{5}$$

Experimental results

A prototype with 400 V/400 V/400 V input/output voltage, 1 kW rated power, and 100 kHz switching frequency was implemented to demonstrate the feasibility of the proposed converter. The specifications and main parameters are listed in Table I. The experimental circuit is shown in Fig. 8.

Table I: Specifications of proposed ripple cancel TAB converter

DC voltage	V_p, V_s, V_t	400V	DC-link capacitor	C_p, C_s, C_t	330 μF electrolytic
Rated power	P	1 kW	Clamping capacitor	$C_1, C_2, C_3, C_4, C_5,$ C_6, C_a, C_b, C_c	2.2 μF PP Film
Turns-ratio	$1:1:n_2:n_2:n_3:n_3$	1:1:1:1:1:1	Switching frequency	f	100 kHz
Transferred power of secondary port	P_s	0.5kW	Inductor	$L_1, L_2, L_3, L_4, L_5, L_6$	35 μH
Transferred power of tertiary port	P_t	0.5kW			

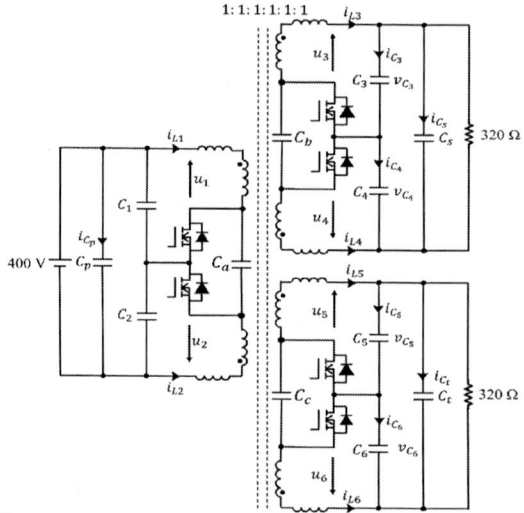

Fig. 8: Experimental circuit

Fig. 9 shows the key waveforms of ripple-canceling characteristic under a full-power 1 kW operation condition. As shown, the currents of clamping capacitors on each port, (i_{C1}, i_{C2}), (i_{C3}, i_{C4}), and (i_{C5}, i_{C6}), have the inversed current with the same amplitude, respectively. Therefore, the ripple current flowing into the DC-link capacitor was significantly reduced. The ripple currents are reduced to 158 and 81.1 mA at the secondary and tertiary ports, respectively. Compared with the simulated results of the conventional TAB converter shown as Fig. 2, the proposed ripple cancel TAB converter reduces ripple by 75% on DC-link capacitors.

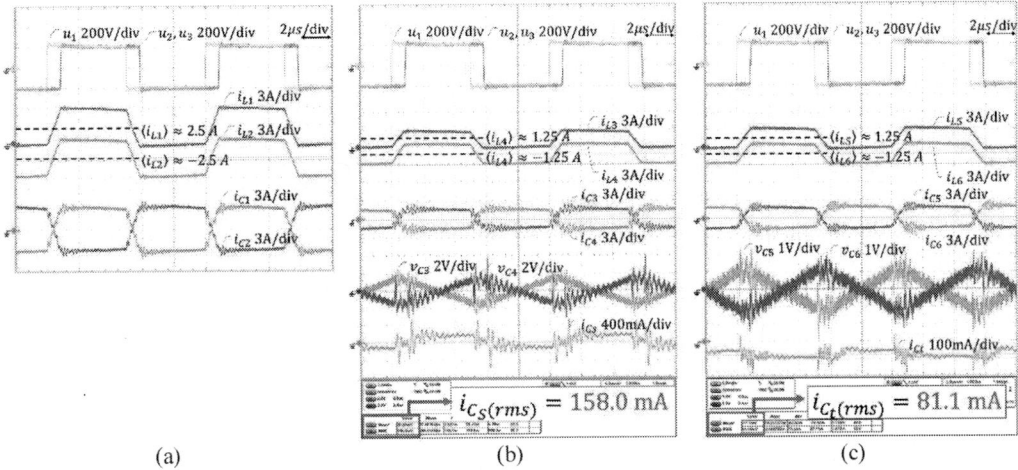

Fig. 9: Key measured waveforms of ripple cancel TAB converter: (a) primary port, (b) secondary port, and (c) tertiary port

Fig. 10 shows the voltage waveform of the main switches at the rated power of 1 kW and the distribution ratio of 50%. As shown in the figure, the gate signals v_{gs} of every switch are turned on after the drain-source voltage v_{ds} is reduced to 0 V. This means that every switch achieves ZVS turn on in this operating condition.

In addition, the drain source voltage v_{ds} of each switch could be well clamped to 400 V by clamping capacitors, C_a, C_b, C_c, without voltage spikes.

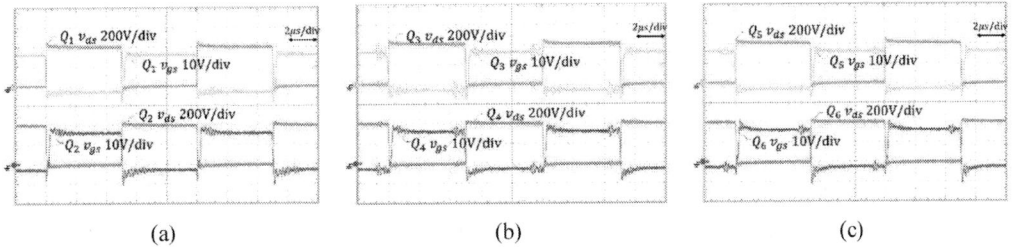

Fig. 10: The voltage waveform at 1 kW rated power. (a) Primary port. (b) Secondary port. (c) Tertiary port

Finally, the efficiency of the proposed converter was measured. Fig. 11 (a) shows efficiency dependence on the operating conditions from 170 W to 1 kW at a 50% distribution ratio. From 380 W to 1 kW, ZVS was achieved. In this output power range, efficiency is over 94%. Fig. 11 (b) shows efficiency dependence on the distribution ratio from 0% to 100% at 1 kW output power. The efficiency is over 92% with any distribution ratio.

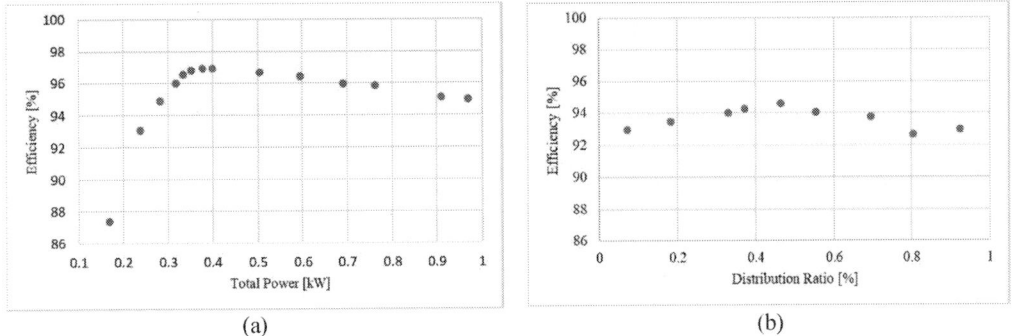

Fig. 11: (a) Efficiency vs. total output power at the distribution ratio of 0.5. (b) Efficiency vs. distribution ratio at the output power of 1 kW

Conclusion

The three-port bidirectional DC/DC converter is considered for a cost-effective and high-efficiency meshed DC distribution microgrid network. To reduce the DC-link capacitor failure ratio in power electronic systems, a bidirectional isolated ripple cancel triple-active-bridge (TAB) converter was proposed in this paper. By applying the proposed topology, the ripple current of the DC-link capacitor is reduced to nearly zero. It also uses half as many semiconductors. In addition, the proposed converter inherits the advantages of the conventional TAB converter such as the same power flow transmission and a wide range of zero voltage switching (ZVS). A prototype with 400 V/400 V/400 V, 1 kW rated power, and 100 kHz switching frequency was implemented. The zero ZVS is found to be from 38% to 100% load conditions. In addition, the voltage stress of switches is well clamped without voltage spikes. The overall 94% efficiency is measured under a 50% load distribution ratio operating condition. In addition, the ripple currents on DC-link capacitors are reduced to 0.16 A and 0.08 A. These features make the proposed converter suitable for a DC distribution microgrid as a high-efficiency and high-reliability power router.

References

[1] A. Kwasinski and P. T. Krein, "A Microgrid-based Telecom Power System using Modular Multiple-Input DC-DC Converters," INTELEC 05 - Twenty-Seventh International Telecommunications Conference, Berlin, 2005, pp. 515-520.

[2] B. G. Dobbs and P. L. Chapman, "A multiple-input DC-DC converter topology," in IEEE Power Electronics Letters, vol. 1, no. 1, pp. 6-9, March 2003.

[3] R. H. Lasseter, "MicroGrids," 2002 IEEE Power Engineering Society Winter Meeting. Conference Proceedings (Cat. No.02CH37309), New York, NY, USA, 2002, pp. 305-308 vol.1.

[4] H. Tao, A. Kotsopoulos, J. L. Duarte and M. A. M. Hendrix, "Family of multiport bidirectional DC-DC converters," in IEE Proceedings - Electric Power Applications, vol. 153, no. 3, pp. 451-458, 1 May 2006.

[5] Y. Kado, D. Shichijo, K. Wada, and K. Iwatsuki, "Multiport power router and its impact on future smart grids," Radio Science, vol. 51, no. 7, pp. 1234–1246, July 2016.

[6] H. Wang, M. Liserre and F. Blaabjerg, "Toward Reliable Power Electronics: Challenges, Design Tools, and Opportunities," in IEEE Industrial Electronics Magazine, vol. 7, no. 2, pp. 17-26, June 2013.

[7] Nippon Chemi-Con Corporation, "Judicious use of aluminum electrolytic capacitors," in Nippon Chemi-Con Corporation of the internert, [Online]. Available: https://www.chemi-con.co.jp/

[8] H. Wang and F. Blaabjerg, "Reliability of Capacitors for DC-Link Applications in Power Electronic Converters—An Overview," in IEEE Transactions on Industry Applications, vol. 50, no. 5, pp. 3569-3578, Sept.-Oct. 2014.

[9] H. Wang, H. S. H. Chung, and W. Liu, "Use of a series voltage compensator for reduction of the dc-link capacitance in a capacitor-supported system," IEEE Trans. Power Electron., vol. 29, no. 3, pp. 1163–1175, Mar. 2014.

[10] R. X. Wang et al., "A high power density single-phase PWM rectifier with active ripple energy storage," IEEE Trans. Power Electron., vol. 26, no. 5, pp. 1430–1443, May 2011.

[11] P.-Y. Huang, T. Ohta, M. Fuji, and Y. Kado, "Bidirectional isolated ripple cancel dual active bridge DC-DC converter", 2019 IEEE International Conference on DC Microgrids (ICDCM2019), Matsue, Japan, pp.

[12] C. Zhao, S. D. Round, and J. W. Kolar, "An isolated three-port bidirectional DC-DC converter with decoupled power flow management," IEEE Trans. Power Electron., vol.23, no. 5, pp. 2443–2453, Sep. 2008."

Design of the Speed Sensorless Field Oriented Control System for Induction Motors Considering Sudden Change of the Rotor Speed

Yoshiki Sakurazawa* , Osamu Yamazaki*, Kazuaki Yuki*, Yosuke Nakazawa*
Kenji Natori**, Keiichiro Kondo***
*Toshiba Infrastructure Systems & Solutions Corporation
1, Toshiba-cho, Fuchu-shi, Tokyo 183-8511, Japan
**Graduate School of Engineering, Chiba University
1-33, Yayoi-cho, Inage-ku, Chiba-shi, Chiba 263-8522, Japan
***Waseda Univerisity
3-4-1, Okubo, Shinjuku-ku, Tokyo 169-8555, Japan
Tel.: +81 / 42-333-2378
E-Mail: yoshiki1.sakurazawa@toshiba.co.jp

Keywords

«Sensorless Control», «Vector control», «Induction motor», «Traction application»

Abstract

This paper studies design of the speed sensorless field oriented control of induction motors considering sudden change of the rotor speed. The controller gains are designed to follow the dynamics of the sudden change of the rotor speed based on an induction motor and current control system model. The primary angular frequency can be estimated properly even when the rotor speed changed suddenly by designing the controller. Also, the time to converge to the command values of the stator current and motor torque is shortened. The effectiveness of the designing method is evaluated by means of simulating slip of the railway vehicle traction.

Introduction

In the field oriented control of induction motors, speed sensors are used to detect the rotor speed. The rotor speed obtained by a speed sensor is used to control the direction of rotor magnetic flux in order to coincide with the direction of the magnetic flux established by the stator magnetizing current in indirect field oriented control. When the speed sensor cannot be used due to malfunction, the primary angular frequency cannot be appropriately controlled and the field oriented control is not achieved. On the other hand, in speed sensorless control, a reliability of the system can be increased and cost can be reduced. Therefore, achieving sensorless control by eliminating speed sensors is desirable.

In speed sensorless field oriented control, it is necessary to estimate the primary angular frequency from the stator current and voltage. Thus, a method that estimates the state of magnetic flux using a magnetic flux observer [1] and a method that controls the primary angular frequency using an error of the q-axis current [2] have been studied. Generally, in these studies, it was assumed that the rotor speed was constant and no sudden changes of the rotor speed occur, so far. In the stability analysis of the speed sensorless field oriented control, some researchers studied motor parameter error [3] [4] and stator voltage error [5]. However, these studies have not considered the error between the stator current and the command value when the rotor speed changes suddenly. Also, in the sensorless control of permanent magnet synchronous motors, the design of the magnetic pole position estimator gain [6] and the stability analysis considering the automatic current regulator [7] have been studied. Therefore, it is required to further improve the stability of the sensorless control in the motor drive systems.

For railway vehicle operations, because the inertial mass of the vehicle is large, the change of the rotor speed is generally small. Therefore, the primary angular frequency estimator gains for the speed sensorless field oriented control has been designed on assumption that the rotor speed does not change suddenly. However, when wheel slip occurs, the rotor speed connected to the driving wheel changes suddenly. In the case of speed sensorless control, it has been reported that error between the current and the command values occurs when the wheel slips [8].

In this paper, a method of designing the primary angular frequency estimator gains of the speed sensorless field oriented control that uses rotor magnetic flux induced voltage [9] is studied. The gains are designed considering the sudden changes of the rotor speed assuming the slip phenomenon of the railway vehicle traction. The model of the induction motor and the automatic current regulator is used for designing the gains. This paper shows that the primary angular frequency can be estimated properly by setting the estimator gains to follow the dynamics of the sudden change of the rotor speed. When the gains are smaller than the designed values, the primary angular frequency estimation is delayed. Then, the field oriented control and the decoupling control in the automatic current regulator are not achieved. As a result, the error between the stator current and the command values occurs when the wheel slips. The effectiveness of the designing method is evaluated by means of simulating slip of the railway vehicle traction.

Speed sensorless field oriented control

Fig. 1 shows a block diagram of a speed sensorless field oriented control system for railway vehicle traction. The rotor of the motor is generally connected to the driving wheel via the gear. In this study, 1 driving wheel model is used as a model of the rolling stock. The details of each block of the motor control system are explained later. In this paper, the designing method of the gains of the primary angular frequency estimator is studied.

Fig. 2 shows vector diagrams of the magnetic flux and the rotor induced voltage in an induction motor. When the field oriented control is achieved, the direction of the rotor magnetic flux ϕ_2 coincides with the direction of the d-axis flux, and the rotor magnetic flux induced voltage E_2 is along the q-axis, as shown in Fig. 2 (a). However, if the direction of the rotor magnetic flux ϕ_2 does not coincide with the d-axis, the rotor magnetic flux induced voltage on d-axis E_{2d} is not zero, as shown in Fig. 2 (b). Therefore, the primary angular frequency is estimated by controlling E_{2d} to become zero. The primary angular frequency estimation formula and equations calculating the rotor magnetic flux induced voltages are described as follows [7]:

$$\hat{\omega}_1 = \frac{E'_{2q}}{\Phi_2^*} + \left(K_{PE2d} + \frac{K_{IE2d}}{s} \right) (E'^{*}_{2d} - E'_{2d}) \tag{1}$$

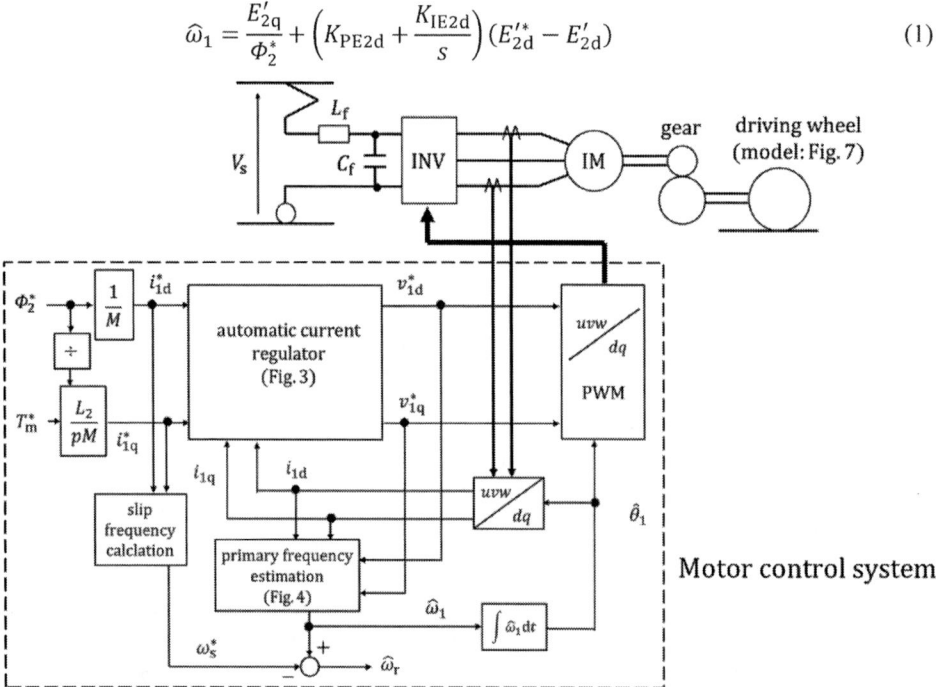

Fig. 1 block diagram of a speed sensorless field oriented control system for railway vehicle traction

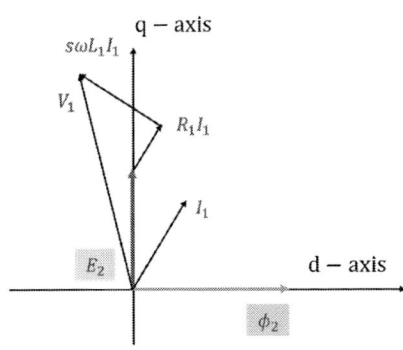

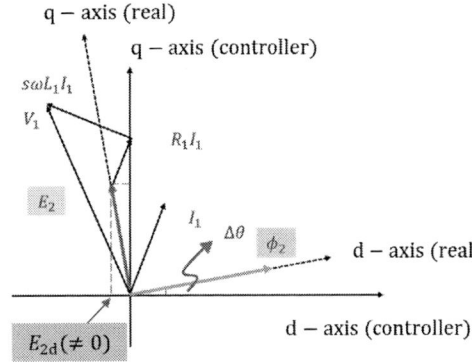

(a) the field oriented control is achieved   (b) the field oriented control is not achieved

Fig. 2 vector illustrations of voltage and magnetic flux

$$E'_{2d} = \left(\frac{M}{L_2}\right)\{v^*_{1d} - (Ri_{1d} - \widehat{\omega}_1\sigma L_1 i_{1q})\} \tag{2}$$

$$E'_{2q} = \left(\frac{M}{L_2}\right)\{v^*_{1q} - (Ri_{1q} + \widehat{\omega}_1\sigma L_1 i_{1d})\} \tag{3}$$

where: $E'_{2d}, E'_{2q}, v^*_{1d}, v^*_{1q}, i_{1d}, i_{1q}$ are the rotor magnetic flux induced voltages converted to the primary side, the stator voltage references and the stator currents in d and q axis, R_1 is the resistance of the stator, L_1, L_2 are the self-inductances of the stator and rotor side, M is the mutual inductance, $\sigma\left(= 1 - \frac{M^2}{L_1 L_2}\right)$ is the linkage coefficient, $\widehat{\omega}_1$ is the estimated primary angular frequency, and $\Phi^*_2(= MI^*_{1d})$ is the rotor flux reference.

A model for designing the primary angular frequency estimator gains

In this section, a model is introduced for designing the primary angular frequency estimator gains. The model includes voltage equations of an induction motor, an automatic current regulator, and the primary angular frequency estimation equations shown in Eq. (1) to Eq. (3). The voltage equation of the induction motor is described as follows:

$$\begin{bmatrix} v_{1d} \\ v_{1q} \\ 0 \\ 0 \end{bmatrix} = \begin{bmatrix} R_1 + p\sigma L_1 & -\omega_1\sigma L_1 & p\dfrac{M}{L_2} & -\dfrac{\omega_1 M}{L_2} \\ \omega_1\sigma L_1 & R_1 + p\sigma L_1 & \dfrac{\omega_1 M}{L_2} & p\dfrac{M}{L_2} \\ -\dfrac{MR_2}{L_2} & 0 & p + \dfrac{R_2}{L_2} & -(\widehat{\omega}_1 - \omega_r) \\ 0 & -\dfrac{MR_2}{L_2} & \widehat{\omega}_1 - \omega_r & p + \dfrac{R_2}{L_2} \end{bmatrix} \begin{bmatrix} i_{1d} \\ i_{1q} \\ \phi_{2d} \\ \phi_{2q} \end{bmatrix} \tag{4}$$

where: p is a differential operator.

Fig. 3 shows the block diagram of the automatic current regulator. The automatic current regulator is designed assuming that the filed oriented control is achieved. If the field oriented control is achieved, the linkage reactance voltage $\sigma L_1\omega_1$ and the rotor magnetic flux induced voltage $\frac{M}{L_2}\omega_1\Phi_2$ are compensated properly. Then, the error of the stator current is controlled by PI controller. In the railway vehicle traction system, the time constant of the automatic current regulator T_{ACR} is generally set to 10 ms.

The motor control system shown in Fig. 1 includes variables such as the primary angular frequency $\widehat{\omega}_1$ and the slip angular frequency ω_s. In order to determine the gains based on the bode diagrams, the equations shown in Eq. (1) to Eq. (3) need to be linearized. In this paper, these equations are linearized

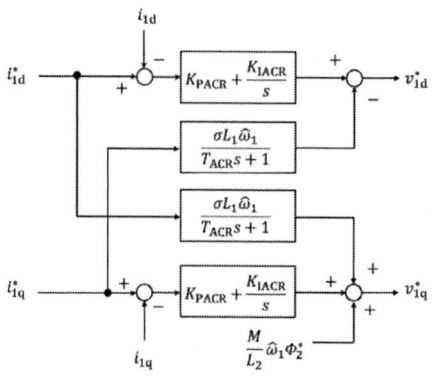

Fig. 3 block diagram of automatic current regulator

for small displacements from an equilibrium point of each variable. The state-space equation for these minute displacements are derived. In this research, the input is a small displacement of the rotor speed $\delta\omega_r$ and the output is a small displacement of the estimated primary angular frequency $\delta\hat{\omega}_1$. The equilibrium points are set to the values when the field oriented control is achieved, and the values are shown in Table 1. The state-space equation of the linearized model is written as follows:

$$
\frac{d}{dt}
\begin{bmatrix}
\delta i_{1d} \\
\delta i_{1q} \\
\delta \Phi_{2d} \\
\delta \Phi_{2q} \\
\delta v_{1d} \\
\delta v_{1q} \\
\delta E'_{2df} \\
\delta E'_{2qf} \\
\delta \hat{\omega}_1
\end{bmatrix}
=
\begin{bmatrix}
-\dfrac{R_1}{\sigma L_1}-\dfrac{R_2(1-\sigma)}{\sigma L_2} & \omega_{10} & \dfrac{MR_2}{\sigma L_1 L_2^2} & \dfrac{\omega_{r0}M}{\sigma L_1 L_2} & \dfrac{1}{\sigma L_1} & 0 \\[2mm]
-\omega_{10} & -\dfrac{R_1}{\sigma L_1}-\dfrac{R_2(1-\sigma)}{\sigma L_2} & -\dfrac{\omega_{r0}M}{\sigma L_1 L_2} & \dfrac{MR_2}{\sigma L_1 L_2^2} & 0 & \dfrac{1}{\sigma L_1} \\[2mm]
\dfrac{MR_2}{L_2} & 0 & -\dfrac{R_2}{L_2} & \omega_{10}-\omega_{r0} & 0 & 0 \\[2mm]
0 & \dfrac{MR_2}{L_2} & -\omega_{10}+\omega_{r0} & -\dfrac{R_2}{L_2} & 0 & 0 \\[2mm]
K_{PACR}\left\{\dfrac{R_1}{\sigma L_1}+\dfrac{R_2(1-\sigma)}{\sigma L_2}\right\}-K_{IACR} & -K_{PACR}\omega_{10} & -\dfrac{K_{PACR}MR_2}{\sigma L_1 L_2^2} & -\dfrac{K_{PACR}\omega_{r0}M}{\sigma L_1 L_2} & -\dfrac{K_{PACR}}{\sigma L_1} & 0 \\[2mm]
K_{PACR}\omega_{10} & K_{PACR}\left\{\dfrac{R_1}{\sigma L_1}+\dfrac{R_2(1-\sigma)}{\sigma L_2}\right\}-K_{IACR} & \dfrac{K_{PACR}\omega_{r0}M}{\sigma L_1 L_2} & -\dfrac{K_{PACR}MR_2}{\sigma L_1 L_2^2} & 0 & -\dfrac{K_{PACR}}{\sigma L_1} \\[2mm]
-\dfrac{MR_1}{T_{E2d}L_2} & \dfrac{\omega_{10}\sigma L_1 M}{T_{E2d}L_2} & 0 & 0 & \dfrac{M}{T_{E2d}L_2} & 0 \\[2mm]
-\dfrac{\omega_{10}\sigma L_1 M}{T_{E2q}L_2} & -\dfrac{MR_1}{T_{E2q}L_2} & 0 & 0 & 0 & \dfrac{M}{T_{E2q}L_2} \\[2mm]
\dfrac{K_{PE2d}MR_1}{T_{E2d}L_2}-\dfrac{\omega_{10}\sigma L_1 M}{\Phi_2^* T_{E2q}L_2} & -\dfrac{K_{PE2d}\omega_{10}\sigma L_1 M}{T_{E2d}L_2}-\dfrac{MR_1}{\Phi_2^* T_{E2q}L_2} & 0 & 0 & -\dfrac{K_{PE2d}M}{T_{E2d}L_2} & \dfrac{M}{\Phi_2^* T_{E2q}L_2}
\end{bmatrix}
\;*
$$

$$
*
\begin{bmatrix}
0 & 0 & i_{1q0} \\
0 & 0 & -i_{1d0} \\
0 & 0 & \Phi_{2q0} \\
0 & 0 & -\Phi_{2d0} \\
0 & 0 & -K_{PACR}i_{1q0} \\
0 & 0 & K_{PACR}i_{1d0} \\
-\dfrac{1}{T_{E2d}} & 0 & \dfrac{\sigma L_1 i_{1q0}M}{T_{E2d}L_2} \\
0 & -\dfrac{1}{T_{E2q}} & -\dfrac{\sigma L_1 i_{1d0}M}{T_{E2q}L_2} \\
\dfrac{K_{PE2d}}{T_{E2d}}-K_{IE2d} & -\dfrac{1}{\Phi_2^* T_{E2q}} & -\dfrac{K_{PE2d}\sigma L_1 i_{1q0}M}{T_{E2d}L_2}-\dfrac{\sigma L_1 i_{1d0}M}{\Phi_2^* T_{E2q}L_2}
\end{bmatrix}
\begin{bmatrix}
\delta i_{1d} \\
\delta i_{1q} \\
\delta \Phi_{2d} \\
\delta \Phi_{2q} \\
\delta v_{1d} \\
\delta v_{1q} \\
\delta E'_{2df} \\
\delta E'_{2qf} \\
\delta \hat{\omega}_1
\end{bmatrix}
+
\begin{bmatrix}
\dfrac{\Phi_{2q0}M}{\sigma L_1 L_2} \\[2mm]
-\dfrac{\Phi_{2d0}M}{\sigma L_1 L_2} \\[2mm]
-\Phi_{2q0} \\[1mm]
\Phi_{2d0} \\[2mm]
-\dfrac{K_{PACR}\Phi_{2q0}M}{\sigma L_1 L_2} \\[2mm]
\dfrac{K_{PACR}\Phi_{2d0}M}{\sigma L_1 L_2} \\[2mm]
0 \\
0 \\
0
\end{bmatrix}
\delta\omega_r
$$

$$(5)$$

Design of a primary angular frequency estimator gains

In this section, gains of the primary angular frequency estimator are determined. In this study, the gains are designed considering the sudden changes of the rotor speed assuming the slip phenomenon of the railway vehicle traction. The mechanical time constant of rolling stock during the slip is generally several hundred milliseconds, and its bandwidth is about 30 rad/s. Therefore, the primary angular frequency estimator gains are designed focusing on around 30 rad/s bandwidth.

Table 1. values at the equilibrium point

parameter	value	parameter	value
R_1	0.0971 Ω	I_{1d0}	120.3 A
R_2	0.0861 Ω	I_{1q0}	146.9 A
L_1	0.03012 H	Φ_{2d0}	3.5 Wb
L_2	0.02993 H	Φ_{2q0}	0.0 Wb
M	0.02910 H	$\widehat{\omega}_{10}$	25.0 rad/s
σ	0.0607	ω_{r0}	21.5 rad/s
T_{ACR}	0.010 s	K_{PACR}	0.183
T_2	0.348 s	K_{IACR}	9.71

First, filter time constants T_{E2d}, T_{E2q} of the calculated rotor magnetic flux induced voltage are determined. Then, the integration time T_{PIE2d} and the proportional gain K_{PIE2d} of the PI controller for E_{2d} compensation are determined.

As shown in Eq. (2) and (3), E'_{2d} and E'_{2q} are calculated from the stator voltage and the stator current. The detected current includes harmonics, which affects the primary angular frequency estimation. In order to reduce the influence, the filtered rotor magnetic flux induced voltage is used for the primary angular frequency estimation. In this study, a first-order low-pass filter is applied. Relational expressions are described as follows:

$$E'_{2df} = \frac{1}{T_{E2d}s + 1} E'_{2d} \tag{6}$$

$$E'_{2qf} = \frac{1}{T_{E2q}s + 1} E'_{2q} \tag{7}$$

where: E'_{2df}, E'_{2qf}, are the filtered rotor magnetic flux induced voltages in d and q axis, respectively.

In this paper, the primary angular frequency estimation equation shown in Eq. (1) is rewritten to Eq. (8). Then, the proportional gain K_{PIE2d} and the integration time T_{PIE2d} are determined.

$$\widehat{\omega}_1 = \frac{E'_{2qf}}{\Phi_2^*} + K_{PIE2d}\left(1 + \frac{1}{T_{PIE2d}s}\right)(E'^*_{2d} - E'_{2df}) \tag{8}$$

Fig. 4 shows the block diagram of the primary angular frequency estimation.

Design of the filter time constants of the rotor magnetic flux induced voltage

The sensorless control using the rotor magnetic flux induced voltage is a method of estimating the primary angular frequency by controlling E_{2d} to be 0. For the quick estimation, the change in E_{2d} need to be followed quickly. The rotor magnetic flux induced voltage E_{2d} is calculated from the stator voltage and the stator current. Therefore, the calculated rotor magnetic flux induced voltage E'_{2d} changes when the stator current changes. The stator current changes in the time constant of the automatic current regulator 10 ms. It is considered that E_{2d} also changes in 10 ms. The filter time constant T_{E2d} is set to 10 ms so that E_{2d} is quickly controlled by following the change in E_{2d}.

$$T_{E2d} = 0.010$$

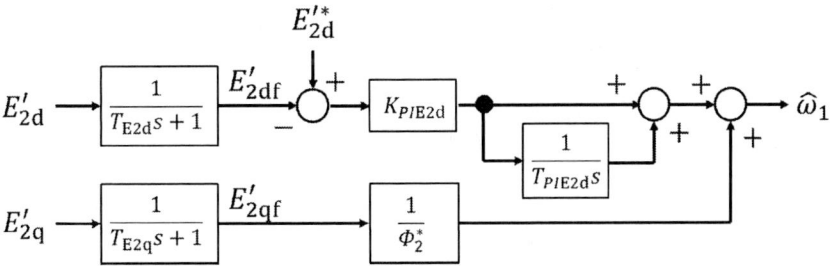

Fig. 4 block diagram of the primary angular frequency estimator

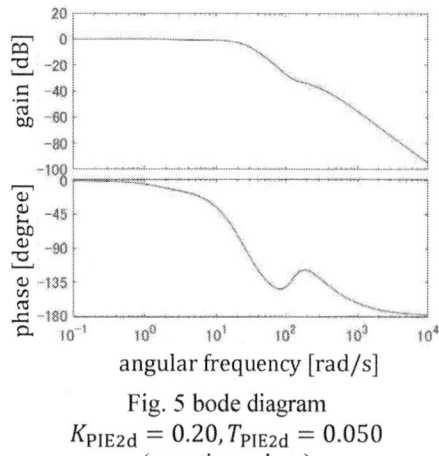

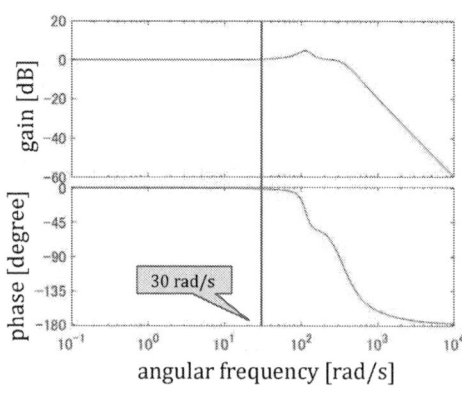

Fig. 5 bode diagram
$K_{\mathrm{PIE2d}} = 0.20, T_{\mathrm{PIE2d}} = 0.050$
(tentative values)

Fig. 6 bode diagram
$K_{\mathrm{PIE2d}} = 12.0, T_{\mathrm{PIE2d}} = 0.050$
(designed gains by the proposed method)

In the sensorless field oriented control using the secondary magnetic flux induced voltage, the primary angular frequency estimation works by using E'_{2q}. Calculated E'_{2q} is the induced voltage by the secondary magnetic flux. The secondary magnetic flux of the induction motor changes depending on the secondary time constant determined by the motor parameters on the rotor side, and the secondary time constant of the induction motor for railway traction vehicle is several hundred milliseconds. Therefore, the filter time constant T_{E2q} of E'_{2q} can be set longer than the filter time constant T_{E2d} of E'_{2d}, and the bandwidth of the estimation using E'_{2d} can be separated from the bandwidth of the estimation using E'_{2q}. The filter time constant T_{E2q} for E'_{2q} is set to about three times of the secondary time constant of the induction motor, and the following value is determined.

$$T_{\mathrm{E2q}} = 1.0$$

Design of the PI controller gains for E_{2d} compensation

First, the integral time T_{PIE2d} shown in Fig. 4 is determined. In a bode diagram of the PI controller, the gain is high at the bandwidth lower than the angular frequency of the integral time ($= 1/T_{\mathrm{PIE2d}}$ [rad/s]). Therefore, the response in the low bandwidth can be improved by setting the integral time short. When the rotor speed changes suddenly, the rotor magnetic flux changes with the secondary time constant of the induction motor, and the time constant of the induction motor is several hundred milliseconds. When the rotor magnetic flux changes, the induced voltage E'_{2d} also changes. Therefore, the integral time T_{PIE2d} should be shorter than the secondary time constant to follow the dynamics. On the other hand, since the time constant of the automatic current regulator is 10 ms, which is the fastest response in the speed sensorless drive system, higher bandwidth than the automatic current regulator bandwidth does not need to be considered. The integral time T_{PIE2d} is determined so that T_{PIE2d} is not close to the time constant of the automatic current regulator and T_{PIE2d} should be shorter than the secondary time constant of the induction motor. At this time, it is considered that the integration time T_{PIE2d} should be set between 50 ms and 100 ms. In this study, the integral time T_{PIE2d} is set to 50 ms in order to estimate the primary angular frequency as quickly as possible. Fig. 5 shows a bode diagram when the proportional gain K_{PIE2d} is 0.20 and the integral time T_{PIE2d} is 50 ms. The proportional gain K_{PIE2d} is set to 0.20 as a tentative value for the design.

Next, the proportional gain K_{PIE2d} shown in Fig. 4 is determined. In this study, the sudden change of the rotor speed when the rolling stock slips is assumed. The mechanical time constant of the rolling stock during slip is several hundred milliseconds, and its angular frequency is about 30 rad/s. When the gain characteristics of the bode diagram exceeds 0 dB at 30 rad/s, the estimated primary angular frequency may become oscillatory. When the gain characteristics is lower than 0 dB, the primary angular frequency cannot be estimated accurately when the rotor speed changes suddenly. Therefore, the gain should be 0 dB at 30 rad/s. In order to satisfy the conditions which is that the gain is 0 dB and the phase delay is almost 0° at 30 rad/s, the proportional gain is designed as $K_{\mathrm{PIE2d}} = 12.0$.

Fig. 6 shows a bode diagram in the case of $K_{PIE2d} = 12.0$. At 30 rad/s, the gain is 0 dB and the phase delay is almost 0 °. When these gains are set, it is expected that the estimated primary angular frequency follows the change in the rotor speed even when the rotor speed changes suddenly.

Evaluation of the determined gains of the estimator

In this section, the determined gains of the primary angular frequency estimator are evaluated by means of simulation of a slip re-adhesion control of the railway vehicle traction. Fig. 7 shows the block diagram of the simulation, and Fig. 8 shows tangential coefficient characteristics between the rail and the wheel for the simulation. The motor torque is set to a torque assuming that the tangential coefficient is about 0.1. When the tangential coefficient characteristics is characteristics A, slip does not occur because the maximum tangential coefficient exceeds the assumed tangential coefficient 0.1. On the other hand, in the case of the characteristics B, the slip occurs because the maximum tangential coefficient is lower than the assumed tangential coefficient 0.1. Slip is simulated by changing the tangential coefficient characteristics at 4 second. When the slip is detected, motor torque is reduced to avoid over slip.

In order to verify the effectiveness of the design method, a simulation study is performed. The case of the designed gains shown in Fig. 6 is compared with the case of the gains shown in Fig. 5 as a comparative example.

Fig. 9 shows simulation results in the case of using the gain shown in Fig. 5 and Fig. 10 shows results in the case of using the determined gains shown in Fig. 6. By designing the estimator gains, the primary angular frequency can be estimated properly even when the rotor speed changes suddenly. In addition, time to converge to command values of the stator current and motor torque is shortened.

Since the field oriented control and the decoupling control in the automatic current control are achieved, the transient tracking errors of the stator current and the motor torque to their command values are reduced when the gains shown in Fig. 6 are used. The automatic current control shown in Fig. 3 is designed on the assumption that the field oriented control is achieved. The assumption is satisfied because the filed oriented control is achieved by estimating the primary angular frequency estimation quickly with the designed gains. Therefore, the error between the motor torque and the command value is improved in the case of using the designed gains shown in Fig. 6. In addition, the rotor speed is also estimated quickly, the slip frequency can be controlled accurately. Therefore, the error between the motor torque and its command value is reduced as shown in Fig. 10. The estimated primary angular frequency is also used for the decoupling control that compensates the leakage reactance induced voltage. Due to the quick estimation, an error between the actual leakage reactance induced voltage $\sigma L_1 \omega_1$ and the output voltage for compensating the leakage reactance induced voltage $\sigma L_1 \hat{\omega}_1$ is reduced. Therefore, the deterioration of the followability of the stator current is improved by using the designed estimator gains.

In addition, when the determined gain values shown in Fig. 6 are used, the wheel re-adheres about 0.2 seconds earlier than the case with the gains shown in Fig. 5. The wheel re-adheres quickly due to the improved followability of the motor torque to its command value. By estimating the primary angular frequency quickly, the motor torque follows the command value even when the wheel slips. As a result, re-adhesion time can be shortened because the motor toque is properly reduced to re-adhere.

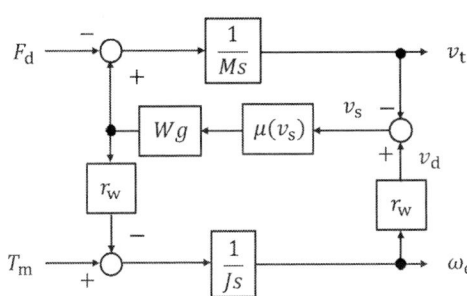

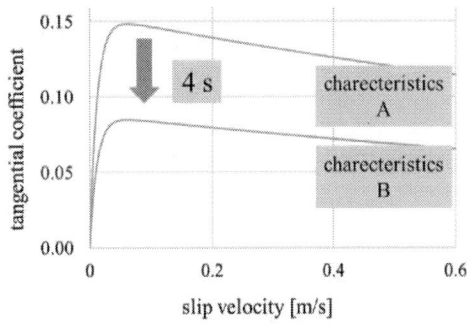

Fig. 7 block diagram of slip simulation

Fig. 8 tangential coefficient characteristics for simulation

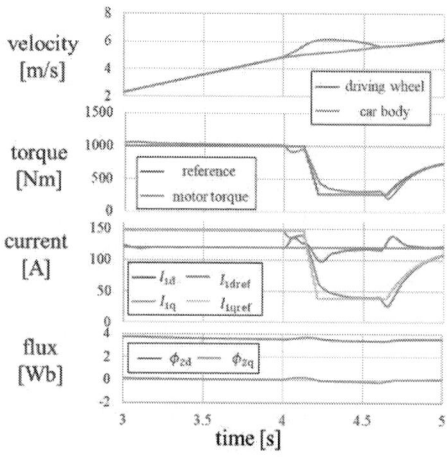

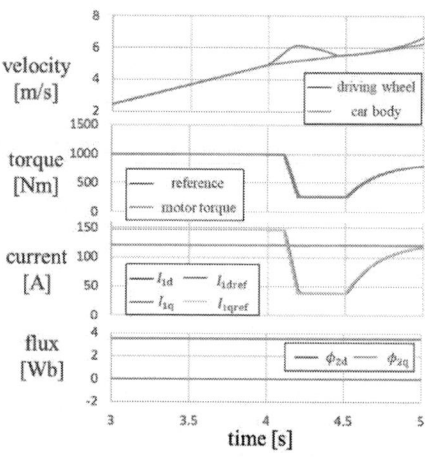

(a) velocity, torque and current

(a) velocity, torque and current

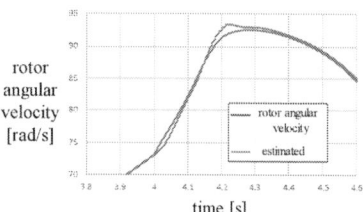

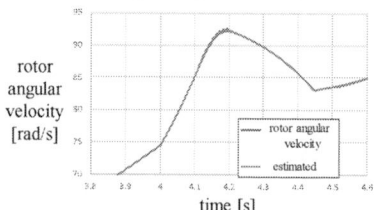

(b) rotor speed and estimated speed

(b) rotor speed and estimated speed

Fig. 9 simulation results of gain values
$K_{\mathrm{PIE2d}} = 0.20, T_{\mathrm{PIE2q}} = 0.050$

Fig. 10 simulation results of designed gain values
$K_{\mathrm{PIE2d}} = 12.0, T_{\mathrm{PIE2q}} = 0.050$

Conclusion

This paper proposed the design method of the primary angular frequency estimator gains of the speed sensorless field oriented control considering the sudden change of the rotor speed. Especially, this study assumed the sudden change of the rotor speed due to the slip phenomenon of the railway traction vehicle. It was confirmed that designed gains enabled to estimate the primary angular frequency properly even when the rotor speed changed suddenly by means of simulating slip of the railway vehicle traction. Also, time to converge to command values of the stator current and the motor torque was shortened.

References

[1] T. Kikuchi, Y. Matsumoto and A. Chiba: Speed Estimation Method for Free-running Induction Motor with High Inertia Load in the Low Speed Range, *2016 IEEE Energy Conversion Congress and Exposition,* pp. 1- 6, 2016.

[2] T. Ogawa, S. Ishida, T. Kojima, T. Sato, H. Taguchi and S. Ohashi: Speed Sensorless Vector Control for Rolling Stock, *2005 European Conference on Power Electronics and Applications*, pp. 1- 7, 2005.

[3] K. Ohyama, K. Shinohara: Small-Signal Stability Analysis of Vector Control System of Induction Motor Without Speed Sensor Using Synchronous Current Regulator, *IEEE Transactions on Industry Applications.*, Vol. 36, No. 6, pp. 1669-1675, 2000.

[4] M. Tsuji, S. Chen, K. Izumi and E. Yamada: Stability Improvement of Speed Sensorless Induction Motor Vector Control System Using q-Axis Flux with Stator Resistance Identification, *PESC 98 Record. 29th Annual IEEE Power Electronics Specialist Conference*, pp. 1587- 1592, 1998.

[5] K. Ide, K. Hazama, T. Tsuji and R. Oguro: Speed Sensorless Field-Oriented Controlled IM with Stator Voltage Error Compensator", *IEEJ Transaction Induction Application*, Vol. 116-D, No.8, pp. 835- 843, 1996. (in Japanese)

[6] D. Lee and K. Akatsu: "Selection Method of Controller Gains for Position Sensorless Control of IPMSM Drives" *IEEJ Journal of Industry Applications*, Vol.8, No.4, pp.720-726. (2019)

[7] S. Maekawa, M. Sugimoto, K. Ishida, M. Nogi and M Kanamori: "Stability Analysis of Sensorless Speed Control for PMSM Considering Current Control System" *IEEJ Journal of Industry Applications*, Vol.8, No.4, pp.736-744. (2019)

[8] K. Kondo, N. Terauchi, K. Yuki and T. Hasebe: Study on the Rotor Flux in Induction Motor in Speed Sensorless Control for Railway Vehicle Traction, in *Proceedings IEEJ Papers Joint Technical Meeting "Semiconductor Power Converter" "Industrial Electrical Application"*, SPC-03-100/ IEA-03-13, 2003. (in Japanese)

[9] K. Kondo, K. Yuki: An application of the induction motor speed sensor less control to railway vehicle traction system, *Conference Record of the 2002 IEEE Industry Applications Conference. The 37th IAS Annual Meeting*, Vol. 3, pp. 2022-2027, 2002.

Efficiency Potential of Solid-State Pulse Modulators using SiC Devices

Spyridon Stathis, Michael Jaritz, Sebastian Blume and Jürgen Biela
Laboratory for High Power Electronic Systems, ETH Zürich, Switzerland
Email: stathis@hpe.ee.ethz.ch

Keywords

"Pulse Transformer", "Resonant Converter", "Silicon Carbide (SiC)", "Power Factor Correction (PFC) Converter", "Solid-State Pulse Modulator", "Efficiency Calculation"

Abstract

In this paper, the efficiency of the CLIC solid-state pulse modulator developed for the CERN research center and the long-pulse modulator system built for the ESS facility is investigated in detail. The analysis shows the current efficiency status of the modulators and how their performance could be improved by using SiC devices.

1 Introduction

Pulse modulator systems are key elements in many application areas such as cancer treatment, particle accelerators and radar systems. The efficiency of these systems becomes more and more important as an increasing number of accelerator facilities try to reduce the greenhouse gas emissions [1–3]. Pulse modulators, which are based on solid-state switches, have been proven to be able to efficiently generate pulses at high power levels [4],[5]. Typically, IGBT modules are used as solid-state switches as they offer high blocking voltages and current ratings, and these have demonstrated a reliable operation in pulsed power applications [6].

The solid-state pulse modulators, typically, consist of several power conversion stages between the grid and the load and each stage contributes to the overall system losses. However, so far, only a few publications have investigated the efficiency of the complete modulator system [1],[3]. Also, the impact of using wide bandgap devices, such as SiC, on the system efficiency has been rarely examined [7]. Hence, in this paper, a comprehensive investigation of the efficiency of two solid-state pulse modulators as well as the efficiency potential using SiC devices are presented.

In the first step, the efficiency of each conversion stage of the CLIC solid-state pulse modulator presented in [8], which is based on Si semiconductors, is calculated and the total efficiency is determined. Afterwards, the efficiency

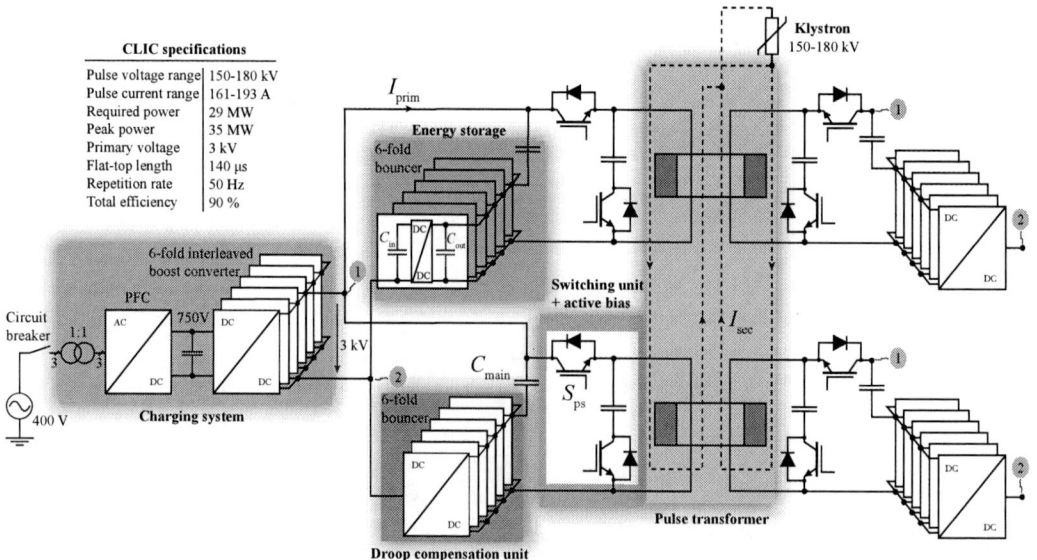

Fig. 1: CLIC pulse modulator schematic overview consisting of: Charging system, droop compensation units along with the main capacitors (energy storage), switching + active bias units, pulse transformer, and klystron load.

improvements by using SiC devices, along with other suggested improvements, is evaluated. Similarly, the efficiency of each sub-system along with the investigation of the impact of using SiC devices is presented for the long-pulse modulator developed for the ESS linear accelerator [9]. The SiC-based systems are not re-optimized in terms of efficiency, but a 1:1 replacement of the power switches is assumed to evaluate the impact of the SiC devices. Further improvements are possible in case an optimization is performed for the SiC-based systems.

In the following section 2, the structure of the CLIC system is explained and the main specifications of the pulse modulator are given. In section 3, the efficiency and the distribution of the losses in each stage are derived for the existing Si-based and for the improved SiC-based modulator system. In section 4, the total efficiency for the Si-based and the SiC-based CLIC modulator is extracted, the contribution of each stage to the total efficiency is presented and in the end the results are discussed. In section 5, the topology and the main specifications of the ESS modulator are shown while in section 6, the loss share of each stage of the current Si-based and the improved SiC-based system is provided. In section 7, the total efficiency is given for the Si-based and the SiC-based ESS modulator and the results are evaluated. Finally, section 8 summarizes the main outcomes of the present work.

2 CLIC Modulator Topology

A schematic overview of the CLIC modulator topology is shown in Fig. 1. The main specifications of the system are also listed in the figure. At the input, a charging system is connected to the 3-phase 400 V AC grid which charges the four main capacitors C_{main} up to 3 kV. The charging system consists of a 1:1 isolation transformer, a power factor correction (PFC) rectifier and a 6-fold interleaved boost converter. Four droop compensation units are connected in series to the four main capacitors in order to compensate the droop during the pulse interval. Each droop compensation unit consists of six interleaved buck-boost modules which are referred as bouncer modules. The four switching units S_{ps}, which also include an active bias circuit to pre-magnetize the core of the pulse transformer, connect the primary winding of the pulse transformer to the main capacitor. A non-linear load, the klystron, is attached to the secondary side of the split two-core pulse transformer [2].

3 Efficiency of CLIC Stages

In the following, the efficiency of each stage is derived for the Si-based and the improved SiC-based modulator. It is noted that the efficiency calculation is based on the maximum junction temperatures that the semiconductors can withstand. The analysis begins with the efficiency calculation of the charging system.

Charging system

The efficiency derivation for the charging system is split into its three sub-systems, namely the isolation transformer, the PFC rectifier and the boost converter.

- **Isolation transformer**: The isolation transformer is a 3-phase 250 kVA unit. The loss distribution for the transformer of the existing modulator can be seen in table I. These values are taken from the data sheet. The core material of the transformer is silicon steel while the windings are made of aluminium. However, if copper windings are used, then the conductivity becomes higher and the winding losses can be reduced by approximately 35%. In addition, if an amorphous core material is used, such as steel with 25 μm ribbon thickness, the core losses can be reduced by 60% [10].

- **PFC rectifier**: For the PFC rectifier, two ACOPOSmulti 8BVP1650 active rectifiers in parallel connection are used [11]. Each active rectifier is rated for 120 kW/750 V DC. The rectifier unit is based on 1.2 kV Si IGBTs and it has an efficiency of 97.5% according to the data sheet. However, even a 99% efficiency is feasible to be achieved for the PFC unit in case SiC technology is adopted [12],[13].

- **6-fold interleaved boost converter**: For charging the main capacitors, a six-fold interleaved boost converter, which operates under boundary conduction mode for achieving zero voltage switching, is used [14]. The output power of a single module is 40 kW, the peak current through the switches is 140 A and the input

Table I: Isolation transformer efficiency & loss distribution

	Existing transformer	**Improved transformer**
Winding losses	2330 W	1515 W
Core losses	700 W	280 W
Conversion efficiency	98.8%	99.2%

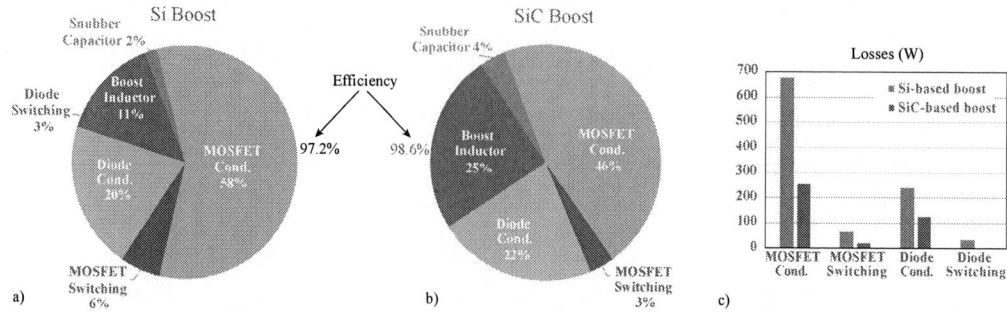

Fig. 2: a) Loss breakdown of the Si-based boost module. b) Loss breakdown of the SiC-based boost module. c) Conduction and switching losses in absolute numbers of each semiconductor for the Si-based and the SiC-based boost module. The efficiency of the Si-based and the SiC-based boost module is 97.2% and 98.6% respectively.

voltage is 750 V. In table II, the device arrangement and the model number of the semiconductors for the Si-based and the SiC-based boost module are listed. The conduction and the switching losses of the switches are calculated according to [15] at 70 kHz switching frequency. The boost inductor is designed considering skin, proximity and core losses [8]. The losses of the snubber capacitors are also considered.

The breakdown of the losses and the efficiency of a 40kW boost module are shown in Fig. 2 a) and b) for the Si-based and the SiC-based boost module respectively. In the SiC-based boost, the conduction and the switching losses of the MOSFETs drop by 12% and 3% respectively whereas the diode switching losses are almost zero due to the ZVS operation and the ultra-fast SiC diodes. The Si-based boost reaches a total efficiency of 97.2% while the SiC-based has an efficiency of 98.6%. The conduction and the switching losses of the semiconductors in absolute numbers are also depicted in Fig. 2 c). In the SiC-based boost module, there is a clear reduction of the conduction and the switching losses compared to the Si-based module.

Table II: Selected boost module devices; (xsyp) for x in series & y in parallel devices

Device	Si-based	SiC-based
MOSFET	IPZ65R019C7 (8s2p)	C3M0016120K (4s2p)
Diode	APT75DQ120B (4s)	IDW40G120C5B (4s)

Droop compensation unit

The CLIC modulator has four droop compensation units and each unit consists of six interleaved active bouncer modules. The topology of a single bouncer module and the main waveforms are depicted in Fig. 3 a) and b) respectively. Moreover, the bouncer specifications along with the devices that are used for the Si-based and the SiC-based bouncer module are shown in table III. In [16], a detailed explanation of the bouncer operation can be found. During the charging time t_{ch}, the high-side switch S_{HS}, and the short-circuit switch S_{SC} are on. Consequently, the inductor current slope is known and the losses of the switches can be calculated analytically. Then, the inductor

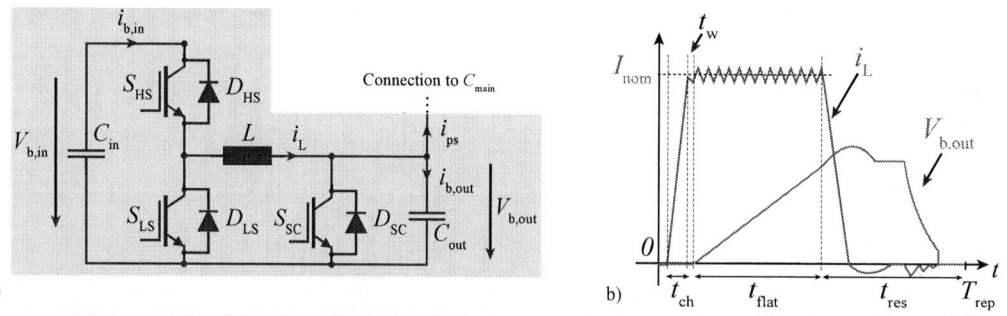

Fig. 3: a) Topology of a bouncer module which consists of a half-bridge S_{HS} and S_{LS}, an inductor L and a short-circuit switch S_{SC}. b) Bouncer module current and voltage during the pulse operation which is divided in four intervals t_{ch}, t_w, t_{flat}, t_{res} over the repetition period T_{rep}.

Table III: Selected bouncer module specifications & devices; (xp) for x in parallel devices

Symbol	Parameter		Device	Si-based	SiC-based
$V_{b,in}$	Input voltage	450 V	S_{HS} (6p)	IKW50N65F5	SCTW90N65G2V
$V_{b,out}$	Output voltage	0-300 V	D_{HS} (6p)	IKW50N65F5	SCTW90N65G2V
I_{nom}	Mean inductor current	628 A	S_{LS} (2p)	IGW50N65H5	SCTW90N65G2V
C_{out}	Output capacitance	66 µF	D_{LS} (4p)	IDW40E65D1	IDW40G65C5
C_{in}	Input capacitance	2 mF	S_{SC} (6p)	IKW50N65F5	SCTW90N65G2V
L	Inductance	31 µH	D_{SC} (6p)	IKW50N65F5	SCTW90N65G2V

current slightly decreases during t_w for ensuring equal current sharing between the interleaved branches [16]. During the pulse interval t_{flat}, the converter operates as a typical buck converter with 100 kHz switching frequency and the output voltage of the bouncer rises linearly from 0 to 300 V. An average duty cycle of 33% is assumed for the calculation of the mean and the RMS current through the switches. The conduction and the switching losses are calculated according to [15] and [17]. For the resonant transition after the pulse interval t_{res}, the components losses are extracted via simulations.

For minimizing the inductor losses, an optimization is performed, which considers skin and proximity losses as well as magnetic core losses [8]. The input and the output capacitor losses are negligible as film capacitors are used.

In Fig. 4 a) and b), the efficiency and the loss distribution are illustrated for the Si-based and the SiC-based bouncer module respectively. It can be observed, that the high side switch contributes 58% (51% switching + 7% conduction) to the total losses for the Si-based module and it is the main loss component. On the other hand, only 20% (12% conduction + 8% switching) of the total losses come from the high side switch in the SiC-based bouncer. The efficiency of the Si-based and SiC-based module are 91% and 95.2% on the bouncer level. However, the power that all the 24 modules (4x6) deliver during the pulse is only 5% of the total 29 MW pulsed power, and this results in the system level efficiencies of 99.4% and 99.7% for the Si-based and the SiC-based bouncer module respectively. In Fig. 4 c), the absolute numbers of the losses for each semiconductor are shown. It is clearly shown that the losses on S_{HS}, which is the main loss contributor, have significantly reduced in the SiC-based bouncer.

Switching unit

For calculating the losses in the pulse switches, the magnetizing current is neglected. The 5SNA1250B450300 IGBT StakPak switching unit is used in the CLIC modulator. In case SiC MOSFETs are used, the CAB450M12XM3 module is assumed for the calculations. 4 in series and 3 in parallel MOSFETs would be arranged. The conduction and the switching losses are computed based on the data sheet parameters. An efficiency of 98.6% is achieved for the Si-based switching unit and 99.5% for the respective SiC-based.

Pulse transformer

In [18], the transformer winding losses due to skin and proximity effect have been discussed and calculated. A trapezoidal shape is assumed for the pulse current which is analyzed with a Fourier series in order to calculate the skin effect losses. Regarding the proximity effect losses, it is assumed that the windings cover the entire height of the core window and the round conductors of the secondary winding are transformed to a sheet conductor

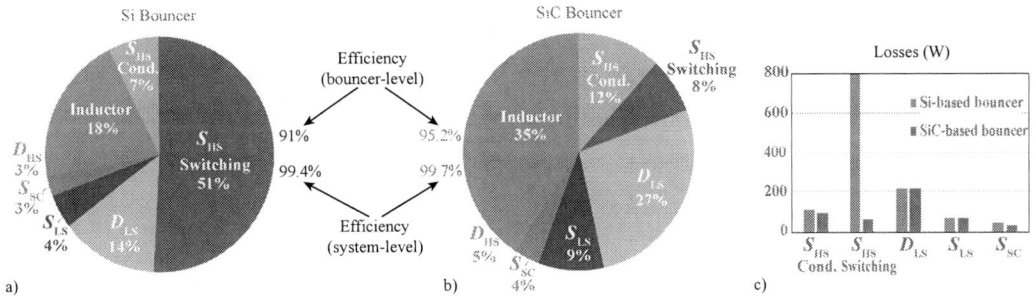

Fig. 4: a) Loss distribution of the Si-based bouncer module. b) Loss distribution of the SiC-based bouncer module. c) Losses in absolute numbers per device for the Si-based and the SiC-based bouncer module. The efficiency of the Si-based and the SiC-based bouncer module is 91% and 95.2% on the bouncer level and 99.4% and 99.7% on the system level respectively.

EPE'20 ECCE Europe

Assigned jointly to the European Power Electronics and Drives Association & the Institute of Electrical and Electronics Engineers (IEEE)

[2]. The core losses are determined by the improved generalized Steinmetz equation by taking into account the pre-magnetization of the core [2].

Another major source of losses in the pulse transformer are the losses due to the pulse shape. To illustrate this, Fig. 5 is provided. Until the flat-top stability (FTS) is reached, energy is lost during the settling time t_{settle}. The same occurs during the fall time t_{fall} of the pulse. In [19], the pulse efficiency has been defined as the ratio between the energy of an ideal rectangular pulse and the used energy of the real pulse during the flat-top period t_{flat}. The real pulse is analyzed in the time domain and, as a result, it is possible to integrate during the settling and fall time to obtain the energy loss.

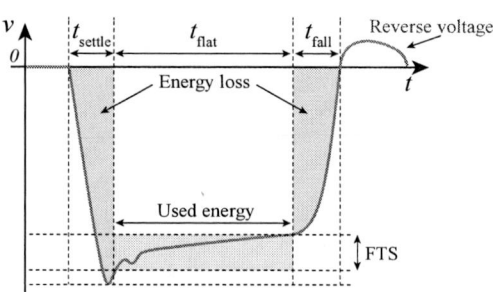

Fig. 5: Pulse shape of CLIC modulator. Once the pulse lies into flat-top stability (FTS) band, the flat-top interval t_{flat} begins.

The conversion efficiency of the split two-core pulse transformer is obtained through an optimization procedure. A detailed explanation of the optimization algorithm steps can be found in [4], but an overview of the algorithm is illustrated in Fig. 6 and it is explained briefly here. The main optimizer sets the lower and upper limits of the optimization variables and initiates the genetic algorithm. In the parameters definition section, the user sets the electrical characteristics of the modulator, such as the desired pulsed power and voltage, the pulse requirements, such as the FTS criteria and maximum permissible overshoot, and other pre-defined geometrical characteristics. Based on these parameters, the volume of the transformer and the tank is obtained through the transformer geometry formulation sub-module. Then, the maximum electric field E_{max} is calculated on the surface of the high voltage conductors by using the charge simulation method [8]. If E_{max} is less than the preset threshold E_{th}, the procedure continues with the analytical calculation of the leakage inductance L_σ and distributed capacitance C_d of the transformer geometry [1]. Finally, the pulse shape is analyzed in the time domain by using the secondary-referred electrical equivalent circuit of the transformer and, if the pulse requirements are met, the

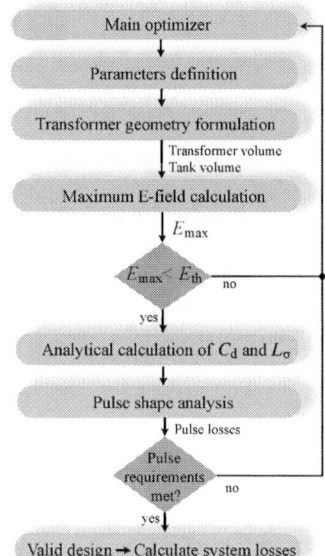

Fig. 6: Flowchart of the pulse transformer optimization.

design is considered valid and the system losses are calculated. The system losses include the winding and the core losses of the transformer as well as the losses due to the pulse shape.

In addition, the length of the high voltage cable, which connects the transformer to the klystron, and the type of the insulating oil affect the transformer efficiency, as they change the total distributed capacitance of the transformer and therefore, the intervals t_{settle} and t_{fall} also change. To visualize this dependency, Fig. 7 is provided. It is observed that the use of mineral oil instead of natural ester oil, can result in higher transformer efficiencies at the same cable length l_{HV}. It is also shown that the higher the cable length, the lower the conversion efficiency of the transformer due to the additional paracitic capacitance of the cable. The selected pareto point for the CLIC system results in a transformer conversion efficiency of 96.7% including winding, core, and pulse shape losses. Approximately 4% are winding losses, 8% core losses, and 88% are losses due to the pulse shape. However, in [4], it has been shown that a 97.5% efficiency could be achieved if a three-core arrangement, mineral oil, and a SiFE with 25 μm lamination core material would be chosen for the CLIC system instead of two-core and natural ester type of oil. However, this solution would lead to a transformer with a significantly higher volume.

4   CLIC Efficiency Results

Combining the results of the previous section, the total efficiency of the existing modulator and the improved SiC-based can be obtained. Using SiC semiconductors along with the other suggested improvements can lead to a substantially higher efficiency for the CLIC modulator system. In Fig. 8 a) and b), the breakdown of the losses for each sub-system of the existing Si-based and the improved SiC-based modulator are shown respectively. The

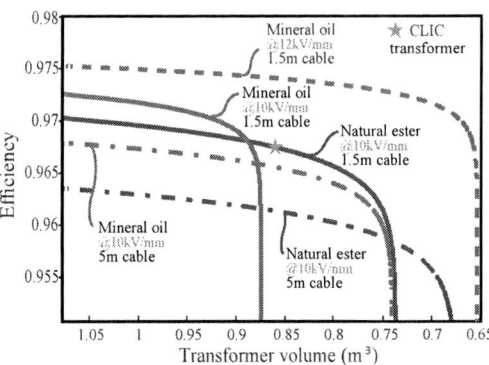

Fig. 7: Pareto limit between conversion efficiency and transformer volume with 1) ester oil, l_{HV} = 5 m with E_{max} = 10 kV / mm, 2) ester oil, l_{HV} = 1.5 m with E_{max} = 10 kV / mm, 3) mineral oil, l_{HV} = 5 m with E_{max} = 10 kV / mm, 4) mineral oil, l_{HV} = 1.5 m with E_{max} = 10 kV / mm, 5) mineral oil, l_{HV} = 1.5 m with E_{max} = 12 kV / mm [2].

improved system shows an efficiency of 93.6% and the existing 88.7%, which means a 4.9% efficiency increase. In the improved system, the pulse transformer losses dominate (39%), which is 11% larger than the existing modulator. In both systems, the pulse transformer constitutes the main loss contributor. In the improved modulator, the PFC, the switching unit and the boost converter have a lower contribution to the total losses compared to the existing modulator system. Moreover, the obtained efficiencies for each conversion stage are shown in the table of Fig. 8 c) while in Fig. 8 d), the total power losses for the two systems are depicted.

5 ESS Modulator Topology

The ESS modulator topology along with the main specifications are presented in [20] and a schematic overview of the system is given in Fig. 9. A PFC charging system is connected to the input 400 V AC mains which forms an 800 V DC-link bus. Next, 2 series/parallel resonant converter basic modules (SPRC-Bm) with a 400 V DC-link voltage are connected in series to the 800 V DC-link bus. A single SPRC-Bm is a Si MOSFET-based full bridge, followed by a resonant tank, a medium frequency transformer and an output rectifier. At the output, the 2 SPRC-Bms are connected in parallel creating an input series, output parallel (ISOP) stack. In order to reach the desired output voltage and power levels, 9 ISOP stacks are connected in parallel at the input and in series at the output. A balancing circuit is also connected to the input of each ISOP stack ensuring an equal voltage sharing across the capacitors C_{DLi} after each pulse. A filter capacitor C_f and the klyston load is attached to the output of the modulator. The converter operates between 100-110 kHz for compensating the droop of the pulse. For the calculations an average switching frequency of 105 kHz is assumed.

6 Efficiency of ESS Stages

In this section, the efficiency of each individual stage of the ESS solid-state pulse modulator is presented for the Si-based and for the improved SiC-based system. The efficiency calculation for the improved SiC-based system are based on the configuration that is illustrated in Fig. 10. In the improved SiC-based modulator, the balancing circuit is not required and the input voltage level for each SPRC-Bm is 800 V.

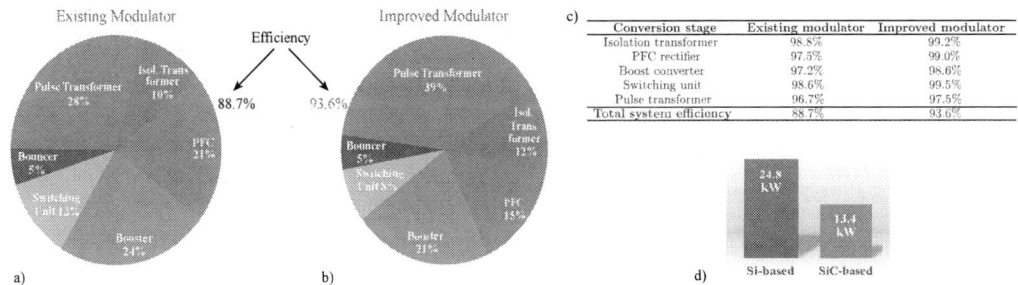

Fig. 8: a) Loss distribution of the existing Si-based CLIC modulator. b) Loss distribution of the improved SiC-based CLIC modulator. c) Conversion stages & total system efficiency results for the CLIC modulator. d) Total losses of the Si-based & the SiC-based CLIC modulator. The existing modulator shows an 88.7% efficiency and the improved modulator has 93.6%.

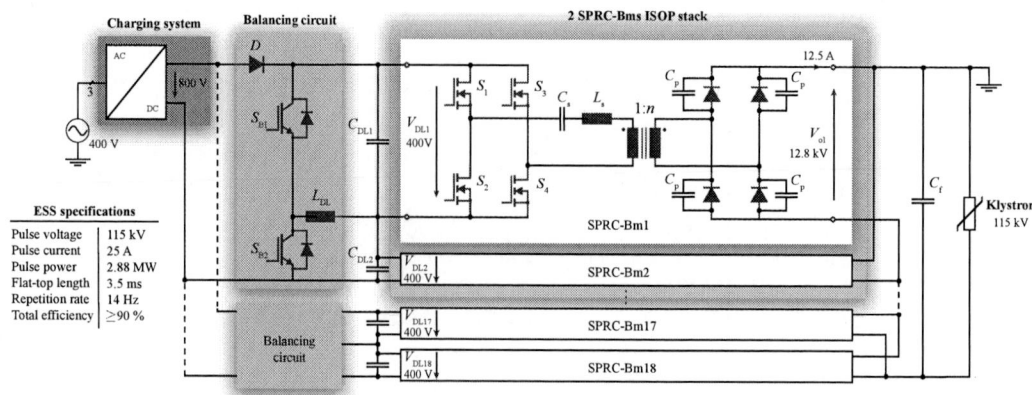

Fig. 9: ESS pulse modulator schematic overview consisting of: Charging system, balancing circuits, two series/parallel resonant converter basic modules (2 SPRC-Bms) in input series output parallel (ISOP) stacks, and klystron load.

Charging system

The input charging system is a standard industry PFC boost converter unit from REVCON with an efficiency of 98% according to the data sheet [21]. A SiC-based solution for the PFC converter could, potentially, have a 99% efficiency as discussed in section 3.

Series/Parallel Resonant Converter-Basic Module

The losses of the isolated DC-DC converter occur on the semiconductors, the passive components and the transformer. The analysis of the efficiency calculation starts with the losses in the power switches. Output filter capacitor losses are not considered as film capacitors are used with very low ESR.

- **Semiconductors losses**: In the first step, the losses of the semiconductors are derived. The selected device arrangement per each branch of the low voltage (LV) and the high voltage (HV) side of the converter is listed in Table IV for the Si-based and the SiC-based module. For the conduction loss calculation, the average and the RMS values of the currents through all the switching components are calculated analytically with the formulas presented in [9]. Then, the conduction losses are computed as proposed in [15]. The MOSFETs and the diodes turn-on under zero voltage and therefore, only turn-off losses occur which are extracted via simulations [9]. Finally, the switching losses of the anti-parallel diodes and the rectifier diodes are considered to be negligible because they turn-off under low di/dt and ultra-fast recovery diodes are used for the rectifier stage.

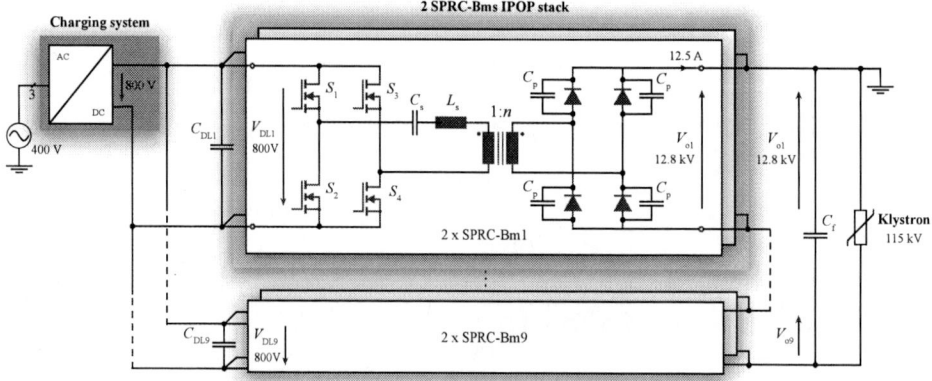

Fig. 10: Improved SiC-based ESS pulse modulator schematic overview consisting of: Charging system, two series/parallel resonant converter basic modules (2 SPRC-Bms) in input parallel output parallel (IPOP) stacks, and klystron load. Comparing to the existing ESS modulator, the balancing circuit is not required and the input voltage level for each SPRC-Bm is 800 V in the improved modulator. The output power and voltage levels of the SiC-based system remain the same.

Table IV: Selected devices for the SPRC-Bm; (xsyp) for x in series & y in parallel devices in each branch.

Device	Si-based	SiC-based
MOSFET (LV)	STY139N65M5 (6p)	IMZ120R030M1H (6p)
Diode (HV)	APT60DQ120SG (36s)	C5D25170H (26s2p)

- **Passive components losses**: At this point, the loss calculation of the series capacitors C_s, the parallel capacitors C_p as well as the series inductance L_s is described. The series capacitor consists of 896 capacitors of 15 nF connected in a series/parallel arrangement. The parallel capacitor of the rectifier has 864 capacitors (216 per diode branch) of 100 nF in a series/parallel connection. The losses for the parallel capacitance are given by the following relation [9]

$$P_{C_p} = D \cdot 2\pi f_s \cdot C_p \cdot \frac{V_{C_{p,1}}^2}{2} \cdot \tan\delta_{C_p} \qquad (1)$$

where D is the duty cycle of the pulse, f_s is the average switching frequency, $V_{C_{p,1}}$ the fundamental component of the voltage across C_p and $\tan\delta_{C_p}$ is the loss factor. The loss factor of the NP0 ceramic material of the capacitor is given in the data sheet. The same formula is applied for the series capacitor.

The required series inductance is an air toroid which offers the advantage of no core losses, no saturation as well as low stray magnetic field [22]. The toroid is made of litz wires and the winding losses calculation is explained in [22]. Given that in the improved SiC-based modulator the input voltage level is twice the voltage level of the existing system and the power level remains the same, this results to half of the current which flows through L_s. Assuming that the DC resistance of L_s can take a value that it is twice the current one, this leads to half of the power losses for the L_s in the improved SiC-based modulator.

- **Transformer losses**: For the primary and the secondary winding of the transformer, litz wire is used and ferrite K2008 as core material. Due to the almost sinusoidal flux density, the standard Steinmetz equation is used for the core losses calculation [23]. However, according to the data sheet, a 20% core losses reduction is feasible in case N97 core material is used [24].

The total winding losses P_{wdg} of the transformer are calculated with the same way as the winding losses of the toroid, except slight differences in the external magnetic field intensity analytic formulas [9], and they are given by the relation

$$P_{wdg} = D \cdot N_s \cdot l_m \left(\frac{N F_s \hat{I}^2}{N_s^2} + \sum_{n=1}^{N} G_s H_{avg,n} \right) \qquad (2)$$

with

$$\hat{H}_{avg,n} = \frac{1}{\pi r_a^2} \int |H|^2 dA_n, \qquad (3)$$

where N_s is the number of strands per bundle, l_m is the mean turn length, N is the number of turns, F_s and G_s are the skin and proximity effect factors respectively, and $\hat{I}$ stands for the peak current of the winding. For the average magnetic field intensity of each turn $H_{avg,n}$, the outer radius of the litz wire bundle r_a as well as the norm of the magnetic field density of each bundle $|H|$ over each turn cross section A_n are needed. In [23], a detailed analysis is presented for calculating the magnitude of the field intensity $|H|$.

In Fig. 11 a) and b), the loss distribution of the Si-based and the SiC-based SPRC-Bm is depicted respectively, excluding the pulse shape losses at this point. The improved SiC-based module has an efficiency of 94.4% which is 1.5% higher than the existing Si-based SPRC-Bm presents. In the improved module, the main power loss factors are the full bridge and the transformer with 40% and 18% respectively, while the contribution of the series inductor dropped from 21.9% to 14%. In addition, the losses in absolute numbers for each component of the Si-based and the SiC-based SPRC-Bm are presented in Fig. 11 c), showing the reduction of the losses in case of a SiC-based SPRC-Bm.

In order to also include the pulse shape losses to the overall system efficiency calculation, Fig. 12 is provided [20]. In Fig. 12 a), the measured output voltage of the built modulator system is shown where V_{out} (blue line) stands for the actual output voltage and $V_{out,avg}$ (green line) is the averaged pulse voltage which is used for computing the rise time t_{rise} and the fall time t_{fall}. Fig. 12 b) illustrates a zoomed version during the rise time (0 to $0.99V_{nom}$) of

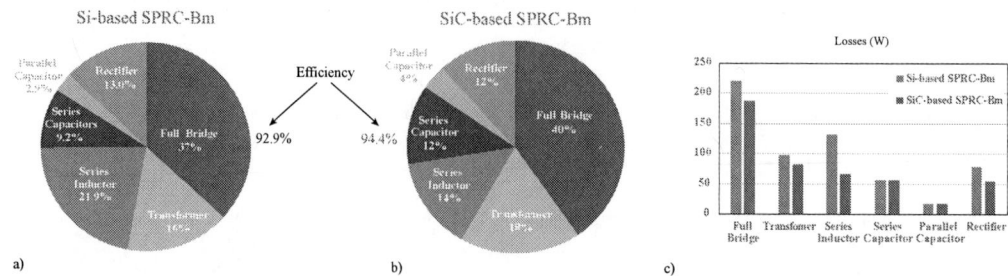

Fig. 11: a) Loss distribution of the Si-based SPRC-Bm. b) Loss distribution of the SiC-based SPRC-Bm. c) Losses in absolute numbers per component for the Si-based and the SiC-based SPRC-Bm. The efficiency of the Si-based and the SiC-based SPRC-Bm is 92.9% and 94.4% respectively. The pulse shape losses are not included in the resonant converter efficiency calculation.

the pulse whereas Fig. 12 c) presents the zoomed version of the pulse during the fall time (V_{nom} to $0.1V_{nom}$). The quantity V_{nom} stands for the nominal output voltage, namely 115 kV. The areas K_1 and K_2 correspond to the lost energy during the rise time and the fall time respectively. The pulse efficiency η_{pulse} is the ratio among the ideal rectangular pulse and the real pulse and it is given by the following relation

$$\eta_{pulse} = \left(\frac{K_{ideal}}{K_{real}} \right) \cdot 100\% \qquad (4)$$

where

$$K_{ideal} = V_{nom} \cdot (t_1 - t_{rise}), \qquad K_{real} = \int_0^{t_2} V_{out}(t)dt \qquad (5)$$

7 ESS Efficiency Results

The analysis from section 6, results in the efficiency values for each conversion stage of the existing Si-based and the improved SiC-based ESS modulator which are listed in Table V. More specific, the improved modulator has an electrical system (PFC + Resonant converter) efficiency of 93.5% which is 2.9% higher than the existing system. Assuming equal pulse losses for both systems, the improved modulator shows an overall efficiency of 91.4% while the existing modulator has 88.6%.

To illustrate the loss distribution for both systems, Fig. 13 is given. In the improved modulator, the main loss contributor are the pulse shape losses with 25.4% whereas, in the existing system, the losses of the full-bridge dominate with 23.7%. In the improved modulator, the losses in the resonant tank drop from 20.1% to 17.8% compared to the existing modulator, which is a 2.3% decrease. The PFC unit contributes 11.4% to the overall efficiency of the SiC-based system while in the existing system, 16.8% of the losses are due to the PFC charging unit. On the contrary, the loss share of the rectifier and the transformer, rise from 10.3% to 14% and from 10.5% to 12.1%, respectively, in the improved modulator system.

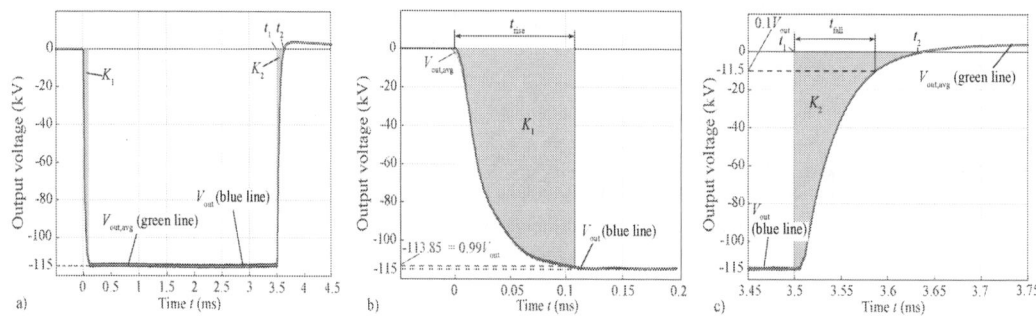

Fig. 12: a) Measured output voltage V_{out} (blue line) and averaged output voltage (green line). b) Zoomed view of the output voltage during the rise time t_{rise}. c) Zoomed view of the output voltage during the fall time t_{fall}. The areas K_1 and K_2 correspond to the lost transferred energy that the klystron cannot use during the rise time and the fall time respectively [20].

Table V: Conversion stages & total system efficiency results for the existing and the improved ESS modulator.

	Existing modulator	**Improved modulator**
PFC rectifier	98.0%	99.0%
Resonant converter	92.9%	94.4%
Pulse shape	97.8%	97.8%
Electrical system efficiency	90.6%	93.5%
Total system efficiency	88.6%	91.4%

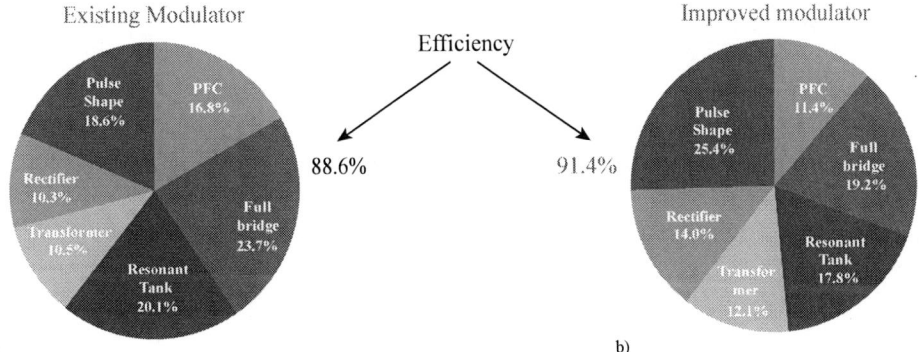

Fig. 13: a) Loss distribution of the existing Si-based ESS modulator. b) Loss distribution of the improved SiC-based ESS modulator. The existing system reaches an efficiency of 88.6% and the improved system has an efficiency of 91.4% including pulse shape losses.

8   Conclusion

In this paper, the loss distribution of all conversion stages of the CLIC and the ESS pulse modulator systems is presented. The analysis leads to the efficiency calculation of each stage and, finally, the total efficiency for the CLIC modulator is obtained as 88.7% while the ESS modulator shows an of efficiency 88.6%, including pulse shape losses for both systems. Furthermore, it is investigated how much SiC devices could improve the efficiency of the existing modulators which are currently based on Si semiconductors. The results indicate that the installation of SiC semiconductors, along with the additional proposed improvements, can lead to an increased overall efficiency which is calculated to be 93.6% for the CLIC system and 91.4% for the ESS system. Essential parameters of the CLIC modulator, which affect the system efficiency and should be considered during the design process, are the type of the insulating oil of the transformer, the length of the high voltage cable, which connects the transformer to the klystron load, as well as the core material and the lamination of the transformer. Finally, further improvements are possible for the performance of the SiC-based modulators in case a new overall optimization would be carried out for both systems.

References

[1] M. Jaritz and J. Biela, "Optimal design of a modular series parallel resonant converter for a solid state 2.88 MW/115-kV long pulse modulator," *IEEE Trans. on Plasma Science*, vol. 42, no. 10, pp. 3014–3022, Oct. 2014.

[2] S. Blume, M. Jaritz, and J. Biela, "Design and optimization procedure for high-voltage pulse power transformers," *IEEE Trans. on Plasma Science*, vol. 43, no. 10, pp. 3385–3391, Oct. 2015.

[3] W. Crewson, M. Lindholm, and K. Elmquist, "Green pulsed power achieved by efficient solid state pulsed power technology," in *IEEE Pulsed Power Conf.*, June 2011, pp. 1471–1473.

[4] S. Blume and J. Biela, "Optimal transformer design for ultraprecise solid state modulators," *IEEE Trans. on Plasma Science*, vol. 41, no. 10, pp. 2691–2700, Oct. 2013.

[5] D. Gerber and J. Biela, "Design of an ultraprecise 127-MW/3- μs solid-state modulator with split-core transformer," *IEEE Trans. on Plasma Science*, vol. 44, no. 5, pp. 829–838, May 2016.

[6] W. Hartmann, R. Fleck, R. Graba, and M. Hergt, "Characterization of commercial IGBT modules for pulsed power applications," in *19th IEEE Pulsed Power Conf. (PPC)*, June 2013.

[7] L. M. Redondo, A. Kandratsyeu, and M. J. Barnes, "Marx generator prototype for kicker magnets based on SiC MOSFETs," *IEEE Trans. on Plasma Science*, vol. 46, no. 10, pp. 3334–3339, Oct. 2018.

[8] M. S. Blume, "Highly efficient pulse modulator system with active droop compensation for linear colliders," Ph.D. dissertation, ETH Zürich, 2016.

[9] M. Jaritz, "Solid-state modulator for generating high voltage pulses in the ms-range with high output power," Ph.D. dissertation, ETH Zürich, 2018.

[10] ABB, "Amorphous core distribution transformers," April 2015.

[11] B&R Industrial Automation GmbH, "ACOPOS Multi User's Manual," April 2018.

[12] A. Stupar, T. Friedli, J. Miniböck, M. Schweizer, and J. W. Kolar, "Towards a 99% efficient three-phase buck-type PFC rectifier for 400 V DC distribution systems," vol. 27, no. 4, April 2012, pp. 1732–1744.

[13] J. Biela, J. W. Kolar, and G. Deboy, "Optimal design of a compact 99.3% efficient single-phase PFC rectifier," in *IEEE Applied Power Electronics Conf. and Exposition (APEC)*, Feb. 2010, pp. 1397–1404.

[14] S. Blume, D. Gerber, and J. Biela, "High precision, low ripple 3kV capacitor charger," in *IEEE Int. Power Modulator and High Voltage Conf. (IPMHVC)*, July 2016, pp. 113–119.

[15] D. Graovac, M. Pürschel, and A. Kiep, "MOSFET power losses calculation using the data-sheet parameters," July 2006, Application Note, V 1. 1.

[16] S. Blume and J. Biela, "Design procedure of an active bouncer for an ultra precise long pulse solid state modulator," in *IEEE Pulsed Power Conf.*, June 2013.

[17] D. Graovac and M. Pürschel, "IGBT power losses calculation using the data-sheet parameters," Jan. 2009, Application Note, V 1. 1.

[18] J. Biela, "Optimierung des electromagnetisch integrierten serien-parallel resonanzkonverters mit eingeprägtem ausgangsstrom," Ph.D. dissertation, ETH Zürich, 2005.

[19] D. Aguglia and E. Sklavounou, "Klystron modulators capacitor chargers design compromises for ac power quality increase of the compact linear collider (CLIC)," in *Int. Symposium on Power Electronics, Electrical Drives, Automation and Motion (PCIM)*, June 2012, pp. 1535–1541.

[20] M. Jaritz and J. Biela, "System design and measurements of a 115-kV/3.5-ms solid-state long-pulse modulator for the european spallation source," *IEEE Trans. on Plasma Science*, vol. 46, no. 10, pp. 3232–3239, Oct. 2018.

[21] REVCON, "Power feed and feedback module REVCON DC(S)," May 2019.

[22] M. Jaritz, S. Blume, D. Leuenberger, and J. Biela, "Experimental validation of a series parallel resonant converter model for a solid state 115-kV long pulse modulator," *IEEE Trans. on Plasma Science*, vol. 43, no. 10, pp. 3392–3398, 2015.

[23] M. Jaritz, S. Blume, and J. Biela, "Design procedure of a 14.4 kV, 100 kHz transformer with a high isolation voltage (115 kV)," *IEEE Trans. on Dielectrics and Electrical Insulation*, vol. 24, no. 4, pp. 2094–2104, 2017.

[24] TDK, "Ferrites and accessories, SIFERRIT material N97," May 2017.

Efficient and scalable power control in multi-port active-bridge converters

Soleiman Galeshi, David Frey, Yves Lembeye
UNIV. GRENOBLE ALPES, CNRS, GRENOBLE INP*, G2ELAB
21 avenue des Martyrs, 38031
Grenoble, France
Tel.: +33 (0)4 76 82 62 99
E-Mail: soleiman.galeshi-mooziraji@g2elab.grenoble-inp.fr
URL: http://g2elab.grenoble-inp.fr

Acknowledgements

This work has been partially supported by the CDP Eco-SESA receiving fund from the French National Research Agency in the framework of the "Investissements d'avenir" program (ANR-15-IDEX-02). The authors also thank the Carnot Institute Energies du Futur for supporting this work.

Keywords

«Converter control», «Modelling», «Estimation technique», «Power converters for EV», «Scalable control».

Abstract

Application of multi-port active-bridge converters is constantly increasing due to their advantages such as bidirectional power transfer, high efficiency, and intrinsic electrical isolation. However, controlling the flow of power in these converters becomes more and more complex as the number of ports increases. Power flow in a multi-port active-bridge converter is controlled through phase shifts and duty cycles. Analytical and Numerical methods, when suitable, often require huge amounts of processing and memory resources for real-time control of the power in converters with three or more ports. This work proposes a method that is computational cost efficient and can be employed for real-time control purposes. It is scalable and can be applied to any number of ports. Simulation and experimental results are provided to validate correct operation of the proposed method.

Introduction

Application of multi-port active-bridge converter (MAB) in energy systems is increasing due to several advantages of this type of converters. The multi-port transformer in MAB provides galvanic isolation and voltage level compatibility. Soft switching of active-bridges allows for increasing the switching frequency and reducing the size of magnetic components. Another advantage is that MAB converters exchange energy between several sources and loads at the same time through the same medium, which is the magnetic core of the transformer. Many researches and experiments have been done on application of triple port active-bridge converter (TAB) [1,2]. Quad-active-bridge converters (QAB) have also been studied and designed for several applications such as interfacing three-phase and dc networks [3,4], electric vehicles [5,6], and micro-grids [7,8]. Two main challenges of MAB converters are the complexity of their design and the associated control. As the number of ports increases, control and design parameters increase as fast and power transfer equations become more and more complex. Considering the complex calculations and high number of parameters in control and switching of a MAB, closed-form analytical solutions become almost impossible to find. Therefore, numerical optimization methods are required to find the optimum possible solution for each power flow. The problem is that not only numerical optimizations take a long time to find a solution, but also, it is not possible to implement them on industrial control chips with limited processing power and memory. Some researches consider special constraints or scenarios that lead to direct solution of the power flow problem [3]. Although they manage to efficiently solve the problem, they still miss the ability to put forward a general control scheme for MAB. Other works employ voltage or current feed-back loops to adjust the phase shifts of each active bridge [5,6]. Employing feed-back loops alone, brings about challenges such as stability issues and the need for correct adjustment of speed and accuracy. Another problem is that large dynamics may occur in those systems, because the flow of power between every two ports are intertwined and the feed-back control loops could affect each other.

Considering the above-mentioned challenges in control of a MAB, this work seeks to propose a method that has acceptable accuracy and dynamic behavior in finding an optimum operating point, while being compatible with limited resources of an industrial control target and assuring stability.

Power flow problem in MAB converters

A multi-port converter, whatever its application is, connects several generators and consumers of energy in one place. The role of the converter is to flow power between its different ports, i.e., the generators and the consumers, with a certain direction and amount. Energy system of a smart building, as a common application of multi-port converters, includes several resources such as photovoltaic panels, wind turbines, storage devices, and several consumers on ac and dc networks [8]. There is a higher-level controller in these systems, which takes into account several inputs (such as solar irradiance, wind speed, state of charge of the batteries, instantaneous demand, and electricity tariffs), and based on predefined strategies, determines how the power must flow in the energy system. The power flow indicates how much power should be fed to or drawn from each port and the converter is responsible for its realization.

The power flow between ports of a MAB can be controlled through applying phase shifts between switching of different legs. Two types of phase shifts can be defined in an active-bridge converter: internal phase shift, which is the phase shift between switching of two legs of a single active-bridge; and external phase shift, which is the phase shift between first harmonics of output voltages of two active-bridges. Fig. 1 shows internal and external phase shifts in a dual-active-bridge converter (DAB). It can be concluded that the control system has two domains of freedom for realization of the desired power flow: internal phase shifts, and external phase shifts. In this study, square-wave voltages (Fig. 2-a) are approximated by their first harmonics, and represented by phasors (Fig. 2-b). Magnitudes of first harmonics of output voltages are controlled through adjusting the internal phase shifts, so they can be considered equivalent to internal phase shifts. In this regard, solving the power flow problem means determining a set of magnitudes (internal phase shifts) of voltage phasors and angles between them (external phase shifts) for of every port, which corresponds to the desired power flow.

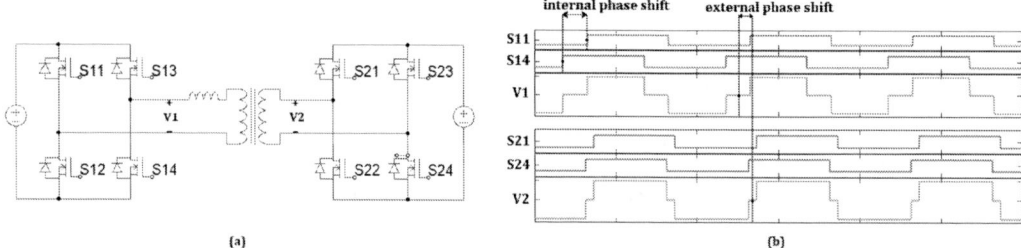

(a)   (b)

Fig. 1: internal and external phase shifts in switching of a DAB. Switches on the same arm are switched complementarily.

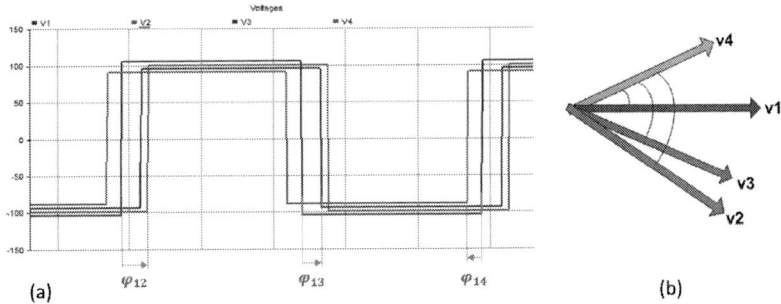

(a) φ_{12} φ_{13} φ_{14} (b)

Fig. 2: voltages and phase shifts in a quad-active-bridge converter: (a) output voltages of active bridges; (b) phasor-space representation.

Using modular matrixes to solve power flow problem

As explained in the previous section, power flow controller must determine magnitude of voltage phasors of every port and the phase difference between them. In an active bridge converter with n active bridges, there are n magnitudes and n-1 phase shifts (one phasor is considered as phase reference) to be determined. Complexity of the power flow problem increases quickly with the number of ports, and there is no direct analytical solution for more than two ports. This work proposes a modular method of solving power flow problem in MAB with any number of ports. Application of modular matrixes for solving power flow problem of a MAB was previously presented in [9] for a special structure of MAB. The current paper presents the solution for a general structure.

A QAB based on general MAB structure is displayed in Fig. 3-a. The first step to determine power transfer is to distribute the inductors between ports. The star-delta transform illustrated in Fig. 3-b is employed to do this. This transform can be generalized for n ports, as

$$L_{i,j} = L_i L_j \sum_{k=1}^{n} \left(\frac{1}{L_k} \right) \tag{1}$$

It is calculated based on superposition theorem. For example, L_{12} in Fig. 3-b would be

$$L_{1,2} = \frac{L_1 L_2 L_3 + L_1 L_2 L_4 + L_1 L_3 L_4 + L_2 L_3 L_4}{L_1 L_2 L_3 L_4} L_1 L_2 = \frac{L_1 L_2 L_3 + L_1 L_2 L_4 + L_1 L_3 L_4 + L_2 L_3 L_4}{L_3 L_4} \tag{2}$$

Through generalizing power transfer of DAB for a MAB with n ports, the power output of port k would be

$$P_k = \frac{V_1.V_k}{L_{1,k}.\omega}.\sin(\varphi_1 - \varphi_k) + \frac{V_2.V_k}{L_{2,k}.\omega}.\sin(\varphi_2 - \varphi_k) + \cdots + \frac{V_n.V_k}{L_{n,k}.\omega}.\sin(\varphi_n - \varphi_k) \tag{3}$$

where, V_k and φ_k are magnitude and phase of first harmonic voltage phasor of port k, respectively; and ω is the angular frequency, $2\pi f$, f being the switching frequency. The equation for power transfer can be simplified using approximate assumption of $sin(\varphi) = \varphi$ as

$$P_k = \frac{V_1.V_k}{L_{1,k}.\omega}(\varphi_1 - \varphi_k) + \frac{V_2.V_k}{L_{2,k}.\omega}(\varphi_2 - \varphi_k) + \cdots + \frac{V_n.V_k}{L_{n,k}.\omega}(\varphi_n - \varphi_k) \tag{4}$$

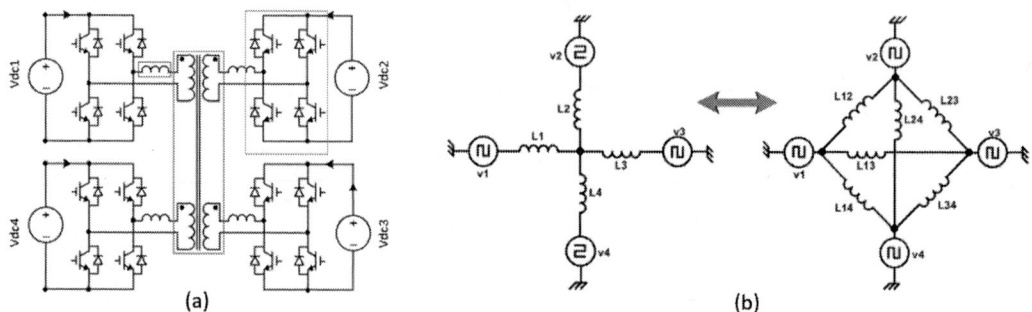

Fig. 3: a quad-active-bridge converter: (a) the structure, with a four-port transformer (the red box), four active bridges (green box), and four inductors (blue box); (b) ac representation and star-delta transform.

Maximum power each port can transfer, P_{max} (the maximum value for power transfer when $sin(\varphi) = 1$), is an important factor in design of MAB and depends inversely on the size of the inductors [10]. If inductors are large, i.e., P_{max} is relatively small, the instantaneous transferred power might often get close to P_{max} and current magnitudes will increase nonlinearly, leading to large losses. Choosing a very small inductor, on the other hand, has its own challenges such as sharp changes in current waveform, higher voltages across the transformer windings, and limited range of soft switching. Considering this trade-off, active-bridge converters are usually designed in a way that their nominal power transfer is around 60% of P_{max} [10]. Under this design condition, the phase shift between the ports will be in the range of ± 37 degrees, and assumption of $sin(\varphi) = \varphi$ in (4) will have maximum 7.3% error, which is acceptable. Rearranging (4) gives

$$P_k = \frac{V_1 . V_k}{L_{1,k}.\omega}.(\varphi_1) + \cdots + \frac{V_2 . V_k}{L_{2,k}.\omega}.(\varphi_2) + \cdots + \frac{V_k . V_k}{L_{k,k}.\omega}.(\varphi_k) \ldots + \frac{V_n . V_k}{L_{n,k}.\omega}.(\varphi_n) - \frac{V_k \varphi_k}{\omega} \sum_{i=1}^{n} \frac{V_i}{L_{n,i}} \tag{5}$$

where, $L_{k,k}$ does not have a real representation and is only added for keeping symmetry in the equation. It will be omitted from calculations in each line of the final matrix. All the power equations can be put into one matrix equation as (6).

$$\omega \begin{bmatrix} P_1 \\ P_2 \\ \cdots \\ P_n \end{bmatrix} = \begin{bmatrix} \frac{V_1 . V_1}{L_{1,1}} - V_1 \sum_{i=1}^{n} \frac{V_i}{L_{1,i}} & \frac{V_1 . V_2}{L_{1,2}} & \cdots & \frac{V_1 . V_n}{L_{1,n}} \\ \frac{V_2 . V_1}{L_{2,1}} & \frac{V_2 . V_2}{L_{2,2}} - V_2 \sum_{i=1}^{n} \frac{V_i}{L_{2,i}} & \cdots & \frac{V_2 . V_n}{L_{2,n}} \\ \cdots & \cdots & \cdots & \cdots \\ \frac{V_n . V_1}{L_{n,1}} & \frac{V_n . V_2}{L_{n,2}} & \cdots & \frac{V_n . V_n}{L_{n,n}} - V_n \sum_{i=1}^{n} \frac{V_i}{L_{n,i}} \end{bmatrix} \times \begin{bmatrix} \varphi_1 \\ \varphi_2 \\ \cdots \\ \varphi_n \end{bmatrix} \tag{6}$$

The matrix of coefficients in (6) may not be reversible under some conditions; for instance, when all inductances and voltage magnitudes are equal. In order to modify this matrix and make it reversible, V_1 can be considered as the reference for phase; therefore, φ_1 will be zero. The first column of the matrix of coefficients can be omitted without changing the outcome, because all of its elements will be multiplied by φ_1, which is zero. Additionally, the last row can be omitted because P_n can be determined as $P_n = -\sum_{i=1}^{n-1} P_i$, based on conservation of power. The final equation would be

$$\omega \begin{bmatrix} P_1 \\ P_2 \\ \cdots \\ P_{n-1} \end{bmatrix} = \begin{bmatrix} \frac{V_1 . V_2}{L_{1,2}} & \frac{V_1 . V_3}{L_{1,3}} & \cdots & \frac{V_1 . V_n}{L_{1,n}} \\ \frac{V_2 . V_2}{L_{2,2}} - V_2 \sum_{i=1}^{n} \frac{V_i}{L_{2,i}} & \frac{V_2 . V_3}{L_{2,3}} & \cdots & \frac{V_2 . V_n}{L_{2,n}} \\ \cdots & \cdots & \cdots & \cdots \\ \frac{V_{n-1} . V_2}{L_{n-1,2}} & \frac{V_{n-1} . V_3}{L_{n-1,3}} & \cdots & \frac{V_{n-1} . V_n}{L_{n-1,n}} \end{bmatrix} \begin{bmatrix} \varphi_2 \\ \varphi_3 \\ \cdots \\ \varphi_n \end{bmatrix} \tag{7}$$

It is now possible to determine phase shifts as

$$\begin{bmatrix} \varphi_2 \\ \varphi_3 \\ \dots \\ \varphi_n \end{bmatrix} = \omega [A]^{-1} \begin{bmatrix} P_1 \\ P_2 \\ \dots \\ P_{n-1} \end{bmatrix} \tag{8}$$

where, $[A]$ is the matrix of coefficients in (7). Equation (8) gives an approximate solution for the power flow problem, i.e., given a desired value of power transfer on each port, (8) can determine a set of phase shifts that if applied to switching of the active bridges, leads approximately to the desired power flow. It should be noted that voltage magnitudes (internal phase shifts) have to be determined before this step. This is usually done by another part of the converter control scheme, whose task is to maintain soft switching conditions. Depending on the variations in dc link voltages and the amount of power on each port, the soft switching control determines the internal phase shifts, and the method proposed in this work uses them for its calculations.

Simulation and experimental results

Fig. 4 shows simulation results of open-loop control of power flow in a QAB. The thick lines are the reference values of powers, changing at 0.2 ms and 0.4 ms moments, and thin lines represent the actual measured powers. The actual powers follow the changes in reference values but with some steady state error. The coupling of powers can be seen on the second port at 0.2 ms and 0.4 ms, where its reference value (P2*) does not change but the actual power (P2) varies. The steady state errors, that are large in some cases, are due to the approximations in the control scheme, the fact that the control is open-loop, and the losses that were not considered in the calculations.

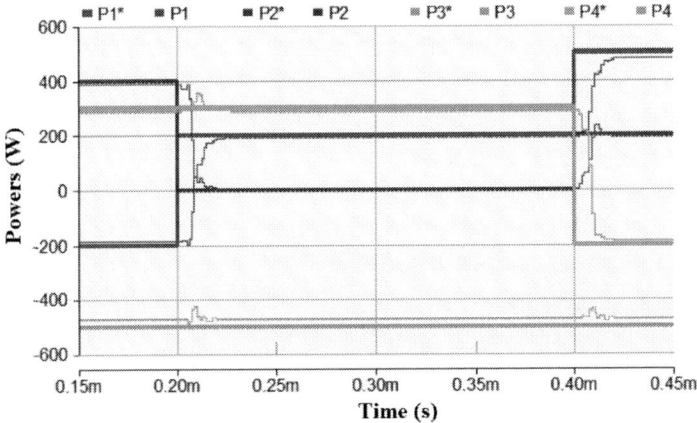

Fig. 4: performance of the proposed power controller when in a QAB. Thick lines show the reference values, and thin lines show actual powers.

Fig. 5 shows the lab-scale prototype that was used to validate the proposed control system experimentally. The details of the converter are provided in Table I. Several experimental tests were performed to validate correct operation of the proposed method. Closed-loop control scheme with PI controllers was employed to validate effective elimination of steady state errors. It should be noted that the reference values of the power sum up to zero due to conservation of power, ignoring the losses in the converter. In other words, it not possible to have all the actual powers equal to their reference value. Therefore, control of one port is always open-loop to allow it to provide the losses.

Fig. 6 shows experimental results. The open-loop results show that the proposed power controller is able to realize the desired power but with some steady state error. The closed-loop results show that the steady state error can be effectively eliminated using PI feedback loops. Due to the limited number of input ports of the oscilloscope and analog outputs of the controller, it was not possible to capture and show all the powers. Measurement of powers on the other ports also showed effective elimination of steady state error, similar to that of the second port in Fig. 6.

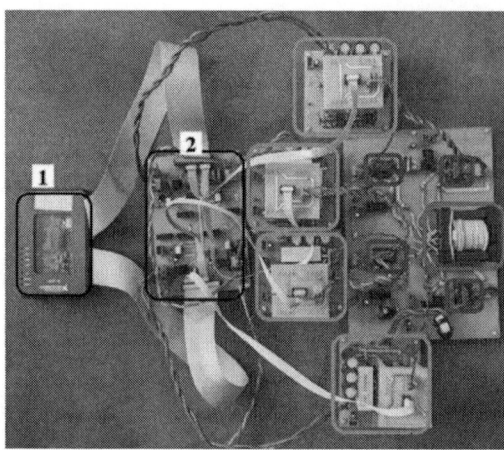

Fig. 5: The quad active-bridge converter that was built in G2ELab for experimental validation of the proposed power control method. The black boxes, numbered 1 and 2, indicate the controller and signal conditioning circuit, respectively. The green and the blue boxes indicate four active bridges and inductors, respectively. The red box is a four-winding transformer. Colors of the boxes correspond to the ones in Fig. 3-a.

Table I: Details of the prototype converter

Parameter	Value
Maximum DC voltage	40 V
Nominal power	250W each port
External inductors	10 µH
Transformer turns	7:7:7:7
Switching frequency	40 kHz

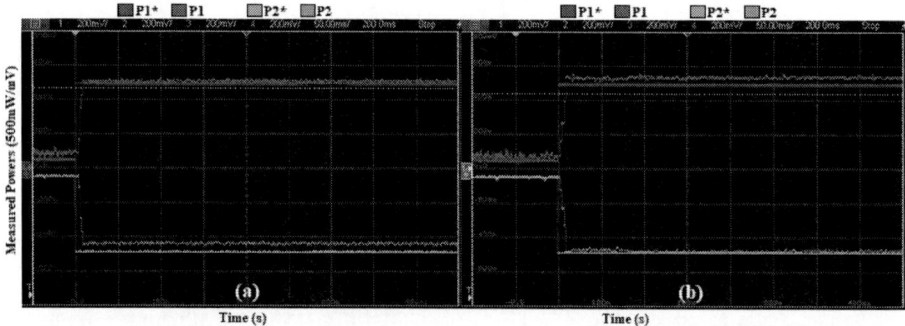

Fig. 6: experimental results of power control in a quad active-bridge converter: (a) open-loop results; (b) closed-loop results. The first port (P1) is always open-loop, to be able to provide the losses. The rest are closed-loop, including P2 that is shown here. Due to the limited number of available analog outputs of the controller and inputs on the oscilloscope, the reference and measured powers of the third and fourth port are not shown.

Performance of the proposed method

An iterative method of solving the problem is presented in [11]. It begins with finding a first approximate solution by assuming the currents to be zero. The solution found this way could be far from the correct solution, but through performing some iterative calculations and updating magnitudes of the current, this method can get very close to the desired power flow. The method presented in [11] and the method presented in this work do the same task: given the desired power flow and voltage magnitudes as inputs, these methods determine a set of phase shifts that, if applied as external phase shifts between switching of the active bridges, will correspond to the desired power flow. The major difference between these methods is the approximations they use for solving the problem. These two methods will be compared in terms of accuracy of their final solution and amount of calculations they require on a controller chip.

Since the reference control signal is the power transfer of each port, a good parameter for comparison of accuracy is the average difference between the desired power transfer and the one measured after applying the phase shifts. Fig. 7 shows the defined error of consecutive iterations of the method proposed in [11] (labeled as "zero currents"), and the error from the method proposed in the current work (Labeled as "matrix"). The "zero currents" method starts with a large error in its first guess, but quickly reduces the error after an additional iteration. The "matrix" method leads to a smaller error, but does not include a mechanism to minimize it. It is noticeable that the error of the proposed method is larger when powers are higher (Fig. 7-a) compared to lower powers (Fig. 7-b). This is due to the $sin(\varphi) = \varphi$ approximation that is used in this method. The phase shifts increase as the powers increase and the difference between $sin(\varphi)$ and φ becomes larger, reaching up to around 7% at the nominal power, where φ is around 37 degrees. Fig. 7 also shows that two iterations of the "zero currents" method is enough for reaching negligible errors.

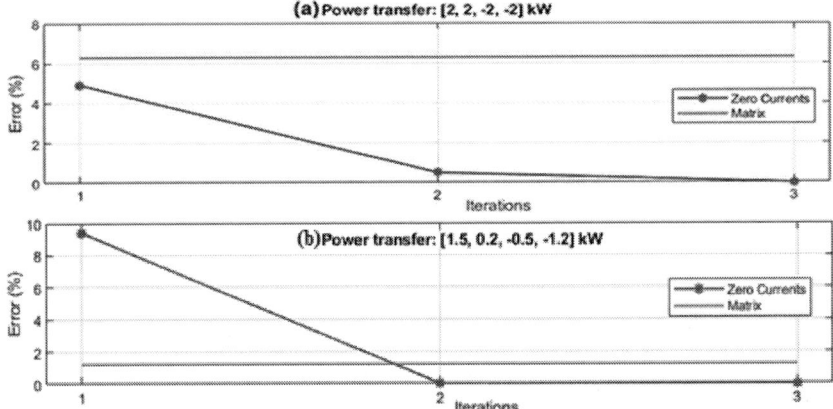

Fig. 7: Comparing convergence of different power control methods in a simulated 2 kW quad active-bridge converter: (a) ideal converter with all powers and voltage at their nominal values; (b) non-ideal converter with ±10% voltage variation.

It should be noted that a feedback PI controller is usually present in control systems that depend on models, such as the one presented in this work. The main goal of using a PI feedback is to eliminate the steady state errors that occur due to the mismatch between the values in the model and the real system. This feedback loop can be counted upon for eliminating the steady state error caused by the approximation of the proposed method, too. Performance of the proposed method with a closed-loop control was illustrated in Fig. 6-b. In this regard, the steady state errors of the proposed method, although larger than those of the "zero currents" method, are within acceptable range and can be eliminated using a feedback loop.

The second comparison is the amount of calculations involved with each method. The final goal of designing a MAB is to build and use it in a real energy system. The control system for this converter should be able to work with limited resources of low-end controller chips used in industrial products. In this regard, it is important to assess performance of the presented methods on a controller and validate that they run correctly. NI myRIO 1900 package, with a 40MHz processor and a Xilinx Z-7010 chip, was used as the controller. Both methods were implemented on the target chip and run on its processor. Fig. 8 illustrates the results. The time required to do the calculations of the proposed method is 4 µs, compared to 127 µs for two iterations of the "zero currents" method of [11]. Considering the overhead processing time of 18 µs, the total runtime would be 22 µs for the proposed method, compared to 145 µs for two iterations of "zero currents". The comparison shows that the proposed method is computational cost efficient and a good candidate for real-time control and optimization purposes.

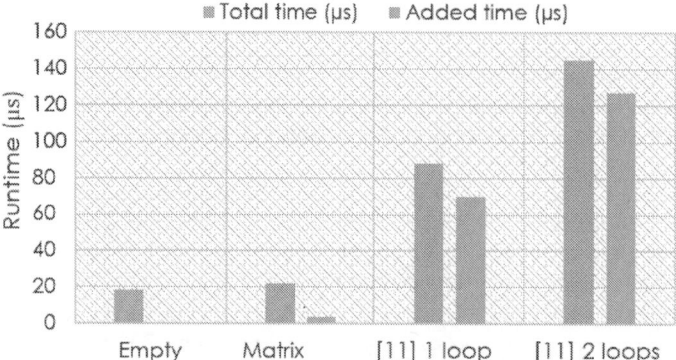

Fig. 8: comparing the time needed to run the proposed method and the method presented in [11]. The first section on the left corresponds to runtime of an empty program and indicates the overhead processing time, which is 18 µs. The second section corresponds to the proposed method, which requires 4 µs to run plus the additional 18 µs overhead processing time. The third and the fourth sections indicate the runtime of the method presented in [11] for one and two iterations, respectively. The total time required for running two iterations is 145 µs.

Conclusions

A modular method is proposed in this paper for control of power flow in multi-port active-bridge converters. Application of these types of converters is increasing, especially in energy systems based on renewable resources. These converters facilitate simultaneous exchange of energy between multiple resources and loads, with lower costs, smaller size, good efficiency, and isolation. The challenge of these converters is their control, which becomes more and more complex when number of their ports increases. The proposed method solves the power flow control problem using an approximate mathematical model of the converter. Compared to a method previously presented, this method has a larger, yet acceptable error due to its approximation. The proposed method acts as a feed-forward loop and it was shown with experimental results that in combination with a feed-back loop, it can effectively control the power flow with no steady state errors. This method requires less mathematical calculations and runs about 7 times faster, compared to a previously proposed method in the literature. This fact makes it applicable for industrial control systems with limited processing resources. Several simulation and experimental results are presented to validate correct performance of the method.

References

[1] N. Delmonte, P. Cova, D. Santoro, A. Toscani and G. Buticchi: Development of a GaN Based Triple-Active-Bridge for DC Nanogrid, 2018 20th European Conference on Power Electronics and Applications (EPE'18 ECCE Europe), Riga, 2018.

[2] K. Katagiri, K. Nishimoto, S. Nakagawa, S. Okutani, Y. Kado and K. Wada: Decoupling Power Flow Control of Triple-Active Bridge Converter with Voltage Difference between Each Port for Distributed Power Supply System, 2018 20th European Conference on Power Electronics and Applications (EPE'18 ECCE Europe), Riga, 2018.

[3] L. F. Costa, G. Buticchi and M. Liserre: Quad-Active-Bridge DC–DC Converter as Cross-Link for Medium-Voltage Modular Inverters, IEEE Transactions on Industry Applications, vol. 53, no. 2, pp. 1243-1253, March-April 2017.

[4] B. J. D. Vermulst, J. L. Duarte, C. G. E. Wijnands and E. A. Lomonova: Quad-Active-Bridge Single-Stage Bidirectional Three-Phase AC–DC Converter With Isolation: Introduction and Optimized Modulation, IEEE Transactions on Power Electronics, vol. 32, no. 4, pp. 2546-2557, April 2017.

[5] A. Chandrasekharan Nair, M. J. Vishal and B. G. Fernandes: A Quad Active Bridge Based on-Board Power Electronic Interface for an Electric Vehicle, 2018 IEEE Energy Conversion Congress and Exposition (ECCE), Portland, OR, 2018, pp. 5941-5947.

[6] M. J. Vishal, A. C. Nair and B. G. Fernandes: Quad-Active-Bridge DC-DC Converter based on-board Power Electronic Interface for Electric Vehicle, 2018 IEEE International Conference on Power Electronics, Drives and Energy Systems (PEDES), Chennai, India, 2018, pp. 1-6.

[7] S. Falcones and R. Ayyanar: LQR control of a quad-active-bridge converter for renewable integration, 2016 IEEE Ecuador Technical Chapters Meeting (ETCM), Guayaquil, pp. 1-6.

[8] S. Galeshi, D. Frey, Y. Lembeye and D. Motte-Michellon: Application of Clustered Multi-port Active-bridge Converters in Microgrids, 2019 21st European Conference on Power Electronics and Applications (EPE '19 ECCE Europe), Genova, Italy, 2019.

[9] B. J. D. Vermulst, J. L. Duarte, E. A. Lomonova and K. G. E. Wijnands: Scalable multi-port active-bridge converters: modelling and optimised control, IET Power Electronics, vol. 10, no. 1, pp. 80-91, 2017.

[10] S. Galeshi, D. Frey, and Y. Lembeye: Design Procedure of DC-DC Multi-Port Active-Bridge Converters, PCIM Europe conference, 2020, in press.

[11] S. Galeshi, D. Frey, Y. Lembeye: Modular Modeling and Control of Power Flow in A Multi-Port Active-Bridge Converter, Symposium de Genie Electrique (SGE 2018), 3-5 Jul. 2018, Nancy, France.

Comparison of Press-Pack and Wire-Bonding Technologies for SiC MOSFETs under Short-Circuit Conditions

Ran Yao[1,2], Francesco Iannuzzo[2], Amir Sajjad Bahman[2], Hui Li[1]

[1]State Key Laboratory of Power Transmission Equipment & System Security and New Technology

Shazhengjie 174, Shapingba, 400044 Chongqing, China

Email: yaoran1234@163.com, cqulh@163.com

[2]Department of Energy Technology

Pontoppidanstræde 111, 9220 Aalborg, Denmark

Email: fia@et.aau.dk, amir@et.aau.dk

Acknowledgements

This work has been carried out under the APETT Project, funded by the Danish Innovation Foundation, 2017-2021, and under the National Key Research and Development Program of China under Grant 2018YFB0905704.

Keywords

«press pack», «silicon carbide», «MOSFET», «short circuit», «finite element modeling».

Abstract

This paper uses the finite element method to analyze the thermal performance of SiC MOSFETs discrete package with different package technologies during short circuit conditions. Firstly, the industrial packaging methods used for power semiconductor devices including SiC MOSFET and Si IGBT are analyzed, which are divided into wire-bonded and press-pack package technologies. Then, the finite element models of wire-bonded and press-pack SiC MOSFET discrete packages are built to analyze the thermal performance with short-circuit conditions. The simulation results show that the press-pack technique performs better with respect to peak temperature under the same electrical conditions.

Introduction

Due to the high switching speed, high breakdown voltage, and low on-resistance, Silicon Carbide (SiC) metal-oxide-field-effect transistors (MOSFETs) devices are promising components for electrical vehicles, motor drives, and high-voltage transmission systems [1]. In fact, SiC MOSFETs can proficiently substitute traditional IGBT technology because of its outstanding compromise with respect to blocking voltage and conduction losses. However, the temperature of the traditional wire-bonded SiC MOSFET modules becomes easily higher than 230 °C under short circuit (SC) conditions [2], and their reliability severely decreases in such a high temperature, which may lead to early catastrophic failures [3]. In applications like modular multilevel converters (MMCs), high-voltage direct current (HVDC), and in general, where switching speed is not as relevant as robustness against short circuit conditions, large benefits are traditionally earned in terms of thermal management with press-pack (PP) technology. In this paper, we investigate press-pack technology for SiC MOSFETs with respect to short-circuit conditions, by comparing it with the wire-bonding approach, for the same part number.

In recent years, many researchers have studied the performance of the SiC MOSFET device during SC tests. H. Du et al. analyzed the degradation indicators of SC tests in SiC MOSFET modules. The results show that a positive shift of threshold voltage, gate leakage current, and on-state resistance increase occurred [4]. G. Romano et al. analyzed the SC failure mechanism of SiC MOSFET, proving that the threshold voltage and carrier mobility change are affected by the high temperature during the SC [5]. Y. Pascal et al. used TCAD to simulate the temperature of a SiC MOSFET, showing that the junction temperature of its cell gets higher than 800 °C with an SC time of 10 μs [6]. However, under the SC conditions, most previous works only analyzed the failure mechanism and related parametric changes but did not propose any enhancement, especially aimed at reducing the junction temperature.

Nowadays, PP technology has widely used to package IGBT chips, due to the advantages of double-side cooling, short-circuit failure mode, and easily connect in series, some researchers try to use this technology to package SiC MOSFET chips [7]. N. Zhu et al. proposed a new PP SiC MOSFET method with some special structures, such as fuzz buttons (flexible press pins made by Gold-plated Beryllium Copper) and low-temperature cofired ceramic (LTCC) microchannel heatsink, which enables the PP module to be used in high power applications [8]. Y. Pascal proposed a PP method by using metal foam to connect the Print Circuit Board (PCB) with the source of the SiC MOSFET chip, which enables the wire-bonded module to perform double-side cooling [9]. However, the practicality and stability of these methods have not been tested, and adopting them may lead to some special failure modes. Therefore, commercial PP technologies are adopted in this paper to package SiC MOSFET chips for short-circuit simulation.

In this paper, firstly, the wire-bonded and press-pack technologies of SiC MOSFETs and IGBT are introduced, respectively. Secondly, for the same part number of C2M0025120D, the finite-element (FE) model of the SiC MOSFET is established with the traditional wire-bonded, WESTCODE, and ABB press-pack approaches. Then, under the SC conditions, by considering the same voltage and current waveform, and the same voltage distribution in the SiC MOSFET chip, thermal field simulation of these models is performed. Finally, the causes of different temperature distribution between these models are discussed.

Different packaging technologies of SiC MOSFET and Si IGBT

Currently, many semiconductor companies use wire-bonded technology to package SiC MOSFET, such as STMicroelectronics and Wolfspeed. Most of them use the TO-247 package for 1200 V SiC MOSFET chip as shown in Fig. 1. [10-11], the SiC die is connected to the base plate by the solder, the source, and the gate connected to the pin by bond wire.

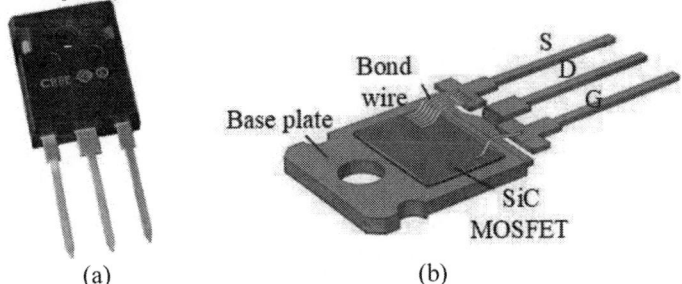

(a)   (b)

Fig. 1: TO-247 package. (a) Wolfspeed C2M0025120D [11]; (B) Internal structure of TO-247 [10].

The PP technology is widely used for IGBT and divided into two approaches, the PP IGBT modules produced by WESTCODE are called press-pack PP, the layout of the multichip PP IGBT module is shown in Fig. 2 (A) [12]. The PP IGBT modules produced by ABB are called StakPak (SP), and the layout of the multichip SP IGBT module is shown in Fig. 2(b) [13]. Comparing PP and SP approaches, in the SP approach, the collector side of the IGBT is soldered to the copper pad and the emitter side is connected to the molybdenum layer by pressure which is applied by the internal spring. But in the PP approach, both sides of the IGBT are connected to the molybdenum layers by the outside clamping force, and there is no solder and spring in the module. Therefore, to analyze the thermal performance of the PP SiC MOSFET module under SC conditions, the PP and SP approach both should be simulated, and the wire-bonded SiC MOSFET module is used as a reference model.

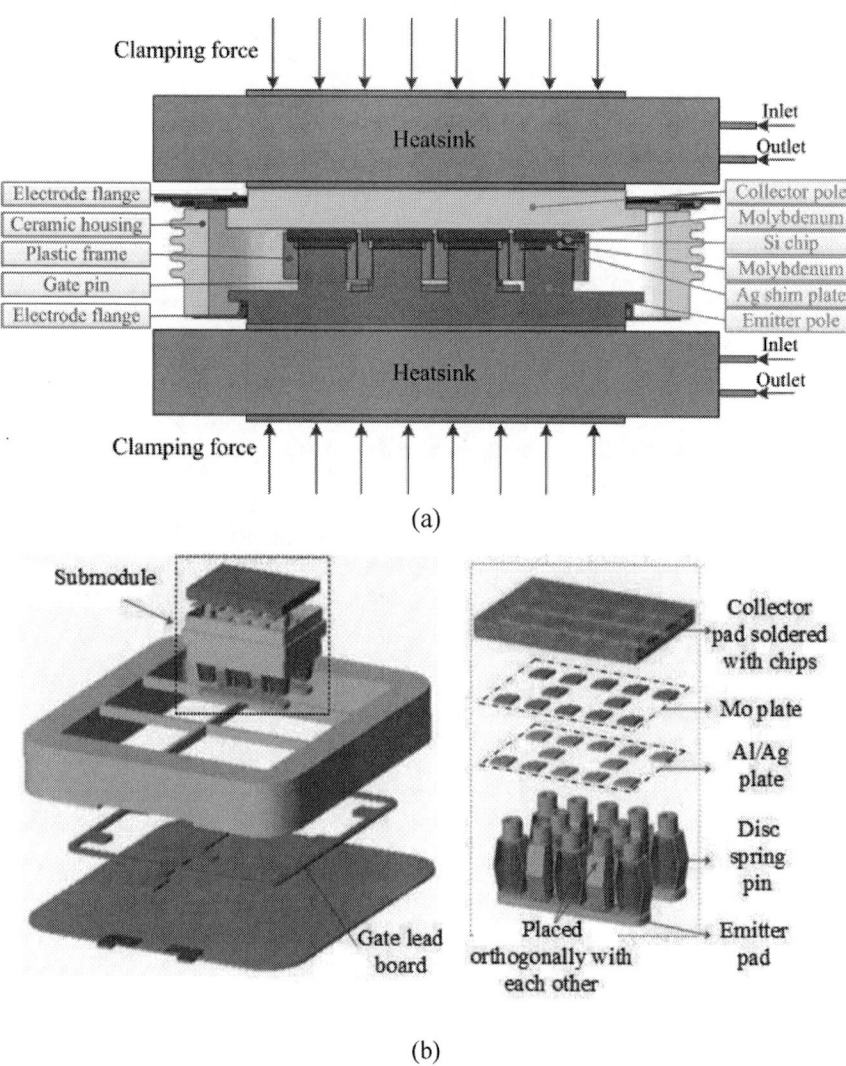

Fig. 2: Two PP approaches. (a) PP IGBT module made by WESTCODE [12]; (b) SP IGBT module made by ABB [13].

FE modeling of wire-bonded, PP, and SP SiC MOSFET discrete packages

The wire-bonded, PP and SP SiC MOSFET discrete packages are built by the COMSOL software with the same SiC MOSFET chip (CPM2-1200-0025B) and shown in Fig. 3 [14]. In Fig. 3(a) shows the internal structure of the wire-bonded discrete package, the shell and resin are hidden. Fig. 3(b) and Fig. 3(c) shows the internal structure of a single chip SiC MOSFET discrete package with PP and SP approaches, the shell, plastic frame, and gate pin are hidden.

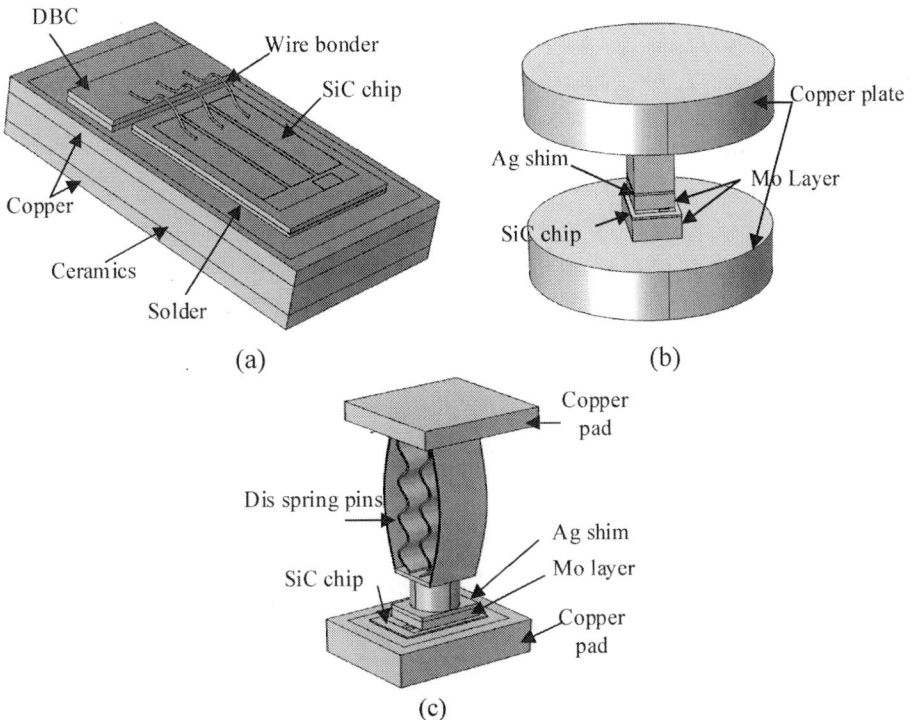

Fig. 3: FE model of the SiC MOSFET discrete packages. (a) Wire-bonded; (b) PP; (c) SP.

For the thermal simulation, to ensure the accuracy of heat transfer, the SiC MOSFET chip, the solder layer, and the molybdenum layers need to by finely meshed. The mesh method of these structures is the maximum element size of 300 μm, the minimum element size of 50 μm, the other structures are meshed by the automatic method setting in the COMSOL, the mesh elements of the wire-bonded, PP and SP SiC MOSFET discrete packages are about 600, 700, 700 thousand, respectively. In the thermal simulation, the boundary conditions of these discrete packages are the same, take the boundary conditions of the PP SiC MOSFET discrete package as an example, as shown in Fig. 4. In Fig. 4 (a), the ambient temperature is set outside the discrete package and fixed at 20 °C, the heat source is applied in the active area of the SiC MOSFET chip, the volume of the active area is about 80% in the chip [15], three metal fingers of aluminum are built on the SiC MOSFET chip. In Fig. 4 (b), during the SC test, the waveform of voltage and current of SiC MOSFET chips (CPM2-1200-0025B) are extracted from the literature and are used to calculate the power loss [9]. The voltage distribution in the SiC MOSFET chip is like a triangle and the maximum value is located about 10 μm underneath the source surface [16]. Such power is applied to the active area, as shown in Fig. 4 (c). Material properties for these discrete packages are shown in Table I [6, 17], which have been used for the simulations. λ is the thermal conductivity, C_p is the heat capacity, ρ is the density.

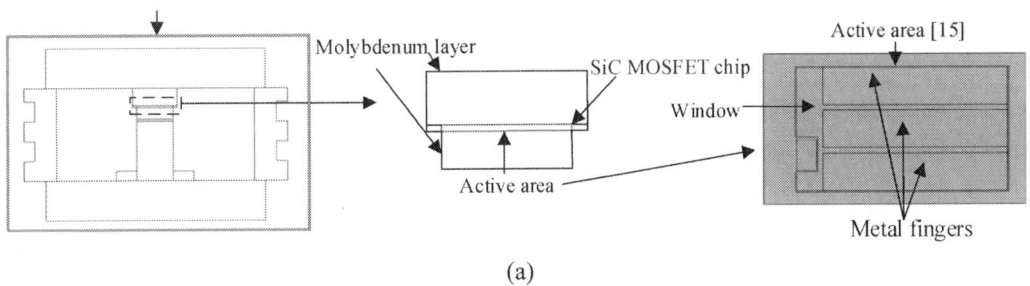

(a)

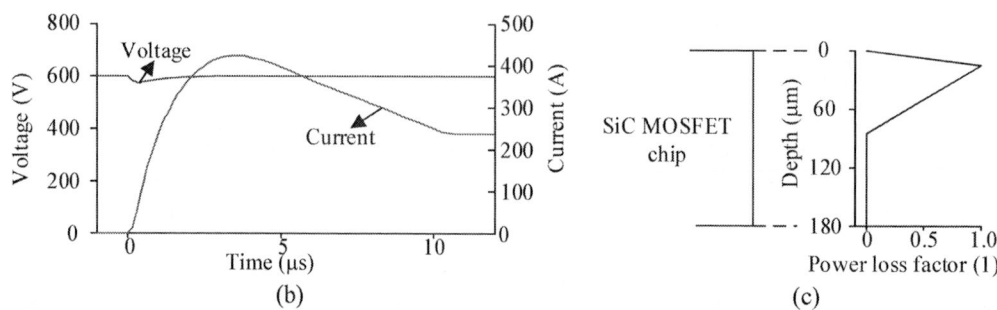

Fig. 4: The setting in the thermal simulation. (a) Boundary conditions; (b) SC test waveform; (c) Voltage distribution in SiC MOSFET chip.

Table I:Material properties in the SiC discrete packages

Material	λ (W/(m$\times$K))	C_p (J/(K$\times$kg))	ρ (kg/m^3)
SiC	353	1031	3211
Copper	400	385	8690
Aluminum	239	910	2700
Molybdenum	138	250	10220
Silver	429	235	10500

Short-circuit simulation results

The wire-bonded, PP and SP SiC MOSFET discrete packages are stimulated with SC time of 10 µs, and as shown in Fig. 4(a), the SiC MOSFET chip have windows between the metal fingers to see the semiconductor temperature, the temperature distribution of these discrete packages are extracted as shown in Fig. 5. In Fig. 5(a), the highest temperature of 186 °C in the wire-bonded discrete package is located through the windows on the SiC MOSFET chip, and the temperature of metal fingers is about 120 °C, the temperature of other structures is 20 °C, this may be caused by the SC time is too short that heat is difficult to propagate to other layers. In Fig. 5(b), the highest temperature of the PP discrete package is located through the windows on the SiC MOSFET chip about 173 °C, the metal fingers is hidden by the molybdenum layer, the other structures is 20 °C. In Fig. 5(c), the temperature distribution of the SP discrete package is the same as that in the PP discrete package. The junction temperature variations of these SiC MOSFET discrete packages are shown in Fig. 5(d), the junction temperature variation trend of the PP and SP SiC MOSFET discrete packages is the same, at 10 µs, the junction temperature is lower than 175 °C. Compering to the PP SiC MOSFET discrete package, the junction temperature variation of the wire-bonded SiC MOSFET discrete package is higher, at 10 µs, the junction temperature is higher than 185 °C. From the datasheet of the CPM2-1200-0025B, the maximum junction temperature limit to a maximum value of 175 °C [14]. Thus, the thermal simulation results indicate that the PP and SP SiC MOSFET discrete packages can be used under SC condition.

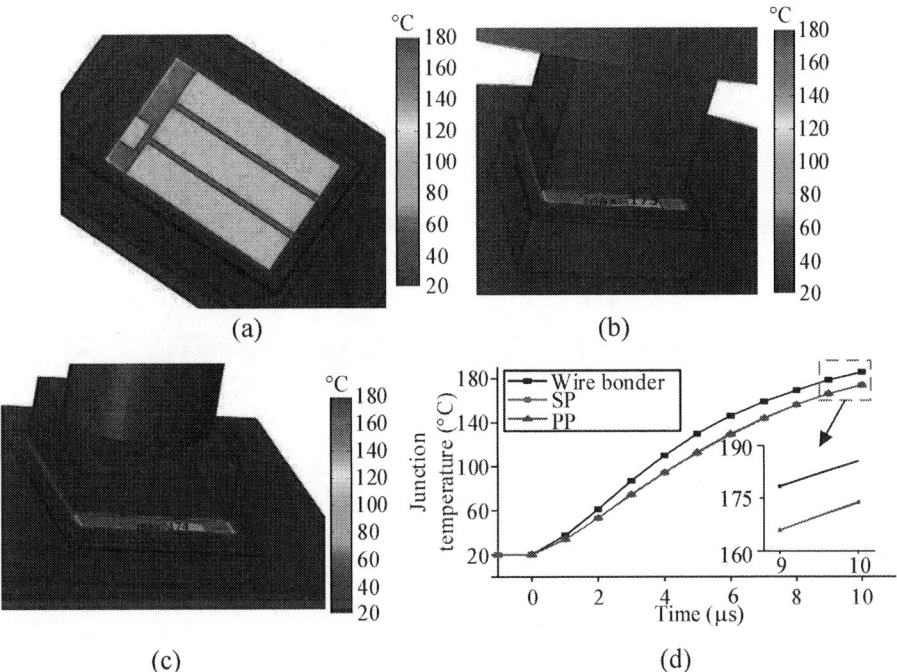

Fig. 5 SC simulation results. (a) Wire-bonded discrete package; (b) PP discrete package; (c) SP discrete package; (d) Temperature comparison.

The surface temperature distribution of the SiC MOSFET chip in the wire-bonded and PP SiC MOSFET discrete packages has been extracted as shown in Fig. 6, because the temperature distribution and junction temperature variation trend of the PP and SP discrete packages are the same, the surface temperature distribution of SiC MOSFET chip in the PP discrete package is extracted as an example for further analysis. In Fig. 6 (a), the temperature on the SiC MOSFET chip is higher than 160 °C, and the temperature in the metal fingers is lower than 120 °C. In Fig. 6 (b), the temperature on the SiC MOSFET chip is higher than 160 °C, and the temperature in the metal fingers is lower than 100 °C. The temperature distribution of the cross-section of geometry under the black line in Fig. 6 (a) and Fig. 6 (b) are extracted as shown in Fig. 6 (c) and Fig. 6 (d), the highest temperature in the SiC MOSFET chip close to the source side, it is due to the triangle voltage distribution in the SiC MOSFET chip. In Fig. 6 (c), the source side is connected to the resin, the thermal conductivity in the resin is lower than 5 W/(m×K), so the heat tends to propagate to underneath layers of the chip, and the highest temperature in the chip under the metal fingers is about 140 °C. In Fig. 6 (d), the source side of the SiC MOSFET chip is connected to the molybdenum layer, this is exactly the opposite case compared to the resin, the thermal conductivity of the molybdenum layer is higher than 130 W/(m×K), the heat tends to propagate to the molybdenum layer, and the highest temperature in the chip is about 120 °C. Thus, the temperature under the metal fingers is lower than other areas on the SiC MOSFET chip. If the metal fingers cover all active area, the junction temperature can be more decreased. Comparing the temperature distribution in Fig. 6 (c) and 6 (d), if the SiC MOSFET chip is connected to the high thermal conductivity material, the junction temperature of the device can be decreased too. So these methods may be used to decrease the temperature of the PP SiC MOSFET module.

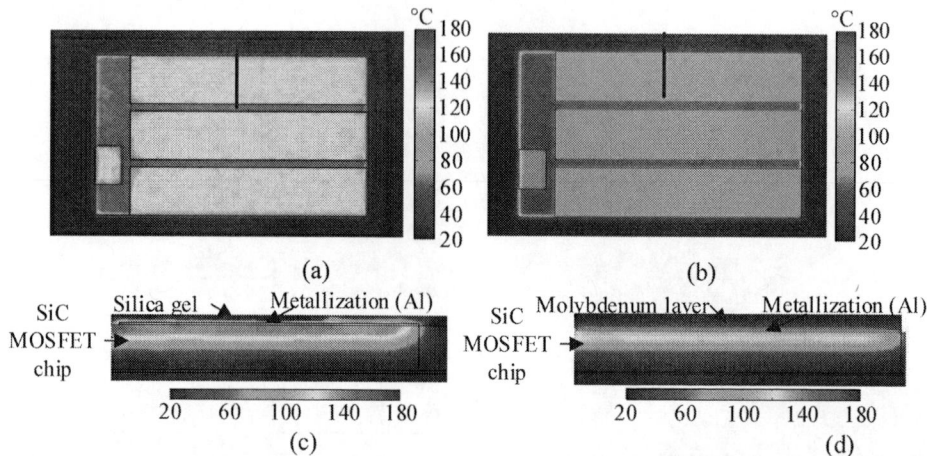

Fig. 6 Temperature distribution on the SiC MOSFET chip. (a) In wire-bonded discrete package; (b) In PP discrete package; (c) Blackline in Fig.6 (a); (d) Blackline in Fig.6 (b)

Conclusion

This paper compares the thermal performance of PP and wire-bonding technologies for SiC MOSFETs under SC conditions. By establishing the FE model of wire-bonded, PP and SP SiC MOSFET discrete packages in thermal simulation, the results show that the PP and SP SiC MOSFET discrete packages have a lower temperature than that of the wire-bonded discrete package. The junction temperature of PP and SP discrete packages is lower than 175 °C with a short circuit time of 10 μm, which satisfies the maximum temperature limit of SiC MOSFET chip listed in the datasheet. Thus, the PP SiC MOSFET discrete package has higher short-circuit tolerance than the wire-bonded discrete package.

References

[1] Dchar, I., Buttay, C., and Morel, H.: SiC power devices packaging with a short-circuit failure mode capability, Microelectronics Reliability, Vol. 76–77 pp 400-404.

[2] Romano, G., Fayyaz, A., Riccio, M., Maresca, L., Breglio, G., Castellazzi, A., and Irace, A.: A comprehensive study of short-circuit ruggedness of silicon carbide power MOSFETs, IEEE Journal of Emerging and Selected Topics in Power Electronics, Vol. 4 no 3 pp 978-987.

[3] Ziemann, T., Tsibizov, A., Kakarla, B., Bort, L., and Grossner, U.: Time-Resolved Short Circuit Failure Analysis of SiC MOSFETs, 2019 31st International Symposium on Power Semiconductor Devices and ICs (ISPSD), 2019.

[4] Du, H., Reigosa, P. D., Iannuzzo, F., and Ceccarelli, L.: Investigation on the degradation indicators of short-circuit tests in 1.2 kV SiC MOSFET power modules, Microelectronics Reliability, Vol 88–90 pp 661-665.

[5] Romano, G., Maresca, L., Riccio, M., D'Alessandro, V., Breglio, G., Irace, A., Fayyaz, A., and Castellazzi, A.: Short-circuit failure mechanism of SiC power MOSFETs, Proceedings of the International Symposium on Power Semiconductor Devices and ICs, 2015.

[6] Pascal, Y., Petit, M., Labrousse, D., and Costa, F.: Thermal simulations of SiC MOSFETs under short-circuit conditions: Influence of various simulation parameters, 2019 IEEE International Workshop on Integrated Power Packaging, IWIPP 2019.

[7] Lai, W., Chen, H. L. M., Kang, S., Ren, H., Yao, R., Pan, L., and Jin, R.: Investigation on the Effects of Unbalanced Clamping Force on Multi-chip Press Pack IGBT Modules, IEEE Journal of Emerging and Selected Topics in Power Electronics, Vol. 7 no 4 pp 2314-2322.

[8] Zhu, N., Mantooth, H. A., Xu, D., Chen, M., and Glover, M. D.: A Solution to Press-Pack Packaging of SiC MOSFETS, IEEE Transactions on Industrial Electronics, Vol. 64 no 10 pp 8224-8234.

[9] Pascal, Y., Labrousse, D., Petit, M., Lefebvre, S., and Costa, F.: Experimental investigation of the reliability of Printed Circuit Board (PCB)-embedded power dies with pressed contact made of metal foam, Microelectronics Reliability, Vol. 88–90 pp 707‑714.

[10] Xu, S. Y., Yang, X., Chen, M. Y., Lai, Wei., Wang, Y. Y., and Li, R.: Analysis and Modeling for the Real-Time Condition Evaluating of MOSFET Power Device Using Adaptive Neuro-Fuzzy Inference System, IEEE ACCESS, Vol. 7 pp 6510–6518.

[11] Datasheet C2M0025120D, https://www.wolfspeed.com/c2m0025120d.

[12] Deng, E., Zhao, Z., Zhang, P., Luo, X., Li, J., and Huang, Y.: Study on the Method to Measure Thermal Contact Resistance Within Press Pack IGBTs, IEEE Transactions on Power Electronics, Vol 34 no2 pp 1509-1517.

[13] Chang, Y., Li, W., Luo, H., He, X., Iannuzzo, F., Blaabjerg, F., and Lin, W.: A 3D thermal network model for monitoring imbalanced thermal distribution of press-pack IGBT modules in MMC-HVDC applications, Energies, Vol. 12 no 7 pp 1-20.

[14] Datasheet CPM2-1200-0025B, https://www.wolfspeed.com/CPM2-1200-0025B.

[15] Pfost, M., Unger, C., Cretu, G., Cenusa, M., Buyuktas, K., and Wahl, U.: Short-Circuit Safe Operating Area of Superjunction MOSFETs, 28th International Symposium on Power Semiconductor Devices and ICs, 2016.

[16] Shenoy, P. M., and Baliga, B. J.: The planar 6H-SiC ACCUFET: A new high-voltage power MOSFET structure, IEEE Electron Device Letters, Vol. 18 no 12 pp 589–591.

[17] Li, H., Yao, R., Lai, W., Ren, H., and Li, J.: Modeling and analysis on overall fatigue failure evolution of press-pack IGBT device, IEEE Transactions on Electron Devices, Vol. 66 no 3 pp 1435-1443.

Error Induced by the Optical Path of a High Accuracy and High Bandwidth Optical Current Measurement System

Stefan Rietmann and Jürgen Biela
Laboratory for High Power Electronic Systems, ETH Zurich, Switzerland
Physikstrasse 3
8092 Zurich, Switzerland
rietmann@hpe.ee.ethz.ch, jbiela@ethz.ch

Keywords

≪Current Sensor≫, ≪Measurement≫, ≪Device modelling≫

Abstract

Birefringence and optical losses in the optical sensor head of a Faraday effect based, high accuracy and high bandwidth current measurement system lead to changes of the polarisation state and orientation. These changes, if not properly determined and calibrated, lead to systematic measurement errors. Therefore, a calibration method for the optical current measurement system focusing on pulse current applications is proposed. The procedure is based on a zero-current measurement and a full (Stokes) polarimeter. It determines the full polarisation state and orientation of a beam propagating through the optical system. The calibration method then allows to determine the parasitic linear birefringence caused by any subsequent optical element in the system. Eventually, an error analysis characterising the potential error reduction by the proposed calibration method is conducted. Further, the additionally introduced error sources caused by the full polarimeter approach are characterised.

1 Introduction

Current measurement systems in high power electronic applications for control and monitoring purposes often require high accuracy and a high measurement bandwidth. A possible solution are magneto-optical current sensors (MOCS) utilizing the Faraday effect for measuring the magnetic field generated by the current. Further advantages of MOCS are their insensitivity to electromagnetic interference and the inherent electrical isolation.

The sensing element, consisting of a magneto-optical (MO) material, is the key element of the current measurement system. Linearly polarised light is incident to this material which turns the plane of polarisation (POP) as function of an applied magnetic field. Choosing the type of material for the sensing element by considering its Verdet constant, optical path length, or optical transmission rate is crucial to achieve the required accuracy and bandwidth as explained in [1].

The integration of the MO material into a sensor head impacts the further behaviour and application spectrum of the current sensor. Intrinsic or all-fiber based sensors have been intensely investigated in the past decades [2, 3, 4, 5]. Due to the long optical path they proved to be rather unsuitable for very high bandwidth applications [1]. Further, the optical fiber needs to be wound around the conductor increasing space demands and impeding practical installation. Extrinsic MOCS systems, as schematically shown in Fig. 1, consist of a dedicated para-, dia- or ferromagnetic bulk material with a short optical path in between the optical fiber guides (up- and downlink).

All MOCS are subject to optically induced errors, in the sensing element, the additional optical fibers and the optical processing elements. These errors arise from linear and circular birefringence, optical losses and environmental influences. Optical error sources in MOCS have been considered in research with special focus on birefringence problems. Solutions to induced and inherent linear birefringence have been addressed especially for all-fiber based MOCS. Potential concepts include the usage of special fibers and fiber treatment, for instance twisted and annealed single-mode (SM) fibers [2, 4], spun high birefringent fibers (SPUN HiBi) [3], and large fiber loop diameters for low bending induced birefringence [5].

Bulk material based extrinsic setups are less sensitive to bending induced birefringence due to their geometric dimensions. The inherent linear birefringence can be determined with zero field / current calibration measurements and subtracted subsequently. Nevertheless, the additional up- and downlink fibers are sensitive to stress induced birefringence. The downlink, as depicted in Fig. 1, carries the modulated signal and therefore needs special care while designing the sensor head. It is important to either preserve the polarisation information carefully or to gather enough information for a valid reconstruction. Proposed solutions for polarisation preservation are using full integration of the optical processing elements into the sensor head [6, 7, 8, 9] with free-space propagation paths as well as downlink optical fibers for signal propagation [10].

The past research of MOCS and its optical errors focused strongly on all-fiber arrangements and complete preservation of the polarisation state after the modulation. An overall consideration of all polarimetric possibilities for a full polarimetric determination of the SOP and the POP is missing. Hence, the polarisation state measurement and the subsequent translation to a magnetic field or current is usually based on the assumption that only linearly polarised light enters and exits the system with no additional and unintended alteration of the polarisation state or orientation. While this assumption is generally valid for systems assuming a low amount of Faraday rotation ($\leq 10°$), it is rendered invalid by the specification in Tab. I and the concept of multiple full rotations of the POP, used in this project and introduced in [1, 11]. In this paper, optical error sources in a bulk material based extrinsic MOCS are discussed with respect to the accuracy and precision targets as given in Tab. I. With a more precise measurement of the polarisation state, error sources can be identified and isolated during calibration. Hence, an analyser configurations (polarimeter), capable of measuring the complete SOP, is used for the calibration procedure. This calibration process allows then for a significant error reduction.

In section 2, a short summary of optical error sources with focus on birefringence is given. Section 3 proposes a calibration method based on the full characterisation of the SOP and a full polarimetric description of the disruptive part of the optical system. In section 4, an error analysis of the potential error reduction by the calibration method is conducted. Further, the additional error sources introduced by the

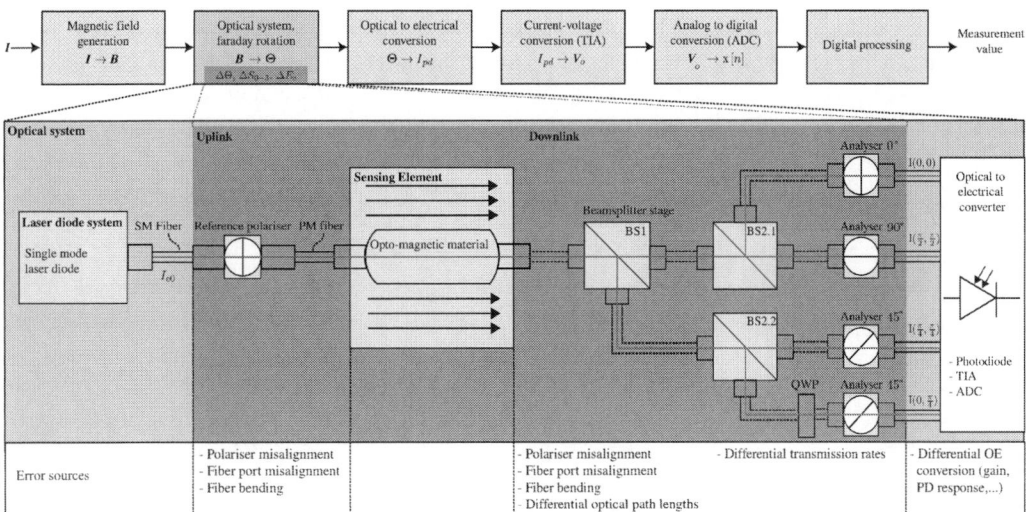

Fig. 1: The optical system of a magneto-optical current sensor is subject to optical losses and birefringence effects. Due to linear and circular birefringence the state of polarisation, the respective optical rotation and the optical intensity of the signal change. Here, the downlink including a full Stokes polarimeter is focused. A calibration procedure reducing the error induced by additional linear birefringence is proposed. Further, an error analysis characterising the possible error reduction but also the additional error introduced by the full polarimeter is conducted.

Table I: Magneto-optical current sensor requirements

	Specification
Current range	100 A ... 10 kA
Frequency bandwidth	DC... $\geq$ 10 MHz
Measurable pulse rise time (5% - 95%)	< 30 ns...ms
Full bandwidth uncertainty (accuracy)	< 0.1%
Reproducibility error	< 25 ppm

full polarimeter approach are considered. Eventually, the paper concludes in section 5.

2 Optical Error Sources

Induced birefringence in MOCS is one of the most significant sources for systematic errors and measurement imprecisions. It causes changes of the polarisation state and orientation. There, any additional change of the polarisation orientation superposes with the desired change by the applied magnetic field. Depending on the chosen analyser setup, any change of the polarisation state leads to a misinterpretation of the measured output signal, as further elaborated in section 3. Thus, these influences need to be calibrated or prevented in order to increase the accuracy (systematic errors) and precision (random errors).

For successfully preventing additional effects in the measurement system, identifying the respective sources is necessary. Since birefringence is present in most of the elements used throughout the optical assembly, it is necessary to specify all elements in terms of their affinity to it.

In the following section, a short definition of birefringence is given. Further, the influence of linear birefringence on the polarisation state and orientation is described with the example of optical fibers.

2.1 Birefringence

2.1.1 Linear Birefringence

Linear birefringence is inherently present in optically anisotropic material, meaning that the material has two different refractive indices (uniaxial case). These materials split an incoming beam into two orthogonally polarised beams, the ordinary and extra-ordinary beam. There, the birefringence of a material is defined as $\Delta n = n_e - n_o$, where n_e and n_o are the refractive indices experienced by the extra-ordinary and the ordinary beam, respectively. Due to the different refractive indices, the phase velocity of the two beams differs, leading to a phase difference described as the linear retardance Δ. As an example, one can assume linearly polarised monochromatic light propagating through birefringent material in z-direction. The polarisation state of the beam is split into the orthogonal polarisation axes as depicted in Fig. 2a. The phase difference between those two orthogonal polarisation components is zero for linearly polarised light. A phase shift is introduced due to the different phase velocities in the x- and y-axes. Hence, the polarisation state is no longer linearly polarised, rather is it elliptically polarised as shown in Fig. 2c

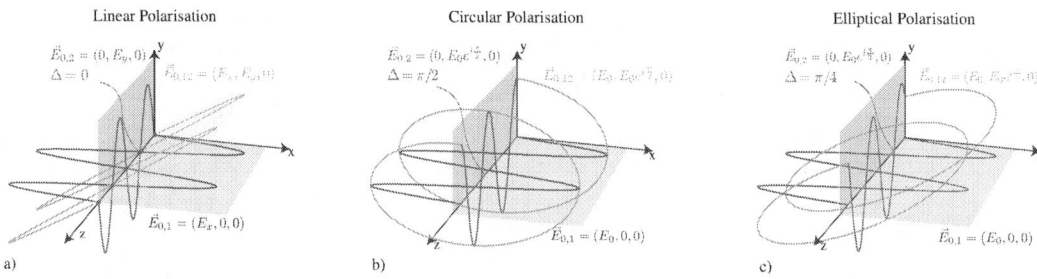

Fig. 2: The polarisation is analytically described with $\vec{E} = \vec{E}_{0,12} \cdot e^{i(kz-\omega t)}$. a) Linear polarisation: The two orthogonal components of the polarisation have a phase difference of $\Delta\varphi = 0$. Depending on the magnitude of the components the orientation of the plane of polarisation is determined. b) Circular polarisation: The two orthogonal components of the polarisation have phase difference of $\Delta\varphi = \pi/2$. c) Elliptical polarisation: The two orthogonal components have phase difference which is not equal to 0 or $\pi/2$.

or, in the special case of a phase shift of $\Delta = \pi/2$, cylindrically polarised as shown in Fig. 2b. As soon as the phase shift reaches a total value of $\Delta = \pi$, the light is linearly polarised again.

2.1.2 Circular Birefringence

Circular birefringence, in opposite to linear birefringence, describes the material induced phase shift between right-hand circular (RHC) and left-hand circular (LHC) polarised waves. Of particular interest for MOCS systems is, that a linearly polarised light beam can be described by the superposition of two counter-rotating circularly polarised waves. The phase difference between those two rotating waves defines the polarisation orientation of the linearly polarised light. For MOCS, inherent circular birefringence can be considered as an offset if present. Hence, calibration of the zero-current field is necessary.

Instead of a constant rotation angle, the Faraday rotation in magneto-optical material is essentially considered a magnetic field modulated circular birefringence. It can be described by the circular birefringence [12] with:

$$\Theta = \Theta_F L = \frac{(n^+ - n^-)\pi L}{\lambda} \tag{1}$$

where Θ describes the rotation of the plane of polarisation, n^+ and n^- the refractive index of the RHC and LHC polarisation and L is the length of the optical path. The term Θ_F is the Faraday effect specific rotation which is mathematically described with $\Theta_F = \mathcal{V} \cdot |\mathbf{H}|$, where $\mathcal{V}$ is the material specific Verdet constant and $\mathbf{H}$ the magnetic field amplitude in the direction of the optical path [1].

2.2 Optical Components

All of the optical components used in the setup shown in Fig. 1 are subject to inherent or induced, linear or circular birefringence. The magneto-optical material is predominantly subject to the desired magnetic field modulated circular birefringence as described in section 2.1.2. Linear birefringence can be observed in most of the optical elements used in this context, most obviously in optical fibers. There are multiple sources of linear birefringence. Most commonly, linear birefringence is caused by mechanical stress due to bending, temperature fluctuations or vibrations. In the following section, optical fibers serve as an example for the impact of induced linear birefringence.

2.2.1 Optical Fibers

In optical fibers linear birefringence is present both induced and inherent. During manufacturing small defects in the optical fiber and slight core ellipticity lead to inherent linear birefringence [13].

Induced linear birefringence is caused by mechanical stress, for example bending, pressure and force put on the optical fiber. It causes the bending induced retardance Δ_b, described in [14] with:

$$\Delta_b = 0.25 \cdot kn^3(p_{11} - p_{12})(1+v)\kappa^2 r^2 \quad (2) \qquad \Delta_b = 7.7 \cdot 10^7 \cdot \left(\frac{r}{R}\right)^2 \left[°/m\right] \tag{3}$$

where n describes the refractive index, p_{ij} are strain optical coefficients, v the poisson count and r is the optical fiber radius. Further, κ describes the curvature $\kappa = 1/R$ of the bend where R is the curvature radius. For silica based fibers [14] specifies Δ_b as in (3). For an optical fiber (e.g. Thorlabs SM600) used at 633 nm with a diameter of 125 μm the minimal long-term bending radius is 25 mm. The bending induced retardance of this particular fiber is around $\Delta_b = 481.3 °/m$ when the fiber is bend to the extent of the maximal allowed radius. A more relaxed bending radius of e.g. 100 mm represents a realistic situation when a fiber is put into a confined space, as for example a box. The bending induced linear retardance for this case is already $\Delta_b \approx 30 °/m$.

The linear retardance, caused by non-ideal optical elements and external influences, changes the overall optical system response. For an accurate measurement a precise description of the influence of linear retardance is necessary.

3 Polarimeteric Signal Processing

Converting the polarisation information of the light beam into an electrical signal is an important part of the overall measurement system but also subject to multiple error sources. In particular, the optical error sources, as described in section 2, influence the polarisation state and orientation of the light beam propa-

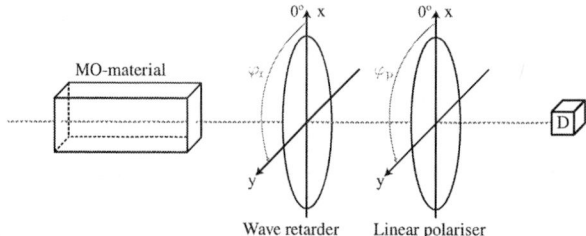

Fig. 3: The schematic figure of the analyser channel shows two rotatable optical elements, a quarter wave plate (QWP) and a linear polariser. The rotation of the relevant axes is indicated by the two parameter φ_r and φ_p, respectively.

gating through the sensor head and the adjacent optical elements. The polarisation state and orientation need to be analysed carefully, in order to identify and isolate the rotation of the plane of polarisation induced only by the current to measure.

In this section, a short introduction to the Stokes and Mueller formalism as well as to basic polarimeteric setups is given. These formalism provide the possibility to describe the polarisation state and orientation based on the beam intensity measurements. For detailed information concerning the theory of polarimetric measurement, please be referred to [15, 16, 17] or [18]. Based on these formalisms, an optical system and a calibration procedure to account for the optical errors are proposed.

3.1 Stokes Vector Measurement

For the determination of the polarisation rotation caused by the Faraday effect, a comprehensive mathematical description of the polarisation state and the polarisation orientation is required. The Stokes vector ($\vec{S}$) in (4) describes the light's polarisation in terms of four parameters. There, S_0 denotes the overall intensity, S_1 the excess of horizontal over vertical polarisation, S_2 the excess of $+45°$- over $-45°$-polarisation and S_3 the excess of right-handed circular (RCP) over left-handed circular (LCP) polarisation. Since only the intensity of a beam can be measured directly, the Stokes vector has to be composed by measured intensities altered through analyser settings (Fig. 3) with different angles. The Stokes vector and the associated intensity measurements $I(\varphi_r, \varphi_p)$ are defined by

$$\vec{S} = \begin{pmatrix} S_0 \\ S_1 \\ S_2 \\ S_3 \end{pmatrix} = \begin{pmatrix} E_{0x}^2 + E_{0y}^2 \\ E_{0x}^2 - E_{0y}^2 \\ 2E_{0x}E_{0y}\cos(\delta) \\ 2E_{0x}E_{0y}\cos(\delta) \end{pmatrix} = \begin{pmatrix} I(0,0) + I(\frac{\pi}{2}, \frac{\pi}{2}) \\ I(0,0) - I(\frac{\pi}{2}, \frac{\pi}{2}) \\ 2 \cdot I(\frac{\pi}{4}, \frac{\pi}{4}) - S_0 \\ S_0 - 2 \cdot I(0, \frac{\pi}{4}) \end{pmatrix} \tag{4}$$

where the angle φ_r represents the orientation of a quarter wave plate (QWP) and the angle φ_p represents the orientation of the transmission axis of a linear polariser. A schematic drawing of single analyser channel, which is used for one of the four $I(\varphi_r, \varphi_p)$ measurements, is given in Fig. 3. According to (4), a minimum number of four different intensity measurements are required to fully characterise the polarisation state and orientation of a beam. Measurement concepts with only two analyser channels, as frequently used in MOCS systems (e.g. [19], [11], [20], [21] or [22]), are only able to determine for example the intensities $I(0,0)$ and $I(\frac{\pi}{2}, \frac{\pi}{2})$ from equation (4). Hence, only the deviation from the horizontal polarisation orientation (S_1) and the overall intensity (S_0) can be determined. Thus, the detectable rotation angle range is $0 \leq \theta_{max} \leq 90°$. However, any polarisation state change towards elliptically polarised states cannot be identified and therefore falsely indicates a change in the polarisation orientation. Hence, a necessary condition for using this measurement setup is the requirement that all light in the system is linearly polarised.

By using an additional analyser channel (e.g. $I(\frac{\pi}{4}, \frac{\pi}{4})$), a full characterisation of the linear plane of polarisation can be achieved. Thus, the detectable rotation angle range enlarges to $0 \leq \theta_{max} \leq 180°$. There, different measurement arrangements exist, as for example Pickering's or Fessenkov's method [16]. These are primarily used in regular polarimetry (e.g. [16, 18] or [23]) in which the aim is to determine the orientation of a linearly polarised beam with no initial reference and a full mapping of the orientation angle.

By using a fourth analyser channel, the full Stokes vector can be characterised. Since the fourth measurement does not add any information to the polarisation orientation, the detectable rotation angle range remains the same as for the measurement concept using three analyser channels. However, by adding the missing information about the beams ellipticity, it fully characterises the polarisation state. A four-channel or full Stokes polarimeter sets the basis for the following optical system description.

3.2 Optical System Description

While it is important to be able to measure the resulting output beam polarisation $\vec{S}_{\text{out}}$, to identify additional changes in the polarisation, a systematic description of the optical effects in the system is required. This description is eventually necessary in order to distinguish the desired Faraday rotation from these additional polarisation changes in the measured output beam.

In this particular application, the most dominant undesired effect is additional linear birefringence. Its effect can be described in form of a variably rotatable linear retarder, as described in section 2. The change in the index of refraction changes the respective phase velocity of one of the orthogonal waves during propagation through a birefringence afflicted medium. The occurring phase shift affects the polarisation state and orientation by manipulating the Stokes parameter S_1-S_3 of the outgoing beam. The Mueller matrix $\mathbf{M}_{\text{linRet}}$ characterises the effects on the polarisation of a beam propagating through a linear retarder. The matrix is given in (5):

$$\mathbf{M}_{\text{linRet}} = \begin{bmatrix} 1 & 0 & 0 & 0 \\ 0 & \cos^2 2\varphi + \cos\Delta\sin^2 2\varphi & (1-\cos\Delta)\sin 2\varphi\cos 2\varphi & -\sin\Delta\sin 2\varphi \\ 0 & (1-\cos\Delta)\sin 2\varphi\cos 2\varphi & \sin^2 2\varphi + \cos\Delta\cos^2 2\varphi & \sin\Delta\cos 2\varphi \\ 0 & \sin\Delta\sin 2\varphi & -\sin\Delta\cos 2\varphi & \cos\Delta \end{bmatrix} \tag{5}$$

where Δ is the retardance induced by the linear birefringence and φ is the rotation of the reference system on which the retardance is acting. With the Mueller matrix, any combination of n series connected optical elements can be described in terms of an input ($\vec{S}_{\text{in}}$) and output ($\vec{S}_{\text{out}}$) polarisation with the following matrix multiplication:

$$\vec{S}_{\text{out}} = \mathbf{M}_n \cdot \mathbf{M}_{n-1} \cdot \ldots \cdot \mathbf{M}_1 \cdot \vec{S}_{\text{in}} \tag{6}$$

Hence, the polarisation related effect in the magneto-optical material and the subsequent optical analyser arrangement can be described with:

$$\vec{S}_{\text{out}} = \mathbf{M}_{\text{linRet}} \cdot \mathbf{M}_{\text{MO}} \cdot \vec{S}_{\text{in}} \tag{7}$$

There, $\mathbf{M}_{\text{MO}}$ describes the Faraday effect in the magneto-optical material with:

$$\mathbf{M}_{\text{MO}} = \begin{bmatrix} 1 & 0 & 0 & 0 \\ 0 & \cos 2\theta_{\text{F}} & \sin 2\theta_{\text{F}} & 0 \\ 0 & -\sin 2\theta_{\text{F}} & \cos 2\theta_{\text{F}} & 0 \\ 0 & 0 & 0 & 1 \end{bmatrix} \tag{8}$$

and $\theta_{\text{F}} = \mathscr{V}|B|L$ describes the Faraday effect ($\mathscr{V}$: Verdet constant, $|B|$: magnetic field flux amplitude, L: optical path length).

Using (7), a description of an optical system influenced by the Faraday effect and by linear birefringence in the magneto-optical material and the subsequent polarimetric analyser is established.

3.3 Linear Birefringence Calibration Procedure

Based on the mathematical model presented in the previous section, a calibration procedure for the MOCS is proposed in the following. Since the considered MOCS is designed for pulse applications, a few basic assumptions can be made:

- The occurrence, time and duration of a pulse event are known.
- The duration of the pulse is finite and relatively short (~µs) compared to the propagation mecha-

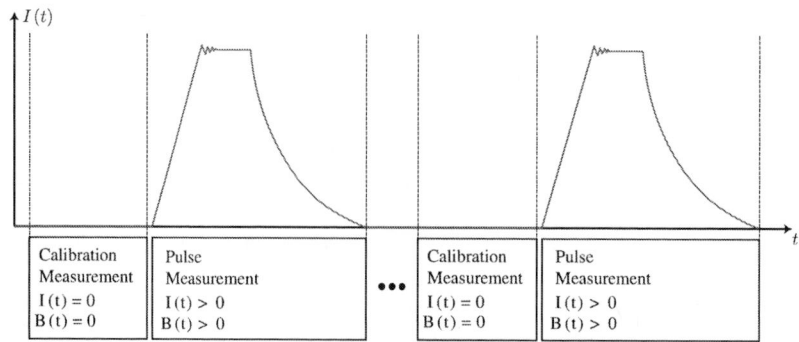

Fig. 4: The calibration measurement allows the detection of the externally applied additional linear birefringence. Assuming pulse current measurements, the environmental influences on the MOCS (movement, temperature, bending) are considered to be constant over the complete measurement time range.

nism of the influences affecting the linear birefringence (movement, temperature, pressure,...).

- The conductor is not conducting any current (i.e. $I_0 = 0$) for a substantial amount of time ($\sim$ μs) between two pulses.

With these assumptions it is possible to calibrate the MOCS prior to each pulse and exclude any environmental influences if the input polarisation state is known and the output polarisation state can be fully measured. There, the assumption that linear birefringence generated by environmental effects does not change in the time range between the calibration and the end of the pulse measurement is essential.

With the Mueller matrix in (5) and the system description in (7), the optical system can be approximately described. This system description incorporates the two most dominant effects to the polarisation state and orientation, the Faraday rotation and the additional linear birefringence. During a calibration measurement the conductor must not carry any current (i.e. $I_0 = 0$), essentially resulting in $|B_0| = 0$. Thus, the Mueller matrix for the magneto-optical material is $\mathbf{M}_{MO} = \mathbb{1}$ and the system description reduces to $\vec{S}_{\text{out}} = \mathbf{M}_{\text{linRet}} \cdot \vec{S}_{\text{in}}$. By knowing the output polarisation $\vec{S}_{\text{out}}$ from the calibration measurement and with the input polarisation $\vec{S}_{\text{in}}$ defined by a reference polariser, the matrix of the additional variably rotatable linear retarder can be determined. With (5), the wanted variables are identified as:

$$
\begin{aligned}
x_1 &= \cos 2\varphi \\
x_2 &= \sin 2\varphi \\
x_3 &= \Delta
\end{aligned}
\tag{9}
$$

Since the first eigenvector of (5) is already determined, which leads to $S_{\text{out},0} = S_{\text{in},0}$ (no depolarisation assumed), the equation system is reduced to a numerically solvable 3×3 system:

$$
\frac{1}{S_{\text{in},0}} \cdot \begin{pmatrix} x_1^2 + x_2^2 \cos(x_3) & [1-\cos(x_3)]x_1 x_2 & x_2 \cdot (-\sin(x_3)) \\ [1-\cos(x_3)]x_1 x_2 & x_2^2 + x_1^2 \cos(x_3) & x_1 \cdot \sin(x_3) \\ x_2 \cdot \sin(x_3) & x_1 \cdot (-\sin(x_3)) & \cos(x_3) \end{pmatrix} \cdot \begin{bmatrix} S_{\text{in},1} \\ S_{\text{in},2} \\ S_{\text{in},3} \end{bmatrix} = \frac{1}{S_{\text{out},0}} \cdot \begin{bmatrix} S_{\text{out},1} \\ S_{\text{out},2} \\ S_{\text{out},3} \end{bmatrix}
\tag{10}
$$

The retardance Δ is derived directly from the solution for x_3 in (10). The rotation of the reference system φ depends on the two conditions x_1 and x_2 from (9).

By knowing the zero-field / zero-current output polarisation, the deviation caused by the additional linear birefringence in the system can be determined. With this information, this influence can be eliminated in a subsequent pulse measurement and the rotation of the polarisation plane can be identified directly from the output polarisation:

$$
\begin{aligned}
\vec{S}_{\text{out}} &= \mathbf{M}_{\text{linRet}} \cdot \mathbf{M}_{\text{MO}} \cdot \vec{S}_{\text{in}} \\
\Rightarrow \mathbf{M}_{\text{linRet}}^{-1} \cdot \vec{S}_{\text{out}} &= \mathbf{M}_{\text{MO}} \cdot \vec{S}_{\text{in}} = \vec{S}_{\text{MO}}
\end{aligned}
\tag{11}
$$

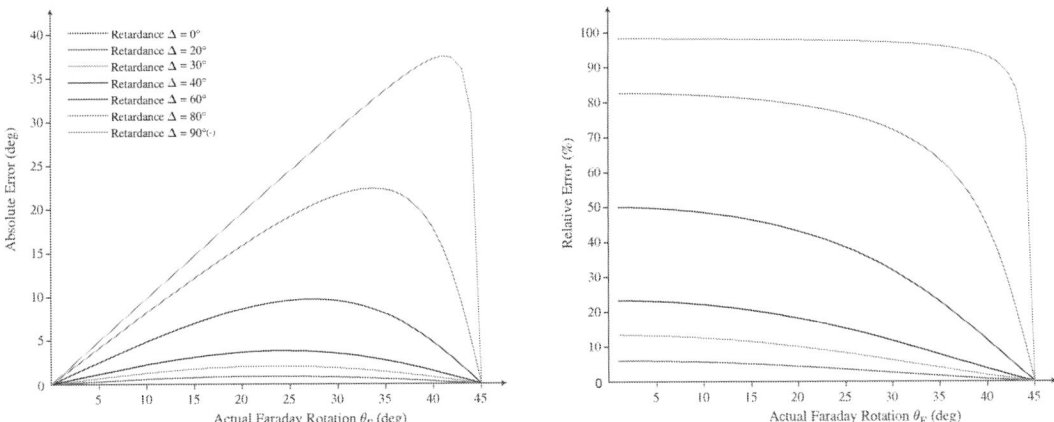

Fig. 5: The potential error range of externally induced additional linear birefringence can lead to a substantial systematic error. The error is periodic over the measurement range of $\theta_F \in \left[0,...,\frac{\pi}{4}\right]$. Hence, for MOCS with measurement ranges larger than a few degrees a calibration method is necessary.

For the inversion of $\mathbf{M}_{linRet}$ it needs to be non-singular.

Eventually, the result $\vec{S}_{MO}$ describes the changes caused by the Faraday effect only. From this result, the orientation of the plane of polarisation can be recalculated utilizing the following equation:

$$\theta_F = -\frac{1}{2}\arctan\left(\frac{S_{MO,2}}{S_{MO,1}}\right) \qquad \text{with } \mathrm{sgn}\left(\cos(2\theta_F)\right) = \mathrm{sgn}\left(S_{MO,1}\right) \tag{12}$$

Due to the range of the principal values of the arctan-function the additional condition in (12) is necessary to determine θ_F over the complete range $0 \le \theta_F \le \pi$.

4 Error Analysis

In this section, two types of errors are considered for an error analysis. First of all, the error reduction due to the calibration procedure is described and justifies the choice of the full polarimeter approach. The effect of the additional linear birefringence on the beams' polarisation is mathematically removed leaving only the effect of the Faraday rotation. However, the calibration method requires more optical evaluation channels, all of which are subject to production and configuration uncertainties. The numerical example refers to the project requirements stated in Tab. I and a maximal magnetic field amplitude of $|\mathbf{B}| = 2\,\mathrm{T}$ applied to the MOCS. This results in an total optimal optical rotation of $\theta_{F,max} = 1980°$. The specification for the magneto-optical material are taken from [1]: $\mathcal{V}_{CdMnTe} = 150°\,\mathrm{T}^{-1}\,\mathrm{mm}^{-1}$, $L = 6.6\,\mathrm{mm}$.

4.1 Error Reduction by Calibration

The system error ε_θ introduced by the additional linear birefringence is determined with the system description given in (7). It is defined as the difference between the actual Faraday rotation θ_F and the measured (altered) rotation of the polarisation orientation $\theta_{F,alt}$.

$$\varepsilon_\theta = \theta_F - \theta_{F,alt} \tag{13}$$

Considering the system description in (7), this error is introduced after the rotation of the polarisation orientation, thus, the measurement result $\theta_{F,alt}$ depends on the Faraday rotation θ_F itself. Figure 5 shows the dependence of the measurement error on the additional linear birefringence (retardance) and the deflection of the plane of polarisation to the reference axis caused by the Faraday rotation. The minimal and maximal error shows at multiples of $\pi/4$. The error pattern is periodic with $\pi/4$ due to the range and symmetry behaviour of the evaluation equation (11).

In section 2.2.1 the magnitude of bending induced linear birefringence is described with an example bending radius of 100 mm. For an example single-mode fiber (e.g. Thorlabs SM600) of 1 m operated at 633 nm, the bending induced retardance is $\Delta \approx 30\,°$. The maximal error for $\theta_F \in [0,...,45]$ is $\varepsilon_\theta = 2.05°$ resulting in a relative error of 9.13 % with respect to the single measurement. Considering the above mentioned project requirements and values, the absolute error introduced by this amount of linear birefringence is 2.02 mT or 0.101 %, already exceeding the maximal allowed uncertainty. Accordingly, measurements with lower maximal field amplitudes experience higher relative errors. Further, note that the choice of the amount of retardance for this example is arbitrary but reasonable. Higher retardance in the range $\Delta \in [0,...,\pi/2]$ will increase the absolute and relative error as well.

By applying the proposed calibration procedure, this error can be eliminated. However, the additional equipment introduces measurement errors, which are taken into account in the following section.

4.2 Polarimeter Error

The uncertainties of commercially available optical elements lead to a deviation of the measured from the expect results. In this analysis, the three channels necessary for the evaluation of θ_F, consisting of three linear polarisers, are considered. The uncertainty of the polariser's orientation influences the intensity measurement in (4), which is necessary for the determination of S_0, S_1 and S_2. Further, the proposed setup implements three parallel channels. There, the intensity measurement is also influenced by the individual transmission rate of the polarisers.

For the analysis of the error induced by the polariser orientation misalignment, the parameter φ_p in $I(\varphi_r, \varphi_p)$ has to be adjusted by the uncertainty $\Delta\varphi_p$ describing the deflection of the polariser from its optimal orientation. Using the Mueller formalism the polarisers are described with:

$$\mathbf{M}_{Pol} = \begin{bmatrix} 1 & \cos 2\varphi_p & \sin 2\varphi_p & 0 \\ \cos 2\varphi_p & \cos^2 2\varphi_p & \sin 2\varphi_p \cos 2\varphi_p & 0 \\ \sin 2\varphi_p & \sin 2\varphi_p \cos 2\varphi_p & \sin^2 2\varphi_p & 0 \\ 0 & 0 & 0 & 0 \end{bmatrix} \tag{14}$$

With (6), the Mueller matrix for the polariser (14) and the Mueller matrix for a linear retarder (5) with $\Delta = \frac{\pi}{2}$ (QWP), the optical system describing one analyser channel is given by:

$$\vec{S}_m = \mathbf{M}_{Pol} \cdot \mathbf{M}_{linRet} \cdot \vec{S}_{MO} \tag{15}$$

with $\vec{S}_m$ being the measured output Stokes vector and $\vec{S}_{MO}$ the Stokes vector transmitted from the before described magneto-optical system. From the right-hand term in (4) and the formulation for a general analyser channel (15), the following general formulation for the measured intensity dependent on the polariser's orientation is derived:

$$I(\varphi_r, \varphi_p) = S_{0,MO} + \cos 2\varphi_r \cdot (S_{1,MO} \cos^2 2\varphi_p + S_{2,MO} \cos 2\varphi_p \sin 2\varphi_p - S_{3,MO} \sin 2\varphi_p) + \tag{16}$$

$$\sin 2\varphi_r \cdot (S_{1,MO} \cos 2\varphi_p \sin 2\varphi_p + S_{2,MO} \sin^2 2\varphi_p + S_{3,MO} \cos 2\varphi_p) \tag{17}$$

Applying the right-hand term of (4) to the formulation for the recalculation of the polarisation orientation in (11) gives a polariser orientation dependent $(\varphi_p, \Delta\varphi_{p,xx})$ formulation (18) for the recalculated polarisation orientation. There, $\Delta\varphi_{p,xx}$ denotes the polariser's individual deviation from the ideal orientation, emphasising that the individual measurement channels are independent from each other.

$$\theta_{F,err} = -\frac{1}{2} \arctan \left(\frac{2 \cdot I(\frac{\pi}{4}, \frac{\pi}{4} + \Delta\varphi_{p,45}) - I(0, 0 + \Delta\varphi_{p,0}) - I(\frac{\pi}{2}, \frac{\pi}{2} + \Delta\varphi_{p,90})}{I(0, 0 + \Delta\varphi_{p,0}) - I(\frac{\pi}{2}, \frac{\pi}{2} + \Delta\varphi_{p,90})} \right) \tag{18}$$

The maximal measurement error ε_M occurs for a maximal deflection between the polarisers measuring $I(0,0)$ and $I(\frac{\pi}{2}, \frac{\pi}{2})$. This can be shown by maximising $\varepsilon_M = |\theta_F - \theta_{F,err}|$.

In this project, as referred to in the beginning of this section, the rotatable polarisers (Thorlabs, FBRP-LPVIS [24]) have a scale allowing for a resolution down to 10 arcmin = 0.167°, hence the reading uncertainty is ±0.083°. The maximal error calculated with (18) is $\varepsilon_M = 0.101°$. With reference to the full measurement range of $[0, ..., 2\,\mathrm{T}]$ the maximal absolute error is $0.1\,\mathrm{mT}$ resulting in a relative error of $0.005\,\%$.

For the analysis of the differential transmission losses in the polarisers, the three intensity measurements defining (11) have to be adjusted each by a separate transmission factor. The material thickness dependent transmission rate T is described by the Beer-Lambert law

$$\frac{I}{I_0} = T = e^{-\alpha \cdot d} \tag{19}$$

where α is the absorption coefficient in (m^{-1}) and d is the material thickness in (m). With (19), the transmission factor uncertainty ΔT can be defined as a function of the uncertainty of the material thickness. The maximal absolute error ε_T occurs for the maximal transmission difference between $\mathrm{I}(\frac{\pi}{4}, \frac{\pi}{4})$ and the two orthogonal measurements, which can be shown again by maximising $\varepsilon_\mathrm{T} = |\theta_\mathrm{F} - \theta_\mathrm{F,err}|$.

The polarisers used for this project (Thorlabs, FBRP-LPVIS [24]) are made up of a nanoparticle film with a nominal thickness of $d = 250\,\mu\mathrm{m}$ and a thickness uncertainty of $\Delta d = \pm 50\,\mu\mathrm{m}$. The maximal absolute error is $\varepsilon_\mathrm{T} = 1.604°$. With reference to the full measurement range of $[0, ..., 2\,\mathrm{T}]$ the maximal absolute error is $1.6\,\mathrm{mT}$ resulting in a relative error of $0.08\,\%$.

5   Conclusion

In this paper, errors introduced by the optical system in magneto-optical current sensor (MOCS) systems are investigated. Additional linear birefringence superimposed with the modulated measurement signal introduces a large change to the state of polarisation. To resolve the error introduced by the additional linear birefringence a zero-current/zero-field calibration method has been proposed. The calibration method is intended for pulse current measurements and assumes negligible fluctuations of the environmental influences changing the linear birefringence of the optical system. This zero-current calibration measurement is then used to recalculate the actual amount of Faraday rotation based on the complete characterisation of the polarisation state and orientation and the mathematical description of the disruptive optical system.

Depending on the amount of additional linear birefringence, the potential error reduction by using the calibration procedure is substantial. Additional linear retardance of $\Delta = 30°$ already results in an maximal measurement error of $0.101\,\%$. With an increasing amount of linear birefringence the error increases, hence adverse environmental conditions exceed the above given exemplary values substantially.

The additional effort necessary to fully characterise the polarisation state and orientation introduces also new potential error sources. Since multiple optical elements, such as wave retarders and linear polarisers are used in parallel, the differential path losses have to be considered in the final measurement. The differential transmission losses between the analyser channels needs to be considered carefully. Since the given example with off-the-shelf products already amounts to a maximal measurement error of $0.08\,\%$, it is necessary to spend considerable effort designing the analysers channel paths with identical characteristics. This includes the analyser elements (linear polariser, wave retarder) as well as the optical path in terms of fibers, connectors or focal lenses. The measurement error by misalignment of the linear polarisers has also a non negligible impact. However, using highly precise adjustable polarisers allows a significant reduction of this error. The example with the given off-the-shelf products reduces the error already to $0.005\,\%$ with respect to the full measurement range.

Acknowledgement

The authors would like to thank CERN for funding the project under contract KE3928/TE. In addition, we would like to thank Dr. M. Barnes and Mr. M.C. Bastos for their valuable scientific input.

References

[1] S. Rietmann and J. Biela, "Sensor Design for a Current Measurement System with High Bandwidth and High Accuracy Based on the Faraday Effect," in *21st European Conf. on Power Electron. and Appl. (EPE / ECCE Europe)*, 2019.

[2] S. Rashleigh and R. Ulrich, "Magneto-optic Current Sensing with Birefringent Fibers," *Appl. Phys. Lett.*, vol. 34, no. 11, pp. 768–770, 1979.

[3] R. I. Laming and D. N. Payne, "Electric Current Sensors Employing Spun Highly Birefringent Optical Fibers," *Journal of Lightw. Technol.*, vol. 7, no. 12, pp. 2084–2094, 1989.

[4] A. Rose, Z. Ren, and G. W. Day, "Twisting and Annealing Optical Fiber for Current Sensors," *Journal of Lightw. Technol.*, vol. 14, no. 11, pp. 2492–2498, 1996.

[5] K. Bohnert, P. Gabus, J. Nehring, H. Brandle, and M. G. Brunzel, "Fiber-optic Current Sensor for Electro-winning of Metals," *Journal of Lightw. Technol.*, vol. 25, no. 11, pp. 3602–3609, 2007.

[6] M. Kanoe, G. Takahashi, T. Sato, M. Higaki, E. Mori, and K. Okumura, "Optical Voltage and Current Measuring System for Electric Power Systems," *IEEE Trans. Power Del.*, vol. 1, no. 1, pp. 91–97, 1986.

[7] A. Cruden, Z. Richardson, J. NcDonald, and I. Andonovic, "Optical Crystal Based Devices for Current and Voltage Measurement," *IEEE Trans. Power Del.*, vol. 10, no. 3, pp. 1217–1223, 1995.

[8] K. Barczak, "Optical Fibre Current Sensor for Electrical Power Engineering," *Bulletin of the Polish Academy of Sciences: Technical Sciences*, vol. 59, no. 4, pp. 409–414, 2011.

[9] A. Brigida, I. M. Nascimento, S. Mendonca, J. Costa, M. Martinez, J. M. Baptista, and P. Jorge, "Experimental and Theoretical Analysis of an Optical Current Sensor for High Power Systems," *Photonic Sens.*, vol. 3, no. 1, pp. 26–34, 2013.

[10] S. Kim, Y.-P. Hong, Y.-G. Kim, and D.-J. Lee, "Field-calibrated Magneto-optic Sensor Based on off-axis Optical probing of Intense Magnetic Fields," *Appl. Opt.*, vol. 56, no. 6, pp. 1701–1707, 2017.

[11] D. Gerber and J. Biela, "High-dynamic and High-precise Optical Current Measurement System Based on the Faraday Effect," *IEEE Trans. on Plasma Sci.*, vol. 43, no. 10, pp. 3550–3554, 2015.

[12] J. P. Castera and T. Suzuki, "Magneto-optical Devices," *Digital Encyclopedia of Appl. Phys.*, 2003.

[13] R. Calvani, R. Caponi, and F. Cisternino, "Polarization Measurements on Single-mode Fibers," *Journal of Lightw. Technol.*, vol. 7, no. 8, pp. 1187–1196, 1989.

[14] R. Ulrich, S. Rashleigh, and W. Eickhoff, "Bending-induced Birefringence in Single-mode Fibers," *Opt. Lett.*, vol. 5, no. 6, pp. 273–275, 1980.

[15] E. Collett, *Field Guide to Polarization.* Spie Bellingham, WA, 2005, vol. FG05.

[16] J. R. Schott, *Fundamentals of Polarimetric Remote Sensing.* Spie Press, 2009, vol. 81.

[17] H. Fujiwara, *Spectroscopic Ellipsometry: Principles and Applications.* John Wiley & Sons, 2007.

[18] M. Bass, C. DeCusatis, J. Enoch, V. Lakshminarayanan, G. Li, C. Macdonald, V. Mahajan, and E. Van Stryland, *Handbook of Optics, Volume II: Design, Fabrication and Testing, Sources and Detectors, Radiometry and Photometry.* McGraw-Hill, Inc., 2009.

[19] B. Chu, Y. Ning, and D. A. Jackson, "Faraday Current Sensor that Uses a Triangular-shaped Bulk-optic Sensing Element," *Opt. letters*, vol. 17, no. 16, pp. 1167–1169, 1992.

[20] U. Holm, H. Sohlstrom, and T. Brogardh, "YIG-sensor Design for Fibre Optical Magnetic Field Measurements," in *2nd Intl Conf on Optical Fiber Sensors*, vol. 514. International Society for Opt. and Photonics, 1984, Conference Proceedings, pp. 333–336.

[21] C. Floridia, J. B. Rosolem, and S. Celaschi, "Mitigation of Output Fluctuations due to Residual State of Input Polarization in a Compact Current Sensor," in *Opt. Fiber Sens.* Optical Society of America, 2018.

[22] P. Mihailovic, S. Petricevic, and R. J, "Improvements in Difference-over-sum Normalization Method for Faraday Effect Magnetic Field Waveforms Measurement," *Journal of Instrumentation*, vol. 1, no. 12, 2006.

[23] P. Hauge, "Survey of methods for the complete determination of a state of polarization," in *Polarized Light: Instruments, Devices, Applications*, vol. 88. International Society for Optics and Photonics, 1976.

[24] Thorlabs Inc. (2020, Jun.) FiberBench Polarization Modules, FBRP-LPVIS. [Online]. Available: https://www.thorlabs.com/newgrouppage9.cfm?objectgroup_id=3101

Analysis of the RMS current stress on the DC link capacitors of the four phase 3-level T-type voltage source converter

Zoran Miletic, Werner Tremmel, Roland Bründlinger, Johannes Stöckl
Austrian Institute of Technology GmbH
Gieffinggasse 2,
1210 Vienna, Austria
+4366488964919/+43505506390
zoran.miletic@ait.ac.at

Petar J. Grbović
University of Innsbruck
Institute of Mechatronics
Techniker Straße 13, 110 (1.OG)
6020 Innsbruck, Austria
+43 664 12 50 951/+43 512 507 62830
petar.grbovic@uibk.ac.at

Keywords: Voltage Source Inverters (VSI), Three-phase system, Converter control, Power quality

Abstract

In the three phase four wire applications, three phases and neutral, often the loads are not well balanced; hence there is a flow of current through neutral conductor. In these applications, the voltage source converters have to handle return or neutral currents into the DC link. This paper presents analysis of the RMS current stress on the DC link capacitors of the four leg 3L T-type voltage source converter caused by unbalanced loads.

Introduction

With evolving tendency to substitute rotating mass synchronous generation with low inertia distributed power converters, the interest in four leg voltage source converters has been increased lately, particularly for three phase four wire applications originated in the micro-grid, off-grid and grid-tied applications on the edge of the distribution grids. In such applications, there is often voltage imbalance [1] or requirement to form a grid and operate under unbalanced load conditions.

In this paper, an impact of the load imbalance or presents of returning neutral current is investigated on a four leg voltage source power converters and especially the RMS current stress on the DC link capacitors.

The DC link is integral part of each voltage source inverter, in terms of operation, size, weight and volume and reliability or life time expectancy. It's usually contacted by electrolytic or film capacitors and laminated DC bus. The DC link operating life time is influenced by the operating DC voltage and the BUS capacitors RMS current. The RMS current, dissipated over the equivalent series resistance of DC link capacitors, heats up DC link capacitors. Increased internal operating temperature of the DC link capacitors increases process of drying up and ageing of capacitors and consequently reduces expected life time of a DC link.

The RMS current stress on DC link is evaluated experimentally by conducting controller hardware in the loop (C-HIL) real time simulation and experimental validation in the lab environment on the four-leg three-level T-type voltage source converter. For the synopsis C-HIL real time simulation data is presented, while for the final paper lab experimental data will be added too.

Typical topologies of three-phase four wire voltage source power converters

In this section, an overview of typical three phase four wire voltage source power converter topologies is presented with outlook on the stress to DC link due to neutral currents flowing into DC link.

In applications with four conductors, the most commonly employed topology is the three-phase split DC link neutral mid-point clamped topology, Fig. 1.

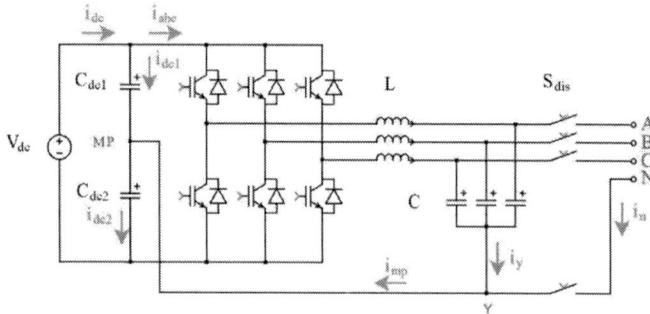

Fig. 1 Three-phase power converter with split DC link neutral mid-point clamped

The neutral current, i_n is given as

$$i_n(t) = i_a(t) + i_b(t) + i_c(t)$$

where, i_a, i_b, i_c represent phase load currents in the time domain. The current flowing into mid-point (MP), i_{mp} is the sum of neutral current i_n and inverter AC filter capacitors currents, i_y, assuming that the AC filter is configured as shown in Fig. 1. The neutral current i_n contains line frequency and low harmonics currents, while the AC filters capacitors current i_y switching harmonics current. Consequently, the RMS current flowing into mid-point (MP) contains fundamental and lower harmonics as well as switching frequency harmonics. According to, [2], the fundamental and lower harmonics impact the DC link with the most weight comparing to higher harmonics including switching harmonics. It should also be noted that the currents flowing into a DC link and its mid-point (MP) affect a DC link capacitor voltage ripple. Increased DC link capacitor voltage ripple reduces effective DC bus voltage and it may trigger the DC link over voltage protection, reduce safe operating range of the semiconductor devices or even result in the failure. [1], [4]

In order to contain the DC link voltage ripple it's required to provide sufficient capacitance rating. Typically, electrolytic capacitors provide storage or capacitance and not so much RMS current rating, contrary to film capacitors with lower capacitance rating for given RMS current rating. Since the zero sequence currents flowing into DC link increase the DC link capacitor voltage ripple it's desirable to have these components reduced.

In order to mitigate DC link RMS currents and capacitor voltage ripple or effectively reduce overall DC link volume, other topologies are being proposed, such as a combination of split DC link neutral clamp with additional phase leg [5] or with series resonant balancers [6], [7]. For the completeness, details of these approaches are going to be detailed in the final paper.

Four leg voltage source power converter

Four leg voltage source converter is well suited for three-phase four-wire applications such as: micro-turbine generators, fuel cell based generators, battery energy storage system, distribution level FACTS; active power filters for a compensation of the harmonic current flowing through the neutral; three phase rectifiers which have to deal with line distortion or imbalance [8] and lately grid forming converters.

Four -legthree-level T-type voltage source converter is shown at Fig. 2

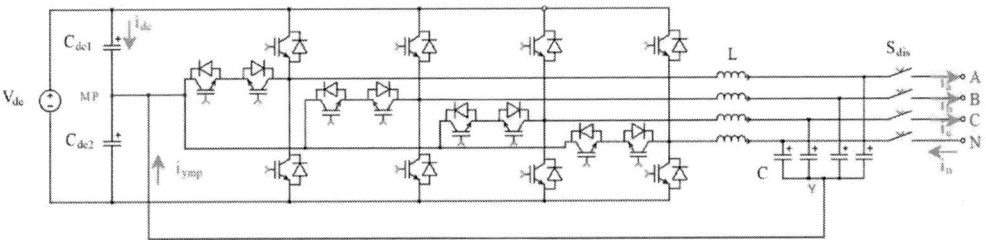

Fig. 2 Four-Leg three-level T-type voltage source power converter

Unlike previously mentioned topologies, at the three-phase four-wire voltage source converter the neutral current doesn't flow into mid-point of the DC link, but it's controlled by 4th leg or neutral leg. In fact, the current through the 4th leg can have arbitrarily direction, to provide path for unbalanced loads or to compensate neutral current. By its nature, four phase voltage source converter is "true" three phase power converter as it can sink or source concurrently positive, negative and zero sequence currents [8].

For case study, the unit under test is a four-phase three-level T-Type voltage source converter with LC filter with nominal current rated to 50A per phase or nominal power rated to 34.5kVA at nominal three phase 400V/50Hz grid

Fig. 3 AIT Smart Grid Converter - four-phaes three-level T-type voltage source converter

The case study was conducted on the controller hardware in the loop (C-HIL) set-up and in the lab environment. The C-HIL set-up comprises of the Typhoon HIL real time simulator and AIT HIL Controller. While lab setup consist of the AIT Smart Grid Converter, Fig. 3, and Regatron DC source and AC grid simulator.

Experimental data

In order to evaluate stress on a DC link of the four phase 3L-Type voltage source converter, the following use cases have been contemplated with the DC bus voltage set to 800V and AC to nominal 230V/50Hz 4 wire connection (ABCN Fig. 2):

- Use case #1 balanced currents: $i_a = 25A/0°, i_b = 25A/-120°$, $i_c = 25A/120°$, $i_n = 0 + $ 3rd harmonic injection
- Use case #2 positive and negative sequence: $i_a = 25A/0°, i_b = 17.625A/-135°$, $i_c = 17.625A/135°$, $i_n = 0 + $ 3rd harmonic injection
- Use case #3 : $i_a = 25A/0°, i_b = 25A/-180°$, $i_c = 25A/180°$, $i_n = 0 + $ 3rd harmonic injection
- Use case #4 : $i_a = 25A/0°, i_b = 25A/-30°$, $i_c = 25A/30°$, $i_n = 0 + $ 3rd harmonic injection
- Use case #5 : $i_a = 25A/0°, i_b = 25A/0°$, $i_c = 25A/0°$, $i_n = 0 + $ 3rd harmonic injection

It should also be noted that 3rd harmonics was injected into modulating signals and dispatched to the PWM modulator, in order to allow better utilization of the DC bus voltage.

The use case # 1 balanced currents is chosen as a baseline, the use case #2, positive and negative sequence without zero sequence currents, and the use cases #3 and #4 as plausible cases with significant presents of neutral or zero sequence currents, while use case #5 represents case of full asymmetry.

The cases #3 and #4 normally cause significant strain on the typical three phase topologies mentioned earlier Fig. 1, [5] [6][7] , due to presents for line frequency currents and its harmonics flowing into a mid-point of DC link.

The four leg three-phase voltage source converter phase currents, i_a, i_b, i_c, i_n, DC link i_{dc} and mid-point DC link i_{ymp}, time domain and harmonic content (lower harmonics only), are shown in Fig. 4 to Fig. 12, while computed broadband RMS currents by applying Parseval theorem in Table 1.

For use cases #1 to #4, as shown in Fig. 4, Fig. 5, Fig. 6 and Fig. 7, the current flowing into mid-point (MP), i_{ymp} contains the switching frequency ripple and only intentionally injected 3rd harmonic while fundamental and other harmonics can be entirely neglected Table 1.

It should also be noted, that for the use cases #1 and #2, i_{dc}, contains insignificant traces of fundamental and lower harmonics, due to absence of zero sequence currents, since $i_n \cong 0$, but it contains switching frequency component Fig. 4, Fig. 5 and Table 1.

For the use cases #3 and #4, there is a measurable present of the fundamental component, as $i_n \neq 0$,as shown in Fig. 6 and Fig. 7, in i_{dc}, however the broadband RMS DC link current, I_{dc}, is not significantly influenced Table 1.

It is also important to evaluate the DC link capacitors voltage ripple, shown in Fig. 13. The ripple depends on DC link capacitance, which is this case is 5.63 mF per each half, and operating conditions. The obtained data correlates with harmonics spectra of i_{dc} for each case. For the use cases with presents of neutral currents, the use case #3 and #4, where the fundamental component existents, the DC link capacitors voltage ripple is impacted the most; In addition it's noticeable that for the worst case scenario, the use case #4, the peak-to-peak DC link voltage imbalance is approx. 20V or 2.5% of the DC bus voltage.

From all the above, it can be concluded the a four leg voltage source converter doesn't require bulky DC link and that the overall volume of DC link can be further reduced and optimised for cost and volume, especially, if it's constructed by film capacitors or by combination of film capacitors, for handling switching frequency content, and electrolytic capacitors, for additional storage and to contain DC link capacitor voltage imbalance.

RMS broadband	Use case #1	Use case #2	Use case #3	Use case #4
I_a	24.54	24.54	24.50	24.84
I_b	24.50	17.57	25.47	23.83
I_c	24.50	17.09	23.83	25.59
I_n	2.05	1.96	24.92	67.44
I_{dc}	13.37	11.11	12.35	18.53
I_{ymp}	9.01	9.04	9.03	9.04

Table 1 Phase currents, i_a, i_b, i_c, i_n, DC link i_{dc} and mid-point DC link i_{ymp} broadband RMS currents

Analysis of the RMS current stress on the DC link capacitors of the four phase 3-level
T-type voltage source converter

MILETIC Zoran

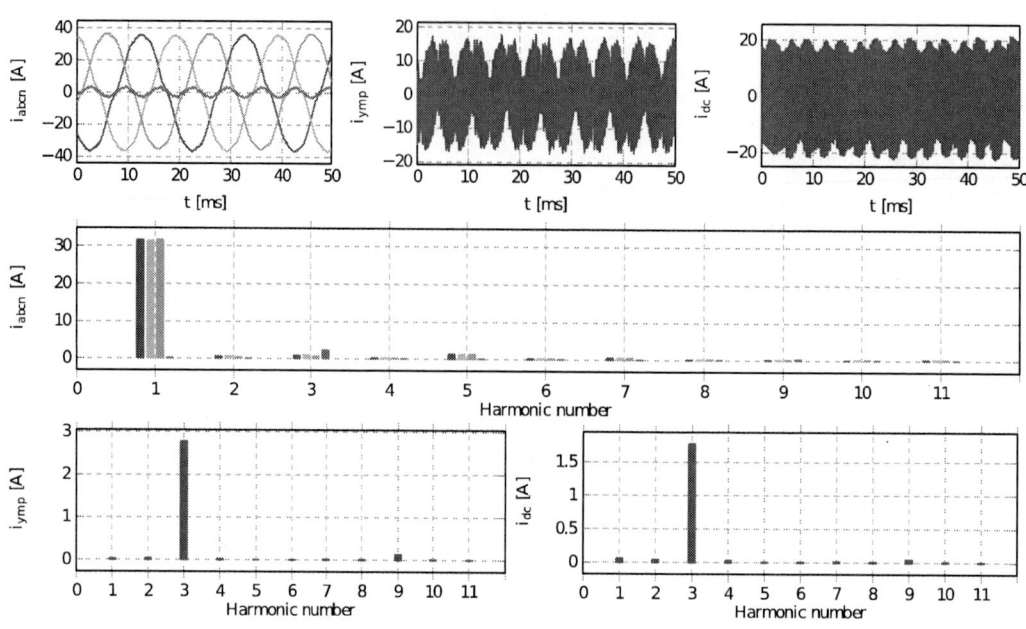

Fig. 4 Use case #1 balanced current: $i_a = 25A/0°, i_b = 25A/{-}120°, i_c = 25A/120°, \quad i_n = 0 + 3^{rd}$ harmonic injection

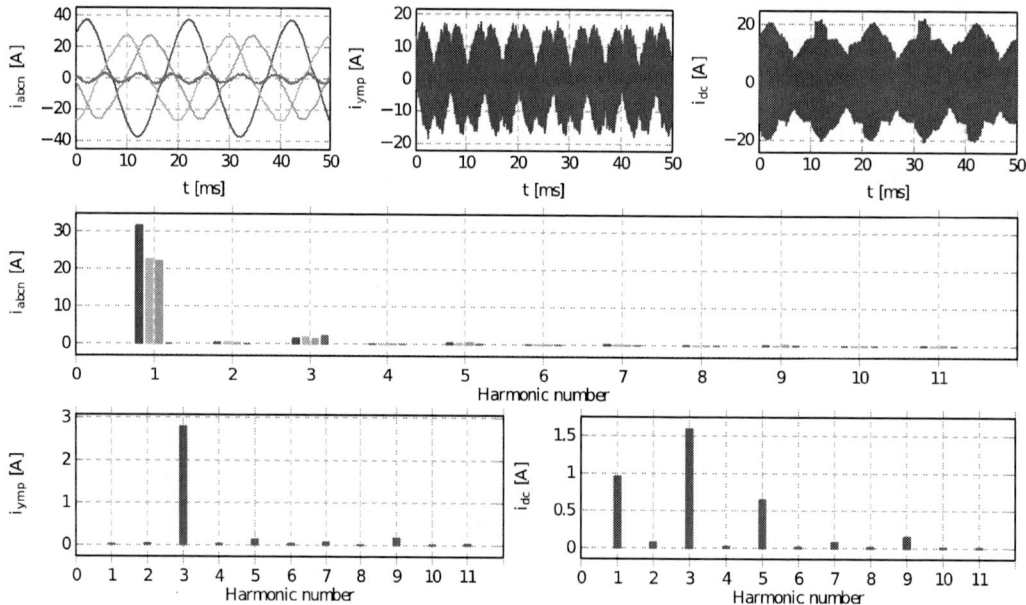

Fig. 5 Use case #2 positive and negative current sequences: $i_a = 25A/0°, i_b = 17.625A/{-}135°, i_c = 17.625A/135°, \quad i_n = 0 + 3rd$ harmonic injection

Analysis of the RMS current stress on the DC link capacitors of the four phase 3-level T-type voltage source converter   MILETIC Zoran

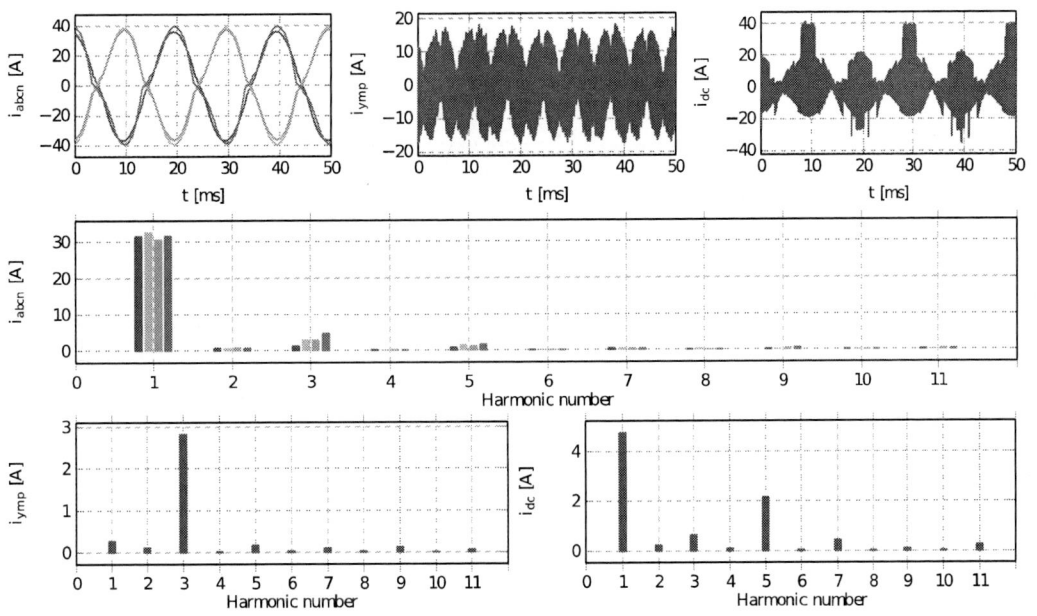

Fig. 6 Use case #3: $i_a = 25A/0°, i_b = 25A/{-}180°,\ i_c = 25A/180°,\ i_n = 0$ + 3rd harmonic injection

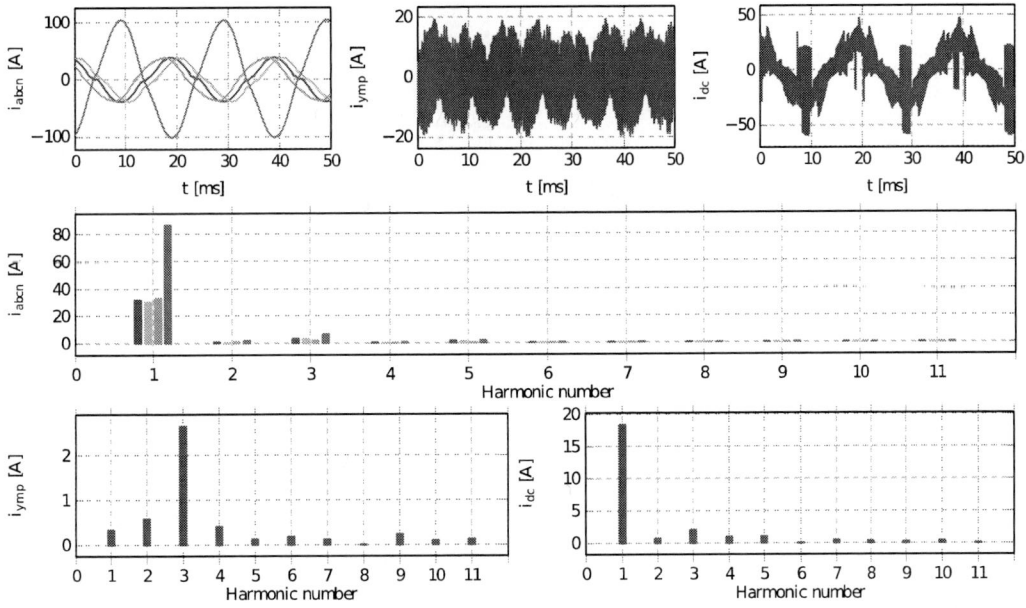

Fig. 7 Use case #4: $i_a = 25A/0°, i_b = 25A/{-}30°,\ i_c = 25A/30°,\ i_n = 0$ + 3rd harmonic injection

Analysis of the RMS current stress on the DC link capacitors of the four phase 3-level T-type voltage source converter — MILETIC Zoran

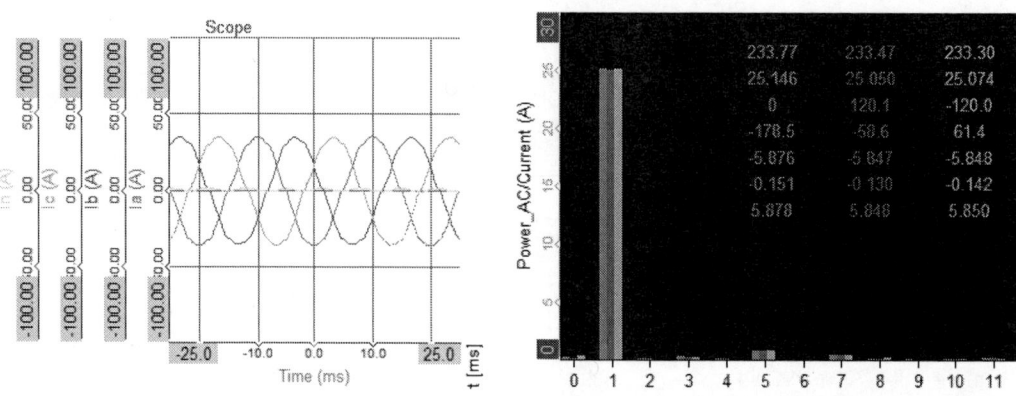

Fig. 8 Use case #1 balanced current: $i_a = 25A/0°, i_b = 25A/-120°$, $i_c = 25A/120°$, $i_n = 0 + 3^{rd}$ harmonic injection – experimental data

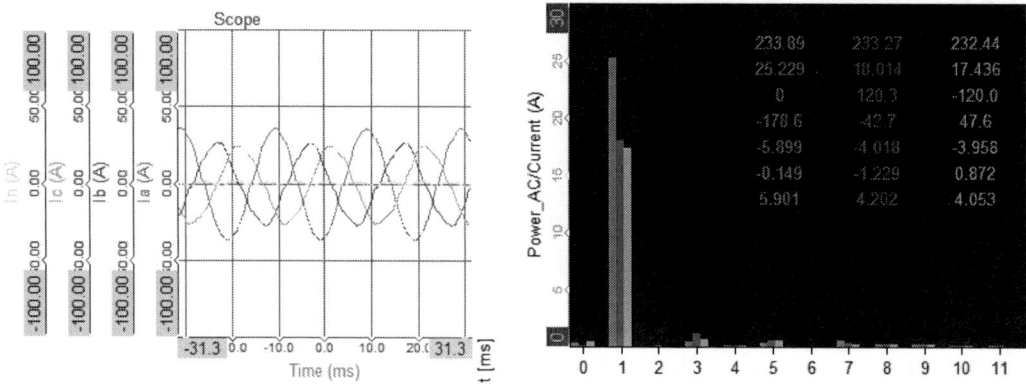

Fig. 9 Use case #2 positive and negative current sequences: $i_a = 25A/0°, i_b = 17.625A/-135°$, $i_c = 17.625A/135°$, $i_n = 0 + 3rd$ harmonic injection – experimental data

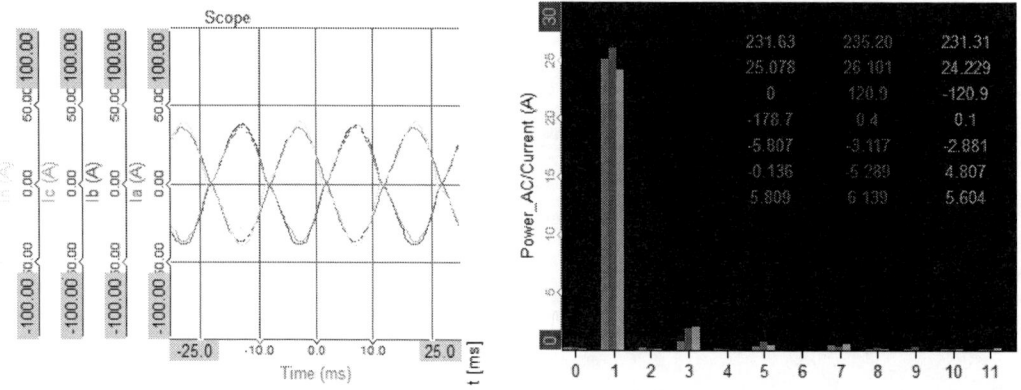

Fig. 10 Use case #3: $i_a = 25A/0°, i_b = 25A/-180°$, $i_c = 25A/180°$, $i_n = 0 + 3rd$ harmonic injection – experimental data

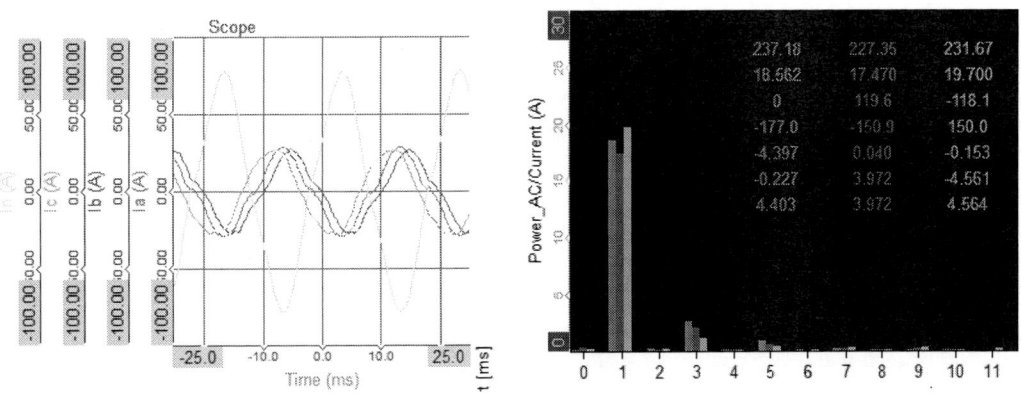

Fig. 11 Use case #4: $i_a = 25A/0°$, $i_b = 25A/-30°$, $i_c = 25A/30°$, $i_n = 0$ + 3rd harmonic injection – experimental data

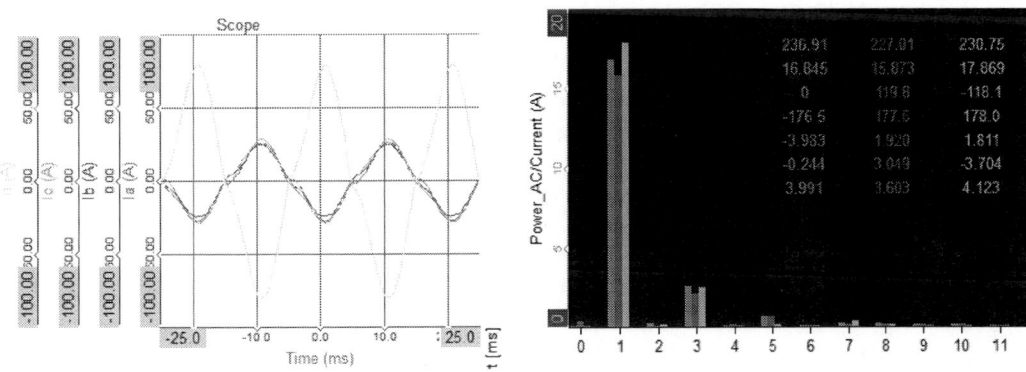

Fig. 12 Use case #5: $i_a = 25A/0°$, $i_b = 25A/0°$, $i_c = 25A/0°$, $i_n = 0$ + 3rd harmonic injection – experimental data

Other benefits of the four phase topology, as well as drawbacks shall be presented and discussed in the follow up papers.

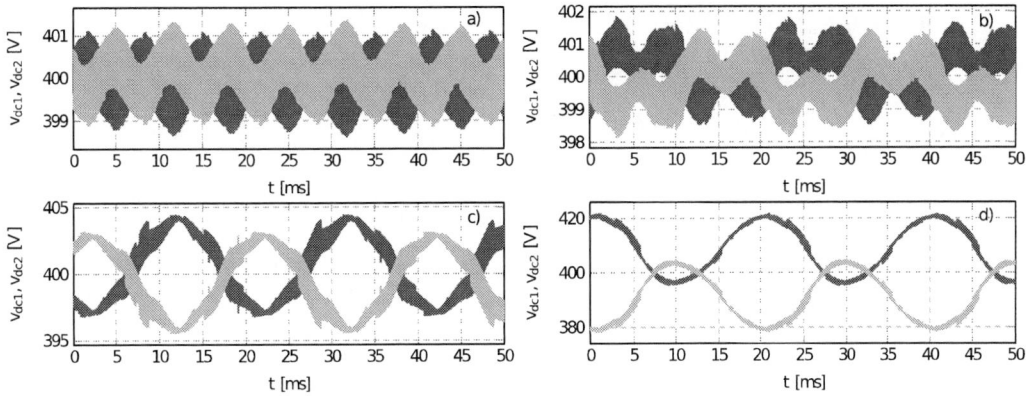

Fig. 13 DC link capacitors voltage ripple a) use case #1, b) use case #2, c) use case #3 and d) use case #4

Conclusion

In this paper, analysis of the RMS current stress on the DC link capacitors of the four-leg 3L T-type voltage source converter is conducted. It's experimentally shown the most of the RMS stress is coming from the switching frequency content and that a compact DC link can be designed even with the case of the load imbalance containing neutral currents. This present one of the major advantages of the four phase voltage sources converter over other typical three phase topologies for applications with four wires, but on the other hand it comes at the expense of having additional 4th phase and more complex control.

References

[1] Z. Miletic, W. Tremmel, R. Bründlinger, J. Stöckl and B. Bletterie, "Optimal control of three-phase PV inverter under Grid voltage unbalance," *2019 21st European Conference on Power Electronics and Applications (EPE '19 ECCE Europe)*, Genova, Italy, 2019, pp. P.1-P.11.

[2] P. J. Grbović, Ultra-capacitors in Power Conversion: Analysis, Design and Applications, Wiley-IEEE Press; 1st edition (December 16, 2013), ISBN-13: 978-1118356265.

[3] M. K. Mishra, A. Joshi and A. Ghosh, "Control schemes for equalization of capacitor voltages in neutral clamped shunt compensator," in *IEEE Transactions on Power Delivery*, vol. 18, no. 2, pp. 538-544, April 2003.

[4] Xu, Xinwei, Midpoint voltage stabilization of split dc bus in three-phase four-wire shunt active power filters: Prototyping of a midpoint voltage balancer for the ECSEL JU funded project CONNECT. Eindhoven : Technische Universiteit Eindhoven, 2019. 116 blz.

[5] Z. Lin, X. Ruan, L. Jia, W. Zhao, H. Liu and P. Rao, "Optimized Design of the Neutral Inductor and Filter Inductors in Three-Phase Four-Wire Inverter With Split DC-Link Capacitors," in *IEEE Transactions on Power Electronics*, vol. 34, no. 1, pp. 247-262, Jan. 2019.

[6] K. Sano and H. Fujita, "A New Control Method of a Resonant Switched-Capacitor Converter and its Application to Balancing of the Split DC Voltages in a Multilevel Inverter," *2007 Power Conversion Conference - Nagoya*, Nagoya, 2007, pp. 122-129.

[7] Benedetto, Marco & Lidozzi, Alessandro & Solero, Luca & Crescimbini, Fabio & Grbovic, Petar. (2018). Five-Level E-Type Inverter for Grid-Connected Applications. IEEE Transactions on Industry Applications. PP. 1-1. 10.1109/TIA.2018.2859040.

[8] Zhang, Rain & Prasad, V. & Boroyevich, Dushan & Lee, F.C.Y.. (2002). Three-dimensional space vector modulation for four-leg voltage-source converters. Power Electronics, IEEE Transactions on. 17. 314 - 326. 10.1109/TPEL.2002.1004239.

An Adaptive Droop Control Method for Interlink Converter in Hybrid AC/DC Microgrids

Mohammad S. Golsorkhi

Department of Electrical and
Computer Engineering,
Isfahan University of
Technology
Isfahan, IR 84156, Iran
golsorkhi@iut.ac.ir

Rasool Heydari

Electrical Engineering Section,
The Mads Clausen Institute
University of Southern
Denmark, Odense, DK-5230,
Denmark
rah@mci.sdu.dk

Mehdi Savaghebi

Electrical Engineering Section,
The Mads Clausen Institute
University of Southern
Denmark, Odense, DK-5230,
Denmark
mesa@mci.sdu.dk

Keywords

« droop control », « microgrid», « distributed power generation », « inverters », « power electronics ».

Abstract

The conventional control scheme of hybrid microgrids (MGs) uses active power-frequency and reactive power-voltage droop control methods to realize proportional active power sharing in ac and dc MGs, respectively. In order to equate the percentage loadings of dc and ac subgrids, the power transfer through the interlink converter is controlled such that the per-unit frequency deviation of ac MG is equated to the per-unit voltage deviation of dc MG. The main drawback of the mentioned strategy is the poor dynamic response caused by the slow dynamics of the conventional droop method as well as the delay associated with frequency measurement in the interlink converter (IC) controller. To enhance the dynamic response, a control scheme based on voltage-current droop characteristics is proposed in this paper. In this method, the d and q axis V-I droop control schemes are adopted for proportional active and reactive power sharing among ac Distributed Energy Resources (DERs), and the dc V-I droop control method is used for coordination of dc DERs. Furthermore, a novel control strategy based on d-axis voltage/dc voltage droop characteristics is proposed for the IC to realize global power sharing with a fast dynamic response. Experimental results are presented to showcase the efficacy of the proposed scheme.

1- Introduction

Throughout the past century, electrical systems have been dominantly based on three-phase ac technology due to its advantage of efficient voltage level transformation and compatibility with fossil fuel-based driven rotating machines. However, the recent increasing penetration of dc-based renewable resources such as photovoltaic (PV) generation and the rising demand for dc power have motivated the introduction of dc grids. The dc grids eliminate the need for dc/ac converters on the generation side and ac/dc conversion on the load side, hence improve the efficiency. In order to take advantage of the high efficiency of dc grids while continuing to utilize ac generations and loads, the concept of hybrid ac/dc microgrid (MG) has been proposed [1]. A hybrid MG is comprised of a dc MG, and an ac MG interconnected through an interlink converter (IC). The ac grid might be connected to the main power grid or be operated in an islanded mode. This paper is focused on the islanded operation mode, in which the power balancing and voltage/frequency control rely on coordinated control of Distributed Energy Resources (DERs).

Two main approaches employed for coordination of DERs in ac and dc MGs are master/slave and droop control methods. While the master/slave control method is easier to implement, the droop control method is widely accepted as the preferred approach due to scalability, high reliability, and ability to operate without communication links [2]. Conventionally, active power-frequency (P-f) and reactive power-voltage (Q-V) droop control schemes are used to realize proportional load sharing among DERs in ac MGs [3] . In dc MGs, active power-voltage (P-f) droop control strategy is commonly employed. The conventional ac and dc droop control strategies can be extended to hybrid MGs as well. By doing so, the electrical load of each of the ac and dc subgrids can be proportionally shared among the DERs in that subgrid. In order to fully exploit the capacity of all DERs in the hybrid ac/dc MG, it is desirable to share the total load of the dc and ac subgrids among all DERs. This task is commonly referred to as global power sharing.

The key for the realization of global power sharing is to control the power flow through IC so as to balance the loading of ac and dc subgrids. The P-f droop characteristics of ac DERs and P-V droop characteristics of dc DERs can be matched by adjusting the IC power flow so that the per-unit frequency deviation in the ac subgrid is equated with the per-unit dc voltage deviations in the dc subgrid. This method is simple to implement and favors global power sharing [4]. However, it features a slow dynamic response and requires large frequency droop coefficients to ensure stability. Several enhancements of the conventional control method have been proposed in the literature, including the extension of the method to hybrid MGs with multiple subgrids [5], the addition of energy storage systems to the IC [6] and enhancement of the dynamics by introducing a feedforward signal into IC's current control loop. However, these improved methods still suffer from large frequency deviations in the ac subgrid. An alternative control method based on active power-frequency-dc voltage droop was proposed in [7] to reduce the frequency deviations in the ac grid. However, this method exhibits a slow dynamic response.

In order to circumvent the issues of slow dynamics and large frequency deviations, a novel control strategy is proposed in this paper. In this scheme, the V-I droop control scheme [8] is adopted in dc and ac subgrids to enable proportional load sharing with zero frequency deviation. Furthermore, a new voltage-voltage droop scheme is proposed for IC to achieve global power sharing with a fast-overdamped dynamic response. The efficacy of the proposed control method is demonstrated by hardware in the loop experimental results.

2- Conventional Control Method of Hybrid AC/DC MG

Fig. 1 shows the schematic diagram of a hybrid ac/dc MG along with the associated control loops. The hybrid MG is comprised of a dc MG, and an ac MG interconnected through an IC. The IC controls the flow of power between the ac and dc MGs. Commonly, the power flowing through IC is controlled in a way that the total ac and dc loads are shared proportionally among the dc and ac DERs, i.e., global power sharing condition is satisfied. If the percentage loading of the dc MG is higher than the ac MG, IC will transfer active power from the ac to dc MG, and vice-versa. This strategy enhances the utilization factor of the DERs in spite of load variations in the ac and dc subgrids [4].

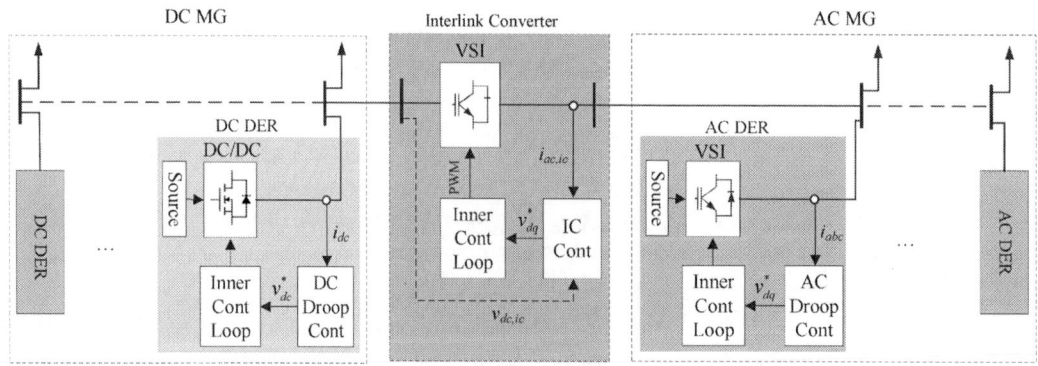

Fig. 1.   Architecture of hybrid ac/dc MG

The conventional control method of hybrid MGs adopts the P-f and Q-E droop control schemes for ac DERs and the P-V droop control method for dc DERs. The P-f and Q-E droop controllers calculate the reference frequency (f) and magnitude (E) of ith ac DER voltages according to the following equations:

$$f_i = f_0 - m_{ac} \frac{P_{ac,i}}{P_{ac,i}^{rated}},$$

(1)

$$E_i = E_0 - n \frac{Q_i}{Q_i^{rated}},$$

(2)

in which $P_{ac,i}^{rated}$, Q_i^{rated}, m_{ac}, n, f_0 and E_0 are the rated active and reactive powers, P-f and Q-E droop coefficients, no-load frequency and no-load voltage of ith ac DER, respectively. The frequency and amplitude of the DER voltage are used to calculate the reference DER voltage, which is then applied to a cascaded voltage-current control loop. The cascaded controller adjusts the pulse width of the PWM signal such that the DER output voltage tracks its reference waveform.

Equations (1) and (2) imply that the change in the frequency/amplitude of each ac DER voltage is proportional with the ratio of the DER's output active/reactive power on the rated values. Under steady-state conditions, the frequency of all DERs is equal, and the DER voltages are close to each other. Therefore, the per-unit active power ($P_{ac,i} / P_{ac,i}^{rated}$) of ac DERs converge to the same value, and the per-unit reactive powers (Q_i / Q_i^{rated}) become almost equal in steady-state. In other words, the ac MG's load is proportionally shared among the ac DERs.

The P-V droop controller of the jth dc DER changes its voltage ($v_{dc,j}$) according to the output power ($P_{dc,j}$), as follows:

$$v_{dc,j} = V_{dc0} - m_{dc} \frac{P_{dc,j}}{P_{dc,j}^{rated}},$$

(3)

where, V_{dc0}, m_{dc} and $P_{dc,j}^{rated}$ are the no-load voltage of dc DERs, P-V droop coefficient, and the rated power of jth dc DER, respectively. Similar to the case of ac MGs, since the voltages of dc DERs have a small difference, (3) implies proportional power sharing among dc DERs.

Global power sharing in the hybrid MG means that the ac and dc DERs have the same per-unit active power outputs. This condition can be expressed as

$$\forall i \in \left\{1, 2, ..., M\right\}, \ \forall j \in \left\{1, 2, ..., N\right\}: \ \frac{P_{ac,i}}{P_{ac,i}^{rated}} = \frac{P_{dc,j}}{P_{dc,j}^{rated}},$$

(4)

in which M and N denote the number of ac and dc DER in the hybrid MG, respectively. By using (1) and (3), and neglecting the voltage variations throughout the dc MG, the condition of (4) can be rewritten as:

$$\frac{f_0 - f}{m_{ac}} = \frac{V_{dc0} - v_{dc}}{m_{dc}}.$$

(5)

According to (1) and (3) the maximum frequency deviation in ac MG and the maximum dc voltage deviation in dc MG are equal to m_{ac} and m_{dc}, respectively. So, (4) implies that global power sharing can be achieved if the per-unit values of ac MG frequency deviation and dc MG voltage deviation are equal. To that end, the reference active power of IC (P_{ic}^*) is controlled as follows:

$$P_{ic}^* = k_p \left(\Delta v_{dc,ic,pu} - \Delta f_{ic,pu} \right) + k_i \int \left(\Delta v_{dc,ic,pu} - \Delta f_{ic,pu} \right) dt$$

(6)

in which

$$\Delta f_{ic,pu} = \frac{f_0 - f_{ic}}{m_{ac}} \tag{7}$$

$$\Delta v_{dc,ic,pu} = \frac{V_{dc0} - v_{dc,ic}}{m_{dc}} \tag{8}$$

The PI controller of (6) adjusts the active power set-point of IC so that the control error, i.e., the difference between per-unit frequency and dc voltage deviations, reaches zero in steady-state conditions. So, as long as the system is stable, this approach guarantees global power sharing under steady-state conditions.

3- Proposed Control scheme

The conventional control scheme of hybrid MG has some major shortcomings. First of all, the P-f droop control method causes frequency deviations in ac MG, which degrade the power quality. Secondly, this method suffers from the poor dynamic response, which can result in large overshoots in the output power of DERs following step load changes. To circumvent these issues, a control scheme based on voltage-current droop characteristics is proposed in this section.

AC DERs Control Strategy

In the ac MG, the V-I droop control method, which was originally proposed in [8], is adopted to enable proportional sharing of the load active and reactive power among DERs. Fig. 2 depicts the schematics of the V-I droop controller of ac DER units. A timing GPS module is used to enable high accuracy time synchronization of local DER controllers. The output signal of the timing GPS is used by the synchronization block to compute the angle of the global synchronous rotating reference frame of ac MG (θ_{global}). By aligning the dq axes of all ac DERs as well as IC to this global frame, the frequency of the ac MG is fixed at the rated value. Furthermore, in order to have proportional load sharing among ac DERs, the d and q components of ith DER's reference voltage are drooped with respect to the d and q components of the current, as follows:

$$v_{d,i}^{*} = E_0 - m_{ac} \cdot g\left(i_{d,i} \Big/ I_{d,i}^{rated}\right), \tag{9}$$

$$v_{q,i}^{*} = -n \cdot h\left(i_{q,i} \Big/ I_{q,i}^{rated}\right), \tag{10}$$

where, $v_{d,i}^{*}, v_{q,i}^{*}, i_{d,i}, i_{q,i}, I_{d,i}^{rated}, I_{q,i}^{rated}$ are the d and q components of the reference voltage, output current and the rated current of the ith ac DER unit. The functions g and h are one-to-one piece-wise linear droop functions, whose slopes increase with the increase in per-unit current, and is the amplitude of the no-load ac voltage.

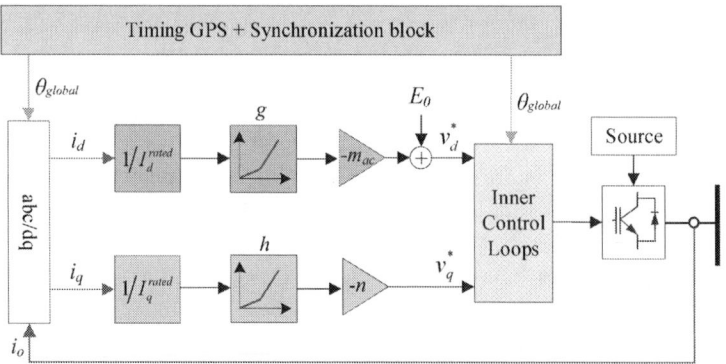

Fig. 2. V-I Droop Control strategy for ac DERs

The calculated reference voltage is applied to the inner control loops which control the PWM signals of the dc/ac converter such that the output voltage tracks its reference.

If the line voltage drops are neglected, the ac DERs output voltages would be the same. Under such condition, (9) implies

$$g\left(i_{d,1}\big/I_{d,1}^{rated}\right) = g\left(i_{d,2}\big/I_{d,2}^{rated}\right) = \ldots = g\left(i_{d,M}\big/I_{d,M}^{rated}\right). \tag{11}$$

Since g is a one-to-one function, it can be inferred that the per-unit values of the DERs' d-axis currents are equal:

$$\frac{i_{d,1}}{I_{d,1}^{rated}} = \frac{i_{d,2}}{I_{d,2}^{rated}} = \ldots = \frac{i_{d,n}}{I_{d,n}^{rated}}. \tag{12}$$

Similarly, by using (10), we have:

$$\frac{i_{q,1}}{I_{q,1}^{rated}} = \frac{i_{q,2}}{I_{q,2}^{rated}} = \ldots = \frac{i_{q,n}}{I_{q,n}^{rated}}. \tag{13}$$

Equations (12) and (13) imply that the d and q components of the load current are proportionally shared among the DERs. According to (9) and (10), the d and q-axis components of the DER voltages are close to the nominal voltage and zero, respectively. So, the active power ($p = v_d i_d + v_q i_q \approx v_d i_d$) is proportional with the d-axis current, and the reactive power ($q = v_q i_d - v_d i_q \approx -v_d i_q$) is related to the q-axis current. Therefore, proportional sharing of the d and q axis currents implies proportional sharing of active and reactive power among the DERs in ac MG.

The main advantage of the V-I droop method compared with the conventional P-f and Q-V droop scheme is the fast dynamic response. Furthermore, the inclusion of the piece-wise linear droop functions (g and h) in the droop control law enhances the accuracy of current sharing. This way, the V-I droop control scheme prevents overcurrent stresses during steady-state and transients.

DC DERs Control Strategy

Fig. 3 illustrates the V-I droop controller of dc DERs. In this case, the reference voltage of ith DER is determined based on the dc voltage- dc current droop characteristics, as follows:

$$v_{dc,i}^{*} = V_{dc0} - m_{dc}g\left(\frac{i_{dc,i}}{I_{dc,i}^{rated}}\right) \tag{14}$$

in which $v_{dc,i}^{*}$, $i_{dc,i}$ and $I_{dc,i}^{rated}$ are the reference voltage, output current, and the rated current of the ith dc DER unit. Furthermore, V_{dc0} is the no-load dc voltage and m_{dc} is the dc droop coefficient.

Similar to the case of ac MG, if the voltage drops across the dc MG lines are neglected, (14) implies that the current sharing among the dc DERs is proportional with their ratings:

$$\frac{i_{dc,1}}{I_{dc,1}^{rated}} = \frac{i_{dc,2}}{I_{dc,2}^{rated}} = \ldots = \frac{i_{dc,n}}{I_{dc,n}^{rated}}. \tag{15}$$

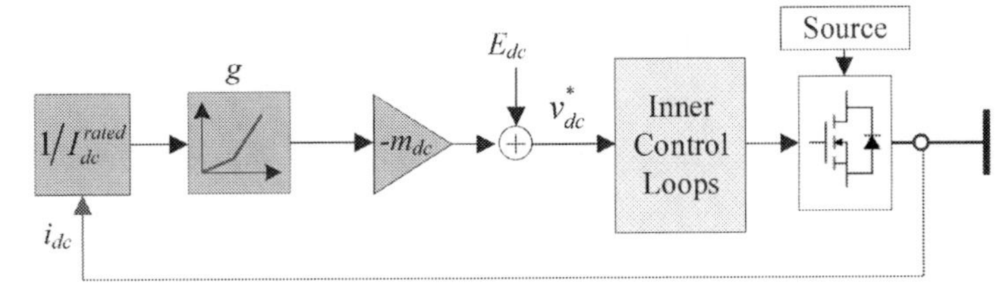

Fig. 3. V-I Droop Control strategy for dc DERs

IC Control Strategy

The schematic diagram of the droop controller for IC is shown in Fig. 4. In order to achieve global power sharing among the ac and dc DERs, the active power flow of IC is controlled by means of a v_{dc}-v_d droop control law, as follows:

$$v^*_{d,ic} = E_0 + \frac{m_{ac}}{m_{dc}}\left(v_{dc,ic} - V_{dc0}\right) + k_d \frac{d}{dt}\left(v_{dc,ic} - V_{dc0}\right) \tag{16}$$

in which $v^*_{d,ic}$ is the d-axis component of the IC ac bus reference voltage, $v_{d,ic}$ is the measured value of the IC dc bus voltage and k_d is the differential droop coefficients. Equation (16) couples the v_d-i_d droop controller in the ac subgrid and the v_{dc}-i_{dc} droop scheme in the dc subgrid. The first term on the right hand side of (16) is the no-load voltage of the ac MG, and the second term is a weighted version of the dc voltage deviation. The third term is included to enhance the dynamic response of the IC with respect to dc load variations.

When the dc load is increased, the dc DERs decrease their voltages according to the v_{dc}-i_{dc} droop characteristics (see (14)). According to (16), IC responds to this change by decreasing the d-axis components of the IC ac bus voltage. As a result, the active power of the IC (dc to ac direction) decreases. Moreover, an increase in the active power of the ac load gives rise to a decrease in the d-axis component of the ac bus voltage (see (9)). Consequently, the active power of the IC (dc to ac direction) increases.

The mechanism of operation of the proposed global power sharing scheme is illustrated in Fig. 5. Three droop characteristics are shown in the figure: 1) v_d-i_d droop of ac DERs, 2) v_{dc}-v_d droop of IC, 3) v_{dc}-i_{dc} droop of dc DERs. It should be emphasized that the droop characteristics of each of the ac (dc) DER with respect to their per-unit currents are identical. Furthermore, if the line voltage drops are neglected, all of the ac (dc) DERs have a common voltage and per-unit current. Consider the case that the ac DERs have a per-unit current of I^{pu}_d. This current is associated with point A on the ac DERs droop characteristics. The v_{dc}-v_d droop of IC, which links the v_d-i_d droop of ac DERs to the v_{dc}-i_{dc} droop of dc DERs, maps the point A to point C. It is observed that at operating point C, the per-unit output current of dc DERs is I^{pu}_{dc}, which is identical with I^{pu}_d. Mathematically,

$$\forall i \in \left\{1,2,...,M\right\}, \; \forall j \in \left\{1,2,...,N\right\} : \; \frac{i_{d,i}}{I^{rated}_{d,i}} = \frac{i_{dc,j}}{I^{rated}_{dc,j}} . \tag{17}$$

Equation (17) can also be driven by combining (9), (14), and (16).

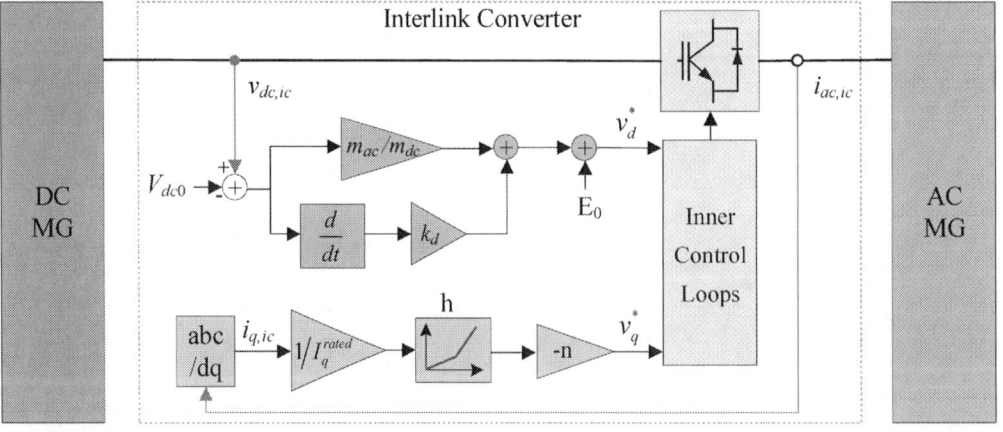

Fig. 4. IC control scheme

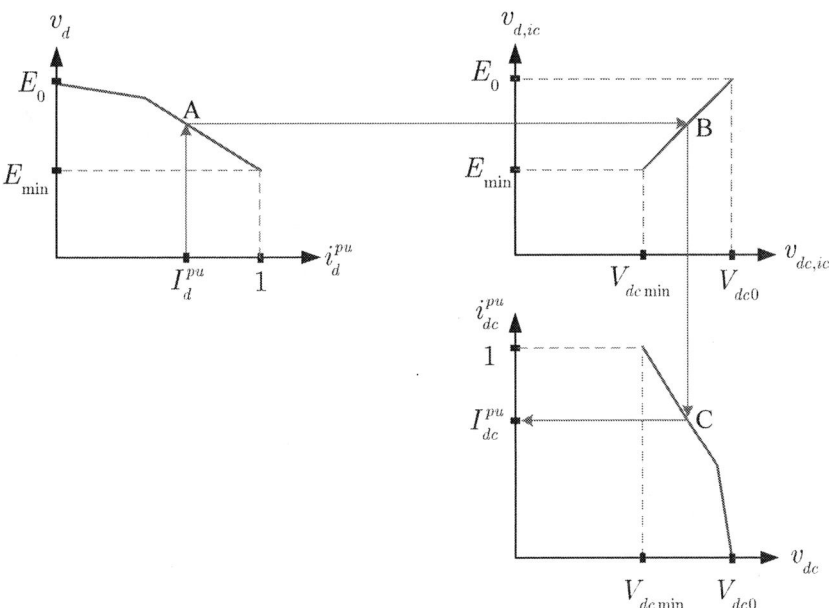

Fig. 5. Mechanism of operation of the proposed global power sharing scheme

To enable IC cooperation with ac DERs in supplying the reactive power of ac load, the q-axis component of the IC ac bus voltage ($v_{q,ic}^{*}$) is controlled based on the following v_q-i_q droop control law:

$$v_{q,ic}^{*} = -nh\left(\frac{i_{q,ic}}{I_{q,ic}^{rated}}\right) \tag{18}$$

It is worth mentioning that in practice, the different DERs in each of ac and dc MGs have different voltages due to the voltage drop across the lines. This results in a power sharing error among the DERs. As shown in [8], this sharing error can be minimized by proper selection of the droop coefficients and droop functions. Furthermore, zero sharing error can be achieved by secondary control schemes [9].

4- Experimental Results

The proposed control strategy was tested in the laboratory-scale MG setup shown in Fig. 6. The MG setup is comprised of three cabinets, each of which includes two 15kW active front end rectifiers and three 15kW inverters. Each of the converters is connected to an FPGA board, which communicates to target PCs via a fast optic fiber communication link. The target PCs, which serve as the main controller unit, receive feedback signals from the FPGA boards and calculate the PWM signals according to the proposed control scheme. Then, the FPGA boards generate PWM pulses, which are applied to the converters' switches. Two of the inverters in the first cabinet were configured as ac DERs, and two of the inverters in the second cabinet were configured as ac loads. Furthermore, one of the inverters in the third cabinet was configured as an IC. The dc DERs and dc loads were emulated in the target PCs. The parameters of the experimental setup are detailed in [10].

In order to demonstrate the efficiency of the proposed control method, the dynamic response of the test MG is studied with respect to ac and dc step load changes. In particular, the ac load is changed from 7kW (PF=0.85) to 14kW(PF=0.85) at t=0s, and the dc load is changed from 3kW to 8kW at t=1s. The performance of the proposed method is compared with the conventional control scheme discussed in Section 2 [4].

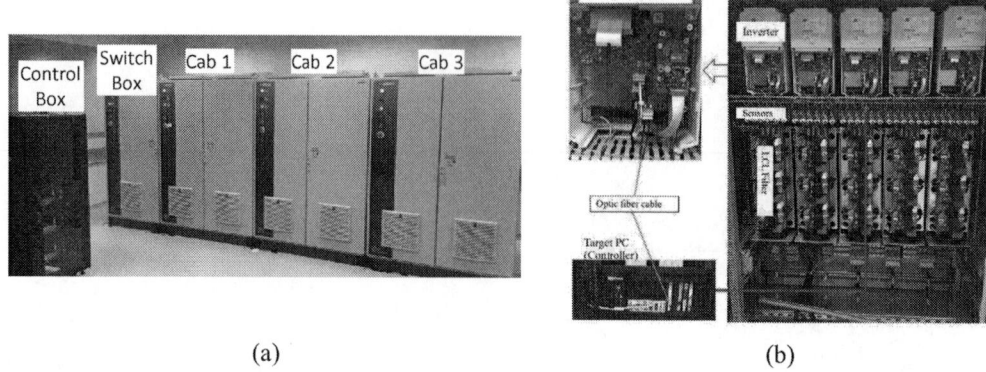

(a)   (b)

Fig. 6. Experimental setup: a) overview of setup, b) components of each cabinet

The experimental results are shown in Fig. 7. Following the step load change in ac MG (@ t=0s), the output power of ac DERs rises. Furthermore, the IC attempts to increase the power flow from dc to ac MG to enable the dc DERs contribute to the ac load. This process is sluggish in the case of the conventional control method, due to the slow dynamic response of IC as well as droop controllers of DERs. As seen in Fig. 7 (a), the step load change is first picked up by ac DERs, and after a few cycles, the dc DERs increase their output powers. Consequently, the ac DERs experience a power overshoot. Similarly, the dc DERs experience overshoots following the step load change in dc MG (@ t=1s). These power overshoots give rise to transient overcurrent stresses in dc and ac DERs. In contrary to the conventional control scheme, the proposed control method features a fast dynamic response, which ensures accurate global power sharing during transients. As shown in Fig. 7(b), the DERs active powers rise smoothly following the load changes, and no overshoot is observed.

In addition to the smooth dynamic response, the proposed control method enhances the voltage regulation of both dc and ac MGs. As seen in Figs. 7 (c) –(f), the maximum voltage deviation of the proposed scheme is almost half of the conventional control method. Such tight voltage regulation has become feasible by adopting piece-wise linear droop functions of (9) and (14).

5- Conclusions

Hybrid ac/dc MGs provide an efficient solution for the integration of ac and dc renewable resources into electrical networks. One key challenge concerning hybrid MGs is the coordination of dc and ac DERs. In this paper, a new adaptive droop control mechanism is presented to enable global power sharing among dc and ac DERs. The proposed method takes advantage of the V-I droop control scheme in ac and dc MGs to decrease the response time of DERs to load changes, hence realizing accurate load sharing during transients. In addition, a new $v_d\text{-}v_{dc}$ droop control scheme is proposed for the IC to guarantee global power sharing among ac and dc DERs with a fast dynamic response. The high damping of the V-I droop characteristics of the dc and ac DERs, on the one hand, and the fast dynamics of the proportional-derivative droop controller of the IC, on the other hand, enable attaining an overdamped dynamic response. Therefore, the proposed method eliminates the transient overcurrent stresses, which could trigger a false trip or shorten the lifetime of the DERs converters. Experimental results showcase the smooth and overshoot-free dynamics of the proposed scheme.

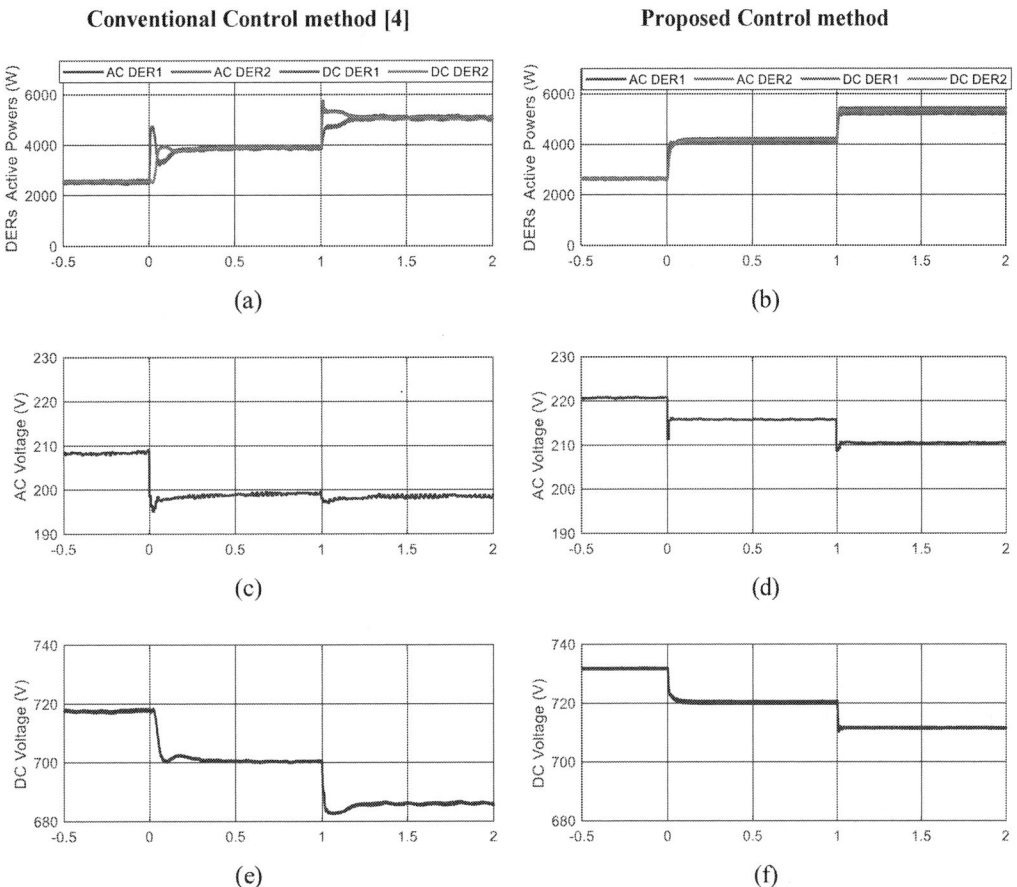

Fig. 7. Comparison between the performance of the conventional and proposed control methods: a,b) DERs active powers, c,d) IC ac bus voltage c) IC dc bus voltage

References

[1] X. Liu, P. Wang, and P. C. Loh, "A hybrid AC/DC microgrid and its coordination control," *IEEE Transactions on smart grid,* vol. 2, no. 2, pp. 278-286, 2011.

[2] A. Gupta, S. Doolla, and K. Chatterjee, "Hybrid AC–DC microgrid: systematic evaluation of control strategies," *IEEE Transactions on Smart Grid,* vol. 9, no. 4, pp. 3830-3843, 2018.

[3] R. Heydari, T. Dragicevic, and F. Blaabjerg, "High-bandwidth secondary voltage and frequency control of vsc-based ac microgrid," *IEEE Transactions on Power Electronics,* vol. 34, no. 11, pp. 11320-11331, 2019.

[4] P. C. Loh, D. Li, Y. K. Chai, and F. Blaabjerg, "Autonomous operation of hybrid microgrid with AC and DC subgrids," *IEEE transactions on power electronics,* vol. 28, no. 5, pp. 2214-2223, 2012.

[5] Y. Xia, W. Wei, M. Yu, X. Wang, and Y. Peng, "Power management for a hybrid AC/DC microgrid with multiple subgrids," *IEEE Transactions on Power Electronics,* vol. 33, no. 4, pp. 3520-3533, 2017.

[6] X. Li, Z. Li, L. Guo, J. Zhu, Y. Wang, and C. Wang, "Enhanced Dynamic Stability Control for Low-Inertia Hybrid AC/DC Microgrid With Distributed Energy Storage Systems," *IEEE Access,* vol. 7, pp. 91234-91242, 2019.

[7] M. Baharizadeh, H. R. Karshenas, and J. M. Guerrero, "Control strategy of interlinking converters as the key segment of hybrid AC–DC microgrids," *IET Generation, Transmission & Distribution,* vol. 10, no. 7, pp. 1671-1681, 2016.

[8] M. S. Golsorkhi and D. D. C. Lu, "A Control Method for Inverter-Based Islanded Microgrids Based on V-I Droop Characteristics," *Power Delivery, IEEE Transactions on,* vol. 30, no. 3, pp. 1196-1204, Jun. 2015.

[9] M. S. Golsorkhi, Q. Shafiee, D. D. Lu and J. M. Guerrero, "Distributed Control of Low-Voltage Resistive AC Microgrids," IEEE Transactions on Energy Conversion, vol. 34, no. 2, pp. 573-584, June 2019.

[10] M. S. Golsorkhi, D. J. Hill and H. R. Karshenas, "Distributed Voltage Control and Power Management of Networked Microgrids," in IEEE Journal of Emerging and Selected Topics in Power Electronics, vol. 6, no. 4, pp. 1892-1902, Dec. 2018.

AUTHOR INDEX

Aarniovuori, Lassi 2829
Abbate, Carmine.................................2802
Abbosh, Amin......................................1006
Abdel-Rahim, Naser 352
Abdelhakim, Ahmed..............................2220
Abdelrahem, Mohamed 900
Abramson, Rose A. 1934
Abusara, Mohammad................................471
Aganza-Torres, Alejandro.........................1813
Aguglia, Davide..................................3330
Ahmad, Bilal.....................................3348
Ahmad, Faheem....................................2987
Ahola, Jero......................................2753
Ait-Ahmed, Mourad................................3289
Aizpuru, Iosu 251, 1205
Alam, M. M. 480, 1551
Alam, Muhammad Farhan.............................416
Alatise, Olayiwola...............................2241
Alawieh, Hadi....................................1685
Albach, Manfred 173, 193
Alexandre, Philippe..............................1905
Ali, Ahmed Ismail M. 1417
Ali, Marwan...............................1118, 2039
Ali, Mohammad...............................2743, 2763
Ali, Waqas.......................................871
Alisar, Ibrahim..................................1205
Alishah, Rasoul Shalchi..........................460
Alkama, Kouceila.................................2564
Allard, Bruno 522, 829, 1470, 1874
Almaksour, Khaled................................1700
Almeida, Bruno F.................................3217
Alonso, C..919
Alqatamin, Moath..................................65
Am, Sokchea......................................3172
Ammann, Ulrich...................................3137
Amrane, Fayssal..................................362
Anders, Erik.....................................944
Andrade, Fabio 1400, 1841, 1850
Andresen, Jan...............................2303, 2545
Anzola, Jon......................................251
Aoustin, Yannick.................................1923
Arandia, Nerea...................................1524
Arazi, M...2881
Arrizabalaga, Antxon.............................1205
Arrozy, Juris....................................1067
Arruti, Asier....................................251
Artiglia, Melissa................................2791
Aríztegui, Raquel González.......................1515

Asllani, Besar...................................1279
Avenas, Yvan................................1685, 2564
Averbukh, Moshe..................................2439
Averous, Nurhan Rizqy.............................153
Azizian, Mohammadreza............................2860
Baburajan, Silpa.................................2573
Bacha, S...406
Bacha, Seddik....................................820
Baërd, H..95
Bagaber, Bakr....................................2613
Bahman, Amir Sajjad..............................2704
Bahrani, Behrooz.................................787
Bai, Wenshuai....................................667
Baker, Erik......................................512
Bakran, Mark-M...........644, 686, 1252, 1533, 1831, 1885,
....2106
Bakri, R...2554
Bakri, Reda......................................2078
Balkowiec, Tomasz................................2029
Barazi, Yazan....................................1057
Barelli, Linda...................................292
Barg, Sobhi......................................416
Barwig, Markus...................................1460
Basic, D. 37, 95
Basic, Duro.................................2938, 2957
Bauer, Pavol...............1224, 1233, 1561, 3422
Bazin, Pascal....................................2467
Beczkowski, Szymon Michal........................2987
Beerten, J.......................................1551
Belhaouane, Moez............................1158, 1756
Bello, Guilherme.................................2049
Benchaib, A......................................745
Benchaib, Abdelkrim.........................820, 1215
Bender, Vitor C..................................3217
Bendfeld, Christian..............................163
Benjamin, Sébastien..............................3156
Benkhoris, Mohamed Fouad.........................3205
Bensebaa, S......................................1363
Bentivegna, N....................................2293
Benzagmout, Abdelhadi............................1905
Beranger, Bruno..................................2467
Berkani, M.......................................1363
Bernet, Steffen.............................124, 1569
Bertele, Felix...................................3137
Bertilsson, Kent............................416, 460
Betto, Kento.....................................3071
Betz, Robert Eric................................927
Bevilacqua, Pascal...............................1279

Beza, Mebtu 969, 1952
Bhajana, V. V. Subrahmanya Kumar 2068
Bidini, Gianni 292
Biela, Juergen 2331, 2583, 2791
Biela, Jürgen 2230, 2409, 2446, 2475, 2513, 2524,
.... 2684, 2712, 2780, 2946
Bier, Anthony 2638
Bikinga, Wendpanga Fadel 2564
Binder, Andreas 2049
Birou, Camille 332
Bissal, Ara 871
Blaabjerg, F. 3237
Blaabjerg, Frede 1, 460, 810, 927, 2088, 2135,
.... 2220, 2393, 2573, 2888, 2898, 2928, 3119
Blanco, Marcos 1076
Blanquez, Francisco R. 1336
Blaquiere, Jean-Marc 1057
Blinov, Andrei 2996
Blume, Sebastian 2684
Böcker, Jan 2341
Böcker, Joachim 1638, 3024
Boersma, S. 745
Bohlen, Oliver 707
Bohnke, M. 1613
Boige, François 1057
Boisaubert, Emile 3172
Bolzan, Thais E. 3217
Bolzoni, A. 1306
Bombois, X. 745
Bongiorno, Massimo 969, 1952
Borcherding, Holger 608
Boulaud, Etienne 3172
Bourennane, Abdelhakim 1096
Bourguet, Salvy 2907
Bouscayrol, Alain 3330
Boutleux, Emmanuel 433
Boutry, Arthur 2366
Bozorg, Mokhtar 3247
Boškovic, N. 2812, 2820
Branca, Xavier 522
Briff, Pablo 9
Bringezu, Thilo 2780
Brockhage, Torben 1766
Brooks, Michael 2773
Bruyere, Antoine 1756
Brückner, Thomas 2938, 2957
Bründlinger, Roland 2723
Büdel, Johannes 1718
Bucher, A. 3403
Bucher, Alexander 203
Budo, Kohei 1450
Buigues, Garikoitz 85

Burgos, Rolando 2366
Burgos-Mellado, Claudio 1354
Busatto, G. 3210
Busatto, Giovanni 2802
Buttay, Cyril 1106, 2265, 2366
Cacciato, M. 909
Cai, Pei 2135
Camail, Philippe 2000, 2265
Camara, M. B. 2881
Camurca, L. 3305
Cardelli, Ermanno 292
Cárdenas, Roberto 1354
Carnielutti, Fernanda 2851
Caron, Hervé 1700
Carpita, Mauro 3247
Carpiuc, Sabin 962
Cascino, S. 2293
Castellazzi, Alberto 2210
Castellini, Simone 292
Castelltort, Arnaud 1747
Castiglia, V. 3237
Catellani, Stéphane 2638
Cavallaro, D. 2293
Chaiba, Azeddine 362
Chakraborty, Sajib 2320, 3111
Cheaito, Hassan 829
Chen, Linglin 1224, 1233
Chen, Qing 542
Chen, Yu 804
Chen, Zhengxin 637
Cheshire, Christoph 3137
Chevalier, Florian 1895
Chillón-Antón, Cristian 853, 1542
Chiumeo, Riccardo 424
Chraye, Hélène 3427
Chrin, Phok 3172, 3376
Chrzan, Piotr J. 3054
Chédot, L. 183
Chédot, Laurent 433, 1106
Ciupageanu, Dana-Alexandra 292
Clerc, Guy 433, 829
Clerici, Alessio 424
Cochelin, Anne-Sophie 3426
Coelho-Medeiros, Rafael 1479
Colak, Ilknur 871
Colas, F. 1579, 2554
Colas, Frédéric 1158
Colmenero, Manuel 1336
Connaughton, Alexander 1866
Cordier, Julien 3272
Corentin, Darbas 1803
Cornea, Octavian 2192

Costa, François 503
Coujard, Clementine 1270
Cravero, Jean-Marc..3330
Crebier, Jean-Christophe 2010
Da Cunha, Julian...2312
Dabbabi, Asma 2907
Dahmen, Christopher..843
Dai, Jing.............................. 820, 1215, 1479
Dakyo, B..2881
Dang, Ziyue..804
Danzer, Michael A..707
Darivianakis, Georgios 998
Davari, Pooya 726, 1006, 2573, 3119, 3295
David, Romain..522
Davidson, Colin 2366, 3425
Davoodi, Amirali...2393
De Doncker, Rik W. 153, 163, 1627
De Jaeger, Jean-Claude 1895
De Jódar, Esther..1658
De La Grandiere, Hubert 3424
De Lauretis, Maria 512
De Mora, Pablo Rodriguez 1533
De Oliveira, Eduardo F....................................2535
De, Dipankar...2210
De-Preville, Guillaume 9
Defrance, Nicolas 1895
Degrenne, N..2865
Delamea, R..2865
Delarue, Philippe 1158, 3330
Delette, G..1613
Delhommais, Mylène.......................................1737
Delpech, F...2554
Demidov, Iurii 2419
Denis, Guillaume...1270
Dennetière, S...2967
Derbey, Alexis...2039
Derkacz, Pawel B..3054
Despesse, Ghislain..503
Despouys, Olivier 1373
Dessante, Philippe...1858
Devos, Guillaume 1858
Di Gregorio, Francesco 1747
Dieckerhoff, Sibylle................ 2274, 2341, 2603
Dierks, Rebecca 3091
Dietz, Armin 1316
Dincan, Catalin 2873
Dinkel, Daniel...1297
Dinulovic, Dragan...2773
Djerioui, Ali..3205
Dong, Dong 2366
Doppelbauer, Martin..1589
Douine, Bruno 1373

Doumiati, Moustapha.......................................2251
Drabek, Pavel..2068
Dragicevic, Tomislav 2393, 2898
Driesen, J. 480
Driesen, Johan..27
Drofenik, Uwe...627
Duarte, J. 2812, 2820
Duarte, Jorge L. 1067, 2656
Duarte, Renan R. 3217
Duerbaum, Thomas 203
Dujic, Drazen 1776, 2486
Dürbaum, Thomas.....................173, 193, 1460
Dworakowski, Piotr...........406, 1106, 2000, 2265, 3006
Džonlaga, Bogdan 1479
Ebersberger, Janine 3340
Ebrahimi, Reza..2860
Eckel, Hans-Günter.....................532, 1666, 2059, 2126, 3282
Eckerle, Richard 765
Ecrabey, Jacques 2467
Egrot, Philippe...1479
Eguia, Pablo 85
Ehlich, Martin 608
El Baghdadi, Mohamed....................................2320
El Jihad, Hamza 1982
Elizondo, Laura Ramirez.................................1561
Ellul, Racquel 1025
Elsabrouty, Ibrahim...871
Elsied, Moataz...3156
Elthokaby, Youssuf...352
Endisch, C. 551
Enjeti, Prasad 1390
Erenler, Yeliz 571
Errigo, F. 183
Escobar, Gerardo 1390
Escofficr, Réne...1470
Esfetanaj, Naser Nourani.........................2860, 3295
Eskandari, Bahman 863
Eslamian, Morteza...3064
Eslampanah, Vahid..2860
Espina, Enrique..1354
Etoz, Burhan..2241
Fadel, Maurice...3101
Fauth, Leon..3081
Fazli, Nastaran...2126
Fehr, Hendrik...1030, 2116
Ferreira, Jan Abraham.......................................892
Ferrieux, Jean-Paul..2039
Finkenzeller, Michael 835
Fischer, Manuel........................1168, 1605, 1709
Fogsgaard, Martin Bendix...............................2258
Foray, Etienne..1874
Forsyth, A. J. 1306

Fort, Jiri 1086
Founier, Etienne 3101
Francois, Bruno 362, 1700, 1756
Frédèric, Poitiers 1803
Frey, D. 406
Frey, David 2695
Freytes, Julián 9
Friebe, Jens 2743, 2763, 3081
Fromme, Christopher 2603
Frost, Damien 223
Fruchier, Olivier 1905
Fu, Siqi 302
Fuchs, Simon 2409, 2513
Fürst, Markus 1885
Galeshi, Soleiman 2695
Gamatié, Abdoulaye 1747
Gandolfi, Chiara 424
Ganjavi, Amir 726, 1006
Gao, Fei 1224, 1233
Gao, Jianbo 1262
Garate, José Ignacio 1524
Garbuio, Lauric 1685
García-Torres, Felix 134
Garnier, Laurent 2467
Gaubert, Jean-Paul 396, 2284
Gauthier, Jean-Yves 590, 736
Gautier, Cyrille 1118, 1858, 2039
Gautier, Maxime 1923
Geng, Zeyang 3198
Gensior, Albrecht 581, 1030, 2116
Gentejohann, Marius 2274
Georges, Didier 820
Gerada, Chris 1944
Gerada, David 1944
Geramirad, Hadiseh 2000
Gerstner, Michael 1316
Geske, Martin 2938, 2957
Geury, Thomas 2320
Geyer, Tobias 2145
Ghamrawi, Ahmad 2284
Ghanes, Malek 3205
Gholami-Khesht, Hosein 3119
Giacomazzo, M. 490
Gierschner, Sidney 2059, 2126
Giewont, William 1016, 1972
Giotakos, Panagiotis I. 2172
Girbau-Llistuella, Francesc 1542
Gireada, Mihaita 2192
Glac, Antonín 3257
Gladen, Marcel 1148
Glasberger, Tomas 3166, 3314
Gleissner, Michael 644, 686

Glushakov, Vasiliy V. 47, 56
Gnärig, Jan Lasse 1030
Golluccio, G. 3210
Golsorkhi, Mohammad S. 2733, 2898
Gomez, Juan S. 1354
Gomis-Bellmunt, Oriol 1542, 2977
Gonzalez, Jose Ortiz 2241
Gonzalez-Torres, J-C. 745
González-Fontderubinat, Paula 1542
Gosses, Kilian 143
Gou, Wanchao 153
Govaerts, G. 480
Gradinger, Thomas B. 627
Grainger, Brandon M. 65
Grbovic, Petar J. 2723
Grecki, Filip 627
Green, Tim C. 276
Green, Tim 2977
Griepentrog, Gerd 1675
Gruson, F. 1579
Gruson, Francois 1158
Gu, Chunyang 1944
Guerrero, Josep. M 3289
Gui, Qiuye 1030
Guichon, Jean Michel 2564
Guillaud, X. 1579
Guillaud, Xavier 1158, 1756, 1952
Guo, Mingzhu 718
Guo, Xuan 317
Gutierrez, A. 919
Gutierrez, Sebastian 598
Gärtner, M. 3403
Götting, Gunther 1289
Hackl, Christoph 900
Hage-Hassan, Maya 1858
Hai, Jie 231
Hallemans, L. 480
Hamid, Muhammad 1289
Hammerer, Horst 2182
Hammes, David 2059
Han, Hua 260, 285, 302, 326
Han, Lubin 370
Han, Weiji 765
Haq, Omer Ikram Ul 3393
Häring, Johannes 644, 686
Harnefors, Lennart 1952
Hartmann, Michael 2258
Harzig, Thibaut 65
Hase, Genki 467, 498
Hasenohr, C. 3403
Hasenohr, Christian 203
Hatori, Kenji 1489

Haug, Martin	2773
He, Maojun	804
He, Yuying	1410, 1435
Hegazy, Omar	2320, 3111
Hein, Lukas	3413
Helle, Lars	2873
Heller, M.	3403
Hénaux, Carole	3101
Henninger, Stefan	3179
Heredero-Peris, Daniel	853, 1542
Herkommer, Christian	1718
Hernandez, Fernando Davalos	598
Herwig, Daniel	1766
Heucke, Sören	2341
Heydari, Rasool	2733, 2898
Hideaki, Yano	1442
Higashihata, Takeshi	1489
Hijazi, A.	183
Hiller, Marc	3366
Hillermeier, Claus	1297
Himker, Niklas	979
Hinkkanen, Marko	2495
Hiraki, Eiji	386
Hirayama, Hiroshi	2376
Hiwatari, Daichi	2376
Hofer, Matthias	2403
Hoffmann, Felix	954
Hofmann, Harald	203
Homann, Michael	307
Hong, Yang	231
Horrein, Ludovic	3330
Horvatic, Iréna	3156
Houari, Azeddine	3205, 3289
Hu, Anliang	2946
Hu, Rui	2594
Huang, Han	1128
Huang, Pin-Yu	2666
Huang, Xingxuan	1972
Huisman, Henk	1067, 1186
Hulea, Dan	2192
Hussain, E. K.	471
Iannuzzo, Francesco	2258, 2704
Ibanez, Federico	598, 3034
Idarreta, Aitor	1205
Idir, Nadir	1793, 1895, 2078
Iman-Eini, Hossein	3305
Ingman, Jonny	563
Inoue, Sadayuki	19
Iraola, Unai	1205
Isaksson, Dan	1016
Ishihara, Hiroki	19
Isobe, Takanori	3358
Itoh, Jun-Ichi	1380, 2200
Jackiewicz, Krzysztof	2029
Jaeger, Johann	3179
Jakob, Roland	2957
Jaritz, Michael	2684
Jasim, Omar	9
Jean-Christophe, Olivier	1803
Jeannin, Pierre-Olivier	3054
Jehle, Andreas	2475
Jelena, Popovic	892
Jeong, Min	2409, 2513
Ji, Shiqi	1972
Jia, Ming	1627
Jiang, Jinhai	637
Jiaqi, Diao	231
Joebges, Philipp	1627
Jonokuchi, Hideki	2376
Jorge, Tenorio	1400
Joryo, Satoshi	3071
Jotwani, Ankit	1138
Joubert, Charles	522
Judge, Paul D.	276
Jun-Ping, He	1599
Junge, Patrick	2613
Juntunen, Raimo	3227
Junyent-Ferré, Adrià	2977
Justin, Elissa Cresenta Anak	2265
Jäppinen, Janne	563
Järvisalo, Heikki	1016
Jørgensen, Asger Bjørn	2987
Kado, Yuichi	2666
Kadri, R.	2554
Kahl, Tino	2603
Kahle, Karsten	1336
Kaiser, Ingmar	1666
Kallfass, Ingmar	1915
Kaminski, Nando	954
Kampen, Dennis	2049, 2153
Kanchan, R. S.	3393
Kang, Yong	797, 804
Karaventzas, Vasilios	2583
Karlsson, Martin	512
Kaszewski, Arkadiusz	2029
Kawabata, Yoshitaka	1177, 1243
Keel, Oliver	2791
Kefer, K.	3403
Kehl, Zdenek	3166
Keller, Christian	2938, 2957
Kennel, Ralph	542, 900, 1262, 2386, 3272
Kersten, Anton	765
Kesbia, Nasreddine	1685
Kestelyn, X.	1579

Ketchedjian, Vasken	1605
Keysan, Ozan	1823
Khanzadeh, Babak	1040
Kharezy, Mohammad	3064
Kikuchi, Naoto	2200
Killeen, Peter	777
Kim, Bunthern	3172, 3376
Kimura, Norihito	105
Kimura, Shota	377
Kindl, Vladimir	1086
Kirchenberger, U.	3403
Kirowitz, Thomas	2403
Kitagawa, Wataru	1425
Kitamura, Taishi	1177
Kiviniemi, Mika	563
Kjaer, Martin Vang	2888
Kjær, Philip	2873
Klass, Stefan	3272
Klier, Samantha	707
Koch, Dominik	880, 1915
Kohlhepp, Benedikt	266, 1460
Kojabadi, Hossein Madadi	2860
Komma, Thomas	835
Komrska, Tomáš	3257
Kondo, Keiichiro	2675
Kone, Lamine	1793
Kopacz, Rafal	2457
Korhonen, Juhamatti	1016
Kosan, Tomas	3166
Kouchaki, Alireza	3015, 3129
Koutroulis, Eftychios	3146
Krall, Felix	1196
Krim, Youssef	1700
Krischan, Klaus	1866
Kroneisl, Michal	1992
Krug, Dietmar	2059
Kucka, Jakub	2020, 3091
Kuder, Manuel	765
Kuebrich, Daniel	203
Kuhlmann, Kai	1718
Kukkola, Jarno	2495
Kumar, Dinesh	726, 1006, 2573
Kuring, Carsten	266
Kusaka, Keisuke	1380, 2200
Kuwana, Kazuki	1425
Kuwata, Akiko	19
Kwasinski, Alexis	2594
Kyyrä, Jorma	3348
Kärkkäinen, Hannu	2829
Kärkkäinen, Tommi J.	563
Kübrich, Daniel	266
Küster, Pierre	571

Labiano, Daniel	1205
Labouré, Eric	1858
Lacarnoy, Alain	654
Lacressonnière, Fabien	332
Ladoux, Philippe	1106, 2265
Lafon, Frederic	1793
Lafoz, Marcos	1076
Lagier, Thomas	1106, 2000, 2265
Lana, Andrey	2419
Langbauer, Thomas	1866
Langmaack, N.	241
Langwasser, M.	3305
Lapassat, N.	37
Larruskain, Marene	85
Lautner, Frank	1831
Lazaroiu, Gheorghe	292
Le Moigne, Philippe	1158, 2078
Le Métayer, Pierre	3006
Le, Hoai Nam	2200
Lee, Seong-Yong	2486
Leedham, Rob	871
Lefebvre, Bruno	1106, 2000, 2366
Lefebvre, S.	1363
Lefebvre, Stéphane	1118
Lehmann, Franziska	944
Lehn, Peter	223
Lembeye, Yves	2010, 2695
Lemmen, Erik	2350, 2358
Leo, Jacopo	75
Leppänen, Joonas	563
Leterme, Willem	276
Letrouvé, Tony	1700
Lexow, Daniel	3282
Li, Boyang	1289
Li, Chi	317
Li, Dingrui	1972
Li, Hui	2704
Li, Jiaqi	9
Li, Jing	1944
Li, Lang	285
Li, Qi	1262
Li, Tao	3044
Li, Weilin	2135
Li, Yongdong	317, 1944
Li, Yu	2386
Li, Zheming	2106
Liang, Chaohui	3044
Liang, Lin	370, 443
Liao, Jianquan	3385
Liao, Yuefeng	1498
Licari, John	1025
Lima, Glauber De Freitas	2010

Lin, Lei 937
Lin-Shi, Xuefang 590, 736
Liserre, Marco 3305
Liu, Cuicui 2505
Liu, Fuxin 1410, 1435
Liu, Libo 1289
Liu, Xudan 804
Liu, Yao 260
Liu, Zhangjie 260, 302, 326
Liukkonen, O. 2630
Llanos, Jacqueline 1354
Llonch-Masachs, Marc 1542
Locment, Fabrice 667, 677
Loisel, Rodica 2907
Loiselay, Florent 1648
Lomonova, E. A. 2812
Lorenz, Malte 2020
Ludois, Daniel C. 777
Lunz, B. 3403
Luo, Fang 370
Luo, Xian 153
Lutz, Josef 3413
Lutze, Marcel 1675
López-Alcolea, Fco. Javier 134
Ma, Yixiao 1435
Mabe, Jon 1524
Machmoum, Mohamed 2907, 3205, 3289
Maekawa, Sari 2838, 2844
Maerz, Martin 1316
Magambo, Jean Sylvio Ngoua Teu 2078
Maharana, Manoj Kumar 2068
Maharana, Suman 2210
Mahr, Florian 3179
Maier, Robert W. 1252, 2106
Mäkelä, Juha 3348
Mallwitz, R. 241, 490
Mallwitz, Regine 213, 1515
Maneiro, Jose 1106, 3006
Mannen, Tomoyuki 3358
Mannerhagen, Felix 3198
Mantellini, Mattia 1076
Mantzanas, Panagiotis 203
Mao, Saijun 892
Marcault, E. 919
Marchesoni, Mario 3321
Marciano, D. 3210
Marciano, Daniele 2802
Margueron, Xavier 2078
Marmolejo, Narciso G. 1589
Marquardt, Rainer 843, 1297
Martin, Christian 1874
Martin, Jérémy 2638

Martinez, Wilmar 598, 3034
Martire, Thierry 1905
Martínez-Gómez, Manuel 1354
März, Martin 143
Mateos, Felix Rodriguez 2583
Mattar, Rita 1118
Mattsson, Aleksi 1016
Maussion, Pascal 3376
Mayorga, John Paul 1658
Mazuela, Mikel 1205
McIntyre, Michael 65
Mehdi, Driss 2284
Meißner, Markus 124, 1569
Mendoza-Araya, Patricio 2917
Meneses, Javiera 2917
Meng, Qingchao 2446
Mercier, Adrien 1858
Mercier, Sylvain 2467
Mertens, Axel 979, 1766, 2020, 2303, 2545, 2613,
........................2743, 2763, 3091, 3340
Mesbahi, Tedjani 3205
Meynard, Thierry A. 654
Mezrag, Bachir 2564
Mezzetti, Margarita 944
Micallef, Alexander 1025
Miceli, R. 3237
Milas, Nikolaos T. 2172
Miletic, Zoran 2723
Millinger, Jonas 512
Minami, Masataka 467, 498
Mirtchev, Alex V. 2429
Mitani, Kohei 1425
Miyauchi, Tsutomu 377
Mizutani, Hiroto 386
Mohamed, Abdalla Hussein 115
Mohamed, Islam 352
Molina-Martínez, Emilio J. 134
Mollov, S. 2865
Molnar, Jan 3166
Monmasson, Eric 1118, 1823
Montero, E. Rodriguez 2098
Montesinos-Miracle, Daniel 853, 1542
Moraes, Tiago José Dos Santos 1693
Morel, Cristina 2251
Morel, F. 183, 406
Morel, Florent 2000, 2366
Morel, Hervè 1279, 1648
Mori, Osamu 386
Morici, Riccardo 1076
Morizane, Toshimitsu 1047, 3071
Mortimer, Benedict 163
Motegi, Shin-Ichi 1786

Mourouvin, Rayane 820
Mourtzis, Dimitris A. 2172
Muehlbauer, Markus 707
Muetze, Annette 1196
Müller, Jan-Kaspar 2763, 3340
Mumtaz, Muhammad Adnan 1326
Munk-Nielsen, Stig 2987
Muñoz, Fredy 3189
Muñoz, Javier 3189
Muntean, Nicolae 2192
Murata, Ryo 386
Musznicki, Piotr 3054
Nabatirad, Mohammadreza 787
Nada, Kaho 19
Nadh, Greeshma 1619
Nair, Durga S. 1619
Najera, Jorge 1076
Najjar, Mohammad 3015, 3129
Nakagaki, Akito 498
Nakamura, Keiichi 1489
Nakashima, Osamu 2376
Nakatani, Shota 1047
Nakazawa, Yosuke 2675
Narula, Anant 969, 1952
Natori, Kenji 2675
Navarro, Gustavo 1076
Navas, Alex F. 1354
Ndagijimana, Fabien 2010
Nee, Hans-Peter 2220
Neumann, Jessica 3101
Nevaranta, N. 2630
Ngoua-Teu, J-S 1613
Nguyen, Ngac Ky 1693
Nguyen, Van Sang 2638
Nicolas, Ginot 1803
Nie, Cheng 1972
Nie, Qingqing 797
Niemelä, M. 2630
Niemelä, Markku 563, 2829
Nikowitz, Mario 2403
Nisch, A. 3403
Nishida, Yasuyuki 1786
Nitzsche, Maximilian 880, 1605, 1709, 1915
Niu, Liyong 342
Norambuena, Margarita 2851
Nymand, Morten 3015, 3129
Obernolte, Urs 608
Oberschelp, Wolfgang 1727
Odriozola, Kepa 654
Oguma, Kenji 377
Ohta, Takahiro 2666
Okamori, Daichi 1047

Okazaki, Akihiro 2838
Okazaki, Yuhei 1040
Oliveira, Joao 1648
Olivier, Jean-Christophe 2251
Omori, Hideki 1047
Omrane, A. 2554
Ordoño, Ander 1524
Orfanoudakis, Georgios I. 3146
Ortega-Perez, Carmen 3330
Ota, Kenji 1489
Ottaviano, Andrea 292
Oumaziz, Amirouche 1096
Pace, Loris 1895
Páez, Juan 1106
Paez, J. D. 406
Palensky, Peter 810, 1561
Pallier, Joris 829
Palm, Herbert 707
Pan, Xuejiao 1128
Passalacqua, Massimiliano 3321
Patarroyo-Montenegro, Juan F. 1841
Patti, Davide 3210
Paulus, Sebastian 2303, 2545
Pavlicek, Vladimir 1086
Pawellek, A. 3403
Pawellek, Alexander 203
Payman, A. 2881
Pei, Xiaoze 342
Peller, Stefan 266
Pelosi, Dario 292
Peltoniemi, Pasi 3227
Peña, R. A. 183
Pendharkar, Ishan 1346
Peng, Han 797, 804
Peng, Hao 804
Peng, Tao 1498
Penin, Carolina 1905
Peralta, Patricio 75
Perenyi, Christian 65
Peretti, Luca 3393
Pérez-Molina, María José 85
Peric, V. 745
Peroutka, Zdenek 3257, 3314
Perriard, Yves 75
Petit, M. 1363
Petit, Mickael 1118, 3054
Petkovic, Marko 1776
Peyghami, Saeed 810, 2573
Pfeifer, Markus 1675
Pfeiffer, Jonas 571, 2535
Phulpin, Tanguy 618
Pidancier, Thomas 3247

Pietrzak-David, Maria	3376
Pilawa-Podgurski, Robert C. N.	1934
Pinheiro, Humberto	2851
Pinomaa, Antti	2419
Pinto, Rafael A.	3217
Pirsto, Ville	2495
Pitel, Ira	1390
Planson, Dominique	1279, 1648
Plaza, Jesus D. Vasquez	1841
Plissonnicr, Marc	1470
Poebl, Monika	835
Polacek, Libor	1086
Pollet, Benjamin	503
Pommier-Petit, Pascal	829
Pool-Mazun, Erick I.	1390
Popuri, Madhuchandra	2068
Pouresmaeil, Edris	863
Pöyhönen, Santeri	2753
Prevost, Thibault	1270
Prieto, Dany	3101
Prieto-Araujo, Eduardo	2977
Pronin, Mikhail V.	47, 56
Puls, Simon	608
Pulvirenti, M.	909, 2293
Pursiainen, Jooa	3227
Putkonen, A.	2630
Pyrhönen, Juha	2829
Pyrhönen, O.	2630
Pyrhönen, Olli	2419
Qashlan, Ziyad H. S.	571
Qiang, Jin	163
Qoria, T.	1579
Qoria, Taoufik	1756
Queval, Loic	1473, 1479
Rabba, Heiko	307
Rabkowski, Jacek	2457, 2996
Radet, Hugo	332
Rahmoun, Yasser	2182
Rahul, Arun S.	1619
Rajput, Shailendra	2439
Ramm, Hannes	307
Ramírez-Scarpetta, J. M.	1850
Rasmussen, Tonny Wederberg	1138
Rathnayake, Hansika	726, 1006
Rault, P.	2967
Raute, R.	989, 1506
Rautio, Juuso	563
Ravyts, S.	480
Ravyts, Simon	27
Rayati, Mohammad	3247
Razzaghi, Reza	787
Rehlaender, Philipp	1638
Reigosa, Paula Diaz	1346, 2258
Reißenweber, Lukas	3263
Rekola, Jenni	3227
Ren, Chunpin	718
Restrepo, Jose Alex	1850
Retianza, Darian V.	1067
Retianza, Darian Verdy	1186
Rezaee, Ali Yahya	460
Richardeau, Frédéric	1057, 1096
Rietmann, Stefan	2712
Rigot, Valentin	618
Risch, Raffael	2230
Riu, Delphine	3156
Roa, Claudio	3189
Robet, Pierre-Philippe	1923
Roboam, Xavier	332
Robyns, Benoit	1700
Rodriguez, Jose	900, 2851
Rodriguez, José Luis	1205
Rokrok, Ebrahim	1756
Roncero-Sánchez, Pedro	134
Roose, T.	1551
Röser, Tobias	3137
Roth-Stielow, Jörg	880, 1168, 1605, 1709
Rouger, Nicolas	1057
Routimo, Mikko	2495
Rouzbehi, Kumars	863
Ru, Yang	231
Ruan, Xinbo	1410, 1435
Rubenbauer, Hubert	3179
Rufer, Alfred	696
Rute, Erwin	1354
Ruthardt, Johannes	1168, 1605, 1709
Saad, H.	2967
Saad, Yamen	3295
Saber, Christelle	1118
Sácz, Doris	1354
Sadarnac, Daniel	618
Saeedian, Meysam	863
Saggio, M.	2293
Sah, Gyanendra Kumar	532
Sahin, Ilker	1823
Saim, Abdelhakim	3289
Sakai, Kazuto	1442
Sakai, Norikazu	1489
Sakai, Yasuhiro	1489
Sakaria, Omar Ahmed	3295
Sakly, Jihen	618
Sakurazawa, Yoshiki	2675
Saleh, Bassem	2648
Salem, Qusay	1289
Sallot, P.	1613

Salvo, L. 909
Sánchez-Sánchez, Enric 2977
Sandelic, Monika 2088
Sandik, Diane -Perle 512
Sangwongwanich, Ariya 1, 2088
Sanseverino, A. 3210
Sanseverino, Annunziata 2802
Santos, Miguel 1076
Sari, A. 183
Sarraute, Emmanuel 1096
Sassatelli, Gilles 1747
Sathik, Mohd. Ali Jagabar 460
Saudemont, Christophe 1700
Savaghebi, Mehdi 2733, 2898
Savarit, Elise 1982
Sawada, Takashi 2623
Sayed, Mahmoud A. 1417
Scarpetta, Jose Miguel Ramirez 1400
Scelba, G. 909
Schafmeister, Frank 1638, 3024
Schanen, Jean-Luc 1685, 2039, 3054
Schiesser, Matthias 962
Schleippmann, Nico 143
Schlesinger, Richard 2524
Schlüter, Michael 2274
Schmidt, Dimitri 1168
Schmitt, Alexander 2182
Schmitt, N. 1363
Schmitt, Stefan 954
Schmitz, Jan 124, 1569
Schobre, Thorben 213, 1515
Schröder, Günter 1727
Schrödl, Manfred 2403
Schulte, Hendrik 1709
Schulz, Matthias 143
Schulz, Nicola 1346
Schulze, Torben A. 307
Schwabe, Christian 3413
Schütt, Michael 532
Sciacca, A. G. 909
Scicluna, K. 989, 1506
Sechilariu, Manuela 667, 677
Segur, R. 745
Seiler, Pascal 2331
Semail, Eric 1693
Sergeant, Peter 27, 115
Shah, Chirag 153
Sharkh, S. M. 471
Sharkh, Suleiman M. 3146
Shi, Guangze 326
Shinoda, Kosei 820, 1215
Shiozaki, Koji 2623

Shirakawa, Tomohide 386
Shousha, Mahmoud 2773
Si-Yuan, Cai 1599
Siala, S. 95
Siala, Sami 1982
Siebke, K. 490
Siemens, Ag 2603
Silventoinen, Pertti 563, 1016
Simola, Aleksi 2753
Singer, Arthur 765
Skala, Bohumil 1086, 3257
Smailus, Erik 1675
Šmídl, Václav 1992
Smit, A. 3403
Snook, Mark 871
Soeiro, Thiago Batista 1224, 1233, 1561
Sokur, Pavel V. 47
Soltau, Nils 1489
Song, Kai 637
Soumaoro, Ousmane 2251
Soupremanien, U. 1613
Sprunck, Sebastian 2535
Stadler, Alexander 3263
Staines, C. Spiteri 989, 1506
Stathis, Spyridon 2684
Staudt, Volker 1148
Stecca, Marco 1561
Steckler, Pierre-Baptiste 736
Stengl, Josef 1086
Stenglein, Erika 173, 193
Štepánek, Jan 3257
Stock, Alexander 1718
Stöckl, Johannes 2723
Stöttner, J. 551
Stotckaia, Anastasiia D. 47, 56
Stras, Andrzej 2029
Streit, Lubeš 3257
Strittmatter, Tobias 1346
Strunk, Robin 979
Su, Guoxing 1410
Su, Mei 260, 285, 302, 326, 1498
Suarez, Camilo 3034
Sugiyama, Kohei 1177
Sun, Jian 1962
Sun, Yao 260, 285, 302, 326, 1498
Svensson, Jan R. 1952
Tadano, Hiroshi 2623
Taheri, Shamsodin 863
Takahara, Takaaki 386
Takahashi, Hirotaka 377
Takahashi, Hiroyuki 1243
Takano, Tomihiro 19

Takeshita, Takaharu	1417, 1425, 1450
Takeuchi, Somi	1243
Talbert, Thierry	1905
Talla, Jakub	3257
Tan, Guoqiang	443
Tanaka, Ami	2844
Tanaka, Miwako	19
Tanaka, Nobuhiko	1489
Tang, Bojin	718
Tang, Houjun	1224, 1233
Tang, Xiaohu	1589
Tannhäuser, Marvin	2603
Tant, Jeroen	27
Tareilus, G.	241
Tarisciotti, Luca	1224, 1233
Tárraga, Sergio	1658
Tatakis, Emmanuel C.	755, 2172, 2429
Taul, Mads Graungaard	927
Tedesco, Davide	2802
Teigelkötter, Johannes	1718
Teirelbar, Ahmed	2648
Tenorio, Jorge	1850
Teramura, Keiko	377
Terbrack, C.	551
Thal, Eckhard	1489
Thiringer, Torbjörn	765, 1040, 3064, 3198
Tibola, Gabriel	2656
Tikhonov, Sergey	1638
Todd, R.	1306
Tolbert, Leon M.	1972
Torres, Alfonso Parreño	134
Torres, Esther	85
Torres, Fernando	3189
Torres, Jorge	1076
Torres, Jose Rueda	810
Touhami, Mustapha	503
Tran, Dai-Duong	2320, 3111
Traoré, Bakou	2251
Tremmel, Werner	2723
Tremouilles, D.	919
Trillaud, Frédéric	1373
Trochimiuk, Przemyslaw	2457
Tröster, Nathan	1168
Tsolaridis, Georgios	2331
Tsoumas, Ioannis	998, 2145, 2163
Turjanica, Pavel	1086
Turki, Faical	307
Twardon, N.	3403
Ufnalski, Bartlomiej	2029
Ulissi, Gabriele	2486
Umetani, Kazuhiro	386
Unruh, Roland	3024

Uwai, Shuto	1177
Vaccaro, Luis	3321
Valtee, Mikko	3227
Van Den Broeck, G.	480, 1551
Van Duivenbode, Jeroen	1186
Van Mierlo, Joeri	2320, 3111
Van Tichelen, P.	480
Vanfretti, L.	745
Vannier, Jean-Claude	1479
Vansompel, Hendrik	115
Vashishtha, Anushruti	1138
Vasquez-Plaza, Jesus D.	1850
Vázquez, Javier	134
Vecchia, Mauricio Dalla	27
Vechiu, Ionel	396
Velardi, F.	3210
Velardi, Francesco	2802
Velazco, Diego	433
Venet, P.	183
Verbelen, Florian	27
Vermeersch, Pierre	1158
Vermulst, Bas	2350, 2358
Viana, Caniggia	223
Videt, Arnaud	1793, 1895
Vienot, Stephane	1793
Vieto, Ignacio	1962
Villarejo, José	1658
Villegas, Carlos	962
Vip, Stephan	2303, 2545
Vitan, Danut	2192
Voborník, Ales	1086
Vogelsberger, M.	2098
Voigt, Matthias	944
Voldoire, Adrien	2039
Vollaire, Christian	2000
Vollmaier, Franz	1866
Vorontsov, Aleksey G.	47, 56
Votava, Martin	3314
Vu, Duc Tan	1693
Wada, Keiji	3358
Wallart, François	433, 736, 1106
Wang, Bo	450
Wang, Dian	677
Wang, Feng	2505
Wang, Fred	1972
Wang, Huai	1, 2888, 3295
Wang, Meiqi	1944
Wang, Qianggang	3385
Wang, Qiwu	1262
Wang, Tianqing	450
Wang, Xuehua	1410, 1435
Wang, Ziyue	443

Wankhede, Yugandhara H.	3081
Wasfi, Amr	2648
Watanabe, Hiroki	1380
Weicker, Martin	2049, 2153
Weimer, Julian	880, 1915
Weinert, Tristan	1727
Weiss, Sébastien	1793
Wernicke, Laurenz	2603
Weyh, Thomas	765
Wickramasinghc, Thilini	1470
Wijnands, Korneel	2350, 2358
Wilk, Andrzej	2265
Will, Frank	944
Winzer, Patrick	2182
Wolbank, T.	2098
Wondrak, W	3403
Wondrak, Wolfgang	644, 686
Wrona, Grzegorz	2457, 2996
Wu, Hailong	1693
Wu, Xiaohua	2135
Wunder, Bernd	143
Xi, Jiawen	342
Xia, Qingping	260
Xiang, Yusheng	1289
Xie, Jian	1289
Xin, Wei	370
Xiong, Weijing	1498
Xu, Chaoqun	718
Xu, Chen	937
Xu, Dianguo	450
Xu, Guo	1498
Xu, Junzhong	1224, 1233
Xu, Lie	1944
Xu, Xinwei	2656
Yakop, Netan	223
Yamada, Shota	1243
Yamashita, Daniela Yassuda	396
Yamazaki, Osamu	2675
Yamdeu, Mathias Tientcheu	3101
Yan, Xiaoxue	443
Yan, Zheng	1326
Yang, Bo	1944
Yang, Guang	637
Yang, Y	3237
Yang, Yongheng	2135, 2393, 2888
Yang, Zhiqing	153, 163
Yano, Takahiro	1177
Yao, Ran	2704
Yao, Wenli	2135
Ye, Shuaichen	2928
Ye, Zichao	1934
Yin, Tianxiang	937

Yu, Qihao	2350, 2358
Yu, Yong	450
Yu, Zhanqing	718
Yuasa, Hiroaki	105
Yüce, Firat	3366
Yuki, Kazuaki	2675
Yuratich, Michael A.	3146
Zacharias, Peter	571, 1813, 2535
Zafar, Talha	2773
Zanetti, E.	2293
Zaoskoufis, Konstantinos	755
Zare, Firuz	726, 1006
Zdanowski, Mariusz	2996
Zehelein, Matthias	880, 1915
Zeller, Valentin	1460
Zeng, Guang	3413
Zeng, Xianwu	342
Zhai, Dongling	718
Zhang, Haibo	1158, 1756
Zhang, He	1944
Zhang, Li	1128
Zhang, Peng	637
Zhang, Xiaokang	590
Zhang, Zhenbin	2386
Zhang, Ziqian	2505
Zhao, Biao	718
Zheng, Zedong	317
Zhou, Dao	2860, 2928
Zhou, Niancheng	3385
Zhu, Chunbo	637
Zhu, Q.	1306
Zhu, Yangming	450
Zhu, Yuanhao	326
Zhuo, Fang	2505
Zi-Fan, Li	1599
Ziegler, Philipp	1168, 1605, 1709
Zinchenko, Denys	2996
Zucuni, Jordan P.	2851
Zuolian, Liu	231

IEEE
445 Hoes Lane
Piscataway, NJ 08854-4141

ISBN 978-1-7281-9807-1